CW00487308

A Dictionary of plants
used by man

By the same author

Dictionary of Botany
Textbook of Practical Biology

# A Dictionary of plants used by man

George Usher
B.Sc., Dip. Agric. Sci.,
D.T.A., M.I. Biol., F.L.S.,
Senior Biology Master, Bedstone School

Constable London

First published in Great Britain 1974
by Constable and Company Ltd
10 Orange Street London WC2H 7EG


ISBN 0 09 457920 2

Set in Montotype Times
Printed in Great Britain by
Tinling (1973) Limited, Prescot, Lancs.
(a member of the Oxley Printing Group Ltd.)

# Preface

The idea for writing this book began over twenty years ago during my student days in the University College of Swansea. It was later developed in Cambridge and in what was the Imperial College of Tropical Agriculture, Trinidad. From the libraries of these three institutions much of the information was gleaned, and to my teachers there I owe a debt of gratitude for the inspiration which has resulted in its final materialization as a dictionary.

It is impossible to mention all the books which have been consulted, but during the later stages "A Dictionary of Economic Plants" by J. C. Th. Uphof was most useful, particularly for reference to the uses made of plants by the North American Indians, and some of the local names of plants from the Far East. From Professor Uphof I have also derived several of the local medicinal uses of plants, and my gratitude is due to him for giving this dictionary a completeness which it would not otherwise have had.

My thanks are due to the staff of the Shropshire Library Service, who have throughout the years been most helpful in producing obscure books with great efficiency.

My greatest debt of gratitude is to my wife for her encouragement, and for living with this book for a great many years.

GEORGE USHER

Bedstone,
Bucknell,
Shropshire.
August, 1971

# Notes on using this dictionary

As this is not a critical classification of economic plants, the classification adopted in "A Dictionary of the Flowering Plants and Ferns" (7th. Edition) by J. C. Willis, revised by H. K. Airy Shaw has been adhered to rigidly. Botanical descriptions of the plants have not been included, as this would make the book far too long.

The main reference to a plant is to be found under its *genus*. This reference includes a brief description of the distribution of the genus, and the total number of species included in it. This will give some idea of the relative economic importance and possible economic potential of the genus.

The main reference is given in *italics*, followed by the authorities for the name, with the vernacular name following in brackets. The country of origin follows; this does not mean that the plant is confined to this area; it may be grown widely or in areas surrounding its country of origin. Any other reference to a species is given without the authorities, and implies that the main article should be consulted.

The whole text is in alphabetical order, except that simple words take priority over compound, e.g. "Dog Senna" will follow "Dog" and come before "Dogbane", and apostrophes and "de", "de la" and similar expressions have been ignored.

The spacing of geographical directions is as follows: No space between the direction and the place means that the two words follow each other directly; a space implies "of". E.g. "N.America" means "North America" while "N. America" means "North of America", thus "N. S.America" means "North of South America."

The term "locally" is used to imply that the plant is grown within the geographical area indicated, on a small scale or simply collected from the wild, and any trading in it is confined to local markets etc.

The term "vegetable" has been used to indicate that the product is eaten after being cooked in some way, usually boiled.

"=" indicates a synonym, but does not indicate any expression of opinion of priority on the author's part.

"see" means that the plant in question has the same distribution and uses as the one referred to.

# A

Aajoo merah – Eugenia rumphii.
Aambeibosse – Teucrium capense.
Aanda – Dichapetalum acuminatum.
Abaca – Musa textilis.
Abag badi – Muncuna poggeri.
Abamule – Laportea podocarpa.
Abba – Grewia flavescens.

***Abbevillea*** Berg. = Campomanesia Ruiz and Pav.

Abele – Populus alba.

***Abelmoschus esculentus*** (L.) Moench = Hibiscus esculentus.

***Abelmoschus moschatus*** Moench = Hibiscus abelmoschus.

***Aberemoa dioica*** Barb. Rodr. = Annona dioica St. Hil.

***Aberia abyssinica*** Clos. = Doryalis abyssinica.

***Aberia caffra*** Harv. and Sond. = Doryalis caffra.

***Aberia gardneri*** Clos. = Doryalis hebecarpa.

ABIES Mill. Pinaceae. 50 spp. extend across N. Temp. regions into Cent. America. The timber is used mainly for lumber and pulp wood, and is generally inferior to spruce or pine.

***A. alba*** Mill. = Abies pectinata.

***A. amabilis*** (Loud.) Forbes (Cascade Fir, Silver Fir. White Fir). W. coast of N. America. The wood is light, close-grained and hard, but not strong. It is used for wood pulp, and interior work in buildings.

***A. balsamea*** (L.) Mill. (Balsam Fir). Newfoundland to Yukon. The wood is mainly used for wood pulp and subsequent paper manufacture, but is also used for crates, barrels etc. A liquid oleoresin (Canada balsam, Balm of Fir, Balm of Gilead, Canada Turpentine) is extracted from the wood. It is used as a cement for microscope slides and lenses, in the production of lacquers, and medicinally in treating urinogenital catarrh. Poultices are made for external use. The leaves, having a pleasant smell, are used for stuffing pillows.

***A. bormuelleriana*** Math. Turkey. Used for wood pulp.

***A. brachyphylla*** Maxim. = A. homolepis.

***A. cephalonica*** Loud (Greek Fir). See A. nordmanniana.

***A. cinerea*** Spreng. = Dichrostachys cinerea.

***A. concolor*** (Gord.) Hoopes (White Fir). Extends from Wyoming to Mexico. The wood is light, coarse, and not strong. It is used for packing cases etc., construction work in houses, and pulp wood.

***A. delavayi*** (Van Tiegh.) Franch. W. China. An important timber locally. It is light, and neither strong nor durable, but is used in general construction work in houses, and for cheap coffins.

***A. excelsa*** Lam. = Picea abies.

***A. fargesii*** Franch. (Frages Fir). C. China. The wood used for construction work, and making coffins.

***A. firma*** Sieb. and Zucc. = A. momi (Japanese Silver Fir, Mommi, Sagamomi). Japan. A light, soft, coarse-grained wood used for building, paper pulp, boxes etc.

***A. grandis*** Lindl. (Grand Fir, Lowland Fir, Lowland White fir). Grows in the valleys from Montana to C. California. The wood is used for interior work in buildings, packing cases etc., and paperpulp. It is light, coarse-grained and not strong.

***A. hirtella*** Lindl. = A. religiosa.

***A. homolepis*** Sieb and Zucc. = A. brachyphylla Maxim. (Nikko Fir, Nikkomomi, Nikko Silver Fir, Takemomi). Japan. The wood is light, soft and coarse-grained. It is used for general building, paper pulp, and logs.

***A. lasiocarpa*** (Hook.) Nutt. (Alpine Fir, Rocky Mountain Fir). Extends along the W. of N. America. The wood is used for pulp. The bark produces an antiseptic gum. Twigs and leaves were used as incense by the local Indians.

***A. magnifica*** Murr. (Red Fir). Sierra Nevada of N. America. The wood is soft and light, and light brown in colour. It is used for wood pulp, packing cases, cheap buildings and fuel.

***A. mariesii*** Masters (Marie's Fir). Mountains of Japan. The heartwood is strong and light, straight-grained, fragrant and polishes well. It is used for buildings, paper pulp, lacquer work and utensils.

***A. momi*** Sieb. = A. firma.

***A. nobilis*** Lindl. (Noble Fir). N. America, Washington, Oregon, and British Columbia. A hard strong light-brown wood, used in building packing cases, aircraft construction, ladder rungs and blinds. Some is used for wood pulp. Often sold as larch.

***A. nordmanniana*** (Stev.) Spach. Caucasus

and Asia Minor. A soft light wood, used for wood pulp and some construction work. It is of rather poor quality. It is sometimes used as Christmas trees. ***A. cephalonica*** is used similarly, but is of better quality timber.

***A. pectinata*** DC.=A. alba. (Silver Fir) Europe and Asia Minor. The wood is yellowish, easily split, elastic and fairly durable. It is used for general construction work, pulp wood, cellulose manufacture, furniture, matches, packing cases etc., masts. It is the principal Christmas Tree in Europe. Strassburg turpentine is distilled from the wood. Silver Pine Needle Oil is distilled from the leaves, and is used in perfumery and toilet preparations on its own or blended with Eau de Cologne and other perfumes. It is used medicinally in the treatment of catarrh and asthma and in cough drops.

***A. pichta*** Forb.=A. siberica.

***A. pindrow*** Royle (Badar, Baludar, Himalaya Silver Fir, Rewari). Himalayas. A soft yellowish wood, not very durable. It is used for inside work in houses, furniture, packing cases, pulp, general carpentry and railway sleepers.

***A. procera*** Rehd.=A. nobilis.

***A. recurvata*** Mast. (Min Shan). China. Used locally for house construction.

***A. religiosa*** (H. B. K.) Schlecht and Cham. =A. hirtella. (Sacred Fir). Mexico. The wood is used locally for building and wood pulp. An oleoresin (Aceite de Palo) is extracted from the wood. It is used medicinally and in paints.

***A. sachalinensis*** (F. Schm.) Mast. Japan and Island of Sakhalin (U.S.S.R.). The wood is used for general building, packing cases etc., wood pulp and ship-building.

***A. shikokiana*** Nakai see A. veitchii.

***A. siberica*** Ledeb.=A. pichta. (Siberian Fir). N. Russia and C. Asia. An oil (Oil of Siberian Fir) which contains bornyl ester is distilled from the fresh leaves. The oil is used in the treatment of bronchitis, as an expectorant, and as a stimulant in treating rheumatism.

***A. squamata*** Mast. (Flaky Fir). Tibet and S. China. The wood is used for house construction in Tibet.

***A. veitchii*** Lindl. (Veitch's Silver Fir). Mountains of Japan. A yellowish, strong elastic wood, used for building, boxes, roof shingles, utensils, machine parts and paper pulp. A. shikokiana is used similarly.

Abietine – Pinus sabiniaria.

Abili – Rhus vulgaris.

Abiurana – Ecclinusa balata.

Abode – Cissus petiolata.

Aboila – Caralluma edulis.

Abrasin Oil – Aleurites montana.

Abrin – Abrus precatorius.

ABROMA Jacq. Sterculiaceae 2 spp. Asia to Australia.

***A. augustum*** (L.) L.f. (Cotton Abroma, Ramie sengat). Malaysia. The bark yields a good cordage fibre, similar to jute. The juice is used in some parts of India to treat indysmenrrhea.

***A. fastuosa*** R. Br.=A. mollis. (Kapas hantoo, Lawe). Malaysia. A shrubby tree, the bark of which yields a fine yarn which is used as twine. It is also used locally to treat itch.

***A. mariae*** Mart.=Theobroma mariae.

***A. mollis*** DC.=A. fastuosa.

ABRONIA Juss. Nyctaginaceae. 32 spp. N. America.

***A. arenaria*** Menz.=A. latifolia.

***A. fragrans*** Nutt. S.W. of N.America.

***A. latifolia*** Esch. (Seacoast Abronia, Yellow Sandverbena) California.

Both these are herbaceous perennials, the roots of which are eaten by the local people. The juice of A. augustum is used in cases of difficult menstruation.

Abrojo – Condalia abtusifolia.

ABRUS Adans. Leguminosae. 12 spp throughout the tropics.

***Abrus abrus*** (L.) Wight.=Abrus precatorius.

***Abrus precatorius*** L.=Abrus abrus (L.) Wight. A vine found in the tropics of the Old and New World. (Crab's Eye, Indian Liquorice, Jequirity, Rosary Pea). The roots which contain glycirrhcin are used as a substitute for liquorice. The seeds, which are black or red, are used as beads; they contain a toxic albumen which can be fatal if it enters the blood stream. It was used in the treatment of trachoma, and by the Chinese to reduce fevers.

***Abrus pulchellus*** Wall. India to S. Australia. The roots are used as a liquorice in Malaya.

Abscess Root – Polemonium reptans.

Absinthe – Artemisia absinthium.

Absinthe, Algerian – Artemisia barrelieri.

Abu Beka – Gardinia lutea.
Abu Beka Resin – Gardinia lutea.
Abui – Pouteria cairnito.
Abura – Mitragyna stipulosa.
ABUTA Aubl. Menispermaceae. 35 spp. S. American tropics.
***A. candellei*** Triana. and Planch.=A. rufescens.
***A. rufescens*** DC=A. rufescens Aubl.
***A. rufescens*** Aubl.=A. candellei Triana. and Planch.=A. rufescens DC. Guiana. A vine the roots of which are a source of white pareira (pereira) which is used medicinally.
ABUTILON Mill. Malvaceae. 100+spp. throughout the tropics and subtropics.
All economic species yield a fibre from the stem which is used in various parts of the world for cordage, sails and other coarse strong fabrics.
***A. angico*** Mart.=Piptadenia rigida.
***A. angulatum*** Mast.=Abutilon intermedium
***A. asiaticum*** G. Don. A tropical shrub.
***A. asiaticum*** Guill and Pierre=A. muticum.
***A. avicennae*** Gaertn.=A. theophrasti Medik. (Button Weed, Butter Print, Chingma, Velvet Weed). An annual herb found throughout the tropics. It is cultivated in China yielding China Jute; sometimes called Indian Mallow.
***A. bedfordianum*** St. Hil. A shrub from Brazil.
***A. graveolens*** Sweet. Tropics. Cultivated especially in Russia for the seeds which yield an edible oil. A. indicum, and A. muticum are used similarly.
***A. indicum*** (L.) Sweet. (Country Mallow, Indian Abutilon). Indian, Malaysia, Philippines. A leaf extract used as a disinfectant on wounds etc. while the bark and roots are diuretic.
***A. intermedium*** Hochst.=A. angulatum. A tropical shrub especially used in Madagascar.
***A. jacquinii*** G. Don.=A. lignosum (Cav.) G. Don. A shrub of the West Indies, Texas and Mexico where it is used especially.
***A. lignosum*** (Cav.) G. Don.=A. jacquinii.
***A. muticum*** Sweet.=A. asiaticum Guill. and Pierre. Small tree. The seeds are eaten by nomads of the Sudan. See A. graveolens.
***A. octocarpum*** F.v. Muell. Australia. The fibre used by the aborigines.
***A. oxycarpon*** F.v. Muell. S. America especially Brazil; Australia. Little cultivated for fibre to manufacture sails.
***A. polyandrum*** G. Don. A shrub, found especially in India and Ceylon.
***A. theophrasti*** Medik=A. avicennae.
***A. triquetrum*** Sweet.=A. trisulcatum.
***A. trisulcatum*** (Jacq.) Urban=A. triquetrum Sweet.=Sida trisulcata Jacq. C. America. Used in Mexico for fibre. The crushed leaves are used against cancer of the mouth, and taken internally against asthma.
***A. venosum*** Lem. See A. oxycarpum.
Abutilon, Indian – Abutilon indicum.
Aby – Talinium cuneifolium.
Abyssinian Banana – Musa ensete.
Abyssinian Finger Grass – Digitaria abyssinica.
Abyssinian Intermediate Barley – Hordeum irregulare.
Abyssinian Mustard – Brassica carinata.
Abyssinian Myrrh Tree – Commiphora abyssinica.
ACACIA Mill. Leguminosae. 750–800 spp. Tropics and sub-tropics.
***A. abyssinica*** Hochst. Trop. Africa. Somali gum from stem.
***A. acuminata*** Benth. (Raspberry Acacia). W. Australia. Wood used for fuel and charcoal, and weapons. Flowers a potential source of perfume.
***A. albicans*** Kunth.=Pithecellobium albicans (Kunt.) Benth.
***A. albida*** Delile. N. Africa. Tree. A gum Arabic (Gomme de Senegal) extracted from the stem. The bark is used for tanning, and the leaves and young shoots fed to camels.
***A. aneura*** F. v. Muell. (Mulga Acacia). S. Australia. The ripe seeds are eaten by the aborigines. The wood is used for making spear throwers, and the leaves are fed to livestock.
***A. angico*** Mart.=Piptadenia rigida Benth.
***A. angustissima*** (Miller) Kuntze.=Mimosa angustissima Miller. (Prairie Acacia). C. America. In Mexico, bark used for tanning, and pods eaten.
***A. arabica*** Willd.=A. nilotica Delile. (Babul Acacia). Tropical Africa. The wood is very hard, durable, and resistant to termites and water. In consequence it has been used since the times of the ancient Egyptians for presses, implement handles, houses and furnitures, and wheels. The pods and bark (Bablah, Neb-Neb) are used for tanning, and dyeing fabrics

yellow, while the young bark yields a fibre. The leaves and fruits are a food for stock. The stem gives a gum (Amrad, Amrawatti, Brown Barbary Gum) which is light brown. The seeds are fermented with dates to give a beverage.

**A. berlandieri** Benth.=A. tephroloba A. Gray. (Huajillo, Matorral, Membre). W.Texas and Mexico. Small shrub. The wood is a fuel, and is used for small carpentry.

**A. bidwilli** Benth. Tree. N. Queensland. The roots are cooked and eaten by the aborigines.

**A. binervata** DC.=A. umbrosa Cunningh. (Black Wattle, Twinvine Wattle). Queensland and New South Wales. The wood is tough and light, and is used for farm implements. The bark is used for tanning.

**A. calamifolia** Sweet. See A. binervata.

**A. catechu** Willd. (Catechu, Dark Catechu). E. India. Tree. The wood is digested with hot water to yield Catechu (Cutch, Black Catechu, Pegu Catechu, Black Cutch) which is used for tanning, dyeing fabric (the original khaki cloth), and in toilet preparations. The presence of tannin and polyphenols make it astringent. It is used in the treatment of diarrhoea and throat infections.

**A. cavenia** Bert. (Cavenia Acacia, Espino Cavan). A small tree of Chili. Cultivated as a source of perfume (Cassie Romaine), especially in S. France. The essential oil is produced by petroleum-ether extraction of the flowers.

**A. cibaria** F. v. Meull. Tree. New South Wales, and W. Australia. Seeds eaten by natives.

**A. cinerea** Spreng.=Dichrostachys cinerea (Spreng.) Wight and Arn.

**A. cochliacantha** Humb. and Bonpl. (Cucharitas, Quisache corteño). Mexico into S. America. Used locally for diseases of the bladder.

**A. concinna** DC. (Soap Pod). Small prickly shrub. India. The pods are used for washing fabrics, and the hair. Sometimes used for cleaning silver.

**A. conferta** Cunningh. Australia. A source of perfume.

**A. coulteri** Benth. Mexico. Shrub or small tree. The hard wood is used for tool handles etc.

**A. cunninghamii** Hook. (Bastard Myal, Korwarkul). Queensland and New South Wales. Tree. The wood used for cabinet-making. It is hard and takes a good polish.

**A. dealbata** Link. (Silver Wattle). S. Australia. Tree. The tree is cultivated in Natal for the bark which is used for tanning. It yields a useful Gum Arabic. The flowers and leaves are the "mimosa" used by florists.

**A. decora** Reichb. (Graceful Wattle Acacia). Australia. Tree. Gum eaten by aborigines.

**A decurrens** (Wendl.) Willd. (Black Wattle, Green Wattle Acacia). Australia. Tree. Bark contains about 40 per cent tannin.

**A. dictyophleba** F. v. Muell. Australia. Tree. Ground seeds eaten by natives.

**A. doratoxylon** Cunningh. (Brigalow, Currawong Acacia, Spear Wood). Tree. Australia. The wood is heavy, close-grained and strong. It is used for spears and boomerangs by the aborigines, and is also used for furniture and farm posts and gates. The flowers are a possible source of perfume. The leaves are used for stock feed.

**A. drepanolobium** Harms. (Iluta lyapi). Small tree. E. and N.E. Africa. Produces East African Gum Arabic, for which it is cultivated, especially in Tanzania.

**A. ehrenbergiana** Hayne. Tropical Africa. shrub. Produces a gum.

**A. elata** Cunningh.=A. dealbata.

**A. etbaica** Schweinf.=***A. nubica*** Benth.

**A. excelsa** Benth. (Brigalow, Ironwood). Tree. Queensland. A very tough, close-grained, scented wood, used for cabinet-making.

**A. falcata** Willd.=A. plagiophylla Spreng. (Burra Acacia). New South Wales. Small tree. Used to intoxicate fish.

**A. farnesiana** (L.) Willd. (Sweet Acacia). Tropics. The wood is hard and used locally in Africa for cabinet work, ship-building and agricultural implements. An inferior gum is produced from the stem. The bark is used for tanning, and a black dye, used for making ink and dyeing leather is extracted from the pods. The flowers (Cassie flowers) are a source of an essential oil (Cassie ancienne) used in perfumery, and for which the tree is cultivated in S. France.

**A. fasciculata** Guill. and Perr.=A. tortilis.

**A. fistula** Schweinf.=A. seyal.

**A. giraffae** Willd. (Giraffe Acacia). S. Africa. Small spiny tree. It produces

Cape Gum, and the bark is used for tanning. The ground pods are fed to livestock.

***A. glaucescens*** Willd. See A. dealbata.

***A. glaucophylla*** Steud. Tropical Africa. Small tree. Somali gum is produced by the stem.

***A. greggii*** Gray. (Cati's Claw). S.W. of N. America, Mexico. Small tree. Locally, the seeds are ground into flour, and a gum from the stem is used.

***A. gummifera*** Willd. (Mogador Acacia). N. Africa. Tree. Produces a gum which is dark, sweet, and semi-soluble in water (Mogador gum, Morocco gum). Much is produced locally.

***A. harpophylla*** F. v. Muell. (Bigalow, Sickle-leaf Acacia). Queensland. Tree. A gum is produced from the trunk, and a perfume from the flowers. A brown dye is extracted from the wood. It is used to colour natural fibres.

***A. hemiteles*** Benth. = A. subcoerulea.

***A. hindsii*** Benth. = A. sinaloensis = A. tropicana. (Cornezuelo, Guisache coreño). Mexico. Small tree. The bark is used locally for treating scorpion stings.

***A. homalophylla*** Cunningh. (Curly Yarran, Gidgee Acacia, Spear Wood). Australia. Tree. The tree produces a gum, and the bark is used for tanning. The dark brown, fragrant wood is used for turning small articles and fancy goods.

***A. horrida*** Willd. = A. karroo Hayne. (Allthorn Acacia, Mimosa Thorn, Sweet Thorn). S. Africa. Tree. Grown for hedges and binding sand. The wood is hard and used internally in buildings. A yellow, brittle gum (Cape gum) is produced from the trunk. The bark and pods are used for tanning, and staining it red-brown. The leaves and young pods are fed to livestock.

***A. jacquemontii*** Benth. (Dhakki, Khumbut). India. Shrub. Produces an inferior gum used in fabric printing and paper manufacture.

***A. karroo*** Hayne = A. horrida.

***A. koa*** Gray (Koa Acacia). Hawaii. The hard wood is used locally for making spears, paddles and ukeleles.

***A. laeta*** R. Br. Tropical Africa. Small tree. Source of Gum Arabic. See A. senegal.

***A. leucophloea*** Willd. (Arinj, Hiwar, Nimbar, Reunja, Safea babul). India and Burma. Medium-sized tree. The red wood is used for agricultural implements, carts, wheels, turnery, oil mills, and is a good fuel. It gives an inferior gum which is used as an adulterant for those of a better quality. The bark is used for tanning (it stains the leather black), as a source of fibre for cordage, nets etc.; it is eaten in times of famine, and is used to flavour alcoholic beverages made from sugar cane and palm juice. The young fruits are edible, and the leaves yield a black dye. A. leucophylla is sometimes included in A. pendula.

***A. longifolia*** Willd. See A. cibaria.

***A. melanoxylon*** R. Br. (Black Sally, Blackwood). Australia, especially S. and W. Tree. The timber is valuable, hard and close-grained, and takes a good polish. It is used for fine cabinet work, tool handles, piano parts, billiard tables, boats and casks. It makes a good veneer.

***A. mellifera*** Benth. Tropical Africa. Small tree. A source of Gum Arabic.

***A modesta*** Wall. N. Punjab to lower Himalayas. Yields a gum (Amritsar gum) used in Indian medicine.

***A. myriadena*** Bert. = Serianthes myriadena.

***A. neboueb*** Baill. E. Africa. Small tree. A source of Gum Arabic.

***A. nigrescens*** Oliv. (Chibunge, Knob Thorn, Kopjesdoorn, Muzoo, Nhlope). C. Africa. Tree. A valuable timber. The hard wood polishes well, and is used a great deal for pit props. The leaves are eaten locally when cooked.

***A. nilotica*** Delile. = A. arabica.

***A. oswaldii*** F. v. Muell. (Umbrella Acacia). Australia. Tree. Seeds eaten by natives.

***A. pallida*** F. v. Muell. Australia. Tree. The roasted young roots eaten by the aborigines.

***A. pendula*** Cunningh. = A. leucophylla. (Boree, True Myall). New South Wales, and Queensland. The timber is close-grained, hard, dark and well-marked. It is used for veneers and small fancy goods. The leaves are eaten by livestock.

***A. penninervis*** Sieb. New South Wales and Queensland. It is used by the natives to intoxicate fish.

***A. pinnata*** Willd. (Aila, Arar, Awal, Shemba). India. S. Africa. Semi-climbing shrub. Bark and fruit pulp used for tanning fish nets, and intoxicating fish in India. In S. Africa bark used for a cordage fibre, and for cleaning mirrors when charred. The twigs and roots are used as toothbrushes.

***A. plagiophylla*** Spreng. = A. falcata.

***A. podalyriaefolia*** Cunningh. New South Wales, Queensland. Bark used for tanning.

***A. pycnantha*** Benth. (Golden Wattle). Australia. Source of gum (Australian gum, Wattle gum), and bark used for tanning.

***A. rehmanniana*** Schinz. Tropical Africa. Small tree. Source of Gum Arabic.

***A. rivalis*** J. M. Black. Australia. Tree. Ground seeds and gum eaten by aborigines.

***A. roemeriana*** Scheele. S. U.S.A., Mexico. Valuable honey plant, especially in Texas.

***A. salicinia*** Lindl. See A. aneura.

***A. saligna*** (Labill.) Wendl. See A. decurrens.

***A. samaryana*** Chev. Tropical Africa. Small tree. A source of Gum Arabic.

***A senegal*** (L.) Willd. = A. verek Guill. and Perr. (Gum Arabic Acacia). Tropical Africa. Small tree. The principle source of Gum Arabic (Ghezirah gum, Gomme Blondes, Gomme Blanche, Kordofan, Prickly Turkey, White Senaar). The gum is harvested from December to June, i.e. when the dry winds are prevalent, and graded according to the tear size. The main trading centres are Senegal and Kordofan. The gum is used in confectionery, the manufacture of ink, in thickening artists' paints, printing fabrics, and shining silks and crèpes.

***A. seyal*** Delile = A. fistula Schweinf. (Seyal Acacia). Tropical Africa. Yields a gum similar to Gum Arabic (Sennarr, Suakim, Talca, Talha, Talki).

***A. sieberiana*** DC. Tropical Africa. Tree. Source of soluble gum.

***A. sinaloensis*** Safford = A. hindsii.

***A. spirocarpa*** Hochst. Tropical Africa. Tree. The bark gives a strong fibre which is used locally.

***A. stenocarpa*** Hochst. (Narrowwing Acacia). E. Africa. Small tree. Source of Gum Arabic known locally as Gomme Salobreda.

***A. stipuligera*** F. v. Muell. See A. dictyophleba.

***A. subcoerulea*** Lindl. F. v. Muell. = A. hemiteles. (Blue-leaved Acacia). W. Australia. Tree. Yields a yellow dye.

***A. subporosa*** F. v. Muell. Victoria and New South Wales. Tree. The tough wood is used for agricultural implements and gun stocks.

***A. suma*** Kunz. E. India. Tree. Source of catechu.

***A. tephroloba*** A. Gray = A. berlandieri.

***A. tortilis*** Hayne. = A. fasciclata Guill. and Perr. (Sejal, Talha). N. Africa. Arabia. Source of a gum (Gomme Rouge).

***A. tropicana*** Safford. = A. hindsii.

***A. umbrosa*** Cunningh. = A. binervata DC.

***A. verak*** Guill. and Perr. = A. senegal.

***A. visco*** Lorentz. Argentina. Tree. Wood used for construction work and packing cases.

***A. wrightii*** Benth. (Uña de gato). California to Mexico. Wood used as fuel.

Acacia, Allthorn – Acacia horrida.

Acacia, Babul – Acacia arabica.

Acacia, Blue-leaved – Acacia subcoerulea.

Acacia, Burra – Acacia falcata.

Acacia, Currawong – Acacia doratoxylon.

Acacia, Gidgee – Acacia homalophylla.

Acacia, Giraffe – Acacia giraffea.

Acacia, Greenwattle – Acacia decurrens.

Acacia, Gum Arabic – Acacia senegal.

Acacia, Koa – Acacia koa.

Acacia, Mogador – Acacia gummifera.

Acacia, Mulgar – Acacia aneura.

Acacia, Narrowwing – Acacia stenocarpa.

Acacia, Praire – Acacia angustissima.

Acacia, Raspberry – Acacia acuminata.

Acacia, Sickle-leaf – Acacia harpophylla.

Acacia, Sweet – Acacia farnesiana.

Acacia, Umbrella – Acacia oswaldii.

ACAENA Mutis. ex L. Rosaceae. 100 spp. in the S. Hemisphere, N. to California and Hawaii.

***A. pinnatifida*** Ruiz. and Pav. Chili. Used locally as a astringent.

***A. sanguisorbae*** Vahl. Australia, New Zealand. A leaf infusion is used in New Zealand for kidney and bladder complaints, stomach-ache, and as a salve for wounds and bruises.

Acajou d'Afrique – Entandrophragma cylindricum.

ACALYPHA L. Euphorbiaceae. 450 spp. extending through tropics and subtropics.

***A. evrardii*** Gagnep. (Trà ring) Laos, Vietnam. The leaves and flowers are used to prepare a diuretic medicine.

***Acalypha indica*** L. (Indian Nettle). Tropical Asia and Africa. The plant contains resin, tannin, an alkaloid, and acalyphin.

It is used, fresh or dried as an emetic.

ACANTHOCEREUS (Berger) Britton and Rose. Cactaceae. 8 spp. extending from S.E. U.S.A. to N.E. Brazil.

***A. pentagonus*** (L.) Britt. and Rose = Cereus pentagonus. (Pitahaya Pitahaya, Naranjada). A large cactus, with edible red fruits.

ACANTHOCHITON Torr. Amaranthaceae. 1 sp. in S.W. U.S.A.

***A. wrightii*** Torr. Leaves and shoot eaten by local Indians, fresh or stored.

ACANTHOPANAX (Decne. and Planch.) Miq. Araliaceae. 50 spp. extending throughout E. Asia, Philippines and Malay Peninsula.

***A. ricinifolius*** (Sieb. and Zucc.) Seem. = Kalopanax pictus. Extends from China through Manchuria and Korea to Japan. The wood is pliable and easily worked, and is used in China to make drums.

***A. spinosus*** (L. f.) Miq. Japan, China. Young leaves used as a vegetable in Japan.

ACANTHOPELTIS Rhodophyceae. Gelidiaceae.

***A. japonica*** Okam. A source of agar in Japan. For main uses of agar, see Gelidium.

ACANTHOPHORA Rhodophyceae. Rhodomeliaceae.

***A. specifera*** (Vahl.) Borg. Indian Ocean and Pacific. Eaten in Java and Philippine Islands.

Acanthorrhiza Wendl. = Crysophilia Blume.

ACANTHOSPERMUM Schrank. Compositae. 8 spp. West Indies, to S. America, Galapagos Islands, and Madagascar.

***A. hispidum*** DC. Brazil and Argentine. A bitter aromatic herb used locally as a diuretic, and to stimulate sweating.

ACANTHOSPHAERA Warb. Moraceae. 2 spp. Amazon and Brazil.

***A. ulei*** Warb. (Balsamo). Tree. The latex is used to reduce fever.

ACANTHOSICYOS Welw. ex Hook. f. Cucurbitaceae. 2 spp. in S. tropical Africa.

***A. horridus*** Welw. ex Hook. f. (Narasplant, Narras). Tropical Africa. Shrub with thorns and very long thick roots. It grows on sand-dunes. The fruit, about the size of an apple has a pleasant semi-acid flavour. The seeds (butter pits) are eaten by the Hottentots. The fruit pulp is used to coagulate the solids of milk which has been heated.

ACANTHOSYRIS Griseb. Santalaceae. 3 spp. temperate S. America.

***A. falcata*** Griseb. Argentine, Bolivia, Paraguay. Small tree. The wood is used for making furniture. The fruits are red, like small plums, and eaten locally. They are also made into a liqueur.

ACANTHUS L. Acanthaceae. 50 spp. extending from S. Europe to subtropical and tropical Asia and Africa. They are xerophytic herbs with spiny leaves.

***A. ebracteatus*** Vahl. (Sea Holly). Tropical Asia. The leaves are used in Malaya to make a cough medicine. They are boiled with molasses, sugar, cinnamon, and flowers of Averrhoa.

***A. mollis*** L. (Bear's Breech). S. Europe. The leaves and roots are used as poultices in parts of S. Europe, and a decoction of them is used to treat diarrhoea.

***A. spinosus*** L. S. Europe. Used as a diuretic and astringent. The leaves of this and possibly ***A. syriacus*** Boiss. were used as models for the carvings on Corinthian columns.

Acaroid, Red – Xanthorroea australis.

Accra Copal – Daniella ogea.

Accra Paste – Carpodinius hirsuta.

Aceito de Palo – Abies religiosa.

Aceitunillo – Aexttoxicon punctatum.

Aceitunillo – Cornus excelsa.

Achillea noire – Achillea atrata.

ACER L. Aceracea. 200 spp. N. Temperate regions. All trees.

***A. campestre*** L. (Hedge Maple). Europe to Asia Minor and Algeria. A light brown tough wood, hard to split and elastic. It is used for agricultural implements, small turnery, tobacco pipes, cutlery, balls, and veneers.

***A. circinatum*** Pursh. (Vine Maple). Pacific N. America. The wood is hard, but not strong, used for tool handles and fishing nets.

***A. crataegifolium*** Sieb. and Zucc. (Hawthorn Maple). Japan. The bark is used to make a paste used in paper-making.

***A. dasycarpum*** Ehrh. = A. saccharinum.

***A. eriocarpum*** Michx. = A. saccharinum.

***A. ginnala*** Max. = A. tataricum L. var. ginnala Max. Japan, China. The leaves are used in Japan to make a tea.

***A. macrophyllum*** Pursh. (Broad-leaved Maple). Pacific N. America. Close-grained, soft light wood, with a rich brown colour. Used for tool handles, furniture, and the interiors of buildings.

***A. mayrii*** Schwer. = A. pictum.

***A. mono*** Max.=A. pictum.

***A. negundo*** L.=Negundo aceroides=N. fraxinifolium. (Box Elder). N. America. The wood is not strong, soft, close-grained and light coloured. It is used for packing cases, etc. cheap furniture, and the interiors of houses. The Indians burn the wood during their ceremonies, and use the bark as an emetic. The sap is an occasional source of sugar.

***A. pictum*** Thunb.=A. mayrii=A. mono. China to Japan. The wood is hard and close-grained, creamy-white in colour. It is used in Japan for furniture, building interiors, ship building and the interior of passenger coaches. The wood is also important as a fuel.

***A. platanoides*** L. (Norway Maple). Europe to the Caucasus and Armenia. The wood is hard, heavy and fine-grained. It is used for furniture, wagons, turning, and the manufacture of rifles.

***A. pseudo-platanus*** L. (Plane Tree Maple, Scotch Maple). Europe to Asia Minor. The yellowish-white wood is hard, heavy and fine-grained. As the wood polishes well, it is used for furniture, wood carving, flooring, shoe lasts, violins, etc.

***A. rubrum*** L. (Red Maple). E. N.America to Texas. The wood is not strong, but close-grained, and is used for small turnery, furniture etc. It is also used for wood pulp.

***A. saccharinum*** L.=A. dasycarpum=A. eriocarpum (Silver Maple, White Maple). E. N.America to Florida. The wood is similar to, and has the same uses as that of A. rubrum. The sap is an occasional source of sugar.

***A. saccharinum*** Wagenh=A. saccharum.

***A. saccharum*** Marsh. (Sugar Maple). E. N.America. The wood is close-grained, tough, hard and heavy. It is reddish-brown in colour, and used for flooring, interior finishing of buildings, furniture, ship-building and small articles. The tree is the source of maple syrup which is obtained by tapping the trees. The syrup is used for sweetening a wide variety of foods, and for the manufacture of sugar.

***A. spicatum*** Lam. (Mountain Maple). N.America especially Labrador and Saskatchewan. The dried bark is used as an antispasmodic locally.

***A. tataricum*** L. var. ginnala Max.=A. ginnala.

ACETOBACTER Bacteria. Bacteriaceae.

***A. aceti*** (Kützing) Beijerinck. Ferments alcohol (ethanol) to acetic acid, the main ingredient of vinegar. The source of the alcohol, e.g. malt, wine, apples etc. gives the vinegar its characteristic flavour. It also produces sorbose by fermentation.

***A. ascendens*** (Henneb.) Berg, et al. Produces ethanol by anerobic fermentation.

***A. melanogenum*** Beijerinck. Produces sorbose by fermentation.

***A. oxydans*** (Henneberg) Berg, et al. Oxidizes glucose to gluconic acid during fermentation.

***A. pasteurianum*** (Hansen) Beijerinck. Produces ethanol by anerobic fermentation.

***A. rancens*** Beijerinck. Produces sorbose by fermentation.

***A. suboxydans*** Kluyver and de Leeuw. Produces sorbose by fermentation, and ethanol under anerobic conditions.

***A. xylinoides*** Henneb. Produces sorbose by fermentation.

***A. xylinum*** (Brown) Beijerinck. Produces sorbose by fermentation.

Acha ajan – Hardwickia binata.

Achar – Mangifera indica.

ACHILLEA L. Compositae. 200 spp. in N. Temperate regions. All are perennial herbs.

***A. ageratum*** L. (Sweet Yarrow). S.Europe. The flowers and fruits used to cure stomach complaints.

***A. atrata*** L.=Ptarmica atrata (Achillea Noire, Schafgarbe, Schwarze Schafgarbe). Alps of S.Austria and Italy. Used to prepare a liqueur (Iva Liqueur).

***A. clavenae*** L.=Ptarmica clavenae. (Assenzio ombrellifera, Steinraute). Alpine regions of Austria, extending into Bulgaria. The leaves (Herba Achilleae clavenae, Bittere Garbe, Speik) are used in a wide variety of home remedies.

***A. lanulosa*** Nutt. (Yarrow). W. N.America. The Indians use a pulp of the leaves to poultice swellings and sores on men and horses. An extract of the roots is used as a local anaesthetic, especially against toothache.

***A. millefolium*** L. (Milfoil, Yarrow). Temperate regions of N.hemisphere. It contains an essential oil, and the alkaloid acilleine. The leaves are sometimes used in salads and soups (Grüdonnerstag-Suppe) in Germany, and smoked as tobacco in Sweden. An infusion of the leaves is used against stomach complaints and complaints of the urinary system, coughs, and as an antiseptic. An infusion of the flower heads is used as a tonic, to encourage sweating, and as an aid to menstruation. The oil (Oleum Millefolii) distilled from the leaves, is a tonic and aromatic.

***A. moschata*** Jacq. (Musk Yarrow). Used in Italy and Switzerland to produce a liqueur (Esprit d'Iva, Iva Liqueur, Iva Wine, Iva Bitter). An infusion of the leaves is used as a tonic, to encourage sweating. Its properties are due to the alkaloids and the essential oil cineol.

***A. nana*** L. see A. atrata.

***A. ptarmica*** L. (Sneezewort). Temperate Europe and Asia. It was formerly grown as a sneezing powder, made from the roots, and the flower heads were used in a wide variety of home remedies.

***A. santolina*** L. E.Africa to Iran. Used in N.Africa. An infusion of the leaves is used for chest complaints, as a tonic, and to aid digestion.

***A. sibirica*** Led. Temperate N.Asia. Has a wide variety of medicinal uses in China.

Achiotillo – Alchornea latifolia.

Achocha – Cyclanthera pedata Schrad.

ACHRAS L. Sapotaceae. 4 spp. in tropical America and W.Indies.

***A. mammosa*** L. = Calocarpum mammosum

***A. nigra*** Poir. = Dipholis nigra.

***A. sapota*** (zapota). L. (Sapodilla). S. Mexico to Honduras. Tree. Cultivated for the edible fruits and gum. The fruits are yellow-brown with a sweet rich flavour. The gum (Chicle) is extracted from the stem by tapping and is used in the manufacture of chewing gum. Much is exported to the United States of America, where it was first introduced as a rubber substitute in 1890. Most is collected from wild trees in Guatemala and Yucatan. The latex is boiled to remove about 67 per cent of the water, moulded, and exported. Chicle is being replaced by synthetics in the manufacture of chewing gum.

ACHYRANTHES L. Amaranthaceae. 100, or 3–5 variable spp. Sub-tropics esp. Africa and Asia. Herbs.

***A. alba*** Eckl. and Zeyh. = Achyropsis leptostachya (E. Mey.) Hook. f. S. and S.E. Africa, Asia. Used in China to aid uterine contraction.

***A. aspera*** L. (Prickly Chaff Flower). Tropical Africa, Asia, Australia. The ground roots are used in India to reduce the pain of scorpion stings; the leaves are eaten in Java; the small branches are used as tooth-brushes in Arabia. The whole plant is extracted as a source of salt in the Chad area of Africa, and the ashes yield an alkali used in dyeing.

ACHYROCLINE Less. Compositae. 20 spp. Tropical Africa, Madagascar, and America. Herbs.

***A. flaccida*** DC. S.America. An extract of the plant is used in Brazil as a tonic and stimulant, to reduce fevers, and to combat intestinal worms.

***Achyropsis leptostachya*** = ***Achyranthis alba*** (E. Mey.)

ACKAMA A. Cunn. Cunoniaceae. 3 spp. in E.Australia and New Zealand. Trees.

***A. australiensis*** (Schl.) C. T. White. = A. quadrivalvis. (Rose Alder). See A. paniculata.

***A. muelleri*** Benth. = A. paniculata.

***A. paniculata*** Engl. = A. muelleri. (Brown Alder). Coastal forests of New South Wales and Queensland. The wood is fine, soft, straight-grained and durable, with a pink-brown colour. It is used for plywood, joinery, interior finishings and small articles.

***A. quadrivalvis*** C. T. White = A. australiensis.

Ackawai Nutmeg – Acrodiclidium camara.

ACNISTUS Schott. Solanaceae. 50 spp. Tropical America.

***A. arborescens*** (L.) Schlecht. (Tree Wild Tobacco). Tropical America. Shrub. Fruits made into jellies.

Acoccanthera P. and K. = Acokanthera G. Don.

ACOKANTHERA G. Don. Apocynaceae. 15 spp. from S.Africa through tropical E.Africa to Arabia.

***A. abyssinica*** D. Don. E.Africa and Somaliland. The wood (Lignum Acocantherae) is used for heart complaints, and, locally as an arrow poison.

***A. ouabaio*** Cathel. Somaliland. Tree. A source of the glucoside ouabain used for heart complaints.

***A. venenata*** G. Don. E.Africa. Tree. A decoction of the leaves, and other parts used as an ordeal poison. They contain ouabai.

Acom – Dioscorea latifolia.

Acomat-batârd – Diphlois nigra.

Aconite, Indian – Aconitum ferox.

Aconite, Japanese – Aconitum fischeri.

ACONITUM L. Ranunculaceae. 300 spp. N.temperate regions. Herbaceous perennials. All contain poisonous alkaloids of the aconitin group in the roots.

*A. **anthora*** L. (Helmet flower). S.Europe. The roots (Radix Anthorae) are used against intestinal worms, and to reduce fevers; formerly they were used as an antidote to poisoning by Ranunculus thora. A decoction of the leaves is diuretic, and induces sweating.

*A. **balfourii*** Stapf. Japan. China. Root used by druggists.

*A. **bodinieri*** Lev. = A. fortunei.

*A. **carmichaelii*** Debx. = A. fischeri.

*A. **chasmanthum*** Stapf. Japan. China. Root used by druggists.

*A. **chinense*** Sieb. = A. fortunei.

*A. **deinorrhizum*** Stapf. Japan. China. Root used by druggists.

*A. **ferox*** Wall. (Indian Aconite). Himalayas. Root used medicinally, and produces bikh poison.

*A. **fischeri*** Forbes and Hemsl. = A. carmichaelii. (Japanese Aconite). Japan. Root a source of drug.

*A. **fortunei*** Hemsl. = A. bodinieri = A. chinense. China, S.E.Asia. Root extract used in China to treat bronchitis. The roots macerated in alcohol are used as a poultice for bruises, fractures etc. in Vietnam. The alcoholic extract is very poisonous.

*A. **hemsleyanum*** E. Pritz. China. The roots, mixed with egg-white, are used to treat boils.

*A. **heterophyllum*** Wall. Himalayas. Eaten locally. Supposed to be a tonic, aphrodisiac, and antiperiodic.

*A. **laciniatum*** Stapf. China, Japan. Root used by druggists.

*A. **lycoctonum*** L. = A. vulparia (Yellow Wolf's Bane). Extends from S.E.Europe to Himalayas and China. It contains aconitin, eosacotin, hypacotin, japacotin, mesacotin, and is used as a narcotic in China.

*A. **napellus*** L. (Monkshood). Central Europe. Cultivated for the roots which when dried are used as a heart and nerve sedative, and as a pain-relieving tincture. The roots are collected during and after flowering. The dried leaves and flowers (Herba Aconiti) are used similarly. An infusion of the plant was used to poison criminals.

*A. **volubile*** Pall. See A. lycoctonum.

*A. **vulparia*** Reichb. = A. lycoctonum.

*A. **wilsonii*** Stapf. ex Mott. China. Cultivated locally, and used as a poultice for boils when mixed with egg white. The boiled roots are used as a cure for coughs, acting as a sedative. This drastic cure can be dangerous.

Acorn – Quercus.

Acorn, Duck – Nelumbo pentapetala.

Acorn Gall – Querus rabur.

ACORUS L. Araceae. 2 spp. N.temperate regions, and sub-tropics. Perennial herbs.

*A. **calamus*** L. (Calamus, Sweet Flag, Sweet Root). N.temperate. Collected wild in England, Germany, the Netherlands, and Russia and U.S.A. and cultivated in Burma and Ceylon. The dried roots (Rhizoma calami), which are collected in the autumn, are used for flavouring, to aid digestion, and as a cure for toothache. The powdered root is used in clothing as an insect repellent, and in tooth powders and hair powders. Calamus oil is distilled from the plant and is used in flavouring a wide range of alcoholic beverages, including gin and bitters. These useful properties are due to the presence of the glucoside acorin.

*A. **gramineus*** Ait. Monsoon regions of Asia. The rhizome (Radix Sanley vel Acori, Sanleo-Calamus), is bitter and was used locally medicinally as A. calamus.

ACRIOPSIS Reinw. ex Blume Orchidaceae. 12 spp. in Indo-China, W.Malaysia, New Guinea, and Solomon Islands. Herbaceous perennials.

*A. **javanica*** Reinw. (Sakat bawang). Malaysia, Indochina. Used locally against fevers.

ACROCARPUS Wight and Arn. Leguminosae. 3 spp From India to Malaysia. Trees.

*A. **fraxinifolius*** Wight and Arn. (Pink Cedar, Red Cedar, Shingle Tree). India to Malaysia. The timber is hard with brown and black stripes. It is used for general building and tea cases.

ACROCERAS Stapf. Graminae. 15 spp. in tropical Africa, Madagascar and Indomalaysia. Grass.

*A. **amplectens*** Stapf. = Panicum zizanoides var angustatum (Kioo, Kittiboe, Peto, Soko). Ivory Coast, Senegal, Sudan. An excellent fodder grass.

*A. **macrum*** Stapf. (Nile Grass). Botswana and N.E. Transvaal. A creeping perennial. Cultivated for hay and pasture, especially in Transvaal. A good permanent pasture.

*A. **sparsum*** Stapf. Malaysia. Perennial grass. Used for livestock fodder in Java.

ACROCOMIA Mart. Palmae. 30 spp. Extend through tropical America and W.Indies. Trees.

*A. **crispa*** (H.B.K.) C. F. Baker. = A. fusiformis. (Corojo Palm). W.Indies. A fine fibre (Corojo fibre) which is rather like flax is extracted from the leaves. This is a cottage industry, and the fibre is made into cloth, and ropes etc. The untreated fibres are made into fly swatters (Plumeros de Pita).

*A. **fusiformis*** (Swartz.) Sweet = A. crispa.

*A. **mexicana*** Karw. (Coyal). Central America. The fruits are edible and sold locally. An edible oil is extracted from the fruits (Oil of Coyal).

*A. **sclerocarpa*** Mart. (Macaja, Micauba, Paraguay Palm). W.Indies and tropical S.America. Oil from fruit pulp and kernel is used for making soap, and can be eaten if refined.

*A. **totai*** Mart. (Mbocaya Palm, Totai Palm). C.E.Paraguay. The whole plant has been used by the local Indians since ancient times. The trunk is used for house construction; the leaves yield a fibre; the leaflets are fed to livestock. The fruit pulp is oily and of good food value for stock, while the kernels give an oil for human consumption. The oil is extracted industrially. The kernel oil is used for soapmaking, and eaten when refined. The pulp oil is used locally for soap making.

*A. **vinifera*** Oerst. C.America. The fruits are eaten locally, and are fermented to produce an alcoholic beverage.

ACRODICLIDIUM Nees. Lauraceae. 30 spp. Trop. Am., W.Indies.

*A. **camara*** Schomb. N.E. S.America. The wood is used for general work. The fruits 'Ackawai Nutmegs' are aromatic. They are used as an antispasmodic, and for digestive complaints, especially against diarrhoea and dysentery.

*A. **mahuba*** Sampaio. Brazil. The seeds are a source of a fat (Mahuba Fat).

*A. **puchury*** Mez. Trop. Am. W.Indies. The fruits (Puchurim nuts) are used medicinally.

ACRONYCHIA J. R. and G. Forst. Rutaceae. 50 spp. Malaysia, Australia, and Pacific Islands. Small trees.

*A. **laurifolia*** Blume. = Cyminosma resinosa DC. = Jambolifera resinosa Lour. Cochin China. The bark is used to caulk fishing boats, and the roots as a fish poison. The leaves contain an essential oil. When young they are used for flavouring, while the older ones are used to perfume baths.

*A. **odorata*** (Lour.) Baill. = Cyminosma odorata = Jambolifera odorata. Trop. Asia. Leaves used as a condiment in Malaysia, and against dengue fever in Annam.

*A. **resinosa*** Forst. Indo-China. Bark used for caulking boats, and the roots for stupifying fish.

ACROSTICHUM L. Pteridaceae. 3 spp. Throughout tropics in mangrove swamps.

*A. **aureum*** L. (Marsh Fern). Throughout tropics. Young leaves eaten by natives of Borneo and area, while the dried fronds are used for thatch.

*A. **meyerianum*** Hook. f. = Stenochlaena tenuifolia.

ACTAEA L. Ranunculaceae. 10 spp. N. Temp.

*A. **racemosa*** Gilib. non L. = A. spicata.

*A. **spicata*** L. = A. racemosa. (Black Cohosh, Black Snakeroot, Bugbane). Temp. Europe and Asia. A tincture is used against pulmonary tuberculosis, muscular rheumatism, dropsy, angina pectoris, whooping cough, hysteria, and chorea.

ACTEPHILIA Blume. Euphorbiaceae. 35 spp. China, Malaysia, and trop. Australia.

*A. **excelsa*** Muell. Arg. Trop. Asia. Small tree. Used locally to prepare a beverage.

ACTINEA Juss. Compositae. 15 spp. N.America.

*A. **biennis*** Gray. S.W. N.America. Herbaceous perennial. Root cortex chewed as a chewing gum substitute by local Indians.

*A. **odorata*** Gray. S.W. N.America and Mexico. An infusion of the flowerheads is drunk locally as a beverage.

*A. **richardsonii*** Hook. S.W. N.America, Mexico. Herbaceous perennial. The root cortex is chewed as a chewing gum substitute by local Indians.

ACTINELLA Pers. = ACTINEA Juss.

ACTINIDIA Lindl. Actinidiaceae. 40 spp. E.Asia. Woody vines.

***A. arguta*** (Sieb. and Zucc.) Max. = A. callosa.

***A. callosa*** Lindl. = A. arguta = A. kolomikta. (Kolomikta Vine). Manchuria, N.China, N.Korea. The fresh fruits (Tara) are the size of a plum and are eaten in China. They are dried (Kismis) in Siberia and eaten during the winter, in cakes, etc.

***A. chinemsis*** Planch. The fruits are the size of a large plum. They are eaten in China, and used in preserves and jams.

***A. kolomikta*** (Max. and Rupr.) Maxim. = A. callosa.

***A. polygama*** (Sieb. and Zucc.) Maxim. (Silver Vine). Japan, China, Manchuria. In Japan, the leaves are eaten boiled, and the fruits are eaten salted.

***A. rubricaulis*** Dunn. W.China. The plants are cultivated for their fruits in China.

ACTINOMYCES Harz. Actinomycetaceae. Widespread.

***A. candidus*** (Krassilnikov) Wakesman. Used in the manufacture of Gruyère cheese.

***Actinophora fragrans*** R. Br. = Schoutinia ovata.

ACTINORHYTIS Wendl. and Drude. Palmae. 2 spp. 1 Malaysia, 1 sp. Solomon Islands. Trees.

***A. calapparia*** (Blume) Wendl. and Drude. = Areca calapparia = Ptychosperma callapparia = Seaforthia callapparia. (Djambe sinagar, Gallappa palm, Pinang calapa, Soonhari). Malayan archipelago. Seeds (Djeboon sarilong) chewed in Java like betel nuts, and are used in Sumatra for the treatment of scurvy.

ACTINOSTEMMA Griff. Cucurbitaceae 1 sp. India to Japan.

***A. lobatum*** Maxim. (Gokizuru). China and Japan. Scrambling vine. Seeds contain 21·5 per cent of an edible, semi-drying oil (Gokizuru oil).

***Actoplanus canniformis*** K. Schum = Donax canniformis.

Adam's Banana – Musa paradisiaca.

Adam's Needle – Yucca filamentosa.

ADANSONIA L. Bombacaceae. 10 spp. Throughout tropics. Trees.

***A. digitata*** L. (Baobab, Cream of Tartar Tree, Ethiopian Sour Gourd, Monkey Bread). Tropical Africa. The fruit pulp is eaten, as it is acid it is used to coagulate rubber latex, and used by the nomadic tribes of E.Africa to coagulate milk. The burned pulp is used by natives to fumigate insects on domestic animals. The pulp is also used as a local remedy against fever and dysentry. The seeds are eaten alone, or mixed to form a gruel with millet. They also contain an oil (Baobab, Fony, Reniala) which is eaten having a pleasant taste. The oil is also used in soap manufacture. The oil has S.G. 0·9198, Sap. Val. 190·5, Iod. No. 67·5.

***A. gregorii*** Muell. (Bottle Tree, Gourd Tree). Australia. The seeds are eaten by the aborigines. The soft wood, and pits at the bases of the branches hold water which is used by the aborigines and animals during the dry season.

***A. madagascariensis*** (Madagascar Baobab). Madagascar. The fruits are eaten locally, and the bark is a source of fibre.

Add-add – Celastrus serratus.

Adder's Tongue – Ophioglossum vulgatum.

Adder's Tongue, Yellow – Erythronium americanum.

Aden Senna – Cassia holosericea.

ADENANDRA Willd. Rutaceae 25 spp. S.Africa. Shrubs.

***A. fragrans*** (Sims.) Roem and Schult. = Diosma fragrans Sims. Cape region of S.Africa. Cultivated by the natives who make a tea from the leaves.

ADENANTHERA L. Leguminosae 8 trop. Tropical Asia, Australia, Pacific. Trees.

***A. microsperma*** Teijsm. and Binn. Java. Cultivated for its durable wood. It is hard, heavy, yellow-orange in colour, and resistant to insects and decay. The wood is used for heavy timbers in houses, bridges etc. and the bark is used for tanning.

***A. pavonina*** L. (Circassian Tree, Sandal Bead Tree, Zumbic Tree). India, S.E.China, and Malaysia etc. The wood (Red Sandal Wood) is hard and red. It is used in India for house timbers and furniture. An extract of the wood yields a red dye. The bark is used for washing clothes and hair. A decoction of the leaves is used against rheumatism and gout. The seeds are made into necklaces etc., and used as a flux for soldering gold ornaments.

ADENIA Forsk. Passifloraceae. 100 spp.

Extend from tropical and sub-tropical Africa to Malaysia. Vines.

***A. cissampeloides*** (Planch.) Harms. = Ophiocaulon cissampeloides. Guinea. The crushed plants are used as a fish poison.

***A. cordifolia*** (Bl.) Engl. = Modecca cordifolia. (Areuj batalingbingan). Indonesia. Stems are used as twine. As they are water resistant, the twine can be used under water.

***A. singaporeana*** (Mast.) Engl. = Modecca singaporeana. (Akar saoot, Saootan). Malaya. In Sumatra the inner bark is made into twine for fishing nets. An extract of the roots is used locally against ringworm.

ADENIUM Roem. and Schult. Apocynaceae. 15 spp. Tropical and sub-tropical Africa; Arabia.

***A. boehmianum*** Schinz. N.Nigeria to S.W.Africa, extending to E.Africa. A decoction of the plant is used locally for heart complaints.

***A. coetaneum*** Schum. See A. boehmianum.

***A. honghel*** A. DC. (Bouron, Honghel, Kaurane). Tropical Africa. Shrub. The whole plant contains the toxic glycoside adeniine, which paralyses the heart. In the east of the area the powdered roots are used to stupify fish, and the juice to poison arrows. In Senegal and the French Sudan, a decoction of the flowers and flower stalks is used as an ordeal poison.

***A. multiflorum*** Klotzsch. Tropical Africa. Herb. The powdered plant is used in parts of the Transvaal to stupify fish.

***A. obesum*** De Wild. (Desert Rose, Mdagu). E.Africa. Stems and roots used to stupify fish and as an arrow poison.

***A. somalense*** A. DC. Used in Somaliland as A. honghel.

***A. speciosum*** Fenzl. Tropical Africa. Herb. Used by natives as an arrow poison.

Aden Myrrh – Commiphora erythraea.

ADENOCALYMMA Mart. ex. Meissn. Bignoniaceae 40 spp. Tropical America.

***A. alliaceum*** (Lam.) Miers. = Bignonia alliacea. Tropical America. Vine. Used against intestinal worms.

***A. punctifolium*** Mart. S.America. Tree. The roots are used locally as a cure for asthma.

***Adenocrepis javanica*** Blume. = Baccaurea javanica.

ADENOPHORA Fisch. Campanulaceae. 60 spp. Temp. Eurasia. Perennial herbs.

***A. communis*** Fischer. Temp. Eurasia. Roots are cooked and eaten in Japan. Cultivated.

***A. latifolia*** Fischer. (Broadleaf Ladybell). Siberia. Roots cooked and eaten.

***A. polymorpha*** Ledeb. China, Russia, Siberia. The roots contain saponin, and are used medicinally in China.

***A. verticillata*** Fisch. Japan. Roots eaten by the Ainu.

***Adenopus*** Benth. = Lagenaria Ser.

Aden Senna – Cassia holosericea.

ADENOSTEMMA J. R. and G. Forst. Compositae. 30 or possibly 5–6 spp. Tropical America, Tropical and S.Africa. 1 spp. pantropical. Herbaceous perennial.

***A. viscosum*** Forst. Tropical Africa and Tropical Asia. A blue dye is extracted from the plant.

ADENOSTOMA Hook. and Arn. Rosaceae. 2 spp. California.

***A. sparsifolium*** Torr. (Redshank Chamise). California. Shrub, or small tree. Wood used by local Indians as arrow heads.

***Adenostylis*** Blume = Zeuxine Lindl.

ADHATODA Mill. Acanthaceae. 20 spp. Tropical Africa and Tropical Asia.

***A. vasica*** Nees. (Malabar Nut Tree). Tropical India. Small shrub. A decoction of the leaves is expectorant, and is used to relieve bronchitis. They are also boiled with sawdust to give a yellow dye. The wood which is carved into beads, also gives a good charcoal for gunpowder.

ADIANTUM L. Adiantaceae. 200 spp. Cosmopolitan, esp. tropical America. Herbaceous, perennials.

***A. aethiopicum*** L. Tropical Africa. Used by natives in S.Africa as a cure for coughs, and an extract of the rhizome is used as an abortive.

***A. capillus-veneris*** L. (Maidenhair Fern). N.temperate. Used to stimulate menstruation, and to prepare a decoction against respiratory complaints.

***A. pedatum*** L. (Maidenhair Fern). N.America and E.Asia. The rhizome contains tannin, a volatile oil, and a bitter principle. It was used by the Indians, and is now used more widely as a demulcent and expectorant.

***A. philippense*** Culantrillo. Philippines. A decoction of the fresh leaves is used for stomach complaints, dysentery, and as a diuretic. A syrup or decoction of the fronds is used to stimulate menstruation.

***A. tenerum*** Sw. Tropical America. A decoction of the plant is used by the natives to stimulate menstruation.

ADINA Salisb. Naucleaceae. 20 spp. Tropical and subtropical Africa and Asia. Trees. All species mentioned here are used for general construction work, especially for the manufacture of native canoes.

***A. cordifolia*** (Willd.) Benth. and Hook. f.=Naudea cordifolia. (Haladra, Haidua, Karam). Moist Himalayas, India, and Pakistan. The juice is used as an insecticide and to kill maggots in sores. An astringent extracted from the roots is used to treat dysentery and diarrhoea.

***A. fagifolia*** (T. and B.) Val.=Nauclea fagifolia T. and B. (Kajoo lasi). Malaysia. Light yellow wood.

***A. microcephala*** Hiern. (Mingerhout, Redwood, Wild Oleander). C. and S.Africa. One of the chief timbers from Malawi. A light wood with a very dark grain

***A. minutiflora*** Val (Geroonggoong, Katoo lobang). Sumatra. A dark honey-yellow wood of soft texture.

***A. polycephala*** Benth. var. ***aralioides*** Miq.= Nauclea aralioides. (Kajoo koonjit, Ki anggrit). Malaysia. A yellow wood. A brew from the bark is used as a stimulant.

***A. rubella*** Hance.=A. rubescens. India, Malaysia. The wood is very durable when it is buried in the ground.

***A. rubescens*** Hemsl. = A. rubella.

ADINANDRA Jack. Theaceae. 80 spp. E. and S.E. Asia, Indonesia.

***A. integerrima*** Anders. Burma, Indo-China, Malacca. Tree. The wood is red and is used locally to make tool handles, and carts.

***Adinobotrys*** Dunn.=Whitfordiodendron.

Adjab Butter – Mimusops djave.

Adjar – Maerua rigida.

Adjeran – Coleus aromaticus.

Adjerem – Haloxylon tetrandus.

Adju Mahogany–canarium mansfeldianum.

ADONIS L. Ranunculaceae. 20 spp. Throughout N.temperate.

Adlay – Coix lachryma – jobi.

***A. aestivalis*** L. (Pheasant's Eye, Summer Adonis).

***A. microcarpa*** L.

***A. vernalis*** L. (Spring Adonis). The 3 spp. are distributed throughout Europe and N.Asia. They contain the glucoside adeniine. The dried plant is used as a heart stimulant.

Adonis, Spring – Adonis vernalis.

Adonis, Summer – Adonis aestivalis.

Adouaiba – Caralluma edulis.

Adzuki Bean – Phaseolus angularis.

AECHMEA Riuz. and Pav. Bromeliaceae. 150 spp. S.America, W.Indies. Epiphytic herbs.

***A. magdalenae*** (André) Standl.=Ananas magdalenae André. C. and S.America. A tough cordage fibre is produced from the leaves.

AEGERITA Pers. ex. Fr Tuberculariaceae 10 spp. Widespread.

***A. webberi*** Faw. (Brown Whitefly Fungus). A fungus which parasitises Citrus Whiteflies.

AEGICERAS Gaertn. Mysinaceae. 2 spp. Throughout Tropics.

***A. corniculatum*** Blanco.=A. majus.

***A. majus*** Gaertn. India through Malaysia to Australia. Shrub. The hard black wood is made into small articles. A fish poison is extracted from the bark.

AEGIPHILA Jacq. Verbenaceae 160 spp. Tropical America, W.Indies.

***A. elata*** Steud. (Guairo Santo). Tropical America. Shrub. The leaves are used to treat dysentery, diarrhoea and ulcers.

AEGLE Corrêa ex. Koen. 3 spp. Indomalaysia. Rutaceae.

***A. barteri*** Hook.=Afraegle paniculata.

***A. marmelos*** (L.) Corr. (Bael fruit, Bengal Quince, Bilva, Siriphal). N.India, Pakistan, lower Himalayas. Tree. The tree is cultivated. The wood is light yellowish, hard, and takes a good polish. It has a wide range of uses from house construction to tool handles and combs. The fruit is hard shelled with a soft aromatic pulp which is eaten and used in making drinks. The pulp is also used to treat dysentery and dyspepsia, and as a soap substitute. A hair oil (Marmelle Oil) is derived from the rind, and a yellow dye used in calico printing. A distillation of the flowers is used as scent. The twigs are used as tooth brushes, and with the leaves yield fodder for livestock. The bark gives an adhesive gum. The tree is sacred to the Hindus.

AEGOPODIUM L. Umbelliferae. 7 spp. Europe, temperate Asia.

***A. podagraria*** L. (Ground Elder, Dwarf Elder, Goutweed). Temperate Europe and Asia. Perennial herb. The leaves are eaten raw, cooked or in soups, esp. in Germany. The leaves (Herba Podagrariae) are used locally as an infusion taken to alleviate gout and sciatica, and as a diuretic and sedative. Externally the leaves are applied as a fomentation.

AEOLANTHUS Mart. Labiatae. 50 spp. Tropical and subtropical Africa. Perennial herbs.

***A. heliotropioides*** Oliv. Herb used to flavour soup locally.

***A. pubescens*** Benth. Cultivated as a soup flavour, and as a cure for intestinal worms.

AEROBACILLUS Prazm. Bacteriaceae. No. of spp. variable.

***A. polymyxa*** (Prazm.) Migula. Used in the production of Buna Rubber. Produces 2,3,-butanediol.

AEROBACTER Prazm Bacteriaceae. No. of spp. variable.

***A. aerogenes*** (Prazm.) Migula. See A. polymyxa.

AERVA Forsk. Amaranthaceae. 10 spp. Temperate and tropical Africa and Asia.

***A. lanata*** Juss. Tropical Africa and Asia. Shrub. Eaten cooked in E.Africa, and as a curry in India.

***A. tomentosa*** Lam. N. Africa, extending into the tropics. The soft hairy spikes are used to stuff pillows and saddles in the Sudan. The roots are used for toothbrushes and as a medicine for horses and camels.

AESCHYNOMENE L. Leguminosae. 150 spp. Tropics and sub-tropics.

***A. americana*** L.=A. glandulosa=A. floribunda. (Hierba rosario, Huevo de rana, Pega-pega). Tropical America. Small shrub. Forage crop.

***A. aspera*** L. (Shola Pith Plant). Tropical Asia. Shrub. The light pith is used in the manufacture of sun helmets.

***A. cristata*** Vatke. See A. elaphroxylon.

***A. elaphroxylon*** (Guill. and Perr.) Taub.= Herminiera elaphroxylon. Shrub. Tropical Africa. The wood is very light and durable. It is used locally as floats of fishing nets.

***A. floribunda*** Mart. and Gal.=A. americana.

***A. glandulosa*** Poir.=A. americana.

***A. grandiflora*** L.=Sesbania grandiflora.

***A. indica*** L. (Kat sola, Kuhilia). India and Pakistan. Small annual shrub. A very light wood used for sun helmets, rafts, small bridges, pads of elephant harness. It is used to make charcoal for gunpowder, and to fire pottery.

***A. spunulosa*** Roxb.=Sesbania aculeata.

***A. uniflora*** E. Mey. S.Africa, India. Perennial herb. Used in India as A. indica.

AESCULUS L. Hippocastanaceae. 1 S.E. Europe, 5 spp. India and E.Asia, 7 spp. N.America. Trees.

***A. arguta*** Buckl. (Ohio Buckeye, Western Buckeye). W. to S.W. N.America. The fruit is ground to produce an emetic by the local Indians, and is used to stupify fish.

***A. californica*** (Spach.) Nutt. (Californian Buckeye). Seeds eaten by local Indians after much boiling to remove the bitter principle.

***A. chinensis*** Bunge. China through Vietnam. Bark used to stupify fish.

***A. glabra*** Willd. (Ohio Buckeye, Foetid Buckeye). E.United States of America. The soft light wood is used for turning and other small articles. Artificial limbs are made from the wood, and it is occasionally used for pulping.

***A. hippocastanum*** L. (Horse Chestnut). Balkans, through N. Middle East, and Caucasus to Himalayas. The tree is cultivated for its timber, which is soft, light but not durable. It is used for furniture, boxes etc. and for charcoal. An extract of the flowers is used as a cure for rheumatism. The seeds are boiled to remove tannins and ground to make a flour which is mixed with wheat or rye flour during scarcity. An extract of the seeds relieves haemorrhoids, backache, rheumatism and neuralgia. An extract of the bark is used to reduce fevers, as a tonic, narcotic and to remove intestinal worms.

***A. indica.*** (Camb.) Hook. (Bankhor, Hanudun) N.India to Himalayas. Oil from the seeds used to cure rheumatism, and to relieve stomach complanits in horses.

***A. octandra*** Marsh. (Sweet Buckeye, Yellow Buckeye). E.United States of America. Uses as A. glabra.

***A. pavia*** L.=Pavia rubra. (Red Buckeye). E.United States of America. A paste of the seeds with flour used to stupify fish.

***A. turbinata*** Blume. (Japanese Horse Chestnut) China, Japan. Soft light, close-grained yellow wood used in Japan for small wooden articles and house interiors. Extract of the seeds used to treat discharging eyes of horses.

AETHUSA L. Umbelliferae. 1 spp. Europe, N.Africa, W.Asia. Herb.

***A. cynapium*** L. (Dog Parsley, Dog Poison, Fool's Parsley). A decoction of the plant is used as a sedative and for a wide range of intestinal infections, especially in children.

***A. mutellina*** (L.) St. Lag.=Ligusticum mutellina.

AETOXICON Ruiz. and Pav. Aetoxicaceae. 1 sp. Chili. Tree. The wood is pale red-brown, and durable. It is used for general carpentry.

Affun Yam – Dioscorea cayennensis.

AFRAEGLE (Sw.) Eng. Rutaceae. 4 spp. W.Africa.

***A. paniculata*** (Schum. and Thonn.) Engl.= Aegle barteri=Limonia warneckei. Seeds source of edible oil rather like palm oil.

AFRAMOMUM K. Schum. Zingiberaceae. 50 spp. Tropical Africa. Perennial herbs. Seeds of all spp. mentioned here are used as condiments.

***A. angustifolium*** Schum E.Africa, and islands of Indian Ocean. Madagascar Cardamon.

***A. hanburyi*** Schum. Cameroons. Cameroon Cardomon.

***A. malum*** Schum. East African Cardamon.

***A. melegueta*** (Rose.) Schum. W.Africa. Grains of Paradise.

African Almond – Brabejum stellatifolium.

African Arrowroot – Canna edulis.

African Bark – Crossopteryx kostschyana.

African Beech – Faurea saligna.

African Blackwood – Dalbergia melanoxylon.

African Bowstring Hemp – Sanseverinia senegambica.

African Breadfruit – Treculia africana.

African Cedar – Juniperus procera.

African Ebony – Diospyros mespiliformis.

African Elemi – Boswellia frereana, Canarium schweinfurthii.

African Grains of Selim – Xylopia aethiopica.

African Grenadille Wood – Dalbergia melanoxylon.

African Kino – Pterocarpus erinaceus.

African Linden – Mitragyna stipulosa.

African Locust – Parkia africana.

African Mahogany – Afzelia africana, Detarium senegalense, Dumoria heckelii, Heretiera utilis, Khaya senegalense Kiggelaria africana, Tarrietia utilis.

African Mammee Apple – Mammea africana.

African Millet – Eleusine coracana.

African Myrrh – Commiphora africana.

African Oak – Chlorophora excelsa, Lophira alata, Oldfieldia africana.

African Padauk – Pterocarpus soyauxii.

African Peach Bitter – Nauclea esculentus.

African Pepper – Xylopia aethiopica.

African Rosewood – Pterocarpus erinaceus.

African Sandalwood – Baphia nitida.

African Sausage Tree – Kigelia africana.

African Starapple – Chrysophyllum africanum.

African Teak – Oldfieldia africana.

African Tragacanth – Sterculia tragacanth.

African Valerian – Fedia cornucopiae.

African Walnut – Lovoa klaineana.

African Whitewood – Enantia chlorantha.

AFROLICANIA Mildbraed. Chrysobalanaceae. 1 sp. Tropical W.Africa.

***A. elaeosperma*** Mildbr. Fruits (Mahogany Nuts) yield an oil (Po-Yak) used to make paints and varnishes.

***Afrormosia*** Harms.=Pericopsis.

Afunu – Laportea aetuana.

AFZELIA Sm. Leguminosae. 14 spp. Tropical Africa and Asia. Trees.

***A. africana*** Sm. (African Mahogany). Tropical Africa. The hard durable timber is used for furniture-making, and locally for construction work. The natives of the Sudan use the burnt pods for soap.

***A. bijuga*** (Colebr.) Gray.=Intsia amboinensis. (Fiji Afzelia, Molucca Ironwood). Pacific Islands. The strong durable wood is used locally for construction work and small carpentry.

***A. bipinaensis*** Harms. (Boanga, Mupwenge). Tropical Africa. The heavy durable wood is orange-brown, and is used locally for general construction work.

***A. cochinchinensis*** Pierre. (Có ca té, Tè Kha). Thailand, Indo-china. A decoction of the bark is used locally as a tonic for horses.

***A. cuanzensis*** Welw. (Mkongomwa, Mubapi, Pod Mahogany). Tropical Africa, esp. S. of the area. The wood is light-brown and streaked. It is used for ornamental work such as panelling and furniture. The seeds are made into necklaces, and the trunks are used for canoes. In Zambia, an infusion of the roots is used for inflammation of the eyelids. The bark and roots are used in performing native magic.

*A.* ***galedupa*** Backer. (Kajoo galadoopa). Celebes, Borneo. A resin from the damaged bark (Galadoopa) is used locally to improve the scent of other resins.

*A* ***grandis*** Hort. ex Loud. = Erythrophleum guineënsis.

*A.* ***javanica*** Miq. (Ki djoolang, Djoolang). Indonesia. The beautiful reddish wood is used locally for handles of spears etc., and walking sticks.

*A.* ***pachyloba*** Harms. (Mkokongo, Sifusifu). Tropical W.Africa. The wood is used for cabinet work.

*A.* ***palembanca*** Baker. = Intsia bakeri. (Ironwood, Malacca Tac). Malaysia. The light-brown wood is hard and durable. It is used for construction work, housebuilding etc. A brown or yellow dye, used for dyeing mats and cloths is extracted from the wood.

***Aganope floribunda*** Miq. = Derris thyrsiflora.

AGAPANTHUS L'Hérit. Alliaceae. Perennial herbs. 5 spp. S.Africa.

*A.* ***umbellatus*** L'Hérit. A decoction of the tubers is used locally to treat intestinal and heart complaints.

AGAPETES D. Don. ex G. Don. Ericaceae. 80 spp. E.Himalayas through to Malay Peninsula.

*A.* ***saligna*** Benth. and Hook. E.India. Shrubs. The leaves are used locally to make a tea.

Agar (Agar-agar) – Ahnfeltia spp., Gelidium spp., Graciliaria spp., Pterocladia spp.

Agar, African – Hypnea spicifera.

Agar, Macassar – Eucheuma muricatum.

Agar, Sacharin – Anfeltia plicata.

AGARICUS L. ex Fr. Agaricaceae. Widespread. The number of spp. is variable. There is some confusion between this genus and the closely-related Psalliota. The fruit-bodies of the species mentioned are eaten locally.

*A.* ***appendiculatus*** Bull. = Psathyrella candolleana.

*A.* ***arvensis*** Schäff. = Psalliota arvensis.

*A.* ***bitorquis*** (Quél.) Sacc. Sahara.

*A.* ***campestris*** L. = Psalliota campestris.

*A.* ***canarii*** Jungh. = Oudemansiella canari.

*A.* ***candicans*** Schäff. = Pholiota praecox.

*A.* ***candidus*** Bres. = Clitocybe candida.

*A.* ***coffea*** Briganti = Pleurotus neopolitanus.

*A.* ***decastes*** Fr. = Clitocybe multiceps = Lycophyllum aggregatus. Europe and Japan.

*A.* ***djamor.*** Fr. = Crepidotus djamor.

*A.* ***fastidiosus*** Pers. = Russula foetens.

*A.* ***geotropus*** Bull. = Clitocybe geotropa.

*A.* ***incrassa*** Sow. = Russula foetens.

*A.* ***infundibuliformis*** Schäff. = Clitocybe infundibuliformis.

*A.* ***luzonicus*** Graff. Philippines.

*A.* ***marginatus*** Batch. = Pholiota marginata.

*A.* ***nebularis*** Batsch. ex Fr. = Clitocybe nebularis.

*A.* ***ostreatus*** Jacq. = Pleurotus ostreatus.

*A.* ***praecox*** Pers. = Pholiota praecox.

*A.* ***pratensis*** Scop. = Psalliota arvensis.

*A.* ***procerus*** Scop. = Lepiota procera.

*A.* ***ostreatus*** Jacq. = Pleurotus ostreatus.

Agarobilli – Caesalpinia brevifolia.

Agassei – Cleome arabica.

AGASTACHE Clayt. in Gronov. Labiatae. 30 spp. N.America, Mexico; C.Asia through to China. Herbs.

*A.* ***anethiodora*** (Nutt.) Britt. (Giant Hyssop). S.W. U.S.A. The leaves are used by the local Indians to make a tea.

*A.* ***neomexicana*** (Briq.) Stand. New Mexico. The leaves are used locally to flavour food.

AGATHIS Salisb. Araucariaceae. 20 spp. Indochina through W.Malaysia to New Zealand. Trees.

*A.* ***alba*** (Lam.) Foxw. = Dammara alba. (White Dammar Pine). Malay Archipelago. Source of a copal (Boed Copal, Pontuik Copal, Macassar Copal, Manilla Copal, Singapore Copal). The copal is obtained from the living tree or from the surrounding soil in a semi-fossilized condition. The fossilized copal is of a better quality. It is used to make varnishes.

*A.* ***australis*** (Lamb.) Steud. (Kauri). New Zealand. The wood is used for general building, especially for marine work. A resin (Kauri Copal, Kauri Gum) is extracted from the trunk, and in a semi-fossilized condition from the ground. The best grade (Range Gum) is from the ground, and the poorest grade (Bush Gum) is extracted from the trunk. The gum is used for making varnish, and as a substitute for amber in making small ornaments etc.

***A. dammara*** Rich. = A. loranthifolia. Malaysia and Burma. A copal is extracted from the trunk, and is used to make varnish. It is variously called Batjan, Boed, Borneo, Celebes, Fidji, Bauuan, Macassar, Malengthet, Menedo, Molucca, Oli, Singapore Copal.

***A. labillardieri*** Warb. (Damar putih Kauri). The yellow-brown wood, which is soft, yet strong, but will not survive contact with the ground. It is used for boat building, furniture etc., and paper pulp. It is the source of a copal (Damar putih).

***A. lanceolata*** Panch. New Caledonia. A resin (Dammar Resin, Kauri Resin) is extracted from the trunk.

***A. loranthifolia*** Salisb. = A. dammara.

***A. microstachys*** Warb. Malaysian Archipelago. Timber used locally for general construction work.

***A. moorei*** (Lindl.) Warb. See A. lanceolata.

***A. ovata*** (Moore) Warb. See A. lanceolata.

***A. palmerstonii*** F. v. Muell. (Kauri). Australia. The timber is used for general construction work.

***A. vitiensis*** (Seem.) Warb. = Dammara vitiensis Seem. Fiji. Source of resin (Resin of Fiji).

***Agathopyllum aromaticum*** Willd. = Ravensara aromatica.

AGATHOSMA Willd. 108 spp. S.Africa.

***A. apiculata*** Mey. S.Africa. Tree. An essential oil is extracted from the leaves. It is used against stomach complaints and to induce perspiring during fever.

***Agati*** Adans. = Sesbania.

***Agatophyllum*** Comm. ex Thou. = Ravensara.

AGAVE L. Agavaceae. 300 spp. S.United States of America to tropical S.America. Large semi-woody perennials. Pulque and aquamiel are fermented beverages made from the juice of various spp. Mescal and tequile are distilled from the fermented stems and leaf bases. The drug mecogenin, which occurs in the leaves of several species can be converted into cortisone.

***A. americana*** L. (Century Plant). Originated in Mexico, now widely cultivated as an ornamental in N.America, Africa, and Europe. The fibres are used for drawnwork in the Azores.

***A. atrovirens*** Karw. = A. latissima. Mexico. Used to make pulque and mescal.

***A. cantala*** (Haw.) Roxb. (Cantala). Cultivated in India, and S.E. Pacific area, although probably originated in Mexico. The fibre from the leaves is softer and weaker than sisal, and is rough due to the retting in sea water. It is used for cordage and is called Bombay Aloë Fibre, Bombay Hemp, Catala Fibre, Cebu Maguey, Manila Maguey.

***A. complicata*** Trel. Mexico. Used to produce pulque and aquamiel. A liqueur (Comiteca) is distilled from a pulp of leaves and fermented raw sugar cane juice.

***A. deserti*** Engelm. S.W. United States of America, and N. Mexico. The leaf bases are roasted as a food by local Indians.

***A. falcata*** Engelm. Mexico. Fibre (Ixtle, Tampico) retted from leaves.

***A. fourcroyoides*** Lemaire. (Henequen Agave). Mexico. Cultivated for fibre (Henequen, Yucatan Sisal). The fibre is coarse, and swells when wet. It is used especially for ropes etc. and binder twine.

***A. funkiana*** Koch and Bouché (Jaumave Loguguilla). Mexico. Cultivated for the light red-yellow fibre which is exported for brush-making.

***A. gracilispina*** Engelm. Mexico. Used for fibre (Ixtle), and pulque.

***A. heteracantha*** Zucc. = A. lecheguilla (Lecheguilla). S. United States of America, Mexico. The fibre from the leaves (Ixtle, Jaumave Ixtle) is used for bag-making, cordage, and brush-making.

***A. kirchneriana*** Berger. A fine fibre (Maguey delgado) is retted from the leaves, and mescal is distilled from it.

***A. latissima*** Jacobi. = A. atrovirens.

***A. lecheguilla*** Torr. = A. heteracantha.

***A. lespinassei*** Trel. See A. zapupe.

***A. letonae*** F. W. Taylor. (Letona). C.America. A fine soft fibre is manufactured from the leaves. It is used for cordage, and sugar and coffee sacks. Cultivated in El Salvador since ancient times. Fibre called Letona, Salvador Henequen, Salvador Sisal.

***A. lophantha*** Schiede. Mexico. Source of fibre (Ixtle de Jamauve, Rula Ixtle).

***A. mapisaga*** Trel. Mexico. Source of pulque.

***A. melliflua*** Trel. Mexico. See A. mapisaga.

***A. palmaris*** Trel. See A. tequilana.

***A. pesmulae*** Trel. See A tequilana.

***A. pseudo-tequilana*** Trel. See A. tequilana.

***A. quiotifera*** Trel. (Maguey ceniso).

Mexico. The flower stalks are chewed like sugar cane. (Quiote). Also made into aquameil.

***A. rigida*** Mill. = A. sisalana.

***A. schottii*** Engelm. (Amole). Arizona. Used as a soap substitute.

***A. sisalana*** Perrine. = A. rigida. (Sisal Agave). Mexico. Cultivated in Old and New World tropics for fibre (Bahama Hemp, Sisal Hemp, Yaxci) which is retted from the leaves. The fibre is rather coarse, and is used for general cordage, which swells when wet.

***A. tecquilana*** Weber. (Chino Azul, Mescal) Mexico. Cultivated for the production of mescal. A fibre of moderate value is also produced, but rarely as a bi-product of mescal production.

***A. utahensis*** Engelm. S.W. United States of America. Leaf bases and roots roasted and eaten by local Indians.

***A. victoriae-reginae*** Moore. S.E. United States of America. Source of a strong, though short fibre (Noa).

***A. virginica*** L. (False Aloë). E.United States of America. An extract of the root was used by the Indians to cure stomach complaints.

***A. weberi*** Celo. (Maguaey liso). Mexico. Used to make aquamiel. A mediocre fibre is extracted and used for cordage by natives.

***A. zapupe*** Trel. (Zapupe Azul, Zapupe Estopier). Mexico. A fine, but weak fibre has been extracted from the plant by the natives for a long time.

Agave, Henquen – Agave fourcroyoides.

Agave, Sisal – Agave sialana.

Agbute – Bosquiea angolensis.

AGELAEA Soland. ex Planch. Connaraceae. 50 spp. Tropical Africa, Madagascar, S.E.Asia, Malaysia. Trees.

***A. emetica*** Plan. Madagascar. Oil extracted from the leaves is used to induce vomiting.

***A. lamarckii*** Plan. Madagascar. An extract of the leaves is used in the treatment of gonorrhoea.

***A. villosa*** Pierre. W.Africa. An extract of the leaves is used against dysentery.

AGERATUM L. Compositae. 60 spp. Tropical America. Perennial herbs.

***A. conyzoides*** L. Tropical America. An oil is distilled from the leaves, and used to treat ulcers, wounds, etc., and to reduce fever. It is also used as a purgative. A root extract is used for stomach complaints.

Aggar Atta – Aquilaria agallocha.

Agil – Dysoxylum malabaricum.

Agiri – Roweopsis obliquifoliata.

AGLAIA Lour. Meliaceae. 250-300 spp. China, Indomalaya, tropical Australia, Pacific Islands. Trees, or shrubs.

***A. acida*** Koord. and Val. (Langsatan). Java. The aromatic fruits are eaten locally.

***A. argentea*** Blume. (Jalongan sesak). Malay Archipelago. The wood is heavy, hard and red. It is used for construction work and furniture.

***A. baillonii*** Pierre. = Lepidaglaia ballonii Pierre. = Epicharis baillonii Pierre. (Huynh dhàn, Santal rouge de Cochin China, Sdau phnom). Cambodia. The scented wood is reddish, and used for general construction work and making small wooden articles. The bitter leaves and flowers are eaten in salads. An extract of the bark is used against intestinal worms.

***A. diepenhorstii*** Miq. = A. odoratissima.

***A. eusideroxylon*** Koord. and Val. (Langsat lootoong, Lootoong). Indonesia. The dark red wood is hard and tough. It is used for heavy constructional work.

***A. ganggo*** Miq. (Ganggo, Kajoo wole, Lotong). Malaysia. Yields a heavy durable hard wood. It is resistant to insects, and is used locally for general construction work, and making household utensils.

***A. harmsiana*** Perk. Philippines. Fruit contains an edible pulp.

***A. odorata*** Lour. S. China to Indo-China. The plant is cultivated for the flowers which are used in China to scent tea; as a perfume for clothing in Java, and as a body perfume.

***A. odoratissima*** Blume. = A. diepenhorstii. (Ooka-ooka, Pantjal kidang, Tangloo). Java. Flowers yield an essential oil used in perfumery.

***A. oligantha*** C. DC. (Manatan). Throughout tropical Asia. Pulp of fruit is edible.

***A. samoënsis*** Gray. Pacific Islands. The flowers are used to scent the coconut oil used to scent the hair and body. They are also used for decoration. The wood is used in constructing native houses.

***A. silvestris*** Merr. (Langsa ootan, Lasa sooroo). Indonesia. The roots are used in the preparation of local soup. The wood is very hard and heavy and is used to make native spears.

Agopia – Pseudospondias microcarpa.

Agota – Laportea podocarpa.

AGONANDRA Miers. ex Benth. and Hook. f. Opiliaceae. 10 spp. Mexico to Tropical S.America. Trees.

***A. brasiliensis*** Benth. and Hook. Brazil. The seeds yield a semi-drying oil (Páo Manfin Oil, Ivory Wood Seed Oil). Semi-hydrogenation of the oil gives a rubber substitute, while sulphonation gives a red dye.

AGOSERIS Rafin. Compositae. 9 spp. N.W. America; 1 sp. temperate S.America.

***A. aurantiaca*** (Hook.) Greene. = Macrorhynchus troximoides (Orange Agoseris). W. N.America. The leaves were eaten by local Indians.

***A. villosa*** Rydb. N.W. United States of America. The latex was used as a chewing gum by the local Indians.

AGRIMONIA L. Rosaceae. 15 spp. N.Temperate. Herbs.

***A. eupatoria*** L. (Agrimony). N.Temperate. An extract of the leaves is a very old remedy for intestinal worms, to encourage menstruation, reduce bleeding, as an astringent, diuretic, and liver complaints. The leaves and stems yield a bright yellow dye.

Agrimony – Agrimonia eupatoria.

Agrimony, Water – Bidens tripartitus.

AGRIOPHYLLUM Bieb. Chenopodiaceae. 6 spp. C.Asia. Herbs.

***A. gobicum*** Bunge. (Soulkhir). Mongolia. Seeds an important local food, and plant as a whole used as forage crop.

***Agrocybe aegerita*** (Bruganti) Kühner = Pholiota cylindrica.

AGROPYRON Gaertn. Graminae. 100–150 spp. Temperate regions.

***A. cristatum*** (L ) Gaertn. (Crested Wheatgrass). Eurasia. Perennial. Makes excellent hay and forage. Introduced into America, especially Rocky Mountains.

***A. intermedium*** (Host.) Beauv. (Intermediate Wheatgrass). S. Europe. An excellent, drought-resistant fodder grass.

***A. repens*** (L.) Beauv. (Couch Grass, Quack Grass, Scutch, Twitch). N.temperate. Perennial. Can be a serious weed in pasture and arable land. The rhizome is used in Central Europe. When dried and roasted it is used as a coffee substitute. It is dried and ground as a flour, and used as a decoction to treat coughs and as a diuretic.

***A. smithii*** Rydb. (Western Wheat Grass). N. and C. Great Plains of America. A hardy, tough, quick-spreading species, yielding excellent forage. It is used in erosion control.

***A. spicatum*** (Pursh.) Scribn. and Sm. (Bluebunch Wheat Grass). N.W. America. Perennial. Drought-resistant, and a good forage grass for range grazing.

***A. trachycaulum*** (Link.) Malte. (Slender Wheat Grass). North America, except S.E. Perennial, though short-lived. A good forage grass, but will not stand close-grazing.

AGROPYRUM Roem. and Schult. = Agropyron.

AGROSTEMMA L. Caryophyllaceae. 2 spp. Eurasia. Annual herbs.

***A. githago*** L. (Corn Cuckle). Europe. The leaves are pickled with bacon, as an emergency food. The seeds are poisonous.

AGROSTIS L. Graminae. 150–200 spp. Cosmopolitan, esp. N.temperate. Perennial grasses.

***A. alba*** L. (Fiorin, Redtop, White Bent). Europe, Asia, and N.America. A good pasture, and hay crop, esp. on acid and wet soils.

***A. tenuis*** Sibth. = A. vulgaris.

***A. vulgaris*** With. = A. tenuis. (Rhode Island Bent). Throughout N. Temperate. An important pasture grass used for pasture and hay.

AGROSTISTACHYS Dalz. Euphorbiaceae. 8–9 spp. India and Ceylon to W.Malaysia.

***A. borneensis*** Becc W.Malaya, N.Borneo. Shrub. The stems are used by the Dyaks to blacken their teeth.

Aguacate – Persea gratissima.

Ague Tree – Sassafras alibidum.

Aguedata – Picramnia pentandra.

Aguire – Astrocaryum aculeatum.

Ahidrendra – Stenotaphrum madagascariense.

Ahila-lowa – Celtis brieryi.

Ahipa – Pachyrrhizus ahipa.

AHNFELTIA Phyllophoraceae. Rhodophycae.

***A. concinna*** J. Ag. (Limu Akiaki). Pacific. Eaten boiled by the natives of Hawaii. Often boiled with octopus.

***A. plicata*** (Huds.) Fries. var ***tobuchiensis*** Kanno and Matsub. N.E. Asia. Source of

Sachalin Agar which is produced commercially in Russia and Japan.
Aiauma – Couroupita peruviana.
Aila – Acacia pinnata.
AILANTHUS Desf. Simaroubaceae. 10 spp. Asia and Australia. Trees.
***A. altissima*** (Mill.) Sw. = A. glandulosa.
***A. fauveliana*** Pierre. (Cam tong houng). Indo-china, Thailand. The resin, from the trunk, and the bark of the root are used locally to treat dysentery, and other abdominal complaints.
***A. glandulosa*** Desf. (Tree of Heaven). China. The wood is off-yellow in colour, hard and takes a good polish. It is used for general construction work, furniture and small articles. A decoction of the bark of both stem and root contains a cardiac depressant, and an astringent. It is used to cure vaginal discharges, dysentery, intestinal worms, epilepsy, asthma, and heart complaints.
***A. glandulosa*** Desf. W. China. The tree is cultivated for the leaves to feed silkworms (Attacus cynthia).
***A. malabarica*** DC. (But, Thanh thât). India, N.Vietnam. Cultivated in N.Vietnam. The leaves yield a black dye used to colour silk and satin. A resin (Mattipaul) is extracted from the trunk and used for incense in Hindu temples. In India, the leaves are used against intestinal worms, and as a tonic.
***A. vilmoriana*** Dode. See A. glandulosa.
Ain – Terminalia tormentosa.
Aini – Artocarpus hirsuta.
AIPHANES Willd. Palmae. 40 spp. Tropical America. Trees.
***A. minima*** (Gaertn.) Burret. (Coyor, Coyose). Tropical America. The fleshy parts of the fruits are eaten locally.
Aipi – Manihot esculenta.
***Aira gigantea*** Steud. = Gynerium sagittatum.
Air Potato – Dioscorea bulbifera.
Aisar – Callicarpa lanata.
AIZOON L. Aizoaceae 15 spp. Africa, Mediterranean, Orient, Australia. Herbs.
***A. canariense*** L. N.Africa. Leaves eaten by desert tribes of N. Africa.
Ajinud – Carum roxburghianum.
Ajooras – Alstonia acuminata.
Ajowan Oil – Carum copticum.
AJUGA L. Labiatae. 40 spp. Throughout temperate areas. Herbs.
***A. chamaeptys*** (L.) Schreber. = Chamaepitya vulgaris Spenn. = Teucrium chamaepitya. (European Bugle, Ground Pine). Europe, W. Asia. An infusion of the leaves is diuretic and is used to stimulate menstruation. It is also used to cure dropsy, intermittent fevers, rheumatism, and gout.
***A. decumbens*** Thunb. E.Asia. Used by the Chinese to relieve pain.
***A. reptans*** L. = A. reptens. (Bugle). Europe, Asia, and introduced sp. in N.America. An infusion of the leaves is a mild narcotic. It is used against rheumatism, bleeding of the lungs, jaundice, and obstructions of the liver and spleen.
***A. reptens*** Host. = A. reptans.
Akala – Rubus hawaiiensis.
Akam Yam – Dioscorea latifolia.
Akarba – Anastatica hierochuntica.
Akar benang tikkoos – Vitis papillosa.
Akar bidji – Roucheria griffithiana.
Akar darik-darik – Vitis lanceolaria.
Akar fatima – Labisia pumila.
Akar gamih – Vitis adnata.
Akar gerip pootih – Parameria barbata.
Akar kapajang – Hodgsonia macrocarpa.
Akar kara – Spihanthes acmella.
Akar kawil-kawil – Uncaria sclerophylea.
Akar kemenjan hantoo – Oldenlandia recurva.
Akar laka – Dalbergia junghuhrui.
Akar maloong – Coptosapelta griffithii.
Akar ritjak – Vitis compressa.
Akar rittom – Parinari laurinum.
Akar saoot – Adenia singaporeana.
AKEBIA Decne. Lardizabaliaceae. 5 spp. E. Asia. Vine.
***A. lobata*** (Houtt.) Decne. Japan. The plant is used locally. The fruits are eaten, and the leaves made into a tea. The young shoots are made into baskets after bleaching.
Akee Apple – Blightia sapida.
Akekenke – Celtis adolphi-frederici.
Akelwak – Rhus vulgaris.
Akoko – Euphorbia lorifera.
Akor-Ogea – Daniellia oliveri.
Akoria – Rhus wallichii.
Al Moulinouquia – Corchorus trilocularis.
ALAFIA Thou. Apocynaceae. 20 spp. Tropical Africa. Tree.
***A. perrieri*** Jum. (Alafy). W. Madagascar. A latex from the stem is used as a soap locally.
Alafy – Alafia perrieri.
Alatani – Coccothrinax martiniensis.
Alambi – Hugonia platysepala.
Alang-Alang – Imperata arundinaceae.

ALANGIUM Lam. Alangiaceae. 17 spp. Tropical Africa, Madagascar, Comoro Islands, China, S.E.Asia, Indomalaya, E.Australia.

***A. salviifolium*** Wangerin. Tropical Asia. Small tree. A decoction of the bark is emetic, and widely used medicinally in India.

Alant – Inula helenium.

ALARIA Alariaceae. N.Atlantic.

***A. esculenta*** (Lynb.) Grev. (Bladder Locks, Murlins). Used as a food in W. Scotland and Ireland.

Alaska Cedar – Chamaecyparis nootkatensis.

Albahia – Ocinum micanthrum.

Albarco – Cariniana pyriformis.

***Albertinia candolleana*** Gardn. = Vanillosmopsis erythropappa.

ALBIZIA Durazz. Leguminosae. 100–150 spp. Warm Old World. Trees.

***A. antunesiana*** Harms. (Kawizi, Masakwe, Muriranyge). Tropical Africa. A durable wood used for general construction work, furniture, plywood. It is resistant to termites. Locally it is used for making drums, mortars etc. The roots are eaten ceremonially in Malawi.

***A. anthelmintica*** Brongn. (Mucenna Albizia). E.Africa. The bark (Cortex Mucennae) is used against intestinal worms in Abyssinia.

***A. brownii*** Walp. Tropical Africa. The hard, dark wood is used for general construction work. The trunk is the source of a gum (Nongo Gum).

***A. fastigiata*** Oliv. (Flat crown Albizia). Tropical Africa, Madagascar. The wood is soft, and used for general construction work, wheels, and general farm implements; it is also used as a source of dye. A gum is extracted from the trunk. The seeds are used to make a sauce.

***A. gummifera*** (Gmel.) C. A. Sm. (Mupambangoma, Mutango). Tropical Africa. A light soft, yellowish wood is used for general construction work.

***A. leblecoides*** Benth. = A. odoratissima.

***A. lebbek*** Benth. (Lebbek). Tropical Asia to Australia. Tree cultivated and timber exported (East Indian Walnut). It is easily worked and durable, and is used for general construction work, furniture and waggons. The bark is used to treat boils, and the leaves and seeds to treat diseases of the eye.

***A. moluccana*** Miq. Malaysia. Wood used for making matches, tea-boxes and paper pulp. The bark is used for tanning. The trees are grown for shade in tea plantations.

***A. odoratissima*** Benth. = A. leblecoides. (Fragrant Albizia). Philippine Islands. Bark used to make a fermented beverage (Basi).

***A. procera*** Benth. (Tall Albizia, Tee-coma). Tropical Asia, N. Australia. The dark, close-grained timber is used for general construction work, esp. heavy agricultural machinery like sugar-cane crushers.

***A. saponaria*** (Lour.) Blume. Philippine Islands and N.Guinea. The bark has detergent properties and is used locally as a soap.

***A. toona*** F. M. Bailey. (Red Siris). Australia. The wood is highly decorative having red and yellow streaks. It is durable and termite resistant, and used for internal woodwork in houses and furniture-making.

***A. versicolor*** Welw. ex. Oliv. (Kibangi, Tidneful). Tropical Africa. Used by natives to make quivers.

***A. welwitschii*** Oliv. Tropical Africa. The heavy light-red wood is tough and strong. It is used for turnery, wagons, inside woodwork for houses and small articles.

Alboquillo de Campo – Thalictrum hernandezii.

Alcaparro – Cassia emarginata.

Alce Wood – Cordia sebestina.

ALCHEMILLA L. Rosaceae. 250 spp. Temp; tropical mountains. Herbs.

***A. arvensis*** (L.) Scop. (Parsley Breakstone, Parsley Piett). Europe, N.Africa, Naturalised in N.America. A decoction of the plant is diuretic and used for kidney and gravel complaints.

***A. kuwuensis*** Engl. (Kurwi hume, Lwandula). Tropical C. and W. Africa. Ash is used in the Congo as a treatment for leprosy.

***A. vulgaris*** L. (Lady's Mantle). Temp. Asia, Europe, N. America. An extract of the plant is used as a diuretic, a stimulant to menstruation, to heal wounds and as an astringent.

ALCHORNEA Sw. Euphorbiaceae 70 spp. Tropics. Shrubs, small trees.

***A. cordata*** Benth. = A. cordifolia.

***A. cordifolia*** Muell. = A. cordata. Tropical

Africa. A black dye is extracted from the twigs.

***A. latifolia*** Sw. (Achiotillo). Tropical America. The wood is light, easily worked, but susceptible to decay and insects. The light-brown wood is used for light bent-wood work, matches, toys, boxes etc.

***A. villosa*** Muell. (Ramie Bukit). Malaysia. A tough cordage fibre is obtained from the bark.

Alcock Spruce – Picea bicolor.

Alcornoca – Bowdichia virgilioides.

Alder, Black – Alnus glutinosa.

Alder, Black – Ilex verticillata.

Alder, Brown – Ackama paniculata.

Alder, Buckthorn – Rhamnus frangula.

Alder, European – Alnus glutinosa.

Alder, Hazel – Alnus rugosa.

Alder, Japanese – Alnus japonica.

Alder, Mountain – Alnus viridis.

Alder, Nodding – Alnus pendula.

Alder, Red – Alnus rubra.

Alder, Red – Cunonia capensis.

Alder, Smooth – Alnus rugosa.

Alder, Speckled – Alnus incana.

Alder, Western – Alnus rubra.

Alder, White – Platylophus trifoliatus.

Ale – Hordeum distichon.

ALECTORIA Usneaceae. Lichen.

***A. fremontii*** Tuckerm. Temp. N.America. Used as a famine food by the Indians of California.

***A. jujuba*** (L.) Ach. (Horsehair Lichen). Temperate areas. Used in England as a source of green to brown dye, and is used in perfumery. The Indians of Pacific N. America boil it with the bulbs of Camass. This mixture is then fermented and baked before eating.

***A. nidulifera*** Fr. See A. ochroleuca.

***A. nigricans*** Fr. See A. ochroleuca.

***A. nitidula*** Fr. See A. ochroleuca.

***A. ochroleuca*** (Ehrh.) Nyl. N.Arctic. One of the chief foods of reindeer.

ALECTRA Thunb. Scrophulariaceae. 30 spp. Tropical America, Africa, and Asia. Herbs.

***A. melampyroides*** Benth. S.Africa. Roots and stems yield a yellow dye.

ALECTRYON Gaertn. Sapindaceae. 15 spp. Malaysia to Polynesia. Tropical Australia and New Zealand. Trees.

***A. excelsum*** Gaertn. New Zealand. The wood is durable and elastic. It is used for agricultural implements, interior woodwork and furniture. An oil extracted from the seeds is used as a perfume locally, and is used to treat sores and weak eyes.

***A. macrococcus*** Radlk. Hawaii. The fruits are eaten locally.

Ale hoof – Glechoma hederaceum.

ALEPIDEA La Roche. Umbelliferae. 40 spp. Tropical, and S.Africa. Herbs.

***A. amatymbica*** Eckl. and Zeyh. S.Africa. A bitter resin extracted from the rhizome is used locally for stomach ache, and as a laxative.

Aleppo Pine – Pinus halepensis.

Alerce – Fitzroya cupressoides.

ALETRIS L. Liliaceae. 25 spp. E.Asia; N.America. Herbs.

***A. farinosa*** L. (Colic Root, Star Grass Unicorn Root). E. N.America. The underground parts are dried and are used as a diuretic. N. Carolina, Virginia and Tennessee are the main supplying areas. The leaves are made into an infusion by the local Indians and used to treat colic.

ALEURITES J. R. and G. Forst. Euphorbiaceae. 45 spp. Tropical Asia, Malaysia, Pacific. The seeds are a source of drying oils which give a hard, acid and alkaline resistant cover. They are used to manufacture paints and varnishes.

***A. cordata*** (Thunb.) R. Br. ex Steud. (Japanese Tung Oil Tree). Japan. Cultivated for the oil which is also used for waterproofing fabrics and paper. It has S.G. 0·943–0·940. Sap. Val. 189–196, Iod. No. 148–160. Unsap.=0·4–0·8 per cent.

***A. fordii*** Hemsl. (Tung Oil Tree). China. Cultivated in Old and New World. It was introduced into S.E. U.S.A. in 1905, now it is grown as an orchard crop, producing 140,000 tons of fruit, and 40 million lb. of oil per annum. The oil has been used to waterproof paper and fabrics by the Chinese since ancient times. It is also used in the manufacture of linoleum, printers' ink, Indian ink, tiling fibres board, and brake linings. It is also combined with cobalt, lead, and manganese to make drying agents. The oil has S.G. 0·939–0·949. Sap. Val. 189–195. Iod. No. 157–172. Unsap. 0·4–0·8 per cent.

***A. moluccana*** (L.) Willd.=A. triloba.

***A. montana*** (Lour.) Wils. China, Indo-China. Cultivated for oil (Abrasin Oil,

Mu Oil). S.G. 0·9360–0·9460. Sap. Val. 190·2–195. Iod. Val. 156–167. Unsap. 0·4–0·5 per cent.

***A. triloba*** Forst.=A. moluccana. (Candlenut Oil Tree). Malaysia, Philippine Islands, Polynesia, India, Australia. Source of oil (Bagilumbang Oil, Candlenut Oil, Lumbang Oil). Also used for painting boats as protection against marine boring worms. Also used for illumination and making soap. S.G. 0·920–0·927. Sap. Val. 190–193. Iod. No. 140–164. Unsap. 0·9 per cent.

***A. trisperma*** Blanco. (Banucalang). Philippine Islands. Cultivated in Philippine Islands, Malaya and E.Indies. Produces oil (Bagilumbang Oil, Banucalang Oil). Also used for painting boats.

Alexanders – Smyrnium dus-atrum.

Alexanders – Ligusticum scothicum.

Alexandrian Laurel – Calophyllum inophyllum.

Alexandrian Senna – Cassia acutifolia.

Alfalfa – Medicago sativa.

Alfalfa, Spanish – Medicago sativa.

Alfalfa, Tree – Cysisus pullilans.

Alfalfa, Yellow-flowered – Medicago falcata.

Algaroba – Caesalpina brevifolia.

Algaroba blanca – Prosopus alba.

Algerian Absinthe – Artemisia barrelieri.

ALHAGI Gagnebin. Leguminosae. 5 spp. Mediterranean, and Sahara to C.Asia and Himalayas. Shrubs.

***A. camelorum*** Fisch. (Camel's Thorn). N.African deserts. A sweet exudate from the stem is used as a sweetmeat. A manna from the pods is used as a laxative. It contains a large amount of sucrose.

***A. graecorum*** Boiss. Greece. Fodder for camels.

***A. maurorum*** Medik. N.African desert. Important camel fodder. A manna exuded from the stems. The roots are eaten in times of famine.

***A. persorum*** Boiss. and Buhse. (Kharechitor). Iran. A source of manna. (Tarandjabine). A laxative.

Alibungog – Ehretia philippensis.

Alim – Aquilaria malaccensis.

ALISMA L. Alismataceae. 10 spp. N.Temp. Australia. Herbs.

***A. plantago-aquatica*** L. (Water Plantain). Cosmopolitan. The lower parts of the plant can be eaten if boiled to remove poison. An infusion of the leaves is diuretic and is used in kidney and bladder complaints.

Aliziergeist – Sorbus torminalis.

Alkajoun – Fumaria vaillantii.

Alkali Grass – Puccinellia nutalliana.

Alkali Grass, Nutall – Puccinellia nutalliana.

Alkali Seepweed – Suaeda fruticosa.

Alkanet – Alkanna tinctoria.

ALKANNA Tausch. Boraginaceae. 25–30 spp. S. Europe, Mediterranean to Persia. Herbs.

***A. tinctoria*** (L.) Tausch. (Alkanna) S.E. Europe, Turkey. Cultivated for a red dye (Alkanna, Alkanet) derived from the roots. Occasionally used for dyeing silks and cotton, ointments and liqueurs. It is a specific stain for fats and oils, but has now been replaced by Sudan III and IV.

Alkanna – Alkanna tinctoria.

Alkanna, Syrian – Macrotomia cephalotus.

Alkanna, Turkish – Macrotomia cephalotus.

Alkekengi – Physalis alkekengi.

ALLANBLACKIA Oliv. Guttiferae. 8 spp. Tropical Africa. Trees.

***A. oleifere*** Oliv. C.Africa. Seeds a source of a fat (Kagné Butter) used as a food locally.

***A. stuhlmannii*** Engl. E.Africa. Seeds produce a fat (Mkani Fat) eaten locally. A red dye is produced from the bark.

***Allamanda***=Allemanda.

Alleghany Barberry – Berberis canadensis.

Alleghany Blackberry – Rubus alleghaniensis.

ALLENMANDA L. Apocynaceae. 15 spp. Tropical S.America. West Indies. Shrubs.

***A. blanchetii*** DC.=Plumeria blanchetii. The sap is a cathartic and emetic. Poisonous.

***A. doniana*** Muell. Arg. Brazil. Sap cathartic and emetic.

Allgood – Chenopodium bonus-henricus.

***Alliaria wasabi*** Prantl.=Eutrema wasabi.

***Alliaria offinalis*** Andr.=Sisymbrium alliaria.

Alligator Juniper – Juniperus pachyphlaea.

Alligator Pear – Persea gratissima

ALLIUM L. Alliaceae. 450 spp. N.Temperate. Herbs.

***A. akaka*** Gmel. Temp. Asia. Eaten as food in Iran.

***A. ampeloprasum*** L. (Levant Garlic). Orient, Mediterranean. Bulb eaten locally.

***A. angulare*** Pall. Siberia. Bulb eaten locally esp. in winter.

***A. ascalonicum*** L. (Eschallot, Shallot). Probable origin in W. Asia. Cultivated widely as a garden vegetable. Used for flavouring and pickling; as they rarely produce seeds, propagation is by setts.

***A. bakeri*** Regel=A. chinense.

***A. canadense*** L. (Canadian Garlic, Meadow Leek, Rose Leek, Wild Garlic). E. N.America. Bulbs eaten, boiled or pickled by local Indians.

***A. cepa*** L. (Onion). Probably originated in Persia, now cultivated in Old and New World. (var. ***bulbiferum*** Bailey=var. ***viviparum*** Metz.) – The Top Onion – produces the bulbils at the top of the flower stalk, as well as one at the base. (var. ***aggregatum*** Don.=var. ***solaninum*** Alef.=var. ***multiplicans*** Bailey) – Potato Onion, Multiplier Onion – has more than one bulblet inside a common outer layer of leaf bases. They are eaten raw, boiled, roasted, etc. or dried as an onion salt. Medicinally onions have a wide use. The juice is mildly antiseptic and so the bulbs are used to treat wounds, boils etc.; it is also diuretic and expectorant, so is used to treat dropsy, when mixed with gin, and alone in cough medicines. The juice is supposed to stimulate hair growth, and the outer skin is a source of a dye which was used to stain fabrics. Contains allyl sulphide which gives the scent.

***A. cernuum*** Roth. (Lady's Leek, Nodding Onion). The bulb is very strongly flavoured, so is used for flavouring. They are pickled occasionally.

***A. chinense*** G. Don.=A. bakeri. (Ch'iao T'ou, Rakkyo). China. Cultivated in China and Japan, where much is exported in pickles, and California.

***A. fistulosum*** L. (Cibol, Stone Leek, Welsh Onion). Native, China and Siberia. Cultis vated in Old and New World for leaves which are used in salads.

***A. kurrat*** Schw. (Kurrat, Kurrat baladi, Kurrat nabati). A cultigen grown in Arabia, and Egypt, may be from the time of the Pharaohs. The leaves used in salads, and as a condiment.

***A. ledebourianum*** Schult. Russia to Japan. Cultivated in Japan for the bulbs and leaves which are eaten raw or cooked.

***A. macleanii*** Baker. (Royal Salep). Iran to Afghanistan. Bulb eaten locally.

***A. nipponicum*** Franch. and Sav. Japan. Bulbs used in salads.

***A. obliquum*** L. (Twisted leaf Garlic). Siberia. May be cultivated. Used as a substitute for garlic.

***A. odorum*** L.=A. tuberosum (Chinese Chive, K'iustai). China, Siberia. Cultivated in China as a salad crop, and for flavouring. Seeds used also in China as a tonic and for heart complaints. The Hindus used the seeds in cases of spermatorrhoea.

***A. oleraceum*** L. (Field Garlic). Europe to Siberia. Wild. Occasionally used as seasoning.

***A. porrum*** L. (Leek). Mediterranean Region. Cultivated throughout Old and New World. Bulb and lower part of the leaves eaten, blanched by covering during growth, and boiled.

***A. rubellum*** M. B.C. Asia. Bulbs eaten locally.

***A. sativum*** L. (Garlic). S.Europe to C.Asia. Cultivated throughout the world for flavouring a wide range of foods. The plant, or an oil extracted from it (Oil of Garlic) is an expectorant, used in the treatment of cough, cold, and similar complaints. It is also used in the treatment of toothache, earache, intestinal worms, and to reduce high blood pressure. The juice is used to mend china and glass. Contains allyl-propyl disulphide, diallyl disulphide which gives the characteristic scent.

***A. schoenoprasum*** L. (Chive). Temperate Europe to E.Asia. Cultivated. Leaves used in salads.

***A. scorodoprasum*** L. (Giant Garlic). C.Europe to Asia Minor. Cultivated locally, and used for flavouring food.

***A. senescens*** L. Temperate Europe and Asia. Bulbs eaten in Japan.

***A. sphaerocephalum*** L. (Ballhead Onion). Europe to temperate Asia. Eaten in parts of Siberia.

***A. splendens*** Willd. Japan. Eaten locally, boiled or pickled in vinegar and saké.

***A. tricoccum*** Ait. (Wild Leek). N.America. Bulbs eaten locally.

***A. tuberosum*** Rottl. ex Spreng.=A. odorum.

***A. ursinum*** L. (Bear's Garlic, Wild Garlic). Temperate Eurasia. Eaten locally. Used medicinally to reduce blood-pressure, to treat asthma, and for digestive complaints.

***A. victorialis*** L. (Long root Onion). Temperate Eurasia. Used in Japan to treat colds.

ALLOPHYLUS L. Sapindaceae. 190 spp. Tropics and sub-tropics. Trees or Shrubs.

***A. africanus*** Beauv. (Banda, Bateli, Danda, Imburuyuta). Senegal to Malawi). Wood used locally for general construction work, and leaves used to treat rheumatism and headaches.

***A. lastoursvillensis*** Pellegr. (Basaso, Bolelombo, Nokanga). Congo. An extract from the leaves used locally to treat various chest complaints and to heal wounds.

***A. subcoriaceus*** Bak. f. (Bateli, Kapiapia, Umungwamagi). C.Africa. Wood used locally for construction work. The juice from the leaves is used to treat headaches.

Allspice (Oil) – Pimenta officinalis.

Allspice, Caroline – Calycanthus floridus.

Allthorn Acacia – Acacia horrida.

***Almeida alba*** St. Hil. – Raputia alba.

Almeindina Rubber – Euphoria rhipzaloides.

Almendero – Laplacea curtyana.

Almendo – Terminalia catappa.

Almond – Amygdalus communis.

Almond, African – Brabeium stellatifolium.

Almond, Bitter – Amygdalus communis.

Almond, Country – Terminalia catappa.

Almond, Dwarf – Amygdalus nana.

Almond, Earth – Cyperus esculentus.

Almondette Tree – Buchababia lanzen.

Almond, Hottentot's – Brabeium stellatifolium.

Almond, Indian – Terminalia catappa.

Almond, Java – Canarium amboinense, C. commune.

Almond-leaved Willow – Salix triandra.

Almond Shorea – Shorea eximea.

Almond, Sweet – Amygdalus communis.

Almond, Tropical – Terminalia catappa.

ALNUS Mill. Betulacea. 35 spp. N.Temperate, S. to Assam and Indochina, and Andes. Trees.

***A. arguta*** (Schlecht.) Spach. Mexico. Bark used locally for tanning.

***A. alnobetula*** (Ehrh.) Koch = A. viridis.

***A. cremastogyne*** Burk. W. China. Planted along waterways. Used as fuel.

***A. fruticosa*** Rupr. Alaska, N.E. Asia. Eskimos use the bark to dye reindeer skins.

***A. glutinosa*** (L.) Gaertn. (Black Alder, European Alder). Europe, S.E.Asia and N.Africa. The soft, light, reddish-brown wood is used to make small articles, e.g. wooden shoes, pumps, woodcarving. The bark is used for tanning, giving a red colour. The bark and leaves are astringent and used to stimulate the appetite.

***A. hirsuta*** (Spach.) Rupr. See A. japonica.

***A. incana*** (L.) Moench. (Spickled Alder). Caucasia through Europe to N.America. The light soft wood is used for making small articles, and the bark is used for tanning.

***A. japonica*** (Thunb.) Steud. (Japanese Alder). Japan, Manchuria, Korea. Used in Japan. The wood is red and used for turning and making gunpowder. The bark is used for dyeing.

***A. maximowiczii*** Call. See A. japonica.

***A. nitida*** (Spach.) E. (Himalayan Black Cedar, Kunis, Sharol, Utis). W. Himalayas, to Kashmir. The wood is light, and reddish-white to grey. It is used in construction work, agricultural carts and wheels etc. The bark is used for tanning and dyeing.

***A. oregana*** Nutt. = A. rubra.

***A. pendula*** Matsum. (Nodding Alder). Japan. Wood used for turning and small articles. Inflorescence used for dyeing.

***A. rubra*** Bongn. = A. oregana. (Red Alder, Western Alder). Entire W. United States of America. The wood is very soft and brittle, and used to some extent for making furniture. The Indians made canoes from the hollowed trunks, and the bark as a source of a red dye for fishing nets. The bark is astringent and used to stimulate the appetite, and to treat stomach complaints. The wood is excellent for smoking salmon and halibut.

***A. rugosa*** (Du Roi) Spreng. = A. serrulata. (Hazel Alder, Smooth Alder). E. N.America. Infusion of the bark was used locally as an astringent and emetic for stomach complaints.

***A. serrulata*** Willd. = A. rugosa.

***A. tenuifolia*** Nutt. W. N.America. Bark used by Indians as a source of orange dye.

***A. viridis*** (Chaix.) DC. = A. alnobetula. (Mountain Alder). N.Temperate. Bark used by Eskimoes to stain leather red.

ALOCASIA G. Don. Araceae. 70 spp. Indomalaysia. Herbs.

***A. cucullata*** (Lour.) Schott. (Giant Taro). E.India. Corms eaten locally.

***A. denudata*** Engl. Malaysia. Extract form corms used by locals as an arrow poison.

***A. indica*** (Rosb.) Schott. India, Malaysia. (Alocasia). Cultivated throughout S.E. Asia and Pacific Islands. Used esp. in India for making curries.

***A. macrorrhiza*** (L.) Schott. (Giant Alocasia). Tropical Asia and Australia. Corms eaten locally in some places.

Alocasia – A. indica.

Alocasia, Giant – A. macrorrhiza.

ALOË L. Liliaceae. 275 spp. Tropical and S.Africa; 42 spp. Madagascar; 12–15 spp. Arabia. Large succulents. The drug aloës is extracted from several spp. The leaves are cut across and the juice extracted. This is dried and the drug extracted from the dried juice with water. The active principle is aloin which is used to treat intestinal worms, to encourage menstruation and as a cathartic. It is also used in chemical test for blood in faeces.

***A. abyssinica*** Lam. E.Africa. In the Red Sea area, Mecca (Moka) aloë is extracted. Sold in the crude form in black cakes.

***A. africana*** Mill. See A. ferox.

***A. barbadensis*** Mill. = A. vera.

***A. candelabrum*** Tod. S.Africa. Yields Natal Aloë.

***A. ferox*** Mill. S.Africa. Crude extract blackish; exported and used in veterinary medicine.

***A. horrida*** Haw. = A. ferox.

***A. latifolia*** Haw. S.Africa. The pulped leaf is used locally to treat boils, sores and ringworm. The juice has been used for tanning.

***A. perryi*** Baker. (Socatria Aloë). Socotra. Yields Socotra (Zanzibar) Aloë. Used medicinally.

***A. saponaria*** (Ait.) Haw. (Soap Aloë, White spotted Aloë). S.Africa. Zulus used an infusion of the leaves to treat scours in calves, stomach ailments in chickens, and to remove hair from hides. The leaf pulp is used to treat ringworm.

***A. spicata*** Baker. See A. ferox.

***A. tenior*** Haw. S.Africa. Natives use an extract of the root to treat tapeworms.

***A. vera*** L. = A. barbadensis (Barbados Aloë, Curaçao Aloë). Mediterranean. Introduced to New World Tropics. Cultivated in various parts, esp. Curaçao. Some varieties produce Indian Aloë or Jaffarabad Aloë. Main source of medicinal aloës.

Aloë, Barbados – Aloë vera.

Aloë, Bombay – Agave cantala.

Aloë, Cape – Aloë ferox.

Aloë, Curaçao – Aloë vera.

Aloë, False – Aloë virginica.

Aloë, Indian – Aloë vera.

Aloë, Jaffarabad – Aloë vera.

Aloë, Manila – Agave cantala.

Aloë, Mecca – Aloë abyssinica.

Aloë, Moka – Aloë abyssinica.

Aloë, Natal – Aloë candelabrum.

Aloë, Soap – Aloë saponaria.

Aloë, Socotra – Aloë perryi.

Aloë Tree – Cordia sebestina.

Aloë, Uganda – Aloë ferox.

Aloë, White spotted – Aloë saponaria.

Aloë Wood – Aquilaria agallocha.

Aloë, Zanzibar – Aloë perryi.

Aloore – Xerospermum noronhianum.

ALOPECURUS L. Graminae. 50 spp. Temperate Eurasia; Temperate S.America. Grass.

***A. pratensis*** L. (Meadow Foxtail). Europe, Asia, Caucasus. Perennial. Used for pasture.

***Aloysia citroidora*** Ort. ex Pers. = Lippia citriodora.

ALPHITONIA Reissek ex Endl. Rhamnaceae. 20 spp. Malaysia, Australia, Polynesia. Trees.

***A. moluccana*** T. and B. (Kajoo daoon babalik). Borneo and area. The bark is used locally to make a scent for person and clothes. Chewing the bark is supposed to clear the throat and improve the voice. A paste of bark and leaves is used to treat freckles.

***A. zizyphoides*** (Hook. and Arn.) Gary = Pomaderris zizyphoides. New Caledonia. The wood is greyish-violet, durable and workable. It is used for cabinet making, and interior panelling.

Alpine Ash – Eucalyptus gigantea.

Alpine Dock – Rumex alpinus.

Alpine Fir – Abies lasiocarpa.

Alpine Timothy – Phleum alpinum.

ALPINIA Roxb. Zingiberaceae. 250 spp. Warm Asia; Polynesia. Herbs.

***A. aromatica*** Aubl. = Renealmia domingensis.

***A. chinensis*** Roscoe. China to Vietnam. Rhizomes used locally in China to stimulate circulation of the blood.

***A. conchigera*** Griff. = Languas conchigera. Malaysia. Used esp. in S.Vietnam as a condiment, and to flavour a spirit distilled from rice.

***A. galanga*** (L ) Willd. = Languas galanga (L.) Stuntz. (Greater Galangal, Langwas). Tropical Asia. Cultivated for an essential oil (Essence d'Amali) extracted from the rhizome, and used in perfumery. Rhizomes used to flavour food, and the flowers are eaten raw or pickled.

***A. globosa*** (Lour.) Horan. = Amomum globosum. = Languas globosa. (Mè trè, Ts'ao k'ou jên, Round Chinese Cardamon). Cultivated in China. The seeds are used as a condiment and have a strong, cool aromatic flavour. The seeds and rhizomes are used to cure stomach ailments esp. sickness and diarrhoea.

***A. magnifica*** Rosc. = Nicolaia magnifica.

***A. malaccensis*** (Burm. f.) Rosc. N.India and Malaysia. Source of Essence d'Amali. Rhizome chewed locally like betel nuts.

***A. officinarum*** Hance. = Languas officinarum. (Lesser Galangal, Small Galangal). E. to S.E.Asia. Cultivated in China for the dried rhizome (Galangal). It is used as a condiment. The rhizome contains the resin galangol, and cinerol, eugenol, and sesquiterpenes which make it a relief to flatulence and indigestion.

***A. pyramidata*** Blume. Java to Philippine Islands. The rhizome is used for digestive complaints. An extract from the leaves is used to perfume stimulant baths. The root is a condiment, like ginger. The sap is boiled with sugar and honey, and fermented to make a beverage in the Philippines.

***A. speciosa*** D. Dietr. = Nicolaia magnifica.

Alprose, Rusty-leaved – Rhododendron ferrugineum.

Alrain Root – Mandragora autumnalus.

Alriba Resin – Canarium strictum.

ALSIDIUM Rhodomeliaceae. Red algae.

***A. helminthochorton*** (de la Tour) Kuetz. (Corcasian Moss) Mediterranean. Used against intestinal worms.

Alsike Clover – Trifolium hybridum.

***Alsodeia*** Thou. = Rinorea Aubl.

ALSODEIOPSIS Oliv. Icacinaceae. 11 spp. Tropical Africa. Shrub.

***A. poggei*** Engl. (Bonsoko, Bungwingwi, Dolo). Congo. The roots used locally as an aphrodisiac.

***A. rowlandii*** Engl. (Bungwingwi). Congo. Forests of Nigeria. Used locally as an aphrodisiac.

ALSOPHILA R. Br. Cyatheaceae. A conglomerate genus containing unrelated spp. segregated from ***Cyathea*** by the lack of an indusium.

***A. australis*** R. B. The pith of the young stem contains starch which is eaten by the natives. The older woody parts are made into small articles, e.g. walking sticks.

***A. myosuroides*** Liebm. C.America to S.Mexico. Used by locals to control bleeding from wounds.

***A. rufa*** Fée. C. and S.America. The starchy pith is eaten by the local tribes in times of shortage.

ALSTONIA R. Br. Apocynaceae. 50 spp. Indomalaysia. Polynesia. Trees.

***A. acuminata*** Miq. (Ajooras, Poole batoo). Bali, Java, Moluccas. The wood is heavy and a beautiful yellow colour. It is used for expensive house pillars and boarding. The roots are roasted and used to give palm wine a bitter flavour.

***A. angustifolia*** Wall. (Mergalang). Malaya to the Philippine Islands. The light yellow wood is hard, but warps easily. It is used for construction work and household utensils. A latex is obtained by tapping. It is used to adulterate Gutta Djelutung.

***A. angustiloba*** Miq. See A. spathulata.

***A. batino*** Blanco. = A. macrophylla.

***A. congensis*** Engl. Tropical Africa. The wood is soft and light. It is used locally for furniture, boats, drums and tool handles. A poor latex is used to adulterate rubber.

***A. constricta*** F. v. Muell. (Bitter Bark, Fever Bark). New South Wales, Queensland. The bark contains alstonin which is bitter. It is used to treat intestinal worms and as a tonic and as bitters to flavour drinks.

***A. eximea*** Miq. Sumatra. A latex (Gutta Djelutung) is tapped from the trunk and used to adulterate Gutta Percha.

***A. grandifolia*** Miq. Sumatra. Source of a latex (Gutta Malaboeai) which is used to adulterate Gutta Percha.

***A. macrophylla*** Wall. = A. batino = Echites trifida. Malaysia, Borneo, New Guinea, Philippines. In the Philippines the dried bark is ground into a powder which is used either alone or as an infusion (possibly in wine) to cure dysentery, fever, to encourage menstruation, control bleeding from wounds and as a tonic.

***A. plumosa*** L. New Caledonia. The source of a rarely used latex.

***A. pneumatophora*** Backer (Basong, Bog, Poolai kapoor). Indonesia. The light yellowish wood is used locally for carving. The bark from the pneumatophores is used like cork. A latex from the trunk is used to adulterate Gutta Djelutung.

***A. polyphylla*** Miq. (Djelutung, Mesenteh, Poolai). Indonesia. The light, soft wood is used to make boxcs, and the trunk yields the latex, Gutta djelutung.

***A. scholaris*** R. Br. (Djelutung, Palmira Alstonia, Pulai, Jelutong). Ceylon to Australia. The wood is used to make coffins. The dried bark contains the alkaloids alstonine, ditamine and echitamine. It has been used since ancient times as a tonic and to treat intestinal complaints, including worms. Sold locally

***A. spathulata*** Blume (Pulai paya). Malaysia. The wood from the roots is very light and is used as cork, and for making sun helmets. The latex is used to adulterate Djelutung, and locally to cure skin complaints. The ground bark is used to treat intestinal worms.

Alstonia, Palmira – Alstonia scholaris.

ALSTROEMERIA L. Alstroemiaceae. 50 spp. S.America. Herbs.

***A. haemantha*** Ruiz. and Pav. Chile. Roots a source of starch eaten locally.

***A. ligtu*** L. Chile. The roots a source of starch (Chuño de Concepción) which is sold locally.

***A. revoluta*** Ruiz. and Pav. See A. haemantha.

***A. versicolor*** Ruiz. and Pav. See A. haemantha.

ALTERNANTHERA Forsk. Amaranthaceae. 200 spp. Tropics and sub-tropics. Herbs.

***A. amoena*** (Lem.) Voss. See A. sessilis.

***A. philoxeroides*** (Mart.) Griseb. See A. sessilis.

***A. sessilis*** R. Br. (Racaba). Tropics. The leaves are eaten, usually with fish, in the Congo; eaten cooked in Malaya; and cooked with rice in Indonesia.

ALTHAEA L. 12 spp. W. Europe to N.E. Siberia. Herbs.

***A. officinalis*** L. (Marsh Mallow). Europe, but introd. to N. and S.America. The root (Radix Althaeae) contains 25–35 per cent mucilage, and is used as a treatment of coughs, catarrh. The plant is cultivated, mainly in Holland, France, and Germany. Pasta Althaeae is made by boiling the root with Gum Arabic, and sugar, and mixing it with white of egg. The root is eaten raw in salads and made into a tea. The leaves are also used as an emollient and demulcent. Fibres extracted from the stems and roots have been used in paper manufacture. Sweetmeats were made from the roots.

***A. rosea*** (L.) Cav. (Hollyhock, Garden Hollyhock). China. Cultivated to W. Europe as an ornamental. Flowers are demulcent and diuretic, and is used in curing chest complaints. The darker varieties yield a dye.

ALTINGIA Nor. Altingiaceae. 7 spp. Assam and S.China, to Malay Peninsula, Sumatra and Java. Trees.

***A. excelsa*** Nor. = Liquidambar altingia. The heavy hard, dark red to black wood is used for general construction work in houses. A yellow scented resin (Getah Mala, Rasamala Resin, Rasamala Wood Oil) is extracted from the stem and used locally in perfumery.

***A. gracilipes*** Hemsl. China to S.Vietnam. A latex tapped from the stem is used by the Chinese to treat respiratory ailments.

Alui – Andrographis paniculata.

Alumbi – Julbernardia sereti.

Alumbi-kibile – Macrolobium coeruleum.

Aluta – Vigna multiflora.

Alva Marina – Zostera marina.

Alyceclover – Alysicarpus vaginalis.

ALYSICARPUS Neck. ex Desv. Leguminosae. 25 spp. Warm Africa to Australia. Herbs.

***A. nummularifolius*** DC. = A. vaginalis.

***A. ovalifolius*** (Schumach.) Léon. Tropical Africa. It is used for forage for goats, sheep and horses, in some parts.

***A. rugosus*** (Willd.) DC. Old World Tropics, Australia, W.Indies. Used in Africa as food for stock.

***A. vaginalis*** DC. = A. nummularifolius. (Alyce clover). Old World Tropics. Cultivated for hay and pasture. A late growing legume used for soil-improvement and establishing terraces to conserve soil.

***A. zeyheri*** Harv. and Sond. Tropical Africa. The roots are used locally in some parts against snake bite.

ALYXIA Banks ex. R.Br. Apocynaceae. 80 spp. Madagascar, Indomalaysia. Woody plants.

***A. flavescens*** Pierre. = Gynopogon flavescen Pierre. (Dây sen). Thailand to S.Vietnam

The root is burned ceremonially by the Buddhists.

***A. lucida*** Wall. Madagascar, Malaysia. An infusion of the leaves is used against intestinal worms in Madagascar, and the bark and leaves are used to flavour locally made rum.

***A. madagascariensis*** DC. (Andriam bavifohy). Madagascar. The leaves and bark are used to flavour locally made rum.

Amacet – Tetragastris balsamifera.

Amache – Hyrangea thunberii.

Amada – Curcuma amada.

Amadou – Calvatia lilacina.

Amadou – Fomes fomentarius.

AMANITA (Pers. ex Fr.) Gray. Agaricaceae. 50 spp. Widespread. Many members of the genus are very poisonous, and although some are edible, they should be picked and eaten with great care.

***A. bulbosa*** Bull.=A. phalloides.

***A. caesarea*** (Scop.) Pers. Warm temperate. Eaten in S. Europe since Roman times.

***A. gemmata*** (Fr.) Gill. See A. rubescens.

***A. muscaria*** (L.) Pers.=Agaricus muscarius (Fly Amanita). Temperate. Poisonous. Cooking reduces or eliminates the poison, so the mushroom is eaten with impunity by some people in France and Italy. The raw mushroom is chewed in N. Siberia. The chewing of the mushroom gives hallucinations, in which the subject becomes exalted, having a feeling of power, then rages and shouts, and finally suffers a contortion of spacial dimensions. The boiled fruit-body, with added sugar is used in several places to kill flies.

***A. nivalis*** Grev. See A. rubescens.

***A. ovoidea*** (Scop.) Quél. Mediterranean area. Fruit-bodies are of good quality and are eaten locally.

***A. phalloides*** Fr.=A. bulbosa. (Deadly Amanita). Deadly poisonous. It takes several hours to have any effect. The toxin effects the liver. Muscarine is the poisonous principle.

***A. rubescens*** Pers. Temperate and subtropics. Fruit bodies eaten, and made into a sauce.

***A spissa*** (Fr.) Quél.=Agarigus spissus Fr. Temperate. Fruit bodies eaten.

***A. vaginata*** (Bull. ex Fr.) Quél. See A. spissa.

***A. verna*** Bull. ex Fr. See A. phalloides.

Amanita, Deadly – Amanita phalloides.

Amanita, Fly – Amanita muscaria.

***Amanitopsis canarii*** (Jungh.) Sacc.=Oudemansiella canarii.

AMARACUS Geld. Labiatae. 15 spp. E. Mediterranean. Shrub.

***A. dictamnus*** (L.) Benth.=Origanum dictamnus L. (Dictame). Woody herb. Crete. Used in Ancient Greece to heal wounds. The plant is a tonic and antispasmodic. It is used to stimulate the uterus, and to treat stomach ache, chest complaints, and scrophulosis.

Amaranth, Green – Amaranthus retroflexus.

Amaranth, Prostrate – Amaranthus blitoides.

Amaranth, Redroot – Amaranthus retroflexus.

Amaranth, Slim – Amaranthus hybridus.

AMARANTHUS L. Amaranthaceae. 60 spp. Temperate and tropical. Herbs.

***A. angustifolius*** Lam. See A. tunbergeii.

***A. blitoides*** S. Wats. (Prostrate Amaranth). W. N.America. The leaves are eaten as a pot-herb, and the seeds as a meal (pinole) by the local Indians.

***A. caudatus*** L. (Inca Wheat, Quihuicha, Quinoa). America, Tropical Africa, E.India. Cultivated in S.America for the seeds which are used to make flour, and for the leaves which are a pot-herb.

***A. cruentus*** L. See A. leucocarpus.

***A. diacanthus*** Raf. See A. blitoides.

***A. gangeticus*** L. Tropics. Cultivated in Old World Tropics. The young plant is cooked and eaten.

***A. graezisans*** L. See A. blitoides.

***A. gracilis*** Desf. See A. thunbergii.

***A. grandiflorus*** J. M. Black. Australia. Seeds eaten by aborigines.

***A. hybridus*** L. (Slim Amaranth). Temperate. Cultivated in India for the seeds which are made into a flour. Leaves used as a vegetable.

***A. leucocarpus*** S. Wats. (Huauhtli). C.America. Cultivated in C.America since ancient times for seeds which were used as grain. Cultivation declined rapidly after the Spanish conquest.

***A. lividus*** L. subsp. ***ascendens*** (Loisel.) Thell. See A. thunbergii.

***A. mangostanus*** L. India. Cultivated especially in Japan for the leaves which are eaten boiled as a vegetable.

***A. palmeri*** S. Wats. See A. blitoides.

***A. paniculatus*** L. Tropical Asia. Cultivated locally as a grain crop.

***A. retroflexus*** L. (Green Amaranth, Pigweed, Redroot Amaranth). Temperate. Seeds eaten by Indians of N.America. Young plants used as a vegetable.

***A. spinosus*** L. Throughout Tropics. Leaves used as a vegetable in Indo-china, Singapore and E.Africa. They are also an emollient. The whole plant is used to encourage fevers and sweating.

***A. thunbergii*** Moq. (Bondue). Tropical Africa. Leaves eaten as a vegetable locally.

***A. torreyi*** Benth. See A. blitoides.

***A. viridis*** L. Tropics. Cultivated in various parts of the tropics for the young shoots which are used as a pot-herb.

Amantillo – Abutilon trisulcatum.

Amargosa – Castela texana.

Amate – Ficus involuta.

Amato blanco – Ficus involuta.

Amato Preito – Ficus cotinifolia.

Amatungula – Carrisa arduina.

Amazon Poison Nut – Strychnos castelnaei.

Ambarella – Spondias dulcis.

Ambari Hemp – Hibiscus cannabinus.

Ambari Jute – Hibiscus cannabinus.

Amber (Baltic) – Pinus succinifera.

Amber de Cuspinole – Hymenaea courbaril.

Amboina Sandalwood – Osmoxylon umbelliferum.

Ambong-ambong – Bidens chinensis.

Ambong-ambong laoot – Premna cordifolia.

Ambon Sandalwood – Osmoxylon umbelliferum.

Ambooloog – Metroxylon rumphii.

Ambrette Seed (Oil) – Hibiscus abelmoschus.

AMBROSIA L. Compositae. 35–40 spp. Cosmopolitan. Herbs.

***A. artemisiifolia*** L. (Ragweed). N.America. The achenes contain 19 per cent oil, which has been suggested as a source of edible oil. (Oil of Ragweed). It contains little linolenic acid.

***A. hispida*** Pursh. (Margarita del Mar). Tropical America. An infusion of the leaves is used in Mexico and Dominica to reduce fever. In Cuba, the leaves are soaked in alcohol and used to treat rheumatism and to encourage sweating.

***A. maritima*** L. Mediterranean. Aromatic, and used to flavour liqueurs. It is used to treat flatulence, and bleeding, especially from the nose.

***A. paniculata*** Michx. S.W. N.America and W.Indies. Cultivated in Dominica and made into an infusion which is used to reduce fever.

***A. peruviana*** Willd. W.Indies and S.America. Used in Cuba as a tonic and astringent. May be of some use against rheumatism.

***A. psilostachya*** DC. (Western Ragwort). N.America. It is used by various Indian tribes to treat sore eyes, heal sores, and to prevent intestinal bleeding. Mixed with sages the plant is burnt in Indian sweat houses.

Ambula – Millettia barteri.

AMBURANA Schwacke and Taub. 3 spp. Leguminosae. Brazil. Trees.

***A. acreana*** (Ducke.) Sm. A coarse-grained yellow wood used for general construction work. The seeds yield an oil used in perfumery.

***A. cearensis*** (Allem.) Sm. = A. claudii. = Torresea cearensis. Brazil, N.Argentine. Wood used for general construction work, crates and barrels. The resin yields a medicinal oil. The seeds and bark are made into a syrup with sugar. This is used in Brazil to treat lung diseases.

***A. claudii*** Scw. and Taub. = A. cearensis.

Amchur – Mangifera indica.

AMELANCHIER Medik. Rosaceae. 25 spp. N.Temperate, esp. N.America. Shrubs.

***A. alnifera*** (Nutt.) N. (Saskatoon Service Berry, Western Service Berry). N.America. Berries eaten by Indians.

***A. arborea*** (Michx. f.) Fern. = A. canadensis.

***A. canadensis*** (L.) Medik (Service Berry, Shadblow Service Berry, Shab Bush). E. N.America. Berries eaten in pies etc. Canned. The wood is very hard, strong and close-grained, and dark brown to red. It is used for small articles needing strength, e.g. tool handles, and for fishing rods (Lance wood).

***A. denticulata*** (H.B.K.) Koch. = Cotoneaster denticulata. = Crataegus minor (Madronillo, Membrillo, Tlaxisqui). Mexico. Wood used to make flexible canes (Varitas de Apizaco).

***A. prunicolia*** Greene. (Service Berry). W.United States of America. Fruit eaten by local Indians.

American Apple Mint – Mentha gentilis.

American Barberry – Berberis canadensis.

American Basswood – Tilia americana.

American Beachgrass – Ammophila breviligulata.

American Beech – Fagus ferruginea.
American Bittersweet – Celastrus scandens.
American Blue Vervain – Verbena hastata.
American Centaury – Sabatia angularis.
American Chestnut – Castanea dentata.
American Christmas Mistletoe – Phoradendron flavescens.
American Columbo – Frasera carolinensis.
American Cotton, Short Staple – Gossypium herbaceum.
American Crane's bill – Geranium maculatum.
American Ebony – Brya ebenus.
American Elder – Sambucus canadensis.
American Elemi – Basera gummifera.
American Ephedra – Ephedra americana.
American Flase Hellebore – Veratrum viride.
American Gingseng – Panex quinquefolia.
American Holly – Ilex opaca.
American Hop Hornbeam – Ostrya virginiana.
American Hornbeam – Carpinus caroliniana.
American Horsemint – Monarda punctata.
American Ipecac – Gillenia stipulata.
American Mastic – Schinus molle.
American Mistletoe – Phorodendron flavescens.
American Nutgalls – Quercus imbricaria
American Oil Palm – Coroza deifera
American Pennyroyal – Hedeoma pulegioides
American Raspberry – Rubus strigosus.
American Red Currant – Ribes triste.
American Red Osier Bark – Cornus amomum.
American Red Raspberry – Rubus strigosus.
American Sea Rocket – Cakile edentula.
American Senna – Cassia marilandica.
American Sloughgrass – Beckmannia syzigachne.
American Smoke Tree – Rhus cotinoides.
American Spikenard – Aralia racemosa.
American Storax – Liquidambar styraciflua.
American Styrax – Liquidambar styraciflua.
American Sweet Gum – Liquidambar styraciflua.
American Sycamore – Platanus occidentalis.
American Upland Cotton – Gossypium hirsutum.
American White Elm – Ulmus americana.
American White Hellebore – Veratrum viride.
American Wild Blackcurrant – Ribes americanum.
American Wild Ginger – Asarum canadense.
American Wild Gooseberry – Ribes cynosbati.
American Wild Plum – Prunus americanus.
American Wild Mint – Mentha canadensis.
American Wormseed – Chenopodium ambrosioides var. anthelminticum.
American Yew – Taxus canadensis.
***Amerimnon*** P. Br. = Dalbergia L.
Amiana – Urera oligoloba.
AMMANNIA L. Lythraceae 30 spp. Cosmopolitan. Herbs.
***A. baccifera*** L. See A. senegalensis.
***A. senegalensis*** Lam. Tropical Africa. Used to produce blisters.
AMMI L. Umbelliferae. 10 spp. Azores, Madeira, Mediterranean and W.Africa. Herbs.
***A. majus*** L. = Apium ammi (Bishop's Weed.) Belgium to Iran. Cultivated from the Middle Ages for the seeds (Semen Ammeos vulgaris) which are used as a tonic and to treat stomach complaints, bronchial asthma, angina pectoris, and as a diuretic.
***A. visnaga*** (L.) Lam. (Khella). Mediterranean. Used since ancient times in folk medicine, locally. An extract of the plant was used to treat asthma and angina pectoris. An infusion of the seeds was used to treat urinary complaints, and the fruit stalks are sold as tooth picks in Egypt.
Ammi – Carum copticum.
AMMOBROMA Torr. Lennonaceae. 1 spp. S.W. N.America, S. N.America. Parasitic herbs, without chlorophyll.
***A. sonorae*** Torr. The underground stems are eaten by the local Indians. They are eaten raw, cooked or made into a flour.
AMMODAUCUS Coss. and Dur. 1 sp. Algeria. Herb.
***A. leucotrichus*** Coss. and Dur. Cultivated and sold across the Sahara. The seeds are used to flavour sauces, and a decoction of the leaves is used against chest complaints.
***Ammodenia*** J. G. Gmel. ex Rupr. = Honkenya = Arenaria
***Ammodenia*** Patrin. = Alsine Scop. pp. = Arenaria L.
Ammoniac, Morroccan – Ferula tingitana.

Ammoniac of Cyrenaica – Ferula marmarica.

Ammoniac of Morocco – Ferula communis.

Ammoniac, North Africa – Ferula tingitana.

Ammoniacum Gum – Dorema ammoniacum.

AMMOPHILA Host. 2 spp. Atlantic N.America, Europe and N.Africa. Herbs.

*A.* ***arenaria*** Link. = A. arundinacea.

*A.* ***arundinacea*** Host. = A. arenaria = Psamma arenaria. (European Beach Grass). Coastal Europe. Grown extensively as a sand binder on sand dunes. The leaves are used for thatching, basket- and chair-making, and making mats, sandals, brooms and brushes.

*A.* ***breviligulata*** Fern. (American Beach Grass). Coast of N.America, Great Lakes. Used as a sand binder.

Amole – Agave schottii.

AMOMUM L. Zingiberaceae. 150 spp. Throughout tropics. Herbs.

*A.* ***aromaticum*** Rosb. (Bengal Cardamon, Nepal Cardamon). E. India. Cultivated as a condiment.

*A.* ***cardamomum*** Willd. (Cardamon). Malaysia. Cultivated. Seeds used as a condiment. Rhizomes ground and used to treat colds.

*A.* ***cevuga*** Seemann. Pacific Islands. Leaves used as thatch and bedding. Rhizome used in Fiji to scent coconut oil used as a body perfume.

*A.* ***clusii*** Sm. (Lambungaly, Mungulu, Ndongo a moaba). Tropical Africa. Seeds used in Gabon as part of an aromatic offering to the dead.

*A.* ***dealbatum*** Roxb. E.India, Malaysia. Used as a substitute for the various cardamons.

*A.* ***globosum*** Lour. = Alpinia globosa.

*A.* ***gracile*** Blume. (Serkkom). Indonesia. The dried fruits are chewed to relieve indigestion. Sold locally.

*A.* ***kepulaga*** Sprague and Burkill (Round Cardamon). Malaysia. Cultivated. Fruits used as a condiment, and to cure coughs and colds. They are also chewed as breath sweeteners.

*A.* ***krervanh*** Pierre. Cambodia. Cultivated. Fruits used as a condiment, and to flavour curries. Exported to Europe where it is used to flavour sausages and cordials.

*A.* ***magnificum*** Benth. = Nicolaia magnifica.

*A.* ***maximum*** Roxb. (Java Cardamon). Malaysia. Cultivated in Java. Used as a condiment locally.

*A.* ***rumphii*** Smith = Hornstedtia rumphii.

*A.* ***subulatum*** Roxb. (Nepal Cardamon). Tropical Asia. Used locally as a condiment.

*A.* ***thyrsoideum*** Gagn. (Krakorso). Tropical Asia espccially Indo-China. Fruit used locally as condiment.

*A.* ***xanthioides*** Wall. (Bastard Siamese Cardamon, Wild Siamese Cardamon). Burma. Cultivated in India as a condiment.

Amontana – Ficus baroni.

AMOORA Roxb. Meliaceae. 25 spp. Indomalaysia. Trees.

*A.* ***aphanamixis*** Roen. and Schukt. = Aphanamixis grandifolia. (Gendis, Goola, Khooleuh). Indonesia. Infusion of the bark used against diseases of the chest by natives of the Celebes.

*A.* ***nitida*** Bentu. Australia. (Incense Wood). Used locally for furniture making.

*A.* ***rohituka*** Wight and Arn. = Aglaia polystachya. Tropical Asia. Seeds a source of scented oil (Rohituka Oil), used for lighting and industrially in Bengal, S. China, and Indo-China.

*A.* ***wallichii*** King. Bengal. Red timber used for furniture and interior work.

AMORPHA L. Leguminosae. 20 spp. N.America.

*A.* ***angustifolia*** Boxton. = A. fruticosa var. A. fragrans.

*A.* ***fruticosa*** L var. ***angustifolia*** Pursh. (False Indigo). Shrub. E. and E.C.United States of America. Stems used by local Indians as bedding and to keep meat fresh.

AMORPHOPHALLUS Blume and Decne. 100 spp. Tropical Africa and Asia. Herbs. Tubers boiled and eaten locally.

*A.* ***campanulatus*** (Roxb.) Blume ex Dcne. (Oroy, Whitespot Giant Arum). Tropical Asia.

*A.* ***harmandii*** Engl. and Gehr. Tonkin.

*A.* ***prainii*** Hook. f. (Begung, Lekir). Malaya, Malacca.

*A.* ***rivieri*** Dur. Tropical Asia. Cultivated in China and Japan. Also made into a flour (Konjaku Flour, Konjaku Powder).

Ampaly – Ficus soroceoides.

Ampapar – Mangifera indica.

Ampedal ajam – Salacia grandiflora.

AMPELOCISSUS Planch. Vitidaceae. 95 spp. Tropics.

***A. martini*** Planch.=Vitis barbata Roxb. var. triloba King=V. labrusca Lour. non L. (Dây dác, Nho rùng). Thailand to Vietnam. Roots eaten locally in Cambodia. Fruits used to make jelly.

AMPELODESMA Beauv. Graminae. 1 spp. Mediterranean, N.Africa.

***A. tenax*** Vahl. Sahara. Leaves used locally to make ropes.

***Ampelopsis arachnoidea*** Planch=Vitis arachnoidea.

***Ampelopsis indica*** Blume=Vitis arachnoides.

***Ampelopsis quinqefolia*** Michx.=Parthenocissus quinquefolia.

AMPELOZIZYPHUS Ducke. Rhamnaceae. 1 sp. Brazil. Tree.

***A. amazonicus*** Ducke. (Saracuramira). Brazil. Scented bark used locally as soap.

Ampet – Cratoxylum formosum.

AMPHICARPAEA Ell. mut. DC. Leguminosae. 24 spp. E.Asia, tropical and N.America, S.Africa. Herbs, with subterranean fruits.

***A. edgeworthii*** Benth. Himalayas to Japan. Seeds eaten in Japan.

***A. monoica*** (L.) Ell.=Falcata comosa. (Ground Peanut, Hog Peanut). S.E. United States of America. Seeds eaten locally by Indians

***Amphilopsis*** Nash=Bothriochloa.

AMPHIMAS Pierre ex Harms. Leguminosae. 4 spp. Tropical W.Africa. Tree.

***A. pterocarpoides*** Harms. Tropical Africa. In Liberia, a red resin is extracted from the bark, and an extract from the bark is used to treat dysentry.

AMPHIPTERYGIUM Schiede ex Standl. Julianiaceae. 4 spp. Mexico. Tree.

***A. adstringens*** (Schlecht.) Schiede. Mexico. The bark is the source of a red dye.

Amrad – Acacia arabica.

Amrawatti – Acacia arabica.

Amrita – Eulophia campestris.

Amritsat Gum – Acacia modesta.

Amsat – Mangifera indica.

Amulu-bala – Urera cameroonensis.

Amumbi – Cellis prantlii.

Amur Cork Tree – Phellodendron amurense.

AMYGDALUS L. 40 spp. Mediterranean to C.China. Trees.

***A. communis*** L.=Prunus amygdalus. (Almond Tree). E.Mediterranean, W.Asia. The Sweet Almond (A. communis var. sativa Ludw.) is cultivated extensively in California and S. Europe. The seeds are eaten ripe, dried, or green, and even salted or roasted. They are also used to make almond paste and marzipan used in confectionery, and as a cleaning agent. Sweet Almond Oil (Oleum Amygdalae Expressum), is non-drying and used to soften tissue and relieve congestion. The Bitter Almond (A. communis var. amara DC.) seeds yield an oil (Oil of Bitter Almond, Oleum Amygdalae Amarae) which is non-drying and used as a sedative, especially in cough medicine. The oil has S.G. 0·9175, Sap. Val. 183–196, Iod. No. 95–102. Much of the oil is produced in S.Europe and N.Africa. Prussic acid is a derivative of a glucoside contained in the kernels.

***A. leiocarpus*** Boiss. E. Mediterranean. Yields a gum, (Cherry Gum, Djedk-i-Ardjin, Kirsch-gummi, Persian Gum) which is eaten locally. Sold. A little is exported.

***A. mira*** (K.) Korsh.=Prunus mira. W.China. Fruits like a typical free-stone peach. Eaten locally.

***A. nana*** L.=Prunus nana. (Swarf Almond). E.Europe, Siberia. Seeds yield an oil similar to but inferior to bitter almond oil. Used locally. Fruits also eaten.

***A. persica*** L.=Prunus persica. (Peach). China. Introduced into Europe in Graeco-Roman times, but had been cultivated in China 200 years previously. The Spaniards introduced it into America. It is now one of the most important orchard crops being cultivated throughout the world where the climate is suitable. The world annual production is about 122 million bushels, with the U.S.A. producing about half this total. A. persica var. scleropersica is the clingstone peach in which the stone remains attached to the pulp; var. aganopersica (D.) R. is the free stone peach, with the stone free from the flesh; var. nectarina Ait. is the nectarine which has a smooth skin; var. compressa Lout. (=var. platycarpa Decne.) is the flat peach, the fruits of which are flattened from the top. The fruits are eaten fresh, cooked, canned, dried. They are used to flavour cordials and liqueurs. The stem yields a gum which is used locally in Asia. Waste from the fruit is used to manufacture oil and paste.

***A. saptoides*** Spach.=P. spartoies (Spach.) Schneid. W.Asia. Yields a gum which is used locally.

***A. tanguatica*** Korsh.=P. tangutica. W. China. Very popular locally; cultivated and kernels eaten raw. The ripe fruit splits open to reveal the stone.

AMYRIS P. Br. Rutaceae. 30 spp. Tropical America; West Indies. Trees.

***A. balsamifera*** L. (Candlewood, Rosewood, Sandalwood, West Indian Sandalwood). C.America, S. N.America. The wood is used for building, and is the source of a scented oil (West Indian Sandalwood Oil, Cayenne Linaoe Oil).

***A. bipinnata*** DC.=Bursera bipinnata.

***A. plumieri*** DC. Mexico, West Indies. A resin is extracted from the bark (Elemi of Mexico, Elemi of Yucatan), which is used in lacquers.

***A. polygama*** Cas=Schinus dependens.

***A. sylvatica*** Jacq. Tropical America, esp. Jamaica. A technical resin is extracted from the wood.

ANABASIS L. Chenopodiaceae. 30 spp. Mediterranean to C.Asia. Shrubs.

***A. aphylla*** L. Caspian Sea Area. Yields an insecticide of minor importance.

***A. articulata*** (Forsk.) Moq. Sahara. Fed to camels Yields an edible gum.

***A. tatarica*** Pall.=A. aphylla.

ANACAMPTIS Rich. Orchidaceae. 1 sp. Europe, N.Africa. Herbs.

***A. pyrimidalis*** (L.) Rich. The tubers are made into a nutritious gruel (salep) used medicinally.

ANACARDIUM L. Anacardiaceae. 15 spp. Tropical America. Trees.

***A. giganteum*** Loud. N.W. S.America. Fleshy fruit stalk eaten locally.

***A. humile*** St Hil.=A. subterraneum. Brazil. Oil from the seeds used like that from A. occidentale.

***A. nanum*** St. Hil. See A. humile.

***A. occidentale*** L. (Cashew). Tropical America. Cultivated in Old and New World. The seed kernels contain 45–47 per cent oil. They are eaten (Cashew Nuts) raw or roasted, and much are exported to Europe and U.S.A. The oil is used as a proofing of timber against termites, in manufacturing varnishes and inks, and electrical insulations. It has a high heat resistance, and is light, making it an excellent lubricant for small electric motors, etc. The thickened fruit stalk (Cashew Apple) is used to make jams, sweetmeats, and a wine (Cajuado), in Brazil. A gum (Chasawa Gum, Gomme d'Acajou) is extracted from the stem and used to preserve wood and book binding against insect attack. The gum is produced mainly in E.Africa, and West Indies.

***A. orientale*** L.=Semccarpus anacardium L. f.

***A. rhinocarpus*** DC. Tropical America. The wood (Esparvie, Esparvie Mahogany) is used as a substitute for mahogany. The bark yields a fibre (Mijugua Fibre) which is used locally.

***A. subterraneum*** Liais.=A. humile.

ANACOLSOA Blume. 1 sp. Tropical Africa; 20 spp. Indomalaysia, and Pacific Islands. Shrubs.

***A. luzoniensis*** Merr. (Galo). Philippine Islands. The nuts are eaten locally. They have a thin shell and a pleasant sweet flavour.

ANACYCLUS L. Compositae. 25 spp. Mediterranean. Herbs.

***A. officinarum*** Hayne. (Bertram). A tincture of the root is used locally against toothache, and for mouth infections.

***A. pyrethrum*** (L.) Link. (Pellitory). Mediterranean. Cultivated in N.Africa. for the essential oil which is used to treat toothache and as a mouthwash. It is also used to flavour liqueurs. The dried roots, collected after flowering, contain pyrethrin, which makes them a potential source of insecticide. The root is used medicinally as a stimulant. It is irritant and stimulates the flow of saliva and causes local reddening of the skin.

Andaman Marble Wood – Diospyros marmorata.

Andaman Zebra Wood – Diospyros kurzii.

ANADENDRUM Schott. Araceae. 9 spp. Indomalaysia. Herbs.

***A. montanum*** Schott. Tropical Asia. Epiphytes. The leaves are used in curries.

ANAGALLIS L. Primulaceae. 28 spp. W.Europe, Africa, Madagascar. 1 sp. pantropic, 2 spp. S. America. Herbs.

***A. arvensis*** L. (Cure All, Scarlet Pimpernel). From N.America to Europe and temperate Asia, N.Africa, Australia. Used in home remedies for lung complaints, gall stones, and cirrhosis of the liver.

ANAGYRIS L. Leguminosae. 2 spp. Mediterranean. Shrubs. The seeds are poisonous, containing the alkaloid anagyrin. They are used as an emetic. The leaves (Herba Anagyris) are used as a purgative.

Ana klûa – Piper lolot.

ANAMIRTA Colebr. Menispermaceae. 1 sp. Indomalaysia. Climber.

***A. cocculus*** (L.) Wight and Arn.=A. paniculata.

***A. paniculata*** Colebr. (Fish Berry, Indian Berry). S.E. Asia, Malabar. The berries are used to stupify fish, and when dried are used medicinally to attack internal parasites. They contain the convulsant poison cocculin.

Anamú – Petiveria alliacea.

ANANAS Mill. Bromeliaceae. 5 spp. Tropical America. Herbs.

***A. comosus*** (L.) Merr=A. sativus (Lindl.) Schult. (Pineapple). Brazil. The species is not known to occur naturally. The fruit is a multiple fruit consisting of the swollen floral axis, sepals and floral bracts, and several hundred fruits. Apart from sugars, citric acid and malic acid, the fruit contains the protease bromelin, and is a valuable source of vitamins A and C. The plants are propagated vegetatively from slips, suckers or crowns. It was discovered by the early colonists in the West Indies, and by the end of the 16th century had spread throughout the world. The world production is about 1¾ million tons, 45 per cent of which is produced in the Hawaiian Islands. The other main producers are Brazil, Cuba, Formosa, Mexico, and the Philippines. Most of the fruit is canned. Other uses include fruit juice, ice-cream, jams and various confections; some is eaten raw. The juice is made into an alcoholic beverage Vin d'Ananas, and into industrial alcohol. The waste is made into vinegar or fed to livestock. The leaves are a source of a fine fibre (Piña Fibre) used to make fabric.

***A. magdalenae*** André=Aechmea magdalenae.

***A. sativus*** (Lindl.) Schult.=A. comosus.

***Ananassa*** Lindl.=Ananas.

Anani – Moronobea coccinea.

ANAPHALIS DC. Compositae. 35 spp. Europe, Asia, America. Herbs.

***A. margaritacea*** (L.) B. and H. (Pearly Everlasting). N.America. The leaves (Herba Anaphalidis) are used as an expectorant, and astringent.

ANAPTYCHIA L. Physciaceae. Lichen Temperate.

***A. ciliaris*** (L.) Kremp.=Parmelia ciliaris. Europe. During the 17th century it was used as a powder to whiten hair and wigs (Cyprus Powder).

ANASTATICA L. Cruciferae. 1 sp. Morocco to S. Persia. Herb.

***A. hierochuntica*** L. (Akarba, Kerchoud, Rose of Jericho). Dried plant sold to tourists.

Anatto – Bixa orellana.

Anauca – Erythrina micropteryx.

Anchietea Rootbark – Anchietea salutaris.

ANCHIETEA St. Hil. Violaceae. 8 spp. Tropical S.America.

***A. salutaris*** St. Hil. (Cipó Suma, Piriguaua). Brazil, Paraguay, Uruguay. The root (Anchietea Rootbark) is emetic, used to treat sore throats, and lymphatic tuberculosis.

Anchovy Pear – Grias cauliflora.

ANCHROMANES Schott. Araceae. 10 spp. Tropical Africa. Herbs. The rhizomes of the spp. mentioned are eaten locally when more palatable foods are scarce. The rhizomes have to be soaked and boiled for a considerable time to make them palatable.

***A. difformis*** (Blume) Engl. Tropical Africa. Rain forests.

***A. hookeri*** Schott. Tropical Africa. Rain forests.

ANCISTROCLADUS Wall. Ancistrocladaceae. 20 spp. Tropical Africa, Ceylon, E. Himalayas to W. Malaya.

***A. extensus*** Wall. China to Malaya. Young leaves used as flavour in Siam, and the roots, when boiled, are used to treat dysentery in Malaya.

ANCISTROPHYLLUM (Mann and Wendl.) Mann. and Wendl. ex Kerchove. Palmae. 6 spp. Tropical Africa. Palms.

***A. opacum*** Drude. W. Africa. Stems used for basket-making and house-building.

***A. secundiflorum*** Mann. and Wendl. Tropical W.Africa. Stems used to build huts and fishing gear.

Anda-assy Oil – Joanessia princeps.

***Anda gomesii*** Juss.=Joanessia princeps.

Andaman Canary Tree – Canarium euphyllum.

Andaman Crape Myrtle – Lagerstroemia hypoleuca.

Andaman Zebrawood – Diospyros kurzii.

Andean Blueberry – Vaccinium floribundum.
Andem – Hydrocotyle sibtorpioides.
Andes Cotton – Gossypium peruvianum.
Anderson Wolfberry – Lycium andersonii.
Andi – Carallia brachiata.
ANDIRA Juss. Leguminoseae. 35 spp. Tropical America and Africa. Trees.
*A.* ***araroba*** Aguiar. (Angelin). Tropical America. A powder (Goa Powder, Bahia Powder) is deposited in the wood It contains the active principle chrysorobin, which is used against external parasites.
*A.* ***excelsa*** H. B. K. = Vouacapoua americana. (Brownheart, Partridge Wood). Tropical America. The reddish-brown wood is hard, durable, heavy. It is used for construction work, furniture, and wheel-making.
*A.* ***galeothiana*** Standl. S.Mexico. Used as shade in coffee plantations. The bark is used against intestinal worms locally in Mexico.
*A.* ***inermis*** (Swartz.) H. B. K. C.America, W.Indies, W.Africa. The wood is hard and durable, varying from yellow to black. It is used in Mexico for construction work, and as a shade tree in coffee plantations. The bark and seeds are used locally to cure intestinal worms, to reduce fevers, and as a purgative.
*A.* ***stipulacea*** Benth. Brazil. The wood is used locally for construction work, and the seeds against intestinal worms.
*A.* ***vermifuga*** Mart. (Brazilian Angelin Tree). The bark, which is poisonous in large doses is used locally against intestinal worms, as a narcotic, emetic and purgative.
Andiroba (Oil) – Carapa guineensis.
Ando – Canarium oleosum.
Andong – Rhodamnia arbrea.
Andong petooroon – Polygala glomerata.
Andrium bavifohy – Alyxia madagascariensis.
ANDROCYMBIUM Willd. Liliaceae. 35 spp. Mediterranean to S. Africa. Herbs.
*A.* ***punctatum*** Baker = Erythrostictus punctatus Schlecht. N. Africa. Used locally as a condiment.
ANDROGRAPHIS Wall. Acanthaceae. 20 spp. Tropical Asia. Herb.
*A.* ***paniculata*** Nees. (Kariyat). India and Ceylon. Roots and leaves source of Alui Halviva a medicine which is used as a tonic, against dysentery, diarrhoea, intestinal worms, stomach complaints and to reduce fevers. It is cultivated locally.
ANDROMEDA L. Ericaceae. 1–2 spp. N. Temperate and sub-arctic.
*A.* ***glaucophylla*** Link. (Bog Rosemary). Shrub. N. and N.E. America. Aromatic leaves used locally by Indians to make a tea.
*A.* ***polifolia*** L. (Bog Rosemary). Temperate to Arctic Asia, Europe and America. Leaves and twigs used for tanning in parts of Russia.
ANDROPOGON L. Graminae. 113 spp. Tropical and subtropical. Perennial grasses.
*A.* ***aciculatus*** Retz. = Chrysopogon aciculatus. Tropics. Used in the Philippines as a cattle food, and used to make hats and mats. The roots are cultivated as Chiendent grenille à brosse which is exported from Vietnam.
*A.* ***bicornis*** L. (Penache). Mexico to Brazil, W. Indies. Used for thatching. Hairs sometimes used for stuffing pillows
*A.* ***candidus*** Trin. = Elionurus candidus
*A.* ***caricosus*** L. Tropical Asia. A good hay.
*A.* ***contortus*** L. = Heteropogon hirtus.
*A.* ***emersus*** Fourn. = A. saccharoides.
*A.* ***flexuosus*** Nees. = Cymbopogon flexuosus.
*A.* ***furcatus*** Muhlend. See A. scoparius.
*A.* ***hallii*** Hack. (Sand Bluestem, Turkey Foot). C. N.America. Used to stabilize inland sand dunes, by sowing in sorghum stubble.
*A.* ***hispidus*** Willd. = Arundinella brasiliensis.
*A.* ***jivarancusa*** Jones = Cymbopogon jivarancusa.
*A.* ***laniger*** Desf. W.Africa, E.Indies. The dried rhizomes are sold in small sachets as a scent. This is possibly the Nara Syriagae used by the Ancients.
*A.* ***littoralis*** Nash. (Seacoast Bluestem). Seashore of E. N.America. Used to stop erosion of sand dunes. Seeded.
*A.* ***muricatus*** Retz. = Vetiveria zizanoides.
*A.* ***nardus*** L. = Cymbopogon nardus.
*A.* ***nigritanus*** Benth. = Vetiveria nigritana.
*A.* ***nilagiricus*** Hochst. = Cymbopogon confertiflorus.
*A.* ***odoratus*** Lieb. (Ginger Grass). Far East. An aromatic oil is extracted from the rhizome.

***A. perforatus*** Trin.=A. saccharoides.

***A. rufus*** Kunth. (Yaragua Grass). Tropical Africa, Brazil. Used for livestock fodder.

***A. ruprechtii*** Hack.=Hyparrahenia ruprechii.

***A. saccharoides*** Sw.=A. emersus=A. perforatus. Tropics. Good pasture grass.

***A. scoparius*** Michx. C. N.America. Used to prevent erosion of inland sand dunes. Good for pasture and hay.

***A. sorghum*** (L.) Brot.=Sorghum vulgare.

***A. squarrosus*** L.f.=Vetivera zizanoides.

***A. virginicus*** L. (Broomsedge). S.United States of America, Mexico. Used to prevent the erosion of sand dunes.

Andzilim – Eurypetalum batesii.

ANEILEMA R. Br. Commelinaceae. 100 spp. Warm areas, esp. Old World. Herbs.

***A. beninense*** Kunth.=Commelina beninensis. Tropical Africa. An infusion of the plant is used as a laxative for children locally.

ANEMARRHENA Bunge. Liliaceae. 1 sp. N.China. Herb.

***A. asphodeloides*** Bunge. Used locally to reduce fever.

ANEMONE L. 150 spp. Cosmopolitan. Herbs.

***A. cernua*** Thunb. (Nodding Anemone). China, Japan. Used in Chinese medcine.i

***A. flaccida*** Schmidt. Japan, China. Leaves eaten locally in Japan.

***A. hepatica*** L.=Hepatica nobilis. (Hepatica). N.America, Europe. The crushed leaves yield anemone camphor which is used medicinally. The whole plant is used as a tonic.

***A. hupehensis*** Lem. (Banate, Kabtabu, Omek, Paloi). China, Formosa, Philippines. Hairs from the fruits used in the Philippines to start fires.

***A. japonica*** Sieb. and Zucc. China, Japan. Cultivated in China for the roots which are used in the treatment of heart complaints.

***A. narcissiflora*** L. N. Temperate. Leaves eaten locally by Eskimos, raw, fermented or in oil.

***A. nemorosa*** L. (Grove Wind-flower, Wood Anemone). Europe, through Caucasus and Korea to Japan. Source of Herba Anemone nemorosa. Used to treat bronchitis, arthritis and pleurisy.

***A. patens*** L. (Spreading Pasqueflower). N.America, Europe to Mongolia. The dried plant, or anemone camphor (see A. hepatica) collected during flowering, used to increase appetite, and restore normal digestion. It is also used to regulate menstruation, as a diuretic and expectorant.

***A. pratensis*** L. (Meadow Pasqueflower). Europe. N.Asia. Used as A. patens.

***A. pulsatilla*** L. (European Pasqueflower). Europe. Used as A. patens. Produced mainly in Czechoslovakia.

***A. ranunculoides*** L. Europe to Japan. Used as A. nemorosa.

Anemone, Nodding – Anemone cernua.

Anemone, Wood – Anemone nemorosa.

ANEMONELLA Spach. Ranunculaceae. 1 sp. E. N.America. Herb.

***A. thalictroides*** (L.) Spach =Syndesmon thalictrodes. (Rue Anemone, Wild Rue, Wild Potato). Starchy roots eaten locally in Pennsylvania.

***Anemonopsis*** Pritz. (sphalm.)=Anemopsis Hook.

ANEMOPAEGMA Mart. ex Meissn. Bignoniaceae. 30 spp. Tropical America. Woody small trees.

***A. mirandum*** Mart. ex DC. (Catuabu). Brazil. The root (Rhizoma Anemopaegmae Catuaba) is used to dilate the capillaries, and is a stimulant. It is also used as an aphrodisiac.

ANEMOPSIS Hook. and Arn. Sauruaceae. 1 sp. S.W. United States of America, Mexico. Herb.

***A. californica*** Hook. and Arn.=Houttuyia californica. Used locally in Mexico to treat indigestion, to cleanse the blood, and to cure colds.

Anenga – Pleurospa linifera.

ANETHUM L. Umbelliferae. 1 sp. N.Africa, 3 spp. W.Asia. Herbs.

***A. graveolens*** L. (Dill). S.Europe,to India. Cultivated in India, Rumania, Germany, England, and U.S.A. Small stems and immature umbels are used to flavour soups, sauces and pickles. An oil (Oil of Dill) is distilled from the plant at the early fruiting stage. It is used for flavouring and as a carminative and stimulant. The dried seeds are used similarly. The main oil-producers are Hungary, Oregon and Idaho.

***A. sowa*** Roxb. ex. Flem. (Satapashpi, Sowa, Suwa). India, and surrounding tropical Asia. Fruits and oil distilled from them are used as a carminative

Angadeo Mastiche – Echinops viscosus.
Angan – Fraxinus floribunda.
Angaroka – Dioscorea ovinata.
Angbolo – Ostryoderris lucida.
***Angelesia*** Korth. = Licania.
ANGELICA L. Umbelliferae. 80 spp. N.hemisphere, New Zealand.
***A. archangelica*** L. (Angelica). Temperate Europe to Himalayas. The leaf stalks are candied and used in confectionary. An oil (Oil of Angelica Root) distilled from the root are used to flavour gin, and liqueurs (Benedictine, Chartreuse), and in perfumery. The leaves are used to flavour fish dishes, and jams, and also to make a tonic infusion. The rhizome (Angelicae Radix), the dried leaves and flowers (Herba Angelicae) and the dried fruits are used as a carminative and stimulant. An infusion of the roots is used to treat indigestion, and chest colds.
***A. atropurpurea*** L. (Purple Angelica). N.W. United States of America, and S.W. Canada. Young shoot eaten raw, or cooked. The roots are candied.
***A. breweri*** A. Gray W. N.America. An infusion of the roots is used by the local Indians to cure chest colds.
***A. decursiva*** Franch. and Savat. = Peuce danum decursivum.
***A. edulis*** Miyabe. Japan. Stems eaten locally.
***A. koreana*** Maxim. China, Korea. Used locally to treat rheumatism.
***A. levisticum*** All. = A. levisitcum Baill. = Levisticum officinale.
***A. polymorpha*** Maxim. = Ligusticum acutilobum. E.Asia. The aromatic root is used in China as a stimulant to blood pressure, diuretic, and to stimulate uterine contractions.
***A. rosaefolia*** Hook. New Zealand. Leaves used locally as a diuretic.
***A. sylvestris*** L. (Wild Angelica, Woodland Angelica). Asia to Europe; introduced into N.America. Stems and leaves eaten cooked, locally. A paste of the fruit is used to kill head lice.
***A. tenuisima*** Nakai. Korea to China. Used as a carminative.
***A. uchiyamai*** Nakai. China. Used medicinally in China.
***A. villosa*** (Walt.) B.S.P. (Hairy Angelica). E. N.America. Used to discourage the smoking of tobacco.
Angelica – Angelica archangelica.
Angelica – Archangelica officinalis.
Angelica, Hairy – Angelica villosa.
Angelicae, Herba – Angelica archangelica.
Angelicae, Radix – Angelica archangelica.
Angelica, Purple – Angelica atropurpurea.
Angelica, Seacoast – Archangelica actaeifolium.
Angelica, Wild – Angelica sylvestris.
Angelica, Woodland – Angelica sylvestris.
Angelim amarillo – Vataireopsis speciosa.
Angelin Tree – Andira araroba.
Angelin Tree, Brazilian – Andira vermifuga.
Angelin Tree, Goa – Andira araroba.
ANGELONIA Humb. and Bonpl. Scrophulariaceae. 30 spp. Tropical America; W.Indies. Herbs.
***A. salicariaefolia*** H. B. K. S. America. Used locally to induce perspiring.
Anghi – Fagara lemairei.
Angico Gum – Piptadenia rigida.
Angin anginan – Chrysophyllum roxburghii.
ANGIOPTERIS Hoffm. Angiopteridaceae. 100? spp. Madagascar, tropical Asia, Polynesia.
***A. evecta*** (Forst.) Hoffm. Polynesia, N.E. Australia. Starchy pith eaten locally in Australia. In Polynesia, an aromatic oil is extracted, which is used to scent the coconut oil which is used as a body oil.
Ang-khak – Monascus purpureus.
Angkol – Garcinia lanessonii.
Angola Calabash – Monodora angolensis.
Angola Weed – Roccella fuciformis.
ANGOPHORA Cav. Myrtaceae. 10 spp. E.Australia. Tree.
***A. intermedia*** DC. = Metrosideros floribunda. S.E. Australia. A brown kino is extracted from the trunk.
Angostura Bark – Cusparia febrifuga.
Angostura braziliensis – Esenbeikia febrifuga.
ANGRAECUM Borg. Orchidaceae. 220 spp. Tropical and S. Africa, Madagascar, Philippines. Herbs.
***A. fragrans*** Thouars. (Faham, Fahon, Fahum). Epiphyte. Mauritius. The leaves are highly scented, and are used to flavour ice creams, custards, etc. They also make a tea of excellent flavour (Thé de Bourbon) but it is too expensive to be of great commercial value.
Angso – Palaquium leiocarpon.
Angso Gutta Percha – Palquium leicarpon.

Angu – Fraxinus floribunda.
***Anguina tricuspidata*** O. Uze. = Trichosanthes tricuspedata.
Angwabala – Celtis meldbraedii.
ANIBA Aubl. Lauraceae. 40 spp C. and tropical S.America. Trees.
***A. canelilla*** (H. B. K.) Mez. = Cryptocarya canelilla. Tropical America. The bark is the source of an aromatic oil (Sandalo Brasileiro). The bark is used to make a stimulating beverage, and a powder of the bark is used to scent bed-linen.
***A. cujumary*** Mart. = Ocotea cujumary.
***A. firmulum*** Nees. (Pichrim Bean). Brazil. Seeds are aromatic and used as a condiment.
***A. floribundum*** Meissn. = Cryptocarya minima. Brazil. Wood used for general construction work locally. Bark used medicinally as a mouthwash.
***A. panurense*** Meissn. Brazil. Source of aromatic oil (Cayenne Linaloe, Bois de Rose).
***A. perutilis*** Hemsl. (Comino). Andes. The wood is satiny in appearance, and used for furniture making.
***A. rosaeodora*** Ducke. (Bois de Rose Femelle). Lower Amazon. The wood is a yellowish brown, and contains an aromatic oil. It is used to make drawers which will contain linen, etc. The oil (Essence de Bois Pose, Huile de Linaloe), is extracted from the wood and used in perfumery.
***A. terminalis*** Ducke. (Pao Rosa). Amazon Basin. An essential oil (Rose Wood Oil) is extracted from the wood and occasionally used in perfumery.
Anijswortel – Annesorrhiza capensis.
Anilanche – Pluchea odorata.
Anil cimarrón – Indigofera lespedezioides.
Anime (Copal) – Trachylobium homemannianum.
Anime de Mexico – Hymenaea courbaril.
Aninga – Pleurospa linifera.
Aniinga Uva – Pleurospa arborescens.
Anise (Oil) (Milk) – Pimpinella anisum.
Anise, Chinese – Illicium verum.
Anise, Japanese Star – Illicium ansatum.
Anise Oil – Illicium paviflorum.
Anise Root – Annesorrhiza carpensis.
Anise, Star – Illicium verum.
Anise, Yellow Star – Illicum parviflorum.
Anisette Liqueur – Pimpinella anisum.
***Anisomeles*** R. Br. = Epimeredi.
ANISOPHYLLEA R. Br. Anisophylleaceae. 30 spp. Tropical Africa and Asia; 1 sp. Tropical S.America. Shrubs.
***A. disticha*** Baill. (Kajoo kantjil, Keribor, Lambai ajam). Coast of W.Malaya. The wood is very hard and is used locally to make arrows etc. An infusion of the leaves is used locally to treat diarrhoea.
ANISOPTERA Korth. Dipterocarpaceae. 13 spp. Assam, S.E. Asia, Malaysia. Tree.
***A. grossivenia*** Van Sloten. (Mersawa kunyit). Malaya, Borneo, Indonesia. The wood is hard and light, but as it is not durable it is of little commercial value. It is used mainly for packing cases. A slow-drying dammar is extracted from the trunk.
***A. laevis*** Ridl. Tropical Asia. Hard, pale yellow wood used for general construction work.
***A. marginata*** Korth. Borneo. Yellow hard wood, used for general construction work.
***A. thurifera*** Blanco. (Mersawa, Palosápis). Philippines. Yellowish wood used for general construction work.
ANISOSPERMA Manso. Cucurbitaceae. 1 sp. Brazil. Vine.
***A. passiflora*** Manso. Oil from seeds used as an emetic and to cure intestinal worms.
***Anisum officinarum*** Moench. = Pimpinella anisum.
***Anisum vulgare*** Gaertn. = Pimpinella anisum.
Anivo – Phloga polystachya.
Anivo – Ravenea robustior.
Anjeli Wood – Artocarpus hirsuta.
Anjangasjang – Elaeocarpus grandiflorus.
Anjang-Anjang – Elaeocarpus grandiflorus.
Annatto – Bixa orellana.
ANNESLEA Wall. Theaceae. 4 spp. China, Formosa to Indomalaysia. Trees.
***A. fragrans*** Wall. Tropical Asia. The wood is greyish-brown, and is used for furniture-making, and fancy goods, because of its markings.
ANNESORRHIZA Cham. et Schlecht. Umbelliferae. 15 spp. S.Africa. Herbs.
***A. capensis*** Cham. and Schlecht. (Anijswortel, Anise Root). S. Africa. Roots eaten locally by natives.
ANNONA L. Annonaceae. 120 spp. Warm, esp. America. Trees. The fruits of all spp. mentioned are eaten raw.
***A. aurantiaca*** Barb.-Rodr. Brazil. Fruit eaten locally.
***A. cacans*** Warm. Brazil. Fruits eaten locally.
***A. cherimolia*** Mill. (Cherimoya, Cheri-

moyer). Tropical and subtropical America. Cultivated. A good dessert fruit, also made into drinks.

***A. cineraria*** Pittier. (Riñon). Venezuela. Good dessert fruit; very sweet.

***A. delabripetala*** Raddi. = Rollinia longifolia St. Hil.

***A. dioica*** St. Hil. = Aberemoa dioica Barb.-Rodr. Brazil. Leaves used locally to relieve rheumatism.

***A. diversifolia*** Safford. C.America. Cultivated locally and in Florida. A good fruit eaten raw.

***A. glabra*** L. (Pond Apple). W.Indies, Florida. Fruit has a poor flavour; occasionally used to make jams.

***A. longiflora*** S. Wats. (Wild Cherimoyer of Jalisco). Mexico. Fruit eaten, and a confection is made by boiling with sugar.

***A. mannii*** Oliv. = Anonidium mannii. Tropical Africa. The fruits are of good flavour and eaten locally.

***A. montana*** Macf. (Cimarrona, Mountain Soursop). W.Indies. A good flavoured fruit, cultivated locally.

***A. muricata*** L. (Corossol, Guanábana, Soursop). Tropical America. Cultivated throughout Old and New World Tropics. The fruit is the largest of the Annona spp. Eaten raw, made into drinks, and into soup (in Java).

***A. nana*** Exell. C.Africa. Root used locally to treat snake-bite.

***A. purpurea*** Moc. and Sessé (Soncoya). C.America. Good fruit, sometimes cultivated, and sold locally.

***A. reticulata*** L. (Bullock's Heart,Corazon, Custard Apple). Tropical America. Cultivated throughout Old and New World tropics. Fruit rather insipid, but of sound quality.

***A. scleroderma*** Safford (Posh Té). C.America. A fruit of excellent quality. The thick skin is a valuable property which may be used in breeding, to make the fruits of the genus more transportable.

***A. senegalensis*** Pers. Tropical W.Africa. Not cultivated. The leaves are used locally to produce perfume.

***A. squamosa*** L. (Anon, Custard Apple, Sugar Apple, Sweet Sop). Tropical America. Cultivated throughout Old and New World tropics. An excellent dessert fruit.

***A. testudina*** Safford. (Anona del Monte). C.America. A good dessert fruit, but not cultivated.

Annunciation Lily – Lilium candidum.

ANODENDRON A. DC. Apocynaceae. 20 spp. Ceylon, Japan and Far East. Woody vines.

***A. candolleanum*** Wight. W.Malaysia, Borneo. Yields a cordage fibre which was important locally.

***A. tenuiflorum*** Miq. Sumatra. Cordage made from bark used for fishing nets.

ANOECTOCHILUS Blume. Orchidaceae. 25 spp. Tropical Asia, Australia, Polynesia. Herbs.

***A. albo-lineatus*** Par. and Reichenb. Malaysian mountains. Used as a pot-herb. Sold locally.

***A. geniculatus*** Ridley. See A. albo-lineatus.

***A. reinwardtii*** Blume. See A. albo-lineatus.

ANOGEISSUS Wall ex. Guillem. and Perr. Combretaceae. 11 spp. Tropical Africa, Arabia, India, S.E.Asia. Trees.

***A. latifolia*** Wall. (Ghatti Tree). E.India and Ceylon. A gum is exuded from the trunk (Ghatti Gum, Indian Gum). The gum is water-soluble and is used to make confectionery and as a mordant. A black dye is prepared from the leaves, and the wood is used for general construction work.

***A. leiocarpus*** Guill. and Perr. Tropical Africa, esp. W.Africa. An insoluble gum from the trunk is eaten locally. The roots are used as chew stick, and the wood ash is used as a mordant.

***A. schimperi*** Hocht. Tropical Africa. The brown wood polishes well and is durable. It is used for general construction work. The leaves yield a yellow dye. The ashes from the wood are used locally as a mordant, and to remove the hair from goats' skin during tanning. The plant used to treat intestinal worms of horses and donkeys.

Anon – Annona squamosa.

Anona del Monte – Annona testudina.

Anongro – Morinda lucida.

***Anonidium mannii*** (Oliv.) Engl. and Diels. = Annona mannii.

ANOTIS DC. Rubiaceae. 30 spp. China, Indomalaysia. 1 sp. S.America. Herbs.

***A. hirsuta*** (L. f.) Miq. = Oldenlandia hirsuta L. f. (Kahitootan, Ranggitan). Java. An infusion of the plant is used locally to treat abdominal pains. It is sometimes eaten with rice.

ANREDERA Juss. Basellaceae. 5–10 spp. S. United States of America, and W. Indies to Argentine, Galápagos Islands. Vines.

***A. baselloides*** H. B. K. (Madeira Vine, Mignonette Vine). Mexico to Chile. Leaves and tubers are eaten cooked, locally.

ANSELLIA Lindl. Orchidaceae. 2 spp. Tropical Africa, Natal. Herbs.

***A. humulis*** Bull. C. Africa. An infusion of the roots is used as an emetic by the Zulus. The leaves and stems are used to treat madness in Zambia.

Ansett – Musa enste.

Antanan goonoong – Pimpinella alpina.

Antanum beirit – Hydrocotyle sibtorpioides.

Antehai – Scaphium affinis.

Antelope Horn, Spider – Asclepias decumbens.

ANTHEMIS L. Compositae. 200 spp. Mediterranean to Persia, and N.India. Herbs.

***A. cotula*** L. (Cotula, Dog Chamomille, Dog Fennel, Mayweed). Europe to N.W. Siberia, introduced to N. and S.America, Australia and New Zealand. It is used to reduce spasms, as a tonic, to control menstruation, cure headaches, as an emetic and as a substitute for insect powder. Leaves recommended for rubbing into bee stings.

***A. nobilis*** L. (Chamomille, English Chamomille, Roman Chamomille). Mediterranean. Cultivated for an essential oil (Oil of Roman Chamomille), which is used to flavour liqueurs and in perfumery. The dried flower heads (Flores Chamomillae Romanae) are used as a carminative and tonic.

***A. tinctoria*** L.=Chamaemelum tinctorium (L.) All. (Dyer's Chamomile, Golden Chamomile). Europe to W.Asia; introduced to N.America. Cultivated for flowers which yield a yellow dye.

***A. wiedemanniana*** Fisch. and Mey. Asia Minor to Afghanistan. Flowerheads used locally in Iran as a carminative and to reduce fever.

***Anthistiria*** L. f.=Themeda.

ANTHOCEPHALUS A. Rich. Naucleaceae. 3 spp. Indomalaysia. Trees.

***A. cadamba*** Miq.=A. indicus.

***A. indicus*** A. Rich.=A. cadamba,=A. morindaefolius=Sarcocephalus cadamba. Tropical Asia. The wood is light and soft. It is used to make canoes, carving etc., match boxes, furniture, house construction, tea chests, and spools for jute mills. The bark is used as a tonic and to reduce fever. The receptacle is eaten by men and stock. The flowers are used in religious festivals.

***A. macrophyllus*** Havil. Indonesia. Wood used to make canoes and house construction.

***A. morindaefolius*** Kunth.=A. indicus.

ANTHOCLEISTA Afzel. Potaliaceae. 11 spp. Tropical Africa; 3 spp. Madagascar. Trees.

***A. nobilis*** G. Don. Tropical Africa. Ashes from burnt leaves used to make soap by natives.

ANTHOLYZA L. Iridaceae. 25 spp. Africa. Herbs.

***A. paniculata*** Klatt. (Ethiopean Madflower). S. Africa. The Zulus used the rhizome to treat dysentery and diarrhoea.

ANTHOXANTHUM L. Graminae 20 spp. N. Temperate, mountains of tropical Africa and Asia. Perennial grasses.

***A. odoratum*** L. (Spring Grass, Sweet Vernal Grass). Europe, Caucasus, Asia, W. N.Africa, N.America, Australia. Grown for forage, usually in mixed pasture. It contains coumarin which gives the grass and hay a characteristic sweet smell.

ANTHRISCUS Pers. emend Hoffm. Umbelliferae. 20 spp. Europe, temperate Asia. Herbs.

***A. cerefolium*** (L.) Hoffm.=Chaerefolium cerefolium (L.) Schinz.=Scandix cerefolium L. (Chervil). Europe through to E.Asia, America, Australia, New Zealand. Cultivated for the leaves which contain an essential oil, and are used for flavouring. The leaves (Herba Cerefolii) are also used as a diuretic.

***A. sylvestre*** (L.) Schims. and Thell.= Myrrhis Sylvestris. (Cow Parsley). Europe to Siberia. N.Africa. The juice is used locally to treat skin complaints.

ANTHURIUM Schott. Araceae. 550 spp. Tropical America. W.Indies. Herbs.

***A. oxycarpon*** Poepp. (Yeury-cumajé). Tropical America, esp. Brazil and Peru. The dried leaves are heavily scented (Folha cheirosa) and mixed with tobacco to give it an aroma.

ANTHYLLIS L. Leguminosae. 50 spp. Europe, N.Africa. W.Asia. Herbs.

***A. vulneraria*** L. (Kidney Vetch). Europe, esp. south, and Caucasus. Cultivated for fodder for sheep and goats.

ANTIARIA Lesch. Moraceae. 4 spp. Tropical Africa, Madagascar, Indomalaysia. Trees.

***A. toxicaria*** Lesch. (Upas Tree). Tropical Asia. The latex contains isomers of the glucoside antiarin. This is a powerful heart poison. Used as an arrow and dart poison. The tree was considered so poisonous as to kill all the vegetation around it.

***A. welwitschii*** Engl. C.Africa. Bark used locally to make cloth.

ANTIDESMA L. Stilaginaceae. 170 spp. Old World tropics esp. Asia. Trees.

***A. amboinense*** Miq.=A. stipulare.

***A. bunius*** (L.) Spreng. (Bignay China Laurel, Chinese Laurel). S.E. Asia, Australia. The red and black berries used in jellies, syrups, and in brandy. Used as a sauce for fish in Malaysia where they are sold.

***A. dallachyanum*** Baill. Queensland. Clusters of acid fruit make excellent jelly.

***A. ghaesembilla*** Gaertn.=A. paniculatum. (Dempool, Koontjir, Onjam). Malayan Archipelago. Leaves used locally to flavour food. Fruits are edible.

***A. paniculatum*** Roxb.=A. ghaesembilla.

***A. platyphyllum*** H. Mann. (Bignay). Hawaii. Fruits used to make jellies, syrup and wine. Hard reddish-brown wood used locally.

***A. stipulare*** Blume.=A. amboinense (Boowah tati kambing, Soolaketan). Indonesia. Hard brown wood used locally for tool handles. The fruit juice is purple-black and is used to dye cloth and to colour rice at ceremonial meals.

***A. venosum*** Mey. S.Africa. Small tree. Fruits eaten raw by natives.

ANTIGONON Endl. Polygonaceae. 8 spp. Tropical America. Vine.

***A. leptopus*** Hook. and Arn. (Coral Vine, Mountain Rose). C. to S.America. The tubers eaten locally.

ANTIRHEA Commers. ex Juss. Rubiaceae. 40 spp. W.Indies, Madagascar, tropical E.Asia to Australia. Trees.

***A. coriacea*** Urb. W.Indies. The strong durable wood is used locally for general construction work and furniture.

***Antirrhoea*** Endl.=Antirhea Commers. ex Juss.

Antong sar – Eurycoma longifolia.

ANTROCARYON Pierre – Anacardiaceae. 8 spp. Tropical W.Africa. Trees.

***A. nannani*** De Wild. Tropical Africa, esp. Congo. An oil (Congo Oil, Gonga Oil) is extracted from the seeds and used for cooking etc. The oil cake is fed to stock. The bark is used for tanning.

Antsikhana – Xerochlamys pilosa.

Ant Tree, Surinam – Triplaris surinamensis.

Anu – Leptadenia lancifolia.

Anubing Gum – Artocarpus cumingiana.

ANVELLEA A. Chev. Compositae. 1 sp. N.Africa. Woody shrub.

***A. australis*** Chev. (Arfedj). Used locally for firewood.

Aoa – Ficus prolixa.

Aoura – Astrocaryum vulgare.

APAMA Lam. Aristolochiaceae. 12 spp. Indomalaya, S.China. Shrubs.

***A. tomentosa*** Kuntze=Bragantia corymbosa. Java. Stems and leaves used locally as a detergent to wash clothes. The leaves are used to treat snake-bite.

Apapa – Bauhinia fassagiënsis.

Apata Wood, Benin – Microdesmis puberula.

Apazoto de zorro – Petiveria alliacea.

Apé – Lerospatha candata.

Apedoo – Eurodia melufolia.

APEIBA Aubl. Tiliaceae. 10 spp. Tropical S.America. Trees.

***A. tibourbou*** Aubl. (Pão de Jangada). Brazil. Oil (Apeiba Oil, Buril Oil) extracted from the seeds is used to treat rheumatism.

Apeiba Oil – Apeiba tibourbou.

Apello Gall – Quercus lusitanica.

***Aperia crinata*** Palis.=Dichelachne crinata.

APHANANTHE Planch. Ulmaceae. 1 sp. Madagascar; 3 spp. India, Ceylon to Japan, Indochina, Philippines, Celebs, Java. E.Australia; 1 sp. Mexico. Trees.

***A. aspera*** Engl. Japan. The hard, rough leaves collected during the autumn are used to sandpaper wood.

APHANIA Blume. Sapindaceae. 24 spp. Indomalaya; 1 sp. W. Africa. Woody shrubs.

***A. senegalensis*** Radlk. (Hever). Senegal. Fruits eaten locally. They are rather acid. The seeds are poisonous.

APHLOIA (DC) Benn. Flacourtiaceae. 1 or 6 spp. Tropical E.Africa, Madagascar. Tree.

***A. mauritana*** Baker.=A. theaformis. Infusion of the leaves drunk locally as a beverage.

Apio Arracacia – Arracacia xanthorhiza.

APIOS Fabr. Leguminosae. 10 spp. E.Asia, N. and C.America. Vines.

***A. americana*** Medik.=A. tuberosa.=Glycine apios. (Ground Nut, Potato Bean). E. and S.E. N.America. Tubers boiled or roasted. They are sweet, and were an important food for the Indians. Occasionally cultivated.

***A. tuberosa*** Moench.=A. americana.

Apitong Gurjung Oil Tree – Dipterocarpus grandiflorus.

Apitong Resin – Dipterocarpus grandiflorus.

APIUM L. Umbelliferae. 1 sp. Europe to India, N. and S.Africa. Herb.

***A. bulbocastanum*** (L.) Caruel.=Bunium bulbocastanum.

***A. carvi*** (L.) Crantz.=Carum carvi.

***A. graveolens*** L. (Celery). Cultivated in Old and New World as a vegetable. The petioles of Blanching Celery (***A. graveolens*** var. ***dulce*** (Miller) Pers.) are whitened by earthing up, and eaten raw or cooked. Celeriac (***A. graveolens*** var. ***rapaceum*** (Mill.) Gaud.) has a tuberous root which is eaten raw or cooked as a vegetable. The curly leaves of ***A. graveolens*** var. ***silvestre*** Presl. are used to flavour cooked dishes. The seeds are used to aid digestion, to regulate menstruation, and as a tonic. When ground they are used as a pain-relieving poultice. An oil (Oil of Celery) extracted from the seeds is used to make celery salt, and to flavour foods and sauces. The oil contains selinene, susquiterpene, and d-limonene.

***A. petroselinum*** L.=Petroselinum crispum Hoffm.

APLECTRUM (Nutt.) Torr. Orchidaceae. 1 sp. temperate N. America. Herb.

***A. hymenale*** (Muhl.) Torr. (Putty Root). The root (Radix Aplectri) is used medicinally.

***Aplopappus*** Cass.=Haplopappus.

APOCYNUM L. Apocynaceae. 7 spp. N.America to Mexico, S.Europe to China. Herbs.

***A. androsaemifolium*** L. (Spreading Dogbane, Wild Ipecac). N. America. The root which contains apocynein, and apocynin is used as a laxative, a diuretic and to induce perspiration. The latex is a possible source of rubber.

***A. cannabinum*** L. (Hemp Dogbane, Indian Hemp). N.America esp. E. A fibre extracted from the bark is used as cordage, sails and fishing nets by the local Indians. The latex, used by the Indians as chewing gum is a potential source of rubber. The rhizome contains apocyamarin, and is a heart stimulant, emetic, expectorant, and induces sweating.

***A. sibiricum*** Pall.=A. venetum.

***A. venetum*** L.=A. sibiricum. S.Europe to China. Fibres (Turka, Turkistan Kendir) used to make cordage, cloth, sails and fishing nets.

APODANTHERA Arn. Cucurbitaceae. 15 spp. Warm America. Vine.

***A. smilacifolia*** Cogn. Brazil. Leaves and roots used locally to treat syphilis.

APODYTES E. Mey. ex Arn. Icacinaceae. 16 spp. S.Africa to Malaysia. Trees.

***A. dimidiata*** E. Mey. (White Pear). S.Africa, S.W. Africa, through to Abyssinia. The hard elastic, white wood is used to make wagons.

APONOGETON L. f. Aponogetonaceae. 30 spp. Old World Tropics, and S.Africa. Herbs, aquatic.

***A. bernierianus*** Hook. See A. cordatus.

***A. cordatus*** Jum. Madagascar. The tubers are eaten locally.

***A. distachyus*** L. f. (Cape Asparagus). Flower heads eaten cooked, or in pickles.

***A. viridis*** Jum. See A. cordatus.

APORUSA Blume corr. Bl. Euphoriaceae. 75 spp. Indomalaysia to Solomon Islands. Small trees.

***A. frutescens*** Blume. Java. Wood used for house construction, and tool handles. Bark used as a mordant for dye derived from Indian Mulberry (Morinda trifolia).

***A. lindleyana*** Meull. (Valuka, Vittil). Tropical Asia, esp. India and Ceylon. An infusion of the roots used locally to treat fevers, jaundice, and insanity.

***A. microcalyx*** Hassk.=Tetraactinostigma microcalyx. (Marane, Pelangas, Peuris). Malayan Islands, Philippine Islands. Wood hard and durable. It is used locally for house construction, furniture and kitchen utensils.

APOSTASIA Blume. Orchidaceae. 10 spp. Tropical Asia, Malaysia, Australia.

***A. nuda.*** R. Br. (Kentjing pelandook, Pelampas boodak). Malayan Peninsula. Herbs used locally to treat diarrhoea.

***Apoterium sulatrii*** Blume.=Calophyllum soulattrii.

Apple – Malus pumila.

Apple, African Mammee – Ochrocarpus africana.
Apple, Akee – Blighia sapida.
Apple, Balsam – Clusia rosea.
Apple, Balsam – Echinocystis lobata.
Apple, Balsam – Momordica balsamina.
Apple, Bell – Passiflora laurifolia.
Apple, Bitter – Solanum supinum.
Apple, Bush – Heinsia pulchella.
Apple, Butter – Cucumis myriocarpus.
Apple, Cashew – Anacardium occidentale.
Apple, Chinese – Malus prunifolia.
Apple, Chinese Crab – Malus baccata.
Apple, Chinese Crab – Malus hupehensis.
Apple, Common – Malus pumila.
Apple, Common Custard – Annona reticulata.
Apple, Crab – Malus angustifolia.
Apple, Crab – Malus coronaria.
Apple, Crab – Schizomeria ovata.
Apple, Custard – Annona squamosa.
Apple, Dead-Sea – Quercus tauricola.
Apple, Elephant – Dillenia indica.
Apple, Elephant – Limonia acidissima.
Apple, Emu – Owenia acidula.
Apple, Fragrant Crab – Malus coronaria.
Apple, Himalayan May – Podophyllum emodi.
Apple, Kangaroo – Solanum oviculare.
Apple, Kei – Doryalis caffra.
Apple, Love – Lycopersicon lycopersicum.
Apple, May – Podophyllum peltatum.
Apple, May – Rhododendron nudiflorum.
Apple, Mess – Blakea trinervis.
Apple Mint – Mentha rotundifolia.
Apple, Monkey – Clusia flava.
Apple, Monkey – Licania platypus.
Apple, Mountain – Eugenia malaccensis.
Apple, Oregon Crab – Malus rivularis.
Apple, Otaheite – Spondias dulcis.
Apple, Pond – Annona glabra.
Apple, Prairie – Psoralea esculenta.
Apple, Rose – Eugenia jambos.
Apple, Sand – Parinari mobola.
Apple, Siberian Crab – Malus baccata.
Apple of Sodom – Calotropis procera.
Apple of Sodom – Solanum sodomeum.
Apple, Sugar – Annona squamosa.
Apple, Wood – Limonia acidissima.
Apple Wood – Feronia lemonia.
Apple Wood – Malus pumila.
Appura – Juniperus macropoda.
Apricot – Ameniaca vulgaris.
Apricot Briançon – Armeniaca brigantina.
Apricot, Japanese – Armeniaca mume.
Apricot Kernel Oil – Armeniaca vulgaris.
Apricot, Mammee – Mammea americana.
Apricot Plum – Prunus simonii.
Apricot, St. Domingo – Mammea americana.
Apricot Vine – Passiflora incarnata.

APTANDRA Miers. Olacaceae. 3 spp. Tropical S.America; 1 sp. tropical W.Africa. Trees.

*A.* ***spruceana*** Miers. Brazil. An aromatic oil, used in perfumery is extracted from the roots. (Sandal Oil, Castanha de Cotia Oil, Sando de Maranhão).

APULEIA Mart. Leguminosae. 2 spp. Brazil. Trees.

*A.* ***leiocarpa*** (Vog.) Macbride. = A. praecox.

*A.* ***praecox*** Mart. S.America. The wood is golden-yellow to coppery. It is tough heavy and durable, and used for construction work, and shafts and wheels of carts.

Aqua Aurantii Florum – Citrus aurantium.
Aquacatillo – Meliosma herbertii.
Aqua Chamomilla – Matricaria discoides.
Aquae Rosae Fortior – Rosa centifolia.
Aquameil – Agave complicata.
Aquameil – Agave melliflua.
Aquamiel – Agave quiotifera.
Aquamiel – Agave weberi.
Aquiboquil – Lardezabala biconata.

AQUILARIA Lam. Thymelaceae. 15 spp. S. China, S.E. Asia, Indomalaysia. Trees.

*A.* ***agallocha*** Roxb. The wood becomes fragrant under pathological conditions (Aloes Wood). It is being used for religious purposes by the Mohammedans, Siamese and Chinese. An essential oil (Aggar Atta, Chuwah) is distilled from the wood in Assam. It is used in perfumery.

*A.* ***bancana*** Miq. = Gonystylus miquelianus.

*A.* ***hirta*** Ridley. See A. malaccensis.

*A.* ***malaccensis*** Lam. (Alim, Karas, Mengkaras). Malaya. Wood is burnt locally as incense.

A. moszkowskii Gilg. See A. malaccensis.

AQUILEGIA L. Ranunculaceae. 100 spp. N.Temperate. Herbs.

*A.* ***formosa*** Fisch. (Sitka Columbine). W. N.America. A mash of the seeds is used to remove head lice.

*A.* ***vulgaris*** L. (Common Columbine). Europe, N.Africa, temperate Asia. Cultivated. Plant used as a diuretic, to induce perspiring, to combat scurvy, and as a tranquilizer. The flowers have been used as a tea substitute.

Arabian Coffee – Coffea arabica.

Arabian Cotton – Gossypium herbaceum.
Arabian Dragon' Blood – Dracaena schizantha.
Arabian Horse-Radish Tree – Moringa aptera.
Arabian Jasmine – Jasminum sambuc.
Arabian Senna – Cassia angustifolia.
Arabian Tea – Catha edulis.
Arabian Wolfberry – Lycium arabicum.
Arabic Gum – Acacia senegal.
Araca – Psidum araca.
ARACHIS L. Leguminosae. 15 spp. Brazil, Paraguay. Herbs.
***A. hypogaea*** L. (Goober, Ground Nut, Pea Nut). Native in S. America but cultivated throughout the world. 85 per cent of world production from India, China, W.Africa and U.S.A. It is grown mainly for the oil (Peanut Oil), but only 10 per cent of the U.S.A. production is used for oil. 50 per cent of the U.S.A. crop is used for peanut butter. The fruits ripen below the soil. The oil is used for cooking, margarine-making, and soap manufacture. It stands heating, so is particularly suitable for frying. The meal left after oil extraction is used as a livestock feed. Its high protein content makes it a valuable food, so the meal is used in ice-cream manufacture, and in peanut milk. The meal is also used to make paints and dyes. The young pods are used as a vegetable. The green parts are used to feed livestock. The hulls are used as fuel, making insulation blocks, and bedding for stock.
***A. prostrata*** Benth. S.America. Used locally as a green manure.
Aradal – Garcinia cambogia.
Aragoadha – Pithecellobium bijeminum.
Arak – Saccharum officinale.
ARALIA L. Araliaceae. 35 spp. Indomalaysia, E.Asia, N.America. Herbs or Shrubs.
***A. chinensis*** L. (Chinese Aralia). Japan. Shrub. Young leaves eaten locally as a vegetable.
***A. cordata*** Thunb. (Udo) Japan. Herb. Cultivated for young stalks which are eaten as a vegetable.
***A. ginseng*** P. and P. = Panax ginseng.
***A. horrida*** Benth. and Hook. W. N.America. Young stems eaten by Eskimos.
***A. papyrifera*** Hook. = Tetrapanax papyrifera. China. Tree. The pith is used for the manufacture of rice paper. This is used in China to paint on, to make toys and artificial flowers. It is also used as a surgical dressing.
***A. racemosa*** L. (American Spikenard). N.America. Herb. Rhizomes collected in the autumn are used as a purgative and to induce sweating.
***A. repens*** Max. = Panax repens.
***A. spinosa*** L. (Devil's Walking Stick, Hercules Club, Prickly Ash, Prickly Elder). E. to S.E. United States of America, Japan, Manchuria. Shrub. The bark contains an essential oil, and is used in Japan as a tonic and to induce sweating. The boiled roots are applied to boils, and a cold infusion of the roots is used on sore eyes locally by the Indians of N.America.
Aralia, Chinese – Aralia chinemsis.
Aramina – Urena lobata.
Arangan Oil – Ganophyllum falcatum.
Arapoca – Raputia alba.
Arapoca amarella – Raputia magnifica.
Arar – Acacia pinnata.
Arar – Tetraclinis articulata.
Ararcanga – Aspidosperma desmanthum.
Arariba – Centrolobium robustum.
***Arariba viridiflora*** Allem – Sickingia viridiflora.
***Arariba viridiflora*** Allem. = Simira viridiflora.
Araroba Powder – Vetaireopsis speciosa.
Araticu – Rollinia laurifolia.
ARAUCARIA Juss. Araucariaceae. 18 spp. E.Australia, New Zealand, Norfolk Islands, New Caledonia, S.Brazil to Chile. Trees.
***A. araucana*** (Molina) Koch. = A. imbricata.
***A. bidwillii*** Hook. (Bunya-Bunya). E. Australia. Seeds eaten locally by natives.
***A. brasiliana*** A. Rich. (Brazilian Pine, Parana Pine). Brazil, N.Argentine, Paraguay. An important timber. The brown to red wood is easy to work, but not resistant to decay. It is used for general construction work, making Venetian blinds, and backing electrotypes. Edible seeds are sold locally.
***A. cookii.*** R. Br. New Caledonia. Light-coloured wood is attacked by borers. It is used locally for general carpentry.
***A. cunninghamii*** Sw. (Hoop Pine, Moreton Bay Pine). New South Wales, Queensland. Durable wood used for general construction work.
***A. imbricata*** Pav. = A. araucana. Coastal areas of Chile. The pale yellow wood is

used for general construction work, house timbering, furniture and boxes. Paper pulp is also made from it. The seeds (piñones) are eaten locally roasted. The resin from the trunk is used medicinally by the local Indians.

***A. klinki*** Lauterb. (Hoop Pine, Klinki). New Guinea. The wood is light yellow-pink with a smooth finish. It is used as house timbers, furniture, turning and dowelling, plywood, and paper pulp.

Araucaria, Oil of – Neocallitropsis araucarioides.

ARAUJIA Brot. Asclepiadaceae. 2–3 spp. S.America. Herbs.

***A. serifera*** Brot. Argentina, Peru. A good fibre is extracted from the stem. It is potentially useful for making fabrics.

Arbol de la Cera – Myrica mexicana.

Arbol del diablo – Monsonia americana.

Arbol del Hule – Castilla elastica.

Arbol del muerto – Ipomoea murucoides.

Arbol del venado – Ipomoea murucoides.

Arbol santo – Guaiacum coulteri.

Arbor-Vitae, Eastern (Oil) – Thuja occidentalis.

Arbor-Vitae, Giant – Thuja plicata.

Arbre à Suif – Pycnanthus kombo.

ARBUTUS L. Ericaceae. 20 spp. N. and C.America. W.Europe, Mediterranean, W.Asia. Small trees.

***A. discolor*** Hook. = Arctostaphylos arguta.

***A. macrophylla*** Mart. and Gal. Mexico. Leaves used as food for silkworm (Eucheiria socialis Westw.).

***A. menziesii*** Pursh. (Madrona). Pacific N.America, Mexico. Hard heavy, close grained wood used for furniture, and charcoal. Bark used for tanning.

***A. unedo*** L. (Strawberry Tree). Mediterranean. Edible sweet mealy berries eaten. Sold locally. The berries are also used to make alcoholic beverages and preserves. The bark, leaves and fruits are used for tanning. The bark contains andromedotoxin which is used in treating diarrhoea.

***A. uva-ursi*** L. = Arctostaphylos uva-ursi.

***A. xalapensis*** H. B. K. See A. macrophylla.

Arbutus, Trailing – Epigaea repens.

ARCANGELISIA Becc. Menispermaceae. 3 spp. Malaysia. Woody vines.

***A. flava*** (L.) Merr. = Anamirta flavescens Miq. = Menispermum flavum L. Is. of S.E. Asia. A yellow dye is extracted from the wood in large quantities. An extract of the bark is used to cure ulcers, itches and as a general germicide. An extract of the roots is used as a tonic, to cure intestinal worms and control menstruation. It is also an abortive.

Arcel – Parmelia caperala.

ARCHANGELICA Hoffm. Umbelliferae. 12 spp. N.Temperate. Herbs.

***A. actaeifolium*** Michx. = Coelopleurum actaefolium (Michx.) Coult. and Rose. (Seacoast Angelica). Coast of N.E. America. Young stems and leaf-stalks cooked and eaten by local Eskimos.

***A. gmelini*** DC. = Coelopleurum gmelini (DC) Ledeb. N. America, N.E. Asia. Stems and leaf-stalks eaten, raw or cooked by Eskimos.

***A. officinalis*** (Moench.) Hoffm. = Angelica archangelica.

Archil – Roccella phycopsis.

Arctic Bramble – Rubus arcticus.

ARCTIUM L. Compositae. 5 spp. Paleotemperate. Herbs.

***A. lappa*** L. = Lappa major Gaertn. (Great Burdock). Temperate Europe and Asia; introduced into N.America. The young roots are dried and used as a diuretic, to encourage the appetite, and to induce sweating. They contain inulin, a bitter compound, and an essential oil. The roots are eaten as a vegetable in Japan.

***A. minus*** (Hill.) Bernh. = Lappa minor Hill (Common Burdock). Europe, Asia, introduced into America. Dried root used as diuretic, to encourage appetite, and to induce sweating.

ARCTOPUS L. Umbelliferae. 3 spp. S.Africa. Herbs.

***A. echinatus*** L. S.Africa. Infusion of the roots used locally as a diuretic, to treat skin diseases, and as a mild laxative.

ARCTOSTAPHYLOS Adans. Ericaceae. 70 spp. W. N. and C. America, 1 sp. N.Temperate and circumpolar. Shrubs.

***A. arguta*** Zucc. = Comarstaphylos arguta = Arbutus discolor. S. Mexico. Extracts of the fruit are used locally as an hypnotic.

***A. manzanita*** Parry. (Manzanita). S.W. United States of America. The berries are used to make jelly. The Indians make an alcoholic beverage from the fruits.

***A. nevadensis*** Gray. S.W. United States of America. The leaves are smoked, mixed with tobacco, by local Indians.

***A. pungens*** H. B. K. (Pointleaf Manzanita). California, Mexico. Fruit eaten locally.

***A. tomentosa*** Pursh. British Columbia to

California. The wood is used in the United States for fine furniture. The sweet, dry, mealy fruit is eaten locally, and made into an alcoholic beverage. The leaves are smoked by the Indians.

***A. uva-ursi*** (L.) Spreng.=Arbutus uva-ursi. (Bearberry). N.Hemisphere. Fruits eaten locally. Leaves used for tanning in Russia and Sweden. The dried leaves contain arbutin and quercetin, and are used as a tonic, astringent and diuretic. The leaves are used to make a beverage (Kutai Tea, Caucasian Tea).

ARDISIA Sw. Myrsinaceae. 400 spp. Warm countries. Shrubs and small trees.

***A. boissieri*** A. DC.=A. squamulosa.

***A. colorata*** Roxb.=A. complanata. (Soompoo, roompooh, Lampeni gede). Malaya. An infusion of the leaves used locally to treat stomach complaints.

***A. complanata*** Wall.=A. colorata.

***A. crenata*** Roxb.=A. crispa.

***A. crenulata*** Lodd.=A. crispa.

***A. crispa*** (Thunb.) A. DC.=A. crenata.=A. Crenulata. (Mata ajam, Popindh). S.E. Asia. Fruits eaten locally in Malaya.

***A. drupacea*** Merr.=A. squamulosa.

***A. fuliginosa*** Blume. Java. Sap boiled with coconut oil is used locally to treat scurvy.

***A. laevigata*** Blume. (Ki mangoo). Java. Leaves eaten locally.

***A. pauciflora*** A. DC.=A. quinquegona.

***A. pumila*** Blume=Labisia pumila.

***A. squamulosa*** Presl.=A. boissieri=A. drupacea. Philippines. Flowers and fruits used to flavour fish.

***Arduina bispinosa*** L.=Carissa arduina.

ARECA L. Palmae. 54 spp. Indomalaysia, Solomon Islands, Bismark Islands, N. Australia. Trees.

***A. calapparia*** Blume=Actinorhytis calapparia.

***A. caliso*** Becc. Philippines. Used locally as a substitute for betel nut.

***A. catechu*** L. (Betelnut Palm). Tropical Asia. Cultivated in S.India, Ceylon, Thailand, Malaysia and Philippines for the nut (Betel Nut). The fruit is about 5 cm. long and contains one seed. The seeds are chewed. They have a mildly narcotic effect. For chewing, the seeds are harvested unripe, and the fibrous husk removed. They may be chewed fresh, or cured by boiling in water and sun-drying. The cured nut has a lower tannin content, is more palatable, and stores better. The seed is chewed after treating with lime, and occasionally cardomom or tumeric, the whole being wrapped in the leaf of the Betel Vine (***Piper betle***). The seeds contain the alkaloids arecoline and arecaidine, and are used to cure tapeworms, to reduce fever and as an astringent. They are also used to cure tapeworms and as a laxative for animals. The nuts are also a source of an inferior latex, and are used in Europe in toothpowders.

***A. hutchinsoniana*** Becc. (Pisa Palm). Philippines. The raw terminal bud is eaten to cure intestinal worms.

***A. ipot*** Becc. See A. caliso.

***A. laosensis*** Becc. See A. triandra.

***A. punicea*** Blume.=Pinanga punicea.

***A. sapida*** Soland.=Rhopalostylis sapida.

***A. triandra*** Roxb. India, Burma, Vietnam, Laos, Cambodia. Cultivated in Cambodia. Fruit. Chewed locally in Cambodia.

***A. vestiaria*** Giseke.=Mischophloeus vestiaria.

ARECASTRUM Becc. Palmae. 1 sp. Brazil.

***A. romanzoffianum*** (Cham.) Becc.=Cocos romanzoffiana. (Pindo Palm). A starchy food (Sagu) is made and eaten locally from the pith. The young buds are eaten locally in oil or vinegar.

Areeta – Sapindus trifoliatus.

Areicha – Grewia flavescens.

ARENARIA L. Caryophyllaceae. 250 spp. N.Temperate. Herbs.

***A. peploides*** L.=Ammodenia peploides. (Seabeach Sandwort). N.Temperate to Arctic. Leaves eaten by Eskimos, raw, pickled or in oil.

***A. rubra*** L.=Spergularia rubra.

***A. serpyllifolia*** L.=Alsine serpyllifolia=A. brevifolia. Temperate Asia, Europe and N.America. Used for bladder complaints by the Chinese.

ARENGA Labill. Palmae. 11 spp. Indomalaysia (except N.Guinea), Caroline Islands, Christmas Islands.

***A. minadoiensis*** Becc.=A. tremula.

***A. pinnata*** (Wurmb.) Merr.=A. saccharifera.

***A. saccharifera*** Lab.=A. pinnata. (Sugar Palm). Malaysia. The sap is extracted from the stem and evaporated to yield palm sugar which is used widely locally. The fermented juice is Palm Wine or Toddy, which on distillation gives the spirit Arrack. The pith is used as a source of starch – a minor source of sago. The leaves yield a fibre, and the trunks are used as water pipes.

***A. tremula*** (Blanco.) Becc. = A. minadoiensis. (Gumayaka Palm). Philippines. The buds are eaten locally to induce intoxication and sleep. The fruits are poisonous.

Areuj batalingbingan – Adenia cordifolia.

Areuj beungbeurootan – Vitis arachnoidea.

Areuj harijang – Vitis repens.

Areuj ki barera – Vitis compressa.

Areuj ki barera – Vitis geniculata.

Areuj ki barere – Vitis papillosa.

Areuj pitjong tjeleng – Hodgsonia macrocarpa.

Areuj tiwook – Trichosanthes oxigera.

Arfedj – Anvellea australis.

Argan – Argania spinosa.

Argan Gum – Argania sideroxylon.

Argan Oil – Argania spinosa.

ARGANIA Roem. and Schult. Sapotaceae. 1 sp. Morocco. Tree.

***A. sideroxylon*** Roem. = A. spinosa.

***A. spinosa*** (L.) Skeels. = A. sideroxylon. A cooking oil (Argan Oil, Huile d'Argan, Oleum Araganiae) is extracted from the seeds, and the husks are fed to cattle. The whole fruit is also used as cattle food. The trunk yields gum (Argan), and the wood is used locally for construction work.

ARGEMONE L. Papaveraceae. 10 spp. W. and E.United States of America, Mexico, W. Indies. Herbs.

***A. glauca*** (Prain.) Degener. (Hawaiian Poppy, Puakala). Polynesia. The juice of the stems is used locally to treat warts. The seeds are eaten in cakes etc.

***A. intermedia*** Sweet. S. N.America. Ashes of leaves used for tattooing by local Indians.

***A. mexicana*** L. (Mexican Prickly Poppy). S.W. United States of America, Mexico. Seeds yield oil (Mexican Poppy Seeds Oil, Prickly Poppy Seed Oil). Used to manufacture soap, and for lighting. Medically it is used as a purgative.

Argentine Jujub – Ziziphus mistol.

***Argithamnia tinctoris*** Millsp. = Ditaxis tinctoria.

ARGYREIA Lour. Convolvulaceae. 90 spp. Indomalaysia; 1 sp. Australia. Vine.

***A. speciosa*** Sweet. (Woolly Asiaglory). Malaysia Leaves used locally as poultices.

***Argyrodendron*** F. Meull. = Tarrietia.

***Aria torminalis*** Beck. = Sorbus torminalis.

ARILLASTRUM Panch. ex Baill. Myrtaceae. 1 sp. New Caledonia. Tree.

***A. gummiferum*** Panch. = Spermolepis gummifera. The red wood is hard and used for interior work, and ship-building. A resin is extracted from the wood (Chene Gom), and a tannin is obtained from the bark.

Arinj – Acacia leucophloea.

Arinja Iba – Philodendron speciosum.

Aringa Imbé – Philodendron speciosum.

ARISAEMA Mart. Araceae. 150 spp. E. Africa, tropical Asia, Atlantic N.America. to Mexico. Herbs.

***A. heterophyllum*** N. E. Brown. E.Asia. Used locally in China to stimulate salivation.

***A. japonicum*** Blume. Japan. Corms eaten locally roasted.

***A. lobatum*** Engl. See A. heterophyllum.

***A. pentaphyllum*** Thunb. See A. heterophyllum.

***A. speciosum*** (Wall.) Mart. (Kiralu, Sampki-Khumb). Himalayas. Root used locally as an antidote to snakebite. It is also used to kill intestinal worms in cattle, and treat sheep for colic.

***A. tortuosum*** (Wall.) Schott. See A. speciosum.

***A. triphyllum*** (L.) Torr. (Indian Turnip, Jack-in-the-Pulpit). E. United States of America. The dried tubers, being expectorant and irritant are used to treat asthma, bronchitis, and whooping cough. They were eaten by the Indians, and are used to some extent in the preparation of cosmetics.

ARISTEA Soland. ex Ait. Iridaceae. 60 spp. tropical and S. Africa, Madagascar. Herbs.

***A. alata*** Baker. Tropical Africa. The sap is mixed with that of the fern Pteridium aquilinum and used for tattooing, locally.

***A. polycephala*** Narms. See A. alata.

ARISTIDA L. Graminae. 330 spp. Temperate and sub-tropical. Grasses.

***A. pungens*** Desf. = Arthratherum pungens. Sahara. Seeds eaten locally in times of famine.

ARISTOLOCHIA L. Aristolochiaceae. 350 spp. Tropical and temperate. Woody vines or trailing shrubs.

***A. albida*** Duch. Tropical W.Africa. The root is used locally to treat nematode infections (Guinea worm), and as a tonic.

**A. *amazonica*** Ule. Brazil. Roots used as a tonic, to reduce fevers, as a diuretic, to regulate menstruation, as an abortive, to aid digestion and as an antiseptic.

**A. *antihysterica*** Mart. Tropical America. The roots are used locally in Brazil as an emetic, to induce sweating, and to control menstruation.

**A. *barbata*** Jacq. = Howardia barbata (Jacq.) Klotsch. S.America. The underground parts are used to treat snake-bite in parts of Brazil.

**A. *bracteata*** Retz. Tropical Africa to India. The bitter roots are used to combat intestinal worms and are purgative. In the Sudan, the roots are used to treat scorpion bites, and with the leaves are used to combat Guinea worm.

**A. *brasiliensis*** Mart. and Zucc. Brazil. The rhizome (Raiz de José Domingo, Papo de Peru) is used locally to treat snake-bite.

**A. *burchelii*** Mart. Brazil. The roots are used locally as a tonic, to aid digestion, an antiseptic, to reduce fever, as a diuretic, to control menstruation, and as an abortive.

**A. *chilensis*** Miers. Chile. Roots used locally to control menstruation

**A. *contorta*** Bunge E. Asia. See A. kaempfrei.

**A. *cordiflora*** Mutis ex H. B. K. Columbia. Rhizome used locally against snake-bite.

**A. *cordigera*** Willd. = Howardia cordigera. Brazil. The roots are used locally to control fevers and to control menstruation.

**A. *crenata*** Mart. See A. amazonica.

**A. *cymifera*** Mart. and Zucc. S.W. S.America. In Brazil, the roots are used to treat asthma, to reduce fever, diarrhoea, and as a diuretic. In the Argentine and Paraguay the roots are used to control menstruation.

**A. *elegans*** Mart. See A. amazonica.

**A. *fragrantissima*** Ruiz. Peru. The rhizome is used locally to treat snake-bite.

**A. *galeata*** Mart. and Zucc. Brazil. The rhizome is used locally to treat snake-bite.

**A. *glaucescens*** H. B. K. = Abuta amara. Columbia. The roots yield (Yellow Pareira) which is used as a diuretic and tonic.

**A. *grandiflora*** Swartz. C.America, W.Indies. The roots are used locally to induce sweating, to treat snake-bites, to control menstruation and as an abortive. The leaves are used locally among the Indians of Mexico to induce sweating and to treat colds and chills.

**A. *kaempferi*** Willd. E.Asia. Roots yield the drug (Ma-tou-ling), used to treat asthma, lung complaints and haemorrhoids.

**A. *longa*** L. E.Mediterranean. Used locally to treat snake-bite, as a tonic, and to relieve stomach ailments.

**A. *lutescens*** Duch. See A. amazonica.

**A. *maurorum*** L. E. Mediterranean. Used locally to heal wounds, and to treat scab in sheep.

**A. *maxima*** L. (Contracapetano, Cuajilote, Guaco de sur). C. and N. S.America. Root used locally to treat snake-bite.

**A. *odoratissima*** L. C.America, W.Indies. The rhizome is used locally to treat snake-bite.

**A. *pallida*** L. Asia Minor. The rhizome has been used since the time of the Ancient Greeks to treat snake bite.

**A. *pardina*** Duchartre. (Bejuco amargo, Guaco, Huaco). Mexico. The leaves are used locally to treat fever, and the stems as cordage.

**A. *philippensis*** Warb. (Barubo). Philippines. Roots used locally for the treatment of stomach complaints, and to control menstruation.

**A. *reticulata*** Nutt. (Texas Snakeroot) S. United States of America. The dried rhizome is aromatic, bitter and a stimulant. It is used to encourage sweating, and locally by the Indians to relieve stomach pains and to treat snake bite. It contains borneol.

**A. *rotunda*** L. E.Mediterranean. The roots are used in Iran to combat intestinal worms, as a diuretic and to control menstruation. It is also a tonic.

**A. *rumicifolia*** Mart. and Zucc. Brazil. Used locally to treat snake bite.

**A. *sericea*** Blanco. Philippine Islands. The upper parts of the plant are used fresh to treat intestinal worms, to relieve digestive complaints, and to control menstruation. The root is an abortive.

**A. *serpentaria*** L. (Virginian Snakeroot). S.E. United States of America. Used as A. reticulata.

**A. *tagala*** Cham. Philippine Islands. The roots are used locally as a tonic to relieve digestive complaints, and control menstruation. A poultice of the root applied to the abdomen of infants relieves distention caused by excess gases.

*A. taliscana* Hook. and Arn. Mexico. Used locally to treat snake bite.

*A. theriaca* Mart. Brazil. Roots used locally to treat snake bite.

*A. turbacensis* H. B. K. Columbia. Rhizome used locally to treat snake bite.

*A. warmingii* Mart. Brazil. Roots used as a tonic, antiseptic, to relieve indigestion, to combat intestinal worms, as a diuretic to control menstruation, and as an abortive.

ARISTOTELIA L'Hérit. 5 spp. E.Australia, Tasmania, New Zealand, Peru to Chile. Small trees or shrubs.

*A. macqui* L'Hérit. (Chilean Wineberry, Maqui). Chile. The fruits are eaten locally.

*A. racemosa* Hook. f. New Zealand. The white wood is used for cabinet work, turning, and charcoal making. The fruit is eaten locally.

*A. serrata* Oliv. New Zealand. A decoction of the leaves is used locally to treat rheumatism, sore eyes, burns and boils.

Arizona Cypress – Cupressus arizonica.

Arizona Poplar – Populus arizonica.

Arizona Yellow Pine – Pinus arizonica.

Arjan Terminalia – Terminalia arjuna.

ARJONA Comm. ex Cav. Santalaceae. 10 spp. temperate S.America. Herbs.

*A. tuberosa* Cav. Patagonia. The sweet tubers (Macachi) are eaten locally.

Arkol – Rhus wallichii.

Armand Pine – Pinus armandi.

ARMENIACA Mume. (Sieb. and Zucc.) Sieb. ex Carr. Rosaceae. 10 spp. Temperate Asia. Trees.

*A. brigantina* (Vill.) Pers. = Prunus brigantina Vill. (Briançon Apricot). S. France. Cultivated for a scented oil (Huile de Marmotte) extracted from the seeds.

*A. mume* (Sieb. and Zucc.) Sieb ex Carr. = Prunus mume Sieb. and Zucc. (Japanese Apricot). Japan. Cultivated in Japan and China for the fruit which is eaten raw, or boiled, or preserved in sugar or salt.

*A. vulgaris* Lam. = Prunus armeniaca L. (Apricot). W. Asia. It was once considered native of the Caucasus. The apricot has been cultivated since ancient times in C. and S.E Asia, S.Europe and N.Africa. It was introduced in to California in the 18th century by the mission fathers. The plant is drought-resistant, and long-lived (30–100 years). It grows well in light loamy soil when protected from late frost. The crop is produced commercially in U.S.A. (about 200,000 tons per annum), Iran, Syria, Spain, France, Italy, and Yugoslavia. The fruit is eaten raw, dried for cooking (70 per cent of the U.S.A. crop is dried). It is canned or made into pulp, or ice-cream. Cordials and liqueurs are prepared from the juice. The seeds yield an oil (Apricot Kernel Oil) which is used in toilet preparations and pharmaceuticals. The oil is semi-drying and edible.

ARMERIA (DC.) Willd. Plumbaginaceae. 80 spp. N.Temperate, 2 spp. Andes. Herbs.

*A. elongata* Koch. = A. maritima.

*A. maritima* (Mill.) Willd. = A. elongata = A. vulgaris = Statice armeria. (Gilly Flower, Sea Pink) Europe, Asia. The leaves (Herba Armeriae maritima) are used for slimming.

*A. vulgaris* Willd. = A. maritima.

ARMILLARIA (Fr.) Quél. Agaricaceae. 40 spp. Cosmopolitan. The fruit bodies of all spp. mentioned are eaten locally, and sold in markets.

*A. bulbigera* (A. and S.) Quél. Sweden.

*A. distans* Pat. Congo.

*A. gymnopoda* Bull. = Armillariella tabescens (Scop. ex Fr.) Sing. = Clitocybe tabenscens (Scop. ex Fr.) Pres. Cosmopolitan. The fruit-body is eaten raw, and has a pleasant flavour.

*A. matsutake* S. Ito and Imai. Japan.

*A. mellea* (Vahl. ex Fr.) Quél. C. and E. Europe. The fruit-bodies are eaten cooked, salted or pickled, as they have an unpleasant taste when raw. The fungus (honey fungus) causes serious root diseases of a wide variety of tree crops.

*A. ventricosa* Peck. Japan, and N.America.

ARMORACIA Gilib. 3 spp. S.E. Europe to Siberia. Herbs.

*A. lapathifolia* Gilib. = A. rusticana. = Cochlearia armoracia. (Horse Radish). W. Asia, Mediterranean, Europe. Cultivated for the roots which are used as a condiment (Horse Radish Sauce, Horse Radish Vinegar, Horse Radish Powder). It is usually ground and mixed with vinegar, cream etc. The flavour is due to allyl isothiocyanate, which is formed by the exposure of myorosin and sinigrin in the roots to the air.

*A. rusticana* Gaertn. = A. lapathifolia.

ARNEBIA Forsk. Boraginaceae. 25 spp. Mediterranean, tropical Africa, Himalayas. Herbs.

***A. hispidissima*** DC.=Lithospermum hispisdissimum Lehm. Tropical and N.E. Africa, N.India. The roots yield a red dye.

ARNICA L. Compositae. 32 spp. N.Temperate, to Arctic. Herbs.

***A. fulgens*** Pursh.=A. monocephala=A. pedunculata. C. and N.W. United States of America. The dried rhizome, roots and flower-heads are used as a tonic, irritant, and to heal wounds. A tincture of the roots is used to treat bruises and strains. The root stock contains a yellow crystalline substance (arnicin), and an essential oil.

***A. monocephala*** Rydb.=A. fulgens.

***A. montana*** L. (Arnica). Europe, Asia, W. N.America. Used as A. fulgens. The root is Arnica Root.

***A. pedunculata*** Rydb.=A. fulgens.

Arnica (Root) – Arnica montana.

Arnotto – Bixa orellana.

Aro – Impatiens burtonii.

AROMADENDRON Blume. Magnoliaceae. 3 spp. Malayan Peninsular, Borneo, Java. Trees.

***A. elegans*** Blume. (Ki lunglung). Malay to Java. The light wood is used for house construction and furniture making.

ARRACACIA Bancroft. Umbelliferae 53 spp. Mexico to Peru. Herb.

***A. xanthorhiza*** Bancroft. (Apio, Arracacia, Arracha). Cultivated locally for the root which is eaten as a vegetable, cooked in a variety of ways.

Arracha – Arracacia xanthorhiza.

Arrack – Arenga pinnata.

Arrack – Cocos nucifera.

Arrack – Saccharum afficinale.

Arrayán – Eugenia foliosa.

Arrayán – Psidium oestedeanum.

Arrayán – Psidium satorianum.

ARRHENATHERUM Beauv. Graminae. 6 spp. Europe, Mediterranean. Grasses.

***A. avenaceum*** Beauv.=A. elatior. (Tall Meadow Oat Grass). Europe. Cultivated as a pasture grass, especially for hay.

Arrowhead – Sagittaria sagittifolia.

Arrowleaf – Sagittaria latifolia.

Arrowleaf Balsamroot – Balsamorhiza sagittata.

Arrowroot – Maranta arundinacea.

Arrowroot, African – Canna edulis.

Arrowroot, Bermuda – Maranta arundinacea.

Arrowroot, Brazilian – Ipomoea batatas.

Arrowroot, Brazilian – Manihot esculenta.

Arrowroot, Chinese – Nelumbium speciosa.

Arrowroot, East Indian – Curcuma angustifolia.

Arrowroot, East Indian – Curcuma leucorhiza.

Arrowroot, East Indian – Tacca pinnatifida.

Arrowroot, False – Curcuma pierrana.

Arrowroot, Fiji – Tacca pinnatifida.

Arrowroot, Florida – Zamia integrifolia.

Arrowroot, Guiana – Dioscorea alata.

Arrowroot, Guiana – Dioscorea batatas.

Arrowroot, Guiana – Musa paradisiaca.

Arrowroot, Guinea – Calathea allouia.

Arrowroot, Hawaiian – Tacca hawainensis.

Arrowroot, Indian – Euonymus atropurpureus.

Arrowroot, Japanese – Pueraria thunbergiana.

Arrowroot, Para – Manihot esculenta.

Arrowroot, Portland – Colocasia antiquorum.

Arrowroot, Purple – Canna edulis.

Arrowroot, Queensland – Canna edulis.

Arrowroot, Sago – Cycas circinalis.

Arrowroot, St. Vincent – Maranta arundinacea.

Arrowroot, Sierra Leone – Canna edulis.

Arrowroot, Tahiti – Tacca pinnatifida.

Arrowroot, Williams' – Tacca pinnatifida.

Arrowwood – Viburnum acerifolium.

ARTABOTRYS R. Br. Annonaceae. 100+ spp. Tropical Africa, Indomalaysia. Trees or vines.

***A. intermedia*** Blume. Indonesia esp. Java. An essential oil is extracted from the flowers and used in perfumery etc.

***A. odoratissimus*** R. Br.=A. uncinatus.

***A. uncinatus*** (Cham.) Merr.=A. odoratissimus. (Climbing Ylang-Ylang). India and Ceylon. The flowers are used to make an aromatic, and stimulating beverage.

ARTEMISIA L. Compositae. 400 spp. N.Temperate, S.Africa, S.America. Herbs.

***A. abrotanum*** L. (Southern Wood). Temperate Asia, and S. Europe. Cultivated for the leaves which are used locally to make a stimulating beverage. The leaves are also used to control menstruation, to combat intestinal worms in children, as an antiseptic and detergent.

***A. absinthium*** L. (Absinthe, Wormwood). Mediterranean, S. Siberia, Kashmir.

Cultivated in Europe, N.Africa, and U.S.A. It is used in the preparation of Vermouth, and Muse Verte (wines), but particularly in the preparation of Absinthe. The spirit contains 68 per cent alcohol (by volume) which is dry and bitter. Many other herbs are used. It was first made by Pernod in 1797, but production was banned in Switzerland in 1908, and in France in 1915. The factory moved to Spain, but was closed during the Civil War (1936–39). Only an imitation absinthe is now made. It contains no Absinthe. The harmful effects (absinthism) is caused by absinthin and thujon which cause delirium, hallucinations and permanent mental illness. The leaves and flower tops are used as a tonic, to cure stomach complaints and to combat intestinal worms.

***A. afra*** Jacq. S. Africa. Used locally as A. absinthium.

***A. annua*** L. Temperate Asia. Used in Indo-China to cure stomach complaints. It is possibly effective against jaundice and some skin complaints.

***A. apiacea*** Nance = A. carvifolia.

***A. barrelieri*** Ben. S. Europe. An essential oil extracted from the plant is used in the manufacture of Algerian Absinthe.

***A. capillaris*** Thunb. Far East. Cultivated in China and Malaya (called Rumput Roman) to make poultices for headaches.

***A. carvifolia*** Wall. = A. apiacea = A. thunbergiana. (Ch'ing hao). India, China, Japan, N.Vietnam. In China, Japan and N.Vietnam the leaves and stems are used as a tonic, and to treat certain psychoses. It is also used for flavouring tea.

***A. cina*** Berg. (Levant Wormseed). Russia, Turkestan, and Far East. Cultivated in Russia and W. U.S.A. for the dried flower heads (Flores Cinae), which contain sanonin and are used to combat intestinal worms.

***A. dracunculoides*** Pursh. (False Tarragon). N.America. Leaves eaten by local Indians when cooked between heated stones. They also eat the oily seeds.

***A. dracunculus*** L. (Tarragon, Estragon). Russia to Himalayas. Cultivated for the leaves which are used to flavour vinegar, sauces, pickles, etc. It contains a volatile oil (Estragon Oil, Oleum Dracunculi) and is used for flavouring, perfumery, as a diuretic, to control menstruation, and combat intestinal worms. It also relieves toothache.

***A. filifolia*** Torr. N.America. Chewed by the Indians in N. Mexico to cure digestive complaints.

***A. frigida*** Willd. N.America, Siberia. Used by local Indians of N.America to flavour corn.

***A. glacialis*** L. non Wulfen. (Genépi des Glaciers, Gletscherreute). European Alps. Leaves used to flavour liqueurs and vermouth.

***A. glacialis*** Wulfen non L. = A. laxa.

***A. glutinosa*** J. Gay. S. Europe. Source of a fragrant essential oil used in perfumery.

***A. gnaphalodes*** Nutt. (Sage Bush, Western Mugwort). W. United States of America. A decoction of the leaves is taken by the local Indians to cure colds, and headaches.

***A. herba-alba*** Asso. = A. sieberi. N. Africa. Used locally as a perfume, to combat intestinal worms and to control menstruation.

***A. judaica*** L. N. Africa, Arabia. Cultivated for seeds (Graines à Vers, Semen Contra) which are used locally as a condiment.

***A. laxa*** (Lam.) Fritsch. = A. glacialis Wulfen non L. = A. mutellina. (Genépi). Alpine. Leaves used in the manufacture of a beverage and liqueur, to reduce fevers and to treat stomach complaints.

***A. maciverae*** Hutch. and Dalz. Tropical Africa, esp. N.E. Powdered leaves (Tazargade) used by Arabs to combat intestinal worms.

***A. maderaspatana*** L. = Grangea maderaspatana.

***A. maritima*** L. Coastal Europe to Mongolia. Leaves (Herba Absinthii maritimi) used locally as a tonic and to cure stomach complaints. It is also used to flavour food.

***A. mendozana*** DC. S. America. A decoction of the plant is used in the Argentine to treat stomach complaints.

***A. mexicana*** Willd. (Mexican Mugwort). S.W. United States of America to Mexico. The leaves are used locally by the Indians to combat intestinal worms, to control menstruation, as a stimulant and they are chewed to cure sore throats.

***A. nova*** A. Nels. See A. gnaphalodes.

***A. pontica*** L. (Roman Wormwood). S.E. Europe, W. Asia. Source of Oil of Wormwood, which is used as A. absinthium.

***A. ramosa*** C. Sm. ex Link. (Barbary Santonica). N. Africa, Canary Islands. Used as A. cina.

***A. sieberi*** Bess. = A. herba-alba.

***A. thunbergiana*** Maxim. = A. carvifolia.

***A. tridentata*** Nutt. (Big Sagebush). N. America. Used locally by Indians. Leaves eaten roasted between hot stones, and boiled to produce an infusion used against colds and headaches. The oily seeds are also eaten. The chewed leaves are also used to treat indigestion.

***A. tripartita*** Rydb. See A. dracunculoides.

***A. vulgaris*** L. (Mugwort). Temperate N.Hemisphere. In Europe the leaves and shoots are used to flavour pork and goose. In the Far East they are used to flavour rice and as a tobacco substitute. The boiled young leaves are eaten in Japan.

***A. wrightii*** Gray. See A. dracunculoides.

***Arthratherum*** Beauv. = Aristida.

ARTHROCNEMUM Moq. 20 spp. Mediterranean to Australia and warm N.America. Shrubs.

***A. glaucum*** (Del.) Ung-Sternb. Mediterranean. Used in Greece to flavour porridge.

***A. indicum*** (Willd.) Moq. Tropical Asia. Eaten raw in India.

***A. subterminale*** (Parish.) Standl. = Salicornia subterminalis. California. Seeds eaten as a meal locally by Indians.

ARTHROPHYTUM Schrenk. Chenopodiaceae. 10 spp. W. and C.Asia. Shrubs.

***A. arborescens*** Litw. (Saxaul). Turkistan. A valuable source of firewood. Cultivated.

ARTHROTHAMNUS Gmel. Laminariaceae. Far East.

***A. bifidus*** (Gmel.) Rupr. Japan and China. Used locally to manufacture Kombu. This is the dried, shredded seaweed, which is added to soups etc.

***A. kurilensis*** Rupr. See A. bifidus.

Artichoke, Chinese – Stachys sieboldii.

Artichoke, Globe – Cyara scolymus.

Artichoke, Japanese – Stachys sieboldii.

Artichoke, Jerusalem – Helianthus tuberosus.

ARTOCARPUS J. R. and G. Forst. Moraceae. 47 spp. S.E. Asia. Trees.

***A. altilis*** (Parkinson) Fosberg. = A. communis = A. incisa. (Breadfruit). Malaysia. The fruit is a valuable food, and the tree is cultivated widely in the tropics. The seedless varieties are of better quality, and are propagated by root cuttings or sideshoots. The fruits have a high starch content, with a good protein content, with a little vitamin A. The fruits are usually cooked by frying or boiling before eating.

***A. altissima*** J. J. Smith. Malaysia. The brown streaked wood is close-grained and used for ship-building.

***A. anisophylla*** Miq. Malaysia. The soft durable wood is used for general construction work. The thick latex is used as bird-lime.

***A. brasiliensis*** Gomez. Brazil. Fruit eaten locally. The pulp is eaten raw, and the seeds are roasted.

***A. camansi*** Blanco (Kamansi). Philippines. Occasionally cultivated. Young fruits are eaten as a vegetable. The seeds which contain a high proportion of starch are boiled or roasted.

***A. champeden*** (Lour.) Spreng. = A. polyphema.

***A. chaplasha*** Roxb. (Chaplash). Himalayan area. The yellow-brown wood is used in India for general construction work, ship-building, wheel making, furniture, and carving.

***A. communis*** Forst. = A. altilis = A. incisa.

***A. cumingiana*** Trée. Philippines. Latex (Anubing Gum), a possible base for chewing gum.

***A. dadak*** Miq. Sumatra. The yellowish wood is very hard and used to make bridges and house supports. It is not attacked by termites. The fruits are small, but edible.

***A. dimorphophylla*** Miq. = A. rigidus.

***A. elastica*** Reinw. (Wild Breadfruit). W. Malaysia. Soft durable wood used for general carpentry, boat and house building. Bark used for cordage, and basket-making. Latex (Gumihan Gum), used for chewing gum, and locally for bird-lime. Seeds roasted for eating.

***A. glauca*** Blume. Java. Wood used for house construction. The fruits are acid and used locally to make jelly.

***A. hirsuta*** Lam. (Aini, Anjeli Wood, Pat phanas, Ran ptanas). S.India. Dark strong light durable wood, used as a teak substitute. Used for general construction work, boat-building, furniture, veneers, turning and barrel-making.

***A. incisa*** L. f. = A. altilis.

***A. integrifolia*** L. f. (Jackfruit). India and Malaysia. Cultivated throughout the Old and New World tropics for the fruit. The fruits are about 40 cm. long and contain a soft white pulp. This is eaten cooked or boiled in milk, or as preserves. The seeds are roasted. The young fruits are eaten in soup. In Java the flowers are made into a sweet dish with syrup and agar.

***A. kemando*** Miq. W. Malaysia. The latex is used locally to make a sauce, which is rather sweet. It produces stomach pains if eaten in excess.

***A. lakoocha*** Roxb. (Lakoocha). E. India and Malaysia. Cultivated in Old and New World for the sub-acid fruits and yellow wood. This is used for boat-building and furniture making.

***A. lanceaefolia*** Roxb. Malaysia. Hard tough yellow wood which polishes well. A good lumber sp. Used for general construction work and furniture.

***A. limpato*** Miq.=Prainea limpato. Sumatra. Hard orange-red wood, not attacked by termites and keeps well under water. Used for bridges and house-making.

***A. odoratissima*** Blanco. (Marang). S. Philippines. Cultivated in New World tropics for the fruit, which are smaller than those of A. integrifolia, but of superior quality.

***A. polyphema*** Pers.=A. champeden. (Champedak). Tropical Asia, esp. W. Malaysia. Cultivated for the dark yellow durable wood which is used for boat-building and furniture. The young fruits are eaten locally in soup.

***A. rigidus*** Blume.=A. dimorphophylla. (Monkey Jack). Malaysia. Cultivated locally for the fruit. The red-brown wood is durable and termite-resistant. It is used for house-building, furniture, and boat-building.

***A. teijsmannii*** Miq. Malaysia. Wood used locally for boat-building. Latex used locally for bird-lime.

Arui – Erythrophleum guineënsis.

ARUM L. Araceae. 15 spp. Europe, Mediterranean. Herbs.

***A. ioscoridis*** Sibth. and Smith.=A. hygrophyllum.

***A. dracunculus*** L.=Dracunculus vulgaris.

***A. guttatum*** Salisb.=Dracunculus vulgaris.

***A. hygrophyllum*** Boiss var. Sintenisii Engl.=Arum dioscoridis. Asia Minor. Rhizomes eaten locally boiled or roasted. The fresh rhizomes are abortive.

***A. maculatum*** L. (Cuckoo Pint, Jack-in-the-Pulpit, Wale Robin). Europe. The plant and tubers are poisonous, but the latter are edible if first boiled to remove the poisonous principles. The rhizome is used as an expectorant, and to treat fevers and paralysis.

Arum, Green Arrow – Peltandra virginica.

ARUNDINARIA Michx. Graminae. 150 spp. Warm. Bamboos.

***A. amabilis*** McClure (Tonkin Bamboo, Tonkin Cane). S.China. Cultivated. The large canes are used for pole vaulting, ski sticks and fishing rods (when split). The smaller canes are used to support plants in gardens etc.

***A. falcata*** Nees. (Himalayan Bamboo). Malaysia. Used for fishing rods, and locally for basket work house ceilings, arrows and hookah tubes.

***A. fastuosa*** Marl.=Semiarundinaria fastuosa.

***A. gigantea*** (Walt.) Chapm. (Large Cane, Southern Cane). E. and S. United States of America. Young shoots eaten as a vegetable, and as food for livestock. The seeds were used by Indians and settlers for bread-making etc.

***A. nitida*** Mitf.=Phyllostachys nitida.

***A. prainii*** Gamble. E. India. Used locally for making huts and baskets.

***A. racemosa*** Munro. Himalayas. Used locally for mats and hut roofs.

***A. simonii*** (Carr.) A. and C. Riv.=Pleioblastus simonii. (Simon Bamboo). China. Used for staking garden plants, fishing rods etc. Young shoots are edible, but not much used.

***A. wightiana*** Nees. Tropical Asia, esp. India. Exported. Made into walking sticks and used for basket work etc.

ARUNDINELLA Raddi. Graminae. 55 spp. Warm. Grasses.

***A. brasiliensis*** Raddi.=Aira brasiliensis=Andropogon hispidus. S. America. Used as a forage grass in Brazil

ARUNDO L. Graminae. 12 spp. Tropical and temperate. Grasses.

***A. conspicua*** Forst.=A. kakao.

***A. donax*** L. (Giant Reed). Mediterranean to tropical Asia. Cultivated in Old and New World. Used in Europe to make wind musical instruments, and in N.America for making mats, screens etc.

***A. festucacea*** (Willd.) Link. (Sprangle Top). N.America. Cut for hay.

***A. kakao*** Steud.=A. conspicua (Kakaho). New Zealand. Ash used locally for treating burns. Stems used for kidney complaints and leaves to combat diarrhoea.

Arunt – Sauropus quadrangularis.

Ary – Lannea welwitschii.

***Arytera macrocarpa*** Miq.=Triomma macrocarpa.

Asa – Impatiens irvingii.

Asafoetida – Ferula assa-foetida.

Asafoetida – Ferula narthax.

Asakusanori – Porphyra.

Asana – Bridelia retusa.

Asarabacca – Asarum europaeum.

ASARUM L. Aristolochiaceae. 70 spp. N.Temperate. Herbs.

***A. arifolium*** Mich. E. United States of America. Infusion of leaves used by local Indians to cure pains in the stomach.

***A. balansae*** Franch. (Tê tân nam). N.Vietnam. An extract of the roots is used to control menstruation, and to encourage the production of saliva.

***A. blumei*** Duch. China and Japan. An extract of the rhizome is used locally to treat coughs.

***A. canadense*** L. (American Wild Ginger). E. N.America. The rootstock contains an essential oil, a resin and asarol, a scented oil. It is used dried as a substitute for ginger, and as a tonic and to relieve stomach pains.

***A. caudatum*** Lindl. W. N.America. A decoction of the leaves is used by the local Indians as a tonic.

***A. europaeum*** L. (European Wild Ginger). C. Europe to Asia Minor Dried rhizome used locally to reduce fever, to stimulate menstruation, and as an emetic, especially after drinking too much wine An ingredient of Schneeberger Snuff

***A. sieboldii*** Miq. (Tê tan). China. The powder or alcoholic tincture of the rhizome is used locally to treat toothache. The powder is used to stimulate sneezing.

Asbarg – Delphinium zalil.

ASCHERSONIA Mont. 135 spp.

***A. aleyrodes*** Webber. (Red Aschersonia). Parasitic on the White flies which feed on the foliage of citrus trees.

***A. goldiana*** Sacc. and Ellis. (Yellow Aschersonia). See A. aleyrodes.

Aschese – Eriocoelum microspermum.

ASCLEPIAS L. Asclepiadaceae. 120 spp. N.America. Herbs.

***A. curassavica*** L. (Bloodflower, Wild Ipecac). Pantropic, originated in Curaçao. Used as a substitute for Ipecac. Used as home remedies for dysentery and piles, and to induce sweating. The root is emetic and an extract of the leaves is used against intestinal worms.

***A. decumbens*** (Nutt.) Gray. (Spider Antelope Horn). W. U.S.A. Latex used by local Indians as chewing gum.

***A. eriocarpa*** Benth. (Woolypod Milkweed). S.W. United States of America. Source of a fibre used by local Indians for cordage. The latex is used for chewing gum.

***A. erosa*** Torr. S.W. United States of America. The latex is a possible source of rubber.

***A. fruticosa*** L.=Gomphocarpus fruticosus.

***A. gigantea*** L.=Calotropis procera.

***A. incarnata*** L. (Swamp Milkweed). N. America. The stem yields a fibre (Ozone Fibre), and the buds are eaten as a vegetable by the local Indians. The root and rhizome are emetic and purgative, diuretic and are used to treat intestinal worms. They are also used to treat rheumatism, catarrh and asthma.

***A. involucrata*** Engelm. (Dwarf Milkweed). S.W. United States of America. The latex is used by the local Indians as chewing gum.

***A. linaria*** Cav. (Plumerillo, Solimán, Torbisco, Venenillo). Mexico. The juice is used locally as a violent purgative.

***A. speciosa*** Torr. (Showy Milkweed). N. America. Buds and young shoots and leaves used as a vegetable by the local Indians. The latex is used as chewing gum.

***A. subulata*** Decn. (Candelillia bronca, Desert Milkweed, Yumete). California to Mexico. The latex is used locally in Mexico as a purgative and emetic. It is also a possible source of rubber.

***A. syriaca*** L. (Common Milkweed). N.America. The young buds and fruits eaten as a vegetable by the local Indians. The latex is used as a chewing gum. It is also a possible source of rubber. The dried root is used to relieve pain.

***A. tuberosa*** L. (Butterfly Weed). N.America. The shoots, pods and roots used locally by the Indians as vegetables. The dried root (Orange Milkweed Root, Pleurisy Root) is used as an expectorant and to reduce fevers.

***Ascelepiodora*** A. Gray=Asclepias L.

ASCOPHYLLUM L. Fucaceae. Brown Algae.

***A. nodosum*** (L.) Le Jolis=Fucus nodosua. (Yellow Wrack, Knobbed Wrack, Sea Whistles). Temperate Atlantic Coasts. Iodine is extracted from the ash. It is used locally as a manure, but it must be used with caution on clay soils as the sodium ions it contains can deflocculate the clay crumbs.

ASCYRUM L. Guttiferae. 5 spp. N.America, W.Indies. Herbs.

***A. hypericoides*** L. (St. Andrew's Cross). E. N.America. Used to treat kidney complaints and stone of the bladder.

Aseru – Dichroa febrifuga.

Ash, Alpine – Eucalyptus gigantea.

Ash, Black – Fraxinus nigra.

Ash, Blue – Fraxinus quadrangulata.

Ash, Blueberry – Elaecarpus grandis.

Ash, Blue Mountain – Eucalyptus oreades.

Ash, Canary – Beilschmeidia bancroftii.

Ash, Cape – Ekebergia carpensis.

Ash, Cape – Harpephyllum caffrum.

Ash, European – Fraxinus excelsior.

Ash, European Mountain – Sorbus aucuparia.

Ash, Flowering – Fraxinus ornus.

Ash, Grey Satin – Eugenia gustavioides.

Ash, Hercules Club Prickly – Zanthoxylum clava-herculis.

Ash, Japanese Prickly – Zanthoxylum piperitum.

Ash, Manna – Fraxinus ornus.

Ash, Martinique Prickly – Zanthoxylum martinicense.

Ash, Mountain – Eucalyptus regans.

Ash, Mountain – Eucalyptus salicifolia.

Ash, Mountain – Sorbus americana.

Ash, New South Wales White – Schizomeria ovata.

Ash, Northern Prickly – Zanthoxylum americanum.

Ash, Oregon – Fraxinus oregana.

Ash, Prickly – Aralia spinosa.

Ash, Prickly – Zanthoxylum clava – herculis.

Ash, Red – Fraxinus pubescens.

Ash, Red – Orites excelsa.

Ash, Red Mountain – Eucalyptus gigantea.

Ash, Senegal Prickly – Zanthoxylum senegalense.

Ash Twig Lichen – Ramalina fraxinea.

Ash, White – Fraxinus americana.

Ash, Wingleaf Prickly – Zanthoxylum alatum.

Ash, Yellow – Cladrastis tinctoria.

Ash, Yellowheart Prickly – Zanthoxylum flavum.

Ash, Yellow wood – Flindersia oxleyana.

Ashanti Blood – Mussaendra erythrophila.

Ashanti Pepper – Piper guineense.

ASHBYA Guillier.=Nematospora.

Ashen Tephrosia – Tephrosia cinerea.

Ashleaf Star Tree – Astronium fraxinifolium.

Asiaglory, Woolly – Argyreia speciosa.

ASIMINA Adans. Annonaceae. 8 spp. E. N.America. Shrubs.

***A. costaricensis*** Don. Smith.=Cymbopetalum costaricensis.

***A. reticulata*** Chapm.=Pictothamnus reticulata (Seminole Tea). S. N.America. A decoction of the leaves is used locally by the Indians to cure kidney complaints.

***A. speciosa*** Nash. See A. reticulata.

***A. triloba*** (L.) Dunal. (Pawpaw). S. and S.E. N.America. The fruits are yellow or green, roundish and sweet. They are sold locally, and are potentially a commercial crop. Bark used locally by fishermen for cordage.

Asna saj – Terminalia torentosa.

ASPALTHUS L. Leguminosae. 245 spp. S. Africa. Small shrubs, many xerophytic.

***A. cedarbergensis*** Bolus. Cape Peninsular. An infusion of the leaves is used locally as a beverage.

ASPARAGOPSIS Harv. Bonnemaisoniaceae. Red Algae.

***A. sanfordiana*** Harv. (Limu kohu). Pacific Ocean. Eaten cold with fish or meat after soaking to remove a bitter principle. It is also mixed with other seaweed to make stews and condiments. Dried and exported to Honolulu.

ASPARAGUS L. Liliaceae. 300 spp. Old World, mostly in dry places. Herbs.

***A. abyssinicus*** Hochst. Abyssinia. Cooked roots eaten locally.

***A. acutifolius*** L. Mediterranean. Stems eaten as a vegetable widely in the region.

***A. adscendens*** Roxb. Iran, Afghanistan, Turkestan to Himalayas. Roots used locally as a stimulant and to induce sweating.

***A. africanus*** Lam. S.Africa. The leaves are made into an ointment by the local women to stimulate the growth of hair.

***A. albus*** L. (White Asparagus). N.Africa. Young shoots eaten as a vegetable (Asperge sauvage) and sold locally in Algeria.

***A. lucidus*** Lindl. (Shiny Asparagus). China, Japan. Tubers eaten as a vegetable in Japan.

***A. officinalis*** L. (Garden Asparagus). Temperate Europe, Asia, N.Africa, N.America. Cultivated widely for the young shoots which are eaten green or blanched as a vegetable. The seeds have been used as a coffee substitute.

***A. pauli-guilelmi*** Solms Laub. Tropical Africa. Tubers cooked as a vegetable by natives of W.Africa.

***A. racemosus*** Willd. Middle East, India, Australia. The roots are applied to relieve irritations. They are also used to treat dysentery, and are diuretic. The fruits are eaten in the Sudan.

***A. sarmentosus*** L. E. India, Ceylon. The roots are eaten in Ceylon. Occasionally candied in China.

Asparagus Bean – Dolichos sesquipedalis.

Asparagus Bean – Psophocarpus tetragonolobus.

Asparagus Bush – Dracaena mannii.

Asparagus, Cape – Aponogeton distachyus.

Asparagus, Garden – Asparagus officinalis.

Asparagus Pea – Psophocarpus tetragonolobus.

Asparagus, Shiny – Asparagus lucidus.

Asparagus, White – Asparagus albus.

Aspen – Populus tremula.

Aspen, European – Populus tremula.

Aspen, Great – Populus grandidentata.

Aspen, Largetooth – Populus grandidentata.

Aspen, Trembling – Populus tremuloides.

Asperge sauvage – Asparagus albus.

ASPERGILLUS Mich. ex Fr. Moniliaceae. 50 spp.

***A. clavatus*** Desm. Produces citric acid by fermentation.

***A. fisheri*** Link=Penicillium javanense. See A. flavus.

***A. flavus*** Link. Used in Japan in the preparation of Sake (Rice Wine) and Soya Sauce (Shoy). Different strains are used for the two processes. Soya sauce is prepared from Soya beans (Glycine max) which contain a high proportion of protein in contrast to the high starch content of rice; both have to be fermented, requiring fungi with differing enzyme complexes. The fungus grows on groundnuts, forming alfatoxins, which are toxic to animals, especially birds fed on the nut meal.

***A. itaconicus*** Kinoshita. A possible source of itaconic acid, which is produced by fermentation. This could be used as a substitute for methacrylic acid, which is used to produce methyl methacrylate resins used in the manufacture of plastics.

***A. niger*** van Tiegh. Produces citric acid by fermentation.

***A. oryzae*** (Ahlb.) Cohn. Used in Japan for the manufacture of Sake and beer.

***A. tomarii*** Kita. Used in the Far East to make Tamari Sauce. This is made by the fermentation of Soya Beans and Rice, or Soya Beans.

***A. versicolor*** (Vuillemin) Tiraboschi. Ferments tartaric acid. It has a possible use in the removal of tartaric acid which forms in grape juice when stored for sale.

***A. wentii*** Wehmer. Used in Java for the production of Soya Sauce and similar products. It also produces citric acid by fermentation.

***Asperula*** L.=Gallium L.

ASPHODELUS L. Liliaceae. 12 spp. Mediterranean to Himalayas. Herbs.

***A. albus*** Mill.=A. macrocarpus. S. and C.Europe. The roots are used occasionally locally as a source of alcohol.

***A. fistulosus*** L. Mediterranean. N.Africa. Bulbs cooked as a vegetable by local Arabs.

***A. luteus*** Reichb. S.Europe. Eaten as a cooked vegetable by the Romans and Ancient Greeks.

***A. macrocarpus*** Parl.=A. albys.

***A. microcarpus*** Viû. Mediterranean. Used locally in Egypt as a source of yellow dye used for dyeing carpets.

***A. pendulinus*** Coss. and Dur. N. Africa. Used locally as a diuretic.

***A. racemosus*** L. S.Europe to Asia Minor Roots used in Turkey as a source of gum used in book-binding.

Aspic Oil – Lavendula officinalis.

***Aspidium*** Sw. For economic spp. See Dryopteris.

ASPIDOSPERMA Mart. and Zucc. Apocynaceae. 80 spp. Tropical and S.America; W. Indies. Trees.

***A. curranii.*** Standl. (Carreto). Columbia to Venezuela. Hard durable wood used for heavy timbers of bridges, houses, and railway sleepers.

***A. cuspa*** (H. B. K.) Blake. Venezuela, Guiana. Bark and leaves used as a cure for intestinal worms in Venezuela.

***A. desmanthum*** (Muell.) Arg. (Araracanga). Brazil. Orange-brown wood used locally for general construction work.

***A. eburneum*** Allem. (Peguia Martin, Pau Setim). Brazil. Wood used for cabinet-making and turning.

***A. excelsum*** Benth. (Paddle Wood, Yaruru). The heart wood is hard, and elastic. They are used for mill rollers. The whole wood is used to make tool handles and paddles. A decoction of the bark is used to treat stomach complaints.

***A. nitidum*** Benth. Guiana, Brazil. The elastic wood is used for tool handles etc.

***A. polyneuron*** (Muell.) Arg. (Peroba Rosa). S. Brazil. The yellow-light red wood, is fine-grained. It is used for general construction work, house building and furniture.

***A. quebracho blanco*** Schlecht. (Quebracho Blanco, White Quebracho). Argentina. Bark used locally for tanning leather.

***A. tomentosum*** Mart. (Woolly White Quebracho). Brazil. The wood (Piquia Peroba) is used for general carpentry.

***A. vargasii*** DC. Venezuela. The yellowish, heavy fine-grained wood takes a good polish and is easily split. It is used for making backs of toilet articles, rulers etc.

ASPILIA Thou. Compositae. 125 spp. Mexico to Brazil, S. tropical Africa, Madagascar. Herbs.

***A. latifolia*** Oliv. and Hiern. (Haemorrhage Plant). S. Africa. The fresh plant applied to wounds stops bleeding.

ASPLENIUM L. Aspleniaceae. 650 spp. Cosmopolitan. Herbs.

***A. macrophyllum*** Sw. (Na Babac, Papak-Lauin). E. Asia, esp. Philippines. Used in the Philippines. An extract of the fronds is a very strong diuretic, used especially in cases of beri-beri.

***A. obtusatum*** Forst. Australasia. In New Zealand, the rhizome is used to treat skin complaints.

Aspou ouar – Maerua rigida.

Assai Palm – Euterpe edulis.

Assam Rubber – Ficus elastica.

Assegai Wood – Curtisia dentata.

Assegai Wood – Terminalia sericea.

Assenzio ombrellifera – Achillea clavenae.

ASTELIA Banks. and Soland. Liliaceae. 25 spp. New Guinea, Australia, Tasmania, New Zealand, Hawaii. Herbs.

***A. grandis*** Hook. f. New Zealand. Leaves yield a soft brown fibre, which is used for a variety of purposes.

***A. nervosa*** Banks. and Soland. New Zealand. The berries are eaten locally by the Maoris.

***A. trinervia*** Kirk. (Kauriegrass). New Zealand. The leaves are used to produce a fibre which has a variety of uses, especially used by the early settlers.

ASTER L. Compositae. 500 spp. America, Eurasia, Africa. Herbs.

***A. amellus*** L.=A. trinervius. Europe to China. The roots are used by the Chinese to cure chest complaints, coughs and internal bleeding.

***A. cantoniensis*** Blume=Boltonia cantoniensis.

***A. macrophyllus*** L. (Rough Tongues). E. N.America. Young leaves are occasionally used locally as a vegetable.

***A. trinervius*** Roxb.=A. amellus.

ASTEROPTERUS Adams. Compositae. 7 spp. Mediterranean, Africa. Shrubs.

***A. gnaphaloides*** L. S. Africa. The leaves are used locally as a tea.

ASTILBE Buch.-Ham. Saxifragaceae. 25 spp. E.Asia, N.America. Herbs.

***A. philippensis*** Henry. Philippine Islands. The leaves are used locally to smoke instead of tobacco.

ASTRAGALUS L. Leguminosae. 2000 spp. Cosmopolitan, except Australia. Herbs or small shrubs.

Many spp. are source of Gum of Tragacanth. The gum is extracted from incisions in the stem. It is used, and has been since the times of the ancient Greeks as a cosmetic, in the manufacture of confectionery and in calico printing. It swells in water, but does not dissolve. This makes it useful in medicine for suspending insoluble powders and emulsifying oils.

***A. adscendens*** Boiss. and Haussk. Iran. Source of Gum of Tragacanth (Persian Manna). Used to make confectionery. Sold locally, and a little is exported.

***A. boëticus*** L. Spain. Cultivated in N.Europe. The seeds are used as a substitute for coffee (Swedish Coffee).

***A. brachycalyx*** Fisch. Iran. Produces a poor grade Gum of Tragacanth.

***A. brachycentrus*** Fisch. Iran. Source of a low grade Gum of Tragacanth.

***A. canadensis*** L. (Canada Milk Vetch). N.America. Roots eaten raw or cooked by local Indians.

***A. caryocarpus*** Ker.-Gwal. W. N.America. Pods eaten cooked or raw by local Indians.

***A. cerasocrenus*** Bunge. (Gommer) E. Middle East. Produces a gum.

***A. creticus*** Lam. Iran. Source of a low grade gum.

***A. cylleneus*** Boiss. and Heldr. Greece. Source of a gum (Morea Tragacanth).

***A. echidnaeformis*** Sirjaev. (Ghavan, Panbeh, Talkeh). Iran. One of the most important sources of Gum of Tragacanth.

***A. edulis*** Dur. N. Africa to Iran. Seeds eaten locally in Iran.

***A. elymaiticus*** Boiss. and Haussk. (Golpanbeh). Iran. Source of Gum of Tragacanth.

***A. fasciculifolius*** Boiss. S.W.Asia. Source of a sweet gum (Sarcocolla). Used locally to cure neuralgia of the face. Used by Parsi to set broken bones. Used locally as a cosmetic to impart a shine to the skin. Exported from Kurdistan.

***A. garbancillo*** Cav.=A. mandoni (Garbancillo). Argentina, Chile and Peru. Used as soap locally by the Indians of Peru.

***A. globiflorus*** Boiss. (Golbanbeh). S.W. Iran. Source of a Gum of Tragacanth.

***A. glycyphyllus*** L. (Milk Vetch). Europe to Siberia. Sometimes cultivated as fodder, and used occasionally as a substitute for tea.

***A. gossypinus*** Fisch. (Ghavan, Panbeh). Iran. Source of a good Gum of Tragacanth.

***A. gummifer*** Labill. (Tragacanth). Asia Minor, S. Europe. The principle source of Gum of Tragacanth.

***A. henryi*** Oliv. China. Gum used locally medicinally.

***A. heratensis*** Bunge. (Indian Tragacanth). C.Asia. Source of gum (Katira Gabina). Used locally for stiffening fabrics. Exported.

***A. hoantchy*** Franch. China. Gum used locally as a medicine.

***A. kurdicus*** Boiss. Asia Minor. Source of Gum of Tragacanth.

***A. lotioides*** Lam.=A. siniais.

***A. mandoni*** Rusby.=A. garbancillo.

***A. microcephalus*** Willd. (Ghuchu). Iran. Source of Gum of Tragacanth.

***A. mongholicua*** Bunge. China. Gum used in Chinese medicine.

***A. myriacanthus*** Boiss. (Ghineh Cheraghee). Iran. Yields a poor quality Gum of Tragacanth.

***A. pictus-filifolius*** Gray. W. N.America. Roots eaten by Indians of Arizona.

***A. prolixus*** Sieb. Middle East to N.India. A possible source of Gum of Tragacanth.

***A. pycnocladus*** Boiss. and Haussk. Iran to W.Russia. Source of a gum (Persian Tragacanth).

***A. sinicus*** L.=A. lotioides. China. Used in local medicine to treat gonorrhoea.

***A. stobiferus*** Royle. Iran. Source of a poor quality Gum of Tragacanth.

***A. stromatodes*** Bunge. Iran. Source of a Gum of Tragacanth.

Astrakan Wheat – Triticum polonicum.

ASTREBLA F. Muell. Graminae. 4 spp. Australia. Grasses.

***A. pectinata*** (Lindl.) F. v. Meull.=Danthonia pectinata. (Mitchell Grass). Australia, esp. S. Drought resistant pasture grass.

***A. triticoides*** (Lindl.) F. v. Meull.=Danthonia triticoides. Australia, esp. S. An excellent pasture grass for fattening cattle.

ASTROCARYUM G. F. W. Mey. 50 spp. Tropical America. Palms. All spp. mentioned, except A. ayri, yield edible oils used for cooking etc. They have similar properties to Coconut Oil.

***A. aculeatum*** G. F. W. Mey. (Aguire). Guyana. Oil from seed and pulp of fruit.

***A. ayri*** Mart.=Toxophoenix aculeatissima. Brazil. Fruits used locally to treat erysepalis.

***A. jauari*** Mart. (Jauari Palm). Brazil. Fruits used locally for fish bait.

***A. murumura*** Mart. (Murumuru). Amazonian Basin. Oil (Murumuru Oil, Murumuri Fat). Used locally and exported.

***A. tucuma*** Mart. (Tucuma). Brazil. Oil called Tucuma Oil.

***A. vulgare*** Mart. (Tucuma, Cumara, Aoura, Awarra). Venezuela, Guyana, Brazil, Peru. Fruits kernels (Tucuma Kernels) exported and source of oil. Edible etc. and used locally for soap-making.

ASTROLOMA R. Br. Epacridaceae. 25 spp. Australia. Shrubs.

***A. humifusum*** R. Br. Australia. Fruits eaten locally by aborigines.

***A. pinifolium*** Benth. Australia. Fruits eaten locally by aborigines.

ASTRONIUM Jacq. Anacardiaceae. 15 spp. C. and tropical S.America; W.Indies. Trees.

***A. graveolens*** Rich. E. Brazil. Wood (Fetia Star Tree, Gonçalo Alves). Heavy hard wood, but rather loose structure. Used for veneers. Polishes well.

***A. urundeuva*** (Allem.) Engl. S.America. Durable, heavy, bright red to red-brown. Used for heavy construction work.

ATALANTIA Corrêa. Rutaceae. 18 spp. Tropical Asia, China, Australia. Trees.

***A. glauca*** (Lindl.) B. and H. f.=Triphasia glauca. (Desert Lemon). S. Australia. Fruits used to make soft drinks and preserves.

***A. monophylla*** DC.=Limonia monophylla. E.India. Oil from fruits used to treat rheumatism.

ATALAYA Blume. Sapindaceae. 8 spp. S.Africa, Indochina, E. Malaysia, Australia. Trees.

***A. hemiglauca*** F. v. Muell. (Cattle Bush, White Wood). S. Australia. Leaves eaten by stock.

Atamasco Lily – Zephyranthes atamasco.

Atemoya – Annona squamosa L.×A. cherimola Mill.

ATHAMANTA L. Umbelliferae. 15 spp. Mediterranean. Herbs.

***A. cervaria*** L.=Peucedanum cervaria.

***A. chinensis*** Lour.=Ligusticum monnieri.

***A. cretensis*** L. (Candy Carrot). S. Europe. The whole plant is used for flavouring liqueurs.

***A. roxburghiana*** (Benth.) Wall.=Carum roxburghianum.

ATHEROSPERMA Labill. Atherospermataceae. 2 spp. Victoria, Tasmania. Trees.

***A. moschatum*** Labill. (Southern Sassafras). Tasmania. S. Australia. A whitish, fine wood, which is bent easily. Used for toy-making, clothes brushes, clothes pegs etc. The bark is used locally to make a tea.

ATHRIXA Ker-Gawl. Compositae. 20 spp. Tropical and S.Africa, Madagascar, Arabia; Australia. Herbs, shrubs.

***A. phylicoides*** DC. S.Africa. Leaves used locally to make a beverage.

ATHROTAXIS D. Don. Taxodiaceae. 3 spp. Australia, Tasmania. Trees.

***A. selaginoides*** D. Don. (King William Pine). W.Tasmania. Light wood, straight-grained, easily worked and bends easily. Pink to Yellowish. Used for window and door frames, boat-building, pattern making, vats, pianoes and violins, and battery separators.

ATHYRIUM Roth. Athyriaceae. 180 spp. Cosmopolitan. Ferns.

***A. esculentum*** Copel. Philippines. Young fronds eaten locally fresh or cooked.

Atil – Maerua rigida.

Atlantic Cedar – Cedrus atlantica.

Atlantic Mountain Mint – Pycanthemum incanum.

Atlantic Yam – Dioscorea villosa.

Atol de Piña – Bromelia karatas.

ATRACTYLIS L. Compositae. 20 spp. W.Mediterranean to Japan. Herbs.

***A. gummifera*** L. Mediterranean. A gum is extracted from the plant, and used as an adulterant for Mastic, used to make a chewing gum.

***A. ovata*** Thunb. E.Asia. The Chinese use an infusion of the whole plant to treat indigestion and as a diuretic.

ATRIPLEX L. Chenopodiaceae. 300 spp. Temperate and sub-tropics. Herbs, or shrubs.

***A. campanulata*** Benth. (Small Salt Bush). S.Australia. Food for livestock.

***A. canescens*** (Pursh.) Nutt. (Fourwing Saltbush). W. N.America to Mexico. Seeds eaten locally by Indians.

***A. halimoides*** Lindl. S.Australia. Food for livestock.

***A. halimus*** L. (Mediterranean Saltbush). N. and S.Africa. Used as a food by desert tribes of N.Africa, and ash used as an alkali in soap manufacture. An important pasture plant in S.Africa.

***A. hortensis*** L. (Garden Orach, Mountain Spinach). Temperate Europe and Asia. Introduced into N.America. Sometimes cultivated. The leaves are cooked as a vegetable.

***A. nummularia*** Lindl. S.Australia. Food for livestock.

***A. semibaccata*** R. Br. S.Australia. Food for livestock.

***A. spongiosa*** F. v. Muell. S.Australia. Food for livestock.

***A. vesicaria*** Heward. S.Australia. Food for livestock.

ATROPA L. Solanaceae. 4 spp. Europe, Mediterranean through C.Asia to Himalayas. Herbs. All parts of the plants contain the poisonous alkaloids atropine and hyoscyamine. The dried roots and leaves are used medicinally as a narcotic, pain-reliever, sedative, and diuretic. Used to dilate the pupils of the eyes.

***A. acuminata*** Royle ex Lindl. (Bentamaka, Indian Belladonna, Mait-brand, Yebrui). N.India.

***A. belladonna*** L. (Belladonna, Deadly Nightshade). Europe, esp. Central and S. Introduced to N.America. The berries used to be used by ladies to dilate the pupils to make them beautiful.

ATTALEA Kunth. Palmae. 40 spp. S.America, W.Indies, Tropical Africa. Trees.

***A. cohune*** Mart. (Cohune Palm). C.America. The fruits are made into sweetmeats and fed to livestock, the seeds are a source of non-drying oil (Cohune Oil) which is used in cooking, soap manufacture and for lighting, locally. The young bud is eaten as a vegetable, the leaves are used for thatching, and the trunk for house timbers.

***A. funifera*** Mart. (Bahia Piassava). Brazil. The leaf stalks yield a firm, durable, moisture-resistant fibre (Piassava Fibre) which in its various grades is used for making brushes and twine.

Attar of Roses – Rosa damascena.

Aturi – Portulaca lutea.

Aubergine – Solanum melongena.

Aubrya Baill. = Sacoglottis.

***Aucklandia costus*** Falc. = Saussurea lappa.

AULOSPERMUM Coulter and Rose. Umbelliferae. 13 spp. N.America. Herbs.

***A. longipes*** (S. Wats.) Coult. and Rose. = Cymopterus longipes S. Wats. (Ribseed). W. N.America. Leaves eaten as a vegetable by Indians.

***A. purpureum*** (S. Wats.) Coult. and Rose. = Cymopterus purpureus S. Wats. W. N.America and Mexico. Leaves eaten locally by Indians, and used to flavour soups.

Aurantii Dulcis Cortex – Citrus sinensis.

Auricula – Primula auricula.

Auricula Tree – Calotropis procera.

AURICULARIA Bull. ex Fr. Auriculariacea. 15 spp. Widespread.

***A. Auricula-Judae*** (L.) Schroet. = Hirneola Auricula-Judae (L.) Berk. (Jew's Ear). Temperate. Eaten widely in China.

***A. brasiliensis*** Fr. Tropics. Eaten in Philippines.

***A. cornea*** Ehrenb. Tropics. Eaten in Philippines.

***A. corrugata*** Sow. = A. mesenterica.

***A. delicata*** (Fr.) Henn. Malaysia. Eaten locally.

***A. mesenterica*** (Dicks. ex Fr.) Pers. = A. tremelloides. = A. corrugata = Telephora mesenterica. Widespread. Collected in New Zealand for export to China. Used in England as a cure for throat infections and dropsy.

***A. moelleri*** Lloyd. Tropics. Eaten in Philippines.

***A. polytricha*** (Mont.) Sacc. = Hirneola polytricha Mont. Far East and Australia. Eaten throughout the Far East. Exported from China. Sold in markets.

Australian Desert Kumquat – Eremocitrus glauca.

Australian Grass Tree – Xanthorrhoea australis.

Australian Gum – Acacia pycantha.

Australian Kino – Eucalyptus camaldulensis.

Australian Mahogany – Dysoxylum fraseranum.

Australian Oak – Eucalyptus regnans.

Australian Sandalwood – Santalum lanceolatum.

Australian Sandarac – Callitris calcarata.

Australian Walnut – Endiandra palmerstoni.

Austrian Pine – Pinus nigra.

Austrian Turpentine – Pinus nigra.

Autumn Mandrake – Mandragora autumnalis.

Avara seringa – Micanda siphonoides.

***Aveledoa*** Pittier = Metteniusa.

Avellano – Gevuina avellana.

AVENA L. Graminae. 70 spp. Temperate and mountains of tropics. Grasses.

***A. fatua*** L. (Wild Oat). Europe, introduced to N.America. Used as grain locally by Indians. Probably one of the parents of A. sativa.

***A. nuda.*** Höjer. (Naked Oat). Origin unknown. A hulless sp. Grown in the highlands of the Himalayan area for grain.

***A. orientalis*** Schreb. (Hungarian Oat, Turkish Oat). Origin possibly S.Europe. Sometimes cultivated in S.E.Europe for grain.

***A. sativa*** L. (Common Oat). Origin uncertain. Cultivated by the ancient European Lake Dwellers, and probably originated from the wild A. fatua. Generally cultivated in the cooler areas of the N.hemisphere which are marginal for wheat or barley cultivation. It is grown mainly for stock feed but is also used to make oatmeal or rolled oats. These are cooked as biscuits, cakes, porridge and breakfast foods. Oats are not suitable for making bread flour as the gluten content is not high enough to make a dough. An oil (Oat Oil) is extracted from the grain and used in the manufacture of breakfast cereals. Funfural, used as a solvent is extracted from the husks. The varieties are classified as spring or winter. The world production is about 4,300 million bushels on 130 million acres. U.S.A. produces about 1,500 million bushels, U.S.S.R. 720 million bushels, and U.K. 187 million bushels.

***A. strigosa*** L. Europe. A weed sp. Formerly cultivated for grain, esp. in England.

Avens, Root – Geum urbanum.

Avens, Water – Geum rivale.

Avens, Water – Geum canadense.

Avens, Yellow – Geum macrophyllum.

AVERRHOA L. Averrhoaceae. 2 spp. Tropical. Origin doubtful, could be Brazilian coast. Trees.

***A. bilimbi*** L. (Bilimbi, Cucumber Tree). Only known in cultivation. Possible origin Malaya. The fruits are yellow, acid, about 7 cm. long. They are used to make jams, jelly etc., sweetmeats and pickles. They also make a pleasant drink.

***A. carambola*** L. (Carambola). Tropical Asia. Cultivated throughout the tropics. The fruit is used for various confections, drinks, and eaten raw. It is yellow and star-shaped in cross-section.

AVICENNIA L. Avicenniaceae. 14 spp. Warm. Trees.

***A. alba*** Blume=A. marina.

***A. marina*** Vierh.=A. alba. Coastal regions of Old World tropics, and Australia. The wood is hard, and used locally for house-building and fuel. The bark is used for tanning. The juice from the leaves is an abortive.

***A. nitida*** Jacq. (Black Mangrove). Shores of tropical America. Bark used for tanning.

***A. officinalis*** L. (Baen, Bina, Kari, White Mangrove). Salt marshes of Indian subcontinent. Wood used for house construction; the ashes as a soap substitute, and as a mordant for artists' pigments. The bark is used for tanning, and the green fruits as poultices for boils.

***A. resinifera*** Forst.=A. tomentosa. Coastal tropics. Wood used for house and pier timbers, and boat-building. The bark is used for tanning. The aborigines of Queensland eat the fruits after cooking them.

Avipriya – Prangos pabularia.

Avocado (Pear) (Oil) – Persia gratissima.

Avocado, Coyo – Persea schliedeana.

Avocado, Mexican – Persea drymifolia.

Awal – Acacia pinnata.

Awar Awar – Ficus septica.

Awarra – Astrocaryum vulgare.

AXONOPUS Beauv. Graminae. 35 spp. S.America. Grasses.

***A. compressus*** (Swartz) Beauv. (Carpet Grass). Central America, W.Indies. Grows well in sandy soil, forming a dense sod. A good pasture for hot countries, and for holding sandy soils. Used for firebreaks in stands of conifers.

Ayahuasca – Banisteriopsis caapi.

Ayanki – Afraegle paniculata.

Ayapana du Tonkin – Eupatorium staechadosmum.

***Aydendron*** Nees.=Aniba.

***A. cujumary*** Mart.=Ocotea cujumary.

AZADIRACHTA A. Juss. Meliaceae. 2 spp. Indomalaysia. Trees.

***A. indica*** A. Juss.=Melia azadirachta. (Margosa, Nimba, Neem Tree). E.India, Ceylon. A non-drying oil (Margosa Oil) is extracted from the seeds. It is used for soap-making and to treat skin diseases, locally. The bark (Cortex Azadirachta, Nim Bark), and the leaf extract (Neem Toddy) are used as a tonic, and to reduce fevers. The leaves are also used as an antiseptic. A powder of the dried rootbark is also used to reduce fevers. The leaves, pressed in the leaves of books, prevent attack by insects.

***Azalia*** Desv.=Rhododendron.

Azara – Gomphocarpus lineolatus.

Azeite de pau de rosa – Physocalymma scaberrimum.

Azerolier – Crataegusa zarolus.

AZOLLA Lam. Azollaceae. 6 spp. Tropical and sub-tropical. Small floating ferns.

***A. pinnata*** R. Br. Tropical Asia. Used in Indochina to feed village stock, e.g. ducks. Also used as a green manure.

***A. rubra*** R. Br. New Zealand. Chewed locally to cure sore throats.

AZORELLA Lam. Hydrocotylaceae. 70 spp. N.Andes to temperate S.America; Falkland Islands, New Zealand, Antarctic Islands. Herbs.

***A. caespitosa*** Lam. (Yareta). Falkland Islands. A resin (Bolax Resin) is extracted from the plant. An extract of the plants is used locally to treat kidney and liver complaints.

***A. diapensiones*** A. Gray. See A. yareta.

***A. yareta*** Hauman. (Timiche, Yareta). W. S.America. A prime source of fuel. It is virtually smokeless, and is used domestically, and industrially. The local Indians use the resin as an absorbent for their medicines.

AZOTOBACTER Beijerinck. Bacteriaceae. All spp. mentioned are found in soil and fix atmospheric nitrogen, adding to the soil fertility.

***A. agile*** Beijerinck.

***A. beijerinck*** Lipm.

***A. chroococcum*** Beijerinck.

***A. vinelandii*** Lipm.

***A. vitreum*** Loehnis and Western.

***A. woodstownii*** Lipm.

Aztec Clover – Trifolium amabile.

Aztec Pine – Pinus teocote.

Aztec Tobacco – Nicotiana rustica.

# B

Babaco – Carica pentagona.

Babadotan lalaki – Synedrella nodifolia.

Babalan slatri – Calophyllum soulattrii.

Babassu (Palm) (Oil) – Orbignya speciosa.

BABIANA Ker-Gawl. Iridaceae. 60 spp. Tropical and S. Africa. Herbs.

***B. plicata*** Ker-Gawl. S.Africa. Roots (Baboon Roots), eaten by early settlers.

Bá bjnh – Eurycoma longifolia.

Bablah – Acacia arabica.

Babni – Eulaliopsis binata.

Baboon Root – Babiana plicata.

Babricon Bean – Canavalia campylocarpa.

Babul Acacia – Acacia arabica.

Baby's Breath – Gypsophila paniculata.

Bacaba – Oenocarpus distichus.

Baccae myrtillorum – Vaccinium myrtillus.

BACCAUREA Lour. Euphorbiaceae. 80 spp, Indomalaysia, Polynesia. Trees.

***B. dulcis*** (Jack) Muell. Arg.=Pierardia dulcis. (Kapoondong, Tjoopa). Malaysia, Indonesia. Cultivated for the fruit which has a pleasant flavour and is eaten locally.

***B. javanica*** (Bl.) Muell. Arg.=B. minahassae=Adenocrepis javanica. (Djreik, Kajoo djali, Poondoong). Indonesia. Hard light brown wood used for house construction.

***B. malayana*** King. Malaysia. Fruit pulp edible. Used locally to make a fermented beverage.

***B. minahassae*** Koord.=Baccaurea javanica.

***B. motleyana*** Muell. Arg. Malaysia. Cultivated locally for the acid-sweet fruits which are eaten cooked or raw, and fermented to make a beverage. A dye is extracted from the bark.

***B. racemosa*** (Reinw.) Muell. Arg. Malaysian Islands. Cultivated locally for the acid-sweet fruits, which are yellow-green in colour.

***B. sapida*** Muell. Arg.=Pierardia sapida. China, India, through to Malaysia. Yellow fruits have a pleasant acid flavour. Cultivated.

BACCHARIS L. Compositae. 400 spp. America. Herbs or shrubs.

***B. alamani*** DC.=B. glutinosa.

***B. calliprinos*** Griseb. Argentina. An infusion of the leaves is used locally to treat colic. An extract of the whole plant yields a yellow dye.

***B. coerulescens*** DC.=B. glutinosa.

***B. farinosa*** Spreng.=B. glutinosa.

***B. genistelloides*** Pers.=B. triptera. S.America. Used locally as a tonic and to treat intestinal worms.

***B. glutinosa*** Pers.=B. alamani, B. coerulescens, B. farinosa. (Hierba del carbonero, Jara dulce, Jaral, Jarilla del rio, Sauce). Mexico. The pounded leaves are used as a poultice for sores, and a decoction is used as an eyewash. The branches are used as house rafters.

***B. tridentata*** Vahl. Brazil. The whole plant is used to combat intestinal worms, and as a diuretic, locally.

***B. triptera*** Mart.=B. genistelloides.

Bachang Mango – Mangifera foetida.

Bach dàng nam – Mallotus anamiticus.

Bach liêm – Cissus modeccoides.

Bâ choût – The costela porlana.

Bach tung – Podocarpus imdricatus.

BACIDIA Sw. Lecudeaceae. Lichens.

***B. muscorum*** (Sw.) Mudd. Temperate. Used, esp. in Europe, as the source of a red dye used for dyeing wool.

BACILLUS Cohn. 33 spp. +. Eubacteriales.

***B. acetoethylicum*** Northtr. Can ferment carbohydrates to form acetone and ethanol. A potential source of industrial production.

***B. butylicus*** Fitz. Can ferment carbohydrates to produce butyric acid and butinol.

***B. freudenreichii*** (Migula) Chester. An important soil microorganism as it converts urea to ammonium carbonate, making the nitrogen more readily convertible for absorption by crop plants.

***B. mecerans*** Schard. Ferments carbohydrates to ethanol and acetone.

BACKHOUSIA Hook. and Harv. Myrtaceae. 7 spp. E.Australia. Trees.

***B. citriodora*** F. v. Muell. (Citron Myrtle, Native Myrtle, Scrub Myrtle). Australia. A possible source of perfume and soap manufacture.

***B. myrtifolia*** Hook. and Harv. (Grey Myrtle, Ironwood). S.E. Australia, near water. Hard tough heavy wood is difficult to work, but as it turns well it is used for tool-handles, wheels, wooden screws, etc.

Bac lá – Croton argyratus.

BACOPA Aubl. Scrophulariaceae. 100 spp. Warm. Herbs.

***B. monniera*** (L.) Wettst. = Herpestis monnieri = Gratiola monniera = Monniera cuneifolia. Through tropics. An infusion of the plant is used as a nerve tonic by the Hindus, especially in cases of epilepsy, and as a diuretic in the Philippines.

BACTERIUM Ehrenb. Number of spp. variable, depending on definition of the genus.

***B. acidi-proprionica*** Scherm. Play a part in the production of Emmenthaler cheese, giving it its characteristic flavour by the production of proprionic acid.

***B. rhamnosifermentans*** Castell. Ferments the sugar rhamnose to produce propyl glycol.

BACTRIS Jacq. Palmae. 180 spp. Tropical America. W.Indies. Trees.

***B. gasipaës*** H.K.B. = Guilelma speciosa.

***B. minor*** Jacq. Columbia. Edible fruits sold locally.

Bacu – Cariniana pyriformis.

Bacury (Kernel Oil) – Platonia insignia.

Bacuru Pary – Rheedia macrophylla.

Bada – Salix acmophylla.

Badar – Abies pindrow.

Badiankohi – Prangos pabularia.

Badoh negro – Ipomoea violacea.

Badra-kema – Ferula galbaniflua.

BAECKEA L. Myrtaceae. 1 sp. China, Malaysia; 1 sp. Borneo; 65 spp. Australia, New Caledonia. Small trees.

***B. frutescens*** L. S.China, Malaysia, Sumatra to Australia. Leaves or flowers used in Far East to make a stimulating tea. It is used to combat lack of vigour, and fevers. The plant produces an essential oil (Essence de Bruère de Tonkin), which is exported to France for soap perfume.

Bael Fruit – Aegle marmelos.

Baen – Avicennia officinalis.

Bafole – Dacryodes edulis.

Baga – Elaeophorbia drupifera.

Bagasse – Saccharum officinale.

Bagberi – Saccharum spontaneum.

Baggar – Eulaliopsis binata.

Bagide – Impatiens irvingii.

Bagilumbang Oil – Aleurites triloba.

Bagilumbang Oil – Aleurites trisperma.

Baguang – Lagenaria leucantha.

Bahama Hemp – Agave sisalana.

Bahama White Wood – Canella alba.

Bahan – Populus euphratica.

Bahia Cotton – Gossypium brasiliensis.

Bahia Grass – Paspalum notatum.

Bahia Piassava – Attalea funifera.

Bahia Powder – Andira araroba.

Bahia Rosewood – Dalbergia nigra.

Bahia, Vanilla of – Vanilla gardneri.

Bahia Wood – Caesalpinia brasiliensis.

Bahotot – Triclisia gelletii.

Baical Skullcap – Scutellaria baicalensis.

Baichandi – Dioscorea hispida.

BAIKIAEA Benth. Leguminosae. 10 spp. Tropical Africa. Trees.

***B. plurijuga*** Harms. (Mkushi, Rhodesian Chestnut, Rhodesian Teak, Umgusi, Zambezi Redwood). Rhodesia and Zambia. A valuable timber, durable, but difficult to work. Used widely for general house construction, railway sleepers, and small canoes.

***B. robynsii*** Ghesq. (Bobombe wantando, Dileko). W.Africa. A decoction using the pulverised bark is used to stain earthenware black.

BAILLONELLA Pierre ex Doubard. Sapotaceae. 1 sp. W. Equatorial Africa. Trees.

***B. ovata*** Pierre = Mimusops pierreana Engl.

The seeds have a high fat content, and are eaten locally with meat. The fat (Beurre d'Orère, Orère Butter), is extracted and used locally as a fat.
Bainoro prieto – Pisonia capitata.
Bains – Salix tetrasperma.
Baira – Cynometra alexandri.
Bai tre nep – Spathoglottis eburnea.
Baja California – Arthrocnemum subterminale.
Bajole – Dacryodes edulis.
Bajri – Pennisetum typhoideum.
Bakan – Calamus aquatilis.
Bakanae-byo – Gibberella fujikuroi.
Bakau – Bruguiera eriopetala.
Bakau akik – Rhizophora conjugata.
Bako – Rhizophora conjugata.
Bakoo bakooan – Wilkstroemia candollerna.
Bak pida – Deinbollia molliuscula.
Bakra – Elaeodendron glaucum.
Bakupari – Rheedia brasiliensis.
Balabodeli – Cissus rubiginosa.
Balambo – Cadaba farinosa.
Balam rambui – Palaquium ridleyi.
Balam seminai – Palaquium stellatum.
BALANITES Delile. Balanitaceae. 25 spp. Tropical Africa to Burma. Shrubs or small trees.
***B. maughamii*** Sprague. E.Africa. An edible oil is extracted from the nuts.
***B. roxburghii*** Planch. (Hingan, Ingudi). Dry parts of India and Pakistan. The wood is used to make small articles like walking sticks, and is burnt as a fuel. The seeds yield an edible oil (Zachun Oil). The fruit pulp contains saponin and is used locally to wash silk.
***Balanocarpus*** Bedd. = Hopea.
BALANOPHORA J.R. and G. Forst. Balanophoraceae. 80 spp. Madagascar, to Japan, Malaysia, Australia, Polynesia. Herbaceous parasites on tree roots.
***B. elongata*** Blume. Java. A wax is extracted from the rhizome, and used locally for lighting.
***B. globosa*** Jungh. See B. elongata.
***B. ungeriana*** Val. See B. elongata.
Balata – Manilkara bidentata.
Balata – Manilkara huberi.
Balata abuirana – Ecclinusa balata.
Balata batârd – Diphlois nigra.
Balata blanca – Manilkara Sieben.
Balata, Panama – Manilkara dariensis.
Balata rosada – Sideroxylon cyrtobotryum.
Balata rosada – Sideroxylon resiniferum.
Balata rouge – Ecclinusa sanguinolenta.
Balata saignant – Ecclinusa sanguinolenta.
Balau bookit – Shorea collina.
Balau resin – Dipterocarpus grandiflorus.
Balau resin – Dipterocarpus vernicifluus.
Balaustre – Centrolobium orinocense.
Balche – Lochocarpus longistylus.
Bald Cypress – Taxodium distichum.
Balderjan – Valeriana capensis.
Balemo – Ficus aspera.
Baleric Myrobalans – Terminalia bellerica.
BALFOURODENDRON Mello ex Oliv. Rutaceae. 1 sp. S.Brazil, Paraguay, N.Argentina. Tree.
***B. riedelianum*** Engl. (Guatambu, Pauliso). A fine timber used locally for interior finishing of houses, furniture, agricultural implements and turning.
BALIOSPERMUM Blume. Euphorbiaceae. 6 spp. India, S.E. Asia, Malaysia, Java. Trees.
***B. axillaea*** Blume. India, Burma through to Indochina. The seeds are poisonous, but an oil extracted from them is used as a drastic purgative in India.
Ball Clover – Trifolium nigrescens.
Ballhead Onion – Allium sphaerocephalum.
Ballot – Exocarpos cupresiformis.
BALLOTA L. Labiatae. 35 spp. Europe, Mediterranean, W.Asia. Herbs.
***B. foetida*** Lam. = B. nigra.
***B. hirsuta*** Schult. See B. nigra.
***B. nigra*** L. (Black Hoarhound). Mediterranean to India. A tincture, which contains an essential oil is used locally to treat hysteria. The plant is also used to adulterate Hoarhound (Marrubium vulgare).
Balm (Oil) (Tea) – Melissa officinalis.
Balm, Bastard – Melittis melissophyllum.
Balm, Bee – Monarda fistulosa.
Balm of Gilead – Abies balsamea.
Balm of Gilead – Populus balsamifera.
Balm, Horse – Collinsonia canadensis.
Balm, Mountain – Eriodictyon glutinosum.
Balm, Oswega Bee – Monarda didyma.
Balm, Pony Bee – Monarda pectinata.
Balm, Western – Monardella parviflora.
Balmony – Chelone glabra.
BALOGHIA Endl. Euphorbiaceae. 13 spp. E.Australia, N.Caledonia, Norfolk Islands. Trees.
***B. lucida*** Endl. = Codiaeum lucidum. (Brush Bloodwood). New South Wales, Queensland, Norfolk Islands. A red dye is extracted from the wood.

***B. pancheri*** Baill. Australia, New Caledonia. Bark used in New Caledonia for tanning.
Balood – Calophyllum amoenum.
Balsa – Ochroma lagopus.
Balsam – Copaifera multijuga.
Balsam, American Styrax – Liquidambar styraciflua.
Balsam, Apple – Echinocystis lobata.
Balsam, Apple – Momordica balsamina.
Balsam, Balsamum Copaivae africanum – Oxystigma mannii.
Balsam, Bourbon Mariae – Caulophyllum tacamahaca.
Balsam, Canada – Abies balsamea.
Balsam, Copahiba angelin – Copaifera multijuga.
Balsam, Copahiba marimary – Copaifera multijuga.
Balsam, Copaiba – Copaifera coriacea.
Balsam, Copaiba – Copaifera lansdorffii.
Balsam, Copaiba – Copaifera reticulata.
Balsam, Copaiba – Dipterocarpus alatus.
Balsam, Copaiba – Dipterocarpus trinervis.
Balsam, Copaiba – Daniella thurfera.
Balsam, Copaiba – Hardwickia pinnata.
Balsam, Copaiba – Daniella oliveri.
Balsam, Duhnual – Commiphora opobalsamum.
Balsam, Fir – Abies balsamea.
Balsam, Friar's – Styrax benzoin.
Balsam, Garden – Impatiens balsamina.
Balsam of Gilead – Abies balsamea.
Balsam of Gilead – Commiphora opobalsamum.
Balsam, Gurjum – Dipterocarpus alatus.
Balsam, Hardwickia – Oxystigma mannii.
Balsamina latifolia Blume = Impatiens platypetala.
Balsam, Indian – Leptotaenia multifida.
Balsam, Jacaré copahiba – Eperua oleifera.
Balsam, Labdanum – Cistus villosus.
Balsam, Levant Styrax – Liquidambar orientalis.
Balsam, Maracaibo – Copaifera officinalis.
Balsam, Maracaibo – Dipterocarpus alatus.
Balsam, Maria – Rheedia acuminata.
Balsam, Mecca – Commophora opobalsamum.
Balsam, Opobalsam – Myroxylon balsamum.
Balsam, Oregon – Pseudotsurga menziesii.
Balsam, Oregon – Pseudotsuga taxifolia.
Balsam Pear – Momordica charantia.
Balsam of Peru – Myroxylon balsamum var. pereira.
Balsam, Pigs' – Tetragastris balsamifera.
Balsam, Poplar (Buds) – Populus balsamifera.
Balsam, Rock – Clusia plukenetii.
Balsam root – Balsamorhiza sagittata.
Balsam Spurge – Euphorbia balsamifera.
Balsam, Tolu – Myroxylon balsamum var. balsamum.
Balsam Tree – Copaifera mopane.
Balsam, West African Copaiba – Oxystigma mannii.
Balsamo – Acanthosphaera ulei.
***Balsamocarpum brevifolium*** (Baill.) Clos. = Caesalpinia brevifolia.
Balsamo de misiones – Schinus terebinthifolius.
***Balsamodendron*** Clos. = Commiphora Jacq.
BALSAMORHIZA Hook. Compositae 12 spp. W. N.America. Herbs. Roots of species mentioned eaten cooked by local Indians.
***B. deltoides*** Nutt. (Puget Balsamroot). W. N.America.
***B. hookeri*** Nutt. (Hooker's Balsamroot). W. N.America.
***B. sagittata*** (Pursh.) Nutt. (Arrowleaf Balsamroot, Balsam root, Oregon Sunflower). W. N.America.
Balsamroot – Balsamorhiza sagittata.
Balsamroot, Arrowleaf – Balsamorhiza sagittata.
Balsamroot, Hooker's – Balsamorhiza hookeri.
Balsamroot, Puget – Balsamorhiza deltoides.
Balsamum Copaivae africanum – Oxystigma mannii.
Balsamum Mariae – Calophyllum inophyllum.
Balsam, Yellow – Impatiens nole-tangere.
Balsu – Ochroma pyrimidalis.
Baltic Amber – Pinus succinifera.
Baludar – Abies pindrow.
Balukanag Oil – Chisocheton cuminginanus.
Baluku – Grewia philippensis.
Bambara Ground Nut – Voandzeia subterranea.
Bambazetto – Chrozophora tinctoria.
Bamboo, Beechey – Bambusa beecheyana.
Bamboo, Berry-bearing – Melocanua bambusioides.
Bamboo, Black – Phyllostachys nigra.
Bamboo, Calcutta – Dendrocalamus strictus.
Bamboo, Castillon – Phyllostachys bambusoiodes.
Bamboo, Choang Ko Chuk – Bambusa tuldoides.

Bamboo, Fa Mei Chunk – Bambusa pervariabilis.
Bamboo, Fishpole – Phyllostachys aurea.
Bamboo, Giant – Dendrocalamus giganteus.
Bamboo, Giant Timber – Phyllostachys bambusioides.
Bamboo, Hairy Sheath Edible – Phyllostachys edulis.
Bamboo, Hardy Timber – Phyllostachys bambusioides.
Bamboo, Henon – Phyllostachys nigra.
Bamboo, Himalayan – Arundinaria bambusioides.
Bamboo, Japanese Timber – Phyllostachys bambusioides.
Bamboo, Madake – Phyllostachys bambusioides.
Bamboo, Male – Dendrocalamus strictus.
Bamboo, Meyer – Phyllostachys meyeri.
Bamboo, Monastery – Thyrostachys siamensis.
Bamboo, Moso – Phyllostachys pubescens.
Bamboo, Nai Chuk – Bambusa pervariabilis.
Bamboo, Narihira – Semiarundinaria fastuosa.
Bamboo, Nai Chuk – Phyllostachys pervariabilis.
Bamboo, Pa koh poo chi – Phyllostachys dulcis.
Bamboo Palm – Raphia vinifera.
Bamboo, Plains – Bambusa balcooa.
Bamboo, Simon – Arundinaria simonii.
Bamboo, Slender – Oxytenanthera stocksii.
Bamboo, Spiny – Bambusa arundinaceae.
Bamboo, Sweetshoot – Phyllostachys dulcis.
Bamboo, Terai – Melocanna bambusoides.
Bamboo, Tonkin – Arundinaria amabilis.
Bamboo, Tufted – Oxytenanthera nigrociliata.
Bamboo, Umbrella-handle – Thyrsostachys siamensis.
Bamboo, Yellowgroove – Phyllostachys aureosulcata.
Bamboong – Oenanthe javanica.
Bambuk Butter – Butyrospermum parkii.
BAMBUSA Schreb. Graminae. 70 spp. Tropical and subtropical Asia, Africa, America.
***B. abyssinica*** Rich. = Oxytenanthera abyssinica.
***B. arundinacea*** (Retz.) Willd. (Spiny Bamboo, Tzou chu). Tropical Asia. Young buds eaten in India and China. Wood used in China for house construction, furniture, ornaments etc. See B. vulgaris.
***B. aspera*** Schult. = Denrocalamus asper.
***B. aurea*** Carr. = Phyllostachys aurea.
***B. balcooa*** Roxb. (Plains Bamboo). Plains E.India. Excellent for building purposes.
***B. beecheyana*** Munro. = Sinocalamus beecheyanus. (Beechey Bamboo). S.E. China. An important source of bamboo shoots which are eaten locally and exported. The young shoots are produced by earthing up the plants to exclude light. Shoots which have become green are exceedingly bitter.
***B. cornuta*** Munro. Java. Cultivated for the young shoots which are eaten locally.
***B. glaucescens*** (Willd.) Sieb. ex Munro = B. nana.
***B. multiplex*** Raeusch. China, Japan. Cultivated. Buds eaten locally. Culms used for making paper and making fishing rods.
***B. nana*** Roxb. = B. glaucescens. China. Cultivated. Stems used for fishing rods.
***B. oldhamii*** Munro. China. Buds eaten locally.
***B. palmata*** Mitf. = Sasa palmata.
***B. pervariabilis*** McClure. (Fa Mei Chuk, Nai Chuk, Yan Chuk). S.China. Stems used much locally for construction work and punting poles.
***B. polymorpha*** Munro. S.Asia, esp. N.W. India. Used for house construction, and possible source for paper-making.
***B. sinospinosa*** McClure. The leaf sheaths are used in the manufacture of sandals.
***B. spinosa*** Roxb. E.India. Boiled young shoots eaten locally as a vegetable. Wood used for general construction work.
***B. textilis*** McClure. (Wong Chuk). S.China. Cultivated. Split stems used for making screens, baskets, hats, and making ropes.
***B. tulda*** Roxb. (Deo bans, Talda bans). S. Asia, esp. Assam, Bengal to Burma. Young buds eaten as a vegetable. Stems used for general construction work etc. Split stems used for screens, baskets, mats etc.
***B. tuldoides*** Munro. (Choang Ko Chuk, Nai Chuk). S.China. Cultivated. Used for general construction work, punting poles, crate making etc. Important locally in China.
***B. vulgaris*** Schrad. Cultivated widely throughout the tropics and subtropics. The young buds are eaten as a vegetable. The fibres in the wood are long and soft, making this a valuable species for paper

making. They have a high length/diameter ratio. The pulp is blended with weaker fibres for making wrapping paper and stationery, or used on its own for book-making. Fine-grained amorphous silica (tabasheer) found in the internodes is used as a catalyst. The white powder on the culms contains a chemical closely related to the female sex hormones. The shoots contain nuclease and deaminase, which makes an aqueous extract better than the conventional media for growing bacteria, and the extract intensifies the response to the skin test when added to tuberculin. The dried leaves are used to deodorize fish oils, and diesel fuel is distilled from the culms.

Bami – Manihot esculenta.

Banana – Musa sapientum.

Banana, Abyssinian – Musa ensete.

Banana, Adam's – Musa paradisiaca.

Banana, Casa – Sicana odorifera.

Banana, Cavendish – Musa nana.

Banana, Chinese – Musa nana.

Banana, Cooking – Musa paradisiaca.

Banana, Date – Musa sapientum.

Banana de Brejo – Caladium striatipes.

Banana de Imbre – Philodendron bipinnatifidum.

Banana de Macaco – Philodendron bipinnatifidum.

Banana, Dwarf – Musa nana.

Banana, Horse – Musa paradisiaca.

Banana Passion Fruit – Passiflora vanvolxemii.

Banana, Plantain – Musa paradisiaca.

Banana porete – Musa oleracea.

Banana, Red – Musa sapientum.

Banana, Rose – Musa sapientum.

Banana Shrub – Michelia fuscata.

Banana, Squash – Cucurbita maxima.

Banana, Vegetable – Musa paradisiaca.

Bananes Cristalicées – Musa sapientum.

Banapu – Terminalia coriacea.

Banate – Anemone hupehensis.

Banda – Allophylus africanus.

Bandakuai – Hibiscus esculentus.

Bandaneke – Vanda roxburghii.

Bandoul pech – Tinospora crispa.

Banga – Lygodium smithianum.

Bangalay Eucalyptus – Eucalyptus botryoides.

Bangikat – Populus ciliata.

Bangkongan – Mischocarpus sundaicus.

Bangka minjak – Rhizphora conjugata.

***Banisteria*** auctt. = Banisteriopsis.

***B. lupuloides*** L. = Gouania lupuloides.

BANISTERIOPSIS C. B. Rob. and Small. Malpighiaceae. 100 spp. tropical America. W.Indies. Woody vines.

***B. caapi*** (Spruce ex Griseb.) Morton. (Caapi). S.America, esp N. and E. Cultivated locally for the leaves and young stems from which a beverage contating hallucinatory drugs (telepathine etc.) is obtained. The beverage is called Ayahuasca, or Yajé.

***B. inebrians*** Morton. See B. caapi.

***B. quitensos*** (Ndz.) Morton. See B. caapi.

Baniti – Garcinia cambogia.

Baniti – Garcinia dulcis.

Bankhor – Aesculus indica.

Bank's Rose – Rosa banksiae.

BANKSIA L. f. Proteaceae. 50 spp. Australia. Trees.

***B. integrifolia*** L. f. = B. spicata. (Beef wood). New South Wales, Queensland, and Victoria. Bark used for tanning.

***B. serrata*** L. f. (Wattung-Urree). The coarse purplish wood is used for general boat building, window frames etc., and furniture.

Bans – Dendrocalamus strictus.

Bans khura – Dendrocalamus strictus.

Bantengan – Vitis lanceolaris.

Banucalang (Oil) – Aleurites trisperma.

Banyan – Ficus benghalensis.

Baobab (Oil) – Adansonia digitata.

Baobab, Madagascar – Adansonia madagascariensis.

Baobotrys indica Roxb. – Maesa indica.

BAPHIA Afzel. Leguminosae. 65 spp. Warm Africa, Madagascar; 1 sp. Borneo. Trees.

***B. nitida*** Lodd. (African Sandalwood). A hard red brown wood (Camwood) is used as a source of dye for dyeing wool. The timber is used for heavy house timbers, pestles and walking-sticks. The leaves are used locally in W.Africa to treat skin diseases.

***B. pubescens*** Hook. f. Tropical Africa. The heavy fine-grained timber is used for general construction work. It develops a red colour on exposure, and has a violet scent. It is used as a source of dye.

BAPTISIA Vent. Leguminosae. 35 spp. N.America. Herbs.

***B. australis*** (L.) R. Br. (False Indigo). E. to C. N.America. An infusion of the leaves is used to stimulate appetite and aid digestion.

***B. leucantha*** Torr. and Gray. S.E.

N.America. Used locally by the Indians as an emetic, laxative and cathartic.
***B. tinctoria*** (L.) R. Br. (Rattle Weed, Yellow Wild Indigo). United States of America, esp. South. Used locally in the south as the source of a yellow dye. The dried root is used as an emetic and cathartic.
Baput – Dolichos malosanus.
Baquina – Pothomorphe umbellata.
Barabanja – Mascarenhasia lanceolata.
Baraka – Cynometra hankei.
Barba de chico – Clematis grossa.
Barba de vejo – Clematis grossa.
Barbados Aloë – Aloë vera.
Barbados Cherry – Malpighia glabra.
Barbados Gooseberry – Perestia aculeata.
BARBAREA R. Br. Cruciferae. 20 spp. N.Temperate. Herbs.
***B. praecox*** (Sm.) R. Br.=B. verna. (Scurvy Grass, Winter Cress). Europe, introduced to N.America. Used as a vegetable in parts of the United States of America, and sometimes cultivated as a winter salad. The seeds yield an edible oil.
***B. verna*** (Mill.) Asch.=B. praecox.
***B. vulgaris*** R. Br. (Yellow Rocket). Temperate. Sometimes used as a winter salad. The leaves (Herba Barbariaea) were used in a balsam used to heal wounds.
Barbary Santonica – Artemisia ramosa.
Barbardine – Passiflora quadrangularis.
Barbasco – Paullinia pinnata.
Barbasco – Serjonia mexicana.
Barbatimao – Stryphnodendron barbatimam.
Barberry – Berberis haematocarpa.
Barberry, Alleghany – Berberis canadensis.
Barberry, American – Berberis canadensis.
Barberry, European – Berberis vulgaris.
Barberry, Indian – Berberis aristata.
Barberry, Magellan – Berberis buxifolia.
Barcelona Nuts – Corylus avellana.
Barcoo Grass – Themeda mambranacea.
Bard Vetch – Vicia monantha.
Bargat – Ficus benghalensis.
Barilla – Salsola soda.
Barrister Gum – Mezoneuron scortechinii.
Bark, Acacia – Acacia arabica.
Bark, Acacia – Acacia decurrens.
Bark, African – Crossopteryx kotschyana.
Bark, American Red Osier – Cornus amomum.
Bark, Angostura – Cusparia febrifuga.
Bark, Autour – Symplocos racemosa.
Bark, Barbatimao – Stryphnodendron barbatimam.
Bark, Batavia Cinnamon – Cinnamomum burmannii.
Bark, Bitter – Pinckneya pubens.
Bark, Black Willow – Salix nigra.
Bark, Butternut – Juglans cinerea.
Bark, Caka – Erythrophleum guineënsis.
Bark, Calisaya – Cinchona calisaya.
Bark, Calunga – Simaba ferruginea.
Bark, Canella – Canella alba.
Bark, Cangerana – Cabralea cangerana.
Bark, Casa Virgindade – Stryphnodendron barbatimam.
Bark, Cascara Amarga – Sweetia panamensis.
Bark, Cascara Sangrada – Rhamnus purshiana.
Bark, Cascarilla – Croton cascarilla.
Bark, Cascarilla – Croton eluteria.
Bark, Cascarilla (substitute) – Croton niveus.
Bark, Cassia – Cinnamomum cassia.
Bark, Ceylon Cinnamon – Cinnamomum zeylanicum.
Bark, Clove – Dicypellium caryophyllatum.
Bark, Conesi – Holarrhena antidysenterica.
Bark, Copalchi – Coutarea latifolia.
Bark, Copalchi – Croton niveus.
Bark, Copalchi – Exostoma caribaeum.
Bark, Copalchi – Portlandia latifolia.
Bark, Cortex Geoffreae – Geoffrea surinamensis.
Bark, Cortex Murure – Urostigma cystopodium.
Bark, Corteza de Ojé – Ficus anthelminthica.
Bark, Cramp – Viburnum opulus.
Bark, Cramp – Acer spicata.
Bark, Crown – Cinchona officinalis.
Bark, Cuprea – Remijia pedunculata.
Bark, Cusparia – Cusparia febrifuga.
Bark, Doom – Erythrophleum guineënse.
Bark, Druggist's – Cinchona succitubra.
Bark, Emblic – Phyllanthus emblica.
Bark, False Winter's – Cinnamodendron corticosum.
Bark, Fever – Pinckneya pubens.
Bark, Flowering Dogwood – Cornus florida.
Bark, Georgia – Pinckneya pubens.
Bark, Granatum – Punica granatum.
Bark, Green Osier – Cornus circinnata.
Bark, Georgia – Pinckneya pubens.
Bark, Guapi – Guarea rusbyi.
Bark, Highbush Cranberry – Viburnum opalus.
Bark, Honduras – Picramnia pentandra.
Bark, Honduras – Sweetia panamensis.

Bark, Indian Madder – Oldenlandia umbellata.
Bark, Iripil – Pentaclethra filamentosa.
Bark, Jamaica – Quassia amara.
Bark, Jamrosa – Terminalia mauritiana.
Bark, Jesuit's – Cinchona officinalis.
Bark, Kajo rapat – Parameria barbata.
Bark, Kurchi – Holarrhena antidysenterica.
Bark, Lace – Lagetta lintearia.
Bark, Ledger – Cinchona ledgeriana.
Bark, Letur – Symplocos racemosa.
Bark, Loxa – Cinchona officinalis.
Bark, Massoia – Cinnamomum massoia.
Bark, Mindanao Cinnamon – Cinnamomum mindanaense.
Bark, Monesia – Chrysophyllum glycyphloeum.
Bark, Nim – Azadirachta indica.
Bark, Oliver's – Cinnamomum oliveri.
Bark, Ordeal – Erythrophleum guineënse.
Bark, Orinoco Simarouba – Quassia amara.
Bark, Pale – Cinchona officinalis.
Bark, Pambotana – Calliandra houstoniana.
Bark, Panama – Quillaja saponaria.
Bark, Peruvian – Cinchona officinalis.
Bark, Quassia – Quassia amara.
Bark, Quassia (substitute) – Quassia paraensis.
Bark, Quebracho Blanco – Aspidosperma quebracho-blanco.
Bark, Quebracho Colorado – Quebrachia lorentzii.
Bark, Quercitron – Quercus tinctoria.
Bark, Quillaja – Quillaja saponaria.
Bark, Quillaja (substitute) – Cupania pseudorhas.
Bark, Quina Blanca – Croton niveus.
Bark, Quina do matto – Cestrum pseudoquina.
Bark, Quino do Rio – Ladenbergia hexandra.
Bark, Red – Cinchona succirubra.
Bark, Red Mangrove – Rhizophora mangle.
Bark, Sasah – Aporosa frutescens.
Bark, Sassafras – Sassafras albidum.
Bark, Sassy – Erythrophleum guineënse.
Bark, Seven – Hydrangea arborescens.
Bark, Sintok – Cinnamomum sintok.
Bark, Slippery Elm – Ulmus fulva.
Bark, Soap – Quillaja saponaria.
Bark, Tangedu – Cassia auriculata.
Bark, Tellicherry – Holarrhene antidysenterica.
Bark Tree, Cosmetic – Murraya paniculata.
Bark, Wahoo – Euonymus atropurpureus.
Bark, West Indian Snake – Colubrina ferruginea.
Bark, White Mangrove – Laguncularia racemosa.
Bark, White Oak – Quercus alba.
Bark, White Sally – Eucryphia moorei.
Bark, White Willow – Salix alba.
Bark, Winter's – Drimys winteri.
Bark, Witch Hazel – Hamamelis virginiana.
Bark, Yellow – Cinchona calisaya.
Bark, Yellow – Cinchona ledgeriana.
Bark, Yohimbe – Pausinystalia yohimba.

BARLERIA L. Acanthaceae. 230 spp. Tropics. Prickly small herbs.

***B. prionitis*** L. Tropical Africa and Asia. Cultivated. The juice is used in India to cure catarrh; in Java the leaves are chewed to relieve toothache; and in Thailand an extract of the roots is used to combat intestinal worms.

Barley, Abyssinian Intermediate – Hordeum irregulare.
Barley, Common – Hordeum vulgare.
Barley, Egyptian – Hordeum trifurcatum.
Barley, Irregular – Hordeum irregulare.
Barley, Six-rowed – Hordeum hexastichum.
Barley, Pearled – Hordeum vulgare.
Barley, Two-rowed – Hordeum distichum.
Barnadine – Passiflora quadrangularis.
Barobo – Diplodiscus paniculatum.
Baroowas – Premna cordifolia.
Barooweh – Premna cordifolia.

BAROSMA Willd. Rutaceae. 20 spp. S.Africa. Shrubs.

***B. betulina*** (Thunb.) Bartl. and Wendle. (Bucco). S.Africa. The dried leaves contain the essential oil diosphenol, and are used locally by the natives as a carminative and diuretic.

***B. creulata*** (L.) Hook. (Buchu). S.Africa. The dried leaves used locally as a carminative and diuretic.

***B. serratifolia*** (Curtis) Willd. (Buchu). S.Africa. Dried leaves used locally as a diuretic and carminative.

Barrel Media – Medicago tribuloides.

BARRINGTONIA J. R. and G. Forst. Barringtoniaceae. 100 spp. Palaeotropics. Small trees or woody plants.

***B. acutangula*** Gaertn. India to Moluccas. Tree. Wood used locally for general purposes. The bark contains a saponin and is used as a treatment for colds and as a fish poison.

***B. calyptrata*** R. Br. (Mangrove tree).

Australia. Tree. Bark used locally as a fish poison.

***B. careya*** F. v. Muell. Australia. Woody shrub to tree. Bark used locally to stupify fish.

***B. insignis*** Miq. (Songgom). Indonesia. Root bark used locally as a fish poison.

***B. racemosa*** Roxb. Malaysia and Pacific. Tree. Leaves eaten locally cooked or raw. An oil from the seeds is used for lighting. The roots and fruits are used to stupify fish.

***B. speciosa*** Forst. Tropical Asia. Tree. The wood is yellow to red and easily worked. It is used locally for furniture. The cooked pods are eaten in Indo-china, and are also used as a fish poison.

***B. spicata*** Blume. (Pootat lemlik). Indonesia. Tree. Leaves are eaten locally, and the wood is used for house construction.

Bartsch – Heracleum sphondylium.

Barubo – Aristolochia philippensis.

Basak – Dichroa febrifuga.

BASANACANTHA Hook. f. Rubiaceae. 20 spp. Tropical America. Small woody plants.

***B. armata*** (Sw.) Hook. f.=Randia armata (Swartz.) DC. (Crucito, Tintero). N. C.America. Fruits eaten locally.

Basaso – Allophylus lastourvillensis.

BASELLA L. Bassellaceae. 2 spp. Tropical Africa; 3 spp. Madagascar, 1 sp. Pantropics. Herbaceous vines.

***B. alba*** L.=B. rubra.

***B. cordifolia*** Lam.=B. rubra.

***B. rubra*** L.=B. alba=B. crodifolia. (Vine Spinach). Tropics. Cultivated for the red or green leaves which are eaten locally as a vegetable. The sap from the fruit is used to colour food and in the same ways as agar.

Basi – Albizia odoratissima.

Basil, Hoary – Ocimum canum.

Basil, Holy – Ocimum sanctum.

Basil, Musk – Moschosma polystachyum.

Basil, Sweet – Ocium basilicum.

Basil Thyme – Satureja calamintha.

Basil, Wild – Calamintha clinopodium.

Basil, Wild – Pycanthemum incanum.

Basket Oak – Quercus prinus.

Basket Willow – Salix viminalis.

Basong – Alstonia pneumatophora.

***Bassia*** Koen. ex L.=Madheua.

Bassona Gum – Acacia leucophloea.

Basswood – Tilia spp.

Basswood, American – Tilia americana.

Bastard Balm – Melittis melissophyllum.

Bastard Box – Eucalyptus polyanthemos.

Bastard Bullet – Houmiri floribunda.

Bastard Cedar – Soymida febrifuga.

Bastard Mahogany – Eucalyptus botryoides.

Bastard Marula – Kirkia acuminata.

Bastard Myal – Acacia cunninghamii.

Bastard Peppermint – Tristania suaveolens.

Bastard Sandalwood – Eremophila mitchelli.

Bastard Sandalwood – Myoporum sandwicensis.

Bastard Siamese Cardamon – Amomum xanthioides.

Bastard Stone Parsley – Sison amomum.

Bastard Teak – Butea frondosa.

Bastard Toadflax – Comandra umbellata.

Bastard Vervain – Stachytarpheta dichotama.

Bastra – Callicarpa lantana.

***Batatas edulis*** Choisy=Ipomoea batatas.

Batate – Ipomoea batatas.

Bataúa Palm – Jessenia bataua.

Batavia Cinnamon – Cinnamomum burmanii.

Batchelor's Button, Yellow – Polygala rugelii.

Bate – Blighia welwitschi.

Batei-fun – Scirpus tuberosus.

Bateli – Allophylus africanus.

Bateli – Allophylus subcoriaceus.

Bati Oil – Ouratea parviflora.

Batiputa Oil – Ouratea parviflora.

BATIS P. Br. Batidaceae. 1 sp. New Guinea, Queensland; 1 sp. Hawaii, S.W. United States of America, W.Indies, and Atlantic coast of S.America. Shrubs.

***B. maritima*** L. (Saltwort). Coasts of Hawaii, tropical and C.America, W.Indies, Florida. The salty leaves are eaten raw locally. They were used as a source of sodium salts in the manufacture of soap and glass.

Batjan Copal – Agathis loranthifolia.

Batjan Manila – Agathis loranthifolia.

Batjarongi – Oenanthe javanica.

Batoong – Drypetes longifolia.

Battinan Crape Myrtle – Lagerstroemia piriformis.

BAUHINIA L. Leguminosae. 300 spp. Warm. Shrubs, small trees, or woody climbers.

***B. acuminata*** L. India to China. Shrub. A root extract is used to treat colds in Java, and the leaves are used in Perak as a poultice to treat ulcers of the nose.

***B. elongata*** Korth. (Sebari, Sobheuri).

Indonesia. Climbing Shrub. Bark is used locally as cordage.

***B. esculenta*** Burch. Tropical Africa. Woody shrub. Pods eaten locally in S.Africa.

***B. fassageensis*** Kotschy ex Schweinf. Java. Climbing shrub. An extract of the root is used locally to treat fevers and coughs.

In N.E. and C.Africa the plant (Apapa, Manghamba) produces seeds which are eaten locally.

***B. hirsuta*** Korth. (Kendajakan pootih). Small shrub. Indonesia. The crushed bark is used in Sumatra to mix with horse fodder to stimulate appetite.

***B. malabarica*** Roxb. Tropical Asia. Small tree. The leaves are used to flavour meat and fish, especially in the Philippines.

***B. petersiana*** Bolle. (Kafumbe, Kitata). Tropical C.Africa. The bark is used locally as fibre, and a dye is extracted from the roots.

***B. purpurea*** L. (Camel's Foot Tree, Keolav, Khairival, Sona). China through to S.India. Tree. Cultivated for the wood which is pink, darkening with age. It is used in India for buildings and agricultural implements. A gum (Semki-gona Gum) is extracted from the trunk. The flowers and buds are eaten locally as a pot-herb.

***B. reticulata*** Roxb. Tropical Africa. Small tree. The roots yield a brown-red dye, and the dye from the pods is blue-black. The seeds are eaten and the ashes are used in soap manufacture. The bark yields a fibre which is used for cordage and to make fabric.

***B. retusa*** Roxb. Himalayas. Source of an inferior gum.

***B. rufescens*** Lam. W.Africa. Tree. The wood is used locally for general carpentry. The bark is the source of a fibre used for cordage. Extracts of the bark are used for tanning and to treat dysentery. An infusion of the leaves is used to treat eye diseases, and the root to treat fever.

***B. thonningii*** Schumach. (Kao, Kifumbi, Lolokendamba). Small tree. Tropical Africa. The wood is used to make household utensils. The leaves are used to treat intestinal worms. The fruits are used as a soap, and an extract of the bark is used to treat colds.

***B. ungulata*** L. Mexico. Small tree. The wood is flexible and used locally in the construction of huts.

***B. vahlii*** Wight ex Arn. India. Tree. The bark yields a fibre used in the manufacture of sails.

***B. variegata*** L. Burma, E.Indies and China. The wood is used to make agricultural implements. The leaves and pods are eaten as a vegetable, and the leaves are used to cover cigarettes. The bark is bitter and astringent and used to increase appetite. It is also used in dyeing and tanning. The tree yields a gum (Semba gona, Senn).

Bavarian Winter Pea – Pisum arvense.
Bawala – Gnetum africanum.
Bayating – Tinomiscium philippense.
Bayberry – Myrica carolinensis.
Bay Fat – Lauris nobilis.
Bayisagugul – Commiphora stocksiana.
Bay, Loblolly – Gordonia lasianthus.
Bay Oil – Pimenta acris.
Bay, Red – Persea borbonia.
Bay, Rose – Rhododendron maximum.
Bay, Swamp – Magnolia glauca.
Bay, Sweet – Magnolia glauca.
Bay, Sweet – Persea borbonia.
Bay, Tan – Gordonia lasianthus.
Bay, True – Laurus nobilis.
Baywood – Pterocarpus soyauxii.
Bazimo – Maytenus senegalensis.
Bdellium – Commiphora africana.
Bdellium – Commiphora hildebrandtii.
Bdellium, Bombay – Commiphora erythraea.
Bdellium, Gubam Myrrh – Commiphora hilbrandtii.
Bdellium, Habbak Daseino – Commiphora hilbrandtii.
Bdellium, Habbak Dunas – Commiphora hilbrandtii.
Bdellium, Habbak Dunkal – Commiphora hilbrandtii.
Bdellium, Habbak Harr – Commiphora hilbrandtii.
Bdellium, Habbak Ilka Adaxai – Commiphora hilbrandtii.
Bdellium, Habbak Tubuk – Commiphora hilbrandtii.
Bdellium, Harobil Myrrh – Commiphora myrrha.
Bdellium, Indian – Commiphora mukul.
Bdellium, Mukul-i-Azrak – Commiphora mukul.
Bdellium, Mukul-i-Yahud – Commiphora mukul.

Bdellium, Opaque – Commiphora spp.
Bdellium, Perfumed – Commiphora erythraea var. glabrescens.
Bdellium, Sakulali – Commiphora mukul.
Bea (Gum) – Caesalpinia praecox.
Beach Clover – Trifolium fibriatum.
Beachgrass, American – Ammophila breviligulata.
Beachgrass, European – Ammophila arundinacea.
Beach Pea – Lathyrus maritimus.
Beach Plum – Prunus maritima.
Bead Tree – Elaeocarpus ganitrus.
Beadtree, Sandal – Adenanthera pavonina.
Beaked Hazelnut – Corylus rostrata.
Beakpod Eucalyptus – Eucalyptus robusta.
Beam, White – Sorbus aria.
Bean, Adzuki – Phaseolus angularis.
Bean, Asparagus – Dolichos sesquipedalis.
Bean, Asparagus – Psophocarpus tetragonolobus.
Bean, Babricon – Canavalia campylocarpa.
Bean, Bengal – Mucnuna atterrina.
Bean, Black – Castanospermum australe.
Bean, Black Mauritius – Muncuna atterrina.
Bean, Broad – Vicia faba.
Bean, Bush – Phaseolus vulgaris.
Bean, Butter – Phaseolus lunatus.
Bean, Butterfly – Clitoria ternata.
Bean, Calabar – Physostigma venenosum.
Bean, Castor – Ricinus communis.
Bean, Cherry – Vigna sinensis.
Bean, Cluster – Cyamopsis psoralioides.
Bean, Common – Phaseolus vulgaris.
Bean, Common Coral – Erythrina corallodendron.
Bean, Coral – Erthina arborea.
Bean, Duffin – Phaseolus lunatus.
Bean, Eastern Coral – Erythrina herbacea.
Bean, Field – Phaseolus vulgaris.
Bean, Fleshy – Muncuna pachylobia.
Bean, Florida Velvet – Muncuna deeringiana.
Bean, French – Phaseolus vulgaris.
Bean, Goa – Psophocarpus tetragonolobus.
Bean, Haricot – Phaseolus vulgaris.
Bean, Horse – Vicia faba.
Bean, Hyacinth – Dolichos lablab.
Bean, Ignatius – Strychnos ignatii.
Bean, Indian – Catalpa bignonioides.
Bean, Indian Coral – Erythrina indica.
Bean, Jack – Canavalia ensiformis.
Bean, Jumping – Leucaena glauca.
Bean, Jumping – Sapium biloculare.
Bean, Jumping – Sebastiana pavoniana.
Bean, Kidney – Phaseolus vulgaris.
Bean, Lima – Phaseolus lunatus.
Bean, Lucky – Erythrina abyssinica.
Bean, Lucky – Thevetia nereifolia.
Bean, Lyon – Muncuna nivea.
Bean, Mat – Phaseolus aconitifolius.
Bean, Mescal – Sophora secundifolia.
Bean, Metcalf – Phaseolus retusus.
Bean, Molucca – Caesalpinia bonducella.
Bean, Moth – Phaseolus aconitifolius.
Bean, Mung – Phaseolus mungo.
Bean, Navy – Phaseolus vulgaris.
Bean, Nickar – Caesalpinia bonducella.
Bean, Ordeal – Physostigma venenosum.
Bean, Pichrim – Aniba firmulum.
Bean, Pichurim – Nectandra puchury-minor.
Bean, Pigeon – Vicia faba.
Bean, Pole – Phaseolus vulgaris.
Bean, Potato – Apios tuberosa.
Bean, Puchury – Nectandra puchury-minor.
Bean, Rice – Phaseolus calcaratus.
Bean, Rice – Phaseolus mungo.
Bean, Sabre – Canavalia ensiformis.
Bean, St. Thomas's – Entada scandens.
Bean, Sarawak – Vigna hosei.
Bean, Scarlet – Phaseolus coccineus.
Bean, Screw – Prosopis odorata.
Bean, Sieva – Phaseolus lunatus.
Bean, Soy – Glycine max.
Bean, Soya – Glycine max.
Bean, Sweet – Gleditsia triacanthos.
Bean, Sword – Canavalia ensiformis.
Bean, Sword – Canavalia gladiata.
Bean, Tepary – Phaseolus acutifolius.
Bean, Tonka – Coumarouna odorata.
Bean, Tonka – Dipteryx odorata.
Bean, Trailing Wild – Strophostyles helvola.
Bean Tree Starch – Castanospermum australe.
Bean, Vanilla – Vanilla phaentha.
Bean, Vanilla – Vanilla planifolia.
Bean, Velvet – Muncuna nivea.
Bean Vine – Phaseolus polystachyus.
Bean, Walnut – Eniandra palmerstonii.
Bean, Wayaka Yam – Pachyrhizus angulatus.
Bean, West African Locust – Parkia filicoides.
Bean, Wild – Phaseolus polystachyus.
Bean, Yam – Pachyrhizus erosus.
Bean, Yam – Pachyrhizus tuberosus.
Bean, Yam – Sphenostylis schweinfurthii.
Bearberry – Arctostaphylos uva-ursi.
Bearberry – Lonicera involucrata.
Bearberry – Rhamnus purshiana.

Bearberry – Vaccinium erythrocarpa.
Bear's Bed – Polytrichum juniperinum.
Bear's Breeches – Acanthus mollis.
Bear's Garlic – Allium ursinum.
Bear Grass – Yucca glauca.
Bear Grass, Common – Xerophyllum tenax.
Bear Huckleberry – Gaylussacia ursina.
Beard Tongue, Large-flowered – Penstemon grandiflorus.
Bearded Usnea – Usnea barbata.
Beardless Wild Rye – Elymus triticoides.
Bearswood – Eriodictyon glutinosum.
Bearswood – Rhamnus purshiana.
***Beatsonia portulacaefolia*** Roxb. = Frankenia portulacaefolia.
***Beaucarnea recurvata*** Lem. = Nolia recurvata.
Beautyleaf, Ceylon – Callophyllum calaba.
Beaverbread Scurfpea – Psoralea castorea.
Beaverwood – Celtis occidentalis.
Bebbeeru – Nectandra rodoiei.
Beccabunga – Veronica beccabunga.
Beccabunga, Brooklime – Veronica beccabunga.

BECCARIOPHOENIX Jumelle and Perrier. Palmae. 1 sp. Madagascar.

***B. madagascariensis*** Jum. and Perr. (Monarana). Madagascar. Palm tree. The sap from the stem is drunk as a sweet drink locally, and the leaves are shredded to make hats.

BECKMANNIA Host. Graminae. 2 spp. N.Temperate.

***B. erucaeformis*** (L.) Host. Temperate Asia and N.America.
Grain eaten in Japan.

***B. syzigachne*** (Steud.) Fern. (American Sloughgrass). Asia, N.America. Grass, used for fodder and hay.

Bedali – Radermachera gigantea.
Bedaru – Cantleya corniculata.
Bede-Bede – Conopharyngia chippi.
Bedstraw, Dye – Galium tinctorium.
Bedstraw, (Our) Lady's – Galium verum.
Bee Balm – Monarda didyma.
Bee Balm – Monarda fistulosa.
Bee Balm, Lemon – Monarda citriodora.
Bee Balm, Pony – Monarda pectinata.
Beech, African – Faurea saligna.
Beech, American – Fagus ferruginea.
Beech, Blue – Carpinus caroliniana.
Beech, Brown – Cryptocarya glaucenscens.
Beech, Brown – Litsea reticulata.
Beech, Cape – Myrsine melanophloea.
Beech, European – Fagus sylvatica.
Beech, Malay Bush – Gmelina arborea.
Beech, Mountain – Nothofagus cliffortoides.
Beech, Myrtle – Nothofragus cunninghamii.
Beech, Nut Seed Oil – Fagus sylvatica.
Beech, Oriental – Fagus orientalis.
Beech, Paddy King's – Flindersia acuminata.
Beech, Red – Flindersia brayeyana.
Beech, Red – Nothofagus fusca.
Beech, Silver – Nothofagus menziesii.
Beech, Tasmanian – Nothofagus cunninghamii.
Beech, Wau – Elmerillia papuana.
Beech, White – Gmelina leichtardtii.
Beechey Bamboo – Bambusa beechyana.
Beefsteak Fungus – Fistulina hepatica.
Beefsteak, Vegetable – Fistulina hepatica.
Beef Tongue – Fistulina hepatica.
Beef Wood – Banksia integrifolia.
Beef Wood – Casuarima equisetifolia.
Beef Wood – Grenvillea striata.
Beef Wood – Stenocarpus salignus.
Beer, Ale – Hordeum vulgare var. distichum.
Beer, Birch – Betula lenta.
Beer, Bock – Hordeum vulgare var. distichum.
Beer, Ginger – Saccharomyces pyriformis.
Beer, Hops – Humulus lupulus.
Beer, Lager – Hordeum vulgare var. distichum.
Beer, Merissa – Calotropis procera.
Beer, Near – Hordeum vulgare var. distichum.
Beer, Porter – Hordeum vulgare var. distichum.
Beer, Root – Cinnamomum mercadoi.
Beer, Root – Sassafras albidum.
Beer, Spruce – Picea mariana.
Beer, Stout – Hordeum vulgare var. distichum.
Beer, Tallas – Rhamnus prinoides.
Beet – Beta vulgaris.
Beet, Garden – Beta vulgaris.
Beet, Red – Beta vulgaris.
Beet, Sea – Beta vulgaris.
Beet, Silver – Beta vulgaris.
Beet, Sugar – Beta vulgaris.
Befelopela – Tachiadenus longifolius.
Beggarweed – Desmodium tortuosum.

BEGONIA L. Begoniaceae. 900 spp. Tropical and subtropical, esp. America. Herbs.

***B. bahiensis*** DC. Brazil. Leaves used locally to reduce fever.

***B. cuullata*** Willd. = B. spathulata. E.Brazil. Leaves used locally as a diuretic.

***B. dominicalis*** DC. Dominica and Guadeloupe. Leaves and flowers used locally to prepare a decoction against colds.

***B. luxurians*** Scheidw. = Scheidweilleria luxurians. Brazil. A decoction of the leaves used locally to reduce fevers.

***B. sanguinea*** Raddi. = Pritzelia sanguinea. Brazil. A decoction of the leaves used locally as a diuretic.

***B. spathulata*** Lodd. = B. cucullata.

***B. tuberosa*** Lam. Moluccas. Leaves eaten raw and cooked locally. They are also made into a sauce.

Begung – Amorphophallus prainii.

BEHAIMIA Griseb. Leguminosae. 1 sp. Cuba. Tree.

***B. cubensis*** Griseb. The wood is strong and used for general construction work.

BEILSCHMIEDIA Nees. Lauraceae. 200+ spp. Tropics, Australia and New Zealand. Trees.

***B. baillonii*** Planch and Sébert. New Caledonia. Strong durable timber used for interior work, and cabinet-making.

***B. bancroftii*** White. (Canary Ash, Yellow Walnut). N.Queensland. Brown-green pliable wood. Not durable or easily worked, but good for veneers, plywood and interior work, as it bends well.

***B. corbisieri*** Robyns and Wilcz. (Bongolu). Congo. Red wood used locally for interior work, furniture-making and carving.

***B. gaboonensis*** (Meissn.) Benth. (Okukuluka). W. and C.Africa. An easily worked wood, used for furniture-making and interior work.

***B. giorgii*** Robyns and Wilcz. (Djombi). Congo. Leaves used locally to prepare a body scent (Ngula).

***B. insularum*** Robyns and Wilcz. (Efufuko, Gongolu). Congo. Flexible wood used locally to make bows.

***B. lanceolata*** Planch. and Sébert. New Caledonia. Greyish wood used for interior work.

***B. leemansii*** Robyns and Wilcz. (Efufuko, Eko). Congo. Brownish wood used for boards, boat-making and oars.

***B. mannii*** Stapf. (Spicy Cedar). W.Africa. Flowers are used to flavour rice. The fruits are used locally to cure dysentery. The seeds are sold locally and eaten raw or roasted, in soups or as a vegetable.

***B. oblongifolia*** Robyns and Wilcz. (Kilundeke, Lihuwe). Congo. Yellow-brown to grey wood used locally for construction work and house interiors.

***B. odorata*** Baill. New Caledonia. Wood used for making furniture.

***B. pendula*** (Sw.) Benth. and Hook f. (Guajón). W.Indies, tropical America. The pink-brown wood has black lines. It is strong, heavy, fairly hard and durable in soil. It is used for decorative work, plywood, ship-building, house construction, furniture-making, etc.

***B. tawa*** Hook. f. = Nesodaphne tawa. New Zealand. Fruits eaten by Maoris.

***B. variabilis*** Robyns and Wilcz. (Monkeru). Congo. Wood used for making small boats and for poles.

Bejuco amargo – Aristolochia pardina.

Bejuco Blanco – Exogonium bracteatum.

Bejuco carey – Tetracera volubilis.

Bejuco deagua – Tetracera sessiliflora.

Bejuco de leche – Funastrum clausum.

Bejuco de panune – Entada polystachia.

Bejuco de pescado – Funastrum cumanense.

Bejuco iasú – Vitis sicyoides.

Bejuco legitimo – Bignonia unguis-cati.

Bejuco tomé – Tetracera volubilis.

Bekaro – Pterygota alata.

***Bekkeropsis uniseta*** K. Shum. = Pennisetum unisetum.

BELAMCANDA Adans. Iridaceae. 2 spp. E.Asia. Herbs.

***B. chinensis*** (L.) DC. (Blackberry Lily, Leopard Flower). China and Japan. Cultivated in Old and New World. The rhizome is used in Chinese medicine for chest and liver complaints, in tonics and as a purgative.

Belam Kedjil – Payena dantung.

***Belangera*** Cambess. = Lamanonia Vell.

Belembe – Xanthosoma brasiliense.

Belian – Palaquium ridleyi.

Bell Apple – Passiflora laurifolia.

Belladonna – Atropa belladonna.

Belladonna, Indian – Atropa acuminata.

Belleric Terminalia – Terminalia bellerica.

Bellflower, Chinese – Platycodon grandiflorum.

Bellflower, Japanese – Platycodon grandiflorum.

BELLIS L. Compositae. 15 spp. Europe, esp. Mediterranean. Herbs.

***B. perennis*** L. (Daisy, English Daisy). The flowers are used to treat chest complaints, including spitting of blood, internal bleeding, blood in the urine; and boils.

***Bellota miersii*** C. Gay = Cryptocarya nitida.

Bellwort – Uvularia sessiliflora.
Belinmbang kera – Sarcotheca monophylla.
Beloha – Cyperus confusus.
Ben Oil – Moringa oleifera.
Ben Seed Oil – Moringa oleifera.
Bene (Oil) – Sesamum indicum.
Benedictine – Angelica archangelica.
Bengal Bean – Muncuna atterina.
Bengal Cardamon – Amomum aromaticum.
Bengal Gambir – Uncaria gambir.
Bengal Gram – Cicer arietinum.
Bengal Kino – Butea frondosa.
Bengal Quince – Aegle marmelos.
Bengor Nut – Caesalpinia bouducella.
Bengue – Strychnos icaja.
Benguet Pine – Pinus insularis.
Benin Apata Wood – Microdesmis puberula.
Benin Ebony – Diospyros crassiflora.
Benin Ebony – Diospyros suaveolens.
Benin Mahogany – Guarea thompsonii.
Benin Mahogany – Khaya grandis.
Benin Mahogany – Khaya puchii.
BENINCASA Savi. Cucurbitaceae. 1 sp. Tropical Africa. Vine.
***B. cerifera*** Savi. = B. hispida.
***B. hispida*** Cogn. = B. cerifera. (Chinese Preserving Melon, Wax Gourd, White Gourd). Tropical Africa, Asia. Cultivated locally and used for preserves and pickles. Eaten as a vegetable in Japan.
Bent – Calamus rotang.
Bent – Salix tetrasperma.
Bent, Rhode Island – Agrostis vulgaris.
Bent, White – Agrostis alba.
Bentamake – Atropa acuminata.
Bentangon – Vitis arachnoidea.
Benteli lalaki – Wrightia javanica.
Benteak – Lagerostroemia lanceolata.
Benzoin – Styrax benzoin.
***Benzoin aestivalis*** (L.) Nees. = Lindera benzoin.
***Benzoin benzoin*** Coult. = Lindera benzoin.
Benzoin, Siam – Styrax tonkinense.
Benzoin, Sumatra – Styrax benzoin.
***Benzoin umbellatum*** Ktze. = Linera umbellata.
BERBERIS L. Berberidaceae. 450 spp. N. and S. America, Eurasia, N.Africa. Shrubs.
***B. aquifolium*** Pursh. = Mahonia aquifolium.
***B. aristata*** DC. (Indian Barberry) N.W. Himalayas. The dried stem is used locally as a tonic and to treat fevers. The stem and root yield a yellow dye.
***B. buxifolia*** Lam. (Magellan Barberry). S. and W. S.America. Fruits eaten locally, and made into preserves.
***B. canadensis*** Mill. (Alleghany Barberry, American Barberry). E. N.America. Fruits used for jams.
***B. darwinii*** Hook. (Michay). Chile. Roots and bark source of a tonic and yellow dye. Used locally.
***B. flexuosa*** Ruiz. and Pav. Peru, Argentina, Chile. Roots source of yellow dye.
***B. haematocarpa*** Woot. (Barberry). S.W. N.America. Fruits used in jellies.
***B. ilicifolia*** Scheele = Mahonia trifoliolatus.
***B. laurina*** Thunb. Brazil. Root a source of a yellow dye.
***B. lutea*** Ruiz. and Pav. Peru, Ecuador. Wood source of a yellow dye.
***B. lycium*** Royle. Himalayas. An extract of the roots (Rasant) is used in local medicine to reduce fevers, to aid digestion, and to cure haemarrhoides.
***B. repens*** Lindl. = Mahonia repens.
***B. ruscifolia*** Lam. Brazil. The root is a source of a yellow dye.
***B. vulgaris*** L. (European Barberry). Europe. The stem is a source of a yellow dye, and the stem and root bark is used to produce a tonic. The red berries are used to make preserves, esp. in France (Confiture d'épine). The fine-grained, hard yellow wood is used for small carving, turning, and mosaic work.
Berchaga – Grewia villosa.
Berenjene – Solanum mammosum.
Bergamot (Oil) – Citrus bergamia.
Bergamot Mint – Mentha aquatica.
Bergamot, Wild – Monara fistulosa.
BERGENIA Moench. Saxifragaceae. 6 spp. C. and E. Asia. Herbs.
***B. bifolia*** Moench. = B. crassifolia.
***B. crassifolia*** (L.) Fritsch. = B. bifolia. N.Mongolia, Siberia. The bark is used for tanning, and the leaves are used to prepare a beverage (Tschager Tea).
***B. purpurascens*** (Hook. f. and Thoms.) Engl. E.Asia. In China, the rhizome is used to control bleeding, and to prepare a tonic.
BERLINIA Soland. ex Hook. f. and Benth. Leguminosae. 15 spp. Tropical Africa. Trees.
***B. acuminata*** Soland. Tropical W.Africa. Hard, pinkish streaked timber is used for general carpentry, furniture-making, turning, and locally for canoes and drums.
***B. globiflora*** Hutch. and B. Davy. S.Africa.

The bark, root, and leaves are used as an ordeal poison in Zambia.

***B. sereti*** De Wild. = Julbernardia sereti.

Bermuda Arrowroot – Maranta arundinacea.

Bermuda Grass – Cynodon dactylon.

Bermuda Red Cedar – Juniperus bermudiana.

Bernang – Protium javanicum.

Berracos – Cyrtocarpa procera.

***Berria*** Roxb. = Berrya.

BERRYA Roxb. mut DC. Tiliaceae. 6 spp. Indomalaysia, Polynesia. Trees.

***B. ammonilla*** Roxb. = B. cordifolia.

***B. cordifolia*** (Willd.) Burret. = B. ammonilla. Ceylon, India, through Malaysia to Queensland. The dark red wood is hard, tough, flexible. It is a valuable timber which is used for house-building, general construction work, boats, and carts.

Berry-bearing Bamboo – Melocanna bambusioides.

Berry, Checker – Gaultheria procumbens.

Berry, Christmas – Photinia arbutifolia.

Berry, Coral – Pymphoricarpus orbicularis.

Berry, One – Paris Quadrifolia.

Berry, Partridge – Gualtheria procumbens.

Berry, Tea – Gaultheria procumbens.

Berry, Water – Syzygium guinense.

Berry, Winter – Gaultheria procumbens.

Berseem – Medicago sativa.

Bersinge – Radermachera xylocarpa.

BERTHOLLETIA Humb. and Bonpl. Lecythidaceae. 2 spp. Tropical S.America, W.Indies. Trees.

***B. excelsa*** H. K. B. (Brazil Nut, Castanha do Pará, Cream Nut, Para Nut). Amazon Basin. The nuts, which are seeds, borne like the segments of an orange in a horny fruit contain about 70 per cent oil. About 30,000 tons are produced annually, most of which are collected from the forest floor. Most of the crop is exported to U.S.A. or Europe, and consumed raw especially at Christmas time. The oil is extracted and used locally for cooking and soap-manufacture. The bark is used locally for caulking boats.

***B. nobilis*** H. B. K. See B. excelsa, but not exported.

Bertram – Anacyclus officinarum.

BERTYA Planch. Euphorbiaceae. 23 spp. Australia, Tasmania. Trees.

***B. cunninghamii*** Planch. Australia, esp. Queensland. A resin is extracted from the trunk.

Bester Cola – Cola acuminata.

Bet – Calaums rotang.

BETA L. Chenopodiaceae. 6 spp. Europe, especially Mediterranean. Herbs.

***B. maritima*** L. (Wild Sea Beet). Coastal regions from Europe to India. Formerly eaten in Europe as a cooked vegetable.

***B. vulgaris*** L. (Beet). A cultigen, derived in Europe from B. maritima. There are four main derivatives, (1) Sugar beet, (2) Red Beet, grown as a garden vegetable, and harvested 8–10 weeks after sowing. (3) Mangel wurzel, (Mangel) grown as an animal food. (4) Swiss Chard, or Leaf Beet – the leaves are eaten as a cooked vegetable. Although the plant is a biennial, it is grown as an annual, and except for the Swiss Chard, it is the swollen hypocotyl which is eaten. Red Beet yields about 200 bu./acre, and Sugar Beet and Mangels about 20 tons/acre, but may be much more. Although the leaves of B. maritima have been eaten since pre-Christian times, the Red Beet was not used in Europe until the 16th. century. The high sugar content of sugar beet (modern varieties may contain 17–19 per cent) was not established until 1802, when it was discovered by the German chemist Marggraf. Interest was shown in Prussia, but the blockade of France during the Napoleonic wars gave the initial impetus to the industry in France. It was to collapse only to be revived throughout Europe during the 1914–18 war and the 1939–45 war. Now the world production is over 160 million tons, most of this being produced in the Soviet Union and Europe. During production of sugar, the beet is cut into strips which are soaked in hot water to remove the sugar. Soluble impurities are then removed by treatment with lime and carbon dioxide. The solution is then boiled under vacuum to produce a syrup from which the sugar is allowed to crystallize. The crystals are separated by centrifuging and purified by further recrystallization. The tops of the plants are fed to livestock either fresh or as silage. The pulp from the factory is dried and fed to stock or used as manure. The impure sugar solution (molasses) is used to produce ethyl alcohol (ethanol).

Betauma – Pouzolzia denudata.

Betel – Piper betel.

Betelnut – Areca catechu.
Beterwok – Melochia umbellata.
Bethroot – Trillium erectum.
Bétis Oil – Madhuca betis.
Betony – Stachys officinalis.
Betoor – Calophyllium wallichianum.
Betoor beroobook – Calophyllum pulchierrimum.
Betoum – Pistacia atlantica.
Betu (Oil) – Balanites aegyptiaca.
BETULA L. Betulaceae. 60 spp. N.Temperate, Arctic. Trees.
***B. acuminata*** Wall.=B. alnoides.
***B. alba*** L. pro parte.=B. pubescens. (European Birch). Europe, Greenland, N.W. temperate Asia. The wood is yellow-pink, elastic, difficult to split and not durable. It is used to make small articles, e.g. spoons, shoes, etc. and waggons. It is sometimes used to make garden furniture. The bark is eaten as a famine food in Lapland. The sap is used in various localities to improve the hair.
***B. alnoides*** Buch.-Ham. ex Don.=B. acuminata. India. The bark was used as paper locally.
***B. bhojpathra*** Wall.=B. utilis.
***B. erminii*** Cham. (Erman's Birch). Japan. The bark is used locally to bandage wounds.
***B. japonica*** Sieb. Temperate Asia. A hard yellow-white wood which polishes well. Used for furniture locally, for wooden pipes and spools. It is important in the Japanese plywood industry.
***B. lenta*** L. (Black Birch, Cherry Birch). E. N.America, to Florida. Heavy, strong dark brown wood. It is used for furniture, utensils, agricultural implements, and is a good fuel. Birch Oil is distilled from the twigs and bark. This resembles, and is sometimes marketed as, Oil of Wintergreen. It is used as flavouring and medicinally to relieve strains. The sap has a high sugar content, and is used to brew Birch Beer.
***B. lutea*** Michx. (Yellow Birch). E. N.America. Strong, heavy, light brown wood. It is used for furniture, wheels, packing cases and fuel.
***B. nigra*** L. (Red Birch, River Birch). E. N.America to Florida, and Texas. Light-brown heavy wood. It is used for furniture, turnery, small wooden articles, e.g. shoes.
***B. papygracea*** Ait.=B. papyrifera.
***B. papyrifera*** Marsh.=B. papygracea. (Canoe Birch, Paper Birch). N.America, Alaska to Washington. The wood is strong, light and close-grained, light brown. It is used for a variety of small articles, e.g. spools, turnery, shoe lasts, tooth-picks. The light coloured bark is resinous and waterproof, so it was used by the Indians to make canoes.
***B. pendula*** Roth.=B. verrucosa.
***B. populifolia*** Marsh. (Gray Birch). E. N.America. The wood is soft, light and not durable. It is used for small wooden articles, e.g. spools, clothes pegs, wood pulp and charcoal.
***B. pubescens*** Ehrh.=B. alba.
***B. utilis*** D. Don.=B. bhojpathra. (Himalayan Birch, Indian Birch). Himalayan region. The inner bark was used locally for paper.
***B. verrucosa*** Ehrh.=B. pendula. (White Birch). Europe, across temperate Asia. Yellowish tough, elastic wood used for small wooden articles, and pipes. It is one of the most useful trees in the U.S.S.R. The bark is waterproof and used for roofing. A wood- and leather-preserving tar is distilled from the wood. Dry distillation of the wood and bark yields an oil which is used medicinally as an antiseptic and counterirritant. A wine (Birch Wine) is prepared from the sap. The leaves, treated with chalk give a yellow dye, treated with alum they yield a green dye. The soot makes a black paint.
Beukenhout – Faurea saligna.
Beurre de Nogo – Nogo nogo.
Beurre d'Orère – Baillonella ovata.
Bezzetta – Chrozophora tinctoria.
Bezetta coerulea – Chrozophora tinctoria.
Bezetta rubra – Chrozophora tinctoria.
Bhabar – Eulaliopsis binata.
Bhander – Ziziphus xylocarpus.
Bhang – Cannabis sativa
Bhankalink – Cyanchum arnottianum.
Bharbur Grass – Ischaemum angustifolium.
Bhasma, Gold – Drosera peltata.
Bheri – Casearia tomentosa.
BHESA Buch.-Ham. ex Arn. Celastraceae. 5 spp. Indomalaya, Pacific. Trees.
***B. paniculata*** Wall. (Kajoo djamboo, Sengajoh, Timpoot). The brown, hard wood is used locally for house-building.
Bhillaura – Trewia nudiflora.
Bhogi – Hopea parviflora.
Bhoma – Glochidion hohenackeri.

Bhutakesi – Corydalis gavaniana.
Bhut kesi – Corydalis gavaniana.
Bhutkis – Corydalis gavaniana.
Bible Frankincense – Boswellia carteri.
Bibu – Holigarna arnottiana.
Bicuhyba (Fat) – Virola bicuhyba.
Bidara goonong – Strychnos ligustrina.
Bidara iaoot – Strychnos ligustrina.
BIDENS L. Compositae. 230 spp. Cosmopolitan. Herbs.
***B. bigelovii*** Gray. S.W. United States of America, and Mexico. Leaves used locally by Indians to make a beverage.
***B. chinensis*** Willd. = B. wallichii. (Ambong-ambong, Harcuga, Kertool). S.E. Asia, Malaysia, Indonesia. The young plants are used in Java to make a soup, and the roots are chewed to cure headaches.
***B. fruticola*** L. = Verbesina fruticosa.
***B. pilosa*** L. China, Japan to Philippines. It is used with half boiled rice and fermented to make a wine (Sinitsit, Tafei) in the Philippines.
***B. scandens*** L. = Salmea eupatoria.
***B. tripartitus*** L. (Murr Marigold, Water Agrimony). N. and W. Asia, through to Europe. A decoction of the leaves is used to treat urinary complaints. It is used especially in cases of gout, blood in the urine, dropsy, and bleeding from the uterus. It is diuretic and astringent.
***B. wallichii*** DC. = B. chinensis.
Bie bie – Radirkojera calodendron.
Biembie – Sorindeia gilletii.
Bigarade – Citrus aurantium.
Big-bud Hickory – Carya tomentosa.
Big-flower Selfheal – Prunella grandiflora.
Big leaf Eucalyptus – Eucalyptus goniocalyx.
Big Marigold – Tagetes erecta.
Big Shellbark Hickory – Carya laciniosa.
Big Tree – Sequioa gigantea.
Bigareaus – Prunus avium.
Bignay – Antidesma platyphyllum.
Bignay China Laurel – Antidesma bunius.
BIGNONIA L. Bignoniaceae. 1 sp. C.America. Woody vine.
***B. acutistipula*** Schlecht. = B. unguis-cati.
***B. alliacea*** Lam. = Adenocalymma alliaceum.
***B. brasiliana*** Lam. = Jacaranda brasiliana.
***B. capreolata*** L. = Tecoma conagera.
***B. caroba*** Vell. = Jacaranda procera.
***B. chica*** Humb. and Bonpl. = Doxantha chica.
***B. copaia*** Aubl. = Jacaranda copaia.
***B. echinata*** Jacq. = Pithecoctenium echinatum.
***B. linearis*** Cav. = Chilopsis lineari.
***B. pentaphylla*** L. = Tabebuia pentaphylla.
***B. quercus*** Lam. = Catalpa longissima.
***B. radicans*** L. = Tecoma radicans.
***B. stans*** L. = Tecoma stans.
***B. unguis-cati*** L. = B. acutistipula. (Bejuco legitimo, Xcanol-ak, Liana unida). Mexico, W.Indies, N. S.America. Used locally in Mexico to treat snake-bites.
Bigroot Geranium – Geranium macrorhium.
Bigstring Nettle – Urtica dioica.
Bihambi – Uapaca staudtii.
Bihbool – Vitex glabrata.
Bihi – Rajania cordata.
Bijágara – Colubrina ferruginea.
Bikal-Bohori – Schizostachym dielsianum.
BIKKIA Reinw. Rubiaceae. 20 spp. Small trees. E.Malaysia, Polynesia
***B. mariannensis*** (Raf.) Brong. = Cormogonum marinensis. (Torchwood). South Sea Islands. The wood burns easily and is used locally to make torches.
Bili devadari – Dysoxylum malabaricum.
Bilimbi – Averrhoa bilimbi.
Billberry – Vaccinium meridioale.
Billberry – Vaccinium myrtillus.
Billberry, Bog – Vaccinium uliginosum.
Billet Wood – Diospyros dendro.
Billian, Borneo – Eusideroxylon zwageri.
Billion Dollar Grass – Echinochloa frumentacea.
Billyweb Sweetia – Sweetia panamensis.
Biloa – Aegle marmelos.
Bilsted – Liquidambar styraciflua.
Bimlau – Grewia elastica.
Bimplipatam (Jute) – Hibiscus cannabus.
Bina – Avicenna officinalis.
Bindweed, Black – Tamus communis.
Binggas Terminalia – Terminalia comintana.
Binjai Mango – Mangifera caesia.
Binong – Sterculia javanica.
Binong – Tetramelis nudiflora.
Bintangoor padi – Calophyllum pulcherrimum.
Bintaos – Wrightia javanica.
Binuang – Octomeles sumatrana.
Binukao – Garcinia binucao.
Binunga Gum – Macaranga tanarius.
Bira tai – Garcinia multiflora.
Birch, Black – Betula lenta.
Birch, Canoe – Betula papyrifera.
Birch, Cherry – Betula lenta.

Birch, Erman's – Betula ermanii.
Birch, European – Betula alba.
Birch, Gray – Betula populifolia.
Birch, Himalayan – Betula utilis.
Birch, Indian Paper – Betula utilis.
Birch, Paper – Betula papyrifera.
Birch, Red – Betula nigra.
Birch, River – Betula nigra.
Birch Tree – Bursera gummifera.
Birch, White – Betula verrucosa.
Birch, White – Schizomeria ovata.
Birch, Yellow – Betula lutea.
Bird Cherry – Prunus avium.
Bird Rape – Brassica campestris.
Bird's foot Trefoil – Lotus corniculatus.
Biribá – Rollinia deliciosa.
Bi-ri-jih – Ferula gallbaniflua.
Bisabol Myrrh – Commiphora erythraea.
Bisabol Myrrh – Commiphora erythraea var. glabrescens.
Bisabol Myrrh – Opopanax chironium.
Bisbirinda – Castela texana.
Biscayne Palm – Coccothrinax argentea.
BISCHOFIA Blume. Bischofiaceae. 1 sp. Chili. Tree. Distribution of the genus seems to be in doubt.
***B. javanica*** Blume. S E. Asia, through to tropical Australia. Heavy, dark, hard, red-brown wood is used for general construction work. The bark yields a red dye, used for dyeing baskets.
Bishkopra – Primula reticulata.
Bishop's Elder – Aegopodium podagraria.
Bishop's Weed – Ammi majus.
Bistort – Polygonum bistorta.
Bitangoor – Calophyllum soulattrii.
Bitangoor boonoot – Calophyllum soulattris.
Bitter, African Peach – Nauclea esculentus.
Bitter Almond (Oil) – Amygdalus communis.
Bitter Apple – Cucumis myriocarpus.
Bitter Apple – Solanum supinum.
Bitter Bark – Alstronia constricta.
Bitter Bark – Pinckneya pubens.
Bitter Bush – Pittosporum phyillyraeoides.
Bitter Candytuft – Iberis amara.
Bitter Cassava – Manihot esculenta.
Bitter Cola – Cola acuminata.
Bitter Cress – Cardamine amara.
Bitter Cress – Cardamine pennsylvanica.
Bitter Dock – Rumex abtusifolius.
Bitter Fennel – Foeniculum vulgare.
Bitter Gentian – Gentiana lutea.
Bitterleaf – Vernonia amygalina.
Bitter Lettuce – Lactuca virosa.
Bitter, Mishmee – Coptis teeta.
Bitter Pits – Acanthosicyos horrida.
Bitter Root – Lewisia rediviva.
Bitter Sneezeweed – Helenium tenuifolium.
Bitter Sweet Orange – Citrus aurantium × C. senensis.
Bitter Vetch – Vicia ervilia.
Bittere Garbe – Achillea clavenae.
Bitternut – Carya cordiformis.
Bitters Tree of Gambia – Veronica senegalensis.
Bittersweet – Solanum dulcamara.
Bittersweet, American – Celastrus scandens.
Bitterwood – Quassia amara.
Bitterwood, Mafura – Trichilia emetica.
Bitterwood, White – Trichilia spondioides.
BIXA L. Bixaceae. 3–4 spp. Tropical America, W.Indies. Small tree.
***B. orellana*** L. (Annatto Tree). Tropical America. Cultivated in Old and New World for the seeds, the coats of which contain a yellow dye used for colouring butter and cheese.
Bizimi – Carpolobia glabrescens.
Black Alder – Alnus glutinosa.
Black Alder – Ilex verticillata.
Black Ash – Fraxinus nigra.
Black Bamboo – Phyllostachya nigra.
Black Bean – Castanospermus australe.
Black Bindweed – Tamus communis.
Black Birch – Betula lenta.
Black Boy – Xanthorrhoea hastilis.
Black Bryony – Tamus communis.
Black Butt – Eucalyptus pilularis.
Black Butt – Eucalyptus piperata.
Black Butt, Pink – Eucalyptus eugenoides.
Black Butt, Western Australian – Eucalyptus patens.
Black Cap – Rubus occidentalis.
Black Carroway – Pimpinella saxifraga.
Black Catechu – Acacia catechu.
Black Cedar – Nectandra pisi.
Black Cohosh – Actaea spicata.
Black Cohosh – Cimicifuga racemosa.
Black Cottonwood – Populus heterphylla.
Black Crottle – Parmelia omphalodes.
Black Cummin – Nigella sativa.
Black Cutch – Acacia catechu.
Black Dammar Resin – Canarium bengalense.
Black Drink – Eryngium synchaetum.
Black Drink – Ilex vomitoria.
Black Ebony – Diospyros dendo.
Black-eyed Pea – Vigna catjang.
Black Gambir – Uncaria gambir.
Black Gram – Phaseolus mungo.

Black Grama – Bouteloua eriopoda.
Black Gum – Nyssa multiflora.
Black Haw – Crataegus douglasii.
Black Haw – Viburnum prunifolium.
Black Hazel – Ostrya virginiana.
Black Henbane – Hyoscyamus niger.
Black Hoarhound – Ballota nigra.
Black Huckleberry – Gaylussacia baccata.
Black Ironwood – Olea laurifolia.
Black Jack Oak – Quercus marylandica.
Black Jack Oak – Quercus nigra.
Black Lecanora – Haematomia ventosum.
Black Locust – Robinia pseudoacacia.
Black Mangrove – Avicennia nitida.
Black Mauritius Bean – Mucuna atterrina.
Black Medic – Medicago lupulina.
Black Mulberry – Morus nigra.
Black Mustard – Brassica nigra.
Black Nightshade – Solanum nigrum.
Black Oak – Quercus velutina.
Black Oak, Western – Quercus emoryi.
Black Olive Tree – Laguncularia racemosa.
Black Pepper – Piper nigrum.
Black Persian Mulberry – Morus nigra.
Black Persimmon – Diospyros texana.
Black Pine – Callitris calcarata.
Black Pine – Podocarpus spicata.
Black Podded Pea – Pisum arvense.
Black Poplar – Populus heterophylla.
Black Poplar – Populus nigra.
Black Raspberry – Rubus occidentalis.
Black Ray Thistle – Centaurea nigra.
Black Root (Indian) – Pterocaulon undulatum.
Black Rosewood – Dalbergia latifolia.
Black Sage – Artemisia tridentata.
Black Sage – Salvia mellifera.
Black Sally – Acacia melanoxylon.
Black Salsify – Scorzonera hispanica.
Black Sapote – Diospyros ebenaster.
Black Sapote – Diospyros ebenum.
Black Sassafras – Cinnamomum oliveri.
Black Seed Juniper – Juniperus saltuaria.
Black Sloe – Prunus umbellata.
Black Snakeroot – Actaea spicata.
Black Snakeroot – Cimicifuga racemosa.
Black Snakeroot – Sanicula marilandica.
Black Spruce – Picea mariana.
Black Stemmed Tree Fern – Cyathea medullaris.
Black Stinkweed – Ocotea bullata.
Black Tang – Fucus vesiculosus.
Black Tea Tree – Melaleuca bracteta.
Black Titi – Cyrilla racemiflora.
Black Tupelo – Nyssa sylvatica.
Black Walnut – Juglans nigra.
Black Wattle – Acacia binervata.
Black Wattle – Acacia decurrens.
Black Willow – Salix nigra.
Blackberry, Alleghany – Rubus allegheniensis.
Blackberry, European – Rubus fruticosus.
Blackberry Lily – Belamcanda chinensis.
Blackberry, Mountain – Rubus allegheniesis.
Blackseed Juniper – Juniperus saltuaria.
Black-stemmed Tree Fern – Cyathea medullaris.

BLACKSTONIA Huds. Gentianaceae. 5–6 spp. Europe, Mediterranean. Herbs.

***B. perfoliata*** (L.) Huds. = Chlora perfoliata. (Yellow Wort). W.Europe to Mediterranean. An extract of the plant is used as a yellow dye.

Blackthorn Sloe – Prunus spinosa.

***Blackwellia fagifolia*** Lindl. = Homalium fagifolium.

***Blackwellia padifolia*** Lindl. = Homalium fagifolium.

Blackwood – Acacia melanoxylon.
Blackwood, African – Dalbergia melanoxylon.
Bladder Campion – Cucubalus baccifer.
Bladder Campion – Silene inflata.
Bladder Fucus – Fucus vesiculosus.
Bladder Kelp – Nereocystis luetkeana.
Bladder Locks – Alaria esculenta.
Bladder Seed – Levisticum officinale.
Bladder Wrack – Fucus vesiculosus.

BLAKEA P. Br. Melastomataceae. 70 spp. C. and S.America, W.Indies. Trees.

***B. quinquenervis*** Aubl. Guyana. Bark used for tanning.

***B. trinervis*** Ruiz. and Pav. (Mess Apple). Guyana. Bark used for tanning.

Blancmange – Chondrus crispus.
Blazing Star – Chamaelirium luteum.
Blazing Star – Liatris punctata.

BLECHNUM L. Blechnaceae. 220 spp. Cosmopolitan. Ferns.

***B. orientale*** L. India to Polynesia and Australia. The Chinese use the rhizome to treat intestinal worms and bladder complaints.

BLECHUM P. Br. Acanthaceae. 10 spp. Tropical America, West Indies. Herbs.

***B. pyrimidatum*** P. Br. Tropical America. The leaves and flowers are used locally as a diuretic, and to treat fevers.

***Bleekrodea*** Blume. = Streblus Lour.

BLEPHARIS Juss. Acanthaceae. 100 spp. Old

World Tropics, Mediterranean, S.Africa, Madagascar. Herbs.

***B. capensis*** Pers. S.Africa. Used locally to treat anthrax. The leaves (Herba Blepharidis) are used by the natives to treat snake bites.

***B. edulis*** Pers. Sahara, Arabia. The seeds are eaten by nomadic tribes, and the leaves are fed to camels.

***B. linarifolia*** Pers. See B. edulis.

BLFPHAROCALYX Berg. Myrtaceae. 25 spp. S.America. Trees.

***B. cisplatensis*** Griseb. Argentine, Chile, Uruguay. An extract of the leaves is used locally as a tonic and astringent.

BLEPHARODON Decne. Asclepiadaceae. 30 spp. Mexico to Chile. Woody climbers.

***B. mucronatum*** (Schlecht.) Decaisne. (Chununa de Cabelle). C.America. Used locally to treat snakebites, and the juice is used as an antiseptic.

Blessed Thistle – Cnicus benedictus.

BLETIA Ruiz. and Pav. Orchidaceae. 45 spp. Tropical America, W.Indies. Herbs.

***B. angustata*** Lindl.=Spathoglottis plicata.

***B. hyacinthina*** R. Br.=Bletilla striata. Systematics doubtful. China, Japan. The plant is used by the Chinese to treat boils. The mucilage is also used to waterproof fabrics.

***B. purpurea*** (Lam.) DC.=B. verecunda.= Limodorum purpureum. Florida, C.America, W.Indies. An infusion of the bulbs was used in the W.Indies to treat poisoning by fish, and the whole bulb was used to treat wounds. The dried bulbs were used as a tonic.

***Bletilla striata*** Rehb f.=Bletia hyacinthina.

BLIGHTIA Koenig. Sapindaceae. 7 spp. Tropical Africa. Trees.

***B. laurentii*** De Wild.=B. welwitschii.

***B. sapida*** Koenig. (Akee Apple). W.Africa. Introduced into Jamaica 1778. Now cultivated in W.Indies. The fruits have arils which are about 7 cm. long and are eaten cooked in oil, or with fish. The ripe fruits are also eaten raw, but if they are unripe or over ripe they may be poisonous. In Africa, the flowers are used to make a cosmetic scented water.

***B. welwitschii*** (Hiern.) Radlk.=B. laurentii =B. wildemanniana. (Bate, Botoko, Ngelakusu). W.Africa. The wood is very hard and is used for cabinet work. The fruits produce a juice which is used to stupify fish.

***B. wildemanniana*** Gilg. ex de Wild.=B. welwitschii.

Blind-Your-Eyes Tree – Excoecaria agallocha.

Blister Buttercup – Ranunculus scleratus.

Blister Umbilicaria – Umbilicaria pustulata.

Blond Plantago Seed – Plantago ovata.

Blood, Dog's – Calderonia salvadoriensis.

Blood Flower – Asclepias curassavica.

Blood Flower – Haemanthus natalensis.

Blood Root – Haemodorum coccineum.

Blood Root – Sanguinara canadensis.

Blood Twig Dogwood – Thelyorania sanguinea.

Blood Wood – Eucalyptus corymbosa.

Blood Wood – Pterocarpus angolensis.

Bloodwood, Brush – Baloghia lucida.

Bloodwood, Brush – Synoum glandulosum.

Bloodwood, Kutcha – Eucalyptus terminalis.

Bloody Spotted Lecanora – Haematomma ventosum.

Blue Ash – Fraxinus quadrangulata.

Blue Beech – Carpinus caroliniana.

Blue Cohosh – Caulophyllum thalictroides.

Blue Ebony – Copaifera bracteata.

Blue Fig – Elaeocarpus grandis.

Blue Flag – Iris versicolor.

Blue Grama – Bouteloua gracilis.

Blue Gum – Eucalyptus botryoides.

Blue Gum – Eucalyptus globosus.

Blue Gum – Eucalyptus gonicalyx.

Blue Gum – Eucalyptus saligna.

Blue Gum – Eucalyptus tereticornus.

Blue Gum – Eucalyptus umbellata.

Blueish Tulip Oak – Argyodendron actinophyllum.

Blue Japanese Oak – Quercus glauca.

Blue-leaved Acacia – Acacia subcoerulea.

Blue Mahue – Hibiscus elatus.

Blue Mallee – Eucalyptus polybractea.

Blue Mallet – Eucalyptus gardneri.

Blue Mountain Ash – Eucalyptus oresdes.

Blue Sage – Salvia mellifera.

Blue Toadflax – Linaria canadensis.

Bluebeard – Salvia viridis.

Bluebell – Scilla non-scripta.

Blueberry – Vaccinium corybosum.

Blueberry – Vaccinium myrtillus.

Blueberry – Vaccinium nitidum.

Blueberry, Andean – Vaccinium floribundum.

Blueberry, Ash – Elaeocarpus grandis.

Blueberry, Columbian – Vaccinium floribundum.

Blueberry, Early Sweet – Vaccinium pennsylvanicum.
Blueberry Elder – Sambucus coerulea.
Blueberry, European – Vaccinium myrtillus.
Blueberry, Ground – Vaccinium myrsinites.
Blueberry, High Bush – Vaccinium corymbosum.
Blueberry, Low Sweet – Vaccinium pennsylvanicum.
Blueberry, Swamp – Vaccinium corymbosum.
Blueberry, Sweet – Vaccinium pennsylvanicum.
Bluebunch Wheatgrass – Agropyron spicatum.
Bluegrass, Canada – Poa compressa.
Bluegrass, English – Festuca elator.
Bluegrass, Fowl – Poa palustris.
Bluegrass, Kentucky – Poa pratensis.
Bluegrass, Pine – Poa scabrella.
Bluegrass, Skyline – Poa epilis.
Bluegrass, Texas – Poa arachnifera.
Bluegrass, Timberline – Poa rupicola.
Bluegrass, Wood – Poa nemoralis.
Bluestem, Little – Andropogon scoparius.
Bluestem, Sand – Andropogon halii.
Bluestem, Seacoast – Andropogon littoralis.
Bluestem, Willow – Salix irrorata.
Bluewood Condalia – Condalia obovata.
BLUMEA DC. Compositae. 50 tropical and S.Africa, Madagascar, India, E.Asia to Australia and Pacific.
***B. balsamifera*** (L.) DC.=Conyza balsamifera. (Ngai Camphor). Himalayas to Philippine Islands, Malaysia and Moluccas. Small shrubs. Strongly camphor-scented. Used and sold locally in Malaysia. The leaves are used to flavour food; Ngai Camphor is obtained from the wood and leaves by distillation; and the leaves and roots are extracted to cure fever.
***B. chinensis*** DC.=B. riparia=Conyza riparia. SE. Asia, to Malaysia and Indonesia. Herbs. Mustard-tasting leaves are used locally to flavour soup, and a decoction of them is used to treat beri-beri, and colic.
***B. lacera*** (Burm. f) DC.=Conyza cappa=Conyza lacera. Tropical Africa and Asia. The leaves are used as a vegetable in some parts of Indonesia. The Hindus use an extract of the leaves as a diuretic, astringent, to reduce fever and combat intestinal worms.
***B. lanceolaria*** Drude=B. myriocephala.
***B. myriocephala*** DC.=B. lanceolaria=B. specabilis=Conyza lanceolaria. Herb. India to Indonesia and Vietnam. Cultivated in Vietnam to use the leaves as a condiment. The leaves are also used to treat asthma. In Malaya the leaves are used to reduce rheumatic pains.
***B. pubigera*** (L.) Merr.=Conyza pubigera. Tropical Asia, esp. S. Vine. Root is used in Malaysia to treat stomach upsets.
***B. spectabilis*** DC.=B. myriocephala.
Boabab (Oil) – Adansonia digitata.
Boabab, Madagascar – Adansonia madagascarriensis.
Boanga – Afzelia bipinaensis.
BOBARTIA Salisb. Iridaceae. 17 spp. S.Africa. Herbs.
***B. indica*** L. S.Africa. The leaves are used locally to make baskets.
***B. spathacea*** Ker-Gawl.=B. indica.
Bobbi – Calophyllum tormentosum.
BOCCONIA L. Papaveraceae. 10 Warm Asia; W.Indies. Small trees.
***B. cordata*** S. Wats. C.America. The bark produces a yellow dye which is used locally. It is also the source of a local anaesthetic used in Mexico.
***B. frutescens*** L. C.America to Peru. The sap is used by local Indians to dye feathers. An infusion of the roots is used to treat jaundice and dropsy in Columbia, while in Mexico it is used to treat warts and opthalmia.
***Bocoa edulis*** Baill.=Isocarpus edulis.
Bodenga – Cissus adenopoda.
Bodiva – Lychnodiscus cerospermus.
Bodumbe – Chytranthus macrobotrys.
Boed Copal – Agathis alba.
Boed Copal – Agathis loranthifolia.
BOEHMERIA Jacq. Urticaceae. 100 spp. Tropical and N. Subtropical. Herbs. All economic species yield stem fibres which are extracted by retting.
***B. caudata*** Sw.=B. flagelliformis. Tropical America. Fibres of a poor quality, used locally in Mexico.
***B. flagelliformis*** Liebm.=B. caudata.
***B. grandis*** (Hook. and Arn.) Heller=B. stipularis (Akoloa, Hawaiian Falsenettle). Hawaii. An inferior fibre, used locally, and improved during cultivation.
***B. macrophylla*** D. Don. N.India. Fibre used for fishing nets.
***B. nivea*** (L.) Gaud. (China Grass, Ramie). China, now cultivated in warm countries of Old and New World. The fibre was

mentioned in China in 2200 B.C. and was used to cover Egyptian mummies. The fibres are the longest in the plant kingdom. They have a tensile strength seven times stronger than silk and eight times that of cotton. This is markedly improved on wetting. Its large scale cultivation is not popular because it is a perennial plant and occupies the ground for a full year, thus preventing the growth of a second crop. The fibres are smooth and this makes spinning difficult. Retting does not leave a clean fibre, and the subsequent removal with acid or lye weakens the fibre. The plants will yield for several years, but the fibre quality deteriorates as the plants exhaust the soil. Propagation is by seed, or more frequently by layering, cuttings or root division. It is cultivated in the Far East where harvesting is by hand, and occasionally in Florida. World production (mainly from China) is 20,000–25,000 tons per annum.

***B. platyphylla*** D. Don. var. ***marquesensis*** F. Brown. (Pute, Vairoa). Marquesas. Fibres used locally for making fishing lines.

***B. puya*** Hassk. = Maoutia puya.

***B. stipularis*** Wedd. = B. grandis.

BOERHAVIA L. Nyctaginaceae. 40 spp. Tropical and subtropical. Herbs.

***B. diffusa*** L. = B. repens.

***B. plumbaginae*** Cav. Mediterranean to tropical Africa. An extract of the leaves is used locally in Africa to treat jaundice and to heal wounds.

***B. repens*** L = B. diffusa. Throughout tropics. An extract of the leaves is used in Angola to treat jaundice, to treat asthma and coughs, and as an emetic. The roots and leaves are eaten by the aborigines of C.Australia.

***B. tuberosa*** Lam. Peru. Roots eaten locally.

BOERLAGIODENDRON Harms. Araliaceae. 30 spp. Formosa, Malaysia, and Pacific Island. Shrubs.

***B. palmatum*** Marms. = Trevesia moluccana. Malaysia. Leaves and stems eaten locally, boiled.

BOESENBERGIA Kuntze. Zingiberaceae. 20 spp. Indomalaya. Herbs.

***B. panduratum*** Ridl. Malaysia, Java. Cultivated locally for the rhizome which is used as a condiment and to treat stomach upsets.

Bofale – Parinari glabra.

Bofidiji – Scorodophloeus zenkeri

Bog – Alstonia pneumatophora.

Bog Bilberry – Vaccinium uliginosum.

Bog Myrtle – Myrica gale.

Bog Rosemary – Andromeda glaucophylla.

Bog Rosemary – Andromeda poliofolia.

Bog Spruce – Picea mariana.

Bogen – Sonneratia ovata.

Bohamba – Pycnanthus kombo.

Bohemian Tea – Lithospermum officinale.

Bois Bande – Richeria grandis.

Bois Blane – Phyllostylon brasiliensis.

Bois Cicerou – Pithecellobium jupunba.

Bois Cochon – Moronobea coccinea.

Bois Cochon – Tetragastris balsamifera.

Bois Côte – Tapura antillana.

Bois d'arc – Maclura aurantiaca.

Bois d'Encens – Protium attenuatum.

Bois de Rose – Aydendron panurense.

Bois de Rose Femelle – Aniba rosaeodora.

Bois de Siam – Cupressus hodginsii.

Bois Diable – Licania terminalis.

Bois fidèle – Citharexlum quadrangulare.

Bois jaune – Ochrosia borbonica.

Bois Violin – Guatteria caribaea.

Bojuco de paloma – Trichostigma octandrum.

Bokaara-gass – Ouratea angustifolia.

Bokakango – Hippocratea myriantha.

Bokalahy – Marsdenia vernicosa.

Bokhara Galls – Pistacia vera.

Bolax Resin – Azorella caespitosa.

***Boldea fragrans*** Gay. = Peumus boldus.

Boldo – Peumus boldus.

***Boldu boldus*** (Mol.) Lyons. = Peumus boldus.

Boleko (Nut) – Ongokea klaineana.

Bolelenge – Cissus planchoniana.

Bolelombe – Allophylus lastoursvillensis.

***Boletopsis rufus*** (Schäff.) P. Henn. = Boletus versipellis.

BOLETUS Dill. ex Fr. Boletaceae. 200 spp. Cosmopolitan. The fruit-bodies of the species mentioned are eaten cooked.

***B. appendiculatus*** Schäff. Europe.

***B. badius*** Fr. Europe.

***B. castaneus*** Bull. = Gyroporus castaneus.

***B. cervinus*** Bull. = Elaphomyces cervinus.

***B. chirurchorum*** Bull. = Polyporus officinale.

***B. chrysenteron*** Bull. Europe.

***B. edulis*** Bull. (Polish Mushroom). Europe. Much esteemed, sold fresh or dried.

***B. granulatus*** L. = Suillus granulata. Associated with roots of conifers. Temperate. Delicious fried in butter.

***B. grenvillei*** Klotzsch. See B. granulatus.

***B. laricis*** Jacq. = Polyporus officinalis.
***B. luteus*** L. ex Fr. = Suillus granulata. See B. granulatus.
***B. oudemansii*** Hartsen. = B. placidus.
***B. placidus*** Bon. = B. oudemansii. See B. granulatus.
***B. scaber*** Bull. See B. edulis.
***B. subtomentosus*** L. Europe and Asia.
***B. sudanicus*** Har and Pat. (Hegba mboddo). C.Africa.
***B. versipellis*** Fr. = Boletopsis rufus = Leccinum aurantiancum. Europe. Eaten widely in Europe and sometimes exported. Pickled in U.S.S.R. (Krassny grib), and used in France as a substitute for truffles.
Boliko – Pachyelasma tessmannii.
Bolivian Black Walnut – Juglans boliviana.
Bolly Gum – Litsea reticulata.
Bolly Silkwood – Cryptocarya oblata.
Bolly Wood – Litsea reticulata.
Bolo Coro Ni – Cussonia nigerica.
Bololo – Pseudospondias microcarpa.
Bolongeta Ebony – Diospyros pilosanthera.
Bolong Gomba – Panaeolus papilionaceus.
Bolongo – Fagara lemairei.
Bolongo – Fagara rubescens.
Bolongolo – Fagaria inaequalis.
BOLTONIA L'Hérit. Compositae. 1 sp. E.Asia; 7 spp. N.America. Herbs.
***B. cantoniensis*** (Blume) Fr. and Sav. = Aster cantoniensis. China and Japan. Young leaves cooked as a vegetable in Japan.
BOLUSANTHUS Harms. Leguminosae. 1 sp. S.Africa. Tree.
***B. speciosus*** (Bolus) Harms. (Rhodesian Wisteria, Wild Wisteria). Tropical Africa. The wood is hard and proof against insects. It is used for house construction and wheel hubs.
Bomanga – Brachystegia laurentii.
BOMAREA Mirb. Alstroemeriaceae. 150 spp. Mexico, tropical America, W.Indies. Herbs. The starchy tubers are eaten as a vegetable, locally.
***B. acutifolius*** Link. and Otto. S.Mexico.
***B. edulis*** Herb. Santo Domingo.
***B. glaucescens*** Baker. Ecuador.
***B. ovata*** Mirb. El Salvador.
***B. salilla*** Herb. Chile.
BOMBACOPSIS Pittier. Bombacaceae. 2 spp. C. and tropical America. Trees.
***B. fendleri*** (Seem.) Pitt. C.America. Light wood is used for construction work.
***B. sepium*** Pitt. (Saquen Saqui). Venezuela. Soft wood is used for making storage vats for rum and tanning.
Bombata – Lannea welwitschii.
BOMBAX L. Bombacaceae. 8 spp. Tropics. Trees. Fruits of all species mentioned give a floss used for stuffing pillows etc.
***B. aesculifolia*** H. B. K. = Ceiba aesculifolia.
***B. buonopnzence*** Beauv. (Silk Cotton Tree). Tropical Africa. A bark extract is used in parts of Africa to control menstruation, or on its own to clean teeth. The mucilaginous calyx is used in soups. The wood is very soft, and is used to make drums, canoes and household utensils.
***B. campestris*** Schum. Small tree. Brazil.
***B. ellipticum*** H. B. K. = B. mexicanum. C.America. Bark and root extract is used locally to treat toothache.
***B. globosum*** Aulb. Tropical America. Used as flights of blow pipe darts.
***B. guinensis*** Thoun. = Ceiba guinensis.
***B. longiflorum*** Schum. Brazil. Bark the source of a tough fibre.
***B. malabaricum*** DC. = Salmalia malabarica. (Rokrosimul, Semul, Purani). Tropical Asia. The wood is a possible source of cellulose. It also produces a resin (Gum of Malabar). The root produces a tonic and is used in an Indian aphrodisiac (Musla-Semul) but gives rise to impotence. The gum is also demulcent and used to treat diarrhoea. The root and bark are used as an emetic.
***B. manguba*** Mart. Brazil.
***B. mexicanum*** Hemsl. = B. ellipticum.
***B. pentandrum*** L. = Ceiba pentandra.
Bombay Aloë – Agave cantala.
Bombay Ebony – Diospyros gardneri.
Bombay Ebony – Diospyros montana.
Bombay Hemp – Agave cantala.
Bombay Mace – Myristica malabarica.
Bombay Mastic – Pistacia mutica.
Bombay Nutmeg – Myristica malabarica.
Bombay Rosewood – Dalbergia latifolia.
Bombondelen – Cassia javanica.
Bombooe – Morinda geminata.
Bomei – Clappertonia minor.
Bonavist – Dalichos lablab.
Bondesobe – Hugonia obtusifolia.
Bondole – Trichoscypha acuminata.
Bonduc Nut – Caesalpinia bonducella.
Boneset – Eupatorium perfoliatum.
Boneset – Kuhia eupatorioides.
Bông lôc – Sclerachne punctata.
Bong nga truât nam – Kaempferia pandurata.

Bongo – Brachystegia laurentii.
Bongo – Cavanillesia platanifolia.
Bongolu – Beilschmiedia corbisieri.
Bon hardi – Morinda angustifolia.
BONNAYA Link. and Otto. Scrophulariaceae. 15 spp. Warm. Herbs.
***B. antipoda*** Druce = Ilysanthes antipoda Merr. Tropical Asia. Extract of leaves and roots is used locally to treat intestinal worms.
Bonsoko – Alsodeiopsis poggei.
Bontoto – Mucuna flagellipes.
Bontseke – Carpolobia glabrescens.
Booloong – Solanum blumei.
Boonga njingin – Tabernaemontana divaricata.
Boonge – Bosqueia angolensis.
Boonijaga – Drypetes longifolia
Boonki – Morina lucida.
BOÖPHONE Herb. Amaryllidaceae. 5 spp. S. and E. Africa. Herbs.
***B. disticha*** Herb. = B. toxicaria = Heamanthus toxicarius. (Candelabra Flower, Gifbol, Sore-eye Flower). S.Africa. The leaves and bulbs are very poisonous. The natives of the Orange Free State use them as a suicide poison, causing severe diarrhoea, and as an arrow poison. They are also used in veterinary medicine.
***B. toxicaria*** Herb. = B. disticha.
Boor – Typha elephantina.
Booratoo – Typha elephantina.
Boorea – Syncarpia laurifolia.
Boorondool – Xerospermum noronhianum.
Boosi – Melochia umbellata.
Boosing – Bruguiera eriopetala.
Boowa asa – Cubilia blancoi.
Boowah tati kambing – Antidesma stipulare.
Borage – Borago officinale.
BORAGO L. Boraginaceae. 3 spp. Mediterranean, Europe and Asia. Herbs.
***B. officinalis*** L. (Borage). Mediterranean. Cultivated widely where climate is suitable. The leaves are used in salads and to flavour drinks, e.g. claret, to which they give a cool, cucumber-like bouquet. The leaves and flowers when dried (Herba et Flores Borraginis) have been brewed to give a refreshing, diuretic drink since the Middle Ages. The flowers are a good source of honey.
BORASSUS L. Palmae. 8 spp. Old World tropics. Trees.
***B. flabellifer*** L. (Palmyra Palm). Cultivated in India and Ceylon. The plant has a great number of uses. The timber is hard and durable, and is used for rafters and general construction work. The leaves have been used as paper (Olas) since ancient times. The whole leaves are used to make thatch, hats, bags, etc., while fibres from the leaf-bases are used to make brushes. The pericarp and stem also yield a coarse fibre. The inflorescence is tapped for the sugary sap which is used to make sugar (Jaggery), or fermented to make palm wine (Toddy) and vinegar. The pericarp is edible, especially when roasted, and the kernels are eaten when young. The seedlings are also eaten, or ground into a flour. The leaves are used as a green manure, and also give salt and potash when burnt.
Bordeaux Turpentine – Pinus pinaster.
Borecole – Brassica oleracea var. viridis.
Boree – Acacia pendula.
Boridischah – Ferula galbaniflua.
Borneo Billian – Eusideroxylon zwageri.
Borneo Camphor – Drybalanops aromatica.
Borneo Copal – Agathis loranthifolia.
Borneo, Dead – Dyera costulata.
Borneo, Getah – Willughbeia firma.
Borneo Ironwood – Eusideroxylon zwgeri.
Borneo Mahogany – Calphyllum inophyllum.
Borneo Manila – Agathis lornanthifolia.
Borneo Rubber – Willughbeia firma.
Borneo Shorea – Shorea aptera.
Borneo Tallow – Shorea aptera.
Borneo Tallow – Shorea stenocarpa.
Borneo White Seraya – Parashorea malaanonan.
Borneol – Dryobalanops aromatica.
Bornyac – Euonymus cochinchinensis.
BORONIA Sm. Rutaceae. 70 spp. Australia. Small woody plants.
***B. megastigma*** Nees. S.W. Australia. Source of an aromatic oil (Oil of Boronia) used in perfumery.
Boronia Oil – Boronia megastigma.
Borrera, Yellow – Theloschistes flavicans.
BORRERIA G. F. W. Mey. Rubiaceae. 150 spp. Warm climates. Herbs.
***B. articularis*** Williams. Tropical Asia. Used locally in China as a poultice on boils etc.
***B. capitata*** (Ruiz. and Pav.) DC. (Poya). S.America. Used locally as an emetic, and astringent.
BORZICACTUS Riccobono. Cactaceae. 17 spp. Andes. Xerophytic shrubs.
***B. sepium*** (H.B.K.) Britt. and Rose. =

Cactus sepium H.B.K. = Cleistocactus sepium (H.B.K.) Weber. Ecuador. Fruits eaten locally.

***Boschia griffithii*** Mast. = Durio griffithii.

BOSCIA Lam. Capparidaceae. 37 spp. Tropical and S.Africa. Small trees.

***B. albitrunca*** (Burch.) Gilg. and Benedict. = Capparis albitrunca (Shepherd's Tree, Witagatboom). S.Africa. Locally the root is used as a coffee substitute.

***B. angustifolia*** Rich. Sahara. Seeds eaten locally.

***B. caffra*** Sond. = Maerua pedunculata.

***B. octandra*** Hochst. Tropical Africa. Fruits (Kursan) eaten locally in the Sudan. An extract of the leaves is used to wash the eyes.

***B. senegalensis*** Lam. Tropical Africa. The leaves and fruits are used to make a soup mixed with cereals. The seeds are used as a coffee substitute.

Bosé Wood – Staudia kemerunensis.

Bosé Wood – Xylopia striata.

Bosipi Wood – Oxystigma mannii.

BOSQUEIA Thou. ex Baill. Moraceae. 15 spp. Tropical Africa, Madagascar. Trees.

***B. angloensis*** (Welw.) Ficalho. (Agbute, Boonge). Tropical Africa, esp. W. C. The wood is pinkish-yellow and is used for general construction work and carpentry, small articles, and for torches. The trunk yields a red latex which is used to adulterate rubber, locally to waterproof fabrics and can be used as indelible ink. The roasted seeds are eaten locally.

***B. phoberos*** Baill. Tropical Africa. The latex is used as a red dye.

BOSWELLIA Roxb. ex Colebr. Buseraceae. 24 spp. Tropical Africa and Asia, Madagascar. Trees.

***B. bhaw-dajiana*** Bindw. See B. papyrifera.

***B. carteri*** Birdw. (Bible Frankinsence). Somaliland and Middle East. The trunk is a source of a gum (Olibanum). This is extracted from incisions in the bark which are allowed to drain for about three months before the true gum is collected. Two types of frankinsence are extracted, the deep yellow-red male frankinsence (Zakana) and the lighter yellow female frankinsence (Kundura Unsa). Frankinsence contains bassarin, 6–8 per cent arabin, and a bitter substance. Because it burns easily, with a pleasant odour, it is used for incense in Christian and Jewish Churches. It was used similarly by the ancient Egyptians, but it was not used for embalming. Although its medicinal value is doubtful, it has been used since ancient times to treat fevers, dysentery, tumours, ulcers, disorders of the breast, boils, and given internally to treat gonorrhoea. In China it was used to treat leprosy. Its modern uses include fumigants, and fixatives for perfumes.

***B. frereana*** Birdw. (Elemi Frankinsence). Tropical Africa. Stem yields a pale yellow, lemon-scented balsam (African Elemi, Loban Maidi, Loban Meti).

***B. neglecta*** Moore. See B. papyrifera.

***B. papyrifera*** Hochst. (Elephant Tree). E.Africa. Yields a fragrant resin burnt as incense. Elephants feed on the leaves.

***B. serrata*** Roxb. (Indian Frankincense, Indian Olibanum Tree) N.W.India. Yields a golden-yellow gum (Salai-gungul), which is used medicinally like Bible Frankincense. The wood is used to make small ornamental articles, and as charcoal.

***B. socotrana*** Balf. f. Socotra. Yields a gum of small commercial value.

Botany Bay Gum – Xanthorrhoea hastilis.

Botaosinho – Croton humulis.

Botendi – Pancovia harmsiana.

BOTHRIOCHLOA Kuntze. Graminae. 20 spp. Warm. Grasses.

***B. intermedia*** Stapf. W.Africa, Barbados, Guyana. A fodder grass for grazing and hay, especially on dry limestone.

***B. pertusa*** Stapf. (Sour Grass). Tropical Africa, Middle East, India, Ceylon, Mauritius. Drought-resistant fodder grass.

Botoko – Blighia welwitschia.

Botoko Plum – Flacourtia ramontchi.

Botor tetragonolobus Adans. = Psophocarpus tetragonolobus.

Bot Tree – Ficus religiosa.

BOTRYCHIUM Sw. Ophioglossaceae. 40 spp. Cosmopolitan. Fern.

***B. ternatum*** S.W. China, Japan. Leaves eaten as a vegetable in Japan.

BOTRYTIS Pers. ex Fr. Moniliaceae. 50 spp. Cosmopolitan. Parasitic fungus.

***B. cinerea*** Pers. ex Fr. = B. preussii = B. vulgaris. A widely distributed sp. In Europe and U.S.S.R., grapes attacked by the fungus (Noble Fungus) are collected separately as they give a special bouquet to wines, especially of Sauterne type.

***B. preussii*** Sacc. = B. cinerea.

***B. vulgaris*** Fr.=B. cinerea.
Bottle Gourd – Lagenaria vulgaris.
Bottle Tree – Adansonia gregorii.
BOUEA Meissn. Anacardiaceae. 1+ spp. S.E.Asia, W.Malaysia, Moluccas. Tree.
***B. burmanica*** Griff. var. ***microphylla*** (Griff.) Engl.=B. microphylla. (Raman). Malaysia. Hard durable wood is used for posts, beams, and agricultural implements. It is insect-proof. The plum-like fruits are edible.
***B. microphylla*** Griff. Malaysia. The yellow plum-like fruits are eaten locally, raw, cooked or in pickles. The young leaves are eaten with rice in Java.
Boulboudi – Fumaria vaillantii.
Boulder Lichen – Parmelia conspersa.
Bourbon Balsamum Mariae – Calophyllum tacamahaca.
Bourbon Lily – Lilium candidum.
Bouron – Adenium honghel.
***Bousinggaultia*** Kunth.=Anredera Juss.
Bouski – Combretum elliotii.
BOUTELOUA Lag. mut. P. Beauv. Graminae. 40 spp. Canada to S.America, esp. S.W. United States of America. Grasses.
***B. curtipendula*** (Michx.) Torr. (Side-Oats Grama). C. N.America. Pasture grass. Sown in sorghum stubble on inland dunes.
***B. eriopoda*** Torr. (Black Grama). S.W. U.S.A., N.Mexico. Pasture grass.
***B. filiformis*** (Fourn.) Griff. (Grama Grass). S.W. U.S.A., Mexico. Pasture grass.
***B. gracilis*** (H.B.K.) Lag. (Blue Grama). N.American plains. Pasture grass.
BOVISTA Pers. Lycoperdales. 15 spp. Temperate. Puffball fungi. The fruit-bodies of species mentioned are eaten as food.
***B. nigrescens*** Pers.
***B. pila*** Berk. and Curt.
***B. plumbea*** Pers.
Bovumara – Hopea parviflora.
BOWDICHIA Kunth. Leguminosae. 3 spp. Tropical S.America. Trees.
***B. virgilioides*** H.B.K. (Alcornococ). Brazil, Guyana, Venezuela. Heavy, tough reddish-brown timber, used in Brazil for cartwheels.
BOWENIA Hook. Zamiaceae. 2 spp. N. and N.E. Australia. Small trees.
***B. specabilis*** Hook. Queensland. Large rhizomes eaten by natives.
BOWIEA Harv. ex Hook. f. Liliaceae. 2 spp. S. and E. Africa. Herbs.
***B. volubilis*** Haw. (Umagaqana). S.Africa. Corms used by various local tribes to treat dropsy, barrenness in women, and as a purgative. Excessive doses are fatal.
Bowihi – Strychnos triclisioides.
Bowman's Root – Gillenia trifoliata.
Bow Wood – Maclura aurantiaca.
Box – Buxus sempervirens.
Box, Bastard – Eucalyptus polyanthemos.
Box, Brush – Tristania confera.
Box, Cape – Buxus macocowanii.
Box, Coast Grey – Eucalyptus bosistoana.
Box Elder – Acer negundo.
Box, Floorbed – Eucalyptus microtheca.
Box, Gippsl and Grey – Eucalyptus bosistoana.
Box, Grey – Eucalyptus hemiphloia.
Box, Grey – Eucalyptus polyanthemos.
Box, Huckleberry – Gaylussacia brachycera.
Box, Marmalade – Genipa americana.
Box, Red – Eucalyptus polyanthemos.
Box, Red – Tristania conferta.
Box, White – Eucalyptus hemiphloia.
Box, White – Tristania conferta.
Box, Yellow – Eucalyptus hemiphloia.
Box, Yellow – Eucalyptus melliodora.
Boxwood – Gonioma kamassi.
Boxwood – Gossypiospermum praecox.
Boxwood – Jacaranda copaia.
Boxwood – Schaefferia frutescens.
Boxwood, Brazilian – Euxylophora paraesis.
Boxwood, Ceylon – Canthium didymum.
Boxwood, Florida – Schaefferia frutescens.
Boxwood, San Domingo – Phyllostylon brasiliensis.
Boxwood Tree – Buxus wallichiana.
Boxwood, West African – Nauclea diderrichii.
Boxwood, West Indian – Phyllostylon brasiliensis.
Boxwood, West Indian – Tabebuia pentaphylla.
Boxwood, Yellow – Sideroxylon pohlmannianum.
Bowoi – Pithecellobium minahassai.
Boysenberry – Rubus strigosus.
BRABEJUM L. Proteaceae. 1 sp. S.Africa. Trees.
***B. stellatifolium*** L. (African Almond, Hottentot's Almond, Wild Chestnut). Seeds eaten locally, and roasted to be used as a coffee substitute. Kernel possibly toxic.
BRACHIARIA Griseb. Graminae. 50 spp. Warm. Grasses. All mentioned are pasture grasses.

***B. distichophylla*** Stapf. Annual. W.Africa.

***B. disticha*** (L.) Stapf = Panicum distachum. N.India to Australia.

***B. humidicola*** Stapf. (False Creeping Paspalum). Tropical Africa, esp. Natal.

BRACHYCLADOS D. Don. Compositae. 4 spp. Temperate S. America. Shrubs.

***B. stuckertii*** Speg. Argentina. Infusion of the leaves (Herba Brachycladi) is used to treat asthma and mountain sickness.

BRACHYLAENA R. Br. Compositae. 23 spp. Tropical and S.Africa, Mascarene Islands. Shrubs.

***B. elliptica*** Less. S.Africa. Leaves used locally to treat diabetes.

***B. hutchinsii*** Hutch. Tropical Africa, esp. Kenya. An oil (Muhugu Oil) is distilled from the plant and used in perfumery, especially in soap-making.

BRACHYSTEGIA Benth. Leguminosae. 30 spp. Tropical Africa. Trees.

***B. boehmii*** Taub. (Mfuti, Mombo, Prince of Wales' Feather Tree). C.Africa. The bark is used for tanning, and yields a fibre used for fishing nets. The wood is used for railway sleepers and for charcoal.

***B. laurentii*** (De Wild.) Louis. = Macrolobium laurentii (Bomanga, Bongo, Tenda). C.Africa. The bark is used in the Congo to make cloth.

***B. spicaeformis*** Benth. Tropical Africa. The wood is hard and heavy. It is used for general construction work, and in Zanzibar for boxes, grain stores, roofing, and matches. A fibre derived from the bark is used to make sacking.

***B. stipulata*** De Wild. C.Africa. The bark is used as cordage.

***B. taxifolia*** Harms. (Ngalati, Ngansa). C.Africa. The bark is used as cordage.

***B. utilis*** Davy and Hutch. C.Africa. Bark is used as cordage.

Bracken – Pteridium aquilinum.

BRACKENRIDGEA A. Gray. Ochnaceae. 10 spp. Old World Tropics. Trees.

***B. zanguebarica*** Oliv. = Ochna alboserrata Engl. Tropical Africa. A yellow dye is extracted from the bark.

Braga – Panicum miliaceum.

***Bragantia corymbosa*** Griff. = Apama tomentosa.

BRAHEA Mart. Palmae. 7 spp. S.United States of America, Mexico. Trees.

***B. dulcis*** (H.B.K.) Mart. Mexico. Fruits (Miche, Michire) are sweet and eaten raw, locally. The timber is used for house construction, and the leaves for thatching.

***B. salvadorensis*** H. Wendl. ex Beccari. C.America. The fruits yield an edible oil which is used for cooking etc.

Brahm – Sarothamnus scoparius.

Brake Root – Polypodium vulgare.

Bramble – Rubus fruticosus.

Bramble, Arctic – Rubus arcticus.

Bramble, Cape – Rubus rosaefolius.

Branda – Chiococca alba.

Brahmadandi – Tricholepsis glaberrimum.

Brasan – Tarenna incerta.

BRASENIA Schreb. Cabomaceae. 1 sp. Tropical America and Africa, India, Temperate E.Asia, Australia. Aquatic herb.

***B. peltata*** Pursh. = B. schreberi.

***B. schreberi*** J. F. Gmel. = B. peltata. (Watershield). The fresh leaves are eaten in Japan, flavoured with vinegar.

Brasilin – Haematoxylum brasileito.

BRASSICA L. Cruciferae. 150+ spp. Europe, Mediterranean, Asia. Many cultivated forms. Herbs.

***B. adpressa*** Bois. = Sinapis incana. Greece, Turkey. Leaves of young plants eaten with lemon juice in Greece.

***B. alba*** (L.) Boiss. = Sinapis alba.

***B. alba*** (L.) Rabenh. = Sinapis alba.

***B. arvensis*** (L.) Boiss. (Charlock). Europe, introduced into N.America. Seeds used, especially in Russia, as a source of mustard (Dakota Mustard).

***B. besseriana*** Andrz. = B. juncea var. eujuncea. Russia, tropical Asia. Seeds used as mustard (Sarepta Mustard).

***B. campestris*** L. (Bird Rape). Temperate Europe, Asia, introduced to N.America. Grown around the Black Sea for the oil (Ravinson Oil), extracted from the seeds. It is used locally as a luminant, lubricant, and in the manufacture of soap.

***B. campestris*** L. var. ***chinoleifera*** L. (Chinese Colza). A cultivar grown in Asia for the seed oil (Chinese Colza Oil) which is semi-drying and used as Rape Oil q.v.

***B. carinata*** A. Br. (Abyssinian Mustard). Grown occasionally in Abyssinia as a potherb.

***B. chinensis*** L. (Chinese Cabbage, Pak-choi). China. Cultivated occasionally in Old and New World as a vegetable.

***B. integrifolia*** (West.) Rupr. (Indian Brown Mustard, Rai). Asia, especially India. Cultivated in India. The leaves are eaten as a vegetable. The oil from the seed is used in cooking. A poultice of the leaves is used locally to relieve lumbago, neuralgia, colic etc. by acting as a rubefacient.

***B. japonica*** (Thb.) Sieb. (Japanese Mustard, Potherb Mustard). Cultivated in Japan for the leaves which are used as a vegetable.

***B. juncea*** (L.) Czern. and Coss. = Synapis juncea. (Indian Mustard, Leaf Mustard). Europe, Asia. Cultivated since ancient times for the leaves which are eaten cooked.

***B. napiformis*** (Paill. and Bois.) L. (Tuberous-rooted Chinese Mustard). Cultigen. Grown for the enlarged roots which are eaten cooked.

***B. napo-brassica*** Mill. (Rutabaga, Swedish Turnip, Russian Turnip, Yellow Turnip, White Turnip). A cultigen, which has been grown for centuries which are eaten as a vegetable, especially in the winter. The root is larger and more oval than the turnip.

***B. napus*** L. (Colza, Rape). Uncertain origin. The annual forms are grown for the oil (Rape Oil, Colza Oil) extracted from the seeds. The crude oil is yellow-green with a decided taste. It is purified by treating with caustic soda or sulphuric acid to give a semi-drying oil which is used for cooking (especially in India) as a lubricant, illuminant, for tempering steel and for soap-making. The oil is called Schmalöl in Germany, and Metah or Sweet Oil in Indonesia. The main oil producers are China, India, Pakistan, Sweden, France, and U.S.S.R. The biennial varieties are used for autumn and winter forage, especially in the U.K. and U.S.A.

***B. narinosa*** Bailey. (Broad-beaked Mustard). Probably Asiatic. Grown for leaves, used as a vegetable locally, rarely in Europe.

***B. nigra*** (L.) Koch. (Black Mustard). Europe, introduced to N.America. Probably originated in Mediterranean region where it has been cultivated for 2000 years. It is referred to in the Bible and was used medicinally by the Ancient Greeks. The leaves are used occasionally as a vegetable and for stock feed. The seeds are used as a condiment and as a source of oil which is formed by hydrolysis of the glucoside sinigrin by the enzyme myrosin. The condiment stimulates the secretion of saliva, and peristalsis of the stomach. The oil is edible and is used as a lubricant and in the manufacture of soft soap. Medicinally the seed is used as an emetic and to cause local reddening of the skin.

***B. oleracea*** L. (Cabbage). This is a generic term which includes derivatives from the Mediterranean sea cabbage. There are many varieties all of which are cultivated in temperate Old and New World. All varieties are more or less frost tolerant, and grow best in mild, cool temperate climates. Var. ***acephala*** L. = var. ***viridis*** (Kale, Collard, Borecole, Marrow Cabbage) are grown as a vegetable and as a stock feed. In the tropics it is grown as a perennial. Var. ***gemmifera*** DC. (Brussel's Sprouts), are grown as a vegetable, the large axillary buds being eaten. Firm sprouts will not develop unless the mean day temperature is less than 12°C. They were probably developed in Belgium during the 13th Century. It is a popular crop in Europe, about 40,000 acres are grown in the U.K. alone, but little grown in U.S.A. Var. ***capitata*** L. (Cabbage) has been cultivated in the coastal regions of Europe for about 8,000 years. There are several types designated by the shape of the head and the curliness of the leaves. They include the Savoy, Drumhead and Red Cabbage. Sauerkraut, a German dish is made by fermenting cabbage in brine, producing lactic acid which preserves the cabbage. Var. ***botrytis*** L. (Cauliflower, Broccoli) – the crowded, undeveloped flower buds are eaten as a vegetable. The term "broccoli" is applied to the later, hardier forms. Var. ***italica*** Plenck. (Asparagus Broccoli, Sprouting Broccoli) – the fleshy inflorescence and young buds are eaten as a vegetable. This vegetable was grown in Mediterranean regions since Roman times, and was introduced into U.K. in 1720. Var. ***gongyloides*** L. = B. caulocarpa Pasq. (Kohlrabi). The enlarged stems are eaten as a vegetable, but more frequently as a winter feed for stock. Var. ***costata*** DC. = var. tronchuda Bail. (Portuguese Kale, Tronchuda Kale). A compact form eaten locally in Portugal as a vegetable.

***B. orientalis*** (L.) Andrz. = Conringia orientalis L.

***B. parachinensis*** Bailey. (Mock Pak-Choi).

China. Cultigen. Eaten locally as a vegetable.

***B. pekinensis*** (Lour.) Rupr. (Chinese Cabbage, Pe-tsai). China. Cultivated in Old and New World. A rather loose head is eaten as a vegetable.

***B. perviridis*** (Bailey) Bailey.=B. rapa var. perviridis B. (Spanish Mustard, Tendergreen). E.Asia. Cultigen. The thickened stem is eaten pickled in the East, and the leaves are eaten as a vegetable in U.S.A.

***B. rapa*** L. (Turnip). Probably originated as a cultigen in C. or E. Asia. Thickened hypocotyl and root eaten as a vegetable and as a winter feed for livestock. It had been cultivated for centuries, but its introduction into British agriculture in the 18th century enabled a crop rotation to be perfected and introduced the modern crop rotation system.

***B. rapa*** L. var. ***perviridis*** B.=B. perviridis.

***B. ruvo*** Bailey. (Italian Turnip Broccoli, Rapa Ruvo, Ruvo Kale). S.Europe. Cultigen. Eaten as a vegetable in Italy.

Brauna – Melanoxylon brauna.

***Brauneria angustifolia*** (DC.) Heller – Echinacea angustifolia.

***Brauneria pallida*** (Nutt.) Britt.=Echinacea pallida.

***Brayera anthelminthica*** Kunth.=Hagenia abyssinica.

Brazil Angelin Tree – Andera vermifuga.

Brazil Arrowroot – Euxylophora paraensis.

Brazil Arrowroot – Manihot esculenta.

Brazil Boxwood – Euxylophora paraensis.

Brazil Cherry – Euglenia michelii.

Brazil Copal – Hymenaea courbaril.

Brazil Guava – Psidium araca.

Brazil Jalep – Operculina pisonis.

Brazil Nut (Oil) – Bertholetia excelsa.

Brazil Nutmeg – Cryptocarya moschata.

Brazil Oak – Posoqueria latifolia.

Brazil Peppertree – Schinus molle.

Brazil Redwood – Brosimum paraense.

Brazil Redwood – Caesalpinia brasiliensis.

Brazil Rhatany – Krameria argentia.

Brazil Rosewood – Dalbergia nigra.

Brazil Rosewood (Oil) – Physocalymma. scaberinum.

Brazil Sassafras Oil – Ocotea sassafras.

Brazil Satinwood – Euxylophora paraensis.

Brazil Tea – Stachytarpheta jamaicensis.

Brazil Tulipwood – Dalbergia cearensis.

Brazil, Vanilla – Vanilla gardneri.

Brazil Wood – Caesalpinia echinata.

Brazil Wood – Caesalpinia sappan.

Brazil Wood – Haematoxylon brasiletto.

Brazil Wood – Peltophorum linnaei.

Brea – Pinus teocote.

Bread, Dika – Irvingia barteri.

Bread, Dika – Irvingia gabonensis.

Bread, Guarana – Paullinia cupana.

Bread, Indian – Poria cocos.

Bread, Monkey – Adansonia digitata.

Bread, St. John's – Ceratonia siliqua.

Breadfruit – Artocarpus altilis.

Breadfruit, African – Treculia africana.

Breadfruit, Nicobar – Pandanus leram.

Breadfruit, Wild – Artocarpus elastica.

Breadnut – Brosimum alacastrum.

Breadnut, Para – Brosimum paraense.

Breadnut, Raman – Brosimum alacastrum.

Breadroot – Psoralea esculanta.

Breadroot, Indian – Psoralea esculenta.

Breastwort, Herniary – Nerniaria glabra.

***Brehmia spinosa*** Harv.=Strychnos spinosa.

BREYNIA J. R. and G. Forst. Euphorbiaceae. 25 spp. China, through S.E. Asia to Australia and New Caledonia. Small shrubs.

***B. cernua*** (Muell.) Arg. (Gagilamo, Gambiran, Gamer, Imer). Malaya. A poultice of the leaves is used locally to relieve swellings of the legs.

***B. rhamnoides*** (Retz.) Arg. Tropical Asia. The astringent bark is used in the Philippines to stop bleeding.

Briançon Apricot – Armeniaca brigantia.

Briar Root – Erica arborea.

BRICKELLIA Ell. Compositae. 100 spp. Warm America, W.Indies. Shrubs.

***B. cavanillesii*** Gray.=Coleosanths squarrosus=Eupatorium squarrosus. Mexico. A decoction of the leaves is used to treat intestinal worms, to reduce fever, and to reduce diarrhoea.

BRIDELIA Willd. corr. Spreng. Euphorbiaceae. 60 spp. Africa and Asia. Shrubs or trees.

***B. ferruginea*** Benth. Tropical Africa. An extract of the bark is used in Nigeria to cure thrush in children. In Uganda, the leaves are used to feed the silkworm ***Anaphe infracta.***

***B. lanceolata*** Kurz.=B. monoica.

***B. loureiri*** Hook. and Arn.=B. tomentosa.

***B. microantha*** (Hochst.) Baill. Tropical Africa. The bark is used for tanning and yields a red and black dye. The leaves are used in S.Africa. and S.Cameroons to feed the silkworms. ***Anaphe panda*** and ***A. reticulata.***

***B. minutiflora*** Hook. Tropical Asia. The fruits are eaten raw in various parts.

***B. monoica*** Merr.=B. lanceolata. (Gandri, Kanjere, Kenidal). Tropical to subtropical Asia, Australia. A decoction of the leaves is used in various parts to treat colic. The wood is used to make small articles, e.g. knife-handles.

***B. montana*** Willd.=B. retusa Spreng.

***B. moonii*** Thw.=B. retusa Baill. Ceylon. The hard durable wood is termite-resistant and olive-brown in colour. It is used for general construction work and agricultural implements.

***B. retusa*** Baill.=B. moonii.

***B. retusa*** Spreng.=B. montana. (Asana, Ekdania, Kassi, Pathor). India, Pakistan. The hard olive-brown wood is used for general construction work and to make agricultural implements. The bark is used for tanning, and the fruits are eaten.

***B. rhamnoides*** Griff.=B. tomentosa.

***B. tomentosa*** Blume.=B. loureiri=B. rhamnoides. China, India, through to Indonesia. The bark is used for tanning in Malaysia, and to treat colic, especially in Java.

Bridonnes – Garcinia indica.

Brier, Cat – Smilex rotundifolia.

Brier, Horse – Smilex rotundifolia.

Brier Root – Smilex china.

Brigalow – Acacia doratoxylon.

Brigalow – Acacia ecelsa.

Brigalow – Acacia harpophylla.

Brimstone Coloured Lepraria – Lepraria chlorina.

Brimstone Tree – Morina critrifolia.

Brindonnes – Garcinia indica.

Bristle Cone Pine – Pinus aristata.

Bristle Greenbrier – Smilax bona-nox.

***Britoa*** Berg.=Campomanesia.

Brittlebrush, White – Encelia farinosa.

Brittle Willow – Salix fragilis.

Broach Cotton – Gossypium obtusifolium.

Broad-beaked Mustard – Brassica narinosa.

Broad Bean – Vicia faba.

Broad Cocklebur – Xanthium stramonium.

Broadleaf Dock – Rumex obtusifolius.

Broadleaf Fig – Ficus platyphylla.

Broadleaf Ironbark – Eucalyptus siderophloia.

Broadleaf Kelp – Laminaria saccharina.

Broadleaf Ladybell – Adenophora latifolia.

Broadleaf Lavender – Lavendula latifolia.

Broadleaf Maple – Acer macrophyllum.

Broadleaf Orchis – Orchis latifolia.

Broadleaf Peppermint Tree – Eucalyptus dives.

Broadleaf Spignel – Peucedanum cervaria.

Broadleaf Water Gum – Tristania suaveolens.

Broccoli – Brassica oleracea var. botrytis.

Broccoli, Asparagus – Brassica oleracea var. italica.

Broccoli, Italian Turnip – Brassica ruvo.

Broccoli, Sprouting – Brassica oleracera var. italica.

BROCHONEURA Warb. Myristiaceae. 10 spp. Tropical Africa, Madagascar. Trees.

***B. dardaini*** Heck. See B. voury.

***B. voury*** Warb. (Vory). Madagascar. Seeds are used locally as the source of an edible oil.

BRODIAEA Sm. Alliaceae. 10 spp. (sens. strict.), 30 spp. (sens lat.) W. N.America. Herbs.

***B. capita*** Benth.=Dichelostemma capitum. W. U.S.A. The bulbs are sweet and eaten locally by the Indians of California and Arizona.

Brome, California – Bromus carinatus.

Brome, Mountain – Bromus marginatus.

Brome, Nodding – Bromus anomalus.

Brome, Smooth – Bromus inermis.

BROMELIA L. Bromeliaceae. 4 spp. Tropical America, W.Indies. Herbs.

***B. ananas*** L.=Ananas comosus.

***B. argentina*** Baker=B. serra.

***B. fastuosa*** Lindl. S.Brazil. A fibre (Caraguata, Gravata, Pita) is made from the leaves and used locally.

***B. karatas*** L. Tropical America. The ripe fruits are used to make the beverage Atol de Piña, and the young inflorescence is eaten as a vegetable in El Salvador.

***B. pinguin*** L. (Piñuela). C.America, Venezuela, W.Indies. The fruits, which are very acid are eaten locally.

***B. serra*** Griseb.=B. argentina=Rhodostachys argentina. S.America, esp. C. The source of a fibre (Caraguata Fibre), used for making sacking, and a possible source of paper.

BROMUS L. Graminae. 50 spp. Temperate, and tropical mountains. Grasses. All species mentioned are grown for forage.

***B. anomalus*** Rupr. (Nodding Brome). W. United States of America, Mexico.

***B. carinatus*** Hook. and Arn. (California Brome). Pacific N. America.

***B. catharticus*** Vahl. (Rescue Grass). Cultivated in S. U.S.A. for winter forage.

***B. inermis*** Leyss. (Smooth Brome). Europe, Asia. Cultivated for hay.

***B. marginatus*** Nees. (Mountain Brome). W. N. U.S.A. to Mexico. Used especially in a mixture of Alfalfa and Sweet Clover as a pasture.

***B. unioloides*** H.B.K. sometimes included in B. catharticus. Argentina.

Brondonko – Millettia laurentii.

Bronze Shield Lichen – Parmelia olivacea.

Brook Euonymus – Euonymus americanus.

Brook Lime (Beccabunga) – Veronica beccabunga.

Brookweed – Samolus valerandi.

Broom – Sarothamnus scoparius.

Broom Corn Millet – Panicum miliaceum.

Broom, Scotch – Sarothamnus scoparius.

Broom, Spanish – Spartium junceum.

Broom, Sweet – Scoparia dulcis.

Broom Tea Tree – Leptospermum scoparium.

Broom, Weaver's – Spartium junceum.

Broomjute Sida – Sida rhombifolia.

Broomsedge – Andropogon virginicus.

Broowan – Synedrella nodiflora.

BROSIMUM Sw. Moraceae. 50 spp. Tropical and Temperate S.America. Trees.

***B. alacastrum*** Swartz. (Breadnut, Ramon Breadnut). Tropical America. The wood is hard, white and fine-grained, and is used for general carpentry. The seeds are eaten roasted, and are sometimes used as a coffee substitute. The leaves are used as livestock fodder.

***B. aubletii*** Poepp. and Endl. = B. guianensis (Aubl.) Huber. = Piratinera guianensis Aubl. Guyana. The wood is hard, heavy and straight-grained with beautiful black markings on a red-brown ground. It is very expensive and is used for making small articles, e.g. fishing rod butts, and cabinet work.

***B. costaricanum*** Liebm. Costa Rica. The seeds are eaten boiled, locally. The leaves are fed to livestock.

***B. galactodendron*** D. = B. utile (Cow Tree, Milk Tree). Tropical America. The bark produces a profusion of latex which is drunk raw. The latex is also used as a base for chewing gum. The bark is used as a cloth for making sails and blankets.

***B. guianensis*** (Aubl.) Huber. = B. aubletii.

***B. paraense*** Hub. (Para Breadnut Tree, Brazil Redwood, Cardinal Wood). Tropical S.America. The wood is exported and used for carpentry and fancy goods.

***B. utile*** (H.B.K.) Pitt. = B. galactodendron.

BROUSSONETIA L'Hérit. ex Vent. Moraceae. 7–8 spp. E.Asia, Polynesia. Trees.

***B. kaempfer*** Sieb. = B. kazinoki.

***B. kazinoki*** Sieb. = B. kaempfer. B. monoica = Morus kaemperi. Korea, Japan. Paper is made from the bark.

***B. monoica*** Hance = B. kazinoki.

***B. papyrifera*** (L.) L'Hérit. ex Vent. (Paper Mulberry). China, Japan. The bark is used in China to make paper and cloth (Tapa Cloth). The bark fibre is also used to make rope.

Brown Barberry Gum – Acacia arabica.

Brown Beech – Cryptocarya glaucescens.

Brown Beech – Litsea reticulata.

Brown Heart – Andira excelsa.

Brown Mahogany – Entandromorpha utile.

Brown Mahogany – Lovoa klaineana.

Brown Mahogany – Lovoa swynnertonii.

Brown Mallet – Eucalyptus astringens.

Brown Oak – Fistulina hepatica.

Brown Padauk – Pterocarpus macrocarpus.

Brown Sarsaparilla – Smilax regelii.

Brown Stringybark – Eucalyptus capitellata.

Brown Strophanthus – Strophanthus hispidus.

Brown Tulip Oak – Argyrodendron trifoliatus.

Brown Walnut – Lovoa klaineana.

Brown Whitefly Fungus – Aegerita webberi.

BRUCEA J. F. Mill. Simaroubaceae. 10 spp. Old World Tropics. Trees.

***B. amarissima*** Desv. = B. sumatrana.

***B. antidysenterica*** J. F. Mill. E.Africa, esp. Abyssinia. The fruits and bark are used locally to treat diarrhoea and fevers.

***B. javanica*** (L.) Merr. E.India through to Philippine Islands. The fruits are used in China to treat dysentery and intestinal worms.

***B. sumatrana*** Roxb. = B. amarissima. India through to N.Australia. The roots are used as an insect repellant, and the fruits (Makassaar Pitjes), which contain the alkaloid brucamarin, are used to treat dysentery.

Bruds – Garcinia hombroniana.

BRUGMANSIA Pers. Solanaceae. 14 spp. Tropical America. Small trees.

***B. amesianum*** R. E. Schultes. (Culebra borrachere, Kwai borrachero). Columbian Andes. A narcotic made from an infusion of the leaves is used in Indian witch-craft. It produces frenzy and ultimate unconsciousness which may last for four days.

BRUGUIERA Lam. Rhizophoraceae. 6 spp. Tropical E.Africa, Asia, Australia, Polynesia. Trees.

***B. gymnorhiza*** Lam. (Msinzi, Mui). Tropics. Wood, which is hard and durable is used in Zanzibar and Pemba for general construction work and telegraph poles. The bark is used for tanning.

***B. parviflora*** Wight. and Arn. Coastal India, Malaysia. Germinating embryos eaten locally in Malaya.

***Brumelia nigra*** Sw. = Diphlois nigra.

BRUNFELSIA L. Solanaceae. 30 spp. Tropical America, W.Indies. Shrubs.

***B. hopeana*** (Hook.) Benth. = Franciscea uniflora (Manaca Rain tree). The root contains a poisonous alkaloid manacine, which acts as a muscular contractant. The root is used locally to relieve rheumatism and to treat syphilis.

Brush Bloodwood – Baloghia lucida.

Brush Bloodwood – Synoum glandulosum.

Brush Box – Tristania conferta.

Brush Mahogany – Weinmannia benthami.

Brussel Witloof – Cichorium intybus.

Brussel's Sprout – Brassica oleracea var. gemmifera.

BRYA P. Br. Leguminosae. 7 spp. C.America. W.Indies. Trees.

***B. ebenus*** DC. (American Ebony, Cocuswood, Ebony, Jamaican Ebony). W.Indies. The heavy dark brown wood polishes well. It is used for general carpentry, and is one of the principal timber trees in parts of the W.Indies.

Bruyère de Tonkin – Baeckea fructescens.

BRYONIA L. Cucurbitaceae. 4 spp. Europe, Asia, N.Africa, Canary Islands. Vines.

***B. alba.*** L. (White Bryony). C.Europe to Russia and Iran. The dried roots are used as a purgative and to treat dropsy.

***B. cretica*** L. = B. dioica (Redberry Bryony). The dried root (Radix Bryoniae) was used as a severe purgative, emetic and diuretic.

***B. dioica*** Jacq. = B. cretica.

***B. lanciniosa*** L. = Kedropsis laciniosa.

***B. sagittata*** Blume = Melothria heteriphylla.

***Bryonopsis*** Arn. = Kedrostis Medik.

Bryony, Black – Tamus communis.

Bryony, Redberry – Bryonia dioca.

Bryony, White – Bryonia alba.

Buanha – Garcinia loureiri.

Buaze Fibre – Securidaca longpedunculata.

Bubalka – Cynometra alexandri.

Bubi-kowa – Garcinia paniculata.

***Bubon galbanum*** L. = Peucedanum galbanum.

Bubulaka – Cynometra hankei.

Bucco – Barosma betulina.

***Bucephalon racemosum*** L. = Trophis racemosa.

BUCHANANIA Spreng. Anacardiaceae. 25 spp. Indomalaysia, tropical Australia. Trees.

***B. arborescens*** Blume. (Otak udang). Malaya, Borneo. A light wood which polishes well, but is not durable. It is pale in colour and used for interior work and cigar boxes.

***B. lanza*** Spreng. (Almondette). E.India, Malaysia. The seeds are rich in oil and are exported to U.K. for use in confectionery.

***B. latifolia*** Roxb. E.India, Burma. The oily seeds have a rich, nutty flavour and are eaten locally and used in confectionery. They are also the source of an edible oil. A gum (Chironji-ki-gond) is also extracted from the seeds and sold in India as a glue.

***B. lucida*** Blume. See B. arborescens.

***B. macrophylla*** Blume = Campnosperma macrophylla.

***B. microcarpa*** Laut. New Guinea. The pinkish wood is easily worked, but is neither strong nor durable. It is used for light construction work, boxes, etc.

***B. oxyrhachis*** Miq. = Campnosperma oxyrhachis.

BUCHENAVIA Eichl. Combretaceae. 20 spp. Tropical S.America, W.Indies. Trees.

***B. capitata*** (Vahl.) Eichl. (Granadillo). W.Indies. An attractive light yellow wood, which has a fine lustre and polishes well. It is fairly hard and seasons well. The wood has a wide variety of uses from furniture, veneers and house construction to railway sleepers and boxes.

BUCHLÖE Engelm. Graminae. 1 sp. W. United States of America. Grass.

***B. dactyloides*** (Nutt.) Engelm. (Buffalo Grass). A sod forming grass which will withstand heavy grazing. It is suitable for erosion control.

BUCHNERA L. Scrophulariaceae. 100 spp. Tropical and subtropical, mostly Old World. Herbs.

***B. leptostachya*** Benth. Tropical Africa, Madagascar. Used locally in Madagascar to stain the teeth.

***B. lithospermifolia*** Benth. (Tronera del Monte). C.America. An extract of the leaves is used to treat headaches.
Buchu – Empleurum ensatum.
Buchu, Long – Barosma serratifolia.
Buchu, Short – Barosma crenulata.
***Bucida buceras*** L.=Laguncularia racemosa.
Buckbean, Marsh – Menyanthes trifoliata.
Buckeye, California – Aesculus californica.
Buckeye, Foetid – Aesculus glabra.
Buckeye, Ohio – Aesculus arguta.
Buckeye, Ohio – Aesculus glabra.
Buckeye, Red – Aesculus pavia.
Buckeye, Sweet – Aesculus octandra.
Buckeye, Western – Aesculus arguta.
Buckeye, Yellow – Aesculus actandra.
***Bucklandia*** R. Br. ex Griff.=Exbucklandia R. W. Brown.
Buckthorn, Alder – Rhamnus frangula.
Buckthorn, Cascara – Rhamnus purshiana.
Buckthorn, Common – Rhamnus catharticus.
Buckthorn, Dahuran – Rhamnus dahurica.
Buckthorn, Gloss – Rhamnus frangula.
Buckthorn, Japanese – Rhamnus japonica.
Buckthorn Plantain – Plantago coronopus.
Buckthorn, Redberry – Rhamnus crocea.
Buckthorn, Sea – Hippophaë rhamnoides.
Buckthorn, Woolly – Bumelia lanuginosa.
Buckwheat – Fagopyrum esculentum.
Buckwheat Bush – Cliftonia monophylla.
Buckwheat, Japanese – Fagopyrum esculentum.
Buckwheat, Siberian – Fagopyrum tataricum.
Buckwheat, Tatary – Fagopyrum tataricum.
Buckwheat, Wild – Eriogonum microthecum.
Budahanarikella – Pterygota alata.
BUDDLEJA L. Buddlejaceae. 100 spp. Tropics and subtropics, esp. E.Asia. Shrubs.
***B. americana*** L.=B. callicarpioides=B. rufescens (Cayolizán, Hierba de la Mosca). Mexico to S.America. The leaves, roots, and bark are used to make a decoction for treating rheumatism and as a diuretic. It is also used as a salve for wounds.
***B. australis*** Vell.=B. brasiliensis.
***B. brasiliensis*** Jacq.=B. australis=B. thapsoides. Brazil. The leaves (Folia Buddleiae) are used to make an expectorant, and are used to stupify fish.
***B. callicarpiodes*** H.B.K.=B. americana.
***B. cambara*** Arech. C. S.America. The leaves are used locally in Brazil as a treatment for chest complaints.
***B. curviflora*** Hook. and Arn. China and Japan. The leaves are used in Japan as a fish poison.
***B. madagascariensis*** Lam. Madagascar. The flowers are used to make a dye for cloth (Zaifotsy), and the branches and leaves are used in the manufacture of rum.
***B. marrubiifolia*** Benth. Mexico. An extract of the whole plant is used as a diuretic, while an extract of the leaves is used to colour butter and vermicelli yellow.
***B. officinalis*** Maxim. China. An extract of the leaves is used locally to treat eye complaints.
***B. rufescens*** Willd.=B. americana.
***B. salviflora*** Lam. S.Africa. The wood is hard and tough. It is used for agricultural implements.
***B. scordioides*** H.B.K. S.W. United States of America, Mexico. An infusion of the leaves is used to treat indigestion in parts of Mexico.
***B. thapsoides*** Desf.=B. brasiliensis.
***B. tucumanensis*** Griseb. Argentina. An extract of the twigs is used locally as an astringent and stimulant.
BUELLIA Buelliaceae. Lichens.
***B. canescens*** (Dicks.) De Not.=Lecidia canescens=Lichen canescens. Used as an antibiotic against tuberculosis (Mycobacterium tuberculosis).
Buffalo Berry, Russet – Shepherdia canadensis.
Buffalo Berry, Silver – Shepherdia argentea.
Buffalo Currant – Ribes aureum.
Buffalo Currant – Ribes odoratum.
Buffalo Grass – Buchloë dactyloides.
Buffalo Grass, Small – Panicum coloratum.
Buffalo Nut – Pyrularia pubera.
Buffalo Thorn – Ziziphus mucronata.
Buffel Grass – Urochloa mosambicensis.
Bugbane – Actaea spicata.
Bugbane – Cimifuga foetida.
Bugis amugis–Koorderisodendron pinnatum
Bugle – Ajuga reptans.
Bugle, European – Ajuga chamaepitys.
Bugle Weed – Lycopus virginicus.
Bugle Weed, Virginian – Lycopus virginicus.
Bugloss – Echium vulgare.
Buhonga – Chartachme microcarpa.
Buklo – Clematis javanica.
Bukonde Fungus – Gibberella fujikuroi.
Buktika – Trichoscypha acuminata.
Bukukute – Lannea edulis.

Bulam Fat – Palaquium walsurifolium.
BULBINE Willd. Liliaceae. 55 spp. Tropical and S. Africa. Herbs.
***B. asphodeloides*** R. and S. = B. caespitosa Baker. (Geel Katstert, Snake Flower, Wild Kapiega). S.Africa. The juice from the leaves and stalks is used locally to treat wounds and ringworm. An extract of the bulb is used to treat stomach upsets and as a urinary antiseptic.
***B. caespitosa*** Baker. = B. asphodeloides.
***B. narcissifolia*** Salm-Dyck. (Kopiefa, Snake Root). The juice is used locally to heal wounds and to treat ringworm. A cold infusion of the leaves is used as a purgative, and an infusion of the bulb is a relief of rheumatism.
Bulbus coronae imperialis – Fritillaria imperialis.
Bull Kelp – Durvillea antarctica.
Bull Oak – Casuarina equisetifolia.
Bull Pine – Pinus sabiniana.
Bullace Plum – Prunus institituia.
Bullet – Manikara bidentata.
Bullet, Bastard – Houmiria floribunda.
Bullock's Heart – Annona reticulata.
Bulloo – Owenia acidula.
Bully – Manilkara bidentata.
BULNESIA C. Gay. Zygophyllaceae. 8 spp. Venezuela through to Argentina and Chile. Trees.
***B. sarmienti*** Lorentz. (Paó Santo, Palo Balsamo, Paraguay Lignum). Argentina and Paraguay. An oil is extracted from the wood by steam distillation. It has a soft rose-like scent and is used in perfumery and soap manufacture, especially to conceal the smell of harsh-scented aromatics. The oil is called Essence de Bois Gaiac, Champaca Wood Oil, Oleum Ligni Guaiaci, Oil of Guaiac Wood.
Bulrush, Great – Scirpus lacustris.
Bulrush Millet – Pennisetum typhoideum.
Bulso – Gnetum gnemon.
Bulungu Resin – Canarium edule.
BUMELIA Sw. Sapotaceae. 60 spp. Warm America, W.Indies. Small trees.
***B. angustifolia*** Nutt. = B. spiniflora.
***B. laetevirens*** Hemsl. Mexico. The fruits produce a latex which is used locally as chewing gum. The immature fruits are pickled in salt and vinegar.
***B. lanuginosa*** (Michx.) Pers. (Woolly Buckthorn, Gum Elastic). E. U.S.A. to Mexico. The fruits are eaten by local Indians. A mucilage is extracted from the ground bark and when dried, is used as a chewing gum.
***B. schottii*** Britton. = B. spiniflora.
***B. spiniflora*** A. DC. = B. angustifolia = B. schottii. (Coma, Coma Resin, Downward Plum, Daffron Plum). Florida, Texas, to Mexico. The black, sweet fruits are edible, and are eaten locally as an aphrodisiac.
Buna – Plantanus orientalis.
Bunag – Garcinia mooreana.
Buna-no-ki – Fagus sieboldi.
BUNCHOSIA Rich. ex Juss. Malpighiaceae. 50 spp. Tropical America, W.Indies. Shrubs.
***B. armeniaca*** (Cav.) DC. Andes. Occasionally cultivated, especially in Ecuador. The green fruits have a creamy sweet pulp, and are eaten locally.
Bundjinji – Isolona thomieri.
Bungwingwi – Alsodeiopsis poggei.
Bungwingwi – Alsodeiopsis rowlandii.
BUNIAS L. Cruciferae. 6 spp. Mediterranean, Asia. Herbs.
***B. erucago*** L. Mediterranean, Asia Minor. The leaves are eaten as a salad or cooked as a vegetable.
***B. orientalis*** L. S.E. Europe, Siberia. The leaves are eaten as a vegetable in Russia and Poland, and occasionally fed to livestock.
BUNIUM L. Umbelliferae. 40 spp. Europe to C.Asia. Herbs.
***B. bulbocastanum*** L. = Apium bulbocastanum = Ligusticum bulbocastanum. Europe. Previously cultivated, now used occasionally and sometimes sold. The tubers are eaten as a vegetable, and used medicinally as an astringent. They are sometimes fed to livestock. The leaves are used as a garnish and flavouring. The seeds are used for flavouring food.
Bunsala – Cissus rubiginosa.
Buntal Fibre – Corypha elata.
Bunya-Bunya – Araucaria bidwillii.
Buphane Herb. = Boöphone.
BUPHTHALMUM L. Compositae. 6 spp. Europe, Asia Minor. Shrubs.
***B. oleraceum*** Lour. Warm Asia. The leaves are used in Annam as a condiment for meat, especially fish.
BUPLEURUM L. Umbelliferae. 150 spp. Europe, Africa, Asia, N.America. Herbs.
***B. falcatum*** L. var. ***octoradiatum*** Tourn. Far East. The leaves are used in China to encourage perspiration.
Bur Clover – Medicago hispida.

Bur Clover, Spotted – Medicago arabica.
Bur Oak – Quercus macrocarpa.
BURASAIA Thou. Menispermaceae. 4 spp. Madagascar. Woody vines.
***B. madagascariensis*** Cand. Madagascar. The fruits are eaten locally. The stem and leaves yield a bitter substance which is used to precipitate the sediment in local beer.
Burdekin Plum – Pleiogynium solandri.
Burdock, Common – Arctium minus.
Burdock, Great – Arctium lappa.
Burdock, Sea – Xanthium strumonium.
Burgundy Pitch – Picea excelsa.
Buril Oil – Apeiba tibourbou.
Buri Palm – Corypha elata.
BURKEA Benth. Leguminosae. 2 spp. W. and S. Africa. Trees.
***B. africana*** Hook. S.Africa. A dark coloured gum is extracted from the stem, and the bark is used for tanning in Tanzania.
Burma Guger Tree – Schima noronhae.
Burmese Lacquer – Melanorrhoea usitata.
Burmese Varnish – Melanorrhoea usitata
Burnet – Sanguisorba officinalis.
Burning Bush – Euonymus atropurpureus.
Burom bac – Mussaendra cambodiana.
Burra Acacia – Acacia falcata.
Burrawang Nut – Macrozamia spiralis.
Burro – Capparis flexuosa.
Burrofat – Isomeris arborea.
Bursa Opopanax – Commiphora kataf.
BURSERA Jacq. ex L. Burseraceae. 80 spp. Tropical America. Small trees.
***B. aloëxylon*** (Schiede) Engl. = Elaphrium aloëxylon (Linaloé Tree). Mexico. See B. delpechiana.
***B. bipinnata*** (DC.) Engl. = Amyris bipinnata = Elaphrium bipinnatum. (Copal amargo, Copal amorgoso, Incienso de pais, Tetlate). Mexico. The resin is used locally to treat wounds.
***B. delpechiana*** Pois. ex Engl. Mexico. (Linaloe Wood, Mexican Linaloe Wood). An essential oil is distilled from the bark and the fruits for use in scents, soaps, and cosmetics. The oil production of the bark increases with age, and trees forty to sixty years old are normally used. Wounding of the bark increases the production of oil. Most of the oil is produced from wild trees though there is some cultivation. Distillation is either carried out locally as a small industry or the wood is exported to the U.S.A. where extraction is carried out by modern methods. The oil from the pericarp is inferior to that from the bark. There is some cultivation in India.
***B. gummifera*** L. = Elaphrium simaruba. A resin (West Indian Elemi, Elqueme, Tacamahaca) is extracted from the bark. It is used as a varnish to preserve canoes, and as a glue to mend china and glass. It is used as a diuretic and purgative, also to reduce fevers, in treating dysentery, yellow fever and dropsy. It was used as an incense by the Mayas. It is produced mainly in the W.Indies.
***B. jorullensis*** (H.B.K.) Engl. = Elaphrium jorullensis. Mexico. A resin is derived from incisions in the bark. The resin (Copal de Penca) makes a very glossy varnish. It is also used locally in the treatment of diseases of the uterus. The bark is used for tanning and dyeing.
***B. microphylla*** (A. Gray) Rose. = B. morelensis = Elaphrium microphyllum (Copal, Torote Copal, Cuajiote colorado). Mexico, S.America. The bark is used for tanning and dyeing.
***B. morelensis*** = B. microphylla.
***B. odorata*** Brandeg. = Elaphrium odoratum. Mexico. A resin is extracted from the bark and is used as a glue for mending pottery. The resin is also used as an expectorant and purgative. It is supposed to cure scorpion stings. The bark is used for tanning.
Burweed – Xanthium canadense.
Burweed – Xanthium strumonium.
Busa – Panicum miliaceum.
Bush Apple – Heinsia pulchella.
Bush Beech, Malay – Gmelina arborea.
Bush, Butter – Pittosporum phillyraeoides.
Bush, Cancer – Sutherlandia frutescens.
Bush Cherry – Euglenia myrtifolia.
Bush, Christmas – Ceratopetalum gummiferum.
Bush Cinquefoil – Potentilla fruticosa.
Bush, Creosote – Larrea mexicana.
Bush, Fever – Garrya elliptica.
Bush, Fever – Ilex verticillata.
Bush, Gum – Eriodictyon glutinosum.
Bush, Hop – Dodonaea lobulata.
Bush Morning Glory – Ipomoea leptophylla.
Bush, Pine – Hakea leucoptera.
Bush, Quinine – Garrya elliptica.
Bush, Quinine – Garrya fremontii.
Bush, Rabbit – Chrysothamnus confinis.
Bush, Rabbit – Chrysothamnus nauseosus.

Bush, Rabbit – Chrysothamnus viscidiflorus.
Bush, Sage – Artemisia gnaphalodes.
Bush, Salt – Kochia aphylla.
Bush, Skunk – Garrya fremontii.
Bush, Smoke – Dalea polydenia.
Bush Tea – Cyclopia vogeli.
Bush Tree – Cyclopia vogeli.
Bushmint, Emory – Hyptis emoryi.
Bussipi Wood – Oxystigma mannii.
Bustic – Dipholis salicifolia.
But – Ailanthus malabarica.
Butam – Palaquium walsurifolium.
Butapala – Elaeodendron glaucum.
Butcher's Broom – Ruscus acueatus.
BUTEA Roxb. ex Willd. Leguminosae. 30 spp. Indomalaya, China. Trees.
***B. frondosa*** Roxb.=B. monosperma=B. superba. (Bastard Teak, Bengal Kino). India, Malaysia. The wood is very durable under water and is used in well-making and for charcoal for gunpowder. A gum (Butea Gum, Bengal Kino) is extracted from incisions in the bark. It is a mild astringent. The bark yields a fibre (Pala Fibre) used locally in sail-making. A decoction of the flowers and leaves is used as a diuretic, astringent, and aphrodisiac. A yellow dye is extracted from the flowers. The pounded seeds are used as a purgative and to treat intestinal worms. They are also used to dilate the blood vessels in the skin.
***Buettneria dasyphylla*** Gray=Rulingia pannosa.
Butolo – Cissus petiolata.
BUTOMUS L. Butomaceae. 1 sp. Temperate Eurasia. Herb.
***B. umbellatus*** L. (Flowering Rush). The rhizomes are eaten locally in Russia.
Butter, Adjab – Mimusops djave.
Butter and Eggs – Linaria vulgaris.
Butter, Bambuk – Butyrospermum parkii.
Butter Bean – Phaseolus lunatus.
Butter, Borneo Tallow – Shorea aptera.
Butter, Cacao – Theobroma cacao.
Butter, Dika – Irvingia gabonensis.
Butter, Djave – Mimusops djave.
Butter, Doumori – Tieghamella heckeli.
Butter Fruit – Diospyros discolor.
Butter, Galam – Butyrospermum parkii.
Butter, Goa – Garcinia indica.
Butter, Illipé – Madhuca butyracea.
Butter, Illipé – Madhuca longifolia.
Butter, Indian – Madhuca butyracea.
Butter, Jaboty – Erisma calcaratum.
Butter, Kagné – Allenblacjia oleifera.
Butter, Kanga – Pentadesma butyracea.
Butter, Kokam – Garcinia indica.
Butter, Kombo – Garcinia indica.
Butter, Lamy – Pentadesma butyracea.
Butter, Mowra – Madhuea longifolia.
Butter, Nogo – Nogo nogo.
Butter, Ochoco – Scyphocephalum ochocoa.
Butter, Orère – Baillonella ovata.
Butter, Orris – Iris florentina.
Butter, Otoba – Myristica otoba.
Butter, Owala – Pentaclethra macrophylla.
Butter, Phulwara – Madhuca butyracea.
Butter Pits – Acanthosicyos horrida.
Butter Print Chingma – Abutilon avicennae.
Butter, Sal – Shorea robusta.
Butter, Shea – Butyrospermum parkii.
Butter, Sierra Leone – Pentadesma butyracea.
Butter, Staudtia – Staudtia kamerrumensis.
Butter, Tallow Mowra – Madhuca logifolia.
Butter Tree – Pentadesma butyracea.
Butter, Ucahuba – Virola surinamensis.
Butter, Ucuiba – Virola surinamensis.
Butterbur, Plamate – Petasites palmata.
Buttercup, Blister – Ranunculus scleratus.
Butterfly Bean – Clitoria ternata.
Butterfly Pea – Clitoria ternata.
Butterfly Weed – Asclepias tuberosa.
Butternut – Caryocar nucifera.
Butternut – Juglans cinerea.
Butterwort – Pinguicula vulgaris.
Button Brush – Cephalanthus occidentalis.
Button Clover – Medicago orbicularis.
Button Snakeroot – Eryngium yuccifolium.
Button Snakeroot – Liatris spicata.
Button Weed – Abutilon avicennae.
Buttonwood – Conocarpus erectus.
Buttonwood – Platanus occidentalis.
Buttonwood, White – Laguncularia racemosa.
Butuan – Musa errans var. botoan.
BUTRYOSPERMUM Kotschy. Sapotaceae. 1 sp. N. Tropical Africa. Tree.
***B. parkii*** (Don.) Kotschy. (Shea Butter Tree). Drier parts of C. and W. Africa. The seeds contain about 50 per cent fat (Shea Butter, Galam Butter, which is extracted by boiling the seeds and collecting the floating fat. The fat is used locally as food, for anointing and as an illuminant. Seeds are exported from W.Africa. The fat is used in Europe to make soap and candles, and the refined oil is used to make margarine, and caçao butter substitutes. The oil cake has a high

carbohydrate content and is fed to stock, The wood is used locally to make buildings, rough furniture and household utensils.

BUXUS L. Buxaceae. 70 spp. Temperate Eurasia, Tropical and S.Africa, Far East. N. and C.America, W.Indies. Small trees. All the species mentioned have a very hard, close-grained wood, which is yellowish in colour. They are all used for making small articles, e.g. musical instruments, mathematical instruments, turning and carving, and furniture.

***B. japonica*** Meull. Arg. Japan.

***B. macowanii*** Oliv. (Cape Box). S.Africa.

***B. sempervirens*** L. (Box). S. and C.Europe, Asia Minor, N. Africa. The leaves are used in some countries as a substitute for palm leaves on Palm Sunday.

***B. wallichiana*** Baill. (Boxwood, Chikri, Papar, Sansada). Himalayas. A decoction of the leaves is used to combat intestinal worms.

BYRSONIMA Rich. ex Juss. Malpighiaceae. 120 spp. C. and S.America, W.Indies. Trees.

***B. coriacea*** (Sw.) DC. (Maricao). Tropical America, W.Indies. Heavy red-brown wood, is very susceptible to termites, but is attractively coloured. It is used for interior house work, furnishing, plywood and veneer.

***B. crassifolia*** H.B.K. Tropical America. Berries eaten, and sold locally in Mexico. Wood is used for charcoal.

***B. cubensis*** Juss. Cuba. Heavy red wood is used for wooden ware.

***B. spicata*** Rich. (Maricao). Tropical America. Yellow-brown wood is used for house construction.

BYTTNERIA Loefl. Sterculiaceae. 70 spp. Tropics. Small shrubs.

***B. aculeata*** Jacq. = B. carthagenensis.

***B. carthagenesis*** Jacq. = B. aculeata. Mexico to S.America. An extract of the roots is used in Venezuela as a substitute for sarsaparilla.

# C

Caa Apé – Lerospatha caudata.

Caapi – Banistiopsis caapi.

Cabbage – Brassica oleracea.

Cabbage, Chinese – Brassica chinensis.

Cabbage, Chinese – Ipomoea aquatica.

Cabbage Gum – Eucalyptus sieberiana.

Cabbage, Kerguelin – Panglea antiscorbutica.

Cabbage, Marrow – Brassica oleracea var viridis.

Cabbage, Pak-Choi – Brassica chinensis.

Cabbage Palm – Livistonia australis.

Cabbage, Palmetto – Sabal palmetto.

Cabbage, Pe-tsai – Brassica pekinensis.

Cabbage, Savoy – Brassica oleracea var. capitata.

Cabbage, Skunk – Symplocarpus foetidus.

Cabbage, Swamp – Symplocarpus foetidus.

Cabbage Tree – Cussonia kirkii.

Cabbage Tree – Pisonia grandis.

Cabbage, Western Skunk – Lysichitum americanum.

Cabbage, Wild – Caulanthus glaucus.

Cabelluda – Eugenia tormentosa.

Cabesa de Viejo – Lophocerus schottii.

Cabeza de Angil – Calliandra anomala.

Cabicia Fibre – Furcraea cabuya.

Cablote – Guazuma ulmifolia.

CABRALEA A. Juss. Meliaceae. 40 spp. Tropical S.America. Tree.

***C. cangerana*** Sald. (Cancharana, Pau de Santo). Tropical S.America. Brittle red wood is used locally for house construction, furniture-making and carving. A water extract of the sawdust yields a red dye. An extract of the bark (Cangerana) is used in Brazil to treat fever.

Cabreu Oil – Myrocarpus fastigiatus.

Cabrewa – Myrocarpus fastigiatus.

Cabriowa Wood Oil – Myrocarpus fastigiatus.

Cabrito – Thevetia thevetioides.

Cabulla Fibre – Furcraea cabuya.

Caburá – Myrocarpus frondosus.

Cabuya (Fibre) – Furcraea cabuya.

***Cacalia cylindriflora*** Wall. = Gynura sarmentosa.

Cacao – Theobroma cacao.

Cacao Blanco – Theobroma bicolor.

Cacao Butter – Theobroma cacao.

Cacao Calabacillo – Theobroma leiocarpa.

Cacao cimarrón – Monsonia americana.
Cacao de Mico – Theobroma purpureum.
Cacao de Mico – Theobroma speciosa.
Cacao de Sonusco – Theobroma angustifolium.
Cacao, Lagarto – Theobroma pentagona.
Cacao, Madre de – Gliricidia sepium.
Cacao Montaras – Theobroma albiflora.
Cacao, Nicaraguan – Theobroma bicolor.
Cacao Silvestre – Theobroma angustifolium.
Cacao Simarron – Theobroma albiflora.
Cacaoti – Theobroma mariae.
***Cacara erosa*** Thour. = Pachyrrhizus erosus.
Cacherahua – Indigofera lespedezioides.
Cachibou – Bursera gummifera.
Cachibou – Calathea discolor.
Cachibou – Calathea lutea.
Cachiman – Rollinia sieberi.
Cachiman Montague – Rollinia sieberi.
Cachiman, Wild – Rollinia mucosa.
Caco mico – Theobroma angustifolium.
Caconier – Ormosia monosperma.
***Cactus sepium*** H.B.K. = Borzicactus sepium.
Cactus Candy – Ferocactus wislizeni.
Cactus, Giant – Carnegiea gigantea.
Cactus, Strawberry – Echinocactus euneacanthus.

CADABA Forsk. Capparidaceae. 30 spp. Warm Africa, Madagascar, S.W. Africa to Ceylon; 1 sp. Java to N.Australia. Woody plants.

***C. farinosa*** Forsk. Tropical Africa, Arabia. The leaves and twigs are pounded into a cake with cereals and eaten locally. It is called Farsa or Balambo.

Cade, Oil of – Juniperus oxycedrus.
Cadillo – Triumfetta lappula.

CAESALPINIA L. Leguminosae. 100 spp. Tropical and Subtropical.

***C. arborea*** Zoll. = Peltophorum pterocarpum.

***C. benguetensis*** Elm. = C. sepiaria.

***C. bonduc*** Roxb. Tropics. Woody climber. Cultivated in India, where the seeds are used to treat stomach complaints, and the fat from the seeds is used as a cosmetic cream. The leaves, fruits and roots are used to make a tonic.

***C. bonducella*** Flem. (Bengor Nut, Bonduc Nut, Molluca Bean, Physic Nut). Tropics. In India the seeds are mixed with black pepper to make a tonic and to reduce fevers. A tonic is also made from the bark. The seeds are used to make necklaces etc.

***C. brasiliensis*** Sw. (Bahia Wood, Brazilian Redwood). S.America. Tree. An extract of the wood is used as a red dye.

***C. brevifolia*** Baill. = Balsamocarpum brevifolium (Baill.) Clos. Chile. Shrub. The fruits (Agarobilli, Algaroba) are used for tanning.

***C. coriaria*** (Jacq.) Willd. (Divi-divi). C.America, W.Indies, N. S.America. Shrub or small tree. The hard, dark coloured wood is used for general carpentry. The pods are used for tanning and as a source of a black dye, used for ink in Mexico.

***C. corymbosa*** Benth. = C. paipae.

***C. crista*** L. Tropics. Woody Climber. Cultivated. See C. bonduc.

***C. dasyrachis*** Miq. = Peltophorum dasyrachis.

***C. digyna*** Rottl. E.India. Woody plant. Roots and fruits (Tari) are used for tanning, locally.

***C. echinata*** Lam. (Brazil Wood, Peach Wood, Pernambuco Wood, Ymira Piranga, Nicaragua Wood, St. Martha's Wood). C.America. The wood is red-brown and dense. It sinks in water. It polishes well, being used for shipbuilding and turnery. A red dye is extracted from the wood which is yellow in acid and purple in alkali.

***C. eriostachys*** Benth. (Iguanero, Palo Alejo). C.America. The bark is used in Mexico to stupify fish.

***C. Gardneriana*** Benth. Brazil. Tree. A yellow dye, used locally is extracted from the bark.

***C. glabrata*** H.B.K. = C. paipae.

***C. kavaiensis*** H. Mann. = Mezonevron kavaiensis.

***C. laevigata*** Perr. = C. nuga.

***C. melanocarpa*** Griseb. S.America. Tree. The leaves (Guajacan), are used locally for tanning.

***C. microphylla*** Mart. Brazil. Tree. The wood is used locally for general carpentry.

***C. nuga*** (L.) Ait. = C. laevigata. Tropical Asia. Tree. The roots are used in India as a diuretic and tonic; the ground leaves are used to tone the uterus after childbirth. A pulp of fruits and stems is used to poison fish.

***C. paipae*** Ruiz. and Pav. = C. glabrata = C. corymbosa. Small tree. Ecuador, Peru.

A black dye, which is used as ink, is extracted from the pod.

***C. praecox*** Ruiz. and Pav. S.America, especially Argentina and Chili. A gum, (Bea) is extracted from the stem.

***C. pulcherrima*** (L.) Sw. (Paradise Flower, Peacock Flower). Tropics. Tree. An extract of the pods and leaves are used in E.India as a laxative. An extract of the roots is used in Angola to treat fevers. Extracts of all parts of the plant are used to control menstruation.

***C. rugeliana*** Urb. Cuba. Tree. The reddish wood is used for inlay work and turnery.

***C. sappan*** L. (Japan Wood, Sappan Wood). Tropics. Tree. The wood is used for cabinet work. The bark, mixed with iron salts gives a black dye which is used in India. An extract of the wood yields a red dye used for colouring matting.

***C. sepiaria*** Roxb. = C. benguetensis Elm. Thorny, woody climber. Tropical Asia, esp. India and Far East. The bark is used in E.India for tanning.

***C. torquato*** Blanco. = Mezonevron latisiliquum.

***C. volkensii*** Harms. Tropical Africa. Liane. The roots yield a red dye.

Café du Sudan – Parkia africana.

Cafta – Catha edulis.

***Cailliea callistachys*** Hassk. = Dichrostachys cinerea.

Caimitillo – Micropholis chrysophylloides.

Caimitillo verde – Micropholis garcinaefolia.

Cainito – Chrysophyllum cainito.

Cairo Morning Glory – Ipomoea cairica.

CAJANUS Adans. mut. DC. Leguminosae. 1–2 Tropical Africa, Asia. Trees.

***C. cajan*** (L.) Millsp. = C. indicus. (Congo Pea, Dhal, Pigeon Pea). A very important pulse crop, especially in India. It is grown throughout the tropics, especially in arid regions, where it is sometimes used to combat soil erosion. The young seeds are eaten as a vegetable, or dried when older and used in soups etc., they are also fed to livestock. The whole plant is fed to stock as hay, grown as a green manure, and occasionally used for thatching. The leaves are fed to silk-worms in Madagascar. It has been grown in the Nile valley and other parts of Africa since ancient times, and where it probably originated. The African varieties are fairly uniform, but the Indian ones show great diversity in morphological characteristics and in time of maturing.

***C. indicus*** Spreng. = C. cajanus.

Cajaput (Oil) – Melaleuca leucadendron.

Cajaput – Anacardium occidentale.

Cake Bark – Erythrophleum guineense.

CAKILE Mill. Cruciferae. 15 spp. Sea coasts and lakesides in N.America, Europe, Mediterranean, Arabia, and Australia.

***C. edntula*** (Bigel.) Hook. (American Sea Rocket). Beaches of N.America. The roots are ground with flour to make bread during times of scarcity. The leaves are sometimes used as a salad or vegetable.

Calabao – Uvaria rufa.

Calabar Bean – Physostigma venenosus.

Calabar Ebony – Diopsyros dendro.

Calabash – Lagenaria vulgaris.

Calabash, Angola – Monodora angolensis.

Calabash Nutmeg – Monodora myristica.

Calabash, Sweet – Passiflora maliformis.

Calabash Tree – Crescentia cujete.

Calabazo – Crescentia cujete.

CALADIUM Vent. Araceae. 15 spp. Tropical S.America. Herbs.

***C. bicolor*** (Ait.) Vent. Tropical S.America. W.Indies. var. ***poecile*** (Schott.) Engl. (Toyoba brava), var. ***vellozianum*** (Schott.) Engl. (Mangara). The boiled rhizomes are eaten locally, while the fresh rhizome gives an extract which is used as an emetic and purgative.

***C. sororium*** Schott. Brazil. The fruits are eaten by the natives of the Amazon Basin.

***C. striatipes*** Schott. (Banana de Brejo, Canna de Brejo). Brazil. The roasted rhizomes are eaten locally by the natives. The crushed rhizome is used to treat dropsy, and an alcoholic extract is used to treat angina catarrhalis.

Calagua – Heliocarpus donnel – smithii.

Calaguala – Polypodium aureum.

Calamander Wood – Diospyros quaesita.

Calambac – Aquilaria agallocha.

Calamint – Satureja calamintha.

CALAMINTHA Mill. Labiatae. 6–7 spp. W.Europe to C.Asia. Herbs.

***C. acinos*** Clairv. = Satureja acinos.

***C. arvensis*** Lam. = Satureja acinos.

***C. clinopodium*** Benth. = Clinopodium vulgare = Melissa vulgaris. (Wild Basil). W.Europe to Asia. Introduced to N.America. An extract of the leaves has been used to control menstruation, to relieve stomach complaints, and to control bleeding. The leaves also yield a brown and yellow dye.

***C. graveolens*** Benth. Mediterranean, S.W.Asia, Caucasus. The seeds are used as a stimulant and aphrodisiac.

***C. officinalis*** Moench. = Satureja calamintha.

Calamodin – Citrus mitis.

CALAMOVILFA Hack. Graminae. 4 spp. N.America. Grasses.

***C. longifolia*** (Hook.) Hack. N.America. An important winter grazing in N. C.United States of America. Also used for a poor quality hay.

CALAMUS L. Palmae. 375 spp. Old World Tropics. Trees. The Rattan Palms. The stems are used for basket-making, mats, chair seats, light furniture, walking sticks and ropes.

***C. acanthophyllus*** Becc. Cambodia, S.Vietnam. Edible fruits are eaten locally.

***C. aquatilis*** Ridl. (Bakan, Rattan). Malaysia.

***C. asperrimus*** Blume. Java, Sumatra.

***C. barteri*** Becc. W.Africa.

***C. caesius*** Blume. Malaysia, Borneo, Sumatra. Exported.

***C. derratus*** Mann. and Wendl. Tropical Africa. Used also for small suspension bridges, and house construction.

***C. dongaiensis*** Pierre. S.Vietnam, Cambodia. Fruits eaten locally.

***C. equestris*** Blume. = C. javanensis.

***C. inopes*** Becc. Celebes. Exported.

***C. javanensis*** Blume. = C. equestris. Malaysia, Borneo, Sumatra. Exported.

***C. luridus*** Becc. S.Malaysia.

***C. manan*** Miq. Malaysia, Sumatra.

***C. minahassa*** Warb. (Rattan Datoo). Celebes. Exported.

***C. oblongatus*** Reinw. = Daemonorops oblongatus. Malaysia.

***C. optimus*** Becc. Borneo.

***C. ornatus*** Blume. (Kalapi, Kelichem, Rattan manau). Sumatra to Philippines. Fruits edible, eaten esp. in Philippines.

***C. ovoideus*** Thw. = C. zeylandicum. (Rattan). Ceylon. Young leaf buds eaten raw or cooked, locally.

***C. palustris*** Greff. Cambodia, S.Vietnam. Fruits eaten locally.

***C. radiatus*** Thw. Ceylon.

***C. rotang*** L. (Bent, Bet, Rattan Cane). Assam, S.India. Used also for making small boats, and stiffening clothes. The shoots and seeds are eaten locally.

***C. salicifolius*** Becc. (Lompeak, Mây sât). Cambodia. S.Vietnam. Fruits eaten locally.

***C. scipionum*** Lour. (Rattan). Malaysia, Sumatra. Young buds eaten locally.

***C. tetradactylis*** Hance. Cambodia. S.Vietnam. Fruits eaten locally.

Calamus – Acorus calamus.

Calamus, False – Iris pseudoacorus.

Calamus Oil – Acorus calamus.

CALANDRINIA Kunth. Portulacaceae. 150 spp. Australia, Canada through to W.Chile. Herbs.

***C. balonensis*** Lindl. = Claytonia balonensis. Australia. The plant was eaten as a vegetable by the early settlers.

***C. menziesii*** Torr. and Grey. California. The twigs and leaves are used locally as a vegetable and a garnish.

***C. polyandria*** Benth. = Claytonia polyandria. W. Australia. Eaten locally by the natives.

CALANTHE R. Br. Orchidaceae. 120 spp. Warm. Herbs.

***C. mexicana*** Reichb. C.America, W.Indies. Powdered flowers are used locally in Mexico to stop nose-bleeding.

***C. vestita*** Lind. = Cytheris griffithii = Cytheris regnieri. Burma, Indonesia, Vietnam. The pseudobulbs are used in Vietnam to treat pains in the legs.

Calarin – Citrus mitis × C. reticulata.

Calashu – Citrus mitis × C. reticulata.

CALATHEA G. F. W. Mey. Marantaceae. 150 spp. Tropical America, W.Indies. Herbs.

***C. allouia*** (Aubl.) Lindl. (Sweet Root Corn). W.Indies. The tubers were eaten cooked by the Caribs. The starch is extracted as a source of Guinea Arrowroot.

***C. discolor*** Mey. (Cachibou). Tropical America, W.Indies. The dried leaves are tough and waterproof, and are used for making baskets.

***C. lutea*** (Aubl.) G. F. W. Meyer. (Cachibou). W.Indies to Amazon Basin. The leaves are used in the W.Indies for making thatch.

***C. macrosepala*** Schum. (Chuflé). C. and S.America. The young flowers are cooked as a vegetable.

***C. violacea*** (Rose) Lindl. C.America to Brazil. The young flowers are cooked as a vegetable.

CALCEOLARIA L. Scrophulariaceae. 300–400

spp. Mexico to S.America. Herbs and Small shrubs.

***C. arachnoidea*** Grah. Chile. The root is used locally as an astringent and in the production of a red dye.

***C. thyrsiflora*** Grah. Chile. Small shrub. The leaves are used locally to treat sore throats, gums, lips and tongue.

Calcutta Bamboo – Dendrocalamus strictus.

Candelabra Flower – Boophone disticha.

Calderon Coytillo – Karwinskia calderonii.

CALDERONIA Standley. Rubiaceae. 2 spp. C.America.

***C. salvadoriensis*** Standl. El Salvador, Guatemala. Source of a dye (Dog's Blood, Sangre de Chucho).

***Calea oppositifolia*** L.=Isocarpha appositifolia.

CALENDULA L. Compositae 20–30 spp. Mediterranean to Persia. Herbs.

***C. arvensis*** L. Mediterranean to N.France. An extract of the leaves is used locally in Spain to induce sweating.

***C. officinalis*** L. (Marigold). S.Europe to Middle East. Cultivated in Old and New World as an ornamental in gardens. The outer (ligulate) florets contain an essential oil, a gum (calendulin) and a bitter principle. They are dried and used medicinally to aid digestion and as a stimulant. They are also used as home remedies against jaundice and to treat intestinal worms. Sometimes they are used to adulterate saffron.

Caliatur Wood – Pterocarpus santalinus.

Californian Buckeye – Aesculus californica.

Californian Brome – Bromus carinatus.

Californian Chia – Salvia columbariae.

Californian Golden Rod – Solidago californica.

Californian Grape – Vitis vinifera.

Californian Hazel Nut – Corylus californica.

Californian Holly – Photinia arbutifolia.

California Holly – Rhamnus ilicifolius.

Californian Incense Cedar – Libocedrus decurrens.

Californian Juniper – Juniperus californica.

Californian Laurel – Umbellularia californica.

Californian Olive – Umbellularia californica.

Californian Pepper Tree – Schinus molle.

Californian Poppy – Eschscholotzia californica.

Californian Red Gum – Liquidambar styraciflua.

Californian Rose – Rosa californica.

Californian Soap Plant – Chenopodium californicum.

Californian Soap Root – Chlorogalum pomeridianum.

Californian White Oak – Quercus lobata.

Calisaya Bark – Cinchona calisaya.

CALLIANDRA Benth. Leguminosae. 100 spp. Madagascar, warm Asia and America. Trees.

***C. anomala*** (Kunth.) Macbr.=Inga anomala. (Cabeza de Angel). C.America. The bark is used for tanning, and the root is added to tepache (the local fermented beverage) to retard fermentation.

***C. formosa*** Benth.=Lysiloma formosa. (Sabicu). Tropical America. The heavy wood is used locally for carpentry.

***C. houstoniana*** (Mill.) Standl.=Mimosa houstonii. Mexico. The bark (Pambotana Bark) is exported to Europe where it is used to control menstruation.

CALLICARPA L. Verbenaceae. 140 spp. Tropical and subtropics. Shrubs.

***C. americana*** L. (French Mulberry). S.E. N.America. The leaves are used as a cure for dropsy.

***C. arborae*** Roxb. (Ghiwata, Khoja, Kojo). N.India. An extract of the bark is used locally as a tonic, to stimulate digestion, and to cure skin complaints.

***C. bodinieri*** Lévi. C. and W. China. Used in China to treat conjunctivitis, caused by gonorrhoea, and to control menstruation.

***C. cana*** L.=C. candicans. (Drusha). Malaysia to Australia. The Hindus use an infusion of the roots, leaves and bark to treat skin diseases. Parts of the plant are used as an arrow poison, and an extract of the leaves is used in the Philippines to poison fish.

***C. candicans*** Hochr.=C. cana.

***C. lanata*** L.=C. tomentosa. (Aisar, Bastra, Massandari). India, esp. S. An extract of the bark is used by the Hindus to reduce fever, treat liver complaints, and skin diseases.

***C. longifolia*** Lam. Malaysia to Australia. Used in Malaya to make poultices for treating fevers and stomach complaints.

***C. tomentosa*** Murr.=C. lanata.

CALLICOMA Andr. Cunoniaceae. 1 sp. E.Australia. Tree.

***C. billardieri*** Don.=Codia montana. The reddish wood is fine-grained and is used for turnery and tool handles.

CALLIRHOË Nutt. Malvaceae. 10 spp. N.America. Herbs.

***C. digitata*** Nutt. (Finger Poppy Mallow). S. U.S.A. The roots were eaten by the Indians.

***C. involucrata*** (Torr. and Gray) A. Gray.= Malva involucrata (Purple Poppy Mallow). C. U.S.A. to N. Mexico. The Indians of Dakota burned the roots and inhaled the smoke to cure colds. A decoction of the roots was used to treat internal pains of various origins.

***C. pedata*** (Torr. and Gray) Gray. (Purple Mallow). W. U.S.A. The roots were eaten by the local Indians.

CALLITRIS Vent. Cupressaceae. 16 spp. Australia. New Caledonia, S.Africa. Trees.

***C. arborea*** Schrad.=Widdringtonia juniperoides. S.Africa. The wood is used for furniture making, and a gum from the bark is used by the natives as a diuretic.

***C. calcarata*** R. Br.=Frenela endlicheri. (Black Pine, Murray Pine). N.Victoria and C.Queensland. The bark yields a resin (Australian Sandarac).

***C. cupressoides*** Schrad.=Widdringtonia cupressoides. (Sapreewood). S.Africa. The light wood is used for making buckets and barrels.

***C. glauca*** R. Br. (Murray Pine, White Cypress Pine). Queensland, New South Wales. The streaked wood is yellow to dark brown, and camphor-scented. It is used in house construction esp. for floors, furniture-making and fences.

***C. quadrivalvis*** Vent.=Tetraclinis articulata.

***Callitropsis*** Compton=Neocallitropsis.

CALLUNA Salisb. Ericaceae. 1 sp. Atlantic N.America to Europe, excluding the S.E., Iceland to Siberia. Small shrub.

***C. vulgaris*** (L.) Hull. Heather. The flowers are an excellent source of honey; when dried (Flores Ericae) they are used as a sedative, astringent and diuretic, used in the treatment of rheumatism and bladder complaints. A tea from the dried leaves (Herba Ericae, Herba Callunae) is used similarly in home remedies. The fresh leaves are occasionally used to flavour beer, and were the source of a yellow dye. The bark is sometimes used for tanning, and the whole plant makes a useful broom.

***Callymenia papulosa*** Mont.=Eucheuma papulosa.

CALOCARPUM Pierre. Sapotaceae. 6 spp. Mexico to Tropical S.America. Trees. The species mentioned produce sweet fruits which are red-brown skinned with a reddish pulp. They are eaten raw or in preserves (Crema de Mamay Colorado, and Guava Cheese). The trees are cultivated, and the fruits sold locally. The bark is a minor source of latex (balata) used for making chewing gum.

***C. mammosum*** Pierre.=Achras Mammosa =Lucuma mammosa. (Marmalade Plum, Mamey Colorado, Sapote). C.America.

***C. viride*** Pitt. (Green Sapote). Guatemala to Costa Rica.

CALOCHORTUS Pursh. Liliaceae. 60 spp. Temperate W. N.America, C.America. Herbs. The bulbs of the species mentioned were eaten cooked by the local Indians.

***C. aureaus*** S. Wats. W. U.S.A.

***C. elegans*** Pursh. W. U.S.A.

***C. gunnisonii*** S. Wats. (Sagebrush Mariposa). W. U.S.A., esp. Montana, New Mexico and Arizona. The dried bulb also used as a flour.

***C. nuttallii*** Torr. and Gray. (Mariposa Lily, Sago Lily). W. U.S.A. esp. Montana to New Mexico.

***C. pulchellus*** Dougl. W. U.S.A., esp. California.

CALODENDRUM Thunb. Rutaceae. 2 spp. Tropical and S.Africa. Trees.

***C. capense*** Thunb. (Cape Chestnut). S.Africa. Tree. The soft white wood is tough, and used for planking, wagons etc. The seeds yield a non-drying oil used for soap-making.

CALONCOBA Gilb. Flacourtiaceae. 15 spp. Tropical Africa. Trees.

***C. echinata*** (Oliv.) Gilg.=Oncoba echinata.

***Calonyction aculeatum*** (L.) House.=Impomaea aculeata.

***Calonyction speciosum*** Choisy.=Ipomaea aculeata.

CALOPHYLLUM L. Guttiferae. 8 spp. Madagascar, Mauritius; 100 spp. Indomalaysia, Indochina, Pacific tropical Australia; 4 spp. tropical America, W.Indies. Trees.

***C. amoenum*** Wall. (Balood, Jajoo pakoi). Malaysia, Indonesia. Heavy hard wood is used for house construction and boards. The sap is used to stupify fish.

***C. bancanum*** Miq.=C. pulcherrimum.

***C. brasiliense*** Camb. Brazil. A yellow oil (Sandal Oil, Sandalo Inglez) is extracted from the bark.

***C. burmannii*** Wight.=C. calaba.

***C. calaba*** L.=C. burmannii (Ceylon Beautyleaf). An oil extracted from the seed is used to treat skin irritations. A reddish resin (Resina Ocuje) is extracted from the bark. The soft reddish wood is used in general carpentry.

***C. diepenhorstii*** Miq.=C. soulattrii.

***C. dryobalanops*** Pierre. Tropical Asia. The hard, durable wood is used for general carpentry, and boat-building.

***C. gracile*** Miq.=C. pulcherrimum.

***C. inophyllum*** L. (Alexandrian Laurel). E.Africa, E.India, Polynesia. The wood (Borneo Mahogany) is hard and polishes well. It is used for furniture-making. boat-building, railway sleepers, and wheel hubs. A resin (Balsamum Mariae, Takamahaca Resin) is extracted from the bark. A non-drying oil (Dilo, Domba Oil) is extracted from the seeds. It is used unpurified locally to treat skin diseases and rheumatism. When purified it is injected to relieve pain in cases of leprosy.

***C. javanicum*** Miq.=C. venulosum.

***C. longifolium*** Willd. S.America. A resin (Maina) is extracted from the bark.

***C. muscigerum*** Boerl. and Koord. Malaysia, Indonesia. The hard wood is used for house construction, and the milky juice is used to stupify fish.

***C. plicipes*** Miq.=C. pulcherrimum.

***C. pulcherrimum*** Wall.=C. bancanum=C. gracile=C. plicipes. (Betoor beroobook, Bintangoor padi). Malaya. The wood is weather resistant and used for masts, beams and flooring.

***C. rekoi*** Standl. (Cedro Cimarrón, Cimarrón). Mexico. The mahogany-like wood is used for general carpentry.

***C. saigonense*** Pierre. Tropical Asia. The hard durable wood is used for general carpentry and boat-building.

***C. soulattrii.*** Burm.=C. diepenhorstii= Apoterium sulatrii (Bitangoor boonoot, Membaloong, Bitangoor). Indonesia. The wood is elastic and is used for masts and house posts. It is exported. The bark (Babalan slatri) is sold locally to treat horses for intestinal complaints.

***C. spectabille*** Willd. Tropical Asia, Pacific Islands. The elastic wood is used for masts and house construction.

***C. tacamahaca*** Willd. (Fooraa, Polamaria). Madagascar, Mascara Islands. The bark is a source of a resin (Tacamahaca Resin, Bourbon Balsamum Mariae).

***C. thorelli*** Pierre. Tropical Asia. The wood is used for masts and furniture-making.

***C. tomentosum*** Wight. (Bobbi, Nagari, Poon Spar Tree, Sirpoon Tree). S. India. The red-brown wood is fairly hard and durable, but is attacked by termites. It is used for masts, poles, furniture, dug-out canoes, bridges, and railway carriages. The seeds yield a reddish oil used locally for lighting.

***C. venulosum*** Zoll. and May.=C. javanicum (Ki sapilan). Java. The wood is used to make oars.

***C. walkeri*** Wight. Ceylon, E.India. The reddish wood is durable and is used for general house construction and panelling.

***C. wallichianum*** Planch and Triana. (Betoor, Teroondjam). Malaysia. The wood is used for house construction, and utensils. A resin from the bark is boiled with coconut milk and used to treat skin complaints.

CALOPLACA Hoffm. Caloplacaceae. Lichens.

***C. murorum*** (Hoffm.) Th. Fr. Temperate, growing on rocks and walls. In Sweden a yellow dye is extracted from it for dyeing wool.

CALPOGONIUM Desv. Leguminosae. 10 spp. C. and S.America; W.Indies.

***C. coeruleum*** Hemsl. (Jicuma). C.America. Vine. Used locally to rub clothes when washing.

***C. mucunoides*** Desv. Herb. Guyana. Grown as a green manure.

***Calosanthes indica*** Blume.=Oroxylum indicum.

CALOTROPIS R. Br. Asclepiadaceae. 6 spp. Tropical Africa, Asia. Small Trees.

***C. gigantea*** (L.) R. Br.=Asclepias gigantea. (Madar, Mudar, Tembega, Wara). Tropical Asia. The stem is the source of a strong fibre, which is spun for fishing nets and lines. The floss from the fruits is used in India for stuffing pillows etc. The leaves are used to poultice boils, sores etc. The Chinese in Java candy the inner parts of the flowers.

***C. procera*** (Ait.) R. Br. (Auricula Tree). Tropical Africa and India. A strong fibre used for fishing lines and ropes is derived from the stem. The floss from the seeds is used for stuffing pillows etc. The fruit is called Apple of Sodom. The juice from the plant is rubber-like (Mudar Gummi) and when mixed with salt is used to remove the hair from hides. It is used to kill children in the Sudan. The leaves are used as part of the ferment to make a beer (Merissa) in W.Africa. The root bark is used to treat leprosy in India.

***Calpidia*** Thou. = Ceodes.

CALTHA L. Ranunculaceae. 20 spp. Arctic and N.Temperate. Herbs.

***C. palustris*** L. Marsh Marigold. N.America, Europe, Temperate Asia. The stems and leaves were eaten in various parts as a vegetable. The flower bud, pickled in vinegar are used as a flavouring instead of capers. The roots are eaten locally in Japan.

Caltrop, Water – Trapa natans.

Calunga Bark – Quassia ferruginea.

CALVATIA Fr. Lycoperdaceae. 10 spp. Temperate. Puff balls.

***C. cakavu*** (Zippel) v. Overeem. Fruitbodies eaten in parts of Asia.

***C. candida*** (Rorsk) Holl. The mature fruitbodies are used to prevent wounds bleeding; the spores are mixed with honey to treat sore throats in China. The young fruitbodies are eaten by Indians in America. The mature fruit bodies are burnt to stupify bees when manipulating hives.

***C. cyathiformis*** (Bosc.) Mor. Fruitbodies eaten by various tribes of American Indians.

***C. gigantea*** (Batch ex Pers.) Lloyd. (Giant Puff Ball). The fruitbodies may be over 3 metres across. See C. candida.

***C. lilacina*** (Berk and Mont.) Lloyd. (Fungus Chirurgorum, Fungus Bovista, Crepitus Lupi). See C. candida.

***C. natalense*** Cook and Mass. Fruitbodies eaten in some parts of Asia.

***C. umbrinum*** Pers. The fruitbodies are eaten in some parts of Asia.

CALYCANTHUS L. Calycanthaceae. 3–4 spp. S.W. and E. United States of America. Shrubs.

***C. fertilis*** Walter. = C. glaucus. E. U.S.A. Bark, leaves and roots are used locally by the Indians to control menstruation.

***C. floridus*** L. (Carolina Allspice). E. U.S.A. The bark is aromatic and is used by the local Indians as a spice.

***C. glaucus*** Willd. = C. fertilis.

***C. praecox*** L. = Chimonanthus praecox.

CALYCOGONIUM DC. Melastomataceae. 30 spp. W.Indies. Trees.

***C. squamulosum*** Cogn. (Camasay Negro, Jusillo). W.Indies, esp. Puerto Rico. The heavy, hard wood is pinkish, brown, streaked with black. It splits easily when nailed and is hard to sand paper and machine. The texture and lustre are good. It is used for furniture, veneer, agricultural implements, flooring, and especially for heavy construction work.

CALYCOPHYLLUM DC. Rubiaceae. 6 spp. W.Indies, S.America. Tree.

***C. candidissimum*** (Vahl.) DC. (Lemonwood, Lancewood). S.Mexico to Columbia and Cuba. Brownish, strong heavy wood. It is used to make bows in U.S.A. Locally it is used to make agricultural implements, house construction and turning. It also makes good charcoal.

***C. multiflorum*** Griseb. (Palo Blanco). S.America, esp. Argentina, Paraguay and Brazil. The fine hard wood is used for cogs, turning, shoe-lasts, rulers, flooring and pulleys.

***Calycopteris floribunda*** Lam. = Getonia floribunda.

***Calycopteris nutans*** Kurz. = Getonia floribunda.

CALYPSO Salisb. Orchidaceae. 1 sp. Cold N.Temperate. Herb.

***C. borealis*** Salis. = C. bulbosa. W. N.America. Tubers eaten by local Indians.

***C. bulbosa*** (L.) Okes. = C. borealis.

CALYPTRANTHES Sw. Myrtaceae. 100 spp. Tropical America, W.Indies. Trees.

***C. aromatica*** St, Hil. S.Brazil. The leaves are used locally as a spice, tasting rather like cloves.

***C. schiedeana*** Beng. Mexico. The leaves are used locally as a spice.

CALYPTROGYNE H. Wendl. Palmae. 10 spp. C.America. Trees.

***C. sarapiquensis*** Wendl. Costa Rica. The leaves are used locally for thatching.

CALYSTEGIA R. Br. Convolvulaceae. 25 spp. Temperate and tropics. Herbs.

***C. sepium*** (L.) R. Br. Tropics and subtropics. Cultivated in India and China. The boiled roots are eaten in China, and the young shoots are eaten in India.

Camagoon Ebony – Diospyros pilosanthera.

***Camarophyllus pratensis*** (Pers.) Karst. = Hygrophorus pratensis.

Camasay Negro – Calcogonium squamulosum.

Camass – Camassia quamash.

CAMASSIA Lindl. Liliaceae. 5–6 spp. N.America. Herbs.

***C. esculenta*** Lindl. (Wild Hyacinth, Indigo Squill). E. and S. U.S.A. The bulbs are eaten by the Indians.

***C. quamash*** (Pursh.) Gr. (Camass, Quamash). W. N.America. The bulbs are eaten cooked by local Indians and by early settlers.

Camdeboo – Celtis kraussiana.

Camel's Foot Tree – Bauhinia purpurea.

Camel Grass – Cymbopogon schoenanthus.

Camel's Thorn – Alhagi camelorum.

CAMELLIA L. Theaceae. 82 spp. Indomalaya, China, Japan. Trees.

***C. drupifera*** Dyer non Lour. = C. kissi.

***C. drupifera*** Lour. Himalayas, Burma, to China. The seeds are the source of an oil.

***C. japonica*** L. Camellia. Japan. Cultivated as an ornamental and the seeds are the source of a non-drying oil (Tsubaki Oil) used to make hairdressings.

***C. kissi*** Wall. = C. drupifera Dyer non Lour. (Diengtyrnembhai, Kissi). E. Himalayas to N.India. The leaves are used as a tea locally, and a cake of the seeds is used to stupify fish.

***C. sasanqua*** Thunb. (Sasanqua Camellia). Assam, China, Japan. The seeds are crushed to give Tea Seeds Oil, used in the silk industry, and in the manufacture of soap.

***C. sinensis*** (L.) Kuntze. = C. thea Link. = Thea sinensis L. (Tea, Chinese Tea, Japanese Tea). China, India, Assam. The beverage has been made in China since early times, from whence it spread into Japan and India. The tea-drinking habit was introduced into Europe by the Dutch in the 1600's, but soon the trade was taken over by the East India Company, making the United Kingdom the prime tea-users in the Western Hemisphere. The world production stands at around 2,000 million pounds per annum, with Ceylon producing 20·6 per cent, India 35·0 per cent. These two countries are the chief exporters. China produces a considerable quantity, but most is used internally. The plants are started in nurseries as seedlings and planted out in fields at 4 ft. intervals. The final quality of the tea depends on the climate and cultural conditions. The crop requires good drainage and heavy dressings of fertilizers. The bushes are top pruned to encourage the growth of side shoots, on which the young leaves are borne. The leaves are picked by hand, the terminal buds and the next two or three leaves being selected. Each bush yields about 2 lb. of leaves a year, production beginning in the third or fourth year. The quality of the final product depends on the age of the leaves. The young buds produce the fine Flowery Pekoe, the next oldest Orange Pekoe and the oldest the coarser Bohea. The fresh-picked leaves are fermented to produce a taste and aroma. They are then rolled to break down the cells, and spread out in cool moist rooms when fermentation continues. The leaves are then dried to about 3 per cent moisture, in dry air. They then turn black and are ready for sale. The product is Black Tea. Green Tea is from the same plants, but is not fermented. Black Tea is the chief type exported, the Green Tea being consumed internally especially in China. Oolong Tea is made from partially fermented leaves, and is most popular in America. It is produced mainly in Formosa. Brick Tea is made from the residue of both Black and Green teas. They are used mainly in Tibet and C.Asia where it is mixed with flour and butter to make a soup. The flavour of the tea is due to essential oils, while the colour and "body" depend on the tannin content. The stimulant it contains is caffeine. China Teas contain the higher caffeine content, while Indian Teas have more tannin. An oil (Tea Seed Oil) is extracted from the seeds. It is used for soap-making, as a textile oil and for lighting.

Camellia – Camellia japonica.

Camellia, Sasanqua – Camellia sasanqua.

Cameroon, Cardamon – Aframomum hanburyi.

Cameroon Copal – Copaifera demeusii.

Cameroon Mahogany – Entandrophragma candollei.

Cameroon Mahogany – Entandrophragma rederi.

Cameroon Mahogany – Khaya eyryphylla.

Camichon – Ficus padifolia.

Camogan Ebony – Diospyros discolor.

Camota Rubber – Sapium tabura.
CAMPANULA L. Campanulaceae. 300 spp. N.Temperate, esp. Mediterranean and tropical Mountains. Herbs.
***C. latifolia*** L. Europe, Middle East. The flowers are sometimes used to induce vomiting.
***C. rapunculus*** L. Rampion. Europe to Siberia, N.Africa. The roots are sometimes eaten in salads. Cultivated locally.
***C. verisolor*** Sibth. Italy, Greece, Turkey. The leaves are sometimes eaten in salads.
Camphire – Lawsonia inermis.
Camphor – Cinnamomum camphora.
Camphor, Borneo – Dryobalanops aromatica.
Camphor Fume – Camphorosoma monspeliaca.
Camphor, Ngai – Blumea balsamifera.
***Camphora officinarum*** Nees. = Cinnamomum camphora.
***Camphorata hirsuta*** Moench = Camphorosoma monspeliaca.
***Camphorata monspeliensium*** Crantz. = Camphorosoma monspeliaca.
CAMPHOROSMA L. Chenopodiaceae. 11 spp. E. Mediterranean; C.Asia. Small shrubs.
***C. monspeliaca.*** L. = Camphorata hirsuta, = Camphorata speliensium. (Camphor Fume, Stinking Ground Pine). Mediterranean, Asia. An extract of the plant is used as a stimulant, diuretic, expectorant, and antiasthmatic. It also induces sweating, and is used to control menstruation. The plant smells of camphor.
Campion, Bladder – Cucubalus baccifer.
Campion, Bladder – Silene inflata.
Campion, Moss – Silene acaulis.
Campion, Red – Melandrium rubrum.
Campion, White – Melandrium album.
CAMPNOSPERMA Thw. Anacardiaceae. 15 spp. Tropics. Trees.
***C. brevipetiolata*** Volk. (Tarentang). Malaya, N.Borneo, New Guinea, Indonesia. The wood is soft and not durable. It is light and easy to work, so is used to make boxes, coffins, etc.
***C. macrophylla*** Hook. f. = Buchanania macrophylla Blume. (Terentanabang). Malaysia. The wood (Meranti) is grey-red in colour, soft and light. It is used locally to make boxes. Some is exported.
***C. minor*** Corner. (Terentang). Malaysia, Borneo. The light wood is not durable. It is used locally for interior work, boxes etc. and matchsticks.
***C. oxyrhachis*** Engl. = Buchanania oxyrhachis. (Paoch lebi). Indonesia. The wood is fine and hard, and is used locally for general construction work. An oil, extracted from the seeds is used for cooking and lighting.
CAMPOMANESIA Ruiz. and Pav. Myrtaceae. 80 spp. S.America. Trees.
***C. acida*** Berg. = Britoa acida. (Para Guava). Brazil. The pleasant flavoured fruits are used to make jellies. The plant is occasionally cultivated.
***C. aromatica*** Griseb. W.Indies, Guyana. The pleasant flavoured fruits taste like strawberries, and are eaten locally.
***C. fenzliana*** Berg. = Abbevillea fenzliana (Guabiroba). Brazil. The orange-coloured fruits are used to make jelly.
***C. guaviroba*** Benth. and Hook. = Abbevillea guaviroba. (Guaviroba). S.Brazil. The Orange-yellow fruits are eaten locally. The plant is cultivated occasionally, locally.
CAMPSIANDRA Benth. Leguminosae. 3 spp. Tropical America. Trees.
***C. laurifolia*** Benth. Brazil. An extract of the leaves is used locally as a tonic, to reduce fevers, and to cleanse ulcers.
***Campylophora hypnoides*** J. Ag. = Ceramium hypnoides.
Cam tong houng – Ailanthus fauveliana.
Camwood – Baphia intida.
Camwood Dye – Pterocarpus erinaceus.
Camwood, Red – Pterocarpus osun.
Canada Balsam – Abies balsamea.
Canada Bluegrass – Poa compressa.
Canada Crookneck – Cucurbita moschata.
Canada Garlic – Allium canadense.
Canada Goldenrod (Oil) – Solidago canadensis.
Canada Lettuce – Lactuca canadensis.
Canada Milk Vetch – Astragalus canadensis.
Canada Pitch – Tsuga canadensis.
Canada Plum – Prunus americana.
Canada Spruce Pine – Picea mariana.
Canada Thistle – Cirsium arvense.
Canada Wild Rye – Elymus canadensis.
Cañafistula – Cassia grandis.
Cañafistula – Peltophorum vogelianum.
Cañahua – Chenopodium canihua.
Canaigre – Rumex hymenosepalus.
CANANGA (DC.) Hook. f. and Thoms. Annonanceae. 2 spp. Tropical Asia to Australia. Trees.
***C. latifolia*** Gin. and Gagnep. = Unona latifolia Hook. f. and Thou. = U. bran-

desiana Pierrs. (Tainghe, Tho shui). Malaysia, and Far East generally. The sweetly scented flowers are used to make a body soap, and scented oils, similar to Ylang-Ylang (see below).

***C. odorata*** Hook. f.=C. odoratum L. (Ylang Ylang). Tropical Asia. Cultivated in Old and New World for the perfumed flowers from which an essential oil (Ylang Ylang) is extracted by distillation. It is used in perfumery, cosmetics and soaps, and is one of the ingredients of Macassar Oil.

***C. odoratum*** L.=C. odorata.

CANARIUM L. Burseraceae. 100 spp. Tropical Africa, Asia, N.Australia, Pacific. Trees.

***C. album*** Reanch. Cochin China. The fruits (Chinese Olives) are eaten locally.

***C. amboinense*** Hochr. (Jalo, Java Almond, Iwar, Kanari, Sangi). Indonesia. The seeds are used in confectionery as a substitute for almonds. They are a minor source of an oil used in confectionery and baking. A resin (Getah kanari), is extracted from the bark. It has a smell of cloves and was used locally as an ointment for wounds, and as an incense in sick rooms. It is occasionally used in varnishes and printing ink, and as an insect repellent in bookcases.

***C. begalense*** Roxb. E.India. The stem is the source of a resin (Black Dammar Resin, East India Copal), used in varnishes for painting enamel.

***C. commune*** L. (Java Almond). Moluccas. Cultivated in Malaysia. The seeds are used in confectionery, especially in rice pastries. The seed oil is used in cooking and for lighting. A resin (Manila Elemi) is extracted from the bark and used as a fixative for perfumes and varnishes. An oil distilled from the resin is used to make soap and cosmetics, and also to make varnishes. The resin is also used as incense.

***C. decumanum*** Gaertn. Malaysia. The seeds are eaten locally.

***C. edule*** Hook. f. Tropical Africa. The purple fruits are edible and are eaten locally. A resin (Bulungu Resin) is extracted from the bark and used for waterproofing calabashes, mending pottery, for smoking, and as a perfume. The resin is also used to treat jiggers and skin complaints.

***C. euphyllum*** Kurz. (Andaman Canary Tree). Andaman Islands. The wood is used locally for general carpentry.

***C. hirsutum*** Willd. Moluccas. The stem yields a very hard resin (Damar Sengai) used to make varnishes.

***C. legitimum*** Miq. Amboina. The bark yields a resin (Damar Itam) which is used locally for torches.

***C. littorale*** Blume. (Ki kanari, Sadjang). Java. The heavy pinkish wood is used locally for house-making. The seeds are eaten locally.

***C. luzonicum*** Miq. (Elemi Canary Tree). Philippines. The bark is the source of a resin (Manila Elemi) used for varnishes, torches and waterproofing boats.

***C. mansfeldianum*** Engl. Tropical Africa. The hard wood (Adju Mahogany, Gabon Mahogany) is used like Mahogany.

***C. microcarpum*** Willd.=C. oleosum.

***C. oleosum*** Engl.=C. microcarpum. (Ando, Kanari minjak, Rasamala). Ambon. The stem produces a resin which thickens on standing, and has a pleasant odour. It is used locally to heal wounds, to treat skin complaints, and, mixed with coconut oil as a hair dressing.

***C. ovatum*** Engl. (Pili). S.Luzon. The seeds produce an oil (Pili Nut Oil) which is used in confectionery. The roasted nuts are also used in confectionery and as an adulterant to chocolate. The uncooked seeds are purgative. The whole fruits are eaten cooked locally.

***C. paniculatum*** (Lam.) Benth. Mauritius. A light resin is extracted from the bark, and the wood (Colophan Wood) is used locally.

***C. patentinervium*** Miq. (Teta toondjoo). Indonesia. The seeds are eaten locally.

***C. polyphyllum*** Schum. Malaysia, New Guinea. The seeds are eaten locally in New Guinea.

***C. pseudo-decumanum*** Hochr. (Kajoo tandikat). Indonesia. A resin is produced, particularly from the older trees. It is used by the natives for waterproofing boats, and when mixed with oil, as bird lime. The seeds are eaten locally, and produce an oil which is used for cooking.

***C. rufum*** Benn. Malacca, Philippine Islands. A resin is extracted from the bark, and used to make varnishes.

***C. samoaese*** Engl. Samoa. The source of a resin, used to scent coconut oil used as a cosmetic locally. The trunks are dug out as canoes.

***C. saphii*** Engl.=Dacryodes edulis.

***C. schweinfurthii*** Engl. (Papo Canary Tree). A resin (African Elemi, Elemi of Uganda) is extracted from the bark and used to make ointments, varnishes and printer's ink. In Angola the powdered bark is used to treat ulcers, and the fruits are used as a condiment.

***C. strictum*** Roxb. (Malaysia). A resin (Alriba Resin) is extracted from the bark and used medicinally as an ointment.

***C. sylvestre*** Gaertn. (Kanari ootan, Kanari). Malaysia. A resin from the stem is used locally for torches. The seeds are fed to pigs.

***C. venosum*** Craib. Vietnam. The roots are used locally and by the Chinese to treat itch.

***C. villosum*** Benth. and Hook. Philippine Islands. A resin (Sahing, Pagsahingin Resin) from the bark is used to waterproof boats, as a fuel, and for lighting.

***C. zeylanicum*** (Retz.) Blume. Ceylon. The wood is used to make tea chests.

***C. zollingeri*** Engl. (Kanang). Kangean. The wood is used locally for boat building, and the resin is used for lighting.

Canary Ash – Beilschmiedia bancroftii.
Canary Grass – Phalaris canariensis.
Canary Grass, Reed – Phalaris arundinacea.
Canary Madrone – Arbutus canariensis.
Canary Tree, Andaman – Canarium euphyllum.
Canary Tree, Elemi – Canarium luzonicum.
Canary Tree, Papo – Canarium schweinfurthii.
Canary Wood – Euxylophora paraensis.

CANAVALIA Adans. mut. DC. Leguminosae. 50 spp. Tropical and sub tropical, esp. Africa and America. Climbing vines.

***C. campylocarpa*** Piper. (Brabicon Bean). W.Indies. Grown as a green manure.

***C. ensiformis*** DC. (Chickaswa Lima, Jack Bean, Sword Bean). W.Indies. Cultivated throughout the Old and New World tropics, as fodder for livestock. The young pods are eaten, but the seeds are considered poisonous. The seeds are sometimes roasted and used as a coffee-substitute. The pods of white varieties are preserved in salt, in Japan.

***C. gladiata*** DC. (Sword Bean). S.E.Asia. Cultivated locally. See C. ensiformis.

***C. obtusifolia*** DC. See C. ensiformis.

***C. polystachya*** (Forsk.) Schweinf.=Dolichos polystachos=D. viscosus. Arabia, India, Madagascar, China. Cultivated as a food plant locally. The unripe fruits and seeds are eaten. The ripe seeds are possibly poisonous.

Cancer Bush – Southlandia frutescens.
Cancer Wort – Southlandia frutescens.
Canchalagua – Centaurium chilensis.
Cancharana – Cabralea cangerana.
Candelabra Flower – Boöpone disticha.
Candelabra Tree – Cassia alata.

CANDELARIELLA Ehrh. Lecanoraceae. Lichens.

***C. vitellina*** (Ehrh.) Muell. Arg. Temperate. Used in Sweden as the source of a yellow dye, to colour wool.

Candelilla – Cestrum lanatum.
Candellila – Euphorbia antisyphilitica.
Candelilla broncq – Asclepias subulata.
Candelilla Wax – Pedilanthus pavonis.
Canelón – Pluchea odorata.

CANDIDA Berkhout. Pseudosaccharomycetaceae.

***C. guillermondii*** (Cast.) Lang. and Guerra. Produces riboflavin during fermentation.

***C. tropicalis*** (Cast.) Berkh.=Monilia candida.

Candillo – Urena lobata.
Candied Peel – Citrus medica.
Candleberry – Myrica cerifera.
Candle Nut Oil (Tree) – Aleurites moluccana.
Candle Nut Oil (Tree) – Aleurites triloba.
Candle Plant – Dictamnus albus.
Candle Tree, Food – Parmentiera adulis.
Candlewood – Amyris babamifera.
Candlewood – Dacryodes excelsa.
Candlewood – Dodonaea viscosa.
Candlewood – Gardenia rothmannia.
Candy Carrot – Athamanta cretensis.
Candytuft, Bitter – Iberis amara.
Candytuft, White – Iberis amara.
Cane, Large – Arundinaria gigantea.
Cane, Southern – Arundinaria gigantea.
Cane, Tonkin – Arundinaria amabalis.
Cane, Sugar – Saccharum officinarum.

CANELLA P. Br. Canellaceae. 2 spp. S.Florida, W.Indies, tropical America. Trees.

***C. alba*** Murr.=C. winterana. (Wild Cinnamon). S.Florida to W.Indies. The bark (Canella, Bahama White Wood, White Cinnamon, Wild Cinnamon) is aromatic, and is used as a spice, for flavouring tobacco, and to relieve stomach upsets. It contains eugenol, cineol, pinene, caryophyllene, and about 8 per cent manitol.

***C. winterana*** (L.) Gaertn. = C. alba.

Canella – Canella alba.

Canelillo – Croton ciliato-glandulosus.

Canelillo – Ocotea veraguensis.

Cangerana – Cabralea cangerana.

Cangoo Mallee Eucalyptus – Eucalyptus dumosa.

Canistel – Lucuma rivicoa.

CANNA L. Cannaceae. 55 spp. Tropical and subtropical America. Herbs.

***C. bidentata*** Bertol. Tropical Africa. The starchy rhizomes are used as a famine food, locally. The seeds are used as beads, and the leaves for wrapping food.

***C. edulis*** Ker-Gwal. (Purple Arrowroot, Queensland Arrowroot). W.Indies. Cultivated commercially in Australia. Grown locally as food crop in Pacific Islands and W.Indies. The large purple rhizomes contain about 25 per cent starch and are eaten cooked. The starch is washed out and used as an invalid food (African Arrowroot, Queensland Arrowroot, Sierra Leone Arrowroot) which is easily digested.

***C. gigantea*** Desf. Brazil. The rhizome is used locally as a diuretic and to reduce fever.

***C. glauca*** L. Mexico to Brazil. The rhizomes are eaten locally in Brazil and W.Indies.

***C. latifolia*** Rose. = C. gigantea.

***C. sanguinea*** Warsz. = C. warszewiczii.

***C. speciosa*** Rose. W.Africa. Cultivated in Sierra Leone. The tubers (African Turmeric) are used as a colouring and in spice powders.

***C. warszewiczii*** Dietr. = C. sanguinea. Tropical America. The rhizomes are used as a diuretic, and the leaves are used as an emollient.

Canna de Brejo – Caladium striatipes.

Canna de Imbé – Dieffenbachia seguina.

CANNABIS L. Cannabidaceae. 1 sp. C.Asia. Herb.

***C. sativa*** L. (Hemp). Asia. The plant is one of the oldest in cultivation. There are three main groups of varieties, producing fibre, producing seeds, and producing Marihuana. The fibres are more stiff than flax but not affected by water. It is used for cordage, sailcloth and caulking boats. The seeds contain about 33 per cent oil which is used in the manufacture of paints, varnishes, soap and when purified it is used for cooking. The drug (Bhang, Ganja, Hashish, Marihuana) is produced mainly in the Middle East and India, but is becoming more widely used. Its use is illegal in many countries as it is thought to be mentally injurious producing mental weakness and a breakdown of moral understanding. The drug is produced from the resin excreted from the flowers and fruits; the best yield being produced from the unfertilised female flowers.

Cannon Ball Tree – Couroupita guianensis.

Canoe Birch – Betula papyrifera.

Cañon Grape – Vitis arizonica.

Cañon Live Oak – Quercus chrysolepis.

Cañon Palm – Washingtonia filifera.

CANSCORA Lam. Gentianaceae. 30 spp. Old World Tropics. Herbs.

***C. diffusa*** (Vahl.) R. Br. = Exacum diffusum = Gentiana diffusa. (Chang Bato). Tropical Africa and Asia through to Australia. The Philippines use an infusion of the leaves as a tonic and to reduce stomach pains.

Cantala – Agave cantala.

CANTHARELLUS Adans. ex Fr. Agaricaceae. 70 spp. Widespread. The fruitbodies of the species mentioned are eaten.

***C. cibarius*** Fr. (Chanterelle). Temperate countries. Dried for winter use.

***C. cinnabarina*** Schw. N.America.

***C. clavatus*** Fr. Europe.

***C. cornucopiodes*** Fr. = Craterellus cornucopoides.

***C. cyanoxanthus*** Heim. Madagascar.

***C. floccosus*** Schw. Europe.

***C. glutinosus*** Pat. Indochina.

***C. madagascarensis*** Pat. Madagascar.

***C. umbonatus*** Fr. (Graylings). S.Canada, U.S.A. Good for stewing. Dried for winter use.

CANTHIUM Lam. Rubiaceae. 200 spp. Old World Tropics. Small trees.

***C. didymum*** Gaertn f. = Plectronia didyma. E. India. The wood (Ceylon Boxwood) is light brown and was used for cutlery.

***C. glabriflorum*** Hiern = Plectronia glabriflora. W.Africa. The brownish wood is fine grained and is used locally for general carpentry.

***C. lanciflorum*** (Benth. and Hook.) Hiern. = Plectronia lanciflora. S.Africa. The plum-sized fruits are eaten locally. They are considered as one of the best local fruits.

***C. ventosum*** Hiern.=Plectronia ventosa. S.Africa. The leaves are used by the Zulus to cure dysentry.
CANTLEYA Ridl. Icacinaceae. 1 sp. W.Malaysia. Tree.
***C. corniculata*** Becc. (Bedaru). Hard, heavy yellow-brown wood is difficult to work and season. It is used for heavy construction work, and occasionally for furniture.
Canton Lemon – Citrus limonia.
Canton Linen – Boehmeria nivea.
***Cantuffa exosa*** Gmel.=Pterolobium lacerans.
Canutillo – Leptocarpus chilensis.
Canyon Live Oak – Quercus chrysolepus.
Caoba – Sweitenia macrophylla.
Caoba Mahogany – Sweitenia mahogani.
Caoutchouc des Herbes Landophilia thollonis.
Capa Blanco – Petitia domingensis.
Capá Prieta – Cordia gerasanthus.
Caparrosa – Neea theifera.
***Capea elongata*** Mart.=Ecklonia bicyclis.
Cape Aloe – Aloë ferox.
Cape Ash – Ekbergia capensis.
Cape Ash – Harpephyllum caffrum.
Cape Asparagus – Aponogeton distachyus.
Cape Barren Tea – Correa alba.
Cape Beech – Myrsine melanophloea.
Cape Box – Buxus macowanii.
Cape Chestnut – Calodendron capense.
Cape Ebony – Euclea pseudebenus.
Cape Gum – Acacia giraffea.
Cape Gum – Acacia horrida.
Cape Homem – Philodendron imbé.
Cape Lancewood – Curtisia dentata.
Cape Mahogany – Ptaeroxylon utile.
Cape Mahogany – Trichilia emetica.
Cape Periwinkle – Vincia rosea.
Cape Saffron – Sutera atropurpurea.
Cape Teak – Strychnos atherstonei.
Cape Thorn – Ziziphus mucronata.
Cape Willow – Salix capensis.
***Capellenia moluccana*** T. and B.=Endospermum moluccanum.
Caper – Capparis spinosa.
Caper, Simulo – Capparis coriacea.
Caper Spurge – Euphorbia lathyris.
Capiro – Sideroxylon capiri.
Capiroto – Conostegia xalaphensis.
Capitaine Bois – Psittacanthus martinicensis.
CAPPARIS L. Capparidaceae. 250 spp. Warm. Woody shrubs.
***C. albitrunea*** Burch.=Boscia albitrunca.
***C. aphylla*** Roth.=C. decidua. Sahara. Fruits eaten locally.
***C. aurantioides*** Pres.=C. horrida.
***C. canescens*** Banks. Australia, esp. W. Fruits eaten by natives.
***C. coriacea*** Burch. (Simulo Caper). S.Africa. Found also in Peru and Chile. The fruit (Fructus Simulo) is used locally to treat epilepsy, hysteria and as a sedative.
***C. corybifera*** E. Mey. S.Africa. The flowerbuds are eaten locally pickled in vinegar.
***C. cynophallophora*** L.=C. flexuosa.
***C. flexuosa*** L.=C. cynophallophora.= Morisonia flexuosa. (Burro, Palo de burro, Pan y agua, Xpayumak). Tropical America. Used medicinally in W.Indies. The leaves are used to treat skin complaints, and infusion of the roots is used to treat dropsy and to control menstruation. The bark is used similarly, and as a diuretic. The fruits are used as a sedative.
***C. horrida*** L. f.=C. aurantioides=C. linearis=C. micrantha. India to Philippines. The leaves are used locally as a poultice to reduce swellings of boils and piles. The root bark is used to relieve stomach complaints, as a sedative, and to reduce perspiring.
***C. linearis*** Blanco=C. horrida.
***C. micrantha*** DC.=C. odorata. Tropical Asia, especially Burma to Philippines. The fruits are eaten locally, but only when fully ripe. The roasted seeds are used to treat coughs. A decoction of the roots is used as a uterine stimulant after childbirth and to relieve stomach pains.
***C. mitchellii*** Lindl. Australia, esp. W. and Tasmania. The fruits are eaten by the natives.
***C. nobilis*** F. v. Muell. See C. mitchellii.
***C. odorata*** Blanco.=C. micrantha.
***C. spinosa*** L. (Caper Bush). Mediterranean. Cultivated for the flower buds (Capers) which are pickled as a flavour and condiment especially for fish. Much is exported, especially from France.
***C. tomentosa*** Lam. Tropical Africa. The leaves are fed to camels in N.Nigeria.
***Capraria crustacea*** L.=Lindernia crustacea.
CAPSELLA Medik. Cruciferae. 5 spp. Temperate and sub-tropics. Herbs.
***C. bursa-pastoris*** (L.) Medic. (Shepherd's Purse). Europe to China, N.America. The leaves were used in Europe and

China (Herba Bursae Pastoris) to treat eye complaints, to reduce fever, as a diuretic and to stop bleeding of wounds.

CAPSICUM L. Solanaceae. 50 spp. C. and S.America. Herbs.

***C. annuum*** L. (Chile, Cayenne Pepper, Green Pepper, Paprika). Tropical America. There is some confusion over the systematics of this genus. Some authorities place the perennial varieties used as a condiment in a separate sp., ***C. fructescens*** L. The plant was discovered by the early settlers in S.America and is now cultivated where the climate is suitable. The pungent taste is due to the phenolic compound capsaicin, which is present mainly in the seeds, peduncle and placenta. It is a valuable source of vitamin C, and both vitamins A and E are present in appreciable quantities. The Sweet Peppers (var. ***grossum*** Sendt.) contain little caspaicin and are used in salads or stuffed with meat. They are usually green in colour. The red forms (Pimento) are used for flavouring cheeses, stuffing olives, flavouring meat, or canning. The other varieties are used to produce the condiments. These include var. ***longum*** Sendt., var. ***cerasforme*** Bailey, var. ***fasciculatum*** Irish, var. ***conoides*** Bailey, var. ***acuminatum*** Fingh. Paprika is produced from the dried fruits from which the stem stalks and placenta have been removed. Much is produced in Hungary and Spain. Chile pepper is made from the whole fruits, as is Tabasco pepper and Cayenne pepper but from different varieties. The leaves and the prepared condiment are used to produce local warming of the skin especially to relieve rheumatism and muscular strains and soreness. Internally peppers encourage gastric secretions, but in excess may cause inflammation. Dwarf varieties are grown for decoration, especially at Christmas time.

Capucine Pea – Pisum arvense.

Capulin – Eugenia acapulcensis.

Capulin – Muntingia calabura.

Capulin – Prunus capollin.

Capulin – Rhus virens.

Caquelin – Clusia plukenetii.

Carabeen, Grey – Sloanea woolsii.

Carabeen, Yellow – Sloanea woollsii.

Caracolillo – Homalium racemosum.

Caracu – Xanthosoma caracu.

CARAGANA Fabr. Leguminosae. 80 spp. C.Asia, Himalayas, China. Small trees.

***C. arborescens*** Lam. (Siberian Pea Shrub). Siberia. The pods are eaten locally as a vegetable, and a fibre is made from the bark.

***C. chamlagu*** Lam. N.China. The flowers are eaten locally.

Carageen – Chondrus crispus.

Caraguata – Bromelia fastuosa.

Caraguata – Bromelia serra.

Caraguata – Eryngium pandanifolium.

CARAIPA Aubl. Guttiferae. 20 spp. Tropical S.America. Trees.

***C. fasciculata*** Camb. Brazil, Guyana. The trunk yields a resin used locally to treat wounds.

***C. lacerdiae*** Barb. Rodr. Brazil. An oil extracted from the seeds is used locally to treat skin diseases and eye infections.

***C. minor*** Hub. See C. lacerdiae.

***C. palustris*** Barb Rodr. See C. lacerdiae.

***C. psidifolia*** Ducke. (Tamaquare). See C. lacerdiae.

Caralla Wood – Carallia brachiata.

Carámano – Cassia grandis.

CARALLIA Roxb. Rhizophoraceae. 10 spp. Madagascar, Indomalaya, N.Australia. Trees.

***C. brachiata*** Merr. = C. integerima. (Andi, Carallia Wood, Karalli, Palamkat, Shengali). N.E.India. Wet forests. The hard reddish wood is used for construction work, furniture-making, agricultural implements, panelling, flooring, and small ornamental articles. A beverage is made from the leaves in Malacca. An extract from the bark is used locally in various parts to treat ulcers in the mouth, and other infections of the mouth and throat, and skin irritations.

***C. calycina*** Benth. Ceylon. The yellow-red to brown wood is beautifully marbled and is used for furniture-making and panelling. It takes a good polish.

***C. integerima*** DC. = C. brachiata.

CARALLUMA R. Br. Asclepiadaceae. 110 spp. Africa, Mediterranean, to Burma. Small succulents. The species mentioned grow in the Sahara region and are eaten locally.

***C. edulis*** A. Chev. (Aboila, Adouaiba).

***C. europaea*** (Guss.) N. E. Br. = Apteranthes gussoneana.

***C. maroccana*** (Hook. f.) Maire.

Carambola – Averrhoa carabola.

Carana – Protium carana.
Caranda – Carissa carandas.
Caranday Palm (Wax) – Copernica australis.
CARAPA Aubl. Meliaceae. 7 spp. Tropics. Trees.
***C. guineensis*** Aubl. (Andiroba, Crabwood). S.America, W.Indies, W.Africa. The reddish wood is easily worked and durable. It is used for furniture-making, house building, instrument cases. The seeds are the source of a non-drying oil (Andrioba, Carapa, Coondi, Craw-wood Oil, Touloucouna), which is used as an insect-repellent, for lighting, and in the manufacture of soap.
***C. moluccensis*** Lam. = Xylocarpus moluccensis.
***C. obovata*** Blume = Xylocarpus granatum.
***C. procera*** DC. (Demarara Mahogany). Tropics. The dark red wood is heavy and durable. It is used for general carpentry and furniture-making. An extract of the bark is used as a tonic and to reduce fevers. The seeds yield an oil (Touloucouna Oil) which is used to treat yaws, mosquito bites and burns. It is used in S.France in the manufacture of soap.
Carapa Oil – Carapa guineensis.
Caraway – Carum carvi.
Caraway, Black – Pimpinella saxifraga.
Carbo Ligni Depuratus – Fagus sylvatica.
Carbo Ligni Pulveratus – Fagus sylvatica.
Carcara Amarga – Picramnia pentandra.
CARDAMINE L. Cruciferae. 160 spp. Cosmopolitan, esp. temperate. Herbs.
***C. amara*** L. (Bitter Cress, Large Bitter Cress). Europe, Asia Minor. Used locally in salads.
***C. hirsuta*** L. See C. amara.
***C. pennsylvanica*** Muhl. (Bitter Cress). E. U.S.A. Occasionally used locally in salads.
***C. pratensis*** L. (Cuckoo Flower). Europe, temperate Asia. Occasionally used locally as a salad.
***C. yesoënsis*** Max. Japan. The young leaves and rhizomes are eaten locally.
Cardamon – Amomum cardamomum.
Cardamon (Seeds) (Oil) – Elettaria cardamonum.
Cardamon, Bastard Siamese – Amomum xanthioides.
Cardamon, Bengal – Amonium aromaticum.
Cardamon, Cameroon – Aframomum hanbury.
Cardamon, Ceylon – Elettaria cardamomum.
Cardamon, Ceylon – Elettaria major.
Cardamon, Cluster – Elettaria cardamomum.
Cardamon, East African – Aframomum malum.
Cardamon, Ellepy – Elettaria cardamomum.
Cardamon, Java – Amomum maximum.
Cardamon, Madagascar – Aframomum angustifolium.
Cardamon, Madras – Elettaria cardamomum.
Cardamon, Malabar – Elettaria cardamomum.
Cardamon, Mangalore – Eletaria cardamomum.
Cardamon, Nepal – Amomum aromaticum.
Cardamon, Nepal – Amomum subulatum.
Cardamon Oil – Elettaria cardamomum.
Cardamon, Round – Amomum kepulaga.
Cardamon, Round Chinese – Amomum globosum.
Cardamon, Siam – Elettaria cardamomum.
Cardamon, Wild Siamese – Amomum xanthioides.
CARDARIA Desv. Cruciferae. 1 sp. Mediterranean, W.Asia. Herb.
***C. draba*** (L.) Desv. = Lepidium draba. Introduced into Europe and N.America. The leaves are used in Afghanistan as a vegetable. The seeds were used as a condiment.
***Cardiaca vulgaris*** Moench. = Leonurus cardiaca.
Cardinal, Red – Erythrina arborea.
Cardinal Wood – Brosimum paraense.
***Cardiogyne*** Bur. = Maclura.
Cardón – Pachycereus pringlei.
Cardón hecho – Pachycereus pecten-aboriginum.
Cardon-pelon – Pachycereus pringlei.
Cardoon – Cynara cardunculus.
CARDUUS L. Compositae. 100 spp. Europe, Mediterranean, Asia. Herbs.
***C. bulbosum*** DC. = Cirsium tuberosum.
***C. edulis*** (Nutt.) Greene. = Cirsium edule.
***C. marianum*** L. = Silybum marianum.
***C. nutans*** L. (Musk Thistle). Europe to Siberia, naturalized in N.America. The pith is eaten when boiled. The flowers are used to curdle milk.

***C. ochrocentrus*** (A. Gray) Greene. = Cirsium ochrocentrum.

***C. oleraceus*** Vill. = Cirsium oleraceum.

***C. undulatus*** Nutt. = Cirium undulatum.

CARDWELLIA F. V. Muell. Proteaceae. 1 sp. Queensland. Tree.

***C. sublimis*** F. v. Muell. (Silky Oak). The pale-pink, silky wood is soft and light, but works well and carves easily. It is used for furniture, veneers, carvings, general building, floors, railway carriages and coachbuilding.

CAREX L. Cyperaceae. 1500–2000 spp. Temperate. Herbs.

***C. acuta*** Ali. = C. riparia

***C. acutifolia*** Ehrh. Europe, Asia, N.America. Used for straw in stables etc.

***C. arenaria*** L. = C. spadicea. Coastal Europe, Asia, N.America. The rhizome (Rhizoma Caricis arenariae, German Sarsaparilla) is used in local medicine as a diuretic and to reduce fever.

***C. atherodes*** Spreng. N.America. Used as a hay grass.

***C. brizoides*** L. (Crin Végétal, Waldhaar). C. and S.Europe. Used, especially in Austria, as a packing material.

***C. buxbaumii*** Wahlenb. Europe, Asia, N.America. Used for straw in stables etc.

***C. dispalatha*** Boott. Japan. Cultivated in the rice fields of Japan and used to make straw hats.

***C. disticha*** Huds. Europe, Asia, N.America. Used as straw for stables.

***C. hirta*** L. Europe, temperate Asia, N.Africa. The rhizome (Rhizoma Caricis hirtae) is used locally as a diuretic.

***C. paniculata*** L. Europe, N.America. Used as straw in stables.

***C. riparia*** Curtis. = C. acuta = C. versicaria. Europe, Caucasia, E.Asia, N.Africa. Used as straw in stables.

***C. rostrata*** Stokes. Europe, Asia, N.America. Used as straw in stables.

***C. spadicea*** Gil. = C. arenaria.

***C. stricta*** Good. Europe, Asia, N.America. Used as straw in stables.

***C. versicaria*** Leers. = C. riparia.

CAREYA Roxb. Barringtoniaceae. 4 spp. Indomalaysia. Trees.

***C. sphaerica*** Roxb. (Kandol, Knung, Vung). India to Indo-China and Thailand. The fruits are eaten locally in India. In the Far East the bark is used to treat diarrhoea, and the leaves and twigs are the source of a red dye used to stain cotton fabrics.

Cargo – Cassia grandis.

Carib Grass – Eriochloa polystachya.

Caribbean Princewood – Exostema caribaeum.

CARICA L. Caricaceae. 45 spp. Warm America. Trees.

***C. candicans*** Gray. Peru. The fruits are fibrous, but pleasant tasting, and are eaten locally.

***C. cestriflora*** Solms. = Vasconcella cestriflora. (Papaya de Terra Fria). Columbia. The fruits are made into preserves.

***C. chrysophylla*** Helib. (Higicho). Ecuador. The fruits are used in preserves. The plant is cultivated.

***Carica papaya*** L. (Papaya, Pawpaw, Melon Tree, Mamão, Fructa bomba, Lechosa, Melon zapote). Tropical S.America, W.Indies. The plant is cultivated throughout the tropics, and is an important crop in Hawaii. The young fruits are eaten as a vegetable, but more usually it is a dessert fruit, used in jellies, preserves of all sorts, candies and eaten raw. The plant produces a latex which is extracted by cutting the bark. It contains the proteolytic enzyme papain. The latex is coagulated in earthenware vessels (metal denatures the enzyme) covered to prevent decolourization. It is then sun-dried or dried over a fire, but must not be overheated, stirring all the while. The latex is a commercial source of the enzyme which is used in the production of digestive medicines, canned meat, leather tanning, and shrink-resisting woollen fabrics. The enzyme is also used to prevent cloudiness in chilled beer. The plants are propagated by seeds (no other method has been yet used successfully) which leads to a great variation between the yield and quality of individual plants. The plants begin to yield in their first year but they yield only for a few years.

***C. peltata*** Hook. and Arn. (Papaya de Mico). C.America. The small fruits are eaten locally.

***C. pentagona*** Heilb. (Babaco). Ecuador. Cultivated locally. The fruits are acid and are eaten cooked or as preserves.

***C. quercifolia*** (St. Hil.) Solms. Laub. (Higuerra del Monte). S.America, esp. N. Fruits as preserves or candied.

CARINIANA Casar. Lecythidaceae. 13 spp. Tropical S.America. Trees.

***C. pyriformis*** Meirs. (Albarco, Bacu). N. S.America. The wood (Columbian Mahogany) is made into veneers. It contains a considerable amount of silica. The bark is made into ropes by the local Indians.

*Caripé* – Moquilea utilis.

CARISSA L. Apocynaceae. 35 spp. Warm Africa and Asia. Small Shrubs.

***C. arduina*** Lam. = C. bispinosa = Arduina bispinosa. (Amatungula). S.Africa. The fruits are eaten locally.

***C. bispinosa*** (L.) Desf. = C. arduina.

***C. brownii*** F. v. Muell. Australia. The fruits are eaten by the aborigines in N.Queensland.

***C. carandas*** L. (Caranda). Small tree. India to Malaysia. The red, plum-like berries are eaten locally and made into jellies and preserves.

***C. grandiflora*** A. DC. (Natal Plum). Natal. Cultivated in Old and New World for the large red berries, which are used to make jellies when unripe, and pies when ripe.

***C. ovata*** R. Br. S.Australia. Fruits eaten locally by the aborigines.

CARLINA L. Compositae. 20 spp. Europe, Mediterranean, Asia. Herbs.

***C. acaulis*** L. (Carline Thistle). Mountainous Europe. An extract of the root was used as a purgative, to relieve stomach upsets, to induce sweating, and to control menstruation.

Carline Thistle – Carlina acaulis.

CARLUDOVICA Ruiz. and Pav. Cyclanthaceae. 3 spp. C. and N.W. tropical S.America. Palm-like plants.

***C. angustifolia*** Ruiz. and Pav. Peru. Leaves used locally for thatching.

***C. divergens*** Drude. Brazil, Peru. The leaves are used as cordage in Peru.

***C. labela*** Schult. Mexico. The leaves are used for thatching locally.

***C. palmata*** Ruiz. and Pav. (Panama Hat Palm, Palmita, Toquilla). C.America to N.W. S.America. The leaves are made into Panama Hats (Toquilla Hats), roofing, matting, brooms, baskets etc. The fibre for the hats has to be of the finest quality, being durable, pliable, waterproof and washable. The production of hats is an important industry in Ecuador. The poorer quality fibres are used for other purposes.

***C. sarentosa*** Sagot. Guiana. Used locally to make brooms.

Carmin – Phytolacca chilensis.

Carnation – Dianthus caryophyllus.

Carna-uba Palm – Copernicia cerifera.

Carnauba, Paraguayan – Copernicia australis.

Carnauba Wax – Copernicia cerifera.

CARNEGIEA Britton and Rose. Cactaceae. 1 sp. S.W. United States of America, Mexico. Cactus.

***C. gigantea*** (Engelm.) Britt. and Rose. = Cereus giganteus. (Giant Cactus, Saguaro). S.Arizona to Mexico. One of the largest cacti. The fruits are eaten by the Indians, raw, cooked or preserved. A syrup from the fruit is made into an intoxicating beverage. The seeds are made into a paste which is used as a butter. The woody ribs from the stems are used to make frames for huts, and lances.

Carnero de la costa – Coccoloba schiedeana.

Caro – Vitis sicyoides.

Caroa Verdadeira – Neoglaziovia variegata.

Carob (Gum) – Caratonia siliqua.

Caroba – Jacaranda porcera.

Carobbe di Guidea – Pistacia terebinthus.

Carolina Allspice – Calycanthus floridus.

Carolina Gromwell – Lithospermum carolinense.

Carolina Horse Nettle – Solanum carolinense.

Carolina Pink – Phlox carolina.

Carolina Tea – Ilex vomitoria.

***Carolinea macrocarpa*** Schlecht and Cham. = Pachira macrocarpa.

Carongang – Cratoxylum arborescens.

Carosella – Foeniculum vulgare.

Carpathian Turpentine – Pinus cembra.

Carpet Grass – Axonopus compressus.

CARPINUS L. Carpinaceae. 35 spp. N.Temperate. Small trees. The wood of all the species mentioned is hard and close-grained, yellow-white.

***C. betulus*** L. (European Hornbeam). Europe to Asia Minor. The wood is used for turning, tools, wagon wheels etc.

***C. caroliniana*** Walt. (America Hornbeam, Blue Beech). N.America to C.America. The wood is used for small articles, tool-handles, etc. and fuel.

***C. cordata*** Blume. (Heartleaf Hormbeam). Japan. Used for turning, etc.

***C. laxiflora*** Blume. Japan, Korea. Used

locally for carving, turning, furniture-making and ski-making.

CARPODINUS R. Br. ex G. Don. Apocynaceae. 50 spp. Tropical Africa. Trees. The species mentioned are sources of rubber, extracted from the bark.

***C. chylorrhiza*** Schum. Tropical Africa.

***C. congonlensis*** Stapf. Tropical Africa.

***C. gracilis*** Stapf. Angola, Congo.

***C. hirsuta*** Hua. W.Africa. Rubber known as Accra Paste, or La Glu.

***C. jumellei*** Pierre. Congo.

***C. lanceolata*** K. Schum. Angola.

***C. landolphioides*** Stapf. Cameroons.

***C. leucantha*** K. Schum. Angola.

***C. maxima*** K. Schum. Congo, Cameroons.

***C. uniflora*** Stapf. Cameroon.

CARPOLOBIA G. Don. Polygalaceae. 10 spp. Tropical W.Africa. Shrubs and trees.

***C. alba*** Auct. non G. Don.=C. glabrescens.

***C. glabrescens*** Hutch.=C. alba. (Bizimi, Bontseke, Kamvula). Congo, Angola. The fruits are eaten locally.

***C. lutea*** G. Don. Tropical Africa, esp. Nigeria. The wood is hard and termite resistant. It is used for house construction and walking sticks.

CARPOTROCHE Endl. Flacourtiaceae. 15 spp. Tropical America. Trees. The following spp. are found in Brazil. The seeds are the source of Carpotroche Oil, Oleo de Sapucainha. The oil is used to treat leprosy, and dermatosis.

***C. amazonica*** Benth.

***C. brasiliensis*** Endl. The main source of the oil.

***C. denticula*** Benth.

***C. grandiflora*** Spruce.

***C. laxiflora*** Benth.

***C. longifolia*** Benth.

Carpotroche Oil – Carpotroche brasiliensis.

Carrageen – Chondrus crispus.

Carageenin – Chondrus crispus.

Carragheen – Gigartina mamillosa.

Carragheen – Gigartina stellata.

Carreto – Aspidosperma curranii.

Carrobeen, Grey – Sloanea woollsii.

Carrot (Oil) – Daucus carota.

Carrot, Candy – Athamenta cretensis.

Carrot, Gargan Death – Thapsia garganica.

Carroway, Black – Pimpinella saxifraga.

Cartagena Ipecac – Cephaelis acuminata.

Cartán – Centrolobium orinocense.

CARTHAMUS L. Compositae. 13 spp. Mediterranean, Africa, Asia. Herbs.

***C. oxycanthus*** Bieb. (Kantiari, Kara, Poli, Wild Safflower). Dry N.India and Pakistan. Cold expression of the seeds gives Poli Oil which is used for cooking, for lighting and in the manufacture of Macassar Hair Oil. It is also used for waterproofing tents, tarpaulins etc. Roghum Oil is extracted by hot compression and is used for the manufacture of waxcloth, and locally for greasing ropes. It cements glass and stone.

***C. tinctoria*** L. (Safflower). Mediterranean, Asia Minor. The main cultivations are in Bombay and Hyderabad. The plant has been introduced into the United States of America, Canada, Australia, Israel and Turkey. It has been cultivated since ancient times for the flowers which yield a yellow or red dye carmanthin, used to colour butter, liqueurs, candles, and when mixed with talc is used in rouge. An oil is extracted from the seeds by dry, hot pressing. It is used in varnishes, paints, and linoleum. The fruits are edible, and are fed raw to poultry. When fried they are used in chutney. The seeds cake is a good cattle food.

Carua Palm – Orbignya spectabilis.

Caruba de Castilla – Passiflora mollissima.

CARUM L. Umbelliferae. 30 spp. Temperate, and subtropics. Herbs.

***C. capense*** Sund. (Venkelwortel). S.Africa. The roots are eaten locally.

***C. carvi*** L.=Apium carvi=Seseli carvi. (Caraway). Mediterranean to Himalayas. Cultivated N. and S. Europe, Morocco. The seeds are used to flavour meat, sausage, bread, cheese etc. An essential oil (Oil of Caraway, Oleum Carvi) which contains carvone, is distilled from the fruits, and is also used for flavouring. The seeds and oil are used medicinally to aid digestion, as a stimulant, diuretic, and to reduce fevers. The roots are edible.

***C. copticum*** (L.) Benth. and Hook.=Ammi copticum. Trachyspermum copticum. (Ammi). India. Cultivated locally. Thymol is extracted from the plant and it is used as a condiment. Ajowan Oil is extracted from the seeds and used as an antiseptic and to aid digestion. Omum water is distilled from the seeds, and is used in local medicine.

***C. gairdneri*** (Hook. and Arn.) Gary. (Yampa). W. N.America. The roots are eaten raw by the local Indians. They are also preserved for winter use.

***C. keloggii*** Gray. See C. gairdneri.

***C. nigrum*** Baill.=Pimpinella saxifraga.

***C. roxburghianum*** Benth.=Athamantha roxburghiana. India. Cultivated in India, Ceylon and Far East for the seeds which are used as a condiment especially in curries. They aid digestion.

***Carumbium amboinicum*** Miq.=Pimelodendron amboinicum.

CARYA Nutt. Juglandaceae. 25 spp. E.Asia, N.America. Trees.

***C. alba*** Nutt.=C. ovata.

***C. alba*** Koch not Nutt.=C. tomentosa.

***C. amara*** Nutt.=C. cordiformis.

***C. cathayensis*** Sarg. (Chinese Hickory). E.China. The nuts are eaten locally, made into confectionery. The tough wood is used locally for tool handles.

***C. cordiformis*** (Wangh.) Koch.=C. amara =Hicoria minima Brit. (Bitternut). N.America. The heavy tough wood is used for yokes and hoops.

***C. glabra*** (Mill.) Sweet.=C. porcina= Hicoria glabra. (Pignut Hickory). E. N.America. The dark, heavy flexible wood is used for making agricultural implements, wagons, tool-handles, and burning. The nuts are edible, but are of variable quality.

***C. illinoensis*** Koch.=C. pecan.

***C. lacinosa*** (Michx. f.) Loud.=C. sulcata= Hicoria laciniosa (Big Shellbark Hickory). N.America. The wood is used for agricultural implements, and tool handles. The nuts are edible.

***C. oliviformis*** Nutt.=C. pecan.

***C. ovata*** (Mill.) Koch.=C. alba=Hicoria ovata. (Shagbark Hickory, Shellbark Hickory). N.America. The hard, tough heavy wood is used to make agricultural implements, tool handle and wagons. The nuts (Hickory Nouts) are edible and are sold commercially.

***C. pecan*** (Marsh.) Engl. and Graebn.=C. illinoensis=C. oliviformis=Hicoria pecan. (Pecan). E. and S. U.S.A. The tree is cultivated commercially, especially in S. U.S.A. for the thin-shelled nuts. They are used in confectionery, cakes, etc. or are eaten raw or salted. The oil (Pecan Oil) extracted from the seeds is edible, but is used mainly in the manufacture of cosmetics, and some drugs. Between 100 million and 200 million pounds of nuts are produced annually in the U.S.A. The coarse, brittle wood is not strong, but is used occasionally to make tool handles, agricultural implements and as a fuel.

***C. porcina*** (Michx. f.) Nutt.=C. glabra.

***C. sulcata*** Nutt.=C. laciniosa.

***C. tomentosa*** (Lam.) Nutt.=C. alba (Bigbud Hickory, Mockernut). E. N.America. See C. ovata.

CARYOCAR Allam. ex L. Caryocaraceae. 20 spp. Tropical America. Trees.

***C. amygdaliferum*** Mutis. S.America. The seeds are eaten locally in Ecuador, and are a source of oil (Sawarri Fat, Souari Fat) which is used locally in cooking.

***C. brasiliensis*** Camb. Brazil. The fruits are eaten with meat by the local Indians.

***C. butrospermum*** Willd. Tropical America. The seeds are oily, and are eaten locally.

***C. glabrum*** Perr. (Soapwood). Guyana. The inner bark is used locally as a soap for washing clothes.

***C. nuciferum*** L. Brazil, Guyana. Cultivated in the W.Indies for the nuts (Butter Nut, Guiana Nut, Paradise Nut, Souari Nut), from which an edible oil is extracted. It is exported. The wood is durable and is used for ship-building.

***C. tomentosum*** Willd. (Souari Tree). Guyana. The nuts (Souari Nuts) are eaten locally.

***Caryophyllum aromaticum*** L.=Eugenia caryophyllata.

CARYOTA L. Palmae. 12 spp. Ceylon, Indomalaysia, Solomon Islands, N.E.Australia. Palms.

***C. mitis*** Lour. Tropical Asia, especially India and Malaysia. The young leaves are eaten locally as a vegetable. The stems are a source of Sago. The fibres from the leaf-bases are used locally to cauterise wounds and as the flights of darts.

***C. sympetala*** Gagnep. (Dûng dinh). The centres of the stems are edible.

***C. urens*** L. (Fish-tail Palm, Kittul Palm). India to Ceylon. The old stems are the source of Sago. The leaves are the source of a coarse fibre used to make brushes. The very young leaves are eaten locally. The juice from the trunk is a local source of sugar and is fermented to make an alcoholic beverage.

Casa Banana – Sicana odorifera.

Casaba – Cucumis melo.

Casabe – Manihot esculenta.

Cascade Fir – Abies amabalis.

Cascara Amarga – Sweetia panamensis.

Cascara Buckthorn – Rhamnus purshiana.

Cascara de Linge – Persea lingue.
Cascara Sagrada – Rhamnus purshiana.
Cascarilla – Cinchona pubescens.
Cascarilla Bark – Croton cascarilla.
Cascarilla Bark – Croton eluteria.
Cascarilla Crespilla – Cinchona officinalis.
Cascarilla Delgada – Cinchona pubescens.
Cascarilla, Industrial – Croton echinocarpus.
Cascarilla Provinciana – Cinchona micrantha.
Cascarilla Verde – Cinchona micrantha.
Cascarilla Verde – Cinchona officinalis.
Cascarille Bark – Croton eluteria.
Cascarille, Industrial – Croton echinocarpus.
Cascarón – Crataeva tapia.
Casea Virgindade – Stryphnodendron barbatimam.

CASEARIA Jacq. Flacourtiaceae. 160 spp. Tropics. Trees.

***C. alba*** Rich. Cuba. Yellow strong wood is used for cabinet making.

***C. dolichophylla*** Standl. (Chililllo). Sinaloa to Mexico and Nicaragua. The branches are used in Sinaloa to make bird cages.

***C. hirsuta*** Sw. W.Indies. The soft wood is used to make furniture.

***C. praecox*** Griseb. = Gossypiospermum praecox.

***C. pringelei*** Brig. (Ciruela, Crementinillo). Sinaloa to Mexico. The arils are pleasantly flavoured, and eaten locally.

***C. singularis*** Eichl. Tropical America. An extract of the roots is used locally to treat leprosy and wounds. An oil extracted from the seeds is used to treat leprosy.

***C. sylvestris*** Sw. Tropical America. An extract of the roots (Guassatonga) is used to treat leprosy and wounds. The seeds are the source of an oil used to treat leprosy.

***C. tomentosa*** Roxb. (Bheri, Chilara, Haili). Himalayan region to C.India. The wood is used for carving and making small articles. The fruits are used as a fish poison. The bark is an adulterant of the red dye Kamela, from Mallotus philippensis.

Cashawa Gum – Anacardium occidentale.
Cashew – Anacardium occidentale.
Cashew Apple – Anacardium occidentale.
Cashew Gum – Anacardium occidentale.
Cashew Marking Nut Tree – Semecarpus anacardium.
Cashew Nut, Oriental – Semecarpus anacardium.
Cashon – Pycnanthus kombo.

CASIMIROA La Llave. Rutaceae. 6 spp. C.America. Trees.

***C. edulis*** La Llave. (White Sapote, Zapote Blanco). C.America. The tree is cultivated locally for the fruit. The fruit has an excellent tart, aromatic flavour and is about the size of an apple. It is eaten raw or used in soft drinks. The plants are propagated from seed, and there is much variation in the quality of the fruit from individual trees.

Casis – Ribes nigrum.
Caspa Tea – Cyclopia subterenata.
Caspic Willow – Salix acutifolia.
Cassada – Diphlois salicifolia.

CASSANDRA D. Don. Ericaceae. 1 sp. N.Temperate. Shrub.

***C. calyculata*** (L.) Moench. N.Temperate. An infusion of the leaves is used as a beverage by some tribes of N.American Indians.

Cassareep – Manihot esculenta.
Cassava – Manihot esculenta.
Cassava, Bitter – Manihot esculenta.
Cassava, Sweet – Manihot esculenta.
Cassena – Ilex cassine.
Cassena – Ilex vomitoria.

CASSIA L. Leguminosae. 500–600 spp. Tropical and warm Temperate. Shrubs and small trees.

***C. absus*** L. Throughout Tropics. Herb. The seeds are used in S.Africa to treat eye-complaints and ringworm. In India they are used as a cathartic.

***C. acutifolia*** Del. Red Sea area. The dried leaves and ripe pods are the source of Alexandrian Senna, used as a laxative. The plant is cultivated in N.Nigeria, French W.Africa, and Timbuctoo. The laxative principles are sennoside A and B.

***C. alata*** L. (Candalabra Bush, Ringworm Senna). Throughout tropics. The leaves contain chrysophanic acid which makes the juice useful in treating skin diseases. An infusion of the roots is used to treat rheumatism and as a strong laxative in Guatemala. The seeds are used to treat intestinal worms.

***C. angustifolia*** Vahl. = C. medicinalis. From Arabia to Punjab. The leaves are a source of senna, used as a laxative. The wild type is known as Arabia or Mecca Senna. The plant is cultivated in S.India when the product is called Tinnevelly Senna.

***C. aphylla*** Cav. Argentina. The stems are used locally in the manufacture of brooms.

***C. arborescens*** Mill. = C. emarginata.

***C. auriculata*** L. Tropical Asia. The bark (Tangedu Bark), is used locally for tanning.

***C. chamaecrista*** Chapm. = Chamaecrista fasciculata. (Partridge Pea). E. U.S.A. Herb. The flowers are a good source of honey.

***C. crassiramea*** Benth. Argentina. The stems are used locally to make brooms.

***C. emarginata*** L. = C. arborescens. (Alcaparro, Brucha macho, Flor de San José, Palo de zorrello). C.America, W.Indies. The wood is used locally as the source of a dye in Jamaica. The leaves are used locally for treating insect stings.

***C. fistula*** L. (Purging Cassia, Pudding Pipe Tree). Tropical Africa. Cultivated in Old and New World Tropics for the pods, the pulp of which is used as a laxative (Cassia Fistula, Cassia Pods, Indian Laburnum, Purging Cassia).

***C. florida*** Vahl. = C. siamea.

***C. fruticosa*** Koen. = C. surattensis.

***C. glauca*** Lam. = C. surattensis.

***C. grandis*** L. f. (Cañafistula, Caramáno, Cargo). C. and S.America, W.Indies. The pulp from the pods is laxative. It is used locally in Mexico to treat fever.

***C. holosericea*** Fresen. Abyssinia. The pods and leaves are used as a laxative (Aden Senna).

***C. javanica*** L. (Bombondelen, Tanggooli). Indonesia, esp. Java. The pods are used as a laxative. The wood is beautifully grained and is used locally for house construction. The bark is used for tanning.

***C. laevigata*** Willd. (Smooth Senna). Throughout Tropics. In Guatemala the seeds are used as a coffee substitute. In Mexico they are used to control menstruation.

***C. leschenaulthiana*** DC. India and Indonesia. Used as a green manure.

***C. marilandica*** L. American Senna. E. U.S.A. The leaves are used as a laxative.

***C. medicinalis*** Bisch. = C. angustifolia.

***C. mimosoides*** L. Tropical Asia and Africa. The leaves are used as a tea substitute in Japan.

***C. moschata*** H.B.K. S.America. The leaves are used as a laxative locally in Columbia.

***C. nodosa*** Ham. (Liring, Siboosook, Sooling). W.Malaysia. Cultivated for the wood which is dark and hard. It is used for posts, etc. and tool handles. The roots are used as a soap for washing clothes.

***C. obovata*** Collad. Tropics. The leaves are used as a laxative (Dog Senna, Italian Senna).

***C. occidentalis*** L. (Coffee Senna). Tropics. The seeds are used as a coffee substitute (Mogdad Coffee, Negro Coffee), and as a medicine to treat a wide variety of complaints. These include stomach upsets, fever, dropsy, rheumatism and ringworm. They are also used as a tonic, to reduce fevers and as a diuretic.

***C. ornithopoides*** Lam. = C. sericea.

***C. oxyphylla*** Kunth. C.America. The leaves are used locally as an emetic.

***C. palellaria*** DC. Tropical America. It is used as a green manure in Indonesia.

***C. pumila*** Lam. (Entjeng-e ent tjeng). Annual. Indonesia. It is used as a green manure, especially in plantation crops.

***C. sericea*** Sw. = C. ornithopoides. Tropical America. The seeds are used as a coffee substitute in Brazil, and the leaves are used as wound-dressings. An extract of the roots is used to treat dropsy.

***C. siamea*** Lam. = C. florida. (Djoowar). India to Indonesia. Cultivated for the timber which is hard and black. It is used for heavy construction work, mine props, etc. and as a fuel.

***C. sieberiana*** DC. (Ratu). Tropical Africa. The heavy timber is termite-resistant and is used for carpentry and house construction. The pods are used to intoxicate fish. The roots are used to make a diuretic in Gambia.

***C. singueana*** Del. (Mukupachiwa). Rhodesia, Malawi, Zambia. An extract of the bark is used by native witch doctors.

***C. sophera*** L. Tropics. An extract of the leaves, roots and bark is used as a laxative. The leaf extract is also used to treat ringworm.

***C. surattensis*** Burm. f. = C. fruticosa. = C. glauca. (Kembang, Kooning). Indonesia. The leaves are eaten locally as a vegetable. An extract of the roots is used locally to treat gonorrhoea.

***C. timoriensis*** DC. (Eheng, Haringhing, Kajoo pelen). Tropical Asia to Australia. The dark tough wood is much used locally to make axe handles.

***C. tora*** L. (Sickle Senna). Old and New World Tropics. In India the leaves are used to treat skin diseases, including ringworm; the plant is cultivated for the seeds which are used as a mordant in dyeing cloth. In Mexico the seeds are used as a coffee substitute.

Cassia (Buds) (Oil) – Cinnamomum cassia.

Cassia Cinnamon – Cinnamomum cassia.

Cassia, Indian – Cinnamomum tamala.

Cassia Pods – Cassia fistula.

Cassia Purging – Cassia fistula.

Cassie Ancienne (Oil) – Acacia farnesiana.

Cassie Flowers – Acacia farnesiana.

Cassie Romaine – Acacia cavenia.

CASSIOPE D. Don. Ericaceae. 12 spp. Polar regions, Himalayas. Shrubs.

***C. tetragona*** (L.) D. Don. Arctic. Used as a fuel by the Eskimoes.

Cassumunar Ginger – Zingiber cassumunar.

CASSYTHA L. 20 spp. Old World Tropics. Semi parasites. Vines.

***C. filiformis*** L. Old World Tropics. An extract of the stems is used as a brown dye in E.Africa.

CASTANEA Mill. Fagaceae. 12 spp. N.Temperate. Trees.

***C. argentea*** Blume. = Castanopsis argentea.

***C. crenata*** Sieb. and Zucc. (Japanese Chestnut). Japan. The wood is hard, heavy and durable. It is used for furniture-making, railway sleepers and ship-building. The nuts are eaten locally.

***C. davidii*** Dode. = C. seguinii.

***C. dentata*** (Marsh.) Borkh. (American Chestnut). E. N.America. The timber is greyish-brown, splits easily and is soft to work. Although it warps during drying it is useful especially for furniture-making. It is also used for railway sleepers, posts, boxes etc. The wood contains about 10 per cent tannin, which is extracted from wood chipping by treatment with hot water. The leaves contain tannin and are used occasionally as a tonic. The fruits are edible.

***C. henryi*** Rhed. and Wils. (Henry Cinquapin). C.China. The wood is much valued locally for building. The fruits are of good eating quality.

***C. javanica*** Blume. = Castanopsis javanica.

***C. mollissima*** Blume. (Chinese Chestnut) N. and W.China. The nuts are small but of good eating quality. It has been introduced to N.America where varieties with larger fruits have been bred.

***C. pumila*** (L.) Mill. (Chinaquapin). E. U.S.A. to Texas. The wood is strong and light and is used for railway sleepers and fencing. The fruits are edible and are sometimes sold locally. The root is astringent and is occasionally used to treat fevers.

***C. sativa*** Mill. = C. vesca. (Spanish Chestnut). Mediterranean to Corsica. It is cultivated mainly for the fruits which are eaten roasted or boiled (Marrons). They are made into a flour (Farine de Châtaignes), made into soup, fried in oil (Paltenta, Pattoni, Nicoi), and used in confectionery (Marron Glacé). The wood is fairly hard and durable. It is used for general carpentry, and railway sleepers. Some is used for the manufacture of cellulose.

***C. seguinii*** Dode. = C. davidii. (Chinese Chinquapin). E. and C.China. Small tree. The fruits (Mae pan-li) are good flavoured and eaten locally.

***C. sumatrana*** Oest. = Castanopsis sumatrana.

***C. vesca*** Gaertn. = C. sativa.

Castanha de Cotia Oil – Aptandra spruceana.

Castanha de Maranháo – Sterculia chicha.

Castanha Oil – Telfairia pedata.

CASTANOPSIS (D. Don.) Spach. Fagaceae. 120 spp. Tropical and subtropical Asia. Trees.

***C. argentea*** A. DC. = Castanea argentea Blume. Burma, Malaya. The wood is strong and durable. It is used locally for building. The bark is the source of a black dye.

***C. boisii*** Hick. and Camus. Tonkin. The nuts are eaten locally.

***C. chrysophylla*** A. DC. (Golden-leaved Chestnut, Golden Chinquapin). Washington to California. The soft reddish wood is sometimes used to make agricultural implements. The nuts are eaten locally by the Indians.

***C. inermis*** Benth. and Hook. = C. sumatrana.

***C. javanica*** A. DC. = Castanea javanica. W. Malaysia. The heavy wood is used locally for building.

***C. philippensis*** Vid. Philippines. The nuts are eaten locally.

***C. sclerophylla*** Schott. and Kotschy. E. and C. China. The nuts are eaten locally.

***C. sumatrana*** A. DC. = C. inermis =

Castanea sumatrana. Malay peninsula, Sumatra. The nuts are eaten locally, especially in Sumatra. They are boiled or roasted.

***C. tibetiana*** Hance. Tibet, China. The nuts are eaten locally.

CASTANOSPERMUM A. Cunn. Leguminosae. 1 sp. Subtropical Australia. Tree.

***C. australe*** A. Cunningh. (Black Bean, Moreth Bay Chestnut). The wood is dark, almost black, durable, fairly hard, resistant to decay, and polishes well. It is used for a wide variety of construction work, plywood and veneer. As it is highly decorative it is used for carving and many forms of fancy work. The beans are eaten by the natives, roasted or made into flour (Bean Tree Starch).

CASTELA Turp. Simaroubaceae. 13 spp. W.Indies, C. and S.America, Galapagos Islands. Shrubs.

***C. nicholsonii*** var. ***texana*** Torr. and Gray= C. texana.

***C. texana*** (Torr. and Gray) Rose. (Amargoso, Bisbirinda, Chaparro amargosa). S.W.Texas. An infusion of the bark is used locally to treat eczema, fevers and digestive disorders.

Casthana do Pará – Bertholletia excelsa.

CASTILLA Cerv. Moraceae. 10 spp. Tropical America. Cuba. Trees. All the species mentioned are good sources of rubber, but are of little commercial importance, since the expansion of Hevea rubber production. The latex is found in latex tubes in the stem from which it is extracted by tapping.

***C. costaricana*** Liebm. Costa Rica.

***C. elastica*** Cerv. (Arbol del Hule, Caucho). S.Mexico to C.America. The rubber is used locally to make raincoats. The bark is sometimes used as a cloth by the natives. It was planted commercially in America, and was the main source of exported rubber. But as the rubber was usually extracted by felling the trees it was soon replaced by Hevea. Some plantations are still operative.

***C. lactiflora*** Pittier. Mexico.

***C. nicoyensis*** Pittier. Costa Rica.

***C. ulei*** Warb. (Cauchu). Amazon Brazil.

***Castilloa*** Endl.=Castela.

Castillian Malva – Malva crispa.

Castor (Oil) (Bean) – Ricinus communis.

CASUARINA Adans. Casuarinaceae. 45 spp. Tropical E.Africa, Mascara, S.E.Asia, Malaysia, Australia, Polynesia. Trees.

***C. collina*** Poisson. New Caledonia. The very hard, close-grained wood is used locally for turning and wagon-making.

***C. cunninghamiana*** Miq. New South Wales, Queensland. The hard wood is used locally for roof shingles, and fencing staves.

***C. deplancheana*** Miq. New Caledonia. The hard yellowish wood is used locally for making wagons.

***C. equisetifolia*** L. (Beef Wood, Bull Oak, Swamp Oak). New South Wales, Queensland, N.Australia, Pacific Islands. Naturalized in Florida. The wood is beautifully marked, but coarsed-grained. It is used for roof shingles, and posting. The tree is cultivated in U.S.A.

***C. junghuhniana*** Miq.=C. montana. Java. The wood is dark, and used for general carpentry.

***C. montana*** Jungh.=C. junghuhniana.

***C. rumphiana*** Miq. Moluccas. The beautifully marked wood is used locally for roof shingles and fencing posts.

***C. stricta*** Ait. (River Oak, Shingle Oak). Australia. The wood is tough but not durable. It is used for furniture, roof shingles, turning and agricultural implements.

***C. suberosa*** Otto and Dietr. (River Black Oak, Swamp Oak). Tasmania, New South Wales, Queensland. The bark is used for tanning.

***C. sumatrana*** Jungh. Malaysia. The hard tough wood is used for tool handles.

***C. torulosa*** Ait. (Forest Oak, River Oak). New South Wales, Queensland. The pleasantly grained wood is used for furniture making, and veneers.

Cat Brier – Smilax rotundifolia.

Cat's Claw – Acacia greggii.

Cat's Claw – Pithecellobium unguis-cati.

Cat's Ear – Hypochoeris maculata.

Cat Mint – Nepeta cataria.

Cat-tail Millet – Pennisetum typhoideum.

Cat-tail, Narrow-leaf – Typha angustifolia.

Cata Grande – Tabernaemontana angloensis.

Catalonian Jasmine – Jasminium grandiflorum.

CATALPA v. Wolf. Bignoniaceae. 11 spp. E.Asia, America, W.Indies. Trees.

***C. bignonioides*** Walt.=C. catalpa. (Common Catalpa, Indian Bean). E. U.S.A. to Florida and Texas. The light brown wood is soft but durable in soil. It is used for fencing and railway sleepers.

***C. catalpa*** (L.) Karst.=C. bignonioides.

***C. longissima*** Sims.=Bignonia quercus. W.Indies. The bark is used locally for tanning.

***C. speciosa*** Warder. (Catawba Tree, Western Catalpa). N.America. The wood is soft and light, but durable in contact with the soil. It is used for railway sleepers, telegraph poles, fence posts, and house interiors.

Catalpa, Common – Catalpa bignoniodes.

Catalpa, Western – Catalpa speciosa.

***Cataria vulgaris*** Moench.=Nepeta cataria.

Catawba Tree – Catalpa speciosa.

Catch Thorn – Ziziphus abyssinica.

Catchweed – Galium aparine.

Catechu – Acacia catechu.

Catechu, Black – Acacia catechu.

Catechu, Pegu – Acacia catechu.

Catena – Heliocarpus donnel-smithii.

CATENELLA Red Alga. Rhabdoniaceae.

***C. impudica*** J. Ag. Burma, esp. Tenasserim. Eaten as a food locally, and sold.

***C. nipae*** Zanan. (Kolongkolong). Pacific Ocean. Mainly on the aerial roots in mangrove swamps. Eaten locally as food.

CATHA Forsk. ex Scop. Celastraceae. 1 sp. Africa, Madagascar, Arabia.

***C. edulis*** Forsk.=Celastrus edulis. (Cafta, Khat, Kât). The plant is cultivated in Arabia, where the leaves are used to prepare a beverage (Arabian Tea). In Abyssinia the leaves are used to flavour a honey wine.

Cathay Walnut – Juglans cathayensis.

Cativo – Prioria copaifera.

Catjang – Vigna catkjang.

Catnip – Nepeta cataria.

Catoseed Oil – Chisocheton cuminginanus.

Cattle Bush – Atalaya hemiglauca.

Cattley Guava – Psidium catteianum.

Catuaba – Anemopaegna mirandum.

Caucasian Truffles – Terfezia claveryi.

Caucasian Wing Nut – Pterocarya fraxinifolia.

Caucho – Castilla elastica.

Caucho Blanco – Sapium stylare.

Caucho Blanco Rubber – Sapium pavonianum.

Caucho Blanco Rubber – Sapium thomsoni.

Caucho Mirado – Sapium stylare.

Caucho Negro – Castilla elastics.

Caucho Virgin Rubber – Sapium thomsonii.

CAULANTHUS S. Wats. Cruciferae. 2 spp. W. United States of America. Herbs.

***C. glaucus*** S. Wats.=Streptanthus glaucus. (Wild Cabbage). Nevada. The leaves are eaten by the local Indians.

CAULERPA Lamouroux. Caulerpaceae. Green Alga. Marine, Tropics and subtropics. The species mentioned are eaten locally as food.

***C. clavifera*** Agardh. in Guam. Indian Ocean and Pacific. Eaten in Malaysia.

***C. freycinetti*** Ag. Indian and Pacific Oceans. Eaten in Malaysia.

***C. laetivirens*** W. V. B. Pacific Ocean. Eaten in Malaysia.

***C. peltata*** Lamour. Pacific Ocean. Eaten in Malaysia.

***C. racemosa*** Agardh. E.Malaysia. Eaten locally.

***C. sertularioides*** (Gmel.) M. A. Howes. (Salsalamagui). Philippines. An important food locally.

CAULOPHYLLUM Michx. Leonticaceae. 2 spp. N.E.Asia, N.America. Herbs.

***C. thalictroides*** (L.) Michx. (Blue Cohosh, Papoose Root, Squaw Root). N.America. The dried root is used locally as a diuretic, to treat fits, and to control menstruation.

Cau ńui – Pinanga duperreana.

Cautivo – Prioria copaifera.

CAVANILLESIA Ruiz. and Pav. Bombacaceae. 3 spp. Tropical America. Trees.

***C. platanifolia*** H.B.K. (Bongo, Ciupo, Hamati). C.America. The wood is coarse and very light. It is used for native canoes and to float rafts of hardwood.

Cavendish Banana – Musa cavandishii.

Cavenia Acacia – Acacia cavenia.

Cay-cay Fat – Irvingia oliveri.

Cây cù dèn – Croton poilanei.

Cây giôe – Garcinia multiflora.

Cây só – Dillenia ovata.

Cay Thành mat – Millettia ichthyochtona.

Cay xoai – Dialium cochinchinensis.

CAYAPONIA S. Manso. Cucurbitaceae. 45 spp. Warm America, 1 sp., tropical W.Africa, Madagascar, 1 sp. Java. Vines.

***C. espelina*** Cogn.=Parianthopodus carijo=P. espelina. Brazil. The root is used locally to treat syphilis, diarrhoea, and as a tonic, diuretic, and purgative.

***C. pedata*** Cogn. Brazil. The root is used locally as a strong purgative.

Cayenne Linaloe – Aniba panurensis.

Cayenne Linaloe Oil – Amyris balsamifera.

Cayenne Pepper – Capsicum annum.
Cayeté Oil – Omphalea megacarpa.
Cayolizán – Buddleja americana.
***Cayratia geniculata*** Gagenp. = Vitis geniculata.
Cazazuate – Ipomoea murucoides.
CEANOTHUS L. Rhamnaceae. 55 spp. N.America. Shrubs or small trees.
***C. americanus*** L. (New Jersey Tea, Red Root). E. N.America, Florida to Texas. The bark of the roots and trunk induce the clotting of the blood, and is a possible treatment of bleeding from the lungs. An infusion of the leaves has been used as a tea.
***C. asiaticus*** L. = Colubrina asiatica.
***C. reclinatus*** L'Hér. = Colubrina reclinata. (Mabi). W.Indies, Porto Rico. The bark is used in Porto Rico to make a beverage.
Ceara Rubber (Oil) – Manihot glaziovii.
Cearo del la Sierre – Cupressus arizonica.
Cebil Colorado – Piptadenia macrocarpa.
Cebil Gum – Piptadenia cebil.
Cebu Maguey – Agave cantala.
CECROPIA Loefl. Urticaceae. 100 spp. Tropical America. Trees.
***C. pachystachya*** Tréc. S.America. In Brazil, a fibre made from the bark is used for sail making.
***C. palmata*** Willd. Guyana, N.Brazil. The trunk is a possible source of rubber.
***C. peltata*** L. Tropical America. In Brazil, the fibre from the bark is used for sail-making, and the wood for paper pulp. The wood is used as tinder and the trunks are hollowed out to make water pipes. The juice from the stem is used in W.Indies and S.America to treat dysentery, and in Mexico to burn warts. A decoction of the leaves is used by local Indians to treat liver ailments and dropsy. The young buds are sometimes used as a vegetable.
Cedar, African – Juniperus procera.
Cedar, Alaska – Chamaecyparis nootkatensis.
Cedar, Atlantic – Cedrus atlantica.
Cedar, Bastard – Soymida febrifuga.
Cedar, Black – Nectandra pisi.
Cedar, Burmuda Red – Juniperus bermudicana.
Cedar, California Incense – Libocedrus decurrens.
Cedar, Eastern Red – Juniperus virginiana.
Cedar Elm – Ulmus crassifolia.
Cedar, Himalayan – Cedrus libani.
Cedar, Himalayan Black – Alnus nitida.
Cedar, Himalayan Pencil – Juniperus macrocarpa.
Cedar, Incense – Libocedrus decurrens.
Cedar, Lebanon – Cedrus libani.
Cedar Lichen – Cetraria juniperina.
Cedar Mahogany – Entandrophragma candollei.
Cedar Mahogany – Entandrophragma cylindricum.
Cedar, Manado – Cedrela celebica.
Cedar, Mlanje – Widdringtonia whytei.
Cedar Oil – Juniperus macropoda.
Cedar, Pencil – Dysoxylum fraseranum.
Cedar, Pencil – Dysoxylum muelleri.
Cedar, Pink – Acrocarpus fraxinifolius.
Cedar, Pink African – Guarea cedrata.
Cedar, Red – Acrocarpus fraxinifolius.
Cedar, Southern Red – Juniperus barbadensis.
Cedar, Southern White – Chamaecyparis thyoides.
Cedar, Spanish – Cedrela odorata.
Cedar, Spicy – Beilschmiedia mannii.
Cedar, Stinking – Torreya taxifolia.
Cedar, West Indian – Cedrela odorata.
Cedar, Western Red – Thuja plicata.
Cedar, Western White – Thuja plicata.
Cedar, White – Chukrasia tabularis.
Cedar, White – Dysoxylum malabaricum.
Cedar, White – Tabebuia pentaphylla.
Cedar, White – Thuja occidentalis.
Cedar Wood Oil – Juniperus virginiana.
Cedar, Yellow – Rhodosphaera rhodanthema.
CEDRELA P. Br. Meliaceae. 6–7 spp. Mexico to tropical S.America, 1 sp. Indonesia. Trees.
***C. brasiliensis*** Juss. = C. fissilis.
***C. celebica*** Koord. (Kajoo ammorang). Indonesia. The coarse light wood is made into cigar boxes, and boards. It is occasionally exported as Manado Cedar.
***C. febrifuga*** Blume. not King. = Toona sureni.
***C. febrifuga*** King. = C. toona.
***C. fissilis*** Vell. = C. brasiliensis. Brazil. The semi-soft wood is used for general carpentry, railway sleepers and pencils. An extract of the bark is astringent and is used as an emetic. It is also used to treat vaginal discharges.
***C. glaziovii*** DC. = C. mexicana. (Cedro, Kulche). S.America, especially Mexico to Brazil. The aromatic, resinous wood is used for general carpentry, furniture and

cigar boxes. The bark is used locally to treat epilepsy and fevers, the seeds are extracted to treat intestinal worms, and the sap is a useful mucilage.

***C. huberi*** Ducke. Brazil. The aromatic wood is used for general construction work.

***C. longipes*** Blake. C.America. In Guatemala the wood is used for house construction and furniture.

***C. mexicana*** M. Roemer=C. glaziovii.

***C. odorata*** L. (Cedro, Cigarbox Cedrela, Spanish Cedar). S.America. The wood is coarse and aromatic. It is used for furniture, cigar boxes, and to make moth-proof clothes chests etc. A decoction of the root bark is used to reduce fevers, and the seeds to treat intestinal worms.

***C. serrata*** Royle=C. sinensis.

***C. serrulata*** Miq.=C. sinensis.

***C. sinensis*** Juss.=C. serrata=C. serrulata =Toona sinensis. China. Introduced widely to other countries, and cultivated locally. The wood is soft, durable and is easy to work. The tree is planted as shade in coffee plantations, and the wood is used to make beams, boards, posts, furniture, etc.

***C. sureni*** Burk. Malaysia. An excellent light, durable timber. It is used for general construction work, furniture and house-building.

***C. toona*** Roxb.=C. febrifuga (Cedrela). India to Australia. An excellent light, durable timber, marked with red. Exported to Europe. It is used for furniture, house building, tea chests, oil casks, and cigar boxes. The flowers are the source of a red and yellow dye.

Cedrela – Cedrela toona.

Cedrela, Cigarbox – Cedrela odorata.

CEDRELOPSIS Baill. Ptaeroxylaceae. 2 spp. Madagascar. Trees.

***C. grevei*** Baill. (Katafa, Katrafy). S.W.Madagascar. The bark is used locally to flavour rum.

Cedro – Cedrela glaziovii.

Cedro – Cedrela odorata.

Cedro – Libocedrus chilensis.

Cedro blanco – Cypressus benthamii.

Cedro Bordado – Panopsis rubescens.

Cedro cimarron – Calophyllum rekoi.

Cedro Macho – Guarea guara.

Cedro Oil – Citrus limon.

Cedron – Quassia cedron.

CEDRONELLA Moench. Labiatae. 1 spp. Canaries, Madeira. Herb.

***C. triphylla*** Moench. The leaves are used locally to make a beverage (Thé des Canares).

CEDRUS Trew. Pinaceae. 4 spp. Trees.

***C. atlantica*** Manetti. (Atlantic Cedar). N.Africa. The wood is used locally for construction work.

***C. deodara*** (Roxb.) Loud. (Deodar). C.Asia, esp. N.India and surrounding area. An important timber tree used locally for buildings, bridges, and railway sleepers.

***C. libani*** Barrel. (Cedar of Lebanon, Himalayan Cedar). From N.Africa to Himalayas. Mentioned in the Bible. An important timber tree. The wood is soft and aromatic, yellowish brown. It is durable and resistant to termites. Locally it is used for buildings and railway sleepers. An oil (Kelon-ka-tel) is extracted from the wood. It is similar to turpentine. An essential oil, used in perfumery is also extracted from the wood.

Cego Maschado – Physocalymna scaberrimum.

CEIBA Mill. Bombacaceae. 10 spp. Tropical America and Africa. Trees.

***C. acuminata*** (S. Wats.) Rose. Mexico. The fibres from the seeds are used locally for stuffing pillows and making candle wicks.

***C. aesculifolia*** (Hook.) Britt. and Baker =Bombax aesculifolia=Eriodendron aesculifolium. (Pochote). The seeds are eaten locally and their floss used for stuffing pillows etc.

***C. guinensis*** (Thonn.) Chev.=Bombax guinensis=Eriodendron guinense. A possible source of fibre.

***C. pentandra*** (L.) Gaertn.=Bombax pentandrum=Eriodendron anfractuosum (Silk Cotton Tree, Kapok Tree). Tropical America. Introduced to Africa and Far East. The fibre from the seeds (Kapok) is not suitable for spinning, but is water-resistant. Because of this it is valuable for filling life-jackets. It is also used for stuffing pillows etc. and as an insulation against temperature changes and sound. Over 90 per cent of the world production comes from Indonesia where the fruits are hand-picked and the fibres removed by hand. An oil is pressed from the seeds and used for making soap and for lighting. The soft light wood is used for packing cases etc.

***C. sumauma*** Schum = Eriodendron samauma. Brazil. The seed fibres (Paina) is used for stuffing pillows etc. and is used by the local Indians to make flights for blow-gun darts.

***C. thonningii*** A. Chev. Tropical Africa, esp. N. and W. The soft wood is used locally to make small boats.

Celandine – Chelidonium majus.

Celandine, Lesser – Ranunculus ficaria.

CELSATRUS L. Celastraceae. 30 spp. Tropics and sub-tropics. Woody Vines.

***C. edulis*** Vahl. = Catha edulis.

***C. obscurus*** Rich. = C. serratus.

***C. paniculata*** Willd. (Kanguni, Malkangni, Valulurai). Lower Himalayas to Madras to Philippines. The juice from the leaves is used as a nerve and brain tonic, to relieve rheumatism, to treat paralysis, and as an antidote to opium poisoning. An extract of the seeds is laxative and emetic, stimulant, and used as an aphrodisiac. The oil crushed from the seeds is a stimulant. The bark is used to induce abortions.

***C. senegalensis*** Lam. = Gymnosporia montana.

***C. serratus*** Hochst = C. obscurus. (Add-add) Abyssinia. Various parts of the plant are used medicinally, locally.

Celebes Copal – Agathis loranthifolia.

Celebes Manila – Agathis loranthifolia.

Celeriac – Apium graveolens.

Celery – Apium graveolens.

Celery Hop Pine – Phyllocladus asplenifolius.

Celery-leaved Crowfoot – Ranunculus scleratus.

Celery Pine – Phyllocladus trichomanoides.

Celery, Turnip-rooted – Apium graveolens.

Cellonia – Posidonia australis.

CELOSIA L. Amaranthaceae. 60 spp. Tropics and subtropics. Climbing herbs.

***C. argentea*** L. Tropics. The young shoots and leaves are used as a vegetable in India.

***C. bonnivarii*** Schinz. Tropical Africa. The leaves are used as a vegetable, and a decoction from them is used to treat skin complaints and intestinal worms.

***C. globosa*** Schinz. Tropical Africa. See C. bonnivarii.

***C. laxa*** Schum. and Thonn. Tropical Africa and Asia. The leaves are eaten locally in India as a vegetable. In parts of Africa a decoction of the leaves is used to treat skin complaints and intestinal worms.

***C. scabra*** Schinz. Tropical Africa. In S.E.Africa, a potion from the leaves and stems is used to treat cancer.

***C. stuhlmanniana*** Schinz. (Rugosi). Congo. An extract of the roots is used locally to treat intestinal worms.

***C. trigyna*** L. Africa, Arabia and Madagascar. The leaves are eaten locally as a vegetable, and an infusion of them is used to treat skin complaints and intestinal worms. The plant is sometimes cultivated.

CELTIS L. Ulmaceae. 80 spp. N.Hemisphere, S.America. Trees.

***C. adolphi-frederici*** Engl. (Akekenke). W.Africa, Congo to Uganda. The yellow wood is fairly hard and is used for general carpentry and house interiors.

***C. anfractuosa*** Liebm. = C. iguanaea.

***C. australis*** L. = C. lutea (European Hackberry, Nettle Tree). Mediterranean to India. The yellow-grey wood is hard, tough and durable. It is used for turning, wagon-building and household utensils. The wood is flexible so the thin shoots are used to make walking sticks, whip handles, and fishing rods. The seeds are edible, and the leaves are used as fodder in India.

***C. brasiliensis*** Planch. = C. flagellaris. Brazil. The wood is soft and flexible, but not durable. It is used for general carpentry, paper pulp, and agricultural implements. An extract of the bark is used to treat fevers.

***C. brieyi*** De Wild. (Ahila-lowa, Diania). Congo. The yellow, very hard wood is used for house-building and general construction work.

***C. cinnamomea*** Lindl. (Djalilan, Kroja, Pinari, Putan, Riwat). Himalayas to S.India, Indonesia. The wood (Kajoo lahi) has a strong scent and is used locally to treat nervous disorders and fevers. The Hindus use an infusion of wood scrapings in lemon juice as a medicine against boils etc. and itch. It is also used to relieve headaches.

***C. flagellaris*** Casar. = C. brasilensis.

***C. iguanaea*** (Jacq.) Sarg. = C. anfractuosa. (Garabato blanco, Granjeno). S.Florida to S.America, W.Indies. The fruits are eaten locally in Mexico. The fruit juice is used to treat eye diseases. An extract of the leaves is put in baths to treat fevers.

***C. kraussiana*** Bernh. (Camdeboo, Stinkwood). Abyssinia to S.Africa. The wood is heavy and strong. It is used for fencing, railway sleepers, planking and agricultural implements.

***C. lutea*** Pers.=C. australis.

***C. mildbraedii*** Engl. (Angwabele, Mongambe). W.Africa to Tanzania and E. S.Africa. The white wood is very hard. It is used for general carpentry, house interiors, and is an excellent firewood.

***C. mississippiensis*** Bosc. N.America. The coarse, heavy wood is not strong, but is used for agricultural implements and cheap furniture, fencing and fuel. The fruits are eaten locally.

***C. occidentalis*** L. (Common Hackberry, Beaver Wood, Sugar Berry). N.America. The wood is soft, coarse and heavy. It is used for agricultural implements, cheap furniture, fencing and fuel. The fruits are eaten locally. The roots are used for dyeing linen.

***C. pallida*** Torr.=Momisia pallida (Garabato, Granjeno, Granjeno huastoco). Arizona to W.Texas. A spiny shrub. The flowers are a good source of honey. The wood is used for fence posts and fuel.

***C. philippensis*** Blanco. (Kaju lulu, Malaiino). Indonesia, Philippine Islands. The whitish wood is very hard and strong, but is susceptible to fungal decay. It is used for interior work.

***C. prantlii*** Priemer ex Engl. (Amumbi, Diti, Kihengia). Congo, Angola and Ivory Coast. The hard white wood is streaked black. It is used for general construction work, interiors of houses and building small boats.

***C. reticulata*** Torr. (Western Hackberry). S.W. N.America. The fruits are eaten by the local Indians.

***C. selloviana*** Miq. S.America. The fruits are eaten locally. The roots yield a brown dye.

***C. tala*** Gill. S.America. An infusion of the leaves is used to treat indigestion in Chile and Argentina.

***C. wightii*** Planch. Himalayas to S.India and Indonesia. See C. cinnamomea.

Celtuce – Lactuca sativa.

CENCHRUS L. Graminae. 25 spp. Tropical and warm temperate. Grasses.

***C. catharicus*** Delile. (Chevral, Karindja, Kramkram). Sahara. The grain is used locally to make a flour, eaten as a porridge.

CENTAUREA L. Compositae. 600 spp. Europe and N.Africa, through N.India to N.China; N. and S. America. 1 sp. Australia. Herbs.

***C. acaulis*** L. N.Africa. A yellow dye is extracted from the roots.

***C. benedicta*** L.=Cnicus benedictus.

***C. calcitrapa*** L. Mediterranean. The young stems are eaten locally in Egypt.

***C. cyanus*** L. (Cornflower). Mediterranean. Widely naturalized. The flower heads are used to make a tonic and stimulant.

***C. erygioides*** Lam. Arabia and around Nile. The leaves are eaten as a salad by the Arabs.

***C. nigra*** L. (Knapweed, Ironweed, Star Thistle, Black Ray Weed). Europe, naturalized in N.America. The leaves are used to prepare a tonic which is diuretic and induces sweating.

***C. perrottetii*** DC. N.African deserts. The leaves are used as camel fodder.

***C. picris*** Pale. N.India. The powdered leaves are used to cure intestinal worms in sheep, and is applied to wounds of sheep.

***C. rhaponticum*** L.=Rhaponticum scariosum. Europe. The roots are used to make a purgative.

CENTAURIUM Hill. Gentianaceae. 40–50 spp. Cosmopolitan, but not tropical and S.Africa.

***C. umbellatum*** Gilib.=Erythraea centaurium (Centaury, Drug Centaurium). Europe to Caucasus, N.Africa. Introduced to N.America. The dried flower heads are used as a tonic and to reduce fevers. An infusion of them is used in Europe to relieve stomach catarrh.

***C. chilensis*** Pers. Pacific S.America. The leaves are used as a bitter stimulant and to relieve digestive complaints.

Centaury – Centaurium umbellatum.

Centaury – Serratula tinctoria.

Centaury, American – Sabatia anglaris.

***Centella asiatica*** Urb.=Hydrocotyle asiatica.

Centella – Hydrocotyle asiatica.

CENTIPEDA Lour. Compositae. 6 spp. Madagascar, Afghanistan, E.Asia, Indomalaysia, New Zealand, Polynesia, Chile. Herbs.

***C. minima*** Kuntze.=C. orbicularis.

***C. orbicularis*** Lour. Asia to Australia and Pacific Islands. An extract of the plant is used in Chinese medicine.

Centepede Grass – Eremochloa ophuroides.

CENTROLOBIUM Mart. ex Benth. Leguminosae. 7 spp. Tropical America. Trees.

***C. orinocense*** (Benth.) Pitt. (Baláustre, Cartán). Venezuela, Columbia. The orange wood is used to make quality furniture. Locally it is also used for house building.

***C. robustum*** Mart. (Arariba, Zebrawood). Guyana, Brazil. The hard wood is difficult to work, but as it is attractively marked in light and dark brown, it is used for furniture, and boat-building.

CENTROSEMA (DC.) Benth. Leguminosae. 45 spp. America. Herbs.

***C. plumierii*** Benth. Tropical America. Climber. Used as a green manure in the tropics.

***C. pubescens*** Benth. Tropical America. Used as a green manure in the tropics.

***Centrotheca affine*** Wall. = Lophatherum gracile.

Century Plant – Agave americana.

CEODES J. R. and G. Forst. 25 spp. Mascarene Islands, Malaysia, Australia, Polynesia. Small trees.

***C. austro-caledonica*** Brogn. and Gris. New Caledonia. The coarse wood is used to make paper.

***C. brunoniana*** (Endl.) Skottsb. = Calpidia brunoniana. = Pisonia brunoniana. (Eva, Papala Kepau, Parpara, Puatea). Tropical Asia, Polynesia, New Zealand. The fruits are viscid and poisonous. They were used to commit suicide, and, in Hawaii, to trap birds.

***C. umbellifera*** Foster = Pisonia grandis.

CEPHAËLIS Sw. Rubiaceae. 180 spp. Tropics. Shrubs, or small trees. The species mentioned are indigenous to S.America. The dried roots and rhizomes are the source of Ipecac which is used as an emetic and to treat amoebic dysentery. The emetic properties are due to the presence of the alkaloids emetine, cephaeline and psychtrine. The main cultivated sources are C. ipecacuanha and C. acuminata, which are cultivated in S.America and Malaysia. The other species are found in the forests of S.America and are of minor economic importance.

***C. acuminata*** Karsten. (Cartagena Ipecac). N.Columbia to Nicaragua. Product called Cartagena, Nicaragua or Panama Ipecac.

***C. barcella*** Muell. Tropical America. Also used locally to treat burns.

***C. emetica*** Pers. = Psychotria emetica. Guatemala to Bolivia. The product is called Striated Ipecac.

***C. ipecacuanha*** (Brotero) Rich. = Psychotria ipecacuanha = Uragoga ipecacuanha. S.America, esp. S.Brazil. Product called Ipecac, Ipecacuanha, Para Ipecac.

***C. tomentosa*** (Aubl.) Vahl. = Psychotria tomentosa.

***Cephalandra indica*** Naud. = Coccinea grandis.

CEPHALANTHUS L. Naucleaceae. 17 spp. Warm Asia and America. Shrubs and small trees.

***C. occidentalis*** L. (Button Bush). N.America. to Mexico, W.Indies, E.Asia. An infusion of the bark is used by the Indians as an emetic, laxative, tonic and astringent.

***C. spathelliferus*** Baker. W.Madagascar. The wood is used locally for house-building.

CEPHALOCROTON Hochst. Euphorbiaceae. 8 spp. Tropical Africa. Shrubs.

***C. cordofanus*** Muell. Arg. Tropical Africa. The seeds (Dingili) are the source of oil (Dingili Oil) which is used in W.Sudan for cooking etc.

CERAMIUM Roth. Ceramiaceae. Red Algae.

***C. boydenii*** Gepp. China, Japan. Coasts. Used in Japan to manufacture agar.

***C. hypnoides*** (J. Ag.) Okam. = Campylophora hypnoides. China and Japan. Coasts. Used in Japan to manufacture agar.

***C. rubrum*** (Huds.) J. Ag. Atlantic and Pacific Coasts. Used in Japan for the manufacture of agar.

CERASTIUM L. Caryophyllaceae. 60 spp. Nearly cosmopolitan. Herbs.

***C. semidecandrum*** L. (Mouse-eared Chickweed). Europe, introduced to N.America. Used locally as a vegetable, before the plant has flowered.

CERATONIA L. Leguminosae. 1 spp. Mediterranean. Tree.

***C. siliqua*** L. (Carob, St. John's Bread). Cultivated in Old and New World. The pods contain about 50 per cent sugar, and protein and form a valuable forage for stock. They are also used for human consumption. These are the "Locusts" eaten by John the Baptist. A gum (Locust Gum, Locust Bean Gum, Carob Gum) is extracted from the pods, 1000 lb. of pods yielding 35 lb. of gum. The interior of the pods is removed by machinery, roasted then water extracted. The liquid is evaporated to give the gum. It is used to thicken sauces, ice cream etc., in toilet preparations, in radio work (Kasbar Cream), and to cream rubber latex. The seeds are used as a coffee substitute (Carob Coffee). A molasses (Pasteli) is made from the beans in Cyprus. The wood is easily worked and is used to make furniture and carts.

CERATOPETALUM Smith. Cunoniaceae. 5 spp. New Guinea, E.Australia. Trees.

***C. apetalum*** D. Don. (Coachwood, Rose Mahogany, Scented Satinwood). Coastal forests of New South Wales. The light brown wood is scented of burnt sugar. It is used for all forms of interior woodwork in houses, turning of all types, and all types of plywood and veneers.

***C. gummiferum*** Sm. (Christmas Bush). Australia. A dark red, tough kino is extracted from the bark.

CERATOPTERIS Brongn. Parkeriaceae. 2 spp. Tropics and subtropics. Fern.

***C. thalictroides*** Brongn. Tropics. The young leaves are eaten as a vegetable in Japan.

CERATOTHECA Endl. Pedalicaceae. 9 spp. Tropical and S.Africa. Herbs.

***C. sesamoides*** Endl. Tropical Africa. The leaves are eaten as a vegetable in parts of the Sudan. The plant is cultivated and sometimes sold locally. The seeds are the source of an edible oil used in cooking, locally. They are also used to make cakes etc.

CERBERA L. Apocynaceae. 6 spp. Coasts of tropical India and W.Pacific. Trees.

***C. manghas*** L. S.W.Asia to Polynesia, tropical Australia. The seeds are a source of an oil (Odolla Oil) which is used locally for lighting, and medicinally to treat intestinal worms. The fruit is used as a suicide poison in the Marquesas, and to stupify fish in the Philippines. The wood makes an excellent charcoal.

***C. parviflora*** Wall. = Ochrosia bonbonica.

***C. tanghin*** Hook. = Tanghinia venenifera.

***C. thevetioides*** H.B.K. = Thevetia thevetioides.

***C. venenifera*** Steud. = Tanghinia venenifera.

CERCIDIPHYLLUM Sieb. and Zucc. Cercidiphyllaceae. 1 sp. Chile, Japan. Tree.

***C. japonicum*** Sieb. and Zucc. The light soft wood is not strong, and is used for interior work in houses, carving etc., in Japan.

CERCIDIUM Tul. Leguminosae. 10 spp. Warm America. Trees.

***C. torreyanum*** (S. Wats.) Sarg. (Palo Verde). S.W.United States of America, Mexico. The ground seeds are made into cakes by the local Indians.

CERIS L. Leguminosae. 7 spp. N.Temperate. Trees.

***C. canadensis*** L. (Red Bud). N.America. The bark was used as an astringent by the Indians. The flowers were eaten as salads or pickles by the early Canadian settlers. The tree is now grown as an ornamental in N.America.

***C. chinensis*** Bunge. China. The bark is used locally as an antiseptic.

CEROCARPUS Kunth. Rosaceae. 20 spp. W. and S.W. United States of America, Mexico. Small trees.

***C. betuloides*** Nutt. = C. parvifolius.

***C. latifolius*** Nutt. (Mountain Mahogany). W. N.America. The wood is used locally by the Indians to make bows.

***C. montanus*** Raf. New Mexico. An infusion of the leaves is used as a laxative by the local Indians.

***C. parvifolius*** Nutt. = C. betuloides. Mexico, California. The wood is used locally to make tool handles.

CEREUS Mill. Cactaceae. 50 spp. S.America. W.Indies. Cacti.

***C. chiotilla*** Weber. = Escontria chiotilla.

***C. conglomeratus*** Berger. = Echinocerus conglomeratus.

***C. dasyacanthus*** Englm. = Echinocereus dasyacanthus.

***C. enneacanthus*** Engelm. = Echinocereus enneacanthus.

***C. geometrizans*** Mart. = Myrtillocactus geometrizans.

***C. giganteus*** Engelm. = Carnegiea gigantea.

***C. grandiflorus*** Mill. = Selenicereus grandiflorus.

***C. griseus*** Haw. = Lemaireocereus griseus.

***C. gummosus*** Engelm. = Machaerocereus gummosus.

***C. lantanus*** DC. = Espostoa lanata.

***C. pecten-aboriginum*** Engelm. = Pachycereus pecten-aboriginum.

***C. pentagonus*** L. = Acanthocereus pentagonus.
***C. pringlei*** S. Wats. = Pachycereus pringlei.
***C. queretaroensis*** Weber. = Lemaireocereus queretaroensis.
***C. quitsco*** Rem. = Trichocereus chiloensis.
***C. schottii*** Engelm. = Lophocereus schottii.
***C. thurberi*** Engelm. = Lemaireocereus thurberi.
***C. undatus*** Haw. = Hylocereus undatus.
Cereus, Night-blooming – Selenicereus grandiflorus.
Ceriman – Monstera deliciosa.
CERIOPS Arn. Rhizophoraceae. 2 spp. Tropical coasts of India and W.Pacific. Small trees.
***C. candolleanum*** Arn. = C. tagal. Mangrove Swamps. The wood is yellow-orange, fairly durable, and is used for house-building and boat-making. The bark is used for tanning.
***C. tagal*** Rob. = C. candolleanum.
CEROPTERIS Link. = Pitrogramma.
CEROXYLON Humb. and Bonpl. Palmae. 20 spp. N.Andes. Palm trees. A wax is extracted from the surface of the leaves of the species mentioned. The wax is used for the manufacture of wax matches and candles.
***C. andicola*** Humb. and Bonpl. (Wax Palm of the Andes). Tropical S.America.
***C. klopstockiae*** Mart. = Klopstockia cerifera. N. S.America.
CERUANA Forsk. Compositae. 1 sp. Egypt, tropical Africa. Herb.
***C. pratensis*** Forsk. Used in Egypt to make brooms.
Cervina Truffle – Hydnotria carea.
CESTRUM L. Solanaceae. 150 spp. Warm America, W.Indies. Shrubs.
***C. dumetorum*** Schlecht. (Huele de dia, Palo hediondo, Potonxihuite, Teponán). Mexico. An infusion of the plant is used locally to treat skin complaints.
***C. lanatum*** Mart. and Gal. (Candelilla, Huile de noche, Zorrillo). C.America. An extract of the wood is used as a cathartic, and to treat fevers. The fruits are the source of a black dye. The wood has an unpleasant smell, and is used to keep vermin from the nests of poultry.
***C. nocturnum*** L. (Galán de noche, Hierba hedionda, Huile de noche, Pipiloxihuitl). C.America, W.Indies. An infusion of the plant is used as an antispasmodic in the treatment of epilepsy.
***C. pseudoquina*** Mart. Brazil. The bark (Quina do Matto), is made into pills for the treatment of fevers, and stomach complaints.
CETRARIA L. Parmeliaceae. Lichen. N.Temperate and Arctic.
***C. aculeata*** (Schreb.) E. Fr. Temperate. Used in Scotland and Canary Island as the source of a brown dye for staining wool.
***C. crispa*** Ach. Food for reindeer.
***C. cucullata*** (Bell.) Ach. Food for reindeer.
***C. fahlunensis*** (L.) Schaer. = Parmelia fahunensis. (Swedish Shield Lichen). Used throughout Europe to stain wool red-brown.
***C. glauca*** (L.) Ach. (Pale Shield Lichen). Used in Europe to stain wool light yellow brown.
***C. islandica*** (L.) Ach. (Iceland Moss). Used in Iceland to stain wool brown. A demulcent, nutrient jelly is made by boiling the lichen, and evaporating the liquor. It is used commercially as a base for ointments, as a laxative and to remove the bitter taste from medicines. The whole plant is made into a bread locally. Much is collected for commercial use in Austria, Germany, Scandinavia and Switzerland. The plant is a valuable food for reindeer.
***C. juniperina*** (L.) Ach. (Cedar Lichen). Used in Scandinavia to dye wool yellow. It is also used to poison wolves.
***C. nivalis*** (L.) Ach. (Snow Lichen). Europe. Used to stain wool purple. An important food for reindeer.
***C. pinastri*** (Scop.) S. Gray. (Pine Lichen). Europe. Used to stain woollens green.
Cevadilla Seed – Schoenecaulon officinale.
Ceylon Beautyleaf – Calophyllu calaba.
Ceylon Bowstring Hemp – Sanseverinea zeylanica.
Ceylon Boxwood – Canthium didymum.
Ceylon Cardamon – Elettaria cardamomum.
Ceylon Cinnamon – Cinamomum zeylandicum.
Ceylon Gooseberry – Doryalis hebecarpa.
Ceylon Mango – Mangifera zeylandica.
Ceylon Moss – Gracilaria lichenoides.
Ceylon Oak – Schleichera trijuga.
Ceylon Olives – Elaeocarpus serratus.
Ceylon Piassava – Caryota urens.
Ceylon Raspberry – Rubus albescens.
Chaché – Pluchea odorata.
Chacoli – Myrciaria cauliflora.

Chacpte – Cordia dodecandra.
CHAENOMELES Lindl. Rosaceae. 3 spp. E.Asia.
***C. oblongata*** Mill. = Cydonia vulgaris = Pyrus cydonia. (Quince). Caucasia to S.E.Arabia. Cultivated in Old and New World for the fruits which are used in various preserves. The fruits are virtually inedible when uncooked. The seeds are very mucilaginous and are used to make a demulcent, used mainly in skin cosmetics. Most of the seeds are produced in Russia, Spain, S.France, Portugal and Iran.
***C. sinensis*** (Thounin.) Koehne. = Chaenomeles sinensis. (Chinese Quince). N.China. The fruits are larger than those of the Quince. They are used locally to scent rooms, and to make sweetmeats.
***Chaerefolium sylvestre*** (L.) Schinz. and Thell. = Anthriscus sylvestre.
CHAEROPHYLLUM L. Umbelliferae. 40 spp. N.Temperate. Herbs.
***C. bulbosum*** L. (Turnip-rooted Chervil). Europe to Caucasus. The roots are eaten as a vegetable, especially in winter. Occasionally cultivated.
***C. sylvestre*** L. = Anthriscus sylvestre.
CHAETACHME Planch. Ulmaceae. 4 spp. Tropical and S.Africa. Madagascar. Shrubs.
***C. microcarpa*** Rendle. (Buhonga, Kekoko). Cameroon to Angola, to Kenya. The light wood is used in W.Africa to make various musical instruments. An infusion of the leaves is used locally as a purgative.
CHAETOCARPUS Thw. Euphorbiaceae. 10 spp. Tropics except E. Malaysia, Australia, Pacific. Trees.
***C. castanocarpus*** Thw. Ceylon and W.Malaya. The wood is hard, close-grained and red-brown. It is used locally for beams, posting etc.
***Chaetochloa italica*** (L.) Scribn. = Setaria italica.
CHAETOMORPHA Kuetz. Cladophoraceae. Green Algae.
***C. antennica*** Kuetz. Marine. Pacific, esp. Hawaii. Eaten locally in salads, and eaten in confectionery.
***C. crassa.*** Kuetz. Pacific. See C. antennica.
CHAETOPTELEA Liebm. Ulmaceae. 1 sp. C.America. Tree.
***C. mexicana*** Liebm. = Ulmus mexicana. The wood is used for general lumber. An extract of the bark is astringent, and is used locally to treat coughs.
***Chaetospermum*** (M. Roem.) Swingle = Swinglea.
***Chaetospermum glutinosum*** (Blanco.) Swingle = Swinglea glutinosa.
Chagual Gum – Puya chilensis.
***Chailletia*** DC. = Dichapetalum Thou.
***Chailletia timoriensis*** DC. = Dichapetalum timoriense.
Chain Cotton – Gossypium brasiliense.
Chal-bih – Engelhardita polystachya.
***Chalas paniculata*** L. = Murraya paniculata.
Chalta – Dillenia indica.
CHAMAECYPARIS Spach. Cupressaceae. 7 spp. N.America, Japan, Formosa. Trees. The wood of the species mentioned is strong, close-grained, and usually fragrant. They have a wide variety of uses, including house-building, railway sleepers, posts, etc.
***C. lawsoniana*** (A. Murr.) Parl. (Lawson Cypress). N.W. N.America. The resin is used as a diuretic.
***C. nootkatensis*** (D. Don.) Spach. (Alaska Cedar, Sitka Cypress, Yellow Cypress, Yellow Cedar). Alaska to Oregon.
***C. obtusa*** (Sieb. and Zucc.) Endl. Japan, Formosa.
***C. pisifera*** (Sieb. and Zucc.) Endl. (Sawara Cypress). Japan.
***C. thyoides*** (L.) Britton, Sterns and Poggenb. (Southern White Cedar, White Cedar). E. N.America.
***Chamaedaphne*** Moench. = Cassandra.
CHAMAEDOREA Willd. Palmae. 100 spp. Warm America. Small reedy palms.
***C. elegans*** Mart. = Collina elegans. C.America. The fruits are eaten locally, as are the cooked young leaves.
***C. gracilis*** Millsp. = C. graminifolia. C.America. The unopened inflorescence is eaten locally as a vegetable (Pacayas).
***C. sartorii*** Liebm. = Eleutheropetalum sartorii. S.Mexico. The young inflorescence is eaten locally as a vegetable.
***C. tepejilote*** Liebm. Mexico. The unopened spathes are eaten as a vegetable locally.
Chamaedrys, Germander – Teucrium chamaedrys.
***Chamaedrys officinalis*** Moench. = Teucrium chamaedrys.
***Chamaedrys scordium*** Moench. = Teucrium scordium.
CHAMAELIRIUM Willd. Liliaceae. 1 sp. E. N.America. Herb.
***C. luteum*** (L.) Gray. (Blazing Star, Unicorn Root). The dried rhizome (Helonias)

contains the glucoside chamaelirin, and is used as a diuretic and to tone the muscles of the uterus.

***Chamaenerion angustifolium*** (L.) Scop.= Epilobium angustifolium.

***Chamaerhodendron ferrugineum*** Bubani= Rhododendron ferrugineum.

CHAMAEROPS L. Palmae. 2 spp. W.Mediterranean. Small palms.

***C. humilis*** L. Mediterranean. The only European palm. The leaves yield a fibre (Crin Végétale) used for cordage. The young leaf buds are eaten locally as a vegetable.

***C. ritchieana*** Griff.=Nannorhops ritchieana.

CHAMAESARACHA (A. Gray) Benth. and Hook. f. Solonaceae. 10 spp. W. N.America, Mexico, Andes.

***C. coronopus*** (Dunal) Gray.=Solanum coronopus. S.W. U.S.A. and Mexico. The berries are eaten locally by the Indians.

Chamise, Redshank – Adenostoma sparsifolium.

Chamomille – Matricaria chamomilla.

Chamomille, Dog – Anthemis cotula.

Chamomille, Dyer's – Anthemis tinctoria.

Chamomille, English – Anthemis nobilis.

Chamomille, Golden – Anthemis tinctoria.

Chamomille, Roman – Anthemis nobilis.

Champaca Wood (Oil) – Bulnesia sarmienti.

Chamedek – Artocarpus polyphema.

Champignon – Psalloita campestris.

Chan sa – Drysoxylum loureiri.

Chanal – Gourliea decorticans.

Chanar – Gourliea decorticans.

Chân chim – Vitex quinata.

Chang-Bato – Canscora diffusa.

Chapote – Diospyros texana.

Chaparrel Tea – Croton crymbulosus.

Chaparro amargosa – Castela texana.

Chaparro prieto – Condalia obtusifolia.

Chaplash – Artocarpus chaplasha.

Chapparral Yucca – Hesperoyucca whipplei.

Char – Xanthorrhoea hastilis.

Charcherquem – Visnea mocanera.

Chari – Grewia betulifolia.

Charlock – Brassica arvensis.

Chartreuse – Angelica archangelica.

***Chasalia*** DC.=Chassalia.

CHASMANTHERA Hochst. Menispermaceae. 2 spp. Tropical Africa.

***C. welwitschii*** Troupin. (Mosimbi, Soolo). Climber. Congo, S. W.Africa. The roots are eaten locally as a vegetable.

CHASSALIA Comm. ex Poir. Rubiaceae. 42 spp. Old World tropics, esp. Madagascar. Shrubs.

***C. rostrata*** Miq.=Psychotria rostrata. N.Malaysia, Borneo, Sumatra. An infusion of the leaves is used locally in Penang to relieve constipation.

Chaste Tree – Vitex aguns castus.

Chaste Tree, Guiana – Vitex divaricata.

Chaste Tree, Molave – Vitex parviflora.

Chaste Tree, Timor – Vitex littoralis.

Chatter Box – Epipactis giganteum.

Chattimandy – Euphorbia cattimandoo.

Chaulmogra (Oil) – Hydnocarpus heterophyllus.

Chaulmoorga – Gynocardia odorata.

Chaulmoogra Tree – Hydnocarpus anthelmintica.

***Chavica auriculata*** Miq.=Piper betle.

Chaw Stick – Gouania lupuloides.

Chawar – Hitchenia caulina.

Chay – Homalium fagifolium.

Chay Root – Oldenlandia corymbosa.

Chayaver – Oldenlandia umbellata.

***Chayota edulis*** Jacq.=Sechium edule.

Chayote – Sechium edule.

Chdari – Rhus oxycanthoides.

Ché – Oldenlandia umbellata.

Checker Berry – Gaultheria procumbens.

Checker Berry – Mitchella repens.

Checker Tree – Sorbus torminalis.

Cheese, Airan – Panus rudis.

Cheese, Camembert – Penicillium camemberti.

Cheese, Caucasian Sheep – Panus rudis.

Cheese, Cheddar – Streptococcus lactis.

Cheese, Dolce Verde – Penicillium expansum.

Cheese, Emmenthaler – Bacterium acidipropriocid.

Cheese, Gorgonzola – Penicillium gorgonzola.

Cheese, Green – Melilotus altissima.

Cheese, Green – Trigonella coerulea.

Cheese, Gruyère – Actinomyces candidus.

Cheese, Guava – Calocarpum mammosum.

Cheese, Guava – Psidium guajava.

Cheese, Herb – Peucedanum oxtruthium.

Cheese, Leiden – Cuminum cyminum.

Cheese, Roquefort – Penicillium rowueforti.

Cheese, Schabzieger – Trigonella coerulea.

Cheese, Serret Ver – Trigonella coerulea.

Cheese, Sheep – Panus rudis.

Cheese, Stilton – Penicillium chrysogenum.

Cheese, Swiss – Propionibacterium shermannii.
Cheese, Swiss – Streptococcus thermiphilus.
Cheesemaker – Withania coagulans.
CHEILANTHES Sw. Sinopteridaceae. 180 spp. Tropics and temperate. Ferns.
*C. **tenuifolia*** (Burm.) Swartz. Himalayas and S.China to S.Asia and Polynesia. A decoction of the rhizome is used by various local tribes to treat illness caused by witchcraft.
CHEIRANTHUS L. Cruciferae. 10 spp. Mediterranean and N.Temperate. Herbs.
*C. **cheiri*** L. (Wallflower). C. and S. Europe. Now cultivated as an ornamental and for an essential oil which is occasionally used in perfumery. The plant contains the glucoside cheiranthin which is a heart poison. The flowers were used to make a medicine which is purgative, used to control menstruation, and is antispasmodic.
CHEIRODENDRON Nutt. Araliaceae. 8 spp. Hawaii and Marquesas Islands. Trees.
*C. **marquesense*** F. Brown (Pimata). Marquesas Islands. The leaves are used locally to make leis.
CHEIROSTEMON Humb. and Bonpl. Sterculiaceae. 1 sp. Mexico. Tree.
*C. **pentadactylon*** Larr. (Manitu de León, Mano de León, Pale de Tuyuyo). Mexico, Guatemala. The flowers are used locally to treat eye infections and haemorrhoids.
Cheken – Eugenia chequen.
Chekiang – Cupressus funebris.
Chekiang – Koelreuteria paniculata.
CHELIDONIUM L. Papaveraceae. 1 sp. Subarctic Eurasia. Herb.
*C. **majus*** L. (Celandine). The yellow latex contains the alkaloids chelidonine, chelerythrine, and protopine. It is used locally to cure skin complaints and freckles. When dried it is used as a diuretic, purgative, expectorant and sedative. It is also used to reduce fevers.
CHELONE L. Schrophulariaceae. 4 spp. E. N.America. Herbs.
*C. **glabra*** L. (Balmony). E. N.America. The leaves are used as a tonic, laxative and to remove intestinal worms.
Chêmpaka – Michelia champaca.
Chene Gum – Spermolepis gummifera.
Chengal – Balanocarpus heimii.
Chengal paya – Hopea pentanervia.

CHENOPODIUM L. Chenopodiaceae. 100–150 spp. Temperate. Herbs.
*C. **album*** L. (Lamb's Quarters). Young plants are used as a vegetable. The seeds are made into cakes and gruel by some American Indians. It is occasionally cultivated, and frequently occurs as a weed.
*C. **amaranticolor*** Coste and Reyn. The leaves are used as a vegetable.
*C. **ambrosioides*** L. (Wormwood). The leaves are used locally in Mexico as a condiment in soup *C. **ambrosiodes*** var. ***anthelmiticus*** L. (American Wormseed) is cultivated, especially in Maryland for the oil (Oil of Chenopodium, Oil of Wormseed), which is used medicinally to treat intestinal worms and enteric amoebae. The oil is obtained by steam distillation of the leaves.
*C. **auricomum*** Lindl. W.Australia, Tasmania. Saltbush. It is used in Australia to feed cattle.
*C. **bonus-henricus*** L.=C. esculentus. (All-good, Good King Henry). Used as a potherb. It is sometimes cultivated.
*C. **californicum*** Wats. (Californian Soap Plant). California. The roots are used locally for washing.
*C. **canihua*** Cook. (Cañahua, Kanjaaya, Quanja peruan). Peru, Bolivia. The seeds are used locally to make bread, soups and gruel.
*C. **capitum*** (L.) Aschers. Europe. It was cultivated as a potherb, and is still used occasionally as an emergency food.
*C. **chilense*** L. Chile. The leaves are used to treat intestinal worms.
*C. **esculentum*** Salisb.=C. bonus-henricus.
*C. **foetidum*** Schrader. E.Africa. The seeds are used locally to kill ants.
*C. **foliosum*** Aschers. Europe and East. It was cultivated as a vegetable, and is still recommended as an emergency food.
*C. **fremontii.*** Wats. W. N.America. The ground, roasted seeds are used as a flour by the local Indians.
*C. **leptophyllum*** Nutt. N.America. The plant is eaten raw or cooked by the local Indians. The seeds are mixed with corn meal and eaten.
*C. **murale*** L. Temperate. Occasionally used as a potherb.
*C. **nuttaliae*** Saff. Mexico. The seeds (Hoautli, Xoxhihuauhtli) are eaten locally and are used in certain religious festivals.

***C. pallidicaule*** Heller. S.America, esp. Bolivia and Peru. The seeds are an important local source of food (Cañagua, Coãihua). There are many cultivated varieties known locally.

***C. procerum*** Hochst. (Doi-doi, Mundjeko). C.Africa and Abyssinia. The plant is used locally to treat migraine.

***C. quinoa*** Willd. Chile and Peru. The seeds are an important local food used to make flour, soups and gruel. They are also fermented to make a beverage (Tschitscha). The plant has been cultivated since prehistoric times.

***C. rhadinostachyum*** F. v. Muell. Australia. The seeds are used by the natives of the interior to make a flour.

***C. rubrum*** L. Temperate. Used locally as a potherb.

***C. virgatum*** (L.) Aschers. Europe. The red berries are used locally as a cosmetic.

***C. vulneraria*** L. Europe, Mediterranean and Caucasus. The leaves are the source of a yellow dye.

Chenopodium Oil – Chenopodium ambrosioides.
Chenrung – Morinda angustifolia.
Chenuet, China – Sphaeranthus senegalensis.
Cherapu – Garcinia praininana.
Cherbet Tokhum – Ocimum basilicum.
Cherimoya – Annona cherimola.
Cherimoya of Jaliso, Wild – Annona logiflora.
Cherry – Prunus avium.
Cherry, Barbados – Malpighia glabra.
Cherry Bean – Vigna sinensis.
Cherry Birch – Betula lenta.
Cherry, Bird – Prunus avium.
Cherry, Brazil – Euglenia michelii.
Cherry, Bush – Eugenia myrtifolia.
Cherry, Capulin – Prunus capollin.
Cherry, Chinese Sour – Prunus cantabrigiensis.
Cherry, Choke – Prunus virginata.
Cherry, Cornellan – Cornus mas.
Cherry, Downy Ground – Physalis peruviana.
Cherry, Downy Ground – Physalis pubescens.
Cherry Elaeagnus – Elaeagnus multiflora.
Cherry, European Bird – Prunus padus.
Cherry, Ground – Physalis heterophylla.
Cherry, Ground – Physalis neo-mexicana.
Cherry Gum – Amygdalus leiocarpus.
Cherry Gum – Prunus puddum.
Cherry, Hard – Prunus avium.
Cherry, Heart – Prunus avium.
Cherry, Jamaica – Muntingia calabura.
Cherry Kernel Oil – Prunus cerasus.
Cherry Laurel – Prunus laurocerasus.
Cherry, Mahaleb – Prunus mahaleb.
Cherry, Manchu – Prunus tormentosa.
Cherry, Marasca – Prunus cerasus.
Cherry, Miyana – Prunus maximowiczii.
Cherry, Mountain – Prunus angustifolia.
Cherry, Plum – Prunus cerasifera.
Cherry, Prairie – Prunus gracilis.
Cherry, Rocky Mountain – Prunus melanocarpa.
Cherry, Rum – Prunus serotina.
Cherry, Sour – Prunus cerasus.
Cherry, St. Lucie – Prunus mahaleb.
Cherry, Surinam – Eugenia uniflora.
Cherry, Sweet – Prunus avium.
Cherry, West Indian – Malphigia punicifolia.
Cherry, Western Sand – Prunus bessyi.
Cherry, Wild Black – Prunus serotina.
Cheru – Holigarna arnottiana.
Chervil – Anthriscus cerefolium.
Chervil, Turnip-rooted – Chaerophyllum bulbosum.
Chervin – Sium sisarum.
Chestnut, American – Castanea dentata.
Chestnut, Bay – Castanospermum australis.
Chestnut, Cape – Calodendron capense.
Chestnut, Chinese – Castanea mollissima.
Chestnut, Earth – Lathyrus tuberosus.
Chestnut, Golden-leaved – Castanopsis chrysophylla.
Chestnut, Horse – Aeculus hippocastanum.
Chestnut, Japanese – Castanea crenata.
Chestnut, Moreth Bay – Castanospermum australe.
Chestnut Oak – Lithocarpus densiflora.
Chestnut Oak – Quercus prinus.
Chestnut, Polynesian – Inocarpus edulis.
Chestnut, Rhodesian – Baikiaea plurijuga.
Chestnut, Spanish – Castanea sativa.
Chestnut, Tahiti – Inocarpus edulis.
Chestnut Tongue – Fistulina hepatica.
Chestnut, Water – Eleocharis tuberosa.
Chestnut, Water – Trapa natans.
Chestnut, Wild – Brabejum stellatifolium.
Chia – Hyptis emoryi.
Chia (Oil) – Salvia chia.
Chia, Californian – Salvia columbariae.
Chian Turpentine – Pistacia terebinthus.
Ch'iao T'ou – Allium chinense.
Chibou – Bursera gummifera.
Chibunge – Acacia nigrescens.

Chica, Vanilla – Selenipedium chica.
Chicaquil – Jatropha multifida.
Chicha – Zizyphus mistol.
Chichicastle – Urera baccifera.
Chichipe – Lemaireocereus chichipe.
Chick Pea – Cicer arietinum.
Chickasaw Lima – Canavalia ensiformis.
Chickasaw Plum – Prunus angustifolia.
Chickling Vetch – Lathyrus sativus.
***Chickrassia*** Wight. and Arn.=Chukrasia.
Chickweed, Common – Stellaria media.
Chickweed, Mouse-eared – Cerastium semidecandrum.
Chicle – Achras sapota.
Chicmu – Trifolium amabile.
Chicory – Cichorium intybus.
Chicozapote – Monsonia americana.
Chieh-k'eng – Platycodon grandiflorum.
Chiendent grenille à brosse – Andropogon aciculatus.
Chieu lieu – Terminalia nigrovenulosa.
Chihli – Koelreuteria paniculata.
Chih mu – Fritillaria verticillata.
Chi krassang tomhom – Polygonum odoratum.
Chikri – Buxus wallichiana.
Chilara – Casearia tomentosa.
Chilca Jarilla – Senecio salignus.
Childrens' Tomato – Solanum anomalum.
Chile Hazel – Gevuina avellana.
Chile Laurel – Laurelia aromatica.
Chilean Clover – Medicago sativa.
Chilean Guava – Myrtus ugni.
Chilean Peppertree – Schinus latifolius.
Chilean Trevoa – Trevoa trinervia.
Chilean Wineberry – Aristotelia maqui.
Chilghoza Pine – Pinus gerardiana.
Chili – Capsicum annuum.
Chilicote – Erythrina flabelliformis.
Chililo – Casearia dolichophylla.
Chiloe Strawberry – Fragaria chiloensis.
CHILOPSIS D. Don. Bignoniaceae. 1 sp. S. United States of America, Mexico. Small tree.
***C. lineari*** (Cav.) Sweet.=C. saligna=Bignonia linearis. (Desert Willow, Flowering Willow, Mimbre). The thin branches are used for basket-making, and the thicker ones for fencing. An infusion of the flowers is used locally as a stimulant in heart complaints, and to cure coughs. The flowers are sold locally.
***C. saligna*** D. Don=C. lineari.
Chilte – Cnidoscolus elasticus.
Chite Blanco – Cnidoscolus elasticus.
Chilte, Highland – Cnidoscolus elasticus.
Chiluchi – Iris nepalensis.
CHIMAPHILA Pursh. Pyrolaceae. 8 spp. Europe, Asia, N. and C.America, W.Indies. Small shrubs
***C. umbellata*** Nutt. (Spotted Wintergreen). Europe, Temperate Asia, N.America. An infusion of the leaves (Folia Chimaphilae) is used to treat stones in the bladder, and to reduce the voiding of urine.
CHIMONANTHUS Lindl. Calycanthaceae. 4 spp. China. Shrubs.
***C. fragrans*** Lindl.=C. praecox.
***C. praecox*** (L.) Link.=C. fragrans=Calycanthus praecox=Meratia praecox. China, Japan. The flowers are used in Japan to scent, clothes, sachets, and to make perfumes.
China brasiliensis – Symplocas racemosa.
China Chenuet – Sphaeranthus sengalensis.
China Crab Apple – Pyrus hupehensis.
China Grass – Boehmeria nivea.
China Jute – Abutilon avicennae.
China Linen – Boehmeria nivea.
China Oil – Shorea robusta.
China Rhubarb – Rheum palmatum.
China Root – Smilax china.
Chinaberry Tree – Melia azedarach.
Chinese Anise – Illicium verum.
Chinese Apple – Malus prunifolia.
Chinese Aralia – Aralia chinensis.
Chinese Artichoke – Stachys sieboldi.
Chinese Banana – Musa cavendishii.
Chinese Bellflower – Platycodon grandiflorum.
Chinese Cabbage – Brassica chinensis.
Chinese Cabbage – Brassica pekinensis.
Chinese Cabbage – Ipomoea aquatica.
Chinese Chinquapin – Castanea seguinii.
Chinese Chestnut – Castanea mollissima.
Chinese Chive – Allium odorum.
Chinese Colza (Oil) – Brassica campestris var. chinoleifera.
Chinese Cotton – Gossypium nanking.
Chinese Crab Apple – Malus baccata.
Chinese Crab Apple – Malus hupehensis.
Chines Date – Ziziphus mauritiana.
Chines Date-plum – Diospyros kaki.
Chinese Dwarf Banana – Musa chinensis.
Chinese Ephedra – Ephedra sinica.
Chinese Galls – Rhus semialata.
Chinese Green – Rhamnus dahurica.
Chinese Green – Rhamnus globosa.
Chinese Haw – Crataegus pentagyna.
Chinese Hemlock – Tsuga chinensis.
Chinese Hickory – Carya cathayensis.
Chinese Jujub – Ziziphus jujuba.

Chinese Larch – Larix potaninii.
Chinese Laurel – Antidesma bunius.
Chinese Liquorice – Glycyrrhiza ralensis.
Chinese Matgrass – Cyperus tegetiformis.
Chinese Mustard – Brassica juncea.
Chinese Olives – Canarium album.
Chinese Olives – Canarium bengalense.
Chinese Olives – Canarium nugrum.
Chinese Pear – Pyrus chinensis.
Chinese Peony – Paeonia albiflora.
Chinese Pepper – Zanthoxylum bungei.
Chinese Pistache – Pistacia chinensis.
Chinese Plum – Prunus salicina.
Chinese Potato – Dioscorea batatas.
Chinese Preserving Melon – Benincasa hispida.
Chinese Quince – Chaenomeles sinensis.
Chinese Root – Pachyma hoelen.
Chinese Sassafras – Sassafras tzumu.
Chinese Soapberry – Sapidus mukorossi.
Chinese Sour Cherry – Prunus cantabrigiensis.
Chinese Sumach – Rhus semialata.
Chinese Sweet Gum – Liquidambar formosana.
Chinese Tallow Tree – Sapium sebiferum.
Chinese Torreya – Torreya grandis.
Chinese Vegetable Tallow – Sapium sebiferum.
Chinese White Pine – Pinus armandi.
Chinese Wolfberry – Lycium chinense.
Chinese Wood Oil – Aleurites fordii.
Chinese Yam – Dioscorea batatas.
Ch'ing hao – Artemisia carvifolia.
Chingma – Abutilon avicennae.
Chingma, Butterprint – Abutilon avicennae.
Chino – Pithecellobium mexicanum.
Chino Azul – Agave tequilana.
Chino Bermejo – Agave palmaris.
Chinottos – Citrus aurantium var. myrtifolia.
Chinquapin – Castanea pumila.
Chinquapin, Chinese – Castanea seguinii.
Chinquapin, Golden – Castanopsis chrysophylla.
Chinquapin, Henry – Castanea henryi.
Chinquapin, Water – Nelumbo pentapetala.

CHIOCOCCA P. Br. Rubiaceae. 20 spp. S.Florida, W.Indies, tropical America. Shrubs.

***C. alba.*** (L.) Hitchc. = C. racemosa. (Branda). Tropical, and sub-tropical America. The leaves are used locally to treat snakebite. In Dominica they are used as an abortive.

***Chiogenes*** Salis. ex Torr. = Gualtheria.

***Chionachne masii*** Bal. = Sclerachne punctata.

CHIONANTHUS L. Oleaceae. 2 spp. E.Asia, E. N.America. Trees.

***C. oblongifolia*** Koord. and Val. = Haloragis oblongifolia.

***C. virginica*** L. (Fringe Tree, Old Man's Beard). E. U.S.A., to Texas. The root bark contains the glucoside chionanthin, and is used medicinally as an alternative in the treatment on catarrhal jaundice. The bark is collected in the autumn. Most of the commercial production is in Virginia and N.Carolina.

Chios Mastic – Pistacia lentiscus.
Chiotilla – Escontria chiotilla.
Chipilin – Crotalaria longirostrata.
Chiquito – Combretum butryosum.

***Chiranthodendron*** Cerv ex Cav. = Cheirostemon.

Chirata – Swertia chirata.
Chirimote – Disterigma margaricoccum.
Chirinda Medlar – Vangueria esculenta.
Chironja – Citrus paradisi × C. sinensis.
Chironji-ki-gond – Buchanania latifolia.

CHISOCHETON Blume. Meliaceae. 100 spp. Indomalaysia, Asia, S.China. Trees. The seeds of the species mentioned are the source of non-drying oils.

***C. cuminginanus*** (DC.) Harms. Philippine Islands. The oil (Balukangag Oil, Catoseeds Oil) is used for lighting and as a purgative, locally.

***C. macrophyllum*** King. (Gendis, Goola). Malaya. The oil is used locally for lighting.

***C. pentandrum*** (Blanco) Merr. Philippines. The oil is used locally as hair-oil.

Chitra – Drosera peltata.
Chittagong Wood – Chukrasia tabularis.
Chive – Allium schoenoprasum.
Chive, Chinese – Allium odorum.
Chive, K'iu rsai – Allium odorum.

***Chlora perfoliata*** L. = Blackstonia perfoliata.

CHLORANTHUS Sw. Chloranthaceae. 15 spp. E.Asia, Indomalaya. Shrubs.

***C. inconspicuus*** Sw. = C. spicatus. (Hoa soi). Indochina, China, Japan. The flowers are used to flavour tea in Indochina, and an infusion of the flowers and leaves is used to treat coughs.

***C. officinalis*** Blume. Malaysia. The leaves have been used since ancient times locally, especially in Java to prepare a beverage.

CHLORELLA Beijerinck. Oocystaceae. Green Alga. Several species are high producers of both lipids and proteins. They have been cultured experimentally as a potential source of human food. One potential disadvantage of Chlorella as a source of food is the production of the antibiotic chlorellin in the cultures.

CHLORIS Sw. Graminae. 40 spp. Tropical and warm temperate. Grass.

***C. gayana*** Kunth. (Rhodes Grass). S.Africa. Cultivated widely in the warmer parts of the world as a forage grass. Much is used in Australia. It withstands grazing and is palatable to livestock.

CHLOROGALUM (Lindl.) Kunth. Liliaceae. 7 spp. California. Herbs.

***C. pomeridianum*** (Ker-Gawl) Kunth. (Californian Soaproot). California. The inner part of the bulb is used as a soap, and the outer part is fibrous.

CHLOROPHORA Gaudich. 12 spp. Tropical America, Africa, and Madagascar. Trees.

***C. excelsa*** (Welw.) Benth. and Hook. f.= Milicia africana. (African Oak, Iroko Fustic Wood, Mbang, Odum). Tropical Africa. The wood is dark brown, heavy and tough. It is termite resistant, and is used locally for furniture, ship-building and wagon-making. The latex is occasionally used as a rubber.

***C. tinctoria*** (L.) Gaud.=Morus tinctoria. Tropical America, W.Indies. The wood has a greenish tinge, and polishes well. It is used for furniture and interiors of houses. The dye Fustic is extracted from the heartwood. It contains the pigments morin, and maclurin and is used to dye wool, yellow, green or brown.

***Chlorophyllum molybdites*** (Mey.) Mass.= Lepiota morganii.

CHLOROSPLENIUM Fr. Heliotiaceae. Fungus. 1 sp. Europe.

***C. aeruginosum*** (Oed.) De Not. The mycelium stains the wood through which it grows. Stained oak wood (Tunbridge Ware) is an attractive dark colour and is much valued. The staining is due to the formation of xylochloric acid.

CHLOROXYLON DC. Flindersiaceae. 1 sp. S.W.India, Ceylon. Tree.

***C. swietenia*** DC. (Satinwood). A much-valued commercial timber used for cabinet work, and furniture. It is used locally for agricultural implements and railway sleepers. The wood contains chloroxylonine, which may cause skin irritations.

Choang ko Chuk – Bambusa tuldoides.

Chochoco – Mahonia chochoco.

Choco – Sechium edule.

Chocolate – Theobroma cacao.

Chocolate, Gaboon – Irvingia barteri.

Chocolate, Nicaraguan – Theobroma bicolor.

Chocolate coloured Nephroma – Nephroma parilis.

Chodhava – Epimeredi malabarica.

CHOIROMYCES Vittad. Tuberaceae. 4 spp. Europe, N.America. Fungi. The fruit-bodies of the species mentioned are eaten locally.

***C. magnusii*** (Matt.) Paol.=Terfeza magnusii. S.Europe.

***C. venosus*** Fr.=Tuber album (Troitskie Truffle, White Truffle). Eaten especially in Russia and Czechoslovakia.

Choke Cherry – Prunus virginana.

Choke Cherry, Gray's – Prunus grayana.

Chloagogue indo – Myriocarpa longipes.

Cholla – Opuntia fulgida.

***Chondodendron*** – Ruiz. and Pav.=Chondrodendron.

CHONDRODENDRON Ruiz. and Pav. corr. Miers. Menispermaceae. 8 spp. Brazil, Peru. Woody Vines.

***C. candicans*** (L. C. Rich.) Sandw. Brazil and Peru. The roots are used locally to make arrow poison.

***C. limacifolium*** (Diels.) Mold. Brazil and Peru. The roots are used locally to make an arrow poison.

***C. platyphyllum*** Miers. Brazil. An extract of the roots is used locally to treat stomach complaints and colic of the uterus.

***C. polyathemum*** Diels. Brazil and Peru. The roots are used locally to make arrow poisons.

***C. tomentosum*** Ruiz. and Pav. (Pariera). Brazil and Peru. The dried root contains alkaloids, e.g. chondoinine and is used medicinally as a diuretic and tonic. It is exported as Pareira Root or Radix pareira brava. The root is also used locally as an arrow poison.

CHONDROPETALUM Rottb. Restioniaceae. 18 spp. S.Africa. Herbs.

***C. tectorum*** (L.) Pillans. Hills in S.Africa. The stems are used locally for thatching.

CHONDRUS Stackh. Gigartinaceae. Red Algae.

***C. crispus*** (L.) Stack. (Carrageen, Irish Moss). Temperate Atlantic coasts. A commercial source of Carrageenin which is used as a stabiliser. The stabilising properties are due to the presence of high molecular weight (100,000 to 500,000), negatively charged polymers. The plant is harvested by hand or dredging during the summer months, and the carrageenin extracted by boiling, filtering and setting in moulds. It is used as a clearing agent for sugar, liqueurs, etc., for stabilizing chocolate and milk preparations, ice cream, cheese, salad dressing, soups, toothpaste, deodorants, cosmetics, casein, paints, and is used in treating leather and in calico printing.

***C. ocellatus*** Holm. Japan. Used locally for sizing cloth, and making plaster for houses.

***C. plotynus*** J. Ag. Japan. The whole plant is ground up to make a glue.

CHONEMORPHA G. Don. Apocynaceae. 20 spp. S.E.Asia, Indo-Malaya. Woody vines.

***C. elastica*** Merr. Philippines. A possible source of rubber.

***C. macrophylla*** G. Don. E.India, Malaya. A water-resistant fibre is made from the bark and used locally to make fishing nets.

Chonqué – Tetracera assa.

CHORISIA Kunth. Bombacaceae. 5 spp. Tropical S.America. Trees.

***C. insignis*** H.B.K. (Palo Borracho). S.America, esp. Peru, Ecuador, and N.Argentina. The pods are a source of a fibre which is light and water resistant. It is used for life-jackets, and heat and sound insulation. The fibre is exported.

***C. speciosa*** St. Hil. Brazil to Argentina. (Paina de soda). The fibres from the seed-pods (Paineira, Painaliferin) is used for stuffing pillows etc.

Chote – Parmentiera edulis.

Choudae Del – Ficus vallis.

Choum-Choum – Oryza sativa.

Ch'-pei-tzu – Rhus potanini.

Christ Thorn – Ziziphus spina-Christi.

Christmas Berry – Photinia arbutifolia.

Christmas Bush – Ceratopetalum gummiferum.

Christmas Rose – Helleborus niger.

CHROZOPHORA Neck. ex Juss. corr. Benth. and Hook. Euphorbiaceae. 12 pp. Mediterranean, tropical Africa to India. Herbs.

***C. pilcata*** Juss. var. ***obliquifolia*** Prain. N.Africa, E.India. The leaves are used locally as a purgative, and the fruits yield a blue dye.

***C. senegalensis*** A. Juss. Tropical Africa, esp. French Sudan. The ground leaves are used locally to combat intestinal worms.

***C. tinctoria*** (L.) A. Juss.=Croton tincturium. (Giradol). S.Europe, Mediterranean, India. The flowers, fruits and sap have been used as the source of red and blue dyes since ancient times. They are used to colour linen, Dutch cheeses, wines and liqueurs. The dye is variously known as Bambazetto, Bezzetta coerulea, Bezetta rubra, Tornasolis, Tournesol.

***C. verbascifolia*** Juss. Mediterranean, Middle East, India. The leaves and fruits are used locally in Iran to treat whooping cough, and to stimulate the appetite. The seeds are used by the Bedouins as a butter substitute.

CHRYSACTINIA A. Gray. Compositae. 4 spp. S.W. United States of America, Mexico. Small shrubs.

***C. mexicana*** Gray. (Damianita). Mexico. An infusion of the leaves is used locally as a diuretic, to induce sweating, to reduce convulsions, and as an aphrodisiac.

CHRYSALIDOCARPUS H. Wendl. Palmae. 20 spp. Madagascar, Comoro Islands. Palms.

***C. ankaizinensis*** Jum. Madagascar. The buds are eaten as a vegetable locally.

***C. fibrosus*** Jum. (Lafa). Madagascar. The buds are eaten locally as a vegetable. The leaves are the source of a fibre, used for making fishing nets.

***C. paucifolius*** Jum. Madagascar. The buds are eaten as a vegetable locally.

CHRYSANTHEMUM L. Compositae. Sensu lato. 200 spp. Europe, Africa, Asia and America, sensu stricto. 5 spp. Eurasia, Mediterranean. Herbs to woody herbs.

***C. balsamita*** L.=Pyrethrum balsamita (Costmary, Mintgeranium). W.Asia. Used locally for flavouring ale.

***C. cinerariifolium*** (Trév.) Vis.=Pyrethrum cinerariifolium. Dalmatia. Cultivated in S.Europe and U.S.A. for the flowers. The plant contains the insecticide pyrethrum, and the dried flower buds are used in the preparation of various insecticides, including Dalmation Insect Powder Pyrethrum.

***C. coccineum*** Willd. = C. roseum.

***C. coronarium*** L. (Garland Chrysanthemum). Mediterranean. Cultivated in Old and New World. The young seedlings are cooked as a vegetable in China and Japan.

***C. indicum*** L. China, Japan. The flower heads are eaten in Japan, pickled in vinegar.

***C. leucanthemum*** L. (Ox Eye Daisy). Europe, temperate Asia. Introduced to N.America. A syrup of the leaves is used locally to treat catarrh. It is also used to make cough pastilles. The leaves are sometimes eaten in salads. It is also cultivated as an ornamental.

***C. marschallii*** Aschers. Iran. A possible source of insecticide powder.

***C. parthenium*** (L.) Bernh. = Matricaria parthenium. (Feverfew Chrysanthemum). Mediterranean to Caucasus. The flower heads are used locally to make a tonic, to control menstruation, to combat intestinal worms, to induce abortion, and as an insecticide. They are also used to flavour wine, to make a tea and to flavour pastries.

***C. roseum*** Adam. = C. coccineum. Middle East. The dried flower heads are used to make Persian Insect Powder.

***C. segetum*** L. Europe to Asia. The young shoots are eaten as a vegetable especially in China where they are called Tung-hao.

***C. sinense*** Sab. = Pyrethrum sinense. China, Japan. The flowerheads are eaten as a vegetable in Japan. The leaves are occasionally used for the same purpose.

***C. tanacetum*** Karsch. = Tanacetum vulgare.

Chrysanthemum, Feverfew – Chrysanthemum parthenium.

Chrysanthemum, Garland – Chrysanthemum coronarium.

Chrysarobin – Andira araroba.

CHRYSOBALANUS L. Chrysobalanaceae. 4 spp. Tropical America, W.Indies, tropical Africa. Shrubs or small trees.

***C. icaco*** L. (Coco Plum, Icaco Plum). Tropical America. The fruits are eaten in preserves.

***C. orbicularis*** Schum. Tropical Africa, esp. Congo, Senegal, and Angola. The fruits are eaten locally. The seeds yield an edible oil, which is also used for making candles. The bark is a source of tannin.

Chrysoplene – Chrysosplenium alternifolium.

CHRYSOPHYLLUM L. Sapotaceae. 150 spp. Tropics, esp. America. Trees. Members of the genus, especially those from S.America are minor sources of balata.

***C. africanum*** A. DC. (African Star Apple). Tropical Africa. The plant is cultivated locally for the sweet-acid fruits (Odara Pears). The seeds yield an oil which is used locally in soap making. The wood is used for turning and carving. It is also used for furniture and railway carriages.

***C. albidum*** G. Don. Tropical Africa. The juice from the crushed leaves is used locally for coagulating rubber, and the balata from the trunk is used as a bird lime.

***C. argenteum*** Jacq. W.Indies. The fruits are eaten locally.

***C. bacanum*** Miq. = C. roxburghii.

***C. buranhem*** Riedel. = Lucuma glycyphloea.

***C. casinito*** L. = C. olivaeforme (Cainito, Star Apple). W. Indies, C.America. The plant is cultivated in the tropics for the fruits, which are eaten raw or as preserves. The fruits are hard and have a pleasant flavour. They are of two main types, white (Cainito Blanco) or purple (Cainito Morado). The wood is hard, heavy and light brown. It is used for furniture-making.

***C. dioicum*** Koord. and Val. = C. roxburghii.

***C. glabrum*** Jacq. W.Indies. The fruits are eaten locally.

***C. glycyphloeum*** Casar. Brazil. The bark (Monesia Bark, Cortex Monesiae) is used medicinally as an expectorant, to treat stomach complaints and as an astringent.

***C. lucumifolium*** Griseb. Brazil to Argentina. The fruits are edible but of rather poor flavour. The wood is elastic and is used locally for making agricultural implements, and boxes. It has some possible use for paper pulp.

***C. macoucou*** Aubl. = Aubletella macoucou. Guyana. The fruits are eaten locally.

***C. magalis-montana*** Sond. S.Africa. The fruits have a good flavour and are eaten locally as preserves.

***C. magnificum*** Chev. = Godoya aurata.

***C. maytenoides*** Mart. Brazil. The wood is hard and elastic. It is used locally for general carpentry.

***C. michino*** H.B.K. Columbia to Peru. The fruits are eaten locally.

***C. microcarpum*** Sw. W.Indies. The fruits are the size of a grape (smaller than those of the other edible species) but are of good flavour, and are eaten locally.

***C. olivaeforme*** L.=C. cainito.

***C. roxburghii*** G. Don.=C. bacanum=C. dioicum. (Angin anginan, Kajoo nasi, Ki laetan, Laket). Malaysia. The wood is tough and elastic. It is used locally for axe handles.

***C. rugosum*** Sw.=Micropholis melinoniana.

***C. sessilifolium*** Panch. and Sebert. New Caledonia. The orange-coloured wood is hard, elastic and is easily worked. It polishes well and is used for general carpentry and making wagons.

***C. wakeri*** Panch and Sebert. New Caledonia. An important lumber species. The wood is yellowish, fine-grained, resistant and hard. It is used for interior work and turning.

***C. welwitschii*** Engl. W.Africa. The seeds are used as ornaments in Angola.

***Chrysopia macrophylla*** Camb.=Symphonia macrophylla.

***Chrysopogon aciculata*** Trin=Andropogon aciculatus.

CHRYSOSPELNIUM L. Saxifragaceae. 55 spp. N.Temperate and Arctic, N.Africa, temperate S.Africa. Herbs.

***C. alternifolium*** L. (Chrysoplene, Golden Saxifrage). Temperate Europe, Asia, and N.America. The leaves are used occasionally as salads.

***C. oppositifolius*** L. See C. alternifolium.

CHRYSOTHAMNUS Nutt. Compositae. 12 spp. W. N.America. Herbs.

***C. confinis*** Greene. New Mexico. The salted flowerbuds are eaten by the local Indians.

***C. nauseosus*** Britt. (Rabbit Bush). W. N.America. The gum was used by the local Indians as chewing gum. The plant is a potential source of rubber.

***C. viscidiflorus*** (Hook.) Nutt. (Rabbit Bush). W. N.America. The latex is used by the local Indians as chewing gum.

Ch'uan peimu – Pleione pogonioides.

Chucan – Trifolium amabile.

Chucum – Pithecellobium albicans.

Chufa (Oil) – Cyperus esculentus.

Chuflé – Calathea macrosepala.

Chuglam, White – Terinalia bialata.

CHUKRASIA Juss. Meliaceae. 1–2 spp. China to Indomalaysia. Tree.

***C. tabularis*** Juss. (Chittagong Wood, Indian Red Wood, White Cedar). C.Asia, Malaysia. The reddish wood is used for general carpentry. The tree yields an excellent gum.

Chùm goi cây dâu – Scurrula gracilifolia.

Chunari – Fouquieria macdougalii.

Ch'ung tsao – Cordyceps sinensis.

Chuño de concepción – Alstroemeria ligtu.

Chununa de Caballe – Blepharodon mucronatum.

Chupire (Rubber) – Euphorbia calycalata.

Chupones – Greigia sphacelata.

Churco Bark – Oxalis gigantea.

Chuwah (Oil) – Aquilaria agallocha.

CHYTRANTHUS Hook. f. Sapindaceae. 30 spp. Tropical Africa. Trees. The fruits of the following species are edible, and are eaten locally.

***C. gilletii*** De Wild.

***C. macrobotrys*** (Gilg.) Exell and Mendonca=Glossolepis macrobotrys. (Bodumbe, Bonsow, Otoko).

***C. mannii*** Hook. f.

***C. mortehanii*** (De Wild.) De Voldere ex Hauman.

***C. stenophyllus*** gilg.

Cibol – Allium fistulosum.

CIBOTIUM Kaulf. Dicksoniaceae. 10 spp. Tropical Asia, America and Hawaii. Ferns.

***C. barometz*** Smith. Assam, S.China to Philippines. The soft hairs on the young leaves and shoot are used to stop bleeding from wounds.

***C. chamissoi*** Kaulf. Hawaii. The scales from the stems are used locally to stuff pillows.

***C. menziesii*** Hook. See C. chamissoi.

CICER L. Leguminosae. 20 spp. N.Africa. Abyssinia, E.Mediterranean to C.Asia, Herbs.

***C. arietinum*** L. (Chick Pea, Gram Pea, Bengal Gram, Garbanzos). It probably originated in S.Europe, or W.Asia, but has been cultivated for centuries in the Middle East, Greece and S.Europe. It is now cultivated in the Old and New World and is one of the principal leguminous crops of India. The crop is grown during the colder parts of the year. The young green pods are eaten, or the dried seeds are eaten in soups etc. or ground into a flour and used for making a bread and sweets. The seeds are sometimes used as a coffee substitute. The crop is also an important food for livestock.

***C. jaquemontii*** Jaub. and Spach.=C. songaricum.

***C. microphyllum*** Royle=C. songaricum.

***C. songaricum*** Steph.=C. jaquemontii=C. microphyllum. Afghanistan to Tibet. Cultivated in the Himalayas for the seeds which are eaten.

CICHORIUM L. Compositae. 9 spp. Europe, Mediterranean, Abyssinia. Herbs.

***C. intybus*** L. (Chickory, Succory). Europe. Cultivated especially in W.Europe. The ground, roasted roots are used as an adulterant or substitute for coffee. The blanched leaves are used in winter salads especially in France when they are called Barbe de Capucin. The tighter headed varieties (Witloef) are preferred in Belgium. The roots are boiled and eaten with butter, or stored to produce the winter salads. The plant is also grown as livestock fodder. The ground root is also used medicinally as a diuretic, tonic, and for stomach complaints. It is also used to treat diseases of the liver and gall bladder.

***C. endivia*** L. (Endive, Escarolle). S.Europe, India, may have been introduced from E.Indies, may be indigenous to Egypt. It is cultivated in the Old and New World as a salad crop. The outer young leaves are eaten after being bleached to remove the bitter principle. The plant has been cultivated in the United Kingdom since the 16th century.

***Cicuta amomum*** Crantz.=Sison amomum.

Cigarbox Cedrela – Cedrela odorata.

Cilician Tulip – Tulipa montana.

Cimarrón – Calophyllum rekoi.

Cimarrona – Annona montana.

CIMICIFUGA Wernischek. Ranunculaceae. 15 spp. N.Temperate. Herbs.

***C. foetida*** L. Europe to China. A decoction of the root is used medicinally in China to induce sweating, and as a diuretic.

***C. racemosa*** (L.) Nutt. (Black Cohosh, Black Snakeroot). E. N.America. The root contains a bitter principle, a warm decoction of which induces sweating and is diuretic. It is said to be a treatment for snakebite. The root decoction is also used medicinally to improve the appetite, as a sedative, and to control menstruation.

Cina – Lophocereus schottii.

Cinar – Dillenia indica.

CINCHONA L. Rubiaceae. 40 spp. Andes. Tree. The bark of the various species contain about 30 alkaloids, the most important of which is quinine. The medicinal value of the bark in curing fevers, especially malaria, was realised by the Spaniards, and was named after the successful cure of the Countess Cinchon, wife of the Spanish ambassador to Peru. The drug was introduced into Spain in 1639. For about 200 years the only source was the natural forests of the Andes, and this source was ruthlessly exploited. The plant was introduced to India and Java during the 19th century, where it was grown to a considerable extent in Ceylon until it was displaced by tea. The main world supplier is now Java, producing over 90 per cent of that used. ***C. ledgeriana*** bark contains the highest concentration of quinine, and it is varieties or hybrids of this species which is most widely grown. The crop is harvested by felling the trees after ten years, and stripping the bark. Increased yield is obtained by breeding and grafting of high-yielding clones. The crop is losing importance as a source of quinine since the introduction of synthetic substitutes.

***C. affinis*** Wedd.=C. micrantha.

***C. calisaya*** Wedd. (Yellowbark Cinchona). Bolivia. Bark called Calisaya Bark, Yellow Bark. The alkaloid salts are used medicinally to reduce fevers and as a tonic.

***C. calysaya*** var. ***ledgeriana*** How.=C. ledgeriana.

***C. grandifolia*** Mut.=C. pubescens.

***C. lancifolia*** Mut.=C. officinalis.

***C. ledgeriana*** Moens.=C. calisaya var. ledgeriana. (Ledgerbark Cinchona). Bark called Ledger Bark, Yellow Bark).

***C. micrantha.*** Ruiz. and Pav.=C. affinis. (Cascarilla Provinciana, Cascarilla Verde).

***C. morado*** Ruiz. and Pav.=C. pubescens.

***C. nitida*** Ruiz. and Pav.=C. officinalis.

***C. officinalis*** L.=C. lancifolia=C. nitida. (Cascarilla Crespilla, Cascarilla, Cascarilla Verde, Ichu). Bark called Jesuit's, Loxa, Peruvian, Crown, Pale, Bark, Countess' Bark.

***C. pitayensis*** Wedd. The bark contains up to 6·5 per cent quinine, which is high for naturally occurring trees.

***C. pubescens*** Vahl. = C. morado = C. grandifolia. (Sascarilla, Cascarilla Delgarda).

***C. succiruba*** Pavon. (Redbark Cinchona). Bark called Red Bark, Druggists' Bark.

Cinchona, Ledgerbark – Cinchona ledgeriana.

Cinchona, Redbark – Cinchonia succirubra.

Cinchona, Wild – Cinchona pubescens.

Cinchona, Yellow Bark – Cinchona calisaya.

***Cineraria chinensis*** Spreng. = Senecio scandens.

***Cineraria maritima*** L. = Senecio cineraria.

***Cineraria salicifolia*** H.B.K. = Senecio salignus.

CINNAMODENDRON Endl. Canellaceae. 7 spp. Tropical S.America, W.Indies. Small trees.

***C. corticosum*** Miers. W.Indies. The bark (False Winter's Bark), is used as a spice.

CINNAMOMUM Schaeffer. Lauraceae. 250 spp. E.Asia, Indomalaysia. Trees.

***C. burmanii*** Blume (Batavia Cinnamon). Java. The tree is cultivated locally. The dried bark is used as a spice which is exported mainly to the Netherlands. The wood is soft and heavy and is used locally for building.

***C. cambodianum*** Hance. (Tep pirou). Cambodia. The bark is used in making joss sticks, and locally for treating stomach complaints. The bark is also used in various forms of medicines to treat digestive complaints, liver ailments, to control menstruation, and as an astringent.

***C. camphora*** (L.) Nees. and Eberm. = Camphora officinarum. (Camphor Tree). China and Japan. The tree is cultivated locally, and has been introduced into the countries where the climate is suitable. The tree is the source of Camphor which is extracted by steam distillation of wood chips, or macerated wood, and occasionally the twigs. The wood may contain 5 per cent of the crude camphor, and a single tree can yield about 3 tons of camphor. The crude oil settles from the distillate, and from this the solid crystals of camphor settle out. The liquor is then redistilled to yield other fractions, the most important of which is Safrole. This is used in the manufacture of heliotropin which is used in perfumery and for flavourings. Camphor was one of the raw products in the manufacture of celluloid. The replacing of celluloid by other plastics, and the artificial synthesis of camphor has greatly reduced the importance of the industry. Camphor is still used in the production of medicines, especially linaments, and in the manufacture of insecticides.

***C. cassia*** Blume. = C. obtusifolium var. cassia. S.E.China. The tree is cultivated locally for the bark which is used as a spice. It is much exported. The bark is thicker than true cinnamon, and the flavour is less delicate. It is much used in Europe for flavouring liqueurs, and chocolate, as well as cooking. The immature fruits (Cassia Buds) are also used as a spice. An essential oil, mainly cinnamic aldehyde, (Oil of Cassia, Oil of Cinnamon, Oleum Cinnamoni), is obtained by steam distillation of the twigs and terminal foliage. It is used for flavouring, and in perfumery, and to aid digestion.

***C. cecicodaphne*** Meissn. (Gonari, Malagiri, Rehu). Himalayas, N.India. The wood is pale brown, scented, and insect repellent. It is used in making furniture, agricultural implements, boards, and for general carpentry.

***C. culilawan*** Blume. Malaysia, China. The bark is aromatic, and is used locally as a condiment. The calyx of the fruit is used locally to make sauces.

***C. iners*** Reinw. Tropical Asia, esp. southern parts. A decoction of the leaves is used locally as a stimulant, to induce sweating, to aid digestion, to encourage lactation. A decoction of the roots is used locally in Malaya as a tonic after childbirth and to treat fevers.

***C. loureirii*** Nees. = C. obtusifolium var. lourierii. (Saigon Cinnamon). China, Japan, Java. The tree is cultivated for the bark which is used locally as a spice. It is used for flavouring, and medicinally to aid digestion and as an astringent.

***C. massoia*** Schew. = Massoia aromatica. New Guinea. The bark (Massoia Bark), is used as a substitute for cinnamon.

***C. mercadoi*** Vid. (Kalinggag). Philippine Islands. An oil (Kalingag Oil) distilled from the bark is used locally as an aid to digestion. It is also used for flavouring local beers.

***C. mindanaense*** Elm. (Mindanao Cinnamon). Philippine Islands. The bark is used as a cinnamon substitute.

***C. obtusifolium*** Nees. var. ***cassia*** Perrot and Eberm.=C. cassia.

***C. obtusifolium*** Nees. var. ***loureirii*** Perrot and Eberm.=C. loureirii.

***C. oliveri*** Bailey. (Oliver's Bark, Black Sassafras). Queensland. The bark is used locally as a cinnamon substitute.

***C. parthenoxylon*** Meissn. Malaysia. The bark is used locally for flavouring.

***C. pedunculatum*** J. S. Presl. S.Japan. The seeds yield a wax which is used locally to manufacture candles.

***C. sintok*** Blume. (Sintok). Malaysia. The bark (Sintok Bark) is used in Indonesia to treat diarrhoea, and to remove intestinal worms.

***C. tamala*** (Buch.-Ham.) Nees. and Eberm. (Indian Cassia). India. The bark is used as a substitute for cinnamon.

***C. tetragonum*** A. Chev. (Quê do, Tu be rou). N.Vietnam, Cambodia. The aromatic wood and leaves are used locally to make a stimulating drink.

***C. zeylanicum*** Nees. (Ceylon Cinnamon). Ceylon, India, Malaysia. Cultivated in Old and New World tropics, mainly in small plantations. The bark is the source of the spice cinnamon which was one of the main spices of the Eastern spice trade. Several thousands of tons are produced annually, mainly from Ceylon. It is used as a spice, and was used for embalming by the ancient Egyptians. The trees are pollarded to encourage the production of young shoots. The bark is removed from two-year old shoots, during the monsoon season. The cambium is then active and facilitates the removal of the bark. The extraneous tissue is removed and the bark dried to give the cinnamon "quills". The small pieces are ground to make cinnamon powder. The bark contains about 1 per cent essential oil, most of which is cinnamic aldehyde. Cinnamon Oil is distilled from the bark chippings and other waste material, and is used medicinally to relieve stomach complaints. Cinnamon Leaf Oil is distilled for the green leaves. It contains about 90 per cent eugenol, and is used as a substitute for clove oil, as a flavouring for sweets, foods, and toothpaste, and in perfumery.

Cinnamon, Batavia – Cinnamomum burmanii.

Cinnamon, Ceylon – Cinnamomum zeylanicum.

Cinnamon Fern – Osmunda cinnamomea.

Cinnamon, Mindanao – Cinnamomum mindanaense.

Cinnamon Oil – Cinnamomum cassia.

Cinnamon, Saigon – Cinnamomum loureirii.

Cinnamon, White – Canella alba.

Cinnamon, Wild – Canella alba.

Cinquefoil – Potentilla reptans.

Cinquefoil, Bush – Potentilla fruticosa.

Cipo Imbé – Philodendron imbé.

Cipo Suma – Anchietea salutaris.

Ciprés de la Cordillera – Libocedrus chilensis.

Ciprés de las Guaytecas – Pilgerodendron uviferum.

Circassian Tree – Adenanthera pavonica.

Cironballi, Yellow – Nectandra pisi.

CIRSIUM Mill. Compositae. 150 spp. N.Temperate. Herbs.

***C. arvense*** (L.) Scop.=Cnicus arvensis. (Canada Thistle). Europe, Asia. Introduced to N.America. The leaves are used locally to coagulate milk.

***C. drummondii*** Torr and Gray.=Cnicus drummondii. W. N.America. The roots are eaten locally by the Indians.

***C. edule*** Nutt.=Carduus edulis=Cnicus edulis. W. N.America. The peeled stems are eaten by the local Indians.

***C. japonicum*** DC.=Cnicus japonicus=Cnicus maackii. (Ta hsiao kiai). China, Japan, Vietnam. The stem and leaves are used in China to prevent bleeding. The leaves are eaten as a vegetable in Japan. The bark and roots are used in Indochina to prevent sepsis.

***C. maackii*** Maxim=C. japonicum.

***C. occidentale*** (Nutt.) Jeps. See C. undulatum.

***C. ochrocentrum*** A. Gray.=Carduus ochrocentrus. (Yellow Spined Thistle). Nebraska to Texas. The roots and stems are eaten by the local Indians.

***C. oleraceum*** (L.) Scop=Cnicus oleraceus=Carduus oleracerus, Europe to Siberia. The young parts of the plant are eaten as a vegetable in parts of Russia and Siberia,

***C. pallidum*** Wooton and Standl. S.W. U.S.A. The fruits are made into a flour by the local Indians.

***C. tuberosum*** (L.) All.=Cardus bulbosum =Cnius tuberosus. Europe. The roots are eaten as a vegetable in some parts of

Europe. They can be stored for winter use.

***C. scopulorum*** (Greene) Cock. See C. undulatum.

***C. undulatum*** (Nutt.) Spreng = Carduus undulatus = Cnicus undulatus. N. America. The cooked roots are eaten by the local Indians.

***C. virginianum*** (L.) Michx. See C. undulatum.

Ciruela – Casearia pringlei.

Ciruela – Sponchas purpurea.

CISSAMPELOS L. Menispermaceae. 30 spp. Tropics.

***C. acuminata*** Benth. = C. pareura.

***C. caapeba*** Vell. = C. fasciculata.

***C. capensis*** Thunb. Cape Peninsula. Vine. The leaves are used by the local farmers to make an emetic and laxative. The natives use the leaves to treat snake-bite.

***C. denuata*** Miers = C. fasciculata.

***C. fasciculata*** Benth. = C. caapeba = C. denuata. Brazil. Vine. The roots are used locally to make an astringent tonic, which is also used to reduce fevers.

***C. pareira*** L. = C. acuminata. Small shrub. Tropics. The natives in parts of tropical America use a decoction of the roots to reduce uterine bleeding, and excess flow of blood during menstruation. It is also supposed to prevent abortion, to reduce fevers, to be a diuretic and expectorant. A poultice of the leaves is also used to treat snake-bite. The plant contains the alkaloid pelosine.

***C. psilophylla*** Presl. = Stephania japonica.

CISSUS L. Vitidaceae. 350 spp. Tropics, rarely sub-tropics. Woody Vines.

***C. adenocaulis*** Steud. ex A. Rich. = C. articulata. Tropical Africa. The leaves are eaten locally as a vegetable. The juice from the leaves is used to treat eye diseases.

***C. adnata.*** Roxb. = Vitis adnata.

***C. afzelli*** Gilg. = C. planchoniana.

***C. aralioides*** Planch. = Vitis aralioides.

***C. articulata*** Guill. and Perr. = C. adenocaulis.

***C. adenopoda*** Sprague. (Bodegnga, Libiabia). Congo, Nigeria, Uganda. The leaves are eaten locally as a vegetable.

***C. bambusti*** Gilg. and Brandt. (Ikobombive, Mumara). Congo. The leaves are crushed and applied as a dressing to wounds.

***C. barbeyana*** De Willd. Congo, Tanzania, Uganda. The crushed leaves are used to treat skin complaints of children, and as a remedy for gonorrhoea.

***C. barteri*** (Bak.) Planch. See C. barbeyana.

***C. compressa*** Blume. = Vitis compressa.

***C. cyphopetala*** Fres. (Eksalita, Jadaonda, Kabera, Kamka). N.E.Africa, Congo. The crushed leaves are used locally to treat wounds. A decoction of the leaves is used to treat bronchitis, and a decoction of the roots is used to combat intestinal worms.

***C. eliptica*** Schlecht. = Vitis sicyoides.

***C. geniculata*** Blume. = Vitis geniculata.

***C. galuca*** Roxb. = Vitis repens.

***C. latifolia*** Miq. = Vitis adnata.

***C. livingstonia*** Welw. = C. rubiginosa.

***C. modeccoides*** Planch. = C. vitiginea Lour. not L. = C. triloba. (Bach liêm, Giây voi). Indochina. A decoction of the roots is used locally to treat headaches.

***C. papillosa*** Blume. = Vitis papillosa.

***C. petiolata*** Hook. f. = C. suberosa. (Abode, Butolo, Sangu). Tropical Africa. The leaves are used locally to treat chest complaints, and scurf of sheep and goats.

***C. planchoniana*** Gilg. = C. afzelii = C. producta.

***C. populnea*** Guill. and Perr. = Vitis pallida.

***C. producta*** Gilg. = C. planchoniana.

***C. quadrangularis*** L. = Vitis quadrangularia.

***C. repens*** Lam. = Vitis repens.

***C. rubiginosa*** (Welw. ex Bak.) Planch. = C. livingstonia. (Balabodeli, Bunsala, Mbala, Musalo). Tropical Africa, especially W. The juice of the leaves is sometimes used to make a beverage. It is also used to treat fever in children, and intestinal complaints. The leaves crushed on the teeth are used to relieve toothache, and a decoction of the crushed roots is used to relieve abdominal pains.

***C. sicyoides*** L. = Vitis discolor.

***C. triloba*** Merr. = C. modeccoides.

***C. vitiginea*** Lour. not L. = C. modeccoides.

CISTANCHE Hoffmegg. and Link. Orobanchaceae. 16 spp. N.W.Africa, through Mediterranean and W.India to N.W.China. Parasitic Herbs.

***C. lutea*** Hoffmegg. and Link. = Phelypaea lutea. N.Africa. The young shoots, as they emerge through the ground are eaten locally as a vegetable.

***Cistanthera*** K. Schum. = Nesogordonia.

CISTUS L. 20 spp. Mediterranean, Canary Islands to Transcaucasus. Shrubs.

***C. albidus*** L. Metiterranean. The leaves are used by the Arabs to make a tea.

***C. glomerata*** Lag. = Halimium glomerata.

***C. ladaniferus*** L. Mediterranean. A gum (Guma Labdanum, Droga de Jara) exuded from the plant during the summer is collected and distilled to produce a heavily scented perfume (Labdanum). This is used in various perfumes, cosmetics, soaps, deodorants and insecticides. It mixes well with a wide variety of other perfumes. The main areas of production are Spain, Morrocco, Corsica, Greece and S. France.

***C. salvifolius*** L. Mediterranean. A decoction of the roots is used by the Arabs to cure bronchitis, and stop bleeding. The leaves are mixed with the fruit coat of Pomegranate and used for tanning.

***C. villosus.*** L. Spain. A balsam (Labdanum Balsam) is exuded from the stem and was used in medicine to cure bronchitis. The balsam is also used as a fixative in perfumes and soaps.

CITHAREXYLUM Mill. Verbenaceae. 115 spp. S.United States of America to Argentina. Trees.

***C. quadrangulare*** Jacq. (Bois fidèle, Fiddlerwood, Pedulo Colorado). W.Indies. The wood is strong and reddish. It is used for general carpentry, house-building, and the natives use it for making guitars.

Citrange – Poncirus trifoliata × Citrus sinensis.

Citangequet – Poncirus trifoliata × Citrus sinensis × Fortunella margarita.

***Citromyces*** Wehmer = Penicillium.

Citron – Citrullus lantanus.

Citron – Citrus medica.

Citron, Jewish – Citrus medica.

Citron, Myrtle – Backhousia citrodora.

CITRONELLA D. Don. Icacinaceae. S. Str. 7 sp. Central and tropical S.America. Trees.

***C. gongonha*** (Mart.) Howard. S.America. The leaves are used as tea in Brazil.

Citronella – Cymbopogon nardus.

CITRULLUS Schrad. ex Eckl. and Zeyh. Cucurbitaceae. 3 spp. Africa, Mediterranean, tropical Asia. Vines.

***C. colocynthis*** (L.) Schrad. (Colocynth). Dry Africa and tropical Asia. It is cultivated in India and Mediterranean regions for the fruit pulp which is used medicinally as a purgative.

***C. lanatus*** (Thunb.) Mansf. = C. vulgaris. (Water Melon). Tropical Africa. Cultivated widely in Old and New World where the climate is suitable. It has been cultivated at least since Egyptian times. The fruits are eaten raw, they are of little food value, containing about 95 per cent water, but they are important thirst-quenchers in the more arid regions. They are a pleasant dessert fruit. In parts of E. Europe, they are used to make a syrup, and sometimes left on the land to feed bees. The seeds contain 20–40 per cent edible oil, and about 30 per cent protein. In parts of Africa they are dried and eaten raw, being an important supplement to the diet. The seeds are sometimes ground into a flour, or the oil extracted from them used in cooking. The Citron has an inedible pulp, but the fruit is made into pickles, or preserved in syrup.

***C. vulgaris*** Schrad. = C. lanatus.

CITRUS L. 12 spp. S.China, S.E.Asia, Indomalaya. Trees.

***C. aurantifolia*** (Christm.) Swingle = C. medica var. acida. (Lime). E.Indies. It was introduced to the Mediterranean from India by the Arabs about 1000 A.D. From Europe it was introduced into the W.Indies by Columbus in 1493. It is now cultivated in the tropics and sub-tropics of the Old and New World. Mexico produces about 2 million boxes (each of 70 lb.) per annum, while Egypt produces about 1 million. One ton of fruit yields 60–90 gallons of juice. The juice is a good source of vitamin C, and is used in drinks, confectionery and other foods. The oil (Oil of Lime) is distilled from the fruit, or pressed from it; the pressed oil having a superior flavour and odour. It is used in flavouring and perfumery. The oil contains citral, methyl anthranilate and sesquiterpene. An oil (Lime Seed Oil) is extracted from the seeds and used in the manufacture of soap. The plants are propagated by seeds. ***C. aurantifolia*** var. ***limetta*** is the Sweet Lime.

***C. aurantifolia*** Swingle × ***Fortunella margarita*** Swingle (Limequat). An artificial hybrid, with a pleasant-tasting fruit. It is of no commercial value.

***C. aurantium*** L. (Bigarade, Seville Orange, Sour Orange). S.E.Asia. It is grown in the subtropics of the Old and New World. The tree was grown in Spain before the introduction of the Sweet Orange, but is now grown mainly for rootstocks for other Citrus spp. The rootstocks are resistant to many diseases to which other species are susceptible. The fruits are very sour and the bulk of the crop is used in the manufacture of marmalade. The peel is an ingredient in the distilling of the orange liqueurs, Curaçao, Cointreau, and Grand Marnier. The dried peel is used as an aid to digestion and for minor stomach upsets. An oil (Oil of Bitter Orange) distilled from the peel is used in perfumery and to aid digestion. Distillation of the leaves yields Oil of Petitgrain used in perfuming toilet preparations and Eau de Cologne. This is produced mainly in S. France and Paraguay. The flowers yield Oil of Neroli which is also used in perfumery and as a flavouring. Orange Flower Water (Aqua Aurantii Florum) is also produced from the flowers.

***C. aurantium*** L. var. ***bergamia*** Risso. = C. bergamia.

***C. aurantium*** L. var. ***myrtifolia*** Ker-Gawl. (Myrtle-Leaved Orange). A mutant of C. aurantium is cultivated in Italy and S. France for the fruits (Chinottos) which are candied.

***C. aurantium*** L. × ***C. sinensis*** Osbeck. (Bitter Sweet Orange). A hybrid produced in Florida. The fruits are flatter at the ends than the Sweet Orange. The pulp is sweet, but the intercarpellary membranes are bitter. The tree is occasionally cultivated but is of no commercial value.

***C. bergamia*** Risso and Poit. = C. aurantium var. bergamia. (Bergamot). Tropical Asia. Cultivated in S.Italy and Sicily for the fruit peel from which Bergamot Oil is extracted. The oil contains L-linalol, L-linalyl acetate, D-limonene, diptene, and L-linalyl. It is used in perfumery, hair preparations and Eau de Cologne.

***C. decumara*** L. = C. grandis.

***C. grandis*** Osbeck. = C. decumara = C. maxima. (Shaddock, Pomelo, Pummelo). Malaya. The largest of the citrus fruits, the skin is coarse and thick, and the flesh is firm and bitter. It is grown on a small scale in the Old and New World tropics and subtropics. The fruit is of little commercial value, though those of the better varieties are used as a dessert fruit in some parts of India. Soaked in brandy the fruits are used to make a liqueur called Forbidden Fruit.

***C. hystrix*** DC. Philippines. A thorny shrub. The fruits are used locally to flavour meat and fish dishes, and occasionally to make drinks. The peel is sometimes candied.

***C. japonica*** Thunb. = Fortunella japonica.

***C. limetta*** Risso. (Sweet Lemon). Tropical Asia. The tree is occasionally cultivated for the fruits which have an insipid lemon flavour, but are sweeter than the Lemon.

***C. limon*** Burmann. (Lemon). The origin is uncertain, but is probably in subtropical Asia. It was cultivated by the Greeks and Romans, and cultivated in the Azores by 1494. The main areas of cultivation are now California and Italy. The trees are propagated by grafting onto other citrus (usually Bitter Orange) rootstocks. They begin to yield 3–5 years after planting, and yield about 1500–2000 fruits per tree each year. The world production is about 2000 million metric tons per year, California producing 30 per cent and Italy 20 per cent of this total. The remainder of the crop is produced in Spain, Argentina, Greece, Algeria, Australia and Mexico. The fruits are a good source of Vitamin C, carotene, and Vitamin $B_1$, but the 5 per cent citric acid they contain make them too bitter for normal dessert use. The juice is extracted and used for fruit drinks, confectionery and flavouring. It is also a commercial source of citric acid. The oil (Oil of Lemon, Cedro Oil) extracted from the skin is used in perfumery, flavouring foods, flavouring liqueurs (Liqueur d'Or), and to aid digestion. It contains limoene, terpinene, pinene, citronellal, phellandrine, and sesquiterpenes. The pith from the skin is a commercial source of pectin. The seeds are the source of an oil (Lemon Seed Oil, Lemon Pip Oil) used in the manufacture of soap.

***C. limonia*** Osbeck. (Canton Lemon, Ninmeng). Subtropical China. The fruits are similar to those of C. limon. C. limonia var. otaitensis Tanaka is the Otahite Orange, and C. limon var. Khatta Tanaka in the Khatta Orange.

***C. longispina*** West. (Kamisan, Tamisan). Philippine Islands. The fruit is mild, juicy and of good flavour. It is used locally to make fruit drinks and as a breakfast fruit.

***C. marginata*** Juss. = Fortunella marginata.

***C. maxima*** (Burm.) Merr. = C. grandis.

***C. medica*** L. (Citron). Subtropical Asia. Cultivated mainly in S. Italy and Corsica for the peel which is used to make candied peel, used in cake-making. The Etrog Citron (Jewish Citron) is used by the Jews during the Feasts of the Tabernacle.

***C. medica*** L. var. ***acidica*** Brandis. = C. aurantifolia.

***C. microcarpa*** Bunge = C. mitis.

***C. mitis*** Blanco. = C. microcarpa. (Calamodin, Musk Lime). Malaysia. The small orange fruits are acid with a musty smell. They are occasionally cultivated and used in marmalade, jellies, drinks, and to flavour tea. C. mitris × C. nobilis Lour. is the Calarin, and C. mitris × C. reticulata var. unshiu is the Calashu. Neither of these hybrids are of any commercial value.

***C. nobilis*** Andr. not Lour. = C. reticulata.

***C. nobilis*** Lour. = C. reticulata var. deliciosa.

***C. paradisi*** Macf. (Grapefruit). Probably originated in W.Indies as a mutant of C. grandis. It is cultivated in subtropics of the Old and New World, but 90 per cent of the world production comes from U.S.A. especially California and Florida. The fruits are a popular breakfast dish, and large quantities are eaten raw, canned or as juice. The waste products from canning, including the peel is dried and fed to livestock. The juice is a good source of Vitamin C, and Vitamin $B_1$. An oil (Grapefruit Seed Oil) extracted from the seeds is used in soap manufacture, and the sulphonated oil can be used for dyeing cotton. The following hybrids produce pleasantly flavoured fruits, but are of little or no commercial value. C. paradisi × C. mitis (Sopomaldin), C. paradisi × C. reticulata (Siamelo), C. paradisi × C. reticulata var. deliciosa (Tangelo), C. paradisi × C. sinensis, (Chironja), C. paradisi × (C. paradisi × C. reticulata var. deliciosa) (Tangelolo), C. paradisi × C. reticulata var. unshiu (Satsumelo).

***C. reticulata*** Blanco. = C. nobilis Andr. not Lour. (Mandarin Orange). Cochin China. Discovered in the Far East during the 18th Century it is cultivated in the subtropics of the Old and New World, especially S.Europe, and S. U.S.A. for the fruit which is of excellent dessert quality. C. reticulata var. deliciosa is the Tangerine. The fruits of both these are grown for the oil (Mandarine Oil, Tangarine Oil) which is extracted from the skin and used for flavourings and liqueurs. The main sources of the oil are Sicily and Florida. C. reticulata var. papillaris is the Tizon Orange, and C. reticulata var. unshui is the Satsuma Orange. C. reticulata var. unshiu × C. sinensis (Oranguma), has a pleasant flavour, but is of no commercial value.

***C. sinensis*** Osbeck. (Sweet Orange). Probably originated in China or Malaysia. It has been cultivated for thousands of years. It spread from India into E.Africa and to the E.Mediterranean. It was cultivated in Italy by 1 A.D. and was introduced to the western hemisphere by Columbus. By the mid 16th century it had reached Florida. It is now widely cultivated in the tropics and subtropics of the Old and New World. The main areas of production are Florida, California, Arizona, S.Africa, Italy, S.America, Israel. The world production is about 352 million boxes (each of 70 lb.) Its main production is as a dessert fruit, but over 40 per cent of the U.S. crop is used for frozen concentrated juice. It is also used in flavouring confectionery, marmalades and jellies. An oil (Oil of Sweet Orange) is extracted from the peel, and is used for flavouring, perfumes and to relieve digestive complaints. It contains terpinol, D-limonene, linalol, citral and decyl aldehyde. An oil used in perfumes and soaps is distilled from the shoot, leaves and flowers. The main producing areas are, Italy, Sicily and the W.Indies. An oil (Orange Seed Oil, Orange Pip Oil) extracted from the seeds is used in the manufacture of soap. The flowers also produce a good honey. The fruit pulp residue is used to feed livestock for which purpose it compares favourably with sugar beet pulp. There are various hybrids, some of which are of commercial importance. C. sinensis × C. paradisi (Orangelo) is of no commercial value. C. sinensis × (C. pararisi × C. reticulata var. deliciosa) (Siamor) is of possible commercial value. C. sinensis×C. reticulata (Temple Orange, Temple Mandarin) originated in Florida. It has a fine flavour and a thin skin, and is of minor commercial importance, but has a commercial potential. C. sinensis×Poncirus trifoliata (Citrange) has a bitter fruit which is used occasionally for flavouring drinks and foods. It is much more frost-resistant than the other citrus fruits, and herein lies its breeding potential. It is occasionally used as a root stock for oranges, grapefruits and lemons.

***C. webberi*** West. (Kalpi). Philippines. The fruit is used locally like lemons.

Cittwodi – Dolichandrone falcata.

Ciupo – Cavanillesia platanifolia.

CLADIUM P. Br. Cyperaceae. 50–60 spp. Tropical and Temperate, esp. Australia; 1 sp. cosmopolitan. Sedges.

***C. effusum*** (Sw.) Torr.=Mariscus jamaicensis (Saw Grass). S. U.S.A., tropical America. The leaves are sometimes used to make cheap paper.

***C. mariscus*** Pohl. (Fen Sedge, Thatching Sedge). Europe, N. Asia, Africa. The stems are used to make house thatching.

CLADONIA L. Cladoniaceae. Lichen.

***C. alpestris*** (L.) Rabenh. Valuable reindeer food.

***C. coccifera*** (L.) Willd. Used in some parts of Europe to dye wool red-purple.

***C. fimbriata*** (L.) Willd. (Trumpet Lichen). Used to dye wool red.

***C. gracilis*** (L.) Willd. A valuable reindeer food. It is also used to dye wool green.

***C. mitis*** Sandst. A valuable reindeer food.

***C. pyxidata*** (L.) Hoffm. (Cup Moss). Used to dye wool green. Mixed with honey it was used to treat childrens' coughs and whooping cough.

***C. rangiferina*** (L.) Web. (Reindeer Moss). An important food for reindeer. It is used in Europe to dye wool rust-red. An essential oil distilled from it has a possible use in perfumery. The lichen contains about 30 per cent mannose, and can be fermented to produce alcohol.

***C. sylvatica*** (L.) Hoffm.=C. tenuis. A valuable food for reindeer. An oil distilled from it is potentially useful in perfumery.

***C. tenuis*** (Flke.) Harm.=C. sylvatica.

***C. uncialis*** Ach. A food for reindeer.

***C. verticillata*** Hoffm. A food for reindeer.

CLADOPHORA Kuetz. Cladophoraceae. Green Algae.

***C. nitida*** Kuetz. Hawaii. Fresh water. It is eaten locally as a dish, with salted freshwater shrimps.

CLADOSTACHYS D. Don. Amaranthaceae. 12 spp. Madagascar, Indo-Malaya. Herbs.

***C. amaranthoides*** (Lam.) Merr. Tropical Asia. The young leafy shoots are eaten as a vegetable with rice in Indonesia.

CLADRASTIS Rafin. Leguminosae. 4 spp. E.Asia; 1 sp. E. N.America. Trees.

***C. amurensis*** Benth.=Maackia amurensis. Japan, Korea, Manchuria. The hard wood is strong, close-grained and dark brown. It is used in Japan for furniture, interior work in houses, agricultural implements, and household utensils.

***C. lutea*** (Michx.) Koch.=C. tinctoria.

***C. tinctoria*** Rafin.=C. lutea. (Yellow Ash, Yellow Wood). E. U.S.A. The heavy strong close-grained wood is yellow and used for gun stocks. A yellow dye is extracted from the heartwood.

Clamath Plum – Prunus subcordata.

Clammy Horseseed Bush – Dodonaea viscosa.

CLAOXYLON A. Juss. Euphorbiaceae. 80 spp. Old World tropics. Trees.

***C. indicum*** Hassk.=C. polot. Tropical Asia. In Java, the leaves are eaten as a vegetable with rice.

***C. polot*** Merr.=C. indicum.

CLAPPERTONIA Meissn. Tiliaceae. 3 spp. Tropical W.Africa. Trees.

***C. ficifolia*** Hook.=Honckenya ficifolia. The bark yields a fibre resembling jute.

***C. minor*** Bech.=Honckenya minor=H. parva (Bomei). Small tree. The bark is the source of a rough fibre, used locally.

***C. polyandra*** (Bech.) Decs. The bark is the source of a rough fibre used locally.

Clary Sage – Salvia sclarea.

Clary Wort – Salvia sclarea.

Clasping Mullein – Verbascum phlomoides.

CLAUSENA Burm. f. Rutaceae. 30 spp. Old World Tropics. Small trees.

***C. anisata*** Hook. f. Tropical Africa. The leaves smell of aniseed and are used locally as an insect repellent.

***C. anisum-olens*** (Blanco) Merr. Philippines. The leaves stuffed into pillows are used locally as a mild narcotic. Baths in which the leaves are soaked are used to treat rheumatism.

***C. inaequalis*** Benth. Tropical Africa. The

Zulus used a decoction of the leaves to treat intestinal worms.

***C. lansium*** Skeels. (Wampi). S. China. The grape-sized fruits are pleasantly flavoured and are eaten locally. The plant has been introduced into America.

***C. willdenowii*** Wight. and Arn. Tropical Asia. The pleasantly flavoured fruits are eaten locally.

CLAVARIA Vaill. ex Fr. Clavariaceae. 150 spp. Widespread. Fungi.

***C. aurea*** Schäff.=Clavariella aurea. The fruitbodies are eaten. They are frequently sold locally in Europe.

***C. botrytis*** Pers. See C. aurea.

***C. cristata*** Homsk ex Fr.=Clavulina cristata. Temperate and subtropics. The fruitbodies are eaten locally.

***C. flava*** Schäff. See C. aurea.

***C. formosa*** Pers.=Clavariella formosa. The fruitbodies are used to treat intestinal worms in children.

***C. mairei.*** Donk. The fruitbodies are used as a mild purgative.

***C. wettsteinii*** Sing. See C. aurea.

***C. zippeli*** Lev. (Majang). Tropical Asia. The fruitbodies are eaten locally, especially in Malaysia.

***Clavariella aurea*** (Schäff.) Karst.=Clavaria aurea.

***Clavariella formosa*** (Pers.) Karst.=Clavaria formosa.

***Clavulina cristata*** (Holmsk. ex Fr.) Schrött. =Clavaria formosa.

Clavel de oro – Turnera ulmifolia.

CLAVICEPS Tul. Hypocreaceae. 10 spp. Cosmopolitan. Parasitic fungi.

***C. purpurea*** (Fr.) Tul. (Ergot). The fungus is parasitic on several grasses, including rye, on which it is cultivated, particularly in Russia and Spain. The fungus sclerotium (Ergot, Ergot of Rye, Secale Cornuti) replaces the developing grain. It contains alkaloids of lysergic acid, including ergonovine, erotamine and ergosterine. These are used to bring about contraction of the smooth muscles of the uterus and bladder, and have been used for the relief of migraine. The hallucinatory drug lysergic acid diethylamide (LSD) was originally isolated from ergot. Eating rye products contaminated with ergot causes ergot poisoning.

CLAYTONIA L. Portulacaceae. 20 spp. E.Siberia, N.America to the Arctic. Herbs.

***C. acutifolia*** Pal. E.Siberia, Alaska. The roots are eaten by the Eskimos.

***C. balonensis*** (Lindl.) F. v. M.=Calandrina balonensis.

***C. caroliniana*** Michx. E. N.America. The roots are edible.

***C. perfoliata*** Don.=Montia perfoliata. (Winter Purslane). N.America. Introduced to Europe. The leaves are eaten as a vegetable.

***C. polyandra*** (Benth.) F. v. M.=Calandrina polyandria.

***C. sibirica*** L. (Miners' Lettuce, Spring Beauty). W. N.America, N.Asia. A decoction of the leaves is used as a diuretic by some Indian tribes.

***C. virginiana*** L. (Rose Elf, Spring Beauty). E. U.S.A. The underground parts are eaten locally by the Indians.

Clearing Nut – Strychnos potatorum.

CLEISTANTHUS Hook. f. ex Planch. Euphoriaceae. 140 spp. Old World tropics. Trees.

***C. collinus*** Benth. E.India. The dried fruits are used to poison criminals, and the seeds are used to narcotise fish.

***Cleistocactus sepium*** (H.B.K.) Weber.= Borzicactus sepium.

CLEISTOPHOLIS Pierre. Annonaceae. 5 spp. Tropical W.Africa. Trees.

***C. patens*** (Benth.) Engl.=Oxymitra patens. Sierra Leone, Gabon, Uganda. The wood is used locally to make drums. The bark provides a coarse fibre used locally to make ropes and mats. A decoction of the leaves is used to reduce fevers. A decoction of the roots is used to combat intestinal worms. The peel from the fruits is used as a condiment for soups in Togoland.

***C. staudtii*** Engl. and Diels.=Oxymitra staudtii. S.Nigeria, Cameroons, Gabon. The wood is used locally for making huts.

CLEMATIS L. Ranunculaceae. 250 spp. Cosmopolitan, mostly temperate. Woody climbers.

***C. biondiana*** Pav. China. Extracts from all parts of the plant are used locally as a diuretic.

***C. dioica*** L. C.America, S.America, W.Indies. An infusion of the leaves and flowers is used locally as a cosmetic. An ointment from the leaves is used to treat skin complaints.

***C. gouriana*** Roxb. var. ***finetii*** Rehd. and Wils. S.W.China. An infusion of the leaves and stems is used locally as a diuretic and to induce sweating.

***C. grossa*** Bent.=C. rhodocarpa (Barba de chico, Barba de vejo, Chilillo). S.Mexico and C.America. The roots are used in Mexico to treat distemper of horses.

***C. hexasepala*** DC. New Zealand. An infusion of the bark and stems is used locally to improve the appetite.

***C. hirsuta*** Guill. and Perr. S.Africa. The leaves are used locally in Senegal to treat skin diseases.

***C. javanica*** DC. (Buklo, Galing, Merangan). New Guinea, Indonesia, Philippines. The bruised leaves are used in Mindanao to treat wounds.

***C. scabiasifolia*** DC.=Clemaopsis scabiosifolia. Tropical Africa. Shrubby. A warm decoction of the leaves is used to treat rheumatism, and coughs. Inhaling the vapour is supposed to relieve migraine. The dried roots are used to light fires.

***C. thunbergii*** Steud.=C. hirsuta.

***C. vitalba*** L. C.Europe to Caucasus. The leaves (Herba Clematidis) are used as a diuretic. The young shoots are eaten locally as a vegetable.

***Clematopsis scabiosifolia*** (DC.) Hutch.= Clematis scabiasifolia.

CLEOME (L.) DC. Clemaceae. 150 spp. Tropics and subtropics. Herbs.

***C. arabica*** L. (Agassei). Sahara. The seeds are used locally for flavouring sauces, and the leaves are used to treat abdominal pains.

***C. brachycarpa*** Vahl. Red Sea, across the drier inland regions of Africa. The leaves are used locally to relieve abdominal pains.

***C. chelidonii*** L. f. Tropical Asia. The roots are used in Indochina to treat intestinal worms.

***C. gigantea*** L. (Mussambé). S.America from Peru to Trinidad. The root and leaves are used locally to treat bronchitis and asthma. They are sold locally as Mussambé.

***C. integrifolia*** Torr.=Peritoma serrulatum. N.America. The local Indians ate the plant as a porridge with cornmeal in the spring, and some tribes use it as the source of a black dye for decorating pottery.

***C. monophylla*** L. Tropical Asia. The powdered roots are used locally to restore consciousness after fainting.

***C. viscosa*** L.=Polanisia icosandra. Tropics. Used in the Far East to stimulate the appetite when added to food. The pods are made into pickles. Sold locally.

CLERMONTIA Gaudich. Campanulaceae. 27 spp. Hawaii. Small trees.

***C. gaudichaudii*** (Gaud.) Hbd. Hawaii. The berries are eaten locally, and the latex is used as bird lime.

CLERODENDRUM L. Verbenaceae. 400 spp. Tropical and subtropical. Shrubs.

***C. blumeana*** Schau.=C. buchananii.

***C. buchananii*** Walp.=C. blumeana. (Kembang boogang, Mata ajam, Titinga). Malaya, Indonesia. The crushed leaves are used locally to treat dysentery and burns. The roots are used to treat snakebite.

***C. glabrum*** E. Mey. S.Africa. The bark is used locally as a purgative for calves.

***C. serratum*** Spreng. E.India, Malaysia. In Java the young leaves and flowers are eaten as a vegetable with rice.

CLETHRA Gronov ex L. Clethraceae. 120 spp. Asia, America.

***C. barbinervis*** Sieb. and Zucc. Japan. The leaves are eaten as a vegetable with rice.

CLIBADIUM Allem. ex L. Compositae. 50 spp. Tropical America, W.Indies. Herbs.

***C. surinamensis*** L. Guyana. The leaves and stems are used locally to stupify fish.

CLIDEMIA D. Don. Melastomataceae. 145 spp. Tropical America, W.Indies.

***C. blepharoides*** DC.=Melastoma coccineum. Brazil. Epiphyte. The leaves are used locally for treating ulcers. The fruits of the following species are eaten locally. They are all small woody plants, found in Mexico and C.America.

***C. chinantlana*** (Naud.) Triana.

***C. dependens*** D. Don.

***C. deppeana*** Steud.=Melastroma petiolare.

***C. hirta*** (L.) D. Don.

***C. naudiniana*** Cogn.

CLIFFORTIA L. Rosaceae. 80 spp. Africa. Woody plants.

***C. ilicifolia*** L. S.Africa. An extract of the leaves (Thorntea) is used locally to soothe coughs and as an expectorant.

CLIFTONIA Banks ex Gaertn. f. Cyrillaceae. 1 sp. S.E. United States of America. Shrub.

***C. monophylla*** (Lam.) Sarg. (Buckwheat Bush, Titi). The flowers are a source of honey.

Chikyeng – Elaedendron glaucum.

CLIMACANTHUS Nees. Acanthaceae. 2 spp. S.China, Indochina, Malaysia. Shrubs.

***C. burmanii*** Nees.=Justicia fulgida=J. nutans. The young leaves are eaten as a vegetable in Annan, the leaves are also used to treat eye diseases.

Climbing Entada – Entada scandens.

Climbing Lily – Gloriosa superba.

CLINOPODIUM L. Labiatae. 10 spp. N.Temperate.

***C. ascendens*** Samp. S.Europe. The leaves are used locally in Portugal to aid digestion.

***C. laevigatum*** Standl. Mexico. Shrub. The leaves are used locally to make a tea.

***C. macrostemum*** (Benth.) Kuntze. (Té del Monte). Mexico. Shrub. The leaves are used locally to make a tea.

***C. vulgare*** L.=Calamitha clinopodium.

CLINOSTIGMA Wendl. Palmae. 5 spp. Samoa, Fiji. Trees.

***C. oncorhyncha*** Beccari. The wood is used in Samoa to make the roofs of huts.

CLITANDRA Benth. Apocynaceae. 1 sp. W. to Central Tropical Africa. Small tree. The systematics of the genus is confused; all the species mentioned may be included in the single species, ***C. orientalis.*** All produce a rubber latex of variable quality.

***C. arnoldiana*** De. Wild. (Kappa). Congo, Uganda.

***C. elastica*** Chev. Ivory Coast.

***C. flavidiflora*** Hall. W.Africa.

***C. nzunde*** De Wild. Congo.

***C. orientalis*** Schum. (Kappa). Tropical Africa, esp. Congo, and Uganda. The rubber is called Kappa, or Noire de Congo.

***C. schweinfurthii*** Stapf. Nile region.

***C. simoni*** Gilg. N.W.Cameroons.

CLINTONIA Rafin. Liliaceae. 6 spp. E.Asia, N.America. Herbs.

***C. borealis*** (Ait.) Rad. (Corn Lily, Cow Tongue, Straw Lily). Labrador to N.Carolina. The young leaves are used locally as a vegetable and salad.

***C. umbellata*** (Michx.) Morong. See C. borealis.

CLITOCYBE (Fr.) Quél. Agaricaceae. 150 spp. Cosmopolitan.

***C. candida*** Bres. A possible source of the antibiotic Clitocybin, used in the treatment of tuberculosis.

***C. castanea*** Beeli. Tropical Asia. The fruitbodies (Neense) are eaten locally.

***C. geotropa*** Quél. A possible source of Clitocybin.

***C. gigantea*** Beeli. A possible source of Clitocybin.

***C. hypocalamus*** Van Overeen. Indonesia. The fruitbodies are eaten especially in Java.

***C. infundibuliformis*** Quél. A possible source of Clitocybin.

***C. multiceps*** Peck.=Agaricus decastes.

***C. nebularis*** (Batsch. ex Fr.) Quél. Temperate. The edible fruitbodies are sold locally in Europe.

***C. tessulata*** (Bull.) Sing. Temperate. Eaten by N.American Indians.

***C. tuberaster*** (Briganti jun.) Sacc. The fruitbodies are eaten.

Clitocybin – Clitocybe spp.

CLITOPILUS (Fr.) Quél. Agaricaceae. 30 spp. N.Temperate.

***C. prunulus*** (Scop.) Quél. Cosmopolitan. The fruitbodies (Mousseron) are eaten in Europe and Asia. They are sold locally.

CLITORIA L. Leguminosae. 40 spp. Tropics and subtropics. Herbs or vines.

***C. cajanifolia*** Barth. Tropics. Used locally as a green manure.

***C. lasciva*** Vah. E.Madagascar. The stem yields a fibre which is used locally for making ropes.

***C. ternata*** L. (Butterfly Bean, Butterfly Pea, Kordofan Pea). Tropics and sub-tropics. A useful forage crop, especially in Australia, where it forms part of the natural pasture. It grows well under irrigation in the Sudan, and although it dies down if water is withheld, it shoots again when rewatered. A useful ground cover against weeds. The pods are eaten as a vegetable in various parts of the tropics. The roots are a strong cathartic, and the seeds are used as a purgative in the Sudan. The flowers are used to colour rice blue in Amboina.

Closed Gentian – Gentiana andrewsii.

CLOSTRIDIUM Beij. Bacillaceae.

***C. acetobutylicum*** McCoy et al. Used in the production of acetone and butanol by the anaerobic fermentation of sugars, usually the waste products of sugar manufacture.

***C. beijerinckii*** Donker. Used to ferment the waste sulphite liquors from sugar manufacture to produce butanol, acetone and isopropanol.

***C. butylicum*** (Beijerinck) Donker. One of the main organisms causing the anaerobic retting of flax and hemp, when they are submerged in water. It is a possible agent for the production of butanol.

***C. felsineum*** (Carbone and Tomb.) Bergey et al. See C. butylcum.

***C. pasteurianum*** Winogr. A soil species capable of fixing atmospheric nitrogen.

***C. toanum*** Baba. Used in the production of butanol, acetone, and isopropanol from sugar syrup and sugar wastes.

Cloth, Tapa – Broussonetia papyrifera.
Clotweed – Xanthium spinosum.
Cloudberry – Rubus chamaemorus.
Clove – Eugenia caryophyllata.
Clove – Syzygium aromaticum.
Clove Bark (Oil) – Dicypellium caryophylatum.
Clove, Madagascar – Ravensara aromatica.
Clove Oil – Eugenia caryophyllata.
Clove Pink – Dianthus caryophyllatus.
Clove Tree – Eugenia caryophyllata.
Clover, Alsike – Trifolium hybridum.
Clover, Alyce – Alysicarpus vaginalis.
Clover, Aztec – Trifolium amabile.
Clover, Ball – Trifolium nigrescens.
Clover, Beach – Trifolium fibriatum.
Clover, Bur – Medicago hispida.
Clover, Button – Medicago orbicularis.
Clover, Chilean – Medicago sativa.
Clover, Cluster – Trifolium glomeratum.
Clover, Crimson – Trifolium incarnatum.
Clover, Egyptian – Trifolium alexandrinum.
Clover, Holy – Onobrychus viciifolia.
Clover, Hop – Medicago lupulina.
Clover, Hungarian – Trifolium pannonicum.
Clover, Japan – Lespedeza striata.
Clover, Kara – Trifolium ambiguum.
Clover, Ladino – Trifolium repens.
Clover, Lappa – Trifolium lappaceum.
Clover, Large Hop – Trifolium campestre.
Clover, Persian – Trifolium resupinatum.
Clover, Purple Prairie – Petalostemon purpureum.
Clover, Red – Trifolium pratense.
Clover, Rose – Trifolium hirtum.
Clover, Slender Prairie – Petalostemon oligophyllum.
Clover, Small-flowered – Trifolium parviflorum.
Clover, Snail – Medicago scutellata.
Clover, Spotted Bur – Medicago arabica.
Clover, Strawberry – Trifolium fragiferum.
Clover, Striata – Trifolium striatum.
Clover, Sub – Trifolium subterraneum.
Clover, Teasel – Trifolium parviflorum.
Clover, White – Trifolium repens.
Clover, White Sweet – Melilotus alba.
Clover, Yellow Sweet – Melilotus officinalis.
Clown's Mustard – Iberis amara.
Club Flower – Cordylanthus wrightii.
Clubmoss – Lycopodium clavatum.
Club Rush – Scirpus lacustris.
Club Wheat – Triticum compactum.

CLUSIA L. Guttiferae. 145 spp. Mostly Warm America, Madagascar, New Caledonia. Trees.

***C. alba*** Choisy. = C. palmicida.

***C. flava*** L. (Fat Pork, Monkey Apple). W.Indies. A gum from the stem (Hog Gummi) is used locally for healing wounds.

***C. fluminensis*** Planch. and Triana. Brazil. A resin is collected from the stem, and used in veterinary medicine.

***C. insignis*** Mart. Brazil. A resin extracted from the flowers is used locally for healing wounds.

***C. minor*** L. C. and S.America. A gum from the stem is used locally to make elastic bandages for binding hernias in children.

***C. palmicida*** Rich. = C. alba. S.America. The source of incense used in local churches.

***C. plukenetii*** Urb. (Caquelin, Pomme Chique). Lesser Antilles. Large epiphyte. A resin (Rock Balsam) extracted from the bark is used medicinally. The sap is used locally as bird lime.

***C. rosea*** Jacq. (Cupey). Tropical America, W.Indies. The wood is heavy and strong, and difficult to work, but is not resistant to decay or insect borers. It is used locally for cheap furniture, farm implements posts, and fuel.

Cluster Bean – Cyamopsis psoralioides.
Cluster Cardamon – Elettaria cardamomum.
Cluster Clover – Trifolium glomeratum.
Cluster Pine – Pinus pinaster.

CLUTIA L. Euphorbiaceae. 70 spp. Africa, Asia. Trees.

***C. hirsuta*** Muell. Arg. S.Africa. A decoction of the plant is used by the natives to treat splenic fever. Mixed with brandy it is used by the farmers to treat stomach complaints.

***C. similis*** Muell. Arg. S.Africa. The natives used the leaves to treat anthrax, and the root to treat snakebite.

***Cluytia*** Ait. = Clutia.

***Clypea rotunda*** Steud. = Stephania rotunda.

CNEORUM L. Cneoraceae. 1 sp. Cuba; 1 sp. W. and C. Mediterranean. Shrubs.

***C. tricoccum*** L. (Spurge Olive). Mediterranean. The leaves and fruits are used as a purgative, and to induce local warmth and reddening of the skin. They contain a great deal of tannin. In S.France the plant is used as a fuel.

CNESTIS Juss. Connaraceae. 40 spp. Warm Africa, Madagascar, Malaysia. Woody shrubs.

***C. corniculata*** Lam. (Furuluga, Oboqui). Tropical Africa, especially West. The leaves are used as a powerful astringent.

***C. ferruginea*** DC. Tropical Africa. In Nigeria the leaves are used as a laxative, and in Sierre Leone the fruits are used to clean the teeth.

***C. polyphylla*** Lam. Madagascar. The fruits are used locally to kill dogs.

***C. volubilis*** Blanco.=Rourea volubilis.

CNICUS L. emend Gaertn. Compositae. 1 sp. Mediterranean. Herb.

***C. arvensis*** Hoffm.=Cirsium arvense.

***C. benedictus*** L.=Cantaurea benedicta=Garbenia benedicata. (Blessed Thistle). Mediterranean to Iran and Caucasus. Widely naturalised. The leaves (Herbs Acathi Germaici, Herba Cardui Benedicti) were widely used in mediaeval medicine, but have lost a great deal of their importance. Their use includes treatment of fevers, intestinal complaints, catarrh, hysteria, and diseases of the liver and lungs. The plant has been cultivated especially in Germany, as an emergency source of oil, which is extracted from the seeds. The seed cake is a good stock food.

***C. drummondii*** Gray=Cirsium drummondii.

***C. edulis*** Gray=Cirsium edule.

***C. japonicus*** Maxim.=Cirsium japonicum.

***C. oleraceus*** L.=Cirsium oleraceum.

***C. tuberosus*** Willd.=Cirsium tuberosum.

***C. undulatus*** Grey.=Cirsium undulatum.

CNIDOSCOLUS Pohl. Euphoriaceae. 75 spp. Tropical America. Small trees.

***C. elasticus*** Lundell. Mexico. The latex contains about 50 per cent rubber (Highland Chilte Rubber) which is coagulated by water.

***C. marcgravii*** Pohl=Jatropha urens.

Coachwhip – Fouquiera splendens.

Coachwood – Ceratopetalum apetalum.

Coral Tree, Hawaiian – Erythrina sandwicensis.

Coast Grey Box – Eucalyptus bosistoana.

Cob Nut – Corylus avellana.

Cob Nut – Omphalea triandra.

Coca, Huanuca – Erythroxylum coca.

Có co té – Afzelia cochinchinensis.

Coca Tree – Erythroxylum coca.

Coca, Truxillo – Erythroxylum nova-granatense.

COCCINIA Wight. and Arn. Cucurbitaceae. 30 spp. Tropical and S.Africa; 1 sp. tropical India and Malaysia. Vines.

***C. grandis*** (L.) Voigt.=C. indica=Cephalandra indica. Tropical Asia to Sudan. The plant is occasionally cultivated. The fruits and young shoots are eaten as a vegetable with rice in India, Indochina and Java. In the Sudan they are eaten raw, cooked or candied. The leaves (Herba Cephalandrae) are a laxative and are used to treat diabetes.

***C. indica*** Wight. and Arn.=C. grandis.

COCCOLOBA R. Br. mut L. Polygonaceae. 150 spp. Tropical and subtropical America. Trees.

***C. grandifolia*** Jacq.=C. pubescens. Tropical America. The reddish hard wood is used in general carpentry.

***C. laurifolia*** Jacq. (Pigeon Plum). W.Indies, S.America, Florida. The dark brown wood is hard and heavy, but rather brittle. It is occasionally used for cabinet-making.

***C. oaxacensis*** Grom.=C. schiedeana.

***C. pubescens*** L.=C. grandifolia.

***C. schiedeana*** Lindau.=C. oaxacensis. (Carnero de la costa, Tepalcahiute). Mexico to Guatemala. The wood is used locally for making cart wheels.

***C. uvifera*** (L.) Jacq. (Sea Grape). Tropical America. The hard, purple-brown wood is used to make furniture. The fruits are made into a jelly, and, in the W.Indies into an alcoholic drink. The bark is the source of Jamaican Kino.

***Coccolobis*** P. Br.=Coccoloba.

COCCOTHRINAX Sargent. Palmae. 50 spp. S.Florida, W.Indies. Small palm trees.

***C. argentea*** (Lodd.) Sarg.=C. jucunda (Biscayne Palm). S.Florida. The leaves are used locally to make hats, baskets, etc.

***C. jucunda*** Sarg.=C. argentea.

***C. martiniensis*** Beccari. (Alatani, Latanier). Dominica, Martinique. The leaves are used locally for thatching, basket-making etc. The wood is made into bows.

COCCULUS DC. Menispermaceae. 11 spp. Tropics and subtropics, excluding S.America. Woody vines.

***C. bakis*** Guill and Perr.=Tinospora bakis.

***C. cebatha*** DC.=Menispermum edule. Arabia. The fruits are used, mixed with raisins to make an alcoholic drink.

***C. crispus*** DC.=Tinospora crispa.

***C. ferrandianus*** Gaudich. Hawaii. The fruits are used to capture fish.

***C. fibraurea*** DC.=Fibraurea tinctoria.

***C. filipendula*** Mart. Brazil. A poisonous plant used locally to control menstruation and to treat snakebites.

***C. laurifolius*** DC. Japan to Java. The bark contains the alkaloids cocculine and coclaurine, which are used medicinally as muscle relaxants.

***C. leaeba*** DC.=C. pendulus.

***C. palmatus*** Hook.=Jateorhiza miersii.

***C. pendulus*** (Forst.) Diels.=C. leaeba. Arabia to Afghanistan. The Arabs use the juice to make an alcoholic drink, while in Afghanistan it is used to treat fevers.

***C. sarmentosus*** (Lour.) Diels. Malaysia. The roots are used locally to make a poison.

***C. thunbergii.*** DC. Japan. The tendrils are used locally to make baskets.

***C. tomentosus*** Colebr.=Tinospora sinensis.

***C. tuberculatus*** L.=Tinospora crispa.

COCHLEARIA L. Cruciferae. 25 spp. N.Temperate, S. to E.Himalayas, Mountains of Java (introduced?).

***C. amoracia*** L.=Armoracia lapathifolia.

***C. officinalis*** L. (Scurvy Grass, Scorbute Grass, Spoonwort). Europe, temperate Asia. Occasionally grown for the treatment of scurvy.

***C. wasabi*** Sieb.=Eutrema wasabi.

COCHLOSPERMUM Kunth. Cochlospermaceae. 15–20 spp. Temperate. Xerophytes, small trees or shrubs.

***C. angolense*** Welw. and Oliv. Angola. The seeds produce a red dye used locally.

***C. gossypium*** DC. E.India. Source of an insoluble gum (Katira Gum), used as Tragacanth (see Astragalus gummifer). The plants are grown near Indian temples for the decorative yellow flowers.

***C. niloticum*** Oliv. Tropical Africa. The tubers are the source of a yellow dye, and are chewed in some parts as a tonic.

***C. planchoni*** Hook. Tropical Africa. The tubers are the source of a yellow dye used in the Sudan. It is mixed with indigo to make a green dye by the Hausas. An extract of the plant is used locally to control menstruation. The bark is the source of a fibre used to make ropes.

***C. vitifolium*** Spreng.=Maximiliana vitifolia. Mexico. The bark is used locally for making ropes.

Cocillana – Guarea rusbyi.

Cocklebur, Broad – Xanthium strumonium.

Cock's foot – Dactylis glomerata.

Cockspur Thorn – Crataegus crus-galli.

Cockur – Ochrolechia tartea.

Coco – Calocasia esculenta.

Coco – Otoba gordoniifolia.

Coco de Chile – Jubaea chilensis.

Coco de Mer – Lodoicea maldivica.

Coco Grass – Cyperus rotundus.

Coco Nain – Rhyticocos amara.

Coco Palm – Cocos nucifera.

Coco Plum – Chrysobalanus icaco.

Cocobola – Dalbergia retusa.

Cocobola – Lecythis costaricensis.

Cocobola, Yama – Platymiscium dimosphandrum.

Cocograss – Cyperus rotundus.

Cocona – Solanum topiro.

Coconut (Oil) (Milk) – Cocos nucifera.

Coconut, Double – Lodoicea maldivica.

Cocora – Grias peruviana.

Cocorite Palm – Maximiliana caribaea.

COCOS L. Palmae. 1 sp. Probably from Asia or Polynesia. Tree.

***C. coronata*** Mart=Syagrus coronata.

***C. nucifera*** L. (Coconut Palm, Coco Palm). Polynesia. Now widely cultivated throughout Old and New World tropics. An important commercial crop, being the sole source of income for some of the smaller islands in the Pacific. The plant is grown commercially for the fruits which yield Coconut Oil. The plants come into bearing after 7 years growth, reaching a maximum in about the 12th year. A single tree may produce 400 fruits a year, yielding about 200 lb. of copra and 10 gal. of oil. The fruits are collected by hand and split open, and the "flesh" (Copra) removed. This is dried in the sun or over fires. It is either exported raw, or the oil is extracted and exported. The copra contains 60–65 per cent oil which is extracted by steam-heating then pressing. The residue (Poonac) is an excellent food for livestock. The oil is refined by the removal of fatty acids by alkali, and deodorized by treating with superheated steam. The oil contains a high percentage of lauric acid which makes it very suitable for the manufacture of quick-lathering soaps. It is used extensively in the manufacture of margarine, cosmetics, and synthetic rubber. About 3·5 million tons of copra are produced annually, mainly from the Far East, but also from Mexico and Brazil. The flesh is also shredded and sold as desiccated coconut which is used in confectionery and cookery. The outer husk of the nut is retted in sea water, and when beaten, the fibre (Coir) is extracted. This is used as a cheap stuffing material, brushes, and for coconut matting; more locally it is used for making salt-resistant rope. The husks are used locally as utensils, and when burnt yield a highly adsorbant charcoal well-suited for gas masks. The liquid from the unripe fruit makes a refreshing drink. The wood is used locally for building construction, utensils and furniture. The outer wood (Porcupine Wood) is exported for cabinet-making. The leaves are woven into sheets (Cadjans) which are used for thatching, and house-building, baskets etc. The leaf-midribs are used as poles for a variety of purposes. The apical bud, "cabbage" is eaten as a vegetable or pickled. The juice from the inflorescence stalk contains sugar (Jaggery) which is obtained by evaporation. The fermented juice is the beverage Toddy, which when distilled yields the highly intoxicating spirit Arrack. Further fermentation of toddy produces a vinegar. In Java a decoction of the roots is used to treat dysentery; in India the hairs from the lower surface of the leaves are used to stop bleeding; in Ghana the bark is used to clean the teeth, as an antiseptic, and the ash is used to treat scabies; in Mexico the "milk" is considered diuretic and is used to combat intestinal worms.

***C. romanzoffiana*** Cham. = Arecastrum romanzoffianum.

***C. yatay*** Mart. Argentina. (A doubtful genus). The fruits are used locally to make an alcoholic drink. The young buds are eaten as a vegetable.

Cocos Wood – Brya ebenus.

Cocoyam – Colocasia antiquorum.

Cocus Wood – Brya ebenus.

Cocuiza Fibre – Furcraea humboldtiana.

***Codia montana*** Labill. = Callicoma billardieri.

***Codiaeum lucidum*** Muell. Arg. = Baloghia lucida.

CODIUM Stackh. Codiaceae. Green Algae. The genera mentioned are eaten locally. Extracts of the plant are effective against intestinal worms.

***C. fragile*** (Sm.) Huit. Japan.

***C. geppii*** O. C. Schmidt. (Tambalang). Pacific Coasts. Eaten in Philippines.

***C. lindenbergii*** Binder. Japan. Eaten fresh or dried and salted.

***C. muelleri*** Kuezt. Pacific Coasts. Eaten in Hawaii, in a dish of peppers and squid.

***C. tenue*** Kütz. Philippines.

***C. tomentosum*** (Huds.) Stackh. Japan, Philippines. It is eaten, in Japan in soup or flavoured with soy sauce and vinegar.

Codo de fraile – Thevetia thevetioides.

CODONOCARPUS A. Cunn. ex Hook. Gyrostemonaceae. 3 spp. Australia. Trees.

***C. cotinifolia*** (Desf.) F. Muell. Arg. (Medicine Tree, Native Poplar). Australia. The roots are eaten as food by the natives.

CODONONOPSIS Wall. Campanulaceae. 30–40 spp. C. and E.Asia, Himalayas, Malaysia. Herbs.

***C. lanceolata*** Benth. China. The roots are the source of a tonic and aphrodisiac, used locally.

***C. tangshen*** Oliv. China. The roots are the source of a tonic and aphrodisiac used locally.

***C. viridiflora*** Wall. C.Asia, Himalayas. See C. lanceolata.

***C. ussuriensis*** Hemsl. Japan, China. The cooked or raw roots are eaten locally in Japan.

COELOCOCCUS H. Wendl. Palmae. 2 spp. Polynesia. Trees.

***C. amicarum*** Warb. (Polynesian Ivory Nut Palm). Carolina Islands. Cultivated in the Philippines for the hard seeds which are used as an ivory substitute in the manufacture of buttons.

***Coelodiscus anamiticus*** Gagnep. = Mallotus anamiticus.

***Coelopleurum gmelinii*** (DC.) Ledeb. = Archangelica gmelinii.

***Coelorhopalon obovatum*** (Berk.) v. Overeem. = Xylaria obovata.

COELSTEGIA Benth. Bombacaceae. 5 spp. W. Malaysia. Trees.

***C. griffithii*** Benth. (Dorian oogeh, Dowren hantos). Malaysia, Indonesia. The bark is used to treat fishing nets in Malaysia.

Coeur kambaur – Tinospora sinensis.

Coeur palmiste – Euterpe dominicana.

COFFEA L. Rubiaceae. 40 spp. Old World tropics, especially Africa. Small trees.

***C. arnoldiana*** de Wild. Congo. A possibility for cultivation as it will tolerate drier climates than C. arabica, and does not require shade.

***C. arabica.*** L. (Arabian Coffee). Highlands of Ethiopia. This plant accounts for about 90 per cent of the world's coffee production, which amount to some 66·5 million tons per year. Of this Brazil produces about 50 per cent, and Columbia about 12 per cent. Other producers include, Ivory Coast, Mexico, Java, Angola, and Jamaica. The plant was first cultivated in Arabia in the 6th century, and the main supplies to Europe came from the Yemen. As the drink became popular in Europe, the Dutch introduced the crop to Ceylon (1685) and Java 1699. In 1727 it was introduced to Brazil, and into the W.Indies about the same time. Ceylon was the main producing country until the end of the 19th century when it was replaced by tea, the coffee having been destroyed by the rust fungus Memileia vastatrix. It is a tropical crop requiring over 60 inches of rain a year, and permanent shade from larger trees. The plants are grown from seeds and are pruned to keep them about 15 ft. tall. Production begins about 3 years after sowing, and continues for 30 to 40 years. The fruit is a drupe containing two seeds which are the coffee beans of commerce. The outer pericarp is removed either by sun-drying and hull, or more commonly by a wet process. In the latter process, the berries are floated to remove unripe ones and debris. The ripe berries, which sink to the bottom of the tanks, are pulped and fermented to remove the pericarp. The seeds are then dried in the sun or artificially, after which they are hulled to remove two skins, the silver skin and the parchment, which still adhere to them. They are then ready for sale. The typical coffee flavour is developed during roasting. The actual flavour depends on the variety of plant used, the cultural conditions, the method of curing and drying, and the age of the beans. The flavour of the beans is due to the oil caffeol, which goes rancid if the beans are kept too long. The stimulating properties are due to the 1–2 per cent caffeine contained in the beans. Besides these the beans contain small amounts of glucose, dextrose, and proteins. Caracolla or Pea Berries are obtained from the fruits at the tips of the branches, which contain only one seed. Coffee is used principally as a beverage, but is used in flavouring foods, confectionery and ice-cream. The liqueurs Tia Maria and Kahlua have coffee as one of the main ingredients. The dried beans are also a commercial source of caffeine, and are used medicinally as a stimulant and diuretic. It stimulates the muscles, kidneys, heart, and central nervous system.

***C. bertrandii*** Chev. Madagascar. The seeds contain no caffeine. A possible source of breeding material to produce caffeine-free coffee.

***C. bonnieri*** Dub. See C. bertrandii.

***C. breviceps*** Hiern. Tropical Africa. A possible commercial source of coffee.

***C. canephora*** Pierre. Congo. There is some confusion between this species and C. robusta. It is occasionally cultivated. The beans are of reasonable quality.

***C. dewevrei*** de Wind. Tropical Africa, esp. Congo. A potential source of cultivated coffee.

***C. excelsa*** Chev. Tropical W.Africa. Occasionally cultivated in Old and New World. The coffee is of good quality, and the trees are good yielders.

***C. gallienii*** Dub. Madagascar. A potential source of caffeine-free coffee.

***C. humboldtiana*** Baill. French Africa. A potential source of caffeine-free coffee.

***C. laurentii*** de Wild. = C. robusta.

***C. liberica*** Hiern. (Liberian Coffee). Tropical W.Africa. Cultivated in W.Indies, Guyana, and many coffee-producing areas of the Old World. The coffee is much inferior to that from C. arabica, but the plants are more hardy, and do not require shade. A source of breeding material, and rootstocks.

***C. maclaudii*** Chev. French Guinea. The source of a reasonably good coffee.

***C. mogenetii*** Dub. Madagascar. A source of caffeine-free coffee.

***C. mozambicana*** DC. = C. racemosa.

***C. quillon*** Wester. Tropical Africa, esp. Congo. A good yielder, and potential source of commercial coffee.

***C. racemosa*** Lour.=C. mozambicana. (Inhambane Coffee). E.Africa, to Mozambique. Occasionally cultivated. A reasonably good coffee.

***C. robusta*** Linden=C. laurentii. (Robusta Coffee, Rio Nunez Coffee). Congo. A lowland type, resistant to Hemileia vastatrix. The quality of the coffee is little inferior to that from C. arabica, and it will grow where C. arabica does not thrive. Most of the Java crop is from this species, and it is also grown in India, parts of C.Africa, and Trinidad.

***C. stenophylla*** G. Don. (Highland Coffee of Sierra Leone). Sierra Leone. Occasionally cultivated in Sierra Leone, India, Ceylon and W.Indies. The beans are small but of good quality.

Coffee, Arabian – Coffea arabica.
Coffee, Congo – Coffea stenophylla.
Coffee, Carob – Ceratonia siliqua.
Coffee, Chestnut – Castanea sativa.
Coffee, Highland – Coffea stenophylla.
Coffee, Inhambane – Coffea racemosa.
Coffee, Kentucky – Gymnocladus dioica.
Coffee, Liberian – Coffea liberica.
Coffee, Mogdad – Cassia occidentalis.
Coffee, Negro – Cassia occidentalis.
Coffee, Rio Nunez – Coffea robusta.
Coffee, Robusta – Coffea canephora.
Coffee, Robusta – Coffea robusta.
Coffee Senna – Cassia occidentalis.
Coffee, Swedish – Astragalus boëticus.
Coffee, Wild – Triosteum perfoliatum.
Coffin-nail – Anacardium occidentale.
Co gao – Sclerachne punctata.
***Cogswellia*** Roem. and Schult.=Lomatium.
Cohoba (Snuff) – Pipadenia peregrina.
Cohosh, Black – Actaea spicata.
Cohosh, Black – Cimicifuga racemosa.
Cohosh, Blue – Caulophyllum thalictroides.
Cohune Oil – Attalea cohune.
Cohune Tree – Attalea cohune
Coir – Cocos nucifera.

COIX L. Graminae. 5 spp. Tropical Asia. Grass.

***C. lacryma-jobi*** (Job's Tears, Adlay). India. Cultivated throughout the tropics. The hard seeds are used as beads, etc., occasionally they are parched and ground as flour, particularly in E.Asia and Philippines. In China the seeds are used as a diuretic, and the parched seeds are used in Japan to make a tea. The foliage is used as food for livestock.

Cojolia – Pithecellobium arboreum.
Cojón de Berraco – Tabernaemontana amygdalaefolia.
Cojón de Gato – Tabernaemontana citrifolia.
Coke – Grewia paniculata.

COLA Schott. and Endl. Sterculiaceae. 125 spp. Africa. Trees.

***C. acuminata*** Schott. and Endl.=Garcinia kola. (Cola). Tropical Africa. The seeds contain 2·4–2·6 per cent caffeine, and theobromine. The trees are grown commercially, on a small scale in Africa and Jamaica. The seeds are collected by hand and sun-dried to give the commercial product (Bitter Cola, Male Cola, False Cola). Much is exported to America and Europe for the manufacture of soft drinks, and for use medicinally. The caffeine acts as a stimulant, to the heart and nerves, and as a diuretic. The natives chew the seeds to enable them to do heavy work over a long period of time.

***C. ballayi*** Cornu. Tropical E.Africa. Sometime used as is C. acuminata.

***C. cordifolia*** R. Br. Tropical Africa. The seeds are eaten locally.

***C. diversifolia*** de Wild. Tropical Africa. The seeds are eaten locally.

***C. nitida*** Schott. and Endl. (Cola) Tropical W.Africa. Closely related to C. acuminata. It is used and cultivated similarly.

Cola – Cola acuminata.
Cola – Cola nitida.
Cola, Bitter – Cola acuminata.
Cola de Mono – Cyathea mexicana.
Cola de zorra – Gymnosperma glutinosum.
Cola, False – Cola acuminata.
Cola, Male – Cola acuminata.
***Colbertia obovata*** Blume=Dillenia aurea.

COLCHICUM L. Liliaceae. 65 spp. Europe, Mediterranean, to C.Asia and N.India. Herbs.

***C. autumnale*** L. (Meadow Saffron). The seeds and corms contain the toxic alkaloid colchicine which is used medicinally in the treatment of gout, rheumatism, eczema and bronchitis. It is diuretic, sedative and increases the appetite. The drug is also used in plant genetics to induce polyploidy. The plant is also grown as an ornamental.

***C. luteum*** Baker. Himalayas to Iran. Contains colchicine. The corms are used locally in Iran for the treatment of rheumatism.

***C. montanum*** L.=C. ritchii Egypt. Used locally to increase weight.

***C. ritchii*** R. Br.=C. montanum.

***Coleosanthus squarrosus*** (Cav.) Balke=Brickellia cavanillensii.

COLEUS Lour. Labiatae. 150 spp. Old World tropics. Herbs.

***C. amboinicus*** Lour.=C. aromaticus.

***C. aromaticus*** Benth.=C. ambionicus=C. carnosus. (Adjeran, Daoon Ajenton). Malaysia. The plant is occasionally cultivated for the leaves. A decoction of the leaves is used for washing clothes and hair, and treating bronchitis and asthma. A poultice of the leaves is used to treat scorpion bites and headaches, and the leaves are used for seasoning meat.

***C. barbatus*** Benth.=C. forskohlii.

***C. carnosus*** Hassk.=C. aromaticus.

***C. dazo*** Chev. Congo and Sudan. Cultivated locally for the tubers which are eaten.

***C. edulis*** Vatke. E.Africa. Cultivated in Ethiopia for the tubers which are eaten.

***C. forskohlii*** (Poir.) Briquet.=C. barbatus.=Ocimum asperum. E.Africa, India. Cultivated locally in India for the tubers which are eaten.

***C. rotundifolius*** Chev. and Perrot.=Plectanthrus rotundifolius.=P. tuberosus. (Hausa Potato). Tropical Africa to the Far East. Cultivated throughout the area for the tubers which are eaten.

Colibah – Eucalyptus microtheca.

Colic Root – Aletris farinosa.

Colic Root – Liatris squarrulosa.

COLLETIA Comm. ex Juss. Rhamnaceae. 17 spp. Temperate and sub-tropical S.America. Trees.

***C. cruciata*** Gill. and Hook. Argentina to Chile. The wood is used locally for house construction, agricultural implements and wagons.

***C. ferox*** Gill. and Hook. S.America. A decoction of the leaves is used locally in various parts to reduce fevers.

***C. treba*** Bert.=Trevoa trinervata.

***Collina elegans*** (Mart.) Liebm.=Chamaedorea elegans.

COLLINSONIA L. Labiatae. 5 spp. E. N.America. Herbs.

***C. canadensis*** L. (Horse Balm, Stoneroot). E. N.America to Florida. The dried rhizome is used as a diuretic, to induce sweating and to treat dropsy and complaints of the gall bladder in children.

COLLYBIA (Fr.) Quél. Agaricaceae. 200 spp. Cosmopolitan. The fruitbodies of the species mentioned are eaten locally.

***C. acerata*** (Fr.) Gill. America, Europe, Japan.

***C. albuminosa*** (Bras.) Petch. Found in termitaria.

***C. boryana*** (Mont.) Sacc.=Lentinus cubensis.

***C. butyracea*** (Bull.) Quél. America, Europe, Japan.

***C. distorta*** (Fr.) Quél. America, Europe, Japan.

***C. euphyllus*** Berk. and Broome=Oudemansiella canarii.

***C. eurhiza*** Berk.=Rajapa eurhiza.

***C. microcarpa*** (Berk. and Br.) Hoehn.=Entoloma microcarpum. Tropical Asia, in termitaria. Eaten particularly in Malaysia.

***C. radiata*** Retham. Tropics. Found in termitaria.

***C. velutipes*** (Curt.) Quél.=Flammula velutipes.

COLOCASIA Schott. Araceae. 8 spp. Indomalaysia, Polynesia. Herbs.

***C. antiquorum*** (L.) Schott. (Cocoyam, Dasheen, Eddo, Malanga isléna, Taro). E.Asia, Polynesia. Cultivated throughout the wet tropics for the tubers, which contain large quantities of small starch grains. The tubers are eaten boiled; the boiling removes the bitterness caused by calcium oxalate crystals. The leaves can also be used as a vegetable after boiling. The tubers are more nutritious than potato, containing more protein, calcium and phosphorus. The plant is also used for the production of Portland Arrowroot.

***C. esculenta*** Schott. (Elephant's Ear, Dasheen). Tropical Asia. Used as C. antiquorum, having smaller and more palatable tubers. Usually called Dasheen in the W.Indies.

***C. indica*** (Lour.) Hassk. (Tolambo). S.W.Malaysia to Java. The juice from the leaves is used locally as an arrow poison.

Colocynth – Citrullus colocynthis.

***Colocynthis*** Mill.=Citrullus.

Colombian Berry – Rubus macrocarpus.

Colombian Blueberry – Vaccinium floribundum.

Colombian Copal – Hymenaea courbaril.

Colombian Mahogany – Cariniana pyriformis.
Colombian Virgin (Rubber) – Sapium thomsonii.
Colombo, American – Frasera carolinensis.
COLONA Cav. Tiliaceae. 30 spp. S.China, S.E.Asia, Indomalaysia.
***C. auriculatum*** Desf. (Nilau kootjing, Dhaloobang tali). Malay Peninsula. Shrub. The bark is a source of cordage, used locally, and exported locally.
***C. javanica*** Blume. Java. Fibre from bark, used locally for fishing nets, and cordage.
Colonche – Opuntia megacantha.
Colophan Wood – Cararium paniculatum.
Colophony – Pinus merkusii.
Colophony – Pinus palustris.
***Colophospermum mopane*** (Kirk. ex Benth.) Kirk. ex J. Leonard. = Copaifera mopane.
Colorado Grass – Panicum texanum.
Colorado, Mamey – Calocarpum mammosum.
Colorin – Erythrina flabelliformis.
Colorines – Rhynchosia pyramidalis.
COLPOÖN Bergius. Santalaceae. 1 sp. S.Africa. Small tree.
***C. compressus*** Berg. S.Africa. The bark is used for tanning.
Coltsfoot – Tussilago farfara.
Coltsfoot, Sweet – Pelastes palmata.
COLUBRINA Rich. ex Brongn. Rhamnaceae. 16 spp. Tropical and subtropical America; 1 sp. Tropical E.Africa and Mauritius; 7 spp. E. and S.E.Asia, Indomalaysia, Queensland, and Pacific Islands. Climbing shrubs, or small trees.
***C. asiatica*** (L.) Brongn. = Ceanothus asiaticus = Rhamnus carolinianus. Tropical Africa, tropical Asia to Australia and Polynesia. Used mainly in Malaysia. The leaves are eaten as salad with fish. The fruit is used to induce abortions, and to poison fish. The bark and roots are used as soap.
***C. ferruginea*** Brongn. (Bijágara). W.Indies. The strong, red, compact wood is used for general carpentry. The bark (West Indian Snake Bark) is used to treat dysentery, and to control bleeding. The leaves are used against intestinal worms and to control menstruation.
***C. oppositifolia*** Brongn. (Kauila). Hawaii. The hard wood was used locally for building huts, making spears, and for making hair-pins.
***C. reclinata*** (L'Hér.) Brongn. = Ceanothus reclinatus.
***C. rufa*** Reiss. Brazil. The bark (Saguaragy) is used to treat fevers. The tree is occasionally cultivated, and some of the bark is exported. The fine, strong wood is used for bridges, fence posts, railway sleepers, ship-building, vehicles, and furniture-making.
***Columbia*** Pers. = Colona.
Colu de mico – Pithecellobium arboreum.
Columbine, Common – Aquilegia vulgaris.
Columbine, Meadow Rue – Thalictrum aquifolium.
Columbine, Sitka – Aquilegia formosa.
Columbo, American – Frasera carolinensis.
Columbus Grass – Sorghum almum.
Colza (Oil) – Brassica napus.
Colza, Chinese – Brassica campestris var. chinoleifera.
Coma – Bumelia spiniflora.
Coma resinera – Bumelia spiniflora.
COMANDRA Nutt. Santalaceae. 5 spp. N.America; 1 sp. S.E. Europe to Asia Minor. Herbs.
***C. umbellata*** (L.) Nutt. (Bastard Toadflax). E. N.America. Parasitic. The sweet fruits are eaten locally by the Indians.
***Comarum*** L. = Potentilla L.
Combee (Resin) – Gardenia gummifera.
Combee (Gum) (Resin) – Gardinia lucida.
Comb Hyptis – Hyptis pectinata.
COMBRETOCARPUS Hook. f. Anisophylleaceae. 1 sp. W.Malaysia, except Malay Peninsula. Tree.
***C. rotundatus*** (Miq.) Dans. (Keruntun). The red-brown wood is fairly hard and denser than water. It is coarse-grained and not durable nor insect-resistant. It is used for heavy construction work, railway sleepers and fuel.
COMBRETUM Loefl. Combretaceae. 250 spp. Tropics, except Australia. Trees.
***C. alatum*** Guill. and Perr. = C. micranthum G. Don. W.Africa. A decoction of the leaves is used locally to cure black water fever.
***C. bracteatum*** Engl. and Diels. W.Africa. A decoction of the leaves is used locally to reduce fevers and as a tonic.
***C. butyrosum*** Tul. W.Africa. The seeds yield a butter-like fat (Chiquito) which is used locally.
***C. coccineum*** Lam. Madagascar. The bark is the source of a fibre, used locally for ropes etc.
***C. confertum*** Laws. C. and W.Africa. The bark is used locally as an ordeal poison.

***C. elliotii*** Engl. and Diels. (Bouski, Diangara, Guiré). Dry Sudan. A gum produced from the stems is used by the nomads to thicken soups and make various confections.

***C. erythrophyllum*** Sond. S.Africa. A gum from the stems is used locally as a substitute for Gum Tragacanth.

***C. ghasalense*** Engl. and Diels. Sudan. The wood is used locally to make incense.

***C. glutinosum*** Guill. and Perr. (Sima Bali). Tropical Africa. In N. W.Africa the roots and bark are used to make a yellow dye. The ash is alkaline, and is used to fix Indigo Blue in cotton fabrics. A decoction of the leaves is used to cleanse wounds.

***C. hartemannianum*** Schweinf. Tropical Africa. The stem is a source of a gum (Mumuye Gum) which is used as a substitute for Gum Arabic. In some parts it is used as a perfume.

***C. hypotilinum*** Diels. W.Africa. The stem yields a gum (Taramniya) used in N.Nigeria to cure toothache. The leaves are used in Gambia as a purgative.

***C. lecananthum*** Engl. and Diels. W.Africa. The stem yields a gum, used locally as chewing gum.

***C. leonense*** Engl. and Diels. Tropical Africa. A source of Mumuye Gum. See C. hartemannianum.

***C. micranthum*** G. Don. = C. alatum.

***C. sokodense*** Engl. W.Africa. A source of Mumuye Gum. See C. hartemannianum. A decoction of the roots is used locally to cure dysentery.

***C. sundaicum*** Miq. Malaysia. A decoction of the leaves has been claimed to be a cure for opium addiction.

***C. tetralophum*** Clarke (Soongoong, Tingting). Malaysia and New Guinea. The fruits are used to cure intestinal worms.

***C. trifoliatum*** Vent. Tropical Africa. The scented wood is used as a source of perfume in the Sudan.

***C. zeyheri*** Sond. (Mufuka, Muruka, Mushende). Central Africa. A decoction of the roots is used locally to relieve haemorrhoids.

Cometa – Pithecoctenium echinatum.

Comezuelo – Acacia hindsii.

Comfrey, Common – Symphytum officinale.

Comfrey, Prickly – Symphytum asperrinum.

Comfrey, Prickly – Symphytum uplandicum.

Comino – Aniba perutilis.

COMMERSONIA J. R. and G. Forst. Sterculiaceae. 1 sp. Tropical E.Asia; 8 spp. Australia. Trees.

***C. dasyphylla*** Andr. = Rulingia pannosa.

***C. echinata*** Forst. New South Wales, Queensland, Malaya. The fibre from the bark is used by the Australian aborigines to make nets.

COMMIPHORA Jacq. Burseraceae. 185 spp. Warm Africa, Madagascar, Arabia to W.India. Trees. All the species mentioned produce a gum resin or bdellium from wounds in the stem. These have a great variety of local names, and some of the species from which the resins are extracted have not been fully elucidated. This applies particularly in Somaliland where there are a number of resins produced under the general name of Habbak.

***C. abyssinica*** (Berg.) Engl. (Abyssinian Myrrh). Somaliland, Abyssinia, Arabia. The resin (Hotai) is sometimes used medicinally as a stimulant and astringent, and to treat stomach complaints. An oil (Oil of Myrrh) distilled from it is used in perfumery.

***C. africana*** Endl. (African Myrrh). Ethiopia, Sudan. Yields a resin (bdellium).

***C. berryi*** Engl. E.India. Resin called Mulu Kilavary.

***C. erythraea*** Engl. (Bisabol Myrrh Tree). Tropical Africa. The resin called Perfumed Bdellium of Bombay, Bisabol Myrrh, Coarse Myrrh of Aden. Used in perfumery.

***C. erythraea*** Engl. var. ***glabrescens*** Engl. (Haddi Tree). Somaliland. The resin (Perfumed Bdellium, Bisabol Myrrh) is a major source of bdellium for the manufacture of perfume. It was used for embalming by the Egyptians etc. The wood (Gafal Wood) is scented and is burnt as incense. The bark is also burnt in houses to give a pleasant odour.

***C. hildebrandtii*** Engl. (Dunkal Tree). Somaliland. Source of a rare bdellium used in perfumery.

***C. kataf*** Engl. = Balsamodendron kataf (Opopanax Myrrh Tree). Tropical Africa. The resin (Bursa Opopanax) is used in perfumery, by the Chinese to increase the milk yield of cattle, and as a size to give a gloss to paint on walls.

***C. mukul*** Engl. (Mukul Myrrh Tree). Arabia to E.India. The bdellium is known as Indian Bdellium, Mukul-i-Azrak, Mukul-i-Yahud, or Sakulali, depending on the colour. It is used in Persia to treat stomach complaints, and rheumatism.

***C. myrrha*** (Nees.) Engl. (Common Myrrh Tree, Harobol Tree). Ethiopia to Somaliland. Resin called Harobol Myrrh. It is used in perfumery and for incense. It was used by the Ancients for embalming.

***C. opobalsam*** Engl.=Balsamodendron gileadense. (Mecca Myrrh Tree). Balsam, which is relatively rare is called Mecca Balsam, Balm of Gilead, Duhnual Balsam. It is used in perfumery and for incense. Locally it is used to treat stomach complaints, to induce sweating and to heal wounds.

***C. pedunculata*** Engl. Tropical Africa. Source of a resin.

***C. pilosa*** Engl. Tropical Africa. The bark is the source of a red-brown dye, used locally for dyeing cloth.

***C. schimperi*** Engl. Abyssinia, S.W.Arabia. Source of a resin.

***C. stocksiana*** Engl. (Bayisagugul, Malaikiluvai). Sind, Baluchistan. The gum is used by the Hindus for treating ulcers.

***C. zanzibarica*** Engl. Central Africa. The fruits are a source of a semi-drying oil.

Common Apple – Malus communis.

Common Banana – Musa sapientum.

Common Barley – Hordeum vulgare.

Common Bean – Phaseolus vulgaris.

Common Bear Grass – Xerophyllum tenax.

Common Betony – Stachys officinalis.

Common Borneo Camphor – Dryobalanops aromatica.

Common Breadroot – Psoralea esculenta.

Common Buckthorn – Rhamnus cathartica.

Common Buckwheat – Fagopyrum esculentum.

Common Burdock – Arctium minus.

Common Butterwort – Pinguicula vulgaris.

Common Camass – Camassis quamash.

Common Catalpa – Catalpa bignonioides.

Common Chaulmoogra Tree – Hydnocarpus anthelmintica.

Common Chickweed – Stellaria media.

Common Clubmoss – Lycopodium clavatum.

Common Columbine – Aquilegia vulgaris.

Common Comfrey – Symphytum officinale.

Common Coral Bean – Erythrina corallodendron.

Common Custard Apple – Annona reticulata.

Common Everylasting – Gnaphalium polycephalum.

Common Germander – Teucrium chamaedrysa.

Common Gorse – Ulex europaeus.

Common Greenbrier – Smilax rotundifolia.

Common Gromwell – Lithospermum officinale.

Common Hackberry – Celtis occidentalis.

Common Heather – Calluna vulgaris.

Common Hyacinth – Hyacinthus orientalis.

Common Ivory Palm – Phytelephas macrocarpa.

Common Jasmine – Jasminom officinale.

Common Jasmin Orange – Murraya exotica.

Common Jujub – Zizyphus jujuba.

Common Juniper – Juniperus communis.

Common Kangaroo Grass – Themedia ciliata.

Common Lespedeza – Lespedeza striata.

Common Licorice – Glycyrrhiza glabra.

Common Lizardtail – Saururus cernuus.

Common Madder – Rubia tinctorium.

Common Mahogany – Swietenia mahagoni.

Common Mallow – Malva rotundifolia.

Common Matrimony Vine – Lycium halamifolium.

Common Mignonette – Reseda odorata.

Common Milkweed – Asclepias syriaca.

Common Moonseed – Menispermum canadense.

Common Motherwort – Leonurus cardiaca.

Common Myrrh – Commiphora myrrha.

Common Oat – Avena sativa.

Common Osier – Salix viminalis.

Common Persimmon – Diospyros virginiana.

Common Pokeberry – Phytolaccoa decandra.

Common Rue – Ruta graveolens.

Common Sassafras – Sassafras alibidum.

Common Smoke Tree – Rhus cotinus.

Common Sweetleaf – Symplocos tinctoria.

Common Tansey – Tanacetum vulgare.

Common Tigerflower – Tigridia pavonia.

Common Tobacco – Nicotiana tabacum.

Common Twig Lichen – Ramalina calicaris.

Common Valerian – Valeriana officinalis.

Common Vetch – Vicia sativa.

Common Wheat – Triticum vulgare.

Common Willow – Salix caprea.

Common Whip Tree – Luehea divaricata.

Common Yucca – Yucca filamentosa.

Compass Plant – Silphium laciniatum.

COMPTONIA L'Hérit ex Ait. Myricaceae. 1 sp. E. N.America. Shrub.

***C. aspleniifolia*** Ait.=C. peregrina. (Sweet Fern). The leaves are used locally as a stimulant and astringent, and to relieve stomach pains and diarrhoea. They were used by the Indians as a condiment and to aid childbirth.

***C. peregrina*** (L.) Coult.=C. aspleniifolia.

CONANTHERA Ruiz. and Pav. Tecophilaeaceae. 5 spp. Chile. Herbs. The bulbs of the following species are eaten locally.

***C. bifolia*** Ruiz. and Pav.

***C. campanulata*** Lindl.=C. simsii.

***C. simsii*** Sweet=C. campanulata=Cumingia campanulata.

***Conchomyces*** Overeem.=Crepidotus.

***Concylocarpus apulus*** (L.) Hoffm.=Tordylium apulum.

CONDALIA Cav. Rhamnaceae. 18 spp. Warm America. Shrubs.

***C. ferrea*** Griseb.=Krugeodendron ferreum.

***C. lycioides*** Gray.=Zyziphus lycioides. (South western Condalia, Crucillo). S.W. U.S.A. and Mexico. The bark is used locally as soap, and local Indians use a decoction of the roots to treat sore eyes.

***C. obovata*** Hook. (Bluewood Condalia) S.W. U.S.A. The fruits are edible and are made into jelly. The wood is the source of a blue dye.

***C. obtusifolia*** (Hook.) Weberh.=Rhamnus obtusifolia. (Abrojo, Chaparro prieto). W. Texas. The roots are used locally as soap, and an extract of the roots is used to treat sores on horses.

Condalia, Bluewood – Condalia obovata.

Condalia, Southwestern – Condalia lycioides.

Condor Vine – Marsdenia reichenbachii.

Condurango Bark – Marsdenia reichenbachii.

Condurango Wine – Marsdenia reichenbachii.

Cone Flower, Purple – Echinacea pallida.

Conehead Thyme – Thymus capitatus.

Conesi Bark – Holarrhena antidysenterica.

Confetti Tree – Maytenus senegalensis.

Confitura – Trophis racemosa.

Confiture d'Épine Vinette – Berberis vulgaris.

Congo Coober – Voandzeia subterranea.

Congo Coffee – Coffea robusta.

Congo Copal – Copaifera demeusii.

Congo Mallee Eucalyptus – Eucalyptus dumosa.

Congo Oil – Antrocaryon nannani.

Congo Pea – Cajanus cajans.

Congo Zebra Wood – Julbernardia sereti.

Congoasa – Mayenus ilicifolia.

CONIUM L. Umbelliferae. 4 spp. N.Temperate Eurasia, S.Africa. Herbs.

***C. maculatum*** L. (Poison Hemlock). N.Temperate Eurasia. Very poisonous, due to the presence of the alkaloid coniine. The dried unripe fruits are used medicinally as a pain-killer and sedative. They were used in the preparation of a criminal poison by the Ancient Greeks.

***Connaropsis*** Planch ex Hook. f.=Sarcotheca.

CONNARUS L. Connaraceae. 130 spp. Tropical Africa, America, and Asia. Trees.

***C. africanus*** Lam. Tropical Africa. On the W.coast, the ground seeds are used as a purgative and to combat intestinal worms.

CONOBEA Aubl. Scrophulariaceae. 7 spp. Warm America, W. Indies.

***C. scoparioides*** Benth. S.America. Used locally in Colombia to treat toothache.

CONOCARPUS L. Combretaceae. 2 spp. Florida to tropical America and W.Indies, N.E.tropical Africa. Trees.

***C. erectus*** L. (Buttonwood). Coastal tropical America. The bark is used as an astringent and tanning leather. The wood is useful in making charcoal.

***Conocybe*** Fayod=Galera.

CONOPHARYNGIA G. Don. Apocynaceae. 25 spp. Africa, Mascarene Islands. Small trees or shrubs.

***C. chippii*** Stapf. Tropical W.Africa. A latex (Bede-Bede) is extracted from the stem.

***C. crassa*** Stapf. W.Africa. The juice is used to coagulate rubber.

***C. elegans*** (Stapf.) Stapf. (Kakope, Toad Tree, Umkadhlu). Rhodesia. The fruits are eaten locally. The latex is used to stop bleeding from wounds, and an extract from the roots is used to treat lung complaints.

***C. pachysiphon*** Stapf.=Tabernaemontana pachysiphon. Tropical Africa. The latex is used as an adulterant for rubber. The bark yields a fibre, used locally for making cloth. The leaves give a black dye, used by the native women for dyeing their hair. The juice from the leaves and shoots is used locally as bird lime.

CONOSTEGIA D. Don Melastomataceae. 50 spp. Tropical America, W.Indies. The fruits of the following species are eaten locally.

**C. arborea** (Schlecht) Schauer. Mexico.

**C. mexicana** Cogn. S.Mexico.

**C. subhirsuta** DC. S.Mexico.

**C. xalapensis** (Bonpl.) D. Don.=Melastroma xalapense. (Capiroto). C.America.

Conquito – Scheelea pteussii.

CONRINGIA Fabr. Cruciferae. 7 spp. Mediterranean. Europe to C.Asia. Herbs.

**C. orientalis** (L.) Andrz.=Brassica orientalis. Mediterranean C.Europe. An oil extracted from the seeds is used locally for cooking.

Constantinople Nut – Corylus colurna.

Contracapetano – Aristolochia maxima.

CONVALLARIA L. Liliaceae. 1 sp. N. temp.

CONVOLVULUS L. Convolvulaceae. 250 spp. Cosmopolitan, mostly temperate. Herbs, some climbing.

**C arborescens** Humb. and Bonpl.=Ipomoea arborescens.

**C. cairicus** L.=Ipomoea cairica.

**C. corymbosus** L.=Rivea cormbosa.

**C. floridus** L. f. Canary Islands. The underground parts are the source of an essential oil (Guadil, Oil of Rhodium).

**C. hederaceus** L.=Ipomoea hederacea.

**C. japonicus** Thunb.=Ipomoea angustifolia.

**C. macranthus** H.B.K.=Ipomoea murucoides.

**C. nil** L.=Ipomoea hederacea.

**C. ovalifolius** sensu Hook. and Arn.=Jacquemontia sandwicensis.

**C. pes-caprae** L.=Ipomoea pes-caprae.

**C. quahutzehuatl** Sessé and Moc.=Ipomoea arborescens.

**C. reptans** L.=Ipomoea aquatica.

**C. scammomia** L. (Scammony). Asia Minor, Mediterranean. A resin (Levant Scammony Resin, Levant Scammony, Resina Scammoniae) is exuded from the roots, and is used medicinally as a cathartic. It causes the evacuation of large quantities of water. The dried root is exported from Syria and Turkey.

**C. sidaefolius** (H.B.K.) Choisy=Rivea corymbosa.

**C. soldanella** L. Europe, temperate Asia, N.Africa, N. and S.America. The leaves (Herba soldanellae) are used as a purgative and diuretic.

**C. soparius** L. f. Canary Islands. The underground parts are a source of an essential oil.

**C. turpethum** L.=Operculina turpethum.

**C. violaceus** Spreng.=Ipomoea violacea.

CONYZA Less. Compositae. 60 spp. Temperate and sub-tropics. Herbs.

**C. balsamifera** L.=Blumea balsamifera.

**C. cappa** Blanco.=Blumea lacera.

**C. chinensis** Lam.=Veronia chinensis.

**C. lanceolaria** Roxb.=Blumea myriocephala.

**C. lacera** Borm. f.=Blumea lacera.

**C. odorata** L.=Pluchea odorata.

**C. persicaefolia** Oliv. and Hiern. Tropical Africa. The leaf sap is used locally in the Congo as a black dye.

**C. pubigera** L. Blumea pubigera.

**C. riparia** Blume.=Blumea chinensis.

Coober, Congo – Voandzeia subterranea.

Coolwort – Tiarella cordifolia.

Coondi Oil – Carapa guineensis.

Coontie – Zamia integrifolia.

Copahiba angelin – Copaifera multijuga.

Copahiba marimary – Copaifera multijuga.

Copaiba (Balsam) – Copiafera coriacea.

Copaiba Balsam – Copaifera landsdorffii.

Copaiba Balsam – Dipterocarpus trinervis.

Copaiba Balsam – Copaifera reticulata.

Copaiba Balsam – Hardwickia pinnata.

Copaiba, West African – Oxystigma mannii.

COPAIFERA L. Leguminosae. 25 spp. tropical America; 5 spp. tropical Africa. Trees.

**C. bracteata** Benth. N. S.America. The wood (Blue Ebony, Marawayana, Violet Wood), is brown-violet in colour, heavy and polishes well. It is used for furniture and fancy work.

**C. coriacea** Mart. (Copaiba). Brazil, Colombia. The resin (Copaiba Balsam) from the trunk is used as a diuretic, stimulant, and to treat infections of the urino-genital tract. It is also used as a disinfectant and to treat dysentery. In art work it is used to remove varnish from oil paintings and to varnish photographic paper.

**C. demeusii** Harms. Congo. The source of a resin (Congo Copal, Cameroon Copal), used for lacquers and varnishes.

**C. glycicarpa** Ducke. Brazil. Source of a resin used for treating skin diseases, wounds and gonorrhoea.

**C. gorskiana** Benth. Tropical Africa. Source of a resin of little commercial value, called Inhambane Copal.

**C. guibourtiana** Benth.=Guibourtia copallifera. Tropical Africa. Source of a resin

(Sierra Leone Copal) used in the manufacture of varnishes.

***C. guyanensis*** Desf. Amazon Basin. Source of a resin used in varnishes etc., and in the treatment of skin complaints, wounds and gonorrhoea.

***C. hymenaefolia*** Moric. W.Indies. The wood is hard and durable. It is used for posts, railway sleepers, and other heavy construction work.

***C. jacquini*** Desf. = C. officinalis.

***C. lansdorffii*** Desf. Brazil. The resin (Copaiba Balsam) is used to treat skin complaints and gonorrhoea, and in perfumery. The wood is used locally for furniture, turning, ship-building, and the bark is used for tanning.

***C. mopane*** J. Kirk ex Benth. = Colophospermum mopane. (Balsam Tree, Mopane, Rhodesian Ironwood, Turpentine Tree). Tropical Africa. The wood is hard, heavy and dark-brown. It is used for pit-props, posts, railway sleepers, heavy construction work, cabinet work and fancy articles. An inhalation of the root is used locally to cure temporary madness. A copal is collected from the trunk.

***C. multijuga*** Hayne. Amazon Basin. The source of a clear, liquid light-scented balsam (Copahiba angelin, Copahiba marimary) used commercially.

***C. officinalis*** L. = C. jacquini. N. S.America. The trunk yields a resin (Maracaibo Resin).

***C. reticulata*** Ducke. Brazil. The resin (Copaiba Balsam, Oleo Vermelho) is used to treat skin diseases and gonorrhoea.

***C. salikounda*** Heckel. W.Africa. The stem yields a brittle resin (Sierra Leone Copal).

Copal – Bursera microphylla.

Copal – Cynometra sessiliflora.

Copal, Accra – Daniella ogea.

Copal, Ako Ogea – Pardaniella oliveri.

Copal, amargo – Bursera bipinnata.

Copal amorgoso – Bursera bipinnata.

Copal, Batjan – Agathis loranthifolia.

Copal, Boed – Agathis alba.

Copal, Boed – Agathis loranthifolia.

Copal, Borneo – Agathis loranthifolia.

Copal, Brazil – Hymenaea courbaril.

Copal, Cameroon – Copaifera demeusii.

Copal, Celebes – Agathis loranthifolia.

Copal, Colombian – Hymenaea courbaril.

Copal, Congo – Copaifera demeusii.

Copal de Penca – Bursera jorullensis.

Copal, East Indian – Canarium bengalense.

Copal, Fidji – Agathis loranthifolia.

Copal, Gum of the Gold Coast – Daniella similis.

Copal, Illorin – Daniella thurifera.

Copal, Indian – Vateria indica.

Copal, Inhambane – Copaifera gorskiana.

Copal, Kauri – Agathis australis.

Copal, Labuan – Agathis loranthifolia.

Copal, Macassar – Agathis alba.

Copal, Macassar – Agathis loranthifolia.

Copal, Madagascar – Hymenaea verrucosa.

Copal, Madagascar – Trachylobium hornemannianum.

Copal, Malengthet – Agathis loranthifolia.

Copal, Manila – Agathis alba.

Copal, Manila – Agathis loranthifolia.

Copal, Menedo – Agathis loranthifolia.

Copal, Molucca – Agathis loranthifolia.

Copal, Mozambique – Trachylobium hornemannianum.

Copal, Niger – Daniellia oblonga.

Copal, Ogea – Daniellia ogea.

Copal, Oli – Agathis loranthifolia.

Copal, Pontuik – Agathis alba.

Copal, Sambas – Agathis loranthifolia.

Copal, Sierra Leone – Copaifera guibourtiana.

Copal, Sierra Leone – Copaifera salikounda.

Copal, Singapore – Agathis alba.

Copal, Singapore – Agathis loranthifolia.

Copal Tree – Daniellia similis.

Copal Tree, Indian – Vateria indica.

Copal, Zanzibar – Hymenaea verrucosa.

Copal, Zanzibar – Trachylobium hornemannianum.

Copalchi – Croton niveus.

Copalchi – Portlandia perosperma.

Copalchi Bark – Exostema caribaeum.

Copalchi Bark – Portlandia latifolia.

Copalquin – Portlandia perosperma.

***Copelandia papilioacea*** (Bull.) Bres. = Panaeolus papilionaceus.

COPERNICIA Mart. Palmae. 30 spp. Tropical America. W.Indies. Palm trees.

***C. australis*** Becc. (Caranday Palm). S.America. The wax from the leaves (Caranday Wax, Paraguayan Carnauba) is hard, and used for gramophone records, candles, wax varnishes, and polishing pastes. Limited amounts of the wax are exported. The wood is tough and durable. The trunks are used as telephone poles and for general construction work. The leaves are used for thatching huts.

***C. cerifera*** Mart. (Carnauba Palm, Wax Palm). S.America, mainly Brazil. The hard wax, which is shaken from the leaves, is used for gramophone records, candles, wax varnishes, and polishing pastes. The wax (Carnauba Wax) has a high melting point. Much of the wax is exported from Brazil.

***C. Sancta Martae*** Becc. Colombia. The trunks are used to build houses, locally, and the leaves are used as thatch.

***C. tectorum*** Mart. See C. Sancta Martae.

***C. textilis*** León. (Yarey). W.Indies. The leaves are used locally in Cuba to make hats.

Copiava, Balsam – Daniellia oliveri.

Copo – Ficus cotinifolia.

Copra – Cocos nucifera.

COPRINUS (Pers. ex Fr.) S. F. Gray. Agaricaceae. 200 spp. Cosmopolitan.

***C. ater*** Copel. Malaysia. Eaten locally in Philippines.

***C. atramentarius*** (Bull.) Fr. Temperate. The black decomposition products of the fruit-bodies is used as an ink. The presence of the spores in the ink, make it valuable in signing documents, and police work, as they make the ink easy to identify.

***C. bryanti*** Copel. Malaysia. Eaten locally in Philippines.

***C. comatus*** (Fl. D.) Gray. (Shaggy Mane). Temperate. The young, freshly-picked fruit-bodies are good to eat, and when they decay they are the source of an inferior black ink.

***C. concolor*** Copel. Malaysia. Eaten locally in the Philippines.

***C. fimetarius*** (L.) Fr. N.America. The fruitbodies are eaten.

***C. macrorhizus*** (Pers.) Rea. Tropical Asia. The fruitbodies are eaten, especially in Malaysia.

***C. micaceus*** (Bull.) Fr. Temperate. The fruitbodies are eaten.

***C. microsporus*** Berk. and Broome. Tropical Asia. The fruitbodies are eaten in Malaysia.

***C. rimosus*** Copel. Malaysia. Eaten locally in the Philippines.

COPROSMA J. R. and G. Forst. Rubiaceae. 90 spp. Malaysia, Australia, Polynesia, New Zealand, Chile. Trees.

***C. australis*** Robinson. New Zealand. An infusion of the leaves is used locally to treat cuts, sores and bruises, fevers and urinary complaints. A poultice of the leaves is used to treat broken limbs. An infusion of the bark is also used to treat similar complaints as well as stomach ailments. An infusion of the young shoots is used to relieve bladder stoppages, while the juice from the plant is used to treat skin complaints.

Copte – Cordia dodecandra.

COPTIS Salisb. Ranunculaceae. 15 spp. N.Temperate and Arctic. Herbs.

***C. anemonaefolia*** Sieb. and Zucc. Japan. The leaves are used locally to relieve stomach upsets.

***C. brachypetala*** Sieb. and Zucc. Japan. The roots yield a yellow dye, used locally.

***C. chinensis*** Wils. China. Used locally as a tonic and to treat indigestion.

***C. occidentalis*** Nutt. Japan. The roots yield a yellow dye, used locally.

***C. teeta*** Wall. Mountains of Assam. The root bark (Mamira, Mishmee Bitter) is used, and sold locally as a bitter tonic. A decoction of the leaves is used to reduce inflammation of the eyes.

***C. trifolia*** (L.) Salib. (Goldthread). N.America. The plant contains the alkaloids berberine and coptin. The whole plant is dried and used locally as a tonic and to relieve stomach complaints. It is also used to treat mouth ulcers.

COPTOSAPELTA Korth. Rubiaceae. 15 spp. Indochina, Malaysia. Climbing shrubs.

***C. griffithii*** Hook. f. (Akar maloong). Malaysia. A decoction of the whole plant is used locally to treat stomach upsets.

Coquitos Palm – Jubaea chilensis.

Coral Bean – Erythrina arborea.

Coral Bean – Erythrina corallodendron.

Coral Bean, Eastern – Eythrina herbacea.

Coral Bean, Indian – Erythrina indica.

Coral Berry – Symphoricarpos arbicularis.

Coral Peony – Paeonia corallina.

Coral Root – Corallorhiza odontorhiza.

Coral Sumac – Metopium toxiferum.

Coral Tree – Erythrina indica.

Coral Tree, Hawaiian – Erythrina sandwicensis.

Coral Vine – Antigonon leptopus.

Coralillo – Pithecellobium arboreum.

Coralina – Erythrina flabelliformis.

CORALLORHIZA Chatelain. Orchidaceae. 15 spp. N.Temperate. Herbaceous saprophytes.

***C. odontorhiza*** Chapm. (Coral Root, Crawley, Dragon's Claw). N.America. The rhizome is used to induce perspiration, and is used locally in the treatment of fevers and pleurisy.

Corazon – Annona reticulata.

CORCHORUS L. Tiliaceae. 100 spp. Warm. Herbs or woody shrubs.

***C. capsularis*** L. (Jute). Native of Africa. Cultivated in Old and New World for the bast fibres. The fibres are coarse, containing 75 per cent cellulose and 11 per cent lignin. They do not stand wetting but are in demand for sacking, twine, carpet-backing etc. The main advantage of the fibre (Gunney, White Jute) is that it is cheap, 90 per cent of the world's 3 million tons annual production coming from India where labour costs are low. Some is produced in Brazil. The fibres are extracted by hand after retting the stems in water, yielding about 1200 lb. of fibre per acre. An increase in labour costs, and the use of substitutes is leading to a reduction in the world production of jute. The short fibres from the base of the stem are used in the manufacture of paper. The leaves are eaten locally as a vegetable.

***C. olitorius*** L. (Daisee, Jew's Mallow, Tossa). Tropics. Cultivated in W.Africa for the leaves which are used as a vegetable. The stem yields a fibre similar to, but of poorer quality than jute.

***C. siliquosus*** L. N. S.America, to S. N.America, W.Indies. The leaves are used locally as a substitute for tea.

***C. tridens*** L. Tropics. The leaves are used locally in tropical Africa as a vegetable.

***C. tridens*** L. (Al Moulinouquia). Senegal through to India. The plant is cultivated locally for the leaves which are used as a vegetable.

Cordate Walnut – Juglans sieboldiana.

Corde Caco – Banisteriopsis longifolia.

Corde Molle – Philodendron oxycardium.

CORDEAUXIA Hemsl. Leguminosae. 1 sp. Tropical Africa. Shrub.

***C. edulis*** Hemsley. (Geeb, Jeheb, Yeheb). Somaliland. The fruits are eaten locally, and the leaves are infused to make a beverage.

Corde de l'Eau – Doliocarpus calineoides.

CORDIA L. Ehretiaceae. 250 spp. Warm. Trees.

***C. alba*** (Jacq.) Roem. Mexico to N. S.America, W.Indies. The yellow strong wood is used locally for general carpentry. An infusion of the flowers is used to induce perspiration, and the fruits are used locally to coagulate indigo.

***C. alliodora*** (Ruiz. and Pav.) Cham. Mexico to S.America, W.Indies. The wood is used locally for house construction. An extract of the leaves is used to treat stomach complaints and as a tonic. The fruits are eaten locally in Mexico, and the ground seeds are used in the W.Indies to treat skin complaints.

***C. boisseri*** DC. S.W. United States of America, Mexico. The fruit pulp and leaves are used locally to treat coughs.

***C. cordifolia*** Wall. = C. grandis.

***C. dodecandra*** DC. (Chacopte, Copte, Siricote). Mexico and Guatemala. The heavy dark wood polishes well and is used for furniture-making and general construction work. The fruits are eaten locally, and the tree is occasionally cultivated for them. The leaves are sometimes used as a substitute for sandpaper, and to scour pots.

***C. gerascanthus*** L. (Capá Prieta, Prince Wood). W.Indies, Porto Rico. The light red wood is used for furniture-making, doors, boats, and sugar mills.

***C. gharaf*** (Forsk.) Ehrenb. ex Aschers. (Tridarent). Dry areas, Senegal to India. The fruits are eaten locally.

***C. grandis*** Roxb. = C. cordifolia. Indochina, across to Assam. The mucilage from the fruit is used locally for glue.

***C. irwingii*** Baker. W.Africa. The wood is very durable and is used locally to make shingles.

***C. marchionica*** Drake. (Makomako). Marquesas. The flowers are used locally to make leis.

***C. myxa*** L. Greece, Turkey. (Khartoum Teak, Sudan Teak, Sebastens). The yellow-brown wood polishes well and is used for furniture-making. The pulp from the fruits is used to treat coughs in the East, and Iran, and more widely as bird lime.

***C. rothii*** Roem. and Schultes. Tropical Africa across to India. The wood is used in India for general building and making agricultural implements. In India, the fruits are eaten pickled, and the bark fibre is used to make ropes. The gum is used to adulterate Gum Arabic.

***C. sebestina*** L. (Aloë Tree, Geiger Tree,

Siricote Blanco, Vomitel). Florida to Mexico, W.Indies. The edible fruit is used locally to treat fevers. The leaves are used locally to treat bronchitis and intestinal complaints.

***C. sebestiana*** Fort. = C. subcordata.

***C. senegalensis*** Juss. = C. gharaf.

***C. subcordata*** Lam. = C. sebestiana. Tropics. The wood is used in Hawaii to make rafts, and the mucilage from the fruits is used to stick fabrics.

Cordgrass, Praire – Spartina pectinata.

Cordial, Godfrey's – Sassafras albidum.

Cordocillo – Thamnosma montana.

CORDYCEPS (Fr.) Link. Hypocreaceae. 100 spp. Cosmopolitan, mostly parasitic on caterpillars.

***C. sinensis*** (Berk.) Sacc. (Ch'ung-tsao). China. Parasitic on the caterpillar of Hepialus sp. It is used locally to cure the craving for opium, and as an antidote to opium poisoning. It is also used as a tonic and stimulant.

CORDYLANTHUS Nutt. ex Benth. Scrophulariaceae. 40 spp. W. N.America. Herbs.

***C. wrightii*** Gray. (Club Flower). Texas, Arizona. The flowers are used locally by the Indians for bleaching cloth.

CORDYLINE Comm. ex Juss. Agavaceae. 15 spp. Tropical warm temperate. Shrubs or trees.

***C. australis*** Hook. f. (Palm Lily). New Zealand. The fibre from the leaves was used as a source of wine, and the whole leaves are used for paper making.

***C. fruticosa*** Goep. = C. terminalis.

***C. hyacinthoides*** W. F. Wight. = C. roxburghiana.

***C. indivisa*** Steud. New Zealand. The leaves yield a fibre which is used by the Maoris to make fabric.

***C. roxburghiana*** (Schultes) Merr. = C. hyacinthoides. Tropical Asia. Occasionally cultivated for the leaf fibres which are used locally for cordage, paper and cloth. The underground parts are used by the Hindus to treat coughs. It may also be a purgative, tonic, and reduces temperature.

***C. terminalis*** Kuenth. = C. fruticosa = Taetsia ferrea. (Palm Lily). India, Australia and Pacific Islands. In Hawaii the underground parts are fermented to make a beverage and the leaves are used to make skirts. In Polynesia the leaves are used for wrapping fish to bake it. In Samoa the leaves are made into skirts.

COREOPSIS L. Compositae. 120 spp. America, tropical Africa, Hawaii. Herbs.

***C. cardaminifolia*** (DC.) Torr. and Gray. S. United States of America, Mexico. A decoction of the plant is used by the local Indians to make a beverage.

***C. leavenworthii*** Torr. and Gray. (Tickweed). S. United States of America. An infusion of the plant is used by local Indians to treat heat exhaustion.

Coriander (Seed) (Oil) – Coriandrum sativum.

CORIANDRUM L. Umbelliferae. 2 spp. Mediterranean. Herbs.

***C. sativum*** L. (Coriander). Cultivated particularly in India, Russia, Hungary and Holland for the seeds. These are used for flavouring curries, liqueurs, sauces, soups, preserved meats etc. The dried fruits are also used as a carminative. The seeds are also the source of an oil (Oil of Coriander, Oleum Coriandri) the active principles of which are coriandrol, d-x-pinene, and z-pinene. This is used for flavouring, a carminative, in perfumery, and soap-manufacture. The young leaves are used for soups locally.

CORIARIA L. Coriariaceae. 15 spp. Mediterranean to Japan, New Zealand, Mexico to Chile. Trees or small woody plants.

***C. myrtifolia*** L. Mediterranean. The bark and leaves are used for tanning.

***C. sarmentosa*** Forst. (Tuhu). New Zealand. The berries are used by the Maoris to make a beverage. The shoots and seeds are poisonous to livestock.

***C. thymifolia*** Humb. and Bonpl. S.America. The berries are used to make a black ink in Colombia.

***Coridothymus capitatus*** Reichb. = Thymus capitatus.

Corigliano – Glycyrrhiza glabra.

Cork Oak – Quercus suber.

Cork Tree – Parinarium mobola.

Cork Tree, Arnur – Phellodendron amurense.

Cork Tree, Sachalin – Phellodendron sachalinense.

Corks – Parmelia omphalodes.

Corkwood – Duboisia myoporoides.

Corkwood, Florida – Leitneria floridana.

Cork Wood Tree – Entelea arborescens.

***Cormigonum mariannensis*** Raf. = Bikkia mariannensis.

***Cormigonum mar*** Raf. = Bikkia mariannensis.
Corn (Oil) – Zea mays.
Corn Cuckle – Agrostemma githago.
Corn Ergot – Ustilago maydis.
Corn, Guinea – Sorghum vulgare.
Corn, Indian – Zea mays.
Corn Lily – Clintonia borealis.
Corn, Pop – Zea mays.
Corn Poppy – Papaver rhoeas.
Corn Root, Sweet – Calathea allouia.
Corn Salad – Valerianella olitoria.
Corn Salad, Italian – Valerianella eriocarpa.
Corn Smut – Ustilago maydis.
Corn Spurrey – Spergularia arvensis.
Corn, Squirrel – Dicentra canadensis.
Corn, Turkey – Dicentra canadensis.
Cornel, Round-leaved – Cornus circinnata.
Cornel, Silky – Cornus amomum.
Cornelian Cherry – Cornus mas.
Cornezuelo – Acacia hindsii.
Cornflower – Centaurea cyanus.
CORNULACA Delile. Chenopodiaceae. 7 spp. Egypt to C.Asia. Small shrubs.
***C. monacantha*** Del. W.African deserts. Valuable camel fodder.
CORNUS (Tourn.) L. (Sens. lat.) 60 spp. C. and S.Europe to Caucasus, E.Asia, N.America. Small trees.
***C. alternifolia*** L. f. (Pagoda Dogwood). N.America. The hard close-grained wood is used for turning.
***C. amomum*** Mill = C. sericea. (Kinnikinnik, Silky Cornel). E. N.America. The bark (American Red Osier Bark) is used to treat indigestion, dropsy and diarrhoea. It is used powdered as a toothpowder.
***C. asperifolia*** Mich. (Rough Leaved Dogwood). E. and S.E. N.America. The Indian used the wood for arrow shafts.
***C. circinnata*** L'Hér. (Green Osier, Round-leaved Cornel). E. N.America. The bark (Green Osier Bark) is used to treat fevers, jaundice, and as a tonic.
***C. declinata*** Sessé and Moc. = C. excelsa.
***C. excelsa*** H.B.K. = C. declinata = C. tolucensis. (Aceitunillo, Tepeacuilotl, Topoza). C.America. The bark is used locally as a tonic.
***C. florida*** L. (Flowering Dogwood). E. N.America. The wood is hard, heavy and dark brown. It is used for small turning, wheels, bearings, tool handles, and engravers' blocks. The root bark is used as a tonic and astringent.
***C. kousa*** Hance var. ***chinensis*** Osborn. China. The fruits (Yangmei) are juicy and are eaten locally.
***C. mas*** L. = C. mascula. (Cornelian Cherry). Europe to Asia Minor. The fruits are sweet and pleasant-tasting. They are eaten raw or made into preserves. In France they are made into an alcoholic drink (Vin de Cornoulle). The hard wood is used for turnery etc.
***C. mascula*** Hort. = C. mas.
***C. nuttallii*** Aud. (Western Flowering Dogwood). Pacific N.America. The hard, close-grained light coloured wood is used for cabinet work, tool-handles etc. The bark is used locally to treat fevers.
***C. sanguinea*** L. = Thelycrania sanguinea.
***C. sericea*** L. = C. amomum.
***C. stolonifera*** Michx. = Thelycrania serica.
***C. suecica*** L. N.America, N.Europe and Asia. Small shrub. The small fruits are eaten by the Eskimos.
***C. tolucensis*** H.B.K. = C. excelsa.
CORNUTIA L. Verbenaceae. 12 spp. Tropical America, W.Indies. Small trees.
***C. pyramidata*** L. C.America, W.Indies. The fruits yield a blue juice which is used as an ink in Dominica, and to dye cloth by the Caribs. If the juice is made alkaline by boiling with lime it becomes red.
Coroba – Scheelea macrocarpa.
Coroba – Scheelea macrolepis.
Corojo Fibre – Acrocomia crispa.
Corojo Palm – Acrocomia crispa.
COROKIA A. Cunn. Escalloniaceae. 6 spp. New Zealand, Polynesia, Rapa Islands. Shrubs.
***C. buddleoides*** A. Cunn. New Zealand. A decoction of the leaves is used locally to treat stomach ache.
Coromandel Ebony – Diospyros hirsuta.
Coromandel Ebony – Diospyros melanoxylon.
CORONILLA L. Leguminosae. 20 spp. Europe, Mediterranean. Herbs.
***C. emerus*** L. S. and C.Europe, S.Russia, Syria. The leaves (Folia Colutae) are used as a diuretic and to treat heart complaints.
CORONOPUS Zinn. 10 spp. Almost cosmopolitan. Herbs.
***C. lepidioites*** Coss. Sahara. The leaves are eaten locally as a vegetable.
Corossol – Annona muricata.
COROZO Jacq. ex Giseke. Palmae. 1 sp. Tropical America. Tree.
***C. oleifera*** Bailey = Elaeis melanocarpa

(American Oil Palm, Corozo, Noli Palm). Tropical America. The fruits and seeds yield an oil (Oil of Corozo) used for lighting, soap manufacture, and machinery. The fat from the seeds is edible, while the fat from the fruit is used in Colombia to treat dandruff, and as a hair tonic.

Corozo – Corozo oleifera.

Coroza – Scheelea excelsa.

Coroza – Scheelea liebmannii.

Coroza – Scheelea lundellii

Coroza – Scheelea preusii.

Coroza zonza – Scheelea zonenis.

CORREA Andr. Rutaceae. 11 spp. Temperate Australia. Tree.

***C. alba*** Andr. = C. cotinifolia. Tasmania. The leaves are used locally to make a tea (Cape Barren Tea).

***C. cotinifolia*** Salis. = C. alba.

CORRIGIOLA L. Caryophyllaceae. 10 spp. Cosmopolitan. Herbs.

***C. littoralis*** L. = C. telephiifolia. Europe, especially Mediterranean. The crushed roots are used to relieve stomach aches, as a tonic, and in Morocco to make perfumes.

***C. telephiifolia*** Pourr. = C. litoralis.

Corsican Moss – Alsidium helminthochorton.

CORTADERIA Stapf. Graminae. 15 spp. S.America. Grasses.

***C. argentea*** Stapf. = Gynerium argenteum. (Pampas Grass). S.America. Cultivated locally for paper manufacture.

Cortex Azadirachta – Azadirachta indica.

Cortex Geoffreae – Geoffroea suinamensis.

Cortex Monesiae – Chrysophyllum glycyphloeum.

Cortex Mucennae – Albizia anthelmintica.

Cortex Mururé – Ficus cystopodium.

Cortex Pereriae – Geissospermum laeve.

Corteza de Ojé – Ficus anthelmintica.

CORTINARIUS Fr. Agaricaceae. 400 spp. especially N.Temperate. The fruitbodies of the following species are edible.

***C. ararmillatus*** Fr.

***C. elatus*** (Pers.) Fr.

***C. emodensis*** Berk. (Onglau). Himalayas.

***C. fulgens*** (A. and S.) Fr.

***C. latus*** (Pers.) Fr.

***C. multiformis*** Fr.

***C. violaceus*** (L.) Fr.

***Cortinellus shiitake*** Henn. = Lentinus edodes.

CORYDALIS Vent. Fumariaceae. 320 spp. N.Temperate; 1 sp. tropics into E.Africa. Herbs.

***C. ambigua*** Cham. and Schlecht. Japan. The tubers are eaten locally.

***C. bulbosa*** DC. Europe, Asia. The boiled tubers are eaten locally by tribes around the Caspian Sea.

***C. gavaniana*** Willd. (Bhut Kesi, Bhutakesi, Bhutkis). W.Himalayas. The root is used by the Hindus to treat skin complaints, syphilis, as a tonic, to improve the appetite, and as a diuretic.

***C. solida*** Swartz. Europe, temperate Asia. The roots are eaten locally in Siberia.

CORYLUS L. Corylaceae. 15 spp. N.Temperate. Shrubs or small trees. The following species are cultivated locally for the nuts, which are generally eaten locally.

***C. americana*** Walt. E. N.America.

***C. avellana*** L. (European Hazel). Europe, temperate Asia. It is cultivated widely within its climatic range for the nuts (Barcelona Nuts, Cob Nuts, Filberts, Hazel Nuts). The main areas of commercial production and export are Turkey (200,000 tons per annum), Spain (20,000 tons per annum), U.S.A. (mainly Oregon) (10,000 tons per annum), and Italy (50,000 tons per annum). The kernels are eaten raw or in oil, or are crushed to extract the oil (Hazel Nut Oil, Filbert Oil), which is used for food, a lubricant, soap-manufacture, perfumery and paints. The wood is soft, light and not durable. It is used for walking sticks, light aircraft, sieves, and as a source of charcoal for making gunpowder. The trees are propagated by budding or grafting onto seedlings, yielding in their fourth year, and giving about $\frac{1}{2}$ ton of nuts per acre. There are several commercial varieties. The leaves are occasionally used as a tobacco substitute.

***C. californica*** (A. DC.) Rose. (Californian Hazelnut). W. N.America. Used mainly by the local Indians. Not cultivated.

***C. colurna*** L. (Turkish Hazel, Constantinople Nut, Filbert of Constantinople). S.E.Europe, S.W.Siberia.

***C. heterophylla*** Fisch. (Siberian Hazelnut). China.

***C. rostrata*** Ait. (Beaked Hazelnut). E. N.America.

***C. tubulosa*** Willd. (Lambert's Filbert). S.Europe. A hybrid with C. avellana is sometimes cultivated.

CORYNANTHE Welw. Rubiaceae. 8 spp. Tropical Africa. Trees.

***C. paniculata*** Welw. Tropical Africa. The hard white wood is used for general building purposes. It is durable, and very dense.

***C. yohimbe*** Schum. = Pausinystalia yohimba.

CORYNOCARPUS R. and G. Forst. Corynocarpaceae. 4–5 spp. New Guinea, Queensland, New Hebrides, New Caledonia, New Zealand. Trees.

***C. laevigata*** Forst. New Zealand. The fruits and seeds are eaten locally by the Maoris, and they use the trunks to make canoes.

CORYNOSTYLIS Mart. Violaceae. 4 spp. Tropical America. Shrubs.

***C. arborae*** (L.) Blake = C. hybanthus = Viola arborea. Mexico. The roots are used locally as an emetic.

***C. hybanthus*** Mart. and Zucc. = C. arborea.

CORYPHA L. Palmae. 8 spp. Ceylon, S.E.Asia, Indomalaya. Palm trees.

***C. australis*** R. Br. = Livistonia australis.

***C. elata*** Roxb. (Buri Palm). Tropical Asia. The leaf petioles are the source of a fibre (Buntal Fibre) used locally to make rope and hats (Balinag Hats, Lacban Hats). A finer fibre from the unopened leaves is made into fancy articles, cloth and hats (Calaseao Hats, Pototan Hats). The mature seeds are made into necklaces and buttons while the kernels of the young seeds are eaten as sweetmeats. The sap is sweet and is used to make a sugar syrup, a fermented beverage (Tuba), and vinegar.

***C. lucuola*** Lam. = Licuala rumphii.

***C. umbraculifera*** L. (Talipot Palm). India and Ceylon, Malaya. In Ceylon, the leaves are used for thatching etc. The seeds are used for necklaces, buttons etc. The crushed fruits are used to stupify fish. The leaves were used by the Buddhists for writing their sacred books.

COSCINIUM Colebr. Menispermaceae. 8 spp. Indomalaya, Indochina, Vines.

***C. fenestratum*** Colebr. E.India. The wood yields a yellow dye, and the roots are used locally as a tonic and to treat stomach complaints.

Cosmetic Bark Tree – Murraya paniculata.

Costa Rican Guava – Psidium friedrichsthalianum.

Costa Rican Sarsaparilla – Smilax relelii.

Costmary – Chrysanthemum balsamita.

COSTUS L. Costaceae. 150 spp. Tropics. Herbs.

***C. afer*** Ker-Gawl. (Ginger Lily). Tropical Africa. The powdered fruits are used locally to treat coughs, and a decoction of the stem is used to treat sickness. Strips of the outer stem are used to make baskets.

***C. lucanusianus*** Braun. and Schum. Tropical Africa, especially W.C. An extract of the plant is sometimes used locally to coagulate the latex from Landophilia owariensis.

***C. spicatus*** (Jacq.) Swartz. C and N. C.America. The sap is used locally as a diuretic.

Costus (Root Oil) – Caussurea lappa.

Côt tzu – Psoralea corylifolia.

***Cotinus americanus*** Nutt. = Rhus cotinoides.

***Cotinus coggygria*** Scop = Rhus cotinus.

COTONEASTER (B. Ehr.) Medik. Rosaceae. 50 spp. N.Temperate. Shrubs.

***C. nummularia*** Fisch. and Mey. = C. racemiflora.

***C. racemiflora*** (Desf.) Koch. = C. nummularia. N.Africa through to N.India. A sweet manna (Shir Khist) is formed on the plants and is used locally in Iran and India.

Cotton Abroma – Aroma augustum.

Cotton, American Upland – Gossypium hirsutum.

Cotton, Andes – Gossypium peruvianum.

Cotton, Arabian – Gossypium herbaceum.

Cotton, Bahia – Gossypium brasiliense.

Cotton, Broach – Gossypium obtusifolium.

Cotton, Chain – Gossypium brasiliense.

Cotton, Chinese – Gossypium nanking.

Cotton, Gallini – Gossypium barbadense.

Cotton Gum – Nyssa aquatica.

Cotton, Kathiawar – Gossypium abtusifolium.

Cotton, Khaki – Gossypium nanking.

Cotton, Kidney – Gossypium brasiliense.

Cotton, Kumpta – Gossypium obtusifolium.

Cotton, Lavender – Santolina chamaecyparissus.

Cotton, Levant – Gossypium herbaceum.

Cotton, Maltese – Gossypium herbaceum.

Cotton, Nanking – Gossypium nanking.

Cotton, Pernambuco – Gossypium brasiliense.

Cotton, Peruvian – Gossypium peruvianum.

Cotton, Red Peruvian – Gossypium microcarpum.

Cotton, Sea Island – Gossypium barbadense.
Cotton Seed Oil – Gossypium herbaceum.
Cotton, Short Staple American – Gossypium herbaceum.
Cotton, Siam – Gossypium nanking.
Cotton, Stone – Gossypium brasiliense.
Cotton, Surat – Gossypium obtusifolium.
Cotton, Syrian – Gossypium herbaceum.
Cotton Tree – Gossypium arboreum.
Cotton Tree – Ochroma limonensis.
Cottonweed – Gnaphalium uligonosum.
Cottonwood, Black – Populus heterophylla.
Cottonwood, Fremont's – Populus fremontii.
Cottonwood, Swamp – Populus heterophylla.
Cotula – Anthemis cotula.

COTYLEDON L. Crassulaceae. 40 spp. S.Africa, S.W. N.America, 1 sp. Eritraea, Arabia. Herbs.

***C. edulis*** (Nutt.) Brewer. California. The leaves are eaten by the local Indians.

***C. lanceolata*** (Nutt.) Brewer and Wats. See C. edulis.

***C. orbiculata*** L. (Pig's Ear, Hondeoor, Konterie). S.Africa. The plant is poisonous, but the juice from the leaves is sometimes used locally to treat epilepsy.

***C. pulverulenta*** (Nutt.) Brewer and Wats. See C. edulis.

Couch Grass – Agropyron repens.

COUEPIA Aubl. Chrysobalanaceae. 58 spp. C. and tropical S.America. Shrubs or small trees.

***C. polyandra*** (H.B.K.) Rose = Hirtella polyandra. Mexico. The fruits (Zapote Amarillo) are eaten locally.

Cough Root – Leptotaenia multifida.

COULA Baill. Olacaceae. 3 spp. Tropical W.Africa. Trees.

***C. edulis*** Baill. Tropical W.Africa. The brown wood is easily worked and is used as a substitute for mahogany. The seeds are eaten locally fresh, or cooked, or fermented as a flavouring for meat.

Coulter Pine – Pinus coulteri.

COUMA Aubl. Apocynaceae. 15 spp. Brazil, Guyana. Trees.

***C. guatemalensis*** Standl. Brazil. The latex is used to make chewing gum.

***C. guianensis*** Aubl. Guyana. The sweet fruits are eaten locally.

***C. macrocarpa*** Barb. Brazil. The latex is used to make chewing gum.

***C. utilis*** Muell. Brazil. The latex, mixed with Castor Oil, is used locally to treat intestinal worms in livestock. The crushed fruits are used to treat lice etc. on men and livestock. A resin from the stem is a varnish for pottery.

COUMAROUNA Aubl. Leguminosae. 13 spp. Tropical America. Trees.

***C. odorata*** (Willd.) Aubl. = Dipteryx odorata. (Tonka Tree). Guyana, Venezuala. The seeds contain about 35 per cent non-drying oil, and 3 per cent coumarin which is used for flavouring, confectionery, to make perfumes, making bitters, as a vanilla substitute, and perfuming tobacco and snuff. Much of the commercial crop is collected from wild trees, although some is grown commercially in Venezuela, W.Indies and Malaya. Much is now replaced by synthetics. The seeds (Tonka Beans) are sometimes called Dutch or English Tonka. In Trinidad, the beans are sometimes cured in rum.

Countess Powder – Cinchona officinalis.
Country Almond – Terminalia catappa.
County Ipecacuanha – Naregamia alata.
County Mallow – Abutilon indicum.
Couranira – Humiria balsamifera.

COURATARI Aubl. Lecythidaceae. 18 spp. Tropical S. America. Trees.

***C. tauari*** Berg. (Tauary). S.America. The bark is used locally by the Indians, to make cloth.

Courbaril – Hymenaea courbaril.

***Courbonia*** Brongn. = Maerua.

COUROUPITA Aubl. Lecythidaceae. 20 spp. Tropical America, W.Indies. Trees.

***C. peruviana*** Berg. (Aiauma). Peru. The wood is soft, and is used for treating skin diseases of livestock by the local Indians.

COURSETIA DC. Leguminosae. 15 spp. S.California to Brazil. Shrubs, or small trees.

***C. mexicana*** S. Wats. = Willardia mexicana.

***C. microphylla*** Gray. S.W. U.S.A., Mexico. The invasion of the plant by Tachardia spp. causes the exudation of a lac which is used by the local Indians to seal jars of syrup.

COUTAREA Aubl. Rubiaceae. 7 spp. Tropical America, W.Indies. Shrubs.

***C. latifolia*** Moc. and Sessé. Mexico. The bark (Copalchi Bark), is used to make a tonic and aromatic bitters. It was used in the treatment of fevers. It was exported to Europe.

***C. pterosperma*** (S. Wats.) Standl. (Copalchi, Copalquin, Palo margo). Mexico. The bark is used locally in the treatment of malaria and other fevers, and for lung complaints.

COUTOUBEA Aubl. Gentianaceae. 5 spp. S.America, W.Indies. Herbs.

***C. alba*** Lam.=C. apicata.

***C. apicata*** Aubl. Tropical America. The extract from the roots is bitter and is used locally to treat stomach complaints, fevers and intestinal worms. It is also used as a tonic.

***Covillea tridenta*** (DC.) Vaill.=Larrea mexicana.

Cowa – Garcinia cowa.

Cowage – Muncuna pruriens.

COWANIA D. Don. Rosaceae. 5 spp. S.W.United States of America, Mexico. Shrubs.

***C. mexicana*** D. Don.=C. stansburiana. California to Mexico. The inner bark is used locally by the Indians to make a fibre used in making ropes, cloth, mats and sandals.

***C. stansburiana*** Torr.=C. mexicana.

Cowberry – Vaccinium vitis-idaea.

Cowitch – Muncuna pruriens.

Cow Lily – Nuphar advena.

Cow Parsley – Chaerefolium sylvestre.

Cow Parsnip – Heracleum lantanum.

Cow Parsnip – Heracleum sphondylium.

Cowpea – Vigna sinensis.

Cowpea, Hindu – Vigna catjang.

Cowslip – Primula veris.

Cow Tamarind – Pithocellobium saman.

Cow Tongue – Clintonia borealis.

Cow Tree – Brosimum utile.

Coxoba Snuff – Piptadenia peregrina.

Coyal – Acrocomia mexicana.

Coyal Real – Scheelea liebmannii.

Coyo Avocado – Persea schiedeana.

Coyor – Aiphanes minima.

Coytillo, Calderon – Karwinskia calderonii.

Coytillo, Humboldt – Karwinskia humboldtiana.

Crab Apple – Malus angustifolia.

Crab Apple – Malus coronaria.

Crab Apple – Schizomeria ovata.

Crabeye Lichen – Lecanora parella.

Crab's Claw – Stratiodes aloides.

Crab's Eye – Abrus precatorius.

Crabwood – Carapa guineensis.

***Cracca virginiana*** L.=Tephrosia virginiana.

CRAIBIA Harms. and Dunn. Leguminosae. 10 spp. Tropical Africa. Trees.

***C. grandiflora*** (Micheli) Harms.=Pterocarpus grandiflorus. (Ifofolo, Zengakiveli). Equatorial Africa. The white wood is used locally for making furniture and general carpentry.

CRAMBE L. Cruciferae. 25 spp. Europe, Mediterranean, N.Atlantic Islands, tropical Africa, W. and C.Asia. Herbs.

***C. maritima*** L. (Sea Kale). Coasts of Europe. The bleached petioles are eaten as a vegetable. The plant is occasionally cultivated, especially in the United Kingdom.

***C. tatarica*** Jacq. (Tatarian Sea Kale). S.E.Europe, W.Siberia, S.Russia. The bleached petioles are occasionally eaten as a vegetable.

Cramp Bark – Viburnum opulus.

Cranberry – Vaccinium macrocarpon.

Cranberry, Mountain – Vaccinium erythrocarpum.

Cranberry, Small – Vaccinium oxycoccus.

Cranberry Tree – Viburnum opulus.

Crane's Bill – Geranium robertianum.

Crane's Bill, American – Geranium maculatum.

Crane's Bill, Nepalese – Geranium nepalense.

Crane's Bill, Wallich's – Geranium wallichianum.

Crane's Bill, Wood – Geranium sylvaticum.

CRANIOLARIA L. Martyniaceae. 3 spp. S.America. Herbs.

***C. annua*** L. N. S.America. A decoction of the roots is used locally as a laxative. When candied they are eaten as a sweetmeat in the W.Indies.

Crape Myrtle, Andaman – Lagerstroemia hypoleuca.

Crape Myrtle, Battinan – Lagerstroemia piriformis.

Crape Myrtle, Queen – Lagerstroemia flosreginae.

CRATAEGUS L. Rosaceae. 200 spp. Mainly N.Temperate. Trees.

***C. aronica*** Bosc.=C. azarolus.

***C. azarolus*** L.=C. aronica. (Azerolier). S.Europe to N. Africa. The yellow red or white fruits are eaten locally, fresh or preserved. Cultivated locally and sold in markets.

***C. crus-galli*** L. (Cock's-Spur Thorn.) E. and S.E. U.S.A. The heavy hard wood is used for small articles, e.g. tool-handles.

***C. cuneata*** Sieb. and Zucc. China. The fruits are used locally to treat stomach complaints and dysentery.

***C. douglassi*** Lindl. (Black Haw). Pacific N.America. The fruits are used locally to make jellies.

***C. crenulata*** Roxb.=Pyracantha crenulata. China. The leaves are used locally as a tea substitute.

***C. flava*** Ait. (Summer Haw). S.U.S.A. The yellow fruits are used locally to make jellies.

***C. hupehensis*** Wils. China. Cultivated locally for the fruits, which are large but have an insipid flavour.

***C. hyplosia*** Koch.=C. mexicana.

***C. mexicana*** Moc. and Sessé=C. hyplosia. Mexico. The fruits are used locally to make jellies.

***C. mollis*** Scheele C. U.S.A. The fruits are used locally to make jellies.

***C. oxycantha*** L.=Mespilus oxycantha. (Hawthorn, May). Europe. The young leaves are used as a tea, which reduces blood pressure. They are also used as a tobacco substitute. The seeds are used as a coffee substitute. The hard, heavy wood is difficult to work. It is used in turnery, making wagons, and for walking sticks.

***C. pentagyna*** Waldst. and Kit.=C. pinnatifida. (Chinese Haw). N.China to Siberia. Cultivated in China. The fruits are eaten cooked, preserved, in jellies, or candied. They are sold locally.

***C. pinnatifida*** Bunge.=C. pentagyna.

***C. stipulosa*** (H.B.K.) Steud. (Manzanilla). C.America. The mealy fruits are used locally to make jellies and preserves. They are also used in syrups and eaten cooked.

***Crataeva*** L.=Crateva.

CRATERELLUS Pers. Thelephoraceae. 20 spp. Widespread.

***C. cornucopioides*** (L. ex Fr.) Pers. (Horn of Plenty). The fruitbodies look unattractive, but are of an excellent flavour.

CRATERISPERMUM Benth. Rubiaceae. 17 spp. Tropical Africa; 1 sp. Seychelles. Shrubs.

***C. laurinum*** Benth. Tropical Africa. The bark is used as the source of a yellow dye in Sierra Leone. It is used for dyeing cloth.

CRATEVA L. Capparidaceae. 9 spp. Tropics, excluding Australia and New Caledonia. Trees.

***C. macrocarpa*** Kurz.=Capparis magna. Tropical Asia. The leaves are used as a vegetable in Cochin China.

***C. nurvala*** Buch.=C. religiosa. Tropical Asia and America. The bark is used in India as a laxative and to stimulate the appetite. In Burma the pickled flowers are used to stimulate digestion.

***C. religiosa*** Forst.=C. nurvala.

***C. tapia*** L. (Cascarón, Manzana de Playa, Palo de guaco, Zapotilla amarillo). Tropical America. The bark is used as a treatment for dysentery, stomach complaints, tonic, and to reduce fevers. The roots are used to blister the skin.

***Cratoxylon*** Blume.=Cratoxylum.

CRATOXYLUM Blume. Guttiferae. 12 spp. Indochina, Indomalaya. Trees.

***C. arborescens*** (Vahl.) Blume. (Carongang). Malaysia. The light pinkisk wood is soft and coarse. It is susceptible to termite attack, but fungus resistant. It is used for general construction work, interior woodwork and plywood.

***C. celebicum*** Blume. (Kajoo arang). Indonesia. The wood is hard and insect resistant. It is used for making boats, and charcoal.

***C. clandestinum*** Blume. See C. celebicum.

***C. formosum*** Dyer.=Tridesmis formosa. (Ampet, Garoonggang, Kemootoon). S.E.Asia, Indonesia. The bark is used locally to treat abdominal complaints. A yellow to black resin from the young branches is used to treat scurvy and wounds. The crushed leaves are used to treat burns.

***C. glaucum*** Korth. See C. arborescens.

***C. polyanthum*** Korth. (Kajoo looloos). W. Malaysia. The hard fine wood is used for general construction work, furniture, and axe handles.

***C. racemusum*** Blume. (Harinmeng gede, Oorangoorangan). Java. The wood is burnt to make charcoal.

Crawfood Plantain – Plantago coronopus.

Crawley – Corallorhiza odontorhiza.

Crawwood Oil – Carapa guineensis.

Cream of Tartar Tree – Adansonia digitata.

Cream Nut – Bertholletia excelsa.

Creeper, Virginian – Parthenocissus quinquefolia.

Creeping Snowberry – Chiogenes hispidula.

Creeping Thyme – Thymus serphyllum.

Crema de Caruba – Passiflora mollissima.

Crema de Mamey Colorado – Calocarpum mammosum.

CREMASPORA Benth. Rubiaceae. 3 spp. Tropical Africa, Madagascar. Shrubs.

***C. africana*** Benth. Tropical Africa. A blue-black dye from the fruits is used in W.Africa to paint the body.

CREMASTRA Lindl. Orchidaceae. 7 spp. E.Asia. Herbs.

***C. wallichiana*** Lindl. (Hakkuri). Japan, China to Himalayas. The roots are used locally in Japan to treat toothache.

Crême de Menthe – Mentha piperita.

Crême de Vanille – Vanilla planifolia.

Crementinillo – Casearia pringlei.

Creosote – Fagus ferruginea.

Creosote – Fagus sylvatica.

Creosote Bush – Larrea mexicana.

CREPIDOTUS (Fr.) Quél. Agaricaceae. 40 spp. Cosmopolitan. The fruitbodies of the following species are eaten locally.

***C. djamor*** (Fr.) v. Overeem. Tropical Asia, especially Malaysia.

***C. verrucisporus*** v. Overeem. = Conchomyces verrucisporus. Malaysia.

CREPIS L. Compositae. 200 spp. N.Hemisphere, tropical and S.Africa. Herbs.

***C. japonica*** Benth. Tropical Asia, Australia. The shoots are eaten as a vegetable in China, known as Huang-hua ts'ai.

Crepitus lupi – Calvatia lilacina.

CRESCENTIA L. Bignoniaceae. 5 spp. Tropical America. Trees.

***C. alba*** H.B.K. = Parmentiera alata.

***C. cujete*** L. (Calabash Tree, Calabazo, Cujete, Jicaro). Tropical America. Cultivated locally. The fruit shells are used locally for cups etc. The seeds are eaten cooked. The fruit pulp is used as a laxative, astringent, expectorant and emollient. The young fruits are eaten pickled, locally. They are rather like pickled walnuts.

Cress, Bitter – Cardamine amara.

Cress, Bitter – Cardamine pennsylvanica.

Cress, Field Penny – Thlaspi arvense.

Cress, Garden – Lepidium sativum.

Cress, Large Bitter – Cardamine amara.

Cress, Para – Spilanthes acmella.

Cress, Penny – Thlaspi arvense.

Cress, Water – Nasturtium officinale.

Cress, Winter – Barbarea praecox.

CRESSA L. Convolvulaceae. 5 spp. Tropics and subtropics. Herbs.

***C. cretica*** L. Tropics. Used in the Sudan as a tonic.

Crested Dog's Tail – Cynosurus cristatus.

Crested Wheatgrass – Agropyron cristatum.

Cretian Dost – Origanum onites.

Crimson Clover – Trifolium incarnatum.

Crin Végétale – Carex brizoides.

Crin Végétale – Chamaerops humilis.

Crin Végétale – Tillandsia usneoides.

Crinkleroot – Dentaria diphylla.

CRINUM L. Amaryllidaceae. 100–110 spp. Tropics and sub-tropics. Bulbous herbs.

***C. asiaticum*** L. = C. giganteum. India to Polynesia. The leaves and bulbs are used locally as an emetic and to induce sweating.

***C. flaccidum*** Herb. Australia. The starch from the bulbs is eaten by the aborigines as a kind of gruel.

***C. giganteum*** Blanco = C. asiaticum.

***C. latifolium*** L. Tropical Asia. The roasted bulbs are used in India to induce warming of the skin to relieve rheumatism.The juice from the leaves is used to relieve ear-ache.

CRITHMUM L. Umbelliferae. 1 sp. Rocky coasts of Europe and Mediterranean. Herbs.

***C. maritimum*** L. (Samphire). The leaves have a salty flavour and are used locally in salads or in pickles.

Croatian Tea – Lithospermum officinale.

CROCOSMIA Planch. Iridaceae. 6 spp. Tropical and S.Africa. Herbs.

***C. aurea*** Planch. = Tritonia aurea. Tropical Africa. The flowers yield a yellow dye which is used as a substitute for saffron.

CROCUS L. Iridaceae. 75 spp. Europe and Mediterranean. Herbs.

***C. sativus*** L. (Saffron Crocus). C.Europe to Iran. The flowers are cultivated in France, Spain and elsewhere for the styles of the flowers which yield the dye saffron. 100,000 flowers are needed to produce 1 kg. of saffron. It is used as a colouring agent due to the presence of the yellow glucoside crocin. Medicinally it is used to control menstruation, and to induce sweating. The Romans used saffron as a dye, medicine, spice and incense, while it was commonly used by wealthy Arabian women as a cosmetic for staining the nails and eyebrows.

Crocus, Saffron – Crocus sativus.

Croisette – Gentiana cruciata.

Crookneck, Canada – Cucurbita moschata.

Crookneck, Winter – Cucurbita moschata.

Crosnes du Japon – Stachys sieboldi.

CROSSANDRA Salisb. Acanthaceae. 50 spp.

Tropical Africa, Madagascar, Arabia. Herbs.

***C. infunduliformis*** (L.) Nees. (Gobbi, Kanagambaram, Priyadarsa). Tropical Asia, especially India, where it is cultivated, especially in the North, as an aphrodisiac.

Cross-leaved Heather – Erica tetralix.

CROSSOPTERYX Fenzl. Rubiaceae. 1 sp. Tropical and S.Africa. Small tree.

***C. kotschyana*** Fenzl. The hard light brown wood is used in Mohammedan areas for tables containing copies of the Koran. The bark (African Bark) is used in Sierra Leone to reduce fevers. The seeds are used in various parts as a fumigant for bark cloth, and when powdered as a body powder.

Crotal – Ochrolechia tartarea.

CROTALARIA L. Leguminosae. 550+ spp. Tropics and subtropics. Herbs.

***C. alata*** Ham. Malaya, Indonesia. An excellent green manure, especially when grown in the shade.

***C. anagyroides*** H.B.K. Venezuela. Used as a green manure and cover crop in warm countries.

***C. burhia*** Buch.-Ham. E.India. The fibre from the stems is used locally for sail-making.

***C. cannabina*** Schweinf. Sudan. Cultivated. The stem fibres are used locally for making ropes.

***C. carmioli*** Polak.=C. guatemalensis.

***C. ferruginea*** Grah. Malaya, Indonesia. An excellent green manure, especially in the shade.

***C. glauca*** Willd. Tropical Africa. The leaves, flowers and pods are eaten locally as a vegetable.

***C. goreënsis*** Guill. and Pur. (Gamba Pea). Tropical Africa. It is used in warm countries as a green manure.

***C. guatemalensis*** Benth.=C. carmioli. C.America. The leaves are used occasionally locally as a vegetable.

***C. intermedia*** Kotschy. Tropical Africa. The fibre from the stem is used locally to make rope and string.

***C. juncea*** L. (Sann Hemp, San Hemp, Sunn Hemp). Of uncertain origin, it is grown on a fairly large scale in India and other Oriental countries for the fibre. The phloem fibres are extracted by retting. They are not as strong as hemp, but more durable than jute. Much is exported from India, and is used for cordage, canvas, etc. The plant is also grown as a green manure and forage crop. In U.S.A. it is considered a potential source of fibre for paper-making.

***C. longirostrata*** Hook. and Arn. (Chipilin). C.America. Used locally as a vegetable. Cultivated.

***C. retusa*** L. Tropics. The stem is a source of a fibre used for cordage and canvas. It is occasionally cultivated.

***C. retzii*** A. Hitchc.=C. spectabilis.

***C. saltiana*** Andr. E.Africa. The stem yields a fibre used locally to make nets for catching game.

***C. spectabilis*** Roth.=C. retzii. Tropical Asia. Widely used as a green manure. The seeds are poisonous to stock.

***C. striata.*** DC. Tropics. Used in tropics and subtropics as a green manure and cover crop.

***C. upembiensis*** Wilczek. (Mufambo). Congo. The roots are eaten locally.

***C. usaramoensis*** Baker. Tropical Africa. It is used as a green manure and cover crop in tropical areas.

CROTON L. Euphorbiaceae. 750 spp. Tropical and subtropical. Shrubs.

***C. alanosanus*** Rosa. Mexico. The source of a resin, used locally to treat toothache.

***C. argyratus*** Blume. (Bac lá, Gây la bac). India to Indochina. The leaves are used in Indochina to make a tea.

***C. campestre*** St. Hil. Tropical America. The roots are used in Brazil to treat skin diseases and urinary disorders.

***C. caudatus*** Geisel. India to Malaya. A decoction of the roots is used locally to relieve constipation.

***C. ciliato-glandulosus*** Orteg.=C. penicillatus. (Canelillo, Cuanaxonaxi, Picosa, Soliman). C.America. The leaves are used locally in Mexico to treat fevers, and as a purgative. They are fed to goats to increase milk production.

***C. cortesianus*** H.B.K. Mexico. The caustic juice is used locally to treat skin complaints.

***C. crymbulosus*** Rottr. S.W. U.S.A. The flower heads are used locally to make a tea (Chaparral Tea).

***C. dioidus*** Chav.=C. elaegnifolius=C. gracilis. (Hierba del zorrillo, Hierba del gato, Rosval, Rubaldo). Mexico. The plant is used locally to treat hysteria, and as a bath to relieve rheumatism.

***C. draco*** Schleicht. (Sangre de Grago). Mexico. The bright red sap is used locally as a dye, and to treat diseases of horses' hooves.

***C. echinocarpus*** Muell. Arg. Brazil. The bark is used locally to make a bitters (Industrial Cascarille) which is used against intestinal worms. The Indians use an extract of the bark to treat eye diseases.

***C. eluteria*** Benn. W.Indies. The bark (Cascarilla Bark) is used to make an aromatic bitter. It is mainly supplied from the Bahamas.

***C. elaegnifolius*** Vahl.=C. dioidus.

***C. floribundus*** Spr. Brazil. The bark is a source of Industrial Cascarille, used as a treatment of intestinal worms, and by the Indians to treat eye diseases.

***C. gracilis*** H.B.K.=C. dioidus.

***C. humilis*** L. Tropical America. The roots (Botaosinho) are used in Mexico to treat urinary diseases and skin diseases.

***C. joufra*** Miq.=C. oblongifolius.

***C. laccifer*** L. India, Ceylon. The source of a lac used in varnishes.

***C. linearis*** Jacq. Florida to W.Indies. The leaves are used locally to make a tea.

***C. macrostachys*** Hochst. Ethiopia. The juice is used locally to treat intestinal worms.

***C. niveus*** Jacq.=C. arboreus=C. pseudo-china. (Copalchi, Quina, Vara blanca). Tropical America. The bark (Copalchi, Quina blanco) is used as a substitute for Cascarilla bark in the treatment of fevers and as a tonic. It was exported to Europe.

***C. oblongifolius*** Roxb.=C. joufra=C. elaeognirifolius. India to Indochina. The powdered seeds are used in India as a violent purgative, insecticide, and to stupify fish.

***C. oligandrum*** Pierre. Tropical Africa. The bark is scented and is used locally as an incense in pagan religious ceremonies. It is also used locally to treat stomach complaints.

***C. poilanei*** Gagnep. (Cây cù dèn, Rang cira). Indochina. The bark is used to treat eye diseases in Cambodia.

***C. pseudo-china*** Schlecht.=C. niveus.

***C. reflexifolius*** H.B.K. C.America. The bark is used locally to reduce fevers.

***C. setigerus*** Hook. (Turkey Mullein). S.W. U.S.A. The leaves and bark are used locally to stupify fish.

***C. texensis*** (Klotzsch) Muell. Arg. (Texas Croton, Skunkweed). E. and S.E. U.S.A. The upper parts of the plant are used by various local tribes of Indians to treat stomach ache, eye-diseases, bathe sick children, and as an insecticide.

***C. tiglium*** L. (Purging Croton). Tropical Asia. Cultivated locally for the seeds which yield an oil (Croton Oil) used as a violent purgative. The whole seeds are used to stupify fish.

***C. xalapensis*** H.B.K. Tropical America. A gum from the trunk is used in Mexico to clean the teeth.

Croton Oil – Croton tiglium.

Croton, Purging – Croton tiglium.

Croton, Texas – Croton texensis.

Crottle – Ochrolechia tartarea.

Crottle, Black – Parmelia omphalodes.

Crottle, Dark – Parmelia physodes.

Crottle, Light – Lecanora parella.

Crottle, Stone – Parmelia caperata.

Crottle, White – Pertusaria corallina.

Crowberry – Empetrum nigrum.

Crowfoot, Celery leaved – Ranunculus sceleratus.

Crowfoot Plantain – Plantago coronopus.

Crown Bark – Cinchona officinalis.

Crown Imperial – Fritillaria imperialis.

Crucecillo de la costa – Randia mitis.

Crucillo – Condalia lycioides.

Crucito – Basanacantha armata.

CRYPTERONIA Blume Crypteroniaceae. 4 spp. Assam; 4 spp. S.E.Asia, Malaysia. Trees.

***C. paniculata*** Blume. (Kajoo tjeleng, Kibanen). Indonesia. The young shoots are eaten locally with cooked rice.

***Cryptocarpus capitatus*** S. Wats.=Pisonia capitata.

CRYPTOCARYA R. Br. Lauraceae. 200–250 spp. Tropics (Excluding C.Africa) and subtropics. Trees.

***C. aromatica*** Becc.=Massonia aromatica=Sassafras goesianum. New Guinea. The bark is used locally as a lotion.

***C. canelilla*** H.B.K.=Aniba canelilla.

***C. erythroxylon*** Maid. (Rose Maple). Coast of New South Wales, and N.Queensland. The wood is coarse and uniform, weathering from pink to yellow. It is used for internal work in houses, furniture, carts etc.

***C. glaucescens*** R. Br. (Brown Beech, Jackwood, Silver Sycamore). Coastal New South Wales and Queensland. The light brown wood is used in general household interiors, furniture, box making, veneers and turning.

***C. latifolia*** Sond. E. S.Africa. The fat extracted from the nut (Nitronga Nuts) is used locally for cooking.

***C. membranacea*** Thw. (Gal-mora). The light yellow, close-grained wood is used to make picture frames and similar articles.

***C. minima*** Metz. = Aydendron floribundum.

***C. moschata*** Nees. and Mart. Brazil. The fruits (Brazilian Nutmeg) are used as a spice.

***C. nitida*** Phil. = Bellota miersii. Chile. The bark is a source of fibre which is used locally.

***C. oblata*** F. M. Baill. (Bolly Silkwood, Tarzali, Tarzali Silkwood). Coasts of Queensland. The coarse pink-brown wood is susceptible to termite attack and not durable. But it has an attractive silky lustre which makes it appreciated as a timber for furniture, internal woodwork and plywood.

***C. patentinervis*** F. v. Muell. Coastal New South Wales and Queensland. The wood is used for inside woodwork, furniture, and carts etc.

***C. peumus*** Nees. = Laurus peumus. Chile. The bark is used locally for tanning.

CRYPTOMERIA D. Don. Taxodiaceae. 1 sp. Japan. Large Tree.

***C. japonica*** D. Don. The close-grained red-brown wood is used for house and ship building, and general construction work, boxes etc. When seasoned in the soil the wood turns dark green. Then it is called Jindai-sugi, and is much valued. The leaves are burnt as incense.

***Cryptoporus volvatus*** (Peck) Shear. = Polyporus volvatus.

CRYPTOSEPALUM Benth. Leguminosae. 15 spp. Tropical Africa. Trees.

***C. pseudotaxus*** Bak. f. (Mmamba, Munienze). C.Africa. The flowers are an important source of honey locally.

CRYPTOSTEGIA R. Br. Periplocaceae. 2 spp. Madagascar. Vines.

***C. grandiflora*** R. Br. Madagascar. Occasionally cultivated as an ornamental.

***C. madagascariensis*** Bojer. Madagascar. The bark is a source of fibre, used locally to make fishing nets. The bark is also used locally for making rum. C. grandiflora and C. madagascariensis both produce a latex containing about 7–10 per cent rubber. The latex occurs in all parts of the plants except the woody stems, and is typically extracted by cutting off the tips of the young shoots, and collecting the exuded globules by hand. Planting were made in C.America during World War II to supplement the supply of Hevea rubber. This project has now been abandoned. The rubber is called Palay Rubber.

CRYPTOTAENIA DC. Umbelliferae. 4 spp. Italy, Transcaucasia, W.Equitorial Africa, E.Asia, N.America. Herbs.

***C. canadensis*** DC. N.America. Cultivated in Japan for the young leaves which are eaten as a vegetable, and the roots which are eaten fried.

CRYSOPHILA Benth. and Hook. f. Palmae. 9 spp. Mexico, C.America. Trees.

***C. warscewiczii*** Wendl. C.America. The leaves are used for thatching.

Crystal Tea Ledum – Ledum palustre.

CTENOLOPHON Oliv. Ctenolophonaceae. 1 sp. tropical Africa; 2 spp. Malaysia. Trees.

***C. parvifolium*** Oliv. Malaya. The hard durable wood is used locally for house-building.

Cuadrado – Manihot carthaginensis.

Cuajilote – Aristolochia maxima.

Cuajilote – Parmentiera edulis.

Cuajiote colorado – Bursera microphylla.

Cuauchichic – Garrya longifolia.

Cuba Hemp – Furcraea hexapetala.

Cuba Lancewood – Duguetia quitarensis.

Cube Gambir – Uncaria gambir.

Cubeb – Piper chubeba.

Cubeba (Oil) – Piper chubeba.

Cubebs – Piper chubeba.

Cubese – Macrolobium coeruleum.

CUBILIA Blume. Sapindaceae. 1 sp. Philippine Islands, Celebes, Moluccas. Tree.

***C. blancoi*** Blume = C. rumphii. The leaves and seeds are eaten locally.

***C. rumphii*** Blume = C. blancoi.

Cuca – Mimosa purpurescens.

Cuanaxonaxi – Croton ciliato-glandulosus.

Cuchamperrito – Funastrum cumanense.

Cuchara caspi – Malouetia tamaquarina.

Cucharillo – Trichilia havanensis.

Cucharitas – Acacia cochliacantha.

Cuckoo Flower – Cardamine pratensis.

Cuckoo Pint – Arum maculatum.

Cucuá – Poulsenia armata.

CUCUBALUS L. Caryophyllaceae. 1 sp. N.Temperate. Herb.

*C. baccatus* Gueld.=C. baccifer.

*C. baccifer* L.=C. baccatus. (Bladder Campion). The leaves used to be applied as an astringent.

Cucubano – Guettarda laevis.

Cucumber (Oil) – Cucumis sativus.

Cucumber, Mandera – Cucumis sacleuxii.

Cucumber Root, Indian – Medeola virginiana.

Cucumber, Squirting – Ecballium elaterium.

Cucumber Tree – Averrhoa bilimbi.

Cucumber Tree – Magnolia acuminata.

Cucumber, Wild – Cucumis myriocarpus.

Cucumber, Wild – Cucumis naudinianus.

CUCUMIS L. Cucurbitaceae. 25 spp. Mostly from Africa, a few from Asia. Vines.

*C. anguria* L. (Gooseberry Gourd, West Indian Gourd). Tropical America. The young fruits are eaten boiled and pickled.

*C. cocomon* Roxb. (Oriental Pickling Melon). China, Japan. The young fruits are eaten in soups, and pickled.

*C. colocynthis* Thunb.=C. myriocarpus.

*C. ficifolius* A. Rich. C. and S.Africa. The roots are used by the natives of Rhodesia to commit murder.

*C. melo* L. (Canteloupe, Melon, Musk Melon, Sweet Melon). Africa. The fruits are eaten raw, as a dessert or breakfast fruit and are occasionally used in preserves. There are several varieties and horticultural groups. The varieties are as follows: var. *cantaloupenis* Naud. (Canteloupes), which usually have a rough warty skin, and are grown mainly in Europe, var. *inodorus* Naud. (Honey Dew, Melon, Winter Melons) the fruits of which ripen late and can be kept through the winter, var. *saccharinus* Naud. (Pineapple Melon, Sucrins), var. *acidulus* Naud. (Cuccumber Melon) which is not edible, var. *flexuosus* Naud. (Snake Melon, Serpent Melon) which have long thin fruits grown as ornamentals and occasionally preserved, var. *dudain* Naud. (Pomegranate Melon, Queen Anne's Pocket Melon) with scented fruits used for preserves, but which are inedible when ripe, var. *chito* Naud. (Lemon Melon, Orange Melon, Lemon Cucumber, Mango Melon) used for making preserves, var. *agrestis* Naud. found wild in Asia and Africa, and eaten locally. The fruits of C. melo have been used from ancient times by the Ancient Egyptians and Romans, and are now grown commercially mainly in America.

*C. metuliferus* E. Mey. Tropical Africa. The fruits are used in salads locally. The plant is cultivated.

*C. myriocarpus* Naud.=C. colocynthis. (Wild Cucumber, Bitter Apple). S.Africa. The fruit pulp is used locally by the natives as a purgative, but it can be fatal. A decoction of the whole plant is used during some native marriage rites.

*C. naudinianus* Sond. (Wild Cucumber, Kololo). S. and C.Africa. The roots are used by some native tribes to commit murder.

*C. prophetarum* L. Tropical Africa. The fruits are bitter, and are used in the Sudan as an emetic.

*C. sacleuxii* Paill. and Bois. (Mandera Cucumber). Zanzibar. The young fruits are used locally as pickles.

*C. sativus* L. (Cucumber). S.Asia. Widely cultivated in the Old and New World for the fruits which are used in salads. The young fruits (Gherkins) are used in pickles. There are many horticultural varieties varying from rough to smooth skinned, and from green to nearly white. An oil, extracted from the seeds, (Cucumber Seeds Oil, Huile de Concombre, Oleum Cucumis sativi) is used for cooking, especially in France.

*C. trigonus* Roxb. Australia. The fruits are eaten locally by the natives.

CUCURBITA L. Cucurbitaceae. 15 spp. America and Asia. Vines.

*C. ficifolia* Bouché.=C. melanosperma. E.Asia. Cultivated in Old and New World for the fruits which are eaten boiled and preserved.

*C. foetidissima* H.B.K. (Wild Gourd). S.W. U.S.A., Mexico. The seeds are eaten locally by the Indians. The whole fruit is used as a soap substitute, and an extract of the ground roots is used as a laxative locally in Mexico.

*C. maxima* Duch. (Gourd, Pumpkin, Squash). America. Cultivated in drier parts of Old and New World. The plants yield quickly without much attention. The fruits are eaten raw or cooked. The fruits are very variable in size and colour, but are always rounded or flattened at the ends. The fruits have been eaten by natives in America since prehistory. The seeds yield a semi-drying oil

(Pumpkin Seed Oil) which is used in cooking and for lighting.

***C. melanosperma*** A. Br.=C. ficifolia.

***C. moschata*** Duch. (Pumpkins). Tropical Asia. (Canada Crookneck, Cushaw, Winter Crookneck, Pumpkin). Tropical Asia. Distinguished from C. maxima by having hairless stems and leaves. The fruits are eaten boiled. Cultivated widely in India and parts of Africa. This species can withstand higher temperatures than other species.

***C. pepo*** L. (Marrow, Vegetable Marrow). Himalayas. Cultivated widely in tropics and subtropics, and in temperate regions as a garden crop. The fruits are eaten boiled or roasted, when unripe. They store well. The fruits vary considerably in shape and size, but are more elongated than those of the other C. spp. The plants are hairy, and the tendrils branch from the base.

Cudbear, (Tinctute of) – Ochrolechia tartarea.

CUDRANIA Tréc. Moraceae. 4 spp. Japan to Australia.

***C. javanensis*** Trecul.=Vaniera cochinchinensis. Spiny herb. S.Japan through Himalayas to Australia. The fruits are eaten raw or preserved in Japan. The wood, treated with alum, yields a yellow dye, and when treated with indigo gives a green dye. The leaves are eaten locally in the Moluccas.

***C. tricuspidata*** (Carr.) Bureau=C. triloba.

***C. triloba*** Hance.=C. tricuspidata=Maclura tricuspidata. Small tree. Korea, through China to Japan. The fruits are pleasantly flavoured and are eaten locally in China.

Cudweed – Graphalium uligonosum.

Cuilón – Mimosa purpurascens.

Cujete – Crescentia cujete.

Cujin – Inga rensoni.

Cuojo (Fat) – Virola venezuelensis.

Culebra borrachere – Brugmansia amesianum.

CULLENIA Wight. Bombacaceae. 2 spp. S.India, Ceylon. Trees.

***C. excelsa*** Wight. (Wild Durian). Ceylon, India. The wood is light brown and easily split. It is used for making packing cases etc. and cigar boxes.

Culver's Root – Veronica virginica.

Cumara – Astrocaryum vulgare.

Cumarú – Amburana acreana.

Cumin (Oil) – Cuminum cyminum.

***Cumingia campanulata*** D. Don=Conathera simsii.

CUMINUM L. Umbelliferae. 2 spp. Mediterranean through the Sudan to C.Africa. Herbs.

***C. cyminum*** L. (Cumin). Mediterranean and Arabia. Cultivated since ancient times in the Mediterranean region. It is now cultivated in the Old and New World, especially in Morocco, Mediterranean region and India. The seeds are used extensively for flavouring, especially liqueurs, as a constituent of curry powders, and Lieden cheese. They contain pinene, dipentene, and cominic aldehyde and are used in home remedies for stomach complaints and as an antispasmodic. They are also used in veterinary medicine. The seeds contain about 10 per cent oil (Cumin Oil) used extensively in perfumery.

Cummin, Black – Nigella sativa.

Cû môt – Stepania rotunda.

Cumquat – Fortunella crassifolia.

Cumquat – Fortunella japonica.

Cumquat – Fortunella margarita.

CUNILA Royen ex L. Labiatae. 15 spp. E. N.America to Uruguay. Herbs.

***C. mariana*** L.=C. origanoides.

***C. origanoides*** (L.) Butt.=C. mariana.= Mappia origanoides (Dittany, Stone Mint). E. N.America to Texas. The leaves were used by local Indians and settlers to make a tea to treat colds and fevers.

CUNNINGHAMELLA Matr. Choanephoraceae. 5 spp. Widespread.

***C. elegans*** Lendner. This species is used to test soils for available phosphorus. The estimate is based on the size of the fungus colonies grown on the soil for 48 hours at 20°C.

CUNNINGHAMIA R. Br. Taxodiaceae. 3 spp. S.China, Formosa. Trees.

***C. lanceolata*** Hook.=C. sinensis.

***C. sinensis*** R. Br. China. The light wood is durable and fragrant. It is used widely in China for reforestation. It is used for building, general carpentry, boat-building and tea chests.

CUNONIA L. Cunoniaceae. 1 sp. S.Africa. Tree.

***C. capenis*** L. (Red Alder). The wood is light, durable, and fairly soft. It polishes well and is used for furniture-making, making wagons, and mills.

Cuôi con chông – Uvaria purpurea.
Cup Moss – Cladonia pyxidata.
Cup Plant – Silphium perfoliatum.
Cup Plant, Indian – Silphium perfoliatum.
Cup, Ragged – Silphium perfoliatum.
CUPANIA L. Sapindaceae. 55 sp. Warm America. Small trees.
***C. glabra.*** Sw. W.Indies. The strong red wood is used for general carpentry.
***C. gracilis*** Panch. and Sebert. New Caledonia. The wood is used locally for turning and cabinet work.
***C. pseudorhus*** A. Rich. = Jagera pseudorhus. Australia. The bark is used by the natives of Queensland as a fish poison, and during World War I, to produce foam on soft drinks.
***C. stipitata*** Panch. and Sebert. New Caledonia. The wood is fine-grained and pale red in colour. It is used locally for cabinet-making.
Cupey – Clusia rosea.
CUPHEA P. Br. 250 spp. America. Lythraceae. Herbs.
***C. glutinosa*** Cham. and Schlecht. Brazil and Argentine. An infusion of the leaves is used locally as a diuretic and purgative.
CUPHOCARPUS Decne. and Planch. Araliaceae. 1 sp. Madagascar. Tree.
***C. inermis*** Baker. The wood is used locally to make musical instruments.
Cuprea Bark – Remijia pedunculata.
CUPRESSUS L. Cupressaceae. 15–20 spp. Mediterranean, Sahara, Asia, N.America. Trees.
***C. arizonica*** Greene. (Arizona Cypress, Cearo de la Sierra). S. N.America, Mexico. The wood is used for general construction work and as a fuel.
***C. benthamii*** Endl. = C. lusitanica.
***C. dupreziana*** A. Camus. (Cyprès du Tassili, Tarout). N.Africa. The wood was used for construction work by the Tuarec.
***C. funebris*** Endl. China. Cultivated in C.China for the timber which is used for general construction work, boat-building, agricultural implements and coffins. The wood is hard, tough, durable, and white.
***C. hodginsii*** Dunn. = Fokienia hodginsii. S.China, Vietnam. The wood is much used by the Chinese for making coffins. An essential oil is distilled from the wood used to make the perfume Bois de Siam.
***C. lustanica*** Mill. = C. benthamii. (Cerdo blanco). Mexico, C.America. The wood is used in Mexico for general construction work. The bark is used locally as an astringent.
***C. macrocarpa*** Hartw. (Monterey Cypress). California. It is widely planted in warm countries as a shade tree and for timber used in general construction work.
***C. sempervirens*** L. (Italian Cypress). S.Europe to W.Asia. An oil (Oil of Cypress) is distilled from the leaves and used in perfumery, soap manufacture and as a treatment for whooping cough. Much of the oil is produced in the Bouches du Rhône of France.
***C. torulosa*** Don. W.China to Himalayas. The wood is used in China for general construction work.
Curaçao Aloe – Aloë vera.
Curaçao (Liqueur) – Citrus aurantium.
CURANGA Juss. Scrophulariaceae. 1 spp. Indomalaysia. Herb.
***C. amara*** Juss. = C. fel-terrae.
***C. fel-terrae*** Merr. = C. amara. A decoction of the leaves is used locally to treat dropsy, to induce sweating, as a diuretic, and to control menstruation.
Curare Poison Nut – Strychnos toxifera.
CURATELLA L. Dilleniaceae. 2 spp. Tropical America. W.Indies. Trees.
***C. americana*** L. (Sandpaper Tree). C.America to N. S.America. The coarse heavy red-brown wood is used for furniture-making and making charcoal. The bark is used for tanning. The leaves are rough due to a high silica content, and are used as a substitute for sandpaper.
Curcas Oil – Jatropha curcas.
CURCULIGO Gaertn. Hypoxidaceae. 10 spp. Tropics. Herbs.
***C. latifolia*** Dryand. India to Malaya. The leaves produce a fibre used to make fishing nets in Borneo.
***C. orchioides*** Gaertn. Tropical Asia. The leaves are used in the Philippines to treat skin diseases.
***C. recurvata*** Dryand. India to Australia. The fibres extracted from the leaves are used by the natives of Luzon to make wigs.
CURCUMA L. Zingiberaceae. 12 spp. Indomalaysia to China. Herbs.
***C. aeruginosa*** Roxb. Burma. The rhizome is a local source of starch.

***C. alismatifolia*** Gagnep. Cambodia, Thailand, Laos. The flowers are eaten locally as a vegetable.
***C. amada*** Roxb. (Amada, Mango Ginger). India. Cultivated locally for the rhizomes which have the flavour of mangoes. They are used in pickles, and in local medicine to relieve digestive upsets and reduce the swellings of bruises etc.
***C. angustifolia*** Roxb. (East Indian Arrowroot). N.India, Himalayas. Cultivated for the rhizomes which yield starch called East Indian Arrowroot, Tik, Tikur, Tikor, or Travancore Starch.
***C. aromatica*** Roxb. (Yellow Zedory, Wild Turmeric). India. Cultivated locally for the rhizomes which yield an orange-red dye. The rhizome is also used in local medicine.
***C. caesia*** Roxb. (Kalihaldi, Narkachura). Bengal. Cultivated locally for the rhizome which is used to treat stomach complaints, as a stimulant and externally on bruises.
***C. domestica*** Loir. = C. longa.
***C. heyneana*** Valeton. Java. The rhizome is a local source of starch.
***C. leucorhiza*** Roxb. (Tikan). India. The rhizomes are a local source of arrowroot sometimes called East Indian Arrowroot.
***C. longa*** L. = C. domestica (Curcuma, Turmeric). S.Asia. Cultivated in India, China, E.Indies and W.Indies. The washed, peeled and powdered rhizome is used as a condiment in curries etc., as an adulterant for mustard and ginger, and as the source of a yellow dye. The dye can be used for colouring cotton, silk, and wool without using a mordant. The colour is due to cucumin which can be oxidized by potassium permanganate to produce vanilla. The dye is also used as a pH indicator, for alkalines. In India and the Far East the juice is used for a wide variety of medicinal purposes, e.g. treating stomach complaints, bruises; fumes from the burning rhizome relieve colds and catarrh, and a paste of the rhizome accelerates the formation of scabs caused by smallpox and chicken pox.
***C. pierreana*** Gagnep. Malaya. Cultivated locally for the rhizome which yields starch (False Arrow Root).
***C. stenchloa*** Gagnep. Cambodia. The powdered rhizome is used locally to treat stomach upsets.
***C. xanthorrhiza*** Roxb. Indonesia. Cultivated locally in Java for the starch which is eaten as a porridge.
***C. zeodoaria*** Roscoe. (Zedoary). India. It has been cultivated for centuries in India and the Far East. The dried rhizome is used as a condiment (Zedory, Zedoaria), and as a source of starch. It contains α-cienol and a resin which are aromatic, making the rhizome a source of perfume used in the manufacture of cosmetics. The rhizome is also used to treat stomach complaints.
Curcuma – Curcuma longa.
Cure All – Anagallis arvensis.
Curled Dock – Rumex crispus.
Curled Mallow – Malva crispa.
Curlycup Grindelia – Grindelia squarosa.
Curly Yarran – Acacia homalophylla.
Currant, American Red – Ribes triste.
Currant, American Wild Black – Ribes americanum.
Currant, Buffalo – Ribes aureum.
Currant, Buffalo – Ribes odoratum.
Currant, European Black – Ribes nigrum.
Currant, Golden – Ribes aureum.
Currant, Indian – Symphoricarpus orbicularis.
Currant, Missouri – Ribes aureum.
Currant, Native – Leptomeria acida.
Currant, Prickly – Ribes lacustre.
Currant, Red – Ribes rubrum.
Currant Rhubarb – Rheum ribes.
Currant, Swamp Red – Ribes triste.
Currant, Wild Black – Ribes americanum.
Currawong Acacia – Acacia doratoxylon.
CURTISIA Ait. Cornaceae. 1 spp. S.Africa. Tree.
***C. dentata*** (Burm.) C. A. Sm. = C. faginea. (Assegai Tree, Cape Lancewood). The hard reddish wood polishes well and is used for furniture making and wheel spokes.
***C. faginea*** Ait. = C. denetata.
Curuba – Passiflora maliformis.
Curuba – Sicana odorifera.
Curupag – Piptadenia macrocarpa.
CUSCUTA L. Cuscutaceae. 170 spp. Tropical and Temperate. Parasitic herbs.
***C. epithymum*** (L.) Murr. (Devil's Guts, Lesser Dodder). Asia, Europe, naturalised in N.America, S.Africa. The leaves are used locally as a laxative.
Cushaw – Cucurbita moschata.
Cush-cush – Discorea trifida.
Cushta – Theobroma angustifolium.

CUSPARIA Humb. Rutaceae. 30 spp. Tropical S.America. Trees.

***C. febrifuga*** Humb.=C. trifoliata Engl. Brazil. The bark (Angostura Bark) is bitter. An extract of the bark is used to treat dyspepsia, dysentery and diarrhoea. It is used locally to treat fevers. The extract affects the spinal motor nerves and is used to treat some forms of paralysis.

Cusparia Bark – Cusparia febrifuga.

Cusso – Hagenia abyssinica.

CUSSONIA Thunb. Araliaceae. 25 spp. Tropical and S.Africa, Madagascar, Mascarene Islands. Trees.

***C. kirkii*** Seem. (Cabbage Tree, Mushonjwa, Mutewetwe, Umbrella Tree). C. Tropical Africa. The wood is used locally in Zambia to make xylophones.

***C. nigerica*** Hutch. (Bolo Coro Ni). Sudan. The leaves are used locally as an aphrodisiac by men, and by women to treat genital diseases.

Custard Apple – Annona squamosa.

Custard Apple, Common – Annona reticulata.

Cutch, Black – Acacia catechu.

Cuxiniquil – Inga preussii.

CYAMOPSIS DC. Leguminosae. 3–4 spp. Tropical and subtropical Africa, Arabia, India. Herbs.

***C. psoralioides*** DC. (Cluster Bean). E.India. Cultivated since ancient times in Asia and later in C.Africa. The seeds and unripe pods are eaten cooked, while the whole plant is valuable fodder and a green manure. A gum from the seeds is used in food manufacture, the paper and textile industries.

***Cyamopsis tetragonolobus*** (L.) Taubert. (Guar). N.India. A drought-resistant plant which has been cultivated in India since ancient times, and has been grown commercially in S.W. U.S.A. on a small scale. A gum (Guar Gum) is extracted from the endosperm. It is a polymer of galactose and mannose. The gum has a high viscosity in low concentrations and is stable over a wide pH range. It is used in cosmetics, in paper manufacture, to depress the appetite, and in the mining industry as a flocculant. The whole plant is a useful fodder for livestock.

CYATHEA Sm. Cyatheaceae. 600 spp. Tropics and subtropics. Tree ferns.

***C. arborea*** (L.) J. E. Smith. Mexico, W.Indies, Venezuela. The dried pith is used by the Caribs as tinder.

***C. canaliculata*** Willd. Madagascar, Réunion, Mauritius. The starch extracted from the pith is used locally.

***C. dealbata*** Swartz. New Zealand. The pith is eaten locally by the natives.

***C. medullaris*** Swartz. (Black-stemmed Tree Fern). Australia. The stem contains starch, which is eaten locally by the natives.

***C. mexicana*** Schlecht and Cham. (Cola de Mono, Ocopetate). C.America. The scales on the fronds are used locally to stop external bleeding.

***C. usambariensis*** Hiern. Tropical E.Africa. The pith and the young fronds are used by the natives to treat intestinal worms.

***C. viellardii*** Mett. New Caledonia. A mucilage from the base of the frond is eaten locally.

CYATHOCALYX Champ. ex Hook. f. and Thoms. Annonaceae. 38 spp. Indomalaya. Trees.

***C. globusus*** Merr. Philippine Islands. The seeds are used locally for chewing.

***C. zeylanicus*** Champ. Ceylon. The soft wood rots quickly, but is used for tea chests etc. The flowers are added to betel nuts for chewing in some parts of Ceylon.

CYBISTAX Mart. ex Meissn. Bignoniaceae. 3 spp. Tropical America. Trees.

***C. donnell-smithii*** (Rose) Seib.=Tabebuia donnell-smithii. C.America. The light yellow-brown wood is coarse, soft and straight-grained. It is used locally for making furniture, and is called Primavera or White Mahogany.

***C. sprucei*** Schum. Peru. Cultivated locally for the leaves which are the source of a blue dye.

CYCAS L. Cycadaceae. 20 spp. Madagascar, E. and S.E.Asia, Indo-Malaya, Australia, Polynesia. Small Trees.

***C. circinalis*** L. India to Philippine Islands. The stem is a minor source of sago, and the seeds are a famine food in parts of the Philippines.

***C. media*** R. Br. Australia. The seeds are poisonous when raw, but are eaten pounded and cooked by the natives of Queensland.

***C. revoluta*** Thunb. Japan. The trunk is a minor source of sago. The large red seeds are eaten locally raw or cooked.

***C. rumphii*** Miq. (Pakis radja, Pakoo padak, Gogopoa, Sajor kalapa). Malayan Archipelago, Australia. The young, unfolded fronds are eaten as a vegetable. The juice from the young leaves is used in Java to treat stomach complaints. The boiled seeds are eaten whole, or made into a flour.

***C. thouarsii*** M. C. Fatra. Madagascar. The fruits are eaten locally.

CYCLAMEN L. Primulaceae. 15 spp. Europe, Mediterranean to Persia. Herbs.

***C. europaeum*** L. (Snowbread). C.Europe. The corm is used as a local treatment for stomach complaints, toothache, bladder pains, as a purgative, and to control menstruation.

CYCLANTHERA Schrad. Cucurbitaceae. 15 spp. Tropical America. Vines.

***C. edulis*** Naud. (Pepino de Comer). Peru. The fruits are used locally for pickles.

***C. pedata*** Schrad. Tropical America. The fruits are eaten locally, especially in Peru, as a vegetable.

CYCLEA Arn. ex Wight. Menispermaceae. 30 spp. Tropical Asia. Herbs.

***C. barbata*** Miers. = C. peltata.

***C. peltata*** Hook. = C. barbata. Malaya. The tubers are used medicinally in Cochin China and Java to treat fevers, as a tonic and to treat stomach upsets. The leaves (Tjintjaoo Idjo) are eaten as a delicacy in Java.

CYCLOLOMA Moq. Chenopodiaceae. 1 sp. N.America. Herb.

***C. atriplicifolium*** (Spreng.) Coult. = Salsola atriplicifolia. (Winged Pignut). The seeds are used by the local Indians to make a flour and a gruel.

***Cyclophorus*** Desv. = Pyrrosia.

CYCLOPIA Vent. Leguminosae. 15 spp. S.Africa. Woody Shrubs. The leaves of the following spp. are used as a substitute for tea.

***C. brachypoda*** Benth. = C. subetrenata.

***C. genistoides*** Vent. S.Africa.

***C. longifolia*** Vogel. S.Africa.

***C. subterenata*** Vog. = C. brachypoda. S.Africa. (Caspa Tea).

***C. tenuifolia*** Lehm. Cape Peninsula.

***C. vogeli*** Harv. S.Africa. (Bush Tea).

***Cyclostemon longifolius*** Blume. = Prypetes longifolia.

CYDONIA Mill. Rosaceae. 1 sp. E.Asia, Caucasus, N.Persia. Tree.

***C. delavayi*** Cardot = Docynia delavayi.

***C. oblonga*** Mill. = C. vulgaris. = Pyrus cydonia. (Quince). Cultivated in warm temperate Old and New World for the fruits which are used for making preserves. The mucilage from the seeds is used commercially for making demulcent lotions. The seeds are produced commercially in Iran, Spain, S.France, Portugal and Russia.

***C. sinensis*** Thouin. = Chaenomeles sinensis.

***C. vulgaris*** Pers. = C. oblonga.

CYMBIDIUM Sw. Orchidaceae. 40 spp. Tropical Asia, Australia. Herbs.

***C. canaliculatum*** R. Br. Australia. The pseudobulbs are eaten locally by the natives.

***C. ensifolium*** Sw. = C. sinense = Limodorum ensatum. India to Indonesia. In Indonesia, a decoction of the flowers is used to treat sore eyes, and the leaves are used as a diuretic.

***C. finlaysonianum*** Lindl. = C. wallichii = C. tricolor. Epiphyte. Coastal regions from Malacca to Celebes. The roots are supposed to keep evil spirits from villages, and when chewed and sprinkled on a sick elephant will cure it.

***C. sinense*** Lindl. = C. ensifolium.

***C. tricolor*** Miq. = C. finlaysonianum.

***C. virescens*** Lindl. Japan. A decoction of the salted flowers is used locally as a beverage. The flowers are also eaten as a preserve in plum vinegar.

***C. wallichi*** Lindl. = C. finlaysonianum.

CYMBOPETALUM Benth. Annonaceae. 11 spp. Mexico to tropical America. Small trees.

***C. costaricense*** (D. Smith) Fries. = Asimina costaricensis. Costa Rica. The petals are used by the local Indians to flavour chocolate.

***C. penduliflorum*** Baill. (Flor de la Oreja). C.America. Cultivated locally for the petals which are used to flavour chocolate.

CYMBOPOGON Spreng. Gramineae. 60 spp. Tropical and subtropical Africa and Asia. Grasses. Many of the species mentioned produce essential oils which are extracted by steam distillation and used in soaps, cosmetics, etc.

***C. caesius*** Stapf. India. Oil of Kachi Grass.

***C. citratus*** (DC.) Stapf. (W.Indian Lemon Grass). Tropical Asia. Cultivated in the Far East, Belgian Congo, W.Indies and C.America. Source of Lemon Grass Oil. The leaves are harvested several times a year and the plants yield economically for three to four years. Citral, extracted from Oil of Lemon Grass is used to synthesise ionone which is used to make synthetic violet perfume and synthesising vitamin A.

*C.* ***coloratus*** Stapf. Fiji.

*C.* ***confertiflorus*** Stapf.=Andropogon nilagiricus. Loas.

*C.* ***excavatus*** (Hochst.) Stapf. S.Africa. The leaves are used locally for thatching. No essential oil is produced.

*C.* ***flexuosus*** Stapf.=Andropogon flexuosus. (East Indian Lemon Grass). S.India. The essential oil (Oil of Inchy, Lemon Grass des Indes orientals, Lemon Grass de Cochin) has the scent of violets and lemons. It is used in the perfume and soap industries. It is used in Vietnam to treat cholera and rheumatism. The oil is derived from red-stemmed varieties grown mainly in India, and to a lesser extent in C.America, W.Indies, and Madagascar. The main constituent of the oil is citral, and it has some value as an insect repellent. Unlike other species this one is usually propagated from seed, the plants yielding for about six years. White stemmed varieties, growing wild in parts of India yield an oil similar to Palmarosa Oil.

*C.* ***giganteus*** Chiov. Nigeria. Oil of Tsauri Grass.

*C.* ***jwarancusa*** Schult.=Andropogon jwarancusa. (Izkhir, Karakusa, Lanjak). Himalayas to Bombay. Used locally to treat stomach complaints, cholera, coughs, rheumatism, gout and fevers.

*C.* ***martini*** (Roxb.) Watts. (Rosha Grass). Drier parts of India. The ***motia*** variety is grown in the drier parts of India and yields Palmarosa Oil (East Indian Geranium Oil) which contains about 90 per cent geranol. The ***sofia*** variety grows in moister areas and yields Gingergrass Oil which is less valuable. Palmarosa Oil is also produced in Java and Sumatra. Both the oils are used mainly as adulterants to more expensive perfumes and in the manufacture of soap.

*C.* ***nardus*** (L.) Rendle=Andropogon nardus. Tropical Asia. Cultivated in Java, Ceylon, Formosa and Guatemala. The oil (Citronella) is used in perfumery, soap manufacture, scenting aerosols, making insecticides, disinfectants and shoe polish. The leaves are used locally for flavouring food and as a tea. Fractional distillation of the oil yields menthol which is used medicinally to relieve coughs.

*C.* ***proximus*** Stapf. Sudan, Ethiopia. The leaves are used locally for thatching. No essential oil.

*C.* ***schoenanthus*** Spreng. (Camel Grass). Middle East. The oil is used in the Punjab as a perfume and medicinally. Formerly esteemed by the Greeks and Romans as a perfume.

*C.* ***senaarensis*** Chiov. Sudan. Oil of Mahareb Grass.

*C.* ***validus*** Stapf. ex Davy. S.Africa. The leaves are used locally for thatching. There is no essential oil.

***Cyminosma*** Gaertn=Acronychia.

CYMOPTERUS Rafin. Umbelliferae. 18 spp. W. N.America. Herbs.

*C.* ***acaulis*** (Pursh.) Rydb. Nevada to California. The roots and leaves are eaten by the local Indians.

*C.* ***bulbosus*** L. S.W. U.S.A. and Mexico. The leaves are used by the local Indians to flavour food.

*C.* ***fendleri*** Gray. Colorado, Utah, New Mexico. The roots are used for flavouring by the local Indians, and the leaves are eaten as a vegetable.

*C.* ***globosus*** S. Wats. Nevada to California. The roots and leaves are eaten by the local Indians.

*C.* ***longipes*** S. Wats.=Aulospermum longipes.

*C.* ***montanus*** Nutt. (Gamote). W. N.America. The roots are eaten by the local Indians as a vegetable.

*C.* ***purpureus*** S. Wats.=Aulospermum purpureum.

CYNANCHUM L. Asclepiadaceae. 5 spp. ***s. str.*** E.Europe, temperate Asia. Herbs, possibly woody.

*C.* ***arnottianum*** Wight. (Bhankalink). N.India. The ground, dried leaves are used locally to treat maggots in domestic animals. A possible source of insecticide.

*C.* ***clausum*** Jacq.=Funastrum clausum.

*C.* ***linare*** Brown. (Kitsanga). C.Madagascar. The roots are eaten locally.

*C.* ***sarcostemoides*** Cost. and Bois. (Falotsy). Madagascar. Latex used locally as bird lime.

*C.* ***vincetoxicum*** B. Bs.=Vincetoxicum officinale. (Swallow's Wort). Europe, N.Africa, to Himalayas. The roots

(Swallow Root, Rhizoma Vincetoxici) are used to treat dropsy, as a diuretic, and to induce sweating.

CYNARA L. Compositae. 14 spp. Mediterranean to Kuristan. Large herbs.

***C. cardunculus*** L. (Cardoon). Mediterranean. Cultivated, especially in Europe for the leaf stalks, which are eaten as a vegetable when blanched.

***C. scolymus*** L. (Globe Artichoke). S. Europe. Cultivated where the climate is suitable. The fleshy receptacle and floral bracts are eaten as a vegetable. The receptacle is pickled or canned. It has been cultivated since ancient times in the Mediterranean region, but previously the blanched foliage was used as a vegetable.

CYNODON Rich. Graminae. 10 spp. Tropics and sub-tropics. Perennial grass.

***C. dactylon*** (L.) Pers. (Bermuda Grass). Africa, possibly India. Cultivated through the tropics and subtropics, as a fodder for livestock. It makes good hay, but is difficult to control and can become a weed especially in gardens etc. It was introduced into Georgia by 1751 and has spread throughout S. U.S.A. where hybrids are used as lawn grasses.

***C. plectostachyus*** Pilger (Star Grass). Tropical Africa. Used for hay and grazing in warm countries. It can become a weed.

CYNOGLOSSUM L. Boraginaceae. 50–60 spp. Temperate and sub-tropics. Herbs.

***C. glochidiatum*** Wall. = C. wallichii.

***C. officinale*** L. (Hound's Tongue). Temperate Europe, Asia, and N.America. The young roots, leaves, contain a resin, tannin, and inulin. A decoction of them has a paralysing effect on the motor nerves, and was used as a sedative. The young leaves are used in Switzerland as a vegetable or salad.

***C. wallichii*** G. Don. = C. glochidiatum. N. India. The Hindus used the juice from the roots to stop sickness in children.

CYNOMETRA L. Leguminosae. 60 spp. Tropics. Trees.

***C. bokalaensis*** De Wild. = C. hankei.

***C. cauliflora*** L. E.India and Malaysia. The young pods have a pleasant sharp flavour and are used pickled with Spanish pepper, soya, etc. or fish.

***C. cubensis*** Rich. Cuba. The dark red wood is used for general carpentry.

***C. hankeri*** Harms. = C. bokalaensis. (Baraka, Bubulaka, Mutchuna). Tropical Africa, especially Congo, Nigeria. The wood is hard and resists decay. It is used locally for general carpentry, railway sleepers, house building and agricultural implements.

***C. pedicellata*** De Wild. (Kabalabala, Mabangu). Tropical Africa. The wood is hard and is used locally for general carpentry, house-building and railway sleepers.

***C. sessiliflora*** Harms. Congo. A copal is extracted from the stem.

***C. spruceana*** Benth. = Troachylobium hornemannianum.

CYNOMORIUM L. Cynomoriaceae. 2 spp. Mediterranean to Mongolia. Shrubs.

***C. coccineum*** L. N.Africa. The ground roots are used as a condiment by the Tuaregs.

CYNORKIS Thou. Orchidaceae. 125 spp. Tropical and S.Africa, Mascarene Islands. Herbs.

***C. flexuosa*** Lindl. Madagascar. The tubers are eaten locally.

***Cynosorchis*** Thou. = Cynorkis.

CYNOSURUS L. Graminae. 3–4 spp. Europe, W.Asia, N. and S.Africa. Annual Grasses.

***C. cristatus*** L. (Crested Dog's tail, Dog's tail Grass). Europe through to Asia Minor, N.America. Grown as a pasture for livestock, and for hay.

CYPERUS L. Cyperaceae. 550 spp. Tropics and warm temperate. Perennial grass-like herbs.

***C. alopecuroides*** Rottb. Tropics. The stems are used to make mats in Egypt.

***C. aristatus*** Rottb. = C. inflexus. N. and S.America. The starchy root tubers are eaten locally in New Mexico.

***C. articulatus*** L. Old and New World. The stems are used locally to make mats. The tubers are pleasantly scented and when dried are put in with clothes to perfume them. Sometimes cultivated.

***C. baroni*** Clarke. Madagascar. The stems and leaves are used locally to make baskets and mats.

***C. cenus*** Presl. Tropical America. In C.America, the stems are used to make baskets, beds etc.

***C. cephalotes*** Vahl. Tropical Asia, Australia. Cultivated in rice fields of Japan for the stems which are made into mats.

***C. confusus*** Cherm. (Beloha). Madagascar.

The stems and leaves are used locally to make baskets.

***C. debilissimus*** Bak. (Forombato). Madagascar. The leaves are used locally to make hats.

***C. esculentus*** L. (Chufa, Earth Almond, Tiger Nut, Yellow Nut Grass). Cosmopolitan. Cultivated widely. In tropical Africa and India the tubers are used as a vegetable, while in India they are a source of a flour and an adulterant for cocoa and coffee. The plant has been found in Egyptian tombs, and is still cultivated in Spain and Italy as a food, and for the juice (Horchata de Chufas) which is pressed from the tubers and drunk locally as a chilled beverage. The tubers are used as a vegetable and food for pigs. They contain 15–20 per cent sucrose, and 22 per cent starch as well as 20–30 per cent non-drying oil (Chufa Oil, Earth Almond Oil, Sedge Oil, Tiger Nut Oil), which is expressed locally and used for cooking.

***C. haspan.*** L. Tropical Africa, India. The burnt plant is used as a source of salt locally in E.Africa.

***C. hexastachyos*** Rottb. = C. rotundus.

***C. iria*** L. Tropics of Old and New World. In India the juice is used as a tonic and to treat stomach complaints.

***C. lenticularis*** Schrad. E.Asia. The juice from the rhizome is used in Chinese medicine to cause contractions of the uterus.

***C. longus*** L. (English Galingale). Europe to E.Asia. The rhizomes are sweet-scented, rather like violets, and an extract of them is sometimes used in perfumery especially as an additive to lavender water.

***C. maculatus*** Boeck. Tropical Africa. The tubers are a local source of perfume sold locally.

***C. madagascariensis*** Roem. and Schwein. Madagascar. The stems and leaves are used locally for making baskets, mats etc.

***C. malaccensis*** Lam. Tropical Asia, Australia. The stems and leaves are used locally for making mats, baskets etc.

***C. natans*** Buch.-Ham. = C. cephalotes.

***C. nudicaulis*** Poir. Madagascar. The stems are used for making hats (Panama Hats of Madagascar).

***C. papyrus*** L. (Papyrus). Nile, Syria, Sicily. The stems were used by the ancient Egyptians to make paper. This was done by splitting the stems and pressing the strips together when they were still wet.

***C. rotundus*** L. = C. hexastachyos. (Cocograss, Nut Grass). A serious weed throughout the tropics. The dried tubers (Soucher) are used in perfumes. It was called Radix Junci by the Romans. Soucher is used particularly in India to perfume hair and clothes.

***C. schweinitzii*** Torr. (Schweinitz Cyperus). Canada and N.Central U.S.A. Used for fattening horses.

***C. stolonifera*** Retz. Tropical Asia, Australia. The sweet-smelling tubers are used locally to treat stomach complaints and as a heart stimulant

***C. tegetiformis*** Roxb. (Chinese Matgrass). Asia. Cultivated in China for making mats etc.

***C. vaginatus*** R. Br. Australia. A fibre extracted from the leaves is used by the natives to make fishing nets, cordage etc.

CYPHOLOPHUS Wedd. Urticaceae. 30 spp. Philippines, Java, Polynesia. Woody herbs.

***C. macropyllus*** Wedd. Samoa, Philippines. A fibre extracted from the bark is used locally to make mats.

CYPHOMANDRA Mart. ex Sendtn. Solanaceae. 30 spp. Central and S.America. W.Indies.

***C. betacea*** Sendtn. (Tree Tomato). Peru, Brazil. Cultivated since ancient times in Peru. Introduced into Old and New World. The orange-coloured fruits are about the size of an egg. They are eaten raw or used for jams and preserves.

***C. hartwegi*** Sendtn. Large perennial herb. S.America. The red berries are eaten cooked in Chile, Colombia and Argentina.

Cyprès du Tassili – Cupressus dupreziana.

Cypress, Arizona – Cupressus arizonica.

Cypress, Bald – Taxodium distichum.

Cypress, Italian – Cupressus sempervirens.

Cypress, Lawson – Chamaecyparis lawsoniana.

Cypress, Mouterey – Cupressus macrocarpa.

Cypress, Oil – Cupressus sempervirens.

Cypress, Sawara – Chamaecyparis pisifera.

Cypress, Sitka – Chamaecyparis nootkatensis.

Cypress, Southern – Taxodium distichum.

Cypress, Summer – Kochia scoparia.

Cypress, Swamp – Taxodium distichum.

Cypress, Swamp – Taxodium mucronatum.

Cypress, Yellow – Chamaecyparis nootkatensis.

CYPRIPEDIUM L. Orchidaceae. 50 spp. N.Temperate. Herbs.

***C. parviflorum*** Salisb. N.America. The dried rhizome is used as a stimulant and antispasmodic.

***C. pubescens*** Willd. See C. parviflorum.

Cyprus Powder– Anaptychia ciliaris.

Cyprus Powder – Usnea barbata.

CYRILLA Garden ex. L. Cyrillaceae. 1 sp. S.E. United States of America to N. tropical S.America. W.Indies. Shrub.

***C. racemiflora*** L. (Black Titi, Ironwood). The flowers are a good source of honey.

CYRTOCARPA Kunth. Anacardiaceae. 2 spp. Mexico. Trees.

***C. procera*** H.B.K. The fruits (Berracos) are eaten locally. The wood which is soft and scented is used for small carvings, trays, souvenirs etc. The bark is used as a soap substitute.

***Cyrtomium*** Presl. = Phanerophlebia.

CYTOSPERMA Griff. Araceae. 18 spp. Tropic, especially New Guinea. Herbs.

***C. edule*** Schott. Pacific. New Guinea. The starchy rhizomes are eaten locally when cooked. Cultivated throughout Polynesia.

***C. merkusii*** (Hassk.) Schott. Philippines. The starchy rhizomes are eaten locally. A decoction of the flowers is used to remove blood clots and to control menstruation.

***Cytheris griffithii*** Wight. = Calanthe vestita.

***Cytheris regnieri*** Reichb. = Calanthe vestia.

CYSTOCLONIUM Harv. Rhodophyllaceae. Red Algae. Marine.

***C. armatum*** Harv. Pacific. Eaten locally in Japan.

CYTINUS L. Rafflesiaceae. 6 spp. Mediterranean, S.Africa, Madagascar. Parasitic herbs.

***C. hypocistis*** L. Mediterranean. Parasitic, especially on Cistus (Broom). A decoction of the aerial parts and fruit (Succus Hypocistidis) is used locally to treat dysentery. The young shoots are eaten as a vegetable.

CYTISUS L. Leguminosae. 25–30 spp. Atlantic Islands, Europe, Mediterranean. Small shrubby trees.

***C. pallidus*** Poir. Canary Islands. Cultivated locally for fodder.

***C. palmensis*** Hutch. See C. pallidus.

***C. prolifer*** Kit. = C. pullilans.

***C. pullilans*** Kit. (Tree Alfalfa, Tree Lucerne). E.Europe. Shrubby herbs. An excellent forage crop, hardy and withstanding drought.

CYTTARIA Berk. Helotales. 6–7 spp. S.Hemisphere. Parasitic on Nothofagus.

***C. dwarwinii*** Berk. S.Temperate and antarctic S.America. A principle food in Patagonia and Fuegia.

# D

Daagon – Itoa stafii.

Dabbada – Chrozophora senegalensis.

Dabe – Erythroxylum mannii.

Dackowar Grasstree – Xanthorrhoea arborea.

DACRYDIUM Soland. Podocarpaceae. 20–25 spp. Indomalaysia, Tasmania, New Zealand. Trees.

***D. cupressinum*** Soland. (Red Pine). New Zealand. The red yellow wood has beautiful markings and is used for interior panelling etc. and cabinet making. It is also used for bridge-building, house-building, and keels of ships. The bark is used for tanning and stains the leather slightly red. The trunk is the source of a resin (Rimu Resin).

***D. elatum*** Wall. Malaya. The wood is light brown heavy and fine-grained. It is used for general construction work, boards, and boxes.

***D. franklinii*** Hook. (Huon Pine), Malaysia, Borneo, New Caledonia, Australia and New Zealand. The yellow to yellow-brown wood is fairly hard, straight-grained, light, soft and strong; it has a pleasant smell. The wood is used for door and window frames, furniture, turning, boat-building, troughs, pattern making and drawing boards. An oil (Huon Pine Oil) is distilled from the wood. It contains methyl eugenol, and is used in the manufacture of toilet waters, medicinal soaps, for preserving case in paints and as a source of vanillin.

***D. intermedium*** T. Kirk. (Yellow Silver Pine). New Zealand. The yellowish, straight-grained wood is firm and resinous. It is used for boat-building.

***D. westlandicum*** T. Kirk. (Westland Pine). New Zealand. The wood is firm, compact, and even-grained. It is valuable and is used for cabinet-work, bridges, wharves etc., house building, and agricultural implements.

DACRYODES Vahl. Burseraceae. 50 spp. Tropics. Trees.

***D. edulis*** (G. Don.) Lam. = Canarium saphii = Pachylobus edulis. Tropical Africa, especially Congo and Nigeria. The fruits (Banfole) are eaten locally.

***D. excelsa*** Vahl. = Pachylobus hexandrus. (Candle Wood Tree, Gommier, Tabonuco). The brown wood has a medium to fine texture, is fairly heavy and has a fine lustre. It works well and stains easily. It is used for furniture-making, building small boats, shingles, crates and veneer. Sometimes it is used as a substitute for mahogany.

***D. hexandra*** Griseb. San Domingo. The trunk produces a resin (Tabonico).

DACTYLIS L. Graminae. 5 spp. Temperate Eurasia. Perennial grasses.

***D. glomerata*** L. (Cock's foot, Orchard Grass). Europe. Cultivated in most temperate countries for pasture and hay. It is usually grown mixed with a legume, e.g. clover or alfalfa as a permanent pasture.

DACTYLOCLADUS Oliv. Crypteroniaceae. 1 sp. Borneo. Tree.

***D. stenostachys*** Oliv. (Jongkong). The wood is soft and light, varying from yellowish-pink to reddish-brown with age. It takes a good finish and is used for joinery, furniture-making, Venetian blinds, boat-building and broom heads.

DACTYLOCTENIUM Willd. Graminae. 10 spp. Warm. Grasses.

***D. aegyptiacum*** Willd. Sahara and Sudan. The seeds are used locally for food. Annual.

***D. australe*** Steud. (Durban Grass). S.Africa, especially Natal. Used to prevent erosion of sea-coast sand dunes.

***Daedalacanthus viscidum*** Anders. = Eranthemum viscidum.

DAEDALEA Pers ex. Fr. Polypodiaceae. 25 spp. Cosmopolitan.

***D. unicolor*** (Bull.) Fr. Europe. A possible source of antibiotics.

DAEMONOROPS Blume ex. Schult. f. Palmae. 100 spp. Indomalaysia. Small palms, Rattan Palms.

***D. didymophyllus*** Becc. Malacca. Source of Dragon's Blood, Resina Draconis, Sanguis Draconis. The resin is derived from the scales on the fruit. It is commercially important being used in varnishes and lacquers to give a mahogany stain. It is also a stimulant and astringent and was used in tooth pastes and mouth washes. It is also used as a protective coating when etching zinc; the parts which are not to be etched are covered with the resin.

***D. draco*** Blume. (Dragon's Blood Palm). Malaya. See D. didymophyllus. This is the major source of the resin.

***D. fissus*** Blume. Borneo. The stems are used for making mats.

***D. forbesii*** Becc. Sumatra. The split stems are used for basket work.

***D. longipes*** Mart. Malaysia. Used to make rattan furniture.

***D. micranthus*** Becc. Penang, Perak, Malacca. See D. didymophyllus.

***D. motleyi*** Becc. Borneo. See D. didymophyllus.

***D. oblongus*** Blume. = Calamus oblongus.

***D. sparsiflorus*** Becc. Borneo. See D. didymophyllus.

***D. trichrous*** Miq. Sumatra. The split stems are used for basket work etc.

***Daenia cordata*** R. Br. = Pergularia tomentosa.

Daffodil – Narcissus pseudo-narcissus.

DAHLIA Cav. Compositae. 20 spp. Mexico, Guatemala. Herbs.

***D. rosea*** Cav. = D. variabilis. (Dahlia, Garden Dahlia). Mexico. The tubers are eaten locally. They contain inulin, and are used to make food preparations for diabetic patients. Originally it was introduced into Europe as a food source, but is now grown as an ornamental.

Dahlia – Dahlia rosea.

Dahomey Rubber – Ficus vogelii.

Dahoon Holly – Ilex cassine.

Dahurian Buckthorn – Rhamnus dahurica.

Dahurian Lily – Lilium dauricum.

Daikon – Raphanus sativus.

Daumyo Oak – Quercus dentata.

Daindaté – Nauclea esculentus.

Daira – Wrightia tomentosa.

DAIS Royen ex L. Thymelaeaceae. 2 spp. S.Africa, Madagascar. Shrubs.

***D. glaucescens*** Decne. Madagascar. The

bark contains a fibre which is used locally for the manufacture of string.

Daisee – Corchorus olitorius.

Daisy (English) – Bellis perennis.

Daisy, Oxeye – Chrysanthemum leucanthemum.

Daisybush, Travers – Olearia traversii.

Daka – Lauro-Cerasus gardneri.

Dakota Mustard – Brassica arvensis.

DALBERGIA L. Leguminosae. 300 spp. Tropics and subtropics. Trees or woody vines. The tree species are the source of the valuable rosewoods. These are rarely planted for timber, but are used as shade trees for tea plantations. The wood is often used for veneers.

***D. baroni*** Baker. Madagascar. Rosewood of Madagascar.

***D. caerensis*** Ducke. Brazil. The wood (Brazil Tulipwood, Kingwood, Violeta) is striped alternately with dark violet and black. It is fragrant and is used for making small articles, inlay work etc.

***D. cochinchinensis*** Pierre. Tropical Asia. Rosewood of Siam.

***D. cubilquitzensis*** (D. Sm.) Pitt. Guatemala, Honduras. The wood (Rosewood of Guatemala) is orange striped with violet. It is heavy, hard, and polishes well. It is used for cabinet work, interior finishes, wheels, axles, and wagon building.

***D. ferruginea*** Roxb.=D. limonensis=D. luzonensis. Borneo, Philippines, New Caledonia. An extract of the wood is used locally to control menstruation and as an abortive. The bark and roots are used as a fish poison.

***D. granadilla*** Standl.=Amerimnon granadillo. (Granadillo). Mexico. The wood is purple striped with purple-black. It is very beautiful and polishes well, and is used for cabinet work.

***D. greveana*** Baill. See D. baroni.

***D. hupeana*** Nance. China. The wood is very strong, and is used locally for making pulley blocks, wheel spokes, oil presses, tool handles etc.

***D. junghuhnii*** Benth.=D. parviflora=D. zollingeriana. (Akar laka). India to Malaya. Woody vine. The heartwood from the stems and roots is scented and is used for joss sticks ('hsiang-chen-hsiang) and incense in temples. The wood (Lacca Lignum) is of minor importance. Oil distilled from wood used locally in medicine.

***D. latifolia*** Roxb. E.India. The wood (Bombay Rosewood, Black Rosewood, East Indian Rosewood, South Indian Rosewood, Malabar Rosewood, Black Rosewood, Javanese Palissander) is deep red to purple, streaked with yellow or black. It is used for making small articles, turning, flooring, and the interiors of houses.

***D. limonensis*** Benth.=D. ferruginea.

***D. luzonensis*** Vogel=D. ferruginea.

***D. melanoxylon*** Guill. and Perr. Tropical Africa. The wood (African Blackwood, African Grenadille Wood, Mozambique Ebony, Senegal Ebony, Unyoro Ebony) is used as a substitute for ebony in making furniture, carving, musical instruments etc. Locally it is used for making hammers and arrows. The roots are used as a local cure for toothache.

***D. miscolobium*** Benth.=Miscolobium nugrum. Brazil. A source of a rosewood of good quality.

***D. multiflora*** Heyne ex. Wall.=D. sympathetica.

***D. nigra*** Allem. Brazil. The wood (Brazilian Rosewood, Bahia Rosewood, Rio Rosewood, White Rosewood) is brown to violet, streaked with black. It is rose-scented and good to work. Its uses include cabinet-making, making pianos, tool-handles, carving, billiard tables etc.

***D. parviflora*** Roxb.=D. junghuhnii.

***D. pinnata*** (Lour.) Prain.=D. tamarindifolia. (Keti, Daman). N.India to Indo-China. The roots are used locally to treat intestinal worms and as a masticatory.

***D. retusa*** Hemsl. C.America. The wood varies in colour, but is generally deep red. It is called Cocobolo, Nicaragua Rosewood, Palo negro, and is used for a wide variety of small articles including cutlery, tool-handles, inlaying, interior work of cars, small boxes and carvings, buttons, musical instruments, etc.

***D. sissoo*** Roxb. (Sissoo of India). Tropical Asia, especially India. The wood is hard, brown and durable. As it takes a good finish it is used for flooring, furniture and carving.

***D. spruceana*** Benth. (Jacaranda do Pará). S.America, especially Brazil. The wood (Rosewood of the Lower Amazon) is hard, durable, heavy and strong. It is used for cabinet work.

***D. stevensoni*** Standl. British Honduras.

The wood (Rosewood of British Honduras) is pink-brown to purple, and is used mainly to make musical instruments.

***D. stipulaceae*** Roxb. = D. ferruginea.

***D. sympathetica*** Nimmo ex. Grah. = D. mutiflora. Hills of W.India. The bark is used locally in Goa to remove pimples.

***D. tamarindifolia*** Roxb. = D. pinnata.

***D. zollingeriana*** Miq. = D. junghuhnii.

DALDINIA Ces. and de Not. Xylariaceae. 13 spp. Cosmopolitan.

***D. concentrica*** (Bolt.) Ces. and de Not. Cosmopolitan. The ground stroma is used in the Moluccas to treat skin diseases.

DALEA L. ex Juss. Leguminosae. 250 spp. America. Shrubs or herbs.

***D. emoryi*** Gray. W. United States of America, Mexico. A yellowish dye extracted from the branches is used to stain deer skins by the local Indians.

***D. enneandra*** Nutt. = Parosela enneandra (Slender Parosela, Za-bà-a). C. to W. U.S.A. Shrub. The tough stems are used by the local Indians to make small arrows.

***D. lanata*** (Spreng.) Britt. = Parosela lanata. W. U.S.A., Mexico. The roots are eaten by the local Indians.

***D. polyadenia*** Torr. (Smoke Bush). W. N.America. Herb. A decoction of the leaves is used by the Indians to treat colds.

Dallis Grass – Paspalum dilatatum.

Dalmatian Insect Powder – Chrysanthemum cinerariifolium.

Dama dere itam – Hopea celebica.

Daman – Dalbergia pinnata.

Damar batoo – Vatica faginea.

Damar Itam – Canarium legitimum.

Damar putih – Agathis labillardieri.

Damar Sengai – Canarium hirsutum.

Damar siap mata – Triomma macrocarpa.

Damaran – Mussaendra afelii.

Damask – Hesperis matronalis.

Damask Rose – Rosa damascena.

Damiana – Turnera diffusa.

Damianita – Chrysactinia mexicana.

Dammar – Anisoptera grossivenia.

Dammar – Anisoptera marginata.

Dammar – Hopea mengarawan.

Dammar Batu – Hopea maranti.

Dammar Daging – Shorea leprosula.

Dammar Hiroe – Vatica papuana.

Dammar Itam – Canarium legitinum.

Dammar Kedemut – Hopea fagifolia.

Dammar Kloekoep – Shorea eximea.

Dammar Kumus – Shorea glauca.

Dammar, Njating Matapoesa – Hopea sangal.

Dammar, Njating Matpleppek – Hopea sangal.

Dammar Penak – Balanocarpus heimii.

Dammar Penak – Balanocarpus maximus.

Dammar Pine, White – Agathis alba.

Dammar Putih – Agathis laballadieri.

Dammar Rasak – Vatica faginea.

Dammar Resin – Agathis lanceolata.

Dammar Resin – Shorea wiesneri.

Dammar, Rock – Hopea odorata.

Dammar, Rose – Vatica rassak.

Dammar, Sal – Shorea robusta.

Dammar Sengai – Canarium hirsutum.

Dammar Temak – Shorea hypochra.

Dammar Tenang – Shorea koodersii.

Dammar Tubang – Shorea eximea.

Dammar, White – Vateria indica.

***Dammara alba*** Lam. = Agathis alba.

Damson – Sorinderia juglandifolia.

Damson Plum – Prunus insititia.

DANAIS Comm. ex Vant. Rubiaceae. 40 spp. Madagascar, Mascarene Islands. Vines.

***D. gerradi*** Baker. Madagascar. A fibre is extracted from the bark and a dye from the roots. Both are used locally.

Dancen – Itoa stapfii.

Danda – Allophylus africanus.

Dandelion (Wine) – Taraxacum officinale.

Dandelion, Leafy-stemmed False – Pyrrhopappus carolinanus.

Dangleberry – Gaylussacia frondosa.

Dango – Willughbeia teneii flora.

Dantong – Payena dantung.

Dantung – Payena stipularis.

Denga – Lygodium smithianum.

***Daniella*** Willis = Daniellia.

DANIELLIA Benn. Leguminosae. 11 spp. Tropical W.Africa. Trees. All the species mentioned produce a copal derived from the roots or the ground around old trees.

***D. oblonga*** Oliv. Tropical Africa. Resin (Niger Copal) also from the stems.

***D. ogea*** Rolfe. W.Africa. Accra Copal, Ogea Copal, used in manufacture of varnishes.

***D. oliveri*** Rolfe = Paradaniella oliveri (Copaiba Tree). Tropical W.Africa. The copal is called Ako-Ogea. A decoction of the roots is used locally to treat skin diseases and gonorrhoea.

***D. similis*** Craib. (Copal Tree). Nigeria,

Ghana. Gum is called Gum Copal of the Gold Coast.

***D. thurifera*** Bennett. Tropical Africa. A frankincense (Liiorin Gum, Illorin Copal, Ogea Gum of Sierra Leone, or Balsam of Copaiba) is used as a perfume. The coarse light wood is used for packing cases, furniture and handles of small articles. It is liable to attack by insects.

***Danthonia pectinata*** Lindl.=Astrebla pectinata.

***Danthonia triticoides*** Lindl.=Astrebla triticoides.

Daoon ajenton – Coleus aromaticus.

Daoon baroo laoot – Mallotus trifoliata.

Daoon dorian – Durio griffithii.

Dapdab harnagan – Fagara rhetsa.

DAPHNE L. Thymelaeaceae. 70 spp. Europe, N.Africa, temperate and sub-tropical Asia, Australia, Pacific. Shrubs.

***D. cannabina*** Wall. Himalayas. The bark is used locally to make paper.

***D. cneorum*** L. S.Europe. In Spain the bark is used to stupify fish.

***D. genkwa*** Sieb. and Zucc. China. The leaves are used locally to treat intestinal worms.

***D. gnidium*** L. Mediterranean. In Sardinia the roots are used to stupify fish, and an extract of the bark is used to induce abortions.

***D. mezereum*** L. (Mezereon). Europe, N.Africa, Asia Minor to Siberia. The dried bark contains a glucoside, daphnin, and a resin mezerin. It is used medicinally as a stimulant, diuretic, and to induce the production of saliva. The bark is produced commercially in Algeria and S.France.

***D. odora*** Thunb. Japan. Cultivated locally for the flowers which are used in perfumed water and sachets.

***D. oleoides*** Schreb. Europe to C. Asia. The roots are used locally in C. Asia as a purgative.

***D. papyrifera*** Sieb.=Edgeworthia papyrifera.

***D. pseudo-mezereum*** Gray. Japan. The bark-fibre is used locally to manufacture paper.

DAPHNIPHYLLUM Blume. Daphniphyllaceae. 10 spp. China, Japan, Formosa, Indomalaya. Shrubs.

***D. humile*** Maxim. Japan. The leaves are smoked locally as a tobacco substitute.

DAPHNOPSIS Mart. and Zucc. Thymelaeaceae. 46 spp. Mexico, W.Indies, to E.Argentina. Shrubs.

***D. brasiliensis*** Mart. and Zucc. Brazil. The bark from young plants is used locally to treat psoriasis and erysipelas.

***D. swartzii*** Meissn. W.Indies. The bark is diuretic, a stimulant, and induces the flow of saliva.

Dara laoot – Strychnos ligustrina.

Darah – Planchonia valida.

Dark Catechu – Acacia catechu.

Dark Crottle – Parmelia physodes.

Darkleaf Malanga – Xanthosoma atrovirens.

Darling Plum – Reynosia septentrionalis.

Darnel – Lolium temulentum.

Darunaj-akrabi – Doronicum roylei.

Daru Urandra – Urandra corniculata.

DARWINIA Rudge. Myrtaceae. 35 spp. Australia. Trees. An essential oil (Oleum Darwiniae aetherum) is distilled from the wood of the following species. It is used in perfume manufacture.

***D. fascicularis*** Benth.

***D. grandiflora*** Benth.

***D. taxifolia*** Muell. Arg.

Dasheen – Colocasia antiquorum.

Dasheen – Colocasia esulenta.

DASYLIRION Zucc. Agavaceae. 18 spp. S.W. United States of America, Mexico. Shrubby herbs.

***D. recurvatum*** Macbride=Nolina recurvata.

***D. simplex*** Trel. (Sotol). Mexico. The leaves are used locally for basket work.

***D. texanum*** Scheele. (Texas Sotol). Texas. The pulp from the apex of the plant is sugary and was used by the local Indians as a food and in the manufacture of a beverage (Sotol).

***D. wheeleri*** S. Wats. (Wheeler Sotol). Arizona, New Mexico. The roasted pulp was eaten by the local Indians and made into a beverage.

***Dasyloma javanica*** Miq.=Oenanthe javanica.

Datah ney – Detarium senegalense.

Date, Chinese – Ziziphus mauritana.

Date Palm – Phoenix dactylifera.

Date Plum – Diospyros lotus.

Date Plum, Chinese – Diospyros kaki.

Dátil – Yucca baccata.

DATISCA L. Datiscaceae. 1 sp. Mediterranean to Himalayas and C.Asia; 1 sp. S.W. United States of America to Mexico. Herbs.

***D. cannabina*** L. C.Asia. Cultivated for the leaves, roots and stems which yield a yellow dye. The dye when mixed with alum was used to dye silk yellow. It was used particularly in China and Japan, and France.

DATURA L. 10 spp. Tropics and warm temperate, especially America. Herbs.

***D. alba*** Nees. and Esenb.=D. fastuosa.

***D. arborea*** L. Peru, Chile. The leaves were smoked by the priests of the ancient religions. The plant is intoxicating and was considered divine. It is often grown as an ornamental.

***D. candida*** (Pers.) Safford. Colombia. The leaves are highly poisonous and narcotic. A decoction of the leaves was drunk during ancient religious rites to induce prolonged delirium.

***D. dolichocarpa*** (Lagerh.) Safford. See D. carpa.

***D. fastuosa*** L.=D. alba. Tropical Africa. The leaves yield a blue or yellow dye used in Zanzibar. They are also smoked as a cure for asthma.

***D. inoxia*** Mill.=D. meteloides (Sacred Datura). S.W. U.S.A., Mexico. The leaves and roots were used by the local Indians to make a drink which dulls the senses.

***D. metel*** L. (Hindu Datura). India. Introduced and cultivated in many parts of the Old and New World. The dried leaves and seeds are used medicinally as a source of daturine an alkaloid with similar properties to atropine, (see Atropa belladonna).

***D. meteloides*** DC.=D. inoxia.

***D. stramonium*** L. (Jimson Weed, Thornapple, Stramonium). N.America, N.Africa, C. and S.Europe. Cultivated as a source of daturine.

Datura, Hindu – Datura metel.

Datura, Sacred – Datura inoxia.

DAUCUS L. Umbelliferae. 60 spp. Europe, Asia, Africa and America. Herbs.

***D. carota*** L. (Carrot). Afghanistan. A biennial grown widely throughout the temperate regions of the world for the roots which are used as a vegetable and occasionally food for livestock. It was grown in the Mediterranean regions in pre-Christian times and introduced into France, Germany and China by the 13th century. It was not accepted in England until the early 17th century. The orange pigment from the roots (carotene) is used in colouring butter. It is also a precursor of vitamin A, making carrots a valuable source of this vitamin in the diet. The roasted roots have been used as a substitute for coffee. An oil from the seeds (Oil of Carrot) contains asarone, carotol, daucol, l-limonene, and pinene, and is used for flavouring, in liqueurs and perfumes. An alcoholic extract of the seeds is used in manufacturing liqueurs in France. A sugar syrup is extracted from the roots. There are a great many horticultural varieties classified on the shape, size and colour of the roots. Some attempt has been made to raise these to the status of botanical varieties.

***D. pussilus*** Michx. (Rattlesnake Weed). N.America to Mexico. The roots are eaten as a vegetable by the local Indians.

DAVIDSONIA F. Muell. Davidsoniaceae. 1 sp. Queensland, New South Wales. Tree.

***D. pruriens*** Muell. Arg. (Do-rog). The plum-shaped berries are used to make preserves.

DAVILLA Vand. Dilleniaceae. 38 spp. Tropical America, W.Indies. Vines.

***D. kunthii*** St. Hil. C.America, Colombia. The tough stems are used locally for tying house-frames. The bark is the source of a black dye.

Dawai-dawai – Ziziphus calophylla.

Dây cau múc – Uncaria tonkinensis.

Dây dác – Ampelocissus martini.

Daylily, Grassleaf – Hemerocallis minor.

Daylily, Tawny – Hemerocallis fulva.

Dây sen – Alyxia flavescens.

Dây vàng giang – Fibraurea recisa.

Dead Borneo – Dyera costulata.

Deadly Amanita – Amanita phalloides.

Deadly Nightshade – Atropa belladonna.

Deadnettle, Purple – Lamium purpureum.

Deadnettle, White – Lamium album.

Dead Sea Apple – Quercus tauricola.

Deathcarrot, Gargan – Thapsia garganica.

DEBREGEASIA Gaudlich. Urticaceae. 5 spp. Ethiopia, Arabia, Afghanistan, Indomalaysia, E.Asia. Shrubs.

***D. edulis*** Wedd. Japan. The yellow fruits are eaten locally.

***D. hypoleuca*** Wedd. W.Himalayas. A fibre from the stems is used locally for cordage.

DECALEPIS Wight and Arn. Periplocaceae. 1 spp. Deccan Peninsular. Herb.

***D. hamiltonia*** Wight and Arn. (Mahali kizhangu). The roots are used locally to stimulate the appetite and as a laxative.

Decan Hemp – Hibiscus cannabinus.

DECASACHISTIA Wight and Arn. Malvaceae. 12 spp. India to Hainan and Malay Peninsular. Shrubs.

***D. parviflora*** Kurz. = D. thorelii. (Phlou méant). Thailand, Cambodia, Laos, S.Vietnam. The roots are eaten locally.

DECASPERMUM J. R. and G. Forst. Myrtaceae. 30 spp. Indomalaya.

***D. fruticosum*** Forst. = Nelitris paniculata = Psidium decaspermum. Small tree. Tropical Asia through Polynesia and Australia. The berries are sweet and are eaten in Java and Sumatra. The leaves are eaten in Java as a condiment.

Dediteh – Wrightia pubescens.

***Deeringia*** R. Br. = Cladostachys.

Deerberry – Mitchella repens.

Deer's Tongue – Saxifraga erosa.

Deer's Tongue – Trilisa odoratissima.

Degendeg – Indigofera arrecta.

***Deguelia scandens*** Aubl. = Derris pterocarpus.

DEINBOLLIA Schumacher and Thonn. Sapindaceae. 40 spp. Warm Africa, Madagascar. Trees.

***D. molliuscula*** Radlk. ex Mildbr. (Bak pida, Dipasa, Dongo). Congo, Angola. The wood is used locally for general construction work.

Deleg – Polyscias nodosa.

***Delima dioica*** Sessé and Moc. = Tetracera sessiliflora.

***Delima mexicana*** Sessé and Moc. = Tetracera sessiliflora.

***Delima sarmentosa*** L. = Tetracera sarmentosa.

***Delesseria coccinea*** Ag. = Plocamium coccineum.

DELPHINIUM L. Ranunculaceae. 250 spp. N.Temperate. Herbs. Many hybrids are grown as ornamentals. The seeds and vegetative parts of the plant contain the alkaloid delphinine which is used as an insecticide.

***D. ajacis*** L. (Rocket Larkspur). Europe, especially Mediterranean. A tincture of the dried ripe seeds is used medicinally as a parasiticide.

***D. brunonianum*** Royle. (Laskar). W.Himalayas, Tibet. The juice from the leaves is used locally to kill ticks on domestic animals.

***D. coeruleum*** Jack. ex Camb. (Dhakangu). Himalayas. An extract of the roots is used locally to treat maggots in goats.

***D. consolida*** L. (Forking Larkspur). Europe, temperate Asia. A tincture from the seeds is used to destroy nits and lice in the hair. The flowers when treated with alum were a source of a green dye.

***D. staphisagria*** L. (Stavisacre). S.Europe, Asia Minor. The seeds are the source of an insecticide.

***D. virescens*** Nutt. (Praire Larkspur). Prairies of C. U.S.A. The local Indians use the seeds in rattles.

***D. zalil*** Aitch and Hemsl. Iran, Afghanistan. The flowers yield a yellow dye (Asbarg) much used locally for dyeing silk and cotton.

Demarara Greenheart – Ocotea radiaei.

Demarara Mahogany – Carapa procera.

Demehi – Vicia calcarata.

Dempool – Antidesma ghaesembilla.

Dempool – Glochidium obscurum.

Dempoollelet – Glochidium borneense.

Dendé – Elaeis guiniensis.

DENDROBIUM Sw. Orchidaceae. 900 spp. Tropical Asia to Polynesia and Australasia. Herbs.

***D. crumentum*** Sw. Tropical Asia. The leaves are used by the Malays to treat nerve and brain complaints, and an extract of the pseudobulbs is used to treat ear ache.

***D. nobile*** Lindl. Tropical Asia. The Chinese used the leaves as a tonic and to treat stomach complaints.

***D. purpureum*** Roxb. E.Malaya. The leaves are used as a poultice for boils, swellings, etc. locally.

***D. salaccense*** Lindl. Malay peninsula, Java. The local women wear the dried leaves in their hair to give it a pleasant smell. The leaves are also used for flavouring rice.

***D. utile*** Smith. Malay Peninsular. The split stems are used locally for basket work.

DENDROCALAMUS Nees. Graminae. 20 spp. China, Indomalaya. Very large bamboo grasses.

***D. asper*** (Schult.) Backer. = Bambusa aspera. E.India, Java. The shoots are edible if dug from the ground before emergence. The stalks are used in Java for house-building etc.

***D. giganteus*** Munro. (Giant Bamboo, Nan Chu). China, Burma, to Thailand. The stems are used for a wide variety of purposes, especially in China, where they are made into rafts, water-buckets and split to make cho-sticks. Cultivated.

***D. hamiltonii*** Nees and Arn. Himalayas to Burma. Used locally for making baskets. The shoots are eaten as a vegetable.

***D. hookerii*** Munro. N.India. The shoots are made into buckets locally.

***D. latiflorus*** Munro=Sinocalamus latiflorus. S.China. Cultivated for the young shoots which are eaten as a vegetable by the Chinese and Japanese. Much is canned for export. The leaf sheaths are used in S.China for lining bamboo casks.

***D. strictus*** Nees. (Male Bamboo, Calcutta Bamboo, Bans, Bans khura, Kopar, Medar). India, Pakistan. The common bamboo. It is used for a wide variety of constructional purposes, mats, baskets, sticks, furniture, poles, water pipes, musical instruments, masts, barrels, paper manufacture, rafts. The dry bamboo is used for torches, and the charcoal for smith's work. The leaves are used for fodder for livestock.

DENDROPANAX Decne. and Planch. Araliaceae. 75 spp. Tropics and subtropics. Trees.

***D. arborae*** Decne and Planch. N. S.America; W.Indies. The wood is strong, yellow, streaked with dark brown. It is used locally for general carpentry.

***Dendropogon usneoides*** (L.) Raf.=Tillandsia usneoides.

***Dendrosma*** Planch and Sebert.=Geijera.

DENTARIA L. Cruciferae. 20 spp. Eurasia, E. N.America. Herbs.

***D. diphylla*** Michx. (Crinkleroot). E. N.America. The roots were eaten as a vegetable by the local Indians.

***D. laciniata*** Muhl. See D. diphylla.

Dentate ohelo – Vaccinium dentatum.

***Dentinum repandum*** (L. ex Fr.) Gray= Hydnum repandum.

Deo bans – Bambusa tulda.

Deodar – Cedrus deodara.

DEPLANCHEA Vieill. Bignoniaceae. 9 spp. Malaysia, Australia, New Caledonia. Trees.

***D. speciosa*** Vieill. New Caledonia. The wood is hard and fine-grained. It is used for turning and cabinet work.

DERMATOCARPON L. Dermatocarpaceae. Lichens.

***D. miniatum*** (L.) Mann. (Sprout Lichen). Temperate. Found on rocks. It is used in Europe to dye wool a sage-green.

DERRIS Lour. Leguminosae. 80 spp. Tropics. Woody vines. The plant contains the insecticide rotenone (together with deguelin, toxicarol, and tephrosin). Commercially the insecticide is important as it has no effect on warm-blooded animals. The plants mentioned are used locally by the natives to stupify fish, which can be eaten without ill-effects. For commercial production, the plants are propagated by stem cuttings. After three years, the plants are harvested and the roots cut up and dried. The dried root may contain up to 20 per cent rotenone. The root may be used as an insecticide direct when dried and ground and diluted with talc; or the rotenone is extracted by solvents and dissolved in oil. (It is insoluble in water.)

***D. elliptica*** Benth. E.India, Malaya to New Guinea. Cultivated in Old and New World tropics. The main commercial source of rotenone.

***D. guienensis*** Benth.=D. pterocarpus.

***D. koolgibberah*** F. M. Bailey. Australia. Used in Queensland for killing fish.

***D. malaccensis*** Prain. Malaya.

***D. polyantha*** Miq. Sumatra.

***D. pterocarpus*** (DC.) Aug. Chev.=D. guienensis=Deguelia scandens (Timbo da Matta). Amazon Basin.

***D. pyrrhothyrsa*** Miq.=D. thyrsiflora.

***D. thyrsiflora*** Benth.=D. pyrrhothyrsa= Aganope floribunda=Milletia thyrsiflora. India through to Indonesia.

***D. uligonosa*** Benth. Africa, Madagascar, Asia, Australia. Derris Root=Derris elliptica.

DESBORDESIA Pierre ex van Tiegh. Ixonanthaceae. 1 sp. Tropical W.Africa. Tree.

***D. insignis*** Pierre. The seeds are used in sauces by the natives.

DESCHAMPIA Beauv. Graminae. 60 spp. Temperate and cool mountains. Grasses.

***D. caespitosa*** (L.) Beauv. (Tussock Grass, Tufted Hair Grass). W. N.America to Mexico. It is used for pasture and hay.

DESCURAINIA Webb. and Benth. Cruciferae. 55 spp. Eurasia, S.Africa, cold to temperate America. Herbs.

***D. halictorum*** Cock. W. N.America. The roots were eaten as a vegetable by the Indians of New Mexico.

***D. incisa*** (Engelm.) Greene. W. N.America. The seeds were parched and eaten as a flour by the local Indians.

***D. parviflora*** (Lam.) Standl. W. N.America. The seeds were used as a flour by the local Indians.

***D. pinnata*** (Walt.) Howell. The leaves were eaten as a vegetable and the seeds used as a flour by the local Indians.

***D. sophia*** (L.) Webb ex Pranol. = Sisymbrium sophia. (Flixweed). Throughout Europe, Asia and N.Africa, introduced to N.America and New Zealand. The seeds are occasionally used locally as a substitute for mustard.

Desert Lemon – Atalantia glauca.

Desert Milkweed – Aclepias subulata.

Desert Pepperwort – Lepidium fremontii.

Desert Ramona – Salvia carnosa.

Desert Rose – Adenium obesum.

Desert Seepweed – Suaeda suffrutescens.

Desert Tea – Ephedra trifurca.

Desert Trumpet – Eriogonum inflatum.

Desert Willow – Chilopsis lineari.

DESMODIUM Desv. Leguminosae. 450 spp. Tropics and subtropics. Herbs or semi-woody shrubs.

***D. discolor*** Vog. Brazil. Grown for fodder in warm countries.

***D. gangeticum*** DC. Old World tropics. In India it is used as a green manure, and a decoction of the plant is used to treat catarrh and to reduce fevers.

***D. gyroides*** DC. Tropical Asia. Grown as a green manure in various parts of the tropics.

***D. heterophyllum*** DC. S.E.Asia to Philippines. Used as a fodder for cattle.

***D. lasiocarpum*** DC. = D. latifolium.

***D. latifolium*** DC. Tropical Africa and Asia. Used for fodder for horses in Nigeria.

***D. oldhami*** Oliv. Japan. The leaves are used locally as a tea substitute.

***D. mauritanum*** (Willd.) DC. = D. ramosissimum.

***D. ramosissimum*** G. Don. = D. mauritanum. Tropical Africa. The leaves are used locally to treat dysentery, to reduce fevers, and to treat eye-diseases.

***D. salicifolium*** (Poiv. ex Lam.) DC. Tropical Africa. Used locally as a green manure.

***D. tortuosum*** (Sw.) DC. = Meibomai purpurea. (Beggarweed). Tropics and subtropics. It is used as a green manure in warm countries, including the U.S.A.

***D. triflorum*** DC. Old and New World tropics. Used as a green manure. A decoction of the leaves is used locally to treat dysentery.

DESMONCUS Mart. Palmae. 65 spp. Tropical America. Palms.

***D. chinantlensis*** Liebm. Mexico. The leaves are used locally to make baskets.

***D. major*** Creug. See D. chinantlensis.

***Desmos chinensis*** Lour. = Xylopia discolor.

Detanve – Pouzolzia denudata.

DETARIUM Juss. Leguminosae. 4 spp. Tropical Africa. Trees.

***D. microcarpum*** Guill. and Perr. = D. senegalense.

***D. senegalense*** Gmel. (Datah ney, Niey datah). W.Africa. The wood (African Mahogany) is grey, hard, insect resistant, and easy to work. It is used for general carpentry. A sweet smelling resin is also extracted from the trunk. The root pulp is sweet and is used locally as a sugar-substitute. The pith from the pods is edible. A decoction from the bark was used as an ordeal poison.

Determa – Ocotea rubra.

Detoo – Morinda geminata.

Deutsche Sarsaparil – Carex arenaria.

DEUTZIA Thunb. Philadelphaceae. 50 spp. Himalayas, E.Asia to Philippines; 1–3 spp. Mexico. Shrubs or small trees.

***D. scabra*** Thunb. Japan. The whitish wood is smooth-grained and is used locally for mosaic work and wooded nails.

Devanula – Lobelia meotinaefolia.

Devilpepper, Java – Rauwolfia serpentina.

Devilpepper, Trinidad – Rauwolfia canescens.

Devil's Bit – Liatris squarrulosa.

Devil's Bit – Succisa pratensis.

Devil's Claw – Martynia parviflora.

Devil's Gut – Cuscuta epithymum.

Devil's Shoestring – Tephrosia virginiana.

Devil's Walking Stick – Aralia spinosa.

Devil's Wood – Osmanthus americana.

Dewberry, European – Rubus caesius.

Dewberry, Northern – Rubus flagellaris.

Dewberry Southern – Rubus trivalis.

Dhak – Butea frondosa.

Dhakungu – Delphinium coeruleum.

Dhakki – Acacia jacquemontii.

Dhal – Cajannus cajans.

Dhalabrauisabia – Cynoglossum wallichii.

Dhaloobang tali – Colona auriculatum.

Dhamani – Grewia tiliaefolia.

Dhamin – Grewia subinaequialis.

Dhamum – Grewia elastica.

Dharb – Saccharum spontaneum.

Dhaval – Lobelia nicoinaefolia.

Dhol asamudrika – Leea macrophylla.

Dholdak – Erythrina suberosa.

Dhooma – Shorea robusta.

Dhoowak manting – Nyssa javanica.
Dhup – Juniperus macropoda.
Dhupa Fat – Vateria indica.
Dial besar – Eugenia acuminatissima.

DAILIUM L. Leguminosae. 1 spp. Tropical S.America; 40 spp. Tropical Africa, Madagascar, Malaysia. Trees.

***D. cochinchinensis*** Pierre. (Cay xoai, Xa meh, Kralanh). Laos, Cambodia, Thailand, S.Vietnam. The bark is chewed locally.

***D. indum*** L. (Tamarind Plun). Malaysia. The wood is hard, very heavy and yellow-brown in colour. It is an important local timber for house construction and boat-building, and is also used to make rulers. The small black fruits are a local delicacy.

***D. maingayi*** Baker. Malaysia. The fruits are eaten locally.

***D. ovoideum*** Thw. Ceylon. The fruits are eaten locally.

***D. platysepalum*** Baker. Malaya. The fruits are eaten locally.

***Dialyanthera*** Warb = Otoba.

DIANELLA Lam. Liliaceae. 30 spp. Tropical Asia, Australia, Polynesia, New Zealand. Herbs.

***D. ensifolia*** Lam. S.W.Asia, Indonesia to Australia. The plant is used locally as a rat poison.

***D. lavarum*** Degener = D. odorata (Ukiuki). Hawaii. The berries are used locally as the source of a blue dye.

***D. nemorosa*** Lam. = D. sandwicensis. Tropical Asia to tropical Australia, Polynesia. The berries are used in Hawaii as the source of a blue dye. The roots are scented and are used in making cosmetics and as a fumigant.

***D. odorata*** Hileb non Blume. = D. lavarum.

***D. sandwicensis*** Hook. and Arn. = D. nemorosa.

Diangara – Combretum elliotii.
Diania – Celtis brieryi.

DIANTHUS L. Caryophyllaceae. 300 spp. Europe, Asia, Africa, especially Mediterranean. Herbs.

***D. caryophyllus*** L. (Carnation, Clove Pink, Picotee). Mediterranean. Widely cultivated as an ornamental in gardens and greenhouses. Large numbers are grown to be sold as cut flowers. The flowers contain an essential oil (eugenol, benzyl benzoate, benzyl salicylate, phenyl ethyl alcohol, methyl salicylate) used in perfumery. About 500 kg. of flowers produce 100 g. of absolute oil. Most of the flowers are produced in S.France and Italy.

***D. prolifera*** (L.) Scop. = Tunica prolifera. (Little Leaf Tunic Flower). C. and S.Europe, N.Africa through to Caucasus. The leaves are sometimes used locally as a tea.

DIATOMS. The sedimented remains of the silicious shells of unicellular green algae. These are mined and sold as Fossil Flour, Keiselghur, Silicious Earth or Infusorial Earth. The white powder has great adsorbtive powers and is used in clarifying a wide variety of fine chemicals. It will absorb about four times its own weight of water and is used to absorb liquids. It is mildly abrasive and is used in fine metal polishes, tooth paste etc. It is used as an insulator against heat loss, and is acid resistant. As it is chemically inert the powder makes a good filler for paints, paper and in the manufacture of dynamite. Mounted on microscope slides, type specimens are used to test the accuracy of lenses.

***Dicalyx odoratissima*** Blume = Symplocos odoratissima.

***Dicalyx tinctorius*** Blume = Symplocos fasciculata.

DICENTRA Borckh. corr Bernh. Fumariaceae. 20 spp. W.Himalayas to E.Siberia, Japan and N.W.China, N.America. Herbs.

***D. canadensis*** (Godd.) Walp. (Squirrel Corn, Turkey Corn). E. N.America. The tubers are used locally as a tonic and to improve the appetite.

***D. cucllaria*** (L.) Bernh. (Dutchman's Breeches). E. N.America. The tubers are used locally as a tonic and to increase the appetite.

DICHAPETALUM Thou. Dichapetalaceae. 150–200 spp. Tropics, especially Africa. Creeping shrubs or vines.

***D. acuminatum*** De Wild. (Aanda, Etudo). W. and Equatorial Africa. The inflorescence is used in the Congo to make arrow heads.

***D. timoriense*** Engl. = Chailletia timoriensis. Java. The leaves are eaten locally as a vegetable, the fruits are cooked with fish, and the bark fibres are used to make fishing nets.

***D. toxicaria*** Engl. = Chailletia toxicaria. W.Africa. The seeds are poisonous and are used locally for killing rats, stupifying fish, and were used as a criminal poison.

DICHELACHNE Endl. Graminae. 5 spp. Australia, New Zealand. Grasses.

***D. crinata*** Hook. f.=D. hookeriana=Apena crinata. (Long Hair Grass). Australia, New Zealand. An important pasture grass in Australia.

***D. hookeriana*** Trin.=D. crinata.

***D. sciurea*** Hook. f.=D. sieberiana=Stipa dichelachne. (Shorthaired Plumegrass). S.Australia, New Zealand. An important fodder grass in Australia.

***D. sieberiana*** Trin.=D. sciurea.

***Dichelostemma capitatum*** (Benth.) Wood.=Brodiaea capitata.

***Dichopsis*** Thw.=Palaquium.

DICHROA Lour. Hydrangeaceae. 13 spp. China, S.E.Asia, Indo-Malaysia. Herbs.

***D. febrifuga*** Lour. (Aseru, Basak). Himalayas. The roots and leafy shoots are used in India and China to reduce fevers and as an emetic.

DICHROSTACHYS (A. DC.) Wight and Arn. Leguminosae. 20 spp. Africa to Australia, especially Madagascar. Shrubs.

***D. cinerea*** Wight and Arn.=Acacia cinerea =Cailliea callistachys.). Tropical Africa. The wood is strong and heavy. It is used for cog-wheels, pegs and fine carvings etc. The bark is used locally for tanning, and a decoction of the roots is used to treat intestinal worms.

DICKSONIA L'Herit. Dicksoniaceae. 30 spp. Malaysia, Australia, New Zealand, New Caledonia, tropical America, St. Helena. Tree ferns.

***D. antartica*** Lab. Australia. The pith of the upper stem contains starch and is eaten roasted by the aborigines.

***D. blumei*** Moore. Malaysia. The scales from the leaves are used locally to stop bleeding.

DICOMA Cass. Compositae. 48 spp. C. E. and S.Africa, Madagascar, Socotra; 1 sp. India. Herbs.

***D. anomala*** Sond. (Maagbessie, Wormbos). S.Africa. A decoction of the roots is used by both Africans and Europeans to treat dysentery, diarrhoea and intestinal worms.

DICRAEANTHUS Engl. Podostemaceae. 2 spp. Tropical Africa. Flattened herbs, looking like liverworts (Bryophytes), living in fast-running water.

***D. africanus*** Engl. (Saule). Congo and Cameroons. The whole plant is eaten locally as a salad.

DICRASTYLIS Drumm. ex Harv. Dicrastylidaceae. 15 spp. Australia. Small trees.

***D. ochrostricta*** F. v. Muell. Australia. The cottony mass at the base of the stem is used by the natives of C.Australia to decorate the body.

Dictame – Amaracus dicatmus.

DICTAMNUS L. Rutaceae. 6 spp. C. and S.Europe to Siberia and N.China. Herbs.

***D. albus*** L.=D. fraxinella=Fraxinella dictamnus. (Dittany, Candle Plant). Temperate and S.Europe, through to N.China. The root contains the alkaloid dictamine and is used medicinally as a uterine stimulant. The leaves are used locally in various parts as a tea.

***D. fraxinella*** Pers.=D. albus.

DICTYOPHORA Desv. Phallaceae. 6 spp. Warm.

***D. indusiata*** (Vent.) Fischer. Tropics. The fruit-bodies are made into a pulp with flowers of Hibiscus rosa-sinensis and used to treat tumours.

***D. phalloides*** Desv. Tropics. Used in Mexico with Psilocybe to produce hallucinations.

DICTYOPTERIS Lam. Dictyotaceae. Brown Alga. Pacific.

***D. plagiogramma*** Mont. (Limu Lipoa). Pacific. Eaten by the natives of Hawaii.

DICTYOTA Lam. Dictyotaceae. Brown Alga. Pacific.

***D. acutiloba*** J. Ag. Eaten by the natives of the Pacific Islands.

***D. apiculata.*** J. Ag. Pacific. Eaten by the natives of the Pacific Islands.

DICYPELLIUM Nees. Lauraceae. 1 sp. Brazil. Tree.

***D. caryophyllatum.*** (Mart.) Nees. The bark (Clove Bark) is sold for flavouring. An oil (Clove Bark Oil) is distilled from the bark and is also used for flavouring.

DIDYMOCARPUS Wall. Gesneriaceae. 120 spp. Tropical Africa, Madagascar, S.E.Asia Indomalaysia, Australia. Herbs.

***D. reptans*** Jack. Sumatra, Malay Peninsula. A decoction of the leaves is used locally to relieve constipation, dysentery, and stomach upsets.

DIDYMOPANAX Decne. and Planch. Araliaceae. 40 spp. Tropical America. Trees.

***D. longipetiolatum*** March. (Mandioqueira). S.Brazil. The wood is fine, straight-grained, easy to work but does not resist decay. It is used locally to make tough packing cases etc.

***D. morototoni*** (Aubl.) Decne. (Matchwood). W.Indies, C.America, Venezuela. The wood is soft, light, straight-grained and brittle. It does not resist decay. It is used locally for general construction work, interiors of buildings, paper pulp, and boxes.

DIEFFENBACHIA Schott. Aracea. 30 spp. Tropical America, W.Indies. Herbs.

***D. seguina*** (L.) Schott. (Canna de Imbé, Dumb Plant). W.Indies, tropical S.America. The rhizome is used locally to treat rashes and itching of the skin. Chewing the rhizome causes temporary dumbness, lasting for several days.

Dieng-tyrnembhai – Camellia kissi.

Diente de culebra – Serjania mexicana.

DIGENEA Ag. Rhodomeliaceae. Red Algae. Tropical Seas.

***D. simplex*** (Wulf.) J. Ag. Tropical seas. Eaten in Japan. It is used in Japan and China to treat intestinal worms and as a laxative for children.

DIGERA Forsk. Amaranthaceae. 2 spp. Old World Tropics. Herbs.

***D. arvensis*** Forsk. Tropical Africa, Asia. The leaves are used in the Sudan as a vegetable and as fodder for livestock.

Digger Pine – Pinus sabiniana.

DIGITALIS L. Scrophulariaceae. 20–30 spp. Europe, Mediterranean, Canary Islands. Herbs.

***D. ferruginea*** L. See D. purpurea.

***D. grandiflora*** Jacq. See D. purpurea.

***D. lanata*** Ehrh. See D. purpurea.

***D. lutea*** L. See D. purpurea.

***D. purpurea*** L. (Foxglove). C. and S.Europe. The leaves contain digitalin, digitoxin, digitalein and digitonin which constitute digitalis. This is used medicinally as a heart stimulant. The plant is cultivated. The leaves are collected during the summer and autumn and dried in the shade. The pulverized leaves may be used direct, or the drugs extracted by solvent extraction. The plant has long been used by herbalists for the treatment of epilepsy. It was introduced into modern medicine during the eighteenth century for the treatment of dropsy, and for several other purposes, but is now used almost exclusively as a heart stimulant.

DIGITARIA Fabr. Graminae. 380 spp. Warm. Grasses.

***D. abyssinica*** (Hochst.) Stapf. (Abyssinian Finger Grass). Tropical Africa. Used for erosion control in S.Africa. It is suitable for grazing, but can become a troublesome weed.

***D. decumbens*** Stent. (Pangola Grass). S.Africa. An excellent forage grass, and used for hay. It is widely cultivated especially in W.Indies and S. U.S.A.

***D. diversinervis*** Stapf. (Richmond Finger Grass). Tropical Africa. An excellent pasture for winter grazing, especially in warm summer rainfall areas.

***D. excilis*** Stapf. = Paspalum exile = P. longiflorum. Tropical Africa. Cultivated for the grains by the Hausa tribes of N. Nigeria.

***D. horizontalis*** Willd. = Panicum sanguinale.

***D. iburua*** Stapf. N.Nigeria. Cultivated locally as a cereal crop.

***D. longiflora*** Pers. = Paspalum brevifolium. Malaysia. Used locally for grazing.

***D. pentzii*** Stent. (Woolly Finger Grass). S.Africa. Used locally as a pasture, but it is palatable to stock only during the autumn and winter.

***D. scalarum*** Chiov. (Dunn's Finger Grass). E.Africa. Used locally and in S.Africa as a pasture, but can become a weed.

***D. smutsii*** Stent. S.Africa. Used locally for grazing and hay.

***D. swazilandensis*** Stapf. (Swaziland Finger Grass). S.Africa. An excellent grazing grass, especially for sheep. It is also used for erosion control.

Dika (Bread) (Nut) – Irvingia barteri.

Dika Bread – Irvingia gabonensis.

Dika Butter (Fat) – Irvingia gabonensis.

Dikkamaly – Gardenia gummifera.

Dileko – Baikiaea robynsii.

Dill – Anethum graveolens.

Dill, East Indian – Peucedanum graveolens.

DILLENIA L. Dilleniaceae. 60 spp. Mascarene Islands, S.E.Asia, Indomalaysia, N.Queensland, Fiji. Trees.

***D. aurea*** Smith. = Colbertia obovata (Djoonti, Sempoor). Indonesia, especially Java. The ash, mixed with clay is used locally to make fire-resistant bricks and pottery. A paste of the bark is used to treat thrush and weakening of the gums. This is sometimes mixed with the juice from Sweet Basil (Ocimum basilicum). The juice from the leaves, rubbed into the scalp is supposed to prevent baldness.

***D. elliptica*** Thunb. (Songo). Indonesia. The

fruits are used locally in the Celebes instead of lemons. They are eaten raw or with salt.

***D. excelsa*** Glid. = Wormia excelsa (Dregel, Kajoo ringon, Segel). Indonesia. The wood is hard, but is not durable, and does not split. It is used locally for posts, bridges etc.

***D. indica*** L. (Chalta, Cinar, Elephant Apple, Chalta). Himalayas, India to Indonesia. The wood is hard, and durable in water. It is red flecked with white. It is used for boat-building and dugout canoes, panelling, inlay work, gunstocks, firewood and charcoal. The fleshy bracts around the fruits are eaten in India as a vegetable or preserves. The fruits are edible and made into curries in Malaya. The rough leaves are used for polishing ivory. The fruit pulp is used to make a hair wash.

***D. megalantha*** M. Bayani. Philippines. The green acid fruits are used for making preserves locally.

***D. mindanaense*** Elm. Philippines. The fruits and their fleshy bracts are eaten locally especially as preserves.

***D. ochreata*** Gilg. = Wormia ochreata (Ngear). N.Celebes. The wood is used locally for making tool handles.

***D. ovata*** Wall. (Cây só, Phlou thom, Xo pho). India to N.Vietnam. The fruits are eaten locally in Cambodia, where the bark is used to treat diarrhoea.

***D. pentagyna*** Roxb. India, Malaya. The leaves are used locally as sandpaper to smooth wood.

***D. philippensis*** Rolfe. Philippines. The fruits are eaten locally as preserves. The bark yields a red dye.

***D. reifferscheidia*** F. Vil. Philippines. The fruits are used locally to make preserves and sauces.

***D. subfruticosa*** (Griff.) Mart. (Simpoh ayer). Malaya, Borneo. The leaves are used in some parts to wrap goods in markets.

Dilo Oil – Calophyllum inphyllum.

DILOBEIA Thou. Proteaceae. 1 sp. Madagascar. Tree.

***D. thouarsii*** Toem. (Tavolohazo, Vivaona). The oil crushed from the seeds is used by the local women as a body oil. The wood is used for torches.

Diloso – Sorindeia claessensii.

DILSEA Stack. Dumontiaceae. Red Algae. N.Temperate. Marine.

***D. edulis*** Stackh. = Iridaea edulis. N.Temperate Atlantic. Eaten locally on the coasts of Ireland S.W.England.

Dimi – Roureopsis obliquifoliata.

DIMORPHANDRA Schott. Leguminosae. 25 spp. Tropical America. Trees.

***D. gonggrijpii*** Sandw. = Mora gonggrijpii. Venezuela, Guyana. The wood is hard, heavy and durable. It is brown with a slight purple tinge, and is used for general construction work, and ship-building.

***D. megistosperma*** Pitt. C.America. The seeds yield a red dye.

***D. mora*** Benth. = Mora excelsa (Mora). Trinidad, Guyana and Venezuela. The wood is resistant to decay and marine borers. It is used for ship-building and general construction work, pavings and railway sleepers.

Dinda – Leea macrophylla.

Dineygama – Sclerocarya birraca.

Dingili (Oil) – Cephalocroton cordojanus.

Dingleberry – Vaccinium erythrocarpum.

DINOCHLOA Buese. Graminae. 20 spp. S.E.Asia, Indomalaya. Large woody grasses.

***D. compactiflora*** Benth. and Hook. f. = Melocalamus compactiflorus. Tropical Asia, especially Bengal and Burma. The large mealy seeds are edible. The culms are made into shoes.

***D. dielsiana*** Pilger = Schizostachyum dielsianum.

Dioko dia mfinda – Ganophyllum giganteum.

DIONYCHA Naud. Melastomataceae. 2 spp. Madagascar. Small trees.

***D. bojeri*** Naud. Madagascar. A black dye is extracted from the bark and is used locally for dyeing silk.

DIOÖN Lindl. corr Miq. Zamiaceae. 3–5 spp. Mexico, S.America. Cycads.

***D. edule*** Lindl. (Chamal, Palma de la Virgin, Palma de Macetas). Mexico. The seeds have a high starch content and are eaten locally as a flour, boiled or roasted. An extract of the seeds is used locally to treat neuralgia.

DIOSCOREA L. Dioscoreaceae. 600 spp. Tropics and subtropics. Vines. This genus contains the yams which are grown throughout the wetter tropics for the starchy tubers. These are the main starchy food for millions of people. The plants are easy to cultivate and are remarkably free from diseases. They are propagated vegetatively from tubers. The tubers are easily stored and in some cases reach over 100 lb. (50 kg) in weight. The starch they contain cannot be extracted for commercial use as the grains are held together by a mucilage. The tubers also contain the poisonous alkaloid dioscorine, but this is present in small amounts in the edible species, and is destroyed by boiling. They also contain steroidal sapogenins which were used in the manufacture of cortisone. The tubers are not exported, but much are sold in local trade. The morphology of the tubers differs in different species. They may arise from the hypocotyl, or from the internode above it, or may include both these parts. The tubers take from eight to ten months to mature after planting, and vary tremendously in size and colour, on which the varietal differences are based mainly.

***D. alata*** L. (Greater Yam, Ten-month Yam, Water Yam, White Yam, Ñame de Agua). S.Pacific. Cultivated throughout the tropics. The source of a poor quality arrowroot (Guiana Arrowroot).

***D. altissima*** Lam. (Ñame Dunguey). W.Indies. It has been used as a food in S.America for centuries.

***D. atropurpurea*** Roxb. (Malacca Yam). Burma, Malaya.

***D. arachnida*** Prain. and Burk. = D. collinsae. Thailand, Vietnam. Eaten locally.

***D. batatas*** Decne. = D. divaricata Blanco. (Chinese Potato, Chinese Yam, Ñame de China). Tropical Asia, Malaysia. Cultivated widely in Old and New World.

***D. bemandry*** Jum. and Perr. Madagascar. Eaten locally.

***D. bulbifera*** L. (Air Potato, Potato Yam). Tropical Asia. It produces tubers on the aerial stem as well as underground. Cultivated widely, it is the only species that is grown widely in the U.S.A.

***D. cayenensis*** Lam. = D. rotundata. (Affun Yam, Yellow Yam, Guinea Yam, Yellow Guinea Yam, Negro Yam, Ñame Amarillo, Ñame Guinea). Tropical America. Cultivated in Old and New World tropics. Also called Twelve Month Yam, the tubers taking a year to mature.

***D. cinnamomifolia*** Hook. = D. teretiuscula = D. tuberosa. Brazil. Cultivated locally.

***D. collinsae*** Prain and Burk. = D. arachidna.

***D. daemona*** Roxb. = Epipremnum giganteum.

***D. deltoidea*** Wall. (Kilari, Kins, Kirta). Himalayas, N.India, S.China. A decoction of the tubers is used locally as a fish poison, and to kill lice. Not edible.

***D. divaricata*** Blanco. Philippines. Eaten locally.

***D. dodecaneura*** Vell. = D. hebantha. Brazil. Eaten locally.

***D. dumetorum*** (Kunth.) Pax. E.Africa. An important local food. Cultivated.

***D. elephantipes*** Salisb. (Hottentot Bread). S.Africa. The huge corky tubers protrude from the ground.

***D. esculenta*** (Lour.) Burk. (Asiatic Yam, Fancy Yam, Lesser Asiatic Yam, Potato Yam, Ñame Papa). Pacific and tropical Asia. Cultivated in India, S.Seas, Vietnam, Burma, Ceylon and W.Africa. The tubers are small, from the base of the stem. They are non-poisonous, and do not store well.

***D. fandra*** Hum. and Perr. Madagascar. Eaten locally.

***D. glandulosa*** Klotzsch. Brazil. Eaten locally.

***D. globosa*** Roxb. E.India. Eaten locally.

***D. hastata*** Vell. S.America. Eaten locally in Brazil.

***D. hastifolia*** Nees. W.Australia. Eaten, and cultivated by the aborigines.

***D. hebantha*** Mart. = D. dodecaneura.

***D. heptoneura*** Vell. = D. lutea.

***D. hirsuta*** Dennst. = D. hispida.

***D. hispida*** Roxb. = D. hirsuta. (Baichandi, Karukanda, Peiperendai). Tropical Asia. The juice from the tubers is used locally as an arrow poison, when mixed with that of the Upas tree (Antiaria toxicaria). Tubers not edible.

***D. hylophila*** Harms. Africa. The stems are used as a fibre by the natives of Usambara.

***D. latifolia*** Benth. (Acom, Akam Yam, Ñame Akam). W.Africa. Cultivated in Old and New World tropics.

***D. lutea*** G. F. W. Mey. = D. heptoneura. Tropical America. The tubers are eaten and also roasted as a coffee substitute.

***D. luzonensis*** Schau. Philippines. Eaten locally.

***D. macabiha*** Jum. and Perr. (Fanganga,

Macabiha). Madagascar. The tubers are poisonous, but are eaten locally.

***D. macroura*** Marms. = D. sansibariensis.

***D. mamillata*** Jum. and Perr. Madagascar. Eaten locally.

***D. ovinala*** Jum. and Perr. Madagascar. Eaten locally.

***D. papuana*** Rich. Pacific Islands. Eaten locally. Cultivated.

***D. poilanei*** Prain. and Burk. Vietnam. The tubers are poisonous, and are used locally as a fish poison.

***D. prazeri*** Prain. and Burk. Himalayas, N.India, S.China. A decoction of the tubers is used locally as a fish poison and to kill lice.

***D. pyrifolia*** Kunth. Malaysia. Wild food plant, eaten locally.

***D. rhipogonoides*** Oliv. (Dye Yam). China, Formosa. The roots are used on Formosa for dyeing and tanning fishing nets, and in parts of China for dyeing cloth. Tubers not edible.

***D. rotundata*** Poir. = D. cayenensis.

***D. sansibariensis*** Pax. = D. macroura = D. welwitschii. Tropical Africa, Madagascar. The tubers are poisonous but have a good flavour. Grown locally, and used for arrow poison and stupifying fish.

***D. seriflora*** Jum. and Perr. Madagascar. Eaten locally.

***D. soso*** Jum. Madagascar. Eaten locally.

***D. spinosa*** Roxb. (Spiny Yam, Wild Yam). E.India, Polynesia. The wild plants are eaten locally.

***D. teretiuscula*** Klotzsch. = D. cinnamomifolia.

***D. tokora*** Mak. China. The tubers are used locally to stupify fish.

***D. transversa*** R. Br. Australia. The tubers are eaten by the aborigines of Queensland.

***D. trifida*** L. f. (Cush-cush, Yampi, Ñame Mapicey). W.Indies, Guyana. Cultivated throughout tropical America.

***D. trifoliata*** H.B.K. S.America. Eaten locally in Brazil.

***D. triloba.*** Kast. = D. trifoliata.

***D. tuberosa*** Vell. = D. cinnamomifolia.

***D. villosa*** L. (Atlantic Yam, Wild Yam). S.E. N.America. The rhizome contains a resin and a saponin which are used medicinally as an expectorant and diaphoretic. The plant is cultivated locally.

***D. welwitschii*** Renole = D. sansibariensis.

DIOSPYROS L. Ebenaceae. 500 spp. Warm. Trees. The heartwood of many species is the ebony of commerce. The ebonies vary in colour but all are hard, close-grained, heavy, taking a good polish. They are used for making small articles, e.g. brushes and combs, turning, piano keys, inlaying, tool-handles and furniture. The sap wood is soft and white. Much of the heartwood is exported to Europe. The fruits of many species are edible.

***D. abyssinica*** Hiern. Ethiopia. The dark wood is used for gun stocks.

***D. atropurpurea*** Gürke. W.Africa. The brown wood is streaked with black. Exported from Cameroons. The fruits are eaten locally.

***D. bipindensis*** Gürke. Tropical Africa. South Cameroon Ebony.

***D. buxifolia*** Hiern. = D. microphyllum.

***D. decandra*** Lour. (Mak châng, Mûong, Thi). Indochina. The fruits are used in Cambodia to treat insomnia and to control menstruation. An infusion of the leaves is used to treat intestinal worms.

***D. cameroonensis*** Gürke. W.Africa. The black wood is used in Ghana and Liberia for house-building, household implements, and tool handles.

***D. canomoi*** A. DC. = D. multiflora.

***D. chinensis*** Blume. = D. kaki.

***D. chrysophyllos*** A. DC. Mascarene Islands. White ebony.

***D. conzattii*** Standl. Mexico. The fruits are eaten locally.

***D. crassiflora*** Hiern. (Benin Ebony). Black wood, exported from Nigeria.

***D. dendo*** Welw. (Billet Wood, Black Ebony, Calabar Ebony, Gabon Ebony). Tropica Africa, especially Cameroons. The wood is black. Exported.

***D. discolor*** Willd. = D. mabola. (Mabola). Malaya, Philippines. The wood is Camogan Ebony. The fruits (Butter Fruit) are cream-coloured with a dry flesh. They are eaten locally.

***D. ebenaster*** Retz. (Black Sapote, Zapte Negro). Mexico. The blackish fruits are eaten with lemon or orange juice. The tree is cultivated.

***D. ebenum*** Koenig. (Ebony, Black Sapote, Sapote Negro). India, Ceylon. The finest commercial ebony.

***D. embryopteris*** Pers. (Gaub). Tropical Africa and Asia. In India the bark is used as the source of a dye for tanning fishing nets. The edible fruits yield a gum when unripe, which is used locally for caulking and preserving boats.

**D. evila** Pierre. Tropical Africa. Gabon Ebony.

**D. flavescens** Gurke. Tropical Africa. Gabon Ebony.

**D. foliosa** Wall. India. The wood is one of the White Ebonies.

**D. frustescens** Blume (Gegentelan). Java. The wood is yellow, and rots in contact with soil. It is used locally for the main pillars of houses and for charcoal.

**D. gardneri** Thw.=D. pyregrina. E.India, Ceylon. Bombay Ebony.

**D. haplostylis** Boiv. Madagascar. Madagascar Ebony.

**D. hirsuta** L. f. Madagascar. Coromandel Ebony.

**D. insignis** L. f. E.India, Ceylon. Bombay Ebony.

**D. kaki** L. f.=D. chinensis. (Chinese Date Plum, Japanese Persimmon). China. The fruits are orange-red, about the size and appearance of a tomato. They are eaten raw or, dried in China (called Ki-kwe). The plant is grown commercially in China and Japan, and was introduced to France and the Mediterranean regions during the 19th century. It was introduced to S. U.S.A. a little later. In all these regions it is grown commercially.

**D. kurzii** Hiern. Nicobar. The wood (Zebra Wood, Andaman Zebra Wood) is black, striped with grey.

**D. lotus** L. (Date Plum). W.Asia to China. Cultivated locally for the fruits which are eaten fresh, over-ripe or dried.

**D. lotus** Blanco=D. multiflora.

**D. loureiriana** G. Don. Tropical Africa. The native of Mozambique used the root to cloud their teeth red.

**D. mabola** Roxb.=D. discolor.

**D. macrocalyx** Guerke. Shrub. Tropical Africa. The roots and bark are used locally to give a black dye.

**D. malacapai** A. DC. Philippines. The wood is called White Ebony.

**D. maritima** Blume. India to Vietnam, Celebes and Australia. The fruits and bark are used in S.Vietnam to stupify fish.

**D. marmorata** Parker. Malaya. The wood is grey, streaked with black, and very beautiful (Andamanese Marble Wood).

**D. melanida** Poir. Philippines. White Ebony.

**D. melanoxylon** Roxb. India, Ceylon. Coromandel Ebony. The leaves are used for wrapping cigarettes.

**D. mespiliformis** Hochst. (Swamp Ebony, West African Ebony, Zanzibar Ebony). Tropical Africa. The wood is dark brown to black. It is used locally for tool handles, small articles etc. Much is exported. The fruit pulp is made into a fermented beverage. The fruit is used in N.Nigeria to make a toffee (Ma'di).

**D. microrhombus** Hiern. Madagascar. Madagascar Ebony.

**D. microphyllum** Bedd.=D. buxifolia. S.India, Malaysia, Africa. Source of an Ebony.

**D. mombuttensis** Guerke. Nigeria. Yoruba Ebony.

**D. montana** Roxb. E.India, Ceylon. Bombay Ebony.

**D. multiflora** Blanco. Philippines. The wood is Camogan Ebony. The juice from the fruit is used as an arrow poison, and the bark as a fish poison.

**D. oppositifolia** Thw. E.India, Ceylon. Coromandel Ebony.

**D. pentamera** F. v. Meull. (Grey Persimmon). Coast of Queensland. The wood is greyish and is used locally for tool-handles, mallets, and shuttles.

**D. perrieri** Jum. Madagascar. Madagascar Ebony.

**D. pilosanthera** Blanco. Philippines. Camagoon Ebony (Bolongeta Ebony).

**D. pseudo-ebenum** Koord. and Val. (Ki areng). Java. An excellent ebony, now practically extinct.

**D. pyregrina** (Gaertn.) Gürke.=D. gardneri.

**D. quaesita** Thw. Ceylon. Calamander Ebony.

**D. ramiflora** Roxb. E.India, Ceylon. Bombay Ebony.

**D. reticulata** Willd. Mauritius. An excellent Ebony.

**D. rubra** Gaertn. Mauritius. Red Mauritius Ebony.

**D. sandwicensis** A. DC. Hawaii. The wood is red-brown and is used locally for building temples.

**D. suaveolens** Gürke. Tropical Africa. Benin Ebony.

**D. sylvatica** Roxb. E.India, Ceylon. Bombay Ebony.

***D. tesselaria*** Poir. Mauritius. Mauritius Ebony.

***D. texana*** Scheele. (Black Persimmon, Chapote). Texas, into Mexico. The wood is an Ebony. The fruits are used in Mexico for dyeing sheepskins.

***D. tupru*** Buch. E.India, Ceylon. Bombay Ebony.

***D. utilis*** Koord. and Val. Moluccas, Celebes. Macassar Ebony.

***D. virginiana*** L. (North American Ebony, Common Persimmon). E. N.America to Florida and Texas. The wood is dark brown. The fruit is edible, but is very astringent until over-ripe. It is occasionally cultivated. Used as a root stock for D. kaki.

***D. wallichi*** King. Thailand. The fruits are used locally as a fish poison.

***Diotis*** Schreb. = Otanthus.

Dipasa – Deinbollia molliuscula.

DIPCADI Medek. Liliaceae. 55 spp. Mediterranean, Africa, Madagascar. Herbs.

***D. cowanii*** (Ridl.) Perr. (Rongolo voalova). Madagascar. The bulbs are used locally for poisoning rats.

***D. erythaeum*** Webb. and Benth. = D. unicolor. Arabia to Egypt. The bulbs are used medicinally as a heart stimulant and expectorant.

***D. unicolor*** Baker. = D. erythaeum.

DIPHLOIS A. DC. Sapotaceae. 20 spp. Florida, W.Indies. Trees.

***D. nigra*** Gris. = Achras nigra = Bumelia nigra. Small tree. W.Indies. The latex is used as a balata substitute (Balâta Batard, Acomat Batârd).

***D. salicifolia*** (L.) A. DC. (Bustic, Cassada). S. Florida to Mexico, W.Indies. The dark brown wood is very hard and heavy. It is used for cabinet work.

DIPHYSA Jacq. Leguminosae. 15 spp. Mexico to C.America. Small trees.

***D. robinioides*** Benth. C.America. The wood is used locally as the source of a yellow dye.

***Diplanthera*** Banks ex Soland. ex Br. = Deplanchea.

DIPLAZIUM Sw. Athyriaceae. 400 spp. Tropics and N.Temperate. Ferns.

***D. asperum*** Blume. Tropical Asia. The leaves are eaten locally as a vegetable.

***D. esculentum*** Sw. Tropical Asia, Malaysia. The leaves are eaten as a vegetable in Java.

DIPLODISCUS Turcz. Tiliaceae. 7 spp. W.Malaysia. Trees.

***D. paniculatus*** Turcz. (Baroba). Philippines. The seeds contain starch and are eaten locally when boiled.

DIPLOGLOTTIS Hook. f. Sapindaceae. 1–2 spp. Australia. Trees.

***D. cunninghamii*** Hook. f. Queensland. The orange-red fruits have a pleasant acid flavour and are used locally for preserves.

***Diplophractum*** Desf. = Colona.

DIPLORHYNCHUS Welw. ex Ficalho. and Hiern. Apocynaceae. 1 sp. Tropical Africa. Trees.

***D. mossambicensis*** Benth. The trunk yields a latex which is used as rubber. It is also used as an eye-wash particularly to relieve the effects of puff adder venom in the eye; spread on the skins of the local drums to improve their tone; and as a bird lime. A decoction of the roots is used by the Europeans to treat black-water fever.

DIPLOSPORA DC. Rubiaceae. 25 spp. China, Indomalaysia, Australia. Trees.

***D. malaccense*** Hook. f. S.Malaysia, Thailand. The leaves are used to make a tea in Perak and Malacca.

DIPLOTAXIS DC. Cruciferae. 27 spp. Europe, Mediterranean. Herbs. The leaves of the species mentioned are used locally in salads.

***D. acris*** Boiss. = D. sieberi.

***D. duveyrierana*** Coss. (Harra, Ranekfait). Sahara. The seeds are used as a poultice to treat galls on camels.

***D. pendula*** DC. Algeria.

***D. sieberi*** Presl. = D. acris. Nile region, Arabia.

DIPLOTROPIS Benth. Leguminosae. 12 spp. Tropical America. Trees.

***D. brachypetalum*** Tul. Guyana. The hard wood polishes well and is used for furniture, house building and boat-building. An extract from the bark is used locally to destroy vermin.

Dips – Vitis vinifera.

DIPSACUS L. Dipsacaceae. 15 spp. Eurasia, Mediterranean, tropical Africa. Herbs.

***D. fullonum*** L. ssp. sativus (L.) Thell. = D. sativus. (Teasel). Europe, Mediterranean, Asia Minor. The mature flower heads, which are strongly hooked, are used by fullers to comb out wool. Occasionally cultivated for the flower heads.

***D. japonicus*** L. China, Japan. In China a decoction of the roots is used to treat haemorrhoids, breast cancer, and to regulate difficult menstruation.

***D. sativus*** Honck. = D. fullonum ssp. sativus.

DIPTEROCARPUS Gaertn. Dipterocarpaceae. 76 spp. Ceylon, India to W.Malaysia. Trees.

***D. alatus*** Roxb. = D. costatus E.India, Burma. The trunk is tapped for a balsam (Gurjum Balsam) which is used locally for caulking boats and as an adulterant for other balsams.

***D. costatus*** Gaetn. = D. alatus.

***D. gracilis*** Blume. Malaysia. Source of a balsam used in paints. The hard heavy wood is used locally for house-construction.

***D. grandiflorus*** Blanco. Philippines. The trunk yields a balsam (Apitong Resin, Baláu) which gives a hard tough finish to varnishes. The red-brown wood is hard and strong. It is used locally for house-construction, furniture, bridges etc., and pavings.

***D. hispidus*** Thw. Ceylon. The red-brown wood is hard and durable. It is used for house-construction and boat building.

***D. kerrii*** King. Malaya. Yields a resin (Minyak Keruong) used locally for caulking boats.

***D. kunstleri*** King. Tropical Asia. The wood is more grey than the other species. It is used for house-building.

***D. lamellantus*** Hook. Tropical Asia. The seeds yield an oil (Gurjum Oil), used by the Chinese to treat dropsy.

***D. marginata*** Kork. (Keroowing) Borneo. The wood is used locally as a source of charcoal.

***D. skinneri*** King. (Keroowing booloo). Malaysia. The wood is durable and easy to work. It is used for boarding. The trunk is a minor source of a resin.

***D. trinervis*** Blume. (Palahlar, Klalar, Pala). Indonesia, especially Java. The bark yields a resin used locally for torches. The wood is used for charcoal.

***D. turbinatus*** Gaertn. E.India, Burma. Source of Gurjum balsam.

***D. vernicifluus*** Blanco. Philippines. Yields a resin (Baláu, Panau Resin).

***D. warburgii*** Brand. (Kaladan). Borneo, Philippines. The wood is used locally for house-construction and making utensils.

DIPTERYGIUM Decne. Cruciferae. 1 sp. Egypt to W.Pakistan. Shrub.

***D. glaucum*** Decne. The plant is used as fodder for camels.

DIPTERYX Schreb. = Coumarouna.

DIRCA L. Thymelaeaceae. 2 spp. N.America. Shrubs.

***D. palustris*** L. (Leather Wood, Swamp Wood). E. N.America. The bark was used as a fibre by the local Indians.

***Diserneston gummiferum*** Jaub. and Spach. = Dorema ammoniacum.

Dishcloth Gourd – Luffa acutangula.

***Disoon*** A. DC. = Myoporum.

DISSOTIS Benth. Melastromataceae. 140 spp. Africa. Herbs or shrubby herbs.

***D. grandiflora*** Benth. Tropical Africa. A juice used for sweetening is extracted from the roots by drying, steaming and beating. It is used mainly in W.Africa where it is made also into a fermented beverage (Biti).

***D. rotundifolia*** Triana. Tropical Africa. The leaves are used locally to combat intestinal worms.

DISTERIGMA Niedenzu ex Drude. Ericaceae. 35 spp. Tropical Andes. Shrubs. The fruits of both species mentioned are pleasant to taste and are eaten raw in Ecuador. They are occasionally sold.

***D. margaricoccum*** Blake. (Chirimote). Ecuador.

***D. popenoei*** Blake. Ecuador.

DISTYLIUM Sieb. and Zucc. Hamamelidaceae. 15 spp. Assam to Japan, S.E.Asia, Malaysia, Java; C.America. Trees.

***D. racemosum*** Sieb. and Zucc. China, Japan. The dark brown wood is very hard and fine-grained. It is used in Japan for making small articles and musical instruments. The ash from the wood is used in the glazing of porcelain.

DITAXIS Vahl. ex A. Jurr. Euphorbiaceae. 50 spp. Warm America, W.Indies. Herbs, or shrubby herbs.

***D. tinctoria*** (Millsp.) Pax. and Hoffm. = Argithamnia tinctoria. C.America to S.Mexico. A pink dye is extracted from the plant.

Ditch Stonecrop – Penthorum sedoides.

Diti – Celtis prantlii.

Dittander – Lepidium latifolium.

Dittany – Cunila origanoides.

Dittany – Dictamnus albus.

***Dittelasma rarak*** Hook. = Sapindus rarak.

Divi-divi – Caesalpinia coriaria.
Djaha keling – Terminalia sumatrana.
Djalilan – Celtis cinnamonea.
Djambe sinagar – Actinorhytis calapparia.
Djamoer rajap – Rajapa euhisa.
Djamoor Gagjih – Oudemansiella canarii.
Djamoor manis – Lentinus djamor.
Djangko-rang – Schefflera aromatica.
Djarong boobookooar – Eranthemum viscidum.
Djati – Tectonia grandis.
Djati sabrang – Peronema canescens.
Djato – Eugeissona insignis.
Djave Butter – Mimuspoes djave.
Djawura – Garcinia lateriflora.
Djeboog sarilong – Actinrhytis calapparia.
Djedari – Rhus oxycanthoides.
Djedk-i-Ardin – Amygdalus leiocarpus.
Djelben – Pisum elatius.
Djelo – Willughbeia tenuiflora.
Djelutong – Alstonia polyphylla.
Djelutong – Dyera costulata.
Djelutong – Dyera laxiflora.
Djelutong badak – Tabernaemontana corymbosa.
Djelutong, Gutta. – Alstonia eximea.
Djelutung – Alsontia scholaris.
Djelutung, Getah – Alstonia polyphylla.
Djending – Rhodamnia arborea.
Djerookan – Glycosmis cochinchinensis.
Djintenan – Gomphostemma phlomoides.
Djirak – Sympolcos javanica.
Djirou – Physostigma venenosum.
Djombi – Beilschmiedia giorgii.
Dhonge areuj – Blumea chinensis.
Djookoot malela – Polygala glomerata.
Djookoot rindik – Polygala paniculata.
Djoolang – Afzelia javanica.
Djoonti – Dillenia aurea.
Djoo-war – Cassia siamea.
Djreik – Baccaurea javanica.
Do – Elaeophorbia drupifera.

DOBERA Juss. Salvadoraceae. 2 spp. S.Arabia, N.W.India, E.Asia. Trees.

***D. roxburghii*** Planch. Tropical Africa, India. An essential oil is extracted from the flowers and used as a perfume in the Sudan.

Dock, Alpine – Rumex alpinus.
Dock, Bitter – Rumex obtusifolius.
Dock, Broad-leaved – Rumex obtusifolius.
Dock, Curled – Rumex crispus.
Dock, Patience – Rumex patientia.
Dock, Spanish Rhubarb – Rumex abyssinicus.
Dockmackie – Viburnum acerifolium.
Dockowar Grass Tree – Xanthorrhoea arborea.
Doctor's Gum – Metopium toxiferum.
Doctor's Gum – Symphonia globulifera.

DOCYNIA Decne. Rosaceae. 6 spp. Himalayas, Burma, W.China. Trees.

***D. delavyi*** (Franch.) Schneid. = Cydonia delavya = Eriolobus delavayi. S.W.China. The fresh fruits (Tao yi) are used to ripen persimmons. The fruits are packed in alternate layers in jars, and after 8 hours the persimmons are ripe (bletted).

Dodder, Lesser – Cuscuta epithymum.

DODECATHEON L. Primulaceae. 1 sp. Arctic, N.E.Asia; 50 spp. Pacific N.America; 1 sp. Atlantic N.America. Herbs.

***D. hendersonii*** A. Gray. (Henderson Shooting Star). W. N.America. The leaves and roasted roots were eaten by the local Indians.

Dodol – Garcinia mangostana.

DODONAEA Mill. Sapindaceae. 60 spp. Tropics and subtropics, especially Australia. Small trees or shrubs.

***D. lobulata*** F. v. Muell. (Hop Bush). Australia. Used as livestock fodder.

***D. madagascariensis*** Radik. Madagascar. The leaves are used for feeding silkworms.

***D. thunbergiana*** Eck. and Zeyh. S.Africa. A decoction of the leaves is used locally as a purgative.

***D. viscosa*** Jacq. (Switchsorrel). Tropics. In India, the hard brown wood is used for making small articles, turning and engraving. In Australia the fruits are used as hops in the manufacture of beer, and more generally a decoction of the bark is used as an astringent, or the bark itself as a poultice.

Dog Chamomille – Anthemis cotula.
Dog Fennel – Anthemis cotula.
Dog Fennel – Eupatorium capillifolium.
Dog Hip – Rosa canina.
Dog Lichen – Peltigera canina.
Dog's Mercury – Mercurialis perennis.
Dog Nettle – Urtica urens.
Dog Parsley – Aethusa cynapium.
Dog Poison – Aethusa cynapium.
Dog Rose – Rosa canina.
Dog Senna – Cassia obovata.
Dog's Tail Crested, Grass – Cynosurus cristatus.
Dog's Tooth – Cynodon dactylon.
Dogbane, Hemp – Apocynum cannabinum.

Dogbane, Spreading – Apocynum androsaemifolium.
Dogberry – Thelycrania sanguinea.
Dogberry – Ilex verticillata.
Dogtooth – Epicampes macroura.
Dogwood, Bloodtwig – Thelycrania sanguinea.
Dogwood, Flowering – Cornus florida.
Dogwood, Jamaica – Piscidia erythrina.
Dogwood, Pagoda – Cornus alternifolia.
Dogwood, Rough-leaved – Cornus asperifolia.
Dogwood, Western Flowering – Cornus nuttallii.
Doi-doi – Chenopodium procerum.
Dok dâm duan – Popowia aberrans.
Dok ka aan – Osbeckia crinita.
Do kom – Gonocaryum subrostratum.
Dok toa pa – Maese balansae.

DOLICHANDRONE (Fenzl.) Seemann. Bignoniaceae. 9 spp. Tropical E.Africa, Madagascar, tropical Asia, Australia. Trees.

***D. falcata*** Seem. (Cittwodi, Hawar, Manchingi). Tropical Asia. A decoction of the bark is used locally as a fish poison. A decoction of the leaves is used to induce abortions.

***Dolicholus*** Medik. = Rhynchosia.

DOLICHOS auctt. (Lam.?). Leguminosae. 150 spp. Warm. Herbs.

***D. ahipa*** Wedd. = Pachyrrhizus ahipa.

***D. biflorus*** L. (Horse Gram). Tropics. Used locally for livestock fodder and as a green manure.

***D. hosei*** Craib. = Vigna hosei.

***D. lablab*** L. = Lablab niger. (Hyacinth Bean, Bonavist, Lablab, Lubia). India. Cultivated in India, parts of Asia, Africa, especially the Sudan. The seeds are eaten, but must be cooked beforehand to destroy a poisonous glucoside. The green parts are used as forage and as a green manure.

***D. malosanus*** Bak. (Baput, Igikindye, Malando Polo). C.Africa. Cultivated locally for the fruits and flowers which are eaten as vegetables.

***D. phaseoloides*** (Sw.) DC. = Rhynochosia pyramidalis.

***D. polystachyos*** Forsk. = Canavalia polystachya.

***D. pruriens*** Jacq. = Mucuna pruriens.

***D. pseudopachyrrhizus*** Herms. Tropical Africa. The leaves and shoots are used locally as an insecticide.

***D. sesquipedalis*** L. (Asparagus Bean). Tropical Asia. The young pods are eaten as a vegetable, and the foliage is used as forage.

***D. sinensis*** L. = Vigna sinensis.

***D. sphaerospermus*** DC. Jamaica. The seeds are eaten locally as a vegetable.

***D. viscosis*** Roxb. = Canavalia polystachya.

DOLIOCARPUS Roland. Dilleniaceae. 40 spp. C. and tropical S.America, W.Indies. Woody vines.

***D. calineoides*** (Eichl.) Gilg. (Corde de l'Eau). W.Indies. The stem contains a lot of free stored water, which is released on cutting. This is used locally as an emergency water supply.

Dolo – Alsodeiopsis poggei.
Dom ong kol – Garcinia lanessonii.
Domba Oil – Caolphyllum inophyllum.

DOMBEYA Cav. Sterculiaceae. 50 spp. Africa; 300+ spp. Madagascar and Mascarene Islands. Trees. Fibres are made locally from the bark of all the species mentioned. The fibres are used for cordage, rough fabrics and bags.

***D. burgessiae*** Gerr. S.Africa. The wood is used locally for rubbing to make fire.

***D. cannabina*** Boj. (Hafotra, Hofotsa). Madagascar.

***D. condensata*** Hoch. (Somangana). Madagascar.

***D. coria*** Baill. Madagascar.

***D. elliptica*** Boj. Madagascar.

***D. floribunda*** Bak. Madagascar.

***D. obovalis*** Baill. Madagascar.

***D. perrieri*** Hoch. Madagascar.

***D. spectabilis*** Bojer. Tropical Africa.

***Donacodes incarnata*** T. and B. = Hornstedia rumphii.

DONAX Lour. Marantaceae. 3 spp. Indomalaysia. Pacific, tropical W.Africa. Herbs.

***D. canniformis*** (Forst. f.) K. Schum. = Actoplanes canniformis = Maranta arundinacea Blanco non L. Tropical Asia, Pacific Islands. The split stems are used locally for basket-work, fish traps and making hats.

***D. cuspidata*** Schum. Tropical W.Africa. Fishing nets are made from the fibre in Ghana, and in N.Nigeria the leaves are used for wrapping kola nuts.

***Dondia*** Adans. = Suadea.

Dongo – Deinbollia molliuscula.
Dongolo – Fagara inaequalis.
Don rang cua – Maesa balansae.

Doodook rajap – Scyphiphora hydrophyllacea.
Doodo songot – Kedropsis laciniosa.
Dooka anggang – Phyllanthus indicus.
Doolangdooloang – Glochidion obscurum.
Doolb – Plantanus orientalis.
Doom Bark – Erythropheleum guineënsis.
DOONA Thw. Dipterocarpaceae. 11 spp. Ceylon. Trees.
***D. zeylanica*** Thw. Ceylon. The wood is hard and durable. It is used locally for house-building, both interior and exterior.
DOREMA D. Don. Umbelliferae. 16 spp. C. and S.W. Asia. Herbs.
***D. ammoniacum*** D. Don.=Diserneston gummiferum. Arabia, across to India. A resin (Gum Ammoniacum) collected from insect punctures in the stem, is used medicinally as an expectorant, antiseptic, and to aid digestion. Exported from Bombay.
Dorian booroong – Durio carinatus.
Dorian oogeh – Coelostegia griffithii.
Dorio papa – Durio carinatus.
Do-rag – Davidsonia pruriens.
DORONICUM L. Compositae. 35 spp. Temperate Eurasia, N.Africa. Herbs.
***D. falconeri*** Hook. f. Himalayas, N.India. The roots are used locally to make a tonic.
***D. pardalianches*** L. (Leopard's Bane, Panther Strangler). W.Europe. The roots are used in the preparation of heart stimulants and nerve tonics. Cultivated as an ornamental.
***D. roylei*** DC. (Darunaj-akrabi). Himalayas, N.India. A decoction of the roots is used locally as a tonic and to prevent giddiness when climbing to great height.
DORSTENIA L. Moraceae. 170 spp. Tropics. Herbs.
***D. brasiliensis*** Lam. S.America. A decoction of the rhizome is used in Brazil as a tonic, purgative, diuretic, emetic, to induce sweating, and to control menstruation.
***D. contrajerva*** L. Tropical America. The dried rhizome is used to flavour cigarettes, and in Costa Rica to reduce fevers.
***D. convexa*** De Willd. (Iteno, Likolo). Congo. The leaves are used locally to heal wounds.
***D. drakena*** L. Mexico. An infusion of the leaves is used locally to relieve a "hangover" from excess alcohol.
***D. klainei*** Pierre. (Molondo, Myemo, Olando). Tropical Africa. The roots are scented, and are used especially in the Gabon as a perfume, salve and gargle. They are also used in ancestor worship.
DORYALIS E. Mey. corr. Warb. 30 spp. Flacourtiaceae. 30 spp. Africa, Ceylon. Bushy shrubs. The fruits of the species mentioned are eaten locally either raw or in various preserves. They are pleasantly flavoured, tart or subacid.
***D. abyssinica*** (A. Eich.) Warb.=Aberia abyssinica. Ethiopia. Fruits orange-yellow.
***D. caffra*** (Hook. f. and Harv.) Warb.= Aberia caffra. (Kei Apple, Umkokolo). S.Africa. Introduced to New World tropics. Fruits green-yellow.
***D. hebecarpa*** (Gardn.) Warb.=Aberia gardneri. (Ceylon Gooseberry, Ketembilla). Tropical Asia, especially India and Ceylon. Introduced throughout Old and New World tropics. Cultivated. Fruits dark purple.
DORYPHORA Endl. Atherospermataceae. 1 sp. New South Wales. Tree.
***D. sassafras*** Endl. (Sassafras). The fragrant wood is used locally for house-building, furniture and insect-proof boxes etc. An essential oil containing saffrole is distilled from the bark and leaves on a small scale. It is used in perfumery.
Dotted Saxifrage – Saxifraga puctata.
Double Claw – Martynia probicidia.
Double Coconut – Lodoicea maldivica.
Douglas Fir – Pseudotsuga menziesii.
Douglas Mulefoot Fern – Marattia douglasii.
Douglas Knotweed – Polygonum douglasii.
Douglas Spruce – Pseudotsuga meniesii.
Doum Palm, Egyptian – Hyphaena thebaica.
Doum Palm, South African – Hyphaene crinita.
Doumori Butter – Tieghemella heckeli.
Doundaké – Sarcocephalus esculentus.
Douo – Elaeophorbia drupifera.
***Dovyalis*** E. May.=Doryalis E. Mey. corr. Warb.
DOXANTHA Miers. Bignoniaceae. 1 sp. Tropical America. Vine.
***D. chica*** Plumb. and Bonpl. The flowers and leaves are the source of a red dye, used by the Indians for colouring their bodies.
Downward Plum – Bumelia spiniflora.
Downy Grape – Vitis cinerea.
Downy Ground Cherry – Physalis peruviana.

Downy Ground Cherry – Physalis pubescens.

Downy Rattlesnake Plantain – Goodyera pubescens.

Downy Rose Myrtle – Rhodomyrtus tomentosa.

Dowora kome – Intsia plurijuga.

Dowren hantos – Coelostegia griffithii.

DRACAENA Vand. ex L. Agavaceae. 150 spp. Warm Old World. Trees.

***D. aurea*** H. Mann. = Pleomele aurea (H. Mann.) N. E. Brown. (Halapele). Hawaii. The wood was used locally for carving idols.

***D. cinnabari*** Balf. (Socotra Dragon's Blood Dracaena). Socotra. A rich red resin (Dragon's Blood, Socotra Dragon's Blood) is collected from the stems. It is used to make varnishes, and medicinally as an astringent to stop bleeding.

***D. congesta*** Ridl. Malaysia. More herbaceous than the other species. A decoction of the roots is used locally to treat rheumatism and to combat intestinal worms.

***D. draco*** L. (Dragon Dracaena, Dragon Blood Tree). Canary Islands. A red resin (Dragon's Blood, Sanguis Draconis, Sang de Dragon) is extracted from the stem and used in varnishes, for pigmenting paper, and for pharmaceutical plasters.

***D. mannii*** Baker. (Asparagus Bush). Tropical Africa. The young shoots are eaten locally as a vegetable by both natives and Europeans. The chopped leaves are eaten cooked with rice by the natives. The wood is the source of a light red dye used locally in S.Nigeria.

***D. schizantha*** Baker. S.Arabia through E.Africa. A red resin (Arabian Dragon's Blood, Socotra Dragon's Blood) is extracted from the stems. It is used for varnishes etc.

***Dracocephalum ibericum*** Bieb. = Lallemantia iberica.

DRACONTIUM L. Araceae. 13 spp. Mexico to tropical S.America. Herbs.

***D. asperum*** C. Koch. (Jacaraca Taia). N.Brazil. The large tubers are used locally to treat snake-bite, and when ground are used to alleviate asthma.

***D. polyphyllum*** L. (Jacaraca Merim). Guyana to N.Brazil. The tubers are eaten by the natives of the Amazon Basin.

DRACONTOMELON Blume. Anacardiaceae. 8 spp. Malaysia to Fiji. Trees.

***D. mangiferum*** Blume. Malaysia into India. The fruits are used in Malaysia to flavour fish, and the flowers are used as a condiment.

DRACUNULUS Mill. Araceae. 2 spp. Mediterranean. Herbs.

***D. vulgaris*** Schott. = Arum dranunculus = A. guttatum. S.Europe, Mediterranean. The burning rhizomes are used locally as a fumigant for parasites on livestock.

Dragon's Blood – Daemonorops draco.

Dragon Blood – Otoba gordoniifolia.

Dragon Blood – Pterocarpus draco.

Dragon's Blood, Arabian – Dracaena schzantha.

Dragon's Blood, Paduak – Pterocarpus draco.

Dragon's Blood Palm – Daemonorops draco.

Dragon Blood Resin – Dracaena cinnabari.

Dragon Blood Resin – Dracaena draco.

Dragon Blood, Socotra – Dracaena cinnabari.

Dragon Blood, Socotra – Dracaena draco.

Dragon Blood Tree – Dracaena draco.

Dragon Blood, West Indian – Pterocarpus draco.

Dragon's Claw – Corallorhiza odontorhiza.

Dragon Dracaena – Dracaena draco.

Dragon Spruce – Picea asperata.

Drakenberg Silky Grass – Pennisetum unisetum.

Dregel – Dillenia excelsa.

***Drepanocarpus isadelphus*** E. Mey. = Machaerium angustifolium.

Drieblaar – Teucrium capense.

DRIMIA Jacq. Liliaceae. 45 spp. Tropical and S.Africa. Herbs.

***D. ciliaris*** Jacq. = Idothea ciliaris. S.Africa. The juice from the bulbs is used locally as an expectorant, emetic and diuretic.

***D. cowanni*** Ridl. Madagascar. The bulbs are used locally as a rat poison.

DRIMYS J. R. and G. Forst. Winteraceae. 70 spp. Borneo, New Guinea, New Caledonia, Australia, Tasmania, New Zealand, S.America. Trees.

***D. aromatica*** F. v. Meull. Australia. The fruits are used locally as a spice.

***D. axillaris*** Forst. New Zealand. The red wood is used for inlay work. A decoction of the aromatic bark is used locally as a tonic and astringent.

***D. winteri.*** Forst. (Wintersbark Drimys).

Mexico to Argentina and Chile. The bark (Winter's Bark) is aromatic and is used in home remedies. It is tonic and antiscorbutic. Locally it is used to treat stomach complaints and dysentery; when powdered it is used as a condiment. The wood is used locally for house interiors and packing cases.

Droga de Jara – Cistus ladaniferus.

Drooping Leucothoë – Leucothoë catesbaei.

Dropseed, Mesa – Sporobolus flexuosus.

Dropseed, Sand – Sporobolus cryptandrus.

Dropwort – Filipendula hexapetala.

Dropwort, Water – Oenanthe stolonifera.

DROSERA L. Droseraceae. 100 spp. Tropical and temperate (except Australia and New Zealand). Herbs.

***D. angelica*** Huds. See D. rotundifolia.

***D. burmanni*** Wahl. (Makha-jali). India. Used locally to cause warming and reddening of the skin.

***D. indica*** L. See D. burmanni.

***D. lunata*** Buch.-Ham. = D. peltata.

***D. longifolia*** L. See D. rotundifolia.

***D. madagascariensis*** DC. = D. ramentacea.

***D. peltata*** Sm. = D. lunata. (Chitra, Mukjajali). Hills of N.India, Japan, Philippines. In India the leaves (sometimes mixed with salt) are used to cause blistering. They are also a constituent of Gold Bhasma used to cure syphilis, to improve the appetite and as a tonic.

***D. ramentacea*** Burch. = D. madagascariensis. Madagascar. The leaves are used locally to rub on the teeth to preserve them. It is also used to treat indigestion and coughs.

***D. rotundifolia*** L. (Sundew). Temperate Europe, Asia and N.America. The dried leaves contain a proteolytic enzyme and are used in the treatment of whooping cough and bronchitis as an antispasmodic.

Drug Centaurium – Centaurium umbellatum.

Drug Eyebright – Euphrasia officinalis.

Drug Lionsear – Leonotis leonurus.

Drug Solomon's Seal – Polygonatum officinale.

Drug Speedwell – Veronica officinalis.

Druggist's Bark – Cinchona succirbra.

Drugsquill, Indian – Urginea indica.

Drummond's Ironweed – Vernonia missurica.

Drusha – Callicarpa cana.

DRYAS L. Rosaceae. 2 spp. Arctic and Alpine. Small shrubs.

***D. octopetala*** L. Arctic and alpine Europe, Asia and N.America. The leaves are used as a tea substitute in the Alps (Kaisertee, Schweizertee).

DRYMOGLOSSUM Presl. Polypodiaceae. 6 spp. Madagascar, tropical Asia, Malaysia. Ferns.

***D. carnosum*** Hook. Nepal, N.India, S.China. A decoction of the frond is used by the Chinese to treat rheumatism and urinary complaints.

***D. heterophyllum*** (L.) Presl. India, Malaysia to Indo-China, Philippines and Australia. The base leaves are used as a cure of eczema and to retard the bleeding from wounds.

***Drymophloeus puniceus*** Becc. = Pinanga punicea.

DRYNARIA (Bory.) J. Sm. Polypodiaceae. 20 spp. Old World tropics. Ferns.

***D. fortunei*** (Kuntz.) S. Sm. China, Formosa. The Chinese use a decoction of the rhizome to treat rheumatism.

***D. quercifolia*** (L.) J. Sm. Malaya, Philippines. A decoction of the rhizome is used as an astringent.

***D. rigidula*** Bedd. Malaya. A decoction of the rhizome is used locally to treat gonorrhoea and dysentery. The young fronds are eaten as a vegetable in the Celebes.

DRYOBALANOPS Gaertn. f. Dipterocarpaceae. 9 spp. Sumatra, Borneo, Malay Peninsula. Trees.

***D. aromatica*** Gaertn. = D. campnhora. (Borneo Camphor Tree). Malaysia. The coarse-grained wood is used locally for general construction work. The wood yields a balsam similar to camphor, from which bornerol is made.

***D. camphora*** Colebr. = D. armoatica.

***D. oblongifolia*** Dyer. (Keladan, Petanang). Malaysia, Indonesia. The wood is used for house-building. The fruits are eaten locally, and the seeds are eaten with rice.

***D. oiocarpa*** v. Sl. Borneo. The wood is resistant to insect attack and is used locally for house-building, ship-building, posts etc.

***D. rappa*** Becc. (Kapur paya). Borneo, Swamps. The wood (Kapur), is very heavy and durable, and is used locally for house-building. The bark is used to make the walls of houses.

DRYOPTERIS Adans. Aspidiaceae. 150 spp. Cosmopolitan. Ferns.

***D. anthelmintica*** Kuntze. The leaves are used by the natives of S.Africa to treat intestinal worms.

***D. barbigera*** (Moore) Kuntze. Himalayan area. The rhizomes are used locally to treat intestinal worms.

***D. blandfordii*** (Hope) C. Chr. Himalayas. The rhizome is used locally to treat intestinal worms.

***D. filix-mas*** (L.) Schott. = Aspidium felix-mas. (Male Fern, Male Shield Fern). The dried rhizome is used medicinally to combat tapeworms.

***D. heterocarpa*** (Blume) Kuntze. Tropical Asia. In Malaya the fronds are rubbed on the skin to treat leucodermia. They are eaten as a vegetable in Penang.

***D. odontoloma*** (Moore) C. Chr. Himalayas. The rhizomes are used locally to treat intestinal worms.

***D. paleacea*** (Sw.) C. Chr. Tropical America. In Colombia the rhizome is used to treat intestinal worms.

***D. pteroides*** Kuntze. Philippines. The stems are used locally to decorate baskets.

***D. schimperiana*** (Hochst.) C. Chr. Himalayas. The rhizomes are used locally to treat intestinal worms.

***D. spinulosum*** (O. F. Muell.) Sw. Temperate Europe, Asia and N.America. The rhizome (Rhizoma Filicis Maris) is used medicinally to treat intestinal worms. They are also eaten cooked on hot stones by the Eskimos.

DRYPETES Vahl. Euphorbiaceae. 200 spp. Tropics, S.Africa, E.Asia. Trees.

***D. bisacata*** Gagnep. = Putranjiva roxburghii. India, Indochina. A decoction of the leaves and fruits is used locally in India to treat liver complaints and fevers and in Thailand to treat rheumatism. The leaves are also used to feed livestock.

***D. longifolia*** Pax. and Hoffm. = Cyclostemon longifolius. Indonesia. The wood is fairly hard and durable. It is used locally for general construction work.

***D. ovalis*** Pax. and Hoffm. = Hemicycla ovalis. (Melamon, Mentaos). Malaysia. The wood is yellow, very hard and heavy, but easy to work. It is used locally for house construction. The older wood darkens and is much used for making walking sticks.

***D. simaluresis*** J. J. Sm. (Lebool fatooh). Indonesia. The heavy wood is immune to attack by insects. It is used for house-building and general construction work.

***D. subsymmetrica*** J. J. Sm. See D. simaluresis.

DUABANGA Buch.-Ham. Sonneratiaceae. 3 spp. Indomalaya. Trees.

***D. moluccana*** Blume. Malaysia. The wood is strong, though light. It is used locally for boards. The bark is ground with that of Mallotus moluccana to make a black dye used locally to stain baskets.

***D. sonneratioides*** Buch.-Ham. India. The soft yellowish wood is used to make boats and canoes and tea boxes.

Duahi – Wrightia tinctoria.

Duahi – Wrightia tomentosa.

DUBOISIA R. Br. Solanaceae. 2 spp. Australia, New Caledonia. Woody shrubs.

***D. hopwoodii*** F. v. Muell. Australia. The leaves and small twigs are chewed by the natives as a stimulant.

***D. myoporoides*** R. Br. (Corkwood, Mgmeo). Australia. The plant contains hyoscine and hyoscyamine of which it is a potential commercial source. It is used locally to stupify fish.

DUCHESNEA Smith. Rosaceae. 6 spp. India, E.Asia. Herbs.

***D. filipendula*** (Hemsl.) Focke. = Fragaria filipendula. China. The fruits (She-p'ao-tzu) are like strawberries and are eaten in China. The plant is cultivated locally.

Duck Acorn – Nelumbo pentapetala.

Duck Potato – Sagittaria latifolia.

Duckweed – Lemna minor.

Duckweed, Tropical – Pistia stratiotes.

Dudhali – Eryngium coeruleum.

Dudoa – Hydnocarpus alcalae.

Duffin Bean – Phaseolus lunatus.

Duggal Fibre – Sarcochlamys pulcherrima.

DUGUETIA A. St.-Hil. Annonaceae. 70 spp. Tropical America, W.Indies. Trees.

***D. quitarensis*** Benth. W.Indies. (Cuban Lancewood, Jamaican Lancewood). The elastic wood is used for whip handles, fishing rods, cart shafts etc.

***D. vallicola*** MacBride. (Yaya). Colombia. The wood is used locally for making tool handles.

Dugulu – Radlkofera calodendron.

Duhnual Balsam – Commiphora opobalsamum.

***Dulgonia laticuspis*** Turcz. = Phyllonoma laticuspis.

Dulce, Pepper – Laurencia pinnatifida.

Dulse – Rhodymenia palmata.

Dumb Plant – Dieffenbachia seguina.

***Dumoria*** A. Chev.=Tieghemella.
Dundee Fibre – Sesbania aculeata.
Dûng dinh – Caryota sympetala.
Dunkal Tree – Commiphora hildebrandtii.
Durban Grass – Dactyloctenium australe.
Durian – Durio zibethinus.
Durian, Wild – Cullenia excelsa.
DURIO Adans. Bombacaceae. 27 spp. Burma to W.Malaysia. Trees.
***D. carinatus*** Mast. (Dorian booroong, Dorio papa). Malaysia. The bark is used locally for roofing. The seeds are eaten.
***D. griffithii.*** Bakh. var. ***heteropyxis*** Bakh. (Daoon dorian). E.Malaysia. The wood is fairly durable and is used locally for house construction.
***D. zibethinus*** Murr. (Durian). Malaysia, Philippines. Cultivated locally for the fruits. Introduced into New World tropics. The fruits taste resinous and sweet, with a cheesey aftertaste. They smell foul and find little favour outside their natural habitat where they are esteemed. The unripe fruits are eaten as a vegetable, or when ripe they are fermented and used as a sauce. The seeds are eaten dried, roasted, in sugar or fried in coconut oil.
Durmast Oak – Quercus petraea.
Durra – Sorghum vulgare.
Durum Wheat – Triticum durum.
DUVILLEA Cham. Fucaceae. Antarctica. Very large Brown Algae.
***D. anartica*** (Cham.) Hariot=D. utilis. (Bull Kelp) Antarctic. Eaten as a vegetable in New Zealand and Australia. It is also used as a manure and by the Maoris for packing mutton birds. The agar contains an effective vermifuge.
***D. utilis*** Bory.=D. anartica.
Dusty Miller – Senecio cineraria.
Dutch Tonka – Coumarouna odorata.
Dutchman's Breeches – Dicentra cucullaria.
***Duvaua dependens*** DC.=Schinus dependens.
Dwarf Almond – Amygdalus nana.
Dwarf Banana – Musa cavendishii.
Dwarf Cape Gooseberry – Physalis peruviana.
Dwarf Cape Gooseberry – Physalis pubescens.
Dwarf Elder – Aegopodium podagraria.
Dwarf Elder – Sambucus ebulus.
Dwarf Huckleberry – Gaylussacia dumosa.
Dwarf Ironwood – Lophira alata.
Dwarf Milkweed – Asclepias involucrata.
Dwarf Pine Needle Oil – Pinus pumilio.
Dwarf Plantain – Plantago virginica.
Dwarf Sumach – Rhus copallina.
***Dyckia glaziovii*** Baker=Neoglaziovia variegata.
Dye Bedstraw – Galium tinctorium.
Dye, Camwood – Pterocarpus erinaceus.
Dye Fig – Ficus tinctoria.
Dye yam – Dioscoria rhipogonoides.
Dyer's Chamomille – Anthemis tinctoria.
Dyer's Greenwood – Genista tinctoria.
Dyer's Indian Mulberry – Morina tinctoria.
Dyer's Weed – Reseda lutea.
Dyer's Weld – Reseda lutea.
Dyer's Woodruff – Galium tinctorium.
DYERA Hook. f. Apocynaceae. 2–3 spp. W.Malaysia. Trees. The species mentioned are grown in Malaysia for the gum juletong which is used in the manufacture of chewing gum. It is obtained by tapping. It is coagulated in air, or more usually by using acids. The latex is plastic when warm but hardens on cooling. Only a few million pounds are produced annually, and it is usually mixed with other gums to make chewing gum. It is also used as an adulterant of other latexes, including rubber, and in the manufacture of linoleum, asbestos, and cellulose.
***D. costulata*** (Miq.) Hook. f. The main source of jelutong, (Djelutong, Dead Borneo, Pontianak Gutta).
***D. laxiflora*** Hook. f.
***D. lowii*** Hook.
DYSOPHYLLA Blume Labiatae. 25 spp. Temperate Asia, Australia. Herbs.
***D. auricularia*** (L.) Blume.=Mentha auriculata=M. foetida. Asia. A paste of the plant with lime is used in Malaya to relieve stomach upsets in children, by applying it externally.
***Dysoxylon*** Bartl.=Dyslxylum.
DYSOXYLUM Blume. Meliaceae. 200 spp. Indomalaysia. Trees.
***D. acutangulum*** Miw. Sumatra. The wood is strong and beautifully grained. It is used for furniture, building and making coffins.
***D. amooroides*** Miq. (Ki tai Ketoodjeuh, Mamalapa). Java. As the wood is not durable, it is used for boxes, and small articles like toothpicks.
***D. decandrum*** (Blanco) Merr. Philippines. The bark is used locally as an emetic.
***D. densiflorum*** Miq.=Epicharis altissima=E. densiflora. (Kapinango, Kheuroch,

Kraminan, Maranginon, Toombawa Wood). Indonesia. The wood is a beautiful pinkish-brown and scented. It is used for general construction work, housebuilding, furniture, wagon wheels.

***D. euphlebium*** Merr. Amboina. The whole plant smells of onions. The whitish wood is used in house construction, and the leaves are eaten locally with fish.

***D. fraseranum*** Benth. (Australian Mahogany, Pencil Cedar, Roho Mahogany, Rosewood). Australia. The reddish, fragrant wood is used for carving, furniture-making, turning and shipbuilding.

***D. glandulosum*** Talb. = D. malabaricum.

***D. lessertianum*** Benth. See D. fraseranum.

***D. loureiri*** Pierre. = Epicharis loureiri. (Chan sa, Huynh duong, Shantal citrin). S.Vietnam, Cambodia. The scented wood is used locally for making coffins and joss sticks.

***D. malabaricum*** Bedd. = D. glandulosum (Agil, Bili devadari, White Cedar). C India. The light brown-grey wood is sweet scented. It is resistant to insect attack and has a pleasant sheen. Its uses include the manufacture of furniture, carriages, panelling, brake blocks, oil casks for coconut oil, and general construction work.

***D. muelleri*** Benth. (Pencil Wood, Turnip Wood). Australia. The dark red wood is used for cabinet work.

***D. spectabile*** Hook. f. New Zealand. The light-red wood is used for cabinet work. A tea made from the leaves is used as a tonic by the natives.

# E

Eagle Fern – Pteridium aquilinum.
Eagle Wood – Aquilaria agallocha.
Early Blue Violet – Viola palmata.
Early Sweet Blueberry – Vaccinium pennsylvanicum.
Earth Almond (Oil) – Cyperus esculentus.
Earth Chestnut – Lathyrus tuberosus.
Earthnut – Carum bulbocastanum.
East African Cardamon – Aframomum malum.
East African Gum Arabic – Acacia doratoxylon.
East African Sandalwood (Oil) – Osyris tenuifolia.
East Indian Arrowroot – Curcuma angustifolia.
East Indian Arrowroot – Curcuma leucorhiza.
East Indian Arrowroot – Tacca pinnatifida.
East Indian Copal – Canarium bengalense.
East Indian Dill – Peucedanum graveolens.
East Indian Kino – Pterocarpus marsupium.
East Indian Lemon Grass – Cymbopogon flexuosus.
East Indian Rhubarb – Rheum palmatum.
East Indian Rosebay – Ervatamia coronaria.
East Indian Rosewood – Dalbergia latifolia.
East Indian Screw Tree – Helicteres isora.
East Indian Walnut – Albizia lebbek.
East Prussian Pea – Pisum arvense.
Easter Lichen – Stereocaulon paschale.
Easter Lily – Lilium longiflorum.
Eastern Arbor-Vitae (Oil) – Thuja occidentalis.
Eastern Coral Bean – Erythrina herbacea.
Eastern Hemlock – Tsuga canadensis.
Eastern Larch – Larix laricina.
Eastern Red Cedar – Juniperus virginiana.
Eastern White Pine – Pinus strobus.
Eau de Créole – Mammea americana.
Ebène Vert – Rhodocolea racemosa.
Ebeno – Pennisetum mollissimum.
***Eberhardtia aurata*** Lecomte = Godoya aurata.
***Eberhardtia tonkinensis*** Lecomte = Godoya aurata.
Eb nembe – Maerua rigida.
Ebony – Brya ebenus.
Ebony – Diospyros ebenum.
Ebony – Diospyros microphyllum.
Ebony – Diospyros reticulata.
Ebony, American – Brya ebenus.
Ebony, Benin – Diospyros crassiflora.
Ebony, Benin – Diospyros suaveolens.
Ebony, Black – Diospyros dendo.
Ebony, Blue – Copaifera bracteata.
Ebony, Bolongeta – Diospyros pilosanthera.
Ebony, Bombay – Diospyros gardneri.
Ebony, Bombay – Diospyros montana.
Ebony, Calabar – Diospyros dendo.
Ebony, Camagoon – Diospyros pilosantha.
Ebony, Camogan – Diospyros discolor.
Ebony, Cape – Euclea pseudebenus.
Ebony Cocus Wood – Brya ebenus.

Ebony, Coromandel – Diospyros hirsuta.
Ebony, Coromandel – Diospyros melanoxylon.
Ebony, Gabon – Diospyros dendo.
Ebony, Gabon – Diospyros flavescens.
Ebony, German – Taxus baccata.
Ebony, Green – Rhodocolea racemosa.
Ebony, Green – Tecoma leucoxylon.
Ebony, Greenheart – Tecoma leucoxylon.
Ebony, Jamaican – Brya ebenus.
Ebony, Macassar – Diospyros utilis.
Ebony, Madagascar – Diospyros haplostylis.
Ebony, Mauritius – Diospyros tesselaria.
Ebony, Mozambique – Dalbergia melanoxylon.
Ebony, North American – Diospyros virginiana.
Ebony, Orange River – Euclea pseudebenus.
Ebony, Red Mauritius – Diospyros rubra.
Ebony, Senegal – Dalbergia melanoxylon.
Ebony, South Cameroon – Diospyros bipindensis.
Ebony, Unyoro – Dalbergia melanoxylon.
Ebony, West African – Diospyros mespiliformis.
Ebony, White – Diospyros malacapii.
Ebony, Yoruba – Diospyros mombuttensis.
Ebony, Zanzibar – Diospyros mespiliformis.

ECBALLIUM A. Rich. Cucurbitaceae. 1 sp. Mediterranean. Vine.

***E. elaterium*** A. Rich. (Squirting Cucumber). The juice of the unripe fruits is used medicinally as a purgative. Cultivated.

ECCILEA (Fr.) Quél. Agaricaceae. 35 spp. Cosmopolitan.

***E. clypeatus*** (L.) Schroet. The fruitbodies are eaten in Europe and Asia.

ECCLINUSA Mart. Sapotaceae. 20 spp. Tropical America, W.Indies. Trees. An inferior balata is tapped from the stems of the following species.

***E. balata*** Ducke. Brazil. The balata, called Abiurana, is exported.

***E. sanguinolenta*** Pierre=Ragala sanguinolenta. Guyana, Amazon. Balata called Balata rouge, Balata saignant, Wapo.

ECDYSANTHERA Hook. and Arn. Apocynaceae. 15 spp. China to Indo-Malaysia. Trees. The following species are sources of excellent rubber, which is tapped from the trunks. They have not been exploited fully commercially.

***E. barbata*** Miq.=Parameria barbata.

***E. godefroyana*** Pierre. Cochinchina.

***E. micrantha*** DC. Burma, Himalayas.

***E. quintareti*** Pierre. Laos.

***E. tournieri*** Pierre. Indochina.

ECHINACEA Moench. Compositae. 3 spp. Atlantic N.America. Herbs.

***E. angustifolia*** DC.=Brauneria angustifolia (Rattlesnake Weed). E. N.America. The ground roots are chewed by the local Indians. This causes an excessive secretion of saliva, which relieves sore throats.

***E. pallida*** Nutt.=Brauneria pallida. (Purple Cone Flower). E. N.America. The dried rhizome is used medicinally to stimulate the appetite and to induce sweating.

ECHINOCACTUS Link. and Otto. Cactaceae. 10 spp. S.United States of America to Mexico. Cacti.

***E. equitans*** Scheidw.=E. horizonthalonius.

***E. hamatocanthus*** Muehlenpf.=Ferocactus hamatocanthus.

***E. horizonthalonius*** Lem.=E. equitans. S.W. U.S.A. The pulp from the centre of the plants is made into confectionery.

***E. williamsii*** Lem.=Lophophora williamsii.

***E. wislizeni*** Engelm.=Ferocactus wislizeni.

ECHINOCEREUS Engelm. Cactaceae. 75 spp. S. United States of America, Mexico. Shrubby cacti. The fruits of all the species mentioned are pleasantly flavoured and are eaten locally in Mexico, raw or as preserves.

***E. conglomeratus*** Forst.=Cereus conglomeratus. (Pitahaya de Agosto).

***E. dasyacanthus*** Engelm.=Cereus dasyacanthus.

***E. enneacanthus*** Engelm.=Cereus enneacanthus. (Strawberry Cactus).

***E. gonocanthus*** Engelm.

***E. stramineus*** (Englm.) Ruempf. (Mexican Strawberry, Pitahaya).

***E. triglochidiatus*** Englm.

ECHINOCHLOA Beauv. Graminae. 30 spp. Warm. Grasses.

***E. colona*** (L.) Link. (Jungle Rice, Shama Millet). A weed species over much of Africa. Cultivated in Old and New World as a fodder for livestock.

***E. frumentacea*** (Roxb.) Link.=Panicum frumentacea. (Japanese Millet, Japanese Barnyard Millet, Sanwa Millet, Billion Dollar Grass). E.Asia. Possibly a polyploid derived from E. crus-galli. Cultivated in U.S.A. as a fodder grass. It is grown in the Far East as a cereal, especially in areas where rice will not grow.

***E. pyramidalis*** Hitche and Chase. (Hoodoo

wendoo, Farka teli, Sil, Samamgate). A swamp species of the Niger and Nile valleys. An excellent fodder species. The seeds are used to make a flour (Kreb) in various parts especially Senegal.

***E. stagnina*** P. Beauv. A swamp species of the Niger and Nile valleys. An excellent fodder species. A sugar extracted from the stem is used to make confectionery by the natives of C.Africa, and a liqueur by the Mohammedans in W.Africa.

ECHINOCYSTIS Torr. and Grey. Cucurbitaceae. 15 spp. America. Vines.

***E. lobata*** (Michx.) Torr. and Gray. (Balsam Apple). E. N.America to Texas. The ground roots are used by the local Indians as a poultice for headaches. A decoction of the roots is used as a tonic and as an aphrodisiac.

***E. macrocarpa*** Nutt. California. The roots are used as a purgative. The seeds yield a red dye used by the local Indians.

ECHINODONTIUM Ellis and Everh. Hydnaceae.

***E. tinctorium*** (Ellis and W.) Ellis and Everh. N.W. N.America. The fruit-bodies yield a dye used by the local Indians.

ECHINOPS L. Compositae. 100 spp. E.Europe, Asia and Africa. Herbs.

***E. dahuricus*** Fisch. China. The leaves are used locally as a demulcent.

***E. viscosus*** DC. Mediterranean. A gum (Angadeo Mastiche) extracted from the plant is used in Greece as a chewing gum.

***Echites hypoleuca*** Benth. = Macrosiphonia hypoleuca.

***Echites suaveolens*** Mart. and Gel. = Macrosiphonia hypoleuca.

ECHIUM L. Boraginaceae. 40 spp. Canaries, Azores, N. and S. Africa, Europe, W. Asia. Herbs.

***E. vulgare*** L. (Bugloss, Viper's Bugloss). Europe to Siberia, Naturalized in N.America. An extract of the leaves is used as a demulcent and expectorant. It is also used to reduce fevers and to treat nerve complaints.

***Eckebergia*** Batsch. = Ekebergia.

ECKLONIA Kjellm. Laminariaceae. Marine. Brown Algae. Pacific. Members of the genus are among the prime producers of alginates in Japan.

***E. bicyclis*** Kjellm. = Capea elongata. Japan. Eaten locally and used in the production of iodine.

***E. cava*** Kjellm. Japan. Used in the production of iodine and potash, and as manure.

***E. kurome*** Okam. Japan and China. Eaten locally.

***E. latifolia*** Kjellm. Japan and China. Eaten locally.

***E. radicosa*** Okam. = Laminaria radicosa. Japan. Eaten locally.

ECLIPTA L. Compositae. 3–4 spp. Warm America, Africa, Asia, Australia. Herbs.

***E. alba*** Hassk. = E. erecta.

***E. erecta*** L. Tropics. The leaves are used in India as a laxative. A decoction of the plant yields a blue-black dye. This is used in India, often mixed with coconut oil to dye the hair. It is also used locally for tattooing. A decoction of the leaves is used in India to encourage the growth of hair on new-born children.

Ecuador Walnut – Juglans honorei.

Eddo – Colocasia antiquorum.

EDGEWORTHIA Meissn. Thymelaeaceae. 3 spp. Himalayas to Japan. Shrubs.

***E. chrysantha*** Matzum. = E. papyrifera.

***E. garneri*** Meissn = E. papyrifera.

***E. papyrifera*** Sieb. and Zucc. = E. chrysantha = E. gardneri = E. tomentosa = Daphne papyrifera. Burma, Himalayas, China to Japan. Cultivated for the bark fibres which are used, particularly in Japan to make a hand-made paper (Nepal Paper, Mitsumata Paper). The fibres are extracted from the twigs which are harvested every other year.

Edible Snake Gourd – Trichosanthes anguina.

Edible Tulip – Tulipa edulis.

***Edwardsia chrysophylla*** Salisb. = Sophora chrysophylla.

Eel Grass – Vallisneria americana.

Efiri – Triclisia gelletii.

Efufuko – Beilschmeidia insularum.

Efufuko – Beilschmiedia leemansii.

Efurungu – Fagara rubescens.

Efwatakala Grass (Oil) – Melinis minutiflora.

Egg Fruit – Lucuma salicifolia.

Egg Plant – Solanum melongena.

Eglantine Rose – Rosa eglanteria.

Egyptian Barley – Hordeum trifurcatum.

Egyptian Clover – Trifolium alexandrinum.

Egyptian Doum Palm – Hyphaena thebaica.

Egyptian Lettuce Seed Oil – Lactuca scariola.

Egyptian Lupin – Lupinus termis.

Egyptian Pistache – Pistacia khinjuk.
Eheng – Cassia timoriensis.
EHRETIA P. Br. Ehretiaceae. 50 spp. Warm. Mostly Old World. Shrubs or small trees.
***E. acuminata*** R. Br. Korea, China, Japan to Australia. The wood is used by the Chinese for carrying poles.
***E. beurreria*** Blanco. = E. philippensis.
***E. buxifolia*** Roxb. = E. microphylla. Philipines. The leaves are used locally as a tea substitute.
***E. elliptica*** DC. (Kanckaway, Sugarberry). W.Texas to Mexico. The wood is used for wheel spokes and axles, yokes and tool-handles.
***E. microphylla*** Lam. = E. buxifolia.
***E. philippensis*** A. DC. = E. beurreria. (Alibungog). Philippines. The leaves are used locally in the cure of diarrhoea, especially when accompanied by the voiding of blood.
EHRHARTA Thunb. Graminae. 27 spp. S.Africa, Mascarene Islands, New Zealand. Grasses.
***E. calycina*** Eck. ex Nees. (Veldtgrass). Cape of Good Hope. Planted in S. U.S.A. to prevent movement of sand dunes.
Ehungai – Parinarium glabra.
Eichel Kaffee – Quercus robur.
Eichenmoss – Evernia prunastri.
EICHHORNIA Kunth. Pontederiaceae. 7 spp. S.E.United States of America to Argentina, W.Indies. Floating water plants.
***E. crassipes*** Solms. = Pairopus crassipes. (Water Hyacinth). Throughout distribution. A possible food for cattle, but of little nutritive value; a possible source of cellulose. The plant can be a serious weed in waterways.
Einkorn Wheat – Triticum monococcum.
Ekanda Rubber – Raphionacme utilis.
Ekdania-Bridelia retusa.
EKEBERGIA Sparrm. Meilaceae. 15 spp. S. and tropical Africa, Madagascar. Trees.
***E. capensis*** Sparrm. (Cape Ash). S.Africa. The wood which is white, soft, and not durable is used for boards etc., furniture, and building waggons.
Eko – Beilschmiedia leemansii.
Ekongo – Sterculia oblonga.
Eksalita – Cissus cyphopetala.
ELAEAGNUS L. Elaeagnaceae. 45 spp. Europe, Asia, N.America. Shrubs.
***E. argentea*** Pursh. (Silverberry). Canada to Utah and S.Dakota. The fruits are eaten by the local Indians.
***E. latifolia*** L. Tropical Asia. The fruits are eaten by the peoples of Nepal and Hindustan.
***E. multiflora*** Thunb. (Cherry Elaeagnus). Japan. The fruits are eaten locally as preserves and made into an alcoholic beverage.
***E. philippensis*** Perk. Philippines. The acid-sweet fruits are used locally to make jellies.
***E. umbellata*** Thunb. Japan. The scalded fruits are eaten locally.
ELAEIS Jacq. Palmae. 1 sp. Tropical America; 1 sp. tropical Africa. Tall Palms.
***E. guineensis*** Jacq. (Oil Palm Dendé). Tropical W.Africa. This plant occurs wild in the areas between the tropical rain forest and the savannahs. The fruits have served as a local source of oil for centuries, and are still used for cooking etc. The oil is a valuable source of vitamin A. The oil was first extracted from the fruits of natural stands, and exports began from W.Africa in 1842. In 1848 the plant was introduced to the Dutch East Indies, but plantation growing here, and in W.Africa began in the early 20th century. The world production is now over 1 million tons a year, mainly from W.Africa and the Congo region, but with appreciable amounts from Indonesia and Malaysia. There are two types of palm oil, the oil from the mesocarp (Palm oil) and the oil from the kernel of the seed (Palm kernel oil). The palm oil is extracted by pressure, and the mesocarp may contain 50–65 per cent oil. This is usually done in the producing country, and the oil is exported for use in making soap, and for protecting cleaned iron surfaces before tin-plating. Some palm oil is used in making margarine and as a lubricant. It has S.G. 0·92, Sap. Val. 196–205, Iod. No. 48–60, Unsap. 0·2–0·5 per cent. The palm kernels contain 43–53 per cent oil which is usually extracted in the importing country and is used mainly in the manufacture of margarine, and soap. The residue is a valuable stock feed. The oil has S.G. 0·87, Sap. Val. 244–255, Iod. No. 16–23. The plants begin bearing 4–6 years after germination. A mature tree bears 2–6 bunches of fruit in a season, each bunch weighing from 10–35 lb. This may give yields of oil up to 2 tons per acre annually. The plants are usually

propagated from seeds, but germination is erratic and the seeds need treating at about 35°C to induce them to germinate. The seedlings are planted out in pots before finally planting in the field. The whole process from sowing the seed to planting in the field may take up to 18 months. There are four main varieties, distinguished by the thickness of the endocarp (Macrocarpa, Dura, Tenera, and Pisifera). Macrocarpa has the thickest endocarp, grading to Pisifera with no visible endocarp. Most of the cultivated palms belong to the first two varieties.

***E. melanocarpa*** Gaertn. = Corozo oleifera.

ELAEOCARPUS L. Elaeocarpaceae. 200 spp. E.Asia, Indomalaysia, Australia, Pacific Islands. Small trees or shrubs.

***E. bancroftii*** F. v. Muell. Queensland. The seeds are eaten locally by the natives.

***E. bifida*** Hook. and Arn. Hawaii. The bark is the source of a cordage fibre, used locally.

***E. calomala*** (Blanco) Merr. (Kumakun). Philippines. The fruits are the size and colour of a purple plum. They are sub-acid and dry. The fruits are eaten locally.

***E. dentatus*** Vahl. New Zealand. The fruits are like small plums and are eaten by the natives. The bark is the source of a blue-black dye used by the natives for dyeing cloth and it is also used for tanning.

***E. floribundus*** Blume. N.India. Himalayas. A decoction of the bark and leaves is used locally to treat inflammation of the gums.

***E. ganitrus*** Roxb. (Bead Tree). India, Malaya. A larger tree than the other species mentioned. The seeds are used to make sacred beads by the Brahmins and Sinyasis. They are also used to make rosaries, buttons, etc.

***E. grandiflorus*** Smith. = E. lanceolata = Monocera lanceolata. (Anjangasjang). S.E.Asia to Indonesia. The crushed bark is used as a poultice for abcesses on the legs; a decoction of the seeds is used as a diuretic; and a decoction of the leaves is used as a mild stimulant.

***E. grandis*** F. v. Muell. (Blueberry Ash, Blue Fig, Silver Quandong). New South Wales, N.Queensland. The wood is light brown to pale straw-coloured. It is straight-grained, easily worked and bends easily. Because of these properties it is used in aircraft construction, furniture making, and house interiors.

***E. hookerianus*** Raoul. New Zealand. The bark yields a blue dye, used by the natives.

***E. kirtoni*** F. v. Muell. See E. grandis.

***E. lanceolatus*** Blume = E. grandiflorus.

***E. mastersii*** King. (Tjoomantoong merah). Malaysia. The wood is light-coloured, fairly strong and heavy. It is used locally for house-construction and fuel.

***E. oppositifolius*** Miq. Java. Cultivated locally for the red, large berries which are eaten raw.

***E. persicifolius*** Brongn. and Gris. New Caledonia. The wood is light and is used locally for making small ships.

***E. serratus*** L. (Veralu). N.E.Himalayas, N.India through to Malaysia. The fruits (Wild olives, Ceylon Olives) are eaten raw, or as a substitute for olives (Olea europea) in curries etc.

***E. valetonii*** Hochr. (Rempoodong). Bangka. The wood is used locally for making houses and boards.

ELAEODENDRON J. F. Jacq. ex Jacq. Celastraceae. 16–17 spp. Tropics and subtropics. Shrubs or trees.

***E. glaucum*** Pers. (Bakra, Butapala, Chikyeng). Lower Himalayas to N.E.India. The light-brown wood is fairly hard and polishes well. It is used for cabinet-work, small wooden articles and fuel. The roots are highly poisonous. A paste of the fresh root bark is used locally to treat swellings and snake bite, and in dyeing cotton ochre. The ground root is used as a snuff to relieve headaches; when burnt, the smoke is used to break hysterical trances; and a cold water extract is used as an emetic.

***E. lycioides*** Baker. Madagascar. The juice is used locally to stain the finger nails red.

ELAEOPHORBIA Stapf. Euphorbiaceae. 5 spp. Tropical and S.Africa. Trees.

***E. drupifera*** (Schum. and Thonn.) Stapf. (Baga, Do, Douo, Tene). W.Africa. A decoction of the plant is used locally as an ordeal poison and as a fish poison.

ELAPHOMYCES Nees-ex Fr. Elaphomyceta ceae. 20 spp. Widespread, especially S.Europe.

***E. cervinus*** (Pers.) Schroet. (Hart's Truffles). C. and E.Europe into Russia. The fruit-bodies (Fungus cervinus, Boletus cervinus, Trufos de Ciervo) are used locally in veterinary medicine as an aphrodisiac for cattle and pigs.

***Elaphrium*** Jacq. = Bursera.

El Assaik – Forsskalea tenacissima.

ELASTERIOSPERMUM Blume. Euphorbiaceae. 1 sp. S.Thailand, W.Malaysia. Tree.

***E. tapos*** Blume. The hard wood is used locally for general carpentry. The seeds are poisonous when raw, but are eaten roasted.

***Elaterium*** Jacq. = Rytidostylis.

Elder, American – Sambucus canadensis.

Elder, Bishop's – Aegopodium podagraria.

Elder, Blueberry – Sambucus coerulea.

Elder, Box – Acer negundo.

Elder, Dwarf – Aegopodium podagraria.

Elder, Dwarf – Sambucus ebulus.

Elder, European – Sambucus nigra.

Elder, Ground – Aegopodium podagraria.

Elder, Mexican – Sambucus mexicana.

Elder, Prickly – Aralia spinosa.

Elder, Yellow – Tecoma stans.

Elderberry – Sambucus gaudichaudiana.

Elderberry, Red – Sambucus callicarpa.

Elecampane (Oil) – Inula helenium.

Elemi, African – Boswellia frereana.

Elemi, African – Canarium schweinfurthii

Elemi, American – Bursera gummifera.

Elemi Canary Tree – Canarium luzonicum.

Elemi Frankincense – Boswellia frereana.

Elemi, Manila – Canarium commune.

Elemi, Manila – Canarium luzonicum.

Elemi Occidental – Protium iciariba.

Elemi of Guiana – Protium guianense.

Elemi of Mexico – Amyris plumieri.

Elemi of Yucatan – Amyris plumieri.

Elemi Oil – Canarium luzonicum.

Elemi, West Indian – Bursera gummifera.

ELEOCHARIS R. Br. Cyperaceae. 200 spp. Cosmopolitan. Herbs.

***E. austro-caledonica*** Vieill. New Caledonia. Used locally to make baskets.

***E. plantaginea*** R. Br. Indonesia. Cultivated locally to make baskets.

***E. plantaginea*** F. v. Muell. = E. sphacelata.

***E. sphacelata*** R. Br. = E. plantaginea F. v. Muell. Australia. The raw tubers are eaten by the aborigines.

***E. tuberosa*** Schult. (Water Chestnut). E.India, China and Japan. The tubers are eaten cooked in a variety of Chinese dishes. The plant is cultivated, and the tubers are exported, usually canned.

Elephant Apple – Dillenia indica.

Elephant Apple – Limonia acidissima.

Elephant's Ear – Colocasia esculenta.

Elephant Foot, Prickly-leaved – Elephantopus scaber.

Elephant Grass – Pennisetum purpureum.

Elephant Grass – Typha elephantina.

Elephant Tree – Boswellia papyrifera.

ELEPHANTOPUS L. Compositae. 32 spp. Tropics. Herbs.

***E. scaber*** L. (Prickly-leaved Elephant's Foot). Throughout tropics. The plant can be a serious weed. It is used to reduce fevers and as a diuretic.

ELEPHANTORRHIZA Benth. Leguminosae. 10 spp. Tropical and S.Africa. Herbs.

***E. burchellii*** Benth. S.Africa. The plant yields a dye.

ELETTARIA Maton. Zingiberaceae. 7 spp. Indomalaya. Herbs.

***E. cardamomum*** (L.) Maton. (Cardamon). India to Malaysia. Cultivated for the seeds which are used as a spice. The main areas of production are India and Ceylon, and to a lesser extent Guatemala and Malaysia. Most of the production is used internally, and it is one of the most important Indian spices. There are two distinct varieties, Malabar and Ceylon. The Malabar is the taller plant with thinner leaves, rounder capsules, with smaller, but more aromatic seeds. The Ceylon variety is shorter, with broader leaves, a three-sided capsule, with more, larger, but less aromatic seeds. Malabar Cardemon is also called Cluster Cardemon, Siam Cardemon, Mysore Cardemon, Magelore Cardemon, Madras Cardemon and Alepy Cardemon. The flavour of the seeds is due to an essential oil (4–7 per cent Oleum Cardamoni) which contains limonene, cinerol, borneol, sabinene and d-a-terpinol. The seeds are used ground, or whole as a spice in flavouring curries, pickles, cakes, cordials, liqueurs, sausages. The oil is used in making Eau de Cologne, to aid digestion, and as a stimulant. The capsules are harvested when they are not fully ripe to retain the aroma of the seeds. They are then sun-dried, bleached and marketed. The seeds are removed at their destination as soon as possible before processing, or using.

***E. major*** Sm. Tropical Asia. The seeds are a source of cardamon.

***E. speciosa*** Blume. Java. Locally, the fruits are used in chutneys and sometimes candied, the leaves are used as a vegetable with rice.

ELETTARIOPSIS Baker. Zingiberaceae. 10 spp. Indomalaya. Herbs.

***E. sumatrana*** Valet. Malaysia to Thailand. The juice is used locally for flavouring and treating scorpion bites.

ELEUSINE Gaertn. Graminae. 9 spp. tropics and subtropics, 1 sp. Temperate S.America. Annual or perennial grasses.

***E. aegyptiaca*** Desf. (Goose Grass). Tropics and subtropics. The seeds are used as a source of flour in various parts, and the whole plant is fed to livestock. Perennial.

***E. coracana*** (L.) Gaertn. (Finger Millet, African Millet, Ragi) India. Cultivated widely in China, S.India and Central Africa as a cereal crop for internal consumption. The grain is used to make flour and for brewing. It stores well and is attacked by only a few pests. Annual.

***E. indica*** Gaertn. (Goose Grass). The probable ancestor of E. coracana. The young seedlings are used as a vegetable in Java where the stems are used to make mats and the whole plant is used as fodder for livestock. Annual.

***E. tocussa*** Fresen. Abyssinia. The grain is a local source of flour. Annual.

***Eleutheropetalum sartorii*** (Liebm.) Liebm.= Chamaedorea sartorii.

ELIONURUS Humb. and Bonpl. ex Willd corr. Kunth. Graminae. 25 spp. Tropics and subtropics. Perennial grasses.

***E. candidus*** Hack.=Andropogon candidus =Lycurus muticus. S.America. Used for binding sand dunes.

Elk Tree Leaves – Oxydendrum arboreum.

Elm, American White – Ulmus americana.

Elm, Cedar – Ulmus crassifolia.

Elm, English – Ulmus campestris.

Elm, European – Ulmus campestris.

Elm, Japanese – Ulmus japonica.

Elm, Mountain – Ulmus montana.

Elm, Red – Ulmus alata.

Elm, Rock – Ulmus thomasii.

Elm, Slippery – Fremontodendron californica.

Elm, Slippery – Ulmus fulva.

Elm, Sweet – Ulmus fulva.

Elm, White – Ulmus americana.

Elm, Winged – Ulmus alata.

ELMERRILLIA Dandy. Magnoliaceae. 7 spp. S.E.Asia, Philippines, New Guinea. Trees.

***E. celebica*** (Koord.) Dandy. Celebes. The hard wood is used locally for house-building.

***E. ovalis*** (Miq.) Dandy. See E. celebica.

***E. papuana*** Sm. (Hui, Wau Beech). New Guinea. The light brown wood is durable and is worked easily. It is used for carving, moulding and cabinet work.

Elmira Resin – Protium heptaphyllum.

Elobwa – Impatiens irvingii.

Elokoloko – Sorindeia gilletii.

Elondo – Fagara pilosiuscula

Elqueme (Gum) – Bursera gummifera.

ELSHOLTZIA Willd. Labiatae. 35 spp. Eurasia, Abyssinia. Herbs.

***E. cristata*** Willd. Europe through temperate Asia. A decoction of the leaves is used locally in Japan to relieve the effects of excess alcohol.

***Elssholtzia*** Garcke=Elsholtzia.

ELYMUS L. Graminae. 70 spp. N.Temperate, S.America. Perennial grasses.

***E. arenarius*** L. Temperate Eurasia. Cultivated locally in Japan and used for making ropes, mats and paper.

***E. canadensis*** L. (Canada Wild Rye). Great Plains of America, Pacific North-West. Used for forage and hay, particularly in mixtures with other grasses. Suitable for covering inland sand dunes.

***E. condensatus*** Presl. (Giant Wild Rye). Occurs widely in C. and S. U.S.A. A good grazing grass when young, and makes a good hay. It tolerates moderate salinity and is fairly drought-resistant. The seeds were used by the Indians to make flour.

***E. giganteus*** Vahl. (Siberian Wild Rye). Siberia. Used for stabilizing inland sand dunes. Does not tolerate salt.

***E. junceus*** Fisch. (Russian Wild Rye). Russia, N.Asia. Used for stabilizing inland sand dunes. Salt tolerant.

***E. triticoides*** Buckl. (Wild Beardless Rye, Wild Wheat, Squaw Grass). W. N.America to Mexico. The seeds were used by the Indians to make flour.

ELYTRARIA Michx. Scrophulariaceae. 7 spp. Tropics and sub-tropics. Herbs.

***E. tridentata*** Vahl.=Tubiflora squamosa. C.America. A decoction of the plant is used locally in El Salvador to treat dysentery.

EMBELIA Burm. f. Mysinaceae. 130 spp. Tropical and subtropical Africa, Madagascar, E.Asia, Indomalaya, Pacific. Shrubs or woody climbers.

***E. benthamii*** Hance=E. laeta.

***E. laeta*** Mez. = E. benthamii = E. obovata = Samara lacta. China to Vietnam. The fruits contain embelin and are used in N.Vietnam to treat tapeworms.

***E. obovata*** Hemsl. = E. laeta.

***E. philippensis*** A. DC. Philippines. The leaves are eaten locally as a vegetable with fish.

***E. ribes*** Burm. f. India, Malaysia. The seeds contain embelin and are used to treat tape worms in children. They are also used to aid digestion and as a tonic. The whole dried fruit is used to adulterate black pepper.

Embiráro – Pterogyne nitens.

Emblic – Phyllanthus emblica.

***Embothrium rubricaule*** Giord. = Stenocarpus salignus.

Emetic Holly – Ilex vomitoria.

Emetic Root – Euphorbia corollata.

EMEX Neck. ex Campderá. Polygonaceae. 2 spp. Mediterranean, S.Africa, Australia. Herbs.

***E. spinosa*** Campderá. Europe, S.Africa. A decoction of the leaves is used by the natives of S.Africa to treat indigestion, and to treat digestive troubles of livestock.

EMILIA Cass. Compositae. 30 spp. Old World Tropics. Herbs.

***E. sonchifolia*** DC. Tropical Asia and Africa. Cultivated in Java for the leaves which are eaten with rice or in soups.

EMINIA Taub. Leguminosae. 6 spp. Tropical Africa. Herbs.

***E. polyadenia*** Hauman. (Monkoo). Congo. The roots are used locally by the natives to make a drink.

EMINIUM (Bl.) Schott. Araceae. 5 spp. E.Mediterranean to C.Asia. Herbs.

***E. spiculatum*** (Blume) Kuntze. Mediterranean to Iran. The tubers are eaten as a vegetable in Egypt.

Emmer (Wheat) – Triticum dicoccum.

Emmer, Kolchic – Triticum georgicum.

Emory Bushmint – Hyptis emoryi.

Emory Oak – Quercus emoryi.

Empang – Euodia latifolia.

EMPETRUM L. Empetraceae. 2 spp. (Good), 15–16 spp. (Vasilev). N.Temperate and Arctic, S.Andes, Falkland Islands, Tristan da Cuhna. Shrubs.

***E. nigrum*** L. (Crow berry). Temperate Europe, Asia and N.America. The berries are edible and are made into a drink with sour milk in Iceland, and are eaten with seal fat in Greenland.

***E. rubrum*** Vahl. Falkland Islands, Chile, Patagonia. The berries are eaten locally.

Empit – Pentaspadon motleyi.

EMPLEURUM Ait. Rutaceae. 2 spp. S.Africa. Shrubs.

***E. ensatum*** (Thunb.) Eckl and Zeyh. = E. serrauatum. S.Africa. The leaves are used locally to relieve digestive upsets and as a diuretic.

***E. serrulatum*** Ait. = E. ensatum.

Emu Apple – Owenia acidula.

Emu Bush – Eremophila oppositifolia.

***Enalus*** Aschers and Guerke = Enhalus.

ENANTIA Oliv. Annonaceae. 9 spp. W.Africa, 1 sp. tropical E.Africa. Trees.

***E. chlorantha*** Oliv. Tropical Africa. The wood (African Whitewood) is soft, yellow and browning with age. It is used for general carpentry, house-building and furniture-making.

***E. kummeriae*** Engl. and Diels. Tropical Africa. The wood yields a red dye.

***E. polycarpa*** Engl. W.Africa. The wood yields a yellow dye, used locally for colouring skins and mats. The bark extract is used to treat sores and ulcers.

ENCELIA Adans. Compositae. 15 spp. W.United States of America, to Chile, Galapagos Islands. Shrubs.

***E. farinosa*** Gray. (White Brittle Brush). S.W. U.S.A. to Mexico. A gum from the plant is used as chewing gum by the Indians of Arizona. A resin is sometimes used as incense in churches.

ENCEPHALARTOS Lehm. Zamiaceae. 30 spp. Tropical and S.Africa. Cycads.

***E. caffer*** Miq. S.Africa. The starch from the stem is used by the Hottentots to make bread (Kaffir Bread).

***E. hildebrandtii*** A. Br. See E. caffer.

***E. miquellii*** F. v. Muell. = Macrozamia miquellii.

***E. poppgei*** Aschers. (Piondo, Tchianda). Congo. The starch from the stem is eaten locally. A glue is made from the sap in the stem.

***E. spiralis*** Lehm. = Macrozamia spiralis.

Encina blanca – Quercus grisea.

Encina colorado – Quercus uruapanensis.

Encina colorado – Quercus eduuardi.

Encina prieta – Quercus grisea.

ENDIANDRA R. Br. Lauraceae. 80 spp. Malaysia, Australia, Polynesia; 1 sp. Assam(?). Trees.

***E. palmerstoni.*** C. T. White. (Australian Walnut, Queensland Walnut, Walnut

Bean, Oriental Wood). Queensland. The wood is streaked with black, green, brown and pink to give a very beautiful effect. It polishes well and is much esteemed for panelling etc. in public buildings, furniture and cabinet work. The wood is much valued and a great amount is exported.

Endive – Cichorium endivia.

Endjin – Lychnodiscus cerospermus.

ENDOMYCES Reess. Endomycetaceae. 2–3 spp. Temperate.

***E. vernalis*** Ludw. This fungus produces large amounts (up to 25 per cent of its dry weight) of fat from carbohydrate. It was used during the Second World War in Germany to manufacture fats by growing the fungus of beer wort.

ENDOSPERMUM Benth. Euphorbiaceae. 10 spp. S.E.Asia, Malaysia, Fiji. Trees.

***E. chinense*** Benth. China, Malaya. The light wood is used to make matches, rough packing cases etc.

***E. malaccense*** Muell. Agr. (Terbulan). Malaya, Borneo. The wood is very soft and light, and not durable outside. It is used locally for the interiors of houses and matches.

***E. moluccanum*** Becc. = Capellenia moluccana. Malaya, Moluccas. The light wood is used for floats for fishing nets. The young leaves are cooked as a vegetable, while the old leaves are used as a laxative.

***E. peltatum*** Merr. Philippines. The wood is used to make matches.

***Endymion nonscriptum*** Garcke. = Scilla non scripta.

***Endymion nutans*** Dum. = Scilla non scripta.

Engambili – Monodora laurentii.

ENGELHARDTIA Leschen and Bl. corr. Bl. Juglandaceae. 5 spp. Himalayas to Formosa, S.E.Asia, Malaysia; 3 spp. C.America. Trees.

***E. polystacha*** Radlk. (Cha-bih, Lewa). Assam. The bark is used locally to stupify fish.

***E. serrata*** Blume. Malaya. The light, yellow-grey wood is used locally to make houses.

***E. spicata*** Blume = Juglans pterococca. China. The bark is used locally for tanning.

Engelmann Spruce – Picea engelmanni.

Engelwasser – Myrtus communis.

Engkabang Nut – Shorea macrophylla.

ENGLEROMYCES P. Henn. Nectriaceae. 1 sp. on bamboo. Africa.

***E. goetzii*** P. Henn. Malawi. The cooked fruit-bodies are used locally against stomach complaints.

English Bluegrass – Festuca elatior.

English Chamomille – Anthemis nobilis.

English Daisy – Bellis perennis.

English Elm – Ulmus campestris.

English Galingale – Cyperus longus.

English Horsemint – Mentha longifolia.

English Ivy – Hedera helix.

English Liverwort – Peltigera canina.

English Oak – Quercus robur.

English Serpentary – Polygonum bistorta.

English Walnut (Oil) – Juglans regia.

English Yew – Taxus baccata.

ENHALUS Rich. Hydrocharitaceae. 1 sp. Indomalaya, Australia. Submerged water-plants.

***E. acoroides*** Steud. = E. koengii. In New Guinea the seeds are eaten boiled or roasted and the fibre from the stems is used to make fishing nets.

***Enhydra*** DC. = Enydra.

ENICOSTEMA Blume. Gentianaceae. 1–3 spp. Tropical America, Africa and Asia. Herbs.

***E. littorale*** Blume. Tropics. The leaves are used in the Sudan to relieve stomach complaints, as a laxative and a tonic.

***E. verticillatum*** Engl. (Rebba bisah). Java. A decoction of the plant is used locally to reduce fevers.

Enset – Musa ensete.

ENTADA Adans. Leguminosae. 30 spp. Warm.

***E. phaseoloides*** (L.) Merr. = E. scandens.

***E. polyphylla*** Benth. (Pashaco). Guyana, Brazil to Peru. A resin from the trunk is used for dyeing leather black. Tree.

***E. polystachia*** (L.) DC. (Bejuco de panune, Guiamol, Parra rosa). Tropical America. The stems are used locally as ropes and when crushed in water they are used as a soap-substitute. Large shrub.

***E. scandens*** Benth. = E. phaseloides. (St. Thomas Bean, Climbing Entada). Tropics. Cultivated on a small scale. The bark is used as a soap substitute in the Far East, especially for washing the hair. The beans, from very large pods are eaten roasted; the leaves are eaten as a vegetable, and a fibre from the stem is used to make cordage for fishing nets, ropes, sails etc. Vine.

ENTANDROPHRAGMA D. DC. Meliaceae. 35 spp. Tropical and S.Africa. Trees. The wood from all the species mentioned is exported for furniture-making etc., as a mahogany substitute. The woods are not particularly heavy or hard, but are scented, have a pleasant brown colour and polish well.

***E. angolense*** (Welw.) A. DC. Africa. Ljebu Mahogany.

***E. angolense*** Auct. non A. DC.=E. cylindricum.

***E. candollei*** Harms. Tropical Africa. Sapele Mahogany, Scented Mahogany.

***E. cylindricum*** (Sprague) Sprague.=E. lebrunii=E. angolense Auct. non. A. DC. Tropical Africa, especially W. and C. Sapele. Acajou d'Afrique.

***E. lebrunii*** Staner=E. cylindricum.

***E. macrophyllum*** Chev. Tiama Mahogany.

***E. rederi*** Harms. Cameroon Mahogany.

***E. utile*** (Dawe and Sprague) Sprague. Brown Mahogany, Heavy Mahogany, Short-Capsuled Mahogany.

ENTELEA R. Br. Tiliaceae. 1 sp. New Zealand. Tree.

***E. arborescens*** R. Br. (Cork Wood Tree). The heavy light wood is used by the Maoris for floats for fishing nets, small boats etc.

ENTEROLOBIUM Mart. Leguminosae. 10 spp. Tropical America. W.Indies. Trees.

***E. cyclocarpum*** (Jacq.) Griseb.=Mimosa cyclocarpa. N. C.America through to W.Indies. The wood is durable in water, and so is used for small boats, water-troughs etc. It is also used for furniture-making. The fruits are eaten as a vegetable and are used as a soap-substitute. They are also the source of a gum (Goma de Caro). The bark yields a syrup used to treat colds and is also used as a soap substitute.

***E. timboüva*** Mart. (Timboüva). S.America, especially the E. side. All parts of the plant except the leaves contain saponin and are used as soap substitutes. The testa and root bark are used in Brazil to treat intestinal worms.

ENTEROMORPHA Link. Ulvaceae. Marine Green Algae. The species mentioned are eaten locally as a vegetable.

***E. complanata*** Kutz. Pacific Islands, Japan.

***E. compressa*** Grev. Pacific Islands, Japan.

***E. intestinalis*** (L.) Link. (Lumot). Philippines, Hawaii, China.

***E. prolifera*** (Muell.) J. Ag. (Limu Eleele). Hawaii.

Entjeng-e ent tjeng – Cassia pumila.

***Entoloma clypeatus*** (L.) Quél.=Hypor-hodius clypeatus.

***Entoloma microcarpum*** (Berk. and Br.) Sacc.=Collybia microcarpa.

ENVOCLADIA Lamour. Rhodophyllaceae. Red Marine Algae.

***E. muricata*** Lamour. Used in California to manufacture agar.

ENYDRA Lour. Compositae. 10 spp. Warm. Water herbs.

***E. fluctans*** Lour.=E. helonchu. India to Thailand, China. The leaves are used in Malaya and Cambodia as a condiment. Occasionally cultivated. In India and Malaya, the leaves are used as a laxative.

Eon – Fillaeopsis discophora.

EPERUA Aubl. Leguminosae. 12 spp. Tropical S.America. Trees.

***E. falcata*** Aubl. Tropical America, especially Guiana. The purple-red wood (Waloba, Wallaba) is valued in Brazil for furniture-making etc. It is also used to make paper pulp, but the yield is low.

***E. grandiflora*** (Aubl.) Benth. Tropical America. The wood is used for paper pulp.

***E. jenmani*** Oliv. Tropical America. The wood is used for paper pulp.

***E. oleifera*** Ducke. Brazil. The stem yields a thick resin (Jacaré copahiba) used in making varnishes.

***E. rubiginosa*** Miq. Tropical America. The wood is used to make paper pulp.

***E. schomburgkiana*** Benth. Tropical America. The wood is used to make paper pulp.

Epazote – Lippia berlandieria.

EPHEDRA L. Ephedraceae. 40 spp. Warm temperate N. and S. America, Eurasia. Shrubs. The plants contain varying amounts of the alkaloid ephedrine which is used medicinally to increase the blood-pressure, to decrease local congestion of the blood and to excite the sympathetic nervous system. It is also used in the treatment of asthma.

***E. americana*** Humb. and Bonpl.=E. andina. (America Ephedra). S.America. An infusion of the stems and roots is used locally as a diuretic and to purify the system.

***E. andina*** Poepp=E. americana.

***E. antisyphilitica*** Berland=E. peduculata. Mexico, W.Texas. A decoction of the stem is used to treat urinogenital diseases.

***E. aspera*** Engelm. Mexico. See E. antisyphilitica.

***E. disticha*** L. = E. vulgaris. Mediterranean, C.Europe to Siberia. A commercial source of ephedrine.

***E. equisetina*** Bunge. Coasts of S.China. Commercial source of ephedrine. Exported.

***E. gerardiana*** Wall. Pakistan, N.India. Source of ephedrine.

***E. intermedia*** Screnk and Mey. Pakistan. Source of ephedrine.

***E. nebrodensis*** Tineo. Pakistan, N. India. Source of ephedrine.

***E. nevadensis*** S. Wats. (Nevada Joint Fir). Nevada to California. The roasted seeds were used as a flour by the local Indians. A decoction of the stem was used to treat urinogenital diseases.

***E. ochreata*** Miers. Argentina. A decoction of the branches is used to treat genital discharges and as a diuretic.

***E. peduculata*** Englm. = E. antisyphilitica.

***E. sinica*** Stapf. (Chinese Ephedra, Ma-Huang). Coasts of S.China. Commercial source of ephedrine. Exported. Its medicinal value has been known to the Chinese for over 5,000 years.

***E. trifurca*** Torr. (Longleaf Ephedra, Joint Fir). S.W. U.S.A., Mexico. An infusion of the branches is a popular beverage locally (Desert Tea, Teamster's Tea).

***E. vulgaris*** Ruch. = E. disticha.

Ephedra, American – Ephedra americana.

Ephedra, Chinese – Ephedra sinica.

Ephedra, Longleaf – Ephedra trifurca.

EPICAMPES J. and C. Presl. Graminae. 10 spp. California to Argentina. Perennial grasses.

***E. macroura*** Benth. (Zakaton Grass). Mexico. The roots (Rice Root, Zakaton, Mexican Whish, Dogtooth) are much used to make brooms etc. Exported to U.S.A., France and Germany.

***E. stricta*** Presl. See E. macroura.

***Epicharis altissima*** Blume = Dysoxylum densiflorum.

***Epicharis densiflora*** Blume = Dysoxylum densiflorum.

***Epicharis loureiri*** Pierre = Dysoxylum loureiri.

EPIDENDRUM L. Orchidaceae. 400 spp. Tropical America. Epiphytic herbs.

***E. cochleatum*** L. Mexico through to C. S.America. W.Indies. Used locally by the natives of Mexico as a source of mucilage.

***E. vitelinum*** Lindl. See E. cochleatum.

EPIGAEA L. Ericaceae. 3 spp. Japan, E. coasts of N.America, E. Asia Minor. Shrubs.

***E. repens*** L. (Ground Laurel). E. N.America. A decoction of the leaves was used by the Indians to treat diarrhoea in children. It was also used as a tonic.

EPILOBIUM L. Onagraceae. 215 spp. N. and S. Temperate, Arctic. Herbs.

***E. angustifolium*** L. = Chamenaerion angustifolium (Rosebay Willowherb, French Willow). Temperate Europe, Asia and N.America. The leaves are used in parts of Russia to make a tea (Kapor Tea, Kapporie Tea, Iwan-yschai).

EPIMEREDI Adans. Labiatae. 7 spp. Indomalaya, Australia. Herbs.

***E. indica*** Kuntze. Tropical Asia. The leaves are used in China, India and Ceylon to aid digestion, as a tonic and to flavour sago cakes.

***E. malabarica*** R. Br. (Chodhava, Karithumba, Vaikantha). Tropical Asia, especially India. The leaves are used by the Hindus to relieve teething troubles in children. They are also used to treat snake-bite and scorpion stings.

***E. salviaefolia*** R. Br. Australia. The flowers and leaves are a possible source of perfume.

EPIPACTIS Zinn. Orchidaceae. 24 spp. N.Temperate, tropical to S.Africa, Thailand, Mexico. Herbs.

***E. giganteum*** Dougl. Hook. (Giant Helleborine, Giant Orchid, Chatterbox). S.W. U.S.A., Mexico. An infusion of the plant is used by the Indians of California to treat insanity.

EPIPREMNIUM Schott. Araceae. 25 spp. Indomalaya. Vines.

***E. giganteum*** (Roxb.) Schott = Pothos giganteus. Malaysia. The sap is used as an arrow poison in Malacca.

***E. mirabile*** Schott = E. pinnatum

***E. pinnatum*** (L.) Engl. = E. .mirabile = Monstera pinnatifida. Tropical Asia, Indonesia to Pacific Islands. In Fiji the ground stem is used to relieve neuralgia.

Epokape – Impatiens giorgii.

Epteba – Fagara lemairei.

EQUISETUM L. Equisetaceae. 23 spp. Cosmopolitan except Australia. Herbaceous cryptogams.

***E. arvense*** L. (Horsetail). Temperate. A strong diuretic, the upper parts of the plant (Herba Equisetii) have been used medicinally since ancient times. They are also used to treat dropsy and as a stypic. The strobili are eaten as a vegetable or pickle in Japan and the rhizomes are eaten by the Indians of N.Mexico.

***E. bogotense*** H.B.K.=E. chilense=E. quitense. Colombia, Peru, Ecuador. Used in Peru as a diuretic and in Venezuala to treat diabetes. In Colombia it is used to treat dysentery, gonorrhoea, to control bleeding and as a diuretic.

***E. chilense*** Presl.=E. bogotense.

***E. debile*** Roxb., (Sumbak). Tropical Asia. Used in Malaya to treat pains in the joints.

***E. giganteum*** H.B.K. See E. bogotense.

***E. heleocharis*** Ehrh. Europe, N.Asia, N.America. Used in the Ukraine to control bleeding.

***E. hiemale*** L. Temperate. The rough stems are used in Japan as a substitute for sandpaper. It was used similarly in Europe.

***E. pratense*** Ehrh. Temperate. The rhizomes are eaten as a vegetable by the Indians of Minnesota.

***E. quitense*** Fée.=E. bogotense.

***E. robustum*** A. Br. N.America. A decoction of the plant is used in parts of Mexico as a diuretic and to treat venereal diseases.

ERAGROSTIS Host. Graminae. 300 spp. Cosmopolitan, especially subtropics. Perennial Grasses.

***E. abyssinica*** Schrad.=Poa abyssinica.

***E. brownei*** Nees. Australia. A good local pasture grass.

***E. curvula*** (Schrad.) Nees. (Weeping Lovegrass). S.Africa. An excellent forage grass which is drought resistant, and tolerates extremes of temperature. It is useful for stabilizing sand-dunes.

***E. cynosuroides*** Beauv.=Poa cynosuroides. Nile. Used locally for making mats.

***E. lugens*** Nees.=Poa microstachya. Tropical America. An excellent forage grass.

ERANTHEMUM L. Acanthaceae. 30 spp. Tropical Asia. Herbs.

***E. viscidum*** Blume.=Daedalacanthus viscidum.=Djarong boobookooat). Malaysia, Indonesia. The juice from the leaves is used locally to treat inflammation of the eyes.

Erdorseille – Ochrolechia tartarea.

EREMOCHLOA Buese. Graminae. 8 spp. India, Ceylon, S.China, S.E.Asia, W.Malaysia, Australia. Perennial Grasses.

***E. ophuroides*** (Munro) Hack. (Centepede Grass). S.E.Asia. Introduced to N.America, especially Carolina. It is used in erosion control and as a lawn grass.

EREMOCITRUS Swingle Rutaceae. 1 sp. N.Australia. Shrub.

***E. glauca*** Swingle. (Australian Desert Kumquat). The fruits are used locally to make jams and drinks.

EREMOPHILA R. Br. Myoporaceae. 45 spp. Australia. Trees.

***E. arborescens*** Cunningh.=E. oppositifolia.

***E. mitchelli*** Benth. (Sandalwood, Bastard Sandalwood, Rosewood). S.Australia. The brown wood is hard, beautifully grained and scented. It is used for veneers and cabinet work.

***E. oppositifolia*** R. Br.=E. arborescens. (Emu Bush). Australia. The bark is used locally for tanning.

EREMOSPATHA (Mann and Wendl.) Mann and Wendl. ex Kerchove. Palmae. 12 spp. Tropical Africa. Small palms.

***E. macrocarpa*** Mann and Wendl. Tropical Africa. The canes are used for making baskets, building houses, small bridges and for making garden furniture.

EREMOTHECIUM Borzi. Endomycetaceae. 2 spp. Widespread.

***E. ashbyii*** Schopf. Is used in the commercial production of the vitamin riboflavin.

EREMURUS M. Bieb. Liliaceae. 50 spp. Alpine, W. and C. Asia. Herbs.

***E. aucherianus*** Boiss. C.Asia. The leaves are eaten locally as a vegetable. The bulbs yield a mucilage which is used to stick leather used in book-binding.

***E. spectabilis*** M. Bieb. C. Asia. The leaves are eaten locally as a vegetable.

Ergot – Claviceps purpurea.

Ergot, Corn – Ustilago maydis.

Ergot, Rye – Claviceps purpurea.

Ergosterine – Claviceps purpurea.

ERIA Lindl. Orchidaceae. 375 spp. Tropical Asia, Australia, Polynesia. Herbs.

***E. pannea*** Lindl. (Karakubong). E.Himalayas, Assam. A decoction of the plant is used locally in baths as a treatment for ague.

ERICA L. Ericaceae. 500+ spp. Europe, Atlantic Islands, N.Africa, Asia Minor, Syria, tropical and especially S.Africa. Shrubs or small trees. The flowers are a good source of honey.

***E. arborea*** L. Mediterranean. The roots

(Briar Roots) are used in making bowls for tobacco pipes. Much is produced from Italy and S.France.

***E. cinerea*** L. See E. tetralix.

***E. tetralix*** L. (Cross-leaved Heath). Temperate Europe. Used in Scotland as the source of a yellow dye.

ERIGON L. Compositae. 200+ spp. Cosmopolitan especially N.America. Herbs.

***E. affinis*** DC. Mexico. The roots are used locally to clean the teeth and treat toothache.

***E. canadensis*** L. (Fleabane). N.America, Europe. A decoction of the leaves is used locally to treat dysentery, diarrhoea, bleeding and venereal diseases.

ERIOBOTRYA Lindl. Rosaceae. 30 spp. Warm Asia. Trees.

***E. japonica*** Lindl. = Photinia japonica (Loquat, Japanese Plum, Japanese Medlar). Central East China. The orange-coloured fruits have a hairy skin and a few large seeds. They are eaten raw, cooked or in jellies and preserves. The plant has been developed by breeding in Japan and is cultivated commercially on a small scale in China, Japan, California and the Mediterranean.

***Eriobroma klaineana*** Pierre = Sterculia oblonga.

***Eriocarpum grindeliondes*** Nutt. = Haplopappus nutallii.

ERIOCEPHALUS L. Compositae. 30 spp. S.Africa. Shrubs.

***E. spinescens*** Burch. S.Africa. Used locally as fodder for livestock.

ERIOCHLOA Kunth. Graminae. 20 spp. Tropical and subtropical. Grasses.

***E. polystachya*** H.B.K. = E. subglabra (Nash) Hitch. Carib Grass. Tropical America. Cultivated locally for fodder.

***E. subglabra*** Hitch. Java. Used locally for fodder for cattle.

***E. subglabra*** (Nash) Hitch. = E. polystachya.

ERIOCOELUM Hook. f. Sapindaceae. 10 spp. W.tropical Africa. Trees.

***E. microspermum*** Radlk. ex de Wild. = E. paniculatum. (Aschese, Kaboli, Mbelabela, Olumba). S.W.Africa, Congo. The bark is used locally to treat coughs, venereal diseases and excess flatulence. The wood is an excellent firewood.

***E. paniculatum*** Auct. = E. microspermum.

***Eriocoma fragrans*** D. Don. = Montanoa tormentosa.

***Eriodendron*** DC. = Ceiba.

ERIODICTYON Benth. Hydrophyllaceae. 10 spp. S.W. United States of America, Mexico. Shrubs.

***E. californica*** (Hook. and Arn.) Torr. (Yerba Santa). The dried leaves contain pentatriacontane and cerotinic acid which are expectorants. They are used medicinally as a tea, locally.

***E. glutinosum*** Benth. (Mountain Balsam, Gum Bush, Bearsweed). California. The leaves are expectorant and are used locally in the treatment of asthma and bronchitis.

ERIOGLOSSUM Blume. Sapindaceae. 1 sp. Indomalaya, Australia. Tree.

***E. edule*** Blume = E. rubigonosum. The fruits are edible.

***E. rubigonosum*** Brand. = E. edule.

ERIOGONUM Michx. Polygonaceae. 200 spp. N.America, especially W. U.S.A. Herbs.

***E. corymbosum*** Benth. W. N.America. The leaves are eaten as a vegetable with corn, by the local Indians.

***E. effusum*** Nutt. = E. microthecium.

***E. inflatum*** Torr. (Desert Trumpet). W. U.S.A. The young stems are eaten raw before flowering.

***E. longifolium*** Nutt. (Indian Turnip, Long-leaved Eriogonium). S.E. U.S.A. The roots are eaten by the local Indians.

***E. microthecium*** Nutt. = E. effusum (Wild Buckwheat). W. N.America. An infusion of the roots and leaves is used by the local Indians to treat coughs caused by tuberculosis.

***Eriolobus delavayi*** Schneid. = Docynia delavayi.

ERIOSEMA (DC.) Desv. Leguminosae. 140 spp. Tropics and subtropics. Herbs or small shrubs. The root tubers of the following species are eaten locally as a vegetable. All the species mentioned are distributed in tropical W.Africa, especially the Congo.

***E. cordifolium*** Hochst. ex A. Rich.

***E. erici-rosenii*** R. H. Fries.

***E. flexuosum*** Staner.

***E. griseum*** Bak. (Kabala). The leaves are used locally as a poultice to reduce swellings.

***E. lebrunii*** Staner and De Craener (Manga).

***E. macrostipulum*** Bak. f.

***E. montanum*** Bak. f. (Kagobozoba). The leaves are used locally as a poultice on wounds and as an infusion to treat intestinal worms and as a purgative.

***E. schoutedenianum*** Staner and De Craener.
***E. verdickii*** De Wild.
ERISMA Rudge. Vochysiaceae. 20 spp. N.Brazil, Guyana. Trees.
***E. calcaratum*** Warm. Throughout range. The seeds yield a fat (Jaboty Butter, Jaboty Tallow) used to make candles and soap.
ERITRICHUM Schrad. Boraginaceae. 65 spp. Temperate. Herbs.
***E. pusillum*** Coss. (Mouhden, Moeddeus). Sahara. Yields a red dye, used locally for colouring the skin.
Erman's Birch – Betula ermanii.
ERODIUM L'Hérit. Geraniaceae. 90 spp. Europe, Mediterranean to C.Asia, temperate Australia, S.tropical S.America. Herbs.
***E. cicutarium*** (L.) L'Hérit. (Stork's Bill). Temperate. The awns on the fruits are very hygroscopic and are used as hygrometers and weather indicators.
***E. hirtum*** Willd. N.Africa. The roots are eaten by the Tuareg.
***E. malacoides*** L'Hérit. Sahara. The leaves are eaten in salads in Morocco.
***Eroteum theaoides*** Sw. = Eurya theoides.
ERUCA Mill. Cruciferae. 6 spp. Mediterranean, N.E.Africa. Herbs.
***E. pinnatifida*** (Desf.) Pomel. var. ***aurea*** Batt. Sahara. Grown at oases as livestock fodder and is used for making sauces.
***E. sativa*** L. = E. versicaria (Rocket, Roquette). C.Europe, Mediterranean, Asia. Introduced to America and Australia. Cultivated in India for the seeds which yield about 32 per cent of a pungent, semidrying oil (Jamba Oil) which is used initially for making pickles. After six months the oil loses its acrid taste and is then used for cooking, as a lubricant and for burning. The seeds are also used in mustard. The young leaves are eaten in salads. They are diuretic and relieve stomach upsets; rubbed on the skin, they cause local reddening.
***E. versicaria*** (L.) Cav. = E. sativa.
ERUCARIA Gaertn. Cruciferae. 9 spp. E.Mediterranean, Arabia, Persia. Herbs.
***E. aleppica*** Gaertn. Greece, Turkey, Arabia. Used locally in Greece as a salad and a vegetable.
ERVATAMIA Stapf. Apocynaceae. 80 spp. Old World tropics. Shrubs.
***E. coronaria*** Stapf. = Tabernaemontana coronaria = T. divaricata. (East Indian Rosebay, Grape, Jasmine). E.India, Malaya. The wood is used in incense and perfumes. It is also used medicinally as a refrigerant. The pulp around the seeds is used as a dye, and in Indonesia, a decoction of the roots is used to treat diarrhoea. The plant is cultivated.
***E. cylindrocarpa*** King and Gamble. Malaya. The leaves are pounded with turmeric and rice and applied as a poultice to skin complaints.
***E. hirta*** King and Gamble. Tropical Asia. In Malacca the leaves are used to treat ulcers of the nose.
***E. malaccensis*** King and Gamble. See E. hirta.
Ervil – Vicia ervilia.
***Ervilia sativa*** Linl. = Vicia ervilia.
***Ervum*** L. = Lens Mill.
ERYNGIUM L. Umbelliferae. 230 spp. Tropics and temperate, except tropical and S.Africa. Herbs.
***E. aquaticum*** L. = E. virginianum. E. U.S.A. The local Indians use a decoction of the roots as an emetic. In smaller quantities, the decoction is used to reduce fevers and as an expectorant.
***E. campestre*** L. (Snakeroot). Europe. A decoction of the root is used to treat urinary complaints. It is also diuretic, expectorant and reduces fevers. Occasionally the roots are eaten candied.
***E. carlinae*** Delav. Tropical America. In C.America an infusion of the leaves is used to treat stomach complaints of children.
***E. coeruleum*** Bieb. (Dudhali, Salelimisri). Himalayas. The roots are used in parts of India as a nerve tonic and aphrodisiac.
***E. foetidum*** L. Tropical America. The roots, which smell vile, are used in soups etc. as a condiment, and give a very agreeable flavour.
***E. graecum*** L. S.E.Europe. The young leaves were used in salads by the Ancient Greeks.
***E. maritimum*** L. (Sea Holly) Mediterranean and Atlantic coasts of Europe. The young tops are occasionally eaten as a vegetable.
***E. pandanifolium*** Cham. and Schlechtd. Subtropical S.America. The leaves are the source of a cordage fibre (Caraguata fibre).
***E. planum*** L. Balkans to E.Africa. The roots

(Radix Eryngii), are occasionally used in Europe as a diuretic.

***E. synchaetum*** (A. Gray) Rose. S. U.S.A. A decoction of the roots is used in ceremonial drinks by some Indian tribes.

***E. ternatum*** Poir. Crete. The leaves are used locally as an aphrodisiac.

***E. virginianum*** Lam.=E. aquaticum.

***E. viride*** Lam. See E. graecum.

***E. yuccifolium*** Michx. (Button Snakeroot, Rattlesnake Master). E. U.S.A. The plant is diuretic, expectorant and reduces fevers; large quantities are emetic. A decoction of the roots is used by the local Indians to treat kidney complaints.

ERYSIMUM L. Cruciferae. 100 spp. Mediterranean, Europe, Asia. Herbs.

***E. canescens*** Roth.=E. diffusum. Europe through to S.Russia. The plant yields erysid, used in Russia, especially the Ukraine, to treat heart complaints.

ERYTHEA S. Wats. Palmae. 10 spp. S.W United States of America to C.America. Palm trees.

***E. edulis*** (Wendl.) S.Wats. Mexico, Guadeloupe. The fruit pulp is eaten and the young buds are cooked as a vegetable.

***Erythraea*** Borekh.=Centaurium.

ERYTHRINA L. Leguminosae. 100 spp. Tropics and subtropics. Trees or small trees.

***E. abyssinica*** Lam.=E. caffa (Kaffir Broom, Lucky Bean Tree). Africa. The white soft, woolly wood is used for carving toys, drums etc. and it is used for brake blocks.

***E. arborea*** (Chapm.) Small.=E. herbacea var. arborae. (Red Cardinal). Florida. The red seeds are used as beads (Coral Beans).

***E. caffra*** Thunb.=E. Abyssinica.

***E. corallodendron*** L. (Common Coral Bean). Tropical America. An extract of the bark is used in Brazil to treat asthma.

***E. edulis*** Triana. N. S.America. The seeds are used as a vegetable in Colombia.

***E. flabelliformis*** Kearney=E. purpusi. (Chilicote, Colorin, Coralina). S.W. U.S.A., Mexico. In Mexico the wood is used for carving and in Durango the seeds are used to treat toothache.

***E. fuseda*** Lour.=E. ovalifolia. (Galala ajer, Tjankring). S.E.Asia, Indonesia. The leaves are eaten cooked or raw in Java and Bali. The roots are used to treat beri-beri.

***E. glauca*** Willd. Venezuela, Guyana. Grown locally as shade for cacao.

***E. herbacae*** L. (Eastern Coral Bean). S.E. U.S.A., Mexico, W.Indies. The seeds are used as rat poison in Mexico and a decoction of the roots was used by the local Indians to reduce fevers.

***E. herbacae*** var. ***arborae*** Chapm.=E. arborea.

***E. hypaphorus*** Boerl.=E. lithosperma.

***E. indica*** Lam.=E. variegata (Indian Cora Bean, Coral Tree). Tropical Asia Australia. The powdered wood is used in Siam as face powder. The leaves are fodder for sheep, goats etc. It is also grown as a shade tree for plantation crops.

***E. lithosperma*** Miq. non Blume.=E. hypaphorus=E. secundiflora. S.E.Asia, Indonesia. Grown as a shade tree for tea and coffee.

***E. micropteryx*** Poepp. (Anauca). Peru. Grown as shade for cacao.

***E. monosperma*** Lam.=E. sandwicensis= Corallodendron monospermum. (Hawaiian Coral Tree, Wiliwile). Hawai. The wood is very light and is made into surf boards, carved stools, fishing net floats and outriggers for canoes. The red seeds are made into necklaces.

***E. mossambicensis*** Sim. S.Africa. The light wood was used by the local Portuguese to make xylophones. It is also used to make fire by rubbing.

***E. ovalifolia*** Roxb.=E. fuseda.

***E. purpusi*** T. S. Brandeg.=E. flabelliformis.

***E. rubinervia*** H.B.K. (Gallito). C.America. The flowers and buds are eaten locally as a vegetable and the leaves are eaten in soups.

***E. sandwicensis*** Degener=E. monosperma.

***E. secundiflora*** Hassk.=E. lithosperma.

***E. senegalensis*** DC. Tropical Africa. A decoction of the leaves is used by the natives to treat genital discharges. A fibre from the bark is used to make scented bracelets etc.

***E. suberosa*** Roxb. (Dholdak, Madar, Pangra). India, Pakistan. The light, soft, yellow-grey wood is used locally to make cheap packing cases, jars, scabbards, small boxes and water troughs. The bark is used as a cork substitute and for making insulation board.

***E. variegata*** L.=E. indica.

***E. versperilio*** Benth. Australia. The wood

was used by the aborigines to make shields.

***E. zeyheri*** Haw. S.Africa. The red seeds are used to make necklaces, bracelets etc.

ERYTHRONIUM L. Liliaceae. 25 spp. S.Europe, temperate Asia, Pacific and Atlantic N.America. Herbs.

***E. americanum*** Ker-Gwol. (Serpent's Tongue, Snowdrop, Yellow Adder's Tongue). E. N.America. The plant is used as a poultice to treat scrophulous tumours, as an emetic and to soothe inflamations.

***E. dens-canis*** L. Temperate Europe to N.Asia. The roots yield a starch used in Japan to make a pastry and cakes. The bulbs are eaten with milk in Siberia and Mongolia. The leaves are used as a vegetable.

***E. grandiflorum*** Pursh. W. N.America. The bulbs are used as a vegetable by the local Indians.

ERYTHROPHLEUM Afzel. ex G. Don. Leguminosae. 17 spp. Africa, Seychelles, tropical E.Asia, Australia. Trees. The bark of all the species mentioned is extremely toxic and is used locally as ordeal poisons.

***E. africana*** G. Don. Tropical Africa. Used in E. Africa.

***E. couminga*** Baill. (Kiminga, Komanga). Madagascar, Seychelles.

***E. fordii*** Oliv. S.China to Indo-China. The wood is used for tanning.

***E. guineënse*** G. Don.=E. judiciale=E. ordale=Afelzia grandis (Arui, Mancona, M'ka, M'kasa, Moavi, Redwater Tree, Sasswood, Sassy Bark Tree). W.Africa, especially Senegal and Cameroons. The bark is widely used in W.Africa as an ordeal poison and an arrow poison. It contains the alkaloid erythrophleine which is extracted and used medicinally as a heart stimulant, as a local anaesthetic in dentistry and to treat dysentery and diarrhoea. The hard wood is used locally for general construction work.

ERYTHROPSIS Lindl. ex Schott. and Endl. Sterculiaceae. 6 spp. Tropical Africa, Madagascar, India, S.E.Asia, Malaysia. Trees.

***E. barteri*** Ridley. W.Africa. The bark fibre is used in Ghana to make cloth and cordage.

***Erythroxylon*** L.=Erythroxylum.

ERYTHROXYLUM P. Br. Erythroxylaceae. 250 pps. Tropics and subtropics. Trees.

***E. acuminatum*** (Arn.) Walp.=E. lucidum. Tropical Asia. The leaf juice is used in Ceylon to treat intestinal worms.

***E. burmannianicum*** Griff.=E. cuneatum.

***E. citrifolium*** St. Hil. Brazil. The bark is used locally as a poultice on wounds.

***E. coca*** Lam. (Coca Tree). Bolivia, Peru. Cultivated in Java and Ceylon. The leaves contain alkaloids, e.g. cocaine and related substances which are narcotics and cerebral stimulants used medicinally in local anaesthetics. The leaves are chewed by the natives as a stimulant. The cocoa from this species comes mainly from Bolivia and is called Huanuca Coca.

***E. cuneatum*** Kurs.=E. burmannianicum (Kajoo mootah, Kajoo oorang). W.Malaysia. The durable wood is used locally for boards and posts.

***E. havanense*** Jacq. (Jiba). W.Indies, Mexico. Small tree. The leaves are used locally to prevent bleeding.

***E. lucidum*** Moon.=E. acuminatum.

***E. mannii*** Oliv. (Dabe, Landa). W. tropical Africa. The brown hard wood is used locally to make wagons etc. and ceilings.

***E. monogynum*** Roxb. E.Indies. The wood yields a tar used for treating boats. An essential oil from the tar is used in perfumery.

***E. myrtoides*** Bojer. Madagascar. The black wood is used locally for making ornaments.

***E. nova-granatense*** (Morris) Hier. (Truxillo Coca). Peru, Bolivia. Cultivated for the leaves which yield a cola (Truxillo Cola). See E. cola.

***E. obovatum*** Macf. W.Indies. The red wood is used for poles of various sorts.

***E. ovatum*** Cav. S.America, W.Indies. The wood is used to make furniture in parts of Argentina.

***E. suberosum*** St. Hil. Brazil. The bark yields a red-brown dye used locally for dyeing wool.

***E. tortuosum*** Mart. See E. suberosum.

Esaka – Entandrophragma cylindricum.

Escarolle – Cichorium endivia.

Eschallot – Allium ascalonicum.

ESCHSCHOLZIA Cham. Papaveraceae. 10 spp. Pacific N.America. Herbs.

***E. californica*** Cham. (Californian Poppy). S.W. U.S.A. The leaves are used as a vegetable by the local Indians. The plant is widely grown as an ornamental in gardens.

***Eschweilera longipes*** (Poit.) Miers.=Lecythis longipes.

Escobilla – Gymnosperma glutinosum.

ESCONTRIA Rose. Cactaceae. 3 spp. Mexico. Xerophytic cacti.

***E. chiotilla*** (Weber) Rose=Cereus chiotilla (Chiotilla). Mexico. The fruits are eaten locally.

ESENBECKIA Kunth. Rutaceae. 38 spp. Tropical America, W.Indies. Trees.

***E. febrifuga*** Juss. Argentina, Paraguay, S.E. Brazil. The bark (Angostura brasiliensis, Quina) is used to make bitters and as a quinine substitute. The wood is used to carve small articles.

Esere – Physostigma venenosum.

Esese – Ficus exasperata.

Esparcette – Onobrychis viciifolia.

Esparcette, Spanish – Hedysarum coronarium.

Esparto – Stipa tenacissima.

Esparto Chino – Fimbristylis spadicea.

Esparto mulato – Fimbristylis spadicea.

Esparvie – Anacardium rhinocarpus.

Espave Mahogany – Anacardium rhinocarpus.

ESPELETIA Mutis. Compositae. 80 spp. Andes. Xerophytes.

***E. nerifolius*** Ernst.=Trixis neriifoilia. Venezuela. The plant yields a resin (Icienso de los Crillos) used to make incense.

Espino Cavan – Acacia cavenia.

Espino rubial – Zanthoxylum martinicense.

ESPOSTOA Britton and Rose. Cactaceae. 11 spp. Tropical S.America. Cacti.

***E. lanata*** (H.B.K.) Britton and Rose.= Pilocereus lanatus=Cereus lanatus. Ecuador, Peru. The sweet fruits (Soroco) are eaten in Peru.

Esprit d'Iva – Achillae moschata.

Essang (Nuts) (Oil) – Ricinodendron heudelotti.

Essence d'Amali – Alpinia galanga.

Essence d'Amali – Alpinia malaccensis.

Essence d'Aspic – Lavendula dentata.

Essence de Bois Gaiac – Bulnesia sarmienti

Essence de Bois Rose – Aniba rosaeodora.

Essence de Bruère de Tonkin – Baekea frutescens.

Essence de Myrte-Myrtus communis.

Estragon (Oil) – Artemisia dracunculus.

Etang – Pycanthus kombo.

Ethiopean Madflower – Antholyza paniculata.

Ethiopean Sour Gourd – Adansonia digitata.

ETHULIA L. f. Compositae. 10 spp. Tropical America, tropical and S.Africa, Mascarene Islands, Assam, Sunda Islands. Herbs.

***E. conyzoides*** L. Tropical Africa and Asia. The leaves are used by the natives of Liberia to prevent abortion and by the Zulus to treat intestinal complaints.

Etu do – Dichapetalium acuminatum.

EUCALYPTUS L'Hérit. Myrtaceae. 500 spp. Australia and Tasmania, 2–3 spp. Indomalaya. Trees. The genus is one of the dominant trees of the Australian flora. The timber of many species is used for a wide variety of constructional work, e.g. house-building, fencing, railway sleepers, heavy beams, planking etc. Several produce essential oils by distillation of the leaves while others exude kinos and gums from the trunks. The timber of other species is used to make paper. To save repetition these will be referred to simply as timber, oil, kino, gum or paper respectively as the description of the species. The various oils are used medicinally as expectorants, in treating colds, as mild antiseptics and to reduce fevers. The kinos are used as astringents in medicine and tanning.

***E. acmenoides*** Schau.=E. triantha.

***E. amygdalina*** Lab.=E. salicifolia.

***E. astringens*** Maid. (Brown Mallet). W.Australia. The bark contains about 50 per cent tannin, which is used commercially and exported.

***E. australiana*** Baker and Smith. New South Wales, Victoria. Oil.

***E. baxteri*** Maid and Blakely. New South Wales, S.Australia. The wood is pale brown. Timber.

***E. blaxlandi*** Maid and Camb. New South Wales, S. Australia. The wood is pale brown. Timber.

***E. bosistoana*** F. v. Muell. (Coast Grey Box, Gippsland Grey Box). New South Wales, Victoria. The wood is light brown to pink. Timber.

***E. botryoides*** Smith. (Bangalay Eucalyptus, Bastard Eucalyptus, Blue Gum). Mainly New South Wales and Victoria. Timber used for ship-building and wheels.

***E. camaldulensis*** Dehn. = E. rostrata (Longbeak Eucalyptus, Red Gum, Australian Kino). Australia. Gum (Red Gum). The wood is very hard and resists termites and teredo worms. Timber, Paper, Wood pulp.

***E capitellata*** Smith (Brown Stringybark). New South Wales, S.Australia. The wood is pale brown. Timber.

***E. citriodora*** Hock. (Lemon-scented Gum, Spotted Gum). New South Wales, Queensland. The wood is light brown to grey-brown. Timber, Oil. The oil contains citronellal and is used in perfumery. The tree is grown for the oil in Spain, Portugal, Brazil, Guatemala, Java, S.Africa and Seychelles.

***E. cladocalyx*** F. v. Muell. = E. corynocalyx (Sugar Gum). Victoria, S.Australia. The wood is heavy and yellow-brown. Timber.

***E. cneorifolia*** DC. S.Australia. Oil.

***E. conoidea*** Benth. = E. leucoxylon.

***E. corymbosa*** Smith = E. gummifera (Blood Wood). New South Wales, Queensland. The wood is pink to red and resists termites. Timber, Red kino.

***E. corynocalyx*** F. v. Muell. = E. cladocalyx.

***E. dalrympleana*** Maid. (Mountain Gum). New South Wales. The whittish-pink wood is lighter than most species and is used for lighter construction work and tool handles.

***E. degulpta*** Smith (Kamerer). New Guinea, New Britain to Philippines. The red-brown wood is used locally for house-building etc.

***E. delegatensis*** R. T. Bak. = E. gigantea.

***E. diversifolia*** F. v. Muell. (Karri). W. to S.W.Australia. The wood is durable in soil and water and is resistant to teredo worms and termites. It is used for construction work and marine work.

***E. dives*** Schauer. (Broad-leaved Peppermint Tree). New South Wales. Oil.

***E. dumosa*** Cunningh. (Cangoo Mallee Eucalyptus, Larap). A small tree. Victoria, New South Wales. Insect attack stimulates the production of a starchy exudate (Lerp) which is eaten by the natives. Oil.

***E. eugenoides*** Sieb. = E. scabra (White Stringybark, Blackbutt). New South Wales, Queensland, Victoria. The wood is pinkish and durable. Timber.

***E. falcata*** Turcz. (White Mallet). W.Australia. The bark is a source of tannin.

***E. fergusoni*** R. T. Bak. = E. paniculata.

***E. fruticetorum*** F. v. Muell. = E. polybractea.

***E. gardneri*** Maid. (Blue Mallet). W.Australia. The bark is a source of tannin.

***E. gigantea*** Hook. f. = E. delegatensis. (Alpine Ash, Red Mountain Ash, Woollybutt, White-top Stringbark). S.E.Australia. The wood is pale brown and open. Timber. The wood is used for a wide variety of purposes, but not very heavy construction work.

***E. globulus*** Lab. (Blue Gum, Tasmanian Blue Gum, Tasmanian Blue Eucalyptus, Victoria Blue Gum, Fever Tree). Australia. Oil (Oil of Eucalyptus, Oleum Eucalypti). The timber is durable and is used for ship-building and agricultural implements. The tree is cultivated for oil production.

***E. gomphocephala*** DC. (Tuart, White Gum). W.Australia. The wood is light yellow and very tough and strong. It is used for wagon-making and wheel hubs.

***E. goniocalyx*** F. v. Muell. (Blue Gum, Grey Gum, Big-leaf Eucalyptus, Spotted Gum). Australia. The wood is yellowish-brown and resists decay in soil. It is used mainly for ship-building and wheel-making.

***E. gummifera*** (Gaertn.) Hochr. = E. corymbosa.

***E. gunnii*** Hook. f. Australia. A stem exudate is eaten by the natives.

***E. haemostroma*** Smith (Scribbly Gum). Australia, mainly New South Wales and Queensland. The bark is a minor source of tannin.

***E. hemilampra*** F. v. Muell. = E. resinifera.

***E. hemiphloia*** F. v. Muell. (Grey Box, Yellow Box, White Box). Australia. Timber.

***E. intertexta*** R. T. Baker. Australia. Small tree. The ground seeds are eaten by the natives.

***E. leucoxylon*** F. v. Muell. = E. conoidea. (White Gum, White Ironbark). Oil, Kino, Timber.

***E. macarthuri*** Deane and Maiden. Australia. Oil. Used locally to denature alcohol in perfumery.

***E. macrorhynchia*** F. v. Muell. (Stringybark). Mainly New South Wales and Victoria. A red kino.

***E. maculata*** Hook. = E. variegata (Spotted Gum). Australia. Timber used for ship-building, wheels, bridges and roofing. Yellow-brown kino.

***E. maideni*** F v. Muell. (Spotted Blue Gum, Maiden's Gum). New South Wales. Wood light yellow. Timber.

***E. mannifera*** Cunning. = E. viminalis.

***E. mannifera*** Mudic. (Manna Eucalyptus). Australia. A stem exudate (Eucalyptus Manna) is eaten by the natives.

***E. marginata*** Smith (Jarrah). W.Australia. The wood is hard, red-brown and though coarse, polishes well. Timber. Much is exported to England (for flooring) Africa and India

***E. melliodora*** A. Cunn. (Yellow Box). Australia. The wood is light pink to yellow-brown. Timber.

***E. microcorys*** F. v. Muell. (Tallow Tree). N.E. New South Wales to S.E. Queensland. The wood is light yellow with a waxy texture. Timber.

***E. microtheca*** F. v. Muell. (Floodbed Box, Colibah). Small tree. The bark is used by the natives of Queensland to intoxicate fish.

***E. muelleriana*** Howitt. (Yellow Stringybark). New South Wales, Victoria. The yellowish wood is termite resistant. Timber.

***E. nitens*** Maid. (Shining Gum, Ribbon Gum). N.E. Victoria to S.E. New South Wales. The white wood is not durable. It is used for building construction, but not in contact with the ground.

***E. obliqua*** L'Hérit. (Stringy Bark, Messmate Stringybark). New South Wales, Victoria, S.Australia. The wood is light and is used for house construction and for pulping. The bark yields a fibre used for mats, ropes etc. The bark is also used for thatching.

***E. oleosa*** F. v. Muell. Australia. Small tree. Oil.

***E. oreades*** R. T. Bak. (Blue Mountain Ash). New South Wales, Queensland. The wood is whitish and is used for house construction furniture-making and small articles.

***E. paniculata*** Smith. = E. fergusoni (Grey Ironbark). New South Wales, Queensland. The wood is very hard and deep red-brown. Timber.

***E. patens*** Benth. (Western Australian Blackbutt). W.Australia. Heavy hard wood is used for construction work, wagons and railway sleepers.

***E. persicifolia*** Lodd. = E. viminalis.

***E. pilularis*** Smith (Blackbutt). Australia. Timber.

***E. piperata*** Smith (Blackbutt, Redwood, Sydney Peppermint Eucalyptus, White Stringy Bark). Australia. Rough timber. Pipertone is distilled from the leaves and is used in the manufacture of menthol and thymol.

***E. polyanthemos*** Scham. (Bastard Box, Grey Box, Red Box). Victoria, New South Wales. The wood is hard and durable. It is used for mining timbers, cogs etc.

***E. polybractea*** F. v. Muell. = E. fruticetorum (Blue Mallet). Small tree. Victoria, New South Wales. Oil.

***E. pulverulenta*** Sims. Australia. A stem exudate is eaten by the natives.

***E. redunca*** Schau var. ***elata*** Benth. (Wandoo, White Gum). S.W. W.Australia. The wood is very hard and durable. Timber. The bark is used for tanning.

***E. regnans*** F. v. Muell. (Australian Oak, Mountain Ash, Swamp Gum). E. Victoria, Tasmania. The wood is light brown and open textured. Timber and wood pulp.

***E. resinifera*** Smith = E. spectabilis = E. hemilampra. (Kino Eucalyptus). New South Wales, Queensland. The wood is durable in soil and water. Timber. Yields a ruby coloured, water-soluble kino.

***E. robusta*** Smith. (Beakpod Eucalyptus, Kimbarra, Swamp Mahogany, White Mahogany). New South Wales. The wood is brittle. Timber.

***E. rostrata*** Schlecht = E. camaldulensis.

***E. salicifolia*** Cav. = E. amygdalina (Mountain Ash, Peppermint Tree, Willowleaf Eucalyptus). Victoria, New South Wales. The timber is relatively light. Timber, wood pulp. The trunk yields a red kino (Ribbon Kino).

***E. saligna*** Smith. (Blue Gum, Saligna Gum, Sydney Blue Gum). E. Australia. The brown wood is hard and heavy. Timber and wood pulp for paper etc.

***E. scabra*** Dum-Cours. = E. eugenoides.

***E. siderophloia*** Benth. (Broadleaf Ironbark, Ironbark, Red Ironbark). Queensland. The wood is durable and tough. Timber. Ruby red kino (Botany Bay Kino).

***E. sideroxylon*** A. Cunn. (Red Ironbark). New South Wales. The wood is dark red and very heavy and tough. Timber.

***E. sieberiana*** F. v. Muell. (Cabbage Gum). New South Wales, Victoria, Tasmania. The wood is used for pulp and paper. The trunk yields a garnet-coloured, water soluble kino.

***E. spathulata*** Hook. (Swamp Mallet). W. Australia. The bark is used for tanning.

***E. spectabilis*** F. v. Muell. = E. resinifera.

***E. staigeriana*** F. v. Muell. (Lemon-scented Ironbark). Australia. Oil, used in perfumery.

***E. stuartiana*** F. v. Muell. Australia. Oil, used in perfumery.

***E. subulatum*** Cunningh. = E. tereticornis.

***E. tereticornis*** Smith. = E. subulatum. (Blue Gum, Flooded Gum, Grey Gum, Red Gum). New South Wales, Queensland. Timber.

***E. terminalis*** F. v. Muell. (Kutcha Bloodwood Eucalyptus). Australia. A gum exuded from the leaves is eaten by the natives of Queensland.

***E. triantha*** Link. = E. acmenoides. (White Mahogany, Yellow Stringybark). E. Australia. The yellowish wood is hard and heavy. It resists decay and burning. Timber.

***E. umbellata*** (Gaertn.) Domin. (Blue Gum, Forest Red Gum). New South Wales, Queensland, Victoria. The wood is reddish, heavy and durable. Timber

***E. variegata*** F. v. Muell. = E. maculata.

***E. viminalis*** Lab. = E. mannifera Cunningh. = E. persicifolia). (Manna Gum, Ribbon Eucalyptus, Swamp Gum, White Gum). Australia. The brown coarse-grained wood is durable in contact with soil. Timber and pulp. The bark exudes a sweet liquid which is eaten by the natives.

Eucalyptus, Longbeak – Eucalyptus camaldulensis.

Eucalyptus, Manna – Eucalyptus mannifera.

Eucalyptus, Willowleaf – Eucalyptus salicifolia.

EUCARYA T. L. Mitch. Santalaceae. 4 spp. S. and E. Australia. Shrubs or small trees.

***E. acuminata*** (R. Br.) Sprague. = Santalum acuminatum = S. preissianum. (South Australian Sandalwood, Quandong). Australia. The hard pleasantly scented wood is much valued for cabinet work. An oil (South Australian Sandalwood Oil) is distilled from the wood and used in perfumery. The fruits are eaten raw or in jellies and preserves. The seeds are eaten by the natives.

***E. spicata*** Sprague and Summ. = Santalum spicatum. (West Australian Sandalwood). S.W. Australia. The wood is much valued for cabinet work. An oil (West Australian Sandalwood Oil) is distilled from the wood and is used as a substitute for Sandalwood Oil. It contains santalol, fusanols and sesquiterpene alcohols, and is used in perfumery, in urinary antiseptics etc. It is also used by the Chinese for making incense.

EUCHEUMA J. Ag. Solieriaceae. Red Alga. Pacific, Indian Ocean.

***E. gelatinae*** (Esp.) J. Ag. Used in China for the manufacture of agar which is eaten, used for mounting paintings, lanterns etc. In Japan it is also used for making size for cloth.

***E. muricatum*** (Gmel.) Web. v. Bosse. = E. spinosum. A valuable source of agar (Macassar Agar) which is used in China as food. Japan and the E.Indies export much to China.

***E. papulosa*** Cotton and Yedo. = Callymenia papulosa. Extends to the coast of Somaliland. China is the main producer. The algae is used in the production of isinglass which is used in food preparations, in refining beers and wines, and as a glue.

***E. speciosa*** J. Ag. = Gigartina speciosa. W.Australia. Used in Australia in the preparation of blancmange, jellies etc.

***E. spinosum*** L. = E. muricatum.

EUCHLAENA Schrad. Graminae. 2 spp. Mexico. Annual Grasses.

***E. mexicana*** Schrad. (Teosinte). Mexico. The grass is grown occasionally in Mexico and S. U.S.A. as a fodder crop. Its chief interest is that it is probably one of the progenitors of cultivated maize (Zea mais).

EUCHRESTA Bennett. Leguminosae. 5 spp. E.Asia; 1 sp. Java. Small shrubs.

***E. horsfieldii*** Bennett. (Palakija, Prandjiwa). Himalayas, through Malaysia to Philippines and Java. The seeds contain the alkaloid cytisine which is poisonous. They are used by the Javanese to treat chest ailments, as a tonic and as an aphrodisiac. The Chinese used the seeds as an aphrodisiac.

EUCLEA Murr. Ebenaceae. 20 spp. Africa, Comoro Islands, Arabia.

***E. natalensis*** A. DC. Shrub. Natal, Transvaal. A dye obtained from the roots is used for staining floors.

***E. pseudebenus*** E. Mey. (Cape Ebony, Orange River Ebony). Tree. S.Africa. The wood is used locally for general carpentry and furniture-making.

EUCOMMIA Oliv. Eucommiaceae. 1 sp. China. Tree.

***E. ulmoides*** Hook. (Tu chung). The bark is used locally as a tonic. All parts of the plant, and the seeds contain a gum which is a potential source of rubber.

EUCRYPHIA Cav. Eucryphiaceae. 5–6 spp. S.E.Australia, Tasmania, Chile. Trees.

***E. billandieri*** Spach.=E. lucida.

***E. cordifolia*** Cav. (Ulmo). Chile. The hard, close-grained wood is used for general carpentry, railway sleepers, telegraph poles, furniture, flooring and by the natives for building canoes. The bark is a source of tanning and the flowers give honey.

***E. lucida*** (Labill.) Baill.=E. billandieri. (Leatherwood). Tasmania. The wood is pink to red-brown, hard and works well. It is used for interior work, plywood, cabinet work etc. and tool handles.

***E. moorei*** F. v. Muell. (White Sally). New South Wales, Victoria. The bark is used for tanning.

EUGEISSONA Griff. Palmae. 8 spp. Malaysia, Borneo. Small palms.

***E. insignis*** Becc. (Djato, Kadjatao). Borneo. The centres of the prop roots are used locally to make baskets, bracelets and walking sticks.

***E. minor*** Becc. (Todjong pipit). Borneo. The brace roots were exported to make umbrella handles.

***E. triste*** Griff. Malaysia. The roots are used to make the floors of native houses. The leaf midribs are used to make the walls and they are also used to make blowpipe darts.

***E. utilis*** Becc. (Kadatoa). Borneo. The plant is semi cultivated by the Dyaks for the starch extracted from the stem and the bright purple pollen which is used to colour rice dishes.

EUGENIA L. Myrtaceae. 1,000 spp. Tropics and subtropics. Trees.

***E. acapulcensis*** Steud.=E. maritima=Myrtus maritima. (Capulin). Mexico. The fruits are eaten locally.

***E. acuminatissima*** Kurz.=Jambosa acuminatissima=J saligna=Syzygium subdecurrens (Dail besar, Hejas, Klampok). Malaysia The red wood is strong, but not termite resistant. It is used locally for building. The bark yields a dye which stains fabrics black.

***E. aquea*** Burm. f. (Watery Rose Apple). Ceylon, Malaysia. The fruits are edible, but rather insipid. They are eaten raw or made into a syrup (Roedjak). The plant is cultivated and the fruits sold in Java.

***E. aromatica*** Baill.=E. caryophyllata.

***E. axillaris*** (Sw.) Willd. (White Stopper). W.Indies, Florida. The hard wood is used locally for cabinet work.

***E. brackenridgei*** Gray. Fiji, New Caledonia. The very dark red wood is very hard and polishes well. It is used for cabinet-making, turning and house interiors.

***E. caryophyllata*** Thunb.=E. aromatica=Caryophyllum aromaticum. (Clove Tree). Moluccas. The plant is the source of the clove of commerce. These are the sun-dried, unopened flower buds. They are picked by hand when green, or turning red. Cloves are one of the more important spices of commerce. They are used whole, or ground and an oil (Oil of Cloves) is obtained by distillation. Their flavour and odour are due to the presence of eugenol in high concentrations (14–19 per cent). The cloves are used as spices, in pickles and sauces. The oil is used as a germicide, in dentistry, for relieving toothache, in perfumery, in the synthesis of vanillin and as a clearing agent in microscope work. The plant originated in the Moluccas, where the Dutch had a monopoly of this important spice, but it has been used by the Chinese since the third century B.C. It was known to the Romans and became of commercial importance in Europe during the Middle Ages. The plant is grown commercially in Java, the former Dutch East Indies, Madagascar and the W.Indies, but 90 per cent of the world production is from Tanzania (Zanzibar and Pemba). The plants are propagated from seeds and come into bearing in 8–9 years, after which they yield for about 60 years.

***E. chequen*** Mol. = Myrtus cheken (Cheken). S.America, especially Chile. The leaves are used as a tonic, diuretic and expectorant, especially for treating colds of older people.

***E. clavimyrtus*** Koord. and Valet. = Jambosa glabrata. Java. The bark yields a dye used locally for dyeing fabrics black. The wood is sometimes used for construction work.

***E. confusa*** DC. (Ironwood, Red Stopper). Florida, W.Indies. The dark red wood is hard and is used locally for cabinet work.

***E. conzattii*** Standl. Mexico. The fruits are eaten locally.

***E. cordata*** Laws. (Water Tree). S.Africa. The dark red wood is hard, elastic and durable. It is used locally for heavy construction work, boat-building and water wheels.

***E. cuprea*** Koord. and Valet. Java. The wood is hard and heavy. It is used locally for house-building.

***E. curranii*** C.B.R. (Lipoti). Philippines. The dark red fruits are eaten raw, made into jellies, or used to make wine. The plant is occasionally cultivated locally.

***E. cymosa*** Lam. = Jambosa tenuicuspis. Malaysian Archipelago, Mauritius. The bark yields a black dye used for staining cloth.

***E. densiflora*** DC. = Jambosa densiflora. Malaysia Archipelago. The fruits and flowers are eaten, especially in Java. The bark yields a brown dye, used locally for dyeing fabric.

***E. dombeyana*** DC. (Grumichama). Peru, Brazil. The fruits are red and small. They are eaten raw or cooked and are used in preserves. The plant is occasionally cultivated locally.

***E. edulis*** Benth. Brazil to Argentina. The fruits are eaten locally and used to make vinegar in the Argentine.

***E. floribunda*** West. (Murta, Rumberry). W.Indies. The dark red fruits are used locally to make jam and to make a rum-based liqueur which is exported, mainly to Denmark.

***E. foliosa*** DC. = Myrtus foliosa (Arrayán). Ecuador. The fruits are eaten locally.

***E. formosa*** Wall. Tropical Asia. The fruits have an insipid flavour, but are eaten locally, especially in Cochin-China, where the plant is cultivated.

***E. gerradi*** Sm. S. Africa. The fruits are eaten by the natives and an infusion of the bark is used by the Zulus to treat chest complaints.

***E. grandis*** Wight. Malaysia, Australia. The wood is not durable, but is used in Java to build boats and houses.

***E. gustavioides*** Bailey (Grey Satin Ash). Queensland. The yellowish wood is not durable in contact with the ground, but is used for construction work, etc.

***E. jambolana*** Lam. = Syzygium jambolanum (Java Plum, Jambolan). Tropical Asia to Australia. The fruits are used to make preserves. The seeds are diuretic and are also used to reduce the blood sugar in cases of diabetes.

***E. jamboloides*** Koord. and Valet = Syzygium racemosum. Java. The wood is used occasionally for building. The bark yields a black dye used for staining fabrics.

***E. jambos*** L. = Jambosa vulgaris. (Rose Apple, Jambos). Old World Tropics. Introduced to New World. Cultivated for the fruits which are used for making jellies and are candied. The flowers are occasionally candied. The bark is used locally for tanning.

***E. javanica*** Lam. = E. mananquil.

***E. klotzschiana*** Berg. (Pera do Campo). Brazil. The fruits are very aromatic and are eaten raw or in jellies. The plant is occasionally cultivated locally.

***E. lancilimba*** Merr. (Unani). Philippines. The fruits are used locally for making jellies.

***E. lineata*** (DC.) Duth. = Jambosa lineata. Malaysia. The wood is used for house-building and the bark for tanning fishing nets.

***E. lucidula*** Miq. = E. polyantha.

***E. luschnathiana*** Berg. (Pitomba). Brazil. The orange-coloured fruits are used locally to make jellies. The tree is occasionally cultivated locally.

***E. macrophylla*** Lam. = E. malaccensis.

***E. maire*** Cunningh. New Zealand. The hard strong wood is used locally for piers etc. and cabinet work.

***E. malaccensis*** L. = E. macrophylla (Mountain Apple). Tropical Asia, especially Malaysia. The tree is occasionally cultivated for the fruit.

***E. mananquil*** Blanco = E. javanica. (Mankil, Samarang Rose Apple). Malaysia and especially Philippines. The large red fruits have a good flavour and are very popular in Java. Cultivated.

***E. maritima*** DC. = E. acapulcensis.

***E. michelii*** Lam. (Brazil Cherry). Tropical America. The fruits are eaten locally.

***E. myrtifolia*** Sims. (Bush Cherry, Red Myrtle). Australia. The fruits are used locally to make jellies. The wood is elastic and is used to make boats and oars. The aborigines use it to make boomerangs and shields.

***E. opaca*** Koord. and Valet. Java. The heavy wood is used locally in house-building.

***E. operculata*** Roxb. = Syzygium modosum. Tropical Asia. The bark is used locally for tanning.

***E. pimenta*** DC. = Pimenta officinalis.

***E. polyantha*** Wight. = E. lucidula. Malaysia. The bark is used locally for tanning fishing nets and staining bamboo. A decoction of the leaves and bark is used to treat diarrhoea in Java.

***E. polycephala*** Miq. = Jambosa cauliflora. India to Borneo. The tree is cultivated locally for the fruits which are used to make jellies.

***E. polycephaloides*** C.B.R. (Maigang). Philippines. The fruits are eaten locally raw or in jellies.

***E. rhombea*** (Berg.) Urban. (Red Stopper). W.Indies, Florida. The hard light brown wood is used locally for cabinet-making.

***E. rumphii*** Merr. (Aajoo merah). Indonesia. The bark is used locally for tanning fishing nets.

***E. selloi*** Berg. = Phyllocalyx edulis. (Pitanga tuba). Brazil. The fruits are used locally to make jam.

***E. spicata*** Lam. = E. zeylanica.

***E. stahlii*** (Kaiersk) Krug. and Urban. (Guayabota). Tropical America, W.Indies. The wood is very susceptible to termites and is difficult to work, but because of its pleasant scent and attractive grey-brown graining it is used for furniture-making, carving, interior work etc. It is also used for agricultural implements and boat-building.

***E. thumra*** Roxb. Malaysia. The wood is red and coarse-grained and is susceptible to termite attack. It is used in Java for building. The bark yields a red dye, used to stain fabrics. The fruits are edible.

***E. tomentosa*** Camb. (Cabelluda). Brazil. The fruits are used locally to make jellies.

***E. uniflora*** L. (Pitanga, Surinam Cherry). Tropical America. Cultivated in the tropics and subtropics for the fruit. These are dark orange to black with a sweet, aromatic flavour. They are eaten raw, or in jams, jellies, sherberts etc. The crushed leaves are spread in barns to repel insects.

***E. uvalha*** Camb. (Uvalha). S.Brazil. Cultivated locally for the fruits which are used to make soft drinks.

***E. zeylanica*** Wight. = E. spicata. Malaysia. The bark yields a black dye, used for staining fabrics.

***Euhymenia papulosa*** Kuetz. = Eucheuma papulosa.

EULALIOPSIS Honda. Graminae. 2 spp. India, China, Formosa, Philippines. Perennial grasses.

***E. binata*** (Retz.) Hubbard. = Pollindium binatum. (Babni, Baggar, Bhabar). N.India, Pakistan, to Central India. Important locally for manufacturing paper. It is also used to make cordage and mats.

EULOPHIA R. Br. ex Lind. Orchidaceae. 200 spp. Throughout the tropics. Herbs.

***E. arenaria*** Don. Tropical Africa. The Zulus use a decoction of the pseudobulbs to cure sterility in men and women.

***E. barteri*** Steud. Tropical Africa, especially Liberia. It is used in Liberia in witchcraft concerned with the breaking of sex taboos.

***E. campestris*** Wall. (Amrita, Salibmisri, Sungmisrie). India. Used by the Hindus as a tonic, to treat heart complaints, coughs, inflammations of the mouth and as an aphrodisiac.

***E. epidendraea*** Fischer = E. virens. N.India. A decoction of the pseudobulbs is used locally to treat intestinal worms.

***E. flaccida*** See E. hains.

***E. herbacea*** Lindl. N.W.India. The pseudobulbs are used locally to make a meal (salep).

***E. hians*** N. E. Br. Tropical Africa. An infusion of the plant is used locally to prevent miscarriages and barrenness of women.

***E. livingstonia*** Reich. Madagascar. The pseudobulbs are eaten locally.

***E. plantagina*** Rolfe. See E. livingstonia.

***E. robusta*** Lindl. See E. hians.

***E. sempervirens*** Rolfe. Tropical Africa. An infusion of the leaves is used locally to treat intestinal worms.

***E. virens*** R. Br. = E. epidendraea.

***Eulophus ambiguus*** Nutt. = Lomatium ambigua.

EUODIA J. R. and G. Forst. Rutaceae. 45 spp. Tropical Africa, Asia, Australia, Pacific Islands. Trees.

***E. amboinensis*** Merr. (Gandaroora besar, Giba). Indonesia. The dried bark is used locally as incense.

***E. fraxinifolia*** Hook.=Philagonia fraxinifolia=P. procera. N.India to N.Vietnam. A yellow aromatic oil distilled from the fruits is used locally as a perfume. The oil distilled from the seeds is used for lighting.

***E. latifolia*** DC. (Empang, Ki sampang, Saoyoo). Indonesia. In Java, a resin from the stem is mixed with coconut oil and used to varnish bamboo which has been coloured with vermillion and lime.

***E. lepta*** Merr.=E. ptelaefolia=E. roxburghiana=Zanthoxylum roxburghianum. China to Philippines. A decoction of the plant is used as a bitter tonic and to control menstruation. A decoction of the leaves is used locally to reduce fevers.

***E. lunu-ankenda*** (Gäertn.) Merr. (Kanalei, Midaumabaphang, Vanashempaga). From E.Himalayas through Malaysia and Indochina. An infusion of the flowers and leaves is used in Malaya to control menstruation and as a tonic. An extract of the root-bark, boiled in oil is used locally as a cosmetic.

***E. melufolia*** Benth. var. celebica (Apedoo, Pendoo). Celebes. The wood is used locally to build houses.

***E. ptelaefolia*** Merr.=E. lepta.

***E. roxburghiana*** Benth.=E. lepta.

***E. ruticarpa*** (A. Juss.) Benth. E.Himalayas through to China. Cultivated in China. where the leaves are used to treat stomach complaints and to aid digestion.

***E. sambucina*** Hook.=Philagonia sambucina. (Kajoo menjawak, Mareh). Indonesia. Locally the crushed leaves are used to treat intestinal worms. They are warmed over a fire and cooled before use.

EUONYMUS L. Celastraceae. 176 spp. Nearly cosmopolitan, but most species in China and Japan. Small trees.

***E. americanus*** L. (Brook Euonymus, Strawberry Bush). E. N.America. The bark causes intestinal irritation and is used by some Indian tribes as a purgative.

***E. atropurpureus*** Jacq. (Burning Bush, Indian Arrow Wood). E. N.America. The bark (Wahoo Bark) is used medicinally as a purgative.

***E. cochinchinensis*** Pierre. (Bornyac, Ko maiy). Indonesia. An alcoholic extract of the bark is used locally as an aperitive.

***E. crenulatus*** Wall. W.Himalayas. The aril of the fruit is used locally as a cosmetic. The bark and aril are used as a purgative.

***E. europaeus*** L. (Spindle Tree). Europe through to W.Siberia. The wood is fine-grained and is used for turning, carving etc. and for charcoal for gunpowder. The fruits yield a yellow dye used for colouring butter and the seeds give an oil which is occasionally used for making soap.

***E. hians*** Koehne. Japan. The wood is hard and heavy. It is used locally for carving, mosaics, turning and printing blocks.

***E. japonicus*** Thunb.=E. pulchellus. S.Japan. Cultivated in Spain as a source of rubber.

***E. oyxphyllus*** Miq. Japan, China to Korea. The fine-grained, hard wood is used in Japan for printing blocks, carving etc.

***E. pendulus*** Wall. W.Himalayas. The bark is used locally as a cosmetic and purgative. The aril of the fruit is used as a cosmetic.

***E. pulchellus*** Carr.=E. japonicus.

***E. sieboldiana*** Blume. Japan. The root latex is a potential source of rubber.

***E. thunbergiana*** Hook. China. A tea is made locally from the flowers.

***E. tingens*** Wall. W.Himalayas. The bark is used locally as a purgative and cosmetic. The arils are used as a cosmetic.

EUPATORIUM L. Compositae. 1,200 spp. Mostly America, a few in Europe, Asia, Africa. Herbs.

***E. aromaticum*** L. (Wild Hoarhound, White Snake root). E. N.America, to Missippi, Florida. The aromatic root is used as a diuretic and antispasmodic.

***E. ayapana*** Vent.=E. triplinerve. Tropical America. The plant is occasionally cultivated for the leaves which are made into a digestive tea.

***E. cannabinum*** L. (Hemp Eupatorium, Water Hemp). Europe, N.Africa, Asia. The roots and leaves (Radix et Herba Cannabis Aquatica) contain the glucoside eupatorin and are used locally to treat dropsy.

***E. capillifolium*** (Lam.) Small. (Dog Fennel). E. N.America. The juice is used locally to treat insect bites and the plants are spread on floors etc. to discourage insects.

***E. celebicum*** Blume.=Vernonia arborea.

***E. chinense*** L. China, Japan. The leaves are used in Chinese medicine as a diuretic.

***E. collinum*** DC. (Yerba del Angel). C.America to Mexico. The leaves are aromatic and are used locally as a substitute for hops in the manufacture of beer and to treat liver complaints.

***E. compositifolium*** Walt. E. N.America. The leaves are used to treat insect bites and are spread on floors etc. to discourage insects

***E. dalea*** L. W.Indies, S.America. The leaves contain courmarin and are used as a substitute for vanilla.

***E. laeve*** DC. Argentina, Paraguay. The leaves yield a blue dye.

***E. indigofera*** Perdoli. See E. laeve.

***E. maculatum*** Just. (Joe-Pye Weed, Smoke Weed). S.E. and Central U.S.A. The leaves, roots and flowers are used locally to treat kidney complaints.

***E. perfoliatum*** L. (Boneset). N.America. The dried leaves and flower heads are used medicinally as a purgative emetic, to reduce fevers and to treat colds.

***E. purpureum*** L. (Gravel Root, Queen-of-the-Meadow Root). E. N.America. The root is used to treat bladder stones and kidney complaints. It is also used as a nerve tonic.

***E. rebaudianum*** Bertoni = Stevia rebaudiana.

***E. squarrosus*** Cav. = Brickellia cavanillesii.

***E. staechadosmum*** Hance. (Ayapana du Tonkin, Mân turói). Vietnam. The leaves smell of lavender and an infusion of them is used locally to aid digestion, as a tonic, as an aphrodisiac and by the women as a hair-dressing. The plant is cultivated locally.

***E. triplinerve*** Vahl. = E. ayapana.

***E. villosum*** Sw. Florida to W.Indies. The leaves are diuretic and are used locally to treat cholera and diarrhoea.

EUPHORBIA L. Euphorbiaceae. 2,000 spp. Cosmopolitan, especially in subtropics and warm tropics. Herbs, shrubs or small trees.

***E. aegyptiaca*** Boiss. Small Shrub. Egypt to Sudan. The plant is used locally as a purgative.

***E. atoto*** Forst. f. = E. dulcis. Tropical Asia, especially India, Indo-China, Malaysia to Australia. In Indo-China the latex is used to control menstruation and to induce abortions.

***E. antiquorum*** L. S.W.Asia, Africa. Xerophytic tree. The latex is poisonous, but the young shoots are made into a sweetmeat by the Chinese in Java. This involves prolonged boiling in sugar.

***E. antisyphilitica*** Zucc. = E. cerifera. (Candelilla). Mexico, S.W. U.S.A. Shrua. The plant yields a wax (Candelilla Wax). The plants are collected from the wild and the wax removed by boiling in water, with added sulphuric acid. The production of the wax is a peasant industry and there is a great danger of over-collecting. The wax is used in leather polishes and water proofing. Mixed with rubber it is used for insulating, dental moulding. It is also used in sealing wax, metal lacqueurs, gramophone records, paint removers, paper sizes, lithographic colours etc. Mixed with paraffin it is used to make candles.

***E. balsamifera*** Ait. (Balsam Spurge). W.Africa, Canary Islands. The young shoots are eaten as a vegetable in Senegal.

***E. calyculata*** H.B.K. (Chipire). Mexico. Yields a rubber (Chupire Rubber), which is of little importance.

***E. candelabrum*** Trém. Tropical Africa. The latex is used as an arrow poison in the Sudan.

***E. cattumando*** W. Elliot = E. trigona. (Chattimandy). India. A cement is made from the latex, which is extracted from the stem by boiling. It is used locally for cementing knives into the handles and in larger quantities is used for moulding fancy articles.

***E. cerifera*** Alc. = E. antisyphilitica.

***E. characias*** L. Mediterranean. The plant is used in Greece to stupify fish.

***E. corallata*** L. E. N.America. The root (Purging Root, Emetic Root) was used in small quantities by the local Indians as an expectorant and to reduce fevers. Larger quantities are emetic.

***E. cotinifolia*** L. C.America to N. S.America. The juice is strongly purging and emetic. It was used by the local Indians as an arrow poison, to stupify fish and as a criminal poison.

***E. cyparisias*** L. Europe, N.Asia, N.America. The juice is used in the Ukraine as a cosmetic to remove freckles.

***E. dendroides*** L. Mediterranean. The crushed plant is used locally to intoxicate fish.

***E. dregeana*** E. Mey. Africa. Tree. Yields a rubber for which the plant has been occasionally cultivated.

***E. dulcis*** Blanco. = E. atoto.

***E. elastica*** Alt. = E. fulva.

***E. fiha*** Dec. (Fiha). Madagascar. Tree. The latex is used to make rubbers and guttas, and sea-resistant varnishes.

***E. fulva*** Stapf. = E. elastica = Euphorbiodendron fulvum. Mexico. Tree. The latex is a minor source of rubber (Palo Amarillo Rubber).

***E. hermentiana*** Lam. (Sango, Soungax). Tropical Africa. Tree. The latex is used locally as an arrow poison etc.

***E. hirta*** L. = E. pilulifera.

***E. hypericifolia*** Gray = E. preslii.

***E. hypericifolia*** L. Tropical America, W.Indies. Annual. A decoction of the plant is used in the W.Indies to relieve pains in the uterus.

***E. intisy*** Drake del Cast. (Herotin, Intisy). Madagascar. Once used as a source of rubber (Intisy Rubber).

***E. ipecacuanha*** L. (Ipecac Spurge). N.America. The plant is used medicinally as an emetic.

***E. kamerunica*** Pax. (Solo). Tree. W.Africa. Cultivated locally for the latex which is used as an arrow poison and for tattooing.

***E. kerrii*** Craid. (Tchidi souan). Burma. The roots are used locally for stupifying fish.

***E. lathyris*** L. (Caper Spurge). Asia, Europe, Introduced to N.America. The seeds are used as a purgative and a coffee substitute.

***E. ligularia*** Roxb. = E. neriifolia.

***E. lorifera*** Stapf. (Akoko, Koko). Hawaii. The latex is a possible source of rubber and for manufacturing chewing gum.

***E. marginata*** Pursh. (Snow-in-the-Mountain). Annual. N.America. The latex is used as a chewing gum by the Indians of New Mexico.

***E. neriifolia*** L. = E. ligularia = E. pentagona. Shrub. Tropical Asia. The latex is used locally as a purgative and to remove warts. The juice is taken to relieve the spasms caused by asthma. The young shoots are eaten as a sweetmeat after prolonged boiling in sugar. The leaves are diuretic and are used locally to stupify fish.

***E. pentagona*** Blanco. = E. neriifolia.

***E. pilulifera*** L. = E. hirta. Tropics. Herb. The whole plant is used as a sedative and to assist the breathing of asthmatics.

***E. pirahazo*** Jum. Tree. Madagascar. The plant yields an inferior rubber.

***E. preslii*** Guss. = E. hypericifolia Gray. Herb. E. N.America. The plant was used by the local Indians to treat dysentery.

***E. primulaefolia*** Baker. Madagascar. The leaves are used locally as a rat poison. They are highly purgative.

***E. resinifera*** Berg. Shrub. N.Africa. The plant yields a resin (Gummi Resina Euphorbium) which is used in paints and in veterinary medicine. It was used medicinally as a drastic purgative.

***E. rhipzaloides*** Welw. W.Africa. Shrub. Yields a rubber (Almeidina Rubber, Kartoffelkautschuk) which was exported from Angola. The latex is used as a waterproof paint for ships' hulls.

***E. sequiriana*** Proch. N.Asia, Europe, America. The latex is used in some parts of Russia as a cosmetic to remove freckles etc.

***E. semivillosa*** Proch. Steppes of Russia. An extract of the dried root in vodka is used to cause vomiting after poisoning. The juice of the plant is used locally as a cosmetic to clear the skin.

***E. stepposa*** Zoz. See E. sequiriana.

***E. thomsoniana*** Boiss. (Hirtiz, Hirivi). Herb. Kashmir. The roots are used locally as a soap substitute to wash the hair and as a purgative.

***E. trigona*** Haw. = E. cattimandoo.

***E. venefica*** Trém. Tropical Africa. Shrub. The latex is used locally as an arrow poison.

***E. virgata*** L. See E. sequiriana.

***Euphorbiodendron fulvum*** Millsp. = Euphorbia fulva.

***Euphoria longana*** Lam. = Nephalium longana.

***Euphoria nephelium*** DC. = Nephelium lappaceum.

EUPHRASIA L. Scrophulariaceae. 200 spp. N.Temperate, Mountains of Malaysia, New Zealand, S.America. Herbs.

***E. officinalis*** L. = E. rostkoviana. (Drug Eyebright). Europe. The plant was commonly used in medicine to treat eye diseases and as a tonic.

***E. rostkoviana*** Hayne. = E. officinalis.

***Eupteron nodosus*** Miq. = Polyscias nodosa.

European Alder – Alnus glutinosa.

European Ash – Fraxinus excelsior.

European Aspen – Populus tremula.

European Barberry – Berberis vulgaris.

European Beachgrass – Ammophila arundinacea.
European Beech – Fagus sylvatica.
European Birch – Betula alba.
European Bird Cherry – Prunus padus.
European Blackberry – Rubus fruticosus.
European Blackcurrant – Ribes nigrum.
European Blueberry – Vaccinium myrtillus.
European Bugle – Ajuga chamaepitys.
European Dewberry – Rubus caesius.
European Elder – Sambucus nigra.
European Goat's Rue – Galega officinalis.
European Goldenrod – Solidago virgaurea.
European Gooseberry – Ribes grossularia.
European Grape – Vitis vinifera.
European Hackberry – Celtis australis.
European Hazel – Corylus avellana.
European Holly – Ilex aquifolium.
European Hop Hornbeam – Ostrya carpinifolia.
European Hornbeam – Carpinus betulus.
European Larch – Larix decidua.
European Mandrake – Mandragora officinarum.
European Misletoe – Viscum album.
European Mountain Ash – Sorbus aucuparia.
European Pasque Flower – Anemone pulsatilla.
European Pennyroyal – Mentha pulegium.
European Privet – Ligustrum vulgare.
European Red Raspberry – Rubus idaeus.
European Turkey Oak – Quercus cerris.
European Wild Ginger – Asarum europaeum.
European Wood Sanicle – Sanicula europaea.
European Yellow Lupin – Lupinus luteus.

EUROSCHINUS Hook. f. Anacardiaceae. 10 spp. New Guinea, N.E. Australia, New Caledonia. Trees.

***E. falcatus*** Hook. (Pink Poplar). N.E.Australia. The pinkish wood is soft, light, not durable, difficult to work and susceptible to insect damage. It is used for cheap furniture, packing cases, coffins, brake blocks etc.

EUROTIA Adans. emend. C. A. Meyer. Chenopodiaceae. 8 spp. N.Africa, C. and E.Europe, temperate Asia, W. N.America, New Mexico. Shrubs.

***E. lanata*** (Pursh.) Moq. (Winter Fat, White Sage). W. N.America, New Mexico. The local Indians use a hot decoction of the leaves to remove head lice.

EURYA Thunb. Theaceae. 130 spp. E.Asia, Indomalaya, Pacific, Mexico to Peru, W.Indies. Trees.

***E. japonica*** Thunb. Indomalaya. The wood is used locally to make waggons and for turning.

***E. ochnacea*** (DC.) Szysz. Tropical Asia. The wood is used locally to build houses and boats.

***E. theoides*** (Sw.) Blume=Erotium theaoides, W.Indies, C.America. The leaves are used locally as a tea substitute.

EURYALE Salisb. Euryalaceae. 1 sp. China, S.E.Asia. Waterplant.

***E. ferox*** Salisb. Used in Japan. The young stems and roots are eaten as a vegetable, as are the seeds which are also dried and used to make a starchy flour.

EURYCOMA Jack. Simaroubaceae. 4 spp. Indomalaya. Trees.

***E. longifolia*** Jacq. (Antong sar, Bá bjnh, Tho nan). Indomalaya. In Vietnam the bark is used to treat indigestion and dullness; the fruits are taken to treat dysentery. In Cambodia the bark is used to counteract intoxication and in Malaya the root-bark is used to treat intestinal worms.

EURYPETALUM Harms. Leguminosae. 3 spp. Tropical Africa. Trees.

***E. batesii*** Bak. f. (Andzilim, Mbusa). Cameroons, Gabon. The hard red wood (Andzilim Wood) is used for cabinetmaking.

EUSIDEROXYLON Teijsm. and Binnend. Lauraceae. 2 spp. Borneo. Trees.

***E. malagangai*** Sym. (Malagangei). Borneo. The heavy reddish brown wood is easily worked and is used for heavy construction work.

***E. zwageri*** Teijsm. and Binnend. (Borneo Billian, Borneo Ironwood). The wood is brown to reddish brown and heavy enough to sink in water. It is very hard and is used for heavy construction work, flooring which takes a lot of wear, street paving, railway sleepers, shingles, telephone poles, etc. The timber is exported.

EUTERPE Gaertn. Palmae. 50 spp. Tropical America, W.Indies. Palm trees.

***E. dominicana*** L. H. Bailey (Manicol). Dominica. The central bud (Coeur palmistre) is eaten locally as a vegetable. The leaves are used for thatching and the prop roots are used to make yokes for carrying buckets etc.

***E. edulis*** Mart. (Assai Palm). Tropical

America, W.Indies. The fruits are edible and a drink is made from them by soaking them in water.

***E. oleracea*** Engel. Tropical America. The central bud is eaten locally as a vegetable and the fruits are made into a beverage.

EUTREMA R. Br. Cruciferae. 15 spp. C. and E.Asia. Arctic; 1 sp. S.W. United States of America. Herbs.

***E. wasabi*** (Sieb.) Maxim.=Alliaria wasabi =Cochlearia wasabi=Wasabi pungens (Kansaë, Tsi, Wasabi). Japan to E.Siberia. The plant is cultivated locally for the roots, twigs and petioles which are used as a condiment, especially with fish.

EUXYLOPHORA Huber. Rutaceae. 1 spp. Amazon, Brazil. Tree.

***E. paraensis*** Huber. (Brazilian Satinwood, Brazilian Boxwood, Pau Amarello, Canary Wood). The hard, heavy yellow wood polishes well and is used for furniture and interior woodwork in houses.

Evening Primrose – Oenothera biennis.

Evergreen Oak – Quercus ilex.

Evergreen Sumach – Rhus sempervirens.

Everlasting, Common – Gnaphalium polycephalum.

EVERNIA L. Usneaceae. Lichens.

***E. furfuracea*** E. Fr.=Parmelia furfuracea.

***E. mesomorpha*** Nyl.=Latheria thamnodes. A possible source of essential oils for perfumery.

***E. prunastri*** (L.) Ach.=Parmelia prunastri =Lobaria prunastri. (Oak Moss). N.Temperate, especially Europe. An oleo-resin is solvent-extracted from the lichen and used in perfumery. It is mainly produced in France, Italy and Czechoslovakia and used particularly in soaps. The whole plant is also used to dye wool purple. The ancient Egyptians used it to leaven bread and it is still used for this purpose by the Arabs and Copts. The lichen is still considered a tonic for stomach weakness.

***E. vulpina*** (L.) Ach. (Waino). Found on Conifers. It is used to stain wool yellow. It was used mainly in Norway and Sweden.

Eve's Thread – Yucca filamentosa.

***Evia amara*** Comm.=Spondias mangifera.

***Evodia*** Scop.=Euodia J. R. and G. Forst.

EVOLVULUS L. Convolvulaceae. 100 spp. Tropics and subtropics, especially America. Herbs.

***E. alsinoides*** L.=E. linifolius. Throughout the tropics. The leaves are used widely to reduce fevers and as a tonic. In India the whole plant is extracted with oil and the oil used as a hair tonic. The leaves are made into cigarettes, smoked to relieve bronchitis and asthma.

***E. linifilius*** L.=E. alsinoides.

***Evonymus*** L.=Euonymus.

***Exacum diffusum*** Willd.=Canscora diffusa.

EXBUCHLANDIA R. W. Brown. Hamamelidaceae. 2 spp. E.Himalayas to S.China, Malay peninsula and Sumatra.

***E. populnea*** R. Br. Tropical Asia, especially India and S.China, Malay peninsula. The wood is used for window frames and the bark is used for tanning.

EXCOECARIA L. Euphorbiaceae. 40 spp. Tropical Asia and Africa. Trees.

***E. agallocha*** L. (Blind-your-Eyes Trees). Tropical S.E.Asia to tropical Australia. The wood is used for incense.

***E. indica*** Muell. Arg.=Sapium indicum.

Exile Tree (Oil) – Thevetia nereifolia.

EXOCARPOS Labill. Santalaceae 26 spp. Indochina, Malaysia, Australia, Pacific to Hawaii. Trees.

***E. cupressiformis*** Lab. (Ballot). Australia. The wood is hard and close-grained. It is used to make tool-handles, gun stocks and for engraving, cabinet work, turning etc.

***E. latifolia*** R. Br.=E. miniata. (Shrub Sandalwood). Australia. The hard wood is coarse-grained and fragrant It is used for cabinet work.

***E. miniata*** Zipp.=E. latifolia.

EXOGONIUM Choisy. Convolvulaceae. 25 spp. Tropical America. Vines.

***E. bracteatum*** (Cav.) Choisy.=Ipomoea bracteata. (Bejuco Blanco, Jicama). Mexico. The succulent roots are eaten locally as a vegetable.

***E. jalapa*** Nutt and Coxe.=E. purga.

***E. purga*** (L.) Benth. (Jalap). Tropical America. The plant is cultivated for a resin used medicinally as a purgative.

EXOSTEMA (Pers.) Rich. Rubiaceae. 50 spp. Mexico to Brazil and Peru, W.Indies. Small trees.

***E. caribaeum*** Roem. and Schult. The bark (Copalchi Bark), is used to reduce fevers. The bright yellow wood is used for cabinet making and turning.

***E. sanctae-luciae*** (Kentisch) Britt. (Quina). Lesser Antilles. The bark is used locally as an emetic and purgative and to reduce fevers.

Extract of Witchhazel – Hamamelis virginiana.

Extractum Scordii Dialysatum – Teucrium scordium.

Eyebright, Drug – Euphrasia officinalis.

EYSENHARDTIA Kunth. Leguminosae. 14 spp. S. U.S.A. to Guatemala. Small trees.

***E. amorphoides*** H.B.K. = E. polystachya.

***E. polystachya*** (Ortega) Sarg. = E. amorphoides. (Palo Guate, Kidney Wood). Mexico, S.W. U.S.A. The wood is fluorescent and is used in Mexico to make ornamental troughs for watering fowl. The flowers are a good source of honey.

# F

Faba St. Ignatii – Strychnos ignatii.

***Faba vulgaris*** Moench. = Vicia faba.

FABIANA Ruiz. and Pav. Solanaceae. 25 spp. S.America. Woody shrub.

***F. imbricata*** Ruiz. and Pav. Peru and Chile. The aerial parts of the plant are used to stimulate the liver and kidneys. The wood (Lignum Fabianae, Lignum Pichi Pichi, Pichi Pichi Wood) is used to treat kidney and bladder ailments. The leaves and twigs are diuretic and are used to treat stomach ailments caused by liver disorders and catarrhal and functional diseases of the kidneys.

FAGARA L. Rutaceae. 250 spp. Tropics. Trees.

***F. angolensis*** Engl. Tropical W.Africa. The wood is light yellow, coarse-grained but tough. It is used for general carpentry, paper pulp, plywood, rough boxes and making canoes.

***F. avicennae*** Lam. = Zanthoxylum avicennae.

***F. becquetti*** G. Gilbert. Ruanda-Urundi. The fairly hard wood is used for general construction work.

***F. capensis*** Thunb. = Zanthoxylum capense.

***F. chalybea*** (Engl.) Engl. (Iguga, Mpopwa, Popioe). Congo to Tanzania. The leaves are used locally as a condiment. The seeds yield an oil used in perfumery. A decoction of the roots is used locally to treat venereal diseases.

***F. coriacea*** Kr. and Urb. = Zanthoxylum emarginatum.

***F. gilletis*** De Wild. See F. leprieuri.

***F. inaequalis*** Engl. = F. iturensis. (Bolongolo, Dongolo, Kanyalumbu, Nzongo). S.Nigeria to Rhodesia. The wood is used locally for house construction and furniture. The bark is used locally to treat toothache and the seeds yield an oil which is used for cooking.

***F. iturensis*** Engl. = F. inaequalis.

***F. lemairei*** De Wild. (Anghi, Bolongo, Epteba). Nigeria to Congo. The wood is used locally for general carpentry. The seeds give an oil which is used locally for cooking.

***F. leprieuri*** (Guill. and Perr.) Engl. (Ikondu, Konko kelekete, Konko nkumama). Congo to Angola, to Uganda. An extract of the bark is used by the natives to kill lice.

***F. pilosiuscula*** Engl. (Elondo, Kanyabumba, Lalanga). Congo to Angola. A decoction of the roots is used locally as an aphrodisiac.

***F. piperita*** Poureita = Zanthoxylum nitidum.

***F. pterota*** Blanco. = Zanthoxylum avicennae.

***F. rhetsa*** Roxb. = Zanthoxylum budrunga (Dapdab harnagan, Ki tana). Java. The wood is fine-grained, yellow and brittle. It is used for the handles of knives and weapons and rifle butts. It is also used for small construction work. The unripe fruits are used locally as a spice.

***F. rubescens*** (Planch.) Engl. (Bolongo, Efurungu, Ilonlolonga). Tropical W.Africa to Kenya. The bark is used by the natives to treat toothache.

FAGONIA L. Zygophyllaceae. 40 spp. S.W. United States of America, Chile, Mediterranean, S.W.Africa, S.W.Asia through to N.W.India. Shrubs.

***F. cretica*** L. = F. sinaica. N. and S. Africa. The leaves are used to feed camels and mules.

***F. sinaica*** Boiss. = F. cretica.

FAGOPYRUM Mill. Polygonaceae. 15 spp. Temperate Eurasia. Herbs.

***F. esculentum*** Moench. (Buckwheat). Originated in C.China, but is cultivated in the Old and New World. The floury endosperm yields a flour which is much used in U.S.A. and Canada to make buckwheat cakes. The whole hulled fruit is used as a breakfast food and for thickening soups etc. It is also grown as a green manure and to smother other weeds. The flowers yield an excellent honey. Its use in the U.K. is virtually confined to feeding pheasants. Its yield is lower than other grain crops, but it is useful on hilly, unproductive land. The name is derived from the German Buchweisen (beech wheat) as the fruits look like small beech fruits.

***F. tataricum*** Gaertn. (Indian Wheat, Siberian Buckwheat, Tartary Buckwheat). Asia. The seeds are used locally as a source of flour.

FAGRAER Thunb. Portaliaceae. 50 spp. Indomalaya, S.E.Asia, N.Australia, Pacific. Small trees or shrubs.

***F. fragrans*** Roxb. = F. peregrina. Burma, Malaya. The yellow wood is tough and durable. It is used locally to make houses, bridges and furniture.

***F. grandis*** Panch. and Sebert. New Caledonia. The yellow wood is much used locally for carving.

***F. peregrina*** Blume = F. fragrans.

***F. racemosa*** Jacq. Malaya. The roots are a popular local tonic, especially after fevers.

FAGUS L. Fagaceae. 10 spp. N.Temperate, Mexico. Trees.

***F. americana*** Sweet. = F. ferruginea.

***F. betuloides*** Mirb. = Nothofagus betuloides.

***F. cliffortoides*** Hook. f. = Nothofagus cliffortoides.

***F. ferruginea*** Dryand. = F. grandiflora = F. americana. (American Beech). E. N.America to Florida and Texas. Where it occurs, this tree is frequently the dominant species. The wood is hard and heavy, varying in colour from white to red-brown. It is used for veneers, barrel-making, boxes etc., paper-making and for carving and turning a wide variety of articles. Destructive distillation of the wood gives a wood tar, creosote, methyl alcohol and acetic acid. The creosote is light-coloured and is used medicinally as an antiseptic, expectorant and antipyretic. Distillation of the creosote gives guiaiacol which is used as an expectorant and intestinal antiseptic. The charcoal is used in art work. The nuts are edible.

***F. fusca*** Hook. = Nothofagus fusca.

***F. grandiflora*** Ehrh. = F. ferruginea.

***F. menziesii*** Hook. f. = Nothofagus menziesii.

***F. obliqua*** Mirb. = Nothofagus obliqua.

***F. orientalis*** Lipsky. (Oriental Beech). Balkans, Asia Minor, N.Iran, Caucasus. See F. sylvatica.

***F. procera*** Popp. and Endl. = Nothofagus procera.

***F. sieboldi*** Endl. (Buna-no-ki). Japan. The wood is hard, tough and polishes well. It is used locally to make furniture and agricultural implements. It is also planted outside Japan as an ornamental.

***F. sinensis*** Oliv. = F. sylvatica var. longipes. China. The wood is used locally to make boats and agricultural implements.

***F. solandri*** Hook. f. = Nothofagus solandri.

***F. sylvatica*** L. (European Beech). Europe to Asia Minor and Central Asia. The tree is the dominant member of the climax vegetation in many areas. The wood is reddish, durable and hard. It is used for furniture, waggons, railway sleepers, ship-building, turning and carving a wide variety of small articles. As it is fairly water-resistant, it was widely used in the construction of small dams. The wood also makes an excellent fuel. Destructive distillation of the wood yields creosote, tar and methyl alcohol. The creosote is used medicinally as an antiseptic and antipyretic. The charcoal (Carbo Ligni Pulveratus) or purified called Carbo Ligni Depuratus is used as an adsorbant in cases of alkali and phosphorus poisoning. The nuts (Beech Mast) are edible and were eaten in times of want. They are fed to stock and pigs, pheasants and poultry. The seeds yield 18 per cent non-drying, edible oil used mainly for making soap and for lighting, but is used for cooking in parts of France. The nuts are eaten roasted. The mast was called bucks which gave the name to the county of Buckinghamshire. The tree is frequently planted as a hedge, as it retains its leaves for a long time. The red-leaved variety (Copper Beech) is frequently planted as an ornamental.

***F. sylvatica*** Oliv. var. ***logipes*** Oliv. = F. sinensis.

Faham – Angraecum fragrans.

Fahon – Angraecum fragrans.

Fahtsai – Nostoc commune var. flagelliforme.

Fahum – Angraecum fragrans.
***Falcata comosa*** (L.) Kuntze – Amphicarpa monoica.
FALLUGIA Endl. Rosaceae. 1 sp. S.W. United States of America, Mexico. Shrub.
***F. mexicana*** Walp. = F. paradoxa.
***F. paradoxa*** (D. Don.) Endl. = F. mexicana = Sieversia paradoxa. The local Indians make the branches into brooms and arrows and use an infusion of the leaves to wash their hair.
Falotsy – Cynanchum sarcostemoides.
Falsch – Populus ciliata.
False Aloë – Agave virginica.
False Arrowroot – Curcuma pierreana.
False Boneset – Kuhnia eupatorioides.
False Calamus – Iris pseudoacorus.
False Chaulmooga Oil – Gynocardia odorata.
False Cola – Cola acuminata.
False Creeping Paspalum – Brachiaria humidicola.
False Hellebore – Veratrum californicum.
False Hellebore, American – Veratrum viride.
False Hellebore, White – Veratrum album.
False Huckleberry – Menziesia ferruginea.
False Indigo – Amorpha fruticosa.
False Indigo – Baptisia australis.
False Pinkroot – Phlox carolina.
False Tarragon – Artemisia dracunculoides.
False Winter's Bark – Cinnamodendron corticosum.
Falsenettle, Hawaiian – Boehmeria grandis
Fame Flower, Orange – Talinum aurantiacium.
Fame Flower, Potherb – Talinum triangulare.
Fa Mei Chuk – Bambusa pervariabilis.
Fancy Yam – Dioscorea esculenta.
Fandrana – Pandanus concretus.
Fanganga – Dioscorea macabiha.
FARAMEA Aubl. Rubiaceae. 120 spp. Tropical S.America, W.Indies. Trees.
***F. odoratissima*** DC. S.America, W.Indies. The light yellow wood is strong and used locally for general carpentry.
Farge's Fir – Abies fargesii.
Farib-buti – Farsetia hamiltonii.
Farinha – Manihot esculenta.
Farihazo – Neodypsis baronii.
Farine de Châtaignes – Castanea sativa.
Farka teli – Echinochloa pyramidalis.
Farsa – Cadaba farinosa.
FARSETIA Turra. Cruciferae. 15 spp. Morocco to N.W. India, Central Africa. Herbs. The species mentioned are used locally to treat rheumatism.
***F. aegyptiaca*** Turra. N.W.India.
***F. hamiltonii*** Royle. (Farib-buti). N.W.India.
***F. jacquemontii*** Hook. f. and Th. N.W.India.
***F. ramosissima*** Hochst. (Tozokamit). Mauritania.
Fat, Bay – Laurus nobilis.
Fat, Bicuhyba – Virola bicuhyba.
Fat, Bulam – Palaquium walsurifolium.
Fat, Cay-cay – Irvingia oliveri.
Fat, Cuojo – Virola venezuelensis.
Fat, Dhupa – Vateria indica.
Fat, Dika – Irvingia gabonensis.
Fat, Gorli – Calocoba echinata.
Fat, Laurel Berry – Laurus nobilis.
Fat, Mahuba – Acrodiclidium mahuba.
Fat, Mandiro – Fevillea cordifolia.
Fat, Maripa – Maximiliana maripa.
Fat, Mkani – Allanblackia stuhlmanni.
Fat, Naim – Lophira alata.
Fat, Njatuo – Palaquium oleiferum.
Fat, Odolla – Cerera manghas.
Fat, Otoba – Otoba otoba.
Fat Pork – Clussia flava.
Fat, Pracachy – Pentaclethra filamentosa.
Fat, Pracaxi – Pentaclethra filamentosa.
Fat, Sawarri – Caryocar amydaliferum.
Fat, Sekua – Fevillea cordifolia.
Fat, Siak – Palaquium oleiferum.
Fat, Souari – Caryocar amygdaliferum.
Fat, Taban Merak – Palaquium oleiferum.
Fat, Tankawang – Shorea balangeran.
Fat, Ucahuba – Virola surinamensis.
Fat, Ucuiba – Virola surinamensis.
Fat, Virola – Virola sebifera.
Fat, Winter – Eurotia lanata.
Fatchoy – Nostoc commune var. flagelliforme.
FATSIA Decne and Planch. Araliaceae. 1 sp. Japan, 1 sp. Formosa. Trees.
***F. papyrifera*** Hook. = Aralia papyrifera = Tetrapanax papyrifera. China. The pith is used to make a rice paper which is used extensively in China to make artificial flowers and for painting. It is also used for surgical dressings.
Fatuhiva – Pelea fatuhivensis.
FAUREA Harv. Proteaceae. 18 spp. Tropical and S.Africa, Madagascar. Trees.
***F. saligna*** Haw. (African Beech, Beukenhout). C. and S.Africa. The wood is hard and durable. It is used locally for building, firewood and charcoal. The bark is used for tanning. The natives use a decoction

of the roots to treat gonorrhoea and dysentery.

***F. speciosa*** Welw. Tropical Africa. The orange coloured wood is used locally for turning and making furniture.

Favas de Santo – Fevillea trilobata.

FAVOLUS Fr. Polyporaceae. 30 spp. mainly tropics.

***F. spathulatus*** (Jungh.) Bresw. Tropical Asia. The fruitbodies are eaten locally in the Sundra Islands.

Feather, Gray – Liatris spicata.

Fécule de Kabija – Tacca pinnatifida.

Fécule de Pia – Tacca pinnatifida.

FEDIA Gaertn. (emend Moench.) Valerianaceae. 3 spp. Mediterranean. Herbs.

***F. cornucopiae*** DC. (African Valerian). Mediterranean. The leaves are used as a vegetable in parts of N.Africa.

FEIJOA Berg. Myrtaceae. 2 spp. Brazil. Small trees.

***F. sellowiana*** Berg. (Fiejoa). S.Brazil to Paraguay. The tree is cultivated in the tropics and subtropics for the fruit which is eaten raw or in jams and preserves.

Feijoa – Feijoa sellowiana.

Fei-tsao-tou – Gymnocladus chinensis.

Fendler Potato – Solanum fendleri.

Fengoky – Caesalpinia adansoniodes.

Fennel – Foeniculum vulgare.

Fennel, Bitter – Foeniculum vulgare.

Fennel, Dog – Anthemis cotula.

Fennel, Dog – Eupatorium capillifolium.

Fennel, Florence – Foeniculum vulgare.

Fennel, Hog – Peucedanum officinale.

Fennel, Hog – Peucedanum ostruthium.

Fennel, Italian – Foeniculum vulgare.

Fennel, Roman – Foeniculum vulgare.

Fennel, Sweet – Foeniculum vulgare.

Fennel, Water – Oenanthe aquatica.

Fen Sedge – Cladium mariscus.

Fenugrec – Trigonella foenum-graecum.

Fenugreek – Trigonella foenum-graecum.

***Ferdinanda eminens*** Lag. = Podachaerium eminens.

FERETIA Delile. Rubiaceae. 2 spp. Tropical Africa, Abyssinia. Woody shrubs.

***F. apodanthera*** Delile. Abyssinia. The seeds are used locally as a coffee substitute.

***F. canthioides*** Hiern. Tropical Africa. In Nigeria, the crushed fruits are mixed with indigo and used to put marking on the face. A decoction of the roots is used to treat gonorrhoea.

Fern, Cinnamon – Osmunda cinnamomea.

Fern, Douglas Mulefoot – Marattia douglasii.

Fern, Eagle – Pteridium aquilinum.

Fern, Maidenhair – Adiantum capillus-veneris.

Fern, Maidenhair – Adiantum pedatum.

Fern, Male – Aspidium filix mas.

Fern, Male Shield – Aspidium filix mas.

Fern, Marsh – Acrostichum aureum.

Fern, Sweet – Comptonia asplenifolia.

Fernleaf Nitta Tree – Parkia filicoidea.

FEROCACTUS Britton and Rose. Cactaceae. 35 spp. S.W. United States of America, Mexico.

***F. hamatocanthus*** (Muehlenpf.) Britton. and Rose. = Echinocactus hamatocanthus. N.Mexico into Texas and New Mexico. The fruits are used locally as a substitute for lemons.

***F. wislizeni*** (Engelm.) Britton and Rose = Echinocactus wislizeni. Mexico, Texas and Arizona. The pulp from the stems is used locally as a sweetmeat.

***Ferolia*** Aubl. = Parinari.

***Feronia elephantum*** Correa = Limonia acidissima.

***Feronia limonia*** Swingle = Limonia acidissima.

***Feronia ternata*** Blanco = Swinglea glutinosa.

FERONIELLA Swingle. Rutaceae. 3 spp. S.E.Asia, Java. Trees.

***F. lucida*** Swingle Java. The fruits are eaten locally.

FERRARIA Burm. ex Mill. Iridaceae. 2 spp. Tropics and S.Africa.

***F. purgans*** Mart. Tropical America. The roots are used as a purgative in many countries of tropical America.

FERULA L. Umbelliferae. 133 spp. Mediterranean to C.Asia. Herbs.

***F. alliacea*** Bois. (Hing, Hingu, Kayam). Persia. A gum from the roots is used locally to treat digestive disorders, hysteria, epilepsy and scorpion stings.

***F. assa-foetida*** L. (Asafoetida). Persia to W.Afghanistan. A gum (Asafoetida) is extracted from incisions in the roots.
It is used locally as a condiment and to flavour sauces. An oil (Oleum Asae Foetidae) is steam-distilled from the gum. Both the oil and the gum are used medicinally in the treatment of convulsions, croup and colic in children, as a laxative and expectorant.

***F. communis*** L. var. ***brevifolia*** Marcz.

Morocco. The roots yield a gum (Ammoniac of Morocco).

***F. foetida*** (Bunge) Regel. Turkestan. Yields a gum. See F. assa-foetida.

***F. galbaniflua*** Boiss. and Buhse. (Galbanum). Persia to Turkestan. A gum (Galbanum, Kasnib, Boridschah, Badra-Kema) is collected from incisions in the stem. It is used medicinally as an expectorant, to aid digestion and to reduce spasms. It is exported mainly from Asia Minor and Bombay.

***F. jaeschkeana*** Vatke Kashmir. A gum from the stems is used locally to treat wounds and bruises.

***F. marmarica*** Asch. and Taub. N.Africa. Yields a gum (Ammoniac of Cyrenaica).

***F. narthex*** Boiss. (Silphium). Baltisthan. Yields asafoetida. See F. assa-foetida. The gum was used as a spice by the ancient Egyptians and Greeks. Extensive healing properties were attributed to it by the Romans.

***F. persica*** Willd. Persia, Caucasus. The stem yields a gum (Sagapenum Gum) used locally to treat lumbago and rheumatism.

***F. rubicaulis*** Boiss. Persia. Yields a gum. See F. assafoetida.

***F. sancta*** Boiss = F. tingitana.

***F. schair*** Borszcz. Turkistan. Yields a Galbanum. See F. galbaniflua.

***F. sumbul*** Hook. f. C.Asia. The root (Mush, Violet Root) is heavily scented and is used in Persia as a perfume and in incense. It is also used medicinally as a nerve tonic and to treat spasms and hysteria.

***F. szowitziana*** DC. (Sagapen). C.Asia. Yields a resin (Saganpen Resin). See F. galbaniflua.

***F. tingitana*** L. = F. sancta. N.Africa to Syria. Yields a gum (North African Ammoniac, Moroccan Ammoniac, Gummi Resina Ammoniacum).

Fescue, Greenleaf – Festuca viridula.

Fescue, Meadow – Festuca elatior.

Fescue, Red – Festuca rubra.

Fescue, Sheep's – Festuca ovina.

Fescue, Thurber – Festuca thurberi

FESTUCA L. Graminae. 80 spp. Cosmopolitan. Perennial grasses.

***F. elatior*** L. (Meadow Fescue, English Bluegrass). Europe. A valuable pasture in both the Old and New World.

***F. ovina*** L. (Sheep's Fescue). Europe. A valuable pasture in both Old and New World, especially on chalk downs etc. for sheep grazing.

***F. rubra*** L. (Red Fescue). Temperate N.Hemisphere. Grown widely as a pasture, especially on poor hill soils.

***F. thurberi*** Vasey. (Thurber Fescue). W. U.S.A. Used as a pasture in sub-alpine regions.

***F. viridula*** Vasey. (Greenleaf Fescue). W. N.America. A valuable pasture in subalpine regions.

Fetia Star Tree – Astronium graveolens.

Fetid Buckeye – Aesculus glabra.

Fetid Marigold – Pectis papposa.

Fetter Bush – Leucothoë catesbaei.

Feua – Metrosideros collina.

***Feuillaea reticulata*** O. Kuntze = Inga feuillei.

Fever Bark – Alstonia constricta.

Fever Bark – Pinckneya pubens.

Fever Bush – Garrya elliptica.

Fever Bush – Ilex verticillata.

Fever Tree – Eucalyptus globosus.

Fever Tree – Zanthoxylum capense.

Feverfew Chrysanthemum – Chrysanthemum parthenium.

FEVILLEA L. Cucurbitaceae. 9 spp. Tropical America. Vines.

***F. cordifolia*** L. S.America. W.Indies. The seeds yield a fat (Mandiro Fat, Sekua Fat) which is used locally to make candles and soap. It is also used as a laxative.

***F. trilobata*** DC. Brazil, W.Indies. The seeds yield an oil (Favas de Santo Ignacio do Brasil) which is used to treat rheumatism and skin diseases.

Few-flowered Psoralea – Psoralea tenuiflora.

FIBRAUREA Lour. Menispermaceae. 5 spp. Assam, Indochina, Philippines, Borneo. Woody climbers.

***F. chloroleuca*** Miers. = F. tinctoria.

***F. recisa*** Pierre. (Dây vàng giang, Nam hoàng liên). Indochina. The roots contain a bitter principle which is much used as a tonic in N.Vietnam. A yellow dye is extracted from the stems.

***F. tinctoria*** Lour. = F. chloroleuca = Cocculus fibraurea = Menispermum tinctorum. (Huang t'eng). Malaysia, through to S.Vietnam. The stem yields a yellow dye used in S.Vietnam. A bitter principle from the roots is used locally as a tonic, diuretic and to relieve constipation. It is used in Malaya to relieve fevers.

FIBRES – The name is used as a noun and

not as an adjective, e.g. "Flax fibre" is mentioned under "Flax" only.

Ficalho – Bosqueia angolensis.

FICUS L. Moraceae. 800 spp. Warm areas, chiefly Indomalaya and Polynesia. The fruit of this genus is a syconium, depending on gall wasps for fertilization.

*F. annulata* Blume. Burma, Malaya. Tree. Occasionally cultivated in Australia and Malacca as a source of rubber.

*F. anomani* Hutch. Tropical West Africa. Tree. The latex is used as a bird lime in Ghana.

*F. anthelmintica* Mart. Peru, Brazil. Tree. The latex (Leche de Ojé) and bark (Corteza de Ojé) are used to treat malaria in Peru.

*F. artocarpoides* Warb. (Likumo). W.Africa. Epiphytic liana. The bark is used locally to make cloth.

*F. aspera* Forst. (Balemo, Noomaie, Rough-leaved Fig). Tree. Victoria and Queensland. The fruit is eaten by the aborigines.

*F. asperifolia* Miq. Tropical Africa. Tree. The leaves are used locally as sandpaper.

*F. auriculata* Lour.=F. roxburghii.

*F. baroni* Bak. (Amontana). Madagascar. Tree. The fibre from the bark is used locally.

*F. benghalensis* L. (Banyan, Bargat, Vov, Wur). India, Pakistan. Large tree. Large aerial roots are produced from the branches. The tree is sacred to the Hindus and much attention is paid to their preservation and cultivation. The wood is fairly hard and durable in water. It is used locally for furniture, house-building, carts etc. The aerial roots are used as straight poles, e.g. for tent poles. The fruits are eaten, fresh or dried. The young leaves and shoots are eaten as a famine food. The fibre from the bark is used to make paper, ropes and slow matches. The latex is a local bird lime and is also used to heal cracks in the feet. The latex is also used to oxidise copper ware in India. One form (F. krishnae) produces cup-shaped leaves known as Krishna's Figs.

*F. bonplandiana* Miq.=F. involuta.

*F. capensis* Thunb. S.Africa. Tree. The fruits are eaten by the natives who also use the wood to make fire by rubbing.

*F. carica* L. (Fig). Mediterranean to India. Cultivated in warmer temperate regions of the Old and New World. The fruits are of considerable economic importance. The plant has been cultivated in the Mediterranean region for at least 5,000 years and is frequently mentioned in the Bible and in Egyptian writings of that period. The fruit is eaten raw or dried and contains about 64 per cent sugar. They are candied, made into jam, brewed into an alcoholic beverage (Fig Wine, Fig Brandy), or canned. A syrup from the fruit is used as a laxative. There are many varieties, some of which require the presence of male flowers (caprifigs) to bring about pollination. The world production is of the order of 1½ million tons annually, of which about 200,000 tons are dried. The main producers are Portugal, Greece, Turkey and Italy, while California is the main producing area of the U.S.A. The plants are propagated by cuttings and many commercial growers have to irrigate. Most of the crop is picked by hand.

*F. cerifera* Blume=F. ceriflua.

*F. ceriflua* Jungh=F. cerifera=F. varigata=F. subopaca. (Gondang, Kondang, Tjoro). Tropical Asia. Tree. The trunk yields a wax (Fig Wax, Gondang Wax) which is used locally for making candles.

*F. complicata* H.B.K.=F. padifolia.

*F. conglomerata* Roxb.=F. cunia.

*F. conraui* Warb. (Likumo). W.Africa to Congo. Small tree. The bark is used locally to make cloth.

*F. consociata* Blume. W.Malaysia. Tree. The bark is used locally to make a cloth used in book-binding.

*F. cotinifolia* H.B.K.=Urostigma glaucum=U. longipes. (Amato Prieto, Copó, Higuerón). Tree. Mexico to Costa Rica. The bark was used by the natives and the early Spanish settlers to make paper. The powdered bark is mixed with the latex and used locally to treat cuts and bruises.

*F. crassinervia* Desf. W.Indies, especially N.islands. An infusion of the leaves is used in Cuba to treat liver complaints.

*F. cunia* Hamil.=F. conglomerata. Himalayas to Malaysia and N.Vietnam. The fruits are eaten locally.

*F. cunninghami* Miq. Trees. Australia. The latex has a limited local use as a source of rubber.

*F. cystopodia* Gasp.=Urositgma cystopodium. Brazil. Tree. The bark (Cortex

Mururé) is used as a laxative. An oil (Oleum Mururé, Mururé Oil) extracted from the fruits is used locally to treat rheumatism.

***F. dicranostyla*** Mildbr. Tropical Africa. Tree. In Togoland, a decoction of the leaves and bark is used to remove the hair from hides.

***F. dob*** Warb. Senegal. Tree. Yields a poor quality rubber.

***F. elastica*** Roxb. (India Rubber Tree, Karet Tree). Burma, India, Malaya. Tree. The latex was an important source of rubber (Assam Rubber, Indian Rubber) before the commercial introduction of Hevea (q.v.). The plant was never planted commercially. It is a popular indoor house plant.

***F. exasperata*** Vahl. (Esese, Yakasa). Tropical Africa, especially the East. The leaves are used locally as sandpaper.

***F. fulva*** Reinw. Burma, Sumatra. Tree. The latex is occasionally used to adulterate other latex.

***F. gibosa*** Blume. W.Malaysia. Tree. The bark is used locally to make a cloth used in book-binding.

***F. glomerata*** Roxb. (Umar). Tropical India, Pakistan. Tree. The light soft wood is not durable, but is relatively water-resistant. It is used to make small houses, furniture, boxes and tea-chests, carvings and other small articles, cart shafts and a poor quality pulp. The fruits are eaten locally and a bird lime is made from the latex. The tree is sacred to the Hindus.

***F. glomerata*** Willd. Australia. Tree. The fruits are eaten by the aborigines.

***F. glumosa*** Delil. E.Africa. Tree. The fruits are eaten locally. The bark fibre is made into a cloth, the latex is used as bird lime and a decoction of the bark is used for tanning.

***F. gnaphalocarpa*** Steud. Tropical Africa. Tree. The fruits are fermented to make a very pleasant spirit. The latex is used to make a gutta percha which has some commercial value.

***F. hypophaea*** Schltr. Climber on tree. News Guinea. A possible source of latex for rubber.

***F. indica*** L.=Urostigma tjiela. India to Philippines. Tree. A decoction of the bark is used locally as a diuretic and tonic. The latex is used as a poultice for bruises, lumbago and rheumatism. A decoction of the roots, sometimes mixed with honey is used to treat gonorrhoea and taken with the fruit, is supposed to be an aphrodisiac. The plant is used as a food plant for the silk worm, Theophila religiosa, in Assam.

***F. involuta*** (Liebm.) Miq.=F. bonplandiana=Urostigma involutum. (Amata, Amanto Blanco, Matapalo). The fruits are eaten locally. The juice is applied externally to relieve stomach pains.

***F. krishnae*** C. DC.=F. benghalensis.

***F. lacor*** Hamilt. Tree. Coasts of Far East, mainly China. The Chinese used the bark to stimulate sweating.

***F. laevigata*** Vahl. (Jagüey Blanco, Shortleaf Fig). W.Indies, S.America. The coarse wood is straight-grained and tough and of limited commercial value. It is used for house interior work, boxes etc., although it is very susceptible to termite damage.

***F. leucantatoma*** Poir.=F. septica.

***F. lingua*** Warb. W.Africa. The bark is used locally to make a cloth.

***F. luteola*** De Wild. See F. lingua.

***F. marquesensis*** F. Br. (Marquesan Fig). S.E.Polynesia. Tree. A fibre (Tapa, Heapo) is extracted from the bark and used to make a good cordage.

***F. melleri*** Bak. (Monoka). Madagascar. Tree. The latex is used locally as a bird lime.

***F. michelsonii*** Bout. and Léon. W.Africa. The bark is used locally to make a cloth.

***F. microcarpa*** Vahl.=F. thonningii. Tree. Tropical Africa. The bark is used locally to make a cloth.

***F. mucuso*** Welw. ex Ficalho. (Liheli, Likuye). W.Africa, east to Tanzania. Tree. The wood is light and is used for the interiors of houses and making packing cases.

***F. natalensis*** Hochst. W. and through Central Africa. The bark is used to make cloth, particularly in Uganda.

***F. nekbudu*** Warb. Congo. Tree. The latex makes an excellent rubber, which becomes red on drying.

***F. obliqua*** Forst. Fiji. Tree. It yields an inferior latex which is used to adulterate those of better quality.

***F. obtrusifolia*** Roxb. E.India to Burma. Tree. The latex is used occasionally as a source of rubber.

***F. padifolia*** H.B.K.=F. complicata=F.

sonorae = Urostigma complicatum. (Camichon, Higuilo, Nacapuli, Palo de Coco). Mexico. The fruits are eaten locally. The bark was used as paper by the original Indians, and by the Spanish settlers.

**F. petiolaris** H.B.K. Mexico. Tree. The gum is acid and used in surgery to repair broken bones and support hernias. The elastic aerial roots are used locally to make suspension bridges. The bark was used to make paper.

**F. platyphylla** Delile. (Broadleaf Fig). Tropical Africa. The latex is a base for chewing gum and is used locally as a bird lime and for cleaning brass. It also gives a rubber (Niger Gutta, Red Kano Rubber, Ogbagha Rubber). The bark is used for tanning and making cordage and cloth.

**F. platypoda** Cunningh. Queensland, Victoria. The fruits are eaten by the aborigines.

**F. preussii** Warb. Cameroons. Tree. The tree is a source of a rubber of some commercial importance.

**F. prolixa** Forst. (Aoa, Otau, Tahitian Fig). S.E. Polynesia. Tree. The bark fibre, especially from the young stems and aerial roots makes an excellent cordage fibre (Tapa, Keapo).

**F. pumila** L.=F. repens. (Okgue). Japan, China, Formosa. Tree. The fruits are edible and are used in Japan to make jellies.

**F. pynaerti** De Wild. (Mwambala). Congo. Climber. The bark fibre is used to make fine ropes.

**F. radula** Willd. S.Mexico to Peru. Tree. The bark is used to make cloth, mats and ropes.

**F. religiosa** L. (Bo Tree, Peepul Tree). Tropical Asia. Tree. A sacred tree to Hindus and Buddhists and is hence widely cultivated. The small fruits are edible, but are used only as a famine food. The latex is used as sealing wax, for mending ornaments etc. and as bird lime. The bark fibres were used in Burma to make paper. In Assam it is used as a food plant for the silk worm, Theophila religiosa.

**F. repens** Rottl.=F. pumila.

**F. retusa** L. Tropical Asia to Australia. Large tree, often epiphytic. In China a powder of the small adventitious roots is mixed with salt and used to treat toothache.

**F. ribes** Reinw. Java. Tree. The bark and leaves are used locally for chewing and to treat malaria.

**F. rigo** Bailey. New Guinea, Australia. Tree, sometimes epiphytic. Occasionally planted as a commercial source of rubber.

**F. roxburghii** Wall.=F. auriculata. India to Indo-China. Tree. The fruits are eaten locally.

**F. rumphii** Blume. Malaysia. Tree. The fruits are eaten locally.

**F. sakalavarum** Baker. Madagascar. Tree. The fruits are eaten locally raw, or in preserves.

**F. schlechteri** Warb. New Caledonia. Tree. The tree yields an inferior rubber.

**F. septica** Burm.=F. leucatatoma (Awar awar, Sirih popar). Malaysia. The roots are chewed with the root of Musa paradisica as an antidote to eating poisonous fish, crabs etc.

**F. soroceoides** Baker. (Ampaly). Madagascar. Tree. The leaves are used locally as sandpaper.

**F. sonorae** S. Wats.=F. padifolia.

**F. stuhlmannii** Warb. (Mukango). Congo to Tanzania. Tree. The fruits are eaten locally.

**F. subopaca** Miq.=F. ceriflua.

**F. sycomorus** L. (Sycomore Fig, Mulberry Fig). N.Africa. Tree. The tree is widely cultivated for the edible fruit. The latex is used to coagulate milk. The wood was used by the ancient Egyptians to make mummy cases.

**F. thonningii** Blume.=F. microcarpa.

**F. tiliaefolia** Baker. (Voara). Madagascar. Tree. The tree is cultivated for the fibres made from the bark.

**F. tinctoria** Forst. (Dye Fig, Mahi). S.E.Polynesia. Tree. The bark from the trunk is used to make a poor quality cordage, while the fibre from the young shoots is of better quality and is used to make fishing nets. The fruit juice is mixed with the juice from the leaves of Cordia subcordata to make a scarlet dye. This is used locally to dye cloth and to paint the faces of warriors etc.

**F. toxicaria** L. Java. Tree. A wax from the plant is used locally for batik work.

**F. trichopoda** Baker. Madagascar. Tree. The latex coagulates in air and is used locally to dress wounds.

***F. tsjaleka*** Burm f. = F. venosa. Himalayan region. Tree. The leaves are used locally to feed the silk worms Ocinara signifera and O. lida.

***F. vallis*** Warb. (Choudae Del, Lowa voluha). Tropical Africa. Tree. The bark is used locally to make cloth.

***F. varigata*** Blume = F. ceriflua.

***F. venosa*** Dryand. = F. tsjaleka.

***F. verrucolosa*** Warb. Tropical Africa. Tree. The fruits are eaten in Angola.

***F. vogelii*** Miq. (Dob). Tropical Africa. The tree yields a good quality rubber (Dahomey Rubber). The bark fibres are used to make cloth and the wood is used locally as a soap.

Fiddlewood – Citharexylum quadrangulare.

Fidji Copal – Agathis loranthifolia.

Field Garlic – Allium oleraceum.

Field Mint – Mentha arvensis.

Field Mushroom – Psalliota arvensis.

Field Pea – Pisum arvense.

Field Penny Cress – Thlaspi arvense.

Fig – Ficus carica.

Fig, Banana – Musa sapientum.

Fig, Blue – Elaeocarpus grandis.

Fig, Broadleaf – Ficus platyphylla.

Fig, Dye – Ficus tinctoria.

Fig, Hottentot – Mesembryanthemum edule.

Fig, Indian – Opuntia ficus-indica.

Fig, Marquesan – Ficus marquesensis.

Fig, Mulberry – Ficus sycomorus.

Fig, Rough-leaved – Ficus aspera.

Fig, Short-leaved – Ficus laevigata.

Fig, Sycomore – Ficus sycomorus.

Fig, Tahitean – Ficus prolixa.

Fig Wax – Ficus ceriflua.

Figwort – Scrophularia nodosa.

Figwort, Maryland – Scrophularia marilandica.

Fiha – Euphorbia fiha.

Fiji Afzelia – Afzelia bijuga.

Fiji Arrowroot – Tacca pinnatifida.

Fiji, Resin of – Agathis vitiensis.

Fiji Sandalwood (Oil) – Santalum yasi.

Fikongo – Brachystelma bingeri.

Filbert – Corylus avellana.

Filbert of Constantinople – Corylus colurna.

Filbert, Lambert's – Corylus tubulosa.

Filbert Oil – Corylus avellana.

Filbert, Siberian – Corylus heterophylla.

FILIPENDULA Mill. Rosaceae. 10 spp. N.Temperate. Herbs.

***F. hexapetala*** Gilib. = Spiraea filipendula = Ulmaria filipendula. (Dropwort). N.Europe to Asia Minor. The roots and flowers and leaves (Radix Filipendula and Herba et Flores Filipendula) were used extensively during the 16th and 17th centuries to treat stones of the kidney and bladder. They were also used to treat epilepsy. They are still used as local remedies for coughs, dropsy, genital discharges and intestinal worms. The roots and young leaves are occasionally used in salads and in Sweden to feed pigs.

***F. ulmaria*** (L.) Maxim. = Spiraea ulmaria. (Queen of the Meadow). Temperate Europe and Asia. Introduced to N.America. An oil, distilled from the flower buds (Oil of Meadowsweet) is used in perfumery. The flowers and stems are used in local remedies as a tonic, diuretic, to control bleeding and as an antispasmodic.

FILLAEOPSIS Harms. Leguminosae. 1 spp. Tropical Africa. Tree.

***F. discophora*** Harms. (Eon, Totum). The light brown wood is very hard and heavy. It is difficult to work, but polishes well. It is used in making houses, furniture and paving blocks.

FIMBRISTYLIS Vahl. Cyperaceae. 300 spp. Tropics and subtropics, especially Indomalaysia and Australia. Herbs. The species mentioned give fibres used to make mats, basketwork etc.

***F. diphylla*** Vahl. Tropical Asia.

***F. globulosa*** Kunth. Tropical Asia. Cultivated.

***F. spadicea*** Vahl. Tropical America. Fibre called Esparto Chino, Esparto Mulato, Gamelotte Fibre.

Finger Grass, Abyssinian – Digitaria abyssinica.

Finger Grass, Dunn's – Digitaria scalarum.

Finger Grass, Richmond – Digitaria diversinervis.

Finger Grass, Swaziland – Digitaria swazilandensis.

Finger Grass, Woolly – Digitaria pentzii.

Fingerleaf Morning Glory – Ipomoea digitata.

Finger Millet – Eleusine coracana.

Finger Poppy Mallow – Callirhoë digitata.

Fingobary – Landolphia dubardi.

Fingotra – Landolphia crassipes.

Fiorin – Agrostia alba.

Fique Fibre – Furcraea macrophylla.

Fir, Alpine – Abies lasiocarpa.

Fir, Balsam – Abies balsamea.

Fir, Cascade – Abies amabilis.

Fir, Douglas – Pseudostuga menziesii.
Fir, Farges – Abies fargesii.
Fir, Flaky – Abies squamata.
Fir, Grand – Abies grandis.
Fir, Himalayan Silver – Abies pindrow.
Fir, Japanese Silver – Abies firma.
Fir, Joint – Ephedra trifurca.
Fir, Lowland White – Abies grandis.
Fir, Marie's – Abies mariesii.
Fir, Nevada Joint – Ephedra nevadensis.
Fir, Nikko Silver – Abies homolepis.
Fir, Noble – Abies nobilis.
Fir, Nordmann – Abies nordmanniana.
Fir, Red – Abies magnifica.
Fir, Red – Pseudotsuga menziesii.
Fir, Rocky Mountain – Abies lasiocarpa.
Fir, Sacred – Abies religiosa.
Fir, Siberian – Abies sibirica.
Fir, Silver – Abies amabilis.
Fir, Silver – Abies pectinata.
Fir, Veitch's Silver – Abies veitchii.
Fir, White – Abies amabilis.
Fir, White – Abies concolor.
Firewheel Tree, Willow – Stenocarpus salignus.

FIRMIANA Marsigli. Sterculiaceae. 15 spp. E.Africa, Indomalaya, S.E. and E.Asia. Trees.

***F. affinis*** Terrac. = Scaphium affinis.

***F. barterii*** (Mast.) Schum. Tropical Africa, especially West. The very light wood is used locally to make floats for fishing nets and household utensils.

***F. plantanifolia*** Brit. = F. simplex.

***F. simplex*** (L.) Wight. = F. plantanifolia = Sterculia plantanifolia. (Phoenix Tree). China, Formosa. The wood is used locally to make furniture and a fibre is obtained from the bark by retting. It is used for cordage.

Fish Berry – Anamirta paniculata.
Fish-tail Palm – Caryota urens.
Fishdeath Tephrosia – Tephrosia toxicaria.
Fishfuddle Tree, Jamaica – Piscidia erythrina.
Fishpole Bamboo – Phyllostachys aurea.

FISSISTIGMA Griff. Annonaceae. 60 spp. Tropical Africa, S.W. China, Indomalaya, N.E.Australia. Trees.

***F. leichhardtii*** Benth. = Unona leichhardtii. (Merangara). Australia. The fruits are eaten by the aborigines.

FISTULINA Bull. ex Fr. Polyporaceae. 5 spp. Europe and N.America.

***F. hepatica*** Fr. (Beefsteak Fungus, Beef Tongue, Chestnut Tongue, Oak Tongue, Vegetable Beefsteak). Temperate, growing mainly on Oak trees. The mycelium stains oak brown (Brown Oak), making the timber particularly valuable. The fruitbodies are edible.

FITZROYA Hook. f. Cupressaceae. 1 sp. Chile. Tree.

***F. cupressoides*** (Molina) Johnst. = F. patagonica. (Alerce, Lahuán). The wood is light and durable. It is used locally for general carpentry, making barrels, cigar boxes, musical instruments and troughs.

***F. patagonica*** Hook. = F. cupressoides.

Fivefinger – Potentilla reptans.
Fiveleaf Grass – Potentilla reptans.

FLACOURTIA Comm. ex L'Hérit. Flacourtiaceae. 15 spp. Tropical and S.Africa, Mascarene Islands, S.E.Asia, Malaysia, Fiji. Shrubs or small trees. The fruits of all the species mentioned have a pleasant sub-acid flavour and are eaten locally, particularly in various preserves.

***F. cataphracta*** Roxb. (Runeala Plum). India, Malaysia. Introduced to tropical America.

***F. euphlebia*** Merr. Philippines.

***F. inermis*** Roxb. (Plum of Martinique). Origin uncertain.

***F. ramontchi*** L'Hérit. (Botoko Plum, Governor's Plum, Madagascar Plum, Ramontschi). Malaysia, Madagascar. Cultivated in Old and New World tropics. The fruits are edible and are also used in parts of Africa to treat jaundice and enlargement of the spleen. The wood is red, hard and durable. It is used for agricultural implements and turning.

***F. rukam*** Zoll. and Mor. (Rukam). Malaysia, Philippines.

***F. sepiaria*** Roxb. Malaysia, Cochin-China, introduced to tropical America.

Fladwood's Plum – Prunus umbellata.
Flag, Blue – Iris versicolor.
Flag Iris – Iris germanica.
Flag, Sweet – Acorus calamus.
Flag, Yellow – Iris pseudoacorus.

FLAGELLARIA L. Flagellariaceae. 3 spp. Tropical Africa, Formosa, Indomalaya, Australia, Pacific. Semi-woody climbers.

***F. indica*** L. E.India to New Caledonia. The tough stems are used in E.Malaysia and Thailand to make baskets etc. The leaves are used as a detergent for washing the hair.

Flaky Fir – Abies squamata.

FLAMMULA (Fr.) Quél. Agaricaceae. 120 spp. Widespread, especially on wood.

***F. velutipes*** (Curt.) Karst. = Collybia velutipes. Temperate. The fruitbodies are eaten.

Flannel Mullein – Verbascum thaspsus.

Flat Pea – Lathyrus sylvestris.

Flatcrown Albizia – Albizia fastigiata.

Flax – Linum usitatissimum.

Flax, New Zealand – Phormium tenax.

Flax, Purging – Linum catharticum.

Flax, Rocky Mountain – Linum lewisii.

Flaxweed – Linaria vulgaris.

Flea Seed – Plantago psyllium.

Fleabane – Erigeron canadensis.

Fleabane, Marsh – Pluchea sericea.

Fleeceflower, Japanese – Polygonium cuspidatum.

***Flemingia*** Roxb. (in Ait.) = Maughania.

Fleshy Pod Bean – Muncuna pachylobia.

Fleur de Dieu – Petrea kohautiana.

***Fleurya*** Gaudich. = Laportea Gaudich.

FLINDERSIA R. Br. Flindersiaceae. 22 spp. Moluccas, New Guinea, E.Australia, New Caledonia. Trees.

***F. acuminata*** White (Paddy King's Beech, Putt's Pine, Silver Silkwood, White Silkwood). Queensland. The wood is light, soft, fairly strong and easy to work. It is used for general building, interiors of houses, furniture and butter churns and pats.

***F. australis*** R. Br. (Flindose, Rasp-Pad). New South Wales, Queensland. The hard durable wood is used for staves.

***F. brayleyana*** F. v. Muell. (Maple Silkwood, Queensland Maple, Red Beech, Silkwood). Queensland. The brownish-pink wood is light, works well and polishes well. An important local timber, being used for all forms of interior work, panelling, furniture-making, tramcars, carriages, boats, aeroplane propellers and rifle stocks.

***F. maculosa*** F. v. Muell. New South Wales, Queensland. The plant yields a gum which contains about 80 per cent arabin.

***F. oxleyana*** F. v. Muell. (Long Jack, Yellow Wood, Yellow Wood Ash). New South Wales, Queensland. The yellow-brown wood is fairly hard, uniform, tough and works easily. It is used for high quality interior work in houses, cabinet work, body and seat frames in railway coaches, wheel-making, tool handles and skis. It is also the source of a yellow dye.

Flindose – Flindersia australis.

Flixweed – Descurainia sophia.

Floodbed Box – Eucalyptus microtheca.

Flooded Gum – Eucalyptus tereticornus.

Flor de dolores – Senecio salignus.

Flor de la Oreja – Cymbopetalum penduliflorum.

Flor de los Muertos – Laelia autumnalis.

Flor de paisto – Hibiscus bifurcatus.

Flor de San José – Cassia emarginata.

Flor de Todos Santos – Laelia speciosa.

Florence Fennel – Foeniculum vulgare.

Florentine Iris – Iris florentina.

Flores Chamomillae Romanae – Anthemis nobilis.

Flores cinae – Artemisia cina.

Flores Coso – Hagenia abyssinica.

Flores Ericae – Calluna vulgaris.

Flores Malvae – Malva sylvestris.

Flores Sambuci – Sambucus nigra.

Flores Syringea – Syringa vulgaris.

Florida Arrowroot – Zamia integrifolia.

Florida Corkwood – Leitneria floridana.

Florida Moss – Tillandsia usneoides.

Florida Torreya – Torreya taxifolia.

Florida Velvet Bean – Mucuna deeringiana.

Florida Woodbox – Schaefferia frutescens.

Floripondio – Datura arborea.

FLOSCOPA Lour. Commelinaceae. 20 spp. Tropics and subtropics. Herbs.

***F. scandens*** Lour. Tropical Asia and Australia. The juice is used to treat sore eyes.

FLOURENSIA DC. Compositae. 30 spp. S.W. United States of America to Argentine. Herbs.

***F. cernua*** DC. (Tarbrush). S. U.S.A. and Mexico. The leaves and flowers (Hojase) are used to treat indigestion.

Flowering Ash – Fraxinus ornus.

Flowering Dogwood – Cornus florida.

Flowering Rush – Butomus umbellatus.

Flowering Usnea – Usnea florida.

Flowering Willow – Chelopsis lineari.

FLUEGGEA Willd. Euphorbiaceae. 3 spp. Old World Tropics. Shrubs.

***F. leucopyros*** Dalz. = F. microcarpa.

***F. melanthesioides*** F. v. Muell. = F. microcarpa.

***F. microcarpa*** Blume. = F. leucopyros = F. melanthesioides = Securinga obovata. Throughout Old World Tropics. The bark is used to stupefy fish.

Fluellin – Linaria elatine.

***Fluggea*** Willd. = Flueggea.

Fly Amanita – Amanita muscaria.

FOCKEA Endl. Asclepiadaceae. 10 spp. Tropical and S.Africa. Woody vines.

***F. mutiflora*** Schum. Angola. Yields a latex used to adulterate rubber.

FOENICULUM Mill. Umbelliferae. 5 spp. Mediterranean, Europe. All the species (or varieties) mentioned are grown for the seeds which yield about 12–18 per cent semidrying oil (Oil of Fennel, Oleum Foeniculi) and 1–6 per cent of a volatile oil. The composition of the oils varies with the species. Oil of Fennel contains approximately 55 per cent anethol and 20 per cent fenchose, chavicol, anisic aldehyde. It is used medicinally to relieve stomach upsets, to flavour foods, candies, liqueurs (especially Absinthe) and in scenting soaps and perfumes. The volatile oil is used similarly in medicine and to scent soaps.

***F. azoricum*** Mill. = F. vulgare var azoricum (Mill.) Thell. (Carosella, Italian Fennel). A cultivar originated in Italy. The broad leaf-bases and stalks are blanched and eaten in salads.

***F. dulce*** Mill. = F. vulgare var. dulce (Mill.) Thell. (Florence Fennel, Sweet Fennel, Roman Fennel).

***F. officinale*** All. = F. vulgare.

***F. piperatum*** Ucr. = F. vulgare var. piperitum (Ucr.) Cout. (Bitter Fennel).

***F. vulgare*** Mill. = F. officinale. (Fennel). Mediterranean. It has been cultivated for thousands of years in Middle East, China and India and has been widely introduced throughout the world. The main areas of cultivation are India, Russia, Europe, Japan.

Folha cheirosa – Anthurium oxycarpon.

Folha de Arubu – Philodendron laciniatum.

Folha de Fonte – Philodendron cordatum.

Folia Andromedae – Oxydendrum arboreum.

Folia Buddleiae – Buddleja brasiliensis.

Folia Colutae – Coronilla emerus.

Folia et Herba Urticae – Urtica dioica.

Folia Jaborandi – Pilocarpus pinnatifolius.

Folia Orthosiphonis – Orthosiphon stamineus.

Folia Rubi fruticosi – Rubus fruticosus.

***Folotsia*** Costantin and Bois. = Cynanchum.

FOMES (Fr.) Kickx. Polyporaceae. 100 spp. Perennial on wood.

***F. applantus*** auct. = Ganoderma applanatus.

***F. auberianus*** Mont. = F. microporus = Ungulina auberiana. Tropics on dead wood. A poisonous species, used by the natives of New Guinea to promote frenzies before battle etc. They also use it as a contraceptive drug and an abortive. It may have some place in modern medicine in treatment of mental illness.

***F. fomentarium*** (L.) Gill. Temperate zone. The fruitbodies used to be a common source of tinder (Amadou). It is now used occasionally to manufacture buttons, flowerpots, smoking caps and slippers. In Siberia it is used as an adulterant for snuff.

***F. hemiterphrus*** Berk. The fruitbodies are used as razor strops in Australia.

***F. igniarius*** (L.) Gill. The fruitbodies yield a brown dye.

***F. microsporus*** (Swartz) Fr. = F. auberianus.

***F. pachyophloeus*** Pat. Tropical Asia. The powdered fruitbodies are eaten in the Philippines.

***Fontanella brasiliensis*** St. Hil. = Quillaja brasiliensis.

Fony Oil – Adansonia digitata.

Food Candle Tree – Parmentiera edulis.

Food Inga – Inga edulis.

Fool Mushroom – Panaeolus papilionaceus.

Fool's Parsley – Aethusa cynapium.

Fooraa – Calophyllum tacamahaca.

Forest Mahogany – Eucalyptus resinifera.

Forest Oak – Casuarina torulosa.

Forest Red Gum – Eucalyptus umbellata.

FORESTIERA Poir. Oleaceae. 15 spp. America, especially S. United States of America and Mexico, W.Indies. Small trees.

***F. acuminata*** (Michx.) Poir. (Swamp Privet). E. U.S.A. The wood is hard and is used for turning.

Forget-me-not – Mysotis palustris.

Forking Larkspur – Delphinium consolida.

Formosa Sweet Gum – Liquidambar formosana.

Forombata – Cyperus debilissimus.

***Forskohlea*** Batsch. = Forsskalea.

FORSSKALEA L. Urticaceae. 6 spp. Canary Islands, S.E.Spain, Africa, Arabia, W.Indies. Trees.

***F. tenacissima*** L. (El Assaik, Harick, Kleasaik). Sahara. The inner bark is used locally to make ropes.

FORSTERONIA G. F. W. Mey. Apocynaceae. 50 spp. Central and tropical S.America, W.Indies. Shrubs. Both species mentioned give a rubber of little commercial value.

***F. floribunda*** Muell. Arg. Brazil, Jamaica.

***F. gracilis*** Muell. Arg. Guiana.

FORSYTHIA Vahl. Oleaceae. 1 spp. S.E.Europe, 6 spp. E.Asia. Shrubs.

***F. suspensa*** Vahl. (Lien ch'iao). China, Japan. A decoction of the fruit coat is used locally to treat boils and other skin infections. It is also used as a diuretic, to treat intestinal worms and to control menstruation. A decoction of the leaves and twigs is used to treat breast cancer.

FORTUNELLA Swingle Rutaceae. 6 spp. E.Asia, Malaysia. Small trees.

***F. crassifolia*** Swingle (Cumquat, Meiwa Kumquat). China, Japan. The sweet fruits are eaten raw. The plant is occasionally cultivated.

***F. japonica*** (Thunb.) Swingle. = Citrus japonica. (Cumquat, Marumi Kumquat). Japan. The fruits are acid and are used occasionally for jellies and preserves.

***F. margarita*** (Lour.) Swingle. = Citrus margarita. (Cumquat, Nagami Kumquat). Japan. Cultivated commercially in China, Japan and Florida. The acid fruits are used for jams and preserves.

Fou-fou – Manihot esculenta.

FOUQUIERIA Kunth. Fouquieriaceae. 10 spp. S.W. United States of America, Mexico. Small trees or shrubs.

***F. macdougalii*** Nash. (Chunari, Jabonecillo). Mexico. The bark is used locally as a soap substitute for washing wool.

***F. splendens*** Engelm. (Ocotillo). S.W. U.S.A. to Mexico. A decoction of the roots is used by the local Indians as a stimulant. The latex is a possible source of rubber.

Fourleaf Loosestrife – Lysimachia quadrifolia.

Four o' Clock – Mirabilis jalapa.

Fourwing Saltbush – Striplex canescens.

Fourwing Sophora – Sophora tetraptera.

Fowl Bluegrass – Poa palustris.

Fox Grape – Vitis labrusca.

Fox Tail, Meadow – Alopercurus pratensis.

Foxberry – Vaccinium vitis-idaea.

Foxglove – Digitalis purpurea.

Foxtail Millet – Setaria italica.

Foxtail Pine – Pinus aristata.

FRAGARIA L. Rosaceae. 15 spp. N.America, Chile, Eurasia to S.India. Herbs. The cultivated strawberries are intercrosses between the four species mentioned below, from which many cultivars have been bred. The main cultivars are derived from F. virginata × F. chiloensis. These plants were introduced to Europe where they were hybridized in 1714.

The strawberry is one of the most popular dessert fruits and is used a great deal in the making of jams and preserves. Many are canned or used to flavour ice-cream. The growing frozen food industry takes an increasing amount of the world production and the demand for fresh strawberries makes air transport of the fruit from one country to another a viable proposition. A home made wine (Crême de Fraise) is also made from the fruit and they are used to flavour syrups used medicinally. A tea substitute is made from the leaves.

***F. chiloensis*** (L.) Duch. (Chiloe Strawberry). Pacific N. and S. America. The plant is cultivated in the Andes. The fruits are eaten locally.

***F. filipendula*** Hemsl. = Duchesnea filipendula.

***F. moschata*** Duch. (Hautbois Strawberry). Temperate Europe. The fruits are small, but of excellent flavour. The plant is sometimes cultivated locally.

***F. vesca*** L. (Strawberry). Europe and temperate Asia. The fruits are eaten locally.

***F. virginiana*** Duch. (Virginian Strawberry). E. N.America, to Texas and Arizona. The fruits are pleasantly flavoured and are eaten locally.

Fragrant Albizia – Albizia odoratissima.

Fragrant Crab Apple – Malus coronaria.

Fragrant Sandalwood – Eucarya spicata.

Fragrant Sumach – Rhus aromatica.

***Franciscea uniflora*** Pohl. = Brunfelsia hopeana.

***Frangula alnus*** Miller = Rhamnus frangula.

FRANKENIA L. Frankeniaceae. 80 spp. Temperate and subtropical seacoasts. Halophytes. Shrubs.

***F. berteroana*** Gay. Chile. Burnt locally as a source of salt.

***F. ericifolia*** C. Sm. Cape Verde Islands, Canary Islands. The plant is used in the Cape Verde Islands as a fish poison.

***F. grandifolia*** Cham. and Schlecht. (Yerba Reuma). California to Mexico. The plant is used locally to treat dysentery and diarrhoea, catarrh and vaginal discharges.

***F. portulacaefolia*** Spreng. = Beatsonia portulaceaefolia. St. Helena. The leaves are used locally as a tea substitute.

Frankincense – Boswellia carteri.

Frankincense, Bible – Boswellia carteri.

Frankincense, Elemi – Boswellia frereana.

Frankincense, Female – Boswellia carteri.
Frankincense, Indian – Boswellia serrata.
Frankincense, Male – Boswellia carteri.
Frankincense, Ogea – Daniella thurfera.

FRASERA Walt. Gentianaceae. 15 spp. N.America. Herbs.

***F. carolinensis*** Walt. (American Columbo). E. N.America. The roots are used to make a tonic.

***Fraxinella dictamnus*** Moench. = Dictamnus albus.

FRAXINUS L. Oleaceae. 70 spp. N.hemisphere, especially E.Asia, N.America and Mediterranean. Trees. The genus includes the Ash trees. The "ash" of the Bible is a translation of the Hebrew "Oren" and probably means the Apello Pine (Pinus halepensis). There is no species of Fraxinus indigenous to Palestine.

***F. americana*** L. (White Ash). E. N.America. An important timber tree. The brown wood is hard, heavy and tough. It is used for the interiors of houses, furniture, agricultural implements, oars and carriage work. The young bark from the branches and roots is used to make a bitter tonic.

***F. chinensis*** Roxb. China. A mealy bug (Coccus pela) feeds on the tree and exudes a white wax (Insect White Wax) which is used in China in the manufacture of candles and to glaze high quality paper. It is also used in China to coat pills and to polish jade, soap stone and furniture.

***F. edenii*** Boerl. and Koord. = F. griffithii.

***F. excelsior*** L. (European Ash). Europe, Caucasus. The yellow-brown wood is fairly hard, heavy and elastic. It is used in general carpentry and especially in agricultural implements, rulers, sports gear and aeroplane parts.

***F. floribunda*** Wall. (Angan, Angu, Kanga). Temperate regions of the Himalayas. A sweet manna is derived from cuts in the stem and is used locally for sweetening and as a laxative.

***F. griffithii*** Clarke. = F. edenii. Himalayas to Java. The leaves are smoked locally as a substitute for opium. The smoke resembles that of opium in scent and taste, but it does not have the same narcotic or habit-forming effects.

***F. latifolia*** Benth. = F. oregana.

***F. longicuspis*** Fr. and Sav. = F. sieboldiana.

***F. mariesii*** Hook. Central China. See F. chinensis.

***F. nigra*** Marsh. = F. sambucifolia (Black Ash). E. N.America. The dark brown wood is coarse-grained, heavy, durable and soft. It is used in house interiors, furniture-making, making barrels and baskets. Veneers (Ash Burl) are made from thin pieces of wood cut across the knots in the trunk and branches.

***F. oregana*** Nutt. = F. latifolia (Oregon Ash). Pacific N.America. The hard brittle, coarse-grained wood is light brown. It is used for interiors of houses, furniture, wagons and carriages. This is a valuable timber tree.

***F. ornus*** L. (Flowering Ash, Manna Ash). S.Europe, Balkans to Asia Minor. See F. floribunda.

***F. oxypylla*** Mar. Morocco, Algeria. The fruits are used in Morocco to flavour foods and as an aphrodisiac.

***F. pennsylvanica*** Marsh = F. pubescens.

***F. pubescens*** Lam. = F. pennsylvanica. (Red Ash). E. N.America. See F. americana.

***F. quadrangulata*** Michx. (Blue Ash). E. N.America. The wood is light yellow, streaked with brown. It is hard, heavy and rather brittle. Its uses include, general construction work, making agricultural implements and wagons and for flooring.

***F. sambucifolia*** Lam. = F. nigra.

***F. sieboldiana*** Blume = F. longicuspis. Japan to Korea. The wood is soft, light and elastic. It is used to make sports gear, furniture and utensils. Thin strips of it give an ornately grained veneer.

***Fremontia*** Torrey = Fremontodendron.

FREMONTODENDRON Corville. Sterculiaceae. 4–6 spp. California to Mexico. Small trees.

***F. californica*** Torr. (Slippery Elm). California to Mexico. A mucilage from the inner bark is occasionally used to make poultices.

Fremont's Cottonwood – Populus fremontii.
French Bean – Phaseolus vulgaris.
French Berries – Rhamnus infectorius.
French Lavender – Lavandula stoechas.
French Mulberry – Callicarpa americana.
French Oil Turpentine – Pinus pinaster.
French Psyllium – Plantago arenaria.
French Rose – Rosa gallica.
French Sorrel – Rumex scutatus.
French Willow – Epilobium angustifolium.
French Willow – Salix triandra.

***Frenela endlicheri*** Parlat. = Callitris calarata.

FREYCINETIA Gaudich. Pandanaceae. 100 spp. Ceylon to New Zealand and Polynesia. Small shrubs.

***F. arborea*** Gaudich. Hawaii. The stem fibres are used locally to make baskets, as cordage and for making fish traps and hats.

***F. banksii*** Cunningh. New Zealand. The leaf fibres are used by the natives to make cloth.

***F. funicularis*** Merr. Vine. Moluccas. The young inflorescence is eaten locally as a vegetable.

Friars' Balsam – Styrax benzoin.

Frijoles – Phaseolus vulgaris.

Frijolollo – Pithecellobium arboreum.

Fringe Tree – Chionanthus virginica.

Fringed Rue – Ruta chalepensis.

FRITILLARIA L. Liliaceae. 85 spp. N.Temperate. Herbs.

***F. camschatcensis*** Ker-Gawl. W. N.America. The small bulbs are eaten by the natives of S.E.Alaska. They are sometimes dried and stored for winter use.

***F. imperialis*** L. (Crown Imperial). Middle East. Cultivated as an ornamental. The bulbs are a minor source of starch. They were previously used medicinally (Bulbus coronae imperialis).

***F. pudica*** (Pursh.) Spreng. W. N.America. The bulbs were eaten by the local Indians as a vegetable, or raw, and they were dried for winter use.

***F. roylei*** Hook. W.China. The ground bulbs are used in Chinese medicine to treat asthma. There is some local trade in the bulbs for this purpose.

***F. verticillata*** Willd. var. ***thunbergii*** Baker. (Chih mu). The bulbs contain fritiline which diminishes the excitability of the respiratory centres and paralyses voluntary movement. They are used in Chinese medicine to combat the effects of opium, to increase lactation and treat breast cancer and in the treatment of haemorrhages, dysentery and fevers.

Fromager inerme – Ceiba guinensis.

Frost Grape – Vitis vulpina.

Frost Weed – Helianthemum canadense.

Fructa de Macaco – Rolliniopsis discreta.

Fructus Myrtilli – Vaccinium myrtillus.

Fructus Simulo – Capparis coriacea.

Fructa bomba – Carica papaya.

FUCHSIA L. Onagraceae. 100 spp. New Zealand, Tahiti, Central and S.America. Shrubs.

***F. excorticata*** L. f. New Zealand. The whole plant is used locally in steam baths and to treat bleeding after childbirth.

***F. macrostemma*** Ruiz. and Pav. Chile. The wood yields a black dye. The leaves and barks are used to reduce fevers and as a diuretic.

***F. magellanica*** Lam. See F. macrostemma.

FUCUS L. Fucaceae. Brown Algae.

***F. fuscatus*** C. Ag. N.Pacific. Eaten in parts of Alaska.

***F. serratus*** L. N.Atlantic. Used in the manufacture of iodine and potash. It is used in many coastal regions as a manure and as a winter feed for livestock. It has been used to treat scurvy, and its high iodine content makes it useful in the treatment of goitre. It is supposed to aid in the treatment of obesity.

***F. vesiculosus*** L. (Bladder Wrack, Lady Wrack, Bladder Fucus, Black Tang, Sea Ware). N.Atlantic. See F. serratus.

Fudo – Heeria insignia.

Fukuti – Millettia barteri.

Fuller's Teazel – Dipsacus fullonum L. ssp. sativus.

FUMARIA L. Fumariaceae. 55 spp. Europe, Mediterranean to Central Asia and Himalayas. Herbs.

***F. media*** Loisel. = F. officinalis.

***F. officinalis*** L. = F. media Loisel. (Fumitory, Earth Smoke). Europe, N.Africa, to temperate Asia. The leaves are used locally as a tonic, to "purify the blood" and to treat minor stomach complaints. The leaves were used to make a yellow and green dye.

***F. paviflora*** Lam. Middle East to Afghanistan. The leaves are used in Iran as a blood "purifier", as a laxative and a diuretic.

***F. vaillantii*** Lois. (Alkafoun, Boulboudi). Oases of the Sahara. The leaves are used locally as a condiment.

Fume, Camphor – Camphorosma monspeliaca.

Fumitory – Fumaria officinalis.

FUNASTRUM Fourn. Asclepiadaceae. 10–15 spp. Tropical America. Shrubs.

***F. clausum*** (Jacq.) Schlech. = Cynanchum clausum = Philibertia crassifolia = P. palmeria (Bejuco de leche, Petaquilla, Mata tórsalo). Florida through to S.America, W.Indies. In Costa Rica, the juice from the leaf is applied to the point of invasion by dipterous larvae in the skin. The juice kills the larva.

***F. cumanense*** (H.B.K.) Schlech. = Sarcostemma cumanense = S. arenarium. (Bejuco de pescado, Cuchamperrito). Mexico, through to S.America. The stems are used in Central America to string fish.

Fungus, Bovista – Calvatia lilacina.

Fungus, Bukonde – Gibberella fujikuroi.

Fungus cervinus – Elaphomyces cervinus.

Fungus Chirurgorum – Calvatia lilacina.

Fungus Chirurgorum – Polyporus officinalis.

Fungus, Milk – Oidium lactis.

***Funkia*** Spreng. = Hosta.

Funori – Gloiopeltis furcata.

FUNTUMIA Stapf. Apocynaceae. 3 spp. Tropical Africa. Trees.

***F. elastica*** Stapf. = Kickxia elastica. (Silkrubber Tree, Lagos Silkrubber Tree, Lagos Rubber Tree). The main rubber-producing species in W.Africa, but has now been mainly replaced by Hevea, except in the Cameroons where production is still carried out. The tree was introduced to the W.Indies, but has been superceded there by Hevea.

FURCRAEA Vent. Agavaceae. 20 spp. Tropical America. Large coarse herbs. All the species mentioned yield fibres from the leaves. They are extracted by retting and are of good quality. The fibres are used to make ropes, hammocks sacks etc.

***F. cabuya*** Trel (Cabuya). Costa Rica to Panama. The fibres are called Cabuya, Cabuia, Cabulla. They are used mainly in local industry.

***F. cubensis*** (Jacq.) Vent. = F. hexapetala.

***F. geminispina*** Jacobi = F. humboldiana.

***F. gigantea*** Vent. (Piteira, Piteira Gigante). Cultivated in Old and New World Tropics, but the only major area of cultivation is Mauritius. The fibre is called Mauritius Hemp or Pitura Fibre.

***F. hexapetala*** (Jacq.) Urb. = F. cubensis. W.Indies, especially the N. islands. The fibre is called Cuba Hemp or Pitre Fibre. It is used locally but not cultivated.

***F. humboldiana*** Trel. = F. geminispina (Maguey de Cozuiza). Cultivated locally. The fibres are called Cocuiza Fibre.

***F. macrophylla*** Baker. (Fique). Columbia. The plant is cultivated locally and there is some small export trade in the fibre.

***F. samalana*** Trel. = F. selloa.

***F. selloa*** Koch = F. samalana. S.America. The fibre is excellent with some small local use.

***F. tuberosa*** Ait. W.Indies. The fibres are used locally.

Furuluga – Cnestis corniculata.

***Fusanus*** R. Br. = Eucarya.

FUSARIUM Link. ex Fr. Tuberculariaceae. 65 spp. Fungi.

***F. aleyrodes*** Petch. (White Fringe Fungus). Attacks White Flies on Citrus.

***F. moniliforme*** Sheld. = Gibberella fujikuroi.

***F. orthoceras*** Sacc. Produces the antibiotic enniantin.

Fustic – Chlorophora tinctoria.

Fustic Tree, Iroko – Chlorophora excelsa.

# G

Gabon Chocolate – Irvingia barteri.

Gabon Ebony – Diospyros dendo.

Gabon Ebony – Diospyros flavescens.

Gabon Mahogany – Canarium mansfeldianum.

Gabon Mahogany – Khaya klainei.

Gadhbhains – Salix acmophylla.

Gadoong tikoos – Tacca palmata.

Gafal Wood – Commiphora erythraea var. glabrescens.

GAILLARDIA Fougeroux. Compositae. 26 spp. N.America; 2 spp. Temperate S.America. Herbs.

***G. pinnatifida*** Torr. S.W. U.S.A. and Mexico. The leaves are used by some Mexican natives as a diuretic.

GALACTIA P. Br. Leguminosae. 140 spp. Tropics and subtropics. Herbs.

***G. villosa*** Reichb. Russian Steppes. An infusion of the young stalks is used locally as an expectorant for treating coughs and sore throats.

Galadoopa – Afzelia galedupa.

Galala ajer – Erythrina fuseda.

Galam Butter – Butyrospermum parkii.

Galancha Tinospora – Tinospora cordifolia.

Galán de noche – Cestrum nocturnum.

Galanga – Kaempferia galanga.

Galangal – Alpinia officinarum.

Galangal, Greater – Alpinia galanga.

Galangal, Lesser – Alpinia officinarum.
Galangal, Small – Alpinia officinarum.
Galappa Palm – Actinorhytis calapparia.
Galbanum – Ferula galbaniflua.
Galbanum – Ferula rubicaulis.
Galbanum – Ferula schair.

GALEGA L. Leguminosae. 3 spp. Mediterranean to Persia; 3 spp. Tropical E.Africa. Herbs.

***G. officinalis*** L. (Galega, European Goat's Rue). Central Europe through to Persia. The dried flower heads are used as a tonic and to stimulate milk production. The whole plant is used as food for livestock, for which it is occasionally cultivated.

***G. toxicaria*** Sw.=Tephrosia toxicaria.

GALEOPSIS L. Labiatae. 100 spp. Temperate Eurasia. Herbs.

***G. dubia*** Leers.=G. ochroleuca.

***G. ochroleuca*** Lam.=G. dubia=Tetrahit longiflorum. (Hemp Nettle). Europe. A decoction of the leaves (Herba Galeopsidis) is used locally to treat lung, intestinal and spleen complaints.

GALERA (Fr.) Quél. Agaricaceae. 50 spp. Cosmopolitan.

***G. siligineoides*** Heim. S.Mexico. Found on trees. The fruit bodies (Ya'nte) are eaten during various ceremonies to produce hallucinations.

GALINSOGA Ruiz. and Pav. Compositae. 4 spp. Mexico to Argentina. Herbs.

***G. parviflora*** Cav. (Galinsoga). Tropical America, but escaped throughout N.America, Europe and Asia. The young plants are eaten as a vegetable in S.E.Asia.

Galinsoga – Galinsoga parviflora.

GALIPEA Aubl. Rutaceae. 13 spp. Central and S.America. Trees.

***G. jasminiflora*** (St. Hil.) Engl. S.America. The bark is used as a quinine substitute in Brazil and a decoction of the roots is used locally to treat warts.

Galipoong booloo – Gomphostemma phlomoides.
Galingale, English – Cyperus longus.

GALIUM L. Rubiaceae. 400 spp. Cosmopolitan. Herbs.

***G. aparine*** L. (Goosegrass, Catchweed). Europe to Siberia, N. and S.America. The dried fruits are sometimes used in Ireland as a coffee substitute.

***G. odorata*** L. (Sweet Woodruff). Europe to Siberia, N.Africa. The plant contains coumarin and is used to perfume stored linen and as a flavouring for wines, liqueurs and snuff. It is particularly used in Germany and Austria to make the drink Maiwein, Maitrank or Maybowl. The whole plant is used as a sedative for children and old people (Herba Matrisilvae, Herba Asperulae orodatae). The plant is also used to treat inflammation of the uterus and vagina and as an antispasmodic.

***G. orizabense*** Hemsl. S.Mexico. The dried leaves are used by the local Indians to treat intestinal parasites and to reduce fevers.

***G. tinctorium*** L. (Dye Bedstraw, Dyer's Woodruff). Europe N.America. The roots yield a red dye, formerly used for dyeing cloth.

***G. umbrosum*** Soland. New Zealand. The whole plant is used in the treatment of gonorrhoea.

***G. verum*** L. (Our Lady's Bedstraw). Europe, Asia Minor to Caucasus. The dried plants were used to fill matresses. The roots give a red dye. The juice curdles milk and was used in some parts of England and Scotland to do this. The leaves are diuretic and purgative, and were used to treat stones of the bladder and kidneys.

Gall, Acorn – Quercus robur.
Gall, Aleppo – Quercus lusitanica.
Gall, American Nut – Quercus imbricata.
Gall, Bokhara – Pistacia vera.
Gall, Chinese – Rhus semialata.
Gall, Ch'-pei-tzu – Rhus potanini.
Gall, Dead Sea Apple – Quercus tauricola.
Gall, Istrian – Quercus ilex.
Gall, Japanese – Rhus semialata.
Gall, Knoppern – Quercus robur.
Gall, Mala Insana – Quercus tauricola.
Gall, Mala Sodomitica – Quercus tauricola.
Gall, May Apple – Azalea nudiflora.
Gall, Morea – Quercus cerris.
Gall, Pistacia – Pistacia terebinthus.
Gall, Rave – Quercus tauricola.
Gall, Sakum – Tamarix gallica.
Gall, Smyrna – Quercus lusitanica.
Gall, Takut – Tamarix articulata.
Gall, Teggant – Tamarix articulata.
Gall, White – Quercus lusitanica.
Gall, Wu-pei-tzu – Rhus semialata.
Gallae turcicae – Quercus lusitanica.
Gallberry – Ilex glabra.

GALLESIA Casaretto. Phytolaccaceae. 2 spp. Peru, Brazil. Trees.

***G. gorazema*** (Vell.) Moq. Brazil. A decoction of the wood is used locally to treat intestinal worms and diseases of the lymphatic system. The wood yields an essential oil used in S.America to treat gonorrhoea.

***G. integrifolia*** (Spreng.) Hrams. (Guaraema). Tropical America. The leaves are used as a soap substitute in Peru and Brazil.

Gallini Cotton – Gossypium barbadense.

Gallito – Erythrina rubinerva.

Gal-mora – Cryptocarya membranacea.

Galo – Anacolosa luzoniensis.

Galoba djantoong – Hornstedtia rumphii.

Gall of the Earth – Prenanthes serpentaria.

Gamba Pea – Crotalaria goreënsis.

Gambia, Bitters Tree of – Veronia senegalensis.

Gambian Tea Bush – Lippia adoënsis.

Gambiran – Breynia cernua.

Gambir, Bengal – Uncaria gambir.

Gambir, Black – Uncaria gambir.

Gambir, Cube – Uncaria gambir.

Gambir ooran – Trigonopleura malayana.

Gambo Hemp – Hibiscus cannabinus.

Gamboge – Garcinia morella.

Gamboge, Pipe – Garcinia hanburyi.

Gamboge, Siam – Garcinia hanburyi.

Gamboge Tree – Garcinia tinctoria.

Gamelotte Fibre – Fimbristylis spadicea.

Gamer – Breynia cernua.

Gamote – Cymopterus montanus.

Gampi – Wilkstroemia canescens.

Gandubikachu – Homalomena rubescens.

Gandaroora besar – Euodia amboinensis.

Gandri – Bridelia monoica.

Gangganan – Tetrameles nudiflora.

Ganggo – Aglaia ganggo.

GANODERMA Karst. Polyporaceae. 50 spp. especially tropics.

***G. amboinense*** (Lam. ex Fr.) sensu Pat. Indonesia and S.Pacific. Used as a drug in Indonesia, but its properties are unknown.

***G. applanatum*** (Wallr.) Karst.=Fomes applantus. See Fomes fomentarius.

***G. cochleara*** (Blume and Nees) Murr. See G. amboinense.

GANOPHYLLUM Blume. Sapindaceae. 1 sp. Tropical W.Africa; 1 sp. Philippines. Trees.

***G. falcatum*** Blume. Malaysia, Philippines. The wood is used in Java for heavy construction work, houses and bridges. The seeds yield a fat (Arangan Oil) used for lighting.

***G. giganteum*** (A. Chev.) Hauman. (Dioko dia mfinda, Kididila, Nzwmbila). Tropical W.Africa. The wood is white and extremely beautiful. It is highly prized for interiors of houses. It is used locally to make charcoal. The fruits are edible.

***Ganua mottleyana*** Pierre=Madhuoa mottleyana.

Gaozaban – Onosma bracteatum.

Garabata – Pisonia aculeata.

Garabato – Celtis pallida.

Garabato blanco – Celtis iguanaea.

Garambullo – Pisonia capitata.

Garamgarum – Roucheria griffithiana.

Garampara – Ouratea castaneidolia.

Garbancillo – Astragalus garbancillo.

Garbanzos – Cicer arietinum.

***Garbenia benedicta*** Benth. and Hook. f.= Cnicus benedictus.

GARCIA Rohr. Euphorbiaceae. 2 spp. Mexico. Trees.

***G. nutans*** Rohr. S.Mexico to Columbia, W.Indies. The seeds contain some 80 per cent of a drying oil which is used for varnishes etc.

GARCINIA L. Guttiferae. 400 spp. Tropics (especially Asia), S.Africa. Trees.

***G. atroviridis*** Griff. (Gelugur). Assam to Malaysia. Sometimes cultivated locally. The fruits are large and the pulps acid. They are eaten raw with sugar, or dried and eaten in soups.

***G. balansae*** H. Bn.=G. morella.

***G. bancana*** Miq. (Kelabang). Malaysia. The fruits are eaten locally.

***G. barrettiana*** Wester. (Kadis). Philippines. The fruits are sub-acid looking like a small orange. They are eaten locally.

***G. benthami*** Pierre. Malaysia, Cochin-China. The fruit pulp is white and pleasantly flavoured. Eaten locally.

***G. binucao*** Choisy (Binukao). E.India to Philippines. The fruit look like small flattened lemons. They are eaten with fish in the Philippines.

***G. cambogia*** Desrouss. (Banti, Aradal, Hila, Goraka). The fruits are the size of an orange and longitudinally ridged. They are dried and eaten with fish locally.

***G. celebica*** L.=G. cornea. Malaysia. The wood is used locally for general building.

***G. cernua*** Bak. (Voahandrintsahona). Madagascar. The fruits are eaten locally.

***G. cochinchinensis*** (Lour.) Choisy. Cochin-China. Cultivated locally for the fruits which are eaten raw. They resemble small oranges.

***G. cola*** Heckel. = Cola acuminata.

***G. conrauana*** Engl. (Ntu). Cameroons. The fruits look like small oranges and are eaten locally.

***G. cornea*** Blume. = G. celebica.

***G. cornea*** L. (Kirasa). Malaysia, Burma. The fruits are purple with a white flesh. They are eaten locally.

***G. cowa*** Roxb. (Cowa, Gambozi). Assam to Burma. The yellowish fruits are ridged and about the size of an orange. They are of good flavour and are eaten locally.

***G. cumingiana*** Pierre. (Malaba). Philippines, N.Luzon. The fruits are eaten locally.

***G. delpyana*** Pierre (Trameng). Indochina. The fruits are yellow and round, about 3 cm. in diameter, with a spongy pericarp. They are eaten locally.

***G. dioica*** Blume. (Tjeuri). Indonesia. The fruits are round, yellow and about the size of a small orange. They have an excellent flavour and are eaten locally.

***G. dives*** Pierre. (Pildis). Philippines. The fruits are eaten locally.

***G. dulcis*** Kurz. (Baniti). Philippines to Java. The fruits are orange-yellow and subacid. They are eaten locally. A green dye is extracted from the bark and used with indigo to give a brown colour. The plant is sometimes cultivated.

***G. gaudichaudii*** Planch. Cochin-China. Occasionally cultivated for a gum extracted from the bark. See G. hanburyi.

***G. gerrardii*** Harvey. (Umbini). Natal. The black fruits are eaten locally.

***G. globulosa*** Ridley. (Kandis). Malaysia. The fruits are orange coloured. They are eaten locally.

***G. hanburyi*** Hook. f. (Gamboge, Siam Gamboge Tree). The bark yields a gum resin (Gamboge, Pipe Gamboge, Siam Gamboge) which is collected from cuts in the trunk. Gamboge is soluble in turpentine but not water. It is used to make water colours, lacquers, varnishes etc. giving them a bright yellow colour. It is also a strong laxative.

***G. harmandii*** Pierre. (Remir). Cambodia. The fruits are eaten locally.

***G. heterandra*** Wall. See G. gaudichaudii.

***G. hombroniana*** Pierre. (Bruds, Mangis Butan). Malaysia, Nicobar Islands. The fruits have an excellent delicate flavour. This species is possible material for hybridizing.

***G. indica*** Choisy. (Kokan, Ktambi). Tropical Asia. Cultivated locally for the fruits (Bridonnes) which are pleasantly sub-acid. They are eaten raw or in syrups and jellies. The seeds give a fat (Goa Butter, Kokam Butter) which is eaten and used medicinally in ointments.

***G. lancaefolia*** Roxb. (Kirindur). Assam, Burma. The fruits are like orange plums and are eaten locally.

***G. lanessonii*** Pierre. (Angkol, Domong kol). S.Vietnam, Laos, Cambodia. Used medicinally in Cambodia. A decoction of the bark is used as a purgative. An extract of the heartwood is laxative and is used to treat constipation and liver complaints. The gum from the trunk is used to reduce fevers and a root extract is used to treat jaundice.

***G. lateriflora*** Blume (Kariis, Djawura). Philippines to Java. The fruits, which are eaten, show a wide variety of shape and flavour, making this a potential source of breeding material. The wood is used to make pestles.

***G. latissima*** Miq. Malaysia. The hard wood is used to make tobacco pipes.

***G. livingstonii*** T. Andres (Gupenja, Imbé, Mwausungulu, Pama). Tanzania to Zambia. The fruits are purple and about 5 cm. long. They are eaten and used locally to make a wine.

***G. loureiri*** Pierre (Buanha). Indochina. The yellow fruits are grooved with a white pulp. They are dried and used locally as a condiment for rice.

***G. macrophylla*** Miq. (Selapan). Sumatra. The fruits are eaten locally.

***G. mangostana*** L. (Mangosteen, Dodol). Moluccas. The plant is grown widely in the Far East. The fruits are purple and about the size of a small orange. They have a superb flavour and their wide commercial use is prevented only by their poor keeping qualities. The plant has been cultivated as a garden crop since early times. Propagation is by seeds and now, more frequently by cuttings, but the young plants take many years to become established due to the slow root growth. Attempts have been made since the 19th century to introduce mangosteens to the W.Indies, but with only limited success.

***G. merguensis*** Wight. Tropical Asia, especially south. The bark yields a brown dye and the resin is used for varnish.

***G. mooreana*** Wester. (Bunag). Philippines. The fruits are eaten locally.

***G. mestoni*** Baill. Queensland. The fruits look like an orange, are slightly sour, but have a pleasant flavour.

***G. morella*** Desv. India. See G. hanburyi.

***G. multiflora*** Champ.=G. balansae=G. tonkinensis. (Cây giôc, Bira tai). Laos, N.Vietnam to Hong Kong). The fruits are lemon-flavoured and are used locally as a lemon substitute. The seeds give an oil used for lighting.

***G. nigro-lineata*** Planch. Malacca. The oval, orange fruits are eaten locally.

***G. oliveri*** Pierre. (Tromeng). Indochina. The red-orange fruits are eaten locally and sometimes preserved in salt.

***G. ovalifolia*** Oliv. Tropical Africa. The wood is used to make canoes in the Congo.

***G. paniculata*** Roxb. (Bubi-kowa). Bengal to E.Himalayas. The orange fruits resemble a small plum. They have a good flavour and are eaten locally.

***G. parvifolia*** Miq. Malaysia. The small yellow fruits are eaten as a condiment with Spanish pepper on fish.

***G. pedunculata*** Roxb. (Tikul). Bengal. The fruits are large and orange. They are used locally to make drinks and in curries.

***G. planchoni*** Pierre. Indochina. The fruits are large, warted and grooved, with a pleasant flavour. They are eaten raw and dried.

***G. praininana*** Kze. (Cherapu). Malaysia. The fruits are round, smooth, about the size of a small orange, but yellow in colour. They are sub acid and are eaten locally.

***G. roxburghii*** Engl. See G. gaudichaudii.

***G. tinctoria*** (DC.) W. F. Wight. (Gamboge Tree, Matau). S.India, Malaya. Cultivated in Old and New World tropics. The fruits are about the size of an orange with a yellow pulp. They have an excellent flavour and are eaten mainly as a breakfast fruit.

***G. terpnophylla*** Thw. Ceylon. The wood is yellow, hard and strong. It is used for beams and posts.

***G. tonkinensis*** Vesque.=G. multiflora.

***G. venulosa*** Choisy. E.India through to Philippines. The fruits are eaten locally.

***G. vilersiana*** Pierre (Prahout, Vàng nhura). Cambodia, Laos, S.Vietnam, Thailand. The fruits are eaten and sold locally. The bark yields a green dye of some commercial importance.

***G. wentzeliana*** Engl. (Mogola). Tropical Africa. Woody climber. The fruits look like grapes. They have a sweet pleasant flavour and are eaten locally.

***G. wightii*** T. Anders. See G. gaudichaudii.

Garden Asparagus – Asparagus officinalis.
Garden Balsam – Impatiens balsamina.
Garden Beet – Beta vulgaris.
Garden Burnet – Sanguisorba officinalis.
Garden Cress – Lepidium sativum.
Garden Gardenia – Gardenia florida.
Garden Hollyhock – Althaea rosea.
Garden Lovage – Levisticum officinale.
Garden Myrrh – Myrrhis odorata.
Garden Orach – Atriplex hortensis.
Garden Pea – Pisum sativum.
Garden Rhubarb – Rheum rhaponticum.
Garden Sage – Salvia officinale.
Garden Sorrel – Rumex acetosa.
Garden Tulip – Tulipa gesneriana.

GARDENIA Ellis. Rubiaceae. 250 spp. Old World Tropics. Trees.

***G. arborea*** Roxb.=G. gummifera.

***G. brasiliensis*** Spreng. Guyana, Brazil. The fruits are eaten by the natives.

***G. brighami*** Mann. Hawaii. The pulp from the fruits is used locally to dye cloth yellow.

***G. campanulata*** Roxb. (Hsaythanpaya). Himalayas, Assam. The fruits are used locally as a laxative and to treat intestinal worms. They are also used as a fish poison and to kill insect larvae in the skin.

***G. erubescens*** Stapf. and Hutch. Tropical Africa. The fruits are used locally as a condiment in soups and sauces. In N.Nigeria the seeds are used as a black cosmetic.

***G. florida*** L. (Garden Gardinia). Tropical China. The flowers are extracted with volatile solvents to give an essential oil used in perfumery. 3,000–4,000 kg. of flowers yield about 500 g. of absolute oil. The oil contains benzyl acetate, linalol, linalyl acetate, terpineol, styrolyl acetate and methyl anthranilate. The flowers are also used for scenting tea and the fruits give a yellow dye used locally.

***G. gummifera*** L. f.=G. arborea=G.

inermis. E.India. A gum (Combee, Dikkamaly) is extracted from the bark. It is antiseptic and is used to treat intestinal worms and minor stomach upsets. It is also used in veterinary medicine to keep insects away from wounds on livestock.

***G. inermis*** Dietr. = G. gummifera.

***G. jasminoides*** Ellis = G. florida.

***G. jovis tonantis*** Hiern. Tropical W.Africa. An extract of the roots is used locally as a tonic. The fruit is used as a fish poison and the seeds yield a black cosmetic stain.

***G. kalbreyeri*** Hiern. Tropical Africa. The seeds yield a black cosmetic stain.

***G. lucens*** Planch. and Sebert. New Caledonia. The light strong wood is used by the natives for carving small articles.

***G. lucida*** Roxb. Burma. The bark yields a resin (Combee Gum, Combee Resin) used to repel insects and worms from attacking the skin of humans and livestock. It is also used to treat skin diseases.

***G. lutea*** Fresen. Abyssinia to Blue Nile. The bark yields a sweet scented resin (Abu Beka). A decoction of the roots with sorghum flour is used locally to treat blackwater fever. The hard yellow wood is used to make knife handles. The fruit is edible.

***G. remyi*** H. Mann. (Remy's Gardinia) Hawaii. The fruits give a dye used locally to dye cloth yellow. The sticky leaf buds are used as a cement.

***G. rothmannia*** L. f. (Candlewood). S.Africa. The very hard wood is used for engraving, tool handles, feloes of wheels and wagon work.

***G. stanleyana*** Paxt. = Randia maculata.

***G. taitensis*** DC. Tahiti. The flowers are worn ornamentally and are used to scent coconut oil. An infusion of the flowers is used to treat headaches.

***G. ternifolia*** Schum. and Thonn. = G. thunbergia.

***G. thunbergia*** L. f. = G. ternifolia. Tropical Africa. The white wood is used locally to make household utensils. The wood ash is used in the manufacture of soap and as a lye for dyeing. An essential oil from the flowers is used by Sudanese women as a perfume. The fruits are used as a black cosmetic. A decoction of the stem and roots in palm wine, mixed with Guinea grains and Butryrospermum roots, are used as a laxative to treat severe constipation.

***G. viscidissima*** S. Moore. W.Africa. The wood is used locally to make a variety of small articles.

***G. vogelii*** Hook. f. The plant gives a dye used locally to paint the body.

Gardenia Oil – Gardenia florida.

Gardenia, Remy's – Gardenia remyi.

Gardon de Candelabro – Trichocereus chiloensis.

Gargan Death Carrot – Thapsia garganica.

Garland Chrysanthemum – Chrysanthemum coronarium.

Garland Flowers – Hedychium coronarium.

Garlic – Allium sativum.

Garlic, Bear's – Allium ursinum.

Garlic, Canada – Allium canadense.

Garlic, Field – Allium oleraceum.

Garlic, Giant – Allium scorodoprasum.

Garlic, Hedge – Sisymbrium allaria.

Garlic, Levant – Allium ampeloprasum.

Garlic, Twisted leaf – Allium obliquum.

Garlic, Wild – Allium canadense.

Garoonggang – Cratoxylum formosum.

Garousse – Lathyrus cicera.

GARRYA Dougl. ex Lindl. Garryaceae. 18 spp. W. United States of America, Mexico, W.Indies. Shrubs.

***G. elliptica*** Dougl. (Fever Bush, Quinine Bush). W. N.America. A decoction of the bark and leaves is used locally to treat fevers.

***G. fremontii*** Torr. (Skunk Bush, Quinine Bush). W. N.America. The leaves are used to make a tonic in California.

***G. gracilis*** Wang. = G. longifolia.

***G. longifolia*** Hartw. = G. gracilis, G. macrocarpa, G. oblonga. (Cuauchichic, Quauhchichic, Ovitano). Mexico. The bitter bark contains garryine and is used locally to treat diarrhoea.

***G. macrophylla*** Hartw. = G. longifolia.

***G. oblonga*** Benth. = G. longifolia.

GARUGA Roxb. Burseraceae. 4 spp. Himalayas to S.China, E.Malaysia, Malaysian Archepelago, N.E.Australia, Pacific Islands. Trees.

***G. abilo*** Merr. = G. mollis. (Kajoo kambing, Ketool). Celebes. The coarse wood resists decay and is used locally to make houses and by the Chinese to make coffins.

***G. mollis*** Turcz. = G. abilo.

***G. pinnata*** Roxb. (Ghogar, Jum, Karak). Tropical Asia, especially India. The juice from the stem is used locally to treat conjunctivitis and cataracts. The leaf juice is mixed with honey as a treatment for asthma.

***Gastrochilus*** Wall. = Boesenbergia.

GASTRODIA R. Br. Orchidaceae. 20 spp. Asia, Indomalaya to New Zealand (? tropical W.Africa). Herbs.

***G. cunninghamii*** R. Br. Malaysia. The tubers are eaten locally when roasted.

***G. elata*** R. Br. See G. cunninghamii.

***G. sesamoides*** R. Br. Tasmania. The tubers are eaten locally.

Gatae – Pisonia grandis.

Gatapa – Hibiscus tiliaceus.

Gaub – Diospyros embryopteris.

Gaucho Blanco – Sapium pavonianum.

Gaucho Blanco – Sapium stylare.

Gaucho Blanco – Sapium thomsoni.

Gaucho Mirado – Sapium stylare.

Gaucho Virgin – Sapium thomsonii.

GAULTHERIA Kalm. ex L. Ericaceae. 200 spp. Circumpacific; 2 spp. E. N.America; 8 spp. E.Brazil. Shrubs.

***G. antipoda*** Forster. Tasmania, New Zealand. The fruits are eaten locally.

***G. cumingana*** Vidal. Philippines, Formosa. The leaves are used to make a tea in the Philippines. The tea is considered to aid digestion.

***G. fragrantissima*** Wall. (Indian Wintergreen). Steam distillation of the leaves yields Indian Wintergreen. See G. procumbens.

***G. fragrantissima*** Wall. var ***punctata*** Smitn. Java, Sumatra. Yields an oil similar to Wintergreen. It is used locally as hair oil and perfume. A decoction of the leaves is used to relieve stomach upsets.

***G. hispida*** R. Br. Australia. The fruits are eaten by the aborigines.

***G. hispidula*** (L.) Torr. and Gray. (Creeping Snowberry, Moxie Plum). E. N.America. The berries are eaten raw, usually with sugar and cream. They are made into a preserve in Newfoundland. The local Indians make a drink from the leaves sweetened with maple sugar. There is some local trade in the fruits.

***G. leucocarpa*** Blume. Java, Sumatra. The plant gives an essential oil. See G. fragrantissima var punctata.

***G. myrsinites*** Hook. W. N.America. The fruits are made into preserves.

***G. procumbens*** L. (Wintergreen, Wintergreen berry, Tea Berry). E. N.America. Steam distillation of the leaves gives an essential oil (Wintergreen Oil, Oil of Gaultheria) which contains a high proportion of methyl salicylate. It is used as an antiseptic and to treat rheumatism, but its main use is in flavouring. The berries are eaten in pies and the leaves are used locally to make a tea (Mountain Tea).

***G. shallon*** Pursh. (Shallon). E. N.America. The dried fruits are used as a winter food by the local Indians.

Gavu – Urginea burkei.

Gây la bac – Croton argyratus.

GAYLUSSACIA Kunth. Ericaceae. 9 spp. N.America; 40 spp. S.America. Shrubs. The following spp. are found in E. N.America. The fruits are eaten raw, but especially in pies.

***G. baccata*** (Wang.) Koch. (Black Huckleberry).

***G. brachycera*** (Michx.) Gray. (Box Huckleberry).

***G. dumosa*** (And.) T. (Dwarf Huckleberry).

***G. frondosa*** Torr. and Gray. (Dangleberry).

***G. ursina*** Curtis. (Bear Huckleberry).

Gaz – Quercus cerris.

Gazenjubeen – Quercus cerris.

Gaza much – Quercus cerris.

Gbakaya – Hugonia platysepala.

Gebangan kipok – Pimpinella alpina.

Gěděběl bengook – Monilia sitophila.

Geeb – Cordeauxia edulis.

Geel katstert – Bulbine asphodeloides.

Gengentelan – Diospyros frutescens.

Geiger Tree – Cordia sebestina.

GEIJERA Schott. Rutaceae. 7 spp. New Guinea, E.Australia, New Caledonia, Loyalty Islands. Trees.

***G. deplanchia*** Planch. and Sebert. New Caledonia. The wood is used for interiors of houses and for general carpentry.

Geiss kappern – Sarothamnus scoparius.

Geissaspis apiculata – Humularia apiculata.

GEISSORHIZA Ker-Gawl. Iridaceae. 65 spp. S.Africa, Madagascar. Herbs.

***G. bojeri*** Baker. Madagascar. The bulbs are used locally to aid digestion.

***Geissois benthamii*** F. v. Muell. = Weinmannia benthami.

***Geissois lachnocarpa*** Maid. = Weinmannia lachnocarpa.

GEISSOSPERMUM Allem. Apocynaceae. 5 spp. Tropical Brazil. Shrubs.

***G. laeve*** Baill. = G. vellosii. Brazil. The stem is used to make a tonic, which is also used to reduce fevers (Cortex pereirde, Pão Piereira).

Gelangan kumus – Shorea laevis.

GELIDIUM Lamour. Gelidiaceae. Red Algae, mainly N.Pacific. This is one of the primary agar-producing genera. Agar occurs in the cell walls of the plant. It is a nitrogen-free gel involving galactose and a sulphate and possibly calcium and magnesium salts. It is extracted from the plants by boiling, after they have been dried and washed. Alternatively, extraction may be by alternate freezing and thawing. The agar is then made into powder, shreds, flakes or bricks. Japan is the main producing country, but it is also produced in U.S.A., U.S.S.R., China, Mexico, S.Africa, New Zealand and Australia. Some 650,000 lb. are used annually in the U.S.A. alone as a culture medium for bacteria, fungi, etc., as a laxative base, in tinned food, ice cream and prepared food powders, in dairy produce, cosmetics, dentistry, activators for sprays, photographic film, sizing paper and fabrics, for various specialised chemical tests and as a lubricant for drawing wire.

***G. arborescens*** Gard. Japan, China etc. Agar.

***G. amansii*** (Lamour) Lamour. Japan, China, S.Africa. Agar.

***G. cartilagineum*** (L.) Gaill. var. ***robustum*** Gardner. (Agarweed). Pacific, N.America. Agar.

***G. divaricatum*** Mast. Pacific. Agar. The plants are eaten in China, usually after drying. They are usually served with sugar or vinegar.

***G. japonicum*** (Harv.) Okam. Japan, China etc. Agar.

***G. linoides*** Kuetz. Japan, China etc. Agar.

***G. nudifrons*** Gardner. Japan, China etc. Agar.

***G. pacificum*** Okam. Japan, China etc. Agar.

***G. rigens*** Mast. Tropics. The dried seaweed is eaten in Japan. It is sold locally.

***G. subcostatum*** Okam. Japan, China etc.

GELSEMIUM Juss. Loganaceae. 1 sp. S.China, Indochina, N.Borneo, Sumatra; 2 spp. S.E. United States of America, N.Mexico. Vines.

***G. elegans*** Benth.=Leptopteris sumatrana =Medicia elegans. Tropical and subtropical Asia. The leaves are used in some parts as a criminal poison and, particularly by women, to commit suicide.

***G. sempervirens*** (L.) Ait. f. (Yellow Jessamine). E. U.S.A. The plant produces toxic alkaloids (gelsemine, gelsemidine) in the roots. These are used medicinally as depressants of the nervous system.

Gelugur – Garcinia atroviridis.

Gembor – Parameria barbata.

***Gendarussa vulgaris*** Ness.=Justicia gendarussa.

Gendavasi – Vitex trifolia.

Gendis – Amoora aphanamixis.

Gendis – Chisocheton marcrophyllum.

Genépi – Artemisia laxa.

Genépi des Glaciers – Artemisia glacialis.

Geneps – Melicoccus bijugatus.

GENIOSTOMA J. R. and G. Forst. Loganiaceae. 60 spp. Madagascar to New Zealand.

***G. ligustrifolium*** A. Cunn. New Zealand. The juice is used locally to treat skin diseases in children and the bark is used to treat itch.

GENIPA L. Rubiaceae. 6 spp. Warm America, W.Indies. Trees.

***G. americana*** L. (Genipa, Marmalade Box). S.America. The fruit is popular locally. It looks like a brown orange, with a rather unattractive-looking brown flesh. It needs to soften before ripening, when it has a distinctive but pleasant flavour. The fruit is used to make the drink Genipapado and is fermented to make Licor de Genipado. The wood is strong and light coloured. It is used, particularly in Brazil for general carpentry. A dye from the stem is used by the local Indians for tattooing.

***G. caruto*** H.B.K. S.America. The fruits are used locally as a laxative.

Genipapado – Genipa americana.

Genépi – Artemisia laxa.

GENISTA L. Leguminosae. 75 spp. Europe, N.Africa, W.Asia. Shrubs.

***G. germanica*** L.=G. villosa. Europe. The plant was used to make a yellow dye.

***G. roetam*** Forsk.=Retama roetam. (Retem-shrub). Deserts of Palestine. The wood is a valuable local source of charcoal. The leaves are poisonous, eaten in excess and are used locally as an abortive.

***G. saharae*** Coss. Algeria. The leaves are food for camels.

***G. tinctoria*** L.=Spartium tinctorium (Dyer's Greenwood). Europe to S.W.Siberia.

The leaves, flowers and twigs are used to make a yellow dye, used for colouring fabrics. When mixed with woad it makes a green dye (Kendal Green). The plant is used locally to make a purgative and diuretic.

Gentian Bitter – Gentiana lutea.

Gentian, Closed – Gentiana andrewsii.

Gentian, White – Lasperitium latifolium.

Gentian, Yellow – Gentiana lutea.

GENTIANA L. Gentianaceae. 400 spp. Cosmopolitan, excluding Africa, mainly alpine. Herbs.

***G. adsurgens*** Cerv. S.Mexico. The roots are used to treat stomach complaints and as a stimulant by the local Indians.

***G. andrewsii*** Griseb. (Closed Gentian). E. N.America. The roots are used by the local Indians to treat snakebites. The white population use a decoction of the roots as a tonic to promote the appetite and digestion.

***G. cruciata*** L. (Croisette). N. and C.Europe to Siberia. The leaves are used locally to stimulate digestion and as a tonic.

***G. diffusa*** Vahl. = Canscora diffusa.

***G. kurroo*** Royle. Kashmir, Himalayas. See G. lutea.

***G. lutea*** L. (Yellow Gentian). Central and S.Europe, Asia Minor. The dried roots are used to stimulate the appetite. They are used in the manufacture of Gentian Bitters, used for the same purpose and to make liqueurs. Usually, the powdered root is mixed with orange peel, cardamon seed, glycerine, alcohol and water. It was used by the Greeks (about 180 B.C.) as an antidote to poisons and to dilate wounds.

***G. pannonica*** Scop. See. G. lutea.

***G. pneumonanthe*** L. Europe through to temperate Asia. The flowers are used to make a blue dye.

***G. punctata*** Gebh. See G. lutea. Used in France and Germany.

***G. purpurea*** L. See G. lutea. Used in France and Germany.

Gentileng – Vitex glabrata.

Gasearpa Groundnut – Kerstingiella geocarpa.

GEODORUM G. Jacks. Orchidaceae. 16 spp. India to Polynesia and Australia. Herbs.

***G. nutans*** (Presl.) Ames. Philippines, Formosa. The mucilage from the plant makes an excellent glue used for sticking the parts of guitars etc.

GEOFFROEA Jacq. Leguminosae. 6 spp. Tropical America, W.Indies. Trees.

***G. superba*** Humb. and Bonpl. S.America. The fruits (Umari), are eaten locally in Brazil.

***G. surinamensis*** Pille. S.America. The bark (Cortex Geoffreae) is used to combat intestinal worms.

GEONOMA Willd. Palmae. sensu latu 240 spp. sensu stricto 150 spp. Tropical America, W.Indies. Palm trees.

***G. binervata*** Oerst. C.America, Mexico. The young inflorescence is eaten as a vegetable in Mexico and the leaves are used for thatching.

***G. dominicana*** L. (Yanga). See G. hodgeorum.

***G. hodgeorum*** L. H. Bailey ex Hodge. Dominica. The leaves are used for thatching.

GEOPHILA D. Don. Rubiaceae. 30 spp. Tropics. Shrubs.

***G. obvallata*** F. Didr. Tropical Africa. The leaves, cooked with the food are used to treat diarrhoea of children in Liberia.

Georgia Bark – Pinckneya pubens.

Gerancine – Oldenlandia corymbosa.

GERANIUM L. Geraniaceae. 400 spp. Cosmopolitan especially temperate. Herbs.

***G. macrorrhizum*** L. (Bigroot Geranium). Central Europe. The roots and leaves are used as an aphrodisiac in Bulgaria.

***G. maculatum*** L. (American Crane's Bill). N.America. The roots and leaves are used to make an astringent, which is used as a styptic. The leaves are used to treat cholera and diarrhoea of children and kidney complaints.

***G. mexicanum*** H.B.K. Central America, Mexico. The leaves are used locally in Mexico to make a tonic for the elderly.

***G. nepalense*** Sweet. (Bhanda, Nepalese Crane's Bill). The roots give a dye similar to that of Rubia cordifolia q.v.

***G. robertianum*** L. (Crane's Bill, Herb Robin, Red Shank). Europe, an escape in N.America. The ground leaves, or a decoction of the leaves is used to treat stones of the kidneys and blood in the stools.

***G. sylvaticum*** L. (Wood Crane's Bill). Europe to Asia. The flowers give a blue dye, used particularly in the Black Forest of Germany.

***G. wallichianum*** D. Don. (Wallich's Crane's Bill, Kao ashud, Lal jahri, Mamiran).

Himalayas. The roots contain about 30 per cent tannin. They are used for tanning and dyeing, particularly in Kashmir.
Geranium, Bigroot – Geranium macrorrhizum.
Geranium Oil – Cymbopogon martini.
Geranium Oil – Pelagonium spp.
Gherkin, West Indian – Cucumis anguria.
German Ebony – Taxus baccata.
German Iris – Iris germanicum.
German Sarsaparilla – Carex arenaria.
German Spearmint Oil – Mentha spicata var. crispa.
Germander, Chamaedrya – Teucrium chamaedrys.
Germander, Common – Teucrium chamaedrys.
Germander, Water – Teucrium scordium.
Germander, Wood – Teucrium scorodonia.
Geroonggoong – Adina minutiflora.
Gerrard Vetch – Vicia cracca.
GESNERIA L. Gesneriaceae. 50 spp. Tropical America, W.Indies. Herbs.
***G. allagorphylla*** Mart.=G. grandis=G. nitida. Brazil. The roots are used locally as a tonic.
***G. grandis*** Hort.=G. allagorphylla.
***G. nitida*** Hort.=G. allagorphylla.
Gethago – Hunteria corymbosa.
Getah Borneo – Willughbeia firma.
Getah dantong – Payene dantung.
Getah djelutung – Alstonia polyphylla.
Getah djintahen – Hunteria corymbosa.
Getah doojan – Palaquium treubi.
Getah kelapang – Palaquium clarkeanum.
Getah kanari – Canarium amboinense.
Getah Mala – Altingia excelsa.
Getah Nalu – Sideroxylon kaernbachianum.
Getah Pootih – Palaquium maingayi.
Getah Pootih – Palaquium treubii.
Getah Soesoe – Willughbeia firma.
GETONIA Roxb. Combretaceae. 1 spp. Indomalaya. Woody vine.
***G. floribunda***=Calycopteris floribunda=C. nutans. (Dok ko deng, Ksuos). The leaves and twigs are used as a tonic in Cambodia.
GEUM L. Rosaceae. 40 spp. N. and S. Temperate, Arctic. Herbs.
***G. nutans*** Crantz non Lam.=G. rivale.
***G. rivale*** L.=G. nutans. (Water Avens). Eurasia, N.America. The roots are used to make a stomach tonic.
***G. urbanum*** L. (Avens Root). Temperate Europe, Asia, N.America and Australia. The rhizome has a taste of cloves and is used to flavour wines, beers and liqueurs, as a spice and particularly in parts of Europe to treat dysentery, diarrhoea, constipation and stomach upsets. It is also a good heart stimulant.
GEVUINA Molina Proteaceae. 3 spp. New Guinea, Queensland, Chile. Trees.
***G. avellana*** Molina (Chile Hazel). Chile. The pale brown wood is light and strong, but not durable in contact with the soil. It is used locally for furniture, turning and roof shingles. The nuts have a pleasant flavour and are eaten locally.
Ghatti Tree (Gum) – Anogeissus latifolia.
Ghavan – Astragalus gossypinus.
Ghavan – Astragalus echidnaeformis.
Ghelaf – Leptadenia lancifolia.
Gherkin – Cucumis sativus.
Gherkin, West Indian – Cucumis anguria.
Gherireh Gum – Acacia senegala.
Ghineh Cheraghee – Astragalus myriacanthus.
Ghineh Zard – Astragalus brachycentrus.
Ghittoe – Halfordia scleroxyla.
Ghiwata – Callicarpa arborea.
Ghogar – Garuga pinnata.
Ghot ber – Ziziphus xylocarpus.
Ghuchu – Astragalus microcephalus.
Giam – Hopea nutans.
Gian tía – Ilex godajam.
Giang – Leea aequata.
Giant Alocasia – Alocasia macrorrhiza.
Giant Arbor Vitae – Thuja plicata.
Giant Bur Reed – Sparganium eurycarpum.
Giant Cactus – Carnegiea gigantea.
Giant Garlic – Allium scorodoprasum.
Giant Granadilla – Passiflora quadrangularis.
Giant Helleborine – Epipactis giganteum.
Giant Hyssop – Agastache anethiodora.
Giant Nettle – Laportea gigas.
Giant Orchid – Epipactis giganteum.
Giant Raffia Palm – Raphia gigantea.
Giant Reed – Arundo donax.
Giant Rye – Triticum polonicum.
Giant Taro – Alocasia cullata.
Giant Timber Bamboo – Phyllostachys bambusioides.
Giant Wild Rye – Elymus condensatus.
Giây chien – Tetracera assa.
Giây voi – Cissus modeccoides.
Giba – Euodia ambointensis.
GIBBERELLA Fr. Nectreaceae. 10 spp. Widespread. Fungus.
***G. fujikuroi*** (Saw.) Wollenw.=Fusarium moniliforme. Parasitic on tropical crops,

especially rice. It was the original source of the plant growth regulating substance gibberellic acid.

Gidgee Acacia – Acacia homalophylla.

Giêng sàng – Ligusticum monnieri.

Gifbol – Boöphone disticha.

GIGANTOCHLOA Kurz. Graminae. 20 spp. Indomalaya. Tall bamboos.

***G. ater*** Kurz. Indomalaya. The buds are eaten as a vegetable locally.

***G. verticillata*** Munro. Java. The buds are eaten as a vegetable, or pickled in vinegar. This is a much-prized vegetable and is cultivated. There are several varieties grown.

GIGARTINA Stackh. Gigartinaceae. Red Algae.

***G. horrida*** Harv. Pacific. Used in the manufacture of agar, particularly in Malaysia. See Gelidium spp.

***G. mamillosa*** (Good. and Woodw.) Ag. N.Atlantic. Used in the manufacture of carragheenin. See Chondrus crispus.

***G. speciosa*** Sind. = Eucheuma speciosa.

***G. stellata*** (Stackh.) Batt. N.Atlantic. Used in the making of Carragheenin. The seaweed is eaten locally in parts of Britain.

***G. teedi*** (Roth.) Lam. Pacific. The plant is eaten locally in Japan.

GILBERTIODENDRON J. Leónard. Leguminosae. 25 spp. Tropical W.Africa. Trees.

***G. grandiflorum*** (De Wild.) Leónard. = Macrolobium grandiflorum (M'Pokusa). W.Africa, especially Cameroons, Congo, Gabon. The wood polishes well and is used locally for cabinet work

***G. ogoouense*** (Pellgr.) Leónard. Tropical Africa, especially Gabon and Congo. The wood is used for house interiors, general carpentry and railway sleepers.

Gilea, Scarlet – Gilia aggregata.

Gilead, Balm of – Populus balsamifera.

GILIA Ruiz. and Pav. Polemoniaceae. 120 spp. Temperate and subtropical America. Herbs.

***G. aggregata*** (Pursh.) Spreng. (Scarlet Gilea). W. N.America. A decoction of the leaves was used by the local Indians as an emetic.

***G. congesta*** Hook W. N.America. A decoction of the leaves was used by the local Indians as a tonic.

GILLENIA Moench. Rosaceae. 2 spp. N.America. Herbs.

***G. stipulata*** (Muhl.) Trel. (American Ipecac). E. N.America. The roots and bark are used to make an emetic and expectorant.

***G. trifoliata*** (L.) Moench. (Bowman's Root). E. U.S.A. A decoction of the plant was used by the local Indians as an emetic.

Gin, Sloe – Prunus spinosa.

Gilly Flower – Armeria maritima.

Ginger – Zingiber officinale.

Ginger, American Wild – Asarum canadense.

Ginger Bread Palm – Hyphaene coriacea.

Ginger Bread Plum – Parinari macrophyllum.

Ginger, Cassumunar – Zingiber cassumuna.

Ginger, European Wild – Asarum europaeum.

Gingergrass Oil – Cymbopogon martini.

Ginger, Japanese – Zingiber moiga.

Ginger Lily – Costus afer.

Ginger, Mango – Curcuma amado.

Ginger, Mioga – Zingiber mioga.

Ginger, Zerumbet – Zingiber zerumbet.

Ginggijang – Leea aequata.

Gingili Oil – Sesamum indicum.

GINKGO L. Ginkgoaceae. 1 sp. E.China. Tree.

***G. biloba*** L. (Gingko, Maidenhair Tree, Ginkyo). Cultivated, particularly in China and Japan as an ornamental. More recently it has become a popular ornamental in temperate regions. The roasted seeds (Sal Nuts) are a delicacy in China, where the seeds are also used medicinally as an expectorant and sedative.

Gingko – Gingko biloba.

Ginkya – Gingko biloba.

Ginseng – Panax schinseng.

Ginseng, American – Panax quinquefolia.

Ginseng de Phú Yên – Hibiscus sagittifolius.

Gipps and Grey Box – Eucalyptus bosistoana.

Giraffe Acacia – Acacia giraffae.

GIRARDINIA Gaudlich. Urticaceae. 8 spp. Tropical Africa, Madagascar, E.Africa. Herbs, or woody herbs. Fibres are made from the stems of the following species. They are used locally for making cloth and sewing.

***G. bullosa*** (Hochst.) Wedd. = Urtica bullosa. (Kibanzoou). Congo to Abyssinia.

***G. condensata*** (Hochst.) Wedd. Congo to Abyssinia.

***G. heterophylla*** Decne. = G. palmata.

***G. palmata*** Gaud. = G. heterophylla. (Nilgiri Nettle). N.W. Himalayas to Malaysia.

GIRONNIERA Gaudlich. Ulmaceae. 15 spp. Indomalaysia, Polynesia. Trees.

***G. subaequalis*** Planch. S.E.Asia, Malaysia. The soft, light wood is used locally for general construction work.

GISEKIA L. Aizoaceae. 5 spp. Tropical and S.Africa to India, Ceylon and Indochina. Herbs.

***G. pharnacioides*** L. Tropical Africa, India. The leaves are eaten as a famine food.

Gitan obah – Hunteria corymbosa.

Gitan palau ninjak – Willughbeia tenuiflora.

GLADIOLUS L. Iridaceae. 300 spp. Canaries, Madeira, W. and Central Europe, Mediterranean to S.W. and Central Asia, tropical and S.Africa. Herbs. The corms are eaten locally usually roasted.

***G. edulis*** Burch. S.Africa.

***G. quartianus*** A. Rich. W.Africa. The corms are also mixed with Guinea Corn to eat or to make a beverage.

***G. spicatus*** Klatt. Tropical Africa.

***G. unguiculatus*** Baker. W.Africa.

***G. zambesiacus*** Baker. E.Africa.

Gladwin – Iris foetidissima.

Glandbearing Oak – Quercus glandifera.

Glasswort, Leadbush – Salicornia fruticosa.

GLAUCIUM Mill. Papaveraceae. 25 spp. Europe, S.W. and Central Asia. Herbs.

***G. flavum*** Crantz. = G. luteum. (Horned Poppy). Mediterranean to Central Europe. The oil from the dried seeds is used for lighting, making soap and for cooking. The juice is used as a purgative, sedative and an adulterant for opium. The plant has been introduced to N.America.

***G. luteum*** Scop. = G. flavum.

GLAUX L. Primulaceae. 1 sp. N.Temperate coasts. Herb.

***G. maritima*** L. (Sea Milkwort). The young shoots are used in salads and as an emergency food.

GLECHOMA L. Labiatae. 10–12 spp. Temperate Eurasia. Herbs.

***G. hederaceum*** L. = Nepeta glechoma (Ground Ivy, Haymaids, Aleroot). Eurasia. Introduced to N.America. A tea made from the leaves was used in home remedies for kidney diseases and indigestion.

GLEDITSIA L. Leguminosae. 11 spp. Tropics and subtropics. Trees.

***G. amorphoides*** Taub. = Gorugandora amorphoides. Brazil. The wood is strong, elastic and durable. It is used locally for general carpentry.

***G. delavayi*** Franch. China. The wood is used locally for general carpentry. The pods are used as a soap substitute and for tanning.

***G. horrida*** Willd. = G. sinensis.

***G. japonica*** Miq. The pod juice is used as a soap substitute. The plant is cultivated locally.

***G. macrantha*** Desf. China. The pods are used locally as a soap substitute.

***G. sinensis*** = G. horrida. E.China. The pods are used as a soap substitute and for tanning. The wood is used for general carpentry.

***G. triacanthos*** L. (Honey Locust, Sweet Bean). E. N.America. The beans are eaten and fed to stock. The wood is hard and durable, resisting rotting in contact with soil. It is used for fence posts, railway sleepers and wheel-making.

GLEICHENIA Sm. Gleicheniaceae. 10 spp. S.Africa, Mascarene Islands, Malaysia, New Zealand. Ferns.

***G. dichotoma*** Hook. = G. hermanni. Australia. The starch from the roots is eaten by the aborigines.

***G. hermanni.*** R. Br. = G. dichotoma.

***G. linearis*** C. B. Clarke. Tropics. In Malaysia the stems are used to make interior walls of houses, chairs, fish traps, pens etc.

Gli – Lichtensteinia pyrethifolia.

Gli, Ognon – Pancratium speciosum.

Glitscherreute – Artemisia glacialis.

***Glinus lotoides*** Loefl. = Mollugo hirta.

GLIRICIDIA Kunth. Leguminosae. 10 spp. Tropical America, W.Indies, Trees.

***G. sepium*** (Jacq.) Steud. = Lonchocarpus maculatus. Central America to N. S.America. Naturalised in Cuba and Philippines. An important shade tree for coffee and cacao plantations. The bark or seeds are ground with rice and used to kill rats.

***Globaria*** Quél. = Lycoperdon.

GLOBBA L. Zingiberaceae. 50 spp. S.China, Indochina. Herbs.

***G. cambodegensis*** Gagnep. Indochina, Vietnam, Malaysia. The plant is used locally to treat rheumatism and fevers.

***G. kingii*** Baker = G. panicoides.

***G. panicoides*** Miq. = G. kingii = G. stenothyras. (Rièng rung). See G. cambodegensis.

***G. stenothyrsa*** Baker = G. panicoides.

Globe Artichoke – Cynara scolymus.

Globemallow, Narrowleaf – Sphaeralcea angustifolia.

GLOBULARIA L. Globulariaceae. 28 spp. Cape Verde Islands, Canary Islands, S.Europe, Asia Minor, Herbs.

***G. alypum*** L. Mediterranean. The leaves are used locally as a purgative, in the treatment of fevers and as an aphrodisiac.

GLOCHIDION J. R. and G. Forst. Euphorbiaceae. 300 spp. mostly in tropical Asia, Polynesia and Queensland, with a few in Madagascar and tropical America. Shrubs or trees.

***G. borneense*** Boerl. (Dempoolelet, Mareme). Borneo, W.Java. The leaves are eaten locally as a vegetable and the hard wood is sometimes used in the making of houses.

***G. dasyanthum*** Kurz. = G. obscurum.

***G. glaucum*** Blume. = G. obscurum.

***G. hohenackeri*** Bed. (Bhoma, Kuluchan, Kalchia). Asia, but especially India. The bark is used in Indian medicine to stimulate the appetite when food is being rejected by the stomach.

***G. llanosi*** Muell. Arg. = Phyllanthus llanosi. Indochina, Philippines. The young shoots are used locally to flavour fish dishes.

***G. marianum*** Muell. Arg. Pacific Islands. The strong, fine-grained wood is used locally to make cart shafts.

***G. obscurum*** Hook. f. = G. dasyanthum = G. glaucum (Dempool, Doolangdooloang, Ki pare). Malaysia, Indonesia. The wood is used locally to make houses.

GLOIOPELTIS J. Ag. Endocladiaceae. Marine Red Algae.

***G. furcata*** (Post. and Rupr.) J. Ag. China, Japan, Pacific N.America. The source of a glue (Funori, Hailo), used for sizing paper, silk and cloth and for binding Chinese paintings. About 40 million lbs. of this material are used each year in Japan alone.

***G. tenax*** (Turn.) J. Ag. See G. furcata.

GLORIOSA L. Liliaceae. 5 spp. Tropical Africa, Asia. Herbs.

***G. superba*** L. (Climbing Lily). Tropical Asia. The roots are used in India and Burma to commit suicide, while the Hindus used them to treat stomach upsets, piles and apply a paste to treat leprosy and skin parasites.

Gloss Buckthorn – Rhamnus frangula.

***Glossolepis macrobotrys*** Gilb. = Chytanthus macrobotrys.

GLOSSONEMA Decne. Asclepiadaceae. 4–5 spp. Tropical Africa, Asia. Shrubs.

***G. boveanum*** Desf. N.Africa, especially Arabia. The fruits are eaten by the Bedouins.

Glossy Privet – Ligustrum lucidum.

GLUTA L. Anacardiaceae. 1 sp. Madagascar, 12 spp. Indomalaya. Trees.

***G. rhengas*** L. (Rengasz). Malaya. The reddish wood is light, and splits easily. It is used for building, furniture, inlay work and fancy articles.

GLYCERIA R. Br. Graminae. 40 spp. Cosmopolitan, especially N.America. Grasses.

***G. fluitans*** (L.) R. Br. = Panicularia fluitans. (Sugar Grass, Manna Grass). N.America, Europe, Asia. The seeds were eaten by the Indians of N.America.

GLYCINE auctt. (Wight. and Arn.?) Leguminosae. 10 spp. Temperate and tropical Africa and Asia. Herbs.

***G. apios*** L. = Apios americana.

***G. hispida*** Max. = G. max.

***G. homblei*** De Wild. var. ***latistipulatum*** Hauman. Congo. The roots are eaten as a vegetable locally.

***G. max*** Mer. = G. hispida = G. soja = Soja hispida. (Soybean, Soyabean). The origin is unknown, but it is presumed to be derived from ***G. ussuriensis,*** which is native in central China. The plant has been cultivated in China since before 2838 B.C. when there are written accounts of its cultivation. It was introduced to Japan before 200 B.C. through Korea. The plant was introduced to Europe in the 17th century and into America in 1804. It was not commercially exploited outside China until the 20th century. Now the world production is some 30 million tons of beans giving 10 million tons of oil per annum. Of this some 58 per cent is produced in U.S.A. and 35 per cent from China. The rest is produced throughout the world in warm temperate regions. The plant is an annual and is grown for the seeds which are occasionally eaten as a green vegetable. Most of the crop is used to produce oil and soya meal. The oil is the chief vegetable oil produced in U.S.A. It is extracted by hydrocarbon solvents, expeller or hydraulic press. The oil has Sp. Gr. 0·922, Sap. Val 189·9–194·3, Iod. No. 103–152, Unsap. 0·50–1·8 per cent. Purified it is used to make margarine, cooking oil, etc. Industrially it is used in the manufacture of inks, paints, synthetic rubber, varnishes, linoleum, soaps, pharmaceuticals and cosmetics. The seeds yield about 17 per cent oil, and 63 per cent meal. The latter contains 40–50 per cent protein including an unusually high proportion of lycine, making it particularly valuable as an animal feed. The meal is used in a wide variety of prepared foods, e.g. ice cream, cakes etc., as a coffee substitute and a commercial source of casein. The flour is used to make an artificial milk in China. In China and Japan the young shoots are used to make a variety of dishes, e.g. Tempé. The plant is useful as a legume in rotations, especially with cereals such as maize. The green parts make good stock feed, silage and green manure.

***G. soja*** Sieb. and Zucc. = G. max.

GLYCOSMIS Corrêa. Rutaceae. 60 spp. Indomalaya. Shrubs.

***G. cochinchinensis*** Pierre. = G. pentaphylla Corr. (Djerookan, Gongseng, Sikatan). Malaya, Indonesia. A decoction of the roots is used locally to treat diseases of the gall bladder, and a decoction of the roots and leaves to treat pains in the abdomen.

***G. pentaphylla*** Corr. = G. cochinchinensis.

***G. pentaphylla*** (Retz.) Corr. = Limonia pentaphylla = Murraya cerasiformis. Indochina, Malaysia. The juice from the leaves is taken to treat fevers, intestinal worms and liver complaints and rubbed on the skin to cure eczema and other skin diseases. The fibrous roots are used as toothbrushes in E.Bengal.

GLYCYRRHIZA L. Leguminosae. 18 spp. Temperate Eurasia, America, N.Africa, S.E.Australia. Herbs.

***G. asperrima*** L. f. Central Asia, Siberia. The leaves are used locally to make a tea.

***G. glabra*** L. (Liquorice, Licorice). Mediterranean. Cultivated in Mediterranean regions, Louisiana and California for the rhizomes and roots (Radix Glycyrrhizae) which yield the liquorice of commerce. The liquorice is made by boiling the pulped roots, evaporating and moulding the residue. The stick liquorice or the juice (Corigliano, Solazzi) is used as an expectorant, demulcent or mild laxative, but more frequently to mask the bitter tastes of the active ingredients in medicines. The root is also used for chewing and in the manufacture of confectionery and chewing tobacco. Spanish Liquorice or Italian Liquorice and Russian Liquorice are varieties of G. glabra.

***G. lepidota*** (Nutt.) Pursh. (Wild Liquorice). N.America. The roots were eaten by the local Indians.

***G. malensis*** Maxim. China. The roots give a good liquorice, but the plant is not exploited commercially.

***G. ralensis*** Fisch. (Chinese Liquorice). China, Siberia. The roots are used locally as an emollient.

GMELINA L. Verbenaceae. 2 spp. tropical Africa, Mascarene Islands; 33 spp. E.Asia, Indomalaya, Australia. Trees.

***G. arborea*** Roxb. = Premna arborae. (Malay Bushbeech). E.India through to Pacific Islands. The wood is used for mining timbers. The roots are used to relieve pain and reduce temperature.

***G. leichthardtii*** F. v. Muell. (Grey Teak, White Beech). New South Wales, Queensland. The wood is a pale grey-brown and works well. It is durable and termite resistant. The wood is used for deck work on ships, carving and pattern-making and floor blocks.

***G. macrophylla*** Benth. See G. leichthardtii.

***G. macrophylla*** Wall. = G. moluccana.

***G. moluccana*** Baker. = G. macrophylla Wall. = Vitex moluccana. (Titi, Toho). Malaya, Indonesia. The light wood is used for small native boats.

GNAPHALIUM L. Compositae. 200 spp. Cosmopolitan. Herbs.

***G. keriense*** A. Cunn. New Zealand. The juice from the leaves is used locally to sooth bruises.

***G. obtusifolium*** L. = G. polycephalum.

***G. polycephalum*** Michx. = G. obtusifolium. (Common Everlasting). N.America. The local Indians used the leaves to treat bruises and used an infusion of them to treat catarrh of the intestine and lungs.

***G. uligonosum*** L. (Cudweed, Cottonweed). Europe through to Caucasus. The leaves are astringent and are used locally to treat quinsy. The plant is naturalized in America.

GNETUM L. Gnetaceae, 30 spp. Indomalaya, Fiji, N.Tropical S.America, W.Tropical Africa. Shrubby plants.

***G. africanum*** Welw. (Bawale, Longonizia). Tropical W.Africa. The roasted seeds are eaten locally. The twigs are used to make traps for animals and as a binding material.

***G. gnemon*** L. (Bulso). Tropical Asia, Malaysia The cooked fruits and the young leaves are eaten as vegetables in the Philippines. The bark yields a fibre. The plant is sometimes cultivated for the leaves.

***G. indicum*** (Lour.) Merr. See G. gnemon.

***G. scandens*** Roxb. E.Himalayas to S.Indo-China. A climber. The fibre from the stem is used locally occasionally.

Goa Bean – Psophocarpus tetragonolobus.
Goa Butter – Garcinia indica.
Goa Ipecacuanha – Naregamia alata.
Goa Powder – Andira araroba.
Goat Nut – Simmondsia californica.
Goat's Rue – Galega officinalis.
Goat's Rue – Tephrosia virginiana.
Goat Willow – Salix caprea.
Gobbi – Crossandra infunduliformis.
Gogopoa – Cycas rumphii.
Gokizuri Oil – Actinostemma lobatum.
Golbanbeth – Astragalus globiflorus.
Gold Bhasma – Drosera peltata.
Gold Edge Lichen – Stricta crocata.
Gold Thread – Coptis trifolia.
Goldband Lily – Astragalus globiflorus.
Golden Chain – Laburnum anagyroides.
Golden Chamomille – Anthemis tinctoria.
Golden Chinquapin – Castanopsis charysophylla.
Golden Club – Orontium aquaticum.
Golden Currant – Ribes aureum.
Golden Evergreen Raspberry – Rubus ellipticus.
Golden Gram – Phaseolus mungo.
Golden-leaved Chestnut – Castanopsis chrysophylla.
Golden Ragwort – Senecio aureus.
Golden Rod, Californian – Solidago californica.
Golden Rod, Canadian – Solidago canadensis.
Golden Rod, European – Solidago virgaurea.
Golden Rod, Leavenworth – Solidago levenworthii.
Golden Rod, Sweet – Solidago odora.
Golden Samphire – Inula crithinoides.
Golden Saxifrage – Chrysoplenium alternifolium.
Golden Seal – Hydrastis canandensis.
Golden Wattle – Acacia pycnantha.
Goldenweed, Nuttall – Aplopappus nuttallii.
Goldmoss Stonecrop – Sedum acre.
Goldthread – Coptis trifolia.
Gollang – Protium javanicum.
Golombi – Stereosperumum kunthianum.
Golpanbeh – Astragalus elymaiticus.
Goma Anime de Mexico – Hymenaea courbaril.
Goma de Caro – Enterolobium cyclocarpum.
Goma de Orore – Pithecellobium hymeneaefolia.
Gomart – Bursera gummifera.
Gombo – Hibiscus esculentus.
Gomme d'Acajou – Anacardium occidentale.
Gomme blanche – Acacia senegal.
Gomme Blondes – Acacia senegal.
Gomme de Benaile – Moringa oleifera.
Gomme de Sénégal – Acacia albida.
Gomme de Sénégal – Sterculia tomentosa.
Gomme rouge – Acacia tortilis.
Gomme Salobreda – Acacia stenocarpa.
Gommer – Astragalus cerasocrenus.
Gommier – Bursera gummifera.
Gommier Blanc – Protium attenuatum.
Gommier l'Incense – Protium attenuatum.
Gomouti Palm – Arenga pinnata.

***Gomphia*** Schreb. = Ouratea.

GOMPHIDIUS Fr. Agaricaceae. 10 spp. N.Temperate. The fruitbodies of the following species are eaten, having a pleasant flavour.

***G. glutinosus*** (Schäff.) Fr. N.Temperate.

***G. oregonensis*** Peck. W. N.America.

***G. roseus*** (Fr.) Karts. Europe to W.Siberia.

***G. subroseus*** Kauffm. W. N.America.

GOMPHOCARPUS R. Br. Asclepiadaceae. 50 spp. Tropical and S.Africa. Small shrubs.

***G. cornutus*** Welw. Madagascar. A decoction of the leaves and roots is used locally to treat asthma and as an emetic.

***G. fruticosus*** R. Br. = Asclepias fruticosa. Africa. The hairs from the seeds coats are used to stuff pillows.

***G. lineolatus*** R. Br. (Azara). W.Africa. A decoction of the leaves is used locally to treat stomach complaints and intestinal worms. The plant contains the glucoside uzarin which is a heart stimulant.

***G. rigidus*** R. Br. S.Africa. The leaves are used locally to treat stomach complaints and diarrhoea.

GOMPHOSTEMMA Wall. Labiatae. 40 spp. India, E.Asia, W. Malaysia. Herbs.

***G. phlomoides*** Benth. (Djintenan, Galipoong booloo, Kopetan). Malaysia, Indonesia. The leaves are used locally to treat wounds and are sold for this purpose.

Gonari – Cumamomum cecicdaphne.

Gonçalo Alves – Astronium graveolens.

Gondang (Wax) – Ficus ceriflua.

Gonggom – Sarcolobus spanoghei.

Gongolu – Beilschmiedia insularum.

Gongseng – Glycosmis cochinchinensis.

***Goniolimon tataricum*** (L.) Boiss. = Statice tataricum.

GONIOMA E. Mey. Apocynaceae. 1 sp. S.Africa. Tree.

***G. kamassi*** Mey. (Kamassiwood). The yellow, heavy, hard wood is exported to Europe as Boxwood, and is used for turning and carving.

GONIOTHALAMUS Hook. f. and Thoms. Annonaceae. 115 spp. Indomalaya. Shrubs.

***G. repevensis*** Pierre. (Romduol si phlé). Cambodia, Laos. The sweet, pleasantly flavoured fruits are eaten locally.

GONOCARYUM Miq. Icacinaceae. 20 spp. Formosa, S.E.Asia, Indomalaya. Small trees.

***G. subrostratum*** Pierre. (Do kom, Seng muang). Indochina. The young leaves are eaten as a vegetable and used in the treatment of beri-beri in Laos. In S.Vietnam they are fermented to make alcohol.

GONOCRYPTA Baillon. Periplocaceae. 1 sp. Madagascar. Woody plant.

***G. grevii*** Baill. (Kompitro). The stem yields a good rubber.

Gonyo Oil – Antrocaryon nannani.

GONYSTYLUS Teijsm. and Binn. Thymelaeaceae. 25 spp. Malaysia, Solomon Islands, Fiji. Trees.

***G. bancanus*** Baill. = G. bancanus.

***G. bancanus*** (Miq.) Kurz. = G. bancanus Baill. = G. miquelianus = Aquilaria bancana. (Ramin). Malaysia. The hard fine-grained wood is used for boards and posts. It is scented and is also used for incense. The oil from the wood is burnt as incense, and to relieve asthma. The oil is also used in perfumery.

***G. miquelianus*** Teijsm and Binn. = G. bancanus.

Goober – Arachis hypogaea.

Goober, Congo – Voandzeia subterranea.

Good King Henry – Chenopodium bonus henricus.

GOODYERA R. Br. Orchidaceae. 40 spp. Temperate Eurasia, tropical Asia, Mascarene Islands, Australia, Polynesia, temperate N.America. Herbs.

***G. pubescens*** (Willd.) R. Br. = Peramium pubescens. (Scrofula Weed, Downy Rattlesnake Plantain). E. N.America. The bulbils are occasionally used locally to treat scrofula

Goola – Amoora aphanamixis.

Goola – Chisocheton macrophyllum.

Goonda – Sphenoclea zeylanici.

Goongum – Podocarpus elata.

Gooseberry, American Wild – Ribes cynosbati.

Gooseberry, Barbados – Pereskia aculeata.

Gooseberry, Ceylon – Doryalis hebecarpa.

Gooseberry, Dwarf Cape – Physalis peruviana.

Gooseberry, Dwarf Cape – Physalis pubescens.

Gooseberry, European – Ribes grossularia.

Gooseberry, Gourd – Cucumis anguria

Gooseberry, Otaheite – Phyllanthus destichus.

Gooseberry, Prickly – Ribes cynosbati.

Gooseberry, Smooth – Ribes oxycanthoides.

Gooseberry, West Indian – Pereskia aculeata.

Gooseberry Wine – Ribes grossularia.

Goosegrass – Eleusine aegyptiaca.

Goosegrass – Eleusine indica.

Goosegrass – Galium aparine.

Goose Tongue – Plantago decipiens.

Goraka – Garcinia cambogia.

Gordoncillo – Piper leucophyllum.

GORDONIA Ellis. Theaceae. 40 spp. China, Formosa, Indomalaya; 1 sp. S.E. United States of America. Trees.

***G. excelsa*** Blume. Malaysia, Java. The hard deep red wood is used for general building, decorative work and rice pestles. The bark yields a black dye used to stain basket work.

***G. integerrima*** T. and B. = Laplacca integerrima.

***G. lasianthus*** (L.) Ellis. (Loblolly Bay, Tan Bay). S.E. U.S.A. The light red wood is soft and close-grained It is used locally for cabinet work The bark is used for tanning.

Gorli (Fat) Oil – Oncoba echinata.

Gorli (Seed) Oil – Oncoba echinata.

Gorse – Ulex europaeus.

***Gorugandora amorphoides*** Griseb. = Gleditsia amorphoides.

GOSSWEILERODENDRON Harms. Leguminosae. 1 sp. Tropical Africa. Tree.

***G. balsamiferum*** Harms. The light yellow wood is strong and polishes well. It is used to make furniture, light building material and general construction work. The bark yields a resin used locally for lighting.

GOSSYPIOSPERMUM (Griseb.) Urb. Flacourtiaceae. 2 spp. Cuba, Tropical S.America. Trees.

***G. praecox*** (Griseb.) Wils. (Boxwood). Cuba to Venezuela. The hard, light yellow wood is used for making mathematical instruments, veneers, engraving blocks, key boards, various spindles, inlay work and turning.

GOSSYPIUM L. Malvaceae. Tropics and subtropics. 20 spp. (sensu Hutchinson, Silow and Stephens), 67 spp. (sensu Prokhanov). Herbs. The genus are large, semi-xerophytic herbs (sometimes reaching the proportions of small trees) with extensive tap-root systems. Many species have hairless or only lightly hirsute testas, while the testas of the cultivated species are very hairy. The economic product is the testa hairs, which are pure cellulose. They owe their importance as a source of fibre to their drying out and twisting, which makes them easy to spin and gives a tough thread. The hairs are of two kinds – the long light coloured lint which can be easily detached from the seed coat and the shorter, more strongly coloured fuzz which is more difficult to detach from the seed coat. Intercrosses between species occur frequently. The basic chromosome number is 13 and broadly speaking the important Old World species (G. herbaceum and G. arboreum) are diploid, while the New World species (G. barbadense and G. hirsutum) are tetraploid. The New World species contained Old World genomes long before the plant was used commercially. How this occurred is not known. The plant has been grown in India since before 3000 B.C. and was used by the pre-Inca civilizations of Peru. Its first introduction into Europe was during the Moorish invasion of Spain – during the 10th century A.D. and although cotton was imported from India, there was no large-scale cotton industry in England until the 16th century. The development of spinning machinery led to the astronomic increase in the demand for cotton and to the development of the American cotton growing. Cotton is grown virtually everywhere where climatic conditions are suitable. The plant needs about 200 frost-free days during the year and an average temperature during the growing season of 15–17°C. The world production is about 12 million tons of cotton and 20 million tons of oil annually. Of these the U.S.A. produces about 30 per cent, U.S.S.R. and China each about 15 per cent, India 9 per cent, United Arab Republic 4·3 per cent, Mexico 4·5 per cent, Brazil 3·9 per cent, Pakistan 3·0 per cent.

The plant is an annual. Apart from U.S.A., where cultivation and harvesting is now mainly mechanised, these processes are carried out by hand. In the drier regions, e.g. Egypt cotton is grown under irrigation. The seeds with their attached fibres are ginned to remove the lint from the seeds, which are further ginned to remove the fuzz. The cotton fibres are then bailed and sent for spinning etc. The seeds are crushed and heated to 110°C before pressing to remove the oil. This is purified by treating with sodium hydroxide, deodourized and bleached. The remaining cake containing 16–20 per cent protein is used mainly as cattle food, but some is used as fertilizer and to make dyes. The oil (Cottonseed Oil) has Sp. Gr. 0·917, Sap. Val. 195, Iod. No. 108·2, Unsap. 0·9 per cent and is semi drying. It is used largely to make cooking oils, margarine, salad oils etc., but some is used to make soap, soap powders, lubricants, etc. The seeds contain gossypol (a phenolic compound) which is poisonous, but it is destroyed during processing by combination with the protein. The hulls from the seeds are used for livestock feed, fertilizer, lining oil wells, explosives and making xylose from which alcohol can be produced. The flowers are a good source of honey and the petals are used in India to make a yellow-brown dye. The root bark is used medicinally to stimulate uterine contractions, to stop haemorrhages and to control menstruation. The fibres are one of the most important fabric fibres in the world. Apart from their use as textiles, they are also used to make cordage, rayon and the interiors of motor tyres. Absorbent cotton wool has had the oily surface removed from the staple and is used for surgical dressing.

***G. arboreum*** L (Tree Cotton) A large plant. India, Africa. Cultivated locally. var. ***sanguinea*** Watt. gives a staple of good quality, while that of var. ***neglecta*** Watt. is much shorter.

***G. barbadense*** L. (Sea Island Cotton, Egyptian Cotton). A tetraploid of New World origin, grown in the W.Indies, but is the main species grown in Egypt (United Arab Republic). It gives a cotton of excellent quality with a fine lustre and the longest staple of any (3–5 cm.). The quality of the Egyptian-grown cotton is slightly inferior to that from the W.Indies.

***G. barbadense*** Oliv. = G. peruvianum.

***G. brasiliense*** Macf. Brazil. Grown locally, giving a fine lint (Chain Cotton, Bahia Cotton, Kidney Cotton, Pernambuco Cotton, Stone Cotton).

***G. herbaceum*** L. Origin unknown. Gives a cotton of good quality. The plant is grown widely to give Arabian Cotton, Maltese Cotton, Levant Cotton, Syrian Cotton and some of the American Short Staple Cotton.

***G. hirsutum*** L. (American Upland Cotton). Nearly all the cotton from the U.S.A. is of the type. It gives American Short Staple Cotton (staple 1–2·5 cm.) and American Long Staple Cotton (staple 2·5–3·5 cm.).

***G. mexicanum*** Tod. (Ichicaxihuitl). C.America. A wild species cultivated locally in C.America for the short staple cotton.

***G. microcarpum*** Tod. (Red Peruvian Cotton). S.America. The fibres are used locally.

***G. nanking*** Meyen. Tropical Asia. Cultivated in Asia and Africa. The fibres are silky and slightly reddish (Chinese Cotton, Nanking Cotton, Siam Cotton, Khaki Cotton). var. ***bani*** Watt. gives a good fibre, while var. ***roji*** Watt. yields a short, rough fibre.

***G. obtusifolium*** Roxb. Grown widely in Nidia giving a coarse, reddish fibre (Broach Cotton, Kathiawar Cotton, Kumpta Cotton, Surat Cotton).

***G. peruvianum*** Cav. = G. barbadense Oliv. S.America. Gives a good quality fibre (Peruvian Cotton, Andes Cotton).

***G. punctatum*** Schum. and Thonn. (Koton bernaoui). Cultivated for local use in the Sudan and into N.Nigeria. The staple is very short and the yield small, but the plant is well-suited to the climate and the yield is sufficient to meet local demand.

GOUANIA Jacq. Rhamnaceae. 70 s p. Tropics and subtropics. Vines.

***G. javanica*** Miq. Indochina, Philippines, Malacca. The crushed roots are used locally as poultices for sores.

***G. leptostachya*** DC. Malaysia. The bark is used locally to wash the hair. In Java the pulped plant is used as a poultice for skin complaints.

***G. lupuloides*** (L.) Urban. = Banisteria lupuloides. (Chaw Stick, Toothbrush Tree). Tropical America. The stem is chewed to harden the gums. The powdered stem is used in toothpowder, and is exported to Europe.

***G. tiliaefolia*** Lam. Philippines. The roots are used locally as a soap substitute.

Goupi Wood – Goupia glabra.

GOUPIA Aubl. Goupiaceae. 3 spp. Guyana, N.Brazil. Trees.

***G. glabra*** Aubl. (Goupi Wood). Guyana. The red-brown, heavy tough wood takes a good polish. It is used for furniture-making, boat-building, railway sleepers and paving blocks.

Gourd – Cucurbita maxima.

Gourd, Bottle – Lagenaria vulgaris.

Gourd, Dishcloth – Luffa acutangula.

Gourd, Edible Snake – Trichosanthes anguinea.

Gourd, Ethiopian Sour – Adansonia digitata.

Gourd, Gooseberry – Cucumis anguria.

Gourd, Japanese Snake – Trichosanthes cucumeroides.

Gourd, Sinkwa Towel – Luffa acutangula.

Gourd, Snake – Trichosanthes anguinea.

Gourd Tree – Adansonia gregorii.

Gourd, Wax – Benincasa hispida.

Gourd, White – Benincasa hispida.

Gourd, Wild – Cucurbita foetidissima.

GOURLIEA Gillies ex Hook. Leguminosae. 1 sp. Temperate S.America. Small tree.

***G. decorticans*** Gill. The fleshy fruits are edible and form an important part of the diet in some parts of Chile. The plant is also fed to livestock when it is called Chanal or Chanar.

Goutweed – Aegopodium podagraria.

Gouty Stem – Adansonia gregorii.

Governor's Plum – Flacourtia ramontchi.

Graceful Wattle Acacia – Acacia decora.

GRACILARIA Greville. Gracilariacea. Marine Red Alga. All species are potential sources of agar.

***G. compressa*** (J. Ag.) Grev. Pacific. Eaten in Japan.

***G. confervoides*** (L.) Grev. Cosmopolitan. Eaten as a food in the Far East and used to make agar in Australia, S.Africa and N.Carolina.

***G. coronopifolia*** J. Ag. Pacific. Eaten in Hawaii (Limu Manauea) with octopus, chicken grated with coconut or in soups.

***G. eucheumoides*** Harv. (Kanot, Kanot). Eaten in the Philippines.

***G. lichenoides*** (L.) Harv. (Ceylon Moss). Indian and Pacific Oceans. Eaten throughout the Far East as a jelly or complete. It is exported to China where it is used medicinally to treat dysentery.

Graines à Vers – Artemisia judaica.

Graines d'Avignon – Rhamnus infectorius.

Grains of Paradise – Aframomum melegueta.

Grains of Selim – Xylopia aethiopica.

Gram – Cicer arietinum.

Gram, Bengal – Cicer arietnum.

Gram, Black – Phaseolus mungo.

Gram, Golden – Phaseolus mungo.

Gram, Green – Phaseolus mungo.

Gram, Horse – Vigna sinensis.

Gram Pea – Cicer arietinum.

Grama, Black – Bouteloua eriopoda

Grama, Blue – Bouteloua gracilis.

Grama Grass – Bouteloua filiformis.

Grama, Side Oat – Bouteloua curtipendula.

Grambula – Myrtillocactus geometrizans.

GRAMMATOPHYLLUM Blume. Orchidaceae. 10 spp. Malaysia, Polynesia. Herbs.

***G. leopardinum*** Rchb. f. = G. scriptum.

***G. rumphianum*** Miq. = G. scriptum.

***G. scriptum*** Blume. = G. leopardinum = G. rumphianum. Moluccas, Amboina. A decoction of the leaves is used locally to treat stomach complaints and intestinal worms. The crushed pseudobulbs are used to heal sores. In the Moluccas the seeds are used as an aphrodisiac.

Granadilla, Giant – Passiflora quadrangularis.

Granadilla Real – Passiflora quadrangularis.

Granadillo – Dalbergia granadillo.

Granadillo – Buchenavia capitata.

Grand Fir – Abies grandis.

GRANGEA Adans. Compositae. 6 spp. Tropical Africa, Madagascar, tropical Asia. Herbs.

***G. maderaspatrana*** (L.) Poir. = Artemisia maderaspatana. Tropical Africa, Asia. The leaves are used in Hindu medicine to treat stomach complaints.

Granjeno – Celtis iguanaea.

Granjeno – Celtis pallida.

Granjeno huastoco – Celtis pallida.

***Granulobacter butylicum*** Beijerinck = Clostridium butylicum.

Grape – Vitis vinifera.

Grape, Californian – Vitis vinifera.

Grape, Canõn – Vitis arizonica.

Grape, Downy – Vitis cinerea.

Grape, European – Vitis vinifera.

Grape, Fox – Vitis labrusca.

Grape, Frost – Vitis vulpina.

Grape Honey – Vitis vinifera.

Grape Jasmine – Ervatamia coronaria.

Grape Mango – Sorindeia madagascariensis.

Grape, Muscadine – Vitis rotundifolia.

Grape, Oregon – Mahonia aquifolium.

Grape, Oregon – Mahonia repens.

Grape, River Bank – Vitis vulpina.

Grape, Sea – Coccoloba uvifera.

Grape, Southern Fox – Vitis rotundifolia.

Grape, Summer – Vitis aestivalis.

Grape, Sweet Winter – Vitis cinerea.

Grapefruit (Seed) (Oil) – Citrus paradisica.

Grapple Plant – Harpagophytum procumbens.

Giant Bamboo – Dendrocalamus giganteus.

Grass, Abyssinian Finger – Digitaria abyssinica.

Grass, Alpine Timothy – Phleum alpinum.

Grass, American Beach – Ammophila breviligulata.

Grass, Bahia – Paspalum notatum.

Grass, Bermuda – Cynodon dactylon.

Grass, Bharbur – Ischaemum angustifolium.

Grass, Billion Dollar – Echinochloa frumentacea.

Grass, Black Grama – Bouteloua eriopoda.

Grass, Buffel – Urochloa mosambicensis.

Grass, Buffalo – Buchloë dactyloides.

Grass, Camel – Cymbopogon schoenanthus.
Grass, Canada Blue – Poa compressa.
Grass, Canary – Phalaris canariensis.
Grass, Carib – Eriochloa polystacha.
Grass, Carpet – Axonopus compressus.
Grass, Centipede – Eremochloa ophuroides.
Grass, China – Boehmera nivea.
Grass, Chinese Mat – Cyperus tegetiformis.
Grass, Coco – Cyperus rotundus.
Grass, Colorado – Panicum texanum.
Grass, Columbus – Sorghum almum.
Grass, Dallis – Paspalum dilatatum.
Grass, Dog's Tail – Cynosurus cristatus.
Grass, Drakenberg Silky – Pennisetum unisetum.
Grass, Dunn's Finger – Digitaria scalarum.
Grass, Durban – Dactyloctenium australe.
Grass, Eel – Vallisneria americana.
Grass, Efwatakala – Melinis minutiflora.
Grass, Elephant – Pennisetum purpureum.
Grass, Elephant – Typha elephantina.
Grass, Esparto – Stipa tenacissima.
Grass, European Beach – Ammophila arundinacea.
Grass, Five-leaf – Potentilla reptans.
Grass, Fowl Blue – Poa palustris.
Grass, Goose – Eleusine indica.
Grass, Goose – Galium aparine.
Grass, Grama – Bouteloua filiformis.
Grass, Green Needle – Stipa viridula.
Grass, Green Sprangletop – Leptochloa dubia.
Grass, Greenleaf Fescue – Festuca viridula.
Grass, Guatemala – Tripsacum laxum.
Grass, Guinea – Panicum maximum.
Grass, Harding – Phalaris tuberosa.
Grass, Hilo – Paspalum congugatum.
Grass, Inchi – Cymbopogon caesius.
Grass, Inchi – Cymbopogon flexuosa.
Grass, Italian Rye – Lolium italicum.
Grass, Johnson – Sorghum halepense.
Grass, Kangaroo – Anthristiria ciliata.
Grass, Kachi – Cymbopogon caesius.
Grass, Kentucky Blue – Poa pratensis.
Grass, Kikuyu – Pennisetum clandestinum.
Grass, Land's – Panicum laevifolium.
Grass, Lemon – Cymbopogon citratus.
Grass Linen – Boehmeria nivea.
Grass, Longhair Plume – Dichelachne crinata.
Grass, Mahareb – Cymbopogon senaarensis.
Grass, Manna – Glyceria fluitans.
Grass, Mauritius – Panicum molle.
Grass, Meadow Fescue – Festuca elatier.
Grass, Meadow Foxtail – Alopecurus pratensis.
Grass, Mitchell – Astrebla pectinata.
Grass, Mutton – Poa fendleriana.
Grass, Napier – Pennisetum purpureum.
Grass, Natal – Tricholaena rosea.
Grass, Needle and Thread – Stipa comata.
Grass, Nile – Acroceras macrum.
Grass, Nodding Brome – Bromus anomalus.
Grass, Nut – Cyperus rotundus.
Grass, Nuttall Alkali – Puciniella nuttalliana.
Grass, Orchard – Dactylis glomerata.
Grass, Pampas – Cortaderia argentea.
Grass, Pangola – Digitaria valida.
Grass, Pangola Giant – Digitaria valida.
Grass, Para – Panicum barbinode.
Grass, Pepper – Lepidium sativum.
Grass, Perennial Rye – Lolium perenne.
Grass, Peavine – Lathyrus sativus.
Grass, Pine Blue – Poa scabrella.
Grass, Porcupine – Stipa vaseyi.
Grass, Quack – Agropyron repens.
Grass, Red Fescue – Festuca rubra.
Grass, Redtop – Agrostis alba.
Grass, Reed – Phragmites communis.
Grass, Reed Canary – Phalaris arundinacea.
Grass, Rescue – Bromus catharticus.
Grass, Rhode Island Bent – Agrotis vulgaris.
Grass, Rhodes – Chloris gayana.
Grass, Rib – Plantago lanceolata.
Grass, Richmond Finger – Digitaria diversinervis.
Grass, Ripple – Plantago major.
Grass, Rocha – Cymbopogon martini.
Grass, Saw – Cladium effusum.
Grass, Scorbute – Cochleria officinalis.
Grass, Scurvy – Barbarea praecox.
Grass, Scurvy – Cochlearia officinalis.
Grass, Shorthair – Dichelachne sciurea.
Grass, Shorthair Plume – Dichelachne sciurea.
Grass, Small Buffalo – Panicum coloratum.
Grass, Smooth Brome – Bromus inermis.
Grass, Sour – Amphilophis pertusa.
Grass, Spring – Anthoxanthum odoratum.
Grass, Squaw – Elymus triticoides.
Grass, Star – Cynodon plectostachyus.
Grass, Sugar – Glyceria fluitans.
Grass, Surf – Phyllospadix scouleri.
Grass, Swatow – Boehmeria nivea.
Grass, Swaziland Finger – Digitaria swazilandensis.
Grass, Sweet – Hierochloë odorata.
Grass, Sweet – Sporolobus indicus.

Grass, Sweet Vernal – Anthoxanthum odoratum.
Grass, Tambookie – Hyparrhena aucta.
Grass, Tangle – Heteropogon hirtus.
Grass, Tape – Vallisneria americana.
Grass, Teff – Poa abyssinica.
Grass, Texas Blue – Poa arachnifera.
Grass, Thatch – Saccharum spontaneum.
Grass, Thurber Fescue – Festuca thurberi.
Grass, Tiger – Thysanolaena maxima.
Grass, Timberline Blue – Poa rupicola.
Grass, Timothy – Phleum pratense.
Grass, Towoomba – Phalaris tuberosa.
Grass Tree, Australian – Xanthorrhoea australis.
Grass Tree, Dockowar – Xanthorrhoea arborea.
Grass Tree Gum – Xanthorrhoea australis.
Grass Tree Gum – Xanthorrhea hastilis.
Grass Tree, Spearleaf – Xanthorrhoea hastilis.
Grass, Tsauri – Cymbopogon giganteus.
Grass, Tufted Hair – Deschampsia caespitosa.
Grass, Tussock – Deschampsia caespitosa.
Grass, Umbrella – Panicum decompositum.
Grass, Vandyke – Panicum flavidum.
Grass, Vasey – Paspalum urvillei.
Grass, Velvet – Holcus lanatus.
Grass Weed – Zostera marina.
Grass, Wood Blue – Poa nemoralis.
Grass, Woolly Finger – Digitaria pentzii.
Grass, Wrack – Zostera marina.
Grass, Yaragua – Andropogon rufus.
Grass, Yellow Nut – Cyperus esculentum.
Grass, Zakaton – Epicampes macroura.
Grassleaf Daylily – Hemerocallis.

GRATELOUPIA J. Ar. Grateloupicaeae. Red Algae. Pacific.

***G. affinis*** (Harv.) Okam. Eaten in Japan. Source of an inferior gum.

***G. filicina*** (Wulf.) J. Ag. Eaten in Japan.

***G. ligulata*** Holmes. Eaten in China.

GRATIOLA L. Scrophulariaceae. 2 spp. N. and S. Temperate. Tropical Mountains. Herbs.

***G. monniera***=Bacopa monniera.

***G. officinalis*** L. (Hedge Hyssop). Central and S.Europe, N. and W.Asia, N.America. The leaves and roots are used as a laxative, emetic and diuretic, particularly in infections of the liver, dropsy and enlargement of the spleen.

Grauna – Melanoxylon brauna.
Gravata – Bromelia fastuosa.
Gravel Root – Eupatorium purpureum.
Gray Birch – Betula populifolia.
Gray's Chokeberry – Prunus grayana.
Graylings – Cantharellus umbonatus.
Greasewood – Larrea mexicana.
Greasewood, Mexican – Sarcobatus vermiculata.
Great Aspen – Populus grandidentata.
Great Bulrush – Scirpus lacustris.
Great Burdock – Arctium lappa.
Great Laurel – Rhododendron maximum.
Great Millet – Sorghum vulgare.
Greater Galangal – Alpinia galanga.
Greater Yam – Dioscorea alata.
Grecian Silk Vine – Periploca graeca.
Greek Juuiper – Juniperus excelsa.
Greek Turpentine – Pinus halepensis.
Greek Valerian – Polemonium coeruleum.
Green Amaranth – Amaranthus retroflexus.
Green Arrow Arum – Peltandra virginica.
Green, Chinese – Rhamnus dahurica.
Green Ebony – Rhodocolea racemosa.
Green Ebony – Tecoma leucoxylon.
Green Gram – Phaseolus mungo.
Green Needle Grass – Stipa viridis.
Green Osier – Cornus circinnata.
Green Pepper – Capsicum annuum.
Green Sapote – Calocarpum viride.
Green Spangletop – Leptochloa dubia.
Green Strophanthus – Strambosnia kombe.
Greenbrier – Smilax china.
Greenbrier, Bristly – Smilax bona-nox.
Greenbrier, Common – Smilax rotundifolia.
Greenbrier, Long-stalked – Smilax pseudochina.
Greenbrier, Saw – Smilax bona-nox.
Greengage – Prunus insititia.
Greenheart – Nectandra rodiaei.
Greenheart, Demarera – Ocotea rodiaei.
Greenheart Ebony – Tecoma leucoxylon.
Greenheart, Surinam – Tecoma leucoxylon.
Greenleaf Fescue – Festuca viridula.
Greenwattle Acacia – Acacia decurens.
Greenwood – Halfordia scleroxyla.
Greenwood, Dyer's – Genista tinctoria.

GREIGIA Regel. 18 spp. Central America to Chile, Juan Fernandez. Herbs.

***G. sphacelata*** Regel. Chile. The berries (Chupones) are eaten locally.

Grenadille de Quijos – Passiflora popenvoii.
Grenadille, Giant – Passiflora quadrangularis.
Grenadilla, Purple – Passiflora edulis.
Grenadilla Real – Passiflora quadrangularis.
Grenadilla, Sweet – Passiflora lingularis.
Grenadille Wood, African – Dalbergia melanoxylon.

Grenadine – Punica granatum.

GREVILLEA R. Br. Proteaceae. 190 spp. E.Malaysia, New Hebrides, New Caledonia. Trees.

***G. gillivrayi*** Hook. New Caledonia. The reddish wood is used for cabinet work.

***G. lineata*** R. Br.=G. striata.

***G. robusta*** Cunn. (Silk Oak, Warragarra). New South Wales, Queensland. The wood is light, but hard. It is used for house interiors, cabinet work and is sometimes grown as shade for coffee.

***G. striata*** R. Br.=G. lineata. (Beef Wood, Turraie). Australia. The hard wood polishes well and is used for cabinet work, furniture etc.

GREWIA L. Tiliaceae. 150 spp. Africa, Asia, Australia, especially in the tropics. Trees.

***G. asiatica*** L.=G. subinaequalis. (Dhamin, Parusha, Phalsa, Palisa). Tropical Asia. Cultivated. The yellowish wood is strong and elastic. It is used in India for yokes, bows and spear shafts. The bark fibre is used to make cordage and the bark is used in clarifying sugar. The fruits have a pleasant sub-acid flavour. They are eaten raw, or made into a fermented beverage. The fruits have some medicinal value in relieving stomach complaints. The bark from the roots is used to treat rheumatism.

***G. betulifolia*** Juss. (Chari. Tarakat). Sahara. The nutty-flavoured fruits are eaten locally.

***G. bicolor*** Juss. (Imigis). Senegal to Sahara. The fruits are eaten locally and the leaves are used for washing clothes.

***G. calvata*** Bak. Madagascar. The bark fibres are used locally to make cordage.

***G. carpinifolia*** Juss. India, tropical Africa. A decoction of the leaves is used in Africa to remove lice from the hair.

***G. celtidifolia*** Juss. (Talok). Indonesia. The elastic wood is used locally for knife handles and waggon parts.

***G. elastica*** Royle.=G. vestita. (Bimlau, Dhamum, Khursi). S.Himalaya, N.India. The very elastic wood is used for tool handles, various forms of shafts, agricultural implements, tennis raquets, bows, furniture etc. The fruit is used to make sherbets and the leaves are fed to livestock.

***G. excelsa*** Vahl. Tropical Africa, Asia. The bark fibre is used locally for cordage.

***G. faucerei*** P. Dang. Madagascar. The bark yields a good cordage fibre (Hafotra).

***G. fibrocarpa*** Mast. (Tjenderai ootan). Malaysia. The fruits are used locally to make soup (Sajoor).

***G. flavescens*** Juss. (Areicha). Sahara, Sudan to Senegal. The fruits (Abba) are eaten locally.

***G. glandolosa*** Wahl. Madagascar. The bark yields a fibre used locally for cordage.

***G. laevigata*** Vahl.=G. oblongifolia (Kola, Ooris ooresan, Torhoi). Tropical Asia to Australia. The bark is used locally for cordage.

***G. macrophylla*** Bak. Madagascar. The bark fibre is used locally for cordage.

***G. mollis*** Juss. Tropical Africa. The bark yields a fibre. It is also eaten in soups, or when dried used as a flour. The fruits are eaten locally. The wood is used in the Sudan to make bows and the wood ash is a source of salt.

***G. oblongifolia*** Blume=G. laevigata.

***G. occidentalis*** L. Tropical Africa. The wood is used by the natives of the Cape Peninsula to make bows.

***G. paniculata*** Roxb. (Co ke, Ktu, Poplear thom). India, Indonesia, Indochina. Distillation of the wood yields acetone. A decoction of the roots is used in Malaya to treat fevers and abdominal pains and in Cambodia and S.Vietnam it is used to relieve coughs.

***G. philippensis*** Blanco. (Baluko). Philippines. The fruits have a pleasant sub-acid flavour and are eaten locally.

***G. populifolia*** Vahl. Tropical Africa. The bark gives a cordage fibre, used locally.

***G. repanda*** Bak. Madagascar. The bark gives a cordage fibre, used locally.

***G. salutaris*** Span (Haoo loonis, Kanila, Pasolder). Malaysia, Indonesia. The rasped bark (Kajoo timor) is made into a paste and applied externally or taken internally to treat internal bruising. The bark is used similarly in other regions to treat external bruises.

***G. stylocarpa*** Juss. (Muling). Philippines. The fruits are pleasantly-flavoured and are eaten locally.

***G. subinaequialis*** DC.=G. asiatica.

***G. tiliaefolia*** Vahl. (Dhamani, Phalsa). Lower Himalayas to S.India. The wood is elastic, durable and polishes well. It is used locally for a wide variety of agricultural implements, yokes, tent poles, masts, sports equipment, barrels, carriage parts, spinning machinery. The bark is used for cordage. The fruits are eaten and the leaves are fed to livestock.

***G. trinervata*** Bak. (Merika). Madagascar. The bark is used locally for cordage.

***G. vestita*** Ball. = G. elastica.

***G. villosa*** Willd. (Berchaga). Senegal to Mauritania. The fruits are eaten locally.

Grey Box – Eucalyptus hemiphloia.

Grey Box – Eucalyptus polyanthemos.

Grey Carrobeen – Sloanea woollsii.

Grey Feather – Liatris spicata.

Grey Gum – Eucalyptus gomiocalyx.

Grey Gum – Eucalyptus tereticornus.

Grey Ironbark – Eucalyptus paniculata.

Grey Myrtle – Backhousia myrtifolia.

Grey Persimmon – Diospyros pentamera.

Grey Satin Ash – Eugenia gustavioides.

Grey Teak – Gmelina leichthardtii.

Greyish Urceolaria – Urceolaria cinerea.

GRIAS L. Lecythidiceae. 15 spp. Central to S.America, W.Indies. Trees.

***G. cauliflora*** L. (Anchovy Pear). W.Indies. Cultivated locally for the fruit.

***G. peruviana*** Miers. (Cocora). Peru, Colombia, Venezuela. The Indians use scrapings from the germinating seeds infused in water as an emetic.

Grijs Appel – Parinari mobola.

GRINDELIA Willd. Compositae. 60 spp. America, excluding Central America. Herbs.

***G. camporum*** Greene (Gumplant). W. N.America. The dried leaves and flower tops are used to relieve asthma, hay fever, whooping cough and to treat burns. They contain 20 per cent resin, saponin, tannin, robustic acid and grindelol.

***G. cuneifolia*** Nutt. See G. camporum.

***G. robusta*** Nutt. (Shore Grindelia) W. U.S.A. The flower heads and leaves are used locally to treat coughs and sore throats, to relieve stomach upsets and to "purify" the blood.

***G. squarosa*** (Pursh.) Dun. (Curlytop Grindelia). W. N.America. See G. camporum.

Grindelia, Curlytop – Grindelia squarosa.

Grindelia, Shore – Grindelia robusta.

GRISELINIA Forst. f. Cornaceae. 6 spp. New Zealand, Chile, S.E.Brazil. Trees.

***G. littoralis*** Raoul. New Zealand. The reddish wood is firm, dense, but slightly brittle. It is used for railway sleepers and ship-building.

***G. lucida*** Forst. New Zealand. The brownish wood is very durable and is used locally for fence posts.

Groats – Fagopyrum esculentum.

Gromwell – Lithospermum ruderale.

Gromwell, Carolina – Lithospermum carolinense.

Gromwell, Common – Lithospermum officinale.

Gromwell, Hoary – Lithospermum canescens.

Gros Terfaz Blanc – Tirmania africana.

Grosse Cérise – Malpighia urens.

Ground Berry – Physalis neo-mexicana.

Ground Blueberry – Vaccinium myrsinites.

Ground Cherry – Physalis heterophylla.

Ground Cherry – Physalis neo-mexicana.

Ground Hemlock – Taxus candensis.

Ground Ivy – Glechoma hederaceum.

Ground Laurel – Epigaea repens.

Ground Lemon – Podophyllum peltatum.

Ground Nut – Apios americana.

Ground Nut – Arachis hypogaea.

Ground Nut, Bambara – Voandziea subterranea.

Groundnut, Geocarpa – Kerskingiella geocarpa.

Groundnut Peavine – Lathyrus tuberosus.

Ground Peanut – Amphicarpaea monoica.

Ground Pine – Ajuga chamaepitys.

Ground Tomato – Physalis neo-mexicana.

Groundsel – Senecio vulgaris.

Grove Windflower – Anemone nemorosa.

Grumichama – Eugenia dombeyana.

***Grumilea*** Gaertn. = Marpouria.

Guabiraba – Campomanesia guaviroba.

Guabiroba – Campomanesia frezliana.

Guacamayo – Trichostigma octandrum.

Guacima – Guazama ulmifolia.

Guaco – Aristolochia pardina.

Guaco – Mikania guaco.

Guaco de sur – Aristolochia maxima.

Guadil (Oil) – Convolvulus floridus.

Guaguasi – Zuelania guidonia.

Guaiac wood (Oil) – Bulnesia sarmienti.

GUAIACUM L. Zygophyllaceae. 6 spp. Warm America, W.Indies. Trees.

***G. coulteri*** A. Gray. = G. planchoni. (Arbol Santo, Matlaquahuitl, Yaga-na). Mexico. The wood is hard and resinous. It is used for firewood and for making strong articles.

***G. officinale*** L. (Lignum Vitae, Lignum sanctum, Guajacan Negro). Tropical America. The gum (Guaiacum) extracted from the wood is mildly laxative. It was introduced to Europe in 1526 when it was credited with many and diverse medical properties, hence the name Lignum Vitae (tree of life). It is now not used much in medicine, but is used as a test for oxygen, and as antioxidant in the commercial preparation of lard. The wood contains some 26 per cent resin which is extracted by heating from holes bored in the logs or by boiling the wood chips in salt water. The wood is greenish-yellow, hard, light and durable. It is used in ship-building and to make small objects where hardness is needed, e.g. hammers.

***G. planchoni*** A. Gray.=G. coulteri.

***G. sanctum*** L. (Lignum Vitae). Tropical America. See G. officinale.

Guaiacum Gum – Guaiacum officinale.

Guaiacum Resin – Guaiacum officinale.

Guairo Santo – Aegiphila elata.

Guajacan – Caesalpinia melanocarpa.

Guajacan Negro – Guaiacum officinale.

Guajillo – Pithecellobium brevifolium.

Guajilote – Parmentiera edulis.

Guajón – Beilschmiedia pendula.

Guanabana – Annona muricata.

Guaniqui – Trichostigma octandrum.

Guano – Ochroma pyramidale.

Guapaque – Ostrya guatemalensis.

Guapi Bark – Guarea rusbyi.

Guar (Gum) – Cyamopsis tetragonolobus.

Guaraguao – Guarea guara.

Guarana – Paullinia cupana.

Guararema – Gallesia integrifolia.

GUAREA Allem. ex L. Meliaceae. 150 spp. Tropical America; 20 spp. Africa. Trees.

***G. africana*** Welw. Tropical Africa. The white wood is light and is used for general carpentry, locally.

***G. cedrata*** (Chev.) Pell.=Trichella cedrata. (Pink African Cedar, Pink Mahogany). Tropical Africa. The wood is pale mahogany-coloured. It is used for furniture-making, boat-building and canoes.

***G. guara*** (Jacq.) P. Wilson.=G. trichiloides =Trichilia guara. (Guaraguao, Yamagua). W.Indies. A resin from the leaves and stems is used in Cuba to control bleeding. The purple wood is used for furniture and inside work on houses, it also yields a reddish oil (Sandalo do Para). The powdered bark is used as an emetic and to control bleeding.

***G. martiana*** C. DC.=G. purgans. Tropical America. The powdered bark is used in Brazil as a depurative mixed in manufactured medicines.

***G. purgans*** St. Hil=G. martiana.

***G. rusbyi*** (Britt.) Rusby. (Cocillana). Bolivia, on the slopes of the Andes. The bark (Guapi Bark), is used locally as an emetic and in medicines as an expectorant. In larger doses it is also used medicinally as an emetic.

***G. thompsoni*** Sprague and Hitch. (Benin Mahogany). Tropical Africa. The wood is used to make furniture. It is exported.

***G. trichiloides*** L.=G. guara.

Guassatonga – Casearia sylvestris.

Gautambi – Balfourodendron riedelianum.

Guatemala Rosewood – Dalbergia cubilquitzensis.

Guatemala Walnut – Juglans mollis.

GUATTERIA Ruiz. and Pav. Annonaceae. 250 spp. S.Mexico to Brazil. Trees.

***G. caribaea*** Urban. (Mato, Bois Violin), Porto Rico and Lesser Antilles. The wood is used locally to make boards and the bark is a source of fibre used to make cordage, sails and for caulking canoes.

***G. cargadero*** Triana and Planch. S.America. The bark is used as a cordage fibre in parts of Colombia.

***G. veneficiorum*** Mart.=Unonopsis veneficiorum.

Guava – Psidium guajava.

Guava, Brazilian – Psidium araca.

Guava, Cattley – Psidium cattleianum.

Guava Cheese – Calocarpum mammosum.

Guava, Costa Rican – Psidium friedrichsthalianum.

Guava, Para – Campomanesia acida.

Guava, Strawberry – Psidium cattleianum.

Guavo – Inga punctata.

Guavo Bejúco – Inga edulis.

Guavo de Castilla – Inga spectabilis.

Guavo de Mono – Inga goldmannii.

Guavo peludo – Inga macuna.

Guavo real – Inga radians.

Guayabi – Patagonula americana.

Guayabi – Patagonula bahiensis.

Guayabillo – Psidium satorianum.

Guayule (Rubber) – Parthenium argentatum.

GUAZUMA Mill. Sterculiaceae. 4 spp. Tropical America. Trees.

***G. grandiflora*** G. Don.=Theobroma grandiflora. Amazon Basin. The fruits are edible and are used in sherberts locally. The plant is cultivated for the fruits.

***G. tomentosa*** H.B.K. Tropical America.

The elastic wood is used to make shoe lasts.

*G. ulmifolia* Lam.=Theobroma guazuma. (Cabolte, Guacima, Tablote). Mexico to tropical America. The wood is elastic, light and coarse-grained. It is used for shoe lasts, barrels, ribs of small boats, charcoal for gunpowder. The bark was used to treat asthma and is used to combat baldness. The juice from the plant is used to clarify syrup.

Guelder Rose – Viburnum opulus.

Guele – Prosopsis africana.

Guérit petit – Justicia gendarussa.

Gueroxiga – Parmentiera edulis.

GUETTARDA L. Rubiaceae. 20 spp. New Caledonia, 60 spp. Tropical America. Trees.

*G. angelica* Mart. Brazil. The roots are used locally to treat diarrhoea of cattle and horses.

*G. laevis* Urban. (Cucubano). W.Indies. The wood is strong, heavy, fine-grained and polishes well. It is susceptible to termites, but is used for furniture, interior work, fancy articles, agricultural implements, tool handles and bridges.

*G. speciosa* L. Tropical Asia, Madagascar. (Zebra Wood). See Connarus guianensis.

Guger Tree, Burma – Schima noronhae.

Guiambe – Philodendron squamiferum.

Guiamol – Entado polystachia.

Guiana Arrowroot – Dioscorea alata.

Guiana Arrowroot – Dioscorea batatas.

Guiana Arrowroot – Musa paradisiaca.

Guiana Chaste Tree – Vitex divaricata.

Guiana Nut – Caryocar nuciferum.

Guiana Symphonia – Symphonia globulifera.

GUIBOURTIA Benn. Leguminosae. 4 spp. Tropical America; 11 spp. Tropical Africa. Trees.

*G. coleosperma* (Benth.) Léon. (Mushi). Tropical Africa, especially Central. The seeds are eaten locally and an oil expressed from them is much used locally for cooking.

*G. copallifera* Benn.=Copaifera guibourtiana.

GUILIELMA Mart. Palmae. 7 spp. Tropical S.America. Palm trees.

*G. insignis* Mart. S.America. The juicy fruits are eaten locally raw or dried.

*G. speciosa.* Mart.=Bactris gasiaës. (Peach Palm). Central and S.America. The peach-like fruits are eaten locally as a vegetable and made into a fermented beverage. The oil from the seeds (Oil of Macanilla) is used locally for cooking. The hard wood is a local building material and is made into bows. The spines are used for tattooing.

*G. utilis* Oerst. Central America. The fruits are eaten as a vegetable in Costa Rica. The wood is used for bows. The plant is occasionally cultivated.

Guimatsu – Larix dahurica.

Guimba – Xylopia obtusifolia.

Guimberana – Philodendron cordatum.

Guinea Arrowroot – Calathea allouia.

Guinea Corn – Sorghum vulgare.

Guinea Grass – Panicum maximum.

Guinea Hen Weed – Petiveria alliacea.

Guinea Pepper – Xylopia aethiopica.

Guinea Yam – Dioscorea cayenensis.

GUIOA Cav. Sapindaceae. 70 spp. Indomalaya, Australia, Pacific. Small trees.

*G. koelreuteria* (Blanco) Merr.=G. perrottetii. Philippines. An oil from the seeds is used locally to treat skin diseases.

*G. perrottetii* Radik.=G. koelreuteria.

*G. pleuropteris* Radik. (Kajoo lentadak). Indonesia. The elastic wood is used locally for the handles of agricultural implements.

Guirambo – Macrosiphonia hypoleuca.

Guiré – Combretum elliotii.

Guisacho coreño – Acacia hindsii.

Guisaro – Psidium molle.

Guisquil – Sechium edule.

GUIZOTIA Cass. Compositae. 12 spp. Tropical Africa. Herbs.

*G. abyssinica* (L. f.) Cass.=G. oleifera. Tropical Africa. The plant is cultivated for the seeds which yield an oil (Niger Seeds Oil, Ramtilla Oil, Werinnua Oil). The plant is cultivated mainly in India, but also in E.Africa, W.Indies and Germany. The oil is used locally for cooking and lighting, but some is exported to Europe where it is used for soap-making and cooking fats. The plant is grown in Rhodesia as a green manure and for silage.

*G. oleifera* DC.=G. abyssinica.

Gulangabin – Rosa eglanteria.

Gul-i-pista – Pistacia vera.

Gullan – Passiflora psilantha.

Gully Root – Petiveria alliacea.

Gulvel – Tinospora malabarica.

Gum, Acajou – Anacardium occidentale.

Gum, Ako Ogea – Daniellia oliveri.
Gum, American Mastic – Schinus molle.
Gum, Ammoniacum – Dorema ammoniacum.
Gum Ammoniac of Cyrenaica – Ferula marmarica.
Gum Ammoniac of Morocco – Ferula communis.
Gum, Amrad – Acacia arabica.
Gum, Amrawatti – Acacia arabica.
Gum, Amritsar – Acacia modesta.
Gum, Anani – Moronobea coccinea.
Gum, Angado Mastiche – Echinops viscosus.
Gum, Angico – Piptadenia rigida.
Gum, Anime – Hymenaea courbaril.
Gum, Anubing – Artocarpus cumingiana.
Gum Arabic – Acacia albida.
Gum Arabic – Acacia dealbata.
Gum Arabic – Acacia greggi.
Gum Arabic – Acacia horrida.
Gum Arabic – Acacia jacquimontii.
Gum Arabic – Acacia leucophloea.
Gum Arabic – Acacia senegal.
Gum Arabic – Acacia seyal.
Gum Arabic – Acacia sieberiana.
Gum Arabic – Acacia stenocarpa.
Gum Arabic, East African – Acacia drepanolobium.
Gum, Argan – Argania spinosa.
Gum, Asafetida – Ferula asafoetida.
Gum, Australian – Acacia pycnantha.
Gum, Babul – Acacia arabica.
Gum, Barister – Mezoneuron scortechinii.
Gum, Bassona – Acacia leucophleoea.
Gum, Bea – Caesalpinia praecox.
Gum, Benzoin – Styrax benzoin.
Gum, Binunga – Macaranga tanarius.
Gum, Bisabol Myrrh – Commiphora erythraea.
Gum, Black – Nyssa multiflora.
Gum, Blue – Eucalyptus botryoides.
Gum, Blue – Eucalyptus globulus.
Gum, Blue – Eucalyptus goniocalyx.
Gum, Blue – Eucalyptus saligna.
Gum, Blue – Eucalyptus tercticornis.
Gum, Blue – Eucalyptus umbellata.
Gum, Bolly – Litsea reticulata.
Gum, Botany Bay – Xanthorrhoea hastalis.
Gum, Broad-leaved Water – Tristania suaveolens.
Gum, Brown Barberry – Acacia arabica.
Gum, Bush – Eriodictyon californica.
Gum, Bush – Eriodictyon glutinosum.
Gum, Butea – Butea superba.
Gum, Cabbage – Eucalyptus sieberiana.
Gum, Californian Red – Liquidambar styraciflua.
Gum, Cape – Acacia giraffae.
Gum, Cape – Acacia horrida.
Gum, Cape – Acacia karroo.
Gum, Carob – Ceratonia siliqua.
Gum, Cashawa – Anancardium occidentale.
Gum, Cashew – Anacardium occidentale.
Gum, Cebil – Piptadenia cebil.
Gum, Chagual – Puya chilensis.
Gum, Chene – Arillastrum gummiferum.
Gum, Cherry – Prunus cerasus.
Gum, Cherry – Prunus puddum.
Gum, Chewing – Achras sapota.
Gum, Chicle – Achras sapota.
Gum, Chinese Sweet – Liquidambar formosana.
Gum, Chironji-ki-gond – Buchanania latifolia.
Gum, Combee – Gardenia gummifera.
Gum, Combee – Gardenia lucida.
Gum Copal of the Gold Coast – Daniella similis.
Gum, Cotton – Nyssa aquatica.
Gum, Dikkamaly – Gardenia gummifera.
Gum, Doctor's – Metopium toxiferum.
Gum, Doctor's – Symphonia globulifera.
Gum, Elastic – Bumelia lanugunosa.
Gum, Elemi – Protium heptaphyllum.
Gum, Elqueme – Bursera gummifera.
Gum, Feronia – Feronia limonia.
Gum, Flooded – Eucalyptus tereticornus.
Gum, Forest Red – Eucalyptus umbellata.
Gum, Formosa Sweet – Liquidambar formosana.
Gum, Galbanum – Ferula galbaniflua.
Gum, Galbanum – Ferula rubicaulis.
Gum, Galbanum – Ferula schair.
Gum, Gamboge – Garcinia morella.
Gum, Gamboge Siam – Garcinia hanburyi.
Gum Ghatti of Bombay – Anogeissus latifolia.
Gum, Ghezireh – Acacia senegal.
Gum, Gomo de Orore – Pithecolobium hymeneaefolia.
Gum, Grass Tree – Xanthorrhoea australis.
Gum, Grass Tree – Xanthorrhoea hastilis.
Gum, Grey – Eucalyptus goniocalyx.
Gum, Grey – Eucalyptus tereticornis.
Gum, Guaiacum – Guaiacum sanctum.
Gum, Guar – Cyamopsis tetragonoloba.
Gum, Gumihan – Artocarpus elastica.
Gum, Hanjigoad – Balanites orbicularis.
Gum, Hog – Clusia flava.
Gum, Hotai – Commiphora abyssinica.
Gum, Hyawa – Protium heptaphyllum.

Gum, Icacia Chichle – Sterculia tomentosa.
Gum, Illorin – Daniella thurifera.
Gum, Indian – Anogeissus latifolia.
Gum, Jingan – Lannea grandis.
Gum, Jingan – Lannea wodier.
Gum, Karya – Sterculia urens.
Gum, Kardofan – Acacia senegal.
Gum, Kateera – Sterculia urens.
Gum, Katira – Cochlospermum gossypum.
Gum, Katira Gabina – Astragalus heratensis.
Gum, Kauri – Agathis australis.
Gum, Kino – Pterocarpus erinaceus.
Gum, Labdanum – Cistus ladaniferus.
Gum, Lemon-scented – Eucalyptus atriodora.
Gum, Lucca – Olea europaea.
Gum, Locust – Ceratonia siliqua.
Gum, Locust Bean – Ceratonia siliqua.
Gum, Maguey – Puya chilensis.
Gum, Maiden's – Eucalyptus maideni.
Gum, Malabar – Bombax malabaricum.
Gum, Mango – Mangifera indica.
Gum, Manna – Eucalyptus viminalis.
Gum, Mesquite – Prosopis juliflora.
Gum, Mogador – Acacia gummifera.
Gum, Morocco – Acacia gummifera.
Gum, Mountain – Eucalyptus dalrympleana.
Gum, Mudar – Calotropis procera.
Gum, Mule – Manihot dichotoma.
Gum, Mulu Kilavary – Commiphora berryi.
Gum, Mumuye – Combretum hartemannianum.
Gum, Mumuye – Combretum leonense.
Gum, Mumuye – Combretum sokodense.
Gum, Nongo – Albizia brownii.
Gum, Ogea Sierra Leone – Daniella thurifera.
Gum, Olibanum – Boswellia carterii.
Gum, Olive – Olea europaea.
Gum, Opopanax – Opopanax chironium.
Gum, Orenburgh – Larix decidua.
Gum, Oriental Sweet – Liquidambar orientale.
Gum, Persian – Amygdalus leiocarpus.
Gum, Pipe Gamboge – Garcinia hanburyi.
Gum, Opium – Papaver somnifera.
Gum, Red – Eucalyptus camaldilensis.
Gum, Red – Eucalyptus tereticornis.
Gum, Red – Liquidambar styraciflua.
Gum, Remanso – Manihot dichotsma.
Gum, Resina de Cuapinole – Hymenaea courbaril.
Gum, Resina lutea – Xanthorrhoea australis.
Gum, Ribbon – Eucalyptus nitens.
Gum, Sagapan – Ferula szowitziana.
Gum, Sagapenum – Ferula persica.
Gum, Salai-gugul – Boswellia serrata.
Gum, Saligna – Eucalyptus saligna.
Gum, Sarcocolla – Penaea sarcocolla.
Gum, Scribbly – Eucalyptus haemastoma.
Gum, Semba gona – Bauhinia variegata.
Gum, Semki-gona – Bauhinia purpurea.
Gum, Sennarr – Acacia seyal.
Gum, Senegal – Acacia senegal.
Gum, Senn – Bauhinia variegata.
Gum, Shining – Eucalyptus nitens.
Gum, Siam Benzoin – Styrax tonkinense.
Gum, Somali – Acacia glaucophylla.
Gum, Sour – Nyssa multiflora.
Gum, Spotted – Eucalyptus citriodera.
Gum, Spotted – Eucalyptus goniocalyx.
Gum, Spotted – Eucalyptus maculata.
Gum, Spotted Blue – Eucalyptus maideni.
Gum, Spruce – Picea mariana.
Gum, Suakim – Acacia seyal.
Gum, Sugar – Eucalyptus cladocalyx.
Gum, Sumatra Benzoin – Styrax benzoin.
Gum, Swamp – Eucalyptus regnans.
Gum, Swamp – Eucalyptus viminalis.
Gum, Sweet American – Liquidambar straciflua.
Gum, Sweet Chinese – Liquidambar formosana.
Gum, Sweet Oriental – Liquidambar orientalis.
Gum, Sydney Blue – Eucalyptus saligna.
Gum, Taramniya – Combretum hypotilinum.
Gum, Tartar – Sterculia cineria.
Gum, Tragacanth – Astragalus adscendens.
Gum, Tragacanth – Astragalus echidnaeformis.
Gum, Tragacanth – Astragalus elymaeticus.
Gum, Tragacanth – Astragalus globiflorus.
Gum, Tragacanth – Astragalus gossypinus.
Gum, Tragacanth – Astragalus gummifera.
Gum, Tragacanth – Astragalus microcephalus.
Gum, Tragacanth – Astragalus myriacanthus.
Gum, Tragacanth – Astragalus prolixus.
Gum, Tupelo – Nyssa aquatica.
Gum, Tupelo – Nyssa ogeche.
Gum, Turpentine – Pinus palustris.
Gum, Velampisini – Feronia limonia.
Gum, Victorian Blue – Eucalyptus globosus.
Gum, Water – Nyssa biflora.
Gum, Wattle – Acacia pycnantha.

Gum, White – Eucalyptus gomphocephala.
Gum, White – Eucalyptus leucoxylon.
Gum, White – Eucalyptus redunca.
Gum, White – Eucalyptus viminalis.
Gum, Yakka – Xanthorrhoea australis.
Gumayaka Palm – Arenga tremula.
Gumbar – Trewia nudiflora.
Gumihan Gum – Artocarpus elastica.
Gummi Acaroides – Xanthorrhoea hastilis.
Gummi, Hog – Clusia flava.
Gummi Peucedani – Peucedanum officinalis.
Gummi Resina Ammoniacum – Ferula tingitana.
Gummi Resina Euphorbium – Euphorbia resinifera.
Gumplant – Grindelia camporum.

GUNDELIA L. Compositae. 1 sp. Asia Minor, Syria, Persia. Herb.

***G. tournfortii*** L. The leaves are eaten locally as a vegetable.

GUNNERA L. Gunneraceae. 50 spp. Malaysia (Except Malaysian Peninsula), Tasmania, New Zealand, Hawaii, Juan Fernandez, Mexico to Chile, tropical S.Africa. Herbs.

***G. chilensis*** Lam. = G. scabra. Chile. The roots are used locally for tanning and the young leaf stalks are eaten as a vegetable.

***G. macrophylla*** Blume. Indonesia. (Harijang gede, Sookmadiloowih). The fruits are used locally as a tonic and stimulant. Sold locally.

***G. perpensa*** L. S.Africa. A decoction of the roots is used by the farmers to treat dyspepsia and a decoction of the roots in brandy is used to treat kidney complaints.

***G. scabra*** Ruiz. and Pav. = G. chilensis.

Gupenja – Garcinia livingstonia.
Gurch – Tinospora malabarica.
Gurjum (Oil) – Dipterocarpus alatus.
Gurjum (Balsam) – Dipterocarpus alatus.
Gurjum Oil – Dipterocarpus lamellantus.
Gutapa – Passiflora pinnatistipula.

GUTIERREZIA Lag. Compositae. 30 spp. N.W. N.America to subtropical S.America. Bushy shrubs.

***G. divaricata*** Nutt. = G. sarothrae.

***G. juncea*** Greene = G. sarothrae.

***G. sarothrae*** (Pursh) Britt. and Rusby. = G. divaricata = G. juncea. W. U.S.A. The twigs are used by the local Indians as brooms.

Gutta Djelutung – Alstonia eximea.
Gutta Malaboeai – Alsonia grandifolia.
Gutta, Niger – Ficus platyphylla.
Gutta Percha – Ficus gnaphalocarpa.
Gutta Percha – Madhuca nottleyana.
Gutta Percha – Palaquium ahernianum.
Gutta Percha – Palaquium gutta.
Gutta Percha – Palaquium leiocarpa.
Gutta Percha – Palaquium maingayi.
Gutta Percha – Palaquium obovatum.
Gutta Percha – Palaquium oxleyanum.
Gutta Percha – Payena dantung.
Gutta Percha – Payena leerii.
Gutta Pontianak – Dyera costulata.

GYMNACRANTHERA Warb. Myristicaceae. 17 spp. Indomalaya. Trees.

***G. canarica*** Warb. = Myristica canarica. S.India. The seeds yield an oil used to make candles.

GYMNARTOCARPUS Boelage. Moraceae 3 spp. Philippines, Sumatra, Java. Trees.

***G. woodii*** Merr. Philippines. The boiled or roasted seeds are eaten locally.

***Gymnelaea sandwicensis*** L. Johns. = Osmanthus sandwicensis.

GYMNEMA R. Br. Asclepiadaceae. 25 spp. Old World Tropics, S.Africa, Australia. Vines.

***Gymnema sylvestre*** R. Br. E.India, Australia, Tropical Africa. The leaves have been used since antiquity to treat diabetes mellitus in India. The leaves have an anti-saccharin effect due to the presence of gymnemic acid. They are chewed or extracted in water.

***G. syringifolium*** Boerl. (Sajor pepe). Malaya. The leaves are eaten locally as a vegetable. The roots are used to treat poisoning from fish and shortness of breath.

GYMNOCLADUS Lam. Leguminosae. 2 spp. Assam, Burma, China; 1 sp. N.America. Trees.

***G. canadensis*** Lam. = G. dioica.

***G. chinensis*** Baill. Central China. The soaked, crushed pods (Fei-tsao-tou) are used as a soap substitute for washing delicate fabrics. The ground seeds are mixed with camphor, musk, cloves, sandalwood and putchuck to make a perfumed toilet soap (P'ing-she fei-tsao).

***G. dioica*** (L.) Koch. = G. canadensis. (Kentucky Coffee Tree). E. U.S.A. The light brown wood is coarse, heavy and durable. It is used for fencing, railway sleepers and furniture-making. The seeds are roasted as a coffee substitute.

GYMNOGONGRUS Harv. Phyllophoraceae. Marine Red Algae. Pacific.

***G. flabelliformis*** Harv. Pacific. Eaten as a food in Japan.

***G. pinnulata*** Harv. Pacific Coasts. Eaten as a food in Japan, where a paste of the thallus is also used to wash the hair.

GYMNOPETALUM Arn. Cucurbitaceae. 3 spp. S.China, Indomalaya.

***G. leucostictum*** Miq. (Kemarogan). Indonesia. Procumbent herb. A decoction of the leaves is used locally to stimulate the appetite after illness.

***G. quinquelobatum*** Miq. (Timpoot poolau). Malaya. Vine. The unripe fruits are eaten locally as a vegetable in soup or with rice.

GYMNOPODIUM Rolfe. Polygonaceae. 3 spp. Central America. Shrubs.

***G. antigonoides*** (Rob.) Blake. Mexico. The flowers produce an excellent honey and the wood makes a fine charcoal.

GYMNOSPERMA Less. Compositae. 1 sp. S. United States of America to Central America. Shrub.

***G. corymbosum*** DC.=G. glutinosum.

***G. glutinosum*** Less.=G. corymbosum= Selloa glutinosa. (Cola de zorra, Escobilla, Jarilla, Mariquita, Tatalencho). A decoction of the plant is used internally to treat dysentery and a decoction of the gum is used externally to treat a wide variety of complaints including rheumatism and ulcers. A popular home remedy in Mexico.

GYMNOSPORIA (Wight and Arn.) Benth. and Hook. f. Celastraceae. 100 spp. Tropics and subtropics, especially Africa. Shrubs.

***G. montana*** Benth.=Celastrus senegalensis. Tropical Africa. A decoction of the stem is used to treat dysentery and stomach complaints of children in Senegal.

***G. senegalensis*** Loes. Tropical Africa, India, Mediterranean. The root bark is used locally to treat dysentery. The ashes are used in the Sudan instead of salt.

***G. senegalensis*** (Lam.) Loes.=Maytenus senegalensis.

GYNANDROPSIS DC. Cleomaceae. 1 sp. Tropics and subtropics.

***G. gynandra*** (L.) Briq. The seeds contain an oil like mustard oil. They are used as a condiment and to treat intestinal worms. The leaves are eaten as a vegetable in India and Nigeria.

GYNERIUM Humb. and Bonpl. Graminae. 1 sp. Mexico to subtropical America. Tall grass.

***G. argenteum*** Nees.=Cortaderia argentea.

***G. asgittatum*** (Aubl.) Beauv.=Aira gigantea. (Wild Cane). The stems are used locally to make matting, baskets, hats, fans etc. The growing buds are used as a soap for washing the hair.

GYNOCARDIA R. Br. Flacourtiaceae. 1 sp. Assam to Burma. Tree.

***G. odorata*** R. Br. (Chaulmoogra, Kushthapa). The seeds produce an oil (False Chaulmoogra Oil) which was used to treat skin complaints, especially leprosy. The fruits are used as a fish poison.

GYNURA Cass. Compositae. 100 spp. Tropical Africa and Madagascar to E.Asia and Malaysia. Herbs.

***G. cernua*** Benth. Tropical Africa. A decoction of the leaves is used in Madagascar to treat a wide variety of complaints. The leaves are eaten as a vegetable in Nigeria.

***G. finlaysoniana*** DC.=G. sarmentosa.

***G. japonica*** Mak.=G. pinnatifida.

***G. pinnatifida*** DC.=G. japonica=Senecio japonicus. (San eh'i, Tien ch'i). Cultivated locally for the roots which are used to treat wounds and haemorrhages

***G. sarmentosa*** DC.=G. finlaysoniana= Cacalia cylindriflora. India, throughout Indonesia. The leaves are used as a condiment in Malaya. A decoction of the leaves is used in Cambodia to treat fevers, in Malacca to treat dysentery and in Java to treat kidney complaints.

GYPSOPHILA L. Caryophyllaceae. 125 spp. Eurasia (especially E.Mediterranean), Egypt; 1 sp. Australia, New Zealand.

***G. arrostii*** L. (White Soup Root). S.Italy. The root (Radix Saponariae albae), contains large amounts of saponin. It is used to treat skin complaints and to induce sweating and as a diuretic.

***G. paniculata*** L. (Baby's Breath). South and Central Europe to Caucasus. The root (Radix Saponariae levanticae), contains saponin and is used as a purgative.

***G. struthium*** Fisch. See G. paniculata. The root is called Radix Saponariae hispanicae.

GYRINOPS Gaertn. Thymelaeaceae. 1 sp. Ceylon; 7 spp. E.Malaysia. Small trees.

***G. cumingiana*** Decne. Philippines, Celebes. The bark and roots are used locally to stop bleeding and the bark and wood are used as a quinine substitute.

***G. walla*** Gaertn. Ceylon. The soft, light white wood is used for cabinet work and inlays.

***Gyrinopsis*** Decne = Gyrinops.

GYROCARPUS Jacq. Gyrocarpaceae. 7 spp. Tropics and subtropics. Trees.

***G. americanus*** Jacq. Tropics. The white wood is soft and light. It is used for making toys and boxes.

***Gyromitra*** Fr. = Helvella.

GYROPHORA Ach. Lichen.

***G. cylindrica*** (L.) Ach. = Umbilicaria cylindrica. Temperate and sub-arctic. It is used in Iceland to make a green-brown dye to colour woollens.

***G. deusta*** (L.) Ach. = Umbilicaria flocculosa. (Rock Tripe). Temperate and subarctic. Used, particularly in Sweden, to make a purple dye for wool. It is also used to make paint (Tousch).

***G. dilleni*** (Tuck.) Arn. N. and S.America, N.Europe. Used to make a jelly like blancmange. It contains a bitter principle which can be removed with soda water. The high starch content makes the plant a useful emergency food.

***G. esculenta*** Miyoshi = Umbilicaria esculenta. Japan. A local delicacy (Iwa-take, Rock Mushroom).

***G. hyperborea*** Ach. See G. dilleni.

***G. muhlenbergii*** Ach. (Rock Tripe, Tripe de Roch). See G. dilleni.

***G. vellea*** (L.) Ach. Temperate. Used in Sweden to make a purple dye for wool, also see G. dilleni.

***Gyroporus*** Quél. = Boletus.

# H

Habbak Daseino Bdellium – Commiphora hildebrandtii.

Habbak Dundas Bdellium – Commiphora hildebrandtii.

Habbak Dunkai Bdellium – Commiphora hildebrandtii.

Habbak Harr Bdellium – Commiphora hildebrandtii.

Habbak Ilka Adaxai Bdellium – Commiphora hildebrandtii.

Habbak Tubuk Bdellium – Commiphora hildebrandtii.

HABENARIA Willd. Orchidaceae. 600 spp. Old and New World, tropics and subtropics. Herbs.

***H. commelinifolia*** Wall. ex Lindl. W.Himalayas. The tuberous roots are used locally to make a gruel.

***H. foliosa*** R. Br. Tropical Africa. The Zulus use an infusion of the roots as an emetic.

***H. menziesii*** (Lindl.) Macoun. = H. orbiculata.

***H. multipartita*** Blume. (Oowi oowi). Java. The roots are eaten locally.

***H. orbiculata*** (Pursh) Torr. = H. menziesii = Orchis orbiculata (Heal All, Large Round leaved Orchid, Moonset). N.America. The leaves are used as a home remedy to treat blisters.

***H. rumphii*** Lindl. Amboina. The tubers are made into a preserve in Indonesia.

***H. sparciflora*** S. Wats. S.W. N.America and Mexico. The plant is used as a famine food by the Indians.

Habhel – Juniperus drupacea.

***Habzelia obtusifolia*** A. DC. = Xylopia obtusifolia.

Hackberry – Celtis occidentalis.

Hackberry, European – Celtis austris.

Hackberry, Western – Celtis reticulata.

Haddi Tree – Commiphora erythraea var. glabrescens.

Hadochoon – Macaranga mappa.

HAEMANTHUS L. Amaryllidaceae. 50 spp. Tropical and S.Africa. Herbs.

***H. coccineus*** L. S.Africa. The bulbs, pickled in vinegar, are used by the farmers as an expectorant to treat asthma and a diuretic to treat dropsy.

***H. natalensis*** Pappe. (Blood Flower, Snake Lily). The root is poisonous and is used by the Zulus as an expectorant and emetic.

***H. toxicarius*** Thunb. = Boophone disticha.

***Haemarthria compressa*** R. Br. = Rottboellia compressa.

***Haemarthria guyanense*** Steud. = Rottboellia compressa.

***Haemarthria fasiciculata*** Kunth. = Rottboellia fasciculata.

HAEMATOMMA Lecanoraceae. Lichen.

***H. ventosum*** (L.) Mass = Lecanora ventosa. (Black Lecanora, Bloody Spotted Lecanora). N.Temperate, on rocks. The lichen is used in Sweden to dye wool a rust brown.

Haematoxylin – Heamatoxylum campechianum.

HAEMATOXYLUM L. Leguminosae. 3 spp.

Mexico, Central America, W.Indies, S.W.Africa. Trees.

***H. brasiletto*** Karst. (Brazilwood, Nicaragua Wood). Tropical America. The bright orange wood is hard. The dye darkens to red on exposure when it is called Brasilin. The wood is exported from Mexico.

***H. campechianum*** L.=H. pechianum. (Lignum Campechianum, Logwood, Palo Campechio). Tropical America. The red wood darkens to brown or purple on exposure. It is used for furniture-making and fancy articles. The heartwood yields the purple dye Haematoxylin which is used for dyeing wool and for making stains for microscope preparations. An extract of the wood is also used to treat dysentery.

***H. pechianum*** L.=H. campechianum.

***Haemocharis integerrima*** Koord and Val.= Laplacea intergerrima.

HAEMODORUM Sm. Haemodoraceae. 20 spp. Australia. Herbs.

***H. apiculatum*** R. Br. Australia. The plants are used by the aborigines to treat dysentery.

***H. coccineum*** Hook. (Blood Root). Australia. The roots are eaten by the aborigines to stimulate the appetite.

***H. paniculatum*** R. Br. Australia. The roots are eaten roasted by the aborigines.

***H. planifolium*** R. Br. Australia. The roots are eaten roasted by the aborigines.

***H. terefolium*** R. Br. Australia. The roots are eaten roasted by the aborigines.

Haemorrhage Plant – Aspilia latifolia.

Hafotra – Dombeya cannabina.

Hafotra – Dombeya perrieri.

Hafotra – Grewia faucerei.

Hagar Ad Tree – Commiphora hildebrandtii.

HAGENIA J. F. Gmel. 1 sp. Abyssinia. Tree.

***H. abyssinica*** J. F. Gmel.=Brayera anthelmintica. (Cusso). Cultivated locally for the pistillate flowers (Flores Koso), which are used to treat intestinal worms.

Hagla – Typha elephantina.

Haiari, White – Lonchocarpus densiflorus.

Haidu – Adina cordifolia.

Hairi – Casearia tomentosa.

Hair Cap Moss – Polytrichum juniperinum.

Hairy Angelica – Angelica villosa.

Hairy Vetch – Vicia villosa.

HAKEA Schrad. Proteaceae. 100 spp. Australia. Trees.

***H. leucoptera*** R. Br. (Needle Bush, Pine Bush). Australia. The wood is soft, but polishes well. It is used for veneers and to make tobacco pipes. The roots are used by the aborigines as an emergency source of water.

***H. rubicaulis*** Colla.=Stenocarpus salignus.

Haladra – Adina cordifolia.

Halapepe – Dracaena aurea.

Hal-dummala (Resin) – Vateria acuminata.

Haleki laoon ooloon – Mallotus trifoliata.

Halfa – Stipa tenacissima.

HALFORDIA F. v. Muell. Rutaceae. 4 spp. New Guinea, E.Australia, New Caledonia. Trees.

***H. scleroxylon*** F. v. Muell. (Ghittoe, Green Wood, Kerosine Wood, Saffron Wood). Queensland. The wood is pale yellow, hard, close-grained and durable. It is used for fishing rods and textile rollers.

Halganne – Palaquium ellipticum.

HALIMIUM (Dunal) Spach emend Willk. Cistaceae. 14 spp. Mediterranean. Shrubs.

***H. glomeratum*** (Lag.) Grosser.=Cistus glomerata=Helianthemum glomeratum =Helianthemum astylum. (Jumita). C.America. A decoction of the plant is used in Mexico to treat stomach upsets and diarrhoea.

HALORAGIS J. and R. G. Forst. Haloragidaceae. 1 spp. Madagascar; 75 spp. E.Asia, Indomalaya through to Australia, Tasmania, New Zealand, Pacific Islands, Juan Fernandez and Chile. Herbs.

***H. erecta*** Eichl. New Zealand, Chile. An infusion of the plant is used locally in New Zealand to treat scrofula.

HALOXYLON Bunge. Chenopodiaceae. 10 spp. W.Mediterranean to Mongolia, S. to Persia, Afghanistan, Burma and S.W.China.

***H. aphyllum*** (Minkw.) M. M. Iljin. Central Asia. Tree. The plant is grown as a sand binder and the wood is used for general carpentry.

***H. articulatum*** Bunge.=H. tamaricifolium.

***H. persicum*** Bunge. See H. aphyllum.

***H. schweinfurthii*** Aschers. Arabia. The plant yields a manna eaten locally.

***H. tamaricifolium*** (L.) Maire.=H. articulatum (Ichufa, Roummef). Sahara. The plant is used as firewood. The ash is mixed with tobacco and used locally as snuff.

***H. tetrandrus*** Moq.=Salsola tetrandra. (Adjerem) Sahara. A valuable local source of firewood.

HALYMENIA Harv. Grateloupiaceae. Red Algae.

***H. formosa*** Harv. Pacific. Used locally as a food in the Philippines.

HAMAMELIS L. Hamamelidaceae. 6 spp. E.Asia, E. N.America. Small trees.

***H. virginiana*** L. (Witch Hazel). E. N.America to Florida and Texas. The bark and leaves are produced commercially in Virginia, S.Carolina and Tennessee. They are used medicinally to reduce swelling, especially in the treatment of bruises, varicose veins and haemorrhoids. They contain gallotannic acid, hamamelitannin and gallic acid. The bark is sold as Witchhazel Bark or Hamamelis Cortex and the decoction as Extract of Witchhazel or Hamamelis Water.

Hamamelis Cortex – Hamamelis virginiana.

Hamamelis Water – Hamamelis virginiana.

Hamati – Cavanillesia platanifolia.

Hammez – Pisum elatius.

Hamperoo – Voacanga foetida.

***Hancea muricata*** Benth. = Mallotus furetianus.

HANCORNIA Gomez. Apocynaceae. 1 sp. Brazil. Small tree.

***H. speciosa*** Gomez. (Mangabeira). The tree is tapped for an inferior rubber (Mangabeira Rubber), but the yield is so small as to warrant its collection only in times of emergency. The fruits are used to make a marmalade in parts of Brazil.

HANGHOMIA Gagnep. and Thénint. Apocynaceae. 1 sp. Indomalaysia. Woody vine.

***H. marseillei*** Gagnep. and Thénint. (Hang kan). The roots are burnt as incense in local pagodas and the pounded roots are used to make a liquid for the ceremonial washing of priests' feet. The root powder is also used to make a cosmetic used by men and women on these ceremonial occasions.

Hang kan – Hanghomia marseillei.

Hanjigoad – Balanites orbicularis.

***Hannoa*** Planch. = Quassia.

***Hansenia*** Lindner = Hanseniospora.

HANSENIOSPORA Zikes. Saccharomycetaceae.

***H. apiculata*** (Rees.) Lindner. Found on Grapes. It is supposed to add a particular taste and bouquet to wines.

Hantap badak – Sterculia javanica.

Hanudun – Aesculus indica.

Haoo loonis – Grewia salutaris.

***Hapalopilus*** Karst. = Polyporus.

HAPLOCLATHRA Benth. Guttiferae. 4 spp. N.Brazil. Trees.

***H. paniculata*** Benth. N.Brazil. The handsome red wood (Mura Piranga), is used to make musical instruments.

HAPLOPAPPUS Cass. corr. Endl. Compositae. 150 spp. W.America. Herbs.

***H. nuttallii*** Torr. and Gray. = Eriocarpum grindelioides. W. N.America. The roots are used by the local Indians to make a beverage and to treat coughs.

***H. spinulosus*** (Pursh.) DC. = Sideranthus spinulosus. S.W. N.America. The leaves and roots are used by the local Indians to treat toothache.

HAPLOPHYTON A. DC. Apocynaceae. 3 spp. S.W. United States of America, Mexico, Cuba. Woody plant.

***H. cimicidum*** A. DC. (Herba de la Cucaracha). S.Arizona, throughout Central America and Cuba. In Mexico, a decoction of the leaves with maize meal is used as an insecticide to kill cockroaches and human parasites.

Harafa – Raphia pedunculata.

Hahahara – Phylloxylon ensifolius.

Harcuga – Bidens chinensis.

Hard Pear – Strychnos henningsii.

Harding Grass – Phalaris tuberosa.

HARDWICKIA Roxb. Leguminosae. 2 spp. India. Trees.

***H. binata*** Roxb. (Acha ajan, Passid). Central and S.India. The dark red wood is very heavy, tough, durable and polishes well. It is used for house- and bridge-building, agricultural implements, bearings of various sorts, looms and carving. The bark is used for tanning and yields a fibre used for cordage, sail-making and paper-making. The stem gives a resin used locally as a wood preservative. The leaves are fed to livestock.

***H. manna*** Oliver = Oxystigma mannii.

***H. pinnata*** Roxb. India. The stem yields a Copaiba Balsam.

Hardy Timber Bamboo – Phyllostachys bambusioides.

Hare's Lettuce – Sonchus oleraceus.

Hari – Sclerocarya birraca.

Harick – Forsskalea tenacissima.

Haricot Bean – Phaseolus vulgaris.

Harijang gede – Gunnera macrophylla.

Haringhing – Cassia timoriensis.

Harinmeng gede – Cratoxylum racemusum.

Harmela – Peganum harmela.

Harobol Myrrh – Commiphora myrrha.
***Haronga*** Thou. = Harungana.
HARPAGOPHYTUM DC. ex Meissn. Pedaliaceae. 8 spp. S.Africa, Madagascar. Herbs.
***H. procumbens*** DC. (Grapple Plant, Kloudoring, Wool Spider). S.Africa. The tubers are used as a medicine by the natives and in some remedies by the Europeans. They are used as an infusion to treat fevers and as a purgative and as an ointment on boils etc.
HARPEPHYLLUM Bernh. ex Krauss. Anacardiaceae. 1 sp. S.Africa. Tree.
***H. caffrum*** Bernh. (Cape Ash, Kaffir Plum). The red fruits have an acid, juicy pulp and are used locally to make jellies. The bark is used by the natives as an emetic and to purify the blood.
HARPULLIA Roxb. Sapindaceae. 37 spp. Indomalaya, tropical Australia, Pacific. Trees.
***H. arborea*** (Blanco) Radlk. Philippines. The bark is used locally as a soap substitute.
***H. pendula*** Planch. (Mogun, Mogun, Tulip Wood). Queensland, New South Wales. The wood is beautifully marked in black to yellow and is used in cabinet work.
Harra – Diplotaxis duveyrierana.
HARRISONIA R. Br. Simaroubaceae. 4 spp. Tropical Africa, S.E.Asia, Indomalaya, tropical Australia. Shrubs.
***H. perfoliata*** (Blanco) Merr. = Paliurus perfoliatus. Tropical and S.E.Asia. The bark is used in the Philippines to treat dysentery, various forms of diarrhoea and cholera.
Hart's Tongue – Phyllitis vulgare.
Hart's Truffles – Elaphomyces cervinus.
Hart's Wort – Peucedanum cervaria.
HARUNGANA Lam. Guttiferae. 1 sp. Tropical Africa, Madagascar, Mauritius. Small tree.
***H. paniculata*** (Pers.) Lodd. The plant yields a red oil which becomes sticky on exposure to air. It is used locally to treat skin diseases and as a bird lime (Nguba).
Hats, Balinag – Corypha elata.
Hats, Calaseao – Corypha elata.
Hats, Lacban – Corypha elata.
Hats, Panama – Carludovica palmata.
Hats, Panama of Madagascar – Cyperus nudicaulis.
Hats, Pototan – Corypha elata.
Hats, Puerto Rican – Sabal causiarum.
Hats, Salokots – Heterospathe elata.
Hats, Sola-Topis – Aeschynomene aspera.
Hats, Sun Helmet – Aeschynomene aspera.
Hats, Torquilla – Carludovica palmata.
Hatsikana – Xerochlamys pilosa.
Hausa Potato – Coleus rotundifolius.
Hautbois Strawberry – Fragaria moschata.
Havra – Eruca pinnatifida.
Haw, Black – Crataegus douglasii.
Haw, Black – Viburnum prunifolium.
Haw, Chinese – Crataegus pentagyna.
Haw, Summer – Crataegus flava.
Hawaiian Arrowroot – Tacca hawaiiensis.
Hawaiian Coral Tree – Erythrina sandwicensis.
Hawaiian Falsenettle – Boehmeria grandis.
Hawaiian Jacquemomtia – Jacquemontia sandwicensis.
Hawaiian Olive – Osmanthus sandwicensis.
Hawaiian Poppy – Argemone glauca.
Hawaiian Sandalwood – Santalum freycinetianum.
Hawar – Dolichandrone falacata.
Hawkbit – Leontodon hispidus.
Hawthorn – Crataegus oxyacanthus (oxyacanthoides).
Hawthorn Maple – Acer crataegifolium.
Haymaids – Glechoma hederaceum.
Hazel Alder – Alnus rugosa.
Hazel, Californian – Corylus californica.
Hazel, Chile – Gevuina avellana.
Hazel, European – Corylus avellana.
Hazel, Turkish – Corylus colurna.
Hazel Nut (Oil) – Corylus avellana.
Hazelcrottle – Lobaria pulmonaria.
Hazelnut, Beaked – Corylus rostata.
Hazelnut, Siberian – Corylus heterophylla.
Hazelraw – Lobaria pulmonaria.
Hazomena – Khaya madagascariensis.
Headache Tree – Premna integrifolia.
Heal-all – Habebaria orbiculata.
Heal-all – Prunella vulgaris.
Heapo – Ficus marquesensis.
Heapo – Ficus prolixa.
Heartleaf Hornbeam – Carpinus cordata.
Heartnut – Juglans sieboldiana.
Heath Tea Tree – Leptospermum ericoides.
Heather – Calluna vulgaris.
Heather, Cross-leaved – Erica tetralix.
Heavy Mahogany – Entandrophragma utile.
Heavy Sapele – Entandrophragma candollei.
***Heckeria*** Kunth. = Pothomorphe Miq.
HEDEOMA Pers. Labiatae. 30 spp. America. Herbs.
***H. incana*** Torr. = Poliomintha incana.
***H. pulegioides*** (L.) Pers. (American Pennyroyal). N.America. The dried leaves and

flower heads are used medicinally and in home remedies to treat colds and digestive complaints.

HEDERA L. Araliaceae. 15 spp. Canary Islands, W. and C.Europe to Mediterranean to Caucasus, W.Himalayas to Korea and Japan, Queensland. Woody vines.

***H. helix*** L. (English Ivy). Europe to temperate Asia. Cultivated as an ornamental. The leaves boiled with soda are used as a soap substitute in washing clothes. The twigs yield a yellow-brown dye. The wood is hard and suitable for carving.

Hedge Garlic – Sisymbrium alliaria.

Hedge Hyssop – Gratiola officinalis.

Hedge Maple – Acer campestre.

Hedge Mustard – Sisymbrium officinale.

Hedge Sison – Sison amomum.

***Hedriostylus corniculatus*** Hassk. = Plukenetia corniculata.

HEDYCARYA J. R. and G. Forst. Monimiaceae. 25 spp. S.E.Australia to Solomon Islands and Fuji. Trees.

***H. angustifolia*** Cunningh. = H. australasica = H. cunninghamiana. (Native Mulberry, Smooth Holly). Victoria, New South Wales, Queensland. The light, tough, close-grained wood is used for cabinet-making.

***H. arborea*** Forst. New Zealand. The leaves are used locally in vapour baths.

***H. australasica*** A. DC. = H. angustifolia.

***H. cunninghamiana*** Tul. = H. angustifolia.

HEDYCHIUM Koen. Zingiberaceae. 50 spp. Madagascar. Indomalaya, S.W.China. Herbs.

***H. coronarium*** Koenig. (Garland Flower). India, Malaya. A possible source of paper pulp.

***H. longicornutum*** Griff. E.Indies, Malaysia. A decoction of the roots is used locally to treat ear ache.

***H. spicatum*** Ham. E.India. The rhizome is used in some parts of tropical Asia in the manufacture of perfume.

***Hedyctis capitellata*** Wall. = Oldenlandia recurva.

HEDYOSMUM Sw. Chloranthaceae. 1 sp. S.China; 40 spp. Tropical America, W.Indies. Shrubs.

***H. bonplandianum*** Mart. = H. brasiliense.

***H. brasiliense*** Mart. = H. bonplandianum. Brazil. A decoction of the leaves is used locally as a diuretic, tonic, aphrodisiac, to induce sweating and to treat stomach complaints.

HEDYSARUM L. Leguminosae. 150 spp. N.Temperate. Herbs.

***H. alpinum*** var. ***americanum*** Britt. (Liquorice Root). U.S.A. and Canada to Arctic. The cooked roots are an important food for the Eskimoes.

***H. coronarium*** L. (Spanish Espercette). Mediterranean. The plant is occasionally cultivated as food for cattle.

***H. mackenzii*** Rich. (Liquorice Root). Central Canada to Alaska. The roots are sweet and are eaten during the Spring by the local Indians.

HEERIA Meissn. Anacardiaceae. 1 sp. S.Africa. Tree.

***H. reticulata*** (Bak. f.) Engl. The leaves are used by the natives to encourage lactation of women and as an aphrodisiac.

***Hegba mboddo*** – Boletus sudanicus.

***Heimia salicifolia*** (H.B.K.) Link. = Nesaea salicifolia.

HEINSIA DC. Rubiaceae. 10 spp. Tropical Africa. Trees.

***H. pulchella*** Schum. Tropical W.Africa. The fruits (Bush Apples) have a pleasant flavour and are eaten locally. The leaves are scented and are used by the local women as a perfume. The wood is elastic and is made into tool handles and spring traps.

Hejas – Eugenia acuminatissima.

HELENIUM L. Compositae. 40 spp. W.America. Herbs.

***H. grandiflorum*** Gilib. = Inula helenium.

***H. puberulum*** Nutt. (Rosilla). W. U.S.A. The leaves are used in home remedies as a tonic, in a snuff for catarrh and as an antiscorbutic.

***H. tenuifolium*** Nutt. (Bitter Sneezewort). N.America. The leaves are used by the local Indians to clear the nose of mucous.

HELIANTHEMUM Mill. Cistaceae. 100 spp. W. and C.Europe, Mediterranean, S. to Cape Verde Islands, Sahara and Somaliland, E. to Central Asia and Persia. Herbs.

***H. astylum*** Dunal. = Halimium glomeratum.

***H. canadense*** (L.) Michx. (Frostweed). The dried leaves are used as a tonic and to stimulate the appetite.

***H. glomeratum*** Lag. = Halimium glomeratum.

HELIANTHUS L. Compositae. 110 spp. America. Herbs.

***H. annuus*** L. (Sunflower). W. N.America. Cultivated mainly in Russia and S.E. Europe and to a lesser extent in India and China. More recently cultivation has increased in Argentina. The plant is grown for the seeds from which Sunflower Seed Oil is extracted. The world production of sunflower seeds exceeds 7 million tons annually. The oil has Sp. Gr. 0·922–0·926, Sap. Val. 189–194, Iod. No. 120–136, Unsap. 0·7–1·20 per cent. It is semi-drying and contains mainly glycerides of oleic, linoleic and palmitic acids. The linoleic acid content is reduced from about 70 per cent to 20 per cent with an increase in the environmental temperature during growth, while the oleic acid content increases from 15 to 65 per cent. The seeds contain between 32 per cent and 45 per cent oil and the meal left after extraction contains about 50 per cent protein, making it a valuable feed for stock. The oil is used as a salad oil, in making margarine, for varnishes and paints and for burning. The seeds are also eaten roasted and are used medicinally as an expectorant and diuretic. The flowers are an excellent source of honey and of a yellow dye used by the local Indians. The stems and leaves are made into silage for livestock and when burnt give a good yield of potassium salts. The pith from the stems is used in cutting sections for making microscope slides.

***H. doronicoides*** Lam. (Oblonghead Sunflower). W. U.S.A. The tubers are eaten by the local Indians.

***H. giganteus*** L. N.America. The seeds are made into a flour by the local Indians.

***H. lenticularis*** Douglas. Texas to Mexico. The seeds are eaten by the local Indians.

***H. maximiliani*** Schrad. Central U.S.A. The swollen roots are eaten by the local Indians.

***H. tuberosus*** L. (Jerusalem Artichoke, Topinambour). N.America. The swollen roots are eaten as a vegetable or in pickles. They contain the carbohydrate inulin and are used in dietary preparations. They were cultivated by the Indians of N.America and were introduced to Europe early in the 17th century. The plant is little cultivated in U.S.A., but is grown in France as a livestock feed.

HELICHRYSUM Mill. corr. Pers. Compositae. 500 spp. S.Europe, tropical and S.Africa, Madagascar, Socotra, S.W.Asia, S.India, Ceylon, Australia. Herbs.

***H. arenarium*** Moench. Europe, Caucasus. The flower heads were used locally to treat intestinal worms, skin diseases and as a diuretic.

***H. cochinchinensis*** Spreng. Cochin-China. The young shoots are used locally in making a rice dish (Bârh Khúc).

***H. serpyllifolium*** Lessing. S.Africa. The leaves are used locally to make a tea (Hottentot Tea).

***H. stoechas*** DC. See H. arenarium.

HELICONIA L. Heliconiaceae. 80 spp. Tropical America. Herbs.

***H. bihai*** L. = H. caribaea. Tropical America. The leaves are used for thatching and are of possible value in paper-making. The young shoots are eaten as a vegetable in the W.Indies.

***H. brasiliensis*** Hook. Brazil. The roots are used locally to treat gonorrhoea and the seeds to treat diarrhoea.

***H. caribaea*** Lam. = H. bihai.

***H. schliedeana*** Klotsch. S.Mexico. The leaves are used as a local thatching material and for wrapping.

HELICTERES L. Sterculiaceae. 60 spp. Tropical Asia and America. Trees.

***H. guazumaefolia*** H.B.K. Central and S.America. W.Indies. The bark fibre is used in Cuba to make sandals (Alpargatas).

***H. isora*** L. (East Indian Screw Tree). Tropical Asia. The bark fibre is used locally for cordage. The pods are used locally in the treatment of dysentery, diarrhoea, flatulence and to improve the appetite. They are sold in bazaars.

HELIOCARPUS L. Tiliaceae. 17 spp. Mexico to Paraguay. Trees. The following species occur mainly in Mexico, where the light wood is used for floats and bottle corks. The beaten bark is used as paper and a decoction of the bark is used as an ointment for sores. The young branches are used as coarse ropes and made into hammocks. The plants are known as Jolocin, Calagula, Catena, Majahua and Xolotzin.

***H. appendiculatus*** Turcz.

***H. donnel-smithii*** Rose.

***H. glanduliferus*** Robin.

***H. tomentosus*** Turcz.

HELIOPHILA Burm. f. ex L. Cruciferae. 75 spp. S.Africa. Herbs.

***H. suavissima*** Burch. (Semamelwana). S.Africa. An infusion of the leaves are used by witch doctors before using divining bones.

HELIOTROPIUM L. Boraginaceae. 250 spp. Tropical and temperate. Herbs.

***H. anchusaefolium*** Poir. Argentina. A decoction of the leaves is used locally to induce sweating.

***H. angiospermum*** Murr. = H. parviflorum. (Nemaax, Rabo de Nico). Tropical America. An infusion of the leaves is used in Mexico as a tonic and to treat dysentery.

***H. elongatum*** Willd. Brazil. A decoction of the leaves is used locally to relieve asthma and as a diuretic.

***H. parviflorum*** L. = H. angiospermum.

***H. tuberculosum*** Boiss. (Jatimisak, Pi-patbuti). N.Indian desert regions. A decoction of the plant is used to treat insect and scorpion bites and to treat the eyes of camels.

Hellebore – Helleborus niger.

Hellebore, American False – Veratrum viride.

Hellebore, American White – Veratrum viride.

Hellebore, False – Veratrum californicum.

Hellebore, White – Veratrum album.

Hellebore, White False – Veratrum album.

Hellebornine, Giant – Epipactis giganteum.

HELLEBORUS L. Ranunculaceae. 20 spp. Europe, Mediterranean to Caucasus; 1 spp. W.China. Herbs.

***H. niger*** L. (Christmas Rose, Hellebore). Europe. The rhizome and roots contain helleborine and helleborein. They are used medicinally as a drastic laxative, as a heart stimulant and to control menstruation.

Helmet Flower – Aconitum anthora.

HELMINTHOSTACHYS Kauf. Ophioglossaceae. 10 spp. Ceylon, Himalayas to Queensland. Ferns.

***H. zeylanica*** Hook. f. Throughout range. The young fronds are eaten as a vegetable and in salads in the Philippines. The rhizome is used to treat malaria and eaten with betel as a tonic. The rhizome is exported to China.

Helonias – Chanaelirium luteum.

HELVELLA L. ex Fr. Helvellaceae. 25 spp. N.Temperate. The fruit bodies of the following species are eaten locally.

***H. crispa*** (Scop.) Fr. Eaten in Asia.

***H. esculenta*** (Pers.) Fr. Europe. The fruit bodies are dried for winter use. They contain helvellic acid which is poisonous and can only be removed by treatment. Their sale is forbidden

***H. gigas*** Krombh. Eaten in parts of Asia.

***H. infulva*** Schaff. Eaten in parts of Asia.

***H. lacunosa*** Afzel. Eaten in parts of Asia.

HELWINGIA Willd. Helwingiaceae. 4–5 spp. E.Himalayas to Japan, Formosa. Shrubs.

***H. rusciflora*** Willd. China, Japan. The leaves are eaten as a vegetable in Japan.

Henero Tea – Polygonum aviculare.

HEMEROCALLIS L. Liliaceae. 20 spp. Temperate Eurasia, especially Japan. Herbs.

***H. fulva*** L. (Tawny Daylily). Europe and Asia. The dried flowers are used as a condiment in China and Japan.

***H. graminae*** Andr. = H. minor.

***H. minor*** L. = H. graminae. China, Japan. The flowers are eaten locally.

***Hemicycla ovalis*** J. J. Sm. = Drypetes ovalis.

HEMIDESMUS R. Br. Periplocaceae. 1 sp. S.India. Woody plant.

***H. indicus*** R. Br. = H. wallichii. The plant is used to make Indian Sarsparilla which is used locally as a diuretic, demulcent and to improve the appetite.

***H. wallichii*** Miq. = H. indicus.

***Hemitelia*** R. Br. = Cyathea.

Hemlock, Chinese – Tsuga chinensis.

Hemlock, Eastern – Tsuga canadensis.

Hemlock, Ground – Taxus canadensis.

Hemlock, Oil – Tsuga canadensis.

Hemlock, Pitch – Tsuga canadensis.

Hemlock, Poison – Conium maculatum.

Hemlock, Western – Tsuga heterophylla.

Hemlock, Yunnan – Tsuga yunnanensis.

Hemp (Seed Oil) – Cannabis sativa.

Hemp, African Bowstring – Sanseverinia senegambica.

Hemp, Ambari – Hibiscus cannabinus.

Hemp, Bahama – Agave sisalana.

Hemp, Bombay – Agave cantala.

Hemp, Ceylon Bowstring – Sansenerinia zeylandica.

Hemp, Deccan – Hibiscus cannabinus.

Hemp, Dogbane – Apocynum cannabinum.

Hemp, Eupatorium – Eupatorium cannabinum.

Hemp, Gambo – Hibiscus cannabinus.

Hemp, Indian – Apocynum cannabinum.

Hemp, Manila – Musa textilis.

Hemp, Mauritius – Furcraea gigantea.

Hemp, Queensland – Sida rhombifolia.
Hemp, Rajmahal – Marsdenia tenacissima.
Hemp, Sisal – Agave sisalana.
Hemp, Sunn (San) – Crotalaria juncea.
Hempnettle – Galeopsis ochroleuca.
Hen and Chickens – Sempervivum tectorum.
Henbane – Hyoscyamus niger.
Henbane, Black – Hyoscyamus niger.
Henderson Shooting Star – Dodecatheon hendersonii.
Henequen – Agave fourcroyoides.
Henequen Agave – Agave fourcroyoides.
Henequen, Salvador – Agave letonae.
Henna – Lawsonia inermis.
Henna Shrub – Lawsonia inermis.
Henon Bamboo – Phyllostachys nigra.
Henry Chinquapin – Castania henryi.
Henweed, Guinea – Petiveria alliaceae.

***Hepatica nobilis*** Schreb. = Anemone hepatica.

***Heptapleurum aromaticum*** Seem. = Schefflera aromatica.

HERACLEUM L. Umbelliferae. 70 spp. N.Temperate. Herbs.

***H. lanatum*** Michx. (Cow Parsnip). W. N.America. The roots young flowers and stems were eaten as a vegetable.

***H. persicum*** Desf. Iran. The seeds are used locally as a condiment in pickles.

***H. sphonodylium*** L. (Cow Parsnip). Europe, Asia, N.America. The boiled leaves and fruits are fermented to make a drink (Bartsch), especially in C.Europe. They are used in France in making liqueurs.

***H. wallichii*** DC. Sikkim, Nepal. The roots are used locally as a tonic and aphrodisiac.

Herb Robin – Geranium robertianum.
Herba absinthii – Artemisia maritima.
Herba Acetosa – Rumex scutatus.
Herba Acanthi Germanici – Cnicus benedictus.
Herba Achilleae Clavenae – Achillea clavenae.
Herba Aconiti – Aconitum napellus.
Herba Anaphalidis – Anaphalis margaritacea.
Herba Anemone nemorosa – Anemone nemorosa.
Herba Angelicae – Angelica archangelica.
Herba Arenaeriae rubra – Spergularia rubra.
Herba Anserinae – Potentilla anserina.
Herba Asperulae odoratae – Gallium odorata.
Herba Auriculae – Primula auricula.
Herba Barbaraea – Barbarea vulgaris.
Herba Brachycladi – Brachyclados stuckertii.
Herba Callunae – Calluna vulgaris.
Herba Cardui benedicti – Cnicus benedictus.
Herba Cephalandrae – Coccinia indica.
Herba Cerefolii – Anthriscus cerefolium.
Herba Clematidis – Clematis vitalba.
Herba Equisetii – Equisetum arvense.
Herba Ericae – Calluna vulgaris.
Herba Erygimi – Sisymbrium officinale.
Herba et Flores Borragines – Borago officinalis.
Herba et Flores Filipendula – Filipendula hexapetala.
Herba et Flores Pilosellae sive Auriculae – Hieracium pilosella.
Herba et Radix Serratulae – Serratula tinctoria.
Herba Galeopsidis – Galeopsis ochroleuca.
Herba Guaco – Mikania guaco.
Herba Lapae minoes – Xanthium strumonium.
Herba Lycii – Lycium halamifolium.
Herba Mari veri – Teucrium marium.
Herba Matrisilvae – Gallium odorata.
Herba Paridis – Paris quadrifolia.
Herba Pirolae Secondae – Pyrola secunda.
Herba Pirolae uniflorae – Pyrola uniflora.
Herba Polygalae – Polygala amara.
Herba prenanthes serpentariae – Prenanthes serpentaria.
Herba Sancta – Verbena officinalis.
Herba Senecionis vulgaris – Senecio vulgaris.
Herba Spergulae rubrae – Spergularia rubra.
Herba Verbebae – Verbena officinalis.
Herba Vinae pervincae – Vinca minor.
Hercules' Club – Aralia spinosa.
Hercules' Club – Zanthoxylum clava-herculis.

***Hericium coralloides*** (L.) Pers. = Hydnum coralloides.

HERITIERA Dryand. Sterculiaceae. 35 spp. W. tropical Africa, Indomalaya, tropical Australia, Pacific. Trees.

***H. javanica*** Blume. Malaysia, Philippines. The heavy red-brown wood is used instead of mahogany in house-building and furniture-making. It is used locally for making canoes and boxes.

***H. littoralis*** Dryand. (Looking glass Trees). Old World Tropics. The wood is durable and strong, and is used for boat-building, wheel spokes and ploughs.

***H. minor*** Lam. (Sundri Tree). India, Mauritius. The wood is used to make charcoal and the seeds are eaten as a famine food.

***H. simplicifolia*** Mast. (Mengkulang). Malaysia. The wood is red-brown hard and heavy. It is not easy to work and not durable in contact with soil. It is used for interiors of houses, furniture-making and plywood.

***H. utilis*** Sprague. (African Mahogany). A good lumber tree, used as mahogany for general carpentry, boards and canoes.

HERMIDIUM S. Wats. Nyctaginaceae. 1 sp. S.W.United States of America. Herb.

***H. alipes*** S. Wats. A decoction of the roots is used by the Indians to treat headaches.

HERMINIERA Guill. and Perr. Leguminosae. 1 sp. Tropical Africa. Tree.

***H. elaphroxylon*** Guill. and Perr. The wood is light tough and durable. It is used widely to make rafts, beds etc.

HERNANDIA L. Hernandiaceae. 20 spp. Central America, Guyana, W.Indies, W.Africa, Zanzibar, Mascarene Islands, Indomalaya, Pacific. Trees.

***H. nukuhivensis*** F. Br. Pacific. The trunks are used locally to make canoes.

***H. ovigera*** L. (Kementing laut). Coasts of Malaysia, Borneo and Philippines. The wood is very soft and light. It works well but rots on contact with soil and is susceptible to termites and borers. It is used for boxes, crates, floats, cheap interior work, cheap plywood, drawing boards etc.

***H. pelata*** Meisen (Jack-in-the-Box). Pacific. The juice from the leaves is used locally for the painless removal of hair.

***H. sonora*** L. (Mago). Tropical America, W.Indies. See H. ovigera.

HERNIARIA L. Caryophyllaceae. 35 spp. Europe, Mediterranean to Afghanistan, S.Africa. Herbs.

***H. glabra*** L. (Rupture Wort, Herniary Breastwort). S.Europe. The leaves are used locally in the treatment of diarrhoea, as a diuretic and expectorant.

Herniary Breastwort – Herniaria glabra.

Herotin – Euphorbia intisy.

***Herpestis*** Gaertn. f. = Bacopa.

***Herrania albiflora*** Goud. = Theobroma albiflora.

***Herrania mariae*** Goud. = Theobroma mariae.

HESPERIS L. Cruciferae. 30 spp. Europe, Mediterranean to Persia, Central Asia, W.China. Herbs.

***H. matronalis*** L. (Damask). Europe to Central Asia. Cultivated locally for the seeds which yield an oil (Honesty Oil, Huile de Julienne, Rotreps Oil) which is used in perfumery.

HESPEROYUCCA Baker. Liliaceae. 1 sp. S.W.United States of America. Large herb.

***H. funifera*** (Koch.) Trel. = H. whipplei.

***H. whipplei*** (Torr.) Baker. = H. funifera = Yucca funifera = Y. whipplei (Chaparral Yucca, Zamandoque). Cultivated in Mexico for the long coarse leaf fibres (Ixtle, Tampico Fibre, Zamandoque Fibre) which is used to make twine and sacks. The flowers are eaten by the local Indians who use the seeds to make a porridge.

HETAERIA Blume. Orchidaceae. 20 spp. Old World Tropics. Herbs.

***H. obliqua*** Blume. Indonesia. An infusion of the leaves is used in Amboina to treat sores.

HETERODENDRUM Desf. Sapindaceae. 5 spp. Australia. Trees.

***H. oleifolium*** Desf. Australia. The fruits are eaten by the aborigines.

HETEROMORPHA Cham. and Schechtd. Umbelliferae. 10 spp. Tropical Africa, Madagascar. Shrubby herbs.

***H. arborescens*** Cham. and Schlechtd. Tropical Africa. The Kaffirs use an infusion of the inner root bark to treat stomach upsets.

HETEROPHRAGMA DC. Bignoniaceae. 2 spp. India, Indochina. Trees.

***H. macrolobum*** Baker = Spathodea macroloba (Soonkai). Indonesia. The wood is dense, fine, durable and brownish-white. It is used for boards and beams.

HETEROPOGON Pers. Graminae. 12 spp. Tropics. Perennial Grasses.

***H. contortus*** (L.) Beauv. = H. hirtus.

***H. hirtus*** Pers. = H. contortus = Andropogon contortus. (Spearhead, Tanglegrass). Old and New World Tropics. Grown as a fodder for livestock. It is used in India for making mats and thatching and in Hawaii for making grass houses.

HETEROPTERYS Kunth. emend Griseb. Malpighiaceae. 100 spp. Tropical America; 1 sp. W. tropical Africa. Lianas.

***H. glabra*** Hook. and Arn. = H. umbellata.

***H. umbellata*** Juss. = H. glabra. Argentina. The stem fibres are used locally for cordage.

HETEROSPATHE Scheff. Palmae. 18 spp. Philippines, New Guinea, Solomon Islands. Palm trees.

***H. alata*** Scheff. Philippines, Malaya. Used in the Philippines. The buds are eaten as a vegetable; the nuts are chewed with betel; the petioles are used in making baskets and the leaflets for making sun hats (Salokots).

HETEROTHALAMUS Less. Compositae. 2 spp. S.America. Shrubs.

***H. brunoides*** Less. = Marshallia brunoides. Argentina to Brazil. The plant is aromatic and is used locally as an excitant and to reduce fevers. The branches are used as brooms and as the source of a yellow dye.

Heua – Metrosideras collina.

HEVEA Aubl. Euphorbiaceae. 12 spp. Tropical America. Trees.

***H. benthamiana*** Muell. Arg. Amazon basin. Occasionally cultivated as a source of rubber.

***H. brasiliensis*** Muell. Arg. (Para Rubber Tree). Amazon Basin. This is the world's most important source of natural rubber, some 2 million tons being produced annually. 90 per cent of this is produced in the Far East, particularly Malaysia and Indonesia. The rest is produced from wild trees and small plantations in Brazil and W.Africa. Early in the 19th century, Macintosh used latex for waterproofing and soon afterwards, Goodyear discovered the process of vulcanizing (used in the manufacture of tyres). This led to a boom in rubber production and a tremendous demand on the natural sources of the Amazon Basin and other rubber producing species. In 1895 Wickham smuggled rubber seed from Brazil. It was germinated in England and shipped to Ceylon. From these trees the original Malaysian plantations were developed. They came into their own after the First World War, when the natural sources were depleted. Ever since the introduction of synthetic rubbers from about 1945, there has been a demand for natural rubber, although it accounts now for only about 20 per cent of world production. The plant needs a rainfall of about 80 inches a year and a temperature of between 75–90°F. The seeds lose their viability quickly and have to be planted immediately. Germination takes about 6 months. The plants begin to yield after 5 years, but reach their peak in about 12 years. Plantations are usually abandoned after 25 years, usually due to the introduction of better varieties. All commercially grown rubber is now propagated by grafting and the development of new high-yielding clones forms an important part of rubber research programmes. A high yielding tree may give up to 1 ton of latex a year. The latex vessels are in the bark and these are tapped by shallow diagonal cuts. These must be made carefully to avoid damaging the underlying cambium. Thin slices are removed from the cut surface, usually on alternate days to ensure an even flow of latex. This may continue for several months after which the trees are rested. The latex is collected from the tree in small cups attached to the trunk. These are emptied into buckets, which are taken to the local factory. Here the latex is coagulated by adding acid, usually acetic, or formic to neutralise the alkalinity of the liquid. The coagulum is then pressed out into sheets and may, or may not, be smoke-cured. The rubber is exported in this form. The seeds contain about 50 per cent of a red semi-drying oil, Para Rubber Seed Oil, which is occasionally used for making soap.

***H. cuneata*** Hub. Amazon Basin. A local source of rubber.

***H. discolor*** Muell. Arg. Amazon Basin. A local source of rubber.

***H. foxii*** Hub. Amazon Basin. A local source of rubber.

***H. pauciflora*** Muell. Arg. Amazon Basin. A local source of rubber.

***H. rigidifolia*** (Benth.) Muell. Arg. Amazon Basin. A local source of rubber.

***Hexagona mori*** Poll. = Polyporus alveolaris.

HEXALOBUS A. DC. Annonaceae. 5 spp. Tropical and S.Africa. Madagascar. Woody plants.

***H. senegalensis*** A. DC. = Uvaria monopetala. Tropical Africa. The whole plant is used in Senegal to treat diarrhoea and as an expectorant.

HIBISCUS L. Malvaceae. 300 spp. Tropics

and subtropics. Woody plants to small trees. Many species are grown as ornamentals.

***H. abelmoschus*** L.=Abelmoschus moschatus. (Muskmallow). Tropics and subtropics. The bark fibre is used locally to make sails. The seeds (Ambrette Seeds) are the source of an oil used in perfumery.

***H. bancroftianus*** Macf. W.Indies. The root is demulcent and is used medicinally in treating coughs and catarrh.

***H. bifurcatus*** Blanco.=H. surattensis.

***H. bifurcatus*** Cav.=H. uncinellus (Flor de paisto). Tropical America. The leaves are eaten as a vegetable in parts of Brazil.

***H. cannabinus*** L. India and Africa. Cultivated for the bark fibres (Deccan Hemp, Kenaf, Bimlipatam Jute, Gambo, Ambari). Although coarser, the fibres compare favourably with those of jute for which the plant is a popular substitute. The roughness is due to irregular lignification of the fibres, which may be 5–10 ft. long. The fibres are extracted by retting, or mechanical decortication. The plant has been cultivated in India for centuries, and on a small scale throughout Africa, but the shortage of jute during World War II stimulated research and production in S. U.S.A., central America, Russia and E.Asia. The plants can be harvested after 90–120 days. The seeds contain about 20 per cent semi-drying oil which can be refined and used for cooking, the resulting cake being used for stock feed. The untreated oil is used for lighting in Africa.

***H. elatus*** Swartz. (Blue Mahue). W.Indies. The wood is used for cabinet work, gun-stocks, carriage poles and fishing rods.

***H. esculentus*** L.=Abelmoschus esculentus. (Okra, Gombo). Central Africa. Cultivated for the fruits which are eaten as a vegetable, particularly in soups etc. They are slimy with a mild flavour, but both sliminess and flavour vary considerably within the numerous varieties. The leaves are also used as a vegetable and the stem fibres are sometimes used locally. The plant has been cultivated for centuries in Central and N.Africa and was probably introduced to the New World during the slave-trading days. It immediately found favour, particularly in S. U.S.A. and W.Indies.

***H. guiensis*** DC.=H. tiliaceus.

***H. hamabo*** Sieb. and Zucc. Japan. The bark fibres are used locally for cordage.

***H. macrophyllus*** Roxb. E.Indies to Java. Cultivated locally. In Java the soft grey-brown wood is used to make houses. In Burma, Malaysia and Thailand the bark fibres are used for cordage.

***H. panduraeformis*** Burm. f. Tropical Australia. The aborigines used the bark fibre to make string etc.

***H. quinquelobus*** G. Don.=H. sterculifolius. Tropical Africa. The bark fibres (West African Jute) are used for cordage. The wood is rather like walnut and is used for general carpentry.

***H. rosa-sinensis*** L. China Japan. The Chinese women use the juice to blacken their eyebrows. The bark is used in China to control menstruation and in Malaya a decoction of the roots is used to treat sore eyes. The plant is widely grown as an ornamental in China and Japan.

***H. sabdariffa*** L. (Roselle, Jamaica Sorrel). India. The plant has been cultivated in India as a fibre plant for centuries. It has similar properties and history to H. cannabinus q.v. The enlarged calyx is used to make jellies and confections, the leaves and shoots are eaten as a vegetable and a fermented beverage is made from the juice.

***H. sagittifolius*** Kurz. (Ginseng de Phú Yên, Nhân Sâm, Sâm nam). Indochina, N.Vietnam. The root is used to treat venereal diseases.

***H. similis*** Blume=H. tiliaceus.

***H. sterculifolius*** Steud.=H. quinquelobus.

***H. surattensis*** L.=H. bifurcatus. Tropical Asia, Africa. The bark fibres are used locally. In Asia the leaves are used locally as a vegetable or in salads. They are also used to relieve coughs.

***H. tiliaceus*** L.=H. guiensis=H. similis=Paritium tiliaceum. (Gatapa, Majagua). Tropics. The bark fibres are produced in various parts on a small scale and exported to Europe. They are used to make sails, fishing nets etc.

***H. uncinellus*** DC.=H. bifurcatus.

***H. unidas*** Lindl. Brazil. The bark fibres are used locally for cordage and textiles.

***H. venustus*** Blume. (Kakapasan, Sekar

waron). Indonesia. The bark fibre is used locally for cordage.
Hibiscus, Rhodesian Tree – Thespesia garckeana.
Hickory, Big-bud – Carya tomentosa.
Hickory, Big Shellbark – Carya laciniosa.
Hickory, Chinese – Carya cathayensis.
Hickory, Pignut – Carya glabra.
Hickory Pine – Pinus pugens.
Hickory, Shagbark – Carya ovata.
Hickory, Shellbark – Carya ovata.
***Hicoria glabra*** (Mill.) Brit. = Carya glabra.
***Hicoria laciniosa*** (Michx. f.) Sarg = Carya laciniosa.
***Hicoria minima*** Brit. = Carya cordiformis.
***Hicoria ovata*** (Mill.) Brit. = Carya ovata.
***Hicoria pecan*** Britt. = Carya pecan.
HIERACIUM L. Compositae. 5,000 microspp. or 1,000 macrospp. Temperate excluding Australasia and tropical mountains. Herbs.
***H. pilosella*** L. (Mouse Ear). Europe, Asia, naturalised in N.America. The leaves, flowers and roots (Herba et flores Pilosellae sive Auriculae) are used in parts of Europe to treat diarrhoea, intestinal worms, bronchial catarrh, fevers and excessive loss of blood during menstruation. A tea from the leaves was used to combat dropsy.
Hierba buena – Lippia geminata.
Hierba de la Cucaracha – Haplophyton cimicidum.
Hierba de la Cucaracha – Macrosiphonia hypoleuca.
Hierba de la Cucaracha – Secondatia stans.
Hierba de las gallinitas – Petiveria alliacea.
Hierba de la Mosea – Buddleja americana.
Hierba de la mula – Lippia umbellata.
Hierba de la pastora – Turnera diffusa.
Hierba de la Perida – Montañoa tomentosa.
Hierba de la viruela – Phyllonoma laticuspis.
Hierba del carbonero – Baccharis glutinosa.
Hierba del gato – Croton dioidus.
Hierba del venado – Turnera diffusa.
Hierba del zorrillo – Croton dioidus.
Hierba de negro – Lippia geminata.
Hierba de Santa Maria – Pluchea odorata.
Hierba dulce – Lippia berlandieria.
Hierba hedionda – Cestrum nocturnum.
HIEROCHLOË (J. G. Gmel.) R. Br. Graminae. 30 spp. Temperate and cold, tropical mountains. Perennial grasses.
***H. odorata*** (L.) Beauv. = Holcus odoratus = Torresia odorata. (Sweet Grass). Europe, Asia, N.America. The Indians of New Mexico burn the grass as incense. It is also used to perfume clothes and to make Sweet Grass baskets.
HEIRONIMA Allem. Euphorbiaceae. 36 spp. Tropical America. W.Indies. Trees.
***H. alchorneoides*** Allem. (Urucurana). Brazil. The roots are used to purify the blood. They are used in patent medicines in S.America and U.S.A.
***H. caribaea*** Muell. Arg. (Tapana). S.America, Trinidad. The red-brown wood is durable and strong. It is used for building, furniture-making, waggons and wheels.
***H. cubana*** Muell. Arg. Cuba. The wood is bright yellow and is used for cabinet work.
High Bush Blueberry – Vaccinium corymbosum.
Highbush Cranberry Bark – Viburnum opulus.
Highland Chilte Rubber – Cnidoscolus elasticus.
Highland Coffee – Coffea stenophylla.
High Mallow – Malva sylvestris.
Higicho – Carica chrysophylla.
Higuera del Monte – Carica quercifolia.
Higuerón – Ficus cotinifolia.
Higuilo – Ficus padifolia.
Hila – Garcinia cambogia.
Hilo Grass – Paspalum conjugatum.
Himalayan Bamboo – Arundinaria falcata.
Himalayan Birch – Betula utilis.
Himalayan Black Cedar – Alnus nitida.
Himalayan Cedar – Cedrus libani.
Himalayan May Apple – Podophyllum emode.
Himalayan Pencil Cedar – Juniperus macropoda.
Himalayan Rhubarb – Rheum emodi.
Himalayan Silver Fir – Abies pindrow.
Himalayan Spruce – Picea smithiana.
Hinahina – Melicytus ramiflorus.
Hindu Cowpea – Vigna catjang.
Hindu Datura – Datura metel.
Hing – Ferula alliaceae.
Hingan – Balanites roxburghii.
Hing-gui-Kalabau – Streptocaulon baumii.
***Hingtsha repens*** Roxb. = Enydra fluctuans.
Hingu – Ferula alliacea.
Hioynh ba – Nauclea officinale.
Hip, Dog – Rosa canina.
Hip, Rose – Rosa pomifera.
HIPPOCRATEA L. Celastraceae. sensu A. C. Smith 1 sp. S.E. United States of America, central Mexico, tropical

S.America, W.Indies; sensu Loesener 120 spp. Tropics. Creeping plants or climbers.

***H. acapulcensis*** H.B.K. Mexico to N. S.America. An extract of the seeds is used locally to kill human body parasites.

***H. myriantha*** Oliv. (Bokakango, Likuko, Nkuba). Equatorial Africa. The stems are used locally to tie oil presses used to express palm oil.

***H. senegalensis*** Lam. = Salacia senegalensis.

***H. verticillata*** Steud. = Salacia senegalensis.

HIPPOCREPIS L. Leguminosae. 20 spp. Europe, Mediterranean to Persia. Herbs.

***H. comosa*** L. (Horseshoe Vetch). Europe. Grown as fodder for livestock.

HIPPOMANE L. Euphorbiaceae. 5 spp. Mexico, W.Indies. Trees.

***H. mancinella*** L. (Manchineel). Tropical America. The wood is used in the W.Indies for cabinet making and interior work. The latex was used by the natives as an arrow poison.

HIPPOPHANË L. Elaeagnaceae. 3 spp. Temperate Eurasia. Trees.

***H. rhamnoides*** L. (Sea Buckthorn). Europe, Asia. The berries are made into jellies, in France they are used to make a sauce for meat and fish dishes, while in Central Asia they are eaten with cheese and milk. The hard, fine-grained wood is sometimes used for lathe work.

HIPPURIS L. Hippuridaceae. 2–3 spp. Almost cosmopolitan. Aquatic herbs.

***H. tetraphylla*** L. = H. vulgaris L.

***H. vulgaris*** L. (Mare's tail). N.hemisphere. The young leaves are eaten by the Eskimoes of Alaska.

HIPTAGE Gaertn. Malpighiaceae. 20–30 spp. Mauritius, W.Himalayas to China, Formosa, Indochina, W.Malaysia, Celebes, Timor, Fiji. Vines.

***H. benghalensis*** Kurz. = H. madablota. India to Burma and Malaya. The leaves are used in India to treat skin diseases. The plant has some insecticidal properties.

***H. madablota*** Gaertn. = H. benghalensis.

***Hirneola*** Fr. = Auricularia.

Hiroong – Nyssa javanica.

***Hirtella polyandra*** H.B.K. = Couepia polyandra.

Hirtiz – Euphorbia thomsoniana.

Hirvi – Euphorbia thomsoniana.

Hisawang – Rhazya stricta.

Hissing Tree – Parinari mobola.

HITCHENIA Wall. Zingiberaceae. 3 spp. India, Malay Peninsular. Herbs.

***H. caulina*** Baker. (Chawar, Nisham). India. The tubers yield a flour which is used to make glue and size and if purified properly it is eaten locally as a gruel.

Hiwar – Acacia leucophloea.

Hoarhound – Marrubium vulgare.

Hoarhound, Black – Ballota nigra.

Hoarhound, Wild – Eupatorium aromaticum.

Hoary Basil – Ocimum canum.

Hoary Gromwell – Lithospermum canescens.

Hoary Milkpea – Galactia mollis.

Hoa soi – Chloranthus inconspicuus.

Hodai Tree – Commiphora hildebrandtii.

HODGSONIA Hook. f. and Thoms. Cucurbitaceae. 1 sp. Indomalaya. Vine.

***H. heteroclita*** Hook. f. and Thoms. = H. macrocarpa.

***H. macrocarpa*** Cogn. = H. heteroclita = Trichosanthes kadam = T. macrocarpa. (Akar kapajang, Areuj pitjong tjeleng). The fruits are about the size of a calabash and are eaten locally. The seeds contain about 36 per cent oil which is extracted and used for cooking and for roasting opium before smoking it.

Hodsale – Palaquium ellipticum.

Hodung – Populus euphratica.

Hoè giáe – Sophora angustifolia.

Hoelen – Poria hoelen.

HOFFMANNSEGGIA Cav. Leguminosae. 45 spp. America, tropical and S.Africa. Herbs.

***H. densiflora*** Benth. S.W. U.S.A., Mexico. The tubers were eaten, roasted by the local Indians.

***H. falcaria*** Cav. See H. densiflora.

Hofotsa – Dombeya cannabina.

Hog Fennel – Peucedanum officinale.

Hog Fennel – Peucedanum ostruthium.

Hog Gum – Metopium toxiferum.

Hog Gummi – Clusia flava.

Hog Millet – Panicum miliaceum.

Hog Peanut – Amphicarpaea monoica.

Hog Plum – Prunus umbellata.

Hog Plum – Spondias lutea.

Hog Plum – Spondias tuberosa.

Hog Plum – Symphonia globulifera.

HOHERIA A. Cunn. Malvaceae. 5 spp. New Zealand. Trees.

***H. populnea*** Cunn. New Zealand. The tough bark is used as cordage. An infusion from the bark is used as a demulcent drink, for treating colds and as a balm for sore eyes and burns. The wood is used for cabinet work and inlaying.

Hoja chegüe – Tetracera volubilis.

Hoja de Cucharilla – Pithecoctenium echinatum.

Hojas de la Pastora – Salvia divinorum.

Hojas de Marua Pastora – Salvia divinorum.

HOLARRHENA R. Br. Apocynaceae. 20 spp. Tropical Africa, Madagascar, India, S.E.Asia, Philippines, Malay peninsular. Small trees.

***H. africana*** A. DC. Tropical Africa. The hairs from the seeds coats are used in Sierra Leone to stuff pillows.

***H. antidysentrica*** Wall. (Tellicherry Bark). Tropical Asia. The seeds are used in local medicine to treat dysentery, fever, diarrhoea, stomach complaints and as a tonic and aphrodisiac. The bark (Conesi Bark, Kurchi Bark) is used to treat diarrhoea, dysentery and as a tonic.

***H. febrifuga*** Klotzsch. (Jasmine Tree, Mwengebushilu). Central Africa. The bark is used locally as an emetic and to reduce fevers. The white wood is used in Malawi to make household utensils.

***H. microteranthera*** Schum. (Piripiri). Tropical Africa. The bark yields a rubber.

***H. wulfsbergii*** Stapf. (Male Rubber Tree). Tropical W.Africa. The bark yields a latex used to adulterate rubber. The bark is used in an infusion of palm wine to treat dysentery. The wood is used locally for carving, tool-handles and combs.

***Holboellia cuneata*** Oliv. = Sargentodoxa cuneata.

HOLCUS L. Graminae. 8 spp. Canaries, N.Africa, Europe to Asia Minor and Caucasus; 1 sp. S. Africa. Perennial grasses.

***H. halepensis*** L. = Sorghum halpense.

***H. lanatus*** L. (Velvet Grass, Yorkshire Fog). Europe and temperate Asia, escaped to America. The grass is grown as a fodder grass.

***H. odoratus*** L. = Hierochloë odorata.

Holiga – Holigarna arnottiana.

HOLIGARNA Buch.-Ham. ex Roxb. Anacardiaceae. 8 spp. Indomalaya. Trees.

***H. arnottiana*** Hook. f. (Bibu, Cheru, Holiga). India. The light grey, soft, satiny wood is used locally for house-building, boats, packing cases and cigar boxes. The bark and rind of the fruit give an acrid juice which is used as a varnish.

Holly, American – Ilex opaca.

Holly, Californian – Photinia arbutifolia.

Holly, Dahoon – Ilex cassine.

Holly, Emetic – Ilex vomitoria.

Holly, European – Ilex aquifolium.

Holly, Longstalk – Ilex pedunculata.

Holly, Sea – Acanthus ebracteatus.

Holly, Sea – Eryngium maritimum.

Holly, Smooth – Hedycarya angustifolia.

Hollyhock – Althaea rosea.

Holm Oak – Quercus ilex.

HOLOCALYX Micheli. Leguminosae. 1 sp. Brazil, Paraguay, N.E.Argentina. Tree.

***H. balansae*** Mich. The yellowish wood is compact, heavy and durable. It is used to make luxury furnishings, carvings etc.

HOLOPTELEA Planch. Ulmaceae. 2 spp. Tropical Africa, Indomalaya. Trees.

***H. grandis*** (Hutch.) Mildbr. (Nembambobolo). Central and W.Africa. The hard yellow wood is used for interior work on houses and for making small boats.

Holy Basil – Ocimum sanctum.

Holy Clover – Onobrychis viciaefolia.

Holy Thistle – Silybum marianum.

Holy Wort – Verbena officinalis.

Hom thom – Ailanthus fauveliana.

HOMALANTHUS A. Juss. corr. Reichb. Euphorbiaceae. 35 spp. Indomalaya, Polynesia. Small trees.

***H. populifolius*** R. Grah. Malaysia to Australia. The bark and leaves yield a black dye used locally for staining basketwork.

HOMALIUM Jacq. Flacourtiaceae. 200 spp. Tropics and subtropics.

***H. caryophyllaceum*** Benth. = H. frutescens = H. foetidum Ridl. Malaysia. The hard light brown wood is used locally for building.

***H. cochinchinensis*** Drude = H. fagifolium.

***H. fagifolium*** Benth. = H. cochinchinensis = Blackwellia fagifolia = B. padifolia. Chay, Song. The bark is used in N.Vietnam for caulking boats. The roots are used to treat venereal diseases. The plant is cultivated.

***H. foetidum*** Benth. Malaysia to Moluccas. The heavy brown wood is used locally for boat-building and for tall buildings.

***H. foetidum*** Ridl. = H. caryophyllaceum.

***H. fructescens*** King = H. caryophyllaceum.

***H. grandiflorum*** Benth. (Kajoo batoo areng, Lamgitlawe). Malaysia. The hard durable wood is used locally for house beams and posts.

***H. racemosum*** Jacq. (Caracolillo). Tropical America, W.Indies. The heavy hard, grey-brown wood is resistant to termites, but difficult to work. It is used for house construction, furniture, agricultural implements, sports goods and heavy construction work.

***H. tomentosum*** Benth. Burma, Malaysia. The wood (Moulmein Lancewood) is used for general lumber and construction work.

***H. vitiense*** Benth. (Oueri). New Caledonia. The light yellow wood is durable and is used for interior work in houses.

HOMALOMENA Schott. Araceae. 140 spp. Tropical Asia, S.America. Herbs.

***H. aromatica*** (Roxb.) Schott. India, S.W.Malaysia to New Guinea. The rhizomes are used in India to make a stimulant.

***H. philippensis*** Engl. Philippines. The rhizomes are used locally to make an embrocation for rheumatism.

***H. rubescens*** Kunth. (Gandubikachu). N.India, Himalayas, Malaysia. The leaves are used by the Malaysians to poison water.

HOMONOIA Lour. Euphorbiaceae. 3 spp. S.China, S.E.Asia, Indomalaya. Small trees.

***H. riparia*** Lour. Himalayas to Malaysia and Philippines. The juice from the leaves is used in Java to blacken the teeth and to firm them in the sockets. In Perak the juice is used to treat skin diseases.

HOMORANTHUS A. Cunn. ex Schau. Myrtaceae. 3 spp. E.Australia. Shrubs.

***H. flavescens*** Cunningh. = H. virgatus.

***H. virgatus*** Cunningh. = H. flavescens. New South Wales. Distillation yields Homoranthus Oil which is used in perfumery.

Homoranthus Oil – Homoranthus virgatus.

***Honckenya*** Willd. = Clappertonia.

Hondcoor – Cotyledon orbiculata.

Honduras Bark – Picramnia pentandra.

Honduras Lancewood – Lonchocarpus hondurensis.

Honduras Mahogany – Swietenia macrophylla.

Honduras Sarsaparilla – Smilax regelii.

Hondureña – Swietenia macrophylla.

Honesty Oil – Hesperis matronalis.

Honey Dew Melon – Cucumis melo.

Honey Locust – Gleditsia triacanthos.

Honey, Palm – Jubaea chilensis.

Honey, Rose – Rosa pomifera.

Honeysuckle – Lonicera caprifolium.

Honeysuckle, Jamaican – Passiflora laurifolia.

Honeysuckle, Maori – Knightia excelsa.

Hongay Oil – Pongamia pinnata.

Honghel – Adenium honghel.

Hoobobbali Tree – Loxopterygium sagotii.

Hoodoo wendoo – Echinochloa pyramidalis.

Hooker's Balsamroot – Balsamorhiza hookeri.

Hoop Pine – Araucaria cunninghamii.

Hoop Pine – Araucaria klinki.

Hooroo mandjel – Laplacea intergerrima.

Hooyoc – Morinda yacatanensis.

Hop (Oil) – Humulus lupulus.

Hop Bush – Dodonaea lobulata.

Hop Clover – Medicago lupulina.

Hop Hornbeam – Ostrya carpinifolia.

Hop, Spanish – Origanum creticum.

Hop, Spanish – Origanum heraclecticum.

Hop Tree – Ptelea trifoliata.

HOPEA Roxb. Dipterocarpaceae. 90 spp. S.China, S.E.Asia, Indomalaya. Trees. The wood of the species mentioned is hard, heavy and strong. It is used for heavy construction work, presses, bridges, house construction, carts, railway sleepers etc. Because of its durability in water it is used extensively locally for boat-building. The species also yield the Dammar resins which are used in the manufacture of varnishes etc.

***H. balageran*** Korth. Malayan Archipelago. Resin called Njating Mahabong.

***H. celebica*** Burck. (Dama dere itam). Indonesia.

***H. dealbata*** Hance. Indo-china, Thailand, Perak.

***H. dryobalanoides*** Miq. Sumatra.

***H. fagifolia*** Miq. Bangka. Resin called Dammar Kedemut.

***H. ferrea*** de Lans. Indo-china.

***H. globosa*** Brandis. Indo-china, Sumatra.

***H. griffithii*** Kurz. Sumatra.

***H. heimii*** King. (Chengal Tree). Malaysia.

***H. intermedia*** King. Sumatra.

***H. lowii*** Foxw. = H. nutans.

***H. maranti*** Miq. Malayan Archipelago. Resin called Dammar Batu.

***H. maximus*** King. Malaysia. Resin called Dammar Penak.

***H. mengarawan*** Miq. Sumatra.

***H. micrantha*** Hook. Sumatra.

***H. multiflora*** Brandis. Indo-china.
***H. myrtiflora*** Miq. Sumatra.
***H. nutans*** Ridl. = H. lowii. (Giam, Nodding Merawan). Malayan Peninsular.
***H. odorata*** Roxb. Burma, Malaya, Perak, Trengganu. An important local timber tree. The resin is called Rock Dammar.
***H. parviflora*** Bedd. (Ironwood of Malabar, Bhogi, Bovumara, Tirpu). Malabar and surrounding area.
***H. pentanerva*** Sym (Chengal paya). Borneo.
***H. sangal*** Korth. Borneo. Resin called Njating from old trees and Njating Matpleppek from young trees.
***H. singkowang*** Miq. = Shorea singkowang.
Horchata de Chufas – Cyperus esculentus.
HORDEUM L. Graminae. 20 spp. Temperate. Annual grasses. The Barleys. The genus has two centres of origin, the highlands of Ethiopia and S.E.Asia. Hordeum has been cultivated since ancient times, certainly since 5000 B.C. in Egypt, 3000 B.C. in N.W.Europe and by the Stone Age Lake Dwellers in Switzerland. It was probably the chief grain crop of the Greeks and Romans and probably throughout Europe as late as the 16th century. The world production now exceeds 100 million tons per year, with Russia being the chief producer and France and the United Kingdom being the main European producers. The crop is grown throughout the world where the climate is suitable. As the plant has a short growing season and is relatively drought-resistant it can be grown as far north as Norway and into the deserts of Africa. It requires a well-drained loam with little nitrogen fertilizers, as these tend to cause lodging and reduce the malting quality. The yield varies from 20–40 bushels an acre and there are many varieties to suit the different conditions. About half the crop is used for livestock feed and about 10 per cent for making beers, ales, lagers and distilling to make whisky. The rest is eaten as flour, pearl barley, as invalid foods and for making malt. Some authorities place all the cultivated forms under H. vulgare, while others reserve this for the six-row barleys, placing the two-row barleys in H. distichum. As with many cereals, there are spring and winter varieties. All the cultivated varieties are diploid with 2n = 14.
***H. distichum*** L. (Two-rowed Barley) = H. vulgare var. distichum.
***H. hexastichum*** L. = H. vulgare var. hexastichum.
***H. irregulare*** Aberg. and Wiebe (Abyssinian Intermediate Barley). A cultigen of intermediate characteristics between H. distichum and H. vulgare. Originated in Ethiopia.
***H. trifurcatum*** Ser. (Egyptian Barley). N.Africa, Asia Minor. Cultivated in N.Africa.
***H. vulgare*** L. (Common Barley, Six-rowed Barley).
***H. vulgare*** L. var. ***distichum*** = H. distichum.
***H. vulgare*** L. var. ***hexastichum*** = H. hexastichum.
Hornbeam, American – Carpinus caroliniana.
Hornbeam, American Hop – Ostrya virginiana.
Hornbeam, European – Carpinus betulus.
Hornbeam, European Hop – Ostrya carpinifolia.
Hornbeam, Heartleaf – Carpinus cordata.
Horn of Plenty – Craterellus cornucopioides.
Horned Poppy (Oil) – Glaucium flavum.
HORNSTEDTIA Retz. Zingiberaceae. 60 spp. Indomalaya. Herbs.
***H. rumphii*** Val. = Amomum rumphii = Donacodes incarnata. (Galoba djantoong, Oona incarnata) Indonesia. The fruits are eaten locally.
Horse Balm – Collinsonia canadensis.
Horse Banana – Musa paradisiaca.
Horse Bean – Vicia faba.
Horse Brier – Smilax rotundifolia.
Horse Chestnut – Aesculus hippocastanum.
Horse Chestnut, Japanese – Aesculus turbinata.
Horse Gram – Dolichos biflorus.
Horse Gram – Vigna sinensis.
Horse Mint – Monarda fistulosa.
Horse Mint – Monarda punctata.
Horse Mint, American – Monarda punctata.
Horse Mint, English – Mentha longifolia.
Horse Nettle, Carolina – Solanum carolense.
Horse Radish – Armoracia lapathifolia.
Horse Radish Tree – Moringa oleifera.
Horse Sugar – Symplocos tinctoria.
Horsetail – Equisetum arvense.
Horsetail Kelp – Laminaria digitata.
Horsehair Lichen – Alectoria jujuba.
Horseheal – Inula helenum.

HORSFIELDIA Willd. Myristicaceae. 80 spp. S.China, S.E.Asia, Indomalaya, N.Australia. Trees.

***H. glabra*** (Blume.) Warb. = Myristica glabra. (Kalapa tijoong, Ki toongila). Indonesia. The bark and leaves are used locally to treat intestinal complaints.

***H. irya*** (Gaertn.) Warb. = Myristica irya. (Kalapa tijoong, Kannarahan). Ceylon through to Indonesia. The wood is olive-green, hard and easy to work. It is used to make boards. The sweet-smelling flowers are used in perfumery.

***H. iryaghedhi*** (Gaertn.) Warb. = H. odorata.

***H. odorata*** Willd. = H. iryaghedhi (Tjempaka selong). Ceylon through to Indonesia. The oil from the seeds is used locally to make candles.

***H. sylvestris*** Warb. E.Malaysia. The oil from the seeds is used locally to make candles. The outer parts of the fruit are used for flavouring.

Hortulana Plum – Prunus hortulana.

HOSLUNDIA Vahl. Labiatae. 2–3 spp. Tropical Africa. Herbs.

***H. opposita*** Vahl. Tropical Africa. An oil (Kamynye Oil) extracted from the plant smells like vanilla and is used in perfumery as a fixative.

HOSTA Tratt. Liliaceae. 10 spp. China and Japan. Herbs.

***H. ovata*** Spreng. (Shibo-shi). Japan. The leaf stalks are eaten as a vegetable locally.

Hotai – Commiphora abyssinica.

Hottentot Almond – Brabejum stellatifolium.

Hottentot Bread – Dioscorea elephantipes.

Hottentot Fig – Mesembryanthemum edule.

Hottentot Tea – Helichrysum serpyllifolium.

Hottentot Tobacco – Tarchonanthus camphoratus.

HOUMIRI Aulb. 3–4 spp. Houmiriaceae. Tropical S.America. Trees.

***H. balsamifera*** St. Hil. (Couranira, Oloroso). Tropical America. The wood is used for general carpentry, locally and for the spokes of wheels.

***H. floribunda*** Mart. (Bastard Bullet). Brazil, Guyana. The red-brown wood is used for house construction, furniture and wheel spokes. The bark is attacked by a fungus which gives a pleasant perfume, used by the local Indians for scenting the hair.

Hound's Tongue – Cynoglossum officinale.

Houp – Montrouziera sphaeraeflora.

Houseleek – Sempervivum tectorum.

HOUTTUYNIA Thunb. Sauruaceae. 1 sp. Himalayas to Japan. Herb.

***H. californica*** Benth. and Hook. = Anemoonpsis californica.

***H. cordata*** Thunb. (Tsi). Cultivated locally in Vietnam where the leaves are used in salads and for treating eye diseases. In Nepal the leaves are eaten in soups, while in China the whole plant is used for treating bladder and kidney complaints and skin diseases.

HOVENIA Thunb. Rhamnaceae. 5 spp. Himalayas to Japan. Small trees.

***H. dulcis*** Thunb. (Japanese Rasin Tree). Himalayas to Japan. The dry sub-acid fruits are eaten locally in China and Japan and are used in China as a diuretic for treating drunkeness.

***Howardia*** Klotzsch. = Aristolochia.

HOYA R. Br. Asclepiadaceae. 200 spp. S.China, S.E.Asia, Indomalaya, Australia, Pacific. Woody vines or epiphytes.

***H. australis*** R. Br. Woody vine, Polynesia, Australia. The flowers are worn ceremonially in Samoa.

***H. coronaria*** Blume. Epiphyte. Malaysia. The leaves are used mixed with those from Capsicum spp. as a digestive stimulant, especially in Java.

***H. latifolia*** G. Don. Woody climber. Malaysia. A decoction of the plant is used in Java to treat stings from fish.

Hsaythanpaya – Gardenia campanulata.

Huaco – Aristolochia pardina.

Huagra Mora – Rubus roseus.

Huajillo – Acacia berladieri.

Huajillo – Acacia coulteri.

Huajillo – Pithecellobium brevifolium.

Huang-hua ts'ai – Crepis japonica.

Huang-nu ya-tzu – Pistacia chinensis.

Huang t'eng – Fibruarea tinctoria.

Huauhtli – Amaranthus leucocarpus.

Huanuca Coca – Erythroxylum coca.

***Hubertia ambavilla*** Bory. = Senecio ambavilla.

Huckleberry – Gaylussacia baccata.

Huckleberry, Bear – Gaylussacia ursina.

Huckleberry, Black – Gaylussacia baccata.

Huckleberry, Box – Gaylussacia brachycera.

Huckleberry, Dwarf – Gaylussacia dumosa.

Huckleberry, False – Menziesia ferruginea.

Huemega – Micranda minor.

Huevos de Gallo – Salpichroa rhomboidea.

HUGONIA L. Linaceae. 40 spp. Tropical Africa, Madagascar, Mascerene Islands, Indomalaya, New Caledonia. Small trees or scrambling vines.

***H. mystax*** L. E.India. The roots are used locally to treat intestinal worms and snake bites.

***H. obtusifolia*** C. M. Eright. (Bondesobe, Ifumbolo, Lombaya lo lowe). Tropical Africa, especially Congo, Cameroons and S.Nigeria. The fruits are eaten locally as a vegetable.

***H. platysepala*** Welw. ex Oliv. (Alambe, Gbakaya, Sikalo). Tropical Africa. The fruits are eaten locally as a laxative and the twigs are used to make musical instruments.

Hui – Elmerrillia papuana.

Huile d'Argan – Argania spinosa.

Huile de Bois – Shorea vulgaris.

Huile de Dia – Cestrum dumetorum.

Huile de Julienne – Hesperis matronalis.

Huile de Linaloë – Aniba rosaeodora.

Huile de Marmotte – Armeniaca brigantina.

Huile de noche – Cestrum lanatum.

Huile de noche – Cestrum nocturnum.

Huile de noche – Pisonia aculeata.

Huili huiste – Karwinskia calderonii.

Huisache – Pithecellobium albicans.

Hulava – Radermachera xylocarpa.

Hulver – Ilex aquifolium.

Humboldt Coytilla – Karwinskia humboldtiana.

Humboldt's Willow – Salix humboldtiana.

***Humiria*** Jaume St. Hil. = Houmiri.

HUMULARIA Duvign. Leguminosae. 40 spp. Temperate and tropical Africa. Woody plants.

***H. apiculata*** (De Wild.) Duvign. = Geissaspis apiculata. (Kobonga). Congo. The dried roots are smoked locally, acting as a strong stimulant.

HUMULUS L. Cannabidaceae. 4 spp. N.Temperate, S. to Indochina, S.W. United States of America. Vines.

***H. lupulus*** L. (Hop). Europe, Asia, N.America. The strobili have long been used as a tonic and sedative and the leaves in salads, but they were not used until the 8th century A.D. for flavouring beer. The world production of dried cones (inflorescences) is about 100,000 tons a year, Germany, United Kingdom and Pacific U.S.A. accounting for about 75 per cent of this. The cones are picked by hand, or machinery and dried from 80 per cent moisture to about 12 per cent. They are added to the wort during beer-making. The wort is boiled with the hops in it. They are then filtered off and the dried remains used as manure. The bitter flavour is caused by lupuline, lupulon, humulon, xanthohomol, cerotic acid and resins produced in glandular hairs on the base of the bracts. These aromatics account for 0·2–0·5 per cent of the weight of the dry strobilus. Oil of Hops is also used in perfumery.

Hungarian Clover – Trifolium pannonicum.

Hungarian Oat – Avena orientalis.

Hungarian Vetch – Vicia pannonica.

Hungarian Turpentine – Pinus cembra.

Hungarian Water – Rosmarinus officinalis.

Hunter's Nut – Omphalea megacarpa.

HUNTERIA Roxb. Apocynaceae. 6 spp. Tropical Africa, S.India, Ceylon, S.China, S.E.Asia, Malay Peninsular. Herbs.

***H. corymbosa*** Roxb. = H. roxburghiana (Gitan obah). Malaya. The latex (Getah ago, Getah djintahan) is used in Malacca to treat yaws.

Huon Pine (Oil) – Dacrydium frankikii.

Hupeh – Cupressus funebris.

Hupeh – Fagus sinensis.

HURA L. Euphorbiaceae. 2 spp. Mexico to tropical S.America, W.Indies. Trees.

***H. crepitans*** L. (Sandbox Tree). Tropical America. The fruits were used as small boxes for holding sand used for drying writing on paper.

***H. polyandra*** Baill. Mexico. The wood is used as telegraph poles. The latex is used for poisoning fish and the seeds for poisoning coyotes. The seeds are also used as a drastic purgative.

Huragi – Osmanthus aquifolium.

Huru tangkalak – Knema laurina.

Husk Tomato – Physalis ixocarpa.

Huynh dhan – Agaia baillonii.

Huynh duong – Dysoxylum loureiri.

Hyacinth (Oil) – Hyacinthus orientalis.

Hyacinth Bean – Dolichos lablab.

Hyacinth, Water – Eichornia crassipes.

Hyacinth, Wild – Cammassia esculenta.

HYACINTHUS L. Liliaceae. 30 spp. Mediterranean, Africa. Herbs.

***H. non scriptus*** L. = Scilla non scripta.

***H. orientalis*** L. (Hyacinth). Mediterranean. Cultivated widely as an ornamental. The sale of bulbs, especially those produced in the Netherlands involves large commercial undertakings. The flowers also produce an essential oil used in perfumery.

HYAENANCHE Lamb. and Vahl. Euphorbiaceae. 1 sp. S.Africa. Tree.

***H. capense*** Thunb. The fruits are used for poisoning hyenas.

HYALOSEPALUM Troupin. Menispermaceae. 8 spp. Tropical Africa, Madagascar. Woody vines.

***H. caffrum*** (Miers.) Troupin. (Kilende, Musimba). Tropical Africa. The fibres are used locally to make mats.

HYBANTHUS Jacq. Violaceae. 150 spp. Tropics and subtropics. Herbs.

***H. album*** St. Hil. Argentina to Brazil. A decoction of the plant is used locally in Brazil as an emetic and purgative.

***H. ipecacuanha*** (L.) Baill. Tropical S.America. The root (Radix Ipecacuanha albae lignosae, White Ipecac, Ipecacuanha branca) is used as a substitute for true Ipecacuanha as an emetic.

Hyawa Gum – Protium heptaphyllum.

HYDNOCARPUS Gaertn. Flacourtiaceae. 40 spp. Indomalaya. Trees. The non-drying oil (Chaulmogra Oil) from the seeds of the following species is used locally to treat leprosy.

***H. anthelmintica*** Pierre. (Chaulmogra Tree), Thailand, Cochin-China.

***H. heterophyllum*** Kurz. = Taraktogenos kruzii. E.India.

***H. wightiana*** Blume. India.

HYDNORA Thunb. Hydnoraceae. 12 spp. Tropical and S.Africa, Madagascar. Leafless parasites.

***H. esculenta*** Jum. and Poir. Madagascar. Parasitic on Acacia. The pleasantly flavoured fruits are about the size of an apple and are eaten locally.

HYDNOTRYA Berk. and Br. Eutuberaceae. 6 spp. Europe, N.America.

***H. carnea*** (Corda) Zobel. Europe. The fruitbodies (Cervena tartofle) are like truffles and are eaten locally, especially in Czechoslovakia.

HYDNUM L. ex Fr. Hydnaceae. 120 spp. Widespread.

***H. coralloides*** Scop. = Hericium coralloides. Temperate and subtropics. The fruit-bodies are eaten in parts of Asia.

***H. fragile*** Petch. Tropical Asia. The fruit-bodies are eaten in Java and Malaysia.

***H. repandum*** L. ex Fr. = Dentinum repandum. Cosmopolitan. The fruit-bodies are eaten widely, especially in Europe.

HYDRANGEA L. Hydrangeaceae. 80 spp. Himalayas to Japan, Philippines to Java, Atlantic N.America, mountains of Central America to Chile. Small trees or shrubs.

***H. arborescens*** L. (Smooth Hydrangea, Mountain Hydrangea, Seven Bark). E. U.S.A. The roots contain the alkaloid hydrangin which is used medicinally as a diuretic. This is extracted from the roots collected in the autumn.

***H. paniculata*** Sieb. Japan. The wood, which is hard and fine-grained is used locally to make umbrella handles, tobacco pipes and walking sticks. The bark is used to make paper.

***H. thunbergii*** Sieb. Japan. The dried, steamed leaves are used locally to make a tea (Amacha).

Hydrangea, Mountain – Hydrangea arborescens.

Hydrangea, Smooth – Hydrangea arborescens.

HYDRASTIS Ellis ex L. Hydrastidaceae. 1 sp. Japan, 1 sp. E. N.America. Herbs.

***H. canadensis*** L. (Golden Seal). E. N.America. The roots contain the alkaloids berberine, canadine and hydrastine, which are used medicinally to reduce bleeding and as a tonic. They are used particularly in the treatment of uterine bleeding and haemorrhoids and as an astringent in the treatment of inflamed mucous membranes. The whole plant was used by the local Indians to make a yellow dye.

HYDROCOTYLE L. Hydrocotylaceae. 100 spp. Tropics and temperate. Herbs.

***H. asiatica*** L. = Centrella asiatica. Perennial herb. Tropical Asia. The leaves are eaten as a salad or vegetable in many parts, particularly Japan and Java.

***H. hirsuta*** Blume. = H. sibtorpioides.

***H. latisecta*** Zell. = H. sibtorpioides.

***H. puncticulata*** Miq. = H. sibtorpioides.

***H. sibtorpioides*** Lam. = H. hirsuta = H. latisecta = H. puncticulata. (Antanum beirit, Andem, Salatoon). Malaysia, Indonesia. The leaves are eaten locally raw or as a vegetable. They are also mixed with sugar or liquorice as a cure for childrens' colds.

HYDRODICTYON Roth. Hydrodictyaceae. Green Algae.

***H. reticulatum*** (L.) Lagerh. Freshwater. Eaten in Hawaii with fresh-water shrimps and salt.

HYDROLEA L. Hydrophyllaceae. 20 spp. Tropical America, Africa and Asia, Herbs.

***H. zeyanica*** Vahl. Tropical Asia. The young leaves are eaten with rice in Java. The plant is cultivated locally.

HYDROPHYLLUM L. Hydrophyllaceae. 10 spp. N.America. Herbs. The leaves and young shoots of the following species were eaten as salads by the Indians and early settlers.

***H. appendiculatum*** Michx. (Waterleaf). E. and S. U.S.A.

***H. occidentale*** Gray (Western Squaw Lettuce). W. U.S.A.

***H. virginicum*** L. (Virginia Waterleaf). E. N.America.

***Hygrocybe*** (Fr.) Karst.=Hygrophorus.

HYGROPHORUS Fr. Agaricaceae. 175 spp. Widespread. The fruitbodies of the following species are eaten in Europe. They are occasionally sold locally.

***H. chrysodon*** (Batsch.) Fr.

***H. hypothejus*** Fr.

***H. lucorum*** Kalchbr.

***H. marzuoles*** (Fr.) Bres.

***H. pratensis*** (Pers.) Fr.=Camarophyllus pratensis.

***H. puniceus*** Fr.

***H. queletii*** Bres.

HYLOCEREUS (A. Berger). Britton and Rose. Cactaceae. 23 spp. Mexico to Peru.

***H. undatus*** (Haw.) Britton and Rose.= Cereus undatus. (Pitahaya, Pitahaya Oregona). Mexico. The fruits are eaten locally. The plants are cultivated and the fruits are sold widely in Mexico.

HYMENACHNE Beauv. Graminae. 8 spp. Tropics. Perennial grasses. The species mentioned are grown in Malaysia as fodder for livestock.

***H. aurita*** Baker.=Panicum javanicum.

***H. interrupta*** Buese.

HYMENAEA L. Leguminosae. 25 spp. Mexico, Cuba, tropical America. Trees. Members of this genus exude a resin around the roots. It is dug out and used to make varnishes.

***H. courbaril*** L. (Courbaril, West Indian Locust). S.Mexico, C.America, W.Indies. The hard wood resembles mahogany and is used for general construction work, ship-building, furniture. The gum is called Brazil Copal, Amber de Cuapinole, Colombia Copal, Goma Anime de Mexico, Resina de Cuapinole. The pulp around the seeds is used locally to make an alcoholic beverage (Atole).

***H. stilbocarpa*** Hayne. Brazil. Yields a resin.

***H. verrucosa*** Gaertn.=Trachylobium hornemannianum. Madagascar, E.Africa, Mascarene Islands. Yields a copal called Madagascar Copal, Zanzibar Copal, Anime Copal.

HYMENOCALLIS Salisb. Amaryllidaceae. 50 spp. Warm America. Herbs. Some species are grown as ornamentals.

***H. guianensis*** Herb.=H. tubiflora.

***H. tubiflora*** Salisb.=H. guianensis=Pancratium guianense. Guyana to Brazil. The bulbs are used locally in Brazil as a diuretic, expectorant and astringent.

HYMENOCARDIA Wall. ex Lindl. Hymenocardiaceae. 9 spp. Tropical Africa; 1 sp. S.E.Asia, Malaysia, Sumatra. Shrubs.

***H. acida*** Tul. Tropical Africa. The bark yields a red dye which is used locally as an antiseptic.

HYMENODICYTON Wall. Rubiaceae. 20 spp. Tropical Africa, Madagascar, Himalayas to the Celebes. Trees.

***H. excelsum*** Wall. Tropical Asia. The soft wood is used locally for carving toys etc., tea boxes. The bark is used locally to reduce fevers.

HYMENOLOPHUS Boerl. Apocynaceae. 1 sp. Sumatra. Tree.

***H. romburghii*** Boerl. A local source of rubber.

HYMENOPAPPUS L'Hérit. Compositae. 10 spp. United States of America, Mexico. Herbs.

***H. filifolius*** Hook. W. N.America. The roots were chewed by the local Indians.

***H. lugens*** Greene. S.E. U.S.A., Mexico. The roots were used by the local Indians as an emetic and to relieve toothache.

HYMENOPHYLLUM Sm. Hymenophyllaceae. 25 spp. Tropics and S.Temperate, N.Temperate Europe and Japan. Ferns.

***H. plumosum*** Kaulf. S.America. A decoction of the leaves is used locally as a diuretic and to induce sweating.

HYOSCYAMUS L. Solanaceae. 20 spp. Europe, N.Africa through the Sahara to S.W. and Central Africa. The leaves contain the alkaloids hyoscyamine and hyocine which are pain-relieving, narcotic and hypnotic.

***H. muticus*** L. Egypt, Arabia to India. The leaves have been used as a sedative since ancient times. They are also used as a substitute for opium.

***H. niger*** L. (Black Henbane). Europe, Asia, N.Africa. Cultivated in Germany, Russia and United Kingdom for their alkaloids which are used medicinally. The leaves are collected during flowering time.

***H. reticulatus*** L. (Kohibhang). Baluchistan. The Hindus use the seeds to relieve toothache.

Hyowana Resin – Protium carana.

HYPARRHENIA Anderss. Graminae. 75 spp. Mediterranean, Africa, Arabia (?). Tufted perennial grasses.

***H. aucta*** (Stapf.) Stent. (Tambookie Grass). Tropical Africa. A rough pasture grass used for reclamation work and for thatching.

***H. dissoluta*** (Nees). E. C. Hubbard. S.Africa. Used locally for thatching.

***H. filipendula*** (Hochst.) Stapf. E. S. and Central Africa, Ceylon, Philippines, E.Australia. Used as fodder in S.Africa.

***H. filipendula*** (Hochst.) Stapf. var. ***pilosa*** (Hack.) Stapf. S.Africa. Used locally for thatching.

***H. hirta*** (L.) Stapf. See H. filipendula.

***H. rufa*** Stapf. Tropical Africa. Used for thatching.

***H. ruprechtii*** Fourn.=Andropogon ruprechtii. Tropical Africa. Used for thatching.

***H. soluta*** Stapf. Tropical Africa. Used for thatching.

***H. subplumosa*** Stapf. Tropical Africa. Used for thatching.

HYPECOUM L. Hypecoaceae. 15 spp. Mediterranean to Central Asia and N.China. Herbs.

***H. pendulum*** L. (Zirgulaki). Baluchistan, Waziristan. The leaves are used locally to make a drink.

HYPERICUM L. Guttiferae. 400 spp. Temperate and tropical mountains. Herbs or small shrubs.

***H. androsaemum*** L. Europe to Caucasus to N.Africa and Iran. The leaves are used in Portugal to treat stomach complaints and as a diuretic.

***H. connatum*** Lam. Argentina, Paraguay, Uruguay. Locally, the tops of the flowerheads are used as a tonic.

***H. japonicum*** Thunb.=H. pusillum. (T'ien chi wang). Tropical and subtropical Asia. The Chinese use the leaves to treat wounds and leech bites.

***H. laxiusculum*** St. Hil. Brazil. The leaves are used locally as an excitant and antispasmodic. They are aromatic and astringent.

***H. perforatum*** L. (Saint John's Wort). Europe to W.Asia, N.Africa, N. and S.America, Australia. The leaves are used to treat a variety of complaints including stomach ache, kidney diseases, intestinal worms and nerve pains. It is used in the making of cosmetics. The leaves and fruits are used to make a tea and a red dye is extracted from the whole plant. An extract containing phytociol is antibiotic.

***H. pusillum*** Choisy=H. japonicum.

***H. teretiusculum*** St. Hil. Brazil. The leaves are used locally as an excitant and to control menstruation.

***Hypestes*** P. and K.=Hypöestes.

HYPHAENE Gaertn. Palmae. 30 spp. Warm Africa, Madagascar, Arabia. Tall palms.

***H. argum*** Mart=Medemia argum. Nubia. The fruits are eaten locally.

***H. coriacea*** Gaertn. (Ginger Bread Palm). Coasts of E.Africa. The fruit pulp is eaten locally and made into a fermented drink. The leaves are used for thatching, baskets etc.

***H. crinita*** Gaertn. (South African Doum Palm). S.Africa. The young leaves are used locally for cordage, the older leaves are used to make hats, baskets etc. The stem sap is fermented to make a drink and the seeds make a vegetable ivory.

***H. shatan*** Cl. (Satra, Satramira). Madagascar. The leaves are used to make mats, baskets etc. and a poor quality paper. The sap from the base of the inflorescence is fermented to make a drink.

***H. thebaica*** Mart. (Egyptian Doum Palm). N.E. and E.Africa. The leaves are used locally for cordage, mats etc. and are made into an inferior paper. The unripe seeds are eaten raw, as is the rind from the seeds, which is also made into sweetmeats and molasses. The ground nuts are used to dress wounds and the nuts are sometimes used as vegetable ivory.

***H. ventricosa*** Kirk. Tropical Africa. The seeds are a source of vegetable ivory.

HYPHOLOMA (Fr.) Quél. Agaricaceae. 80 spp. Temperate, especially on wood.

***H. appendiculatum*** (Bull.) Karst. Temperate. The fruitbodies are eaten. They have a good flavour which is retained on drying.

***H. candolleanum*** (Fr.) Quél.=Psathyrella candolleana.

***H. perplexum*** Peck. (Perplexing Hypholoma). N.America. The rather tasteless fruitbodies are eaten cooked with seasoning, bacon etc.

***H. sublateritium*** (Schäff.) Karst. Temperate. The top of the fruitbody is eaten in Japan, Italy and N.America, when soaked in vinegar and cooked.

HYPNEA J. Ag. Sphaerococcaceae. Red Algae. Subtropical Seas. The following species are eaten locally.

***H. cenomyce*** J. Ag. Pacific and Indian Oceans. Eaten in Indonesia.

***H. cervcornis*** J. Ag. Pacific. Eaten in China.

***H. musciformis*** (Wulf.) J. Ag. Pacific. Eaten in China. Also a source of agar agar.

***H. nidifica*** J. Ag. Pacific. Eaten in Hawaii where it is called Limu Huna.

***H. spicifera*** J. Ag. Coast of S.Africa. Gives African Agar Agar.

HYPOCHOERIS L. Compositae. 100 spp. Cosmopolitan, especially S.America. Herbs.

***H. maculata*** L. (Cat's Ears). Europe to Siberia. The leaves are eaten locally in salads.

HYPOESTES Soland. ex R. Br. Acathaceae. 150 spp. Old World tropics, especially Madagascar.

***H. verticillaria*** R. Br. Tropical Africa. The roots are used in the Sudan for dyeing mats.

HYPOMYCES (Fr.) Tul. Hypocreaceae. 30 spp.

***H. lactifluorum*** Fr. N.America. Parasitic on the agaricacious fungus Lactarius piperatus. Has a good flavour and is eaten locally.

***Hyporhodius*** (Fr.) Schroet=Eccilia.

***Hyporhodius cervinus*** (Schäff) Henn.= Pluteus cervinus.

HYPOXIS L. Hypoxidaceae. 100 spp. America, Africa, E.Asia, Indomalaya, Australia. Herbs.

***H. aurea*** Lour.=H. flava=H. franquevillei. Tropical Asia. Used as an aphrodisiac and rejuvenating tonic.

HYPTIS Jacq. Labiatae. 400 spp. Warm America, W.Indies. Herbs.

***H. albida*** H.B.K. Mexico. The leaves are used locally to flavour food.

***H. breviceps*** Poir. Tropical America. Introduced to Java where it is used as a balm on wounds.

***H. emoryi*** Torr.=H. lanata (Emory Bushmint). S.W. U.S.A. The seeds (Chia) are eaten by the Mexican Indians.

***H. fasciculata*** Benth. S.America. The leaves are used locally to aid digestion, to induce sweating and to relieve catarrh.

***H. lanata*** Torr.=H. emoryi.

***H. laniflora*** Benth. (Salvia). Mexico. A decoction of the leaves is used locally to treat fevers.

***H. mutabilis*** (Rich.) Briq. Central America. A decoction of the leaves is used locally to treat stomach pains.

***H. pectinata*** (L.) Poir.=Nepeta pectinata (Comb Hyptis). Throughout tropics. A decoction of the leaves is used in the W.Indies to treat stomach pains. The leaves are used in Madagascar to make rum and a resin from the plant is burnt as incense in parts of Africa.

***H. spigera*** Lam. Tropical Africa. The plant is cultivated in E. and W.Africa for the seeds which contain about 30 per cent of a good drying oil. The oil is superior to linseed oil in many ways, but the smallness of the seeds make the commercial extraction on a large scale difficult. Small scale extraction does take place in parts of Africa and the oil is used as a substitute for linseed oil and for cooking. The seeds are eaten widely in Africa as a paste, in stews etc.

***H. suaveolens*** Poir.=Marrubium indicum. Throughout tropics. A tea is made from the leaves in W.Africa. In the Philippines a decoction of the leaves is used as an antispasmodic and the leaves are put in baths to relieve rheumatism and induce sweating. The branches are also put in beds and chairs to discourage bugs.

Hyptis, Comb – Hyptis pectinata.

Hyptis spicigera Oil – Hyptis spicigera.

Hyssop – Hyssopus officinalis.

Hyssop, Biblical – Origanum maru.

Hyssop, Giant – Agastache anethiodora.

Hyssop, Hedge – Gratiola officinalis.

Hyssop Oil – Hyssopus officinalis.

Hyssop, Wild – Verbena hastata.

HYSSOPUS L. Labiatae. 15 spp. S.Europe to Central Asia. Herbs.

***H. officinalis*** L. (Hyssop). Mediterranean, Balkans, Asia Minor, Iran. Occasionally cultivated for the oil from the leaves (Oil of Hyssop) which is used for flavouring liqueurs. A decoction of the leaves is used in home remedies on bruises and mixed with honey it is taken to relieve coughs, colds etc. and to treat stomach complaints.

HYSTERANGIUM Vittad. Polypodiaceae. 30 spp. Temperate.

***H. lapidescens*** Horan. The sclerotia are used in China to treat nerve-complaints of children, including epilepsy.

# I

IBERIS L. Cruciferae. 30 spp. Europe. Herbs.

***I. amara*** L. (Bitter Candytuft, Clown's Mustard, White Candytuft). Europe through to Crimea and Algeria. The leaves are used locally in the treatment of rheumatism and as an anti-scorbutic. The seeds are used as a substitute for mustard and in a wide variety of home remedies.

Iboga Root – Tabernanthe iboga.

Ibota Privet – Ligustrum ibota.

Icacia Chiche – Sterculia tomentosa.

ICACINA A. Juss. Icacinaceae. 5 spp. Tropical W.Africa. Herbs.

***I. senegalensis*** Juss. Tropical W.Africa. The tubers are eaten as a vegetable, locally. The starch from the seeds is sometimes used instead of cereal flour in times of want.

***I. trichantha*** Oliv. Tropical W.Africa. The tubers are eaten locally in soups.

Icaco Plum – Chrysobalanus icaco.

Ice Plant – Mesembryanthemum crystalinum.

Iceland Moss – Cetraria islandica.

Iceland Poppy – Papaver nudicaule.

Ichicaxihuitl – Gossypium mexicanum.

***Ichthyomethia piscipula*** L. = Piscidia erythrina.

Ichufa – Haloxylon tamaricifolium.

***Icica*** Aubl. = Protium.

Idar – Leptadenia lancifolia.

***Idothea ciliaris*** Kunth. = Drimia ciliaris.

Ifofolo – Craiba grandiflora.

Ifumbolo – Hugonia obtusifolia.

Igikindye – Dolichos malosanus.

Igisura – Urtica massaica.

Ignacio do Brasil – Fevillea trilobala.

Ignatius Beans – Strychnos ignatii.

Iguanero – Caesalpinia eriostachys.

Iguga – Fagara chalybea.

Ije – Premna tomentosa.

Ijebu Mahogany – Entandrophragma angolense.

Ikobambive – Cissus bambusati.

Ikon-du – Fagara leprieuri.

Ilama – Annona diversifolia.

ILEX L. Aquifoliaceae. 400 spp. Cosmopolitan, except N.America. Trees. The leaves of many species contain caffeine and are of local importance as a tea.

***I. amara*** Perodi. Argentina, Paraguay. The leaves used locally to make a tea.

***I. aquifolium*** L. (European Holly, Hulver). Europe to Caucasus and N.Africa. The white wood is strong and polishes well. It is used for carving, inlay work, veneer, handles of kettles etc., walking sticks and mathematical instruments. The berries are violently emetic. The twigs, with leaves and berries are used extensively in Europe for Christmas decorations and there is a large seasonal trade in them.

***I. asiatica*** Ham. = I. godajam.

***I. capensis*** Harv. and Song. (Water Tree). S.Africa. The white mottled wood is used locally for wheel spokes.

***I. cassine*** L. (Cassena, Dahoon Holly). S. and S.E. N.America. The leaves are used locally as a tea and sold to a limited extent. See I. aquifolium.

***I. conocarpa*** Reiss. The leaves are used occasionally as a tea. They are also used as a tonic, diuretic and to treat stomach complaints.

***I. gigantea*** Bonpl. = I. theezans.

***I. glabra*** (L.) Gray. (Gallberry, Inkberry). S.W. N.America. The flowers are a good source of honey.

***I. godajam*** Colebr. = I. asicatica (Gian tía, Rút). Tropical Asia. The bark is used to make a diuretic and is used to treat diarrhoea.

***I. integra*** Thunb. Japan. The ground bark is used locally as birdlime.

***I. macropoda*** Miq. Japan, Korea. The wood is tough, white and hard. It is used for turning, mosaic work, utensils and matches.

***I. malabarica*** Bedd. = I. wightiana S.India, Ceylon. The white wood is soft and is used to make tea chests etc.

***I. medica*** Reiss. Brazil. A decoction of the leaves is used locally as a diuretic and to treat stomach complaints.

***I. opaca*** Ait. (American Holly). E. N.America to Texas. The light wood is not strong, but tough. It is used for house interiors, cabinet-making and turning. See I. aquifolia.

***I. paraguensis*** A. St. Hil. (Paraguay Tea, Yerba Maté). S.American especially N.Argentina, Paraguay, Uruguay to Brazil. The plant is cultivated locally for the leaves which are made into a tea. Although the tea is of great local importance, little is exported. The leaves contain caffeine and has a pleasant smell but a slightly bitter taste. It is usually drunk with lemon and sugar, but rarely milk. the leaves are collected from a plant every two or three years. They are dried over a fire, ground down and are then ready for use. The wild plants have been used for tea for centuries and cultivation was begun by the Jesuits.

***I. peduculata*** Miq. (Longstalk Holly). Japan. The boiled leaves yield a brown dye, used locally.

***I. pseudobuxus*** Reiss. Brazil. The leaves are used locally to make a tea and the wood is used in general carpentry.

***I. sebertii*** Panch. New Caledonia. The yellow-white wood is hard and durable. It is used locally for house interiors and planks.

***I. theezans*** Mart.=I. gigantea. Brazil. The leaves are used locally as a tea; they are also used as a diuretic, stimulant and to treat stomach complaints.

***I. verticillata*** (L.) Gray. (Black Alder, Dogberry, Fever Bush, Winterberry). E. N.America. The leaves are used for tea.

***I. vomitoria*** Ait. (Carolina Tea, Cassena, Emetic Holly). S. and S.E. N.America. The leaves are used to make a tea (Black Drink) drunk by the Indians on ceremonial occasions. They are also used as a tea for normal use and are sometimes sold. The berries are emetic. See I. aquifolia.

***I. wallichii*** Hook. f. (Bui). S.Indochina. A decoction of the bark is used locally to reduce fevers and as a tonic.

***I. wightiana*** Dalz.=I. malabarica.

ILLICIUM L. Illiciaceae. 42 spp. India, E.Asia, W.Malaysia, Atlantic N.America, Mexico, W.Indies. Trees or shrubs.

***I. anisatum*** L.=I. religiosum. (Japanese Star Anise, Mang tsao, Shikimi). China, Japan. Cultivated. The seeds are very poisonous and are used as a fish poison. They are also used to relieve toothache and to treat skin diseases. The oil from the seeds is given to children to relieve stomach upsets.

***I. cambodiana*** Sarg. Tonking. See I. verum.

***I. parviflorum*** Sarg. (Yellow Star Anise). S.E. U.S.A. The oil from the bark (Anise Oil) is used for flavouring.

***I. religiosum*** Sieb.=I. anisatum.

***I. verum*** Hook. f. (Chinese Anise, Star Anise). S.E.Asia. Cultivated widely in the Far East and Jamaica for the seed oil which is used in flavouring. The oil is exported. The dried fruits are used to relieve indigestion

***Illipé betis*** Merr.=Madhuca betis.

Illipé Butter – Madhuca longifolia.

Illipé Butter Tree – Madhuca butyracea.

Illipé Nuts – Madhuca latifolia.

Illipé Nuts – Shorea gysbertsiana.

Illipé Nuts – Shorea macrantha.

Illipé Nuts – Shorea martiana.

Illipé Nuts – Shorea palembanica.

Illipé Nuts – Shorea seminis.

Illipeé Nuts – Shorea squamata.

Illipé Nuts – Shorea stenoptera.

Illorin Gum – Daniellia thurifera.

Iloko toko – Cissus planchoniana.

Ilonlolongo – Fagara rubescens.

Iluta lyapi – Acacia drepanolobium.

***Ilysanthes antipoda*** Merr.=Bonnaya antipoda.

Imbatabata – Lannea edulis.

Imbé – Garcinia livingstonii.

Imbu – Spondias tuberosa.

Imburana de Cheiro – Amburana acrean a.

Imburral – Oldenlandia umbellata.

Imburuyta – Allophylus africanus.

Imenga – Crytosperma senegalense.

Imigis – Grewia bicolor.

Immersion Oil – Juniperus virginiana.

IMPATIENS L. Balsaminaceae. 600–700 spp. Tropical and temperate Eurasia and Africa, especially Madagascar and the mountains of India and Ceylon. Herbs.

***I. burtonii*** Hook. f.=I. emini=I. kerchhovena. Central Africa. The leaves are used locally to treat wounds.

***I. emini*** Warb.=I. burtonii.

***I. giorgii*** de Wild. (Epokape). Congo. The leaves are eaten locally as a vegetable.

***I. irvingii*** Hook. f. ex Oliv.=I. kirkii=I.

thonneri. (Asa, Bagide, Elobwa). Central Africa. The juice from the plant is used locally as an antiseptic on cuts. The plant is also used as a source of salt.

***I. kerchhovena*** de Wild. = I. burtonii.

***I. kirkii*** Hook. f. ex Oliv. = I. irvingii.

***I. noli-tangere*** L. (Quick in Hand, Touch me not, Yellow Balsam). Europe to China and Japan. The plant is used as a diuretic, laxative and antiseptic on wounds.

***I. platypetala*** Lindl. = Balsamina latifolia. (Patjar leuwung). Malaya, Indonesia. The leaves are used locally to treat skin diseases.

***I. thonneri*** de Wild. = I. irvingii.

IMPERATA Cyr. Graminae. 10 spp. Tropics and subtropics. Perennial grasses.

***I. arundinaceae.*** Cyr. (Alang-alang, Lalang). Temperate. The leaves are used to make paper in China, for mats and raincoats for farmers in Japan and for thatching in the Philippines. The stems are used in the Philippines to make hats. The starch from the rhizomes is fermented to make a drink in Malaya.

***I. brasiliensis*** Trin. = I. caudata = I. sape = Saccharum sape. Tropical America. Used for livestock feed in Brazil.

***I. caudata*** Chapm. = I. brasiliensis.

***I. cylindrica*** (L.) Beauv. (Cogon Grass). S.Africa. Used for reclamation work but is unpalatable to livestock.

***I. sape*** Anders. = I. brasiliensis.

***Imperatoria ostruthium*** L. = Peucedanum ostruthium.

Inaola-a-liwye – Tiliacora bequaertii.

Inca Wheat – Amaranthus caudatus.

Incense Cedar – Libocedrus decurrens.

Incense Wood – Amoora nitida.

Incienso – Myrocarpus frondosus.

Incienso de pais – Bursera bipinnata.

Inchi Grass, Oil of – Cymbopogon martini.

Inchi Lemon Grass des Indes orientales – Cymbopogon flexuosus.

Indian Abutilon – Abutilon indicum.

India Coral Bean – Erythrina indica.

India Drug Squill – Urginea indica.

Indian Gum – Anogeissus latifolia.

India Madder – Rubia cordifolia.

India Sissoo – Dalbergia sissoo.

India Rubber – Marsdenia tenacissima.

Indian Aconite – Aconitum ferox.

Indian Almond – Terminalia catappa.

Indian Aloë – Aloë vera.

Indian Arrow Wood – Euonymus atropurpureus.

Indian Balsam – Leptotaenia multifida.

Indian Barberry – Berberis aristata.

Indian Bdellium – Commiphora mukul.

Indian Bean – Catalpa bignonioides.

Indian Belladonna – Atropa acuminata.

Indian Berry – Anamirta paniculata.

Indian Black Root – Pterocaulon undulatum.

Indian Blue Pine – Pinus excelsa.

Indian Bread – Poria cocos.

Indian Brown Mustard – Brassica integrifolia.

Indian Butter – Madhuca butyracea.

Indian Cassia – Cinnamomum tamala.

Indian Copal – Vateria indica.

Indian Corn – Zea mays.

Indian Cucumber Root – Medeola virginiana.

Indian Cup Plant – Silphium perfoliatum.

Indian Currant – Symphoricarpos orbicularis.

Indian Fig – Opuntia ficus-indica.

Indian Frankincense – Boswellia serrata.

Indian Hemp – Apocynum cannabinum.

Indian Jalap – Ipomoea turpethum.

Indian Juniper – Juniperus macropoda.

Indian Kale – Xanthosoma violaceum.

Indian Laburnum – Cassia fistula.

Indian Liquorice – Abrus precatorius.

Indian Madder – Oldenlandia umbellata.

Indian Mallow – Abutilon avicennae.

Indian Mulberry – Morinda trifolia.

Indian Mustard – Brassica juncea.

Indian Nettle – Acalypha indica.

Indian Olibanum – Boswellia serrata.

Indian Olive – Olea cuspidata.

Indian Paper Birch – Betula utilis.

Indian Percha Tree – Palaquium ellipticum.

Indian Plantago Seed – Plantago ovata.

Indian Poke – Veratrum viride.

Indian Red Wood – Chukrasea tabularis.

Indian Red Wood – Soymida febrifuga.

Indian Rennet – Withania coagulans.

Indian Rhubarb – Saxifraga peltata.

Indian Rice – Zizania aquatica.

Indian Rubber – Ficus elastica.

Indian Sarsaparilla – Hemidesmus indicus.

Indian Silver Grey Wood – Terminalia bialata.

Indian Tobacco – Lobelia inflata.

Indian Tragacanth – Astragalus heratensis.

Indian Turnip – Arisaema triphyllum.

Indian Turnip – Eriogonum longifolium.

Indian Turpentine – Pinus longifolia.

Indian Wheat – Fagopyrum tataricum.
Indian Willow – Salix tetrasperma.
Indian Wintergreen (Oil) – Gaultheria fragantissima.
Indian Yellow – Mangifera indica.
Indigo – Indigofera anil.
Indigo – Indigofera leptostachya.
Indigo – Indigofera tinctoria.
Indigo, False – Amorpha fruticosa.
Indigo, False – Baptisia australis.
Indigo, Mysore Pala – Wrightia tinctoria.
Indigo Squill – Cammassia esculenta.
Indigo, West African – Lonchocarpus cyanescens.
Indigo, Yellow Wild – Baptisia tinctoria.
Indigo, Yoruba – Lonchocarpus cyanescens.
INDIGOFERA L. Leguminosae. 700 spp. Warm. Herbs. The genus is the principle source of the blue dye indigo. The dye was an important commercial commodity during the 17th and 18th centuries, much being imported from the Far East. It has now been replaced largely by the coal-tar dyes. The plants are cut just before flowering and cut into small pieces. The pieces are soaked in water when a yellow solution forms. This oxidises in air to a blue precipitate. Further oxidation is prevented by heating. The precipitate is dried and pressed into blocks before exporting.
***I. anil*** L.=I. suffruticola. (Indigo Plant). Tropical America. Cultivated in Old and New World. It was the principle source of indigo.
***I. arrecta*** Hochst. (Degendeg). E.Africa. An important source of indigo in Abyssinia. Grown as a green manure.
***I. diphylla*** Vent. W.Africa. A local source of blue dye.
***I. hendecaphylla*** Jacq. Africa, Asia. Grown as a green manure in rubber plantations.
***I. lespedezioides*** H.B.K. (Anil cimarró, Cacherahua). S.Mexico to S.America, W.Indies. The plant is used in Guatemala to treat stomach complaints.
***I. pascuorum*** Benth. Tropical America. A decoction of the roots is used locally to treat fevers and as a stimulant.
***I. pauciflora*** Delil. Asia, Africa. Grown as fodder for camels. A decoction of the roots in milk is used in the Sudan as a purgative.
***I. simplifolia*** Lam. Tropical Africa. The roots are used in Nigeria as an arrow poison.
***I. suffruticosa*** Mill.=I. anil.
***I. sumatrana*** Gaertn.=I. tinctoria.
***I. tinctoria*** L.=I. sumatrana. (Indigo Plant). Malaysia. It was the main source of indigo from the Far East.
***I. trita*** L. f. (Kandaram, Torementi, Vekharo). Tropical Asia, especially India. The Hindus used the seeds as a tonic.
Industrial Cascarille – Croton echinocarpus.
INGA Mill. Leguminosae. 200 spp. Tropical and subtropical America, W.Indies. Trees.
***I. anomala*** Kunth.=Calliandra anomala.
***I. feuillei*** DC.=I. reticulata=Feuillaea reticulata. Peru. Cultivated. The fruits have been eaten locally since ancient times.
***I. goldmanii*** Pittier. (Guavo de Mono). Costa Rica, Peru. Grown as shade in coffee plantations.
***I. edulis*** Mart. (Food Inga, Inga Cipo). Mexico through Panama southwards. The trees are grown as shade in coffee plantations. The fruits are eaten locally.
***I. leptoloba*** Schlecht. (Quamo). S.Mexico through to S.America. Grown locally as shade in coffee plantations.
***I. macuna*** Walp. and Duch. (Guavo Peludo). Panama. The fruits are eaten locally.
***I. paterno*** Harms. (Paterno). Mexico to Costa Rica. The pods are eaten locally.
***I. pittieri*** Mich. Costa Rica, Panama. Grown as shade in coffee plantations.
***I. preusii*** Harms. (Cuxiniquil). El Salvador. Grown locally as shade in coffee plantations.
***I. pterocarpa*** DC.=Peltophorum pterocarpum.
***I. punctata*** Willd. (Guavo). S.Mexico to Costa Rica. Grown locally as shade in coffee plantations.
***I. radians*** Pitt. (Guavo Real). S.Mexico to S.America. The pods are eaten locally.
***I. rensoni*** Pitt. (Cujin). El Salvador. The pods are eaten locally.
***I. reticulata*** Spr.=I. feuillei.
***I. ruiziana*** G. Don. (Toparejo). Nicaragua to Panama. The fruits are eaten locally.
***I. spectabilis*** Willd. (Guavo de Castilla). Costa Rica to Panama. The fruits are eaten locally.
***I. spuria*** Humb. and Bonpl. (Nacaspilo). Mexico through to S.America. The fruits are eaten locally.
Inga Cipo – Inga edulis.

Inga Food – Inga edulis.
Inglewane – Bulbine aloides.
Ingo – Ongokea kamerunensis.
Ingo-sha-hush – Neonelsonia acuminata.
Ingoo – Toona sureni.
Ingudi – Balanites roxburghii.
Inhambane Coffee – Coffea racemosa.
Inhambane Copal – Copaifera gorskiana.
Inkberry – Ilex glabra.
Inkcap – Coprinus atramentarius.
Ink Pipe – Monotropa uniflora.
Ink Root – Statice limonium.

INOCARPUS J. R. and G. Forst. Leguminosae. 4 spp. Pacific Islands.

***I. edulis*** Forst. = Bocoa edulis (Polynesian Chestnut, Tahiti Chestnut). Pacific Islands. The seeds form an important food in Samoa. They are eaten raw or cooked.

INOCYBE (Fr.) Quél. Agaricaceae. 150 spp. Widespread.

***I. cutifracta*** Petch. Tropical Asia. The fruitbodies are eaten.

***Inodes*** O. F. Cook = Sabal.
Inomena – Asparagopsis sanfordiana.
***Ionidium ipecacuanha*** L. = Hybanthus ipececuanha.
***Inophloeum*** Pittier = Poulsenia.
Insect Powder, Dalmatian – Chrysanthemum cinerariifolium.
Insect Powder, Persian – Chrysanthemum roseum.
Intermediate Wheatgrass – Agropyron intermedium.
Intisy – Euphorbia intisy.

INTSIA Thou. Leguminosae. 9 spp. Islands off tropical E.Africa, Madagascar, tropical Asia, Malaysia. Trees.

***I. amboinensis*** Thou. = Afzelia bijuga.

***I. bakeri*** Prain. = Afzelia palembanica.

***I. palembanica*** Miq. (Ipil, Kwila, Marbau, Moluccan Ironwood). Indonesia, Philippines, Borneo, New Guinea. The dark red-brown wood is hard and heavy, it is durable and not very susceptible to decay or attack by termites etc. It is used for heavy construction work, bridges, piles etc.

***I. plurijuga*** Harms. (Dowora kome, Tos lelake). Celebes to New Guinea. The hard yellow-brown wood is used for heavy construction work in the Celebes.

INULA L. Compositae. 200 spp. Eurasia, Africa. Herbs.

***I. crithmoides*** L. (Golden Samphire). Coasts of Europe, Mediterranean. The leaves are used as a salad occasionally in England.

***I. helenium*** L. = Helenium grandiflorum (Alant, Elecampane, Horseheal, Scabwort, Yellow Starwort). S. and C.Europe, Balkans, Central Asia. The roots contain an essential oil (Elecampane Oil). The plant is cultivated for the roots which are used in flavours for sweets, desserts etc. The oil is used in vermouths and medicinally in treating coughs, colds, asthma and chest complaints. It is said to kill the tuberculosis bacterium. The root also yields a blue dye.

***I. viscosa*** Ait. Mediterranean. A yellow dye from the root is used locally in Greece.

***Ionidium*** Vent. = Hybanthus.
Ioobai – Myrica nagi.
Ipecac – Cephaëlis ipecacuanha.
Ipecac, American – Gillenia stipulata.
Ipecac, Cartegena – Cephaëlis acuminata.
Ipecac, Farinaceous – Richardsonia pilosa.
Ipecac, Nicaragua – Cephaëlis acuminata.
Ipecac, Panama – Cephaëlis acuminata.
Ipecac, Para – Cephaëlis ipecacuanha.
Ipecac Spurge – Euphorbia ipecacuanha.
Ipecac, Striated – Cephaëlis emetica.
Ipecac, Undulated – Richardsonia pilosa.
Ipecac, White – Hybanthus ipecacuanha.
Ipecac, White – Polygala angulata.
Ipecac, Wild – Apocynum androsaemifolium.
Ipecac, Wild – Asclepias carassavica.
Ipecacuanha – Cephaëlis ipecacuanha.
Ipecacuanha branca – Hybanthus ipecacuanha.
Ipecacuanha, Country – Naregamia alata.
Ipecacuanha, Goa – Naregamia alata.
Ipecacuanha, Portuguese – Naregamia alata.
Ipil – Intsia palembanica.

IPOMOEA L. Convolvulaceae. 500 spp. Tropical and warm temperate. Vines or creepers.

***I. aculeata*** (L.) Kuntze = Calonyction aculeatum = C. speciosum. Old World and New World Tropics. The juice is used to coagulate the latex of Castilla elastica in Mexico and Central America.

***I. altissima*** Mart. = Operculia altissima. Brazil. The very poisonous roots are used locally in small quantities as a violent emetic.

***I. angustifolia*** Jacq. = Convolvulus japonicus. Japan. The root is eaten locally when roasted.

***I. aquatica*** Forsk.=I. reptans=Convolvulus reptans. Tropics. Cultivated in S.China. The leaves and stems are eaten locally as a vegetable (Chinese Cabbage). The stems are sometimes pickled. The Burmese use the juice as an emetic in cases of opium and arsenical poisoning.

***I. arborescens*** (Humb. and Bonpl.) Son.= Convolvulus arborescens=C. quahutzehuatl. (Palo blanco, Palo bolo, Quauhzahuatl). Mexico, El Salvador. Tree. The bark is used locally for treating snakebites and diseases of the spleen.

***I. batatas*** Poir.=Batatas edulis. (Batate, Sweet Potato). Tropical America, W.Indies. The starchy tubers are an important food. There are hundreds of varieties, many of which are called yams, especially in S. U.S.A. The tubers contain large amounts of starch and sugars, and are important source of vitamins A, B and C. The tubers are also canned, used as a source of starch (Brazilian Arrowroot), fermented to produce alcohol and used to make glucose. The cultivated types are hexaploid 2n=90. This is possibly a reason why seeds rarely set, although the plants flower infrequently in cultivation. Propagation is exclusively by cuttings, new varieties being produced by crossing in controlled conditions. Exact figures for world production are unobtainable, as most of the crop is produced on small holdings and enters only into local commerce. It was grown in tropical America for thousands of years and introduced to China. It reached Europe before the common potato and is now grown throughout the world. Japan is one of the leading producers.

***I. bracteata*** Cav.=Exogonium bracteatum.

***I cairica*** (L.) Sw.=Convolvulus cairicus. (Cairo Morning-glory). Tropics, especially Africa and Asia. The stems are used in Hawaii as rough cordage and the tubers are used there as a famine food.

***I. digitata*** L. (Fingerleaf Morning-glory). Tropics, especially India, Ceylon, Malaya. The roots are used as a tonic to improve the appetite and as an aphrodisiac.

***I. gigantea*** Choisy=I. palmato-pinnata.

***I. hederacea*** (L.) Jacq.=Convolvulus hederaceus=C. nil. Throughout the tropics. In India the roasted seeds are used as a purgative, while in China they are used to treat intestinal worms and as a diuretic.

***I. leptophylla*** Tor (Bush Morning-glory). W. N.America. The roots were used as a famine food by the Indians. Some tribes burnt the roots, the smoke was a remedy for nervousness and bad dreams. The ground roots dusted on the body were used to relieve pain and to revive after fainting.

***I. macrantha*** Doh.=I. murucoides.

***I. mammosa*** Chois.=Merremia mammosa. Philippines. The roots are eaten as a vegetable locally.

***I. murucoides*** Roem. and Schult.=I. macrantha=Convolvulus macranthus. (Arbol del muerto, Arbol del venado, Cazazuate, Palo del muerto). Trees. S.Mexico, Guatemala. A decoction of the wood is used in Mexico as a treatment for paralysis. The ash is used as a soap in Guatemala.

***I. orizabenis*** Ledenois. (Mexican Scammony). Mexico. The tubers are used locally to make a gruel (Ipomea, Mexican Scammony, Orizaba Jalap). The roots are used medicinally as a purgative.

***I. ovalifolia*** Choisy.=Jacquemontia sandwicensis.

***I. palmato-pinnata*** Benth. and Hook.=I. gigantea. S.America. The ground roots are used locally as a purgative.

***I. pes-caprae*** (L.) Sweet.=Convolvulus pescaprae. (Seaside Morning Glory). Throughout the tropics. The roots are eaten as a famine food; the seeds are used as a purgative. The leaves are used in ritual baths by the Caribs of the W.Indies.

***I. purga*** Hayne.=Exogonium jalapa. (Jalap). Mexico. Cultivated in Mexico, W.Indies and E.Indies for the roots which are used medicinally as a purgative.

***I. reptans*** (L.) Poir.=I. aquatica.

***I. simulans*** Hanbury. Mexico. Yields Tampico Jalap. See I. orizabensis.

***I. tiliacea*** Choisy. S.America, W.Indies. The tubers are fed to pigs in the W.Indies.

***I. tricolor*** Cav.=I. violacea.

***I. turpethum*** R. Br.=Operculina turpethum.

***I. violacea*** L.=I. tricolor=Convolvulus violaceus. Mexico. The seeds contain the hallucinogenic drugs d-lysergic acid (LSD) amide and d-isolysergic acid;

they are called Badoh Negro and are used ceremonially by the local Indians.

Ipomea – Ipomoea orizabensis.

Iramomi – Picea bicolor.

Iré Rubber – Funtumia elastica.

IRESINE P. Br. Amaranthaceae. 80 spp. Australia, America, Galapagos Islands.

***I. arbuscula*** Uline and Bray. Central America. Tree. The ash is used in Guatemala to make soap.

***I. calea*** (Ibanez) Standl. Shrub. Mexico. A decoction of the plant is used locally as a diuretic and to induce sweating.

***I. celosia*** L. (Jubas Bush). Herb. W.Indies. An infusion of the leaves is used to treat stomach complaints in Cuba.

***I. paniculata*** (L.) O. Ktze. Herb. Tropical America. A decoction of the leaves is used to reduce fevers in parts of Mexico.

***Iriartea durissima*** Oerst. = Socratea durissima.

***Iridaea edulis*** (Stackh.) Harv. = Dilsea edulis.

IRIDOPHYCUS Gigartinaceae. Marine Red Algae. Pacific coast of N.America.

***I. flaccidus*** Setch. and Gud. Coast of California. Used to produce on non-colloidal meal (Kelp Meal) which is made into Kelp tablets for human consumption. It is also mixed with alfalfa for animal feeding.

Iripil Bark – Pentaclethra filamentosa.

IRIS L. Iridaceae. 300 spp. N.Temperate. Herbs.

***I. caroliniana*** Wats. E. N.America. The dried rhizomes are used as an emetic, purgative and diuretic.

***I. ensata*** Thunb. (Irisa Marjul). W.Himalayas. The Hindus use the roots to improve the appetite and in the treatment of dropsy and liver ailments.

***I. filifolia*** Boiss = I. juncea. Mediterranean. The rhizomes are eaten as a vegetable by the Arabs.

***I. florentina*** L. (Florentine Iris). S.Europe. The plant is cultivated for the rhizome which is dried and ground to form Orris. This is used in toilet powders and toothpastes and powders. The oil from the rhizome (Orris Butter) is used in perfumery as a fixative, especially in perfumes with a violet base. The plant is cultivated mainly in France, Italy and Germany. It is harvested after flowering.

***I. foetidissima*** L. (Gladwin, Stinking Gladwin, Spurgewort). Central and S.Europe to Afghanistan, N.Africa. The rhizome is used as a pain-reliever and purgative.

***I. germanica*** L. (German Iris, Flag Iris). Mediterranean. Cultivated extensively. The rhizome (Radix Iridis, Rhizoma Iridis, Verona Orris Root) is used as an emetic, purgative, diuretic and expectorant. It is also used to make toothpowders and sachets. Small pieces of the rhizome (Rhizoma Iridis pro infantibus) are given to children to chew during teething. It is used in some parts of Italy to make rosary beads.

***I. juncea*** Brot. = I. filifolia.

***I. missouriensis*** Nutt. (Wild Iris). W. N.America. The rhizome is chewed by the local Indians to relieve toothache.

***I. nepalensis*** D. Don. (Chiluchi, Shoti, Sosan). Himalayas. The rhizome is used in Hindu medicine as a purgative and diuretic, particularly in treating obstructions of the bile duct.

***I. pseudacorus*** L. (False Calamus, Yellow Flag). Europe, Asia, N.Africa. The rhizome (Radix Acori palustris) is used to relieve toothache, to treat diarrhoea, to aid menstruation and to treat discharges from the vagina. The rhizome also is used in tanning and with the addition of iron salts to make blue and black dyes. The seeds are used to relieve stomach upsets and aid digestion and as a coffee substitute.

***I. setosa*** Pall. Alaska, Siberia. The seeds are used as a coffee substitute by the Eskimos.

***I. vesicolor*** L. (Blue Flag). E. N.America. The dried rhizomes are used as a purgative, emetic and diuretic.

Iris, Flag – Iris germanica.

Iris, Florentine – Iris florentina.

Iris, Wild – Iris missouriensis.

Irisa – Iris ensata.

Irish Moss – Chondrus crispus.

Iroko Fustic Tree – Chlorophora excelsa.

Ironbark – Eucalyptus siderophloia.

Ironbark, Broadleaf – Eucalyptus siderophloia.

Ironbark, Grey – Eucalyptus paniculata.

Ironbark, Lemon-scented – Eucalyptus staigeriana.

Ironbark, Red – Eucalyptus siderophloia.

Ironbark, Red – Eucalyptus sideroxylon.

Ironbark, White – Eucalyptus leucoxylon.

Ironoak – Quercus obtusiloba.

Ironweed – Centaurea nigra.

Ironweed, Drummond's – Veronia missurica.
Ironweed, Kinka Oil – Veronia anthelmintica.
Ironweed, Nigerian – Veronia nigritana.
Ironwood – Acacia excelsa.
Ironwood – Afzelia palembanica.
Ironwood – Backhousia myrtifolia.
Ironwood – Cistanthera papaverifera.
Ironwood – Cyrilla racemiflora.
Ironwood – Eugenia confusa.
Ironwood – Mesua ferrea.
Ironwood – Olea paniculata.
Ironwood – Sideroxylon mastichodendron.
Ironwood, Black – Olea laurifolia.
Ironwood, Borneo – Eusideroxylon zwageri.
Ironwood, Dwarf – Lophira alata.
Ironwood, Malabar – Hopea parviflora.
Ironwood, Molucca – Afzelia bijuga.
Ironwood, Molucca – Intsia palembanica.
Ironwood, Red – Reynosia septentrionalis.
Ironwood, Rhodesian – Copaifera mopane.
Ironwood, Sonora – Olneya tesota.
Ironwood, Uganda – Cynometra alexandri.
Ironwood, White – Toddalia aculeata.
Irregular Barley – Hordeum irregulare.

IRVINGIA Hook. f. Ixononthaceae. 10 spp. Tropical Africa, Malayan Peninsular, Indochina, Borneo. Trees.

***I. barteri*** Hook. f. (Dika Bread, Dika Nut, Gabon Chocolate). W.Africa. The fruits are eaten. The seeds are ground to make a flour. The fat extracted from the seeds is used to make candles and soap.

***I. gabonensis*** Baill. (Wild Mango). Tropical Africa. The fat from the seeds (Dikka Butter, Dikka Fat) is eaten locally. The kernels are used to make a flour. The red-yellow wood is hard and tough. It is used for street paving.

***I. malayana*** Oliv. (Kalik karsik, Paooh bajan). Malaya, Indonesia. The fat from the seeds is used to make candles.

***I. oliveri*** Pierre. Asia, especially Cambodia and Cochin-China. The seeds yield a non-drying fat (Cay-Cay Fat) used to make candles.

***I. smithii*** Hook. f. Tropical Africa. An oil from the seeds is used locally in making perfumes.

ISACHNE R. Br. Graminae. 60 spp. Tropics and subtropics. Perennial grasses.

***I. australis*** R. Br. Australia. Used for livestock feed in the tropics.

Isatan – Isatis tinctoria.

ISATIS L. Cruciferae. 45 spp. Europe, Mediterranean to S.W. and E.Asia. Herbs.

***I. japonica*** Miq. Japan. The leaves yield a green dye used locally.

***I. tinctoria*** L. (Isatan, Woad). Europe to W.Asia. It was an important source of blue dye when the cost of indigo was high and before the invention of the analine dyes. The dye was used by the Ancient Britons to stain the skin and later to dye fabrics. Woad is extracted from the ground leaves, but the blue colour does not develop until oxidation. See Commiphora.

Isafghol – Plantago amplexicaulis.
Isano (Nut) – Ongokea klaineana.

ISCHAEMUM L. Graminae. 50 spp. Tropics and subtropics. Perennial grasses.

***I. angustifolium*** Hack. (Bharbur Grass). N.India, Afghanistan. The plant is used to make cordage, sails, paper and mats.

***I. timorense*** Kunth. Malaysia. Used as food for livestock.

ISCHNOSIPHON Kvern. Marantaceae. 35 spp Tropical America, W.Indies. Herbs.

***I. aruma*** (Aubl.) Koernicke. S.America to the Amazon, Trinidad and Tobago. The leaves are used by the natives to make baskets.

ISERTIA Schreb. Rubiaceae. 25 spp. Central and tropical S.America. Trees.

***I. pittieri*** Standl. S.America. The leaves are used locally as a soap substitute.

Isinglass – Gelidium cartilagineum.
Islio – Physostigma venenosum.
Islay – Prunus ilicifolia.

ISOCARPHA R. Br. Compositae. 10 spp. S. United States of America to tropical S.America, W.Indies. Herbs.

***I. oppositifolia*** (L.) R. Br. = Calea oppositifolia. W.Indies, Mexico to S.America. The leaves are used in Cuba to make a tonic (Manzanella de la Tierra).

ISOETES L. Isoetaceae. 75 spp. Temperate and tropics. Herbaceous cryptogams.

***I. martii*** A. Braun. Brazil. Used locally to treat snake-bite.

ISOLONA Engl. Annonaceae. 20 spp. Tropical Africa, Madagascar. Small trees.

***I. thonneri*** (De Wild. and T. Dur.) Engl. (Bundjinji, Konadala). Congo. The very hard wood is occasionally used for construction work, its main use being for making xylophones.

***Isomeria*** D. Don. ex DC. = Veronia.

***Isonandra gutta*** Hook. f. = Palaquium gutta.

ISOPTERA Scheff. ex Burck. Dipterocarpaceae. 3 spp. W.Malaysia. Trees.

*I. borneensis* Scheef. Borneo. The wood is used locally for building. It is heavy and water-resistant.

*I. sumatrana* van Slooten. (Kedawang, Sengkawang). Sumatra. The heavy wood is used locally for making bridges.

ISOTOMA Lindl. Campanulaceae. 10 spp. Australia; 1 sp. Central and S.America, W.Indies. Herbs.

*I. longiflora* Presl. Tropical America. A decoction of the plant is used locally to treat asthma.

Ispaghul – Plantago amplexicaulis.
Ispaghul – Plantago ovata.
Istrian Galls – Quercus ilex.
Itabo – Yucca elephantipes.
Italian Corn Salad – Valerianella eriocarpa.
Italian Cypress – Cupressus sempervirens.
Italian Fennel – Foeniculum vulgare.
Italian Jasmine – Jasminum grandiflorum.
Italian Liquorice – Glycyrrhiza glaba.
Italian Millet – Setaria italica.
Italian Rye Grass – Lolium italicum.
Italian Senna – Cassia obovata.
Italian Turnip Broccoli – Brassica ruvo.
Itauba Branca – Nectandra rodivei.
Itendale – Rhynchosia monobotrya.
Iteno – Dorstenia convexa.

*Ithyphallus impudicus* Ed. Fischer=Phallus impudicus.

ITOA Hemsl. Flacourtiaceae. 2 spp. China. Trees.

*I. stapfii* van Slooten=Poliothyrsis stapfii=Mesaulosperma stapfii. (Daagon, Dancan). Celebes. The yellow wood is hard and fairly durable. It is used locally in house construction.

IVA L. Compositae. 15 spp. N. and Central America. W.Indies. Herbs.

*I. axillaris* Push. (Poverty Seed). W. N.America. A decoction of the plant is used by the local Indians to relieve stomach pains and cramps, especially in children.

Iva Bitter – Achillea moschata.
Iva Liqueur – Achillea atrata.
Iva Liqueur – Achillea moschata.
Iva Wine – Achillea moschata.
Ivory Coast Khaya – Khaya ivorensis.
Ivory Coast Raffia Palm – Raphia hookeri.
Ivory-like Ramalina – Ramalina scopulorum.
Ivory Nut Palm, Polynesian – Coelococcus amicarum.
Ivory Palm, Common – Phytelephas macrocarpa.
Ivory Palm, Seemann – Phytelephas seemanni.
Ivory Wood – Siphonodon australe.
Ivorywood Seed Oil – Agonandra brasiliensis.
Ivy, English – Hedera helix.
Ivy, Ground – Glechoma hederaceum.
Ivy, Poison – Rhus toxicodendron.
Iwa-take – Gyrophora esculenta.
Iwan-tschai – Epilobium angustifolium.
Iwar – Canarium ambionense.

IXONANTHES Jack. Ixonanthaceae. 10 spp. Himalayas, S.China, Indochina, Philippines, Borneo; 1 spp. New Guinea Trees.

*I. icosandra* Jack. (Jajoo booloos, Pagar anak). Malaysia. The juice from the bark is used locally to coagulate Djelutung. See Dyera costulata.

IXORA L. Rubiaceae. 400 spp. Tropics. Small trees.

*I. concinna* R. Br. E.Indies. The wood is used locally to make walking sticks.

*I. fulgens* Roxb. Malaya. The twigs are chewed as a remedy for toothache.

*I. longituba* Boerl. Indonesia. A decoction of the roots and twigs is used locally to stain basket work a brown colour.

Ixtle – Agave gracilispina.
Ixtle – Agave heteracantha.
Ixtle – Hesperoyucca whipplei.
Ixtle de Jamauve – Agave lophanthe.
Ixtle de Jaumave – Agave funkiana.
Ixtle Fibre – Agave falcata.
Ixtle, Jaumave – Agave heteracantha.
Ixtle, Palma-Samuela carnerosana.
Ixtle, Rula – Agave lophantha.
Izkhir – Cymbopogon jwarancusa.
Izote – Yucca elephantipes.

# J

Jabonecillo – Fouquieria macdougalii.
Jaborandi – Pilocarpus microphyllus.
Jaborandi, Paraguay – Pilocarpus pinnatifolius.
Jaborandi Pepper – Piper longum.
Jaborandi, Pernambuco – Pilocarpus jaborandi.

Jaboticaba – Myrciaria jaboticaba.
Jaboticaba – Myrciaria trunciflora.
Jaboty Butter – Erisma calcaratum.
Jaboty Tallow – Erisma calcaratum.
Jacana – Pouteria multiflora.
Jacaraca Merim – Dracontium polyphyllum.
Jacaraca Taia – Dracontium asperum.
JACARANDA Juss. Bignoniaceae. 50 spp. Central and S.America, W.Indies. Trees.
***J. brasiliana*** Pers. = Bignonia brasiliana. Brazil. The durable wood is used locally for general carpentry.
***J. caroba*** DC. = J. procera.
***J. caroba*** DC. var. ***oxyphylla*** Bur = J. oxyphylla.
***J. coerulea*** Juss. W.Indies. The leaves are used in Cuba as an external treatment for eczema. A decoction of the bark is used as an emetic and purgative and the roots are used to induce sweating.
***J. copaia*** (Aubl.) D. Don. = Bignonia copaia (Boxwood, Palo de Buba). Central America to Brazil. The white wood is coarse, light and firm. It is used for cheap construction work, boxes, matches etc. The pulp is used in paper manufacture. The bark is used locally as an emetic and purgative.
***J. micrantha*** Cham. Brazil. The wood is used locally for house interiors and general carpentry.
***J. mimosaefolia*** D. Don. = J. ovalifolia. Tropical S.America. The wood is durable, but breaks easily. It is used for general carpentry.
***J. obovata*** Mart. = J. subrhombea.
***J. ovalifolia*** R. Br. = J. mimosaefolia.
***J. procera*** Spreng. = J. caroba = Bignonia caroba (Caroba). Guyanas. The leaves are used as a sedative, diuretic, to induce sweating and to stimulate the appetite.
***J. subrhombea*** DC. = J. obovata. Brazil. A decoction of the bark is used locally to induce sweating.
Jacaranda do Pará – Dalbergia spruceana.
JACARATIA A. DC. Caricaceae. 8 spp. Tropical America. Africa. Trees.
***J. mexicana*** A. DC. (Papaya Orejona). Mexico through Central America. The fruits are eaten locally in salads, as a vegetable or as sweetmeats.
Jack Bean – Canavalia ensiformis.
Jack-in-the-box – Hernandia peltata.
Jack-in-the-pulpit – Arisaema triphyllum.
Jack Oak – Quercus marylandica.
Jack Pine – Pinus banksiana.
Jackfruit – Artocarpus integrifolia.
Jackwood – Cryptocarya glaucescens.
Jacob's Ladder – Polemonium coeruleum.
JACOBINIA Nees. ex Moric. Acanthaceae. 50 spp. Mexico to tropical S.America. Shrubs.
***J. spicigera*** (Schlecht.) Bailey = Justica spicigera. Mexico. The leaves develop a blue colour when put in water. They are used in Mexico to whiten clothes.
***J. tinctoria*** (Oerst.) Hemsl. Central America. The leaves yield a blue dye.
JACQUEMONTIA Choisy. Convolvulaceae. 120 spp. Tropics. Herbs.
***J. sandwicensis*** A. Gray. = Convolvulus ovalifolius = Ipomoea ovalifolia. (Hawaiian Jacquemontia, Pauohiika). Hawaii. The plant was used by the natives for a variety of purposes. The roots were eaten, the leaves were eaten with coconut, used as a tea, chewed to treat babies with thrush and used as a purgative.
Jacquemontia, Hawaiian – Jacquemontia sandwicensis.
JACQUINIA L. corr. Jacq. Theophrastaceae. 50 spp. Warm America, W.Indies. Shrubs. The crushed fruits are used locally to poison fish.
***J. aurantiaca*** Bart. = J. umbellata.
***J. pungens*** Gray.
***J. seleriana*** Urb. and Loes.
***J. umbellata*** A. DC. = J. aurantiaca.
Jadaonda – Cissus cyphopetala.
Jaffarabad Aloe – Aloe vera.
***Jagera pseudorhus*** Radlk. = Cupania pseudorhus.
Jaggery – Borassus flabellifer.
Jaggery – Cocos nucifera.
Jaggery – Phoenix sylvestris.
Jagüey blanco – Ficus laevigata.
Jajoo booloos – Ixonanthes icosandra.
Jajoo pakoi – Calophyllum amoenum.
Jakobee – Senecio cineraria.
Jalap – Ipomoea purga.
Jalap, Brazilian – Operculina pisonis.
Jalap, Indian – Ipomoea turpethum.
Jalap, Orizaba – Ipomoea orizabensis.
Jalap, Tampico – Ipomoea simularis.
Jalamala – Salix tetrasperma.
Jalo – Canarium amboinense.
Jalocote – Pinus teocote.
Jalongan – Agaia argentea.
Jamaica Bark – Quassia amara.
Jamaica Dogwood – Piscidia erythrina.
Jamaican Ebony – Brya ebenus.

Jamaican Fishfuddle Tree – Piscidia erythrina.
Jamaican Honeysuckle – Passiflora laurifolia.
Jamaican Kino – Coccolobus uvifera.
Jamaican Lancewood - Duguetia quitarensis.
Jamaican Nectandra – Nectandra coriacea.
Jamaican Quassia – Picrasma excelsa.
Jamaican Sarsaparilla – Smilax ornata.
Jamaican Sorrel – Hibiscus sabdariffa.
Jamba Oil – Eruca sativa.
Jambolan – Eugenia jamolana.
***Jambolifera odorata*** Lour. = Acronychia odorata.
***Jambolifera resinosa*** Lour. = Acronychia laurifolia.
Jambos – Eugenia jambos.
***Jambosa acuminatissima*** Hassk. = Eugenia acuminatissima.
***Jambosa cauliflora*** Miq. = Eugenia polycephala.
***Jambosa densiflora*** DC. = Eugenia densiflora.
***Jambosa glabrata*** DC. = Eugenia clavimyrtus.
***Jambosa lineata*** DC. = Eugenia lineata.
***Jambosa tenuicuspis*** Miq. = Eugenia cymosa.
***Jambosa vulgaris*** DC. = Eugenia jambos.
James Whitkowwort – Paronychia jamesii.
Jamrosa Bark – Terminalia mauritiana.
Janatsi – Debregeasia edulis.
Japan Clover – Lespedeza striata.
Japan Tallow – Rhus succedanea.
Japan Wood – Caesalpinia sappan.
Japanese Aconite – Aconitum fischeri.
Japanese Alder – Alnus japonica.
Japanese Apricot – Armeniaca mume.
Japanese Arrowroot – Pueraria thunbergiana.
Japanese Artichoke – Stachys sieboldi.
Japanese Barnyard Millet – Echinochloa frumentacea.
Japanese Bellflower – Platycodon grandiflorum.
Japanese Buckthorn – Rhamnus japonica.
Japanese Buckwheat – Fagopyrum esculentum.
Japanese Chestnut – Castanea arenata.
Japanese Elm – Ulmus japonica.
Japanese Fleeceflower – Polygonum cuspidatum.
Japanese Gall – Rhus semialata.
Japanese Ginger – Zingiber moiga.
Japanese Horse Chestnut – Aesculus turbinata.
Japanese Kelp – Laminaria japonica.
Japanese Lacquer – Rhus vernicifera.
Japanese Linden – Tilia japonica.
Japanese Medlar – Eriobotrya japonica.
Japanese Millet – Echinochloa frumentacea.
Japanese Mint Oil – Mentha arvensis.
Japanese Mustard – Brassica japonica.
Japanese Pagoda Tree – Sophora japonica.
Japanese Pepper – Zanthoxylum piperatum.
Japanese Persimmon – Diospyros kaki.
Japanese Plum – Eriobotrya japonica.
Japanese Plum – Prunus salicina.
Japanese Poplar – Populus maximowiczii.
Japanese Prickly Ash – Zanthoxylum piperatum.
Japanese Privet – Ligustrum japonicum.
Japanese Rasin Tree – Hovenia dulcis.
Japanese Raspberry – Rubus parvifolius.
Japanese Silver Fir – Abies firma.
Japanese Snake Gourd – Trichosanthes cucmeroides.
Japanese Star Anise – Illicium anisatum.
Japanese Staunton Vine – Stauntonia hexaphylla.
Japanese Stone Pine – Pinus pumila.
Japanese Timber Bamboo – Phyllostachys bambusioides.
Japanese Torreya – Torreya nucifera.
Japanese Tung Oil – Aleurites cordata.
Japanese Turpentine – Larix dahurica.
Japanese Wing Nut – Pterocarya rhoifolia.
Japanese Yew – Taxus cuspidata.
Japanese Zelkova – Zelkova serrata.
Jara dulce – Baccharis glutinosa.
Jaral – Baccharis glutinosa.
JARDINEA Steud. Graminae. 3 spp. Tropical Africa. Perennial grasses.
*J.* ***congoensis*** Franch. Tropical Africa, mainly W. to central. The stems are used locally for making mats, baskets etc.
JARILLA Rusby. Caricaceae. 4 spp. Mexico. Shrubs.
*J.* ***caudata*** (Brandeg.) Standl. (Jarilla). Mexico. The fruits are used locally to make preserves.
Jarilla – Gymnosperma glutinosum.
Jarilla – Jarilla caudata.
Jarilla del no – Baccharis glutinosa.
Jaropa de Mora – Rubus glaucus.
Jarrah – Eucalyptus marginata.
Jasmine, Arabian – Jasminum sambac.
Jasmine, Catalonian – Jasminum grandiflorum.
Jasmine, Common – Jasminum officinale.
Jasmine, Grape – Ervatamia coronaria.
Jasmine, Italian – Jasminum grandiflorum.
Jasmine Tree – Holarrhena febrifuga

JASMINUM L. Oleaceae. 300 spp. Tropics and sub-tropics, except N.America. Shrubs.

***J. auriculatum*** Vahl. (Juhi). Tropical Asia, especially Ceylon, Mauritius, Thailand. The flowers yield an essential oil used in perfumery.

***J. frutescens*** L.=J. luteum=J. syracum. S.Europe, N.Africa. The rhizome is used as an adulterant to Yellow Jessamine (Gelsemium sempervirens).

***J. grandiflorum*** L. (Catalonian Jasmine, Italian Jasmine). S.Europe. Cultivated for the flowers which yield an essential oil used in perfumery.

***J. lanceolatum*** Roxb. var ***puberulum*** Hmesl. China. The roots and twigs are used locally as a pain reliever.

***J. luteum*** Gueld.=J. frutescens.

***J. niloticum*** Gilg. Tropical Africa. The essential oil from the flowers is used as a perfume in the Sudan.

***J. odoratissimum*** L. Madeira. Cultivated locally for the flowers which yield an essential oil used in perfumery.

***J. officinale*** L. (Common Jasmine). Central Asia. Scrambling Shrub. Cultivated widely as an ornamental. Also cultivated for the flowers which yield an essential oil used in perfumery.

***J. paniculatum*** Roxb. China. The flowers are used locally for scenting tea.

***J. sambac*** Ait (Arabian Jasmine). Tropical Asia. The flowers are used for scenting tea.

***J. syriacum*** Boiss. and Gaill.=J. frutescens.

Jatamansi (Oil) – Nardostachys jatamansi.

JATEORHIZA Miers. Menispermaceae. 2 spp. Tropical Africa. Woody vines.

***J. columba*** (Roxb.) Miers.=J. palmata.

***J. macrantha*** (Hook. f.) Excell. and Medonça. (Malengwia, Ngodia). W. and Central Africa. The roots are eaten locally as a famine food.

***J. miersii*** Oliv.=J. palmata.

***J. palmata*** (Lam.) Miers.=J. columba,=J. meirsii=Cocculus palmata. Tropical Africa. Cultivated in Mauritius for the roots (Radix Columba) which contains the alkaloids columbamine, jateorhizine, palamatine. The roots are used medicinally as a tonic, to relieve digestive complaints and in the treatment of diarrhoea and dysentery.

Jatimisak – Heliotropium tuberculosum.

JATROPHA L. Euphorbiaceae. 175 spp. Tropics and subtropics, N.America, S.Africa. Shrubs or small trees.

***J. aconitifolia*** Mill. C.America. The leaves have been used as a vegetable in Mexico since ancient times.

***J. cinerea*** (Ortze) Muell. Arg. Mexico. A decoction of the leaves is used locally as a mordant for dyeing.

***J. curcas*** L. (Physic Nut, Purging Nut). Tropics. Cultivated in the tropics for the seeds which yield Curcas Oil. This is used medicinally as a strong purgative, for lighting, making candles and soap, as a lubricant and in the woollen industry.

***J. gaumeri*** Greenm. (Pomolche). Mexico. A decoction of the roots is used locally for treating snakebite.

***J. gossypifolia*** L. Throughout the tropics. The leaves are used in the W.Indies to reduce fevers, in Ghana as a purgative, in Venezuela as a purgative and to relieve stomach complaints, and in the Philippines as a poultice to reduce swellings of the breast. In Venezuela the roots are used to treat leprosy and the latex to treat ulcers. The seed oil is used in Ghana as a purgative.

***J. heudelotii*** Baill.=Ricinodendron heudelotii.

***J. multifida*** L. (Chicaquil, Tortora, Yucca cimarrona). Tropical America. Naturalized in Old World Tropics. Cultivated locally. In Brazil, the sap is used for treating wounds. The roasted seeds are used to treat venereal diseases and fevers. The seeds are strongly purgative and may cause poisoning.

***J. spathulata*** (Oreg.) Muell Arg. (Snagre de Grado, Terote). Mexico to S.W. U.S.A. The bark is used for tanning and yields a red dye. The stems are used to make whips and baskets.

***J. urens*** L.=Cnidoscolus marcgravii. (Pendo Tree). Tropical America. The seeds are eaten locally and yield a light oil used for cooking.

***J. zeyheri*** Sond. S.Africa. The rhizomes are used locally for tanning.

Jau – Tamarix dioica.

Jauary Palm – Astrocaryum jauri.

Jaujá – Suaeda ramosissima.

Jaumave Ixtle – Agave heteracantha.

Jaumave loguguilla – Agave funkiana.

Java Almond – Canarium ambionense.

Java Almond – Canarium commune.

Java Cardamon – Amomum maximum.

Java Devilpepper – Rauwolfia serpentina.

Java Nato Tree – Palaquium javanense.

Java Plum – Eugenia jambolana.
Java Tea – Orthosiphon stamineus.
Javanese Long Pepper – Piper retrofractum.
Javanese Palissander – Dalbergia latifolia.
JEFFERSONIA Bart. Podophyllaceae. 2 spp. E.Asia, N.America. Herbs.
***J. binata*** B. S. Barton = J. diphylla.
***J. diphylla*** (L.) Pers. = J. binata. (Twinleaf, Rheumatism Root). N.E. U.S.A., S.W. through central states. The roots are used as a stimulant and tonic, to induce perspiring and to relieve stomach pains.
Jeheb – Cordeauxia edulis.
Jelly, Iceland Moss – Cetraria islandica.
Jelutong – Alstonia scholaris.
Jelutong – Dyera costulata.
Jenny Stone Crop – Sedum reflexum.
Jequie Rubber – Manihot dichotoma.
Jequirity – Abrus precatorius.
Jering – Pithecellobium jiringa.
Jerusalem Artichoke – Helianthus tuberosum.
Jerusalem Rye – Triticum polonicum.
Jessamine, Yellow – Gelsemium sempervirens.
JESSENIA Kartst. Palmae. 6 spp. Trinidad, tropical S.America. Palm trees.
***J. batua*** (Mart.) Burrett. (Batauá Palm, Patauá Palm, Seje). Orinoco and Amazon Basins. The pericarp yields an excellent cooking oil (Patauá Oil) which is used locally. There is 18–24 per cent oil in the pericarp, which is extracted by boiling in water.
***J. polycarpa*** Karst. S.America. The pericarps are eaten locally and the kernels yield an oil used locally for cooking.
Jesuit's Bark – Cinchona officinalis.
Jew's Ear – Auricularia auricula-judae.
Jew's Mallow – Corchorus olitorius.
Jew's Plum – Spondias dulcis.
Jewish Citron – Citron medica.
Jhau – Tamarix gallica.
Jhuri – Osyris arborae.
Jiba – Erythroxylum havanense.
Jicama – Exogonium bracteatum.
Jicana – Pachyrrhizus palmatilobus.
Jicaro – Crescentia cujete.
Jicuma – Calopogonium coeruleum.
Jimgan – Lannea grandis.
Jimson Weed – Datura stramonium.
Jindai-Sugi – Cryptomeria japonica.
Jingan Gum – Lannea grandis.
Jingan Gum – Lannea wodier.
Jingan-ki-gona – Lannea grandis.
Jiki – Marsdenia tenacissima.
JOANNESIA Vell. Euphorbiaceae. 3 spp. Venezuela, N.Brazil. Trees.
***J. princeps*** Vell. = Anda gomesii. Brazil. The timber is used locally for construction work. The seeds yield a thick oil (Andaassy Oil) which is used to treat skin diseases and is strongly laxative.
Job's Tears – Coix lacryma-jobi.
Jobo – Spondias lutea.
Joconoxtla – Opuntia imbricata.
Jocote – Spondias purpurea.
Joe-Pye-Weed – Eupatorium maculatum.
Joggobee – Senecio cineraria.
Johnson Grass – Sorghum halepense.
Joint Fir – Ephedra trifurca.
Jojoba Oil – Simmondsia californica.
Jolocin – Heliocarpus donnel-smithii.
***Jonesia asoca*** Roxb. = Saraca indica.
***Jonesia pinnata*** Willd. = Saraca indica.
Jongkong – Dactylocladus stenostachys.
Jonquil (Oil) – Narcissus jonquilla.
Joshua Tree – Yucca arborescens.
Josswood – Mitragyna africana.
Joyapa – Macleania popenoei.
JUBAEA Kunth. Palmae. 1 sp. Chile. Palm tree.
***J. chilensis*** (Molina) Baill. = J. spectabilis.
***J. spectabilis*** H.B.K. = J. chilensis. (Coco de Chile, Coquito Palm, Honey Palm, Wine Palm, Wine Palm of Chile). The sap from the trunk is sweet and when concentrated is sold as palm honey. It is also fermented to make palm wine. The seeds yield an edible oil. They are also sold locally for eating, when they are called Coquitos. The whole fruits are sometimes candied as sweetmeats. The leaves are used for baskets etc.
Jubas Bush – Iresine celosia.
JUGLANS L. Juglandaceae. 15 spp. Mediterranean to E.Asia, Indo-China, N. and Central America, Andes. Trees.
***J. ailantifolia*** Carr. = J. sieboldiana.
***J. boliviana*** Dode. (Bolivian Black Walnut). Bolivia. The seeds (nuts) are eaten locally.
***J. cathayensis*** Maxim. (Cathay Walnut). China. The nuts (seeds) are eaten locally.
***J. cinerea*** L. (Butternut) E. N.America. The wood is coarse, light brown and not strong. It is used for furniture and house interiors. The nuts are eaten and the fruit pericarp yields a yellow-orange dye. The sap is a minor source of sugar. The inner root-bark (Butternut Bark) is used to treat fevers and as a mild purgative.

***J. duclouxiana*** Dode. Mountains of Asia. Cultivated in China. The fruits are eaten.

***J. honorei*** Dode. (Ecuador Walnut, Nogal). Highland of Ecuador. The strong wood is beautifully marked and is used for furniture making etc. The nuts are edible and the sweet kernels are made into a local sweetmeat (Nogada de Ibarra). The plant is sometimes cultivated.

***J. insularis*** Griseb. (Nogal, Palo del Nuez). Cuba. The leaves are used locally in astringent baths to treat skin diseases of the heads of children.

***J. kamaonia*** Dode. = J. regia var. kamaonia. Himalayas. The nuts (seeds) are eaten by the local Chinese.

***J. major*** (Torr.) Heller. = J. rupestris var, major. (Nogal silvestere). Arizona to New Mexico, Mexico. The leaves are used as an astringent in Mexico.

***J. mandschurica*** Maxim. (Manchurian Walnut). Manchuria. The hard heavy brown wood is used locally for general carpentry.

***J. mexicana*** S.Wats. = J. mollis.

***J. mollis*** Engelm. = J. mexicana. (Guatemala Walnut). Mexico. The wood is used for turning, for which it is valued. A decoction of the leaves is used locally to treat rheumatism and the fruit pericarp yields a brown dye.

***J. nigra*** L. (Black Walnut). N.America. The wood is highly valued, particularly for veneer work on furniture, car facias etc. It is dark brown hard and durable, the complete timbers being used for boatbuilding, interior work and gun stocks. The timber is exported to Europe for making gun stocks. The nuts (seeds) are edible and are much-used in fragments in confectionery, ice-cream etc. The collecting of the nuts from wild trees forms a small but important local industry in parts of U.S.A. This was one of the first New World trees to be established in Europe.

***J. pterococca*** Roxb. = Engelhardtia spicata.

***J. regia*** L. (English Walnut, Persian Walnut). The plant has been cultivated so long that its origins are obscure. They are probably from S.E.Europe to the Himalayas and China. It may have been introduced to the United Kingdom by the Romans, but the introduction may have been as late as the 16th century. The early settlers may have introduced the tree to U.S.A. but it was not successfully established until 1923. The timber is hard, heavy and beautifully grained and is much valued for veneer, furniture and gun-stocks. The tree is grown mainly for the fruits, the nuts (seeds) being eaten raw, salted or in confectionery. California and Oregon produce some 100,000 tons of nuts a year, France about 30,000 tons, Italy 20,000 tons, Balkans about 11,000 tons, Turkey 9,000 tons, with China, India, Chile and Australia producing significant amounts. The trees are usually grown from seed, but they are grafted to J. nigra in U.S.A. An oil from the seeds is also used in food, soap and paints. A decoction of the leaves, bark and pericarp is mixed with alum and used to stain wood a dark brown. A decoction of the leaves is used as a tea and as an insect repellent. A decoction of the pericarp is used as a tonic and to treat intestinal worms.

***J. regia*** L. var. ***kamaonia*** DC. = J. kamaonia.

***J. rupestris*** Engelm. (Texas Walnut). S.W. U.S.A. and Mexico. The nuts (seeds) are eaten by the local Indians.

***J. rupestris*** var. ***major*** Torr. = J. major.

***J. sieboldiana*** Maxim. = J. ailantifolia. (Cordate Walnut, Heartnut, Siebold Walnut). Japan, China. The wood is soft, light and dark brown. It is used in Japan to make gun stocks, furniture etc. The pericarp and bark are used to make a brown dye.

Juhi – Jasminium auriculatum.

Jujub – Ziziphus jujuba.

Jujub, Argentine – Ziziphus mistol.

Jujub, Chinese – Ziziphus jujuba.

Jujubbread – Ziziphus jujuba.

JULBERNARDIA Pellegr. Leguminosae. 11 spp. Tropical Africa. Trees.

***J. sereti*** (De Wild.) Troupin. = Berlinia serti. (Alumbi, Kua). Tropical Africa, especially Central and W. The wood (Congo Zebra Wood) is used for inside work in houses. It is exported.

Jum – Garuga pinnata.

Jumping Bean – Sapium biloculare.

Jumping Bean – Sebastiania pavoniana.

Jumpy Bean – Leucaena glauca.

JUNCUS L. Juncaceae. 300 spp. Cosmopolitan, rare in the tropics.

***J. acutus*** L. = Temperate. The stems are used to make mats.

***J. communis*** E. Mey. = J. effusus.

***J. effusus*** L.=J. communis. Temperate. Cultivated for the stems which are used to make mats. The pith is used to make candle wicks, especially in Japan.

***J. glaucus*** Sibth. (Rush). Central Europe through to temperate Asia, N. and S. Africa. The stems are used to make mats etc.

***J. procerus*** E. Mey. Chile. The stems are used locally to make cordage.

Jungle Rice – Echinochloa colonum.

Juniper – Juniperus communis.

Juniper, Alligator – Juniperus pachyphlaea.

Juniper Berry Oil – Juniperus communis.

Juniper, Blackseed – Juniperus saltuaria.

Juniper, Californian – Juniperus californica.

Juniper, Greek – Juniperus excelsa.

Juniper, Indian – Juniperus macropoda.

Juniper, Mexican – Juniperus mexicana.

Juniper Mistletoe – Phoradendron juniperinum.

Juniper, Prickly – Juniperus oxycedrus.

Juniper, Sierra – Juniperus occidentalis.

Juniper, Utah – Juniperus utahensis.

Juniper Wood Oil – Juniperus communis.

JUNIPERUS L. Cupressaceae. 60 spp. N.Hemisphere. Trees or shrubs.

***J. barbadensis*** L. (Bermuda Red Cedar, Southern Red Cedar). W.Indies. Florida. The wood was used to make pencils.

***J. californica*** Carr. (Californian Juniper). California to Mexico. The ground fruits were used as a flour by the local Indians.

***J. communis*** L. (Common Juniper). Temperate Europe, Asia and N.America. The fruits contain 0·5–1·5 per cent essential oils, 10 per cent resin and 15–30 per cent dextrose. After they have matured for about two years they are used to flavour gin, liqueurs etc. The oil (Oil of Juniper) obtained by steam distillation of the fruits is used as a diuretic and urogenital stimulant. The wood and young shoot tips are also used as a diuretic. The oil distilled from the wood is used in veterinary medicine. The best quality berries are obtained from Italy, Czechoslovakia, Hungary and Yugoslavia. The bark is used as cordage in Lapland and the plant is widely grown as an ornamental.

***J. drupacea*** L. Asia Minor. The fruits (Habhel) are eaten locally.

***J. excelsa*** Bieb. (Greek Juniper). Afghanistan. The fruits and oil distilled from them are used to treat indigestion, vaginal discharges and as a diuretic.

***J. macropoda*** Boiss. (Indian Juniper, Appura, Dhup, Pencil Cedar). Malaysia, N. India. The reddish wood is fragrant, hard and durable. It is used locally for building, making water channels, buckets etc. and walking sticks and for fuel. The better quality wood is used in India and Pakistan to make pencils. The distilled shavings and sawdust give Cedar Oil, used in perfumery. The residue from distillation is made into linoleum.

***J. mexicana*** Spreng. (Mexican Juniper). Texas to Mexico, Guatemala. The wood is brown and hard. It is used for construction work, telegraph poles and railway sleepers.

***J. occidentalis*** Hook. (Sierra Juniper). Pacific N.America. The fruits are eaten by the local Indians. The wood is used as fuel and for fencing.

***J. oxycedrus*** L. (Pricky Juniper). Mediterranean. Distillation of the heartwood yields an oil (Oil of Cade) used as an antiseptic and to kill external parasites. It is also used as immersion oil for microscope work.

***J. pachyphlaea*** Torr. (Alligator Juniper). S.W. U.S.A., Mexico. The fruits are eaten by the local Indians.

***J. procera*** Hochst. ex A. Rich. (African Cedar). W.Africa, through to Malawi. The wood is termite resistant and is much used locally for building. It is also used to make pencils. The wood is exported from Kenya.

***J. pseudosabina*** Fisch. and Mey. Turkeystan. The wood is used as incense.

***J. rigida*** Sieb. and Zucc. (Muro, Nezu). Japan, Korea, Manchuria. The tree is cultivated in Japan for the wood which is valued for house construction, fencing, poles and agricultural implements. The oil distilled from the fruit is used locally as a diuretic.

***J. sabina*** L. (Savan, Savin). S.Europe to Caucasus, Central Asia and N.Asia. The leaves are used locally against body lice and to keep moths out of stored clothes. The oil from the twigs (Oil of Savin) is used medicinally to combat intestinal worms, to control menstruation and to induce sweating. It is also an abortive. The oil is also used in perfumery.

***J. saltuaria*** Rehd. and Wils. (Black Seed

Juniper). N.W.China. The wood is used locally for building.

***J. utahensis*** (Engelm.) Lemm. (Utah Juniper). W. U.S.A. The fruits were eaten as a flour by the local Indians.

***J. uvifera*** Don. = Pilgerodendron uviferum.

***J. virginiana*** L. (Eastern Red Cedar). The dull red, light timber is fragrant, but not strong. It is used to make clothes cupboards, chests etc., the smell repelling insects. It is also used in house interior work, buckets and particularly in making pencils. The oil, distilled from the heartwood (Red Cedar Wood Oil) is used in insecticides, moth-repellents etc. and a fixative for perfumes in soap and perfumes and as immersion oil for microscope work. It is also used to some extent in shoe-polish. The oil is a dangerous abortive. The trees are used as Christmas trees in S.E. U.S.A. and the leaves were used as incense by the Indians.

Jura Turpentine – Picea excelsa.

Jurosse – Lathyrus cicera.

Jurubeba – Solanum insidiosum.

Jusillo – Calycogonium squamulosum.

JUSTICIA L. Acanthaceae. 300 spp. Tropics and subtropics. Shrubs.

***J. dahona*** Buch. = J. gendarussa.

***J. fulgida*** Blume = Climanacanthus burmanii.

***J. gendarussa*** L. = J. dahona = J. salicina = Gendarussa vulgaris. India and Far East. A decoction of the roots are used in China to treat Rheumatism and a decoction of the leaves are used similarly in India. In Réunion the whole plant (Guérit petit) is used as an emetic in the treatment of colic. It is used in the Antilles to reduce fevers.

***J. nasuta*** L. = Rhinacanthus nasuta.

***J. nutans*** Burm. = Climacanthus burmanii.

***J. salicina*** Vahl. = J. gendarussa.

***J. spicagera*** Schlecht. = Jacobinia spigigera.

Jute – Corchorus capsularia.

Jute, Ambari – Hibiscus cannabinus.

Jute, Bimplipatam – Hibiscus cannabinus.

Jute, China – Abutilon avicennae.

Jute, West African – Hibiscus quinquelobus.

Jute, White – Corchorus capsularis.

# K

Kabala – Eriosema griseum.

Kabalabala – Cynometra pedicellata.

Kabera – Cissus cyphopetala.

Kaboli – Eriocoelum microspermum.

Kaboo-kaboo – Santinia tomentosa.

Kabumbo – Lannea discolor.

Kabuteng higunte – Tricholoma spp.

Kachaso – Zizyphus abyssinica.

Kachi Grass, Oil of – Cymbopogon coloratus.

Kachlora – Pithecellobium bigeminum.

Kaddam – Mitragyna parvifolia.

Kadis – Garcinia barrettiana.

Kadjangrah – Knema laurina.

Kadjatao – Eugeissona insignis.

KADSURA Juss. Schisandraceae. 22 spp. India, China, Japan, S.E.Asia, W. Malaysia, Moluccas. Climbers.

***K. chinensis*** Hance. = K. coccinea.

***K. coccinea*** A. C. Smith = K. chinensis = Schizandra hanceana. (Xunh xe). China, Vietnam. The Chinese use the plant to make a tonic, aphrodisiac called Ngu vi tu, which is also used to treat chest complaints.

Kadukkodi – Pachygone ovata.

KAEMPFERIA L. Zingiberaceae. 70 spp. Tropical Africa to India, W.Malayasia and S.China. Herbs.

***K. aethiopica*** (Solms.) Benth. Tropical Africa especially W. The tubers are used as a spice in E.Africa.

***K. angustifolia*** Rose. Tropical Asia. The tubers are used locally for chewing and treating colds.

***K. galanga*** L. Tropical Asia. The plant is cultivated throughout the Far East for the rhizomes which are used as a condiment, (Galanga). A poultice of the rhizome in oil is used in the Philippines to hasten healing and to bring boils to a head. In China and India the rhizome is used to wash the hair.

***K. pandurata*** Roxb. (Bong nga truât nam, Ngo shut). India to Indonesia. The rhizomes are used in Indonesia to treat stomach complaints.

***K. rotunda*** L. Tropical Asia. The rhizome is used widely to treat stomach complaints and as an ointment with coconut oil to heal wounds. The leaves and rhizomes

are used in Java as a condiment. The plant is cultivated for the oil distilled from the rhizome and used in cosmetics.
Kafferslangen Wortel – Polygala serpentaria.
Kaffir Boom – Erythrina abyssinica.
Kaffir Bread – Encephalartos caffer.
Kaffir Marvola Nut – Sclerocarya caffra.
Kaffir Plum – Harpephyllum caffrum.
Kaffir Plum – Lannea caffra.
Kaffir Potato – Plectranthus esculentus.
Kaffir Potato – Plectranthus floribundus.
Kafumbe – Bauhinia petersiana.
Kafunbafunda – Rumex maderensis.
Kagné Butter – Allenblackia oleifera.
Kagobozoba – Eriosema montanum.
Kahika – Podocarpus dacrydioides.
Kahitootan – Anotis hirsuta.
Kaim – Mitragyna parvifolia.
Kaisertee – Dryas octapetala.
Kait bensi – Uncaria sclerophylea.
Kajingo – Vigna maranguensis.
Kajo rapat – Parameria barata.
Kajoo ajer perampooan – Leea aequata.
Kajoo ammorang – Cedrela celebica.
Kajoo anjang – Elaeocarpus grandiflorus.
Kajoo arang – Cratoxylum celebicum.
Kajoo batoo areng – Homalium grandiflorum.
Kajoo daging – Vatica bancana.
Kajoo daoon babalik – Alphitonia moluccana.
Kajoo djali – Baccaurea javanica.
Kajoo djamboo – Bhesa paniculata.
Kajoo galadoopa – Afzelia galedupa.
Kajoo gambir – Trigonopleura malayana.
Kajoo kambing – Garuga abilo.
Kajoo kantjil – Anisophyllea disticha.
Kajoo koongit – Terminalia sumatrana.
Kajoo koonjit – Adina polycephala.
Kajoo laboo – Tetrameles nudiflora.
Kajoo lahi – Celtis cinnamomea.
Kajoo lasi – Adina fagifolia.
Kajoo lemah – Haloragis oblongifolia.
Kajoo lentadek – Guioa pleuropteris.
Kajoo looloos – Cartoxylum plyanthum.
Kajoo malaka – Tetramerista glabra.
Kajoo menjawak – Euodia sambucina.
Kajoo mootoh – Erythroxylum cuneatum.
Kajoo nasi – Chrysophyllum roxburghii.
Kajoo oorang – Erythroxylum cuneatum.
Kajoo pelen – Cassia timoriensis.
Kajoo ringoo – Dillenia excelsa.
Kajoo sepat – Macaranga diepenhorstii.
Kajoo sepat – Macaranga triloba.
Kajoo seriawan – Symplocos odoratissima.
Kajoo tandikat – Canarium pseudo-decumanum.
Kajoo tjeleng – Crypteronia paniculata.
Kajoo wole – Aglaia ganggo.
Kaju lulu – Celtis philippensis.
Kaju bedarah – Knema glauca.
Kakakora – Vigna multiflora.
Kakapasan – Hibiscus ventustus.
Kaki – Diospyros kaki.
Kakope – Conopharyngia elegans.
Kakweshe – Tessmannia africana.
Kaladan – Dipterocarpus warburgii.
Kalakudi – Wrightia tinctoria.
Kalapa tijoong – Horsfieldia glabra.
Kalapa tijoong – Horsfieldia irya.
Kalapi – Calamus ornatus.
Kalchia – Glochidion hohenackeri.
Kale – Brassica oleracea var. acephala.
Kale, Indian – Xanthosoma vioaceum.
Kale, Ruvo – Brassica ruvo.
Kale, Sea – Crambe maritima.
Kale, Tatarian Sea – Crambe tatarica.
Kalihaldi – Curcuma caesia.
Kalik karsik – Irvingia malayana.
Kalingag (Oil) – Cinnamomum mercadoi.
KALLSTROEMIA Scop. Zygophyllaceae. 7 spp. N. and N.E.Australia; 16 spp. S.United States of America, W.Indies to Argentina. Herbs.
***K. maxima*** (L.) Torr. and Gray. S.W. U.S.A. to Central America. The young plants are sometimes used as a vegetable in Central America.
KALMIA L. Ericaceae. 8 spp. N.America, Cuba. Shrubs.
***K. latifolia*** L. (Laurel, Mountain Laurel). E. N.America. The heavy hard brittle wood is used for tool handles, turning and fuel. The roots are used to make tobacco pipes. The leaves are narcotic and in excess poisonous. They are used medicinally as a sedative, to treat heart complaints, to improve the appetite, to sooth coughs, to treat blood disorders, diarrhoea and dysentery.
Kaloempang Oil – Sterculia foetida.
Kalombo – Lannea antiscorbutica.
Kalong – Sterculia javanica.
Kalongan – Sterculia javanica.
Kalpakku – Pithecellobium bigeminum.
Kalpi – Citrus webberi.
Kamahi – Weinmannia racemosa.
Kamala (Tree) (Powder) – Mallotus philippensis.
Kamansi – Artocarpus camansi.
Kamassiwood – Gonioma kamassi.

Kamba – Cissus cyphopetala.
Kambing-kambing – Sacrolobus globosus.
Kambul – Rhus wallichii.
Kambur – Momordica subangulata.
Kamerer – Eucalyptus deglupta.
Kamilo – Cubilia blancoi.
Kamisan – Citrus longispina.
Kampas – Koompassia malaccensis.
Kamukungo – Lannea antiscorbutica.
Kamvula – Carpolobia glabrescens.
Kamynye Oil – Hoslundia opposita.
Kan Kalagoua – Setaria verticillata.
Kana – Saccharum munja.
Kanagambarum – Crossandra infunduliformis.
Kanalei – Euodia lunu-ankenda.
Kanalong – Tacca fatsiifolia.
Kanang – Canarium zollingeri.
Kanari – Canarium amboinense.
Kanari – Canarium sylvestre.
Kanari minjak – Canarium oleosum.
Kanari ootan – Canarium sylvestre.
Kandaram – Indigofera trita.
KANDELIA (DC.) Wight and Arn. Rhizophoraceae. 1 sp. E.Asia and W.Malaysia. Tree.
***K. rheedei*** Wight. and Arn. The bark is used for tanning in Tonkin.
Kandis – Garcinia globulosa.
Kandoeen – Symplocos javanica.
Kandol – Caereya sphaerica.
Kandu – Eryngium coeruleum.
Kane polo – Godoya firma.
Kanga – Fraxinus floribunda.
Kanga Butter – Pentadesma butyracea.
Kangaroo Apple – Solanum oviculare.
Kangaroo Grass, Common – Themeda ciliata.
Kanguni – Celastrus paniculata.
Kanila – Grewia salutaris.
Kanjaaya – Chenopodium canihua.
Kanjere – Bridelia monoica.
Kanker Bos – Sutherlandia frutescens.
Kankonde – Tessmannia africana.
Kankora – Ziziphus abyssinica.
Kanna – Sceletium anatomicum.
Kanno – Ahnfeltia plicata.
Kannarahan – Horsfieldia irya.
Kanot-Kanot – Gracilaria eucheumoides.
Kansaë – Eutrema wasabii.
Kansu – Koelreuteria paniculata.
Kantiari – Carthamus oxycanthus.
Kanyabumba – Fagara pilosiuscula.
Kanyalumbu – Fagara inaequalis.
Kao – Bauhinia thonningii.
Kao ashud – Geranium wallichianum.
Kapas hantoo – Abroma fastuosa.
Kapas ooton – Thespesia lambas.
Kapasan – Thespesia lambas.
Kapiega, Wild – Bulbine asphodeloides.
Kapinango – Pysoxylum densiflorum.
Kapok – Bombax buonopozence.
Kapok (Tree) – Ceiba pentandra.
Kapoondong – Baccaurea dulcis.
Kapor Tea – Epilobium angustifolium.
Kappa – Clitandra orientalis.
Kapporie Tea – Epilobium angustifolium.
Kapur – Dryobalanops rappa.
Kapur paya – Dryobalanops rappa.
Kara Clover – Trifolium ambigium.
Kara – Carthamum oxycanthus.
Karak – Garuga pinnata.
Karakubong – Eria pannea.
Karakusa – Cymbopogon jwarancusa.
Karalli – Carallia brachiata.
Karam – Adina cordifolia.
Karamanni Wax – Symphonia globulifera.
Karas – Aquilaria malaccensis.
KARATAS Mill. Bromeliaceae. 20 spp. Mexico to Argentina. Herbs.
***K. plumieri*** Morr. W.Indies, Martinique, Panama. The pleasantly flavoured fruits are eaten locally.
Karaya Gum – Sterculia urens.
Karet – Ficus elastica.
Kari – Avicennia officinalis.
Kariis – Garcinia lateriflora.
Karithumba – Epimeredi malabarica.
Kariyat – Andrographis paniculata.
Karkalia – Mesembryanthemum aequilaterale.
Karo Pittosporum – Pittosporum crassifolium.
Karri – Eucalyptus diversicolor.
Karukanda – Dioscorea hispida.
KARWINSKIA Zucc. Rhamnaceae. 14 spp. S.W. United States of America to Bolivia, W.Indies. Trees.
***K. calderonii*** Standl. (Calderon Coyotillo, Huili huiste.) El Salvador. The dull red, heavy, hard wood is strong and durable. It is used for railway sleepers, wheel hubs, pestles and mortars, shuttles and bowls.
***K. humboldtiana*** (Roem. and Schult.) Zucc. (Humboldt Coytillo). S.W. U.S.A., Mexico. The fruit pulp is edible, but the seeds are poisonous. They are used locally in Mexico to treat convulsions.
Kasa – Pachyelasma tessmannii.
Kasari – Lathyrus sativus.
Kasbar Cream – Ceratonia siliqua.
Kashmal – Lannea grandis.

Kasnih – Ferula galbaniflua.
Kassi – Bridelia retusa.
Kat – Catha edulis.
Kat sola – Aeschynomene indica.
Katabu – Anemone hupehensis.
Katafa – Cedrelopsis grevei.
Kateera Gum – Sterculia urens.
Kateeragum Sterculia – Sterculia urens.
Kathiawar Cotton – Gossypium obtusifolium.
Katio Oil – Bassia mottelyana.
Katio Oil – Madhuca mottleyana.
Katira Gum – Cochlospermum gossypum.
Katjee – Teucrium capense.
Katji pot – Salacia macrophylla.
Katjondong – Tacca pinnatifida.
Katoka – Treculia perrieri.
Katoo lobang – Adina minutiflora.
Katook badak – Sauropus rhamnoides.
Katrafy – Cedrelopsis grevei.
Kattar – Pistacia integerrima.
Kaurane – Adenium honghel.
Kauri – Agathis australis.
Kauri – Agathis labillardierii.
Kauri – Agathis palmerstonii.
Kauri Copal – Agathis australis.
Kauri Gum – Agathis australis.
Kauri Resin – Agathis lanceolata.
Kava Pepper – Piper methysticum.
Kawaka – Lebrocedrus doniana.
Kawizi – Albizia antunesiana.
Kayam – Ferula alliacea.
Kayu-galu Oil – Sindora inermis.
Kbuar – Phoenix sylvestris.
Kea – Mezonevron kavaiensis.
Kedaboo – Sonneratia ovata.
Kedaung – Parkia javanica.

KEDROSTIS Medik. Cucurbitaceae. 35 spp. Tropics and subtropics. Africa, Madagascar, tropical Asia and Malaysia. Vines.

***K. laciniosa*** (L.) Naud. = Bryonia lacinosa. (Doodo songot, Korek korek). Indonesia. The young fruits are eaten in soups locally, especially in Java. The leaves are also made into a paste as a poultice for boils.

Kefir – Saccharomyces kefir.
Kegr – Salvadora persica.
Kei Apple – Doryalis caffra.
Kekoko – Chaetachme microcarpa.
Keladan – Dryobalanops oblongifolia.
Kelaki – Madhuca nancifolia.
Kelabang – Garcinia bancana.
Kelang patejan – Haloragis oblongifolia.
Kelapa bout – Hydnocarpus polypetala.
Kelili jong-kong – Xylopia ferruginea.
Kelimparan tooli – Labisia pumila.
Kelo-ka-tel Oil – Cedrus lebanii.
Kelp, Bladder – Nereocystis luetkeana.
Kelp, Broad leaf – Laminaria saccharina.
Kelp, Bull – Durvillea antartica.
Kelp, Horsetail – Laminaria digitata.
Kelp, Japanese – Laminaria japonica.
Kelp Meal (Tablets) – Iridophycus flaccidum.
Kemang – Mangifera caesia.
Kemarogan – Gymnopetalum leucostictum.
Kembang – Cassia surattensis.
Kembang boogang – Clerodendrum buchananii.
Kembang mentega – Tabernaemontana divaricata.
Kembo – Ostryoderris lucida.
Kementing laut – Herandia ovigera.
Kemootoon – Cratoxylum formosum.
Kenaf Hibiscus (Seed Oil) – Hibiscus cannabis.
Kendajakan pootih – Bauhinia hirsuta.
Kendir – Apocynum venetum.
Kenidai – Bridelia monoica.
Kenooar – Shorea lepidota.
Kentjing perlandook – Apostasia nuda.
Kentucky Bluegrass – Poa pratensis.
Kentucky Coffee Tree – Gymnocladus divica.
Keolav – Bauhinia purpurea.
Keranthai merah – Santiria laevigata.
Kerchoud – Anaststica hierochuntica.
Kerguelen Cabbage – Pringlea antiscorbutica.
Keribor – Anisophyllea disticha.

KERMADECIA Brong. and Gris. Proteaceae. 10 spp. N.E.Australia, New Caledonia, Fiji. Trees.

***K. rotundifolia*** Brong. and Gris. New Caledonia. The wood is used locally for general carpentry and house interiors.

Kermes Oak – Quercus cocifera.
Keroowing – Dipterocarpus marginata.
Keroowing booloo – Dipterocarpus skinneri.
Kerosene Wood – Halfordia scleroxyla.

KERSTINGIELLA Harms. Leguminosae. 1 sp. Tropical W.Africa. Herbs.

***K. geocarpa*** Harms. (Geocarpa Groundnut). The seeds are eaten in some parts of Nigeria.

Kerukup – Shorea pachyphylla.
Keruntun – Combretocarpus rotundatus.

KETELEERIA Carr. Pinaceae. 4–8 spp. E.Asia, Indochina. Trees.

***K. davidiana*** Beissn. China. The soft, light wood is used locally for house building.
Ketembilla – Doryalis hebecarpa.
Keti – Dalbergia pinnata.
Ketjapi kera – Sandoricum emarginatum.
Ketool – Bidens chinensis.
Ketool – Geruga abilo.
Khairival – Bauhinia purpurea.
Khaki Cotton – Gossypium nanking.
Kharassan Wheat – Triticum orientale.
Khare-chitor – Alhagi persorum.
Khasia patchouli – Microtonea cymosa.
Khasia Madder – Rubia khasiana.
Khasia Pine – Pinus khasya.
Khat – Catha edulis.
Khatta Orange – Citrus limonia var. khatta.
KHAYA A. Juss. Meliaceae. 9 spp. Tropical Africa, Madagascar. Trees. The woods of the species mentioned resemble Mahogany and are used for the same purposes.
***K. anthoteca*** C. DC. (White Mahogany). Tropical W.Africa. The wood is exported. The bark is used in Angola to treat fevers.
***K. euryphylla*** Marhs. (Cameroon Mahogany). Tropical W.Africa. Exported.
***K. grandifoliola*** C. DC. (Lagos Mahogany). Tropical Africa.
***K. grandis*** Stapf. (Benin Mahogany). Tropical W.Africa.
***K. ivorensis*** Chev. (Ivory Coast Khaya, Red Mahogany). Tropical W.Africa.
***K. klainei*** Pierre. (Gabon Mahogany). Tropical W.Africa.
***K. madagascariensis*** Jum. and Perr. (Hazomena, Madagascar Mahogany). Madagascar. Wood exported. The trunk yields a yellowish gum.
***K. nyasica*** Stapf. (mululu, Muwawa, Mururu, Red Mahogany). Central Africa. An important local timber, unaffected by termites. It is used as Mahogany. The oil from the seeds is used locally to kill head lice.
***K. senegalensis*** Juss. (African Mahogany, Senegal Khaya). Tropical Africa. The wood is exported.
Khella – Ammi visnaga.
Kheuroch – Dysoxylum densiflorum.
Khirni – Manilkara kauki.
Khoja – Callicarpa arborea.
Khooleuh – Amoora aphanamixis.
Khumbut – Acacia jacquemontii.
Khursi – Grewia elastica.
Khus – Saccharum spontaneum.
Khus-khus – Vetiveria zizanoides.
Ki anggrit – Adina polycephala.
Ki areng – Diospyros pseudo-ebinum.
Ki bengang – Neesia altissima.
Ki beureum laoot – Toona sureni.
Ki beusi – Rhodammia arborea.
Ki djoolang – Afzelia javanica.
Ki kanari – Canarium litorale.
Ki-kive – Diospyros kaki.
Ki laetan – Chrysophyllum roxburghii.
Ki langil – Polyscias nodosa.
Ki lunglung – Aromadendron elegans.
Ki mangro – Ardisia laevigata.
Ki pare – Glochidion obscurum.
Ki saat – Neonauclea excelsa.
Ki sampang – Euodia latifolia.
Ki sapilan – Calophyllum venulosum.
Ki tai ketoodjeuh – Dysoxylum amooroides.
Ki tana – Fagara rhetsa.
Ki toongila – Horsfieldia glabra.
Ki toowa – Leea indica.
Kibanen – Crypteronia paniculata.
Kibangabanga – Heeria insignis.
Kibangi – Albizia versicolor.
KIBATALIA G. Don. Apocynaceae. 25 spp. Tropical Africa, W.Malaysia. Trees.
***K. blancoi*** (Rolfe) Merr.=Kickxia arborea =K. blancoi. Philippines. The bark and leaves are used locally as a fish poison and the bark and roots, to induce abortions.
Kibonga – Humularia apiculata.
Kibutiga – Lannea edulis.
***Kickxia*** Blume=Kibatalia.
***Kickxia elastica*** Preuss=Funtumia elastica.
***Kickxia elatine*** (L.) Dum.=Linaria elatine.
Kidamar – Drypetes longifolia.
Kididila – Ganophyllum giganteum.
Kidney Bean – Phaseolus vulgaris.
Kidney Cotton – Gossypium brasiliensis.
Kidney Vetch – Anthyllis vulneraria.
Kidney Wood – Eysenhardtia polystachya.
KIELMEYERA Mart. Guttiferae. 20 spp. Brazil. Trees.
***K. coriacea*** Mart. (Pau Camp). Brazil. The bark is used as a substitute for cork.
Kieriemoor – Mesembryanthemum stellatum.
Kifoko – Pycnoneurum junciforme.
Kifumbi – Bauhinia thonningii.
Kifuria – Leptderris nobilis.
KIGELIA DC. Bignoniaceae. 3 spp. Warm Africa. Madagascar. Trees. The bark of the following species is used locally to treat dysentery.

***K. acutifolia*** Engl. Cameroons.
***K. africana*** Benth. (African Sausage Tree). S.Africa.
KIGGELARIA L. Flacourtiaceae. 4 spp. Tropical and S.Africa. Trees.
***K. africana*** L.=K. dregeana.
***K. dregeana*** Turez. (Natal Mahogany). S.Africa. The pinkish wood is used for furniture and boards.
Kihengia – Celtis prantlii.
Kikeujeup – Tarenna incerta.
Kikuyu Grass – Pennisetum clandestinum.
Kik-we – Diospyros kaki.
Kilari – Dioscorea deltoidea.
Kilende – Hyalsepalum caffrum.
Kilong lagong – Melothria heterophylla.
Kilundeke – Beilschmiedia oblongifolia.
Kim giao – Podocarpus blumei.
Kimbarra – Eucalyptus robusta.
Kiminga – Erythrophleum couminga.
Kingwood – Dalbergia cearensis.
King William Pine – Athrotaxis selaginoides.
Kinka Oil Ironweed – Vernonia anthelmintica.
Kinnikinnik – Cornus amomum.
Kino, African – Pterocarpus erinaceus.
Kino, Australian – Eucalyptus camaldulensis.
Kin, Bengal – Butea frondosa.
Kino, East Indian – Pterocarpus marsupium.
Kino Eucalyptus – Eucalyptus resinifera.
Kino, Jamaican – Coccolobis uvifera.
Kino, Malabar – Pterocarpus marsupium.
Kino, Ribbon – Eucalyptus salicifolia.
Kins – Dioscorea deltoidea.
Kioo Peto – Acroceras amplectens.
Kiotsenga – Urtica massaica.
Kiraj – Metroxylon rumphii.
Kiralu – Arisaema speciosum.
Kirindur – Garcinia lancaefolia.
KIRKIA Oliv. Ptaeroxylaceae. 8 spp. Tropical and S.Africa. Trees.
***K. acuminata*** Oliv. (Bastard Marula, Mutuwa, Nsena). Central Africa. The wood is used locally for furniture and plywood. It is used in the villages for building kraals etc.
Kirsch-gummi – Amygdalus leicarpus.
Kirschwasser – Prunus cerasus.
Kirta – Dioscorea deltoidea.
Kisambila – Maytenus senegalensis.
Kisariawan – Symplocos odoratissima.
Kishanula – Rhus vulgaris.
Kismis – Actinidia callosa.
Kiso – Muncuna flagellipes.
Kissi – Camellia kissi.
Kissoumpo – Menabea venenata.
Kitan – Linum humile.
Kitata – Bauhinia peteriana.
Kiterong – Schoutenia burmani.
Kitjenkeh – Urophyllum arboreum.
Kitsanga – Cynanchum lineare.
Kitsangana – Pycnoneurum junciforme.
Kittiboe – Acroceras amplectens.
Kittul Fibre – Caryota urens.
K'iustai – Allium odoratum.
KLAINEDOXA Pierre. Ixonanthaceae. 10 spp. Tropical Africa. Trees.
***K. gabonensis*** Pierre. W.Africa. The heavy hard, gold-brown wood is used locally for decking on Congo steamers and as poles etc. in Liberia. The seeds are eaten locally, raw or roasted or as a flour.
Klalar – Dipterocarpus trinervis.
Klamath Plum – Prunus subcordata.
Klampok – Eugenia acuminatissima.
Klanting – Schefflera aromatica.
Kleasaik – Forsskålea tenacissima.
KLEINHOVIA L. Sterculiaceae. 1 sp. Tropical Asia. Tree.
***K. hospita*** L. The bark is used locally as cordage and the juice from the leaves as an eye-wash.
KLEINIA Mill. Compositae. 50 spp. Tropical and S.Africa, Arabia. Herbs.
***K. pteroneura*** DC.=Senecio pteroneurus. N.Africa. The juice is used to treat intestinal complaints and the small branches are used locally in the treatment of rheumatism.
Klepoo ketek – Neonauclea excelsa.
Klinki – Araucaria klinki.
***Klopstockia cerifera*** Humb. and Bonpl.= Ceroxylon klopstockiae.
Kloudoring – Harpagophytum procumbens.
Klue tani – Musa bulbisina.
Knackaway – Ehretia elliptica.
Knapweed – Centaurea nigra.
Kneeholm – Ruscus aculeatus.
Kneipp Tea – Sambucus ebulus.
KNEMA Lour. Myrsiticaceae. 60 spp. S.China, S.E.Asia, Indomalaya. Trees.
***K. bicolor*** Rat.=K. corticosa.
***K. corticosa*** Lour.=K. bicolor. (Mâu chó, Muscadier à suif). Burma through to Thailand. The aromatic oil from the seeds is used in medicinal soaps for treating skin diseases. The oil is used for the same purpose in Vietnam.
***K. glauca*** (Blume) Warb.=Myristica glauca.

(Kaju bedarah). Malaysia, Java, Celebes and Thailand. The oil from the seeds was used locally for lighting.

***K. hookeriana*** (Wall.) Warb.=Myristica hookeriana. Malaysia, Sumatra, Borneo. The wood is not durable, but is used in Sumatra for housing and boards.

***K. laurina*** (Blume) Warb.=Myristica laurina. (Pianggu pipit, Huru tangkalak, Kadjangrah). Burma, Java, Sumatra. The wood is used locally for housebuilding.

***K. palembanica*** (Miq.) Warb. Sumatra. The wood is used locally for carving small articles.

KNIGHTIA R. Br. Proteaceae. 2 spp. New Zealand, New Caledonia. Trees.

***K. excelsa*** R. Br. (Maori Honeysuckle, Rewa, Rewarewa). New Zealand. The red beautifully figured wood is much valued for cabinet-making.

Knob Thorn – Acacia nigrescens.

Knobbed Wrack – Ascophyllum nodosum.

Knobwood – Zanthoxylum capense.

Knotweed, Douglas – Polygonum douglasii.

Knotweed, Prostrate – Polygonum aviculare.

Knotweed, Sachalin – Polygonum sachalinense.

KNOXIA L. Rubiaceae. 15 spp. Indomalaya. Herbs. The following species are used locally to stimulate the fermentation of rice to alcohol, particularly in Cambodia.

***K. brachyata*** R. Br.

***K. corymbosa*** Willd.=K. sumatrensis=K. umbellata.

***K. sumatrensis*** Retz.=K. comrymbosa.

***K. umbellata*** Banks.=K. comrymbosa.

Knung – Careya sphaerica.

Ko maiy – Euonymus cochinchinemsis.

Koa Acacia – Acacia koa.

Koal – Licuala rumphii.

Kobitisondolo – Tacca umbrarum.

Kobus Magnolia – Magnolia kobus.

KOCHIA Roth. Chenopodiaceae. 90 spp. Central Europe to temperate Asia, N. and S.Africa, Australia. Herbs or shrubs.

***K. aphylla*** R. Br. (Salt Bush). Australia. Important as livestock feed during times of drought.

***K. scoparia*** (L.) Schrad. (Summer Cypress). Central Europe to temperate Asia, escaped to N.America. Cultivated in Japan and China where the young shoots are used as a vegetable. The seeds are made into a flour. The whole plants are used as brooms.

Koda Millet – Paspalum scrobiculatum.

***Koellia*** Moench.=Pycnanthemum.

KOELREUTERIA Laxm. Sapindaceae. 8 spp. China, Formosa, Fiji. Trees.

***K. apiculata*** Rehd. and Wils.=K. paniculata var. apiculata. Korea, China, Japan. The flowers are used in China as the source of a yellow dye.

***K. paniculata*** Laxm. China. The flowers yield a yellow dye and the seeds are used locally to make necklaces.

***K. paniculata*** var. ***apiculata*** Rehd. and Wils. =K. apiculata.

Koenig Akee – Blighia sapida.

Kohibhhang – Hyoscyamus reticulatus.

Kojo – Callicarpa arborea.

Kok-saghys – Taraxacum koksaghyz.

Kok toung ka – Strychnos nux-blanda.

Kokam Butter – Garcinia indica.

Kokan – Garcinia indica.

KOKIA Lewton. Malvaceae. 5 spp. Hawaii. Trees.

***K. rockii*** Lewton. Hawaii. The juice from the bark is used locally for dyeing fish nets.

Koko – Euphorbia lorifera.

Kokomba – Mascarenhasia geayi.

KOKOONA Thw. Celastraceae. 5 spp. Ceylon, Burma, W.Malaya. Trees.

***K. zeylanica*** Thw. (Kokjum, Wana pota). Ceylon. The oil from the seeds (Pota-etatel) is used locally as a leech repellant.

Kola – Grewia laevigata.

Kola bitter – Garcinia cola.

Kola male – Garcinia cola.

Kololo – Cucumis naudinianus.

Kolomikta Vine – Actinidia callosa.

Kolongkolong – Catenella nipae.

Komanga – Erythrophleum couminga.

Kombo – Pycnathus kombo.

Kombo-kombo – Musanga smithii.

Kombu – Laminaria japonica.

Kompitro – Gonocrypta grevii.

Konadala – Isolona thonneri.

Kon – Sceletium anatomicum.

Konal – Prangos pabularia.

Kondang – Ficus cerifiua.

Kondapake – Pinanga dicksonii.

Kongo nkumama – Fagara leprieuri.

Königsberger Pea – Pisum arvense.

Konjaku Flour (Powder) – Amorphophallus rivieri.

Konko kelekete – Fagara leprieuri.

Konterie – Cotyledon orbiculata.

Koolim – Scorodocarpus borneensis.
Koolit seriawan – Symplocos odoratissima.
Koomaroko – Zanthoxylum celebicum.
Koomis ootjing – Tacca palmata.
KOOMPASSIA Maingay. Leguminosae. 4 spp. Malaysia, Borneo, New Guinea. Trees.
***K. excelsa*** (Becc.) Taub. (Tapang). Malaya, Borneo, Philippines. The reddish, chocolate coloured wood is heavy and strong. It is difficult to work and not durable in contact with soil, but polishes well. It is used for making heavy furniture and for firewood and charcoal.
***K. malaccensis*** Maing. (Kampas). Malaya, Borneo, New Guinea. The very hard, heavy wood is red streaked with yellow-brown. It is used for heavy construction work, flooring etc. When preserved with creosote, which it absorbs readily it is used for railway sleepers. The root wood is used to make tables.
Kooning – Cassia surattensis.
Koonsboisie – Teucrium capense.
Koontjir – Antidesma ghaesembilla.
Koopooi – Sarcotheca griffithii.
KOORDERSIODENDRON Engl. Anacardiaceae. 1 sp. Philippines, Celebes, New Guinea. Tree.
***K. pinnatum*** Merr. (Bugis amugis, Rangu). The hard red-brown wood is beautifully marked. It is hard and strong, but is not durable in contact with the soil. The wood is used for interior work and furniture.
Kooroo – Toonia sureni.
Kopar – Dendrocalamus strictus.
Kopetan – Gomphostemma phlomoides.
Kopiefa – Bulbine narcissifolia.
Kopjesdoorn – Acacia nigrescens.
Korakan – Eleusine coracana.
Kordofan – Acacia senegal.
Kordofan Pea – Clitoria ternala.
Korean Pine – Pinus koraiensis.
Korek kotek – Kedropsis laciniosa.
Kori – Tamarix gallica.
Korwarkul – Acacia cunninghamii.
Koso – Kagenia abyssinica.
KORTHALSIA Blume. Palmae. 35 spp. Indomalaya. Rattan Palms.
***K. scaphigera*** Mart. Malaysia. The stems are used locally to make baskets.
Korwarkul – Acacia cunninghamii.
KOSTELETZYA C. Presl. Malvaceae. 30 spp. N.America, Russia, Mexico, tropical and S.Africa, Madagascar. Herbs.
***K. pentacarpa*** Led. Russia. See Althaea officinalis.
Kougoed – Sceletium anatomicum.
Krabas prey – Mallotus anamiticus.
Krakorso – Amomum thyrsoideum.
Kralanh – Dailium cochincinensis.
KRAMERIA L. ex Loefl. Krameriaceae. 25 spp. S. United States of America to Chile. Shrubs.
***K. argentea*** Mart. Brazil. The dried root (Brazilian Rhatany, Pará Rhatany) contains tanning. It is used medicinally as an astringent and tonic. It is also used for tanning.
***K. ixina*** L. Colombia, Brazil, Guyana. The dried root (Savanilla Krameria) is used medicinally as a tonic.
***K. parvifolia*** Benth. (Range Rhatany). W. U.S.A. The local Indians use an infusion of the twigs to treat sore eyes and a dye from the roots to stain wool.
***K. tomentosa*** St. Hil. Mexico to Colombia and N.Brazil, W.Indies. The root (Savanilla Ratanhia) is used for tanning.
***K. triandra*** Ruiz. and Pav. (Peruvian Krameria). Bolivian, Peru. The dried root (Peruvian Rhatany) is used as an astringent and tonic. It contains some 10 per cent tannin and is used for tanning. The local women used the root as a tooth preservative.
Kraininan – Dysoxylum densiflorum.
Kreb – Echinochloa pyramidalis.
Krervanh – Amomum krervanh.
Krim Sagiz – Taraxacum hybernum.
Krobonko – Telfairia occidentalis.
Kroja – Celtis cinnamomea.
KRUGEODENDRON Urb. Rhamnaceae. 1 sp. W.Indies, Mexico. Small tree.
***K. ferreum*** (Vahl.) Urb. = Rhamnus brandegiera = R. ferreus = Condalia ferra. The bark and roots are chewed to relieve toothache in Mexico, where a decoction of the roots is used as a purgative.
Ksopo – Menabea vebebata.
Ktu – Grewia paniculata.
Ktambi – Garcinia indica.
Kua – Julbernardia sereti.
Kudzu – Pueraria thunbergiana.
Kudzu Vine, Thunberg – Pueraria thunbergiana.
Kuhilia – Aechynomene indica.
KUHNIA L. Compositae. 7 spp. United States of America, Mexico. Herbs.
***K. eupatorioides*** L. (False Boneset). E. U.S.A. A decoction of the leaves is used locally as a tonic and to induce sweating.
Kui – Larix dahurica.

Kulche – Cedrela mexicana.
Kuluchan – Glochidion hohenackeri.
Kumakum – Elaeocarpus calomala.
Kumiss – Lactobacillus casei.
Kümmel – Carum carvi.
Kummelbranntwein – Carum carvi.
Kumpta Cotton – Gossypium obtusifolium.
Kumquat, Australian Desert – Eremocitrus glauca.
Kumquat, Marumi – Fortunella japonica.
Kumquat, Meiwa – Fortunella crassifolia.
Kumquat, Nagami – Fortunella marginata.
Kunan – Sclerocarya birraca.
Kundura Unsa – Boswellia carteri.
Kunis – Alnus nitida.
Kurakkan – Eleusine coracana.
Kurchi Bark – Holarrhena antidysenterica.
Kuromoji Oil – Linera umbellata.
Kurrat – Allium kurrat.
Kurrat baladi – Allium kurrat.
Kurrat nabati – Allium kurrat.
***Kurrimia*** Wall ex Thwaites = Bhesa.
Kurupum – Pimenta citrifolia.
Kursingh – Radermachera xylocarpa.
Kurwini Mango – Mangifera odorata.
K'u shên – Sophora angustifolia.
Kush-kum-tsagicks – Fritillaria camschatiensis.
Kushthapa – Gynocardia odorata.
Kussum Oil – Schleichera trijuga.
Kussum Tree – Schleichera trijuga.
Kusta – Parinari laurinum.
Kustarak – Saccharomyces vordermanni.
Kutai Tea – Arctostaphylos uva–ursi
Kutche Bloodwood Eucalyptus – Eucalyptus terminalis.
Kulot – Laurencia setaculosa.
Kuttatuvera – Mundulea sericea.
Kutwi hume – Alchemilla kuwuensis.
Kuyonu – Afraegle paniculata.
Kwai borrachero – Brugmansia amesianum.
Kwila – Intsia palembanica.
KYDIA Roxb. Malvaceae. 3 spp. E.Himalayas to S.E.Asia. Small trees.
***K. calycina*** Roxb. Dry India and Burma. The bark is used for cordage, locally in India and the mucilage from the stem is used to clarify sugar.
KYLLINGA Rottb. Cyperaceae. 60 spp. Tropics and sub-tropics, especially Africa. Herbs.
***K. triceps*** Rottb. Tropics. A decoction of the roots is used locally to treat vaginal discharges and as an antispasmodic.

# L

Labdanum Balsam – Cistus villosus.
Labdanum Gum – Cistus landaniferus.
LABISIA Lindl. Myrsinaceae. 9 spp. Malaysia. Shrubs.
***L. pothoina*** Lindl. = L. pumila.
***L. pumila*** Benth. and Hook. = L. pumila = Ardisia pumila. (Akar fatima, Kelimparan tooli). Malaysia. A decoction of the roots is used locally to treat gonorrhoea.
Lablab – Dolichos lablab.
***Lablab niger*** Medik. = Dolichos lablab.
Labrador Tea – Ledum groenlandicum.
Labuan Copal – Agathis loranthifolia.
Labuan Manila – Agathis loranthifolia.
LABURNUM Fabr. Leguminosae. sens str. 2 spp. Central Europe; sens lat. 4 spp. Europe, N.Africa, W.Africa. Trees.
***L. anagyroides*** Medik. = L. vulgare. (Golden Chain). Central Europe. The hard wood is used for turning and carving. The plant, especially the seeds, are poisonous.
***L. vulgare*** Griseb = L. anagyroides.
Laburnum, Indian – Cassia fistula.
Lac – Croton laccifer.
Lac, Mirzapore – Schleichera trijuga.
Lacca lignum – Dalbergia junghuhnii.
Lace Bark – Lagetta lintearia.
Lace Bark – Plagianthus betulinus.
LACHNATHES Ell. Haemodoraceae. 1 sp. N.America. Herb.
***L. carolina*** (Lam.) Dandy = L. tinctoria. (Paintroot, Red Root, Spiritweed, Wool Flower). An extract of the roots and leaves is used to treat coughs and pneumonia and to make a red dye.
***L. tinctoria*** Ell. = L. carolina.
***L. tinctorum*** (J. F. Gmel.) Sprague = L. carolina.
***Laciniaria*** Hill = Liatris.
Lacktree, Malay – Schleichera trijuga.
Lacquer, Burmese – Melanorrhoea usitata.
Lacquer, Japanese – Rhus vernicifera.
LACTARIUS Pers. ex S. F. Gray. Agaricaceae. 130 spp. Mainly temperate. The fruit-bodies of the following species are eaten locally and frequently are sold in markets.
***L. camphoratus*** L. N.Temperate. The dried fruits are used for flavouring fruits and salads.
***L. congolensis*** Beeli. Congo.
***L. deliciosus*** (L.) Gray. Temperate.

***L. flavidulus*** Imai. Japan.

***L. helvus*** (Fr.) Fr. N.Temperate. The dried fruits are used for flavouring soups and salads.

***L. hygrophoroides*** B. and C. E.Asia, America.

***L. lignyotus*** Fr. Temperate.

***L. luteolus*** Peck. N.America.

***L. piperatus*** (Scop.) Grey. Temperate. The fruitbodies have an acrid taste, but are eaten widely. The Chinese used them as a drug and the French used them as a medicine to treat conjunctivitis associated with gonorrhoea.

***L. repraesentaneus*** Britz. Temperate. Eaten pickled in Russia.

***L. rufus*** (Scop.) Fr. Temperate, Eaten pickled in Russia.

***L. torminosus*** Schaff. Sweden, Russia.

***L. scrobiculatus*** Scop. Temperate. Eaten pickled in Russia.

***L. subpurpureus*** Peck. N.America.

***L. vellereus*** Fr. Temperate. Eaten pickled in Russia.

***L. volemus*** (Fr.) Fr. Europe.

LACTOBACILLUS Bacillaceae.

***L. acidophilus*** (Moro) Holland. Used in acidifying milk by the respiration of lactose to lactic acid.

***L. bulgaricus*** (Leurssen and Kühn) Holland. The main organism involved in the manufacture of Yoghurt from milk.

***L. casei*** Rogers. Used particularly in Russia to make the drink Kumiss by the fermentation of cow's or mare's milk. It also plays a part in the preparation of Swiss cheeses.

***L. cucumeris*** Berg et al. With L. palntarum performs the primary fermentation in the production of Sauerkraut. L. pentoaceticus then carries out a second fermentation.

***L. pentoaceticus*** Fred. Peterss. and Davenp. See L. cucumeris.

***L. pentosus*** Fred, Peters and Anders.=L. plantarum (O. Jensen) Holland. Used in the production of lactic acid from sulphite waste liquors from the refining of sugar.

***L. plantarum*** Holland. See L. cucumeris.

***L. plantarum*** (O. Jensen) Holland.=L. pentosus.

LACTUCA L. Compositae. 100 spp. Chiefly temperate Eurasia. Herbs.

***L. amurensis*** Regel.=L. indica.

***L. angustana*** Vilm=L. sativa var. asparagina.

***L. canadensis*** L. (Canada Lettuce). N.America, W.Indies. The young leaves are eaten locally as a vegetable.

***L. denticulata*** Maxim. China, Japan. The leaves are eaten locally in China as a vegetable.

***L. indica*** L.=L. amurensis=L. laciniata=L. saligna. India and Far East. Cultivated as a vegetable in China and Japan and is a commercial crop in these countries. There are several varieties. The leaves are thought to be a tonic and to aid digestion.

***L. laciniata*** Makino=L. indica.

***L. perennis*** L. (Perennial Lettuce). Central and S.Europe. The blanched leaves are used locally in salads.

***L. quercina*** L. Europe, especially central and E.Russia. The plant is cultivated especially in France for the latex (Lactucarium) which is a sedative, used in cough mixtures etc. as an opium substitute.

***L. romana*** Gars.=L. sativa var. longifolia.

***L. saligna*** Lour=L. indica.

***L. sativa*** L. (Lettuce). A cultigen, probably derived from the wild L. scariola of Asia Minor. It is recorded as a food by the Persians in the 6th century B.C. and apparently spread. It was cultivated in China by the 5th century A.D. and was in common use by 16th century. The plant is used mainly for salads, but sometimes as a vegetable. It is cultivated widely as a garden and commercial crop, both in the open and under glass. There are several varieties including var. ***capitata*** L. (Cabbage Lettuce, Head Lettuce) with dense cabbage-like heads; var. ***crispa*** L. (Curled Lettuce) with a looser head, with the margin of the leaves crinkled; var. ***longifolia*** Lam.=***L. romana*** Gars. (Cos Lettuce, Romaine Lettuce) with a loose head of narrow upright leaves; var. ***asparagina*** Bailey=L. ***angustana*** Vilm. (Asparagus Lettuce) which have a thick edible stem without a distinct head.

***L. scariola*** L. (Prickly Lettuce). Europe, Asia, introduced to N.America. Grown in Egypt for the oil from the seeds (Egyptian Lettuce seed Oil) which is used in the preparation of foods.

***L. taraxaciflora*** (Willd.) Schum.=Sonchus taraxcifolius. (Langue de Vache, Wild Lettuce). Tropical Africa, especially W.

Grown locally as a vegetable in W.Africa; also eaten in salads.

***L. thungergii*** Maxim. Mongolia, China, Japan. Used medicinally in China.

***L. virosa*** L. (Bitter Lettuce, Lettuce Opium). Central and S.Europe. Cultivated in Germany, France and U.K. as a source of Lactucarium. See L. quercina.

Lactucarium – Lactuca quercina.

Lactucarium – Lactuca virosa.

Laddo – Prosopsis africana.

LADENBERGIA Klotzsch. Rubiaceae. 30 spp. S.America. Trees.

***L. hexandra*** Wedd. S.America. The bark contains astringent alkaloids, extracted as Quino do Rio and used as a substitute for quinine.

Ladino Clover – Trifolium repens.

Ladybell, Broadleaf – Adenophora latifolia.

Lady's Bedstraw – Galium verum.

Lady's Comb – Scandis pecten-veneris.

Lady's Leek – Allium cernuum.

Lady's Mantle – Alchemilla vulgaris.

Lady's Seal – Tamus communis.

Lady Wrack – Fucus vesiculosus.

LAELIA Lindl. Orchidaceae. 30 spp. Mexico, C.America, E. tropical S.America. Epiphytic orchids.

***L. autumnalis*** Lindl. (Flor de los Muertos). Mexico. The mucilage from the pseudobulbs is used locally to make small images used during religious festivals.

***L. grandiflora*** Lindl. = L. speciosa.

***L. speciosa*** Schltr. = L. grandiflora. (Flor de Todos Santos). See L. autumnalis.

***L. thomsoniana*** Schltr. Central America, W.Indies. The hollow pseudobulbs are used as tobacco pipes locally in the W.Indies.

Lafa – Chrysalidocarpus fibrosus.

LAFOENSIA Vand. Lythraceae. 12 spp. Tropical America. Trees.

***L. punicaefolia*** DC. Tropical America. A yellow dye is extracted from the wood.

Lafu – Neodypsis tanalensis.

LAGENARIA Ser. Curcubitaceae. 1 sp. Throughout the tropics; 1 sp tropical Africa, Madagascar; 4 spp. tropical Africa. Vines.

***L. breviflorus*** Benth. Tropical Africa. The fruits are used in W.Africa for removing the hairs from hides.

***L. leucantha*** Roxb. (Baguang, Sikay, Upo). Tropical Africa, Asia. The plant is cultivated locally for the fruits which are eaten raw.

***L. siceraria*** (Molina) Stand. = L. vulgaris. (Bottle Gourd, Calabash). Probably originated in Africa, but has spread throughout the world. The hard fruit shells have been used as utensils since ancient times by many peoples. The young fruits are eaten as a vegetable in parts of Africa and Asia. The pulp from around the seeds is used as a purgative in India. The plant has been cultivated since ancient times.

***L. vulgaris*** Ser. = L. siceraria.

LAGERSTROEMIA L. Lythraceae. 50 spp. Old World tropics. Trees.

***L. flos-reginae*** Retz. = L. speciosa Pers. (Queen Crapemyrtle). Malaysia. The reddish wood is insect resistant and is used for house building, flooring, bridges and railway sleepers.

***L. hypoleuca*** Kurz. (Andaman Crapemyrtle). Amdaman Islands. The timber (Pyinma Andaman) is used for general carpentry.

***L. lanceolata*** Wall. (Benteak, Bili nandi, Nanan Wood). India. The red-brown timber is elastic, durable and water resistant. It is very valuable and used for a wide variety of purposes, including shipbuilding, furniture, poles, agricultural implements, boxes of various sorts, turning, casks. The bark is used for tanning.

***L. piriformis*** Koehne. (Battinan Crapemyrtle). Philippines. The wood (Banabu) is used for general carpentry.

***L. reginae*** Roxb. = L. speciosa. (L.) Pers.

***L. speciosa*** Pers. = L. flos-reginae.

***L. speciosa*** (L.) Pers. = L. reginae Roxb. Tropical Asia, Australia. A decoction of the leaves is used in the Philippines to treat diabetes mellitus and the seeds are used as a narcotic.

LAGETTA Juss. Thymelaeaceae. 4 spp. W.Indies. Small trees.

***L. lintearia*** Lam. (Lace Bark Tree). W.Indies. The inner bark, removed by maceration is like lace and is used locally for making dresses, etc.

La Glu – Carpodinus hirsuta.

LAGOECIA L. Umbelliferae. 1 sp. Mediterranean. Herb.

***L. cuminoides*** Occasionally used as a substitute for Cuminum cymimum q.v.

Lagondi – Vitex trifolia.

Lagos Mahogany – Khaya grandifoliola.

Lagos Silk Rubber – Funtumia elastica.

LAGUNCULARIA Gaertn. f. Combretaceae.

2 spp. Tropical America, tropical W.Africa. Trees.

*L. **racemosa*** Gaertn. f.=Bucida buceras. (Black Olive Tree, Oxhorn Bucida, White Buttonwood, White Mangrove). Coastal regions of tropical America, Florida and W.Indies. The heavy hard wood is a beautiful green-brown. It is termite resistant and withstands contact with the soil. Its uses include heavy construction work, railway sleepers, flooring, fencing and carts.

Lahuán – Fitzroya cupressoides.

Lakambing – Sarcolobus spanoghei.

Laket – Chrysophyllum roxburghii.

Lakoom – Vitis geniculata.

Lal jahri – Geranium wallichianum.

Lalang – Imperata arundinaceae.

Lalanga – Fagara pilosiuscula.

LALLEMANTIA Frisch. and Mey. Labiatae. 5 spp. Asia Minor to Central Asia and Himalayas. Herbs.

*L. **iberica*** (Bieb.) Fisch. and Mey.=L. sulphurea=Dracocephalum ibericum. (Lallemantia). Dry regions of Asia Minor and the Middle East. The plant is grown for the seeds which contain about 30 per cent of a drying oil (Lallemanntia Oil) which is used locally for cooking and lighting. The leaves are eaten as a vegetable in Persia.

*L. **royleana*** Benth. Persia to Himalayas. The seeds are used in Persia to treat coughs, as a heart stimulant and as an aphrodisiac.

*L. **sulphurea*** C. Koch.=L. iberica.

Lallemantia (Oil) – Lallemontia iberica.

Lâ lot – Piper lolot.

LAMANONIA Vell. Cunoniaceae. 10 spp. S.Brazil, Paraguay. Trees. The bark of the species mentioned is used as a tonic locally.

*L. **speciosa*** Camb.

*L. **tomentosa*** Camb.

Lamb's Quarters – Chenopodium album.

Lambai ajam – Anisophyllea disticta.

Lambert's Filbert – Corylus tubulosa.

Lambinana – Nuxia verticillata.

Lamedor de Moca – Visnea mocanera.

Lamgitlawe – Homalium grandiflorum.

LAMINARIA Lam. Laminariaceae. Large, marine, brown Algae.

*L. **angustata*** Kjellm. See L. japonica.

*L. **cichorioides*** Miyabe. See L. japonica.

*L. **cloustoni*** Edm. See L. digitata.

*L. **digitata*** (L.) Edmonson. (Kelp, Sea Girdles, Seastaff, Tangle, Horsetail Kelp). N.Temperate Atlantic. Used widely in U.S.A. and U.K. to manufacture algin and ammonium, calcium and sodium alginate. It was formerly used as a commercial source of iodine and is used locally as a manure. Algin and its salts are colloids used in the stabilization of foods, confectionery, especially ice-cream, dehydrated and prepared foods, cosmetics and medicinal creams, jellies etc.; paints, water softeners, and the making of dental plates. The young plants are eaten locally.

*L. **japonica*** Aresch. (Japanese Kelp). Coasts of temperate E.Asia. This or a similar species has been recorded as a food in China as long ago as 600 B.C. This is the principle source of Kombu, a food widely eaten in Japan. It is made from the shredded dried seaweed, which may be powdered or coloured with malachite green. It is eaten with meat, as a vegetable, pickled, or candied as a sweetmeat. Sometimes it is used as a tea substitute, or made into a curd with beans. Kombu is called Haiti in China.

*L. **potatorum*** Lab. Temperate oceans. Eaten locally by the Australian aborigines.

*L. **radicosa*** Kjellm=Ecklonia radicosa.

*L. **religiosa*** Miyabe. See L. japonica.

*L. **saccharina*** (L.) Lamour (Broadleaf Kelp, Sweet Tangle). See L. digitata.

*L. **stenophylla*** (Kuetz.) J. Ag. See L. difitata.

LAMIUM L. Labiatae. 40–50 spp. Europe, Asia, tropical Africa. Herbs.

*L. **album*** L. (White Deadnettle). Europe, Asia, introduced to N.America. A decoction of the flowers is used as a home remedy for catarrh, dropsy and vaginal discharges. An infusion of the roots in wine is a home remedy for stones in the kidney. The young stems are used occasionally as a vegetable.

*L. **purpureum*** L. (Purple Deadnettle). Europe to Mediterranean. Introduced to N.America. The leaves are used as a diuretic and purgative and used to stop bleeding.

Lamkis – Palaquium microphyllum.

Lambrisco – Rhus virens.

Lampeni gede – Ardisia colorata.

Lamwick Plant – Phlomis luchnitus.

Lamy Butter – Pentadesma butyracea.

Lanamar – Posidonia australis.

Lanceleaf Sandalwood – Santalum lanceolatum.
Lance Tree – Lonchocarpus capassa.
Lancewood – Amelanchier canadensis.
Lancewood – Calycophyllum candidissimum.
Lancewood – Cardwellia sublimis.
Lancewood – Oxandra lanceolata.
Lancewood, Cape – Curtisia dentata.
Lancewood, Cuba – Duguetia quitarensis.
Lancewood, Honduras – Lonchocarpus houdurensis.
Lancewood, Jamaica – Duguetia quitarensis.
Lancewood, Moulmein – Homalium tomentosum.
Landa – Erythroxylum mannii.
Land's Grass – Panicum laevifolium.
LANDOLPHIA Beauv. Apocynaceae. 55 spp. Tropical and S.Africa, Madagascar, Mascarene Islands. Vines. The species mentioned are minor sources of rubber, which is obtained by tapping the stems and roots. Considerable quantities were produced in W.Africa, from wild plants of L. heudelotii and L. oweniensis, but the habit of the plants makes them unsuitable for large scale cultivation.
***L. comorensis*** Benth. Tropical Africa.
***L. crassipes*** Radl. Madagascar. Rubber called Fingotta.
***L. dawei*** Stapf. E.Africa.
***L. droogmansiana*** de Wild. Congo.
***L. dubardi*** Pierre (Fingobary). Madagascar.
***L. gentilii*** de Wild. Congo. Rubber called Rouge de Kassai.
***L. heudelotii*** A. DC. W.Africa, Sudan.
***L. hispidula*** Pierre. (Fingomainty). Madagascar.
***L. kilmandjarica*** Stapf. = Clintandra kilmandjarica. Tropical Africa.
***L. kirkii*** Dyer. E.Africa. Important locally. The rubber is called Mosambique Blanc, Mosambique Rogue, Nyassa Rubber, Pine Rubber.
***L. klainei*** Pierre. W.Africa.
***L. madagascariensis*** Benth. and Hook. Madagascar. Rubber called Madagascar Rouge.
***L. mandrianambo*** Pierre. (Mandrianambo). Madagascar.
***L. owariensis*** Beauv. Tropical Africa, especially West. The rubber is called Rouge de Congo, Rouge du Kassai.
***L. perieri*** Jum (Piralahi). Madagascar Rubber called Majunra Rouge.
***L. richardiana*** Pierre. Madagascar.
***L. senegalensis*** Kotschy and Peyr. W.Africa.
***L. sphaerocarpa*** Jun. Madagascar. The rubber is called Reiabo Rubber.
***L. stolzii*** Busse E.Africa.
***L. subsessilis*** Pierre. Madagascar.
***L. thollonii*** Dewevre. Congo, Angola. The rubber is tapped from the roots and is called Caoutchouc des Herbes.
***L. ugandensis*** Stapf. Uganda.
Langa Marca – Myrsine grisebachii.
Langsa ootan – Aglaia silvestris.
Langsat – Lansium domesticum.
Langsat lootoong – Aglaia eusideroxylon.
Langsatan – Aglaia acida.
Langset lootoong – Aglaia eusidetoxylon.
LANGSDORFFIA Mart. Balanophoraceae. 1 spp. Mexico to tropical S.America. Parasitic herbs.
***L. hypogaea*** Mart. The plants yield a wax (Siejas) which is used locally to make candles.
***Languas*** Koenig = Alpinia.
Langue de Vache – Lactuca taraxaciflora.
Langwas – Alpinia galanga.
Lanishah – Reaumurea hypericoides.
Laujak – Cymbopogon jwarancusa.
LANNEA A. Rich Anacardiaceae. 70 spp. Tropical Africa; 1 sp. Indomalaya. Trees or shrubs.
***L. acida*** A. Rich. Tropical Africa. The subacid fruits are eaten locally. The stem yields an edible gum and the powdered bark is used by the natives to paint their faces.
***L. amaniensis*** Engl. and Krause. Tropical Africa. The bark yields a red dye which is used locally for dyeing cloth.
***L. antiscorbutica*** (Hiern.) Engl. (Kalombo, Kamukungo, Mubumbo) Congo to Zambia. A decoction of the bark is used locally to treat scorbutic ulcers and scurvy. The hard wood is used during religious ceremonies.
***L. caffra*** Sond. (Kaffir Plum). S.Africa. The heavy elastic wood is used for construction work and furniture.
***L. discolor*** (Sond.) Engl. (Kabumbu, Live Long, Mushamba). Tropical Africa, particularly Rhodesia and Zambia. The soft wood is used locally to make floats for fishing nets and brake blocks. A decoction of the bark is used to treat diarrhoea.
***L. edulis*** (Sond.) Engl. = L. velutina. (Bukukute, Imbatabata, Kibunga).

Tropical Africa. The fruits are eaten locally.

***L. fruticosa*** Hochst. Abyssinia. A gum from the bark is used to adulterate Gum Arabic.

***L. gossweileri*** Exell. and Mendonca. Tropical Africa. The fruits are eaten locally.

***L. grandis*** (Dennst.) Engl. (Jigman, Kashmal, Wodier Wood, Wuda). India, Malaya. The wood (Wodier Wood) is brown, fairly hard and termite resistant. It is used locally for a wide variety of purposes, mainly making small household articles, carving, small boats, packing cases, furniture etc. The bark is used for tanning and dyeing silk a golden brown, it is also powdered as a dentifrice and strips of it are placed on the backs of elephants as a protective saddle. The gum from the trunk (Jingan Gum) is used for sizing paper, calico printing and mixing with lime for plastering. The gum is also used in making confectionery.

***L. gummifera*** Blume Java. The gum has commercial possibilities.

***L. stuhlmannii*** (Engl.) Engl. Central Africa. The bark is used locally as the source of a red dye and by the local fishermen to preserve their nets.

***L. velutina*** Auct. = L. edulis.

***L. welwitschii*** (Hiern.) Engl. (Ayr, Bombata, Mumbu ya dito). Central Africa. The light wood is used locally to make small boats and for light interior work in houses.

***L. wodier*** Roxb. E.India, Burma. The trunk yields a gum (Jingan Gum) used locally for calico printing, for making whitewash and preserving fishing nets.

Lanoot – Mischophloeus vestiaria.

LANSIUM Corrêa. Meiliaceae. 6–7 spp. Indomalaya. Trees.

***L. domesticum*** Jack. (Langsat). Tropical Asia, especially Malaysia and Indochina. The plant is cultivated in the Old and New World tropics for the fruits which are eaten raw. They are about 5 cm. long with a leathery skin and a whitish aromatic pulp. The fruit peel is used as incense in parts of Java.

LANTANA L. 150 spp. Tropical America, W.Indies, tropical and S.Africa. Shrubs.

***L. camara*** L. Tropical America. A decoction of the leaves is used locally as a tonic and stimulant.

***L. lippioides*** Hook. and Arn. = Lippia geminata.

***L. microphylla*** Mart. S.America. A decoction of the leaves is used in Brazil to treat rheumatism and the fruits are used to make a tonic.

***L. pseudo-thea*** St. Hil. = Lippia pseudo-thea.

Lapachillo – Sweetia elegans.

LAPLACEA Kunth. Theaceae. 30 spp. Malaysia, tropical America, W.Indies. Trees.

***L. curtyana*** A. Rich. (Almendero). W.Indies, especially Cuba. The hard wood is used for general carpentry and furniture-making.

***L. intergerrima*** Miq. = Gordonia integerrima = Haemocharis integerrima. (Hooroo mandjel). Indonesia. A rare tree, having been cut nearly to extinction. The wood is hard and durable. Much prized locally for furniture-making.

***L. semiserrata*** Camb. Brazil. A decoction of the seeds is used locally as a diuretic and aphrodisiac.

LAPORTEA Gaudich. Urticaceae. 23 spp. Tropics and subtropics, temperate E.Asia, E. N.America, S.Africa, Madagascar. Herbs or woody plants.

***L. aestuans*** (L.) Gaudich. = Urtica aestuans. (Afudu, Lusala, Tokodole). Congo, Angola, Ethiopia to Mozambique. The leaves are eaten locally as a vegetable and the stem fibres are used to make cordage.

***L. bulbifera*** Wedd. Japan. The young shoots are used as a vegetable locally.

***L. canadensis*** (L.) Gaudich. (Wood Nettle). E. N.America. The best fibres make a very strong fibre which was used by the local Indians after extraction by retting.

***L. gigas*** Wedd. = Urtica gigas. (Giant Nettle). New South Wales, Queensland. The bast fibres make an excellent fibre, used locally.

***L. meyeniana*** (Walp.) Warb. Philippines. A decoction of the leaves and roots is used locally as a diuretic.

***L. photiniphylla*** = Urtica photiniphylla. (Small-leaved Nettle). New South Wales, Queensland. The bast fibres make a good cordage used by the aborigines.

***L. podocarpa*** Wedd. (Abamule, Agota, Makwapusa). Tropical Central Africa. The leaves are used locally as a vegetable and diuretic.

Laqueur, Burmese – Melanorrhoea usitata

***Lappa major*** Gaertn.=Arctium lappa.
***Lappa minor*** Hill.=Arctium minus.
Lappa Clover – Trifolium lappaceum.
Larap – Eucalyptus dumosa.
***Larch*** (as used for lumber) – Abies nobilis.
Larch, Chinese – Larix potaninii.
Larch, Eastern – Larix laricina.
Larch, European – Larix decidua.
Larch, Western – Larix occidentalis.
LARDIZABALA Ruiz. and Pav. Lardizabalaceae. 2 spp. Chile. Shrubs.
***L. biternata*** Ruiz. and Pav. Chile to Peru. The fruits (Aquiboquil) are sweet and have a pleasant flavour. They are eaten locally. A fibre is extracted from the stem.
LARETIA Gill. and Hook. Hydrocotylaceae. 2 spp. Chilean Andes. Herbs.
***L. acaulis*** Gill. and Hook. Chilean Andes. The resin from the stem (Resina Laretae) is used as a substitute for galbanum.
Large Bitter Cress – Cardamine amara.
Large Cane – Arundinaria gigantea.
Large Flowered Beard Tongue – Penstemon grandiflorus.
Large Hop Clover – Trifolium campestre.
Large Round Leaved Orchid – Habenaria orbiculata.
Large Tooth Aspen – Populus grandidentata.
LARIX Mill. Pinaceae. 10–12 spp. Europe, N.Asia, N.America. Trees.
***L. americana*** Michx.=L. laricina.
***L. dahurica*** Turcz. (Guimatsu, Kui). E.Siberia to Japan. The hard wood is used in Japan for a wide variety of heavy construction work, mine props, railway sleepers, water pipes and ship building. A turpentine (Japanese Turpentine) and tannin are extracted from the bark.
***L. decidua*** Mill.=L. europaea. (European Larch). Central Europe to Siberia. The elastic wood is used for a variety of construction work and general carpentry. The wood yields an oleoresin (Venetian Turpentine, Terebinthina Laricina, Terebithina Ventea) and a gum (Orenburgh Gum). A sweet substance (Manna of Briançon) was extracted from the wood and used to treat bladder and bronchial catarrh. It was also used in ointments etc. Much used in reforestation in the U.K.
***L. europea*** DC.=L. decidua.
***L. laricina*** (Du Roi) Koch.=L. americana. (Eastern Larch, Tamarack). E. N.America to Rocky Mountains. The heavy hard wood is used for telegraph poles, railway sleepers and boat building. An extract of the bark is a laxative and diuretic. It is used to treat jaundice and liver obstructions, rheumatism and skin diseases.
***L. occidentalis*** Nutt. (Tamarack, Western Larch). The heavy strong wood is reddish and durable in contact with the soil. It is used for railway sleepers, posts, house interiors and furniture-making.
***L. potaninii*** Batalin. (Chinese Larch). China. The wood is used in China as the other spp. mentioned.
Larkspur, Forking – Delphinium consolida.
Larkspur, Prairie – Delphinium virescens.
Larkspur, Rocket – Delphinium ajacis.
Larouman – Ischnosiphon aruma.
LARREA Cav. Zygophyllaceae. 2 spp. Temperate S.America. Shrubs.
***L. mexicana*** Moric.=Covillea tridentata. (Creosote Bush, Greasewood). S.W. U.S.A. and Mexico. An infusion of the leaves and twigs is used locally as an antiseptic. The flower buds are pickled and used as a condiment. A decoction of the leaves is also used in Mexico as a bath to treat rheumatism.
***L. nitida*** Cav. Argentina and Chile. A decoction of the leaves is used in Chile to aid digestion and to control menstruation.
Lasa sooroo – Aglaia silvestris.
LASERPITIUM L. Umbelliferae. 35 spp. Canaries, Mediterranean through to S.W.Asia. Herbs.
***L. aquilegiaefolium*** Jacq.=Siler trilobum.
***L. latifolium*** L. (Laserwort, White Gentian, Woundwort). Europe. The fruits are used locally to relieve stomach complaints and to flavour beers.
***L. montanum*** Lam.=L. siler.
***L. prutenicum*** L.=L. selenioides. Europe. The root contains a resin which makes it valuable as a local remedy for treating skin complaints. They are used in France to make healing dressings (Thaspia Plasters).
***L. selenoides*** Crantz.=L. prutrnicum.
***L. siler*** L.=L. montanum=Siler montanum. Central and S.Europe. The fruits and roots are used locally as a condiment and to treat toothache. The fruits are used in Austria to flavour liqueurs.
***L. trilobum*** L.=Siler trilobum.
Laserwort – Laserpitium latifolium.

LASIA Lour. Araceae. 3 spp. Indomalaya. Herbs.

***L. aculeata*** Lour. = L. spinosa. Indomalaya. The young leaf stalks and leaves are eaten locally as a vegetable in curries or with rice.

***L. spinosa*** Thw. = L. aculeata.

LASIANTHERA Beauv. Icacinaceae. 1 sp. Tropical W.Africa. Shrub.

***L. africana*** P. B. (Mbe-mundju, Mundja). Archways of the branches are erected in front of villages to ward off evil spirits.

LASIOSIPHON Fresen. Thymeleaceae. 50 spp. Tropical and S.Africa, Madagascar through to W.India and Ceylon. Shrub.

***L. krausii*** Mels. W.Africa. The leaves and roots, which are poisonous are used locally as a drastic purgative, fish poison and criminal poison.

Laskar – Delphinium brunonianum.

***Lastrea*** Bory = Thelypteris.

Latanier – Coccothrinax martiniensis.

***Latheria thamnodes*** (Flot.) Hue. = Evernia mesomorpha.

LATHRAEA L. Orobanchaceae. 7 spp. Temperate Eurasia. Herbaceous root parasites.

***L. squamaria*** L. (Toothwort). Europe through to Himalayas. The underground parts were a home remedy for colic, epilepsy and convulsions.

LATHYRUS L. Leguminosae. 130 spp. N.Temperate and mountains of tropical Africa and S.America. Herbs.

***L. alatus*** Sibth. and Sm. = L. clymenum.

***L. cicera*** L. (Garousse, Jurosse, Mochi). Canaries, through the Mediterranean to Transcaucasus. Cultivated locally in Spain, S.France and Italy as a food plant, but the pods, if not cooked properly can be poisonous.

***L. clymenum*** L. = L. alatus = L. purpureus. Mediterranean. Cultivated locally for fodder.

***L. decaphyllus*** Pursh. = L. polymorphus.

***L. hirsutus*** L. (Rough Pea). Mediterranean. Introduced into many countries as a forage crop, although there is some possibility of livestock being poisoned.

***L. maritimus*** (L.) Bigelow. (Beach Pea, Sea Pea). N.Polar regions. The seeds are used as a coffee substitute by the Eskimos.

***L. maritimus*** Torr. = L. vestitus.

***L. montanus*** (L.) Bernh. (Tuberous Bitter Vetch). Europe. The tubers are eaten roasted locally, especially in U.K.

***L. ochroleucus*** Hook. W. N.America. The seeds were eaten by the local Indians.

***L. ochrus*** DC. S.Europe. Grown for livestock feed in Greece.

***L. odoratus*** L. (Sweet Pea). S.Europe. A very popular garden flower, grown commercially for the flowers from which an essential oil is extracted and used in perfumery.

***L. ornatus*** Nutt. W. N.America. The local Indians eat the pods as food.

***L. polymorphus*** Nutt. = L. decaphyllus. W. N.America. The pods are eaten by the local Indians.

***L. purpureus*** Desf. = L. clymenus.

***L. sativus*** L. (Chickling Vetch, Grass Peavine, Kasari). Mediterranean. Cultivated in the Middle East and N.India for human food and for livestock, especially in poorer areas. The uncooked seeds contain an alkaloid which causes lathyrism (a paralysis of the lower limbs), so that the seeds have to be well-cooked before eating.

***L. sylvestris*** L. (Flat Pea). Europe to W.Asia. Used locally for livestock feed.

***L. tingitanus*** L. (Tangier Pea). Mediterranean. Grown locally and in W. and S.W. U.S.A. as winter pasture for stock.

***L. tuberosus*** L. (Groundnut Peavine, Earth Chestnut). Europe through to W.Asia. Introduced to N.America. Cultivated for the tubers which are eaten locally as a vegetable. The flowers were formerly used for perfume.

***L. vestitus*** Nutt. = L. maritimus Torr. W. N.America. The seeds are used as a coffee substitute by the Eskimos.

LATIPES Kunth. Graminae. 1–2 spp. Senegal to Sind. Grasses.

***L. senegalensis*** Kunth. N.W. Africa. The seeds are eaten as a cereal by the desert tribes.

LAUNAEA Cass. Compositae. 40 spp. Canaries, Mediterranean to E.Asia, tropical and S.Africa. Herbs.

***L. glomerata*** Hook. f. Deserts of Africa and Asia. The young leaves are eaten locally as a salad.

***Launea*** Endl. = Launaea.

Laurel – Kalmia latifolia.

Laurel – Lauris nobilis.

Laurel, Alexandrian – Calophyllum inophyllum.

Laurel Amarillo – Nectandra sintenisii.

Laurel Avispello – Nectandra coriacea.

Laurel Berry Fat – Laurus nobilis.
Laurel, Bignay China – Antidesma bunius.
Laurel, Californian – Umbellularia californica.
Laurel, Cherry – Prunus laurocerasus.
Laurel, Chile – Laurelia aromatica.
Laurel, Chinese – Antidesma bunius.
Laurel de la Sierra – Litsea neesiana.
Laurel Geo – Ocotea leucoxylon.
Laurel, Great – Rhododendron maximum.
Laurel, Ground – Epigaea repens.
Laurel, Mountain – Kalmia latifolia.
Laurel, Oak – Quercus imbricata.
Laurel Prieto – Nectandra membranacea.
Laurel Sabino – Magnolia splendens.

LAURELIA Juss. Atherospermataceae. 1 sp. New Zealand; 1 sp. Chile. Trees.

***L. aromatica*** Juss. (Chile Laurel, Peruvian Nutmeg). Chile, Peru. The seeds are used locally as a spice.

***L. novae-zealandiae*** A. Cunn. New Zealand. The wood is soft but very strong. It is used locally for boat-building and furniture-making. An infusion of the bark, applied externally or taken internally as appropriate is used locally as a tonic, to treat sore throats, boils, ulcers, tuberculosis, gonorrhoea and to control menstruation.

***Lauro-Cerasus*** Duham. = Prunus.

LAURUS L. Lauraceae. 2 spp. Mediterranean, Canaries, Madeira. Trees.

***L. nobilis*** L. (Bay, True Bay, Laurel). Mediterranean. The leaves are used as a condiment. A fat (Bay Fat, Laurel Berry Fat) extracted from the seeds is used to make soap.

***L. peumus*** Domb. = Cryptocarya peumus.

***L. pyrifolia*** Willd. = Ocotea squarrosa.

Lavandin – Lavandula officinalis × L. latifolia.

LAVANDULA L. Labiatae. 28 spp. Atlantic Islands, Mediterranean to Somaliland and India. Small shrubs.

***L. dentata*** L. (Toothed Lavender). Mediterranean to Afghanistan. An infusion of the leaves is used in Persia to wash wounds and to treat catarrh.

***L. latifolia*** Medic. = L. spica Cav. (Broad leaved Lavender). Mediterranean. A scented oil (Oil of Spike, Oleum Spicae, Essence d'Aspic) is extracted from the leaves and flower heads for use in perfumery and porcelain painting. The leaves and flower heads are also hung with clothes etc. to scent them and are used in local remedies to control menstruation and to induce abortion. The plant which is cultivated in S.France is a good source of honey.

***L. officinalis*** Chaix. = L. vera = L. spica L. S. Europe. Cultivated locally for the flowers and leaves from which Oil of Lavender is extracted. This is used in perfumery, to aid digestion, for flavouring medicines etc., in insecticide sprays and in painting china. The flowers and leaves are also hung with clothes to give them a pleasant smell.

***L. officinalis*** Chaix × ***L. latifolia*** Cav. (Lavandin). A cultivated hybrid, producing a better yield of oil than either of the parents.

***L. spica*** Cav. = L. latifolia.

***L. spica*** L. = L. officinalis.

***L. stoechas*** L. = Stoechas officinarum. (French Lavender). Mediterranean. An oil (Stoechas Oil) distilled from the plant is used in Spain to treat lung diseases, asthma and cramp.

***L. vera*** DC. = L. officinalis.

LAVATERA L. Malvaceae. 25 spp. Canary Islands to Mediterranean, through to Himalayas, Central Asia and E.Siberia; Australia and California. Herbs.

***L. behriana*** Schl. = L. plebeia.

***L. kashmiriana*** Bois. (Resha Katmi). Persia, N.Pakistan and Kashmir. All parts of the plant are purgative. An infusion of the roots sweetened with sugar is used to treat intestinal and bladder complaints and coughs.

***L. plebeia*** Sims. = L. behriana (Tree Mallow). Australia. The roots are eaten by the aborigines.

Lavender (Oil) – Lavendula officinalis.
Lavender Cotton – Santolina chamaecyparissum.
Lavender, Broadleaved – Lavendula latifolia.
Lavender, French – Lavendula stoechus.
Lavender, Sea – Limonium limonium.
Lavender, Toothed – Lavendula dentata.
Laver – Porphyra vulgaris.
Laver, Lettuce – Ulva latuca.
Lawe – Abroma fastuosa.
Lawson Cypress – Chamaecyparis lawsoniana.

LAWSONIA L. Lythraceae. 1 spp. Old World Tropics. Shrub.

***L. alba*** Lam. = L. inermis.

***L. inermis*** L. = L. alba. (Camphire, Henna

Shrub). The plant has been grown since ancient times, particularly in N.Africa, Arabia to India. The dried leaves yield a green powder which is used to dye the hair and nails orange-brown (henna) and to dye horses' coats and fabric. Henna-dyed linen has been found wrapping Egyptian mummies. The bark is used by the Arabs to treat jaundice and nervous complaints. The flowers yield a scented oil (Mehndi) used by the Africans as a perfume and by the Arabs in religious feasts.

Lay Xut – Dioscorea arachidna.

Layason – Musa silvestris.

Leadbush Glasswort – Salicornia fruticosa.

Leaf Mustard – Brassica juncea.

Leafcup, Yellow – Polymnia uvedalia.

Leafy-stemmed False Dandelion – Pyrrhopappus carolinianus.

Leather-root Scurfpea – Psoralea macrostachya.

Leatherwood – Dirca palustris.

Leatherwood – Eucryphia lucida.

Leavenworth Goldenrod – Solicago leavenworthii.

Leavenworth Vetch – Vicia leavenworthii.

Lebbek – Albizia lebbek.

Lebool fatooh – Drypetes simalurensis.

LECANIODISCUS Planch. ex Benth. Sapindaceae. 3 spp. Tropical Africa. Trees.

***L. cupanioides*** Planch. Tropical Africa. The flowers are used locally to make perfume.

LECANORA (Lecanoraceae). Lichens.

***L. calcarea*** (L.) Nyl. Temperate, on rocks. Used in Sweden to dye wool a terracotta colour.

***L. esculenta*** Everson (Manna Lichen). Eaten by the desert tribes, mixed with flour. It could be the manna of the Bible.

***L. parella*** Mass. (Crabeye Lichen, Light Crottle). Yields a dye (Orseille d'Auvergne) which was used in France and U.K. to dye wool purple.

***L. tartarea*** Mass. Temperate on rocks etc. The lichen is soaked in stale urine for about three weeks and then dried over peat fires. The resulting blue-black mass can be kept indefinitely and is used in Sweden and Scotland for dyeing wool red.

***L. ventosa*** Ach. = Haematomma ventosum.

Lecanora, Black – Haematomma ventosum.

Lecanora, Bloody Spotted – Haematomma ventosum.

***Leccinum aurantiacum*** (Bull.) Gray = Polyporus versipellis.

Leche de Ojé – Ficus anthelmintica.

Leche Maria – Trophis racemosa.

Lecherillo – Tabernaemontana citrifolia.

Lecheros – Sapium aucuparium.

Lechosa – Carica papaya.

Lechuguilla – Agave heteracantha.

***Lecidia canescens*** (Dicks.) Ach. = Buellia canescens.

LECONTEA A. Rich. Rubiaceae. 3 spp. Madagascar and Brazil. Vines or trees.

***L. amazonica*** Ducke. Brazil. The root yields a yellow aromatic oil (Oleo Sandal do Norte).

***L. bojeriana*** A. Rich. Madagascar. The stem yields a black dye used locally for colouring cloth.

LECYTHIS Loefl. Lecythidaceae. 50 spp. Tropical America.

***L. costaricensis*** Pitt. Central and S.America. The wood (Cocobola Wood) is yellow to red-brown, very hard and heavy. It is used for carving and making small articles, e.g. knife handles.

***L. grandiflora*** Aubl. Guyana. The orange-red wood is very hard and is used for wheel spokes. The seeds are edible and yield an oil, used for lighting and making soap.

***L. grandiflora*** L. Tropical America. The seeds yield an oil used locally to make soap and for lighting. They are also edible.

***L. longipes*** Poit. = Eschweilera longipes (Toledo Wood). N. S.America. The green-brown to red-brown wood is hard, heavy and durable. It is resistant to marine borers and is used for marine construction work.

***L. ollaria*** L. (Monkey Pod). The red-grey wood is resistant to marine borers. It is used for piers etc. The seeds are edible, yielding an oil (Sapucaja Oil) which is used for lighting and soap manufacture.

***L. paraensis*** Ducke. Brazil. The nuts (Para Nuts) are edible and of excellent flavour.

***L. sagotiana*** Miers. See L. longipes.

***L. subglandulosa*** (Steud.) Miers. See L. longipes.

***L. urnigera*** L. Tropical S.America. The seeds are edible and yield an oil used for lighting and the manufacture of soap.

***L. usitata*** Miers. Brazil. The nuts are edible, having a very pleasant flavour.

***L. zabucajo*** Aubl. Tropical America. The nuts (Paradise Nuts, Sapucaia Nuts) are of excellent flavour and are eaten locally.

Ledger Bark – Cinchonia ledgeriana.

Ledgerbark Cinchona – Cinchona ledgeriana.

LEDUM L. Ericaceae. 10 spp. N.Temperate to Arctic. Shrubs.

***L. decumbens*** (Ait.) Lodd. = L. palustre var. decumbens. N.America. The leaves are used as a tea substitute by the local Eskimos.

***L. goenlandicum*** Oedr. (Labrador Tea). Arctic N.America. The leaves were used as a tea substitute.

***L. palustre*** L. (Crystal Tea Ledum). N. Subarctic and Arctic. The leaves are used by the Eskimos and locally in Japan as a tea. They are also used in an infusion as a diuretic and to treat chest complaints, coughs etc.

***L. palustre*** L. var. ***decumbens*** Ait. = L. decumbens.

Ledun, Crystal Tea – Ledum palustre.

LEEA Royen ex L. Leeaceae. 70 spp. Old World Tropics. Shrubs.

***L. aequata*** L. (Giang, Ginggijang, Kajoo ajer perampooan). Malaysia, Indonesia. A mixture of the wood shavings with ginger is used as a poultice to treat paralysis, while the ground wood and leaves are used to poultice infected wounds.

***L. curtisii*** King. (Mali-mali). Malay Peninsula. The leaves, pounded with tobacco, are used locally as a pomade to preserve the hair.

***L. guinensis*** G. Don. Tropical Africa, Vine. A decoction of the leaves is used locally in the Cameroons to treat stomach complaints.

***L. indica*** Merr. = L. sambucina (Ki toowa, Mali-mali hantco, Soolangkar). S.E.Asia, Indonesia. The crushed leaves are used locally as a poultice to treat headaches.

***L. macrophylla*** Roxb. ex Hornem. (Dhol asamudirka, Dinda). Tropical Asia, especially India. The Hindus use the crushed roots as a poultice to treat ringworm and heal obstinate sores. An infusion of the roots is used to treat guinea worms.

***L. rubra*** Blume. Tropical Asia, especially Indochina and Java. A decoction of the leaves is used locally to treat intestinal worms.

***L. sambucina*** Willd. = L. indica.

Leek – Allium porrum.

Leek, Lady's – Allium cernuum.

Leek, Meadow – Allium canadense.

Leek, Rose – Allium canadense.

Leek, Stone – Allium fistulosum.

Leek, Wild – Allium tricoccum.

LEITNERIA Chapm. Leitneriaceae. 1 spp. S.E. United States of America. Tree.

***L. floridana*** Chapm. (Florida Corkwood). The wood has a lower density than cork and is used for fishing net floats.

Lekir – Amorphophallus prainii.

Leko – Licuala rumphii.

Leliso – Xylocarpus moluccensis.

LEMAIREOCEREUS Britton and Rose. Cactaceae. 25 spp. Central America to N. S.America, W.Indies. Large xerophytic cacti. The fruits of the following species are edible, of good flavour and are frequently sold locally.

***L. chichipe*** (Goss.) Britton and Rose. (Chichipe). Mexico.

***L. griseus*** (Haw.) Britton and Rose. = Cereus griseus. Venezuala to Guyanas. The ash is used locally as a potassium fertilizer.

***L. queretaroensis*** (Weber) Safford = Cereus queretaroensis. (Pitahaya). Mexico.

***L. thurberi*** (Emgelm.) Britton and Rose = Cereus thurberi. (Pitahaya, Pitahay dulce). Mexico to Arizona.

LEMNA L. Lemnaceae. 15 spp. Cosmopolitan. Floating water plants. The following species are cosmopolitan and serve as food for geese and ducks.

***L. minor*** L. (Duckweed).

***L. oligorrhiza*** Kurz.

***L. trisulca*** L.

Lemon (Seed) (Oil) – Citrus limon.

Lemon, Bee Balm – Monarda citriodora.

Lemon, Canton – Citrus limonia.

Lemon, Desert – Atlantia glauca.

Lemon Grass de Cochin, Oil of – Cymbopogon flexuosus.

Lemon Grass Oil – Cymbopogon citratus.

Lemon Grass Oil, East Indian – Cymbopogon flexuosus.

Lemon Grass Oil, West Indian – Cymbopogon citratus.

Lemon, Ground – Podophyllum peltatum.

Lemon scented Gum – Eucalyptus citroidon.

Lemon scented Ironbark – Eucalyptus staigeriana.

Lemon, Sweet – Citrus limetta.

Lemon Verbena – Lippia citriodora.

Lemon Vine – Pereskia aculeata.

Lemon, Water – Passiflora laurifolia.

Lemonade Berry – Rhus integrifolia.

Lemonade Sumach – Rhus typhina.
Lemonade Tree – Rhus typhina.
Lemonwood – Calycophyllum candidissimum.
Lenglnegan – Leucas lavandulifolia.
Lenkeng – Nephelium longana.
LENS Mill. Leguminosae. 10 spp. Mediterranean, W.Asia. Herbs.
***L. culinaris*** Med. = L. esculenta.
***L. esculenta*** Moench. = L. culinaris = Ervum lens. (Lentil). Probably originated in S.W.Europe and temperate Asia. The plant has ben cultivated since prehistoric times; it has been found in Bronze Age villages in Switzerland and is probably the "potage" referred to in Genesis 25:34. It is now widely cultivated in the Middle East and India, as well as in France and Germany. The seeds are used mainly to make soups or a flour which is used in India to make dal and also to make the invalid food Revalenta Arabica. The young pods are sometimes eaten as a vegetable and the green or dried plants make excellent forage for livestock.
Lentil – Lens esculenta.
LENTINUS Fr. Agaricaceae. 50 spp. Widespread, especially in the tropics. The fruitbodies of the following species are eaten locally.
***L. araucariae*** Pat. China, Indo-China, Philippines.
***L. connatus*** Berk. Tropics, especially Far East.
***L. cubensis*** Berk. and Curt. = Collybia boryana. New World tropics.
***L. djamor*** Fried. = Crepidopus djamor = Pleurotus flabellatus. (Djamoor manis). Tropics, especially Indonesia.
***L. edodes*** (Berk.) Sing. = Cortinellus shiitake. E.Asia. The fungus (Shiitake) is cultivated in China and Japan and is sold in local markets and canned.
***L. exilis*** Klotz. Philippines.
***L. goossensiae*** Beeli. Congo.
***L. leucotrichous*** Lév. Tropics, especially Far East.
***L. lividus*** Beeli. Congo.
***L. piperatus*** Beeli. Congo.
***L. rudis*** (Fr.) Henn. = Panus rudis.
***L. sajor-caju*** (Fr.) Fr. (Sajor caju). China, Indochina, Philippines, E.Indies.
***L. squarosulus*** Mont. Tropics, especially Far East.
***L. tuber – regium*** Fr. Tropics. The sclerotia (Pachyma tuber-regium) are eaten in Africa and in Malaya are used to treat fevers.
Lentisk Pistache – Pistacia lentiscus.
LEONOTIS (Pers.) R. Br. in Ait. Labiatae. 40 spp. Tropical and S.Africa, 1 sp. Throughout the tropics. Herbs.
***L. leonurus*** R. Br. (Drug Lion's Ear). S.Africa. A decoction of the leaves is used locally to treat snake-bites and to control menstruation. A decoction of the leaves and flowers is used to treat skin diseases and to remove tapeworms.
LEONTICE L. Leonticaceae. 8–10 spp. S.E.Europe to E.Asia. Herbs.
***L. leontopetalum*** L. (Rakaf). Middle East, Asia Minor). The roots are used to treat leprosy in Palestine, and as an antidote to the effects of opium. They are also used in Kashmir to remove stains from cloth.
LEONTODON L. Compositae. 50 spp. Temperate Eurasia, Mediterranean to Persia. Herbs.
***L. carolinianum*** Walt. = Pyrrhopappus carolinianus.
***L. hispidus*** L. (Hawkbit). Europe, Caucasus, Persia, Asia, Asia Minor. The roasted roots are used as a coffee substitute.
***L. taraxacum*** L. = Taraxacum officinale.
LEONURUS L. Labiatae. 14 spp. Temperate Eurasia. Herbs.
***L. cardiaca*** L. (Motherwort). Europe, Asia. A decoction of the leaves is used as a nerve tonic, a heart stimulant, to treat female hysteria, to control menstruation and tone the membranes of the uterus and as an aphrodisiac. The leaves also yield a green dye used in Central Europe.
***L. sibiricus*** L. (Siberian Motherwort). Siberia, Mongolia to China. Used in China as L. cardiaca.
Leopard's Bane – Doronicum pardalianches.
Leopard Flower – Belamcanda chinensis.
Leopard Lily – Belamcanda chinensis.
LEOPOLDINIA Mart. Palmae. 4 spp. Brazil. Tall palm trees.
***L. major*** Wall. Brazil. The ash from the burnt fruits is used locally as a salt substitute.
***L. piassaba*** Wall. (Para Piassaba). Tropical America. The leaf fibres (Piassaba Fibre) is used locally to make heavy cordage and brushes.
***Lepargyrea canadensis*** Greene = Shepherdia canadensis.

LEPIDAGATHIS Willd. Acanthaceae. 100 spp. Tropics and sub-tropics. Herbs.

***L. alopecuroides*** (Vahl.) R. Br. Tropical America, W.Indies. The leaves are used locally, in Dominica to make a soothing beverage used to calm frightened children.

LEPIDIUM L. Cruciferae. 150 spp. Cosmopolitan. Herbs.

***L. draba*** L.=Cardaria draba.

***L. fremontii*** S. Wats. (Desert Pepperweed) S.W. U.S.A. The seeds were used by the local Indians as a food and for flavouring.

***L. latifolium*** L.=Nastursium latifolium (Dittander). Europe, N.Africa, temperate Asia. The leaves were used by the Ancient Greeks as a salad. It was cultivated. The roots were used medicinally with the leaves (Radix et herba Lepidii) to treat abdominal complaints and scurvy. The Sudanese used the leaves to treat diseases of camels.

***L. meyenii*** Walpers. (Maca). Andes. Cultivated widely locally as a vegetable.

***L. oleraceum*** Forsk. New Zealand. The leaves are used as a vegetable on the South Island.

***L. rotundum*** DC. Australia. The aborigines of S.Australia use the leaves as a vegetable.

***L. sativum*** L. (Garden Cress, Pepper Grass). Temperate. Cultivated widely in temperate regions as a salad plant. The leaves are used medicinally as an antiscorbutic and in Ethiopia an edible oil is extracted from the seeds. The roots are sometimes used as a condiment.

LEPIDOPETALUM Blume. Sapindaceae. 6 spp. Pacific Islands. Trees.

***L. perrottetti*** (Camb.) Blume. Philippine Islands. The ground seeds are used locally to poison wild dogs.

LEPIOTA (Pers. ex Fr.) S. F. Gray. Agaricaceae. 150 spp. Widespread. The fruit-bodies of the following species are eaten locally.

***L. congolensis*** Beeli. Tropical Africa. Eaten in the Congo.

***L. excoriata*** Sch. Madagascar.

***L. madagascariensis*** Duf. Madagascar.

***L. mastoidea*** (Fr.) Quél. Temperate.

***L. morganii*** Peck.=Chlorophyllum molybdites. Warm regions. Eaten in Madagascar and Guinea, but sometimes considered poisonous.

***L. procera*** (Scop. ex Fr.) S. F. Gray= Agaricus procerus. (Parasol Mushroom). Temperate.

***L. rachodes*** (Vitt.) Quél. Temperate. The mushroom Ola bala of excellent eating quality and found in Madagascar is probably a spp. of Lepiota.

LEPIRONIA Rich. Cyperaceae. 1 sp. Madagascar, tropical Asia, Australia, Polynesia. Herb.

***L. mucronata*** Rich. Cultivated in Indonesia and China where the leaves are used for baskets, mats, sails and as a packing material.

LEPISANTHES Blume. Sapindaceae. 40 spp. Tropical Asia. Trees.

***L. cuneata*** Hiern. Malaysia. A decoction of the leaves and stems is used locally to treat coughs.

LEPRARIA Lichen.

***L. chlorina*** (DC.) Ach. (Brimstone-coloured Lepraria). Temperate. Used in Scandinavia as the source of a brown dye used to colour wool.

***L. flava*** (L.) Ach. Europe. A possible source of antibiotics.

***L. iolithus*** (L.) Ach. See L. chlorina.

Lepraria, Brimstone-coloured – Lepraria chlorina.

LEPTACTINIA Hook. f. Rubiaceae. 25 spp. Tropical and S.Africa. Shrubs.

***L. densiflora*** Hook. f. Tropical Africa. The leaves yield an essential oil used in perfumery.

***L. senegambica*** Hook. f. See L. densiflora.

LEPTADENIA R. Br. Asclepiadaceae. 4 spp. Tropical and S.Africa. Woody vines or shrubs.

***L. lancifolia*** Decne. (Anu, Ghelaf, Idar). Dry N.Africa. The leaves are used locally to make a sauce and the fruits are used as tinder.

***L. pyrotechnica*** Decne.=L. spartum.

***L. spartum*** Wight.=L. pyrotechnica. N.Africa. The fruits are eaten by the Bedouins and the wood is used as fuel.

***Leptandra virginica*** (L.) Nutt.=Veronica virginica.

LEPTOCARPUS R. Br. Restionaceae. 12 spp. S.E.Asia, Australia, Tasmania, New Zealand, Chile. Herbs.

***L. chilensis*** Mast.=Schoenodum chilense. (Canutillo). Chile. The leaves are used locally to make mats.

***Leptochlaena*** Spreng.=Leptolaena.

LEPTOCHLOA Beauv. Graminae. 27 spp. Tropics and subtropics. Grasses.

***L. capillacea*** Beauv.=L. chinensis. Africa, Australia. Used as a fodder grass in Africa. The seeds are used as a famine food.

***L. chinensis*** Nees.=L. capillacea.

***L. dubia*** (H.B.K.) Nees. (Green Spangletop). S. U.S.A. to Argentina. An important local grazing and hay grass.

LEPTODERRIS Dunn. Leguminosae. 38 spp. Tropical Africa. Trees.

***L. nobilis*** Welw. ex Baker. (Kifuria, Mufundi). Congo to Angola. A decoction of the roots is used by the local natives as a dressing for wounds.

LEPTOLAENA Thou. 12 spp. Madagascar. Trees.

***L. pauciflora*** Baker. Madagascar. The wood is used locally for house-building.

LEPTOMERIA R. Br. Santalaceae. 15 spp. Australia, Tasmania. Small shrubs. The acid fruits are used locally in Australia to make jellies etc.

***L. acida*** R. Br. (Native Currant).

***L. aphylla*** R. Br.

***L. billardieri*** R. Br.

LEPTONYCHIA Turez. Sterculiaceae. 30 spp. Tropical Africa, S.India, Burma, W.Malaysia, New Guinea. Trees.

***L. chrysocarpa*** Schum. Sudan. The wood is used locally for carving small ornaments etc.

***Leptopteris sumatrana*** Blume.=Gelsemium elegans.

LEPTOSPERMUM J. R. and G. Forst. Myrtaceae. 50 spp. Malaysia, Australia, New Zealand, Caroline Islands. Trees.

***L. citratum*** Chill. E.Australia. An oil distilled from the plant is used in the commercial manufacture of citral. This is used in perfumery, flavouring etc. The plant is grown commercially in E.Africa and Guatemala.

***L. ericoides*** Rich. (Heath Tea Tree). New Zealand. The wood is hard and is used to make small piers and wheels. A decoction of the fruits is used locally to treat stomach complaints, diarrhoea and they are applied as a poultice to sores and wounds. The gum from the capsules is used to treat burns, sucked to relieve coughs and constipation in young children. A decoction of the bark is also used locally to treat diarrhoea, epidermal infections and as a sedative. An infusion of the leaves is also used to treat internal pains of various sorts, bruises etc., dysentery and urinary complaints. The sap combats bad breath and purifies the blood.

***L. flavescens*** Sm.=L. thea (Yellow Tea Tree). Australia. The leaves are used locally as a tea substitute.

***L. thea*** Willd.=L. flavescens.

***L. scoparium*** Forst. (Broom Tea Tree). New Zealand. The red, strong, elastic wood is used for inlay work and cabinet-making. The Maoris used it for paddles and spears. The bark was used by the natives for roofing huts. The leaves are used locally as a tea substitute. See L. ericoides.

LEPTOTAENIA Nutt. Umbelliferae. 16 spp. N.America. Herbs.

***L. multifida*** Nutt. (Cough Root, Indian Balsam). W. N.America. The local Indians use the boiled dry roots as a tea substitute. The root is sold as Balsmea and is used medicinally as an infusion to relieve chest coughs, bronchitis etc.

LEPTOTES Lindl. Orchidaceae. 6 spp. Brazil. Herbs.

***L. bicolor*** Lindl.=L. glaucophylla=Tetramicra bicolor. Brazil. The fruits are used locally for flavouring food, especially ice-cream.

***L. glaucophylla*** Hoffmsgg.=L. bicolor.

Lerp – Eucalyptus dumosa.

LESPEDEZA Michx. Leguminosae. 100 spp. Himalayas to China and Japan, Australia, temperate N.America. Herbs. All the species mentioned are used for livestock feed and soil conservation.

***L. bicolor*** Turcz. (Shrub Lespedeza). China, Korea.

***L. cyrtobotrya*** Miq. Japan.

***L. sericea*** Benth. Himalayas through Korea to Japan.

***L. striata*** Hook. (Japan Clover, Lespedeza). E.Asia.

Lespedeza – Lespedeza striata.

Lespedeza, Shrub – Lespedeza bicolor.

Lesser Asiatic Yam – Dioscorea esculenta.

Lesser Celandine – Ranunculus ficaria.

Lesser Dodder – Cuscuta epithymum.

Lesser Galangal – Alpinia officinarum.

***Letharia thamnodes*** (Flot.) Hue=Evernia mesomorpha.

***Letharia vulpina*** (L.) Waino=Evernia vulpina.

Letona – Agave letonae.

Letterwood – Brosimum aubletii.

Lettuce – Lactuca sativa.

Lettuce, Bitter – Lactuca virosa.

Lettuce, Canada – Lactuca canadensis.
Lettuce, Hare's – Sonchus oleraceus.
Lettuce Laver – Ulva lactuca.
Lettuce, Miner's – Claytonia sibirica.
Lettuce, Mountain – Saxifraga erosa.
Lettuce, Opium – Lactuca virosa.
Lettuce, Perennial – Lactuca perennis.
Lettuce, Prickly – Lactuca scariola.
Lettuce, Sea – Ulva lactuca.
Lettuce Tree – Pisonia alba.
Lettuce, Water – Pistia stratiotes.
Lettuce, Western Squaw – Hydrophyllum occidentale.
Lettuce, White – Prenanthes altissimus.
Lettuce, Wild – Lactuca taraxciflora.

LEUCADENDRON R. Br. Proteaceae. 73 spp. S.Africa. Trees.

***L. argenteum*** R. Br. (Silver Tree). Cape Peninsula. The soft wood is sometimes used for boxes etc. The leaves are silky and are sold as curios and fancy articles.

LEUCAENA Benth. Leguminosae. 50 spp. tropical America; 1 sp. throughout the tropics; 1 sp. Polynesia. Trees. Several spp. are grown for shade in tea plantations.

***L. esculenta*** (Moc. and Sessé) Benth.= Mimosa esculenta. Mexico. The salted seeds are eaten locally.

***L. glauca*** Benth=Mimosa glauca. (Jumpy Bean). Tropics. The leaves, flower buds and young fruits are eaten with rice in Java. The brown seeds are used throughout the tropics to make necklaces etc. and the leaves are used as a green manure.

LEUCAS R. Br. Labiatae. 100 spp. Tropical America, W.Indies, tropical and S.Africa, Arabia, S.China, Indomalaya. Herbs.

***L. aspera*** Spreng.=L. plukenetii=L. dimidiata=Phlomis aspera. Tropical Asia, Philippines. The leaves are used in Vietnam to treat insect stings, snake bites and as a condiment. In the Philippines they are used to treat skin infections.

***L. bancana*** Miq.=L. zeylanica.

***L. dimidiata*** Benth.=L. aspera.

***L. lavandulifolia*** Smith.=L. linifolia. (Lenglengan, Patji-patji). Malaya, Indonesia. The leaves, crushed with lime and tobacco are used by the natives to treat wounds on domestic animals. A decoction of the leaves is also used internally and externally to treat Saccharomycosis of horses. Pillows stuffed with the dried leaves are supposed to cure nervous people who sleep on them.

***L. linifolia*** Spreng.=L. lavandulifolia.

***L. martinicensis*** R. Br. (Wild Tea Bush). Tropics. A decoction of the leaves is used to treat intestinal complaints. They are also burnt as an insect repellent.

***L. plukenetii*** Benth.=L. aspera.

***L. zeylanica*** (L.) R. Br.=L. bancana= Plomis zeylanica. Tropical Asia. In Malaya the pounded leaves are used as a poultice for wounds and sores. The boiled whole plant is used in Malaya to treat skin diseases, in S.Vietnam to treat vaginal discharges and in Cambodia they are used to cure snake bites. In Bali the leaves are used as a condiment.

LEUCONOSTOC Lactobacilliaceae. Several species including ***L. dextranicum*** and ***L. citrovorum*** ferment lactose and are instrumental, with other organisms in souring cream and imparting flavour to butter during its manufacture. ***L. mesenteroides*** is used commercially in the manufacture of dextrans from sugar and in the production of lactic acid in making sauerkraut. Dextrans are used as blood plasma extenders, used during blood transfusions, in foods and as an additive to oil-well drilling muds.

LEUCONOTIS Jack. Apocynaceae. 10 spp. W.Malaysia. Trees.

***L. gigantea*** Boerl. var. ***ovalis*** Boerl. Borneo. The latex yields a very hard rubber.

***L. griffithii*** Hook. f. Malacca. Yields a good rubber, but has not been commercially exploited.

LEUCOPOGON R. Br. Epacridaceae. 150 spp. Malaysia, Australia, New Caledonia. Trees.

***L. vieillardii*** Brongn. and Gris. New Caledonia. The hard, close-grained wood is used for inlay work.

LEUCOSPERMUM R. Br. Proteaceae. 40 spp. S.Africa. Small trees.

***L. conocarpum*** R. Br. S.Africa, especially Cape Peninsula. The bark is very astringent and is used for tanning.

LEUCOSYKE Zoll. and Mor. Urticaceae. 35 spp. Malaysia, Polynesia. Small trees.

***L. alba*** Zoll. and Mor.=L. capitellata.

***L. capitellata*** (Poir.) Wedd.=L. alba.= Urtica capitellata. Java, New Guinea, Formosa to Philippines. The bark fibres are used locally to make cordage. A decoction of the roots is used locally to treat headaches, coughs, pulmonary tuberculosis and stomach pains.

LEUCOTHOË D. Don. Ericaceae. 4 spp. E.Asia; 40 spp. America. Shrubs.

***L. catesbaei*** (Walt.) Gray. (Drooping Leucothoë, Fetter Bush). E. U.S.A. The local Indians use the plant to clear the head from colds. It promotes the discharge of mucous from the mucous membranes of the nose.

Leucothoë, Drooping – Leucothoe catesbaei.

LEUROCLINE S. Moore. Boraginaceae. 3 spp. N. tropical Africa. Small shrubs.

***L. chazaliei*** Boiss. Mauritania. Used locally as firewood.

Levant Cotton – Gossypium herbaceum.

Levant Garlic – Allium ampeloprasum.

Levant Madder – Rubia peregrina.

Levant Scammony (Scamonny) (Resin) – Convolvulus scammonia.

Levant Storax – Liquidambar orientalis.

Levant Worm Seed – Artemisia cina.

Leverwood – Ostrya virginiana.

LEVISTICUM Hill. Umbelliferae. 3 spp. S.W.Asia, Europe, cultivated and naturalized in America. Herbs.

***L. officinale*** (Baill.) Koch. = Angelica levisticum. (Bladder Lovage, Garden Lovage, Lovage angelica). S.Europe. The bleached leaves are sometimes eaten as a vegetable. The fruits are used for flavouring foods, confectionery and liqueurs. Oil of Lovage, used for flavouring is distilled from the roots. A decoction of the roots is used medicinally as a stimulant, to aid digestion, relieve stomach pains, to induce sweating and control menstruation.

Lewa – Engelhardtia polystachya.

LEWISIA Pursch. Portulacaceae. 20 spp. W. N.America. Herbs.

***L. rediviva*** Pursh. (Bitter Root). W. N.America. The roots are eaten by the local Indians.

***Leysera*** L. = Asteropterus.

***Leyseria*** Neek. = Asteropterus.

***Leyssera*** Batch. = Asterocarpus.

Liana unida – Bignonia unguis-cati.

LIATRIS Gaertn. ex Schreb. Compositae. 40 spp. N.America. Herbs.

***L. punctata*** Hook. = Laciniaria punctata. N.America. The roots are eaten by the local Indians of New Mexico.

***L. spicata*** Willd. = Laciniaria spicata. (Button Snakeroot, Gray Feather). N. America, especially E. U.S.A. The roots are used as a diuretic and stimulant.

***L. squarrulosa*** Michx. = Laciniaria scariosa. (Colic Root, Devil's Bit). E. U.S.A. The root is used as a tonic and diuretic.

***Libanothamnus*** Ernst. = Espeletia.

Liberian Coffee – Coffea liberica.

Libiabia – Cissus adenopoda.

LIBOCEDRUS Endl. Cupressaceae. 3 spp. New Caledonia; 2 spp. New Zealand, Andes, W. United States of America. Trees. The wood of the species mentioned is soft, easily worked but durable.

***L. bidwillii*** Hook. f. New Zealand. The wood is red and is used locally for fencing, telegraph poles, bridges, railway sleepers etc.

***L. chilensis*** (Don.) Endl. (Cedro, Ciprés de la Cordillera). Andes. The wood is yellowish and durable. It is used for house-building and interiors.

***L. decurrens*** Torr. (Californian Incense Cedar). W. U.S.A. The light red wood is used for fencing, laths, roof shingles and interiors of buildings.

***L. doniana*** Endl. (Kawaka). New Zealand. The wood is dark red, streaked with deeper red. It is used for general building, posts and fences and roof shingles.

LICANIA Aubl. Chrysobalanaceae. 1 sp. Malaysia; 135 spp. S.E. United States of America to tropical S.America and W.Indies. Trees.

***L. arborea*** Seem. (Mexican Oiticia). Central America, Mexico. The oil from the seeds (Mexican Oiticica Oil) is used locally to make candles, soap and grease.

***L. floribunda*** Benth. Tropical America. The strong wood is used for general carpentry.

***L. micrantha*** Miq. Tropical America. The strong wood is used for general carpentry.

***L. microcarpa*** Hook. f. Brazil. The wood is used for general carpentry and the bark as an astringent.

***L. platypus*** (Hemsl.) Pitt. (Monkey Apple, Sansapote, Zapote Cabillo). Central to S.America. The fruit pulp has a rather unpleasant flavour, but is eaten locally.

***L. rigida*** Benth. Brazil. A drying oil (Oiticica Oil) is extracted from the seeds. It is used in varnishes etc.

***L. sclerophylla*** Mart. Tropical America. The wood is strong and is used for general construction work.

***L. splendens*** Korth. (Sampaluan). Malaysia. The wood is hard, heavy, durable, reddish, but difficult to work. It is used for railway sleepers, piling and tool handles.

***L. terminalis*** Hook. f. (Bois Diable). Lesser

Antilles. The wood is used locally for posts etc. In Dominica it is an important source of good charcoal for domestic fires.

***L. turiuva*** Cham. and Schlecht. Tropical America. The strong wood is used for general carpentry.

LICARIA Aubl. Lauraceae. 45 spp. Central and tropical S.America. Trees.

***L. anacardioides*** Benth. Brazil. The bark is used locally as an abortive.

***L. duckei*** Samp. See L. anacardioides.

***Lichen canescens*** Dicks. = Buellia canescens.

Lichen, Ash Twig – Ramalina fraxinea.

Lichen, Boulder – Parmelia conspersa.

Lichen, Bronze Shield – Parmelia olivacea.

Lichen, Cedar – Cetraria juniperina.

Lichen, Common Twig – Ramalina calicaris.

Lichen, Crabeye – Lecanora parella.

Lichen, Dog – Peltigera canina.

Lichen, Eastern – Stereocaulon paschale.

Lichen, Gild Edge – Stricta crocata.

Lichen, Horsehair – Alectoria jujuba.

Lichen, Manna – Lecanora esculenta.

Lichen, Map – Rhizocarpon geographicum.

Lichen, Mealy Blister – Physcia pulverulenta.

Lichen, Midnight – Parmelia stygia.

Lichen, Pale Shield – Cetraria glauca.

Lichen, Pine – Cetraria pinastri.

Lichen, Powdered Swiss – Nephroma parilis.

Lichen, Puffed Shield – Parmelia physodes.

Lichen, Ring – Parmelia centrifuga.

Lichen, Rose and Gold – Stricta aurata.

Lichen, Smoky Shield – Parmelia omphalodes.

Lichen, Snow – Cetraria nivalis.

Lichen, Sprout – Dermatocarpon miniatus.

Lichen, Swedish Shield – Cetraria fahlunensis.

Lichen, Trumpet – Cladonia fimbricata.

Lichen, Warty Leather – Lobaria scrobiculata.

Lichen, Wrinkled Shield – Parmelia caperata.

LICHTENSTEINIA Cham. and Schlechtd. Umbelliferae. 7 spp. S.Africa, St. Helena. Herbs.

***L. interrupta*** Cham. and Schlechtd. S.Africa. The root (Radix Lichtensteiniae) is used by the natives to reduce fevers.

***L. pyrethrifolia*** Cham. and Schlechtd. S.Africa. The Hottentots brew an intoxicating beverage (Gli) from the plant.

Licor de Genipado – Genipa americana.

Licorice. See Liquorice.

LICUALA Wurmb. Palmae. 100 spp. S.E.Asia, Indomalaya, Bismark Islands. Small Palms.

***L. acutifida*** Mart. (Palas padi). Malaya. Used to make walking sticks (Panang Lawers).

***L. elegans*** Blume (Seredik itam). Sumatra. The young, unfolded leaves are used locally for wrapping tobacco.

***L. pumila*** Blume. (Sandand oopool). Sumatra. The cut young leaves are used locally with opium to make it burn better. The young leaves are also used to wrap tobacco.

***L. rumphii*** Blume = Corypha lucuola. (Koal, Leko, Wala). Indonesia. The split leaves are used locally to flavour opium.

Licuri – Syagrus coronata.

Lidah a jam – Polygala glomerata.

Lidah kerbaoo po tih – Alangium ebenaceum.

Lien ch'iao – Forsythia suspensa.

Life Root – Senecio aureus.

Ligala – Monodora laurentii.

Lige – Loranthus discolor.

Light Crottle – Lecanora parella.

Lignum Acocantherae – Acokanthera abyssinica.

Lignum Campechianum – Haematoxylum campechianum.

Lignum Fabianae – Fabiana imbricata.

Lignum Muira-Puama – Liriosma avata.

Lignum, Paraguay – Bulnesia sarmientii.

Lignum Pichi Pichi – Fabiana imbricata.

Lignum Quassiae Jamaicense – Picramnia excelsa.

Lignum Surinamense – Quassia amara.

Lignum sanctum – Guaiacum officinale.

Lignum Santali Rubrum – Pterocarpus santalinus.

Lignum Vitae – Guaiacum officinale.

Lignum Vitae – Guaiacum sanctum.

Lignum Vitae, Maracaibo – Bulnesia arborea.

LIGUSTICUM L. Umbelliferae. 60 spp. N.Hemisphere. Herbs.

***L. acutilobum*** Sieb. and Zucc. Japan and China. A decoction of the dried aromatic roots is used in Japan to combat dizziness and is taken to ease labour of women in childbirth.

***L. bulbocastanum*** (L.) Cr. = Bunium bulbocastanum.

***L. hultenii*** Fern. Alaska and W.Canada.

The leaves are used by the Eskimos to flavour fish.

***L. monnieri*** Calest. = Athamantha chinensis = Selinum monnieri. (Giêng sàng, Xà sàng, She ch'uang tzu). E.Europe through to China and Vietnam. The leaves are used as a condiment in Vietnam. The plant is sometimes cultivated.

***L. mutellina*** (L.) Crantz. = Aethusa mutellina = Meum mutellina. Central and S.Europe. The leaves make excellent forage for livestock in mountainous regions. They are also used as a substitute for parsley and when dried as a tea which is also used to relieve stomach pains.

***L. scoticum*** L. (Alexanders, Scotch Lovage). N.Europe, E. N.America. The blanched leaves are sometimes used as a vegetable on the W. coast of Scotland and in the Hebrides. They are sometimes eaten raw as celery. The young shoots are sometimes candied.

LIGUSTRUM L. Oleaceae. 40–50 spp. Europe to N.Persia; E.Asia; Indomalaya to New Hebrides. Shrubs or small trees.

***L. ibota*** Sieb. (Ibota Privet). Japan, China. The roasted seeds are used as a coffee substitute in Japan. Insect feeding on the stems stimulates the production of a wax which is used industrially.

***L. japonicum*** Thunb. (Japanese Privet). Japan. The roasted seeds are used locally as a coffee substitute. The plant is occasionally cultivated.

***L. lucidum*** Ait. (Glossy Privet). China. Insects feeding on the stems stimulate the production of a wax which is used industrially.

***L. nepalense*** Wall. Japan. The roasted seeds are used locally as a coffee substitute.

***L. vulgare*** L. (European Privet). Europe, W.Asia, N.Africa, introduced to N.America. The hard wood is used for small tools and makes a good charcoal. The plant is grown widely as hedges around gardens, the variegated form being particularly popular. The bark yields a yellow dye used for colouring wool. The fruits yield a green and black dye and an ink. The twigs are flexible and are used to make baskets etc.

Liheli – Ficus mucuso.

Lihurwe – Beilschmiedia oblongifolia.

Likir – Tacca pinnatifida.

Likolo – Dorstenia convexa.

Likuko – Hippocratea myriantha.

Likumo – Ficus artocarpoides.

Likuye – Ficus mucuso.

Lilac – Syringa vulgaris.

LILAEA Humb. and Bonpl. Liliaceae. 1 sp. Rocky Mountains, Mexico to the Andes. Herb.

***L. subulata*** H.B.K. The leaves are used for thatching and making brooms by the natives of S.America.

LILIUM L. Liliaceae. 80 spp. N.Temperate. Herbs.

***L. auratum*** Lindl. (Goldband Lily). Japan. The large bulbs are eaten locally as a vegetable.

***L. avenaceum*** Fisch. = L. maculatum.

***L. brownii*** F. E. Brown = L. candidum Lour. non L. = L. odorum. (Pat ho). S.China to Vietnam. The bulbs are eaten as a vegetable in Vietnam, where the leaves preserved in oil or alcohol are used as a dressing for bruises.

***L. candidum*** L. (Annunciation Lily, Bourbob Lily, Madonna Lily). Mediterranean to S.W.Asia. Cultivated for the flowers which yield an essential oil used in perfumery.

***L. candidum*** Lour. non L. = L. brownii.

***L. columbianum*** Hans. W. N.America. The bulbs are eaten as a vegetable by the local Indians.

***L. cordifolium*** Thunb. Japan. The plant is cultivated for the bulbs from which the starch is extracted and used as food. The young leaves are eaten as a vegetable.

***L. dauricum*** Ker-Gawl. (Dahurian Lily). Japan. The bulbs are eaten by the Ainu.

***L. glehni*** F. Schmidt. (Oba-ubayuri, Turep). Japan. The bulbs are eaten by the Ainu.

***L. lancifolium*** Thunb. Japan. The bulbs are eaten locally.

***L. longiflorum*** Thunb. (Easter Lily). Mediterranean. The flowers contain an essential oil of possible use in perfumery.

***L. maculatum*** Thunb. = L. avenaceum. Japan. The bulbs are eaten by the Ainu.

***L. martagon*** L. (Martagon Lily). Temperate Asia, Europe. The bulbs are eaten in Mongolia and parts of Russia, often dried and eaten with reindeer or cow's milk.

***L. odorum*** Planch. = L. brownii.

***L. parviflorum*** (Hook.) Holz. W. N.America. The bulbs are eaten by the local Indians.

*L.* ***philadelphicum*** L. (Wood Lily). N.America. The bulbs are eaten by the local Indians.
*L.* ***pomponicum*** L. (Pompon Lily). Siberia. The bulbs are eaten locally as a vegetable.
*L.* ***sargentiae*** Wils. (Sargent Lily). China. The flowers are eaten locally.
*L.* ***spectabilis*** Link. China, Japan. The bulbs are eaten locally.
*L.* ***superbum*** L. (Turkcap Lily). E. U.S.A. The bulbs were eaten in soups by the local Indians.
*L.* ***tenuifolium*** Fisch. China, Japan. The bulbs are eaten as a vegetable locally.
*L.* ***tigrinum*** Ker-Gawl. (Tiger Lily). China, Japan. The plant is cultivated locally for the bulbs which are eaten as a vegetable.
*L.* ***wallichianum*** Schult. f. (Findora). W.Himalayas, Nepal. The bulb leaves are used in Hindu medicine as a demulcent in treating chest complaints.
Lily, Annunciation – Lilium candidum.
Lily, Atamasco – Zephyranthes atamasco.
Lily, Blackberry – Belamcanda chinensis.
Lily, Bourbon – Lilium candidum.
Lily, Climbing – Gloriosa superba.
Lily, Corn – Clintonia borealis.
Lily, Cow – Nuphar advena.
Lily, Dahurian – Lilium dauricum.
Lily, Easter – Lilium longiflorum.
Lily, Ginger – Costus afer.
Lily, Goldband – Lilium auratum.
Lily, Madonna – Lilium candidum.
Lily, Mariposa – Calochortus nuttallii.
Lily, Martagon – Lilium martagon.
Lily of the Valley – Convallaria majalis.
Lily, Palm – Cordyline australis.
Lily, Palm – Cordyline terminalis.
Lily, Pompon – Lilium pomponicum.
Lily, Sago – Calochortus nuttallii.
Lily, Sargent – Lilium sargentiae.
Lily, Snake – Haemanthus natalensis.
Lily, Straw – Clintonia borealis.
Lily, Tiger – Lilium tigrinum.
Lily, Turkcap – Lilium superbum.
Lily, White – Pancratium speciosum.
Lily, Wood – Lilium philadelphicum.
Lima Bean – Phaseolus lunatus.
Lima, Chickasaw – Canavalia ensiformis.
Lima Weed – Roccella fuciformis.
Limaó do Matto – Rheedia edulis.
Limber Pine – Pinus flexilis.
Lime (Oil) – Citrus aurantifolia.
Lime – Tilia cordata.
Lime, Musk – Citrus macrocarpa.
Lime, Ogeeche – Nyssa ogeche.
Lime, Queensland Wild – Microcitrus inodora.
Lime, Russell River – Microcitrus inodora
Lime, Spanish – Melicoccus bijugatus.
Lime, Sweet – Citrus aurantifolia var limetta.
Limequat – Citrus aurantifolia × Fortunella margarita.
Limestone Urceolaria – Urceolaria calcarea
***Limnanthemum*** S. G. Gmel. = Nymphoides
LIMNOPHILA R. Br. Scrophulariaceae. 36 spp. Tropical Africa, Madagascar, India to Indochina, Java, Timor. Herbs or small shrubs.
*L.* ***aromatica*** Merr. = L. villosa. India to Australia. In India a decoction of the leaves is used as an expectorant and in Malaysia they are used as a poultice for sores etc.
*L.* ***roxburghii*** G. Don. Tropical Asia and Pacific Islands. The leaves are used locally to flavour dishes and to perfume the hair.
*L.* ***villosa*** Blume = L. aromatica.
***Limodorum ensatum*** Thunb. = Cynbidium ensifolium.
***Limodorum purpureum*** Lam. = Bletia purpurea.
***Limonia glutinosa*** Blanco = Swinglea glutinosa.
***Limonia monophylla*** Roxb. = Atalantia monophylla.
***Limonia pentaphylla*** Retz. = Glycomis pentaphylla.
Limoncillo – Trichilia havanensis.
LIMONIA L. Rutaceae. 1 sp. India to Java. Small tree.
*L.* ***acidissima*** L. = Feronia elephantum = F. limonia. (Elephant Apple, Wood Apple). The plant is cultivated locally for the aromatic fruits which are eaten raw or in sherberts and jellies. They are used to treat indigestion as are the leaves. In Thailand the juice is used as a yellow ink for writing on palm leaves. A water soluble gum (Feronia Gum, Velampisini) is used in glues, water paints and varnishes.
Limu Akiaki – Ahnfeltia cocinna.
Limu Eleele – Enteromorpha prolofera.
Limu Huna – Hypnea nidifica.
Limu Kala – Sargassum echinocarpum.
Limu Kohn – Asparagopsis sanfordiana.
Limu Lipahapaha – Ulva fasciata.
Limu Lipeepee – Laurencia papillosa.
Limu Lipoa – Dictyopteris plagiogramma.

Limu Luan – Porphyra leucostricta.
Limu Manauea – Gracilaria coronopifolia.
Lin – Linum humile.
Linaloé (Oil) (Seed Oil) – Bursera aloëxylon.
Linaloé, Cayenne – Aydendron panurense.
Linaloé Wood, Mexican – Bursera depechiana.

LINARIA Mill. Scrophulariaceae. 150 spp. N.Hemisphere outside the tropics, especially Mediterranean. Herbs.

***L. canadensis*** (L.) Dum. (Blue Toadflax). N.America. A decoction of the leaves is used as a diuretic and laxative. An ointment made from the leaves is used to treat haemorrhoids.

***L. elatine*** L.=Kickxia elatine. (Fluellin). Central to S.Europe, N.America, W.Asia. A decoction of the leaves is used to treat internal bleeding and nose-bleeding.

***L. vulgaris*** Mill. (Butter-and-Eggs, Flaxweed, Yellow Toadflax) Europe, Asia, Escape in N.America. A decoction of the leaves is diuretic and is used to treat bladder and liver complaints, as well as skin complaints and scrofula. It is also a laxative. An ointment of the leaves is used to treat haemorrhoids.

Linden, African – Mitragyna stipulosa.
Linden, Japanese – Tilia japonica.
Linden, Small-leaved – Tilia cordata.
Linden, Tuan – Tilia tuan.
Lidenleaf Sage – Salvia tiliaefolia.

LINDERA Thunb. Lauraceae. 100 spp. Himalayas, E.Asia, W.Malaysia. Shrubs or trees.

***L. benzoin*** (L.) Meissn.=Benzoin aestivalis =B. benzoin. (Spice Bush). Introduced to S.E. U.S.A. where the leaves are sometimes used as a tea substitute and the dried berries as a substitute for Allspice. The aromatic bark is occasionally used to reduce fevers.

***L. membranacea*** Maxim=L. umbellata.

***L. praecox*** Blume. Japan. An oil from the berries is used locally for lighting.

***L. umbellata*** Thunb.=L. membranacea= Benzoin umbellatum. Japan and China. An oil (Kuromoji Oil) is extracted from the seeds and used for lighting.

LINDERNIA All. Scrophulariaceae. 80 spp. Warm, especially Asia and Africa. Herbs.

***L. crustacea*** (L.) F. Muell.=Capraria crustacea. Tropics, mainly tropical Asia. The leaves are used in Tahiti to treat ear ache.

***L. japonica*** Thunb.=Mazus rugosus.

Linen – Linum usitatissimum.
Linen, Canton – Boehmeria nivea.
Linen, China – Boehmeria nivea.
Linen, Grass – Boehmeria nivea.
Lingkersap – Schefflera aromatica.

LINOCIERA Sw. ex Schreb. Oleaceae. 80–100 spp. Tropics and subtropics. Trees.

***L. oblongifolia*** Koord.=Chionanthus oblongifolia. (Kajoo lemah, Kalang patejan). Java. The fruits are used in a wide variety of local medicines, but are of no proved medicinal value.

Linseed Oil – Linum usitatissimum.

LINUM L. Linaceae. 230 spp. Temperate and subtropics, especially Mediterranean. Herbs. The stems of all the species mentioned yield a fibre.

***L. catharticum*** L. (Purging Flax). Europe, N.Africa, Iran, to Caucasus and W.Asia. The leaves were used as a diuretic.

***L. humile*** Mill. (Lin, Kitan). A cultivar from N.Africa. It is closely related to L. usitatissimum and was known to the Ancient Egyptians. It is grown for the seed which is the source of an edible oil.

***L. lewisii*** Pursh. (Rocky Mountain Flax). W. N.America. The bast fibres are used by the local Indians to make cordage for ropes, etc., nets, baskets. They sometimes eat the seeds.

***L. marginale*** Cunningh. Australia. The seeds are eaten by the aborigines, who use the bast fibres for cordage, fishing nets etc.

***L. usitatissimum*** L. (Flax). Temperate Europe and Asia. The plant is grown for the bast fibres (linen) and the seed oil (Linseed Oil). The cultivated plant probably originated from L. angustifolium of the Mediterranean regions. The fibres have been found in Swiss Stone Age dwellings and were widely used by the ancient Egyptians. The fibres are particularly flexible and tough (particularly when wet) and linen was one of the most important fabrics until the use of cotton and then man-made fibres decreased its popularity. Now some half million tons is produced annually, mainly in Russia, France, Poland and Czechoslovakia. Other N.European countries, Japan and some of the Near East countries produce a little, while virtually none is produced in U.S.A. The plant does best in cool moist climates. It is harvested by hand or machinery and the stems are retted, either in natural waters, dew or in controlled temperature baths, the latter giving fibres of the best quality. The dried stems are scutched to remove the extraneous tissue, carded, combed, bleached and spun into yarn. The yarn is then used to make thread, or is woven into cloth, or is sometimes used for canvas, nets etc. The shorter fibres and and those from flax grown for oil are used for making cigarette paper, as they burn with little undesirable odours, and to make other fine paper such as is used to make Bibles.

Linseed oil is produced from the seeds. Some 12 per cent of the word production of 3½ million tons annually comes from plants grown for fibre; the rest is produced from special varieties. The main producing countries are U.S.A. Argentine, India, Russia and Canada. The seeds contain 33–43 per cent oil which is extracted from the heated seeds by pressing. A little cold-pressed oil is produced and used for cooking. The hot pressed oil is a drying oil Sp. Gr. 0·927–0·931; Sap. Val. 189–196 Iod. No. 170–204. Un sap. 0·5–1·6 per cent. It keeps indefinitely out of contact with air, but forms a hard film readily on exposure. The mucilages from the seeds were allowed to precipitate for several years before the oil was used, but now they are removed by treating with sulphuric acid. The drying properties of the oil are improved by heating with the oxides of lead, cobalt or manganese. The oil is used in paints and varnishes (although this use is declining with the introduction of plastic paints), linoleum, printer's ink, soap, plumbing and lithographic varnishes.

Seed residues from hot-pressing contain 33–43 per cent protein and can be used for cattle feed. Those from cold pressing containing cyanogenic glucosides which make them poisonous. The whole seeds were used as food by the Greeks and Romans and are used medicinally as a demulcent laxative. They are applied as a poultice to burns and scalds.

Lion's Foot – Prenanthes serpentaria.

Lionsear, Drug – Leonotis leonurus.

Lipoti – Eugenia curranii.

LIPIA L. Verbenaceae. 220 spp. Tropical America, Africa. Shrubs. Several spp. introduced to America from the Mediterranean by the early colonists are used as savoury herbs in cooking under the name of Oregano.

***L. adoensis*** Hochst. (Gambian Tea Bush). W.Africa. The leaves are used as a tea substitute by the natives and Europeans.

***L. berlandieria*** Schauer. (Epazote, Hierba dulce, Oregano). Mexico. Used locally to flavour food and as a stimulant, to control menstruation and as a demulcent.

***L. chaipensis*** Loese. = L. umbellata.

***L. citriodora*** Kunth. = Aloysia citriodora (Chile, Argentine, Uruguay). An oil distilled from the leaves (Verbena Oil) is used in perfumery. The leaves are used locally as a tea substitute.

***L. dulcis*** Trév. Tropical America. The dried leaves contain camphor and lippiol. They are used medicinally to soothe coughs.

***L. geminata*** H.B.K. = Lantana lippioides (Hierba buena, Hierba de negro, Mirto, Te de pais). Mexico through to S.America, W.Indies. The plant is used as a relaxant, to induce sweating, to treat stomach complaints and to control menstruation.

***L. graveolens*** H.B.K. (Té de Pais). Mexico, Guatemala. The leaves are used in Mexico to flavour food and as a tonic and expectorant.

***L. ligustrina*** (Lag.) Britt. S.W. U.S.A. and Mexico. The leaves are used in Mexico as a relaxant, to control menstruation and to treat bladder complaints. The oil from the leaves is used in S.Europe to make perfumes.

***L. lycioides*** Steud. N. and S.America. An infusion of the flowers is used locally to treat colds.

***L. pringlei*** Briq. = L. umbellata.

***L. pseudo-thea*** Schau. = Lantana pseudo-thea. S.America. The leaves are used locally to make a tea.

***L. scaberrima*** Sond. S.Africa. The dried leaves are used to stop bleeding.

***L. substrigosa*** Turcs. = L. umbellata.

***L. umbellata*** Cav. = L. chiapensis = L. pringlei = L. substrigosa. (Hierba de la mula, Salvia, Nacare, Tobaquilla, Topozana). Mexico. The leaves are used locally to treat stomach complaints.

LIQUIDAMBAR L. Altingiaceae. 6 spp. Atlantic N.America, S.W.Asia, S.E.China, Indochina, Formosa. Trees.

**L. formosana** Hance. (Chinese Sweet Gum, Formosa Sweet Gum). China. The leaves are used locally to feed silk worms and the wood is used to make tea chests.

**L. orientalis** Mill. (Levant Storax, Oriental Sweet Gum). The trunk yields a solid or semi solid gum (Levant Storax) which is used as an expectorant in pastilles, to scent soaps and to make oriental perfumes. It is also used as a parasitic against scabies and other skin diseases and as a fumigant.

**L. styraciflua** L. (American Sweet Gum, Bilsted, Red Gum). The wood (Satin Walnut, Californian Red Gum) is red-brown heavy, but not strong. It is used for building, furniture, especially cabinets and cupboards and pulp wood. The trunk yields a gum (American Storax, Styrax) which is used in cough pastilles, for treating skin diseases and as a fumigant. An oil (Oil of Styrax) extracted from the gum is used in perfumery and to scent soap.

Liquorice – Glycyrrhiza glabra.

Liquorice, Chinese – Glycyrrhiza ralensis.

Liquorice, Indian – Abrus precatorius.

Liquorice, Italian – Glycyrrhiza glabra.

Liquorice Root – Hedysarum alpinum.

Liquorice Root – Hedysarum mackenzii.

Liquorice, Russian – Glycyrrhiza glabra.

Liquorice, Spanish – Glycyrrhiza glabra.

Liquorice, Wild – Glycyrrhiza lepidota.

LIRIODENDRON L. Magnoliaceae. 1 sp. E. N.America; 1 sp. China, Indochina. Trees.

**L. chinensis** Sarg. China, Indochina. The inner bark is used locally as a stimulant.

**L. tulipifera** L. (Tulip Tree, Yellow Poplar). E. N.America. The brittle, soft light wood is used for house interiors, roof shingles, boxes and cabinets. The inner bark contains tulipiferine, a heart and nerve stimulant, and so it is used locally as a stimulant.

LIRIOPE Lour. Liliaceae. 6 spp. E.Asia. Herbs.

**L. graminifolia** (L.) Baker. = L. spicata = Ophiopogon spicatus. E.Asia, especially Japan, Manchuria, Hongkong, Formosa and Philippines. The candied tubers are used locally as a stimulant and aphrodisiac. As a decoction they are used in China to treat lung complaints and coughs, dysentery and fevers.

**L. spicata** Lour. = L. graminifolia.

LIRIOSMA Poepp. and Endl. Olacaceae. 15 spp. Tropical S.America. Trees.

**L. ovata** Miers. (Muirapuama). Brazil, especially Amazon. The wood (Lignum Muira-Puama, Muira-Puama Wood, Radix Muira-Puama) is used as a tonic and aphrodisiac.

**Lissochilus** R. Br. = Eulophia.

Litchi – Nephelium litchi.

LITHARIA Usnaceae. Lichens.

**L. vulpina** (L.) Waino. On tree trunks in the Arctic and on high mountains. It used to be used for poisoning wolves and may have some antibiotic properties.

LITHOCARPUS Blume. Fagaceae. 300 spp. E. and S.E.Asia, Indomalaya, California. Trees.

**L. densiflora** (Hook. and Arn). Rehd. = Pasania densiflora = Quercus densiflora. (Chestnut Oak, Tanbark Oak). California, Oregon. The bark is an important source of tannin.

LITHOSPERMUM L. Boraginaceae. 60 spp. Temperate. Herbs.

**L. canescens** Lehm. (Hoary Gromwell). W. N.America. The roots yield a red dye used by the local Indians as a face paint and dye.

**L. carolinense** (Walt.) G. F. Gmel. (Carolina Gromwell). N.America. A dye from the roots is used by the local Indians as a face paint.

**L. erythrorhizon** Sieb. and Zucc. Japan. Cultivated locally for the roots which yield a purple dye.

**L. hispidissimum** Lehm. = Arnebia hispidissima.

**L. officinale** L. (Common Gromwell). Europe, W.Asia, Middle East, introduced to N.America. The leaves are used locally to make a tea (Bohemian Tea, Croatian Tea).

**L. ruderale** Dougl. (Gromwell, Stoneseeds). W. N.America. A decoction of the roots is used by the local Indians to treat diarrhoea.

Litmus – Ochrolechia tartarea.

Litmus – Rocella fuciformis.

Litmus – Rocella tinctoria.

**Litrea molle** Griseb. = Schinus latifolius.

LITSEA Lam. Lauraceae. 400 spp. Warm Asia to Korea and Japan, Australia, America. Trees or shrubs.

**L. calicaris** Kr. New Zealand. The firm elastic wood is used to make barrels, wheels, coach shafts, ships' blocks, etc.

***L. citrata*** Blume = L. cubeba.

***L. cubeba*** Pers. = L. citrata = Tetranthera polyantha. China to Java. The bark is used locally to treat colic and fevers. The scented young fruits are used in Java as a condiment (Sambal) used to flavour goat's meat. The plant yields an essential oil.

***L. ferruginea*** Benth. New South Wales, Queensland. The light, pale brown wood was used for carving, veneer, plywood, turning, general carpentry, furniture and coffins.

***L. glauca*** Sieb. China, Japan. The tree yields an oil used locally for burning and soap-making.

***L. neesiana*** (Schauer) Hemsl. (Laurel de la Sierra). Mexico. The leaves are used by the local Indians to treat stomach pains.

***L. novoleontes*** Bartlett. Mexico. The leaves are used locally to make a tea.

***L. reticulata*** Benth. (Bolly Gum, Bolly Wood, Brown Beech). New South Wales, Queensland. See L. ferruginea.

***L. sebifera*** Blume. = L. tetranthera.

***L. tetranthera*** Mirb. = L. sebifera. Tropical Asia, especially India and Malaya. The plant is cultivated locally as the source of an oil used by the Chinese to make soap. The bark is used in the Philippines to treat intestinal catarrh.

Little Bluestem – Andropogon scoparius.

Little Flower Lagerstroemia – Lagerstroemia parviflora.

Little Leaf Tunic Flower – Dianthus prolifera.

Little Millet – Panicum sumatrense.

Livelong – Lannea discolor.

Live Oak – Quercus virginiana.

Live Oak, Canyon – Quercus chrysolepis.

Liverwort, English – Peltigera canina.

LIVISTONA R. Br. Palmae. 30 spp. Indomalaya, Australia. Palms.

***L. australis*** Mart. = L. inermis = Corypha australis. (Cabbage Palm). Australia. The young leaves and central bud are eaten as a vegetable by the aborigines. The leaves are used for thatching.

***L. chinensis*** Mart. Philippines. The leaves are used for thatching.

***L. cochinchinensis*** Blume. Cochin-China. The fruits are eaten locally and the leaves are used for thatching.

***L. inermis*** Wendl. = L. australis.

***L. jenkinsiana*** Griff. N.E.India. The leaves are used for thatching.

***L. sariba*** Merr. Tropical Asia, especially Cochin-China, Philippines and Penang. The leaves are used for thatching, mats, hats, etc. The midribs are used for blowpipe darts. The endosperm, pulped in vinegar is eaten locally.

***L. speciosa*** Kurz. Malaya. The leaves are used for thatching.

Lizardtail – Sarurus cernuus.

Ljebu Mahogany – Entandrophragma angloense.

Lobak – Raphanus sativus.

Loban Maidi – Boswellia frereana.

Loban Meti – Boswellia frereana.

LOBARIA Stictaceae. Lichens.

***L. calicaris*** Hoffm. = Ramalina calicaris.

***L. fraxinea*** Hoffm. = Ramilaria fraxinea.

***L. prunastri*** Hoffm. = Evernia prunastri.

***L. pulmonaria*** (L.) Hoffm. (Hazelcrottle, Hazelraw, Lungwort, Rage). Subalpine woods of Europe. Used in Scandinavia and the United Kingdom to dye wool orange-brown. Yields an essential oil used in perfumery and is used occasionally for tanning. The plant is also used in home remedies to treat lung complaints, when it is called Muscus pulmonarius.

***L. scrobiculata*** (Scop.) DC. (Oak Rage, Warty Leather Lichen). Temperate areas on trees and rocks. Yields a brown dye, used in the United Kingdom to stain wool brown.

LOBELIA L. Campanulaceae. 200–300 spp. Cosmopolitan. Herbs.

***L. cardinalis*** L. N.America. A poisonous plant; the leaves contain the alkaloids lobeline and lobelanidine and the essential oil lobelianin. The leaves and young shoots are used medicinally as an emetic, expectorant and nauseant.

***L. excelsa*** Lesch. E.India. The leaves are used locally as a tobacco substitute.

***L. inflata*** L. (Indian Tobacco). N.America. Cultivated in E. N.America. See L. cardinalis.

***L. laxiflora*** H.B.K. Mexico. The roots are used locally as an emetic, expectorant and to treat asthma.

***L. nicotinaefolia*** Heyne. (Devanula, Dhaval). India. The roots are used locally to treat scorpion stings.

***L. pinifolia*** L. S.Africa. The roots are used locally as a stimulant and to induce sweating. They are also used to treat rheumatism and skin diseases.

***L. succulenta*** Blume. Java. The leaves are eaten locally with rice.
***L. syphilitica*** See L. cardinalis.
Loblolly Bay – Gordonia lasianthus.
Loblolly Pine – Pinus taeda.
Locust, African – Parkia africana.
Locust (Black) – Robina pseudo-acacia.
Locust Gum – Ceratonia siliqua.
Locust, Honey – Gleditsia triacanthos.
Locust, West Indian – Hymenaea courbaril.
Lodebark – Symplocos racemosa.
Lode Tree – Symplocos sumuntia.
Lodgepole Pine – Pinus murrayana.
Lodh – Symplocos sumuntia.
LODOICEA Comm. ex J. St. Hil. Palmae. 1 sp. Seychelles. Palm.
***L. callipyge*** Comm. = L. maldivica.
***L. maldivica*** Pers. = L. callipyge = L. seychellarum. (Coco de Mer). The old leaves are used for thatching and the young ones to make hats. The large shells of the fruits are used for bowls etc.
***L. seychellarum*** Pabill. = L. maldivica.
LOESELIA L. Polemoniaceae. 17 spp. California to Venezuela. Shrubs.
***L. mexicana*** (Lam.) Brand. Mexico. The leaves contain the alkaloid loeseline. A decoction of the leaves is used locally to reduce fevers, as an emetic, diuretic and purgative. Applied externally it prevents the hair from falling out.
Loganberry. See Rubus strigosus.
Logwood – Haematoxylum campechianum.
Lokaka – Roureopsis obliquifoliata.
Loki – Takka pinnatifida.
Lole – Salacia soyauxii.
LOLIUM L. Graminae. 12 spp. Temperate Eurasia. Grasses.
***L. italicum*** R. Br. (Italian Rye Grass). Europe. Cultivated in Old and New World as a fodder grass, usually in short-term leys.
***L. perenne*** L. (Perennial Rye Grass). Europe. Cultivated in permanent pastures in Old and New World.
***L. temulentum*** L. (Darnel, Tare, Wray). Temperate. The seeds contain the alkaloid temuline, which may be produced by a fungus. This makes them poisonous and a dangerous contaminant of cereals. The plant is probably the "tares" of the Bible. The seeds are sometimes mixed with barley to increase the intoxicating effect of beer and are used in treating rheumatism and neuralgia.
Lolo – Piper saigonense.
Lolokendamba – Bauhinia thonningii.
LOMANTIA R. Br. Proteaceae. 12 spp. E.Australia, Tasmania, Chile. Trees.
***L. hirsuta*** (Lam.) Diels. (Palo Negro). Chile. The brown wood is used locally for furniture and the brown sap from the bark as a dye.
LOMATIUM Rafin. Umbelliferae. 80 spp. W. N.America. Herbs. The roots of the following species were used by the local Indians to make flour.
***L. ambigua*** (Nutt.) Jones.
***L. canbyi*** (Coult. and Rose) Jones.
***L. cous*** (S. Wats.) Jones.
***L. farinosa*** (Hook.) Jones.
***L. foeniculacea*** (Nutt). Coult. and Rose.
Lombaya lo lowe – Hugonia obtusifolia.
Lombiro – Cryptostegia madagascariensis.
Lompeak – Calamus salicifolius.
LONCHOCARPUS Kunth. Leguminosae. 150 spp. Tropical America, W.Indies, Africa, Australia. Trees.
***L. capassa*** Rolfe. (Lance Tree, Rain Tree). Tropical Africa. The heavy, hard, light yellow wood is used locally to make grain mortars, tool handles etc. The natives of Zambia used the bark to make "hunger Belts", worn in times of famine.
***L. cyanescens*** Benth. (West African Indigo, Yoruba Indigo). Tropical Africa. Vine. The leaves and young roots yield a blue dye, the characteristic colour of the clothes worn by Yoruba women in Nigeria. The roots have been used to treat leprosy.
***L. densiflorus*** Benth. (White Haiari). Guyana. Vine. The roots are used locally as a fish poison.
***L. floribundus*** (Benth.) Killip. = L. nicou.
***L. goosenii*** Hauaman. (Mbongum bongi). Congo. An infusion of the wood is used locally to reduce fevers.
***L. hondurensis*** Benth. (Honduras Lance Wood). Central America. The wood (Rosa Morada) is used for general carpentry.
***L. latifolius*** H.B.K. = L. oxycarpus. W.Indies. The wood is used for general construction work, waggons and marquetry.
***L. longistylus*** Pitt. Central America, Mexico. The bark is made into a fermented beverage in Mexico. The Mayas made a similar drink (Balche) by fermenting the bark in honey.
***L. maculatus*** DC. = Gliricidia sepium.

***L. nicou*** DC.=L. floribundus. Vine. Brazil, Peru. The plant was cultivated for the roots to use as a fish poison. It is now cultivated vegetatively as a source of the insecticide rotenone. It is a useful insecticide as it has no apparent effect on mammals. The plants are dug up after 2–3 years and the roots dried and pulverized to a powder which is sold as an insecticide. The dried root contains about 20 per cent rotenone and the fresh root 0·75–1 per cent. Commercial rotenone is the oil-soluble fraction from the root, containing rotenone, deguelin, phrosin and toxicarol.

***L. oxycarpus*** DC.=L. latifolius.

***L. sericeus*** H.B.K. Tropical Africa and America. The hard wood is used for general construction work and turning.

***L. urucu*** Killip. Climber. S.America. Cultivated as a source of rotenone.

***L. utilis*** Smith. Peru. Cultivated as a source of rotenone.

Longan – Nephelium longana.

Longbeak Eucalyptus – Eucalyptus camaldulensis.

Long Buchu – Barosma serratifolia.

Longhair Plume Grass – Dichelanche crinita.

Long Jack – Flindersia oxleyana.

Longleaf Ephedra – Ephedra trifurea.

Long-leaved Eriogonum – Eriogonum longifolium.

Long-leaved Pine – Pinus palustris.

Longonizia – Gnetum africanum.

Long Pepper – Piper chaba.

Long Pepper – Piper longum.

Longroot Onion – Allium victorialis.

Long-stalked Greenbrier – Smilax pseudochina.

Longstalk Holly – Ilex pedunculata.

Longyen – Nephelium longana.

LONICERA L. Caprifoliaceae. 200 spp. N.America to Mexico, Eurasia to N.Africa, Himalayas, Philippines and S.W. Malaysia. Shrubs, small trees or vines.

***L. angustifolium*** Wall. Himalayas. The sweet berries are eaten locally.

***L. caprifolium*** L. (Honeysuckle). Europe. A decoction of the dried flowers is used as an expectorant and laxative and the flowers are used in a syrup to treat chest complaints.

***L. coerulea*** L. var. ***edulis*** Reg. Siberia, N.Caucasus. The berries are eaten locally in Siberia.

***L. involucrata*** (Rich.) Banks. (Bearberry, Twinberry). N.America. The berries are eaten by the local Indians.

Looking Glass Tree – Heritiera littoralis.

Loo roos – Peronema canescens.

Loosestrife – Lysimachia vulgaris.

Loosestrife, Fourleaf – Lysimachia quadrifolia.

Loosestrife, Purple – Lythrum salicaria.

Loosestrife, Spiked – Lythrum salicaria.

Lootoong – Aglaia eusideroxylon.

Lopez Root – Toddalia aculeata.

LOPHATHERUM Brong. Graminae. 2 spp. E.Asia, Indomalaya, tropical Australia. Grasses.

***L. elatum*** Joll.=L. gracile.

***L. gracile*** Br.=L. elatum=L. japonicum=L. lehmannii=Centrotheca affine. (Tan chu). E.Asia, Indomalaya. The Chinese use the leaves to treat fevers and to make a digestive tea.

***L. japonicum*** Steud.=L. gracile.

***L. lehmannii*** Nees.=L. gracile.

LOPHIRA Banks ex Gaertn. Ochnaceae. 2 spp. Tropical W.Africa. Trees.

***L. alata*** auctt., non Banks ex Gaertn.=L. lanceolata.

***L. alata*** Banks ex Gaertn. (African Oak, Dwarf Ironwood, Niam Tree). Tropical W.Africa. The red-brown hard heavy wood sinks in water; it is easy to work and is used for bridges, paving blocks, railway sleepers, flooring, furniture and turning. An oil (Niam Fat, Niam Oil, Meni Oil) is cold pressed from the seeds and is used locally for cooking, soap-making and as a hair oil.

***L. lanceolata*** Van Tiegh. ex Keay=L. alata auctt., non Banks ex Gaertn. See L. alata Banks ex Gaertn.

LOPHOCEREUS (A. Berger) Britton and Rose. Cactaceae. 4 spp. S.W. United States of America, Mexico. Cacti.

***L. schottii*** (Englem.) Britt. and Rose=Cereus schottii. (Cina, Cabesa de Viejo). California, Arizona, Mexico. The fruits are eaten locally.

LOPHOPETALUM Wight ex Arn. Celastraceae. 4 spp. Indochina, W.Malaysia. Trees.

***L. toxicum*** Loher. Philippines. The bark is used as an arrow poison.

LOPHOPHORA Coult. Cactaceae. 1–3 spp. S.United States of America, Mexico. Cacti.

***L. williamsii*** (Lem.) Coulter=Echinocactus williamsii. (Peyote). S.Texas and Mexico. The dried crowns (Mescal Buttons) have been chewed by the local Indians during ceremonies for centuries. The plant contains the hallucinatory drug anhalonin, an alkaloid similar to LSD. The plant may have some medicinal value.

LOPHOPYXIS Hook. f. Lophopyxidaceae. 2 spp. Malaysia, Solomon Islands. Shrubs.

***L. pierrei*** Boerl. (Tali sesawi). Malaya. The root is used for treating stings and bites, especially from poisonous fish. The crushed leaves are used to dress wounds and ulcers.

***Lophostemon arborescens*** Schott.=Tristania conferta.

Loquat – Eriopotrya japonica.

Loquat, Wild – Uapaca kirkiana.

LORANTHUS Jacq. Loranthaceae. 600 spp. Tropical and subtropical Old World, a few temperate Eurasia and Australasia. Small woody parasites.

***L. calyculata*** DC.=Psittacanthus calyculatus.

***L. chinensis*** Benth.=Scurrula gracilifolia.

***L. cochinchinensis*** Lour.=Macrosolen cochinchinensis.

***L. discolor*** Engl. (Lige, Mpoa). Congo. The fruits are eaten locally.

***L. divaricatus*** H.B.K. S.America. The sticky sap is used locally as a bird lime.

***L. globosus*** Doh.=Macrosolen cochinchinensis.

***L. gracilifolius*** Schult.=Scurrula gracilifolia.

***L. grandiflorus*** King. Malaysia, Java, Thailand. The leaves mixed with turmeric and rice are used in Malaysia as a treatment for ringworm.

***L. nummulariaefolius*** Chev. Tropical Africa. In Somaliland the bark is used for tanning.

***L. zeyheri*** Harv. S.Africa. The sticky sap is used locally as bird lime.

Loroco – Urechites karwinskii.

Lotong – Aglaia ganggo.

LOTUS L. 100 spp. Temperate Europe, Asia, Africa. Herbs.

***L. corniculatus*** L. (Bird's-foot Trefoil). Europe, Asia, Africa. Grown in the Old and New World as the legume in mixed pastures. It is slow to establish, but is particularly useful on poorer soils.

Lotus – Nelumbo nucifera.

Lotus – Nelumbo pentapetala.

Lotus – Ziziphus lotus.

Louisiana Moss – Tillandsia usneoides.

Love Apple – Lycopersicon lycopersicum.

Lovage (Oil) – Levisticum officinale.

Lovage, Scotch – Ligusticum scoticum.

Lovage angelica – Levisticum officinale.

Lovegrass, Weeping – Eragrostis curvula.

Lovishohe – Polygala ruwenzoriensis.

LOVOA Harms. Meliaceae. 11 spp. Tropical Africa. Trees.

***L. klaineana*** Pierre and Sprague. (African Walnut, Brown Mahogany, Tiger Wood Lovoa). Tropical W.Africa. The light soft walnut-brown wood is used for furniture, house interiors, carvings etc.

***L. swynnertonii*** Baker f. Tropical Africa. The wood (Brown Mahogany) is heavy and durable. It is used for furniture, carvings etc.

Lovoa, Tiger Wood – Lovoa klaineana.

Low Sweet Blue berry – Vaccinium pennsylvanicum.

Lowa voluha – Ficus vallis.

Lower Amazon Rosewood – Dalbergia spruceana.

Lowland Ribbon Wood – Plagianthus betulinus.

Lowland White Fir – Abies grandis.

Loxa Bark – Cinchona officinalis.

LOXOPTERYGIUM Hook. f. Anarcardiaceae. 5 spp. Tropical S.America. Trees.

***L. balanse*** Griseb. See L. lorentzii.

***L. lorentzii*** Griseb=Schinopsis lorentzii. (Quebracho Colorado). Argentine and Paraguay. The wood and bark are a commercial source of tannin.

***L. sagotii*** Hook. (Hooboobali Tree). Guyana. The wood (Hooboobali) is used for general carpentry.

Lubam Meti – Boswellia frereana.

Lubanga – Lygodium smithianum.

Lubia – Dolichos lablab.

Lucca Gum – Olea europaeae.

Lucerne – Medicago sativa.

Lucerne, Tree – Cytisus pallilans.

Lucerne, Wild – Stylosanthes mucronata.

Lucky Bean – Erythrina abyssinica.

Lucky Seeds (Beans) – Thevetia nereifolia.

LUCUMA Molina. Sapotaceae. 100 spp. Malaysia, Australia, Pacific Islands. tropical America. Trees. The fruits of the following species are about the size of an apple and have a pleasant flavour when ripe. They are eaten raw, locally where they are sold in markets etc. The trunks are potential sources of balata gums, used in making chewing gum etc.

***L. arguacoensium*** Karst. Columbia.

***L. bifera*** Molina. Chile, Peru.

***L. caimito*** Roem.=Pouteria caimito. (Abui). Brazil, Peru. Cultivated.

***L. capiri*** A. DC.=Sideroxylon carpiri.

***L. glycyphloea*** Mart. and Eichl.=Chrysophyllum buranhem=Pradosia lactescens. Brazil. The bark is used locally as a tonic, astringent and to reduce bleeding.

***L. mammosa*** Gaertn.=Calocarpum mammosum.

***L. nervosa*** A. DC.=L. rivicoa.

***L. obovata*** H.B.K. (Lucumo). Chile, Peru. Cultivated.

***L. procera*** Mart.=Urbanella procera. (Macarandiba). Brazil. Cultivated.

***L. rivicoa*** Gaertn. f.=L. nervosa. Central America, W.Indies. The fruit is also eaten as sherberts etc. and the plant yields a spice (Canistel) used in Brazil.

***L. salicifolia*** H.B.K. (Egg Fruit, Yellow Sapote, Zapote Amarillo). Central America.

Lucumo – Lucoma obovata.

LUDWIGIA L. Onagraceae. 75 spp. Cosmopolitan, especially tropical America. Herbs.

***L. repens*** Sw. Tropics. The tips of the stems are eaten as a vegetable in Cochin-China.

LUEHEA Willd. Tiliaceae. 20 spp. Tropical America, W.Indies. Trees.

***L. divaricata*** Mart. (Whiptree) S.Brazil, Argentine. The brown, heavy hard wood is not durable. It is used for general carpentry, furniture, shoes, saddles, soles of shoes and wooden ware.

LUFFA Mill. Cucurbitaceae. 6 spp. Tropics. Vines.

***L. acutangula*** Roxb. (Dishcloth Gourd, Sinkwa Towel Gourd). Tropical Asia. Cultivated in China and Japan for the young fruits which are eaten as a vegetable or in soups.

***L. aegyptiaca*** Mill.=L. cylindrica.

***L. cylindrica*** (L.) Roem.=L. aegyptiaca. (Suakwa, Vegetable Sponge). Tropical Asia, Africa. Cultivated in Old and New World. The young fruits are eaten in soups etc. The vascular tissue of the old fruits is used as a scrubbing pad.

***L. operculata*** Cogn.=L. purgans.

***L. purgans*** Mart.=L. opercukata=Momordica operculata. Tropical America. The vascular tissue of the fruit is used as a toilet sponge.

Luffa – Luffa cylindrica.

LUISIA Gaudich. Orchidaceae. 30 spp. Tropical Asia to Japan and Polynesia. Herbs.

***L. teres*** Blume. China, to Indochina. The Chinese and Vietnamese used an extract from the roots to treat ulcers and wounds.

Lumbang Oil – Aleurites molucca.

Lumbang Oil – Aleurites triloba.

Lumbungaly – Amomum clusii.

LUMNITZERA Willd. Combretaceae. 2 spp. E.Africa to Malaysia, N.Australia and Pacific Islands. Trees.

***L. coccinea*** Wight. and Arn.=L. littorea. (Red-flowered Mangrove). Salt water swamps. Malaysia, Polynesia. The heavy fine-grained wood is used for boat-building in Polynesia.

***L. littorea*** Voight.=L. coccinea.

***L. racemosa*** Willd.=Petaloma alba. Tropical Asia, Australia, Polynesia. The wood is used for poles, heavy construction work, ship-building, paving, bridges, wharves. The bark is used for tanning. The sap from the bark is used in the Philippines in a mixture with coconut oil to treat skin diseases, especially herpes and itches.

Lumot – Enteromorpha intestinalis.

Lumot kahoi – Usnea philippina.

LUNARIA L. Cruciferae. 3 spp. Central and S.E.Europe. Herbs.

***L. annua*** L. (Penny Flower). Europe. The plant is occasionally cultivated for the young roots, which are eaten.

LUNASIA Blanco. Rutaceae. 10 spp. Philippines, Borneo to E.Malaysia. Small trees.

***L. armara*** Blanco.=L. costulata=Mytilococcus quercifolia. (Maitan, Pintan). E.Malaysia, Java. Alkaloids contained in the bark cause paralysis of the heart. They are mixed with alum and used as a poultice to reduce swelling of the limbs.

***L. costulata*** Miq.=L. amara.

Lungwort – Lobaria pulmonaria.

Lungwort – Pulmonaria officinalis.

Lungwort – Sticta aurata.

Lungwort – Sticta pulmonaria.

Lupin(e), Egyptian – Lupinus terminis.

Lupin(e), Shore – Lupinus littoralis.

Lupin(e), White – Lupinus albus.

Lupin(e), Yellow (European) – Lupinus luteus.

LUPINUS L. Leguminosae. 200 spp. America, Mediterranean. Herbs. The seeds and foliage of the genus contain a bitter principle which causes toxicity – lupinosis. Excess foliage must not be fed to stock and if the seeds are used for human consumption they should be boiled and strained first.

***L. albus*** L. = L. sativus. (White Lupin(e)). Mediterranean. The plant is grown as cattle fodder and as a green manure. The roasted seeds are eaten in Turkey, or in Egypt and are mixed with wheat flour. They are sometimes used as a coffee substitute.

***L. littoralis*** Dougl. (Shore Lupin(e)). W. N.America. The roots were eaten by the local Indians.

***L. luteus*** L. ((European) Yellow Lupin(e)). S.Europe. The plant is grown as a green manure on poor soils. The roasted seeds have long been used as a coffee substitute.

***L. sativus*** Gaertn. = L. albus.

***L. terminis*** Forsk. (Egyptian Lupin(e)). Mediterranean. Cultivated on the saline soils of the Nile delta and in other parts of the tropics for the seeds which are eaten as a flour.

***L. varius*** L. Spain, Portugal, Balearic Islands. The seeds are roasted and used locally as a coffee substitute. The plant is sometimes cultivated.

Lusala – Laportea aestuans.

Lushamla – Vigna maranguensis.

Lusitanian Oak – Quercus lusitanica.

LUVUNGIA Buch.-Ham. ex Wight. and Arn. Rutaceae. 12 spp. Indomalaya. Shrubs.

***L. scandens*** Buch.-Ham. Malaya. A poultice of the roots and fruits is used locally to treat scorpion stings. A perfumed oil extracted from the fruits is used by the Hindus to scent medicines.

Luzon Pine – Pinus insularis.

Lwandula – Alchemilla kuwensis.

***Lychnis diurna*** Sibth. = Melandrium album.

LYCHNODISCUS Radlk. Sapindaceae. 8 spp. Tropical Africa. Trees.

***L. cercospermus*** Radik. = Pancovia lijai. (Bodiva, Endjin, Mobembe, Tshihangi). Congo to Sudan. The wood is used locally for making mortars for pounding grain etc.

LYCIUM L. Solanaceae. 80–90 spp. Temperate and subtropics. Shrubs.

***L. andersonii*** Gray. (Anderson Wolfberry). California, Arizona. The berries are eaten raw, or in soups by the local Indians.

***L. arabicum*** Schwein. (Arabian Wolfberry). N.Africa. The fruits are eaten by the local Arabs.

***L. barbarum*** Ait. non L. = L. halamifolium.

***L. berlandieri*** Gray. Arizona, California. The berries are eaten raw or in soups by the local Indians.

***L. chinense*** Mill. (Chinese Wolfberry). E.Asia. The leaves are eaten locally as a vegetable.

***L. fremontii*** Gray. Arizona to California. The berries are eaten raw or in soups by the local Indians.

***L. halamifolium*** Mill. = L. barbarum. (Common Matrimony Vine). Introduced from China to Mediterranean region where the leaves (Herba Lycii) are used as a purgative and diuretic.

***L. pallidum*** Miers. (Rabbit Thorn). S.W. N.America, into Mexico. The berries are eaten raw or cooked by the local Indians.

***L. sandwicense*** Gray. Hawaii. The berries are eaten locally.

LYCOPERDON Pers. Lycoperdaceae. 50 spp. Cosmopolitan. The Puff Balls. The young fruits bodies of the following species are eaten locally.

***L. fuligineum*** Berk. and Curtis. Tropical Asia.

***L. gemmatum*** Batsch. Eaten locally by the Indians of N.America.

***L. giganteum*** Huss. The fruitbodies are burnt to stupify bees when harvesting honey from hives. They are also used to stop bleeding.

***L. piriforme*** Schaff. Tropical Asia.

***L. perlatum*** Pers. Cosmopolitan.

***L. pratense*** Pers. Tropical Asia.

***L. umbrinum*** Pers. Cosmopolitan. Eaten in Asia.

LYCOPERSICON Mill. = Lycopersicum. 7 spp. Pacific, S.America, Galapagos Islands. Herbs.

***L. lycopersicum*** (L.) Karst. = Lycopersicum esculentum Mill. = Solanum lycopersicum L. (Tomato, Love Apple). Peru, Bolivia, Ecuador. The plant is now widely grown in the Old and New World for the fruits which are used in salads, for making ketchup, pickles, soups, pastes and juice. The plant is first recorded in Italy in 1554. It was first grown in Europe as an ornamental and the fruits were considered poisonous, but they were being eaten generally by the end of the sixteenth century. The plant was not grown commercially in America until the late eighteenth century. The world production is of the order of 18 million tons a year, with the U.S.A. producing 4–5 million tons, three quarters of which is used to make prepared foods and the rest is grown for salads. In northern latitudes the crop is grown under glass, being one of the main market garden crops. In more southern latitudes it is grown as a field crop, making Italy, Spain, Brazil, Japan and Mexico among the world's leading producers. The crop is harvested by hand, but machine harvesting is becoming increasingly dominant, especially in California. There are a great many varieties varying in shape, size and colour of fruit and in leaf form, but the fruit used for canning, etc. must be of uniform size for processing. The seed left after processing yield a semi-drying oil (Tomato Seed Oil) which is used for food and soap manufacture. The pulp is made into cattle cake or fertilizer.

***Lycopersicum*** Hill. = Lycopersicon.

***Lycopersicum esculentum*** Mill. = Lycopersicon lycopersicum.

LYCOPODIUM L. Lycopodiaceae. 450 spp. Tropics and temperate. Small herbs.

***L. annotinum*** L. See L. clavatum.

***L. cernuum*** L. Tropics. The dried plant is sometimes used to stuff pillows etc.

***L. clavatum*** L. (Clubmoss). Temperate. The spores are used as a fine light powder used medicinally for dusting sores etc., as an absorbent, to coat pills and lubricate suppositories. They are also used scientifically for measuring fine vibrations.

***L. complanatum*** L. See L. clavatum.

***L. selago*** L. See L. clavatum.

LYCOPUS L. Labiatae. 14 spp. N.Temperate. Herbs.

***L. asper*** Greene. = L. lucidus var. americana. N.America. The boiled roots are eaten by the local Indians.

***L. lucidus*** Turcz. Japan, China. The boiled rootstock is used as a vegetable in Japan.

***L. lucidus*** Turez. var. ***americanus*** Gray. = L. asper.

***L. uniflorus*** Michx. = L. virginicus.

***L. virginicus*** L. = L. uniflorus. (Virginian Bugle Weed). N.America. The rootstock are eaten as a vegetable by the Indians of British Columbia. They are mildly sedative and are used to treat persistent coughs.

LYCORIS Herb. Amaryillidaceae. 10 spp. E.Himalayas to Japan. Herbs.

***L. radiata*** Herb. China. The bulbs are used locally as an emetic and expectorant.

***Lycurus muticus*** Spreng. = Elionurus candidus.

LYGEUM Loefl. ex L. Graminae. 1 sp. Mediterranean especially N.Africa. Grass.

***L. spartum*** Loefl. ex L. The leaves are used to make a fibre used to make ropes, mats, chair seats, sails etc.

LYGODESMIA D. Don. Compositae. 12 spp. N.America. Herbs.

***L. grandiflora*** (Nutt.) Torr. and Gray. W. N.America. The leaves are eaten boiled as a vegetable by the local Indians.

***L. juncea*** (Pursh.) D. Don. (Rush Skeleton Weed). Central Plains of N.America. The latex is used as a chewing gum by the local Indians.

LYGODIUM Sw. Schizaeaceae. 4 spp. Tropics and subtropics. Climbing ferns.

***L. circinatum*** Sw. Tropical Asia, especially Malaysia. The stems are used to make baskets.

***L. japonicum*** (Thunb.) Swartz. Philippines, Australia, Japan through to Korea and India. A decoction of the leaves is used in China to treat bladder complaints, vaginal discharges and as an expectorant.

***L. scandens*** Swartz. Old and New World tropics. The stems are used to make small baskets in the Bismark Islands and the Philippines. They are also used to make hats.

***L. smithianum*** Presl. (Banga, Denga, Lubunga, M'banga). W. and Equatorial Africa. The stems are used locally to make fishing nets and animal traps.

Lym Bean – Muncuna nivea.

***Lyperia atropurpurea*** Benth. = Sutera atropurpurea.

***Lyperia crocea*** Eckl. = Sutera atropurpurea.

***Lyophyllum aggregatus*** Schaff. = Agaricus decastes.

Lyre-leaved Sage – Salvia lyrata.

LYSICHITUM Schott. Araceae. 1 sp. E.Siberia to Japan; 1 sp. Pacific N.America. Herbs.

***L. americanum*** Hulten and St. John. (Western Skunk Cabbage). Pacific N.America. The rhizomes are eaten as a vegetable by the local Indians and are used by some tribes to purify the blood. The leaves are used as poultices.

LYSILOMA Benth. Leguminosae. 35 spp. S.W. Unites States of America to tropical S.America. W.Indies. Small trees.

***L. bahamensis*** Benth. S.Florida, W.Indies. The wood is used for boat-building in the W.Indies. It is heavy and durable.

***L. candida*** T. S. Brandeg. (Palo blanco). Baja California. The bark, which is exported is used locally for tanning.

***L. divaricata*** (Jacq.) MacBride. Mexico. The bark is used locally for tanning.

***L. formosa*** (Benth.) Hitchc. = Calliandra formosa.

***L. latisiliqua*** Benth. = L. bahamensis (Wild Tamarind).

***L. sabicu*** Benth. Mexico, Cuba, Bahamas. The wood is heavy, fine-grained and durable in water. It is used for boat-building and cabinet work.

LYSIMACHIA L. Primulaceae. 200 spp. Cosmopolitan, especially E.Asia and N.America. Herbs.

***L. candida*** Lindl. = L. obovata.

***L. clethroides*** Duby. China, Japan. The leaves are used as a flavouring for food in China.

***L. foenum-graecum*** Hance. China. The sweet scented leaves are used by the Chinese women to perfume their hair and to correct halitosis.

***L. fortunei*** Max. China, Japan. The leaves are used for flavouring food in China.

***L. nemorum*** L. (Wood Loosestrife, Wood Pimpernel, Yellow Pimpernel). Europe to Caucasus. The leaves (Herba Anagallides lutea) are used as an astringent for healing wounds.

***L. nummularia*** L. (Moneywort). Europe, introduced to N.America. The leaves and flowers are sometimes used to make a tea. The leaves are also used to heal wounds.

***L. obovata*** Buch.-Ham. = L. candida. India. The leaves are used locally as a vegetable.

***L. quadrifolia*** L. (Fourleaf Loosestrife). E. N.America. The local Indians use a decoction of the plant to treat stomach aches.

***L. vulgaris*** L. (Loosestrife). Europe to temperate Asia, introduced to N.America. A decoction of the leaves is used locally as an astringent to treat wounds, bleeding from the nose or mouth and as a gargle.

LYTHRUM L. Lythraceae. 35 spp. Cosmopolitan. Herbs.

***L. salicaria*** L. (Purple Loosestrife, Spiked Loosestrife). Europe, introduced to N.America. The leaves can be eaten as a vegetable and are used as an emergency food.

# M

***Maackia amurensis*** Rupr. and Maxim. = Cladrastis amurensis.

Maagbessie – Dicoma anomala.

***Maba*** J. R. and G. Forst. = Diospyros.

Mabangu – Cynometra pedicellata.

MABEA Aubl. Euphorbiaceae. 50 spp. Central to tropical S.America, Trinidad. Trees.

***M. piriri*** Aubl. Guyana. A potential source of rubber, but of no commercial importance.

Mabi – Ceanothus reclinatus.

Mabolo – Diospyros discolor.

Maca – Lepidium meyenii.

Macabiha – Dioscorea macabiha.

Macacauba – Playtmiscium ulei.

Macachi – Arjona tuberosa.

MACADAMIA F. v. Muell. Proteaceae. 1 sp. Madagascar; 1 sp. Celebes; 5 spp. E.Australia; 3 spp. New Caledonia. Trees.

***M. ternifolia*** F. v. Muell. (Macadamia Nut, Queensland Nut). E.Australia. The seeds are edible and have a delicious flavour when roasted and salted The plant has been introduced into various parts of the tropics and is grown commercially in Hawaii.

***M. tetraphylla*** L. Johnson. (Queensland Nut, Macadamia Nut, Nuez de Australia). E.Australia. A rough-shelled nut used as M. ternifolia.

Macadamia Nut – Macadamia terifolia.

Macadamia Nut – Macadamia tetraphylla.

Macaja – Acrocomia sclerocarpa.

Macarandiba – Lucuma procera.

MACARANGA Thou. Euphorbiaceae. 280 spp. Tropical Africa, Madagascar, Indomalaya, Australia and Pacific Islands. Small trees or shrubs.

***M. bancana*** Muell. Arg. = M. triloba.

***M. diepenhorstii*** Muell. Arg. (Kajoo sepat, Sape). Sumatra. A gum produced in the phloem is used locally as a wood glue for small articles.

***M. gigantea*** Muell. Arg. Tropical Asia, especially Malaya, Java, Sumatra. The sap is used locally as a glue for wood. A decoction of the root bark is used in Malaya to treat dysentery.

***M. grandifolia*** (Blanco) Merr. Philippines. A gum from the stem is astringent and is used locally to treat mouth ulcers.

***M. henricorum*** Hems. (Man lâu). Vietnam. The bark and roots are chewed with betel nuts.

***M. hypoleuca*** Muel. Arg. Malaysia, Sumatra, Borneo. A decoction of the bark is used locally to reduce fevers, as a mild sedative and expectorant.

***M. involucrata*** Baill. Moluccas. A decoction of the bark mixed with cinnamon and liquorice is used to sooth tonsilitis.

***M. mappa*** Muell. Arg. (Hadochoon, Sapia). Indonesia. The gum from the bark is used locally for treating fishing nets.

***M. tanarius*** Muell. Arg. S.China, Malaysia. The bark yields a gum (Binunga Gum) used in the Philippines for glueing guitars etc. An alcoholic beverage is also made by fermenting the bark.

***M. triloba*** Muell. Arg.=M. bancana=Pachystemon bancanus. (Kajoo sepat, Parake oodang, Tootoop antjoor). Malaysian Archipelago. A decoction of the leaves is used in Java to treat abdominal complaints. The soft white wood is sometimes used to make knife handles.

Macaroni Wheat – Triticum durum.

Macassar Agar – Eucheuma muricatum.

Macassar Copal – Agathis alba.

Macassar Copal – Agathis loranthifolia.

Macassar Ebony – Diospyros utilis.

Macassar Manila – Agathis alba.

Macassar Oil – Cananga odoratum.

Macassar Oil – Schleichera trijuga.

Macawood, Quira – Platymiscium pinnatum.

Mace (Oil) – Myristica fragrans.

Mace, Bombay – Myristica malabarica.

Maceron – Smyrnium olus-atrum.

MACHAERUIM Pers. Leguminosae. 150 spp. Mexico to tropical S.America and W.Indies. Trees.

***M. acutifolium*** Vog. Brazil. The fruits are used locally as a diuretic.

***M. angustifolium*** Vog.=Drepanocarpus isadelphus. Brazil. A resin from the bark is used locally to treat snake-bite.

***M. caudatum*** Ducke. Brazil. The fruits are used locally as a diuretic.

***M. macrocarpum*** Ducke. See M. acutifolium.

***M. macrophyllum*** Benth. See M. acutifolium.

MACHAEROCEREUS Britt. and Rose. Cactaceae. 2 spp. Lower California. Cacti.

***M. gummosus*** (Engelm.) Britt. and Rose.=Cereus gummosus. (Pitahaya, Pitahaya agria). Lower California. The acid fruits are eaten locally. The juice from the stems is used to stupify fish.

***Machilus*** Rumph.=Neolitsea.

***Machrochloa tenacissima*** (L.) Kunth.=Stipa tenacissima.

MACLEANIA Hook. Ericaceae. 45 spp. Central to W. tropical S.America. The fruits of the following species are sweet and pleasantly flavoured. They are eaten locally.

***M. ecuadoriensis*** Horold. Ecuador.

***M. popenoei*** Blake. Ecuador.

MACLURA Nutt. Moraceae. 12 spp. Warm America, Africa, Asia. Trees.

***M. africana*** Bur. Tropical Africa. The wood yields a yellow dye.

***M. aurantiaca*** Nutt.=Toxylon pomiferum. (Bois d'Arc, Bow Wood, Osage Orange). S. U.S.A. The bright yellow wood is hard, strong and very flexible. It is used for railway sleepers, fencing and wheel-making and was used by the local Indians for clubs etc. The root bark and trunk yield a yellow dye, used for dyeing fabrics. The fruits contain an antioxidant, which could be valuable in preventing oil-containing foods, cosmetics etc. from becoming rancid.

***M. tricuspidata*** Carr.=Cudrania triloba.

MACODES (Bl.) Lindl. Orchidaceae. 10 spp. Malaysia, Solomon Islands, Herbs.

***M. petola*** Blume. Java, Malacca. The natives drop the juice into the eyes to increase close vision.

Macqui – Aristotelia macqui.

Macrae Raspberry – Rubus maeraei.

***Macrochloa tenacissima*** (L.) Kunth.=Stipa tenacissima.

MACROCYSTIS C. A. Agardh. Laminariaceae. Brown Alga.

***M. pyrifera*** (L.) C. A. Agardh. Pacific coast of America. Harvested especially off the coast of California and used to manufacture sodium alginate. This is used to thicken a wide range of pharmaceutical and toilet creams, especially greaseless preparations.

MACROLENES Naud. Melastomataceae. 20 spp. Siam, Malaysia. Vines.

***M. muscosa*** Blume. Java. The young shoots are eaten locally as a vegetable with rice. The berries are made into preserves or drinks and the juice from the stem is used to treat eye infections.

MACROLOBIUM Schreb. Leguminosae. 100 spp. Tropical America, Africa. Trees.

***M. coeruleum*** (Taub.) Harms.=Vouapa coerulea (Alumbi-kibile, Lubese). Tropical Africa. The wood is used for general construction work, railway sleepers and house interiors.

***M. grandifolium*** De Wild.=Gilbertiodendron grandiflorum.

***M. laurentii*** De Wild.=Brachystegia laurenti..

***Macrorhynchus troximoides*** Torr. and Gray =Agoseris aurantiaca.

MACROSIPHONIA Muell. Arg. Apocynaceae. 10 spp. S.W. United States of America to tropical S.America. Xerophytic shrubs.

***M. hypoleuca*** (Benth.) Muell. Arg.=M. lanugonosa=Echites hypoleuca=E. suaveolens. (Guirambo, Hierba de la Cucaracha, Rosa de San Juan). S. N.America, Mexico. The roots are used locally ground with sugar to kill cockroaches.

MACROSOLEN Blume. Loranthaceae. 40 spp. S.E.Asia, Malaysia. Parasitic shrubs.

***M. cochinchinensis*** v. Tiegh.=Loranthus cochinchinensis=L. globosus. Indochina. The leaves are used in Vietnam as a tea substitute and the fruits are used as a balm against itch.

MACROTOMIA DC. Boraginaceae. 6 spp. Mediterranean to Himalayas. Herbs.

***M. cephalotus*** DC. Greece to Caucasus. The root (Radix Alkannae syricae, Syrian Alkanna, Turkish Alkanna) is used to make a dye.

MACROZAMIA Miq. Zamiaceae. 14 spp. Extratropical Australia. Trees. The ground seeds of the species mentioned are eaten by the aborigines.

***M. miquelii*** A. DC.=Encephalartos miquelii. Australia.

***M. spiralis*** Miq.=Encephalartos spiralis= Zamia spiralis. (Burrawang Nut). New South Wales, Queensland. The seeds also make a good arrowroot.

Macuatus – Rosa californica.

Madagascar Baobab – Adansonia madagascariensis.

Madagascar Cardamon – Aframomum angustifolium.

Madagascar Clove – Ravensara aromatica.

Madagascar Copal – Hymenaea verrucosa.

Madagascar Ebony – Diospyros haplostylis.

Madagascar Mahogany – Khaya madagascariensis.

Madagascar Nutmeg – Ravensara aromatica.

Madagascar Olive – Noronhia emarginata.

Madagascar Periwinkle – Vinca rosea.

Madagascar Plum – Flacourtia ramontchi.

Madagascar Raffia Palm – Raphia pedunculata.

Madagascar Rouge – Landolphia madagascariensis.

Madar – Erythrina suberosa.

Madder – Rubia tinctorum.

Madder, Indian – Rubia cordifolia.

Madder, Indian – Oldenlandia umbellata.

Madder, Khasia – Rubia khasiana.

Madder, Levant – Rubia peregrina.

Madder, Sikkim – Rubia sikkimensis.

Madeira Vine – Anredera baselloides.

Madflower, Ethiopean – Antholyza paniculata.

MADHUCA J. F. Gmel. Sapotaceae. 85 spp. Indochina, Indomalaya, Australia. Trees.

***M. betis*** (Blanco) Merr.=Illipé betis= Payena betis Philippines. The hard wood is termite-resistant and is used to build houses, bridges, wharves etc. An oil (Betis Oil) extracted from the seeds is used for lighting.

***M. butyracea*** Gmel. (Illipé Butter Tree). Tropical Asia, especially India. A non-drying fat (Indian Butter, Phulwara Butter) extracted from the seeds is used locally for cooking.

***M. lancifolia*** Lam.=Payena lancifolia (Kelaki). Indonesia. An oil extracted from the seeds is used locally for cooking.

***M. latifolia*** Macbr. N.India, Indochina, Malaysia. The flowers are rich in nectar and are eaten locally or fermented to make a beverage (Mahua Spirit). They are also used to produce acetone commercially. The seeds yield a fat which is used locally for cooking.

***M. longifolia*** (Koen.) H. J. Lam. (Mowra Butter Tree). S.India. The seeds yield a fat (Illipe Butter, Tallow Mowra Butter) which is used for soap, candles and treating skin diseases. It is also used locally as a butter substitute. The seeds are exported to Europe. The residue after pressing is used as a fertilizer.

***M. malabarica*** Bedd. E.India. The flowers are sweet and are eaten locally. The seeds are a minor source of edible fat.

***M. macrophylla*** Lam. = Payena macrophylla. (Pasra). Java. The wood is used locally for house construction and the seeds are a minor source of edible fat.

***M. mottleyana*** Clarke. = Ganua mottleyana. Malaysia. A poor quality gutta percha is extracted from the trunk. The oil (Katio Oil) is used in making foods. The wood is used locally for house construction etc.

***M. obovatifolia*** (Merr.) Merr. (Manik). Philippines. The fruits are eaten locally.

***M. stenophylla*** Lam. = M. utilis.

***M. utilis*** Lam. = M. stenophylla = Payena utilis (Seminai). The seeds yield an oil which is used locally for cooking.

Ma'di – Diospyros mespiliformis.

Madi – Madia sativa.

MADIA Molina Compositae. 20 spp. Pacific America. Herbs.

***M. glomerata*** Hook. (Tarweed). Pacific N.America. The oily fruits are eaten by the local Indians.

***M. sativa*** Molina. (Tarweed, Madia Oil Plant). S.America. Cultivated locally and in France and Germany for the seeds which yield the edible Madia Oil. This is used as a substitute for Olive Oil. The leaves make a pasture for sheep.

Madia Oil – Madia sativa.

Madonna Lily – Lilium candidum.

Madre de Cacao – Gliricidia sepium.

Madrona – Arbutus menziesii.

Madronillo – Amelanchier denticulata.

Madweed – Scuellaria laterifolia.

Mae pan-li – Castanea seguinii.

MAERUA Forsk. Capparidaceae. 100 spp. Tropical and S.Africa to India. Trees or shrubs.

***M. angolensis*** DC. Tropical Africa, especially W. The yellow heavy hard wood is used for general carpentry. The leaves are used locally as a purgative and sometimes as a vegetable.

***M. crassifolia*** Forsk. = M. rigida.

***M. pedunculosa*** Vahl. = Neibuhria pedunculosa = Boscia caffra. S.Africa. The roots are used as a famine food by the natives.

***M. rigida*** R. Br. = M. crassifolia. (Adjar, Aspou ouar, Atil). N.Sudan. The fruits (Eb nembe, Mil) are eaten locally.

***M. virgata*** Brongn. Red Sea area. The fruits are eaten locally. The ash from the burnt stems and leaves is extracted for salt.

MAESA Forsk. 200 spp. Old World tropics. Shrubs.

***M. balansae*** Mez. (Dok toa pa, Don rang cua). Vietnam. The leaves are used locally to treat itch.

***M. coriacea*** Champ. = M. japonica.

***M. doraena*** Blume. = M. japonica.

***M. floribunda*** Scheff = M. ramentacea.

***M. glabra*** A. DC. = M. ramentacea.

***M. indica*** Wall. M. morsha = Baobotrys indica. China to Vietnam. The leaves are used locally to flavour food.

***M. japonica*** Mor. = M. coriacea = M. doraena. China, Japan to N.Vietnam. A decoction of the leaves is used locally to stop coughing and a decoction of the roots is used to treat vaginal discharges.

***M. leptobotrys*** Hance = M. ramentacea.

***M. morsha*** Hamilt. = M. indica.

***M. ramentacea*** Wall. = M. floribunda = M. glabra = M. leptobotrys. Tropical and subtropical Asia. A poultice of the leaves is used locally, especially in Malaya to treat skin diseases.

***M. tetrandra*** A. DC. Malaysia. A decoction of the roots is used locally to reduce fevers.

Mafura Bitterwood – Trichilia emetica.

Mafura Oil – Trichilia emetica.

Mafura Tallow – Trichilia emetica.

Magellan Barberry – Berberis buxifolia.

MAGNOLIA L. Magnoliaceae. 80 spp. Himalayas to Japan; Borneo and Java; E. N.America to Venezuela; W.Indies. Trees.

***M. acuminata*** L. = Tulipastrum acuminata. (Cucumber Tree, Mountain Magnolia). E. U.S.A. The soft light wood is durable, but not strong. It is used for flooring and furniture-making.

***M. blumei*** Prantl. = Manglietia glauca.

***M. glauca*** L. = M. virginiana. (Sweet Bay, Swamp Bay). E. U.S.A. The soft light wood is used for small household articles, e.g. broom handles.

***M. grandiflora*** L. (Southern Magnolia). S.E. and S. U.S.A. A decoction of the bark is used as a tonic and to induce sweating.

***M. javanica*** Koord. and Val. Java, Sumatra.

The wood is used locally for the handles of weapons.

***M. kobus*** DC. (Kobus Magnolia). Japan, Korea. The soft wood is used in Japan to make matches, engraving and kitchen utensils. The bark is used locally to cure colds.

***M. macrophylla*** Michx. See M. acuminata.

***M. mexicana*** DC. = Talauma mexicana.

***M. obovata*** Thunb. Japan to central China. In Japan the soft wood is used in making furniture, engraving and kitchen utensils.

***M. officinalis*** Rehd. and Wils. Central China. A decoction of the bark is used locally as a tonic, aphrodisiac and to treat colds.

***M. schiedeana*** Schlecht. Mexico. A decoction of the flowers is used locally to treat scorpion stings.

***M. splendens*** Urban. (Laurel Sabino). Puerto Rico. The wood is an olive-green, darkening to brown on exposure. It is easy to work and pleasantly grained, but decays readily. It is used locally for furniture, inlays, carving, interiors and exteriors of housing, planking and plywood.

***M. virginiana*** L. = M. glauca.

Magnolia, Kobus – Magnolia kobus.

Magnolia, Mountain – Magnolia acuminata.

Magnolia, Southern – Magnolia grandiflora.

MAGONIA A. St. Hil. Sapindaceae. 2 spp. Brazil, Bolivia. Trees.

***M. pubescens*** A. St. Hil. = Phaeocarpus campestria. Brazil. A decoction of the bark is used as a general tonic in parts of Brazil. The stem, leaves and fruits are used to poison fish.

Maguey, Cebu – Agave cantala.

Maguey ceniso – Agave quiotifera.

Maguey de Cozuiza – Furcraea humboldtiana.

Maguey delgado – Agave kirchneriana.

Maguey Gum – Puya chilensis.

Maguey liso – Agave weberi.

Maguey, Manila – Agave cantala.

Maguey manso – Agave crassispina.

Mahalatsaka – Piper pachyphyllum.

Mahaleb Cherry – Prunus mahaleb.

Mahali kizhangu – Decalepis hamiltonii.

Mahareb Grass, Oil of – Cymbopogon senaarensis.

Mahavoahavana – Plectaneia microphylla.

Ma-há-wa-soo – Vaupesia cataractarum.

Mahi – Ficus tinctoria.

Mahobohobo – Uapaca kirkiana.

Mahoe – Melicytus ramiflorus.

Mahogany – Swietenia mahagoni.

Mahogany, Adju – Canarium mansfeldianum.

Mahogany, African – Afzelia africana.

Mahogany, African – Detarium senegalense.

Mahogany, African – Heretiera utilis.

Mahogany, African – Khaya senegalensis.

Mahogany, African – Tieghemella heckeli.

Mahogany, Australian – Dysoxylum fraseranum.

Mahogany, Bastard – Eucalyptus botryoids.

Mahogany, Benin – Guarea thompsoni.

Mahogany, Benin – Khaya punchii.

Mahogany, Borneo – Calophyllum inophyllum.

Mahogany, Brown – Entandrophragma utile.

Mahogany, Brown – Lovoa klaineana.

Mahogany, Brown – Lovoa swynnertonii.

Mahogany, Brush – Weinmannia benthami.

Mahogany, Cameroon – Entandrophragma rederi.

Mahogany, Cameroon – Khaya euryphylla.

Mahogany, Caoba – Swietenia mahagoni.

Mahogany, Cape – Ptaeroxylon utile.

Mahogany, Cape – Trichilia emetica.

Mahogany, Columbian – Cariniana pyriformis.

Mahogany, Espave – Anacardium rhinocarpus.

Mahogany, Forest – Eucalyptus resinifera.

Mahogany, Gabon – Canarium mansfeldianum.

Mahogany, Gabon – Khaya ivoriensis.

Mahogany, Honduras – Swietenia macrophylla.

Mahogany, Lagos – Khaya grandifoliola.

Mahogany, Ljebu – Entandrophragma angolense.

Mahogany, Madagascar – Khaya madagascariensis.

Mahogany, Manubi – Trichilia emetica.

Mahogany, Mexican – Swietenia humulis.

Mahogany, Mexican – Swientenia macrophylla.

Mahogany, Mountain – Cercocarpus latifolia.

Mahogany, Natal – Kiggelaria dregeana.

Mahogany, Natal – Trichilia emetica.

Mahogany Nuts – Afrolicania elaeosperma.

Mahogany, Pink – Guarea cedrata.

Mahogany, Pod – Afzelia cuanzensis.

Mahogany, Red – Eucalyptus resinifera.

Mahogany, Red – Khaya nyasica.
Mahogany, Red – Khaya ivorensis.
Mahogany, Roho – Dysoxylum fraseranum.
Mahogany, Rose – Ceratopetalum apetalum.
Mahogany, San Domingo – Swietenia mahagoni.
Mahogany, Sapele – Entandrophragma candollei.
Mahogany, Scented – Entandrophragma candollei.
Mahogany, Short-Capsuled – Entandrophragma utile.
Mahogany, Smooth-barked African – Khaya anthotheca.
Mahogany, Swamp – Eucalyptus botryoides.
Mahogany, Swamp – Eucalyptus robusta.
Mahogany, Tabosa – Swietenia mahagoni.
Mahogany, Tiama – Entandrophragma candollei.
Mahogany, Venezuela – Swietenia candollei.
Mahogany, West African – Mitragyna macrophylla.
Mahogany, West Indies – Swietenia mahagoni.
Mahogany, White – Cybistax donnell-smithii.
Mahogany, White – Eucalyptus robusta.
Mahogany, White – Eucalyptus triantha.
Mahogany, White – Khaya anthotheca.

MAHONIA Nutt. Berberidaceae. 70 spp. Himalayas to Japan and Sumatra; N. and Central America. Shrubs.

***M. aquifolium*** (Lindl.) Don. = Berberis aquifolium (Oregon Grape). W. N.America. The rhizome and roots contain the alkaloids berberine, berbadine and oxyacanthine, so that these parts are used medicinally to improve the appetite and as a tonic.

***M. chochoco*** Fedde. (Chochoco). Mexico. The wood yields a yellow dye and is used locally for tanning.

***M. ganpinensis*** Fedde. China. All parts of the plant are used, to make a poultice, as an antiseptic and for toothache.

***M. nepalensis*** DC. = M. philippensis.

***M. philippensis*** Takida = M. nepalensis. Luzon. The stems and leaves yield a yellow dye.

***M. repens*** (Lindl.) G. Don. = Berberis repens = Odostemon aquifolium. (Oregon Grape). W. N.America. The ripe fruits are eaten raw and are used to make jellies, wines and unfermented drinks. The early settlers used the bark as a tonic, to treat mountain fever, kidney complaints and to reduce fevers.

***M. trifoliatus*** (Moric.) Heller. = Berberis ilicifolia. Texas to Mexico. The fruit is acid, but is used locally to make preserves. The wood is the source of a yellow dye. It is also used for tanning and to make ink.

Mahua Spirit – Madhuca latifolia.
Ma-Huang – Ephedra sinica.
Mahuba Fat – Acrodiclidium mahuba.
Mahue, Blue – Hibiscus elatus.
Maiden's Gum – Eucalyptus maideni.
Maidenhair Fern – Adiantum capillus-veri-eris.
Maidenhair Tree – Ginko biloba.
Maigang – Eugenia polycephaloides.
Maina Resin – Calophyllum longifolium.
Mais des Esprits – Cyrtosperma senegalense.
Maitan – Lunasia amara.
Mait Brand – Atropa acuminata.
Maize – Zea mays.
Maizurrye – Nannorhops ritchieana.
Majagua – Hibiscus tiliaceus.
Majhua – Heliocarpus donnel-smithii.
Majang – Clavaria zippeli.

***Majorana*** Mill. = Origanum.

Majunga noir – Mascarenhasia arborescens.
Majunga rouge – Landophilia perieri.
Majura Oil (Tallow) – Trichilia emetica.
Makassar Pitjes – Brucea sumatrana.
Mak châng – Diospyros decandra.
Makhi-jali – Drosera burmanni.
Makino – Semiarundinaria fastuosa.
Makita – Parinari laurinum.
Makomako – Cordia marchionica.
Makwapusa – Laportea podocarpa.
Mala insana – Quercus tauricola.
Mala sodomitica – Quercus tauricola.
Malaba – Garcinia comingiana.

MALABAILA Hoffm. Umbelliferae. 10 spp. E.Mediterranean through to Persia and Central Asia. Herbs.

***M. pumila*** Boiss. Sahara region. The roots are eaten by the Bedouins.

***M. sekakul*** (Russ.) Boiss. (Sekakul). Mediterranean Africa. Yields a strongly perfumed oil (Sekakul) used as an aphrodisiac locally.

Malabar Ironwood – Hopea parviflora.
Malabar Kino – Pterocarpus marsupium.
Malabar Nut Tree – Adhatoda vasica.
Malabar Rosewood – Dalbergia latifolia.
Malabar Tallow Tree – Vateria indica.

Malaboeai, Gutta – Alstonia grandifolia.

MALACANTHA Pierre. Sapotaceae. 1 sp. Tropical W.Africa. Shrubby tree.

***M. warnekeana*** Engl. The berries have a sweet pulp and are eaten locally.

Malacca Poison Nut – Strychnos malaccensis.

Malacca Teak – Afzelia palembanica.

Malacca Yam – Dioscorea atropurpurea.

MALACHRA L. Malvaceae. 6 spp. Warm America, W.Indies. Herbs.

***M. capitata*** L. Warm America, W.Indies. Cultivated for the excellent stem fibres, which have a limited local use. They are similar to jute.

Malagiri – Cinnamomum cecicodaphne.

Malaiino – Celtis philippensis.

Malaikluval – Commiphora stocksiana.

Malakabuyan – Swinglea glutinosa.

Mala mujer – Urera caracasana.

Malando polo – Dolichos malosanus.

Malanga amarilla – Xanthosoma sagittifolium.

Malanga blanca – Xanthosoma sagittifolium.

Malanga isleña – Colocasia antiquorum.

Malanga, Primrose – Xanthosoma violaceum.

***Malaxis longibracteata*** Reichb. = Oberonia longibracteata.

Malay Brushbeech – Gmelina arborea.

Malay Gutta Percha Tree – Palanguium gutta.

Malay Lacktree – Schliechera trijuga.

Male Bamboo – Dendrocalamus strictus.

Male Cola – Cola acuminata.

Malenghet Copal – Agathis loranthifolia.

Male Rubber Tree – Holarrhena wulfsbergii.

Mal hombre – Urera caracasana.

Mali-mali – Leea curtisii.

Mali-mali hantco – Leea indica.

Malia pootih – Strichnos ligustrina.

Malkangni – Celastrus paniculata.

Mallee, Blue – Eucalyptus polybractea.

Mallee, Cangoo – Eucalyptus dumosa.

Mallet, Blue – Eucalyptus gardneri.

Mallet, Brown – Eucalyptus astringens.

Mallet, Swamp – Eucalyptus spathulata.

Mallet, White – Eucalyptus falcata.

MALLOTUS Lour. Euphorbiaceae. 140 spp. E. and S.E.Asia, Indomalaya to New Caledonia, Fiji and N.E.Australia; 2 spp. Tropical Africa and Madagascar. Shrubby trees.

***M. anamiticus*** O. Ktze. = Coelodiscus anamiticus. (Bang dàng nam, Krabas prey). Vietnam, Cambodia. The leaves are used locally as a tea substitute.

***M. apelta*** Muell.-Arg. China, Indonesia. The seeds are used locally to kill fish.

***M. cochinchinensis*** Lour. = M. paniculatus. S.China to N.Australia. The light wood is used in Indo-China to make matches and packing cases.

***M. discolor*** F. v. Muell. = Rottlera discolor. Australia. The capsules are used to produce a bright yellow dye.

***M. floribunda*** Muell.-Arg. Tropical Asia especially the New Guinea area. The male flowers are used to scent rice flour when it is used as a toilet powder.

***M. furetianus*** Muell.-Arg. = M. maclurei = Hancea muricata. India, Indo-China, Indonesia. The scented leaves are used in Indo-China as a tea substitute.

***M. maclurei*** Merr. = M. furetianus.

***M. paniculatus*** Muell.-Arg. = M. cochinchinensis.

***M. philippensis*** Muell.-Arg. (Kamala Tree). Himalayas to Australia. The hairs on the fruits yield a golden-yellow dye (Kamala Powder) which is much esteemed for dyeing silks. The dye is also used to treat intestinal worms and skin complaints.

***M. trifoliata*** Muell.-Arg. (Daoon baroo laoot, Haleki laoon ooloo, Taan). Indonesia. The fine, light, white wood is used locally for small joinery.

Mallow – Malva rotundifolia.

Mallow – Lavatera plebeia.

Mallow, Castilian – Malva crispa

Mallow, Country – Abutilon indicum.

Mallow, Curled – Malva crispa.

Mallow, Finger Poppy – Callirhoë digitata.

Mallow, High – Malva sylvestris.

Mallow, Indian – Abutilon avicennae.

Mallow, Jew's – Corchorus olitorius.

Mallow, Marsh – Althaea officinalis.

Mallow, Narrowleaf Globe – Sphaeralcea angustifolia.

Mallow, Purple – Callirhoë pedata.

Mallow, Purple Poppy – Callirhoë involucrata.

Mallow, Tree – Lavatera plebeia.

Maloot – Palaquium microphyllum.

MALOUETIA A. DC. Apocynaceae. 25 spp. Central and tropical S.America; W.Indies, tropical Africa. Trees.

***M. nitida*** Spruce. Tropical S.America. The sap is used locally as an arrow poison.

***M. tamaquarina*** Spruce. (Cuchara caspi,

Spoon Tree). Amazon Basin. The local Indians add the leaves to a narcotic drink made from Banisteriopsis caapi. This can cause severe digestive disorders and muscular abnormalities.

MALPIGHIA L. Malpighiaceae. 35 spp. Tropical America, W.Indies. Trees or shrubs. The fruits of the following species are sub-acid and refreshing. They are eaten raw or in preserves and jellies.

***M. glabra*** L. (Barbados Cherry). S.Texas to tropical America, W.Indies. Cultivated locally.

***M. punicifolia*** L. (West Indian Cherry). Tropical America, W.Indies.

***M. urens*** L. Tropical America, W.Indies. Fruits called Grosse Cérise.

Malte – Agropyron trachycaulum.

Maltese Cotton – Gossypium herbaceum.

Maluko – Pisonia alba.

MALUS Mill. Rosaceae. 35 spp. N.Temperate. Trees. The genus is included by some authorities in ***Pyrus*** L. The distinction retained here is the presence of stone cells in the fruits of Pyrus spp. They are absent from the fruits of Malus spp.

***M. angustifolia*** (Ait.) Michx.=Pyrus angustifolia. (Crab Apple). E. N.America. The bitter-sweet fruits are used locally to make jellies, preserves, etc. and cider. The wood is hard and heavy. It is used for tool-handles etc.

***M. baccata*** (L.) Borkh.=Pyrus baccata. (Chinese Crabapple, Siberian Crabapple). N.China to Siberia. The fruits are eaten locally. They are often preserved by drying.

***M. communis*** DC.=M. pumila.

***M. coronaria*** (L.) Mill.=Pyrus coronaria (Fragrant Crabapple, Crabapple). E. N.America. The fruits are used locally for jellies, cider and vinegar. The wood is heavy, but not strong. It is used for small domestic articles, tool-handles etc.

***M. hupehensis*** (Pamp.) Rehd.=M. theifera =Pyrus hupehensis. (Chinese Crabapple) China to Assam. The leaves are used in China to make a tea.

***M. praecox*** (Pall.) Borkh.=Pyrus praecox. Middle East. Cultivated locally and in the Mediterranean area for the small sweet fruits which are eaten raw.

***M. prunifolia*** (Willd.) Borkh.=Pyrus prunifolia. (Chinese Apple). E. temperate China. The fruits are eaten raw or in preserves. There are several local varieties which have flesh colours varying from red to white.

***M. pumila*** Mill.=M. communis=Pyrus malus. (Apple). The plant originated in the Caucasus but have been cultivated for over 2,000 years, gradually spreading from the Mediterranean until next to the grape it is the most important fruit crop. It is cultivated widely in both the N. and S.Temperate zones where the winter temperatures do not fall below −7°C. The fruits keep well under well-ventilated conditions, making the crop available for sale throughout the year. Propagation is by grafting onto selected stocks, and cultivations include pruning and seasonal spraying against pests and diseases. The trees will continue to bear for 100 years, but by this time they have become commercially unprofitable. The world production is about 20 million tons annually, with Europe being the chief producer. About half the crop is consumed as raw fruit while most of the rest is used to make cider. Some is used to make soft drinks and vinegar. Pectin is a by-product from the manufacture of cider. The wood is hard and strong. It is used for turning, tool handles, etc. There are some 2,000 varieties, which fall into three main classes, (1) Cider which are bitter-sweet with more than 0·45 per cent tannin and more than 0·2 per cent sugar, (2) Cooking or culinary, which are large moderately acid, with a firm flesh and (3) dessert—which have a low tannin and acid content and a high sugar content. The varieties are determined by colour, size, flavour, aroma and texture of the flesh.

***M. rivularis*** Roem.=Pyrus rivularis Doulg. (Oregon Crabapple). W. N.America to Alaska. The fruits are eaten raw or in jellies. They are sometimes stored in seal oil for winter use.

***M. theifera*** Rhed.=M. hupehensis.

MALVA L. Malvaceae. 40 spp. N.Temperate. Herbs.

***M. crispa*** L. (Castillian Malva, Curled Mallow). Origin unknown. The leaves are used, especially in Mexico to treat bruises, boils and sore throats. They are also used as a garnish for food.

***M. involucrata*** Torr. and Gray.=Callirhoë involucrata.

***M. meluca*** Graebn. Cool S.America. Sometimes cultivated, especially in Russia as a fodder crop and for the jute-like fibres.

***M. parviflora*** L. China. The young shoots are used locally as a vegetable.

***M. rotundifolia*** L. (Common Mallow). Temperate. The dried leaves contain mucilage and tannin. They are used to soothe inflammed tissues.

***M. sylvestris*** L. (High Mallow). Europe, Asia. Introduced to N. and S.America and Australia. The dried leaves are used as a tea substitute and as they contain a mucilage, they are used medicinally as an expectorant. A decoction of the flowers (Flores Malvae) is used for gargling and as a mouthwash.

***M. verticillata*** L. China, introduced to S.Asia and S.Europe. The young shoots are used as a vegetable in China.

Malva, Castillian – Malva crispa.

Mamèro – Carica papaya.

Mamekh – Paeonia emodi.

Mamey Colorado – Calocarpum mammosum.

Mamira – Coptis teeta.

Mamira – Thalictrum foliosum.

Mamiran – Geranium wallichianum.

Mamiran – Thalictrum foliolosum.

MAMMEA L. Guttiferae. 1 sp. tropical America, W.Indies; 1 sp. tropical Africa; 20 spp. Madagascar; 27 spp. Indomalaya and Pacific Islands. Trees.

***M. africana*** (Don.) Oliv. (African Mammee Apple). Tropical Africa. The seeds are eaten locally.

***M. americana*** L. (Mammee, Mammee Apple, Mammee Apricot, St. Domingo Apricot) N. S.America, W.Indies. The plant is cultivated for the fruit which are eaten raw, or in preserves and jams (Mammee Preserve). The seeds are used locally to treat intestinal worms. A liqueur (Eau de Creole) is distilled from the flowers in the W.Indies. A resin (Resina de Mammee) from the stem is used by the natives to treat chigoes in the feet.

***M. longifolius*** Benth. and Hook. f. (Nag kesar, Surgi, Surungi). India. The reddish grey wood is hard and smooth. It is used for shipbuilding and boards. The flowers yield a red dye, used for dyeing silk and an essential oil used in perfumery.

Mammee – Mammea americana.

Mammee Apple – Mammea africana.

Mammee Apple – Mammea americana.

Mammee Apricot – Mammea americana.

Mammoth Tree – Sequoia gigantea.

Mamoneillo – Mellicoccus bijugatus.

Man Kinh – Vitex quinata.

Man lâu – Macaranga henricornum.

Mân turói – Eupatorium staechadosmum.

Manaca Raintree – Brunfelsia hopeana.

Nanado Cedar – Cedrela celebica.

Manatan – Aglaia oligantha.

Manchineel – Hippomane mancinella.

Manchingi – Dolichandrone falcata.

Manchu Cherry – Prunus tomentosa.

Manchurian Walnut – Juglans mandschurica.

Mancona – Erythrophleum guineensis.

Mandakaki – Tabernaemontana divaricata.

Mandan Wild Rye – Elymus canadensis.

Mandarin Orange (Oil) – Citrus reticulata.

Mandarin, Temple – Citrus sinensis × C. reticulata.

***Mandavilla potosina*** T. S. Brandeg. = Urechites karwinskii.

Mandera Cucumber – Cucumis sacleuxii.

Mandihoba – Manihot glaziovii.

Mandim bepos – Pandanus papuanus.

Mandioca – Manihot esculenta.

Mandioca doce – Manihot dulce.

Mandioqueira – Didymopanax longipetiolatum.

Mandiro Fat – Fevillea cordifolia.

MANDRAGORA L. Solanaceae. 6 spp. Mediterranean to Himalayas. Herbs. Both species mentioned contain the alkaloids mandragornie and hyoscyamine which are narcotic.

***M. autumnalis*** Spr. (Autumn Mandrake). Mediterranean. The root (Alrain Root, Mandragora Root) is used to dilate the eye pupils.

***M. officinarum*** L. (European Mandrake, Mandragora). The root was used as a narcotic, purgative and emetic and sometimes as a potential to aid pregnancy. It is now used in the treatment of asthma, hay-fever and coughs.

Mandragora – Mandragora officinarum.

Mandragora Root – Mandragora autumnalis.

Mandrianambo – Landophilia mandrianambo.

Mandrake – Phyllostachys bambusioides.

Mandrake – Podophyllum peltatum.

Mandrake, Autumn – Mandragora autumnalis.

Mandrake, European – Mandragona officinarum.

MANETTIA Mutis ex L. Rubiaceae. 130 spp. Central and tropical S.America, W.Indies. Trees.

***M. cordifolia*** Mart. = M. ignita. Tropical S.America. The root (Radix Ipecacuanahe striata minor) is used as an adulterant for true Radix Ipecacuanha.

Mang ram bo – Stixis flavescens.

Mang tsao – Illicium anisatum.

Manga – Eriosema lebrunii.

Mangabiera (Rubber) – Hancornia speciosa.

Mangea – Pancovia harmsiana.

Mangel Wurzel – Beta vulgaris.

Manghamba – Bauhinia fassagëensis.

MANGIFERA L. Anacardiaceae. 40 spp. S.E.Asia, Indomalaya Trees.

***M. altissima*** Blanco. (Paho, Pahutan Mango). Philippines. The fruits are used locally to make pickles.

***M. caesia*** Jack. (Kemang). Malaysia. The tree is cultivated locally for the fruits.

***M. foetida*** Lour. (Bachang Mango). Malaysia. The fruits have a very turpentine flavour, but are eaten locally in lime juice or syrup. The young fruits (Sambal) are eaten with fish on Cayenne Pepper.

***M. indica*** L. (Mango). India, Burma, Assam. Cultivated throughout the Old and New World Tropics. The plant has been cultivated since ancient times in its native localities and plays a part in the Hindu religion. It was introduced into Brazil in 1700 and into the W.Indies in 1740. It was not introduced to the U.S.A. (Florida) until 1889. The fruits are eaten raw and are used, particularly in India to make various condiments, e.g. chutney (Murrabala), Mango Pulp (Amsat, Ampapar), Mango Pickle in oil (Achar), Mango Preserve (Murrabala) and Mango Powder (Amchur). The wood is used in India for house-building, boats, furniture, plywood, etc. A gum (Mango Gum) from the trunk mixed with opium and egg white is used locally to treat dysentery. The fruit skins and seeds are used in Malaya to treat intestinal worms, haemorrhoides and uterine haemorrhages. The seeds are eaten during famine. A yellow dye (Indian Yellow, Jaune Indien, Monghyr Piuri) is made in India from the urine of cows fed on the leaves. It is used by artists in making oil paints.

There are some 1,000 cultivars varying considerably in the flavour of the fruit; the poorest fruits have a turpentine flavour and are very fibrous, while the best ones are less aromatic, more sweet and contain very little fibre. The local plantings are usually from seedlings, but when the fruit is grown on a commercial scale propagation is by budding, grafting, or stem layering. Commercial growing is found mainly in India and sometimes under irrigation. The plants bear 4–6 years after planting, but flowering is intermittent and on older trees only about 0·1 per cent of the flowers set fruit. A method has not yet been developed for storing the ripe fruit so consumption is mainly local, the excess fruit being fermented to wine, spirits or vinegar.

***M. laurina*** Blume (Monjet Mango). Malaysia. The fruits are eaten locally.

***M. odorata*** Grif. (Kurwini Mango). Malaysia. The fruits are eaten locally.

***M. rumphii*** Pierre. Malaysia. The fruits are eaten locally, raw and as preserves.

***M. zeylanica*** Hook. Ceylon. The fruits are eaten locally.

Mangis butan – Garcinia hombroniana.

MANGLIETIA Blume. Magnoliaceae. 30 spp. S.E.Asia to Sumatra. Trees.

***M. glauca*** Blume = Magnolia blumei. Java. The brown wood is strong, fine-grained and durable. It is used locally for tools, house-building and general construction work.

Mango – Mangifera indica.

Mango, Bachang – Mangifera foetida.

Mango, Ginger – Curcuma amada.

Mango, Grape – Sorindeia madagascariensis.

Mango, Kurwini – Mangifera odorata.

Mango, Monjet – Mangifera laurina.

Mango, Pahutan – Mangifera altissima.

Mango, Wild – Irvingia gabonensis.

Mangold – Beta vulgaris.

Mangold, Fetid – Pectis papposa.

Mangosteen – Garcinia mangostana.

Mangrove, Black – Avicennia nitida.

Mangrove, Red – Rhizophora mangle.

Mangrove, Red-flowered – Lumnitzera coccinea.

Mangrove, White – Avicennia officinalis.

Mangrove, White – Laguncularia racemosa.

Mangure – Terminalia serica.

MANICARIA Gaertn. Palmae. 4 spp. Tropical America, W.Indies. Palm trees.

***M. saccifera*** Gaertn. (Monkey Cap Palm,

Sleeve Palm, Temiche Palm, Turuy, Ubussu). Central and S.America. The leaves are used widely for thatching and the fibres from the leaf stalks are used to make mats. An edible oil (Ubussu Oil) is extracted from the seeds and used locally.

Manicoba – Manihot glaziovii.

Manicoba de Jequie – Manihot dichotoma.

Manicoba de Piauhy – Manihot piauhyensis.

Manicoba da São Francisco – Manihot heptaphylla.

Manicol – Euterpe dominicana.

MANIHOT Mill. Euphorbiaceae. 170 spp. S.W. United States of America to tropical S.America. Shrubs or small trees. The species which yield commercial rubber are not grown extensively although the product is of good quality. The plants do not lend themselves to extended periods of tapping.

***M. aipi*** Pohl. = M. dulcis.

***M. carthaginensis*** (Jacq.) Mill. (Cuadrado, Xaché, Yuca del Monte, Yuquilla). Mexico to Venezuela. The starchy roots are eaten locally and an oil from the seeds is used as an emetic and purgative.

***M. dulcis*** (J. F. Gmel.) Pax. = M. aipi. (Cuacamote dulce, Guhyaga, Yuca dulce). S.America. The plant is cultivated for the roots which are eaten locally as a vegetable, and from which the starch is sometimes extracted. This species is sometimes considered to be a variety of M. esculenta, having a lower hydrocyanic acid content.

***M. esculenta*** Crantz. = M. utilissima. (Cassava, Manioc, Mandioca, Yuca, Aipi). S.America. The plant is grown throughout the tropics, with a world production of about 80 million tons annually. It is a very important staple source of starch in many areas, especially where the soil is poor. The plant yields 10 tons of tubers per acre under poor soil conditions and with minimum cultivations. The tubers contain some 30 per cent starch, but very little protein. There are numerous varieties varying in the time they take to mature (1–2 years) and the amount of cyanogenic glucosides they contain. The "sweet" varieties contain little cyanogenic glucosides, while the "bitter" varieties contain a higher proportion. These are removed by heating, or prolonged exposure to the sun. The root is eaten boiled as a vegetable, pounded into a dough and cooked as cakes, or dried and then cooked. These products are variously called bami, casabe, fou-fou, cassava bread, conaqui farinha. The tubers are also fermented to make local beers and are used for livestock feed. The ground dried roots are sold as Brazilian Arrowroot. The evaporated juice (Cassareep) is used in making sauces for meat; a similar sauce (West Indian Pepperpot) is made from the tubers. The plant is used locally, mainly, but some countries, especially Brazil grow it for the production of commercial starch which is used to make Tapioca, glue, sugar, syrup and acetone. Tapioca is made by precipitating the starch grains from the pulped tubers and dropping the grains on to hot plates. This cooks the starch into the small "pearls" and hydrolyses some of it to sugars. The tapioca is exported and used in temperate countries to make puddings, soups etc. The plants are propagated vegetatively by stem cuttings.

***M. glaziovii*** Muell. Arg. (Manicoba, Mandihoba). Brazil. Cultivated in various parts of the tropics for the rubber (Ceara Rubber). The seeds yield an oil (Ceara Rubber Oil, Manihot Oil).

***M. heptaphylla*** Ule. (Manicoba da São Francisco). Brazil. Yields São Francisco Rubber.

***M. palmata*** Muell. Arg. Brazil. Yields a rubber.

***M. paiuhyensis*** Ule. (Manicoba de Piauhy). Brazil. Gives Piauhy Rubber.

***M. utilissima*** Pohl. = M. esculenta

Manihot Oil – Manihot glaziovii.

Manil – Moronobea coccinea.

Manila Aloë – Agave cantala.

Manila, Batjan – Agathis loranthifolia.

Manila, Borneo – Agathis loranthifolia.

Manila Copal – Agathis alba.

Manila Copal – Agathis loranthifolia.

Manila Elemi – Canarium commune.

Manila Elemi – Canarium luzonicum.

Manila Hemp – Musa textilis.

Manila, Labuan – Agathis loranthifolia.

Manila, Macassar – Agathis alba.

Manila, Maguey – Agave cantala.

Manila, Malenghet – Agathis loranthifolia.

Manila, Menado – Agathis loranthifolia.

Manila, Molucca – Agathis loranthifolia.

Manila, Pontiak – Agathis alba.

Manila, Singapore – Agathis alba.
Manila Tamarind – Pithecellobium dulce.
MANILKARA Adans. Sapotaceae. 70 spp. Tropics.
***M. bidentata*** (DC.) Chev. = Mimusops globosa = Mimusops bidentata. (Balata Tree, Bullet, Bully, Purgio, Quinilla). S.America. The latex (Balata) is used commercially for the soles of shoes and machine belts and as a substitute for chicle in the manufacture of chewing gum. The heavy dark wood is used for building, posts, bridges, railway sleepers, roof shingles and wheel spokes.
***M. dariensis*** (Pitt.) Standl. = Mimusops dariensis. Panama. Yields a balata (Panama Balata).
***M. elengi*** (L.) Chev. Malaysia. Cultivated locally. The flowers are used to scent linen etc. and yield an essential oil used in perfumery. A decoction of the bark or roots is used to reduce fevers and as a mouthwash.
***M. fasciculata*** H. J. L. and G. Mass. (Sauh, Sner). Indonesia, New Guinea. The orange, heavy tough wood is used for heavy construction work and bridge building.
***M. huberii*** (Ducke). Chev. Brazilian Amazon. A balata is extracted from the stem.
***M. kauki*** Dub. = Mimusops kauki. (Khirni, Palai, Talawrinta). Malaysia. The heavy reddish wood is used for beams, furniture and tool handles. The powdered seeds are used in India to treat intestinal worms, leprosy, delirium and eye diseases. The powdered roots and bark, mixed with honey and water are used to treat diarrhoea of children.
***M. sieberi*** (DC.) Dubard = Mimusops sieberi. Trinidad. The trunk yields a balata.
***M. specabilis*** (Pitt.) Standl. = Mimusops spectabilis. Costa Rica. The heavy wood is used for railway sleepers.
Mañio – Saxe Gothaea conspicua.
Manioc – Manihot esculenta.
Manitu de León – Cheirostemon pentadactylon.
Manketti Nut Oil – Ricinodendron rautanenii.
Mankil – Eugenia javanica.
Mankil – Eugenia mananquil.
Manna – Alhagi spp.
Manna – Tamarix mannifera.
Manna Ash – Fraxinus ornus.
Manna of Briançon – Larix decidua.
Manna Eucalyptus – Eucalyptus mannifera.
Manna Grass – Glyceria fluitans.
Manna Gum – Eucalyptus viminalis.
Manna Lichen – Lecanora esculenta.
Manna Oak – Quercus persica.
Manna, Persian – Astragalus adscendens.
MANNIOPHYTON Muell. Arg. Euphorbiaceae. 1 sp. Tropical Africa. Liane.
***M. africanus*** Muell. Arg. The bark fibres are used in the Congo to make fishing nets and cordage.
Mano de Zopilote – Pothomorphe umbellata.
Mano lango – Agave palinaris.
Mano de león – Cheirostemon pentadactylon.
Manubi Mahogany – Trichilia emetica.
Manuku – Leptospermum ericoides.
Manzana de palya – Crataeva tapia.
Manzanella de la Tierra – Isocarpha oppositiflora.
Manzanilla – Crataegus stipulosa.
Manzanillo – Sapium laurocerasus.
Manzanita – Arctostaphylos manzanita.
Manzanita, Pointleaf – Arctostaphylos pungens.
Maori Honeysuckle – Knightia excelsa.
Maoutia Wedd. Urticaceae. 15 spp. Indomalaya, Polynesia. Herbs.
***M. puya*** Wedd. = Boehmeria puya = Urtica puya. Himalayas to Burma. The bark fibres are used locally for making cloth and sails.
Map Lichen – Rhizocarpon geographicum.
Maple, Broad-leaved – Acer macrophyllum.
Maple, Hawthorn – Acer crataegifolium.
Maple, Hedge – Acer campestre.
Maple, Mountain – Acer spicatum.
Maple, Norway – Acer platanoides.
Maple, Plane Tree – Acer pseudoplatanus.
Maple, Queensland – Flindersia brayteyana.
Maple, Red – Acer rubrum.
Maple, Rose – Cryptocarya erythroxylon.
Maple, Silkwood – Flindersia brayleyana.
Maple, Silver – Acer saccharinum.
Maple, Sugar – Acer saccharum.
Maple Syrup – Acer saccharum.
Maple, Vine – Acer circinatum.
Maple, White – Acer saccharinum.
***Mappia origanoides*** (L.) House = Cunila origanoides.
MAPOURIA Aubl. Rubiaceae. 170 spp. Tropics. Shrubs.

***M. psychotrioides*** DC. Tropical Africa. The wood yields a red dye, used in Sierra Leone for dyeing cloth.
Maracaibo Lignum Vitae – Bulnesia arborea.
Maracaibo Resin – Copiafera officinalis.
Marachino – Prunus cerasus.
Maracuja – Passiflora alata.
Maracuja de Serra – Passiflora amethystina.
Maracuja mirim – Passiflora warmingii.
Marane – Aporusa microcalyx.
Marang – Artocarpus odratissima.
Maranháo Nuts – Sterculia chicha.
Maranginon – Dyoxylum densiflorum.
MARANTA L. Marantaceae. 23 spp. Tropical America. Herbs.
***M. arundinacea*** L. (Arrowroot). W.Indies. Cultivated in Old and New World Tropics, but mainly in St. Vincent, Jamaica and Bermuda. The root tubers contain some 80 per cent starch and 1 per cent protein. They are very fibrous so are not grown as a vegetable, but for the commercial production of arrowroot (Arrowroot Starch, St. Vincent Arrowroot, Bermuda Arrowroot). The plants are harvested just before dormancy and ground in water. The starch grains which settle to the bottom are washed repeatedly and then dried to form the arrowroot of commerce. It is used mainly as an invalid food, to feed children and is used to make biscuits and pastries. The plants are propagated by pieces of tuber or stem, or allowed to regenerate from pieces of tuber left in the ground after harvesting.
***M. arundinacea*** Blanco non L. = Donax canniformis.
Marara, Rose – Weinmannia lachnocarpa.
Mararie – Weinmannia lachnocarpa.
Mararo – Protium carana.
MARASMIUS Fr. Agaricaceae. 150 spp. Cosmopolitan. Many species cause diseases of plants. The species mentioned below are used to flavour foods. Their excessive use may lead to poisoning.
***M. alliatus*** (Schäff.) Schr. = M. scorodonius.
***M. caryophylleus*** (Schäff.) schr. = M. oreades.
***M. collinus*** Fr. Switzerland.
***M. crinus-equi*** Kalchbr. Indonesia. The rhizomorphs are used locally for binding jewellery.
***M. dulcis*** Beeli. Tropical Africa.
***M. esculentus*** (Wulf.) Karst. Austria.
***M. helvelloides*** Henn. E. Central Europe.
***M. oreades*** (Bolt.) Fr. = M. caryophylleus. Europe.
***M. piperatus*** Belli. Tropical Africa.
***M. scorodonius*** (Fr.) Fr. = M. alliatus. Temperate Europe.
***M. venosus*** Henn. and Nym. Europe.
Maratitige – Spilanthes acmella.
MARATTIA Sw. Marattiaceae. 60 spp. Hawaii, parts of Australia, New Zealand. Ferns.
***M. alata*** Hook. and Arn. = M. douglasii.
***M. douglasii*** (Presl.) Baker. = M. alata. (Douglas Mulefoot Fern, Pala). Hawaii. Used locally as a famine food and for making a drink used to treat bronchitis and diarrhoea.
***M. fraxinea*** Smith. = M. salicina. New South Wales, Queensland. The starch from the stem pith is eaten by the aborigines.
***M. salicina*** Smith. = M. fraxinea.
Marawayana – Copaifera bracteata.
Marbau – Intsia palembanica.
Marble Wood – Olea paniculata.
Marble Wood, Andamanese – Diospyros marmorata.
Mareh – Euodia sambucina.
Mareme – Glochidion borneense.
Mareposa – Pithecoctenium echinatum.
Mare's tail – Hippuris vulgaris.
Marganita – Arctostaphylos tormentosa.
Margarita del Mar – Ambiosia hispida.
Margosa (Oil) – Azadirachta indica.
MARGYRICARPUS Ruiz. and Pav. Rosaceae. 10 spp. Andes. Small shrubs.
***M. setosus*** Ruiz. and Pav. (Pearl Fruit). Andes. A decoction of the leaves is used locally in Chile as a diuretic.
Maria Balsam – Rheedia acuminata.
Marianne Yellow Wood – Ochrosia mari annensis.
Maricao – Bysonima coriacea.
Maricao – Bysonima spicata.
Marie's Fir – Abies mariensii.
Marigold (Oil) – Calendula officinalis.
Marigold – Tagetes erecta.
Marigold, Big – Tagetes erecta.
Marigold, Fetid – Pectis papposa.
Marigold, Marsh – Caltha palustris.
Marigold, Murr – Bidens tripartitus.
Marigold, Sweet – Tagetes lucida.
Marihuana – Cannabis sativa.
Marilópez – Turnera ulmifolia.
Maripa Fat – Maximiliana maripa.
Mariposa Lily – Calochortus nuttallii.
Mariposa, Sagebrush – Calochortus gunnisonii.

Mariquita – Gymnosperma glutinosum.
MARISCUS Vahl. Cyperaceae. 200 spp. Tropics and subtropics. Herbs.
***M. jamaicensis*** (Crantz.) Britt. = Cladium effusum.
***M. sieberianus*** Nees. Throughout tropics. The plant is used to treat intestinal worms in parts of Sumatra.
***M. umbellatus*** Vahl. Old World tropics. The rhizomes are eaten roasted in parts of Africa, where the plant is cultivated.
Marjoram Oil – Origanum majorana.
Marjoram, Oil of – Thymus mastichina.
Marjoram, Pot – Origanum vulgare.
Marjoram, Sweet – Origanum majorana.
Marjoram, Winter – Origanum heraclecticum.
Marjul – Iris ensata.
MARKHAMIA Seem. ex Baill. Bignoniaceae. 12 spp. Tropical Africa, Asia. Shrubs or trees.
***M. acuminata*** (Klotzsch) Schum. (Musikinyati, Umpetawhale). Tropical Africa. The roots are used locally in Malawi to treat backache.
***M. lanata*** Schum. Tropical Africa. The yellowish wood is used locally for general construction work, bridges and furniture.
Marking Nut, Cashew – Semecarpus anacardium.
Marmalade Box – Genipa americana.
Marmalade Plum – Calocarpum mammosum.
Marmelle Oil – Aegle marmelos.
Maroñon de la Maestre – Talauma plumieri.
Marquesan Fig – Ficus marquesensis.
Marquesas – Pelea fatuhivensis.
Marrons Glacées – Castanea sativa.
Marrow, Vegetable – Cucurbita pepo.
MARRUBIUM L. Labiatae. 40 spp. Temperate Eurasia, Mediterranean. Herbs.
***M. indicum*** Blanco = Hyptis suaveolens.
***M. vulgare*** L. (Hoarhound). Europe, introduced to N.America. The leaves and tops of the plant are used in home remedies, e.g. syrups, teas, etc. to treat colds, sore-throats etc. The plant is cultivated in U.K. and used commercially for this purpose. The leaves also yield an essential oil used in making liqueurs.
MARSDENIA R. Br. Asclepiadaceae. 5–10 spp. Tropical Africa, Asia, America. Climbers.
***M. cundurango*** Reich. f. = M. reichenbachii.
***M. reichenbachii*** Triana = M. cundurango. (Condorvine). Ecuador, Colombia. The dried bark (Cundurango Bark) is used to make a wine (Cundurango Wine) which is used to treat stomach ailments.
***M. tenacissima*** Wight. and Arn. (Rajmahal Hemp, Jiti, Tomgus). Sub-Himalayas, Bengal. The bark fibres are strong and elastic. They are used for cordage, nets and bow-strings. The latex is used to make India Rubbers for removing pencil marks and is sometimes mixed with other latexes.
***M. tinctoria*** R. Br. Malaysia. The plant is cultivated locally for the deep blue dye extracted from it.
***M. tomentosa*** Morr. and Decne. Japan. The stems are used locally to make bow-strings and cordage.
***M. verrucosa*** Decne. (Bokalahy). Madagascar. Yields a good quality rubber.
***M. zimapanica*** Hemsl. (Tequampatli). Mexico. The root is mixed with meat as a poison for coyotes.
Marsh Buckbean – Menyanthes trifoliata.
Marsh Fern – Acrostichum aureum.
Marsh Fleabane – Pluchea sericea.
Marsh Mallow – Althaea officinalis.
Marsh Marigold – Caltha palustris.
Marsh Rosemary – Statice limonium.
Marsh Samphire – Salicornia herbacea.
Marsh Woundwort – Stachys palustris.
***Marshallia aliena*** Spreng. = Heterothalamus brunoides Less.
MARSILEA L. Marsiliaceae. 60 spp. Tropics and temperate. Herbs.
***M. drumondii.*** A. Br. Australia. The sporocarps are eaten by the aborigines and the leaves are eaten by livestock.
***Marsippospermum grandiflorum*** Hook. = Rostkovia grandiflora.
Martagon Lily – Lilium martagon.
Martinique Plum – Flacourtia inermis.
Martinique Prickly Ash – Zanthoxylum martinicense.
MARTYNIA L. Martyniaceae. 3 spp. Mexico and surrounding area. Herb. The systematics of this genus are confused.
***M. louisiana*** Mill. = M. probocidea.
***M. montevidensis*** Cham. Argentina to Brazil. The seeds are used to make soothing poultices in Argentina.
***M. parviflora*** Wooten (Devil's Claw). S.W. N.America, Mexico. The fruits are used by the local Indians to weave designs into their baskets.
***M. probocidea*** Glox. = M. louisiana. (Double Claw, Ram's Horn, Unicorn

Plant). N.America to Mexico. The young fruits are used in pickles. The plant is sometimes cultivated.

Marula, Bastard – Kirkia acuminata.

Marumi Kumquat – Fortunella japonica.

***Marumia*** Blume = Mcarolenes.

Maruwa – Melia excelsa.

Marvel of Peru – Mirabilis jalapa.

Marvey – Olea paniculata.

Maryland Figroot – Scrophularia marilandica.

Masa – Tetragastris balsamifera.

Masakwe – Albizia antunesiana.

Mascada de Huyaquil – Otoba gordoniifolia.

MASCARENHASIA A. DC. Apocynaceae. 10 spp. Tropical E.Africa, Madagascar. The species mentioned are minor sources of a good rubber.

***M. arborescens*** A. DC. Madagascar. Rubber called Majunga noir.

***M. elastica*** Schum. E.Africa. Mgoa Rubber.

***M. geayi*** Cost. and Poir. (Kokomba). Madagascar.

***M. lanceolata*** Decne. (Barabanja). Madagascar.

***M. lisianthiflora*** A. DC. Madagascar.

***M. longifolia*** Jum. Madagascar.

***M. mangorensis*** Jum. and Pierre. Madagascar.

Massandari – Callicarpa lanata.

***Massoia aromatica*** Becc. = Cinnamomum massoia.

Massoia Bark – Cinnamomum massoia.

Masson Pine – Pinus massoniana.

Mastic – Sideroxylon mastichodendron.

Mastic, American – Schinus molle.

Mastic, Bombay – Pistacia mutica.

Mastic, Chios – Pistacia lentiscus.

Mastic Thyme – Thymus mastichina.

MASTOCARPUS Kuetz. Gigartinaceae. Red Algae.

***M. klenzianus*** Kuetz. Malaysia. The plant yields agar. They are eaten locally.

Masterwort – Peucidanum ostruthium.

Mart Bean – Phaseolus aconitifolius.

Mata ajam – Ardisia crispa.

Mata ajam – Clerodendrum buchananii.

Mata tórsalo – Funastrum clausum.

Matapalo – Ficus involuta.

MATAYBA Aubl. Sapindaceae. 50 spp. Warm America. Trees.

***M. apetala*** Radlk. = Ratonia apetala. W.Indies. The light-coloured wood is used for house doors and window frames.

***M. domingensis*** (DC.) Radlk. (Ngera Lora). W.Indies. The pinkish heavy wood is used for house floors, furniture, agricultural implements, tool handles, carts and waggons.

***M. purgans*** Radlk. Amazon Basin. The seeds (Semen Mataybae) produce an oil which is used as a laxative.

Matchwood – Didymopanax morototoni.

Maté, Yerba – Ilex paraguensis.

Matgrass, Chinese – Cyperus tegetiformis.

Matico, Hachogue – Piper palmeri.

MATISIA Humb. and Bonpl. Bombacaceae. 35 spp. Tropical S.America. Trees.

***M. cordata*** Humb. and Bonpl. (Sapote). S.America. The fruit pulp has a pleasant flavour and is eaten locally.

Matitana – Neodypsis tanalensis.

Matlaquahitl – Guaiacum coulteri.

Mato – Guatteria caribaea.

Matorral – Acacia berlandieri.

Ma-tou-ling – Aristolochia kaenferi.

MATRICARIA L. Compositae. 40 spp. Europe, Mediterranean; 10 spp. S.Africa, 2 spp. W. N.America. Herbs.

***M. chamomilla*** L. (Chamomille). Europe to Afghanistan. The plant is cultivated in the Old and New World. The leaves are used to make a tea, used in home remedies for intestinal worms and as a mild sedative. The flower heads are used medicinally as a nerve tonic, to induce sweating and to treat stomach upsets. They are also used as a hair rinse to give the hair a golden sheen. An essential oil is extracted from the plant and is used in making perfumes, scenting shampoo powders, making liqueurs and scenting tobacco. The oil from Germany is used to dissolve platinum chloride during the process of coating glass and china with platinum.

***M. discoidea*** DC. Asia, introduced to Europe and America. Used as M. chamomilla. The oil is called Oleum Chamomilla or Aqua Chamomilla.

***M. multiflora*** Fenzl. = Tanacetum multiflorum. S.Africa. The powdered plant, or an infusion of it is used locally as a tonic and to treat intestinal worms.

***M. parthenium*** L. = Chrysanthemum parthenium.

Matrimony Vine – Lycium halamifolium.

Matsu – Garcinia tinctoria.

Matsu take – Armillaria matsutake.

Matsub – Ahnfeltia plicata.

Mattipaul – Ailanthus malabarica.

Mâu cho – Kaema corticosa.

MAUGHANIA Jaume St. Hil. Leguminosae. 35 spp. Indomalaya, N.Australia. Herbs.

***M. congesta*** Roxb. Tropical Asia. Yields a bright orange dye (Waras) used for dyeing silk.

Maul Oak – Quercus chrysolepis.

MAURITIA L. f. Palmae. 6 spp. Tropical America, W.Indies. Palm trees.

***M. carana*** Wall. Tropical America. The young leaves produce a fibre used locally for cordage. The fruits are made into a fermented beverage and the trunks are used for building.

***M. flexuosa*** L. f. (Moriche). S.America. The leaves produce a fibre used for cordage and for thatching. The leaf sheaths are used to make sandals. The fruits are extracted for an edible oil and are made into a drink. The sap is fermented into a beverage and the starch from the stem made into a sago. The stems are used to make bridges and posts, while the seeds are used as buttons etc.

***M. vinifera*** Mart Brazil. The leaves produce a fibre used locally for cordage. The fruits are fermented to make a wine and the fruit pulp is eaten as a side dish.

Mauritius Ebony – Diospyros tesselaria.

Mauritius Grass – Panicum molle.

Mauritius Hemp – Furcraea gigantea.

Mauritius Raspberry – Rubus rosaefolius.

Mauritius Terminalia – Terminalia mauritiana.

Mavokeli Rubber – Pentopetia elastica.

MAXIMILIANA Mart. Palmae. 10 spp. Tropical S.America, W.Indies. Palm trees.

***M. caribaea*** Griseb. (Cocorite Palm). W.Indies, N. S.America. The leaves are used for thatching.

***M. maripa*** Drude. Guyana. An edible fat (Maripa Fat) is extracted from the seeds and used locally.

***M. regia*** Mart. Tropical America, especially Brazil. The seeds yield an edible oil.

***M. vitifolia*** Krug. and Urb. = Cochlospermum vitifolium.

MAXWELLIA Baill. Bombacaceae. 1 sp. New Caledonia. Tree.

***M. lepidota*** Baill. The yellow flexible wood is used for tool handles.

May – Crataegus oxyacantha.

May Apple – Podophyllum peltatum.

May Apple – Rhododendron nudiflorum.

Mayberry. See Rubus strigosus.

Mayflower – Epigaea repens.

May Pop – Passiflora incarnata.

Mây sât – Calamus salicifolius.

Mayteno – Maytenus ilicifolia.

MAYTEUS Molina. Celastraceae. 225 spp. Tropics. Trees.

***M. boaria*** Molina = M. chilensis. Chile. An infusion of the leaves is used locally to reduce fevers.

***M. chilensis*** DC. = M. boaria.

***M. ilicifolia*** Mart. = Nemopanthus andersonii. (Congoasa, Mayteno). S.America. An infusion of the leaves is used as a tonic and to treat stomach and intestinal complaints.

***M. obtusifolia*** Mart. Brazil. The hard wood is not durable but is used for general construction work and carpentry.

***M. senegalensis*** (Lam.) Exell. = Celastrus senegalensis = Gymnosporia senegalensis. (Bazimo, Confetti Tree, Kisambila, Mmoza, Umiviesa). Africa, except N. and W. The buds are used locally to treat vaginal discharges, an infusion of the roots is used as an antiseptic on wounds and in Zambia, rootchips in beer are used as an aphrodisiac. The branches are rubbed together to make a fire.

Mayweed – Anthemis cotula.

MAZUS Lour. Scrophulariaceae. 20 spp. E. and S.E.Asia, Indomalaya, Australia. Herbs.

***M. bicolor*** Benth. = M. rugosus.

***M. rugosus*** Lour. = M. bicolor = Lindernia japonica. Throughout range. A decoction of the plant is used in Cambodia as a tonic and to reduce fevers. In Java the decoction is used to treat snakebites.

Mazzard – Prunus avium.

Mba-y abambo – Cratosperma senegalense.

Mbala – Cissus rubiginosa.

Mbang – Chlorophora excelsa.

M'banga – Lygodium smithianum.

Mbe-mundju – Lasianthera africana.

Mbelabela – Eriocoelum microspermum.

Mbocaya Palm (Oil) – Acrocomia totai.

Mbongum bongi – Lonchocarpus goosenii.

Mbusa – Eurypetalum batesii.

Mbwetomasa – Polygonum salicifolium.

Mdagu – Adenium obesum.

Mè trè – Alpinia globosa.

Meadow Fescue – Festuca elatior.

Meadow Foxtail – Alopecurus pratensis.

Meadow Leek – Allium canadense.

Meadow Pasque Flower – Anemone pratensis.

Meadow Rue, Columbine – Thalictrum aquifolium.
Meadow Saffron – Colchicum autumnale.
Meadow Sage – Salvia pratensis.
Meadowsweet, Oil of – Filipendula ulmari.
Mealy Blister Lichen – Physcia pulverulenta.
Mealy Ramalina – Ramalina farinacea.
Meanti hitan – Parashorea lucida.
Mecca Aloë – Aloë abyssinica.
Mecca Balsam – Commiphora opobalsamum.
Mecca Myrrh – Commiphora opobalsamum.
Mecca Senna – Cassia angustifolia.
Medang poodi – Teijmanniodendron pteropodum.
Medang seroo – Schima bancana.
Medar – Dendrocalamus strictus.

MEDEMIA Württemb. ex H. Wendl. Palmae. 2 spp. N. tropical Africa, 1 sp. Madagascar. Palm trees.

***M. argun*** Benth. and Hook. = Hyphaene argun.

***M. nobilis*** Drude. (Satrabe, Satranabe). Madagascar. The leaves are used locally to make baskets etc. and paper. The starch from the stem is eaten locally.

MEDEOLA L. Trilliaceae. 1 sp. N.America. Herb.

***M. virginiana*** L. (Indian Cucumber Root). The rhizomes were eaten by the local Indians.

Medic, Barrel – Medicago tribuloides.
Medic, Black – Medicago lupulina.
Medic, Purple – Medicago sativa.
Medic, Tree – Medicago arborea.

MEDICAGO L. Leguminosae. 100 spp. Temperate Eurasia, Mediterranean, S. Africa. Herbs. The species mentioned below are used as pasture and hay for feeding livestock.

***M. arabica*** (L.) All. (Spotted Bur Clover). S.Europe, N.Africa to Asia.

***M. arborea*** L. (Tree Medic). Mediterranean, N.Africa to Asia.

***M. falcata*** L. (Yellow-flowered Alfalfa). Europe, to N.Africa into Asia. Cultivated in Old and New World.

***M. hispida*** Gaertn. (Bur Clover). Mediterranean.

***M. elupulina*** L. (Hop Clover, Nonsuch, Black Medic). Temperate Europe, to Asia and N.Africa. Cultivated also in America.

***M. orbicularis*** (L.) All. (Button Clover). Mediterranean.

***M. platycarpa*** Trau. E.Europe. Drought and cold resistant, useful in dry areas.

***M. ruthenica*** Trau. Central Russia through Korea to China.

***M. sativa*** L. (Alfalfa, Lucerne, Purple Medic, Chilean Clover, Spanish Alfalfa, Berseem). Temperate Europe, Asia, N.Africa. This is one of the major forage crops of the world, some 60 million acres being sown annually and about half of this in America. It yields about 6 tons per acre under good conditions (about 25 per cent of this is protein), but as it is drought-resistant, it is grown under drier conditions than can be tolerated by other legumes. Alfalfa is often grown under irrigation or as a mixed pasture with grasses. It is one of the oldest forage crops, being grown as such by the Medes and Persians who introduced it to Greece. It was taken to America by the Spaniards but spread slowly in use until it was introduced to California from Chile during the gold rush (1858). There are several varieties. Although grown primarily for fodder, the leaves are also used as a commercial source of chlorophyll, the flowers yield a good honey and the seeds contain a drying oil which could be used in the manufacture of paints.

***M. scutellata*** (L.) Willd. (Snail Clover). N.Africa into Asia Minor.

***M. tribuloides*** Desv. (Barrel Medic).

***Medicia elegans*** Gardn. = Gelsemium elegans.

Medicinal Rhubarb – Rheum officinale.
Medicine Tree – Codonocarpus cotinifolia.

MEDINILLA Gaudich. Melastomataceae. 400 spp. Tropical Africa, Madagascar, Indomalaya, Pacific Islands. Vines.

***M. hasseltii*** Blume. Malaysia. In Sumatra the young leaves are eaten as a vegetable with rice and the fruits are eaten with fish.

Mediterranean Saltbush – Atriplex halimus.
Medlar – Mespilus germanica.
Medlar, Chirinda – Vangueria esculenta.
Medlar, Japanese – Eriobotrya japonica.
Medlar, White – Vangueriopsis apiculata.
Medlar, Wild – Vangueriopsis lancifolia.

***Megabarea trillesii*** Pierre = Spondianthus preussii.

***Megastroma*** Coss. and Durieu. = Eritrichium.

Mehndi (Oil) – Lawsonia alba.

***Meibomia purpurea*** (Mill.) Vail. = Desmodium tortuosum.

Meiwa Kumquat – Fortunella crassifolia.

Melcoacha – Opuntia megacantha.

MELELEUCA L. Myrtaceae. 1 sp. Indomalaya; 100 spp. Australia, Pacific Islands. Shrubs or small trees. Many species yield oils by steam distillation of the stems and leaves.

***M. alternifolia*** Cheel. E.Australia. The oil is very germicidal and is used in medicine and dentistry.

***M. bracteata*** F. v. Muell. (Black Tea Tree). New South Wales, Queensland. The light oil is used in perfumery and as an insect repellent.

***M. ericifolia*** Smith = M. heliophila = M. nodosa. Australia. The tree is planted to prevent the erosion of muddy shores.

***M. leucadendron*** L. (Cajaput Tree). Malaysia to Australia. The oil (Cajatput Oil) smells like camphor and is used medicinally to aid digestion, as an antiseptic and to treat intestinal worms. The hard mottled wood is used for posts and shipbuilding.

***M. linariifolia*** Sm. Australia. The oil is used for antiseptic soaps.

***M. linariifolia*** var. ***trichostachys*** Sm. Australia. The shrub is planted to prevent erosion of coastal sands.

***M. nodosa*** Link. = L. ericifolia.

***M. parviflora*** Lindl. = M. preissiana var. leiostachya.

***M. preissiana*** Schan. var. ***leiostachya*** Schan. Australia. The plant is used to prevent erosion of coastal sands.

***M. suaveolens*** Gaertn. = Tristania suaveolens.

***M. uncinata*** R. Br. (Tea Tree). Australia. The leaves are chewed by the natives to relieve catarrh.

***M. viridiflora*** Brogn. and Gris. New Caledonia. The oil (Niaouli Oil) is produced mainly in New Caledonia and is used medicinally in treating coughs, rheumatism and neuralgia.

***M. wilsonii*** F. v. Muell. Victoria, S.Australia. The oil is used as that from M. leucadendron.

Melamon – Drypetes ovalis.

MELANDRIUM Rochl. Caryophyllaceae. 100 spp. N.Hemisphere. Herbs.

***M. album*** (Mill.) Garcke. = M. dioicum = M. diurnum. = Lychnis diurna. (White Campion). Europe to N.Africa, Asia Minor and Siberia. The roots were used locally for washing clothes.

***M. dioicum*** Schinz. = M. album.

***M. diurnum*** Fries. = M. album.

***M. rubrum*** Garcke. (Red Campion). See M. album.

MELANORRHOEA Wall. Anacardiaceae. 20 spp. S.E.Asia, Malaysia, Borneo. Trees.

***M. inappendiculata*** King. Malaysia to India. The resin is used as a black ink and the leaves are sometimes used as a criminal poison.

***M. laccifera*** Pierre. Indo-China. Yields a lacquer. See M. usitata.

***M. usitata*** Wall. (Burmese Varnish Tree, Theetsee). Burma, Thailand, Manipur. The sap is extracted by tapping the stem. This blackens on exposure to give the valuable Burmese Lacquer, used as a varnish on woodwork and to cement Burmese glass mosaics. The varnish is exported.

MELANOXYLON Schott. Leguminosae. 3 spp. Tropical S.America. Trees.

***M. amazonicum*** Ducke = Recordoxylon amazonicum.

***M. brauna*** Schott. (Brauna, Grauna). Brazil. The hard dark wood is heavy, tough and strong. It is used for bridges, beams, posts, railway sleepers etc. The bark is used for tanning.

MELASTOMA L. Melastomataceae. 70 spp. S.China, Indomalaya, Pacific. Shrubs.

***M. albicans*** Swartz = Miconia albicans.

***M. argentea*** Swartz = Miconia argentea.

***M. coccinium*** Vell. = Clidemia blepharoides.

***M. malabathricum*** L. (Singapore Rhododendron). E.India, Malaya to Australia. In Malaya the powdered leaves are sprinkled on small pox sores to prevent marking. The wood tar is used to blacken the teeth and the foliage is used to feed silkworms.

***M. petiolare*** Schlecht. = Clidemia deppeana.

***M. xalapense*** Bonpl. = Conostegia xalapensis.

Melebekan – Shorea lepidota.

MELIA L. Meliaceae. 2–15 spp. Old World tropics and subtropics. Trees.

***M. azadirachta*** L. = Azadirachta indica.

***M. azedarach*** L. (Chinaberry Tree, Umbrella Tree). S.W.Asia, Persia and Asia Minor. Cultivated in Old and New World tropics. In the Middle East the juice from the leaves is used as a diuretic, to treat intestinal worms and to control menstruation. The bark is also used to treat intestinal worms. The ground fruits are used as an insecticidal powder and the seeds are made into necklaces etc. The wood is used for making furniture.

***M. dubia*** Cav. Malaysia. The red-brown wood is used for building, furniture-making and tea chests.

***M. excelsa*** Jack. (Maruwa, Surian bawang). Indonesia, New Guinea. The medium hard pinkish wood is insect resistant and easily worked. It is used for general carpentry, furniture and carving.

MELIANTHUS L. Melianthaceae. 6 spp. S.Africa. Shrubs or trees.

***M. comosus*** Vahl. S.Africa. The root-bark and leaves are used locally for treating snake-bites.

***M. major*** L. S.Africa. A decoction of the leaves is used locally to wash wounds.

MELICOCCUS P. Br. Sapindaceae. 2 spp. Tropical America, W.Indies. Trees.

***M. bijugatus*** Jacq. (Mamoneillo, Spanish Lime). Tropical America. The fruits are juicy with a pleasant sub-acid flavour. They are sold locally. A decoction of the bark is used locally to treat dysentery.

MELICOPE J. R. and G. Forst. Rutaceae. 70 spp. Indomalaya, warm Australia, New Zealand, Pacific Islands. Trees.

***M. ternata*** Forst. New Zealand. The trunk yields a gum, used by the natives as chewing gum.

MELICYTUS J. R. and G. Forst. Violaceae. 5 spp. New Zealand, Fiji, Norfolk Islands. Trees.

***M. ramiflorus*** Forst. (Hinahina, Mahoe). New Zealand. The charcoal is used to make gun powder and the bark for treating burns.

MELILOTUS Mill. Leguminosae. 25 spp. Temperate and subtropical Eurasia, Mediterranean. Herbs.

***M. alba*** Desr. (White Sweet Clover). Europe, Asia. Cultivated occasionally as food for livestock. It was introduced to U.S.A. about 1700 where it is used, especially in Texas as a livestock food and soil-builder.

***M. altissima*** Thuill. Europe, temperate Asia. Sometimes used locally as fodder. The leaves are used to flavour Swiss Green Cheese and to scent tobacco and cosmetics.

***M. bungiana*** Boiss. = M. suaveolens.

***M. elegans*** Sal. Mediterranean, through Abyssinia to E.Africa. The leaves are sometimes used locally to flavour butter and are used by Abyssinian women to scent their hair.

***M. graveolens*** Bunge. = M. suaveolens.

***M. macrorhiza*** Pers. Europe to Asia. The leaves are used as a vegetable (Yeh-hua tsen) in China, where the seeds are used to treat colds.

***M. officinalis*** (L.) Lam. (Yellow Sweet Clover). Europe, temperate Asia, N.Africa. Introduced into N.America in about 1700 where it is used as a fodder crop and as a soil improver. It contains coumarin which makes it less palatable to livestock than most other forage legumes. The leaves are used in Europe for flavouring cheese, to scent tobacco, snuff and as a moth repellent in clothes. The flowers are a good source of honey.

***M. ruthenica*** Ser. = M. wolgica Poir. S.Russia. The roots are eaten locally.

***M. suaveolens*** Ledeb. = M. bungiana = M. graveolens. E.Asia to Indochina. A decoction of the whole plant is used in Vietnam to treat eye-diseases.

***M. wolgica*** Poir. = M. ruthenica.

MELINIS Beauv. Graminae. 1 sp. tropical S.America, W.Indies; 17 spp. tropical and S.Africa, Madagascar. Grasses.

***M. minutiflora*** Beauv. Tropical America. Used as fodder for livestock in Australia and Rhodesia. The plant yields an oil (Oil of Efwatakala Grass) which is used as a repellent for mosquitoes.

MELIOSMA Blume. Meliosmaceae. 100 spp. Warm Asia, America. Trees.

***M. buchananiifolia*** Merr. S. China. The bark is used as incense by the Chinese.

***M. herbertii*** Rolfe. (Aquacatillo). Tropical America. The wood is moderately hard and easy to work with a pleasant brown orange colour. It is susceptible to termites and decay, but can be used successfully for interiors of houses, boxes etc. furniture and decorative work.

MELISSA L. Labiatae. 3 spp. Europe through to Persia and Central Asia. Herbs.

***M. clinopodium*** Benth. = Calamintha clinopodium.

***M. officinalis*** L. (Common Balm). Mediterranean through to S.W.Persia. The plant is cultivated for the leaves which yield the aromatic Oil of Balm, used in perfumery, ointments and furniture creams.

The leaves are used to flavour liqueurs, salads, soups and vinegar; they are also used to make Balm Tea, a household remedy for headaches and toothache.

***M. vulgaris*** Trev. = Calamintha clinopodium.

MELITTIS L. Labiatae. 1 sp. Europe. Herbs.

***M. grandiflora*** Sm. = M. melissophyllum.

***M. melissophyllum*** L. = M. grandiflora = M. sylvestris. (Bastard Balm). A decoction of the leaves is used in a wide variety of home remedies for purifying the blood, to heal wounds, a diuretic, to control menstruation, a mild sedative and to treat diarrhoea.

***M. sylvestris*** Lam. = M. melissophyllum.

***Melocalamus*** Benth. = Dinochloa.

MELOCANNA Trin. Graminae. 2 spp. Indomalaya. Bamboo-like grasses.

***M. bambusioides*** Trin. (Berry-bearing Bamboo, Metunga, Muli, Terai Bamboo). E.Bengal, Burma. The stems are used for building and making mats. The fruits are the size of an apple and are eaten roasted locally.

MELOCHIA L. Sterculiaceae. 75 spp. Tropics. Shrubs or trees.

***M. arborea*** Blanco = M. umbellata.

***M. corchorifolia*** L. The stem yields a cordage fibre. The leaves are eaten as a vegetable in India and in parts of Africa the root bark is chewed to heal sore lips.

***M. indica*** Grey. = M. umbellata.

***M. umbellata*** Stapf. = M. arborea = M. indica = Visenia indica (Boosi, Betenook, Tangkal bintenoo, Wesnoo). Malaysia, Java. The bark yields a cordage fibre. The wood is not durable and is used locally for house-building and fuel, also tea chests. The leaves are fermented with a mixture of coconut cake and maize grain to make Tempe, which is eaten locally. It can cause poisoning.

MELODINUS J. R. and G. Forst. Apocynaceae. 50 spp. Indomalaya, Australia, Pacific. Trees.

***M. ovalis*** Boerl. Borneo. The latex is used as rubber.

***M. pulchrinervius*** Boerl. Borneo. The latex is used to adulterate gutta-percha.

***Melodorum*** Hook. f. and Thoms. = Fissistigma.

Melon – all forms of Melon are Cucumis melo, except

Melon, Chinese Preserving – Benincasa hispida.

Melon, Oriental Pickling – Cucumis conomon.

Melon Tree – Carica papaya.

Melon zapote – Carica papaya.

MELOTHRIA L. Cucurbitaceae. 10 spp. New World. Vines.

***M. fluminensis*** Gardn. Brazil. The seeds are used locally to treat digestive complaints in cattle.

***M. heterophylla*** Cogn. = Zehneria connivens.

***M. maderaspatana*** Cogn. = Mukia scabrella.

Membaloong – Calophyllum soulattrii.

Membre – Acacia berlandieri.

Membrillo – Amelanchier denticulata.

MEMECYLON L. Memecylaceae. 300 spp. Tropical Africa, Asia, Australia, Pacific Islands. Trees.

***M. costatum*** Miq. Malaysia. The tough wood is durable and is used locally for furniture and house construction.

***M. edule*** Roxb. = M. tinctorium. Coast of Indian Ocean, Sumatra. The heavy hard wood is used locally for house construction. The fruit pulp is eaten and the leaves yield a yellow dye.

***M. polyanthemos*** Hook. f. Tropical Africa. The wood is used in Liberia to make furniture and for house construction.

***M. tinctorium*** Koenig = M. edule.

MENABEA Baill. Periplocaceae. 1 sp. Madagascar. Tree.

***M. venenata*** Baill. (Kissoumpo, Ksopo). The root is used locally as a criminal ordeal poison.

Menedo Copal – Agathis loranthifolia.

Mengilas – Parastemon urophyllum.

Mengkaras – Aquilaria malaccensis.

Menkulang – Heretiera simplivifolia.

Meni Oil – Lophira alata.

MENISPERMUM L. Menispermaceae. 3 spp. Temperate E.Asia, Atlantic N.America, Mexico. Vines.

***M. canadense*** L. (Common Moonseed). E. N.America. The roots are used medicinally as a tonic to improve the appetite and as a diuretic.

***M. crispum*** L. = Tinospora crispa.

***M. edule*** Vahl. = Cocculus cebatha.

***M. japonicum*** Thunb. = Stephania japonica.

***M. rimosum*** Blanco = Tinospora rumphii.

***M. tinctorum*** Spreng. = Fibraurea tinctoria.

Mentaos – Dypetes ovalis.

Mentaos – Wrightia pubescens.

MENTHA L. Labiatae. 25 spp. N.Temperate, S.Africa, Australia. Herbs.

***M. aquatica*** L. = M. citrata. (Bergamot

Mint). S.Europe to N.Africa, Asia. Yields a lemon-scented essential oil used in perfumery.

***M. arvensis*** L. var. ***piperascens*** Malin. Japan, China. A variety of Field Mint (M. arvensis) found in temperate Europe, Asia and America, which yield an essential oil containing 80–90 per cent menthol. This is extracted and used in pharmaceuticals, toothpastes and flavouring cigarette tobaccos. The plant is grown commercially in China, Japan and Brazil.

***M. auriculata*** L. = Dysophylla auricularia.

***M. canadensis*** L. (American Wild Mint). N.America. The leaves were eaten by the local Indians and are now used for flavouring. The plant is also cultivated for the commercial production of thymol and pulegone which are isolated from the essential oil distilled from the leaves.

***M. citrata*** Ehrh. = M. aquatica.

***M. crispa*** auct. non L. = M. spicata var. crispa.

***M. cunninghamia*** Benth. New Zealand. A decoction of the plant is used occasionally to induce sweating.

***M. foetida*** Burm. f. = Dysophylla auriculata.

***M. gentilis*** L. (American Apple Mint). Cultivated hybrid between M. spicata and M. arvensis. It is sometimes seen growing as an escape in Europe and America. The plant is used for flavouring.

***M. longifolia*** (L.) Huds. = M. sylvestris (English Horsemint, Horsemint). Temperate Europe to N.Africa, introduced to N.America. A decoction of the leaves is used to treat stomach complaints. It is antiseptic and is used to wash wounds. The decoction is also taken to relieve pains, especially headaches. This is probably the "mint" of the Bible; it is cultivated extensively in the Middle East.

***M. piperata*** L. (Peppermint). Possibly a hybrid between M. aquatica and M. spicata. It is cultivated in Europe and N.America for the essential oil (Peppermint Oil) which is used for flavouring confectionery, chewing gum and the liqueur Crême de Menthe. The oil and the dried leaves are used medicinally to treat stomach complaints and as a stimulant. It is also a source of menthol.

***M. pulegium*** L. (European Pennyroyal). Europe, Mediterranean to Middle East. The plant is cultivated for the oil (Oil of Pennyroyal) which is used medicinally to treat stomach complaints. The dried leaves were used in home remedies for the same purpose. The oil is also used to scent soaps and to manufacture menthol.

***M. rotundifolia*** (L.) Huds. = M. spicata L. var. rotundifolia. (Apple Mint, Woolly Mint). Europe, introduced to N.America. The leaves are sometimes used for flavouring. The plant is occasionally cultivated.

***M. spicata*** L. = M. viridis (Spearmint). Europe, N.America. Cultivated in both these areas for the essential oil (Spearmint Oil) which is used for flavouring sweets, chewing gum, toothpastes etc. The dried leaves are used medicinally to treat stomach complaints and as a stimulant.

***M. spicata*** L. var. ***crispa*** Schrad. = M. crispa auct. non L. Europe. It is cultivated for the essential oil (German Spearmint Oil). See M. spicata.

***M. spicata*** L. var. ***rotundifolia*** L. = M. rotundifolia.

***M. sylvestris*** L. = M. longifolia.

***M. viridis*** L. = M. spicata.

MENTZELIA L. Loasaceae. 70 spp. Tropical and subtropical America, W.Indies. Herbs.

***M. albicaulis*** Dougl. W. to S.E. U.S.A. The seeds are made into a flour by the local Indians.

MENYANTHES L. Menyanthaceae. 1 sp. N.Temperate. Herb.

***M. cristata*** Rosb. = Nymphoides cristata.

***M. trifoliata*** L. (Marsh Buckbean). The leaves contain the glucoside menyanthin. They are used to flavour beer in Scandinavia, as an emergency food in Russia, and when dried are used as a tea substitute and a tonic.

MENZIESIA Sm. Ericaceae. 7 spp. N.Temperate Asia, America. Shrubs.

***M. ferruginea*** Sm. (False Huckleberry). N.W. America. The berries are eaten by the local Indians.

Merang – Pithecellobium minahasse.

Meranti cheriak – Shorea pauciflora.

***Meratia praecox*** Rehd. and Wils. = Chimonanthus praecox.

Merawan, Nodding – Hopea nutans.

MERCURIALIS L. Euphorbiaceae. 8 spp. Mediterranean, temperate Eurasia to N.Siam. Herbs. The seeds of this genus contain an oil which dried more rapidly than linseed oil and could be of commercial importance.

***M. leiocarpa*** Sieb. and Zucc. Japan. The leaves yield a blue dye, used locally for cloth printing.

***M. perennis*** L. (Dog's Mercury). Europe, N.Africa, Middle East. The leaves yield a good yellow dye.

MERENDERA Ram. Liliaceae. 10 spp. Mediterranean to Abyssinia and Afghanistan. Herbs.

***M. persica*** Boiss. and Kotschy. Persia, Afghanistan. The corms are used to treat rheumatism in Persia.

Merendon Virola – Virola merendonis.

Mergalang – Alstonia angustifolia.

Merika – Grewia trinervata.

Merissa – Calotropis procera.

MERISTOTHECA J. Ag. Solieriaceae. Red Algae.

***M. papulosa*** (Mont.) J. Ag. Indian and Pacific Oceans. Eaten in Japan and China. Imported from E.Indies and Japan.

Merkus Pine – Pinus merkusii.

***Merremia mammosa*** Hall. = Ipomoea mammosa.

Mersawa Kunyit – Anisoptera grossivenia.

MERTENSIA Roth. Boraginaceae. 50 spp. N.Temperate, South into Mexico and Afghanistan. Herbs.

***M. maritima*** (L.) S. F. Gray. N. and temperate N.hemisphere. The rhizomes are eaten by the Eskimos of Alaska.

Mesa – Microphlois chrysophylloides.

Mesa Dropseed – Sporobolus flexuosus.

Mesakhi Fibre – Villebrunea integrifolia.

***Mesaulosperma*** van Slooten = Itoa.

Mescal Beans – Sophora secundiflora.

Mescal Buttons – Lophophora williamsii.

MESEMBRYANTHEMUM L. Senso lato 1000 spp. S.Africa, mainly, but possibly of a wider distribution. Succulent herbs or shrubs.

***M. acinaciforme*** L. S.Africa. Introduced along the Mediterranean coast. The fruits are eaten by the Hottentots.

***M. aqeuilaterale*** Haw. = M. glaucescens (Karkalia). Australia, S.W.America. The aborigines of Australia eat the fruits raw and the leaves when roasted.

***M. anatomicum*** Haw. = Sceletium anatomicum.

***M. angulatum*** Thunb. S.Africa. Cultivated around oases of N.Africa, for the leaves which are eaten as a vegetable and in salads.

***M. chilense*** Mol. Chile. The fruits are eaten locally.

***M. crystalinum*** L. (Ice Plant). S.Africa Canary Islands, Mediterranean. The leaves are eaten locally as a vegetable and in salads.

***M. edule*** L. (Hottentot Fig). S.Africa. The fruits are eaten by the Hottentots.

***M. floribundum*** Haw. S.Africa. Used as a pasture for sheep etc.

***M. forskalei*** Hochst. E.Africa, in region of the Nile. The Bedouins use the crushed fruits to make a bread.

***M. glaucescens*** Haw. = M. aequilaterale.

***M. micranthum*** Haw. S.Africa. The Hottentots use the ash as a source of soda for washing.

***M. stellatum*** Will (Kieriemoor). S.Africa. The plant is an intoxicant and can induce delirium. It is used locally in making beer.

***M. tortuosum*** L. S.Africa. The leaves are slightly narcotic and are chewed by the Hottentots.

Mesentah – Alstonia polyphylla.

MESOGLOIA Suring. Mesogloiaceae. Brown Algae.

***M. crassa*** Suring. Coasts of Japan. Eaten locally when fresh.

MESONA Blume Labiatae. 12 spp. Himalayas to Formosa and Philippines, Java. Herbs.

***M. palustris*** Blume. Java. The leaves are used locally to make a non-intoxicating drink.

***Mespilodaphne*** Nees. = Ocotea.

MESPILUS L. (emend Medik.) Rosaceae. 1 sp. S.E.Europe through to Central Asia. Tree.

***M. aria*** Scop. = Sorbus aria.

***M. aucuparia*** All. = Sorbus aucuparia.

***M. germanica*** L. = Pyrus germanica. (Medlar). The plant is sometimes found in hedgerows in U.K. but is not native. It is cultivated occasionally for the fruits which are only edible when over-ripe (bletted) and slightly frosted. The fruits are sometimes fermented to make a cider.

***M. oxycantha*** Crantz. = Crataegus oxyacantha.

Mesquite – Prosopis juliflora.

Mesquite Mistletoe – Phorodendron californicum.

Mess Apple – Blakea trinervis.

***Messerschmidia argentea*** (L. f.) Johnston = Tournfortia argentia.

Messmate Stringybark – Eucalyptus obliqua.

MESUA L. Guttiferae. 3 spp. Tropical Asia. Trees.

***M. ferrea*** L. (Ironwood, Na). Tropical Asia, India, Malaysia. The very hard wood is used to make lances and walking sticks. The flower buds are used to scent medicines and cosmetics and locally the stamens are used to fill pillows to give them a sweet smell.

***Metagonia penduliflora*** Nutt. = Vaccinium dentatum.

METAPLEXIS R. Br. Asclepiadaceae. 6 spp. E.Asia. Vines.

***M. stauntoni*** Schult. China, Japan. The roots are eaten locally in Japan.

Metcalf Bean – Phaseolus retusus.

***Methysticodendron*** R. E. Schult. = Brugmansia.

METOPIUM P. Br. Anacardiaceae. 3 spp. Florida, Mexico, W.Indies. Small trees.

***M. toxiferum*** (L.) Krug. and Urb. = Rhus metopium. (Coral Sumac, Poison Wood). S.Florida and Mexico. The stem yields a resin (Doctor's Gum, Hog Gum) which is used as a violent purgative. The sap is a severe skin poison and the wood is used to adulterate Quassia amara.

METROSIDEROS Banks ex Gaertn. Myrtaceae. 60 spp. S.Africa, E.Malaysia, Australia, New Zealand, Polynesia, Chile. Trees or climbers.

***M. albiflora*** Soland. ex Gaertn. New Zealand. An infusion of the bark is used locally to treat wounds, pains and bleeding.

***M. collina*** (Forst.) Gray var. ***toviana*** F. Br. (Heua, Feua). Polynesia. The wood is used locally to build houses and boats.

***M. excelsa*** Soland. ex Gaertn. New Zealand. An infusion of the bark is used locally to treat diarrhoea and the honey to treat sore throats.

***M. glomulifera*** Smith = Syncarpia laurifolia.

***M. lucida*** Rich. (Rata). New Zealand. The tough pale red wood is used locally for general carpentry and ship-building.

***M. perforata*** Rich. = M. scandens.

***M. polymorpha*** Gaud. = M. villosa.

***M. procera*** Salisb. = Syncarpia laurifolia.

***M. robusta*** Cunningh. (Northern Rata). New Zealand. The hard heavy durable red wood is used locally for telegraph poles, railway sleepers, bridge-building, carriages and wheel-making.

***M. scandens*** Soland = M. perforata (Rata Vine). New Zealand. The inner stem-bark is used as a dressing for sores and to stop bleeding of wounds.

***M. tomentosa*** Rich. New Zealand. The hard, heavy red wood is strong and resistant to the ship worm (Teredo). It is used for marine timber work and ship-building, boarding etc. An infusion of the bark is used by the aborigines to treat dysentery.

***M. vera*** Lindl. = Nania vera. Malaysia. The wood is very hard and cannot be used for building etc., but as it is resistant to Teredo it is used to make parts of rudders and anchors.

***M. villosa*** Sm. = M. polymorpha. Hawaii. The wood is used locally for carving and making, spears, mallets and canoe outriggers.

METROXYLON Rottb. Palmae. 15 spp. Siam to New Guinea. Palm trees.

***M. elatum*** Mart. = Pigafettia alata.

***M. rumphii*** Mart. (Sago Palm, Prickly Sago Palm, Sagoulier, Ambooloong, Kiraj, Pohon sago). Malaysia, Indonesia. Cultivated for the starch contained in the pith of the trunk, which is used to make Sago. Much is exported especially from Borneo. The plants take up to 15 years to mature and are harvested just before flowering. The pith is grated and kneaded, then soaked in water to separate the fibres from the starch grains. These are then sun-dried for local use, but if for export they are made into a paste and forced through a sieve to give them a uniform round shape. A single tree yields 600–800 lb. of starch. The extracted pith (èla) is fed to pigs. In some parts the pith is roasted whole and eaten. The sago is used locally to make bread, cakes and a kind of porridge, while that which is exported is used mainly to make milk puddings.

***M. sagu*** Rottb. (Sago Palm). Malaya, Moluccas. Cultivated and used as M. rumphii. The sago is used in Malaya as a poultice for shingles and the sap from the fruit, which is poisonous is used as a criminal poison.

METTENIUSA Karst. Alangiaceae. 3 spp. N.W. Tropical America. Trees.

***M. nucifera*** Pitt. Venezuela. The fruits are eaten locally.

Metunga – Melocanna bambusioides.

MEUM Mill. Umbelliferae. 1 sp. Europe. Herb.

***M. athamanticum*** Jacq. (Signel). The plant was cultivated in N. U.K. The roots were eaten as a vegetable in Scotland. The leaves were used in a wide variety of home remedies as a diuretic, to control menstruation and uterine complaints, to treat catarrh, hysteria and stomach ailments.

***M. mutellina*** Gaertn. = Ligustricum mutellina.

Mexican Avocado – Persea drymifolia.

Mexican Elder – Sambucus mexicana.

Mexican Greasewood – Sarcobatus vermiculata.

Mexican Juniper – Juniperus mexicana.

Mexican Linaloe Wood – Bursera delpechiana.

Mexican Mahogany – Swietenia humulis.

Mexican Mahogany – Swietenia macrophylla.

Mexican Mock Orange – Philadelphus mexicanus.

Mexican Mugwort – Artemisia mexicana.

Mexican Oiticica (Oil) – Licania arborea.

Mexican Pinõn – Pinus cembroides.

Mexican Prickly Poppy – Argemone mexicana.

Mexican Sarsparilla – Smilax aristolochiaefolia.

Mexican Scammony – Ipomoea orizabensis.

Mexican Strawberry – Echinocereus stramineus.

Mexican Sycomore – Plantanus mexicana.

Mexican Tea – Chenopodium ambrosioides.

Mexican Whish – Epicampes macroura.

Meyer Bamboo – Phyllostachys meyeri.

Mezcal – Agave kirchneriana.

Mezcal – Agave tequilana.

Mezcal Blanco – Agave pseudo-tequilana.

Mezcal Cucharo – Agave pseudo-tequilana.

Mezcal de Pulque – Agave atrovirens.

Mezereon – Daphne mezereum.

MEZONEVRON Desf. Leguminosae. 30 spp. Tropical Africa and Madagascar to Australia and Pacific Islands. Trees.

***M. glabrum*** Desf. = M. latisiliquum.

***M. kauiense*** Drake = M. kavaiensis.

***M. kavaiensis*** (H. Mann) Hilleb. = M. kauiense = Caesalpinia kavaiensis. (Kea, Uhiuhi). Hawaii. The very hard, black wood is used for spear shafts, house frames and sledge runners.

***M. latisiliquum*** (Cav.) Merr. = M. glabrum = Caesalpinia torquato. (Sapnit). Philippines. The leaves are eaten locally as a salad and are used in an infusion to treat asthma.

***M. scortechinii*** F. v. Muell. Australia. The trunk yields a gum (Barister Gum) which is similar to Gum Tragacanth.

Mfuti – Brachystegia boehmii.

Mgmeo – Duboisia myoporoides.

Mgoa Rubber – Mascarenhasia elastica.

Mgottmerkua – Talinum cuneifolium.

Micauba – Acrocomia sclerocarpa.

Miche – Brahea dulcis.

MICHELIA L. Magnoliaceae. 50 spp. Tropical Asia, China. Trees.

***M. celebica*** Koord. Celebes. The yellow-brown wood is light and durable. It is used for general carpentry, house interiors and coffins.

***M. champaca*** L. (Chêmpaka). India, Malaysia. The flowers contain an essential oil and are used in Thailand to scent perfumes, body oils, clothes etc. The wood is used for general house-building, tea-chests etc. The leaves are fed to silk worms and a decoction of the bark is used locally to reduce fevers.

***M. figo*** Spreng. = M. fuscata.

***M. fuscata*** Blume = M. figo. (Banana Shrub) China. Cultivated in many warm countries for the banana scented oil extracted from the flowers. It is used in scenting hair-oils.

***M. montana*** Blume. Malaysia. The dark brown wood is light and durable. It is used locally to build bridges and houses.

***M. tsiampaca*** L. = M. velutina. Java, Amboina. The light yellow wood is strong and durable. It is used locally for house-building and furniture.

***M. velutina*** Blume = M. tsiampaca.

Michire – Brahea dulcis.

MICONIA Ruiz. and Pav. Melastomataceae. 700 spp. Tropical America, W.Indies; 1 sp. W.Africa. Small trees or shrubs.

***M. albicans*** (Swartz.) Triana = Melastoma albicans. Tropical America. The fruits are eaten locally in Mexico.

***M. argentea*** (Swartz.) DC. = Melastoma argentea. S.Mexico and central America. The wood yields a good charcoal.

***M. impetiolaris*** (Sw.) D. Don. Tropical America, W.Indies. In Chile the leaves are used to make aromatic baths.

***M. liebmannii*** Cogn. Mexico. The fruits are eaten locally.

***M. willdenowii*** Klotzsch. Brazil. A decoction of the bark is used locally to treat swamp fever and the leaves are used as a tea substitute.

MICRANDA Benth. Euphorbiaceae. 14 spp. Tropical S.America. Trees.

***M. minor*** Benth. (Huemega). Peru. The tree yields a latex sometimes used locally to adulterate that from Hevea brasiliensis.

***M. siphonoides*** Benth. (Avara Seringa). Amazon. The latex yields a good rubber, but it is not exploited commercially.

MICRECHITES Miq. Apocynaceae. 20 spp. S.China, Indomalaya. Trees.

***M. napensis*** Quint. Cochin-China. Yields a good rubber which is coagulated from the latex by lemon juice.

MICROCITRUS Swingle. Rutaceae. 5 spp. E.Australia. Trees.

***M. inodora*** Swingle. (Queensland Wild Lime). Queensland. The fruits (Russell River Limes) are used as those of the true Lime (Citrus aurantifolia).

MICRODESMIS Hook. f. Pandaceae. 10 spp. Tropical Africa, S.E.Asia, W.Malaysia. Trees.

***M. puberula*** Hook. f. Tropical Africa. The hard, flexible wood (Benin Apata Wood) is used for hoe-handles, knife handles, combs, spoons, walking sticks and animal traps.

MICROMERIA Benth. Labiatae. 100+ spp. Cosmopolitan. Herbs.

***M. chamissonis*** (Benth.) Greene (Yerba Buena). W. N.America. The dried leaves are used as a tea substitute by the Californian Indians.

***M. douglasii*** Benth. See M. chamissonis.

MICROPHOLIS (Griseb.) Pierre. Sapotaceae. 40 spp. Tropical America, W.Indies. Trees. The wood contains silica which makes it difficult to saw.

***M. chrysophylloides*** Pierre. (Caimitillo). W.Indies, especially Puerto Rico. The yellow-brown wood is hard and strong with a pleasing lustre, but it splits when nailed. It is used for furniture, flooring and other interior work, agricultural implements and general carpentry.

***M. garcinaefolia*** Pierre. (Caimitillo verde). W.Indies, especially Puerto Rico. See M. chrysophylloides.

***M. melinoniana*** Pierre = Stephanoluma rugosa = Chrysophyllum rugosum. Guyana. The trunk yields a balata (Balata Blanc).

MICROSERIS D. Don. Compositae. 14 spp. W. N.America; 1 sp. Chile, 1 sp. Australia, New Zealand. Herbs.

***M. forsteri*** Hook. = Scorzonera scapigera. Australia. The tubers are eaten by the aborigines.

***Microstemon*** Engl. = Pentaspadon.

***Microtanea*** Hemsl. = Microtoena.

MICROTIS R. Br. Orchidaceae. 10 spp. E.Asia, Malaysia, Australia, New Zealand, Polynesia. Herbs.

***M. porrifolia*** R. Br. New Zealand. The tubers are eaten by the natives.

MICROTONEA Prain. Labiatae. 20 spp. Himalayas, W.China, Java. Woody plants.

***M. cymosa*** Prain. (Khasia Patchouli). India, Assam. An oil distilled from the leaves is used in China and Indochina as a perfume, especially to scent soap and fabrics, e.g. carpets.

***M. robusta*** Hemsl. See M. cymosa.

Midaumabaphang – Euodia lunu-ankenda.

Midnight Lichen – Parmelia stygia.

Miel de Tuna – Opuntia megacantha.

Mignonette (Oil) – Reseda odorata.

Mignonette Vine – Anredera baselloides.

Mijugua Fibre – Anacardium rhinocarpus.

MIKANIA Willd. Compositae. 250 spp. Tropical America, W.Indies; 2 spp. S.Africa. Herbs.

***M. guaco*** Humb. and Bonpl. (Guaco). N. S.America. A decoction of the leaves (Herba Guaco) is used locally to treat cholera and diarrhoea, cramp and nervous complaints.

Mil – Maerua rigida.

Milfoil – Achillea millefolium.

***Milicia africanum*** Sim. = Chlorophora excelsa.

MILIUSA Leschen ex A. DC. Annonaceae. 40 spp. Indomalaya, Australia. Trees.

***M. horsfieldii*** Benn. Java. The tough wood is used for lance handles.

Milk Parsley – Peucedanum palustre.

Milkpea, Hoary – Galactia mollis.

Milk Thistle – Silybum marianum.

Milk Tree – Sapoum aucuparium.

Milk Vetch – Astragalus glycyphyllus.

Milkweed – Sonchus oleraceus.

Milkweed, Common – Asclepias syriaca.

Milkweed, Desert – Asclepias subulata.

Milkweed, Dwarf – Asclepias involucrata.

Milkweed Root, Orange – Asclepias tuberosa.

Milkweed, Showy – Ascelepias speciosa.

Milkweed, Swamp – Asclepias incarnata.

Milkweed, Woollypod – Asclepias eriocarpa.

Milkwort, Sea – Glaux maritima.
Milky Tassel – Sonchus oleraceus.
Millet, African – Eleusine coracana.
Millet, Broom Corn – Panicum miliaceum.
Millet, Bulrush – Pennisetum typhoideum.
Millet, Cat-tail – Pennisetum typhoideum.
Millet, Finger – Eleusine coracana.
Millet, Foxtail – Setaria italica.
Millet, Great – Sorghum vulgare.
Millet, Hog – Panicum miliaceum.
Millet, Italian – Setaria italica.
Millet, Japanese (Barnyard) – Echinochloa frumentacea.
Millet, Koda – Paspalum scrobiculatum.
Millet, Little – Panicum sumatrense.
Millet, Native – Panicum decompositum.
Millet, Pearl – Pennisetum typhoideum.
Millet, Pearl – Setaria glauca.
Millet, Proso – Panicum miliaceum.
Millet, Sanwa – Echinochloa frumentacea.
Millet, Shama – Echinochloa colonum.
Millet, Small – Panicum sumatrense.
Millet, Spiked – Pennisetum typhoideum.
Millet, Texas – Panicum texanum.

MILLETTIA Wight and Arn. Leguminosae. 180 spp. Tropics and subtropics (only a few in America). Trees or vines.

***M. atropurpurea*** Benth. Malaysia. The roots and twigs contain rotenone, so that the plant is a potential source of insecticide. The twigs and roots are used locally to stupify fish.

***M. auriculata*** Baker. India, Himalayas. The roots and twigs contain rotenone, a potential source of insecticide. They are used locally to kill parasites on cattle and as a fish poison. The bark yields a poor quality cordage fibre.

***M. barteri*** (Benth.) Dunn. (Ambula, Fukuti). Tropical Africa, especially Congo and West. Vine. The bark is used locally to stupify fish and as cordage in hut-building etc.

***M. caffa*** Meisn. S.Africa. The heavy, hard wood is durable in contact with the soil. It is used locally for wheel spokes and bearings. The fruits are used by the natives to treat intestinal worms.

***M. eriantha*** Benth. = Whitfordiodendron erianthum.

***M. ichthyochtona*** Drake. (Cay Thành mat). Possibly a cultigen, as it is only known in cultivation in Tonkin where the fruits are used to stupify fish.

***M. laurentii*** De Wild. (Brondonko, Ntoka). Tropical Africa, especially the Congo and West. The hard, heavy, dark brown, marbled wood (African Palisander) is insect resistant and is used locally to make small boats.

***M. pachycarpa*** Benth. Tonkin. The roots and fruits are used locally as a fish poison.

***M. piscidia*** Wight. Tonkin. The roots are used locally as a fish poison.

***M. sericea*** Wight. and Arn. = Pongamia sericea. Burma, Malaysia. The roots are used locally to stupify fish.

***M. taiwaniana*** Hayata. Formosa. The roots are used locally to stupify fish.

***M. thyrsiflora*** Benth. = Derris thyrsiflora.

Milo – Sorghum vulgare.
Mimbo – Raphia hookeri.
Mimbre – Chilopsis lineari.

MIMETES Salisb. Proteaceae. 16 spp. S.Africa. Trees.

***M. lyrigera*** Knight. S.Africa. The bark is used locally for tanning.

MIMOSA L. Leguminosae. 450–500 spp. Tropical and subtropical America, few in Africa and Asia. Small scrubby herbs to trees.

***M. arborea*** L. = Pithecellobium arboreum.

***M. barbatimam*** Vell. = Stryphnodendron barbatimam.

***M. colubrina*** Vell. = Piptadenia colubrina.

***M. cyclocarpa*** Jacq. = Enterolobium cyclocarpum.

***M. esculenta*** Moc. and Sessé = Leucaena esculenta.

***M. flexicaulis*** Benth. = Pithecellobium flexicaule.

***M. glauca*** L. = Leucaena glauca.

***M. guianensis*** Aubl. = Stryphnodendron guianense.

***M. houstonii*** L'Hér. = Calliandra houstoniana.

***M. invisa*** Mart. Tropical America. A possible green manure.

***M. purpurascens*** Robinson (Cuca, Cuilón, Iguano). Mexico. The bark is used locally for tanning and is chewed to harden the gums.

***M. stipitata*** Robinson. See M. purpurascens.

Mimosa Thorn – Acacia horrida.

MIMUSOPS L. Sapotaceae. 57 spp. Tropical Africa; 1 sp. Malaysia to Pacific Islands. Trees.

***M. bidentata*** DC. = Manilkara bidentata.

***M. dariensis*** Pitt. = Manilkaro dariensis.

***M. djave*** (Lam.) Engl. (Moumgou). Tropical Africa, especially West. The seeds yield a fat (Adjab Butter, Djave Butter) which is eaten locally and used in Europe to make soap. The cake, left after extraction and the fruits which are used as an ordeal poison are poisonous, containing a poisonous glucoside.

***M. globosa*** Gaertn. = Manilkara bidentata.

***M. henriquesii*** Engl. and Warb. Mozambique. A poor quality gutta-percha is extracted from the trunk.

***M. kauki*** L. = Manilkara kauki.

***M. pierreana*** Engl. = Baillonella ovata.

***M. sieberi*** A. DC. = Manilkara sieberi.

***M. spectabilis*** Pitt. = Manilkara spectabilis.

Minando – Palaquium ahernianum.

Miner's Lettuce – Claytonia sibirica.

Mindanao Cinnamon – Cinnamomum mindanaense.

Mingerhout – Adina microcephala.

Mint, American Apple – Mentha gentilis.

Mint, American Wild – Mentha canadensis.

Mint, Apple – Mentha rotundifolia.

Mint, Atlantic Mountain – Pycanthemum incarnum.

Mint, Bergamot – Mentha aquatica.

Mint, Cat – Nepeta cetaria.

Mint, Emorybush – Hyptis emory.

Mint, Fern – Monarda fistulosa.

Mint, Field – Mentha arvensis.

Mint Geranium – Chrysanthemum balsamita.

Mint, Horse – Monarda fistula.

Mint, Pepper – Mentha piperata.

Mint, Spear – Mentha spicata.

Mint, Stone – Cunila origanoides.

Mint, Virginian Mountain – Pycanthemum virginianum.

Mint, Woolly – Mentha rotundifolia.

Minyadotana – Zschokkea foxii.

Minyak Keruong – Dipterocarpus karrii.

***Minyranthus heterophylla*** Turez. = Sigesbeckia orientalis.

Mirabella Plum – Prunus instititia.

MIRABILIS L. Nyctaginaceae. 60 spp. America.

***M. jalapa*** L. (Four o'Clock, Marvel of Peru). In Japan the powdered seeds are used as a cosmetic and a decoction of the roots is used to treat dropsy. In China a red dye made by infusion of the flowers with water is used to colour foods made from seaweed and confectionery.

Mirasol – Tithonia scaberrima.

Mirim – Rollinia laurifolia.

Mirto – Lippia geminata.

Mirzapore Lac – Schleichera trijuga.

***Misanteca*** Cham. and Schlechtd. = Licaria.

MISCHOCARPUS Blume. Sapindaceae. 25 spp. China, S.E.Asia, Indomalaya, N.E.Australia. Small trees.

***M. sundaicus*** Blume = Cupania lessertiana. (Bangkongan, Poolas laoot). S.E.Asia. A good source of charcoal.

MISCHOPHLOCUS Scheff. Palmae. 1 sp. Moluccas. Palm.

***M. paniculata*** Scheff. = M. vestiaria.

***M. vestiaria*** Merr. = M. paniculata = Areca vestiaria = Seaforthia vestiaria. (Lanoot, Sarewow). The epidermis of the young leaves is used locally to make a fibre from which cloth is woven.

***Miscolobium nigrum*** Mart. = Dalbergia miscolobium.

Mishmee Bitter – Coptis teeta.

Mispero del Monte – Pouteria tovarensis.

Misran – Pedicularis pectinata.

Missouri Currant – Ribes aureum.

Mistletoe – Viscum album.

Mistletoe, American Christmas – Phoradendron flavescens.

Mistletoe, Juniper – Phoradendron juniperinum.

Mistletoe, Mesquite – Phorandendron californicum.

Mitabid – Tricholepis glaberimum.

Mitchell Grass – Astreble pectinata.

MITCHELLA L. Rubiaceae. 2 spp. N.E.Asia, N.America. Herbs.

***M. repens*** L. (Checkerberry, Deerberry, Partridgeberry, Squaw Vine). N.America. A decoction of the leaves is used to treat diarrhoea, dropsy and disorders of the uterus.

Miterwort – Tiarella cordifolia.

MITRACARPUM Zucc. Rubiaceae. 40 spp. Tropical S.America, W.Indies, tropical and S.Africa. Herbs.

***M. scabrum*** Zucc. Tropical Africa. The dried leaves are used locally to heal ulcers and as an antidote to arrow poison.

MITRAGYNA Korth. Naueleaceae. 12 spp. Tropical Africa, Asia. Trees.

***M. africana*** Korth. W.Africa. The light brown wood is used locally for house-building, carving and tools. The bark yields a yellow dye.

***M. macrophylla*** Hiern. W.Africa. The light brown wood (West African Mahogany) is used for house-building, canoes, furniture. Some is exported. The boiled roots are used to treat stomach complaints in Sierra Leone.

***M. parvifolia*** Korth. = Stephegyne parviflora. (Kaddam, Kaim). India, Pakistan. The pink-brown, hard light wood is used for building, furniture, agricultural implements and carving a wide variety of articles. The bark yields a good cordage fibre.

***M. speciosa*** Korth. Malaysia, Siam. The leaves are smoked as an opium substitute.

***M. stipulosa*** Kuntze. (Abura, African Linden). W.Africa. The yellow-brown wood is soft. It is used for house-building, drums, boxes, paddles, canoes and barrels. A decoction of the leaves is used locally to treat coughs and to reduce fevers.

***Mitranthes satoriana*** Berg. = Psidium sartorianum.

***Mitrophora*** Lév. = Morchella.

Miyana Cherry – Prunus maximowiczii.

M'ka – Erythrophleum guiniënsis.

Mkani Fat – Allanblackia stuhlmanni.

Mkokongo – Afzelia pachyloba.

Mkongomwa – Afzelia cuanzensis.

Mkushi – Baikiaea plurijuga.

Mlanje Cedar – Widdringtonia whytei.

M'loundu – Strychnos icaja.

Mmamba – Cryptosepalum pseudotaxus.

Moavi – Erythrophleum guiniensis.

Mobembe – Lychnodiscus cercospermus.

Mobola Plum – Parinarium mobola.

Mocha Coffee – Coffea arabica.

Mochi – Lathyrus cicera.

Mocitahyba – Zollerina ilicifolia.

Mock Orange – Philadelphus mexicanus.

Mock Oyster – Pleurotus ostreatus.

Mock Pak-Choi – Brassica parachinensis.

Mockernut – Carya tomentosa.

Moeddeur – Eritrichium pusillum.

Mofomadoki – Eriosema montanum.

Mogador Acacia – Acacia gummifera.

Mogador Gum – Acacia gummifera.

Mogda Senna – Cassia occidentalis.

Mogun-Mogun – Harpullia pendula.

Mohave Yucca – Yucca mohavensis.

MOHRIA Sw. Schizaeaceae. 3 spp. Tropical and S.Africa. Ferns.

***M. thurifraga*** Sw. Cape Peninsula. A paste of the dried leaves and fat is used by the natives to treat burns.

Moiga Ginger – Zingiber moiga.

Moka Aloë – Aloë abyssinica.

Mokolingo – Muncuna flagellipes.

Mokula – Pterocarpus angolensis.

Molave Chaste Tree – Vitex parviflora.

MOLLUGO L. Aizoaceae. 20 spp. Tropics and sub-tropics.

***M. hirta*** Thunb. = Glinus lotoides. Tropics and subtropics. The leaves are eaten as a vegetable in the Sudan. In parts of India they are used as a purgative, to treat diarrhoea and skin diseases.

***M. oppositifolia*** L. = M. spergula = Polycarpaea frankenioides. Tropical Africa, Asia, Australia. The leaves are eaten in some parts as a vegetable and in others the leaf-juice is used to treat skin complaints.

***M. pentaphylla*** L. = M. sumatrana. India, Malaya, to Japan, Philippines. The leaves are used by the Hindus as a vegetable and in parts of India as an antiseptic, to treat stomach complaints and to control menstruation. In Malaya a poultice of the leaves is used to treat sore legs.

***M. spergula*** L. = M. oppositifolia.

***M. sumatrana*** Gandog. = M. pentaphylla.

Molondo – Dorstenia klainei.

Molucca Bean – Caesalpinia bonducella.

Molucca Copal – Agathis loranthifolia.

Molucca Ironwood – Afzelia bijuga.

Molucca Ironwood – Intsia palembanica.

Molucca Manila – Agathis loranthifolia.

Moluks Ijzerhout – Intsia palembanica.

Mombin, Red – Spondias purpurea.

Mombin, Yellow – Spondias lutea.

Mombo – Brachystegia boemii.

Mommi – Abies firma.

MOMORDICA L. Cucubitaceae. 45 spp. Old World Tropics. Vines.

***M. charantia*** L. (Balsam Pear). Old World Tropics. Cultivated throughout the tropics. The fruits are eaten raw in salads or boiled or fried, especially in Peru. The sap from the leaves is used to treat stomach complaints and intestinal worms.

***M. chochinchinensis*** Spreng. Tropical Asia. The fruits are eaten as a vegetable in the Philippines etc. and the roots are used as a soap substitute.

***M. operculata*** L. = Luffa purgens.

***M. schimperiana*** Naud. Tropical Africa. The plant is used locally as an insecticide.

***M. subangulata*** Blume. (Kambur). Indonesia. The shoots and unripe fruits are eaten locally in a soup (Sajoor).

Monajembe – Scorodophloeus zenkeri.

Monantha Vetch – Vicia monanthos.

Monarana – Beccariophoenia madagascariensis.

MONARDA L. Labiatae. 12 spp. N.America to Mexico. Herbs.

***M. didyma*** L. (Oswego Bee Balm, Oswego Tea). E. N.America. The plant yields an essential oil (Oil of Thyme) used in perfumery and hair tonic and thymol. The local Indians used the leaves as a vegetable and as a hair dressing.

***M. fistulosa*** L. = M. menthaefolia. (Bee Balm, Horse Mint, Wild Bergamot) See M. didyma.

***M. menthaefolia*** Grah. = M. fistulosa.

***M. pectinata*** Nutt. (Pony Bee Balm). W. N.America. The Indians of New Mexico use the leaves to flavour food.

***M. punctata*** L. (American Horsemint, Horsemint). N.America, especially E. The leaves are used to treat stomach complaints, as a stimulant, diuretic and to control menstruation.

MONARDELLA Benth. Labiatae. 20 spp. W. N.America. Herbs.

***M. parviflora*** Greene. (Western Balm). W. N.America. A decoction of the leaves is used by the local Indians to treat colds and indigestion.

MONASCUS van Tiegh. Eurotiaceae. 3 spp. Widespread.

***M. purpureus*** Went. When grown on rice the fungus produces a bright red dye (Ang-khak). The dried grains, produced in China, are used by the Chinese throughout the Far East for colouring food and drinks and sauces. The dye contains slight traces of arsenic which gives it a faint garlic flavour.

Monastery Bamboo – Thyrsostachys siamensis.

***Moesia uniflora*** Gray = Pyrola uniflora.

Monesia Bark – Chrysophyllum glycyphloeum.

***Moneses reticulata*** Nutt. = Pyrola uniflora.

Moneywort – Lysimachia nummularia.

Mongambe – Celtis mildbraedii.

Mongola – Garcinia wentweliana.

Mongolian Oak – Quercus mongolica.

MONILIA Pers. ex Fr. Moniliaceae. 40 spp. Widespread.

***M. candida*** Bon. = Candida tropicalis. This fungus has been used to ferment a variety of waste products etc. to produce animal feedstuffs.

***M. martinii*** Ell. and Sacc. = M. sitophila.

***M. sitophila*** (Mont.) Sacc. = M. martinii = Oidium aurantiacum. The fungus is used in Java to ferment groundnuts to make Ontjom beurum, or the seeds of Muncuna pruriens to make Gěděběl bengook, both of which are easily digestible high protein foods.

Monjet Mango – Mangifera laurina.

Monkeru – Beilschmiedia variabilis.

Monkey Apple – Clusia flava.

Monkey Apple – Licania platypus.

Monkey Bread – Adansonia digitata.

Monkey Cap Palm – Manicaria saccifera.

Monkey Jack – Artocarpus rigidus.

Monkey Pod – Lecythis ollaria.

Monkey Puzzle – Araucaria auracana.

Monkoo – Eminia polyadenia.

Monkshood – Aconitum napellus.

Monk's Rhubarb – Rumex alpinus.

***Monniera cuneifolia*** Michx. = Bacopa monniera.

***Monocera lanceolata*** Hassk. = Elaeocarpus grandiflorus.

MONOCHORIA C. Presl. Pontederiaceae. 3 spp. N.E.Africa to Manchuria and Australia. Water herbs.

***M. hastata*** (L.) Solms. India, Malaysia. The roots are used in India to make a tonic and in Malaysia to feed pigs.

***M. vaginalis*** C. Presl. Indonesia. The roots are used in Java to treat stomach complaints, liver diseases and toothache. A decoction of the leaves is used to reduce fevers.

MONODORA Dunn. Annonaceae. 20 spp. Tropical Africa, Madagascar. Trees.

***M. angolensis*** Welw. (Angola Calabash). Tropical Africa. The seeds are used locally as a condiment.

***M. laurentii*** De Wild. (Engambili, Ligala). Congo. The hard wood is used for making spears. A decoction of the bark is used to treat stomach complaints.

***M. myristica*** Dunal. (Calabash Nutmeg). Tropical Africa, especially W. The ground, dried endosperm is a popular local condiment. The whole seeds are also used as beads.

Monoka – Ficus melleri.

MONOLEPIS Schrad. Chenopodiaceae. 6 spp. Central and N.E. Arctic, Asia, N.America and Patagonia. Herbs.

***M. nuttaliana*** (Schult.) Greene. W. N.America. The salted fried roots were eaten by the local Indians and the seeds were used as grain.

MONOTROPA L. Monotropaceae. 5 spp. N.Temperate. Herbs.

***M. uniflora*** L. (Ink Pipe, Wood Snowdrop)

N.America. A decoction of the root is used as a mild sedative.

Monoza – Maytenus senegalensis.

MONSONIA L. Geraniaceae. 40 spp. Africa, S.W.Asia, N.W.India.

***M. ovata*** Cav. (Monsonia). S.W.Africa. A decoction of the whole plant is used locally to treat dysentery.

MONSTERA Schott. Araceae. 50 spp. Tropical America, W.Indies. Woody Vines.

***M. cannaefolia*** Kunth. = Spathiphyllum cannifolium.

***M. deliciosa*** Liebm. = Philodendron pertusum. (Ceriman). Central America. The fruits have a very pleasant flavour, but contain raphides which make them unpleasant to eat. They are pulped and used to make drinks and ices.

***M. pinnatifida*** Schott. = Epipremnum pinnatum.

MONTAÑOA Cerv. Compositae. 50 spp. Mexico to Colombia. Shrubs.

***M. tomentosa*** Cerv. = Eriocomo fragrans. (Hierba de la Perida). Mexico. A decoction of the leaves is used locally to help women in childbirth and the ground roots mixed with water are used to treat dysentery.

Monterey Cypress – Cupressus macrocarpa.

Montezuma Pine – Pinus montezumae.

MONTIA L. Portulacaceae. 50 spp. N. and S.America, temperate Eurasia, mountains of tropical Africa. Herbs.

***M. fontana*** L. Central Europe. The leaves are eaten as a salad, especially in France.

***M. perfoliata*** (Willd.) How. = Claytonia perfoliata.

***Montrichardia*** Creug. = Pleurospa.

MONTROUZIERA Planch. ex Planch. and Triana. Guttiferae. 5 spp. New Caledonia. Trees.

***M. sphaeraeflora*** Planch. New Caledonia. The red-brown wood is durable and easily worked. It is used locally for general carpentry.

Moonseed, Common – Menispermum canadense.

Moonset – Habenaria orbiculata.

Mooseberry Viburnum – Viburnum pauciflorum.

Mopane – Copaifera mopane.

***Mora excelsa*** Benth. = Dimorphandra mora.

***Mora gonggrijpii*** (Kleinh.) Sander. = Dimorphandra gonggrijpii.

Mora – Dimorphandra mora.

Morabukea – Dimorphandra gonggrijpii.

Mora Común – Rubus adenorichos.

Mora de Castilla – Rubus glaucus.

Mora de Rocoto – Rubus roseus.

MORCHELLA Dill. ex Fr. Helvellaceae. 15 spp., especially temperate. The Morels. The fruitbodies of the species mentioned are eaten locally. They are of good flavour and very popular.

***M. conica*** Pers. ex Pers. Temperate.

***M. elata*** Fr. Temperate.

***M. esculenta*** (L.) Pers. ex Fr. Eaten widely in Europe. The fruitbodies are collected extensively and sold locally. They are cultivated in parts of France.

***M. hybrida*** (Sow. ex Fr.) Boud. Temperate.

***M. rimosipes*** DC. ex Fr. Temperate.

Morea Galls – Quercus cerris.

Morea Tragacanth – Astragalus cylleneus.

Moreth Bay Chestnut – Castanospermum australe.

Moreton Bay Pine – Araucaria cunninghamii.

Moriche – Mauritia flexuosa.

MORINDA L. Rubiaceae. 80 spp. Tropics. Small trees or shrubs.

***M. bracteata*** Roxb. Malaysia. A red dye is extracted from the roots. It is used locally to dye fabrics and basketwork.

***M. geminata*** DC. (Bombooe, Ditoo, Ooanda, Soorooghan). West Africa. The orange roots are used locally as a mordant for indigo when dyeing.

***M. lucida*** Benth. (Anongro, Boonko, Nafoorbogin, Zooagoo). W.Africa. The roots are used locally to treat yellow fever.

***M. speciosa*** Roxb. Malaysia. An infusion of the roots is used locally to treat rheumatism.

***M. tinctoria*** Roxb. (Dyer's Indian Mulberry). The roots yield a red dye used to colour wool and linen.

***M. trifolia*** L. (Indian Mulberry). Malaysia, Indonesia. The leaves yield a red dye and the roots a yellow dye. The plant is cultivated in Java where the dyes are used for batik work. The leaves are eaten locally as a vegetable and an infusion of the bark, roots and fruit is used to dress wounds. The wood (Togari Wood of Madras) is insect resistant.

***M. yucatanensis*** Greenm. (Hooyoc, Piñuela). Mexico. The leaves are purgative and diuretic. They are used locally to treat digestive upsets.

MORINGA Adans. Moringaceae. 12 spp.

N.E. and S.W.Africa, Arabia, Madagascar, India. Trees.

***M. drouhardi*** Jum. (Maroserano). Madagascar. See M. oleifera.

***M. hildebrandti*** Bak. Madagascar. See M. oleifera.

***M. oleifera*** Lam. = M. pterygosperma. (Horseradish Tree). India. Cultivated in Old and New World tropics for the oil (Ben Oil) extracted from the seeds. The oil which is sometimes called Gomme de Benoiele, is used by artists in mixing paints.

***M. pterygosperma*** Gaertn. = M. oleifera.

MORISONIA L. Capparidaceae. 4 spp. W.Indies, S.America. Small trees.

***M. americana*** L. (Arbol del diablo, Cacao cimmarrón, Chicozapote). S.Mexico to N. S.America. Lesser Antilles. An infusion of the flowers is used in the W.Indies to improve the appetite as a mild sedative and to treat intestinal worms.

***M. flexuosa*** L. = Capparia flexuosa.

Morning Glory, Bush – Ipomoea leptophylla.

Morning Glory, Cairo – Ipomoea cairica.

Morning Glory, Fingerleaf – Ipomoea digitata.

Morning Glory, Seaside – Ipomoea pescaprae.

Moroccan Ammoniac – Ferula tingitana.

Morocco Gum – Acacia gummifera.

MORONOBEA Aubl. Guttiferae. 7 spp. Tropical S.America. Trees.

***M. coccinea*** Aubl. (Manil, Ohori). Tropical S.America. The trunk yields a gum (Anani), used by the Indians as a glue for making arrows, fixing knife-handles and a wide variety of household uses. The wood (Bois Cochon) is used locally for general building purposes.

***M. riparia*** (Spruce) Planch. and Triana. Tropical S.America. The latex is used as a glue for household uses, by the local Indians and the resin is burnt in torches.

***M. rupicola*** R. E. Schultes. Tropical S.America. The resin is used by the local Indians for making torches.

Morphine – Papaver somniferum.

Mortinà – Vaccinium mortinia.

Mortiño – Vaccinium floribundum.

MORUS L. Less than 10 ? spp. E. Temperate N.America, S.W. U.S.A. to the Andes, tropical Africa, S.W.Asia, to Japan and Java. Trees.

***M. alba*** L. (White Mulberry). China. Cultivated widely for the leaves on which the silkworm Bombyx mori feed.

***M. cucullata*** Poir. = M. latifolia.

***M. kaempferi*** Ser. = Broussonetia kazinoki.

***M. latifolia*** Poir. = M. cucullata = M. multicaulus (Pai-sang). China, Japan. Grown locally as a food for silk worms.

***M. multicaulis*** Perrot. = M. latifolia.

***M. nigra*** L. (Black Mulberry). China, Japan. Cultivated in Europe and the Near East for the fruit which is eaten raw or cooked. A syrup (Sirupus Mororum) is used medicinally to soothe throats etc.

***M. rubra*** L. (Red Mulberry). E. N.America. The light coarse-grained wood is used locally for ship-building, fences and barrels.

***M. tinctoria*** L. = Chlorophora tinctoria.

MOSCHOSMA Reich. Labiatae. 7 spp. Tropical Africa to Queensland. Herbs.

***M. polystachyum*** Benth. (Musk Basil, Sangat). Malaysia, Indonesia. The leaves are mildly narcotic. In Java they are rubbed on sprains etc. and rubbed on people suffering from shock. A decoction of the leaves is also taken internally for nerve complaints.

Mosimbi – Chasmanthera welwitschii.

MOSQUITOXYLON Krug. and Urb. Anacardiaceae. 1 sp. Jamaica. Tree.

***M. jamaicense*** Krug. and Urb. (Mosquito Wood). The wood is used locally for general building.

Mosquito Wood – Mosquitoxylon jamaicense.

Moss Campion – Silene acaulis.

Moss, Ceylon – Gracilaria lichenoides.

Moss, Corsican – Alsidium helminthochorton.

Moss, Cup – Cladonia pyxidata.

Moss, Florida – Tillandsia usneoides.

Moss, Hair Cup – Polytrichum juniperinum.

Moss, Iceland – Cetraria islandica.

Moss, Irish – Chondrus crispus.

Moss, Louisiana – Tillandsia usneoides.

Moss, Oak – Evernia prunastri.

Moss, Reindeer – Cladonia rangiferina.

Moss, Spanish – Tillandsia usneoides.

Moss, Sphagnum – Sphagnum cymbifolium.

Mossy Cup Oak – Quercus macrocarpa.

Moss Stonecrop – Sedum acre.

MOSTUEA Didr. Loganiaceae. 1 sp. Brazil; 7 spp. Tropical Africa, Madagascar. Shrubs.

***M. stimulans*** Chev. W. Tropical Africa.

The roots are chewed locally in the Gabon as a stimulant and aphrodisiac during night-long festive dances etc.
Motes prey – Tabernaemontana sralensis.
Moth Bean – Phaseolus aconitifolius.
Motherwort – Leonurus cardiaca.
Motherwort, Siberian – Leonurus sibiricus.
Motillo – Sloanea berteriana.
Motta karmal – Dillenia indica.
Mouhden – Eritrichium pusillum.
Moulmein Lancewood– Homalium vitiense.
Moumgou – Mimusops djave.
Mount Atlas Pistache – Pistacia atalantica.
Mountain Alder – Alnus viridis.
Mountain Apple – Eugenia malaccensis.
Mountain Ash – Eucalyptus regians.
Mountain Ash – Eucalyptus salicifolia.
Mountain Ash – Sorbus americana.
Mountain Ash – Sorbus aucuparia.
Mountain Balm – Eriodictyon glutinosum.
Mountain Beech – Nothofagus cliffortoides.
Mountain Blackberry – Rubus allegheniensis.
Mountain Brome – Bromus marginatus.
Mountain Cherry – Prunus angustifolia.
Mountain Cranberry – Vaccinium erythrocarpum.
Mountain Elm – Ulmus montana.
Mountain Gum – Eucalyptus dalrympleana.
Mountain Hydrangea – Hydrangea arborescens.
Mountain Laurel – Kalmia latifolia.
Mountain Lettuce – Saxifraga erosa.
Mountain Magnolia – Magnolia acuminata.
Mountain Mahogany – Cercocarpus latifolia.
Mountain Maple – Acer spicatum.
Mountain Nutmeg – Myristica fatua.
Mountain Papaya – Carica cundinamarcensis.
Mountain Pine – Pinus montana.
Mountain Rose – Antigonon leptopus.
Mountain Rye – Secale montanum.
Mountain Sorrel – Oxyria digyna.
Mountain Sorrel – Rumex paucifolius.
Mountain Soursop – Annona mantana.
Mountain Spinach – Atriplex hortensis.
Mountain Tea – Gaultheria procumbens.
MOURIRI Aubl. Memecylaceae. 50 spp. Central and tropical S.America, W.Indies.
***M. domingensis*** Spach. W.Indies. The fruits are pleasantly flavoured and are eaten locally.
***M. pusa*** Gardn. Brazil. The fruits (Pusa) are pleasantly flavoured and are eaten locally
***Mouriria*** Juss. = Mouriri.
Mouse Ear – Hieracium pilosella.
Mouse Ear – Myosotis palustris.
Mouse-eared Chickweed – Cerastium semidecandrum.
Mousse Chêne – Evernia prunastri.
Mousseron – Cletopilus prunulus.
Moxie Plum – Chiogenes hispidula.
Mowra Butter Tree – Madhuca longifolia.
Mozambique Blanc – Landolphia kirkii.
Mozambique Copal – Tracylobium hornemannianum.
Mozambique Ebony – Dalbergia melanoxylon.
Mozambique Rouge – Landolphia kirkii.
Mozote de caballo – Triumfetta lappula.
Mpesa – Muncuna poggei.
Mpoa – Loranthus discolor.
M'Pokusa – Gilbertodendron grandiflorum.
Mpopwa – Fagara chalybea.
Msala – Oxyanthus speciosum.
Msinzi – Bruguiera gymnorhiza.
Mtata – Phyllogeiton discolor.
Mu Oil – Aleurites montana.
Mubapi – Afzelia cuanzensis.
Mubsamaropa – Pterocarpus angolensis.
Mubondo – Protea angolensis.
Mubulu – Pseudospondias microcarpa.
Mubumbo – Lannea antiscorbutica.
Mubungu – Pericopsis angolensis.
Mucenna Albizia – Albizia anthelmintica.
Much-good – Peucedanum cervaria.
MUCOR Mich. ex Fr. Mucoraceae. 42 spp. Cosmopolitan.
***M. dubius*** Wehmer. Takes part in the fermentation of the Javanese beverage Ragi.
***M. javanicus*** Wehmer. See M. dubius.
***M. piriformis*** A. Fisch. Possibly useful in the industrial manufacture of citric acid.
MUCUNA Adans. Leguminosae. 120 spp. Tropics and subtropics. Vines.
***M. alterrima*** (Piper and Tracy) Holland. Tropical Asia. Used in warm countries as a cover crop, green manure and forage.
***M. deeringiana*** (Bort.) Small. = Stizolobium deeringianum. (Florida Velvet Bean). Possibly a variety of M. pruriens. Malaysia, S.Asia. See M. alterrima.
***M. flagellipes*** Vogel ex Benth. (Bontoto, Kiso, Mokolingo). Congo. The crushed bark and leaves are used locally to make a blue dye.
***M. gigantea*** DC. Malaysia. The seeds are eaten locally.
***M. jonghuniana*** Baker. Java. The beads are

used locally as a charm against diseases.

***M. monosperma*** DC. (Peddadulagondi, Sonsgaravi). Assam, E.Himalayas. A decoction of the seeds is used by the Hindus as an expectorant and as a sedative lotion.

***M. nivea*** DC. (Lyon Bean). E.India. The plant is used as a green manure and cover crop, while the pods are eaten locally as a vegetable.

***M. pachylobia*** (Piper and Tracy) Rock.= Stizolobium pachylobium. India. Cultivated locally for the pods which are eaten as a vegetable.

***M. poggei*** Taub. (Abagabadi, Mpesa, Sepe). Central Africa, Congo. The leaf-juice is used in the Congo as a black ink, and the twigs are used to stupify fish.

***M. pruriens*** (L.) DC.=Dolichos pruriens. (Cowage, Cowitch). E.Indies. See Monila sitophila. The seeds are used locally to treat intestinal worms.

Mudar gummi – Calotropis procera.

MUEHLENBECKIA Meissn. Polygonaceae, 15 spp. New Guinea, Australia, New Zealand, W. S.America. Shrubs.

***M. adpressa*** Meissn.=M. gunii. Australia. The berries are used locally for making pies, etc.

***M. gunii*** Hook. f.=M. adpressa.

***Muehlenbergia*** Schreb.=Muehlenbergia.

Mufambo – Crotalaria upembiensis.

Mufuka – Combretum zeyheri.

Mufundi – Leptoderris nobilis.

Mugengo – Rhus vulgaris.

Mugharokoko – Rhynochosia monobotrya.

Mugwort – Artemisia vulgaris.

Mugwort, Mexican – Artemisia mexicana.

Mugwort, Western – Artemisia gnaphalodes.

Muhaeba – Parinari mobola.

MUHLENBERGIA Schreb. Graminae. 100 spp. Himalayas to Japan, N.America to Andes. Grasses.

***M. huegelli*** Tren. Java. Used locally as feed for livestock.

Muhugu Oil – Brachylaena hutchinsii.

Muhumbu Oil – Brachylaena hutchinsii.

Mui – Brugeuera gymnorhiza.

Muiki – Talinum cuneifolium.

Muira Puama – Ptychopetalum olacoides.

Muira Puama Wood – Liriosma ovata.

Muirapinima Preta – Zollernia paraensis.

Muirapuama – Liriosma ovata.

Mukango – Ficus stuhlmannii.

Mukarata – Heeria insignis.

MUKIA Arn. Cucurbitaceae. 4 spp. Old World Tropics. Vines.

***M. scabrella*** Arn.=Melothria maderaspatana. Tropical Africa, Asia. A decoction of the plant is used in W.Africa to treat flatulence and induce sweating.

Mukjajali – Drosera peltata.

Mukul Myrrh – Commiphora mukul.

Mukumbi – Parinari glaba.

Mukupachiwa – Cassia singueana.

Mukute – Syzygium guinense.

Mukwa – Pterocarpus angolense.

Mulberry, Black – Morus nigra.

Mulberry, Black Persian – Morus nigra.

Mulberry Fig – Ficus sycomorus.

Mulberry, French – Callicarpa americana.

Mulberry, Indian – Morinda trifolia.

Mulberry, Native – Hedycarya angustifolia.

Mulberry, Paper – Broussonetia papyrifera.

Mulberry, Red – Morus rubra.

Mulberry, White – Morus alba.

Mule Gum – Manihot dichotoma.

Mulga Acacia – Acacia aneura.

Muli – Melocanna bambusioides.

Muling – Grewia stylocarpa.

Mulingo – Pycnanthus kombo.

Mullein, Flannel – Verbascum thaspsus.

Mullein, Turkey – Croton setigerus.

Mulu Kilavary – Commiphora berryi.

Mululu – Khaya nyasica.

Mumara – Cissus bambusati.

Mumbu ya dito – Lannea welwitschii.

Mumuye Gum – Combretum hartmannianum.

Mumuye Gum – Combretum leonense.

Mumuye Gum – Combretum sakodense.

Mundjeko – Chenopodium procerum.

MUNDULEA Benth. Leguminosae. 30 spp. Madagascar; 1 sp. tropical Africa, S.India, Ceylon.

***M. sericea*** (Willd.) Greenw.=M. suberosa. (Kattutuvera, Supti). S.India, Ceylon. The seeds are used locally to stupify fish.

***M. suberosa*** Benth.=M. sericea.

Mung Bean – Phaseolus mungo.

Mungulu – Amomum clusii.

Munienze – Cryptosepalum pseudotaxus.

MUNTINGIA L. Elaeocarpaceae. 3 spp. Tropical S.America, W.Indies. Trees.

***M. calabura*** L. (Capulin). Tropical S.America. The fruits are edible with a pleasant flavour and the bark fibres are used locally to make ropes.

Mûong – Diospyros decandra.

Mupambangoma – Albizia gummifera.

Muavi – Parkia bussei.

Mupundu – Perinari mobola.
Mupwenge – Afzelia bipinaensis.
Mundja – Lasianthera africana.
Mura Piranga – Haploclathra paniculata.
Mureré Oil – Ficus cystopodium.
Muriranyge – Albizia antunesiana.
Murlins – Alaria esculenta.
Muro – Juniperus rigida.
Murr Marigold – Bidens tripartitus.
Murrabla – Mangifera indica.
Murray Pine – Callitris calcata.
Murray Pine – Callitris glauca.
MURRAYA Koen. ex L. mut Murr. Rutaceae. 12 spp. E.Asia, Indomalaya, Pacific Islands. Trees.
***M. cerasiformis*** Blanco. = Glycosmis pentaphylla.
***M. koenigii*** Kurz. Himalayas. A decoction of the bark, leaves and root is used locally as a tonic.
***M. paniculata*** (L.) Jacq. = Chalcas paniculata. (Cosmetic Bark Tree). Malaysia. The hard light yellow wood which darkens to light brown (Satinwood) is used to make cutlery handles and walking sticks in Java. The leaves are used to flavour curries and the bark is used as a cosmetic.
Murta – Eugenia floribunda.
Muruka – Combretum zeyheri.
Murumuru (Oil) (Fat) – Astrocaryum murumura.
Mururu – Khaya nyasica.
MUSA L. Musaceae. 35 spp. (?). Old World Tropics. Large herbaceous plants. The systematics of the genus is far from clear. It seems fairly certain that the edible species, which are triploid are derived mainly from M. acuminata, with some possible contribution from M. balbisiana. M. paradisiaca, M. sapientum and possibly M. nana are considered to be varieties of M. acuminata.
***M. acuminata*** Colla. Java, New Guinea. The fruit (Pisang Jacki) are eaten locally.
***M. balbisini*** Colla. (Klue Tani, Pisang Bau). India, Malaysia, Burma, Thailand. Sometimes cultivated locally. The fruits are pickled and the male buds are eaten as a vegetable.
***M. basjo*** Sieb. and Zucc. Japan. The fibres from the pseudostem are used locally to make sails and cloth.
***M. cavendishii*** Lam. ex Paxt. = M. nana.
***M. chinensis*** Sweet. = M. nana.
***M. corniculata*** Rumph. Réunion. The large fruits are eaten locally.
***M. discolor*** Horan. New Caledonia. The fruits are eaten locally and the pseudostem fibres are used for cordage and fabrics.
***M. ensete*** Gmel. (Ansett, Abyssinian Banana, Enset). Central and E.Africa. Cultivated locally. The endosperm of the seeds is eaten locally and starch is extracted from the leaf sheaths. A fibre is extracted from the bases of the stems and young shoots.
***M. errans*** (Blanco) Teodoro. (Botoan). Philippines. The young flowerheads are eaten locally as a vegetable or raw. The ripe fruits are fermented to make vinegar. The young leaves are used as a poultice to treat chest ailments and the juice from the base of the plant is used as an urethral injection to treat gonorrhoea.
***M. fehi*** Bert. (Aiori). New Caledonia, Tahiti. The fruits are eaten locally.
***M. holstii*** K. Schum E.Africa. The pseudostem is a potential source of a good fibre.
***M. nana*** Lour = M. cavendishii = M. chinensis. (Cavendish Banana, Chinese Banana, Dwarf Banana). E.tropical Asia. Grown, but not extensively in the Old and New World tropics, but particularly in the Canary Islands. The fruits are small, but of good flavour and keeping qualities and can be eaten raw. Sometimes considered a dwarf variety of M. sapientum.
***M. oleracea*** Vieill. (Banana Porete). New Caledonia. The rhizomes are eaten locally as a vegetable.
***M. paradisiaca*** L. (Adam's Banana, Horse Banana, Plantain). Tropical Asia. Distinguished from the Banana by having a higher starch content in the fruit. The plant is cultivated widely in the tropics with many local varieties. The high starch content of the fruits and its slightly bitter flavour make it unpalatable to eat raw, but they are eaten widely when roasted or boiled when peeled. The flour from the fruit (Guiana Arrowroot) is an excellent invalid food.
***M. paradisiaca*** L. var. ***sapientum*** Kuntze. = M. sapientum.
***M. sapientum*** L. = M. paradisiaca L. var. sapientum. (Banana). Probably originated in S.E.Asia, Malaysia. It is distinguished from M. paradisiaca by its higher sugar

content in the fruits, but such a distinction is of doubtful taxonomic validity. There are some 200 forms in cultivation throughout the humid tropics. They are grown solely for the fruits which are eaten raw, fried or caramelized. In S.America, especially Venezuela the fruit is used for making vinegar. The plants are propagated entirely from suckers as the fruits do not produce viable seeds. They need a rich well-drained soil, with a minimum temperature of 10°C. The fruits contain some 22 per cent carbohydrate, several vitamins and a little protein. The world production is about 23 million tons per year and about 75 per cent of this is produced in Latin America, where Brazil is the leading producer. In many parts of S.America and Africa the fruit is a staple source of carbohydrate. There is a large export trade (about 4 million tons a year) to temperate countries. For this trade the fruit is picked about three-quarters ripe and kept in cool-storage ships (14°C.) being ripened on arrival at 21°C. The plant yields about 1½ tons an acre, but this figure varies widely. Bananas were introduced to the Americas from the Canary Islands soon after the discovery of America and the banana industry was developed in 1870–80. The crop had been grown for local consumption in the Far East for thousands of years and in Africa a little later. (The name "banana" originated in Sierra Leone). The most important plantation variety is the triploid Gros Mechele, but this is being replaced by Valery, which being rather thin-skinned is box-packed for export, rather than being exported on the stem.

***M. silvestris*** Lamarie. (Layason). Philippines, Moluccas. Yields a fibre which is used locally. See M. textilis.

***M. textilis*** Neé. (Abacá, Manila Hemp). Philippines. The plant is grown for the fibres extracted from the pseudostem. They are very flexible and water-resistant. This makes them valuable for making marine cordage, but they are used locally in the Philippines for making textiles and hats. The poorer quality fibres are used to make paper. The plant is cultivated in the Old and New World tropics giving a world production of some 100,000 tons of fibre a year. 90 per cent of this is produced in the Philippines. The plants are propagated from suckers and there are several varieties. The pseudostem is cut down and de-corticated by hand or more recently by machine. The fibres are washed dried and graded. The stronger coarser fibres are produced in the outer tissues of the stem and the weaker, finer ones within.

***M. tikap*** Warb. Caroline Islands. The leaves yield a fibre which is used locally.

***M. ulugurensis*** Warb. See M. holstii.

Musalo – Cissus rubiginosa.

MUSANGA C. Sm. ex R. Br. Urticaceae. 1 sp. Tropical Africa. Tree.

***M. smithii*** R. Br. (Kombo-kombo). The natives make a drink from the leaves.

Musau – Ziziphus abyssinica.

Muscadier à suif – Knema corticosa.

Muscadine Grape – Vitis rotundifolia.

MUSCARI Mill. Liliaceae. 60 spp. Europe, Mediterranean, W.Asia. Herbs.

***M. comosum*** Mill. Central and S.Europe. The bulbs are eaten locally in Greece.

Muscus pulmonarius – Lobelia pulmonaria.

Mushamba – Lannea discolor.

Mushende – Combretum zeyheri.

Mushi – Guibourtia coleosperma.

Mushingo – Syzygium guinense.

Mushonjura – Cussonia kirkii.

Mushroom, Field – Psalliota arvensis.

Mushroom, Fool – Panaeolus papilionaceus.

Mushroom, Rock – Gyrophora esculenta.

Mushroom, White – Psalliota campestris.

Musimba – Hyalosepalum caffrum.

Musise – Tessmannia dewildemaniana.

Musk Basil – Moschosma polystachyum.

Musk Lime – Citrus microcarpa.

Musk Mallow – Hibiscus abelmoschus.

Muskmelon – Cucumis melo.

Musk Thistle – Carduus nutans.

Musk Yarrow – Achillea moschata.

Musala-Semul – Bombax malabaricum.

Musombo – Syzgium guinense.

MUSSAEINDOPSIS Baill. Rubiaceae. 2 spp. W.Malaysia. Trees.

***M. beccariana*** Baill. W.Malaysia. The heavy resistant, yellow-brown wood is used locally for house construction and bridge-building.

MUSSAENDRA L. Rubiaceae. 200 spp. Old World Tropics. Climbers or shrubs.

***M. afzelii*** Oliv. (Damaran). W.Africa. The juice from the leaves is used locally to treat inflamed eyes.

***M. cambodiana*** Pierre=M. frondosa Lour non L. Indochina. (Buróm bac). The flowers are used locally as an infusion to treat coughs, asthma and fevers. The decoction is also applied to skin diseases.

***M. erythrorphila*** Oliv. (Ashanti Blood). Tropical Africa. The roots are chewed locally to stimulate the appetite.

***M. frondosa*** L. Malaysia. The Japanese use the juice to treat eye infections and a decoction of the leaves to combat intestinal worms.

***M. frondosa*** Lour non L.=M. cambodiana.

***M. theifera*** Pierre. Indochina. A decoction of the fruits is used locally to reduce fevers.

***M. variabilis*** Hemsl. Malaysia. A decoction of the roots is used locally to treat coughs and a decoction of the leaves is used to reduce fevers.

Mussambé – Cleome gigantea.
Mustard, Abyssinian – Brassica carinata.
Mustard, Black – Brassica nigra.
Mustard, Broad-beaked – Brassica narinosa.
Mustard Bush – Salvadora persica.
Mustard, Clown's – Iberis amara.
Mustard, Dakota – Brassica arvensis.
Mustard, Hedge – Sisymbrium officinale.
Mustard, Indian – Brassica juncea.
Mustard, Indian Brown – Brassica integrifolia.
Mustard, Japanese – Brassica japonica.
Mustard, Leaf – Brassica juncea.
Mustard, Potherb – Brassica japonica.
Mustard, Sarepta – Brassica besseriana.
Mustard, Spinach – Brassica perviridis.
Mustard, Tub-rooted Chinese – Brassica napiformis.
Mustard, White – Sinapsis alba.
Mutaku – Gilbertodendron ogoouense.
Mutango – Albizia gummifera.
Mutchuna – Cynometra hankei.
Mutewetwe – Cussonia kirkii.
Mutsikingate – Markhamia acuminata.
Mutton Grass – Poa fendleriana.
Mutuge – Pycanthus kombo.
Mutuwa – Kirkia acuminata.
Muwanga – Pericopsis angolensis.
Muwawa – Khaya nyasica.
Muyini – Pancovia harmsiana.
Muzoo – Acacia nigrescens.
Mwambala – Ficus pynaerti.
Mwausungulu – Garcinia livingstonii.
Myall, Bastard – Acacia cunninghamii.
Myall, True – Acacia pendula.

MYCENA (Fr.) S. F. Gray. Agaricaceae. 150 spp. Cosmopolitan.

***M. fasciculata*** Beeli. The fruitbodies are eaten in parts of the Congo.

Myemo – Dorstenia klainei.

***Mylitta*** Fr.=Hysterangium.

MYOPORUM Banks and Soland. ex Forst. Myoporaceae. 32 spp. Mauritius, E.Asia, New Guinea, Australia, New Zealand. Trees.

***M. debile*** R. Br. Australia. The fruits are eaten locally.

***M. laetum*** Forst. New Zealand. The bark is used locally to treat ulcers and toothache. The leaves are used to dress wounds, bruises and baby eczema and the twigs and leaves are added to steam baths.

***M. platycarpum*** R. Br. Australia. The fruits are eaten locally and a sweet exudation from the stem is a popular local delicacy, especially in S.Australia.

***M. sandwicensis*** Gray. (Bastard Sandalwood). Hawaii. The wood is sold as a substitute for true Sandalwood.

***M. serratum*** R. Br. Australia. The fruits are eaten locally.

MYOSOTIS L. Boraginaceae. 50 spp. Temperate Eurasia, mountains of tropical Africa, S.Africa, Australia, New Zealand. Herbs.

***M. palustris*** Roth.=M. scorpioides. (Forget-me-not, Mouse Ear). N.Hemisphere. An infusion of the leaves is used to treat lung complaints, e.g. bronchitis and whooping cough.

***M. scorpioides*** L. emend Hill.=M. palustris.

***Myrcia*** Soland. ex Lindl.=Pimenta.

MYRICARIA Berg. Myrtaceae. 65 spp. Tropical S.America, W.Indies. Trees. The fruits of the following species are of good flavour and are eaten locally.

***M. cauliflora*** Berg. S.Brazil.

***M. jaboticaba*** Berg. (Jaboticaba). S.Brazil. The fruits are eaten in other parts of the tropics and are made into jellies and a fermented beverage.

***M. tenella*** Berg. S.Brazil.

***M. trunciflora*** Berg. Brazil.

MYRIANTHUS Beauv. Urticaceae. 12 spp. Tropical Africa. Trees.

***M. arborea*** Palisot. Congo. The fruits are eaten locally.

MYRICA L. Myricaceae. 35 spp. Cosmopolitan, but excluding N.Africa, Central and S.E.Europe, S.W.Asia, Australia. Small trees or shrubs.

***M. acris*** DC. = Pimenta acris.

***M. aspleniifolia*** L. = Comptonia asplenii-folia.

***M. carolinensis*** Mill. (Bayberry). E. and S. U.S.A. A wax from the surface of the fruits is used to make candles.

***M. cerifera*** L. (Wax Myrtle). E. and S. U.S.A. The wax from the surface of the fruit is used to make candles. The root bark (Wax Myrtle Bark) is sold as a tonic.

***M. cordifolia*** L. (Wax Berry, Wax Bush). S.Africa. The wax (Berry Wax) is removed from the fruit in boiling water and usəd to make candles.

***M. gale*** L. (Sweet Gale). Temperate Europe, Asia and N.America. The aromatic leaves were used medicinally to treat dysentery and as a moth repellent. A decoction of the leaves was also used to treat skin diseases. They are sometimes used to increase the head on beer.

***M. javanica*** Blume. Java. The seeds are eaten locally.

***M. mexicana*** Willd. (Arbol de la Cera) Mexico. The wax from the fruits is used locally to make candles and is taken as a treatment for diarrhoea and jaundice. An infusion of the root bark is used as an astringent and emetic.

***M. nagi*** Thunb. = M. rubra (Ioobai, Yama Momo.) Tropical and subtropical Asia. The tree is cultivated in China for the seeds which are eaten locally.

***M. pringlei*** Greenm. Mexico. The wax from the fruits is used locally to make candles.

***M. pubescens*** Willd. Cool Andes. The wax from the fruits is used to make candles.

***M. rubra*** Sieb. and Zucc. = M. nagi.

Myrica Oil – Pimenta acris.

YROCARPA Benth. Urticaceae. 15 spp. Central to tropical S.America. Shrubs.

***M. colipensis*** Liebm. = M. longipes.

***M. longipes*** Liebm. = L. colipensis. (Chola-gogue indo). Central America, S.Mexico. A decoction of the plant is used locally in Menico to treat malaria.

MYRISTICA Gronev. Myristicaceae. 120 spp. Old World Tropics. Trees. The various nutmegs are the sun-dried seeds of the various species, while mace is made from the flattened sun-dried arils. They are used as condiments in flavouring pickles, sauces, puddings etc.

***M. argentea*** Warb. New Guinea. The nutmegs (Macassar Nutmeg, Papua Nutmeg) are particularly pungent and smell of wintergreen.

***M. canarica*** Bedd. = Gymnacanthera can-arica.

***M. corticosa*** Hook. f. = Knema corticosa.

***M. discolor*** Merr. = M. simianrum.

***M. fatua*** Houtt. (Mountain Nutmeg). Moluccas. The nutmegs are only used occasionally.

***M. fragrans*** Houtt. (Nutmeg). Moluccas. Cultivated in the Old and New World tropics, but the bulk of the world supplies still come from the Moluccas. Both the mace and the nutmeg must be used with caution as they contain some 4 per cent myristicin which is poisonous. The plant is grown commercially under shade. The spice has been known in Europe since the 12th century, when it was used as a condiment and fumigant. A fat (Nutmeg Butter) is derived from the seeds unsuitable for commerce. This is the source of Oleum Myristicae which is aromatic and used to perfume tooth-pastes and tobacco and is used in per-fumery. It is used medicinally to aid digestion and in embrocations to treat rheumatism. The waste fat is used to make candles.

***M. glabra*** Blyme = Horsfieldia glabra.

***M. glauca*** Blume = Knema glauca.

***M. hookeriana*** Wall. = Knema hookeriana.

***M. iners*** Blume. Java. The wood is used locally to perfume cloth.

***M. irya*** Gaertn. = Horsfieldia irya.

***M. kombo*** Baill. = Pycnanthus kombo.

***M. laurina*** Blume = Knema laurina.

***M. malabarica*** Warb. S.India. The nutmeg (Bombay Nutmeg, Wild Nutmeg) is only slightly flavoured and has little commer-cial value. It is used as an adulterant of true nutmeg and mace.

***M. simianrum*** A. DC. = M. discolor (Tanghas). Philippines. The oil from the seeds or bark is used locally to treat skin diseases.

***M. succedanea*** Blume. Moluccas. The nutmeg and mace are used occasionally.

Myrobolan – Phyllanthus emblica.

MYROCARPUS Allem. Leguminosae. 4 spp. S.Brazil, Paraguay. Trees.

***M. fastigiatus*** Allem. Brazil. An oil (Cabreu Oil) distilled from the tree is rose-scented and is used in perfumery. The trunk also yields a balsam like the Balsam of Peru.

The hard dark red wood is used for general carpentry and construction work.

***M. frondosus*** Allem. (Incienso). See M. fastigiatus.

MYROXYLON L. f. Leguminosae. 2 spp. Tropical America. Trees.

***M. balsamum*** (L.) Harms. = M. toliuferum. (Balsam of Tolu). Venezuela, Colombia, Peru. The trunk yields a balsam (Balsam of Tolu, Opobalsam) which is used in ointments as an antiseptic and in cough syrups as an expectorant. It is also used as a fixative for perfumes.

***M. balsamum*** (L.) Harms. var. ***pereirae*** (Royle) Baill. = M. periera. (Balsam of Peru). Central America. The balsam is produced by wounding the stem. The plant is grown commercially especially in El Salvador. The wood is used commercially as Mahogany. The balsam is used as Balsam of Tolu.

***M. periera*** (Royle) Klotzs. = M. balsamum var. pereirae.

***M. toliuferum*** H.B.K. = M. balsamum.

Myrrh – Commiphora myrrha.

Myrrh – Commiphora schimperi.

Myrrh, Abyssinian – Commiphora abyssinica.

Myrrh, Aden – Commiphora erythraea.

Myrrh, African – Commiphora africana.

Myrrh, Bisabol – Commiphora erythraea var. glabrescens.

Myrrh, Garden – Myrrhis odorata.

Myrrh, Herabol – Commiphora myrrha.

Myrrh, Mecca – Commiphora opobalsamum.

Myrrh, Mukul – Commiphora mukul.

Myrrh, Opopanax – Commiphora kataf.

Myrrh, Sweet Scented – Myrrhis odorata.

MYRRHIS Mill. Umbelliferae. 2 spp. Europe, W.Asia. Herbs.

***M. odorata*** Scop. (Garden Myrrh, Sweet Scented Myrrh). Europe, Caucasus. The leaves are used in home remedies for "Cleaning the blood" and as an expectorant. The roots and fruits are used to flavour brandy-based drinks.

***M. sylvestris*** Spreng. = Anthriscus sylvestre.

MYRSINE L. Myrsinaceae. 7 spp. Azores, Africa to China. Trees.

***M. australis*** (A. Rüch.) Allan. New Zealand. An infusion of the leaves is used locally to treat toothache.

***M. capitellata*** Wallich. Tropical Asia, Himalayas. The fruits are eaten locally.

***M. grisebachii*** Hiern. (Polo Blanco, Langa Marca). Argentina. The wood is used locally for furniture, turning and carts.

***M. melanophloea*** R. Br. (Cape Beech). S.Africa. The attractive red-brown wood is used for making wagons. A decoction of the leaves is used as an astringent.

***M. semiserrata*** Wall. Tropical Asia, Himalayas. The fruits are eaten locally.

***M. urvillea*** DC (Matsam) New Zealand. The bark is used for tanning.

MYRTILLOCACTUS Console. Cactaceae 4 spp. Mexico, Guatemala. Small tree-like cacti.

***M. geometrizans*** (Mart.) Console. = Cereus geometrizans. (Grambula). S.Mexico, Guatemala. The berries are popular locally and are eaten raw or dried.

Myrtle (Oil) – Myrtis communis.

Myrtle – Nothofagus cunninghami.

Myrtle Beech – Nothofagus cunninghami.

Myrtle, Citron – Backhousia citriodora.

Myrtle, Downy Rose – Rhodomyrtus tomentosa.

Myrtle, Grey – Backhousia myrtifolia.

Myrtle-leaved Orange – Citrus aurantum var. myrtifolia.

Myrtle, Native – Backhousia citriodora.

Myrtle, Red – Eugenia myrtifolia.

Myrtle, Scrub – Backhousia citriodora.

Myrtle, Tasmanian – Nothofagus cunninghami.

Myrtle Wax – Myrica cerifera.

MYRTUS L. Myrtaceae. 100 spp. Tropics and subtropics especially America.

***M. bullata*** Soland ex A. Cunn. New Zealand. The leaves are used locally as a poultice for bruises.

***M. communis*** L. (Myrtle) Mediterranean. The leaves are used in Jewish festivals and were used by the Ancient Egyptians, Greeks and Romans. The fruits are sometimes used as a condiment and to relieve stomach upsets. An oil distilled from them (Myrtle Oil, Oleum Myrti, Essence de Myrte, Engelwasser) is used in S.Europe, the Middle East and India as an aromatic tonic. The hard wood is used for furniture, tool handles and walking sticks.

***M. cheken*** Spreng. = Eugenia chequen.

***M. obcordata*** Hook. f. New Zealand. An infusion of the bark and berries is used locally to aid menstruation.

Mysore Pala Indigo – Wrightia tinctoria.

Mysore Raspberry – Rubus albescens.

***Mytilococcus quercifolia*** Zoll = Lunasia amara.

MYXOPYRUM Blume. Oleaceae. 12 spp. Indochina, Indomalaya. Climbers.

***M. nervosum*** Blume. Malaysia. A decoction of the roots is used in Java to relieve pains in the joints and the bark is used as a rough cordage.

# N

Na – Mesua ferrea.

Na Babac – Asplenium macrophyllum.

Nacapuli – Ficus padifolia.

Nacare – Lippia umbellata.

Nacaspilo – Inga spuria.

***Naematoloma*** Karst. = Hypholoma.

Nafoorbogin – Morinda lucida.

Nagami Kumquat – Fortunella margarita.

Nagari – Calophyllum tormentosum.

Nagkesar – Mammea longifolius.

Nahuatl – Rhynchosia pyrimidalis.

Nai Chuk – Bambusa pervariabilis.

Nai Chuk – Bambusa tuldoides.

Nailroot, Silver – Paronychia argentea.

NAJAS L. Najadaceae. 50 spp. Cosmopolitan. Aquatic herbs.

***N. flexilis*** (Willd.) Rostk. and Schmidt. Old World, N.America, W.Indies. The whole plants are harvested locally and used as manure, or sometimes they are dried and used as a packing material.

***N. major*** All. Temperate and tropics. The leaves are eaten as a salad in Hawaii, especially with shell-fish.

Naked Oat – Avena nuda.

Nam hoàng liên – Fibraurea recisa.

Name Akam – Dioscorea latifolia.

Name Amarillo – Dioscorea cayenensis.

Name de Agua – Dioscorea alata.

Name de China – Dioscorea batatas.

Name Dunguey – Dioscorea altissima.

Name Guinea – Dioscorea cayenensis.

Name Mapicey – Dioscorea trifida.

Name Papa – Dioscorea esculenta.

Nan Chu – Dendrocalamus giganteus.

Nanan Wood – Lagerstroemia lanceolata.

Nanga – Cynometra alexandri.

***Nania vera*** Miq. = Metrosideros vera.

Nanking Cotton – Gossypium nanking.

NANNORHOPS H. Wendl. Palmae. 4 spp. Arabia to N.W.India. Palms.

***N. ritchieana*** H. Wendl. = Chamaerops ritchieana. (Fees, Mazurrye, Peer putta). N.W.India. The leaves are used locally to make mats, fans, baskets, etc. The young leaves, flowers and fruits are used as a vegetable and the leaf fibres are made into ropes. The seeds are used as beads.

Nanny Berry – Viburnum cassinoides.

Nanu – Gardenia remyi.

Napa huite – Trichilia hirta.

Napier Grass – Pennisetum purpureum.

NAPOLEONA Beauv. Napoleonaceae. 15 spp. Tropical W.Africa. Trees.

***N. heudelotii*** Juss. W.Africa. In Sierra Leone the pulped fruits are used to treat hernia.

***N. leonensis*** Hutch. and Dalz. In Liberia the bark is chewed with cola nuts, or eaten as a condiment with rice.

Naranjillo – Solanum quitoense.

Narasplant – Acanthosicos horridus.

Narbonne Vetch – Vicia narbonensis.

NARCISSUS L. Amaryllidaceae. 60 spp. Europe, Mediterranean, W.Asia. Herbs.

***N. jonquilla*** L. (Jonquil). Mediterranean. Cultivated for the essential oil used in perfumery and as an ornamental. The perfume has a heavy odour and contains benzyl benzoate, methyl benzoate, methyl cinnamate, indol and linalol.

***N. poeticus*** L. (Poet's Narcissus). Europe. Cultivated as an ornamental and for the essential oil used in perfumery.

***N. pseudo-narsiccus*** L. (Daffodil). Europe. Cultivated widely as an ornamental and commercially for the florists' trade.

Narcissus Oil – Narcissus poeticus.

Narcissus, Poet's – Narcissus poeticus.

NARDOSTACHYS DC. Valerianaceae. 2 spp. Himalayas, W.China. Herbs.

***N. grandiflora*** DC. See N. jatamansi.

***N. jatamansi*** DC. (Jatamansi). Himalayan region. An essential oil distilled from the roots (Nardus Root, Spikenard of the Ancients) was used as a perfumed cosmetic by the aristocracy of ancient Rome. It is now used as a black hair dye and hair tonic and to treat nervous disorders.

Nardus Root – Nardostachys jatamansi.

NAREGAMIA Wight. and Arn. Meliaceae. 1 sp. S.W. tropical Africa; 1 sp. India. Shrubs.

***N. alata*** Wight. and Arn. India. The root (Country Ipecacuanha, Goa Ipecacuanha, Portuguese Ipecacuanha). Is used locally to treat rheumatism and dysentery. It is also used as an emetic.

Nargusta Terminalia – Terminalia obovata.

Naranjillo – Solanum quitoense.

Narihira Bamboo – Semiarundinaria fastuosa.

Narihiradake – Semiarundinaria fastuosa.

Narkachura – Curcuma caesia.

Narras – Acanthosicyos horrida.

Narrenschawamm – Panaeolus papilionaceus.

Narrowleaf Cat Tail – Typha angustifolia.

Narrowleaf Globe Mallow – Sphaeralcea angustifolia.

Narrowleaf Vetch – Vicia angustifolia.

Narrow-wing Acacia – Acacia stenocarpa.

Narve – Premna tomentosa.

Nasi sedjook – Salacia grandiflora.

NASTURTIUM R. Br. Cruciferae. 6 spp. Throughout Europe to Central Asia, Afghanistan, N.Africa, mountains of tropical E.Africa and N.America. Herbs.

***N. apetalum*** DC.=N. indicum var. apetala.

***N. indicum*** DC. var. ***apetala*** Gagnep.=N. apetalum=Sisymbrium apetalum. India to China and Indochina. The seeds are used in China to treat asthma. The whole plant is used locally as a diuretic and to treat scurvy. Seeds produced in Vietnam are exported to China.

***N. latifolia*** Crantz.=Lepidium latifolia.

***N. officinale*** R. Br.=Rorippa nasturtium-aquaticum. (Water Cress). Europe, N.America and temperate Asia. The leaves and stems are eaten in salads. The plant is cultivated. The whole plant is used in home remedies to treat scurvy, to treat kidney complaints, to reduce fevers and "clean the blood".

***N. palustre*** DC. S.Europe. The leaves are used as a salad in S.France.

Nasturtium – Tropaeolum majus.

Nasturtium, Tuber – Tropaeolum tuberosum.

Natal Aloë – Aloë candelabrum.

Natal Grass – Pennisetum unisetum.

Natal Grass – Tricholaena rosea.

Natal Mahogany – Kiggelaria dregeana.

Natal Mahogany – Trichilia emetica.

Natal Plum – Carissa grandiflora.

Natal Slangkop – Urginea macrocentra.

Native Currant – Leptomeria acida.

Native Millet – Panicum decompositum.

Native Mulberry – Hedycarya angustifolia.

Native Myrtle – Backhousia citriodora.

Native Olive – Osmanthus americana.

Native Poplar – Codonocarpus cotinifolia.

Nato – Palaquium obtusifolium.

Nato Tree, Java – Palaquium javanense.

NAUCLEA L. Naucleaceae. 35 spp. Tropical Africa, Asia, Polynesia. Trees.

***N. diderrichii*** De Wild. (West African Boxwood). W.Africa. The hard bright yellow wood is used for boards, furniture, bridges, mortars and canoes.

***N. esculentus*** Afzel. (Daindaté, Doundaté). W.Africa. The sweet juicy fruits are eaten locally and the flowerheads are used as a vegetable. The roots (Peach Root) are used locally as a decoction to treat fevers and indigestion. A yellow dye from the roots is used to stain Morocco leather.

***N. excelsa*** Blume.=N. mollis. (Ki saat, Klepoo ketek). Java. The heavy hard pink-yellow wood is used locally to make houses and household utensils.

***N. mollis*** Blume.=N. excelsa.

***N. officinale*** Pierre. (Huynh ba). S.Vietnam, Cambodia. A decoction of the yellow wood is bitter and is used locally to reduce fevers.

***N. undulatus*** Miq. (Bangal perampocan). Borneo, Malacca, Moluccas. The bitter leaves are occasionally eaten locally, especially as a famine food.

***Nauclea*** Korth. non L.=Neonauclea.

Ndaku – Gilbertodendron ogoouense.

Ndele – Talinum cuneifolium.

Ndongo a moaba – Amomium clusii.

Neb-Neb – Acacia arabica.

NECTANDRA Roland ex Rottb. Lauraceae. 100 spp. Central to sub-tropical S.America. Trees.

***N. cinnamomoides*** Nees. Ecuador. The tree is cultivated locally for the flower calyx which is used as a spice.

***N. coriacea*** (Siv.) Griseb. (Jamaica Nectandra, Laurel Avispello). W.Indies. The pinkish wood is light, easy to work, but susceptible to termites. It is used for house interiors, furniture, carving, toy-making, boats and boxes.

***N. membranacea*** (Sw.) Griseb. (Laurel Prieto). Tropical America, W.Indies. The yellow-brown wood is soft, easy to work but susceptible to termites. It is used for furniture, house interiors, boxes, crates and general light carpentry.

***N. pisi*** Miq. (Black Cedar, Yellow Ciroua-balli). Tropical America, especially

Guyana. The brown wood is used for ship-building and harbour works.

***N. puchury-minor*** Nees. and Mart= Ocotea puchury-minor. Brazil. The seeds (Pichurim Beans, Puchury Beans) are used in local medicine.

***N. rodioei*** (Schomb.) Hook. Tropical America. The wood (Bebeeru, Greenheart, Itauba Branco) is heavy, tough and elastic. It is used for ship-building and harbour works.

***N. sintenisii*** Mez. (Laurel Amarillo). Tropical America, W.Indies. The green-yellow wood is soft, light and susceptible to termites. Used as N. membranacea.

Nectandra, Jamaica – Nectandra coriaceae.

Nectarine – Amygdalus persica.

Nedun – Pericopsis mooniana.

NEEA Ruiz. and Pav. Nyctaginaceae. 80 spp. Mexico to tropical S.America, W.Indies. Small shrubs.

***N. parviflora*** Poepp. and Endl. Tropical S.America. The leaves (Yana múco) are chewed by the Peruvian Indians to preserve and blacken the teeth.

***N. theifera*** Oest. (Caparrosa). Brazil. The leaves are used locally as a tea substitute and the source of a black dye.

Needle and Thread – Stipa comata.

Needle Grass, Green – Stipa viridula.

Needle Pine – Hakea leucoptera.

Neem Tree (Toddy) – Azadirachta indica.

Neensa – Clitocybe castanea.

NESSIA Blume. Bombacaceae. 8 spp. W.Malaysia. Trees.

***N. altissima*** Blume. (Ki bengang). Indonesia. The whitish wood is insect-resistant and is used locally for houses and small boats. A decoction of the fruit skins is diuretic and is used locally to treat gonorrhoea.

***N. malayana*** Bakh. (Sebongang). Indonesia. The dark brown wood is used locally for pillars and beams.

Negra Lora – Matayaba domingensis.

Negro Coffee – Cassia occidentalis.

Negro Pepper – Xylopia aethiopica.

Negro Yaw – Dioscorea cayenensis.

***Negundo aceroides*** Moench.=Acer negundo.

***Negundo fraxinifolium*** Nutt.=Acer negundo.

***Nelitris paniculata*** Lindl.=Decaspermum fruticosum.

***Nelumbium*** Juss.=Nelumbo Adans.

NELUMBO Adans. 1 sp. E. United States of America to Colombia; 1 sp. Asia and N.E.Australia. Aquatic herbs.

***N. lutea*** Pers.=N. pentapetala.

***N. nucifera*** Gaertn.=N. speciosa. Asia, N.E. Australia. (Lotos, Lotus). The rhizomes are eaten as a vegetable, or preserved in sugar in the Far East. They are also ground as a starch (Lotus Meal). The seed kernels are also used as a source of starch, or eaten dry. The leaves are boiled as a vegetable. The stamens are used to flavour tea in Indo-China. The flowers are sacred to the Buddhists. This could be the sacred lotus of the Nile, which is no longer found there.

***N. pentapetala*** (Walt.) Fernald.=N. lutea. (Duck Acorn, Lotus, Water Chinquapin, Water Nut). E. U.S.A. to Colombia. The starchy rhizomes, leaves and seeds were eaten by the local Indians.

***N. speciosa*** Willd.=N. nucifera.

Nemaax – Heliotropium angiospermum.

NEMALION Targioni-Tozzetti. Helminthocladiaceae. Red Algae.

***N. lubricum*** Duby. Pacific. Eaten in Japan.

***N. xiphoclada*** Baker. Pacific. Eaten in Japan.

NEMATOSPORA Peglion. Saccharomycetaceae. 3 spp. Parasitic.

***N. gossypii*** (Ashby and Nowell). Guill. Parasitic in cotton bolls, but potentially capable of producing riboflavin industrially.

Nembambobolo – Holoptelea grandis.

Nemocá – Ocotea spathulata.

***Nemopanthus andersonii*** Hart.=Maytenus ilicifolia.

Neninga – Polygonum salicifolium.

***Neobaronia*** Baker=Phylloxylon.

NEOCALLITROPSIS Florin. Cupressaceae. 1 sp. New Caledonia. Tree.

***N. araucarioides*** Compton. A viscid rose-scented oil (Oil of Araucaria) is distilled from the wood. It is used as a fixative in scenting cosmetics, soaps, creams etc.

NEODYPSIS Baill. Palmae. 15 spp. Madagascar. Palm trees.

***N. baronii*** Jem. (Farihazo). E.Madagascar. The buds are eaten as a vegetable locally.

***N. tanalensis*** Jum. and Perr. (Lafu, Matitana). S.E.Madagascar. The leaves yield a fibre used locally.

NEOGLAZIOVIA Mez. Bromeliaceae. 2 spp. Brazil. Herbs.

***N. variegata*** (Arrunda da Cam.) Mez.= Dyckia glaziovii. (Caroa Verdadeira).

The leaf fibres (which resemble Agave) are used locally for making nets, packing tobacco and making paper.

NEOLITSEA (Benth.) Merr. Lauraceae. 80 spp. E. and S.E.Asia, Indomalaya. Trees.

***N. nanmu*** Hemsl. India, China. The wood is used by the Chinese for coffins.

***N. odoratissima*** Nees. India, China. The wood is used locally for house-building and to make tea chests.

***N. pauhoi*** Kanch. India, China. A mucilage washed from the wood shavings (Pau-Hoi) is used by the Chinese women to stick down their hair.

NEONELSONIA Coulter and Rose. Umbelliferae. 1 sp. Mexico. Herb.

***N. acuminata*** (Benth.) Coulter and Rose. (Ingo-sha-hush). An extract of the leaves and stems is used locally in Colombia to treat intestinal inflammations and "to prevent the death" of women after childbirth.

Neosia Pine – Pinus gerardiana.

***Neottia diuretica*** Willd. = Spiranthes diuretica.

***Neotyphonia integrifolia*** Shafer = Rhus integrifolia.

***Neotyphonia ovata*** Abrams. = Rhus ovata.

Neou Oil – Parinari macrophyllum.

***Neowashingtonia filifera*** Sudw. = Washingtonia filifera.

Nepal Cardamon – Amomum subulatum.

Nepal Paper – Daphne cannabine.

Nepalese Crane's Bill – Geranium nepalense.

NEPENTHES L. Nepenthaceae. 67 spp. Ceylon, Assam, S.China, Indochina, Malaysia, N.Queensland, New Caledonia. Woody climbers.

***N. distllatoria*** L. (Pitcher Plant). Ceylon. The stems are used locally to make small baskets, etc.

***N. reinwardtiana*** Miq. Malaysia. The stems are used locally to make cordage.

NEPETA L. Labiatae. 250 spp. Temperate Eurasia, N.Africa, mountains of tropical Africa. Herbs.

***N. cataria*** L. = Cataria vulgaris. (Catmint, Catnip). Europe, introduced to N.America. A decoction of the dried leaves and flower heads are used to treat stomach upsets, especially in young children.

***N. glechoma*** Benth. = Glechoma hederaceum.

***N. pectinata*** L. = Hyptis pectinata.

NEPHELIUM L. Sapindaceae. 35 spp. Burma to Indochina, W.Malaysia. Trees.

***N. chryseum*** Blume. Malaya. The fruits are eaten locally.

***N. lappaceum*** L. = Euphoria nephelium (Rambutan). Malaysia. Cultivated in tropical Asia for the fine flavoured fruits which are eaten raw. The seeds are eaten roasted and yield a fat (Rambutan Tallow) used for candles.

***N. litchi*** Camb. (Litchi). S.China. The fruit (Litchi) is a favourite among the Chinese. The plant is cultivated in the Old and New World. The dried fruits (Litchi Nuts) are exported. There are numerous varieties.

***N. longana*** (Lam.) Cam. = Euphoria longana. (Longan). S.China. The plant is cultivated locally for the fruits (Longan, Longyen, Lenkeng), which are eaten raw or dried.

***N. mutabile*** Blume (Pulasan). Malaysia. A fruit of excellent flavour, it is widely cultivated in the Far East and China. The fruits are eaten fresh.

NEPHROMA Lichen. Peltigeracea.

***N. arcticum*** (L.) E. Fr. Arctic. Eaten as food by reindeer.

***N. parillis*** Ach. (Chocolate-coloured Nephroma, Powdered Swiss Lichen). Subarctic to temperate. An extract is used in Scotland to dye wool blue.

Nephroma, Chocolate-coloured – Nephroma parilis.

NEPTUNIA Lour. Leguminosae. 15 spp. Tropics and subtropics. Waterplants.

***N. oleracea*** Lour. = N. prostrata. Tropics. The young shoots are sometimes eaten as a vegetable.

***N. prostrata*** Baill. = N. oleracea.

NEREOCYSTIS Postels and Ruprecht. Laminariaceae. Large brown seaweeds.

***N. luetkeana*** (Mert.) Postels and Ruprecht. Pacific N.America. The plant is eaten by the Orientals in N.America. The powdered seaweed is used to make mineral supplement pills for humans. The long stipes are used as fishing lines by the Indians.

NERIUM L. Apocynaceae. 3 spp. Mediterranean to Japan. Shrubs. The plants contain the cardiac poisons neriodrin and neriodorein.

***N. indicum*** Mill. = N. odorum. Tropical Asia. A decoction of the root is used in India and Malaya to induce abortion and for suicide. A poultice of the root is used against ringworm. The flowers are used for perfume and produce a good honey. The plant is cultivated.

***N. odorum*** Soland.=N. indicum.

***N. oleander*** L. (Oleander). Mediterranean. Cultivated in the Old and New World as an ornamental. The roots are used in criminal poisoning and to exterminate rats.

Neroli Oil – Citrus aurantium.

NESAEA Comm. ex Juss. Lythraceae. 50 spp. Tropical and S.Africa, Madagascar, Ceylon, Australia, S.America. Shrubs.

***N. salicifolia*** H.B.K.=Heimia salicifolia. S.W. U.S.A., to Mexico, Jamaica and S.America. A decoction of the plant is used locally as a pleasant intoxicant.

Nesco – Willardia mexicana.

***Nesodaphne tawa*** Hook. f.=Beilschmiedia tawa.

NESOGORDONIA Baill. Sterculiaceae. 17 spp. Tropical Africa, Madagascar. Trees.

***N. papaverifera*** Chev. Tropical W.Africa. The wood (Ironwood) is red-brown and very hard. It is used for railway sleepers and ship bulkheads. Locally it is used for axe handles etc. and mortars.

***Nestegia sandwicensis*** (A. Gray) Degen. and L. Johns=Osmanthus sandwicensis.

Netleaf Oak – Quercus reticulata.

Nettle, Bigstring – Urtica dioica.

Nettle, Carolina Horse – Solanum carolinense.

Nettle, Dog – Urtica urens.

Nettle, Hemp – Laportea gigas.

Nettle, Horse – Solanum carolinense.

Nettle, Indian – Acalypha indica.

Nettle, Nilgiri – Girardinia palmata.

Nettle Root – Urtica dioica.

Nettle Seed – Urtica dioica.

Nettle, Small – Urtica urens.

Nettle, Small-leaved – Laportea photiniphylla.

Nettle, Stinging – Urtica dioica.

Nettle Tree – Celtis australis.

Nettle, Wood – Laportea canadensis.

NEURADA L. Neuradaceae. 1 sp. E.Mediterranean to Indian deserts. Shrub.

***N. procumbens*** L. The plant is fed to camels.

NEUROSPORA Shear and Dodge. Phaeosporae. 4 spp. Widespread.

***N. crassa*** Shear and Dodge. Mutant strains are used for the bioassay of amino acids, vitamins etc.

Neusi – Passiflora organensis.

Nevada Joint Fir – Ephedra nevadensis.

New Caledonian Sandal Wood – Santalum austro-caledonicum.

New Jersey Tea – Ceanothus americanus.

New South Wales White Ash – Schizomeria ovata.

New Zealand Flax – Phormium tenax.

New Zealand Passion Flower – Passiflora tetrandra.

New Zealand Spinach – Tetragonia tetragonioides.

NEWBOULDIA Seem. ex Bur. Bignoniaceae. 1 sp. Tropical W.Africa. Tree.

***N. laevis*** Seem. A decoction of the leaves is used in Ghana to treat dysentery, stomach complaints and fever, and in Nigeria as a salve for sore eyes.

Nezu – Juniperus rigida.

Ngai Camphor – Blumea balsamifera.

Ngâi tuong – Stephania rotunda.

Ngalati – Brachystegia taxifolia.

Ngân chày – Polyalthia thorelli.

Ngansa – Brachystegia taxifolia.

Ngantho – Palaquium obtusifolium.

Ngear – Dillenia ochreata.

Ngelakusa – Blighia welwitschii.

N'Ghat Oil – Plunkenetia conophora.

Ngilimana – Pancovia harmsiana.

Ngo shut – Kaempferia pandurata.

Nguba – Harunguana paniculata.

Ngubi – Pseudospondias microcarpa.

Nhâm sâm – Hibiscus sagittifolius.

Nhlopi – Acacia nigrescens.

Nho – Vitis pentagona.

Nho rùng – Ampelocissus martini.

Niam (Oil, Fat) Tree – Lophira alata.

Niaouli Oil – Melaleuea viridiflora.

Nicaraguan Cacao Tree – Theobroma bicolor.

Nicaragua Ipecac – Cephaëlis acuminata.

Nicaragua Rosewood – Dalbergia retusa.

Nicaragua Wood – Caesalpinia echinata.

Nicaragua Wood – Haematoxylum brasiletto.

Nickar Bean – Caesalpinia bonducella.

Nicobar Breadfruit – Pandanus leram.

NICOLAIA Horan. Zingiberaceae. 25 spp. Indomalaya. Herbs.

***N. hemisphaerica*** (Blume.) Schum. Malaysia. The leaves have a potential use in making paper.

***N. magnifica*** Schum.=Alpinia magnifica=Amomum magnificum. Malaysia. The roots are used for flavouring food and the young flowering shoots are used in curries. The plant is cultivated locally.

NICOTIANA L. Solonaceae. 21 spp. Australia,

Polynesia; 45 spp. N. and S.America, warm temperate. Herbs.

***N. alata*** Link. and Otto. S.America. The leaves are smoked and chewed by the local Indians.

***N. attenuata*** Torr. S.W. U.S.A., Mexico. The leaves are smoked locally.

***N. glauca*** Graham (Tree Tobacco). S.America, introduced as far N. as S.W. U.S.A. Used to make insecticides. It contains the alkaloid anabasine which is particularly effective against aphids.

***N. palmeri*** Gray=N. trigonophylla.

***N. quadrivalvis*** Pursh. W. U.S.A. Cultivated and used by the local Indians for smoking.

***N. rustica*** L. (Aztec Tobacco). Mexico. Cultivated and used extensively for smoking by the E. coast Indians of U.S.A. It is probably the first tobacco to be introduced to Europe by Jean Nicot (hence the name Nicotiana) to Lisbon in 1558. It is now grown on a small scale in Europe and the Far East. The plant has a higher nicotine content than N. tabacum and is used mainly for producing nicotine for medicinal purposes and insecticides. A little is also smoked.

***N. tabacum*** L. (Tobacco). Tropical America. The plant is not now found in the wild and is probably a fortuitous cross between N. sylvestris and N. otophora. It is tetraploid. The cured and dried leaves are used to make tobacco, snuff and are a source of nicotine for the manufacture of insecticides and nicotine sulphate. The smoking habit was introduced to the U.K. by Sir Walter Raleigh in 1585 (see N. rustica) and into Indian and China by the Portuguese, from whence it reached Ceylon and Turkey by 1610. In the same year the first commercial plantings were made in Virginia by John Rolfe. The plant is well adapted to climate, but requires a medium rich soil and is very demanding on labour. The annual world production is about 8,660 million pounds. Of this the U.S.A. produce 21 per cent, China 20 per cent, India 7 per cent, U.S.S.R. 6 per cent, Brazil, Japan, Turkey and Rhodesia about 3 per cent and other counties including Indonesia, E.Africa and the Balkans produce the rest.

When the plants reach a suitable size, the apical buds are removed. This encourages branching and an increase in size of the remaining leaves. The leaves also ripen more evenly and the nicotine content is increased. The first leaves are ready for harvesting about four months after planting. They are usually harvested from the base of the plant first. The type of tobacco produced will depend on the variety of plant, the area where it is grown and the type of curing. Curing involves a slow drying of the plant. During the process the starch and protein contents are greatly reduced, the volatile nicotine is driven off and the organic acid (mainly citric acid) content increases. Curing may be by air, fire or in flues. Air curing entails hanging the leaves in heated, but ventilated barns of controlled humidity for 1–2 months. Fire curing is in barns with fires on the floor, in which the leaves are placed on racks and allowed to come in contact with the smoke during 3–10 weeks. Flue curing takes only 4–6 days and in this process the leaves are hung in smaller barns, heated by pipes (Flues). After curing the leaves are fermented in heaps for 4–6 weeks during which time the smell and taste develop due to the release of essential oils and other chemical changes. The time and conditions of curing depend on the use to which the final product will be put. The final product may be treated with molasses, sugar, rum, apple juice and is sometimes scented. There are a large number of varieties and the nomenclature is further confused by the trade names given to the different types of end-product.

***N. trigonophylla*** Dunal.=N. palmeri. Mexico, S.W. U.S.A. The leaves are smoked by the local Indians.

Nicuri – Syagrus coronata.

***Niebuhria pedunculosa*** Hochst.=Maerusa pedunculosa.

Niepa Bark Tree – Quassia indica.

Niey datah – Detarium senegalense.

Nigaki – Picrasma quassioides.

NIGELLA L. Ranunculaceae. 20 spp. Europe, Mediterranean to Central Asia. Herbs.

***N. sativa*** L. (Black Cummin). Central Europe to N.Africa and W.Asia the plant is cultivated for the seeds which are used as a seasoning for foods. They are sometimes mixed with bread. They are also used to treat intestinal worms and jaundice.

Niger Copal – Daniellia oblonga.
Niger Gutta – Ficus platyphylla.
Niger Seed Oil – Guizotia abyssinica.
Nigerian Ironweed – Vernonia nigritiana.
Night Blooming Cereus – Selenicereus grandiflorus.
Nightshade, Black – Solanum nigrum.
Nightshade, Deadly – Atropa belladonna.
Nightshade, Silverleaf – Solanum elaeagnifolium.
Nikau Palm – Rhopalostylis sapida.
Nikkomomi – Abies homolepis.
Nikko Silver Fir – Abies homolepis.
Nilau kootjing – Colona auriculatum.
Nile Grass – Acroceras macrum.
Nilfiri Nettle – Girardinia palmata.
***Nima quassioides*** Buch-Ham. = Picrasma quassioides.
Nimba – Azadirachta indica.
Nimbar – Acacia leucophloea.
Nim Bark – Azadirachta indica.
Nin meng – Citrus limonia.
***Nipa*** Thunb. = Nypa.
Nipa Palm – Nypa fruticans.
Niri batoo – Xylocarpus moluccensis.
Nisham – Hitchenia caulina.
Nistalmal – Paullinia pinnata.
NITRARIA L. Zygophyllaceae. 7 spp. Sahara through to S.Russia, Afghanistan and E.Siberia; 1 sp. S.E.Australia. Shrubs.
***N. retusa*** (Forsk.) Asch. N.Africa, Middle East. The fruits are mildly narcotic and are eaten by the local Arabs. Sodium carbonate is extracted from the ash of the plant for local use.
***N. schoberi*** L. S.E.Australia. The fruits are eaten by the aborigines.
***N. tridentata*** Desf. Sahara region. The fruits are eaten locally.
NITROBACTER Sack. Bacteriaceae. Several species take part in the nitrogen cycle, oxidising nitrites in the soil to nitrates.
Nitronga Nuts – Cryptocarya latifolia.
NITROSOMONAS Sack. Bacteriaceae. Several species take part in the nitrogen cycle, oxidising ammonium ions to nitrite.
Nitta Tree, Fern-leaved – Parkia filicoidea.
Njating Mahabong – Hopea balagernan.
Njating Matpeleppek – Hopea sangal.
Njating Matapoesa – Hopea sangal.
Njatoh labar – Godoya firma.
Njatok takis – Palaquium microphyllum.
Njatoo doojan – Palaquium treubii.
Njatuo Fat (Tallow) – Palaquium oleiferum.
Njooroo – Xylocarpus moluccensis.
Nkombo – Pycnanthus kombo.
Nkuba – Hippocrayea myiantha.
Nkundu – Salacia soyauxii.
Noa – Agave victoriae-reginae.
Noble Fir – Abies nobilis.
Nodding Alder – Alnus pendula.
Nodding Anemone – Anemone cernua.
Nodding Brome – Bromus anomalus.
Nodding Merawan – Hopea nutans.
Nodding Onion – Allium cernuum.
Nogada de Ibarra – Juglans honorei.
Nogal – Juglans honorei.
Nogal – Juglans insularis.
Nogal silvestre – Juglans major.
NOGO Baehni. Sapotaceae. 1 sp. Tropical Africa.
***N. nogo*** Baehni. The seeds yield a fat (Beurre de Nogo), used locally in Gabon for cooking.
Noire de Congo – Clitandra orientalis.
Nokanga – Allophylus lastoursvillensis.
Noli Palm – Corozo oleifera.
NOLINA Michx. Agavaceae. 30 spp. S.W. United States of America, Mexico. Large monocotyledonous shrubby herbs. The leaves of the following species are used in Mexico for thatching, mats, baskets, brooms and hats.
***N. inermis*** (S. Wats.) Rose.
***N. longifolia*** (Schult.) Hemsl. (Zacate).
***N. recurvata*** Hemsl. = Beaucarnea recurvata = Dasylirion recurvatum.
***N. stricta*** Lem.
Nonay Spruce – Picea excelsa.
Nonda Tree – Parinari nonda.
Nongo Gum – Albizia brownii.
Nonno – Tacca fatsiifolia.
Nonsuch – Medicago lupulinus.
Nov-maie – Ficus aspera.
Nopal – Opuntia cochenillifera.
Nopal – Opuntia ficus-indica.
Nopal – Opuntia megacantha.
Nopal Cardamon – Amomum aromaticum.
Nopal chamacuero – Opuntia dejecta.
Nopal Cordón – Opuntia streptacantha.
Nopal cyotillo – Opuntia azurea.
Nopal Durazillo – Opuntia leucotricha.
***Nopalea*** Salm-Dyck. = Opuntia.
Nopalillo – Opuntia azurea.
Nopalillo – Opuntia karwinskiana.
Nordmann Fir – Abies nordmanniana.
Norino tsukudani – Porphyra vulgaris.
NORONHIA Stadm. ex Thou. Oleaceae. 40 spp. Madagascar, Mauritius, Comoro Islands. Trees.
***N. emarginata*** (Lam.) Stadtm. = Olea emarginata. (Madagascar Olive, Noronhia).

Madagascar. The fruits are eaten locally and the plant is occasionally cultivated.
Noronhia – Noronhia emarginata.
North African Ammoniac – Ferula tingitana.
North American Ebony – Diospyros virginiana.
Northern Dewberry – Rubus flagellaris.
Northern Prickly Ash – Zanthoxylum americanum.
Northern Rata – Metrosideros lucida.
Norway Maple – Acer platanoides.
Norway Spruce – Picea abies.
NOSTOC Vaucher. Nostocaceae. Blue-green algae, fresh-water.
***N. commune*** Vaucher var. ***flagelliforme*** (Berk. and Cart.) Born and Flah. Eaten locally in China where it is called fahtsai or fatchoy.
NOTHOFAGUS Blume. Fagaceae. 35 spp. New Guinea, New Caledonia, Australia, New Zealand, temperate S.America. Trees.
***N. betuloides*** Blume=Fagus betuloides. S.Chile. The wood is hard and heavy and is used locally for furniture, waggons, agricultural implements, ship-building.
***N. cliffortoides*** Oertst.=Fagus cliffortoides. (Mountain Beech). New Zealand. The wood is used for general carpentry, piers and telegraph poles.
***N. cunninghami*** (Hook.) Oerst. (Myrtle, Myrtle Beech, Tasmanian Beech, Tasmanian Myrtle). Tasmania, S.Australia. The pink-brown wood is light, strong, hard and fairly elastic. It is used locally for furniture, all forms of bent work, turning, flooring, panelling and veneers.
***N. fusca*** Oerst.=Fagus fusca (Red Beech). New Zealand. The wood is used locally for house interiors, piers and bridges and railway sleepers.
***N. menziesii*** Oerst.=Fagus menziesii (Silver Beech). New Zealand. The tough red wood is elastic and is used for various forms of barrels, buckets etc. It is also used for house-building and railway sleepers.
***N. obliqua*** Blume.=Fagus obliqua (Roble Pellin). Chile. The dark red heavy, hard wood is used for general construction work and railway sleepers.
***N. procera*** Oerst.=Fagus procera (Rauli). Chile. The bright red wood is used for furniture, barrels etc. A popular wood for furniture.
***N. pumilo*** Reiche. Andes. The wood is used for furniture and general construction work.
***N. solandri*** Oerst.=Fagus solandri. New Zealand. The wood is light red, with a dark heartwood. It is strong and durable and is used for outside construction work, bridges and railway sleepers.
***Nothopanax*** Miq. emend Harms.=Polyscias.
***Nothopanax edgerleyi*** Hook. f.=Panax edgerleyi.
***Nothoprotium sumatranum*** Miq.=Pentaspadon motleyi.
Nsena – Kirkia acuminata.
Ntoka – Millettia laurentii.
Nuez de Australia – Macadamia tetraphylla
Nuez Moscada – Ocotea moschata.
Nunye – Phyllogeiton discolor.
NUPHAR Sibth. and Sm. Nymphaceae. 25 spp. N.Temperate. Water herbs.
***N. adventa*** Ait. f.=Nymphaea adventa (Cow Lily, Yellow Pond Lily). E. U.S.A. The starchy rhizome is eaten raw, or as a vegetable by the local Indians. The seeds are eaten in soups.
***N. luteum*** Sm. Europe to Asia. The starchy rhizomes are a possible emergency food.
***N. polysepala*** Engelm. W. N.America. The seeds are eaten by the local Indians.
Nut, Amazon Poison – Strychnos castelnaei.
Nut, Barcelona – Corylus avellana.
Nut, Bengor – Caesalpinia bonducella.
Nut, Boleko – Ongokea klaineana.
Nut, Bonduc – Caesalpinia bonducella.
Nut, Brazil – Bertholletia excelsa.
Nut, Buffalo – Pyrularia pubera.
Nut, Burrawang – Macrozamia spiralis.
Nut, Butter – Caryocar nucifera.
Nut, Cashew – Anacardium occidentale.
Nut, Clearing – Strychnos potatorum.
Nut, Cob – Corylus avellana.
Nut, Cob – Omphalea triandra.
Nut, Constantinople – Corylus colurna.
Nut, Curare Poison – Strychnos toxifera.
Nut, Dika – Irvingia barteri.
Nut, Earth – Carum bulbocastanum.
Nut, Enkabang – Shorea macrophylla.
Nut, Essang – Ricinodendron heudelotii.
Nut, Grass – Cyperus rotundus.
Nut Grass, Yellow – Cyperus esculentus.
Nut, Ground – Apios americana.
Nut, Guiana – Caryocar nuciferum.
Nut, Hazel – Corylus avellana.
Nut, Hunter's – Omphalea megacarpa.
Nut, Illipé – The following Shorea spp.:

S. macrantha, S. martiana, S. palembanica, S. squamata, S. stenoptera, S. seminis.

Nut, Isano – Ongokea klaineana.

Nut, Kaffir Marvola – Sclerocarya caffra.

Nut, Macadamia – Macadamia ternifolia.

Nut, Macadamia – Macadamia tetraphylla.

Nut, Mahogany – Afrolicana elaeosperma.

Nut, Malacca Poison – Strychnos malaccensis.

Nut, (Oil) Manketti – Ricinodendron rautaneii.

Nut, Maranhão – Sterculia chicha.

Nut, Nitronga – Cryptocarya latifolia.

Nut, Oil – Pyrularia pubera.

Nut, Ojok – Ricinodendron heudelotii.

Nut, Pará – Bertholletia excelsa.

Nut, Paradise – Caryocar nuciferum.

Nut, Paradise – Lecythis zabucajo.

Nut, Physic – Caesalpinia bonducella.

Nut, Physic – Jatropha curcas.

Nut, Pig – Omphalea triandra.

Nut, Pignolia – Pinus pinea.

Nut, Pine – Pinus cembroides.

Nut, Pine – Pinus edulis.

Nut, Pine – Pinus monophylla.

Nut, Pine – Pinus paryana.

Nut, Pistachio – Pistacia vera.

Nut, Promotion – Anacardium occidentale.

Nut, Purging – Jatropha curcas.

Nut, Queensland – Macadamia ternifolia.

Nut, Queensland – Macadamia tetraphylla.

Nut, Sal – Ginkgo biloba.

Nut, Sanga – Ricinodendron heudelotii.

Nut, Sapucaia – Lecythis zabucajo.

Nut, St. Ignatius Poison – Strychnos ignatii.

Nut, Siak Illipé – Palaquium burekii.

Nut, Singhara – Trapa bispinosa.

Nut, Souari – Caryocar nucifera.

Nut, Souari – Caryocar tomentosum.

Nut, Water – Nelumbo pentapetala.

Nutmeg (Oil) – Myristica fragrans.

Nutmeg, Ackawai – Acrodiclidium camara.

Nutmeg, Bombay – Myristica malabarica.

Nutmeg, Brazilian – Cryptocarya moschata.

Nutmeg, Calabash – Monodora myristica.

Nutmeg, Madagascar – Ravensara aromatica.

Nutmeg, Mountain – Myristica fatua.

Nutmeg, Peruvian – Laurelia aromatica.

Nutmeg, Wild – Myristica malabarica.

Nuttall Alkali Grass – Puciniella nuttaliana.

Nuttall Goldenweed – Aplopappus nuttallii.

Nux Vomica – Strychnos nux-vomica.

NUXIA Comm. ex Lam. Buddlejaceae. 40 spp. Tropical Africa, Madagascar, Mascarene Islands. Trees. The wood of the following species is used locally for house-building.

***N. capitata*** Baker. Madagascar.

***N. sphaerocarpa*** Baker. Madagascar.

***N. terminaloides*** Baker. Madagascar.

***N. verticillata*** Lam. (Lambinana). Madagascar. The leaves are used to flavour the local beverage Betsabetsa.

Nyassa Rubber – Landolphia kirkii.

NYCTANTHES L. Verbenaceae. 2 spp. India to Thailand, Java and Sumatra. Shrubs.

***N. arbor-tristis*** L. India. The leaves yield a bright yellow dye.

NYMPHAEA L. Nymphaeaceae. 50 spp. Tropics and temperate. Water herbs.

***N. advena*** Ait. = Nuphar advena.

***N. alba*** L. (White Waterlily). Europe to temperate Asia. The starchy rhizomes are a potential emergency food.

***N. calliantha*** Con. S.Africa. The seeds are eaten locally by the natives.

***N. capensis*** Thunb. S. and S.E.Africa. The rootstocks are eaten locally.

***N. coerulea*** Sav. Tropical Africa. The rhizomes and fruits are eaten locally.

***N. gigantea*** Hook. Australia. The tubers and peeled stalks are eaten by the aborigines.

***N. lotus*** L. Tropical Africa and Asia. The powdered roots are used locally to treat stomach complaints and dysentery. The seeds are used as an emergency food in India and when powdered are used in the Sudan to treat skin diseases.

***N. stellata*** Willd. Tropical Africa and Asia. The rhizomes and seeds are used locally as famine foods.

NYMPHOIDES Seguier. Menyanthaceae. 20 spp. Tropical and temperate. Water plants.

***N. crenata*** F. v. Muell. Australia. The roasted tubers are eaten by the aborigines.

***M. cristata*** Roxb. = Villarsia cristata. Tropical Asia. The plant is eaten as a vegetable in China and the seeds are used to treat intestinal worms. In India the stems and leaves are ground with oil and used as a poultice for ulcers and insect bites.

NYPA Steck. Nypaceae. 1 sp. Ceylon, Ganges delta, Malaysia to Marianne and Solomon Islands, Australia. Palm.

***N. fruticans*** Thunb. (Nipa Palm). The leaves are used locally for thatching, baskets, hats, etc., while the leaf-petioles were used to make arrows. Sugar is extracted from the inflorescence and from it alcohol and vinegar are made. The Chinese eat the fruits as a sweetmeat, or as a candied preserve.

NYSSA Gronov. ex L. Nyssaceae. 10 spp. Himalayas, E.Asia, E. N.America. Trees.

***N. aquatica*** L. (Cotton Gum, Tupelo Gum, Water Tupelo). E. U.S.A. to Texas. The light brown, soft tough wood is used for crates, shoes and broom handles. The wood from the roots is sometimes used for net floats. The flowers yield a good honey.

***N. biflora*** Walt. (Water Gum, Water Tupelo). The soft tough light yellow wood is used for pulp, crates, veneer, shoes, wheel-hubs, piles, yokes and rollers in glass factories.

***N. javanica*** Wang. = N. sessiliflora. (Dhoowak manting, Hiroong, Wooroo gading). W. Malaysia. The light brown wood is hard and fairly heavy. It is used locally for house construction.

***N. multiflora*** Wang. = N. sylvatica.

***N. ogeche*** Marsh. (Ogeeche Lime, Sour Tupelo, Tupelo Gum). S.Carolina and Florida. The fruits (Ogeeche Limes) are eaten locally and the flowers are a good source of honey.

***N. sessiliflora*** Hook. f. and Th. = N. javanica.

***N. sylvatica*** Marsh. = N. multiflora. (Black Gum, Black Tupelo, Pepperidge, Sour Gum). E. U.S.A. See N. biflora.

Nzembila – Ganophyllum giganteum.

Nzeng – Quassia gabonensis.

Nzongo – Fagara inaequalis.

# O

Oak, African – Chlorophora excelsa.

Oak, African – Lophia alata.

Oak, African – Oldfieldia africana.

Oak, Australian – Eucalyptus regnans.

Oak, Black Jack – Quercus marylandica.

Oak, Black Jack – Quercus nigra.

Oak, Blue Japanese – Quercus glauca.

Oak, Blueish Tulip – Argyrodendron actinophyllum.

Oak, Brazilian – Posoqueria latifolia.

Oak, Brown – Fistulina hepatica.

Oak, Brown Tulip – Argyrodendron trifoliatum.

Oak, Bull – Casuarina equisetifolia.

Oak, Bur – Quercus macrocarpa.

Oak, Californian White – Quercus lobata.

Oak, Canyon Live – Quercus chrysolepis.

Oak, Ceylon – Schleichera trijuga.

Oak, Chestnut – Quercus densiflora.

Oak, Chestnut – Quercus prinus.

Oak, Cork – Quercus suber.

Oak, Daimyo – Quercus dentata.

Oak, Durmast – Quercus petraea.

Oak, Emory – Quercus emoryi.

Oak, English – Quercus pedunculata.

Oak, European Turkey – Quercus cerris.

Oak, Evergreen – Quercus ilex.

Oak, Forest – Casuarina torulosa.

Oak, Glandbearing – Quercus glandulifera.

Oak, Holly – Quercus ilex.

Oak, Holm – Quercus ilex.

Oak, Iron – Quercus obtusiloba.

Oak, Jack – Quercus marylandica.

Oak, Kermes – Quercus coccifera.

Oak, Laurel – Quercus imbricata.

Oak, Live – Quercus virginiana.

Oak, Lung – Stricta pulmonata.

Oak, Lusitanian – Quercus lusitanica.

Oak, Manna – Quercus cerris.

Oak, Manna – Quercus persica.

Oak, Maul – Quercus chrysolepis.

Oak, Mongolian – Quercus mongolica.

Oak Moss – Evernia prunastri.

Oak, Mossy Cup – Quercus macrocarpa.

Oak, Netleaf – Quercus reticulata.

Oak, Oregon White – Quercus garryana.

Oak, Oriental – Quercus variabilis.

Oak, Overcup – Quercus lyrata.

Oak, Patana – Careya arborea.

Oak, Pin – Quercus palustris.

Oak, Portuguese – Quercus lusitanica.

Oak, Post – Quercus obtusiloba.

Oak, Pubescent – Quercus pubescens.

Oak, Quebec – Quercus alba.

Oak Rage – Lobaria scrobiculata.

Oak, Red – Quercus borealis.

Oak, Red – Quercus digitata.

Oak, Red Southern – Quercus texana.

Oak, Red Tulip – Argyrodendron perabatal.

Oak, River – Casuarina stricta.

Oak, River – Casuarina torulosa.

Oak, River Black – Casuarina suberosa.

Oak, Scarlet – Quercus coccinea.

Oak, Shin – Quercus gambelii.

Oak, Shin – Quercus undulata.
Oak, Shingle – Casuarina stricta.
Oak, Shingle – Quercus imbricata.
Oak, Silk – Grenvillea robusta.
Oak, Silky – Stenocarpus salignus.
Oak, Sily – Orites excelsa.
Oak, Southern Red – Quercus texana.
Oak, Spanish – Quercus digitata.
Oak, Spanish – Quercus pagoda.
Oak, Swamp – Casuarina equisetifolia.
Oak, Swamp – Casuarina suberosa.
Oak, Swamp Spanish – Quercus pagoda.
Oak, Swamp Spanish – Quercus palustris.
Oak, Swamp White – Quercus bicolor.
Oak, Tanbark – Lithocarpus densiflora.
Oak, Texas – Quercus texana.
Oak, Tongue – Fistulina hepatica.
Oak, Turkey – Quercus cerris.
Oak, Valley – Quercus lobata.
Oak, Water – Quercus nigra.
Oak, Water White – Quercus lyrata.
Oak, Wavyleaf – Quercus undulata.
Oak, Western Black – Quercus emoryi.
Oak, Western White – Quercus lobata.
Oak, White – Quercus alba.
Oak, Willow – Quercus phellos.
***Oakesia sessiliflora*** (L.) Wats. = Uvularia sessiliflora.
Oat, Common (Oil) – Avena sativa.
Oat, Hungarian – Avena orientalis.
Oat, Naked – Avena nuda.
Oat, Sea – Uniola paniculata.
Oat, Wild – Avena fatua.
Oatgrass, Tall Meadow – Arrhenatherum avenaceum.
Oba-ubayuri – Lilium glehni.
OBERONIA Lindl. Orchidaceae. 330 spp. Old World Tropics. Herbs.
***O. anceps*** Lindl. Malacca. The crushed plant is used locally as a poultice for boils.
***O. longibracteata*** Lindl. = Malaxis longibracteata. Ceylon to Cambodia. In Cambodia the crushed leaves are used as a poultice for scorpion bites.
Oblonghead Sunflower – Helianthus doronicoides.
Obon-dji – Urera cameroonensis.
Oboqui – Cnestis corniculata.
Oca – Oxalis crenata.
OCHANOSTACHYS Mast. Olacaceae. 1 spp. W. Malaysia. Tree.
***O. amentacea*** Mast. The heavy strong, durable wood is insect-resistant and is used locally for house-building, furniture and telephone poles.
Orchid, Giant – Epipactis giganteum.
OCHLANDRA Thw. Graminae. 12 spp. Madagascar, India, Ceylon. Bamboos.
***O. travancorica*** Benth. S.India. The plant is used locally in making paper.
***Ochna alboserrata*** Engl. = Brackenridgea zanguebarica.
Ochoco (Butter) – Scyphocephalium ochocoa.
***Ochocoa gaboni*** Pierre = Scyphocephalium ochocoa.
***Ochrocarpus*** Thou. = Mammea.
OCHROLECHIA Lecanoraceae. Lichen.
***O. tartarea*** (L.) Mass. (Cockur, Crotal, Crottle). Temperate. Yields a purple dye (Cudbear, Tincture of Cudbear) and a pH sensitive pigment (Erdorseille) similar to litmus.
OCHROMA Sw. Bombacaceae. 1 sp. S.Mexico to Bolivia, W.Indies. Tree.
***O. lagopus*** Sw. = O. pyrimidale.
***O. limonensis*** Rowlee. = O. pyrimidale.
***O. pyrimidale*** (Cav. ex Lam.) Urb. = O. lagopus = O. limonensis (Balsa, Cork Wood, Cotton Tree, Guano). This is one of the lightest woods. It is readily attacked by insects, decays easily and warps. It is used locally for rafts etc., for fruit boxes and to make toys. Its main commercial use is in making life-jackets and as a heat insulator in refrigerators.
OCHROSIA Juss. Apocynaceae. 30 spp. Madagascar to Australia, Pacific Islands. Trees.
***O. borbonica*** Gmel. = Cerbera parviflora (Bois Jaune). Adaman Islands, Annam, Ceylon, Vietnam, Malaya, Réunion. The bark (Quinquina du pays) is used in Réunion and Vietnam to reduce fevers and the leaves are used to make a tonic.
***O. mariannensis*** A. DC. (Marianne Yellow Wood). Pacific Islands. The fine yellow wood is used to make furniture in Guam.
***O. oppositifolia*** Schum. (Oopas laki laki, Songga langit). Java. A decoction of the roots is used locally to counteract the effects of eating poisonous fish and crabs and to treat abdominal pains.
***O. sandwicensis*** A. DC. Hawaii. The bark and roots yield a yellow dye used locally to dye cloth.
OCIMUM L. Labiatae. 150 spp. Tropics and warm temperate, especially Africa. Herbs.
***O. americanum*** L. W.Africa. An infusion of the leaves is used locally to reduce

fevers and the whole leaves are used to flavour food.

**O. aristatum** Blume=Orthosiphon aristatus.

**O. basilicum** L. (Sweet Basil). India, S.E.Asia, N.E.Africa. A weed in the Sudan Gezira. The plant is cultivated, throughout its range, but particularly in Morocco and Réunion for the essential oil which is used in perfumery, soap-making, to flavour liqueurs and sauces. The plant is used as a vegetable in India and in some parts of the Mediterranean coasts the seeds are used to make a drink (Cherbet Tokhum).

**O. canum** Sims. (Hoary Basil). Old World Tropics. The leaves are eaten as a vegetable in India, while in the Sudan a paste of the leaves is used as a poultice for skin diseases.

**O. grandiflorum** Blume=Orthospihon stamineum.

**O. gratissimum** L. (Ramtulsi, Vriadhatulasi). Tropical Asia. Cultivated, especially in India where the leaves are used in steam baths for the treatment of rheumatism and as a decoction to treat venereal diseases.

**O. graveolens** A. Br. Abyssinia. The leaves are used locally to flavour foods.

**O. kilmandscharicum** Gurke. E.Africa. The leaves contain some 60 per cent camphor. The plant was cultivated as a commercial source of camphor during World War II, but has not been used much since.

**O. micranthum** Willd. (Albahaca). Mexico to N. S.America, W.Indies. The leaves are used in El Salvador to relieve ear ache.

**O. sanctum** L. (Holy Basil, Tulsi). Old World Tropics. The plant is sacred to the Hindus and is grown in front of temples. The leaves are used as a condiment.

**O. viride** L. W.Africa. A decoction of the leaves is used locally to reduce fevers.

Ocopetate – Cyathea mexicana.

Ocote – Pinus teocote.

OCOTEA Aubl. 3–400 spp. Tropical and subtropical America, a few in tropical and S. tropical Africa. Trees.

**O bullata** E. Mey. (Black Stinkwood). S.Africa. The dark brown hard wood is used for furniture-making, house-building, wagons, planking and gun stocks. The bark is used for tanning.

**O. caudata** Mez. Tropical S.America. The essential oil safrole is distilled from the wood. It is used in perfumery and to make flavouring.

**O. cujumary** Mart.=Aydendron cujumary =Oreodaphne floribunda. Brazil. The wood is used locally to make boats and for general carpentry.

**O. indecora** Schott.=Mespilodaphne indecora. Guyana, to Brazil. A decoction of the rootbark is used locally in Brazil to treat rheumatism.

**O. leucoxylon** (Sw.) Maza. (Laurel Geo). Tropical America, W.Indies. The fairly soft honey-coloured wood is used for general carpentry, cheap furniture, light construction work, boxes and plywood.

**O. moschata** (Meissn.) Metz. (Nuez Moscada). Tropical America, W.Indies. The dark brown streaked wood is attacked by termites, but is used to make furniture, turning, carving, construction work, bridges.

**O. puchury-minor** Mart.=Nectandra puchy-minor.

**O. rodiaei** (Schomb.) Mez. (Demarera Greenheart). Guyana. The hard olive-green to black wood is durable and strong. It is used for marine construction work, paving and boarding.

**O. rubra** Mez. (Determa, Wane). Guyana, Brazil. The light, red-brown wood is insect-resistant and works well. It is used for interior work, furniture, boats and sugar boxes.

**O. sassafras** Mez.=Mespilodaphne sassafras. Brazil. Brazilian Sassafras Oil (Safrole) is distilled from the wood and used in perfumery, soap-manufacture, deodorants and flavourings. Locally the roots, bark and leaves are used as a diuretic, to induce sweating and to treat rheumatism.

**O. spathulata** Mez. (Nemocá). W.Indies, especially Puerto Rico. The variegated wood varies in colour from a pinkish to greenish brown. It is very hard and heavy but polishes well. It is used for furniture, interior work, agricultural implements, heavy construction work, boat-building, mill work and making novelties.

**O. squarrossa** Mart.=Laurus pyrifolia. Brazil. A decoction of the bark is used locally as a tonic.

**O. usambarensis** Engl. See O. caudata.

**O. veraguensis** (meissn.) Mez.=Sassafridium veraguense. (Canelillo, Palo colorado, Sigua canelo). S.Mexico to

Panama. The hard wood is used locally for making furniture and general construction work.

Ocotillo – Fouquiera splendens.

OCTOMELES Miq. Tetramelaceae. 1–2 spp. Malaysia, but not the Malay Peninsula, Java or Lesser Sundra Islands. Trees.

***O. sumatrana*** Miq. (Binuang). Indonesia, Borneo, Philippines. The soft, light, pale yellow wood decays readily, but is used for interior work, plywood, veneers, matchboxes, floats and coffins.

Odara Pear – Chrysophyllum africanum.

***Odina*** Roxb. = Lannea.

Odolla Fat – Cerbera manghas.

***Odostemon aquifolium*** Rydb. = Mahonia repens.

Odum – Chlorophora excelsa.

***Odendea*** (Pierre) Engl. = Quassia.

OEMLERIA Reichb. Rosaceae. 1 sp. Pacific N.America. Tree.

***O. cerasiformis*** (Torr. and Gray) Greene. The fruits are eaten by the local Indians.

OENANTHE L. Umbelliferae. 40 spp. Temperate Eurasia, mountains of tropical Africa. Herbs.

***O. aquatica*** (L.) Poir. = O. phellandrium (Water Fennel). Europe through to Siberia and Iran. The fruits are diuretic and expectorant. They are used medicinally in treating laryngitis, asthma and catarrh.

***O. crocata*** L. S.Europe. The crushed roots are used to stupify fish in Sardinia.

***O. javanica*** DC. = O. laciniata = O. stolonifera = Dasyloma javanica. (Water Dropwort, Bamboong, Batjarongi, Pampoong, Piopo). Indochina, China, Japan, Malaysia. Cultivated. The leaves and shoots are eaten as a vegetable and used as a flavouring locally.

***O. laciniata*** Zoll. = O. javanica.

***O. phelandrium*** Lam. = O. aquatica.

***O. sarmentosa*** Presl. (Water Parsley). Pacific N.America. The sweet tubers were eaten as a vegetable by the local Indians.

***O. stolonifera*** Wall. = O. javanica.

OENOCARPUS Mart. Palmae. 16 spp. Tropical S.America. Palms.

***O. bacaba*** Mart. Guyana to Brazil. The fruits yield an oil, used locally for cooking.

***O. bataua*** Mart (Tooroo). Guyana to Brazil. The juice from the flesh of the fruit is used locally to make a drink.

***O. distichus*** Mart. (Bacaba). S.America. The fruits are used to make a popular local drink (Bacaba Branca). A darker coloured similar drink (Bacaba Vernelha) is also drunk on its own or with sugar or manioc. An oil from the fruits is used locally for cooking.

OENOTHERA L. Onagraceae. 80 spp. America (especially temperate) W.Indies. Herbs.

***O. biennis*** L. = Onagra biennis. (Evening Primrose). Europe and N.America. The roots are sometimes eaten as a vegetable and the leaves in salads. The plant is sometimes cultivated.

Ofbit – Succisa pratensis.

Ogbagha Rubber – Ficus platyphylla.

Ogea Copal – Daniellia ogea.

Ogea Gum of Sierra Leone – Daniellia thurifera.

Ogeeche Lime – Nyssa ogeche.

Ognon Gli – Pancratium speciosum.

Ohelo – Vaccinium berbifolium.

Ohelo (Dentate) – Vaccinium dentatum.

Ohio Buckeye – Aesculus arguta.

Ohio Buckeye – Aesculus glabra.

Ohori – Moronobea coccinea.

OIDIUM Link ex Fr. Moniliaceae. 25 spp. especially tropical.

***O. aurantiacum*** Lev. = Monilia sitophila.

***O. lactis*** Fresen. (Milk Fungus). The fungus yields gums which can be moulded and hardened, when grown on glucose, mannose or glycerol. These gums may have some commercial value in making small moulded articles.

***Oils.*** The term "of" has been omitted throughout.

Oil, Abrasin – Aleurites montana.

Oil, Ajowan – Carum copticum.

Oil, Ako-Ogea – Paradaniella oliveri.

Oil, Algerian – Ruta montana.

Oil, Allspice – Pimenta officinalis.

Oil, Ambrette – Hibiscus abelmoschus.

Oil, American Wormseed – Chenopodium ambrosioides.

Oil, Anda-assay – Joannesia princeps.

Oil, Andiroba – Carapa guineensis.

Oil, Angelica Root – Angelica archangelica.

Oil, Anise – Illicium parviflorum.

Oil, Anise – Pimpinella anisum.

Oil, Apeiba – Apeiba tibourbou.

Oil, Apricot Kernel – Armeniaca vulgaris.

Oil, Arangan – Ganophyllum falcatum.

Oil, Araucaria – Neocallitropsis araucarioides.

Oil, Arbor Vitae – Thuja occidentalis.

Oil, Argan – Argania spinosa.
Oil, Asa foetida – Ferula asa-foetida.
Oil, Aspic – Lavandula officinalis.
Oil, Avocado – Persea americana.
Oil, Azeite de pau de rosa – Physocalymma scaberrimum.
Oil, Babassu – Orbignya speciosa.
Oil, Bacury Kernel – Platonia insignis.
Oil, Bagilumbang – Aleurites triloba.
Oil, Bagilumbang – Aleurites trisperma.
Oil, Balm – Melissa officinalis.
Oil, Balukanag – Chisocheton cuminginanus.
Oil, Banucalang – Aleurites trisperma.
Oil, Baobab – Adansonia digitata.
Oil, Bati – Ouratea parviflora.
Oil, Batiputa – Ouratea parviflora.
Oil, Bay – Pimenta acris.
Oil, Beech Nut – Fagus sylvatica.
Oil, Ben – Moringa oleifera.
Oil, Bene – Sesamum indicum.
Oil, Bergamot – Citrus bergamia.
Oil, Betis – Madhuca betis.
Oil, Betu – Balanites aegyptoaca.
Oil, Bitter Almond – Amygdalus communis.
Oil, Bitter Orange – Citrus aurantium.
Oil, Black Mustard – Brassica nigra.
Oil, Boronia – Boronia megastigma.
Oil, Brazil Nut – Bertholletia excelsa.
Oil, Brazilian Rosewood – Physocalymma scaberrimum.
Oil, Brazilian Sassafras – Ocotea sassafras.
Oil, Buril – Apeiba tibourbou.
Oil, Cabreu – Myrocarpus fastigiatus.
Oil, Cabriowa Wood – Myrocarpus fastigiatus.
Oil, Cade – Juniperus oxycedrus.
Oil, Cajaput – Melaleuca leucadendron.
Oil, Canadian Golden Rod – Solidago canadensis.
Oil, Candle Nut – Aleurites moluccana.
Oil, Candle Nut – Aleurites triloba.
Oil, Carapa – Carapa guineensis.
Oil, Cardamon – Elettaria cardamomum.
Oil, Carpotroche – Carpotroche brasiliensis.
Oil, Carrot Seed – Daucus carota.
Oil, Cassia – Cinnamomum cassia.
Oil, Cassie Ancienne – Acacia farnesiana.
Oil, Castanha – Telfairia pedata.
Oil, Castanha de Cotia – Aptandra spruceana.
Oil, Castor – Ricinus communis.
Oil, Catoseed – Chisocheton cuminginanus.
Oil, Cayenne Linaloe – Amyris balsamifera.
Oil, Cayeté – Omphalea megacarpa.
Oil, Ceara Rubber – Manihot glaziovii.
Oil, Cedar – Juniperus macropoda.
Oil, Cedro – Citrus limon.
Oil, Celery – Apium graveolens.
Oil, Chamomille – Matricaria chamomilla.
Oil, Champaca Wood – Bulnesia sarmienti.
Oil, Chaulmogra – Hydnocarpus heterophyllus.
Oil, Chenopodium – Chenopodium ambrosioides.
Oil, Cherry Kernel – Prunus cerasus.
Oil, Chia – Salvia chia.
Oil, Chinese Colza – Brassica campestris var. chinoleifera.
Oil, Chinese Wood – Aleurites fordii.
Oil, Chua – Shorea robusta.
Oil, Chufa – Cyperus esculentus.
Oil, Churwah – Aquilaria angallocha.
Oil, Cinnamon – Cinnamomum cassia.
Oil, Citronella – Cymbopogon nardus.
Oil, Clary – Salvia sclarea.
Oil, Clove – Eugenia caryophyllata.
Oil, Coconut – Cocos nucifera.
Oil, Cohune – Attalea cohune.
Oil, Colza – Brassica napus.
Oil, Congo – Antrocaryon nannani.
Oil, Coondi – Carapa guineensis.
Oil, Copaiba – Copaiba spp.
Oil, Coriander – Coriandrum sativum.
Oil, Corn – Zea mays.
Oil, Corozo – Corozo oleifera.
Oil, Costus Root – Saussurea lappa.
Oil, Cottonseed – Gossypium herbaceum.
Oil, Coyal – Acrocomia mexicana.
Oil, Crawwood – Carapa guineensis.
Oil, Croton – Croton tiglium.
Oil, Cubeba – Piper cubeba.
Oil, Cucumber – Cucumis sativa.
Oil, Cumin – Cuminum cyminum.
Oil, Curcas – Jatropha curcas.
Oil, Cypress – Cupressus sempervirens.
Oil, Dill – Anethum graveolens.
Oil, Dilo – Calophyllum inophyllum.
Oil, Dingili – Cephalocroton cordofanus.
Oil, Domba – Calophyllum inophyllum.
Oil, Dwarf Pine Needle – Pinus pumilio.
Oil, Earth Almond – Cyperus esculentus.
Oil, East African Sandalwood – Osyris tenuifolia.
Oil, East Indian Lemon Grass – Cymbopogon flexuosus.
Oil, Efwatakala Grass – Melinis minutiflora.
Oil, Egyptian Lettuce Seeds – Lactuca scariola.
Oil, Elecampane – Inula helenium.

Oil, Elemi – Canarium luzonicum.
Oil, Emmersion – Juniperus virginiana.
Oil, English Walnut – Juglans regia.
Oil, Essang – Ricinodendron heudelotii.
Oil, Estragon – Artemisia dracunculus.
Oil, Eucalyptus – Eucalyptus globosus.
Oil, Exile – Thevetia nereifolia.
Oil, False Chaulmogra – Gynocardia odorata.
Oil, Fennel – Foeniculum vulgare.
Oil, Fiji Sandalwood – Santalum yasi.
Oil, Filbert – Corylus avellana.
Oil, Flax Seed – Linum usitatissimum.
Oil, Fony – Adansonia digitata.
Oil, Gardenia – Gardenia florida.
Oil, Garlic – Allium sativum.
Oil, Gaultheria – Gaultheria procumbens.
Oil, Geranium – Pelagonium radula.
Oil, German Spearmint – Mentha spicata.
Oil, Gingergrass – Cymbopogon martini.
Oil, Gokizuru – Actinostemma lobatum.
Oil, Gonyo – Antrocaryon nannani.
Oil, Gorli – Caloncorba echinata.
Oil, Grapefruit Seed – Citrus paradisi.
Oil, Guadil – Convolvulus floridus.
Oil, Guaiac Wood – Bulnesia sarmienti.
Oil, Gurjum – Dipterocarpus lamellanthus.
Oil, Hazelnut – Corylus avellana.
Oil, Hemlock – Tsuga canadensis.
Oil, Hemp Seeds – Cannabis sativa.
Oil, Homoranthus – Homoranthus virgatus.
Oil, Honesty – Hesperis matronalis.
Oil, Hongay – Pongamia pinnata.
Oil, Hop – Humulus lupulus.
Oil, Horned Poppy – Glaucium flavum.
Oil, Huon Pine – Dacrydium franklinii.
Oil, Hyacinth – Hyacinthus orientalis.
Oil, Hyptis spicigera – Hyptis spicigera.
Oil, Hyssop – Hyssopus officinalis.
Oil, Inchi Grass – Cymbopogon martini.
Oil of Inchy Lemon Grass des Ides orientales – Cymbopogon flexuosus.
Oil, Indian Wintergreen – Gaultheria fragantissima.
Oil, Inoy Kernel – Poga oleosa.
Oil, Ivory Wood Seed – Agonandra brasiliensis.
Oil, Jamba – Eruca sativa.
Oil, Japanese Mint – Mentha sativa.
Oil, Japanese Tung – Aleurites cordata.
Oil, Jatamansi – Nardostachys jatamansi.
Oil, Jojoba – Simmondsia californica.
Oil, Jonquil – Narcissus jonquilla.
Oil, Juniper Berry – Juniperus communis.
Oil, Juniper Wood – Juniperus communis.
Oil, Kachi Grass – Cymbopogon coloratus.
Oil, Kalingag – Cinnamomum mercadoi.
Oil, Kaloempang – Sterculia foetida.
Oil, Kamynye – Hoslundia opposita.
Oil, Katio – Bassia mottelyana.
Oil, Kayu-galu – Sindora inermis.
Oil, Kelon-ka-tel – Cedrus lebanii.
Oil, Kenaf Seed – Hibiscus cannabinus.
Oil, Kuromoji – Lindera umbellata.
Oil, Kussum – Schleichera trijuga.
Oil, Lallemantia – Lallemantia iberica.
Oil, Lavender – Lavandula officinalis.
Oil, Lemon – Citrus limon.
Oil, Lemongrass – Cymbopogon citratus.
Oil of Lemongrass de Cochin – Cymbopogon flexuosus.
Oil, Lime – Citrus aurantifolia.
Oil, Linaloé (Seed) – Bursera aloë xylon.
Oil, Linseed – Linum usitatissimum.
Oil, Lovage – Levisticum officinale.
Oil, Lumbang – Aleurites molucca.
Oil, Lumbang – Aleurites triloba.
Oil, Macanilla – Guilielma speciosa.
Oil, Macassar – Cananga odorata.
Oil, Macassar – Schleichera trijuga.
Oil, Mace – Myristica fragrans.
Oil, Madia – Madia sativa.
Oil, Mafura – Trichilia emetica.
Oil, Mahareb Grass – Cymbopogon senaarensis.
Oil, Mandarin – Citrus reticulata.
Oil, Manihot – Manihot glaziovii.
Oil, Manketti Nut – Ricinodendron rautanenii.
Oil, Margosa – Azadirachta indica.
Oil, Marigold – Calendula officinalis.
Oil, Marjoram – Origanum majorana.
Oil, Marjoram – Thymus mastichina.
Oil, Marmelle – Agele marmelos.
Oil, Meadowsweet – Filipendula ulmaria.
Oil, Mehndi – Lawsonia alba.
Oil, Meni – Lophira alata.
Oil, Mexican Oiticica – Licania arborea.
Oil, Mexican Prickly Poppy Seed – Argemone mexicana.
Oil, Mignonette – Rededs odorata.
Oil, Mihugu – Brachylaena hutchinsii.
Oil, Mu – Aleurites montana.
Oil, Muhumbu – Brachylaena hutchinsii.
Oil, Mureré – Ficus cystopodium.
Oil, Murumuru – Astrocaryum murmura.
Oil, Myrica – Pimenta acris.
Oil, Myrrh – Commiphora abyssinica.
Oil, Myrtle – Myrtus communis.
Oil, Naim – Lophira alata.
Oil, Narcissus – Narcissus poeticus.
Oil, Neou – Parinarium macrophyllum.

Oil, Neroli – Citrus aurantium.
Oil, N'Ghat – Plukenetia conophora.
Oil, Niam – Lophira alata.
Oil, Niaouli – Melaleuca viridiflora.
Oil, Niger Seed – Guizotia abyssinica.
Oil Nut – Pyrularia pubera.
Oil, Nutmeg – Myristica fragrans.
Oil, Oat – Avena sativa.
Oil, Olive – Olea europaea.
Oil, Oiticica – Licania rigida.
Oil, Orange – Citrus sinensis.
Oil, Orange Flower – Citrus aurantium.
Oil, Oreoseleni – Peucedanum oreoselinum.
Oil, Ouricuru Plam Kernel – Syagrus coronata.
Oil, Owala – Pentaclethra macrophylla.
Oil Palm – Elaeis guineensis.
Oil, Palm – Elaeis guineensis.
Oil, Palmarosa – Cymbopogon martinii.
Oil, Palosapsis – Anisoptera thurifera.
Oil, Pào Manfin – Agonandra brasiliensis.
Oil, Para Rubber Seed – Hevea brasiliensis.
Oil, Pataúa – Jessenia bataua.
Oil, Patchouly – Pogostemon patchouly.
Oil, Peanut – Arachis hypogea.
Oil, Peach Kernel – Amygdalus persica.
Oil, Pecan – Carya pecan.
Oil, Pennyroyal – Mentha pulegium.
Oil, Peppermint – Mentha piperita.
Oil, Perilla – Perilla frutescens.
Oil, Petitgrain – Citrus aurantium.
Oil, Pili Nut – Canarium ovatum.
Oil, Pimenta – Pimenta officinalis.
Oil, Pine Needle – Pinus montana.
Oil, Pine Needle – Pinus sylvestris.
Oil, Piririma – Sygrus cocoides.
Oil, Pistachio Nut – Pistacia vera.
Oil, Poli – Carthamus oxycanthus.
Oil, Pongam – Pongamia pinnata.
Oil, Poppy Seed – Papaver somniferum.
Oil, Pota-eta-tel – Kakoona zeylandica.
Oil, Po-yak – Afrolicania elaeosperma.
Oil, Pracachy – Pentaclethra filamentosa.
Oil, Pracaxy – Pentaclethra filamentosa.
Oil, Prickly Poppy Seed – Argemone mexicana.
Oil, Pumpkin Seed – Cucurbita pepo.
Oil, Ragweed – Ambrosia artemisiifolia.
Oil, Rantilla – Guizotia abyssinica.
Oil, Rape – Brassica rapa.
Oil, Rasamala Wood – Altingia excesla.
Oil, Ravinson – Brassica campestris.
Oil, Red Cedar Wood – Juniperus virginiana.
Oil, Reniala – Adansonia digitata.
Oil, Reseda – Reseda odorata.
Oil, Rhodium – Convolvulus floridus.
Oil, Rice – Oryza sativa.
Oil, Roghum – Carthamus oxycanthus.
Oil, Rohituka – Amoora rohituka.
Oil, Roman Chamomille – Anthemis nobilis.
Oil, Rose – Rosa Damascena.
Oil, Rosemary – Rosmarinus officinalis.
Oil, Rosewood – Aniba terminalis.
Oil, Rue – Ruta graveolens.
Oil, Safflor – Carthamus tinctoria.
Oil, Sage – Salvia officinalis.
Oil, Sandal – Guarea guara.
Oil, Sandal – Lecontea amazonica.
Oil, Sandel – Calophyllum brasiliense.
Oil, Santal – Santalum album.
Oil, Sapucaja – Lecythis ollaria.
Oil, Sassafras – Sassafras alibidum.
Oil, Savin – Juniperus sabina.
Oil, Savory – Satureja hortensis.
Oil, Sedge – Cyperus esculentus.
Oil, Sesame – Sesamum indicum.
Oil, Siberian Fir – Abies sibirica.
Oil, Silver Pine Needle – Abies pectinata.
Oil, Simson – Trianthema salsoloides.
Oil, Sioerno – Xanthophyllum lanceolatum.
Oil, South Australian Sandalwood – Eucarya acuminatus.
Oil, Spanish Origanum – Thymus capitatus.
Oil, Spearmint – Mentha spicata.
Oil, Spike – Lavandula latifolia.
Oil, Star Anise – Illicium verum.
Oil, Stillingia – Sapium sebiferum.
Oil, Stoechas – Lavandula stoechas.
Oil, Styrax – Liquidambar styraciflux.
Oil, Suir – Xanthophyllum lanceolatum.
Oil, Sunflower Seeds – Helianthus annuus.
Oil, Supa – Sindora supa.
Oil, Sweet Almond – Amygdalus communis.
Oil, Sweet Basil – Ocimum basilicum.
Oil, Sweet Birch – Betula lenta.
Oil, Sweet Orange – Citrus sinensis.
Oil, Sweet Orange Pip – Citrus sinensis.
Oil, Sweet Pea – Lathyrus odoratus.
Oil, Tangarine – Citrus reticulata.
Oil, Tansy – Tanacetum vulgare.
Oil, Tarragon – Artemisia dracunculus.
Oil, Tea Seed – Camellia sasanqua.
Oil, Teel – Sesamum indicum.
Oil, Thyme – Monarda didyma.
Oil, Thyme – Thymus vulgaris.
Oil, Tiger Nut – Cyperus esculentus.
Oil, Tigli – Croton tiglium.
Oil, Tomata Seed – Lycopersicon lycopersicum.
Oil, Touloucouna – Carapa guineensis.
Oil, Touloucouna – Carapa procera.

Oil, Tsauri Grass – Cymbopogon giganteus.
Oil, Tsubaki – Camellia japonica.
Oil, Tuberose Flower – Polianthes tuberosa.
Oil, Tucma – Astrocaryum tucuma.
Oil, Tucuma – Astrocaryum vulgare.
Oil, Tulip Wood – Pysocalymna scaberrimum.
Oil, Tung – Aleurites fordii.
Oil, Turkey Red – Agonandra brasiliensis.
Oil, Turkey Red – Ricinus communis.
Oil, Ubussu – Manicaria saccifera.
Oil, Verbena – Lippia citriodora.
Oil Vine, Zanzibar – Telfairia pedata.
Oil, Werinnua – Guizotia abyssinica.
Oil, West Australian Sandalwood – Eucarya spicata.
Oil, West Indian Lemon Grass – Cymbopogon citratus.
Oil, West Indian Sandalwood – Amyris balsamifera.
Oil, White Cedar – Thuya occidentalis.
Oil, White Mustard Seed – Sinapis alba.
Oil, Wild Marjoram – Thymus mastichina.
Oil, Wild Olive Seeds – Ximenia americana.
Oil, Wintergreen – Gaultheria procumbens.
Oil, Wormseed – Chenopodium ambrosioides.
Oil, Wormwood – Artemisia pontica.
Oil, Ylang-ylang – Canangium odoratum.
Oil, Zachun – Balanites roxburghii.
Oil, Zit-el-Harmel – Peganum harmala.
Oiticica Oil – Licania rigida.
Ojitos de Picho – Rhychosia phaseoloides.
Ojo de cangnejo – Rhynchosia phaseoloides.
Ojod – Vitis lanceolaria.
Ojok Nut – Ricinodendron heudelotii.
Oklahoma Plum – Prunus gracilis.
Okgue – Ficus pumila.
Okote – Sorindeia gilletii.
Okra – Hibiscus esculentus.
Okukuluka – Beilschmiedia gaboonensis.
Okwa – Treculia africana.
Ola Bala – Lepiota spp.
Olas – Borassus flabellifer.

OLAX L. Olacaceae. 55 spp. Tropical Africa, Madagascar, Indomalaya, Australia. Trees.

***O. wightiana*** Wall. = O. zeylandica. Ceylon. The leaves are eaten locally as a vegetable.

***O. zeylanica*** Wall. = O. wightiana.

Old Maid – Vinca rosea.
Old Man's Beard – Chionanthus virginica.
Old Man's Beard – Clematis vitalba.
Old Man's Beard – Usnea barbata.

OLDENLANDIA L. Rubiaceae. 300 spp. Warm. Small shrubs and vines.

***O. corymbosa*** L. (Chay Root). Asia, especially India and Ceylon. The roots yield a purplish dye, which turns green when treated with sulphuric acid, when it is called Gerancine. Both dyes are used for colouring cloth.

***O. recurva*** Miq. = Hedyctis capitellata. (Akar kemenjan). Borneo, Malacca. An extract of the roots is used locally to treat dysentery and rheumatism. A paste of the leaves is used as a poultice on snakebites.

***O. umbellata*** L. (Chayaver, Ché, Imburral, Indian Madder, Turbuli). Tropical Asia, especially India, Ceylon. Burma. The plant is cultivated for the roots which yield a red dye, used for colouring cloth. The dye is made from the yellow root extract by treating it with alkali.

OLDFIELDIA Benth. and Hook. f. Euphoriaceae. 4 spp. Tropical Africa. Trees.

***O. africana*** Benth. and Hook. f. (African Oak, African Teak). Tropical Africa. The hard heavy wood sinks in water and is used for boat-building.

OLEA L. Oleaceae. 20 spp. Mediterranean, through Africa to E.Asia, E.Australia, New Zealand and Polynesia. Trees.

***O. cunninghamiana*** Hook. f. New Zealand. The strong, heavy dark brown wood is used for furniture, turning, railway rolling stock, bridges and piers.

***O. cuspidata*** Wall. = O. ferruginea. (Indian Olive). Himalayas. The wood is used for general carpentry.

***O. emarginata*** Lam. = Noronhia emarginata.

***O. europaea*** L. (Common Olive). Mediterranean. Cultivated locally since antiquity for the fruits which yield Olive Oil. The plant has definitely been grown in Crete since 3500 B.C. It was introduced into America in 1769 and later into S.America and Australia. The world production is about 10 million tons of fruit annually, of which 90 per cent is used for oil production. Spain produces about 38 per cent of the world crop, followed by Italy 20 per cent, Greece 13 per cent and Turkey, Morocco, California and Australia. For table use there are two types of olive, green and black. The green olives are harvested unripe, treated with sodium hydroxide to neutralize a bitter principle and pickled in brine. The black olives are harvested ripe and allowed to undergo a lactic acid fermentation for 1–6 months before treatment with sodium hydroxide. For oil production, the fruit is harvested ripe by shaking from the trees. The berries are crushed to extract the oil, which is washed to remove the bitter principle, allowed to age for a short time, filtered through a diatomaceous earth and bottled. The fruits contain 20–30 per cent fresh weight of non-drying oil, Sp. Gr. 0·9148–0·9191, Sap. Val. 185–200, Iod. No. 77–94, Unsap. 0·6–1·5 per cent. The oil is used mainly in cooking, salad oils, as a laxative, emolient and demulcent. It is also used to make soap and as a lubricant. The bulk of the oil is used in the country of production. The oil is used for anointing in the Christian Churches. The residue of the fruit after pressing is used as a fuel in California and further chemically extracted in Europe where the resulting oil is used to make soap, or mixed with the edible oil. Olive Gum (Lucca Gum) was used medicinally and is still used in Italy during the manufacture of perfumes. The hard heavy wood is used for turning and carving.

There are hundreds of cultivars and propagation is by budding, cuttings or layering. The better cultivars, under good cultivation will yield half a ton of fruit per acre (the Spanish average is about $\frac{1}{3}$ ton per acre). The plants do not come into full bearing for 15–20 years after planting.

***O. ferruginea*** Royle = O. cuspidata.

***O. fragrans*** Thunb. = Osmanthus fragrans.

***O. laurifolia*** Lam. (Black Ironwood). S.Africa. The heavy, hard dark brown wood is used for waggons, poles and agricultural implements.

***O. paniculata*** R. Br. (Ironwood, Marble wood, Marvey). Australia. The hard tough wood is used for turning and staves.

***O. sandwicensis*** A. Gray = Osmanthus sandwicensis.

***O. thozetii*** Panch. and Sebart. Australia, New Caledonia. The strong heavy wood is used to make furniture.

***O. verrucosa*** Link. (Warty Olive). S.Africa. The heavy hard wood is used for waggons, mill parts, agricultural implements and furniture.

Oleander – Nerium oleander.

Oleander, Wild – Adina microcephala.

Oleander, Yellow – Thevetia nereifolia.

OLEANDRA Cav. Oleandraceae. 40 spp. Tropics. Herbs.

***O. colubrina*** (Blanco) Copel. Terrestrial or epiphytic fern. A decoction of the stipe is used to control menstruation.

OLEARIA Moench. Compositae. 100 spp. New Guinea, Australia, New Zealand. Shrubby trees.

***O. colensoi*** Hook. f. New Zealand. The hard light brown satiny wood is used for ornamental work.

***O. traversii*** F. v. Muell. (Traver's Daisybush). New Zealand. The hard satiny wood is used for cabinet work.

Oleo de Sapucainha – Carpotroche brasiliensis.

Oleo oreoseleni – Peucedanum oreoselinum.

Oleo Sandalo do Norto – Lecontea amazonica.

Oleo Vermelho – Copaifera reticulata.

Oleum Amygdalae Expressum – Amygdalus communis.

Oleum Arganiae – Argania spinosa.

Oleum Asae Foetidae – Ferula assa-foetida.

Oleum Cardamoni – Elettaria cardamomum.

Oleum Caryophylli – Euglena caryophyllata.

Oleum Cedri Folii – Thuja occidentalis.

Oleum Chamomilla – Matricaria discoidea.

Oleum Cinnamomi – Cinnamomum cassia.

Oleum Coloncohae – Coloncoba echinata.

Oleum Coriandri – Coriandrum sativum.

Oleum Darwinae aetherum – Darwinia fascicularis.

Oleum Eucalyptii – Eucalyptus globosus.

Oleum Foeniculi – Foeniculum vulgare.

Oleum Ligni Guaiaci – Bulnesia sarmieti.

Oleum Millefolii – Achillea millefolium.

Oleum Mururé – Ficus cystopodium.

Oleum Myriciae – Pimenta acris.

Oleum Myristicae – Myristica fragrans.

Oleum Myrti – Myrtus communis.

Oleum Pini Pumilionis – Pinus montana.

Oleum Rosemarini – Rosmarinus officinalis.

Oleum santali – Santalum album.

Oleum Sciadopitys – Sciadopitys verticillata.

Oleum Spicae – Lavandula latifolia.

Oleum Terebinthinae – Pinus palustris.

Oleum Thymi – Thymus vulgaris.

Oli Copal – Agathis lorathifolia.

Olibanum – Boswellia carterii.

Olibanum, Indian – Boswellia serrata.
OLIGOCERAS Gagnep. Euphorbiaceae. 1 sp. Indochina. Tree.
***O. eberhardtii*** Gagnep. The fruits are eaten locally.
Olive (Oil) (Gum) – Olea europaea.
Olive, Californian – Umbellularia californica.
Olive, Ceylon – Elaeocarpus serratus.
Olive, Chinese – Canarium album.
Olive, Hawaiian – Osmanthus sandwicensis.
Olive, Indian – Olea cuspidata.
Olive, Madagascar – Noronhia emarginata.
Olive, Native – Osmanthus americana.
Olive, Spurge – Cneorum tricoccum.
Olive Tree, Black – Laguncularia racemosa.
Olive, Warty – Olea verrucosa.
Olive, Wild – Elaeocarpus serratus.
Olive, Wild – Sideroxylon mastichodendron.
Oliver's Bark – Cinnamomum oliveri.
Olivier Yellow – Terminalia obovata.
OLNEYA A. Gray. Leguminosae. 1 sp. California, Mexico. Tree.
***O. tesota*** A. Gray. (Sonora Ironwood, Tesota). The seeds are eaten by the local Indians.
Ololiuqui – Rivea corymbosa.
Olombe – Solanum pierreanum.
Olona – Touchardia latifolia.
Olondo – Dorstenia klainei.
Oloroso – Houmiri balsamifera.
Olumba – Eriocoelum microspermum.
Olupua – Osmanthus sandwicensis.
Omek – Anemone hupehensis.
OMPHALEA L. Euphorbiaceae. 20 spp. Tropical America, Africa, Madagascar, Indochina, W.Malaysia, Celebes. Climbers or trees.
***O. diandra*** L. W.Indies. The fruits are eaten locally.
***O. megacarpa*** Hemsl. S.America, W.Indies. The plant is cultivated. The seeds (Hunter's Nuts) are eaten locally. An oil (Cayeté Oil) expressed from them is used as a cathartic.
***O. triandra*** L. Tropical America. The fruits (Cob Nuts, Pig Nuts) are eaten locally and the plant is a potential source of latex. It is sometimes cultivated.
OMPHALOCARPUS Beauv. Sapotaceae. 5 spp. Tropical W.Africa. Trees.
***O. anocentrum*** Pierre. Tropical W.Africa. An oil extracted from the seeds is used locally for cooking.
***Omphalogonus*** Baill. = Parquetina.
Omum Water – Carum copticum.
ONCINOTIS Benth. Apocynaceae. 25 spp. Tropical and S.Africa, Madagascar. Trees.
***O. hirta*** Oliv. Congo. A potential emergency source of latex.
ONCOBA Forsk. Flacourtiaceae. 5 spp. Tropical Africa. Trees.
***O. echinata*** Oliv. = Caloncoba echinata. Tropical W.Africa. Cultivated in Central America for the seeds which yield an oil (Gorli Oil, Gorli Seeds Oil) used in the treatment of leprosy and other skin diseases.
***O. spinosa*** Forsk. (Kpoe, Krutu). Tropical Africa. The ground fruits are used locally as snuff and the wood is used for inlays.
ONCOCARPUS A. Gray. Anacardiaceae. 6 spp. Philippines, New Guinea, Fiji. Palm-like trees.
***O. vitiensis*** A. Gray. Fiji. The natives make a drink from the fruits.
ONCOSPERMA Blume. Palmae. 5 spp. Ceylon, Indochina, W.Malaysia. Palms.
***O. filamentosa*** Blume. Malaysia. The wood is water-resistant and is used locally for piles for lake dwellings and flooring for houses. The unopened leaves are used locally as a vegetable.
***O. horridum*** Scheff. See O. filamentosa.
One Berry – Paris quadrifolia.
One-flowered Vetch – Vicia articulata.
ONGOKEA Pierre. Olaceae. 2 spp. W. Equatorial Africa. Trees.
***O. gore*** Pierre = O. klaineana.
***O. kamerunensis*** Engl. (Ingo). The yellow wood is used locally for building houses.
***O. klaineana*** Pierre = O. gore (Boleko, Isano). The seeds (Boleko Nuts, Isano Nuts) yield an oil which is used locally for cooking.
Onion – Allium cepa.
Onion, Baldhead – Allium sphaerocephalum.
Onion, Longroot – Allium victorialis.
Onion, Nodding – Allium cernuum.
Onion, Sea – Urginea maritima.
Onion, Welsh – Allium fistulosum.
Onjam – Antidesma ghaesembilla.
ONOBRYCHIS Mill. Leguminosae. 120 spp. Europe, Mediterranean to Central Asia. Herbs.
***O. sativa*** Lam. = O. viciifolia.
***O. viciifolia*** Scop. = O. sativa. (Esparcette, Holy Clover, Sainfoin). Europe, W.Asia, introduced to N.America. Sometimes grown as fodder in Europe.

ONONIS L. Leguminosae. 75 spp. Canary Islands, Mediterranean, Europe to Central Asia. Herbs.

***O. spinosa*** L. (Restharrow). Europe to Turkestan. A decoction of the stem and leaves is used in Central Europe to treat bladder complaints, gout, gall stones and skin diseases. An infusion of the flowers is used to "clean the blood".

***Onopordon*** Hill.=Onopordum.

ONOPORDUM L. Compositae. 40 spp. Europe, N.Africa, W.Asia. Herbs.

***O. acanthium*** L. Europe. The flower heads are used to adulterate saffron.

ONOSOMA L. Boraginaceae. 150 spp. Mediterranean to Himalayas and China. Herbs.

***O. bracteatum*** Gaud. (Gaozaban). Iran to W.Himalayas. An infusion of the flowers is used locally as a tonic and to improve the appetite.

***O. echinoides*** L. Central Europe to Himalayas. The roots yield a red dye (Orsanette) used in India to dye fats and wool, in place of Alkanna.

Ontjom beurum – Monilia sitophila.

***Onychium*** Reinw.=Lecanopteris.

Ooka-ooka – Aglaia odoratissima.

Oona nepa – Hornstedtia rumphii.

Oopas laki-laki – Ochrosia oppositifolia.

Oorang-oorangan – Cratoxylum racemusum.

Ooris ooresan – Grewia laevigata.

Oowi oowi – Habenaria multipartia.

OPERCULINA S. Manso. Convolvulaceae. 25 spp. Tropics. Vines.

***O. altissima*** Meisn.=Ipomoea altissima.

***O. pisonis*** Mart. Brazil. The root yields Brazilian Jalap, which is used medicinally as a purgative.

***O. turpethum*** (L.) S. Manso.=Convolvulus turpethum=Ipomoea turpethum. Throughout the tropics. The roots contain 4–10 per cent Turpenthin Resin which is used medicinally in an alcoholic tincture as a drastic purgative.

***Ophioderma falcatum*** Degener.=Ophioglossum pendulum.

OPHIOGLOSSUM L. Ophioglossaceae. 30–50 spp. Tropics and temperate. Ferns.

***O. ovatum*** Bory. S.Madagascar. The leaves are eaten locally as a vegetable.

***O. pendulum*** Hook. and Arn.=Ophioderma falcatum. Hawaii. A decoction of the plant is used locally to treat coughs.

***O. reticulatum*** L. Malaysia. The young leaves are eaten as a vegetable in the Moluccas.

***O. vulgatum*** L. (Adder's Tongue). Temperate Asia, through Europe to E. N.America. A decoction of the plant is emetic and is used to treat sickness, dropsy and hiccoughs. A poultice of the leaves is a treatment for scrofulous ulcers.

OPHIOPOGON Ker-Gawl. Liliaceae. 20 spp. Himalayas to Japan and Philippines. Herbs.

***O. japonicus*** (L. f.) Ker-Gawl. Japan. The tubers are eaten locally.

***O. spicatus*** Ker-Gawl.=Liriope graminifolia.

***O. spicatus*** Kunth. China. The plant is cultivated locally for the roots which are used as an aphrodisiac, to improve the milk secretion and to relieve dyspepsia.

***Ophiurus appendiculatus*** Steud.=Rottboellia exaltata.

Opium (various types) – Papaver nudicaule and P. somniferum.

Opium Poppy – Papaver somniferum.

***Oplismenus frumentaceus*** Kunth.=Panicum crus-gali.

Opok – Vitis lanceolaria.

OPOPANAX Koch. Umbelliferae. 3 spp. Persia to Balkans. Herbs.

***O. chironium*** (L.) Koch. (Opopanax). Mediterranean. The resin Opopanax is extracted from incisions in the roots and used in perfumery. It was used medicinally.

Opopanax – Opopanax chironium.

Opopanax Myrrh – Commiphora kataf.

OPUNTIA Mill. Cactaceae. 250 spp. America, Galapagos Islands. Xerophytic shrubs or trees.

***O. azurea*** Rose. (Nopal cyotillo, Nopalillo). Mexico. The fruits are eaten locally

***O. basiliana*** Engelm. and Bigel. S.W. N.America. The young stems, flower buds and flowers are eaten as a vegetable by the local Indians. They are cooked by steaming in pits.

***O. bigelow*** Engelm. S.W. N.America. The stems are fed to livestock.

***O. camanchica*** Engelm. S.W. N.America, Mexico. The fruits are eaten by the local Indians.

***O. castillae*** Griffith.=O. megacantha.

***O. cholla*** Weber. S.W. N.America, Mexico. The stems are fed to livestock.

***O. chochenillifera*** Mill. (Nopal). Mexico.

The plant is cultivated locally as food for the cochineal insect from which a red dye is extracted. The dye is used for colouring foods. The fruits of the plant are eaten locally and the stems are used to poultice bruises, boils etc.

***O. clavata*** Engelm. New Mexico. The fruits and stems are eaten, roasted by the local Indians.

***O. dejecta*** Salm-Dyck. (Nopal chamacuera). Mexico. The fruits are eaten raw and the stems as a vegetable.

***O. engelmannii*** Salm-Dyck. S.W. N.America. The fruits are eaten locally raw or cooked and the stems are eaten as a vegetable.

***O. ficus-indica*** Mill. (Indian Fig, Prickly Pear, Tuna de Castilla). Mexico. The plant is cultivated in the tropics and subtropics for the fruits which are eaten fresh, dried or cooked. They are very popular locally and an appreciable amount are shipped from Mexico to the U.S.A.

***O. fulgida*** Engelm. (Cholla). S.W. N.America, Mexico. The stem yields a gum which is used in Mexico as a size and for stiffening fabrics. The stems are fed to livestock.

***O. humifusa*** Raf. S.W. N.America. The fruits are eaten locally, raw, cooked or dried for the winter.

***O. imbricata*** (Haw.) DC. (Joconoxtla, Tuna). S.W. N.America, Mexico. An extract from the fruits is used to set the dye prepared from the cochineal insect. The woody parts of the stem are used to make canes.

***O. karwinskiana*** (Salm-Dyck) Schum. (Nopalillo). Mexico. A decoction of the roots is used locally to treat dysentery.

***O. leucotricha*** DC. (Nopal Durazillo). Mexico. The fruits are eaten locally.

***O. megacantha*** Salm-Dyck = O. castillae. (Nopal, Tuna). Mexico. The plant is cultivated locally for the fruits, which are considered the best of the Opuntia fruits. There are several local varieties. They are eaten raw or cooked. Cakes of the dried fruit (Queso de Tuna) are sold locally. A syrup from the fruits (Miel de Tuna) is used raw or boiled to a paste (Melcoacha). It may be fermented to make a drink (Colonche) to which pulque may be added (Nochote). The young shoots are eaten as a vegetable or used as a poultice for inflammations. The juice from the stems is sometimes mixed with tallow in making candles.

***O. spinosior*** Engelm. S.W. N.America, Mexico. The young shoots are fed to livestock.

***O. strepacantha*** Lem. (Tuna Cardona, Nopal Cordón). Mexico. The plant is cultivated locally for the fruit.

***O. versicolor*** Engelm. S.W. N.America, Mexico. The fruits are eaten by the local Indians.

Orach, Garden – Atriplex hortensis.

Orange, Bitter – Citrus aurantium.

Orange, Bitter-sweet – Citrus aurantium × C. sinensis.

Orange, Blood – Citrus sinensis.

Orange, Calarin – Citrus mitis × C. reticulata var. deliciosa.

Orange, Calashu – Citrus mitis × C. reticulata var. unshiu.

Orange, Chinotto – Citrus aurantium var. myrtifolia.

Orange Flower Oil – Citrus aurantium.

Orange, Khatta – Citrus limonia var. khatta.

Orange, King – Citrus reticulata.

Orange, Mandarin – Citrus reticulata.

Orange, Mexican Mock – Philadelphus mexicanus.

Orange, Murcott – Citrus sinensis × C. reticulata.

Orange, Myrtle-leaved – Citrus aurantium var. myrtifolia.

Orange Oil – Citrus sinensis.

Orange, Orangelo – Citrus sinensis × C. paradisi.

Orange, Oranguma – Citrus reticulata var. unshiu × C. sinensis.

Orange, Osage – Maclura aurantiaca.

Orange, Otahite – Citrus limonia var. otaitensis.

Orange River Ebony – Euclea pseudoebenus.

Orange, Satsuma – Citrus reticulata var. unshiu.

Orange, Satsumelo – Citrus paradisi × C. reticulata var. unshiu.

Orange, Seville – Citrus aurantium.

Orange, Siamelo – Citrus paradisi × C. reticulata.

Orange, Siamor – Citrus sinensis × C. paradisi × C. reticulata var. deliciosa.

Orange, Sopomaldin – Citrus paradisi × C. mitis.

Orange, Sour – Citrus aurantium.

Orange, Sweet – Citrus sinensis.
Orange, Tangarine – Citrus reticulata var. deliciosa.
Orange, Temple – Citrus sinensis × C. reticulata.
Orange, Tizon – Citrus reticulata var. papillaris.
Orange Fame Flower – Talinum aurantiacum.
Orange Milkweed Root – Asclepias tuberosa.
Orange Tree, Wild – Toddalia aculeata.
Oranguma – Citrus reticulata var. unshiu × C. sinensis.
ORIGNYA Mart. ex Endl. Palmae. 25 spp. Brazil, Bolivia. Palms.
***O. martiana*** Barb. S.America. Hevea rubber is coagulated by the smoke from the burning seeds.
***O. speciosa*** Berk. (Babassu Palm). S.America, especially Brazil. The fruits yield an oil (Babassu Oil) which is widely used in the local soap industry. The fruits are harvested from wild stands. The plant is not cultivated.
***O. spectabilis*** Mart. (Carua Palm). S.America, especially N. The seeds yield an oil which is used locally for cooking etc.
Orcein – Rocella spp.
Orcella – Roccella fuciformis.
Orcella Weed – Roccella montagnei.
Orchard Grass – Dactylis glomerata.
Orchid, Large Round-leaved – Habenaria orbiculata.
***Orchipeda foetida*** Blume = Voacanga fortida.
ORCHIS L. Orchidaceae. 35 spp. Temperate Eurasia to India and S.W.China. Herbs. The tubers of the following species are ground into a white demulcent and nutritious powder which is sometimes given to convalescent children. In India it is used to make sweetmeats and in Greece and Turkey it is eaten with honey. In the East the powder is used as an aphrodisiac and to treat intestinal catarrh.
***O. latifolia*** L. (Broad leaved Orchis).
***O. mascula*** L.
***O. militaris*** L.
***O. morio*** L.
***O. orbiculata*** Pursh. = Habenaria orbiculata.
Orchis, Broad-leaved – Orchis latifolia.
Ordeal Bark – Erythrophleum guineensis.
Ordeal Bean – Physostigma venenosum.
Oreganillo – Weinmannia pinnata.
Oregano – Lippia berlandieria.
Oregon Ash – Fraxinus oregana.
Oregon Balsam – Pseudotsuga menziesii.
Oregon Crab Apple – Malus rivularis.
Oregon Grape – Mahonia aquifolium.
Oregon Grape – Mahonia repens.
Oregon Sunflower – Balsamorhiza sagittata.
Oregon White Oak – Quercus garryana.
Oreja de coyote – Turnera ulmifolia.
Orenburgh Gum – Larix decidua.
***Oreodaphne floribunda*** Benth. = Ocotea cujumary.
***Oreodoxa oleracea*** Mart. = Roystonea oleracea.
***Oreodoxa regia*** H.B.K. = Roystonia regia.
Oreoseleni Oil – Peucedanum oreoselenium.
Orère Butter – Baillonella ovata.
Oriental Beech – Fagus orientalis.
Oriental Cashew Nut – Semecarpus anacardium.
Oriental Oak – Quercus variabilies.
Oriental Plane – Plantanus orientalis.
Oriental Prickly Melon – Cucumis cocomon.
Oriental Sesame (Oil) – Sesamum indicum.
Oriental Sweet Gum – Liquidambar aorientalis.
Oriental Sycamore – Platanus orientalis.
Oriental Wood – Endiandra palmerstonii.
Origano – Origanum vulgare.
ORIGANUM L. Labiatae. 15–20 spp. Europe through Mediterranean to Central Asia. Herbs.
***O. compactum*** Benth. N.Africa. The leaves are used locally as a condiment.
***O. creticum*** Sut. S.Europe. The dried leaves are used as a condiment (Spanish Hops) and as a substitute for thyme.
***O. dictamnus*** L. = Amaracus dictamnus.
***O. glandulosum*** Desf. N.Africa. The dried leaves are used locally as a condiment.
***O. heraclecticum*** L. = O. vulgare var. prismaticum. (Winter Marjoram). Mediterranean to U.K. The leaves are used to flavour soups and salads and to make a herbal tea. They are sometimes used as a thyme substitute (Spanish Hops).
***O. majorana*** L. = Majorana hortensis. (Sweet Marjoram). Mediterranean, S.Europe. Cultivated in the Old and New World for the leaves which are used to flavour meat dishes and canned meats. The oil (Oil of Marjoram) is obtained from the leaves by steam distillation and used for the same purposes. A decoction

of the dried leaves is used medicinally to relieve stomach upsets.

***O. maru*** L. Asia Minor. Probably the Biblical "Hyssop".

***O. onites*** L. (Pot Majoram, Cretian Dost). S.Europe, Asia Minor. The leaves are used as a condiment.

***O. vulgare*** L. (Origano, Pot Marjoram). Europe. Used as a condiment and introduced to N.America by the early colonists. It is not of such a good quality as Sweet Marjoram. A decoction of the leaves is used in home remedies for stomach upsets, bronchitis and toothache.

***O. vulgare*** L. var. ***prismaticum*** Gaudn.=O. heraclecticum.

Orinoco Scrap – Sapium jenmani.

Orinoco Simaruba Bark – Quassia amara.

ORITES P. Br. Proteaceae. 9 spp. Temperate E.Australia, Andes. Trees.

***O. excelsa*** R. Br. (Red Ash, Sily Oak). Australia. The grey wood is durable and is used locally for agricultural implements.

***Orithia oxypetala*** Gray=Tulipa edulis.

Orizaba Jalep – Ipomoea orizabensis.

ORMOSIA G. Jacks. Leguminosae. 50 spp. Tropics. Trees.

***O. calavensis*** Azaol. Malaysia. A decoction of the leaves is used locally to treat stomach upsets.

***O. krugii*** Urban. (Palo de Matos). W.Indies, especially Puerto Rico. The strong, coarse-grained pinkish wood is susceptible to termites and marine borers. It is used locally for general carpentry, housebuilding, furniture, mill work.

***O. monosperma*** (Sw.) Urban. W.Indies. The hard coarse wood is used locally for building.

***O. sumatrana*** Prain. Java, Sumatra, Moluccas. The hard wood is used for interior work in native houses.

ORNITHOGALUM L. Liliaceae. 150 spp. Old World. Herbs.

***O. narbonensis*** L. Europe. The bulbs are sometimes eaten.

***O. pyrenaicum*** L. S.France, Spain. The young flowering shoots are sometimes eaten as a vegetable.

***O. ubellatum*** L. Europe. The bulbs are sometimes eaten as a vegetable.

ORNITHOPUS L. Leguminosae. 10 spp. Subtropical S.America, tropical Africa, Mediterranean, W.Asia. Herbs.

***O. sativus*** Link. (Seradella, Serratella). S.Europe. Sometimes grown locally as a green manure and cover crop.

OROBANCHE L. Orobanchaceae. 140 spp. Temperate and subtropics. Parasitic herbs. The underground parts of the following species are eaten by the local Indians of W. N.America.

***O. californica*** Cham. and Schlecht.

***O. fasciculata*** Nutt.

***O. ludoviciana*** Nutt.

***O. tuberosa*** (Gray) Heller.

***Orobus lathyroides*** L.=Vicia unijuga.

ORONTIUM L. Araceae. 1 sp. Atlantic N.America. Herb.

***O. aquaticum*** L. (Goldenclub). The underground parts and seeds are eaten by the local Indians. They need boiling several times to remove the bitter principle.

OROXYLUM Vent. Bignoniaceae. 2 spp. S.China, S.E.Asia, Indomalaya. Trees.

***O. indicum*** Vent.=Calosanthes indica. India, Malaysia. The bark is used in Java to treat stomach complaints.

Orris (Butter) – Iris florentina.

Orris Root, Verona – Iris germanica.

Orsanette – Onosma echinoides.

Orseille – Roccella tinctoria.

Orseille d'Auvergne – Lecanora parella.

Ortega – Urera caracasana.

Ortega de caballo – Urera baccifera.

ORTHOSIPHON Benth. Labiatae. 30 spp. Tropical Africa, Madagascar; 20 spp. E.Asia, Indomalaya. Herbs or half-shrubs.

***O. aristicus*** (Blume) Miq.=Ocimum aristatum. Tropical Asia to Australia. A decoction of the leaves is used throughout the area to treat kidney and bladder complaints.

***O. grandiflorus*** Bold. Asia, Malaysia. See O. aristicus. The plant is cultivated locally.

***O. rubicundus*** Benth.=Plectranthus tuberosus. Tropical Africa. The plant is cultivated locally for the starchy tubers which are eaten as a vegetable.

***O. stamineus*** Benth.=Ocium grandiflorum. S.E.Asia to Australia. Cultivated in Java for the leaves (Java Tea, Folia Orthosiphonis) a decoction of which is used medicinally in Java and the Netherlands to treat urinary complaints.

ORYZA L. Graminae. 25 spp. Tropics. Grasses.

***O. barthii*** A. Chev. (Wild Rice). Tropical Africa. Used locally as green manure and livestock feed. The grains are sometimes eaten.

***O. glaberrima*** Steud. W. tropical Africa. The grain is sometimes eaten locally. Occasionally cultivated.

***O. longistaminata*** Chev. and Roehr. Central Africa. Cultivated locally as a grain crop.

***O. minuta*** C. presl. Malaysia, Philippines. A possible ancestor of the small-grained rice varieties.

***O. sativa*** L. (Rice). The origins are obscure, but the plant is probably indigenous to Africa, India and Indonesia. The genetics of the group is obscure and the species may be a block name for several different ones. The diploid number of chromosomes is 24, while some American varieties are tetraploid. There are thousands of cultivars which seem to fall into at least two groups the "japonica" group originating in Japan and Korea and the "indica" group from China, India and Java. From the point of view of use they are classified into long, medium and short grained. The long-grained take longer to mature and are less starchy and glutinous when cooked, while the short-grained mature more quickly and are sticky and starchy when cooked. The long-grained types are grown in the warmer climates and the short-grained in the cooler, e.g. Japan. Considered from the cultivation aspects there are two types, the lowland and upland types. The lowland types are grown in seed beds and transplanted to flooded paddy fields (in S.E.Asia during the monsoon season of November to January), and the fields are drained some four months later when the crop is harvested by hand. The upland types are grown on higher ground as a dry crop. The yield of the upland types is usually lower than that of the lowland types. Apart from the tropical countries the crop is grown under irrigation in the sub-tropics, e.g. America, Spain, Italy and the yields are usually higher under these conditions.

Rice is the prime cereal crop, being the basic diet of over half the world's population. The world production is some 500,000 million lb. per annum and there are about 250 million acres under rice production yielding 1,450 lb. per acre. The main producers are China (35 per cent), India (21 per cent), Japan (7 per cent), Pakistan (6 per cent), Indonesia (6 per cent), Thailand (3 per cent) and Burma (3 per cent). The crop is grown locally in Africa and America where it is of local importance. The U.S.A. produce about 1 per cent of the world total. The grain is used to make a wide variety of dishes and as stock food. It is fermented to make Rice Wine (Saki). The straw is used for stock feed and to make hats etc., Rice Oil is extracted from it. Milling rice removes the valuable aleurone layer and embryo, making it deficient in proteins and vitamin $B_1$. The extensive and exclusive use of polished rice leads to the development of beri-beri. This could be reduced by parboiling the rice before milling and then undermilling. The husks are extracted for Tikitiki Extract which has a high vitamin $B_1$ content and is used in the treatment and prevention of beri-beri.

ORYZOPSIS Michx. Graminae. 50 spp. N.Temperate and sub-tropics. Grasses.

***O. hymenoides*** Roem. and Schult. W. N.America. The grain is eaten by the local Indians.

Osage Orange – Maclura aurantiaca.

OSBECKIA L. Melastromataceae. 100 spp. Tropical Africa to Australia. Shrubs.

***O. crinita*** Roxb. = O. stellata = O. septemnervia. (Dok ka aan). N.Vietnam, Laos, Thailand. A decoction of the leaves is used locally to treat toothache.

***O. septemnervia*** Ham. = O. crinita.

***O. stellata*** D. Don. = O. crinita.

Osier – Salix viminalis.

Osier, Green – Cornus circinnata.

Osier Willow – Salix viminalis.

Osier Willow, Purple – Salix purpurea.

OSMANTHUS Lour. Oleaceae. 15 spp. E. and S.E.Asia, America.

***O. americana*** (L.) Gray. (Devil Wood, Native Olive) E. N.America. The heavy, hard dark brown wood is used occasionally for general carpentry.

***O. aquifolium*** (Sieb. and Zucc.) Benth. Japan. (Hiiragi). The wood is used locally to make furniture and fancy articles.

***O. fragrans*** Lour. = Olea fragrans. China. The scented flowers are used by the Chinese to scent tea. The fruits are edible.

***O. sandwicensis*** Knobl. = Gynmelaea sandwicensis = Nestigis sandwicensis = Olea sandwicensis. (Hawaiian Olive, Olupua,

Pua). Hawaii. The hard wood is used locally to make axe handles and fish spears.

***Osmaronia*** Greene=Oemleria.

OSMELIA Thw. Flacourtiaceae. 1 sp. Ceylon; 3 spp. Malaysia, except Java and Sundra Islands. Trees.

***O. celebica*** Koord. Celebes. The strong wood is used locally for house-building.

OSMORHIZA Rafin. Umbelliferae. 15 spp. Caucasus to Himalayas and Japan, N.America, Andes. Herbs.

***O. claytoni*** (Michx.) Clarke. W. N.America. A decoction of the roots is used by the local Indians to help to put on weight

***O. occidentalis*** (Nutt.) Torr. (Sweet Root). W. N.America. A decoction of the roots is used by the local Indians to treat colds. The root is sometimes smoked for the same purpose.

OSMOXYLON Miq. Araliaceae. 2 spp. Malaysia. Trees.

***O. umbelliferum*** Merr. (Ambon Sandalwood, Rozemarijinhout, Sasooroo). Indonesia. The reddish scented wood is burnt as incense.

OSMUNDA L. Osmundaceae. 10 spp. Tropical and temperate. Ferns.

***O. cinnamomea*** L. (Cinnamon Fern). E. N.America. The young fronds are eaten as a vegetable by the local Indians.

***O. claytonia*** L. See Aspidium felix-mas.

***O. regalis*** L. (Royal Fern). Temperate. The Japanese use the hairs from the leaves to mix with wool to make cloth and raincoats.

Osoberry – Oemleria cerasiformis.

OSTRYA Scop. Carpinaceae. 10 spp. N.Temperate (S. to Central America). Trees.

***O. carpinifola*** Scop. (European Hop Hornbeam). Central to S.Europe, Asia Minor. The orange, tough hard wood is used for general carpentry and is made into charcoal.

***O. guatemalensis*** (Winkl.) Rose.=O mexicana. (Guapa ךue). Mexico, Central America. The hard tough wood is used for railway sleepers. The bark is used for tanning.

***O. mexicana*** Rose=O. guatemalensis.

***O. virginiata*** Koch. (American Hop Hornbeam). E. N.America. The hard, tough, light brown wood is used for fencing, tool handles etc.

OSTRYODERRIS Dunn. Leguminosae. 7 spp. Tropical Africa. Vines.

***O. lucida*** (Welw.) Baker. f. (Angbolo, Kembo). Congo, Angola. The leaves yield a dye which is used locally.

Oswega Tea – Monarda didyma.

Oswega Bee Balm – Monarda didyma.

OSYRIS L. Santalaceae 6–7 spp. Mediterranean and Africa to India. Trees.

***O. arborea*** Quall. (Jhuri, Popli, Tamparal). Lower Himalayas into India. The leaves are used locally as an emetic.

***O. tenuifolia*** Engl. E.Africa. The brown wood (East African Sandalwood) yields an oil (East African Sandalwood Oil). See Santalum album.

Otaheite Apple – Spondias dulcis.

Otaheite Gooseberry – Phyllanthus distichus.

Otahite Orange – Citrus limonia var. otaitensis.

Otak udang – Buchanania arborescens.

OTANTHUS Hoffmgg. and Link. Compositae. 1 spp. W.Europe, Mediterranean to Caucasus. Herb.

***O. maritimus*** (L.) Hoffmgg. and Link.= Diotis candidissima. A decoction of the leaves is used by the Arabs to reduce fevers and control menstruation.

Otau – Ficus prolixa.

Otó – Xanthosoma violaceum.

OTOBA (DC.) Karst. Myristicaceae. 9 spp. Central and tropical S.America.

***O. gordoniifolia*** (DC.) Warb. (Coco, Mascada de Huyaquil). Peru. The trunk yields a red varnish (Sangre de Drago, Dragon's Blood) which is used locally.

***O. otoba*** (Benth. and Hook.) Warb. (Otoba). Central America to Peru. The seeds are used in Colombia to treat external parasites and the fat (Otoba Fat) extracted from them is used in making soap.

Otoba (Fat) – Otoba otoba.

OTOPHORA Blume. Sapindaceae. 30 spp. Indochina, W.Malaysia.

***O. alata*** Blume. Borneo, Java. The fruits are eaten locally.

***O. specabilis*** Blume. Malaysia. The fruits are eaten locally. The tree is sometimes cultivated.

Ottar (Otto) of Roses – Rosa damascena.

Ouabain – Acocanthera oubaio.

OUDEMANSIELLA Speg. Agariaceae. Occur widely in Indonesia on tree trunks. The following species are much esteemed as food.

***O. canarii*** (Junghuhn) von Höhnel.=

Agaricus canarii. = Amanitopsis canarii = Collybia euphyllus.

***O. mucida*** (Schrad ex Fr.) v. Hoehn.

***O. radiata*** (Relh. ex Fr.) Sing.

OUGEINIA Benth. Leguminosae. 1 sp. India. Tree.

***O. dalbergioides*** Benth. (Sannan, Tinnus). The light-red-brown wood is hard and tough. It is used for building, furniture and agricultural implements.

Ouôlo – Terminalia avicennioides.

OURATEA Aubl. Ochnaceae. 300 spp. Tropics. Trees.

***O. angustifolia*** (Vahl.) Baill. Tropical Asia, especially India and Ceylon. The wood (Bokaara-grass) is used for building in Ceylon. A decoction of the leaves is used as a tonic and to relieve stomach complaints in Malabar.

***O. castaneidolia*** (DC.) Engl. (Garampara). Brazil. The kernels of the fruits yield an oil which is used to make soap, while the oil from the shells is fast drying and suitable for varnishes.

***O. sumatrana*** Gilg. Malaysia, Indonesia. The hard heavy wood is used locally to make boats and pumps. A decoction of the leaves and roots is used to treat dysentery.

Ouricuru Palm (Kernel Oil) – Syagrus coronata.

Our Lady's Bedstraw – Galium verum.

Overcup Oak – Quercus lyrata.

Overtop Palm – Rhyticocos amara.

Ovinala – Dioscorea ovinala.

Ovitano – Garrya longifolia.

Ovoga – Poga oleosa.

Owala Oil (Butter) – Pentaclethra macrophylla.

OWENIA F. v. Muell. Meliacae. 6 spp. Queensland, New South Wales. Trees.

***O. acidula*** F. v. Muell. (Bulloo, Emu Apple). Australia. The fruits are eaten locally.

***O. cerasifera*** F. v. Muell. Queensland. The fine hard wood is used for turning and carving.

OXALIS L. Oxalidaceae. 800 spp. Cosmopolitan, chiefly Central and S.America, S.Africa. Herbs.

***O. acetosella*** L. (Wood Sorrel). Europe, N.America. The leaves are sometimes eaten locally as a vegetable.

***O. cernua*** Thunb. S.Africa. The underground parts are sometimes eaten as a vegetable in S.France and N.Africa.

***O. corniculata*** L. See O. acetosella.

***O. crenata*** Jacq. Peru. The small tubers are an important local vegetable. They have been introduced to Europe where they are called Oca.

***O. deppei*** Lodd. Mexico. The tubers are eaten locally and were introduced to France and Belgium.

***O. strica*** L. (Upright Yellow Wood Sorrel). W. N.America to Mexico. The leaves are eaten as a vegetable by the local Indians.

OXANDRA A. Rich. Annonaceae. 22 spp. Central and tropical S.America. W.Indies. Trees.

***O. lanceolata*** (Sw.) Baill. = Uvaria lanceolata. W.Indies, especially Cuba, Jamaica, Puerto Rico. The wood (Lancewood) is used for general carpentry.

Oxeye Daisy – Chrysanthemum leucanthemum.

Oxhorn Bucida – Laguncularia racemosa.

OXYANTHUS DC. Rubiaceae. 50 spp. Africa. Trees.

***O. speciosum*** DC. (Msala). E.Africa. The wood is used for general carpentry.

***Oxycoccus macrocarpon*** (Ait.) Pursh. = Vaccinium macrocarpon.

OXYDENDRUM DC. Ericaceae. 1 sp. E. United States of America. Tree.

***O. arboreum*** (L.) DC. (Sorrel Tree, Sour Wood). The leaves (Folia Andromedae, Elk Tree Leaves) contain andromedotoxin which is used medicinally to treat heart complaints. As they are bitter, the leaves are also chewed to allay thirst.

***Oxymitra patens*** Benth. = Cleistopholis patens.

***Oxymitra staudtii*** Engl. and Diels. = Cleistopholis staudtii.

OXYRIA Hill. Polygonaceae. 1 sp. N. Arctic and subarctic, mountains of Eurasia and California. Herb.

***O. digyna*** (L.) Hill. (Mountain Sorrel). The leaves are eaten by the Indians of Alaska.

OXYSTELMA R. Br. Asclepiadaceae. 4 spp. Old World Tropics. Vines.

***O. esculentum*** R. Br. = O. zippeliana = Periploca esculenta. Tropical Asia. In Vietnam the fruits are eaten, the roots are fed to livestock and are used for treating jaundice. The latex is used as a poultice for ulcers.

***O. zippeliana*** Blume = O. esculentum.

OXYSTIGMA Harms. Leguminosae. 8 spp. Tropical Africa. Trees.

***O. mannii*** Harms.=Hardwickia mannii. Tropical W.Africa. The wood (Bosipi Wood, Bussipi Wood) is light and light red in colour. It is used for furniture and cigar boxes. The plant also yields a balsam (West African Copaiba, Balsamum Copaivae africanum, Hardwickia Balsam). See Copaiba.

OXYTENANTHERA Munro Graminae. 20 spp. Tropical Africa, E.Asia, Indomalaya. Woody Bamboos.

***O. abyssinica*** Munro.=Bambusa abyssinica=Schizostachyum abyssinicum. Tropical Africa, especially E. The stems are used to make boats and the smaller branches to make arrows.

***O. nigrociliata*** Munro. (Tufted Bamboo). Tropical Asia, especially India. The stems are used locally to make small boats.

***O. stocksii*** Munro. (Slender Bamboo). Tropical Asia, especially India. The stems are used locally in making boats.

***Oxythece*** Miq.=Neoxythece.

Oyster Nut Tree – Telfairia occidentalis.

Oyster Plant – Tragopogon porrifolius.

Oyster Plant, Spanish – Scolymus hispanicus.

Oyster, Vegetable – Traopogon porrifolius.

Ozone Fibre – Asclepias incarnata.

# P

Pacayas – Chamaedorae graminifolia.

PACHIRA Aubl. Bombacaceae. 2 spp. Tropical America. Trees.

***P. aquatica*** Aubl. (Provision Tree). S.Mexico to S.America. The roasted seeds are eaten locally.

***P. longifolia*** Hook.=P. macrocarpa.

***P. macrocarpa*** (Schlecht. and Cham.) Walp.=P. longifolia=Carolinea macrocarpa. (Apompo, Sapotón). Mexico and Central America. The leaves are used locally to treat eye inflammations and the seeds are used as a substitute for cocoa.

PACHYCEREUS (A. Berger) Britton and Rose. Cactaceae. 5 spp. Mexico. Large Cacti.

***P. pecten-aboriginum*** (Engelm.) Britton and Rose=Cereus pecten-aboriginum. (Cardón hecho Hecho). The natives make a flour from the seeds.

***P. pringlei*** (S. Wats.) Britton and Rose.=Cereus pringlei. (Cardón, Cardón-pelon). The seeds and fruits are made into a flour which is used locally and the stems are used to make houses.

PACHYCORMUS Corville. Anacardiaceae. 1 sp. Lower California. Tree.

***P. discolor*** (Benth.) Corville=Rhus veitchiana. The bark is used for tanning, some is exported to Europe.

PACHYELASMA Harms. Leguminosae. 1 sp. W.Africa. Tree.

***P. tessmannii*** Harms. (Boliko, Kasa). The fruits and seeds are used locally to stupify fish.

PACHYGONE Miers. Menispermaceae. 12 spp. China, S.E.Asia, Indo-Malaya, Australia, Pacific Islands. Vines.

***P. ovata*** (Poir) Miers. ex Hook. f. and Th. (Kadukkodi). Tropical Asia, or seashore. The dried fruits are used locally to stupify fish and exterminate vermin.

***Pachylobus edulis*** G. Don=Dacryodes edulis.

***Pachylobus hexandrus*** (Hamilton) Engl.=Dacryodes excelsa.

***Pachyma*** Fr.=Poria.

PACHYPODANTHIUM Engl. and Diels. Annonaceae. 4 spp. Tropical W.Africa. Trees.

***P. staudtii*** Engl. and Diels. Tropical W.Africa. The tough wood is used locally for building houses, boarding etc.

PACHYRHIZUS Rich. ex DC. Leguminosae. 6 spp. Tropics. Herbs. The tubers of the following species are eaten locally as a vegetable. In some parts they are of great local importance.

***P. ahipa*** (Wedd.) Parodi.=Dolichos ahipa. (Ahipa). Bolivia, Argentine. Cultivated in Bolivia. It could be a cultivar.

***P. angulatus*** Rich. (Wayaka Yambean). Malaya.

***P. bulbosus*** Spreng. (Yambean). Probably from tropical Asia. Cultivated in Old and New World tropics. The pods are also eaten as a vegetable.

***P. erosus*** Rich.=Cacara erosa. Mexico. Cultivated in Old and New World tropics.

***P. palmatilobus*** Benth. and Hook. (Jicana). Tropical America. Cultivated in Mexico and Central America.

***P. tuberosus*** Spreng. (Yambean). S.America, W.Indies. Cultivated locally. The pods are also eaten as a vegetable. The starch from the tubers is used to make custards and puddings.

***Pachystemon bancanus*** Miq. = Macaranga triloba.

Pacific Plum – Prunus subcordata.

Pacific Yew – Taxus brevifolia.

Padauk, African – Pterocarpus soyauxii.

Padauk, Brown – Pterocarpus macrocarpus.

Padauk, Dragonsblood – Pterocarpus draco.

Padauk, Red – Pterocarpus macrocarpus.

Padauk, Sandalwood – Pterocarpus santalinus.

Padauk, Vengai – Pterocarpus marsupium.

***Padbruggea pubescens*** Craib. = Whitfordiodendron pubescens.

Paddlewood – Aspidosperma excelsum.

Paddlewood – Uvaria buesgenii.

Paddy – Oryza sativa.

Paddy King's Beech – Flindersia acuminata.

Paduak Dragon's Blood – Pterocarpus draca.

***Padus avium*** Mill. = Prunus padus.

PAEDERIA L. Rubiaceae. 50 spp. Tropics. Climbing Shrubs.

***P. foetida*** L. China, Malaysia, Philippines. A decoction of the powdered leaves is used in Java and Malaya to treat intestinal ailments.

PAEONIA L. Paeoniaceae. 33 spp. Temperate Eurasia, W. N.America. Herbs.

***P. albiflora*** Pall. = P. lactiflora. (Chinese Peony). China to Siberia. The roots are eaten as a vegetable in Mongolia and in China a decoction of the plant is used to treat nervous complaints.

***P. anomala*** L. Siberia. A decoction of the roots is used locally to treat stomach complaints and dysentery and to stop bleeding from the uterus.

***P. corallina*** Retz. (Coral Peony). N.W.Africa. A decoction of the plant is used in Morocco to treat spasms and hysteria.

***P. emodi*** Wall. (Udslap, Mamekh). W.Himalayas. The Hindus use a decoction of the roots to treat hysteria and convulsions, colic, obstruction of the bile duct and diseases of the uterus.

***P. foemina*** Garsault. = P. officinalis var. foemina. (Peony, Piney). S.Europe to Asia Minor. A decoction of the roots is used to treat convulsions, spasms, epilepsy etc.

***P. lactiflora*** Wall. = P. albiflora.

***P. moutans*** Sims. (Tree Peony). China. The roots are used locally as a treatment for spasms etc. The flowers are eaten as a vegetable in Japan.

***P. officinalis*** L. See P. moutans.

***P. officinalis*** L. var. ***foemina*** L. = P. foemina.

Pagoda Dogwood – Cornus alternifolii.

Pagoda Tree, Japanese – Sophora japonica.

Pagsahingin Resin – Canarium villosum.

Pah koh poo chi – Phyllostachys dulcis.

Paho – Mangifera altissima.

Pahoorie – Platonia grandiflora.

***Pahudia*** Miq. = Afzelia.

Pahutan Mango – Mangifera altissima.

Pai-fu-long – Smilax china.

Pai ho – Lilium brownii.

Pail l'ou weng – Polycarpaea corymbosa.

Paina – Ceiba sumauma.

Paina de soda – Chorisia speciosa.

Painaliferin – Chorisia speciosa.

Paina-paina – Plukenetia corniculata.

Paineria – Chorisia speciosa.

Paint Root – Lachnanthes carolina.

Pai-sang – Morus latifolia.

Paissava fibre – Attalea funifera.

Pak-choi – Brassica chinensis.

Pak-Choi, Mock – Brassica parachinensis.

Pakis radja – Cycas rumphii.

Pakoo padak – Cycas rumphii.

Pala – Dipterocarpus trinervis.

Pala – Wrightia tomentosa.

Pala Fibre – Butea frondosa.

Palagaste – Tithonia scaberrima.

Palahlar – Dipterocarpus trinervis.

Palai – Manilkara kauki.

Palakija – Euchresta horsfieldii.

Palamkat – Carallia brachiata.

PALAQUIUM Blanco. Sapotaceae. 115+ spp. Formosa, S.E.Asia, Indomalaya, Solomon Islands. Trees. Various members of the genus produce Gutta Percha, a balata or gum which is pliable when hot but hardens and darkens on exposure to air. It is an excellent heat and electrical insulator and impervious to water. It was used for marine cables, denture plates, golfball centres, transmission belts and acid resistant containers. A maximum world production was reached during the first World War. This was about $\frac{3}{4}$ million pounds annually, but this has gradually declined with the increasing use of synthetic plastics. The original and main source was P. gutta, but as the extraction involved the felling of the trees, other species were soon involved and the stem was tapped for the latex. Now the Gutta

Percha is extracted from the leaves and twigs by macerating them in cold water and collecting the gum floating on the surface. It is stored under water to prevent hardening.

***P. ahernianum*** Merr. Philippines. Yields Philippine Gutta Percha.

***P. burckii*** Lam. Sumatra. The wood is used to make boards and boats. The seeds (Siak Ilipé Nuts) yield an edible fat (Tallow Siak) which is used locally. The seeds and the fruit pulp is eaten locally.

***P. clarkeanum*** King and Gamble. Malay Peninsular. (Getah kelapang). Yields an inferior Gutta Percha.

***P. ellipticum*** Engl.=Dichopsis elliptica (Halaganne, Hodsale, Indian Percha Tree, Palla). India. The red-brown, durable, fairly hard wood is used locally for house-building, furniture, cooperage, masts, mine works, chests, carts and spears. The seeds yield an oil used to make soap and for lighting.

***P. gutta*** Burck.=Isonandra gutta. (Malay Gutta Percha Tree). Malaysia. The main source of Gutta Percha. Cultivated.

***P. hexandrum*** Engl.=P. clarkeanum.

***P. javanense*** Burck. (Java Nato Tree). Java. The seeds yield a fat which is used locally for cooking and lighting.

***P. krantziana*** Pierre. Cochinchina. Yields an inferior Gutta Percha.

***P. leiocarpon*** Boerl. Borneo. Gutta Percha called Angso.

***P. macrocarpon*** Burck. Sumatra, N.Celebes, Molucca. The hard heavy brown wood is used locally to make houses.

***P. maingayi*** King and Gamble. Malacca. A good Gutta Percha (Getah Pootih).

***P. microphyllum*** King and Gamble. (Lakis, Maloot, Njatok takis). Malaysia. The hard wood is used locally for posts, boards and household utensils.

***P. obovatum*** Engl. var. ***occidentale*** Lam. Malaysia. Yields a good Gutta Percha. The red hard wood does not decay in water and is not attacked by termites. It is used locally for piers, etc. and house-building.

***P. obtusifolium*** Burck. (Nato, Ngantho, Tofiri modjioo). Celebes, Bali, Moluccas. The wood is used locally for boats, houses etc.

***P. oleiferum*** Blanco. Malaya. The seeds yield a fat (Siak, Njatuo Fat, Taban Merak Fat, Njatuo Tallow, Taban Merak Tallow), which is used locally for cooking, or when refined used as a substitute for cacao butter.

***P. oxleyanum*** Pierre. Malaysia. Yields a Gutta Percha.

***P. ridleyi*** King and Gamble. (Balam rambui, Belam). Malaysia, Indonesia. The red wood is hard and insect resistant. It is used locally for beams and posts.

***P. rostratum*** Burck. W.Borneo, Sumatra. The reddish wood is durable and is used for house-building and canoes.

***P. semarum*** Lam. (Samaram). Sumatra. The durable wood is used locally to make houses, boarding, boats and posts. It is exported to Singapore. The seeds yield an edible fat.

***P. stellatum*** King and Gamble. (Balam seminai). Malaya. The wood is brown, durable and easy to work. It is used locally for house-building, posts, boxes and canoes.

***P. treubii*** Burck. (Getah pootih, Njatoo doojan). Malaya, Borneo. The Gutta Percha is used to adulterate others of better quality.

***P. walsurifolium*** Pierre. (Butam). Malaya, Sarawak. Yields a Gutta Percha used as an adulterant for those of better quality. The seeds give a fat (Bulam Fat) which is bitter and is used locally for lighting. The wood is attacked by insects, but is used locally for boards.

Palas padi – Licuala acutifida.

Palas Tree – Butea frondosa.

Palata – Persea gratissima.

Palay Rubber – Cryptostegia grandiflora.

Palay Rubber – Willughbeia martabanica.

Pale Bark – Cinchona officinalis.

Pale de Tayuyo – Cheirostemon pentadactylon.

Pale Shield Lichen – Cetraria glauca.

Palisa – Grewia asiatica.

Palisander, African – Milletia laurentii.

Palisander, Javanese – Dalbergia latifolia.

***Paliurus perfoliatus*** Blanco.=Harrisonia perfoliata.

Palla – Palaquium ellipticum.

Palm, American Oil – Corozo oleifera.

Palm, Assai – Euterpe edulis.

Palm, Babussa – Orbignya speciosa.

Palm, Bamboo – Raphia vinifera.

Palm, Bataúa – Jessenia batana.

Palm, Betelnut – Areca catechu.

Palm, Bircayne – Coccothrinax argentea.

Palm, Buri – Corpha elata.

Palm, Cabbage – Livistonia australis.
Palm, Cabbage – Roystonea oleracea.
Palm, Canon – Washingtonia filifera.
Palm, Caranday – Copernicia australis.
Palm, Carnauba – Copernicia cerifera.
Palm, Carua – Orbignya spectabilis.
Palm, Coconut – Cocos nucifera.
Palm, Cocorite – Maximiliana caribaea.
Palm, Cohune – Attalea cohune.
Palm, Common Ivory – Phytelephas macrocarpa.
Palm, Coquitos – Jubaea chilensis.
Palm, Corojo – Acrocomia crispa.
Palm, Corozo – Corozo oleifera.
Palm, Corozo – Scheelea spp.
Palm, Date – Phoenix dactylifera.
Palm, Egyptian Doum – Hyphaene thebaica.
Palm, Fish-tail – Caryota urens.
Palm, Galappa – Actinorhytis calapparia.
Palm, Giant Raffia – Raphia gigantea.
Palm, Ginger Bread – Hyphaene coriacea.
Palm, Gomuti – Arenga pinnata.
Palm, Gumayaka – Arenga tremula.
Palm, Honey – Jubaea chilensis.
Palm, Ivory – Phytelephas macrocarpa.
Palm, Ivory – Phytelephas seemanni.
Palm, Ivory Coast Raffia – Raphia hookeri.
Palm, Ivory Nut – Coelococcus amicarum.
Palm, Jauary – Astrocaryum jauari.
Palm Kernel Oil – Elaeis guineensis.
Palm Lily – Cordyline australis.
Palm Lily – Cordyline terminalis.
Palm, Madagascar Raffia – Raphia pedunculata.
Palm, Mbocaya – Acrocomia totai.
Palm, Monkey Cap – Manicaria saccifera.
Palm, Nikau – Rhopalostylis sapida.
Palm, Nipa – Nypa fruticans.
Palm, Noli – Corozo oleifera.
Palm Oil – Elaeis guineensis.
Palm, Overtop – Rhyticocos amara.
Palm, Palmyra – Borassus flabellifer.
Palm, Panama Hat – Carludovica palmata.
Palm, Paraguay – Acrocomia sclerocarpa.
Palm, Patauá – Jessenia bataua.
Palm, Paxiuba – Socratea exrrhiza.
Palm, Peach – Guilielma speciosa.
Palm, Pindo – Arecastrum romanzoffianum.
Palm, Pisa – Areca hutchinsoniana.
Palm, Polynesian Ivorynut – Coelococcus amicarium.
Palm, Puerto Rican Hat – Sabal causiarum.
Palm, Royal – Roystonea regia.
Palm, Rumf Sago – Metroxylon rumphii.
Palm, Sago – Metroxylon rumphii.
Palm, Sago – Metroxylon sagu.
Palm, Saw Palmetto – Serenoa serrulata.
Palm, Seemann Ivory – Phytelephas seemanni.
Palm, Sleeve – Manicaria saccifera.
Palm, South African Doum – Hyphaena crinita.
Palm, Sugar – Arenga pinnata.
Palm Sugar – Cocos nucifera.
Palm Sugar – Phoenix sylvestris.
Palm, Tagua – Phytelephas macrocarpa.
Palm, Talipot – Corypha umbraculifera.
Palm, Temiche – Manicaria saccifera.
Palm, Thatch – Thrinax parviflora.
Palm, Totai – Acrocomia totai.
Palm, Tucuma – Astrocaryum vulgare.
Palm, Turury – Manicaria saccifera.
Palm, Ubussu – Manicaria saccifera.
Palm, Urucury – Cocos coronata.
Palm, Wax – Ceroxylon andicola.
Palm, Wax – Copernicia cerifera.
Palm, Wild Date – Phoenix sylvestris.
Palm Wine – Arenga pinnata.
Palm Wine – Caryota urens.
Palm Wine – Jubaea chilensis.
Palm Wine – Mauritia vinifera.
Palm Wine – Raphia hookeri.
Palm, Wine Raffia – Raphia vinifera.
Palm, Yaray – Sabal causiarum.
Palma Barreta – Samuela carnerosana.
Palma Blanca – Washingtonia sonorae.
Palma Christi – Ricinus communis.
Palma colorado – Washingtonia sonorae.
Palma criollo – Yucca macrocarpa.
Palma de la Virgin – Dioon edule.
Palma de Macetas – Dioon edule.
Palma de Vino – Scheelea excelsa.
Palma istlé – Samuela carnerosana.
Palma Ixtle – Samuela carnerosana.
Palma negra – Washingtonia sonorae.
Palma oriente – Yucca australis.
Palma Real – Sceelea zonensis.
Palmarosa (Oil) – Cymbopogon martinii.
Palmate Butterbur – Petasites palmata.
Palmate Violet – Viola palmata.
Palmetto, Cabbage – Sabal palmetto.
Palmetto, Saw – Serenoa serrulata.
Palmira Alstonia – Alstonia scholaris.
Palmita – Carludovica palmata.
Palmita – Yucca elephantipes.
Palo Alejo – Caesalpinia eriostachys.
Palo Amarillo Rubber – Euphorbia fulva.
Palo Balsamo – Bulnesia sarmenti.
Palo Blanco – Calycophyllum multifolium.
Palo Blanco – Ipomoea arborescens.
Palo Blanco – Lysiloma candida.

Palo Blanco – Myrsine grisebachii.
Palo Bolo – Ipomoea arborescens.
Palo Borracho – Chorisia insignis.
Palo Campechio – Haematoxylum campechianum.
Palo Chino – Pithecellobium mexicanum.
Palo Colorado – Ocotea veraguensis.
Palo de arco – Acacia coulteri.
Palo de Buba – Jacaranda copaia.
Palo de Burro – Capparis flexuosa.
Palo de Coco – Ficus padifolia.
Palo de Cotorra – Randia mitis.
Palo de Guaco – Crataeva tapia.
Palo de Leche – Sapium pavonianum.
Palo de Zorrello – Cassia emarginata.
Palo del Muerto – Ipomoea murucoides.
Palo del Nuez – Juglans insularis.
Palo hediondo – Cestrum dumetorum.
Palo Margo – Portlandia perosperma.
Palo de Matos – Ormosia krugii.
Palo mitas – Pithecoctenium echinatum.
Palo Muerto – Aetoxicon punctatum.
Palo Negro – Dalbergia retusa.
Palo Negro – Lomatia hirsuta.
Palo Pangue – Gunnera chilensis.
Palo piojo – Willardia mexicana.
Palo verde – Cercidium torreyanum.
Palo zopilote – Swietenia humulis.
Paloi – Anemone hupehensis.
Palosápsi Mersawa – Anisoptera thurifera.
Palosápsis Oil – Anisoptera thurifera.
Palta – Persea gratissima.
Paluch – Populus ciliata.
Pambotana Bark – Calliandra houstoniana.
Pampas Grass – Cortaderia argentea.
Pampoong – Oenanthe javanica.
Pan y agua – Capparis flexuosa.
PANAEOLUS (Fr.) Quél. Agaricaceae. 30 spp. Cosmopolitan.
***P. campanulatus*** L. var. ***sphinctrinus*** (Fr.) Bres. Europe, N. and S.America. The fruitbody contains hallucinogenic drugs. It is used during religous ceremonies by the Mazatecs of S.Mexico, who call it The Flesh of the Gods (Teonanactl).
***P. papiliomaceus*** Bull. = Copelandia papilionacea. Cosmopolitan (Fool's Mushroom, Narrenschwamm). Eating the fruitbodies causes mental derangement. It was supposed to have been used by Hungarian wise women in love philtres.
Panama Balata – Manikia dariensis.
Panama Bark – Quillaja saponaria.
Panama Hat Palm – Carludovica palmata.
Panama Ipecac – Caphaëlis acuminata.
Panama Tree – Sterculia carthaginensis.
Panau Resin – Dipterocarpus vernicifluus.
PANAX L. Araliaceae. 8 spp. Tropical and E.Asia, N.America. Herbs.
***P. edgerleyi*** Hook. f. = Nothopanax edgerleyi (Raukawa). New Zealand. The Maoris extract a perfumed oil from the leaves.
***P. fruticosum*** L. = Nothopanax fruticosum.
***P. ginseng*** May. = P. schinseng.
***P. murrayi*** F. v. Muell. Australia. Tree. The plant yields a gum similar to Gum Acacia, for which it is sometimes used as a substitute.
***P. quinquefolia*** L. (American Ginseng). E. N.America. The root (Radix Ginseng, Radix Ninsi) is used in America, Europe and China as an aphrodisiac and by the Chinese as a panacea for restoring vigour to young and old. The plant is cultivated locally and exported to Hong Kong. It contains the glucoside panquillon. There is no real evidence of its value as a drug.
***P. repens*** Max. = Aralia repens. (San chi, Shen yeh, Tam that). China, Indochina. The Chinese use the roots as a tonic. The plant is cultivated locally.
***P. schinseng*** Nees. = P. ginseng = Aralia ginseng. N.China. The plant is cultivated locally. See P. quinquefolia.
Panbeh – Astragalus echidnaeformis.
Panbeh – Astragalus gossypinus.
PANCHERIA Brongn. and Gris. Cunoniaceae. 25 spp. New Caledonia. Trees. The purple-red wood of both these species is durable and polishes well. It is used for turning and cabinet work.
***P. obovata*** Brongn. and Gris.
***P. ternata*** Brongn. and Gris.
Pan chu – Phyllostachys pubescens.
PANCOVIA Willd. Sapindaceae. 10–12 sp. Tropical Africa. Trees.
***P. harmsiana*** Gilg. (Botendi, Mangea, Muyini, Ngilimana). Congo, Gabon. The hard wood is used locally for construction work and handles of implements. The fruits are eaten as an emergency food.
***P. lujai*** De Wild. = Lychnodiscus cerospermus.
PANCRATIUM L. Amaryllidaceae. 15 spp. Mediterranean to tropical Asia, tropical Africa.
***P. guianense*** Ker-Gawl. = Hymenocallis tubiflora.
***P. speciosum*** L. f. (Ognon Gli, White Lily). W.Indies. The juice is used locally as an emetic.

PANDA Pierre. Pandaceae. 1 sp. Tropical W.Africa. Tree.

***P. oleosa*** Pierre.=Porphyranths zenkeri. The seeds yield an oil which is used locally for cooking.

PANDANUS L. f. Pandanaceae. 600 spp. Old World Tropics. Trees.

***P. amboinensis*** Warb. Malaysia. The leaves are used locally for making mats.

***P. aquaticus*** F. v. Muell. Australia. The seeds are eaten roasted by the aborigines.

***P. atrocarpus*** Griff. Malaysia. The leaves are used to make mats, sails etc.

***P. bidur*** Jungh. Malaysia. In the Moluccas the fibres from the proproots are used to make cordage and fishing nets etc.

***P. concretus*** Baker. (Fandrana). Madagascar. The leaves are used locally to make sacks.

***P. copelandi*** Merr. Philippines. The leaves are used locally to make mats, baskets etc.

***P. dubius*** Spreng. Amboina. The fibre from the proproots is used locally to make cordage for hammocks, beds etc.

***P. edulis*** Thou. Madagascar. The fruits are eaten locally.

***P. furcatus*** Roxb. E.India, Malaysia. The leaves are used locally to make mats.

***P. heudelotianus*** Balf. Tropical Africa. The leaves are used locally to make mats and the fibres from the proproots to make cordage.

***P. houlletii*** Car. (Screw-pine). Tropical Asia, especially Malaysia. The fruits are sweet with a pleasant flavour, rather like pineapple. They are eaten locally.

***P. kerstingii*** Warb. Coast of W.Africa. The leaves are used locally to make mats.

***P. leram*** Jones. (Nicobar Breadfruit). Malaysia. The cooked fruits are eaten locally.

***P. odoratissimus*** L.=P. tectorus. (Thatch Screw Pine). Seychelles, Malaysia, Australia, Pacific Islands. The plant is cultivated and there are several varieties. The leaves are used for thatching, mats etc. and are made into sugar bags. The seeds are edible. The male flowers produce an essential oil used in perfumery.

***P. odorus*** Thunb. Malaysia. The young leaves are eaten locally, especially in Java where they are eaten with beans.

***P. papuanus*** Solms-Laubach. (Mandim bepos, Tabalooko. New Guinea. The fibres from the proproots are used locally to make fishing nets.

***P. polycephalus*** Lam. Malaya. The young leaves are eaten raw.

***P. simplex*** Merr. Philippines. The leaves are used locally to make mats, bags, slippers and fancy goods.

***P. sparganoides*** Baker. Madagascar. The leaves are used locally to make sacks.

***P. spurius*** Miq. Malaysia. The leaves are used locally for mats, baskets etc. Cultivated in Java.

***P. tectorus*** Soland=P. odoratissimus.

***P. togoensis*** Warb. Togoland. The leaves are used locally to make mats.

PANGIUM Reinw. Flacourtiaceae. 1 sp. Malaysia. Tree.

***P. edule*** Reinw. (Pangi). The seeds are eaten locally after cooking to destroy the hydrocyanic acid they contain. The crushed raw seeds are used to stupify fish and also yield an oil used locally for lighting.

Pangola Grass – Digitaria decumbens.

Pangra – Erythrina suberosa.

***Panicularia fluitans*** Kuntze=Glyceria fluitans.

Panicgrass – Panicum amarum.

PANICUM L. Graminae. 500 spp. Tropics and warm temperate. Grasses. The grain species grow under drier conditions and in poorer soils. They are not grown on very large acreages, but are of extreme local importance to peasant communities.

***P. amarum*** Ell. (Panicgrass). Shores of E. and S. U.S.A. Planted to prevent erosion of seashores.

***P. ambiguum*** Trin. Tropical Asia, Pacific. Used locally for livestock fodder.

***P. barbinode*** Trin. (Para Grass). S.America. Cultivated widely in warm countries as livestock pasture and for hay.

***P. brizoides*** Jacq.=P. flavidum.

***P. burgii*** Chev.=P. stagninum.

***P. coloratum*** L. (Small Buffalo Grass). S.Africa. An excellent pasture and hay species.

***P. crus-ardeae*** Willd.=P. elongatum=P. sulcatum. S.America. Used as a pasture grass in Brazil.

***P. crus-galli*** L.=Oplismenus frumentaceus. Temperate. The seeds are used as a cereal in Japan, for making porridge etc.

***P. decompositum*** R. Br.=P. laevinode=P. proliferum. (Native Millet, Umbrella Grass). Australia. Grown locally as a pasture grass. The seeds are used as a cereal by the aborigines.

***P. distachum*** L.=Brachiaria disticha.

***P. elongatum*** Poir.=P. crus-ardeae.

***P. flavidum*** Retz.=P. brizoides. (Vandyke Grass). Australia. A good pasture grass, used particularly for wintering stock.

***P. frumentacea*** Roxb.=Echinochloa frumentacea.

***P. laevifolium*** Hack. (Land's Grass). S.Africa. An excellent forage grass, particularly for hay and silage. Used mainly in Botswana.

***P. laevinode*** Lindl.=P. decompositum.

***P. maximum*** Jacq. (Guinea Grass). Tropical and subtropical Asia, Africa and America. A very important fodder grass, particularly in the warm, humid tropics.

***P. microspermum*** Fourn.=P. tricanthum. Tropical America. The sweet-smelling rhizome is used in Brazil as a diuretic and mild stimulant.

***P. miliaceum*** L. (Broom Corn Millet, Hog Millet, Proso Millet). India. Cultivated in Old and New World, especially India, China, Japan and Russia. The flour is used for making bread etc. and the grain is fed to livestock. The seeds are highly nutritious, containing some 10 per cent protein and 4 per cent fat, as well as starch. An alcoholic beverage (Braga, Busa) is made from the seeds in Central and E.Europe and Asia Minor.

***P. miliare*** Lam.=P. sumatrense.

***P. molle*** Sw. (Mauritius Grass). Tropical America. Grown locally as fodder for livestock.

***P. obtusum*** H.B.K. (Vine mesquite). W. U.S.A., Mexico. The grain is used as flour by the local Indians.

***P. proliferum*** F. v. Muell.=P. decompositum.

***P. sanguinale*** L.=Digitaria horizontalis. Cosmopolitan. Used as fodder for livestock.

***P. sarmentosum*** Roxb. Tropical Asia, Malaysia. Used locally as fodder for livestock.

***P. stagninum*** Retz.=P. burgii. Tropical Africa, especially the Sudan. A syrup from the stems is used locally to make sweetmeats and drinks.

***P. sulcatum*** Bert.=P. crus-ardeae.

***P. sumatrense*** Roth. ex Roem. and Schult.= P. miliare. (Little Millet, Small Millet). India. Smaller than the other grain millets, it is grown extensively locally on the poorer land.

***P. texanum*** Buckl. (Texas Millet). Texas. Sometimes grown locally as livestock fodder.

***P. tricanthum*** Nees.=P. microspermum.

***P. turgidum*** Forsk. N.Africa. The seeds are used locally for flour etc.

***P. virgatum*** L. (Switchgrass). N. through to Central America. The plant makes a good fodder and is used extensively to prevent erosion of inland dunes and conservation work.

PANOPSIS Salisb. Proteaceae. 25 spp. Tropical America. Trees.

***P. rubescens*** (Pohl.) Pitt. (Cedro Bordado, Yolombo). The hard red-brown wood is used locally for furniture-making and fancy goods.

Pansy – Viola tricolor.

Panther Strangler – Doronicum pardalianches.

Pantjal kidang – Aglaia odoratissima.

Pantjal kidang – Phyllanthus indicus.

PANUS Fr. Agaricaceae. 20 spp. Widespread on wood.

***P. rudis*** Fr.=Lentinus rudis. Cosmopolitan. The fruit bodies are used in N.Caucasus to flavour sheep's milk, cheese and are occasionally eaten.

Páo de Jangada – Apeiba tibourbou.

Páo d'Embira – Xylopia carminativa.

Páo Manfin Oil – Agonandra brasiliensis.

Páo Piereira – Geissospermum laeve.

Páo Rosa – Aniba terminalis.

Páo Santo – Bulnesia sarmienti.

Paoch lebi – Campnosperma oxyrhachis.

Paooh bajan – Irvingia malayana.

Papain – Carica papaya.

Papaja ootan – Polyscias nodosa.

Papak lauin – Asplenium macrophyllum.

Papar – Buxus wallichiana.

PAPAVER L. Papaveraceae. 100 spp. Europe, Asia, S.Africa, Australia, America. Herbs.

***P. nudicaule*** L. (Iceland Poppy). Arctic, E. and W.Hemispheres. The seeds contain some opium and are used locally to relieve pain.

***P. rhoeas*** L. (Corn Poppy, Red Poppy). Europe to temperate Asia. Introduced to N.America. The flowers are used in home medicine as an expectorant. The red dye from the flowers is used to colour wines and medicines.

***P. somniferum*** L. (Opium Poppy). Mediterranean to India. The plant is cultivated in the Middle and Far East for the latex Opium. The latex is the source of the alkaloids morphine, codeine which are used medicinally as pain-relievers and as sedatives. Crude opium also contains laudenine, nicotine and papaverine. The addictive drug heroine is also made from opium. Opium is smoked extensively in the East, being known as Chinese, Egyptian, Indian, Persian or Turkish Opium. There are several varieties of the plant, grouped into two main classes; "Glabra" (Druggist's or Manufacturer's Opium) used mainly for the extraction of medicinal drugs and "Alba" (Soft or Shipping Opium) used for smoking. The crude opium is extracted from the plant from longitudinal incisions in the unripe seed capsule. This is done ten to twenty days after the capsule is formed. The seeds, which contain no opium, have some 44–50 per cent oil (Poppy Seed Oil). They are used locally in India in making curries and sweetmeats, or are spread on bread. Much seed is exported from India to Europe where the drying oil is used to make paints and varnishes and soap. It is also used to make salad dressings, especially in France and Germany. The oil is also used in India for lighting, as well as in preparing foods. White seeds are preferred to the darker ones for the extraction of oil. The oil seed cake is a valuable cattle fodder.

Papaya – Carica papaya.

Papaya de Mico – Carica peltata.

Papaya de Terra Fria – Carica cestriflora.

Papaya, Mountain – Carica cundinamarcensis.

Papaya Orejona – Jacaratia mexicana.

Paper Birch – Betula papyrifera.

Paper Mulberry – Broussonetia papyrifera.

Papita – Cordia globosa.

Papo Canary Tree – Canarium schweinfurthii.

Papo de Peru – Aristolochia brasiliensis.

Papoose Root – Caulophyllum thalictroides.

PAPPEA Eckl. and Zeyh. Sapindaceae. 3 spp. Tropical and S.Africa. Small trees.

***P. capensis*** Eckl. and Zeyh. S.Africa. The seeds yield an oil which is used as a lubricant and to make soap.

***P. ugandensis*** Baker f. (Umumena). Central Africa. The oil from the seeds is used by the natives for cooking food.

Paprika – Capsicum annuum.

Papua Nutmeg – Myristica argentea.

Papyrus – Cyperus papyrus.

Para Arrowroot – Manihot esculenta.

Para Breadnut – Brosimum paraense.

Para Cress – Spilanthes acmella.

Para de Mula – Agave pesmulae.

Para Grass – Panicum barbinode.

Para Guava – Campomanesia acida.

Para Ipecac – Cephaëlis ipecacuanha.

Para Nut – Bertholletia excelsa.

Para Piassaba – Leopoldinia piassaba.

Para Rhatany – Krameria argentea.

Para Rubber (Seed Oil) – Hevea brasiliensis.

Para Sarsaparilla – Smilax spruceana.

PARABARIUM Pierre. Apocynaceae. 20 spp. S.China, Indochina. Vines. Both species mentioned are potential sources of commercial rubber.

***P. dindo*** Pierre. Indochina.

***P. dinrang*** Pierre. Indochina.

***Paradaniella*** Willis = Daniellia.

Paradise Flower – Caesalpinia pulcherrima.

Paradise Nut – Caryocar nuciferum.

Paradise Nut – Lecythis zabucajo.

Paraguay Jaborandi – Pilocarpus pinnatifolius.

Paraguay Lignum – Bulnesia sarmienti.

Paraguay Palm – Acrocomia sclerocarpa.

Paraguay Tea – Ilex paraguensis.

Paraguayan Carnauba – Copernicia australis.

Parake oodang – Macaranga triloba.

PARAMERIA Benth. Apocynaceae. 6 spp. Indochina Malaya. Vines.

***P. barbata*** Schum. = P. glandulifera = Ecdysanthera barbata. (Akar gerip pootih, Gembor). Indochina, Malaya. The stem yields a latex which coagulates in warm water and is used locally. A decoction of the bark (Kajo rapat) is used locally in Java to heal wounds and to contract the uterus. In Bali, a paste of the bark is used to treat dysentery.

***P. glandulifera*** Benth. = P. barbata.

***P. philippensis*** Radlk. Philippines. The bark fibres are used locally for cordage.

Parana Pine – Araucaria brasiliana.

PARASHOREA Kurz. Dipterocarpaceae. 11 spp. S.E.Asia, W.Malaysia. Trees.

***P. lucida*** Kurz. = Shorea lucida. (Meanti hitan, Sooranto). Sumatra. The dark, heavy wood is used locally for construction work and boat-building.

***P. malaanonan*** (Blanco) Merr. (Borneo White Seraya, Seraya puteh). The heavy wood works and polishes well, but is attacked by termites. It is an important local wood and is exported. It is used for general construction work, furniture-making, flooring, decking, railway carriages and car body work.

Parasol Mushroom – Lepiota procerea.

Parasol Pine – Sciadopitys verticillata.

PARASTEMON A. DC. Chrysobalanaceae. 2 spp. Malaysia. Trees.

***P. urophyllum*** A. DC. (Mengilas). Malaysia. The hard, heavy red-brown wood finishes well and is used for general construction work, bridges and piers and boat keels. It burns well as a firewood.

Pareira Root – Chondrodendron tomentosum.

Pareira, Yellow – Aristolochia glaucescens.

Pariera, White – Abuta rufescens.

PARIANA Aubl. Graminae. 34 spp. Tropical S.America.

***P. lunata*** Nees. Colombia to Brazil. In Colombia the leaves are used to wrap gold and platinum dust.

***Parianthopodus carijo*** Manso = Cayaponia espelina.

PARIETARIA L. Urticaceae. 30 spp. Temperate and tropics. Herbs.

***P. officinalis*** L. (Pellitory of the Wall. Paritory). Central and S.Europe. A decoction of the leaves is diuretic and laxative. It is used to treat bladder complaints, stones and suppression of urine.

PARINARI Aubl. Chrysobalanaceae. 60 spp. Tropics. Trees.

***P. annamense*** Hance. China, Malaysia. The flowers are used in Thailand to make a cosmetic oil.

***P. campestre*** Aubl. Guyana. The fruits are eaten locally.

***P. curatellaefolium*** Planch. Tropical Africa. The fruits are eaten locally and the wood is used for general carpentry.

***P. emirnense*** Baker. (Vandevenona). Madagascar. The fruits are eaten locally.

***P. excelsum*** Sab. Tropical Africa. The wood is used for general construction work, furniture and household utensils. The ashes are used in the Congo to cure skins.

***P. glabra*** Oliv. (Bofale, Ehungai, Mukumbi). W.Africa, Congo. The hard red wood is used for general carpentry and posts.

***P. guayanensis*** Aubl. (Satinwood) Guyana. The red-brown wood is beautifully figured and is used to make furniture.

***P. laurinum*** A. Gray. (Akarrittom, Kusta, Makita). Pacific islands. A semisolid fat is extracted from the seeds and is used to make a very quick-drying varnish.

***P. macrophyllum*** Sab. (Ginger-bread Plum). Tropical Africa. The wood is used for general building purposes and canoes. The fruits are eaten locally and their skins are used to scent ointments. A decoction of the bark is used locally to treat inflamed eyes and a similar preparation from the bark and leaves is used as a mouthwash. An oil (Neou Oil) is extracted from the seeds.

***P. mobola*** Oliv. (Cork Tree, Grijs Appel, Hissing Tree, Mobola Plum, Muhacha, Mupundu). Tropical Africa. The hard, light brown wood is not durable, but is used locally for beams, poles, railway sleepers, canoes and mortars. The bark is used for tanning, darkening the leather. A decoction of the roots is used locally to treat cataract. The pleasantly flavoured fruits are very popular especially in S.Africa and Rhodesia. They are eaten raw, used to make a drink (Luzwazhi) and a fermented beverage. They are also eaten in a kind of soup. The seeds are added to fish dishes.

***P. nonda*** F. v. Muell. (Nonda Tree). Australia. The fruits are eaten by the aborigines.

***P. oblongifolium*** Hook. f. Malacca. The dark orange, heavy durable wood is used locally for general construction work.

***P. polyandrum*** Benth. Tropical Africa. The wood is used to build houses, particularly in rural areas and to make charcoal.

***Parinarium*** Juss. = Parinari Aubl.

PARIS L. Trilliaceae. 20 spp. N.Temperate Old World. Herbs.

***P. quadrifolia*** L. (One Berry). N.Europe to Siberia and Asia Minor. A decoction of the leaves (Herba Paridis) is used to treat headaches, neuralgia, rheumatism and gout.

***Paritium tiliaceum*** (L.) Britt. = Hibiscus tiliaceous.

Paritory – Parietaria officinalis.

PARKIA R. Br. Leguminosae. 40 spp. Tropics.

***P. africana*** R. Br. = P. biglobosa. (African Locust). Tropical Africa. The sweet fruit pulp is eaten raw. The parched seeds are eaten in cakes, used to make a condiment, or roasted to make a coffee (Café du Sudan). They are also used to treat stomach upsets. The bark is used for tanning leather, staining it red and in Gambia a decoction of the bark is used for treating toothache. The wood is used locally for light carpentry.

***P. biglobosa*** Benth.=P. africana.

***P. bussei*** Hrams. (Muavi). Tropical Africa. The seeds yield Muavi Poison, a criminal poison used in Malawi.

***P. filicoidea*** Welw. (Fernleafed Nitta Tree, West African Locust Bean). Tropical Africa. The bark is used in the Sudan to tan leather a dark brown. The dried powdered seeds are used locally as a condiment.

***P. javanica*** Merr. (Kedaung). Indonesia, Borneo. The soft wood is not durable, but easy to work. It is used locally for light-weight construction work. The whole pods are used as a condiment.

***P. speciosa*** Hassk. Malaysia. The raw seeds are eaten locally and the whole pods are used as a condiment.

PARKINSONIA L. Leguminosae. 2 spp. Tropical America, S.Africa. Trees.

***P. microphylla*** Torr. S.Arizona through to Mexico. The seeds are eaten by the local Indians, raw, or as a flour.

PARMELIA L. Parmeliaceae. Lichens.

***P. acebatulum*** Duby. Ireland. Used locally to dye woollens an orange-brown.

***P. caperata*** (L.) Ach. (Arcel, Stone Crottle, Wrinkled Lichen). Yields a brown-yellow to bright yellow dye, used to dye wool on the Isle of Man.

***P. conspersa*** (Boulder Lichen, Sprinkled Parmelia). The red-brown dye is used in England to dye woollens.

***P. centrifuga*** (L.) Ach. (Ring Lichen). Yields a red-brown dye, used in England to colour wool.

***P. ciliaris*** Ach.=Anaptychia ciliaris.

***P. fahlunensis*** Ach.=Cetraia fahlunensis.

***P. furfuracea*** (L.) Ach.=Evernia furfuracea.

***P. kamtschadalis*** (L.) Ach. Tropical Asia. Yields a light red dye, used in India to colour calico.

***P. olivacea*** (L.) Ach. (Bronze Shield Lichen). Used in U.K. to colour woollens brown.

***P. omphalodes*** (L.) Ach. (Black Crottle, Corks, Smoky Shield Lichen). Yields a purple-red dye used in Scandinavia and Scotland to dye wool.

***P. physodes*** (L.) Ach. (Dark Crottle, Puffed Shield Lichen). The brown dye is used in Scotland and Scandinavia to dye wool. The lichen is eaten in soup by the Indians of Wisconsin.

***P. pollinaria*** Ach.=Ramalina pollinaria.

***P. prunastri*** Ach.=Evernia prunastri.

***P. saxatilis*** (L.) Ach. Yields an orange to red-brown dye used in N.Ireland and Scotland to dye wools. The dye gives Harris tweed its characteristic smell.

***P. stygia*** (L.) Ach. (Midnight Lichen). The brown dye is used in U.K. to dye woollens.

Parmelia, Sprinkled – Parmelia conspersa.

PARMENTIERA DC. Bignoniaceae. 8 spp. Mexico to Colombia. Trees.

***P. alata*** Miers=Crescentia alata. Old World Tropics, Mexico, Central America. In Nicaragua a drink is made from the seeds, the shells are used for cups and the wood is used to make carts.

***P. cerifera*** Seem. Central America. The plant is used locally to provide fodder for livestock.

***P. edulis*** DC. (Chote, Cuajilote, Guajilote, Gueroxiga, Food Candle Tree). Central America. The fruits are eaten locally, raw, cooked, or in preserves and pickles. A decoction of the roots is diuretic and is used to treat dropsy.

Parna – Garcinia livingstonii.

PARONCHIA Mill. Caryophyllaceae. 50 spp. Cosmopolitan. Herbs.

***P. argentea*** Lam. (Silver Nailroot). Mediterranean. The leaves are used in Morocco to make an aphrodisiac.

***P. jamesii*** Torr. and Gray. (James Whitkow Wort). W. U.S.A. The leaves are used by the local Indians to make a tea.

***Parosela*** Cev.=Dalea.

Parosela, Slender – Dalea anneandra.

PARQUETINA Baill. Periplocaceae. 1 sp. W. Equatorial Africa.

***P. calophyllus*** Baill. Yields a latex, possibly useful as emergency supply of rubber.

Parra rosa – Entada polystachia.

Parsley – Petroselenium hortense.

Parsley, Bastard Stone – Sison amomum.

Parsley, Breakstone – Alchemilla arvensis.

Parsley, Cow – Chaerefolium sylvestre.

Parsley, Dog – Aethusa cynapium.

Parsley, Fool's – Aethusa cynapium.

Parsley, Milk – Peucedanum palustre.

Parsley Piert – Alchemilla arvensis.

Parsley, Turnip Rooted – Petroselenium hortense.

Parsley, Water – Oenanthe sarmentosa.

Parsnip – Pastinaca sativa.

Parsnip, Cow – Heracleum lanatum.
Parsnip, Cow – Heracleum sphondylium.
Parsnip, Water – Sium cicutaefolium.
Parsnip, Water – Sium thunbergii.

PARSONIA R. Br. Apocynaceae. 100 spp. S.China, S.E.Asia, Indomalaya, Australia New Zealand, Polynesia. Trees.

***P. helicandra*** Hook. and Arn. (Pe-nuli-valli). Tropical Asia. Tropical Asia, especially W.India. The Hindus used the juice from the plant as a medicine to treat insanity.

PARTHENIUM L. Compositae. 15 spp. America, W.Indies. Herbs, shrubs or vines.

***P. argentatum*** Gray. (Guayule). S.W. U.S.A., Mexico. Shrub. The plant produces a latex which is of value as a rubber. It is extracted by maceration of the whole plant. It was used commercially during World War II, but since then production has greatly declined.

***P. hysterophorus*** L. N. and S.America. The local Indians use the boiled roots to treat dysentery.

***P. integrifolium*** L. E. U.S.A. The local Indians use the leaves as a poultice for burns.

PARTHENOCISSUS Planch. Vitidaceae. 15 spp. Temperate Asia, America. Vine.

***P. quinquefolia*** (L.) Planch. = Ampelopsis quinquefolia. (Virginia Creeper). S. N.America. The local Indians use the dye from the fruits to stain the skin and war feathers. A decoction of the bark is used as a tonic and to treat dropsy and the local Indians use a decoction of the roots to treat diarrhoea.

Partridge Berry – Gaultheria porcumbens.
Partridge Berry – Mitchella repens.
Partridge Pea – Cassia chamaecrista.
Partridge Wood – Andira excelsa.
Parusha – Grewia asiatica.
***Pasania costata*** Gamble – Quercus costata.
***Pasania cyclophora*** Gamble – Quercus cyclophora.
***Pasania densiflora*** Oerst. = Lithocarpus densiflora.
Pashaco – Entada polyphylla.
Pasionaria – Passiflora ciliata.
Pasolder – Grewia salutaris.

PASPALUM L. Graminae. 250 spp. Warm. Grasses.

***P. brevifolium*** Flueg. = Digitaria longiflora.

***P. conjugatum*** Berg. (Hilo Grass). Tropics. Grown as fodder for livestock.

***P. dilatatum*** Poir. (Dallis Grass). S.America. An important fodder grass in S. U.S.A. and S.Africa.

***P. exile*** Kipp. = Digitaria exilis.

***P. longiflorum*** Vhev. = Digitaria exilis.

***P. mandiocanum*** Trin. Brazil. An important local fodder grass.

***P. notatum*** Flueg. (Bahia Grass). Mexico to S.America, W.Indies. Grown locally as a fodder grass and in S. U.S.A.

***P. scrobiculatum*** Lam. (Koda Millet). Old and New World. A decoction of the underground parts is used in the Philippines to aid childbirth. The plant is sometimes grown for fodder.

***P. urvillei*** Steud. (Vasey Grass). S.America. A coarse grass usually grown for silage.

Paspalum, False Creeping – Brachiaria humidicola.
Pasque flower, European – Anemone pulsatilla.
Pasque flower, Meadow – Anemone pratensis.
Pasque Flower, Spreading – Anemone patens.

PASSERINA L. Thymelaeaceae. 15 spp. S.Africa. Shrubs.

***P. filiformis*** L. S.Africa. The bark is used to make cordage, used especially to make whips and tying thatch.

Passid – Hardwickia binata.

PASSIFLORA L. Passifloraceae. 500 spp. Mainly in America, but a few in Asia and Australia; 1 sp. Madagascar. Vines. The fruits of the species mentioned have a tart agreeable flavour. They are eaten raw, or more frequently made into drinks and sherbets.

***P. alata*** Dryand (Maracuja). Brazil, Peru. Cultivated locally.

***P. amethystina*** Mikan. (Maracuja se Serra). Brazil.

***P. antioquiensis*** Karts. = Passiflora vanvolxemii.

***P. cearensis*** Barb. (Peora). Brazil. Cultivated locally.

***P. ciliata*** Dryand. (Pasionaria). Mexico. The juice of the plant is used as a sedative in the treatment of hysteria and insomnia, especially of children.

***P. edulis*** Sims. (Purple Grenadilla, Passion Fruit). Brazil. Cultivated in Old and New World Tropics. The fruit juice is produced commercially in some countries.

***P. incarnata*** L. (Apricot Vine, May-Pop). E. N.America. The dried flower heads are used to treat insomnia, neuralgia, diarrhoea and discharges from the vagina.

***P. laurifolia*** L. (Bell Apple, Jamaican Honeysuckle, Water Lemon). Tropical America. Cultivated locally in the W.Indies.

***P. ligularis*** Juss. (Sweet Grenadilla). Tropical America. Cultivated locally.

***P. maliformis*** L. (Curuba, Sweet Calabash). Tropical America. Cultivated locally.

***P. mollissima*** (H.B.K.) Bailey. (Caruba de Castilla, Tasco). Ecuador, Colombia. Cultivated locally, especially in Ecuador and made into a drink called Crema de Caruba.

***P. organensis*** Gardn. (Nensi). Brazil. Cultivated locally. The fruits are also used in pastries.

***P. parahybensis*** Sims. (Perluxo). See P. organensis.

***P. pinnatistipula*** Cav. (Gutupa, Tasco). Ecuador, Colombia.

***P. popenovii*** Killip. (Grenadille de Quijos). Ecuador.

***P. psilantha*** (Sodiro) Killip. Ecuador. Cultivated.

***P. quadrangularis*** L. (Barbadine, Giant Granadilla, Granadilla Real). Tropical America. Cultivated throughout the tropics. The unripe fruits are also eaten as a vegetable.

***P. salvadorensis*** D. Smith. El Salvador. The leaves are used locally as a diuretic.

***P. tripartita*** (Juss.) Poir. (Tasco). Ecuador. Cultivated locally.

***P. van-volxemii*** Triana and Planch = P. antioquiensis. = Tacsonia van-volxemii (Banana Passion Fruit). Colombia. Cultivated locally and in New Zealand.

***P. villosa*** Mast. See P. organensis.

***P. warmingii*** Mast. (Maracuja Mirim). Brazil.

Passion Fruit – Passiflora edulis.

Passion Fruit, Banana – Passiflora van-volxemii.

Pasta Althaeae – Althaea officinalis.

Pasteli – Ceratonia siliqua.

PASTINACA L. Umbelliferae. 15 spp. Temperate Eurasia. Herbs.

***P. sativa*** L. = Peucedanum sativum. (Parsnip). Eurasia. Cultivated in Old and New World temperate regions for the starchy roots which are eaten as a vegetable. The starch in the roots hydrolyses on exposure to cold, making the root sweet. It has been cultivated in Europe since Roman times and was introduced to America during the 17th century.

Pastorata – Turnera diffusa.

Pat phanes – Artocarpus hirsuta.

PATAGONULA L. Ehretiaceae. 2 spp. Brazil, Argentine. Trees.

***P. americana*** L. (Guayabi). Brazil, Argentine. The dark brown wood yields a dye used to stain lighter woods the colour of walnut or mahogany.

***P. bahiensis*** Moric. (Guayabi, Ope branco). Brazil, Argentine, Paraguay and Uruguay. The dark heart wood is used to make furniture, interiors of buildings and turning. The sap wood is used for agricultural implements.

Patana Oak – Careya arborea.

Patashiti – Theobroma bicolor.

Pataua (Oil) – Jessenia batana.

Patchouli, Khasia – Microtanea cymosa.

Patchouly (Oil) – Pogostemon cablin.

Paterno – Inga paterno.

Pathor – Bridelia retusa.

Patience Dock – Rumex patientia.

Patjar leuweung – Impatiens platypetala.

Patji-patji – Leucas lavandulifolia.

Patolang gabat – Trichosanthes quinquangulata.

***Patrisia*** Rich. = Ryania.

Patti – Cannabis sativa.

Pau Amarello – Euxylophora paraensis.

Pau Camp – Kielmeyera coriacea.

Pau de Santo – Cabralea cangerana.

Pau-Hoi – Neolitsea pauhoi.

Pau-liso – Balfourodendron riedeilanum.

Pau Setim – Aspidosperma eburneum.

PAULLINIA L. Sapindaceae. 180 spp. Warm America. Trees, or shrubs.

***P. asiatica*** L. = Toddalia aculeata.

***P. cupana*** Kunth. (Guarana). S.America. The seeds contain 4·88 per cent caffeine and a paste of them is used locally to make a stimulating drink to combat fatigue. An alcoholic beverage is also made from them when mixed with cassava. The seeds are used medicinally to treat diarrhoea. The plant is cultivated, particularly in Brazil.

***P. mexicana*** L. = Serjania mexicana.

***P. pinnata*** L. = Serjania curassavica. (Barbasco, Nistamal). Mexico, central America, W.Indies. The stems are used as cordage. The sap from the plant is used locally as a criminal poison and to stupify fish and to poison arrows. The seeds are used for the same purposes.

***P. yoco*** Schult. and Killi. (Yoco). Tropical America, especially Colombia. The bark contains some 3 per cent caffeine and is used locally to make a stimulating drink.

PAULOWNIA Sieb. and Zucc. Scrophulariaceae. 17 spp. E.Asia. Trees.

***P. imperialis*** Sieb. and Zucc. = P. tomentosa. China. The wood is used in China and Japan to make musical instruments and for boxes, carving clogs etc. It gives a good charcoal suitable for making gunpowder.

***P. tomentosa*** Koch. = P. imperialis.

Pauohiika – Jacquemontia sandwicensis.

PAUSINYSTALIA Pierre ex Beille. Rubiaceae. 13 spp. Tropical W.Africa. Trees.

***P. yohimba*** Pierre = Coryanthe yohimbe. Tropical W.Africa. The bark (Yohimbe Bark) is used as an aphrodisiac in the Cameroons.

***Pavia rubra*** L. = Aesculus pavia.

PAVONIA Cav. Malvaceae. 200 spp. Tropics and subtropics. Shrubs or herbs.

***P. bojeri*** Baker. Madagascar. The bark fibres are used locally to make cloth.

***P. dasypetala*** Turzc. Central America. The bark fibres are used by the local Indians to make cloth.

***P. hirsuta*** Guill. and Perr. Tropical Africa. The natives add the mucilagenous roots to milk to accelerate the production of butter during churning.

***P. schimperiana*** Hochst. Tropical Africa. The bark fibres are used locally for cordage.

***P. urens*** Cav. Tropical Africa. The bark fibres are used locally for cordage.

Pawpaw – Asimina triloba.

Pawpaw – Carica papaya.

Paxiuba (Palm) (Wood) – Socratea exrrhiza.

PAYENA A. DC. Sapotaceae. 16 spp. S.E.Asia, W.Malaysia. The species mentioned yield a latex which is elastic at normal temperatures, but becomes plastic on heating. They were of some local importance.

***P. betis*** F.-Vill. = Madhuca betis.

***P. dantung*** Lam. (Belam kedjil, Dantong). Indonesia. The latex called Getah dantong.

***P. lancifolia*** Burck. = Madhuca lancifolia.

***P. leerii*** Kurz. Burma, Malaysia.

***P. macrophylla*** Burck. = Madhuca macrophylla.

***P. stipularis*** Burck = P. sumatrana. Sumatra, celebes.

***P. sumatrana*** Miq. = P. stipularis.

***P. utilis*** Ridl. = Maduca utilis.

Pea, Asparagus – Psophocarpus tetragonolobus.

Pea, Bavarian Winter – Pisum arvense.

Pea, Beach – Lathyrus maritimus.

Pea, Black-eyed – Vigna catjang.

Pea, Black Podded – Pisum arvense.

Pea, Butterfly – Clitoria ternata.

Pea, Capucine – Pisum arvense.

Pea, Chick – Cicer arietinum.

Pea, Congo – Cajanus cajan.

Pea, Cow – Vigna sinensis.

Pea, East Prussian – Pisum arvense.

Pea, Field – Pisum arvense.

Pea, Flat – Lathyrus sylvestris.

Pea, Gambia – Crotalaria goreënsis.

Pea, Garden – Pisum sativum.

Pea, Gram – Cicer arietinum.

Pea, Grass – Lathyrus sativa.

Pea, Hindu Cow – Vigna catjang.

Pea, Konigsberger – Pisum arvense.

Pea, Kordofan – Clitoria ternata.

Pea of the Oasis – Pisum elatius.

Pea, Partridge – Cassia chamaecrista.

Pea, Pigeon – Cajanus cajan.

Pea, Rosary – Abrus precatorius.

Pea, Rough – Lathyrus hirsutus.

Pea, Sand – Pisum arvense.

Pea, Sea – Lathyrus maritimus.

Pea, Shrub – Caragana arborescens.

Pea, Siberian – Caragana arborescens.

Pea, Smyrna – Pisum arvense.

Pea, Sweet – Lathyrus orodatus.

Pea, Tangier – Lathyrus tingitanus.

Peach – Amygdalus persicae.

Peach Bitter, African – Nauclea esculentus.

Peach Kernel Oil – Amygdalus persica.

Peach Palm – Guilielma speciosa.

Peach Root – Nauclea esculentus.

Peach Wood – Caesalpinia echinata.

Peach Wood – Caesalpinia sappan.

Peacock Flower – Caesalpinia pulcherrima.

Peanut (Oil) – Arachis hypogaea.

Peanut, Ground – Amphicarpaea monoica.

Peanut, Hog – Amphicarpaea monoica.

Pear – Pyrus communis.

Pear, Alligator – Persea gratissima.

Pear, Anchovy – Grias cauliflora.

Pear, Avocado – Persea gratissima.

Pear, Balsam – Momordica charantia.

Pear, Chinese – Pyrus chinensis.

Pear, Hard – Strychnos henningsii.

Pear, Odara – Chrysophyllum africanum.
Pear, Prickly – Opuntia ficus-indica.
Pear, Thorn – Scolopia zeyheri.
Pear, White – Apodytes dimidiata.
Pearl Fruit – Margyricarpus setosus.
Pearl Millet – Pennisetum typhoideum.
Pearl Millet – Setaria glauca.
Pearly Everlasting – Anaphalis margaritacea.
Peavine, Grass – Lathyrus sativus.
Peavine, Groundnut – Lathyrus tuberus.
Pecan (Oil) – Carya pecan.

PECTIS L. Compositae. 70 spp. S.United States of America through to Brazil, W.Indies, Galapagos Islands. Herbs.

***P. leonis*** Rydb. W.Indies, especially Cuba. A decoction of the leaves is used locally to treat intestinal complaints, asthma and bronchitis.

***P. papposa*** Harv. (Foetid Marigold). S.W. U.S.A. The flowers are used by the local Indians as a condiment with meat.

PEDALIUM Royen ex L. Pedaliaceae. 1 sp. Tropical Africa, Madagascar, tropical Asia. Herb.

***P. murex*** L. The leaves are eaten as a vegetable in Africa. The seeds are diuretic and are used medicinally to treat urinary disorders.

Peddadulagondi – Muncuna monosperma.

PEDICULARIS L. Scrophulariaceae. 500 spp. N.Temperate. Herbs.

***P. lanata*** Willd.=P. langsdorffi.

***P. langsdortfii*** Fisch=P. lanata. Arctic Canada. The roots are eaten locally as a vegetable.

***P. pectinata*** Wall. (Misran). W.Himalayas, Kashmir. The Hindus used the dried leaves to treat spitting of blood.

PEDILANTHUS Neck. ex Poir. Euphorbiaceae. 14 spp. Florida through to tropical S.America, W.Indies. Shrubs.

***P. pavonis*** (Klotzsch. and Garcke) Boiss. Mexico. A decoction of the leaves is used locally as a purgative and to control menstruation. The plant yields a wax (Candelilla Wax) which has a high melting point. It is used for varnishes, gramophone records and candles.

Pedulo Colorado – Citharexylum quadrangulare.
Peepul Tree – Ficus religiosa.
Peer putta – Nannorhops ritchieana.
Pegah – Santinia tomentosa.

PEGANUM L. Zygophyllaceae. 5–6 spp. Mediterranean to Mongolia, S. United States of America to Mexico. Shrubs.

***P. harmala*** L. (Harmela Shrub). Mediterranean to Central Asia. The seeds are narcotic. In some parts of the East they are burnt to inhale the smoke which gives a feeling of exaltation. They were used to treat eye diseases, as an aphrodisiac and to stimulate the appetite. In parts of Russia the roots are used to treat nervous complaints and rheumatism. A red dye (Turkey Red) is extracted from the fruits and used in Turkey to dye the tarbooshes. An oil (Zit-el-Harmel) is extracted from the seeds.

Pega-pega – Triumfetta lappala.
Pegar arrak – Ixonanthes icosandra.
Pegu Catechu – Acacia catechu.
Peguia Marfin – Aspidosperma eburneum.
Peiperendai – Dioscorea hispida.
Peladjan – Pentasperon motleyi.

PELAGONIUM L'Hérit. Geraniaceae. 250 spp. Tropics, especially S.Africa; 1 spp. each in Canaries, St. Helena, Tristan da Cunha, E.Mediterranean, S.Arabia, S.India, Australia, New Zealand. Shrubs.

The oil producing species are derived from S.Africa. They are grown commercially in Algeria, Madagascar, Réunion, Ceylon and E.Africa and have been introduced to the U.S.A. The plants are propagated by stem cuttings and last for several years. The young shoots are collected three times a year and the oil distilled from them. The oil (Geranium Oil) is used in perfumery and soap manufacture. The main principle in the oil is geraniol, but the oils from different species have different odours.

***P. antidysentericum*** Kostel. S.Africa. The roots boiled in milk are used locally to treat dysentery.

The oil-producing species include:

***P. capitatum*** Willd.
***P. crispum*** L.Hérit.
***P. exstipulata*** Willd.
***P. fragrans*** Willd.
***P. fulgidum*** Ait.
***P. glutinosum*** L'Hérit.
***P. graveolens*** L'Hérit.
***P. odoratissimum*** Ait.
***P. quercifolium*** Baum.
***P. radula*** L'Hérit.=P. roseum.
***P. roseum*** Willd.=P. radula.
***P. vitifolium*** L'Hérit.

Pelampas boodak – Apostasia nuda.

Pelangas – Aporusa microcalyx.

PELEA A. Gray. Rutaceae. 75 spp. Pacific, mostly Hawaii. Trees.

***P. anisata*** Mann. Hawaii. The sweet-scented fruits are strung in garlands.

***P. fatuhivensis*** F. Br. Marquesas. The leaves are used locally to scent coconut oil, used as a body oil.

PELECYPHORA Ehrenb. Cactaceae. 2 spp. Mexico. Low growing cacti.

***P. aselliformis*** Ehrenb. (Peyotillo). The fruits are used locally to treat fevers.

Peler kambing – Sarcolobus spanoghei.

Pellitory – Anacyclus pyrethrum.

Pellitory of the Wall – Parietaria officinalis.

PELTANDRA Rafin. Araceae. 4 spp. S.United States of America and Atlantic N.America. Swamp herbs.

***P. virginica*** (L.) Kunth. (Green Arrow Arum, Virginian Tukahoe). E. U.S.A. The corms, spadix and fruits were eaten as vegetables by the local Indians.

PELTIGERA L. Lichens.

***P. aphthosa*** (L.) Hoffm. Temperate. Used as a source of d-galactose and d-mannose.

***P. canina*** (L.) Hoffm. (Dog Lichen, English Liverwort). An infusion of the whole plant is used to treat liver complaints. It is slightly laxative and a tonic. It is used in some parts of Europe to dye wool a rust-red.

***Peltiphyllum peltatum*** Engl. = Saxifrage peltata.

PELTOGYNE Vogel. Leguminosae. 25 spp. Tropical S.America, especially Guyana and Brazil. The wood of the following species is called Purple Heart. It is a dark purple, heavy, hard and durable and is used for interior work, carving etc.

***P. catingae*** Ducke.

***P. densiflora*** Spruce.

***P. excelsa*** Ducke.

***P. gracilipes*** Ducke.

***P. lecointei*** Ducke.

***P. maranhensis*** Ducke.

***P. paniculata*** Benth. The main source of the wood.

***P. paradoxa*** Ducke.

PELTOPHORUM (Vogel) Walp. Leguminosae. 12 spp. Tropics. Trees.

***P. adnatum*** Griseb. W.Indies. The purple wood is used for general carpentry, house-building, carving etc.

***P. dasyrachis*** Kurz. = Caesalpinia dasyrachis. Sumatra. Grown as shade for coffee and cacao.

***P. linnaei*** Benth. (Brazil Wood). Tropical America. The orange-yellow wood is used for furniture-making and wheel spokes.

***P. pterocarpum*** Baker. = Caesalpinia arborea = Inga pterocarpa. (Soga). S.E.Asia to Australia. The plant is cultivated for the bark and is grown as shade for coffee. The powdered bark is used in tooth-powders and in water to treat opthalmia, dysentery and ulcers. It is used for tanning and in Java is extracted to yield a brown dye used to stain Batik work.

***P. vogelianum*** Walp. (Cañafistula). Paraguay and N.E.Argentine. The reddish wood is durable and is used locally for general carpentry, furniture and turning.

PEMPHIS J. R. and G. Forst. Lythraceae. 2 spp. Tropical Africa and Madagascar to Pacific. Small trees.

***P. acidula*** Forst. Old World tropics. The very hard wood is used for nails, tool handles and boat-building. It is used in Polynesia for clubs etc. and fish hooks.

Penache – Andropogon bicornis.

PENAEA L. Penaeaceae. 6 spp. S.Africa. Small shrubs. The following species yield a gum (Sarcocolla) which tastes like liquorice and is similar to Gum Tragacanth.

***P. mucronata*** L.

***P. sarcocolla*** L.

Pencil Cedar – Dysoxylum fraseranum.

Pencil Cedar – Dysoxylum muelleri.

Pendo Tree – Jatropha urens.

Pendoo – Euodia melufolia.

***Penicellaria arabica*** A. Braun. = Pennisetum typhoides.

***Penicellaria mosambicensis*** Kl. ex A. Braun. = Pennisetum typhoides.

Penicillin – Penicillium notatum.

PENICILLIUM Link. ex Fr. Moniliaceae. 137 spp. Cosmopolitan.

***P. camemberti*** Thom. Used in the manufacture of Camembert cheese.

***P. chrysogenum*** Thom. Takes a part in the formation of Stilton cheese. It is also a potential commercial source of gluconic acid.

***P. citrinum*** Thom. A potential commercial source of citric acid.

***P. expansum*** Link. Takes part in the production of Dolce Verde cheese.

***P. glaber*** Wehm. A potential commercial source of citric acid.

***P. glaucum*** Link. Yields a glucose oxidase which is used to remove glucose from eggs before drying them for sale as powdered eggs. The enzymes could be of use in clarifying cider.

***P. gorgonzola*** Biourge. One of the several organisms which takes part in the production of Gorgonzola cheese.

***P. javanicum*** van Beijma. This fungus synthesises fat. It was used in Germany during World Wars I and II for this purpose.

***P. luteum*** Thom. A potential commercial source of citric acid.

***P. luteum-purpurogenum*** Thom. A potential industrial source of gluconic acid.

***P. notatum*** West. Produces the antibiotic penicillin. This is used widely to treat diseases of man and animals which are caused by Gram-positive bacteria (particularly pneumonia and gonorrhoea). Penicillin is ineffective against Gram-negative bacteria.

***P. patulum*** Bainer.=P. iurticae. Produces gentisyl alcohol.

***P. pfeifferanus*** Wehm. A possible commercial source of citric acid.

***P. roquefortii*** Thom. Takes part in the production of Roquefort cheese.

***P. urticae*** Bainer.=P. patulum.

Penjari – Thalictrum foliolosum.

PENNANTIA J. R. and G. Forst. Icacinaceae. 4 spp. Australia, New Zealand, Norfolk Islands. Trees.

***P. corymbosa*** Forst. New Zealand. The very hard wood is used for furniture, tool handles and turning. The Maoris use it for making fires by rubbing.

PENNISETUM Rich. Graminae. 130 spp. Warm. Grasses.

***P. clandestinum*** Hochst. ex Chiov. (Kikuyu Grass). Tropical Africa, especially E. Grown for pasture and hay.

***P. compressum*** R. Br. Australia. Grown as a fodder grass.

***P. cylindricum*** Sw. ex Trin.=P. typhoides.

***P. glaucum*** (L.) R. Br.=P. typhoideum.

***P. mollissimum*** Hochst. (Ebeno). Sahara. The grass is grown as forage and the grain is used locally as a cereal.

***P. purpureum*** Schum. (Elephant Grass, Purple Grass, Napier Grass). Tropical Africa. A good fodder grass and is also used to make paper.

***P. typhoideum*** Rich. (Pearl Millet, Spiked Millet, Bulrush Millet, Cat's Tail Millet, Bajri). Cultivated throughout tropical Africa and the Indian subcontinent as a grain crop. The plant does not resist drought as well as Sorghum, nor has the grain such a high nutritional value. It is also grown for livestock fodder.

***P. typhoides*** (Burm. f.) L. C. Rich.=P. cylindricum=Penicellaria arabica=Penicellaria mosambicensis). Cultivated in N.E., E. and Central Africa and India as a pasture grass. The grain is used in Kenya to make beer.

***P. unisetum*** (Nees.) Benth.=Bekkeropsis uniseta. (Drakenberg Silky Grass, Natal Grass). N. and E. Africa. Cultivated as fodder for livestock.

Penny Cress – Thlaspi arvense.

Penny Flower – Linaria annua.

Pennyroyal – Satureja rigidus.

Pennyroyal, American – Hedeoma pulegioides.

Pennyroyal, European – Mentha pulegium.

PENSTEMON Schmid. Scrophulariaceae. 1 sp. N.E.Asia; 250 spp. N.America; 1 sp. Central America. Herbs.

***P. grandiflorus*** Nutt. (Large Flowered Beard Tongue). Central N.America. The local Indians use an extract of the roots to treat toothache.

PENTACE Hassk. Tiliaceae. 25 spp. S.E.Asia, W.Malaysia. Trees.

***P. polyantha*** Hassk. (Sageung). Java. The wood is used locally for house-building and bridges.

PENTCHLETHRA Benth. Leguminosae. 3 spp. Tropical America, Africa. Trees.

***P. filamentosa*** Benth. (Iripil Bark Tree, Pracaxi). Tropical S.America. The hard tough wood is used locally for house-building and furniture. The seeds contain a fat (Pracachy, Pracaxi Fat) which is used for making candles and soap. Some is exported. The seeds are poisonous and are used in Brazil as an emetic; a paste of the seeds is used locally to treat ulcers and snake-bite.

***P. macrophylla*** Benth. (Owala Oil Tree). Tropical Africa. The hard wood is used to make wheels, turning and general carpentry. The seeds are used locally to make a flour and produce a fat (Owala Oil, Owala Butter) which is used to make candles and soap. A decoction of the bark is used locally in treating ulcers.

***Pentacme scamensis*** Kurz=Shorea siamensis.

PENTADESMA Sabine Guttiferae. 4 spp. Tropical Africa, Seychelles.

***P. butyracea*** Sabine. (Butter Tree, Tallow Tree). Tropical Africa. The seeds produce a fat (Kanga Butter, Lamy Butter, Sierra Leone Butter) which is used locally for cooking and commercially in the manufacture of candles, soap and margarine. The seeds are sometimes used as a substitute for Cola nuts.

Pentagul – Dalbergia sympathetica.

PENTASPADON Hook. f. Anacardiaceae. 5 spp. Malaysia, Solomon Islands. Trees.

***P. motleyi*** Hook. f. = Nothoprotium sumatranum. (Empit, Peladjan). Malaysia. The soft, white wood is attacked by insects. It is used for floor-boards. The seeds are eaten raw or cooked and in Sarawak the oil from them is used in cooking. The latex from the stem is used locally to treat ringworm.

***P. velutina*** Engl. Tropical Asia. The stem latex is used locally to treat ringworm.

PENTHORUM Gronov. ex L. Penthoraceae. 1–3 spp. E.Asia, Indochina, Atlantic N.America. Herbs.

***P. sedoides*** L. (Ditch Stonecrop, Virginia Stonecrop). Atlantic N.America. A decoction of the leaves is laxative and demulcent. It is used to treat child cholera, haemorrhoids and other intestinal complaints.

PENTOPETIA Decne. Periplocaceae. 10 spp. Madagascar. Shrubs.

***P. elastica*** Jum. and Perrier. Madagascar. Yields a rubber (Mavokeli Rubber), which is used locally.

***Pentstemon*** Ait. = Penstemon.

PENTZIA Thunb. Compositae. 35 spp. S.Africa, a few tropical and N.Africa. Small shrubs.

***P. incana*** O. Kuntze. = P. virgata.

***P. virgata*** Less. = P. incana. S.Africa. Used locally as fodder for livestock.

Pe-nuli-valli – Parsonia helicandra.

Penyau – Upuna borneensis.

Peony – Paeonia foemina.

Peony, Chinese – Paeonia albiflora.

Peony, Coral – Paeonia anomala.

Peony, Tree – Paeonia moutan.

Peora – Passiflora cearensis.

PEPEROMIA Ruiz. and Pav. Peperomiaceae. 1000+ spp. Tropics and subtropics, especially America. Herbs.

***P. leptostachya*** Hook. and Arn. Tropics. In the Pacific Islands, the natives use the juice from the leaves to treat skin diseases, burns and eye infections.

***P. vividispica*** Trel. Central and S.America. The raw leaves are eaten locally.

Pepino – Solanum muricatum.

Pepino de Comer – Cyclanthera edulis.

Pepper – Piper nigrum.

Pepper, Alleppy – Piper nigrum.

Pepper, Ashanti – Piper guineense.

Pepper, Bell – Capsicum annuum.

Pepper, Black – Piper nigrum.

Pepper, Cayenne – Capsicum annuum.

Pepper, Cherry – Capsicum annuum.

Pepper, Chili – Capsicum annuum.

Pepper, Chinese – Zanthoxylum bungeri.

Pepper, Cone – Capsicum annuum.

Pepper Dulce – Laurencia pinnatifida.

Pepper Grass – Lepidium sativum.

Pepper, Green – Capsicum annuum.

Pepper, Guinea – Xylopia aethiopica.

Pepper, Jaborandi – Piper longum.

Pepper, Japanese – Zanthoxylum piperatum.

Pepper, Javanese Long – Piper retrofractum.

Pepper, Kava – Piper methysticum.

Pepper, Long – Piper chaba.

Pepper, Long – Piper longum.

Pepper, Madagascar – Piper nigrum.

Pepper, Negro – Xylopia aethiopica.

Pepper, Red – Capsicum annuum.

Pepper, Saigon – Piper nigrum.

Pepper, Sweet – Capsicum annuum.

Pepper Tree, Brazilian – Schinus molle.

Pepper Tree, Californian – Schinus molle.

Pepper Tree, Chilean – Schinus latifolius.

Pepper Tree, Peru – Schinus dependens.

Pepper, Wall – Sedum album.

Pepper, White – Piper nigrum.

Pepperidge – Nyssa multiflora.

Peppermint (Oil) – Mentha piperita.

Peppermint, Bastard – Trisania suaveolens.

Peppermint Eucalyptus, Sydney – Eucalyptus piperata.

Peppermint Tree – Eucalyptus salicifolia.

Peppermint Tree, Broad-leaved – Eucalyptus dives.

Pepperwort, Desert – Lepidium fremontii.

Pera do Campo – Eugenia telotzschiana.

***Peramium pubescens*** (Willd.) Salisb. = Goodyera pubescens.

Percha Tree, Indian – Palaquiuim ellipticum.

Perennial Lettuce – Lactuca perennis.

Perennial Rye Grass – Lolium perenne.

Perepat – Sonneratia caseolaria.

Perepat lanang – Scyphiphora hydrophyllacea.

PERESKIA Mill. Cactaceae. 20 spp. Mexico to tropical S.America, W.Indies. Vines.

***P. aculeata*** Mill.=P. pereskia. (Barbados Gooseberry, West Indian Gooseberry, Lemon Vines). Mexico. The fruits are eaten locally, raw or preserved and the leaves are used as a vegetable. The plant is cultivated.

***P. bleo*** (H.B.K.) DC. S.America. The leaves are used as a vegetable in parts of Colombia.

***P. pereskia*** (L.) Karst.=P. aculeata.

PERESKIOPSIS Britton and Rose. Cactaceae. 17 spp. Mexico, Central America. Cacti.

***P. porteri*** (Brandeg.) Britton and Rose. Mexico. The fruits are eaten locally.

PEREZIA Lag. Compositae. 90 spp. S.United States of America to Patagonia. Herbs.

***P. multiflora*** Less. Brazil and Argentina. A decoction of the leaves is used locally to induce sweating.

***P. nana*** Gray. S.W. U.S.A. to Mexico. Pipitzahoic acid is extracted from the roots for use as a pH indicator on the alkali side.

***P. wrightii*** Gray. See P. nana.

PERGULARIA L. Ascelpiadaceae. 3–5 spp. Africa and Madagascar to India. Vines.

***P. africana*** N. E. Br. Tropical Africa, especially W. and Central. The juice is used to adulterate Dragon's Blood.

***P. minor*** Anders.=Telosma cordata.

***P. tomentosa*** L. Dry regions of N.W.Africa. The stems are soaked in the milk to start the curdling process during cheese making.

PERIANDRA Mart. ex Benth. Leguminosae. 8 spp. Central America, W.Indies, Brazil. Shrubs.

***P. dulcis*** Mart. Brazil. The roots (Raiz Doco) are used as a substitute for liquorice.

PERICOPSIS Thw. Leguminosae. 5 spp. Tropical Africa; 1 sp. Ceylon, 1 sp. Palau and Caroline Islands. Trees.

***P. mooniana*** Thw. Ceylon. The brown mottled wood (Nedun) is hard and much esteemed for making furniture.

PERILLA L. 4–6 spp. India to Japan. Herbs.

***P. arguta*** Benth. China, Japan. In Japan the seeds are eaten raw or as a vegetable, the leaves and flowers are used as a condiment, or eaten salted. The leaves are used to dye the salted fruits of Prunus mume a light purple and the cotyledons are dried in salt to use as a condiment. The plant is cultivated.

***P. fructescens*** (L.) Britt.=P. ocimoides. India. The plant is cultivated in India, Japan and Korea for the seeds from which Perilla Oil is extracted. The leaves are used as a condiment in China and Japan. The oil is very quick-drying, Sp. Gr. 0·930–0·937, Sap. Val. 188–197, Lod. No. 185–205, Unsap. 0·6–1·3 per cent. It gives a hard tough finish and is used to make varnishes, printing ink and paints and also to waterproof paper. In U.S.A. it is mixed with soy-bean oil to make protective paints.

***P. ocimoides*** L.=P. frutescens.

Perilla Oil – Perilla frutescens.

PERIPLOCA L. Periplocaceae. 10 spp. N. and tropical Africa, to E.Asia. Woody vines.

***P. calumpitensis*** Llanos.=Streptocaulon baumii.

***P. canescens*** Afz. Ivory Coast. The stem yields a potentially useable latex.

***P. esculenta*** Roxb.=Oxystelma esculentum.

***P. graeca*** L. (Grecian Silk Vine). S.Europe to W.Asia. The alkaloid periplocin is extracted from the bark. It is used medicinally as a heart stimulant. See digitalin under Digitalis.

PERISTROPHE Nees. Acanthaceae. 30 spp. Warm Africa to E.Malaysia. Shrubs.

***P. bivalvis*** (L.) Merr. Tropical Asia. The stems and leaves yield an orange dye.

***Peritoma serrulatum*** DC.=Cleome integrifolia.

Periwinkle–Vinca minor.

Periwinkle, Cape – Vinca rosea.

Periwinkle, Madagascar – Vinca rosea.

Pernambuco Cotton – Gossypium brasiliensis.

Pernambuco Jaborandi – Pilocarpus jaborandi.

Pernambuca Wood – Caesalpinia echinata.

Peroba Rosa – Aspidosperma polyneuron.

Perobinha – Sweetia elegans.

PERONEMA Jack. Verbenaceae. 1 sp. Burma, Malaysia, Thailand. Tree.

***P. canescens*** Jacq. (Djati sabrang, Soongkai, Loo roos). The wood is valued in Sumatra for house and bridge-making. An extract of the leaves is used locally to treat toothache.

PERROTTETIA Kunth. Celastraceae. 20 spp. E.Asia, Malaysia, Australia, Hawaii, Mexico to Colombia. Trees.

***P. sandwicensis*** Gray. Hawaii. The natives use the wood for rubbing to make fires.

PERSEA Mill. Lauraceae. 150 spp. Tropics. Trees.

***P. americana*** Mill. = P. gratissima.

***P. borbonia*** (L.) Raf. (Red Bay, Sweet Bay). S. and S.E. U.S.A. The hard, heavy, reddish, brittle wood is used for house interiors and furniture. The leaves are used as a condiment, especially for soups and stuffing fowl.

***P. drymifolia*** Cham. and Schlect. (Mexican Avocado). Mexico. Frequently considered a variety of P. gratissima. The plant is hardier than the Avocado, but has smaller fruits with thinner skins and aromatic leaves.

***P. gratissima*** Gaertn. = P. americana (Avocado Pear, Alligator Pear, Aguacate, Palta). Native of Mexico, Central America and N. S.America. It has been cultivated for the fruits since before the Spanish conquest. The plant is grown throughout the tropics for the fruits which are eaten mainly as a salad dish. If one includes P. drymifolia (q.v.), there are three main types, the other two being the Guatemalan and the West Indian. These may have originated by hybridization, as there are many local varieties resulting from hybridization between the three main groups. The Guatemalan type is thick-skinned and less hardy than the Mexican (P. drymifolia), while the W.Indian is least hardy, leathery-skinned and more suited to tropical conditions. The trees are grown commercially in California, Florida, S.Africa, Australia, Brazil, Cuba, Israel and the Pacific Islands. Although the fruits are exported, this is difficult and costly, as they do not keep well. Yields vary considerably but are about 100 lb. per tree per year. Propagation is by budding and grafting of selected varieties. The fruits have a high nutritional value, containing 5–25 per cent oil, a high proportion of protein and appreciable amounts of the vitamin B complex. The oil from the seeds (Avocado Oil) is non-drying and is used as a salad dressing in Hawaii and to make soap in Guatemala. It is of potential use in the cosmetics industry. Locally various medicinal claims are made for the plant. The bark and leaves are used to treat stomach and chest complaints and to control menstruation. The seeds are used to treat dysentery and diarrhoea and the fruit skins to remove intestinal parasites.

***P. lingue*** Nees. S.America. The bark (Cascara de Lingue) is used for tanning.

***P. meyeniana*** Nees. S.America, Chile. The bark is used locally for tanning.

***P. schiedeana*** Nees. (Coyo Avocado). Central America. The fruits are eaten locally in salads.

Persian Berries – Rhamnus infectorius.

Persian Clover – Trifolium resumpinatum.

Persian Gum – Amygdalus leiocarpus.

Persian Insect Powder – Chrysanthemum roseum.

Persian Manna – Astragalus adscendens.

Persian Tragacanth – Astragalus pycnocladus.

Persian Walnut – Juglans regia.

Persian Wheat – Triticum carthlicum.

Persimmon – Diospyros virginiana.

Persimmon, Black – Diospyros texana.

Persimmon, Grey – Diospyros pentamera.

Persimmon, Japanese – Diospyros kaki.

Persio – Roccella tinctoria.

PERTUSARIA L. Lichen. Pertusariaceae.

***P. corallina*** (L.) Fr. (White Crottle). Temperate. Used in Scotland to dye wool a red-purple.

***P. pseudocorallina*** (Sw.) Arn. Subarctic. Used in Scandinavia to dye wool a purple-red.

Peru Balsam – Myroxylon pereirae.

Peru, Marvel of – Mirabilis jalapa.

Peru Peppertree – Schinus dependens.

Peruvian Bark – Cinchona officinalis.

Peruvian Cotton – Gossypium peruvianum.

Peruvian Cotton, Red – Gossypium microcarpa.

Peruvian Krameria – Krameria triandra.

Peruvian Nutmeg – Laurelia aromatica.

Peruvian Rhatany – Krameria triandra.

***Petaloma alba*** Blanco. = Lumnitzera racemosa.

PETALOSTEMON Mich. mut. Pers. Leguminosae. 50 spp. N.America. Herbs.

***P. oligophyllum*** (Torr.) Rydb. (Slender Prairie Clover). Prairies of W. U.S.A. to Mexico. The leaves are eaten as a vegetable by the Indians of New Mexico.

***P. purpureum*** (Vent.) Rydb. (Purple Prairie Clover). W. U.S.A., Mexico. The local Indians make a tea from the leaves and use the roots for chewing.

Petananang – Dryobalanops oblongifolia.

Petaquilla – Funastrum clausum.

Petaquillas – Pithecoctenium echinatum.

PETASITES Mill. Compositae. 5 spp. Europe to Central Asia. Herbs.

***P. frigidus*** (L.) Fries. W. N.America. The Alaskan Eskimos eat the leaves as a vegetable.

***P. japonicus*** F. Schmidt. (Sachalin). Japan. The leaf-stalks and flower buds are eaten as a vegetable. The stalks are sometimes preserved and the flower buds are used as a condiment.

***P. palmata*** Gray. (Palmate Butterbur, Sweet Coltsfoot). N.America. The burnt plants are used as salt by the local Indians.

***P. speciosa*** (Nutt.) Piper. W. N.America. The local Indians use the ashes as salt.

Petitgrain Oil – Citrus aurantium.

PETITIA Jacq. Verbenaceae. 2 spp. W.Indies. Trees.

***P. domingensis*** Jacq. (Capa Blanco). The light brown wood is heavy, hard but easy to work. It is fairly durable and insect-resistant. It is used for posts, etc. house interiors, furniture, carving, turning, rollers in coffee-hulling mills.

***P. poeppigii*** Scheuer. W.Indies. The strong wood is used to make ships.

PETIVERIA L. Phytolaccaceae. 1 sp. Tropical America, New Zealand. Shrub.

***P. alliacea*** L.=P. octandra. (Anamú, Apazote de zorro, Guinea Hen Weed, Gully Root, Hierba de las gallinitas, Zorrillo). The plant has a strong characteristic smell. It is used locally in S.America to treat nervous complaints, intestinal worms, scorpion stings and toothache. It is also used to control menstruation and induce abortions. The leaves are placed with woollen cloth to deter insects.

***P. octandra*** L.=P. alliacea.

PETREA L. Verbenaceae. 30 spp. Tropical America, W.Indies.

***P. kohautiana*** Presl. (Fleur de Dieu). Lesser Antilles. A decoction of the flowers is used in Dominica to induce abortions.

PETROSELENIUM Hill. Umbelliferae. 5 spp. Europe, Mediterranean. Herbs.

***P. crispum*** (Mill.) Nym. ex A. W. Hill.=P. hortense=P. sativum=Apium petroselinum. (Parsley). S.Europe, Asia Minor. The leaves have been used as a condiment since ancient times. The plant is cultivated in Old and New World. The fruits are used medicinally as a diuretic and to relieve minor stomach upsets. The leaves yield an oil (Oil of Parsley) which is used for flavouring foods, sauces etc. The Tuberous Rooted Parsley has enlarged roots which can be eaten as a vegetable.

Pe-tsai – Brassica pekinensis.

PEUCEDANUM L. Umbelliferae. 120 spp. Temperate Eurasia, tropical and S.Africa. Herbs.

***P. altissimum*** Desf.=P. officinale.

***P. araliaceum*** Benth. and Hook. f. Tropical Africa. The leaves are used locally for scenting clothes.

***P. aucheri*** Boiss. Baluchestan. The seeds are used locally to treat minor stomach upsets.

***P. cervaria*** (L.) Lapeyr.=Athamanta cervaria=Selinium cervaria. (Broad-leaved Spignel, Hart's Wort, Much-Good). C. to S.Europe. The roots and fruits (Radix et Semen Cervariae) were used in preparations to treat stomach aches, gout, dropsy, fevers, and to control menstruation. The plant was cultivated.

***P. decursivum*** Maxim.=Angelica decursive =Prophyroscias decursiva. China, Korea, Japan, N.Vietnam. An alcohol extract of the root is used as a linament to treat rheumatism in N.Vietnam.

***P. dhana*** Ham. Himalayas. The roots are used locally as a tonic.

***P. fraxinifolium*** Hiern. Tropical Africa. The scented leaves are used as a diuretic and in Sierra Leone to treat intestinal worms. They are also used to make scented body lotions.

***P. galbanum*** Benth.=Bubon galbanum. S.Africa. A decoction of the leaves is used locally as a diuretic.

***P. graveolens*** Benth. and Hook.=P. sowa. (East Indian Dill). Tropical Asia. The dried fruits are used as a condiment and to relieve stomach upsets.

***P. magaliesmontanum*** Houd. S.Africa. The natives use the leaves in stimulating baths.

***P. nagpurense*** Prain. India. A decoction of the roots is used locally to treat stomach upsets.

***P. officinale*** L.=P. altissimum=Selinum officinale. (Hog's Fennel, Sulphur Root). Europe, Asia. A decoction of the root is diuretic, induces sweating, is antiscorbutic and controls menstruation. It is also used in veterinary practice. A resin from the plant (Gummi Peucedani) is similar to Gum Ammoniac.

***P. oreoselinum*** (L.) Munch. Europe to Caucasus. The roots, leaves and seeds were used medicinally as a diuretic (Radix, herba, et Semen Oreoseleni). The plant yields an essential oil (Oleo Oreoseleni).

***P. ostruthium*** (L.) Koch.=Imperatoria ostruthium=Selenium ostruthium. (Hog Fennel, Master Wort). Europe, Asia. An alcoholic extract of the plant is used to treat stomach upsets. The roots were used to treat bronchial catarrh and the leaves were used to flavour cheeses.

***P. palustre*** (L.) Moench=Selenium palustre =Selenium sylvestre. (Milk Parsley). Europe, Asia. In S.E.Europe, the roots are used as a substitute for ginger.

***P. sativum*** S. Wats.=Pastinaca sativa.

***P. sowa*** Kurz.=P. graveolens.

PEUMUS Molina emend Pers. Monimiaceae 1 sp. Chile. Tree.

***B. boldus*** Molina.=Boldea fragrans= Boldu boldus. The leaves contain an essential oil and the alkaloid boldine. They are used medicinally as a diuretic and stimulant.

Peuris – Aporusa microcalyx.

Peyote – Lophora williamsii.

Peyotillo – Pelecyphora aselliformis.

Phá côt chí – Psoralea corylifolia.

PHACELIA Juss. Hydrophyllaceae. 200 spp. N.America, Andes. Herbs.

***P. tanacetifolia*** Benth. (Tansy Phacelia). W. U.S.A. Produces a good honey for which the plant is sometimes grown commercially.

Phacelia, Tansy – Phacelia tanacetifolia.

***Phaeanthus cumingii*** Vidal.=Polyalthia suberosa.

***Phaeocarpus campestris*** Mart.=Magnolia pubescens.

***Phaeomeria*** Lindl. ex K. Schum.=Nicolaia.

PHALARIS L. Graminae. 20 spp. N. and S.Temperate. Grasses.

***P. arundinacea*** L. (Reed Canary Grass). Temperate Europe, Asia. Introduced to America in 19th century. Grown in Old and New World for pasture and hay.

***P. canariensis*** L. (Canary Grass). Mediterranean. Grown in the Old and New World for bird seeds, which are sometimes used for human consumption.

***P. tuberosa*** L. (Harding Grass, Towoonba Grass). Mediterranean. Grown in warm countries, especially S.Africa as winter pasture.

PHALLUS Pers. Phallaceae. 10 spp. Widespread. Stinkhorns.

***P. impudicus*** (L.) Pers.=Ithyphallus impudicus. N.Hemisphere. A poisonous fungus which is used in home remedies for treating gout.

Phalsa – Grewia asiatica.

Phalsa – Grewia subinaequialis.

Phalsa – Grewia tiliaefolia.

PHANEROPHLEBIA Presl.=Aspidiaceae. 20 spp. Tropical Asia, Hawaii, tropical America, S.Africa. Ferns.

***P. forunei*** J. Sm. China, Japan, Korea. The Chinese use the rhizome to prevent bleeding.

PHASEOLUS L. Leguminosae. 200–240 spp. Tropics and subtropics, especially America. Herbs. Members of the genus are grown mainly for the fruits, which are used for human food and feeding livestock. The seeds are nutritious, containing about 21–25 per cent protein, 58–64 per cent carbohydrate and 1–1·5 per cent fat. The plants also produce root nodules, in association with bacteria (***Rhizobium*** spp.). These add nitrogen-containing substances to the soil, which make the plants useful in crop rotations. There are several varieties of each species, varying in form from climbing vines to dwarf forms and also varying in pod shape and size and in the shape, size and colour of the seeds.

***P. aconitifolius*** Jacq. (Mat Bean, Moth Bean). Tropical Asia. Cultivated in India for the seeds which are used as human and livestock food.

***P. acutifolius*** Gray. (Tepary Bean). S. U.S.A., Mexico. Grown locally for the seeds. The plant has been grown by the local Indians since ancient times. A very drought-resistant species.

***P. adenanthus*** E. Mey.=P. rostratus Wall. Tropics. The Hindus eat the tuberous roots as a vegetable.

***P. angularis*** Wight. (Adzuki Bean). Tropical Asia. The plant has been grown in China, Japan and Korea since ancient times. The seeds are eaten as a vegetable or used as a flour.

***P. aureus*** Roxb.=P. mungo.

***P. calcaratus*** Roxb. (Rice Bean). Tropical Asia. Cultivated throughout the Far

East for the seeds which are usually eaten with rice. It is also grown for forage for livestock.

***P. coccineus*** L. = P. multiflorus. (Scarlet Runner Bean). Central America. Cultivated widely in temperate and subtropical regions, mainly as a garden crop. The immature pods are eaten as a vegetable.

***P. diversifolius*** Pers. S. U.S.A. The local Indians eat the roots as a vegetable.

***P. helvola*** (L.) Britt. (Trailing Wild Bean). E. N.America. Recommended for soil conservation of sand-dunes.

***P. limensis*** Macf. = P. lunatus.

***P. lunatus*** L. = P. limensis. (Lima Bean, Butter Bean, Sieva Bean). Central America. The plant has been grown locally since ancient times for the seeds. There are climbing and dwarf varieties. It is particularly important in W.Africa and U.S.A. (mainly California) where it accounts for 10 per cent of the total dry bean production. It is also used for canning and freezing.

***P. metcalfei*** Woot and Standl. = P. retusus.

***P. multiflorus*** Willd. = P. coccineus.

***P. mungo*** L. = P. aureus. (Mung Bean, Green Gram, Golden Gram, Black Gram, Urd, Woolly Pyrot). Tropical Asia. Grown throughout the Far East, India, Asia Minor and Mediterranean Europe. It has been grown in India since ancient times and is still the most widely grown pulse crop in the country. The seeds are eaten as a vegetable and are spouted as bean sprouts used by the Chinese as a vegetable or in salads. The plant is also grown as a cover crop and green manure and the straw is fed to livestock.

***P. polystachyus*** (L.) B.S.P. (Wild Bean, Bean Vine). E. and S. N.America. The dried seeds are used as a vegetable, especially by the local Indians.

***P. retusus*** Benth. = P. metcalfei. (Metcalf Bean). S. U.S.A., Mexico. Used as a forage crop in dry areas, the seeds are sometimes eaten.

***P. trilobus*** L. (Pillepesary). Tropical Asia. Sometimes cultivated in India for the seeds, which are used as a vegetable and for livestock forage.

***P. vulgaris*** L. (Common Bean, French Bean, Haricot Bean, Kidney Bean). N. S.America. The plant was grown by the Incas and is now grown in the Old and New World as a vegetable. The seeds (Frijoles) are important locally in Latin America. Some varieties are grown for the seeds and others for the pods (Snap Beans). There are a large number of varieties adapted to various climatic conditions and uses, e.g. dry seed or for freezing and canning. Brazil, China and U.S.A. produce over half the world supply of dry beans.

Pheasant's Eye – Adonis aestivalis.

PHELIPAEA Desf. Orobanchaceae. 4 spp. S.W.Asia. Herbs.

***P. lutea*** Desf. N.Africa. The roots are eaten by the nomads as an emergency food.

***P. violacea*** Desf. See P. lutea.

***Phelbodium aureum*** (L.) J. Smith. = Polypodium aureum.

***Phellinus*** Quél. = Fomes.

PHELLODENDRON Rupr. Rutaceae. 10 spp. E.Asia. Trees.

***P. amurense*** Rupr. (Amur Cork Tree). China, Japan. A decoction of the bark is used in Japan to treat skin diseases and the berries are used as an expectorant.

***P. sachalinense*** Sargt (Sachalin Cork Tree). Korea, China, Japan. The bark is used as buoys for fishing nets etc., and the unthickened bark yields a yellow dye. The hard red-brown wood is used in Japan for house interiors, railway sleepers and furniture.

PHELLOPTERUS (Torr. and Gray) Coult. and Rose. Umbelliferae. 5 spp. S.W. United States of America. Herbs.

***P. montanus*** Nutt. S. U.S.A. The baked roots are used as a flour by the local Indians.

***Phelypaea*** D. Don = Phelipaea.

***Phelypaea lutea*** Desf. = Cistanche lutea.

PHILADELPHUS L. Philadelphaceae. 75 spp. N.Temperate, especially E.Asia. Shrubs.

***P. mexicanus*** Schlecht. (Mexican Mock Orange). The flowers and branches are distilled to give a scented water which has been used as a perfume since ancient times.

***Philagonia frazinifolia*** Hook. = Euodia fraxinifolia.

***Philagonia procera*** DC. = Euodia fraxinifolia.

***Philagonia sambucina*** Blume = Euodia sambucina.

***Philibertia crassifolia*** Hemsl. = Funastrum clausum.

Philippine Gutta Percha – Palaquium ahernianum.

PHILLYREA L. Oleaceae. 4 spp. Madeira, Mediterranean to N. Persia. Trees.

***P. latifolia*** L. (Tree Phillyrea). Mediterranean. The wood is used in Greece for turning and making saddles. A decoction of the leaves is used as a diuretic and to control menstruation. It is also used as a mouthwash.

***P. media*** L. See P. latifolia.

PHILODENDRON Schott. Araceae. 275 spp. Warm America, W.Indies. Vines.

***P. bipinnatifidum*** Schott. (Banana de Macaco, Banana de Imbé). S.Brazil. The fruits are eaten locally, raw or in jellies. The seeds are used to treat intestinal worms.

***P. cordatum*** Kunth. (Guimberana). S.Brazil. The leaf juice (Folha de Fonte) is used in a soap paste to treat skin diseases of livestock in parts of Brazil.

***P. imbe*** Schott. = P. sellowianum. (Cape Homem, Cipo Imbé) S.America. A poultice of the leaves is used locally to treat rheumatism, boils, oedema.

***P. laciniatum*** (Vall.) Engl. Brazil. The leaves (Folha de Arubu) are used with oil as a poultice for rheumatism.

***P. ochrostemon*** Schott. See P. imbé.

***P. oxycardium*** Schott. (Corde Molle). Central America, W.Indies. The stems are used locally for making baskets.

***P. pertusum*** Kunth. = Monstera deliciosa.

***P. radiatum*** Schott. Central America. The leaves are used locally in baths to treat rheumatism and rickets. The aerial roots are used to make baskets.

***P. sagittifolium*** Liebm. S.Mexico. The aerial roots are used locally to make baskets.

***P. selloum*** Koch. Brazil. The fruits are eaten raw or in jellies. The seeds are used locally to treat intestinal worms and a decoction of the roots is used as a purgative.

***P. sellowianum*** Kunth. = P. imbé.

***P. speciosum*** Schott. (Aringa Iba). Brazil. The leaves are used locally as a poultice for boils and rheumatism and the seeds are used to treat intestinal worms

***P. squamiferum*** Poepp. (Guiambe). Brazil, Guyana. A poultice of the leaves is used in Brazil to treat dropsy and oedema.

***P. warsewiczii*** Schott. See P. radiatum.

PHLEUM L. Graminae. 15 spp. Temperate Eurasia, N.America to Mexico, temperate S.America. Grasses.

***P. alpinum*** L. (Alpine Timothy). Alpine Europe, Asia and N.America. A good forage grass for mountain pastures.

***P. pratense*** L. (Timothy Grass). Temperate Europe, Asia. One of the most important pasture and hay grasses in temperate regions. Introduced to E. U.S.A. in 1747. The grass is often grown in a pasture mixture with red clover.

PHLOGA Nor. ex Hook. Palmae. 2 spp. Madagascar. Palm trees.

***P. polystachya*** Nor. (Anivo, Tsiriky). Madagascar. The leaves are used locally as a source of cordage fibre and to make hats. The juice from the stem is fermented to make a drink.

PHLOMIS L. Labiatae. 100+ spp. N.Temperate Old World. Herbs.

***P. lychnitus*** L. (Lamwick Plant). S.Europe. The leaves are sometimes used to adulterate sage.

***P. zeylanica*** L. = Leucas zeylanica.

Phlou méant – Decaschistia parviflora.

Phlou thom – Dillenia ovata.

PHLOX L. Polemoniaceae. 66 spp. N.America, Mexico; 1 sp. N.E. Asia. Herbs.

***P. carolina*** L. (Carolina Pink, Thick-leaf Phlox). Coasts of S.E. U.S.A. The roots (False Pinkroot) are used to adulterate Pinkroot.

Phlox, Thick-leaf – Phlox carolina.

***Phoberos mundtii*** Presl. = Scolopia mundtii.

***Phoberos zeyheri*** Presl. = Scolopia zeyheri.

PHOENIX L. Palmae. 17 spp. Warm Africa, Asia. Palm trees.

***P. acaulis*** L. N.India. The stem is a local source of sago.

***P. dactylifera*** L. (Date Palm). Asia Minor. One of the staple food trees of N.Africa. It was introduced to Spain and from there to Mexico during the 18th century by the Spanish missionaries. It is grown commercially in California, but N.Africa and Asia Minor are by far the most important producing areas. The trees are grown primarily for the fruits, but the leaves are used for thatching and fuel, and the stems are used for house-building; even the seeds are fed to camels. The plants are propagated by suckers and come into bearing after about five years, reaching full bearing in about fifteen years, when a tree will produce 100 lb. of fruit a year. The plants are dioecious, so pollination is often carried out artificially in commercial plantings. The fruits contain some 60–70 per cent sugars and 2 per cent fat and 2 per cent protein. They are eaten fresh or dried, mixed with milk, or fermented to make an alcoholic beverage (Arrack). In temperate countries they are used in jams, cakes and confectionery. There are a large number of varieties of date, but they can be grouped into three main types, (1) soft dates with a low sugar (60 per cent) content. These are used for local consumption and eaten fresh or pressed into cakes (Agwa). (2) Semi-soft, which contain a little more sugar and are harder. They are usually harvested before they are fully ripe and make up the bulk of the exported dates. (3) Dry dates, these have the highest sugar content and dry out on the tree. These keep well and are usually stored for use when the other types are not available.

***P. pusilla*** Gaertn.=P. zeylanica. (Wild Date). E.India, Ceylon. The leaves are used to make baskets and mats.

***P. sylvestris*** Roxb. (Wild Date Palm). India. Cultivated in N.E.India. The sap (Toddy) is used to produce Palm sugar (Jaggery). The trunks are tapped when the trees are about five years old and collection continues from October to February. A single tree yields some 350 lb. of toddy a year making 35 lb. of sugar.

***P. zeylanica*** Trin.=P. pusilla.

Phoenix Tree – Firmiana simplex.

PHOLIDOCARPUS Blume. Palmae. 7 spp. Indochina, W.Malaysia, Moluccas. Palms.

***P. kingiana*** Ridl. Malaysia. The midribs are used locally to make blow-pipe darts.

PHOLIOTA (Fr.) Quél. Agaricaceae. 120 spp. especially temperate. The fruitbodies of the following species are eaten locally.

***P. candidans*** (Schäff) Schrot.=P. praecox.

***P. caperata*** (Bers.) Karst. Temperate.

***P. cylindrica*** DC.=Agrocybe aegerita. Mediterranean.

***P. marginata*** (Batsch.) Quél.=Agaricus marginatus. N.America.

***P. mutabilis*** (Schäff.) Quél. Europe.

***P. nameko*** (T. Ito) S. Ito and Imai. Japan.

***P. praecox*** Pers.=P. candicans=Agaricus candicans=Agaricus praecox. Europe.

***P. squarrosa*** (Muell.) Karst. N.America.

***P. squarrosoides*** (Peck.) Sacc. Japan.

***P. terrestris*** Overh. Japan.

PHORADENDRON Nutt. Viscaceae. 190 spp. America, W.Indies. Parasitic shrubs.

***P. californicum*** Nutt. (Mesquite Mistletoe). Parasitic on Acacia, Cercidium, Prosopis in S.W. U.S.A. The dried fruits are eaten by the local Indians.

***P. flavenscens*** (Pursh.) Nutt. (American Christmas Mistletoe). N.America. The Indians use a decoction of the plant to treat post-childbirth bleeding.

***P. juniperinum*** Englem. (Juniper Mistletoe). S.W. U.S.A., Mexico. The local Indians use the seeds as a coffee substitute.

PHORMIUM J. R. and G. Forst. Agavaceae. 2 spp. New Zealand, Norfolk Islands. Large herbs.

***P. tenax*** J. R. and G. Forst. (New Zealand Flax, New Zealand Hemp). New Zealand. Grown for the leaf fibres which are used for cordage, sacking etc. The fibres are not particularly strong. The plant has been grown by the Maoris for cloth and cordage for a long time. It is now grown commercially on a fairly small scale in New Zealand, S. U.S.A., Mauritius, Central Africa, Ceylon, St. Helena, Argentina and Chile.

PHOTINIA Lindl. Rosaceae. 60 spp. Himalayas to Japan and Sumatra, N.America. Shrubs.

***P. arbutifolia*** Lindl. (Californian Holly, Christmas Berry). California to Mexico. The berries are eaten raw, or cooked by the local Indians.

***P. japonica*** Thunb.=Eriobotrya japonica.

PHRAGMITES Adans. Graminae. 3 spp. Cosmopolitan. Grasses.

***P. communis*** Trin. (Reed Grass). Cosmopolitan. The stems are used to make mats. The young shoots are eaten as a vegetable in parts of Japan. Some tribes of American Indians eat the rhizomes as a vegetable, or ground into a flour after roasting.

PHRYNIUM Willd. Marantaceae. 30 spp. Tropical Africa, Indomalaya. Herbs.

***P. confertum*** (Benth.) Schum. Tropical Africa. The stems are used locally to make baskets and mats.

PHTHIRUSA Mart. Loranthaceae. 60 spp. Tropical America. Parasitic shrubs. The following species are potential sources of rubber.

***P. pyrifolia*** (H.B.K.) Eichl.

***P. theobromae*** (Willd.) Eichl.

Phulwara Butter – Madhuca butyracea.

PHYLLANTHUS L. Euphorbiaceae. 600 spp. Tropics and subtropics except Europe and N.Asia. Trees, shrubs or Herbs.

***P. acutifolius*** Mart. Tree. Brazil. A decoction of the roots is used locally and in patent medicines to treat stones in the bladder.

***P. alatus*** Blume = P. urinaria.

***P. cantoniensis*** Hornem. = P. urinaria.

***P. corcovadensis*** Muell. Arg. See P. acutifolius.

***P. distichus*** (L.) Muell. Arg. (Otaheite Gooseberry). India, Malaysia. Cultivated throughout the tropics for the fruits which are used in pickles and preserves.

***P. echinatus*** Wall. = P. urinaria.

***P. emblica*** L. (Emblic, Myrobolan). Shrub. Tropical Asia, India. Cultivated throughout the tropics for the acid fruits which are used in jellies and preserves. They are sometimes eaten raw. The bark is used for tanning.

***P. engleri*** Pax. S.Africa. Herb. The root bark is used as a suicide poison.

***P. indicus*** Muell. Arg. = Prosorus indica (Dooka anggang, Pantjal kidang). Tree. S.E.Asia. The hard wood is used locally to build houses and to make wheels. The seeds are poisonous.

***P. kirganelia*** Blanco. = P. niruri.

***P. llanosi*** Muell. Agr. = Glochidion llanosi.

***P. niruri*** L. = P. kirganelia. Shrub. Throughout the tropics. In S.India a decoction of the plant is used to treat dysentery and as a diuretic; in the Philippines it is used to treat stomach complaints and in Reunión to treat dropsy, diarrhoea and vaginal discharges. In San Domingo and Puerto Rico the leaves and roots are used to treat fevers and in India and Ghana the leaves are used as a treatment for gonorrhoea.

***P. reticulatus*** Poir. Tropics. The leaves are used locally as a diuretic and the roots yield a red dye.

***P. urinaria*** L. = P. alatus = P. cantoniensis = P. echinatus. Tropical Asia. Herb. A decoction of the leaves is used locally to reduce fevers and as a tonic. In Cambodia it is also used in the treatment of liver disorders and diarrhoea.

PHYLLITIS Ktz. Scytosiphonaceae. Brown Algae.

***P. fascia*** Ktz. Pacific Ocean. The dried seaweed is eaten in coastal areas of Japan.

***Phyllocalyx edulis*** Berg. = Eugenia selloi.

PHYLLOCLADUS Rich. Phyllocladaceae. 7 spp. Philippines, Borneo, Moluccas, New Guinea, Tasmania, New Zealand. Trees.

***P. aspelenifolius*** (Labil.) Hook. = P. rhomboidales. (Celery Hop Pine). Tasmania. The heavy, light brown wood bends well and is resistant to decay and insect attack. It is used for coach- and boat-building, furniture, flooring, barrels etc.

***P. rhomboidales*** L. C. Rich. = P. asplenifolius.

***P. trichomanoides*** D. Don. (Celery Pine). New Zealand. The heavy, strong, light-coloured wood is used for pit props, railway sleepers and piling. The bark is used for tanning kid for gloves and contains a red dye.

PHYLLOGEITON (Weberb.) Herzog. Rhamnaceae. 2 spp. Tropical and S.Africa. Trees.

***P. dicolor*** (Klotzsch.) Herzog. (Mtata, Nunye, Umzingila). Central Africa. The fruits are eaten locally.

PHYLLONOMA Willd. ex Schult. Dulongiaceae. 8 spp. Mexico to Peru. Small trees.

***P. laticuspis*** (Turcz.) Engl. = Dulongia laticuspis. (Hierba de la viruela). Mexico. A decoction of the leaves is used locally to treat smallpox.

PHYLLOPHORA L. Phyllophoraceae. Red Algae.

***P. rubens*** (L.) Grev. Baltic and Black Sea. Used in S.Russia as a source of agar and iodine.

PHYLLOSPADIX Hook. Zosteraceae. 2 spp. Japan, Pacific N.America. Marine herbs.

***P. scouleri*** Hook. (Surf Grass). Pacific N.America. The local Indians eat the roots as a vegetable.

PHYLLOSTACHYS Sieb. and Zucc. Graminae. 40 spp. Himalayas to Japan. Bamboo grasses. The stems of the species mentioned are used for fishing rods, plant stakes, walking sticks etc., the young shoots of many are eaten as a vegetable.

***P. aurea*** (Carr.) A. and C. Riv. = Bambusa aurea. (Fishpole Bamboo). China. Cultivated in China and Japan. The shoots are eaten.

***P. aureosulcata*** McClure. (Yellow Groove Bamboo). China. Cultivated. The shoots are edible, but the bamboo is not of a very good quality.

***P. bambusioides*** Sieb. and Zucc. (Giant Timber Bamboo, Hardy Timber Bamboo,

Japanese Timber Bamboo). China. One of the largest and most important timber bamboos. It is widely cultivated in China and Japan. The larger stalks are used for construction work. The shoots are edible but rather bitter when raw. The Castillion Bamboo is a smaller variety producing smaller culms, but sweeter edible shoots.

***P. dulcis*** McClure. (Sweetshoot Bamboo, Pah koh poo chi). China. The shoots are of excellent eating quality, but the wood is inferior to the other bamboos.

***P. flexuosa*** A. and C. Riv. China. The shoots are of medium eating quality and the wood is useful for lighter work.

***P. heteroclada*** Steud. China, India. The plant is used mainly for making paper.

***P. meyeri*** McClure (Mayer Bamboo). China. Yields a bamboo of excellent quality and edible shoots.

***P. nigra*** (Lodd.) Munro. (Henon Bamboo). China. Cultivated in China and Japan for the edible shoots. They are rather bitter before cooking, but the bitter principle is removed with the cooking water.

***P. nitida*** (Mitf.) Nakai. = Arundinaria nitida. N.China. One of the more cold resistant species. The wood is used locally for baskets and light fencing.

***P. nuda*** McClure. China. A hardy species. The shoots are edible.

***P. pubescens*** Mazel ex Lehaie. (Pan chu). Cultivated in China and Japan for the wood and edible shoots which are canned and exported.

***P. viridis*** (Young) McClure. China. The shoots are edible and the wood is used locally.

***P. viridi-glaucescens*** (Carr.) A. and C. Riv. China. The shoots are edible and the wood is used for the usual purposes.

PHYLLOSTYLON Capanema ex Benth. and Hook. f. Ulmaceae. 3 spp. W.Indies to Paraguay. Trees.

***P. brasiliensis*** Capanema. (San Domingo Boxwood, West Indian Boxwood, Bois Blanc, Sabonero). W.Indies to Argentine. The light yellow wood polishes well. It is frequently stained black to resemble ebony and is used for knife handles.

PHYLLOXYLON Baill. Leguminosae. 4 spp. Madagascar, Mauritius. Trees.

***P. ensifolius*** Baill. (Harahara, Tsiavango). Madagascar. The resin from the trunk is used locally to stupify fish.

***P. perrieri*** Drake. See P. ensifolius.

***P. phyllanthoides*** Baker. Madagascar. The hard wood is used locally for spade handles.

***P. xiphocladia*** Baker. See P. phyllanthoides.

Phyllyrea – Phyllyrea latifolia.

PHYSALIS L. Solanaceae. 100 spp. Cosmopolitan, especially America. The fruits of the following species are eaten locally, usually in sauces etc. and by the local natives.

***P. alkekengi*** L. (Alkekengi, Strawberry Tomato, Winter Cherry). Central and S.Europe. Cultivated in Old and New World for the fruits which are sometimes eaten, but are used mainly for treating fevers and urinary disorders, gout and rheumatism.

***P. fendleri*** Gray. N.America.

***P. heterophylla*** Nees. N.America.

***P. ixocarpa*** Brot. (Husk Tomato, Tómatl, Uiltōmatt). Mexico. The fruits are eaten locally. The plant is cultivated.

***P. lanceolata*** Michx. N.America.

***P. minima*** L. (Sunberry). Tropics. The fruits are eaten as a vegetable.

***P. neo-mexicana*** Rydb. (Ground Berry, Ground Tomato, Tomate de Campo). Closely related to P. pubescens. S. N.America.

***P. peruviana*** L. S. N.America to N. S.America, W.Indies. Europe, Asia. The berries are used for sauces and preserves.

***P. pubescens*** L. (Downy Ground Cherry, Dwarf Cape Gooseberry, Strawberry Tomato). S. N.America, to N. S.America, W.Indies, Europe, Asia. The berries are used in sauces and preserves.

***P. virginiana*** Mill. N.America.

***P. viscosa*** L. N.America.

PHYSCIA Schreb. Lichen. Physciaceae.

***P. pulverulenta*** (Schreb.) Nyl. (Mealy Blister Lichen). Temperate. Used in Europe as the source of a yellow dye used to colour wool.

Physic Nut – Jatropha curcas.

PHYSOCALYMNA Pohl. Lythraceae. 1 sp. Tropical S.America. Tree.

***P. floribundum*** Pohl. = P. scaberrimum.

***P. floridum*** Pohl. = P. scaberrimum.

***P. scaberrimum*** Pohl. = P. floribundum. = P. floridum. (Brazilian Rosewood, Maschado, Tulip Wood). The heavy, hard, red wood is used for furniture, turning and fancy articles. An essential oil

(Azeite de pau de rosa, Brazilian Rosewood Oil, Rosewood Oil, Tulip Wood Oil) is distilled from the wood and used in perfumery.

PHYSOSTIGMA Balf. Leguminosae. 5 spp. Tropical Africa. Vines.

***P. cylindrospermum*** Balf. See P. venenosum.

***P. venenosum*** Balf. (Djirou, Esere, Islio). Tropical W.Africa. Introduced to India and Brazil. The seeds (Calabar Bean, Ordeal Bean) are used locally as a criminal poison. Medicinally they are a source of physostigmine salicylate, used to stimulate secretions from glands and peristalsis.

PHYTELEPHAS Ruiz. and Pav. Palmae. 15 spp. Tropical America. Palms. The hard white seeds of the following species are used as a source of vegetable ivory, used to make buttons, knobs, chessmen etc.

***P. macrocarpa*** Ruiz. and Pav. (Common Ivory Palm, Tagua Palm).

***P. seemanni*** Cook. (Seemann Ivory Palm).

PHYEUMA L. Campanulaceae. 40 spp. Mediterranean, Europe, Asia. Herbs.

***P. comosum*** Vill. = P. orbiculare.

***P. orbiculare*** L. = P. comosum. (Rampion). Europe. The leaves and roots are occasionally eaten as a vegetable or in salads.

PHYTOLACCA L. Phytolaccaceae. 35 spp. Tropics and subtropics. Herbs.

***P. abyssinica*** Hoffm. E.Africa. The leaves and stems are eaten locally as a vegetable and the berries yield a red dye.

***P. acinosa*** Roxb. Tropical Asia, China, Japan. The leaves are used locally as a vegetable. Cultivated in parts of India.

***P. americana*** L. = P. decandra.

***P. chiliensis*** Miers. (Carmin). Chile. The berries yield a red dye, used locally.

***P. decandra*** L. = P. americana (Common Pokeberry). N.America. The dried roots are used medicinally as an emetic and purgative. The berries give a red dye which is used for colouring sweets, etc., wine and was used as ink.

***P. rivinoides*** Kunth. and Bouché. (Venezuela Pokeberry). Central and S.America. The leaves and shoots are eaten locally as a vegetable and the roots are used as a soap substitute.

Pia – Tacca pinnatifida.

Pianggu pipit – Knema laurina.

***Piaropus crassipes*** (Mart.) Britt. = Eichornia crassipes.

Piassaba – Raphia gigantea.

Piassaba, Bahia – Attalea funifera.

Piassaba, Ceylon – Caryota urens.

Piassaba, Pàra (Fibre) – Leopoldinia piassaba.

Piauhy Rubber – Manihot piauhyensis.

PICEA A. Dietr. Pinaceae. 50 spp. N.Temperate, especially E.Asia. Trees. These are very important lumber species. The wood is generally light-coloured, soft and not strong. It is used for paper pulp, general light construction work, house interiors, boxes and musical instruments. It is sometimes used for boats and barrels.

***P. abies*** (L.) Karst. = P. excelsa. (Norway Spruce). Europe, N.Asia, Balkans. A very important lumber species in Europe. The resin is a source of Jura Turpentine and Burgundy Pitch which is used medicinally as a salve. An essential oil is distilled from the needles and used in perfumery.

***P. alcockiana*** Carr. = P. bicolor.

***P. alba*** Link. = P. glauca.

***P. asperata*** Mast. (Dragon Spruce). W.China.

***P. bicolor*** (Maxim.) Mayr. = P. alcockiana (Alcock Spruce, Iramomi). Japan.

***P. canadensis*** (L.) B.S.P. = P. glauca.

***P. complanata*** Wils. China.

***P. engelmanni*** (Parry) Engelm. (Engelmann Spruce). W. N.America. The wood is also used for charcoal and the bark for tanning.

***P. excelsa*** Link. = P. abies. = Abies excelsa.

***P. glauca*** (Moench.) Voss = P. alba. = P. canadensis. (White Spruce). N.America.

***P. glennii*** Mast. (Sakhalin Spruce). Japan. The wood is beautifully grained.

***P. jezoensis*** Carr. (Yeddo Spruce, Yezomatsu Spruce). E.Asia. Used for most purposes, especially in Japan, where the resin is used locally to heal wounds.

***P. mariana*** (Mill.) B.S.P. = P. nigra. (Black Spruce, Bog Spruce). Bogs of N.America. Used for pulping. The gum from the branches is used for chewing (Canada Spruce Gum) and a beer (Spruce Beer) is made from boiling the branches.

***P. nigra*** (Ait.) Link. = P. mariana.

***P. orientalis*** (L.) Link. Asia Minor.

***P. purpurea*** Mast. (Purple Cone Spruce). China.

***P. rubra*** Link. = P. rubens. E. N.America. Also yields a gum used for chewing.

***P. rubens*** Sarg. = P. rubra.

***P. sitchensis*** (Bong.) Carr. (Sitka Spruce). Pacific N.America. An important local timber.

***P. smithiana*** (Wall.) Boiss. (Himalayan Spruce). Himalayas.

Pichiché – Psidium satorianum.

Pichichio – Solano mammosum.

Pichi Pichi Wood – Fabiana imbricata.

Pichrim Beans – Aniba firmulum.

Pichurim Beans – Nectandra puchury-minor.

Pickerel Weed – Pontederia cordata.

Pico de Gallo – Xylopia obtusifolia.

Picosa – Croton ciliato-glandulosus.

Picotee – Dianthus caryophyllus.

PICRALIMA Pierre. Apocyanaceae. 1 sp. W. Equatorial Africa. Tree.

***P. klaineana*** Pierre=P. nitida. The bark and seeds are used locally to treat fevers and combat intestinal worms.

***P. nitida*** Th. and Hel.=P. klaineana.

PICRAMNIA Sw. Simaroubaceae. 55 spp. Mexico to tropical S.America, W.Indies. Small trees.

***P. excelsa*** (Sw.) Planch. (Jamaican Quassia). W.Indies. The yellow-white soft wood (Lignum Quassiae Jamaicense) is bitter to taste and is used medicinally as a substitute for quassia and to combat intestinal worms.

***P. pentandra*** Sw. Florida to W.Indies, Bahamas. The bark (Honduras Bark, Carcara Amarga) is diuretic and used to treat skin diseases. In Cuba a decoction of the leaves, roots and bark is used to reduce fevers. The moist leaves are used by the Caribs to stain violet baskets made from the roots of Monstera deliciosa.

PICRASMA Blume. Simaroubaceae. 6 spp. W.Himalayas to Japan, Malaysia. Fiji. Trees.

***P. quassioides*** Benn.=Nima quassioides=Rhus ailanthoides. (Nigaki, Shurni). Himalayas, Korea, China, Japan. The bitter bark and wood, used as a substitute for quassia, reduces fevers and is used as an insecticide. The hard yellow wood is used in Japan for household utensils and mosaic work.

PICRORHIZA Poyle ex Benth. Scrophulariaceae. 2 spp. W.Himalayas. Herbs.

***P. kurroa*** Royle. China. The leaves are used in China as a cathartic.

Pié de Mula – Agave pesmulae.

Pie Plant – Rheum rhaponticum.

***Pierardia dulcis*** Jack.=Baccaurea dulcis.

***Pierardia sapida*** Roxb.=Baccaurea sapida.

Pig's Balsam – Tetragastris balsamifera.

Pig's Ear – Cotyledon orbiculata.

Pig Nut – Omphalea triandra.

PIGAFETTIA (Mart.) Becc. Palmae. 3 spp. E.Malaysia. Palm trees.

***P. elata*** Wendl.=Metroxylon elatum. Celebes. The leaf fibres from young leaves are used locally as thread and made into mats.

Pigeon Bean – Vicia faba.

Pigeon Pea – Cajanus cajan.

Pigeon Plum – Coccoloba laurifolia.

Pignolia Nut – Pinus pinea.

Pignut – Simmondsia californica.

Pignut Hickory – Carya glabra.

Pignut, Winged – Cycloloma atriplicifolium.

Pigweed – Amaranthus retroflexus.

Pilay – Rubus niveus.

Pilchi – Tamarix dioica.

Pildis – Garcinia dives.

PILERODENDRON Florin. Cupressaceae. 1 sp. S.Chile. Tree.

***P. uviferum*** (Don.) Florin.=Juniperus uvivera=Thuja tetragona. (Ciprés de las Guaytecas). The hard durable wood is used locally for house-building, furniture and poles.

Pili (Nut Oil) – Canarium ovatum.

Pillesesara – Phaseolus trilobus.

PILOCARPUS Vahl. Rutaceae. 22 spp. Tropical America, W.Indies. Trees. The leaves of the following species contain the alkaloid pilocarpine, which is used medicinally in the treatment of diabetes, asthma, to cause contraction of the eye muscles and in large doses as an emetic.

***P. jaborandi*** Holmes. Brazil. Leaves called Pernambuco Jaborandi.

***P. microphyllus*** Stapf. (Jaborandi). N.E.Brazil.

***P. pinnatifolius*** Lem. Tropical America. Leaves are called Folia Jaborandi, Paraguay Jaborandi.

***Pilocereus lanatus*** Weber.=Espostoa lanata.

Pilohe pimenta – Reynoldsia marchionensis.

Pimata – Cheirodendron marquesense.

PIMELODENDRON Hassk. Euphorbiaceae. 6–8 spp. Malaysia. Trees.

***P. amboincum*** Hassk.=Carum amboinicum Moluccas, Amboina. The latex is used locally to varnish wood and the bark is used as a purgative.

PIMENTA Lindl. Myrtaceae. 18 spp. Tropical America, W.Indies. Trees.

***P. acris*** (Sw.) Kostel.=Myrcia acris. W.Indies. Grown commercially in Ceylon, Kenya and W.Africa, but mainly in Jamaica for the leaves. The leaves contain Oil of Bay, Oil of Myrcia, Oleum Myrciae, the main constituents of which are eugenol, chavieol, myrcine and methyl eugenol. The oil is distilled from the leaves with alcohol (originally rum) and used in the preparation of soaps, hair-dressing (Bay Rum) and perfumes.

***P. citrifolia*** (Aubl.) Urb. (Kurupum). W.Indies. The wood is insect resistant and is used locally for house-building.

***P. officinalis*** Lindl.=Eugenia pimenta. (Allspice, Pimenta, Jamaican Pepper). W.Indies. Cultivated in Jamaica for the unripe fruits which are the spice Allspice. The spice is used widely in pickles, sauces, baking, mincemeat and sausages. It was first imported to Europe in 1601, probably as a substitute for caramom. An essential oil (Pimenta Oil, Oil of Allspice) is distilled from the leaves and is used in perfumery, soaps. The fruits are also used medicinally to aid digestion. The oil contains 65–80 per cent eugenol.

Pimenta (Oil) – Pimenta officinalis.

Pimenta de Macaco – Xylopia carminativa.

Pimenta omoa – Reynoldsia marchionensis.

Pimentron – Capsicum annuum.

Pimiento – Capsicum annuum.

Pimpernel, Wood – Lysimachia nemorum.

Pimpernel, Yellow – Lysimachia nemorum.

PIMPINELLA L. Umbelliferae. 150 spp. Eurasia, Africa; 1 sp. Pacific N.America. Herbs.

***P. alpina*** Koord.=P. pruatjan. (Antanan goonoong, Gebangan kipok). Indonesia. The roots are used in Java as a diuretic and aphrodisiac.

***P. anisum*** L.=Anisum officinarum=A. vulgare. (Anise.) E.Mediterranean. Cultivated in S.Europe, S.Russia, Near East, N.Africa, India, Pakistan, China, Mexico, Chile, U.S.A. for the seeds which are used for flavouring foods, medicines, beverages (Anisette liqueur, Anise Milk) and tooth powders. The seeds are used medicinally to relieve stomach upsets and to induce perspiring. An essential oil (Oil of Anise), distilled from the seeds is used for the same purposes. The oil is produced mainly in S.Spain, Bulgaria and S. Russia.

***P. diversifolia*** DC. Himalayas. The leaves are used locally to relieve stomach upsets.

***P. pruatjan*** Molk.=P. alpina.

***P. saxifraga*** L.=Carum nigrum. (Black Carroway). N.Africa. The seeds are used for flavouring food locally and in Mediterranean countries.

Pimple Mallow – Callirhoë pedata.

Pin Oak – Quercus polustris.

Pinang calapa – Actinorhytis calapparia.

PINANGA Blume. Palmae. 115 spp. S.E.Asia, Indomalaya. Palm trees.

***P. banaensis*** Mag. Indonesia. The seeds are used locally as a substitute for betel nuts and the pith of the stem is eaten as a vegetable in Cambodia.

***P. dicksonii*** Blume. (Kondapake). Tropical Asia, especially India. The seeds are used locally as a substitute for betel nuts.

***P. duperreana*** Pierre. (Cau núi, Sla condor). See P. banaensis.

***P. punicea*** Merr.=Areca punicea=Drymophloeus puniceus. Indonesia. The split stems and leaf fibres are used locally for weaving baskets etc. and the leaf fibres are used in the Celebes for making fabrics.

Pinari – Celtis cinnamomea.

PINCKNEYA Michx. Rubiaceae. 1 sp. S. U.S.A. Small tree.

***P. pubens*** Michx. (Fever Tree). The bark (Bitter Bark, Georgia Bark) contains cinchona and is sometimes used to make tonics and to treat malaria.

Pindo Palm – Arecastrum romanzoffianum.

Pine, Apello – Pinus halepensis.

Pine, Arizona Yellow – Pinus arizonica.

Pine, Armand – Pinus armandi.

Pine, Austrian – Pinus nigra.

Pine, Aztec – Pinus teocote.

Pine, Benguet – Pinus insularis.

Pine, Black – Callitris calcarate.

Pine, Black – Podocarpus spicata.

Pine Bluegrass – Poa scabrella.

Pine, Brazilian – Araucaria brasiliensis.

Pine, Bristle Cone – Pinus aristata.

Pine, Bull – Pinus sabiniana.

Pine Bush – Hakea leucoptera.

Pine, Celery – Phyllocladus trichomanoides.

Pine, Celery-Hop – Phyllocladus asplenifolius.

Pine, Chilghoza – Pinus gerardiana.

Pine, Chinese White – Pinus armandi.
Pine, Cluster – Pinus pinaster.
Pine, Coulter – Pinus coulteri.
Pine, Digger – Pinus sabiniana.
Pine, Eastern White – Pinus strobus.
Pine, Foxtail – Pinus aristata.
Pine, Ground – Ajuga chamaepitys.
Pine, Hickory – Pinus pugens.
Pine, Hoop – Araucaria cunnunghami.
Pine, Hoop – Araucaria klinki.
Pine, Huon – Dacrydium franklinii.
Pine, Indian Blue – Pinus excelsa.
Pine, Jack – Pinus banksiana.
Pine, Japanese Stone – Pinus pumila.
Pine, Khasia – Pinus khasya.
Pine, King William – Athrotaxis selaginoides.
Pine, Korean – Pinus koraiensis.
Pine Lichen – Cetraria pinastri.
Pine, Limber – Pinus flexis.
Pine, Loblolly – Pinus taeda.
Pine, Lodgepole – Pinus murrayana.
Pine, Long-leaved – Pinus palustris.
Pine, Luzon – Pinus insularis.
Pine, Masson – Pinus massoniana.
Pine, Merkus – Pinus merkusii.
Pine, Montezuma – Pinus montezumae.
Pine, Moreton Bay – Araucaria cunninghamii.
Pine, Mountain – Pinus montana.
Pine, Murray – Callitris calcarata.
Pine, Murray – Callitris glauca.
Pine, Needle – Hakea leucoptera.
Pine, Needle Oil – Pinus montana.
Pine, Needle Oil – Pinus sylvestris.
Pine, Neosia – Pinus gerardiana.
Pine, Nut – Pinus cembroides.
Pine, Nut – Pinus edulis.
Pine, Nut – Pinus monophylla.
Pine, Nut – Pinus parryana.
Pine, Parana – Araucaria brasiliana.
Pine, Parasol – Sciadopitys vertillata.
Pine, Pitch – Pinus rigida.
Pine, Pond – Pinus serotina.
Pine, Ponderosa – Pinus ponderosa.
Pine, Putt's – Flindersia acuminata.
Pine, Radiate – Pinus radiata.
Pine, Red – Dacrydium cupressinum.
Pine, Red – Pinus resinosa.
Pine, Rocky Mountain White – Pinus flexilis.
Pine Rubber – Landolphia kirkii.
Pine, Sand – Pinus clausa.
Pine, Scotch (Scots) – Pinus sylvestris.
Pine, Screw – Pandanus spp.
Pine, Short-leaved – Pinus echinata.
Pine, She – Podocarpus elata.
Pine, Short-leaved – Pinus echinata.
Pine, Slash – Pinus caribaea.
Pine, Southern – Pinus palustris.
Pine, Stinking Ground – Camphorosma monospeliaca.
Pine, Stone – Pinus pinea.
Pine, Sugar – Pinus lambertiana.
Pine, Swamp – Pinus caribaea.
Pine, Swiss Stone – Pinus cemba.
Pine, Table Mountain – Pinus pugens.
Pine, Torrey's – Pinus torreyana.
Pine, Umbrella – Sciadopitys verticillata.
Pine, Western White – Pinus montana.
Pine, Western Yellow – Pinus ponderosa.
Pine, Westland – Dacrydium westlandicum.
Pine, White Bark – Pinus albicans.
Pine, White Cypress – Callitris glauca.
Pine, White Dammar – Agathis alba.
Pine, White – Podocarpus elata.
Pine, Yellow – Pinus echinata.
Pine, Yellow Silver – Dacrydium intermedium.
Pineapple – Ananas comosus.
Piney – Paeonia foemina.
Piney Tallow – Vateria indica.
PINGUICULA L. Lentibulariaceae. 35 spp. N.Hemisphere, outside the tropics, Andes, Antarctic. Insectivorous herbs.
***P. vulgaris*** Sm. (Butterwort). N.Hemisphere. The leaves are used locally to coagulate milk.
Pinie – Pinus pinea.
Pink – Dianthus caryophyllus.
Pink African Cedar – Guarea cedrata.
Pink Blackbutt – Eucalyptus eugenoides.
Pink, Carolina – Phlox carolina.
Pink Cedar – Acrocarpus fraxinifolius.
Pink Mahogany – Guarea cedrata.
Pink Poplar – Euroschinus falcatus.
Pink, Sea – Armeria maritima.
Pinkroot – Spigelia marilandica.
Pinkroot, False – Phlox carolina.
Pino triste – Pinus lumholtzii.
Piñon – Pinus cembroides.
Piñon – Pinus edulis.
Piñon – Pinus monophylla.
Piñon – Pinus parryana.
Pintan – Lunasia amara.
Piñuela – Bromelia pinguin.
Piñuela – Morinda yucatanensis.
PINUS L. Pinaceae. 70–100 spp. N.Temperate, mountains of N. tropics. Trees. Many of the species are important for lumber and several yield turpentine which is used in making paints and polishes and

as a solvent for resins and waxes. Medicinally it is used in ointments and linaments for rheumatism and chest colds, etc. and internally to treat stomach upsets, cystitis and venereal diseases. The resin left after distilling the turpentine (Rosin, Yellow Resin, Colophony) is used for varnishes, paints, inks, polishes, soap, sealing wax, plastics, floor covering, waterproofing cardboard and fireworks. An oil (Rosin Oil, Rosinol, Retinol) obtained by dry distillation of the rosin is used industrially to make carbon black and printer's ink, dyes, axle greases, varnishes and brewer's pitch. Medicinally it is used to treat skin diseases and gonorrhoea. An oleoresin is extracted from the trees by tapping. This is distilled to give the turpentine, leaving the rosin as a residue. About 30 million pounds of turpentine are produced annually in U.S.A. some by the process described above, some by the distillation of pine stumps left after felling and about one third during the sulphur paper-making process. Russia, Greece, Yugoslavia, Austria, India and Indonesia also produce some turpentine.

***P. albicaulis*** Englm. (White Bark Pine). Pacific N.America. The seeds are eaten by the local Indians.

***P. aristata*** Englm. (Bristle Cone Pine, Fox-tail Mine). Arizona. The wood is sometimes used locally for lumber. It is not strong. Sometimes used for pit-props.

***P. arizonica*** Engelm. (Arizona Yellow Pine). The light soft wood is sometimes used locally for lumber.

***P. armandi*** Franch. (Chinese White Pine, Armand Pine). China, Formosa. The soft wood is sometimes used locally for building and making cheap furniture.

***P. australis*** Michx. f. = P. palustris.

***P. banksiana*** Lam. = P. divaricata (Jack Pine). E. N.America. The light wood is soft and coarse. It is used for fuel, wood pulp and making boxes.

***P. caribaea*** Maorelet. (Slash Pine, Swamp Pine). S.E. U.S.A., Central America, Bahamas. The hard heavy strong wood, used for general construction work, railway sleepers, is a source of turpentine and resin.

***P. cembra*** L. (Swiss Stoen Pine). Austria to N.W. Russia. The wood is soft and close-grained. It is used for making furniture, turning, constructing Alpine houses. The leaves yield a turpentine (Carpathian or Hungarian Turpentine) and the seeds are eaten locally in milk-foods and pastries, in Norway and Russia.

***P. cembroides*** Zucc. (Mexican Piñon, Nut Pine, Piñon). S.W. U.S.A., Mexico. The nuts are eaten locally and are sold as Pine Nuts.

***P. clausa*** (Engelm.) Sarg. (Sand Pine). S.E. U.S.A. The soft light wood is not strong, but is occasionally used for masts of small ships.

***P. contorta*** var. ***murrayana*** Englm. = P. murrayana.

***P. coulteri*** D. Don. (Coulter Pine). California. The seeds are eaten by the local Indians.

***P. divaricata*** Dum. Cours. = P. banksiana.

***P. echinata*** Mill. (Short-leaved Pine, Yellow Pine). E. and S. U.S.A. The yellow wood is heavy, hard and coarse. It is used as lumber.

***P. edulis*** Engelm. (Nut Pine, Piñon). S.W. U.S.A., Mexico. The wood is brittle and soft and is used for posts and fuel. The seeds are eaten locally and sold in the markets.

***P. excelsa*** Wall. (Indian Blue Pine). Himalayas to Afghanistan. The wood is used locally for construction work. The tree yields a turpentine and tar.

***P. flexilis*** James. (Limber Pine, Rocky Mountain Pine, White Pine). W. N.America. The soft, close-grained, light wood is sometimes used for general carpentry.

***P. gerardiana*** Wall. (Neosia Pine, Chilghoza Pine). Himalayas to Afghanistan. The seeds are eaten locally.

***P. halepensis*** Mill. (Aleppo Pine). S.Europe, W.Asia. Source of Greek Turpentine.

***P. insignis*** Doug. = P. radiata.

***P. insularis*** Endl. (Benguet Pine, Luzon Pine). Philippines. The source of a turpentine.

***P. khasya*** Poyle. (Khasia Pine). Burma. Yields a turpentine and a resin.

***P. koraiensis*** Sieb. and Zucc. (Korean Pine). Korea. The seeds are eaten locally.

***P. lambertiana*** Dougl. (Sugar Pine). Pacific U.S.A. The light soft wood is used for house interiors and shingles. The sweet liquid exuded from wounds is used as a laxative.

***P. laricio*** Poir.=P. nigra.

***P. longifolia*** Roxb. N. and E. India. The wood is used locally for construction work and yields a turpentine (Indian Turpentine). The charcoal is used by the Chinese for fireworks.

***P. lumholtzii*** Robins and Fern. (Pino triste). Mexico. The wood is used locally to make musical instruments and a decoction of the leaves to treat stomach upsets.

***P. macrophylla*** Lindl.=P. montezumae.

***P. maritima*** Lam.=P. pinaster.

***P. massoniana*** D. Don. (Masson Pine). S.China. The wood is fairly hard, close-grained and durable. It is used locally for general construction work, furniture etc.

***P. merkusii*** Jungh. (Merkus Pine). Burma, N.Sumatra. The wood is durable and is used for posts and house construction. The tree is a valuable source of turpentine and colophony.

***P. monophylla*** Torr. and Frem. (Nut Pine, Piñon). S.W. U.S.A., Mexico. The nuts are eaten by the local Indians and the charcoal is used for smelting.

***P. montana*** Mill.=P. mungo. (Mountain Pine). Alpine Europe. Small tree. Steam distillation of the leaves gives Pine Needle Oil, which is used medicinally as an antiseptic and an inhalent for chest coughs. The twigs yield Oleum Pini Pumilionis, used for the same purposes. The wood is used locally to make wooden shoes.

***P. monticola*** Doug. (Western White Pine). W. N.America. The soft wood is used for general construction work and interiors of houses etc.

***P. montezumae*** Lamb.=P. russeliana=P. macrophylla. (Montezuma Pine). Mexico, Central America. The resin is used locally for dressing wounds.

***P. mugo*** Turra.=P. montana.

***P. mugo*** var. ***pumilio*** (Haenke) Zenari=P. pumilio.

***P. murrayana*** Balf.=P. contorta var. murrayana. (Lodgepole Pine). W. N.America. The soft light wood is not durable, but is sometimes used for pit props, railway sleepers and fuel.

***P. nelsoni*** (Shaw.) Mexico. The seeds are eaten locally.

***P. nigra*** Arn.=P. laricio. (Austrian Pine). Austria to Balkans. Yield a turpentine (Austrian Turpentine) which is used mainly in ointments and plasters.

***P. palustris*** Mill=P. australis. (Long-leaved Pine, Southern Pine). S. and S.E. U.S.A. A very important lumber tree. The wood is called Pitch Pine or Southern Pine. It is hard, heavy, tough and durable, although rather coarse-grained. The wood is used for general building, house interiors, pulping, fences, railway sleepers, boxes, masts, fuel and charcoal. The tree also gives turpentine and rosin.

***P. parryana*** Engelm. (Nut Pine Piñon) California, Mexico. The nuts are eaten locally.

***P. parviflora*** Sieb. and Zucc.=P. pentaphylla.

***P. pentaphylla*** Carr.=P. parviflora. Japan. The soft, light wood is used locally for house-building, house interiors, furniture, ship-building, matches, wooden tiles, carving and wooden water pipes.

***P. pinaster***=P. maritima (Cluster Pine). Mediterranean. Yields a turpentine (Boreaux Turpentine, French Oil Turpentine).

***P. pinea*** L. (Stone Pine, Pinie). S.Europe. Cultivated for the edible fruits (Pignolia Nuts) which are exported.

***P. ponderosa*** Laws. (Ponderosa Pine, Western Yellow Pine). W. N.America, California to Mexico. The hard strong wood is used for railway sleepers, pit props, general construction work, posts and fuel. It is an important lumber tree.

***P. pumila*** Englm. (Japanese Stone Pine). Japan. The wood is used locally for charcoal.

***P. pumilio*** Haenke.=P. mugo var. pumilio. Central Europe. Steam distillation of the leaves gives Dwarf Pine Needle Oil, which is used in perfumery, soap and air-freshening aerosols. It is also used medicinally for treating chest and bladder complaints, rheumatism and skin diseases.

***P. pungens*** Michx. (Table Mountain Pine). E. U.S.A. The wood is used locally for charcoal.

***P. radiata*** D. Don.=P. insignis. (Radiate Pine). California, Mexico. A soft light wood which is not decay or insect resistant, but can be worked easily. Cultivated widely as a forest tree in Australia and New Zealand. It is used for house interiors, furniture, matches, plywood, matchboxes, toys, turning, brush handles etc.

***P. resinosa*** Ait. (Red Pine). E. N.America. The hard light wood is used for general building, masts, spars and bridges. The bark is used for tanning.

***P. rigida*** Mill. (Pitch Pine). S.E. N.America. The wood is used for fuel and charcoal, sometimes for lumber.

***P. russeliana*** Lindl.=P. montezumae.

***P. sabinaria*** Doug. (Digger Pine, Bull Pine). California. Distillation of the wood produces the aromatic Abietine. The seeds are eaten by the local Indians.

***P. serotina*** Michx. (Pond Pine). N.Carolina to Florida. A local source of turpentine. The wood is sometimes used for lumber.

***P. strobus*** L. (Eastern White Pine). The light wood is not strong, but is used to make furniture, house interiors, matches, masts, laths and wooden ware.

***P. succinifera*** Conw. A fossil tree found in Baltic areas of Europe. A source of Amber (Baltic Amber, Succinite) which is carved to make fancy articles and sometimes used in lacquers.

***P. sylvestris*** L. (Scotch Pine). N.Europe, N.Asia, Central Europe. An important European forest species, widely cultivated. The wood is not strong, but durable. It is used for general carpentry, furniture, railway sleepers, masts. Distillation of the wood yields tar, pitch and the wood is used to make cellulose. The leaves are a source of Pine Needle Oil, used in perfumery and medicine.

***P. taeda*** L. (Loblolly Pine). E. and S. U.S.A. The brittle wood is used for general construction work.

***P. teocote*** Cham. and Schlecht. (Aztec Pine, Jalocote, Ocote). Mexico. Distillation yields a turpentine (Ocotzol, Trementina de Ocote) and a tar (Brea) which is used to make soap and for torches.

***P. tonkinensis*** Chev. (Thông). Tonkin. The wood is used locally for building, furniture etc. The resin is used locally for making a soap.

***P. torreyana*** Carr. (Torrey's Pine). California. The seeds are eaten locally.

Piondo – Encephalartos poppgei.

Piopo – Oenanthe javanica.

Pi-pat-buti – Heliotropium tuberculosum.

Pipe Gamboge – Garcinia hanburyi.

PIPER L. Piperaceae. 2000 spp. Tropics. Vines or scrambling shrubs.

***P. aduncum*** L. Tropical America. A decoction of the leaves is used as a diuretic and stimulant.

***P. angustifolium*** Ruiz. and Pav. Central America, W.Indies, Mexico. The leaves (Matico) which contain stearoptene matico camphor are used medicinally as a urino-genital antiseptic and to prevent bleeding. The dried leaves are exported.

***P. attenuatum*** Ham. N.India, Himalayas. A decoction of the root is used locally as a diuretic.

***P. auritum*** H.B.K. Mexico, Central America. The leaves are used locally for seasoning food.

***P. bantamense*** Blume. Malaysia. The bark is used to scent clothes during washing.

***P. betle*** L.=Chavica auriculata. (Betel, Sirih). Malaysia, Indochina. The plant is cultivated widely in the Far East for the leaves. These are wrapped around Betel Nuts (Areca catechu) after treating with lime and cutch and possibly cloves, etc. and used for chewing. The effect is mildly stimulatory and apparently without ill effects. The effect of the juice is to blacken the teeth. The chewing of betel nuts has been practiced in the Far East for centuries. The oil from the leaves (Betel Oil) is used to treat sore throats and laryngitis and as an inhalant for diphtheria. The Chinese use the fruits, leaves and roots to relieve stomach upsets and in India a mixture of the root with Black Pepper (Piper nigrum) is used to produce sterility in women.

***P. caducivracteum*** C. DC. (Vitne). Malaysia. Used locally as P. betle.

***P. capense*** L. f. S.Africa. A decoction of the fruits was used by the early settlers to relieve minor stomach upsets.

***P. chaba*** Hunter=P. officinarum. Malaysia, Philippines. The seeds (Long Pepper) are used as a condiment.

***P. clusii*** C. DC. W.Africa. The seeds (Ashanti Pepper) are used locally as a spice. The fruits yield an oil (Oil of Cubeb). See P. cubeba.

***P. cubeba*** L. (Cubeba). Malaya. Cultivated, especially in Java and Singapore, for the berries (Cubeb Berries), from which is distilled Oil of Cubeb. The oil is used medicinally as an antiseptic, diuretic, expectorant and to relieve digestive upsets. It is also used in flavouring foods, bitters and some cigarette tobaccos.

***P. excelsum*** Forst. Australia, New Zealand, Pacific. In New Zealand, the plant has a variety of medicinal uses. A decoction of the leaves is used to treat bladder and kidney complaints, kidney trouble, skin diseases, stomach pains and toothache. The leaves and bark are used for stomach ache and to dress wounds. The roots relieve toothache and kidney and bladder troubles, while the fruits and seeds are used to relieve constipation, stimulate the salivary glands and treat kidney trouble. The seeds are also considered aphrodisiac.

***P. fragile*** Benth. Moluccas. The bark is used locally to treat yaws.

***P. guineense*** Schum. and Thom. (Ashanti Pepper). The fruits (Ashanti Pepper, Poivre du Kissi) are a popular local spice. Some are exported and there is some local cultivation.

***P. hederaceum*** Cunningh. New South Wales. The flowers are an important local source of honey.

***P. leucophyllum*** (Miq.) C. DC. (Gordoncillo). Mexico. A decoction of the plant is used locally to treat fevers and as a wash to remove head parasites.

***P. lolot*** C. DC. (Ana klûa tâo, Lâ lot, Poivre lolot). Indochina. The leaves and fruits are used locally as a condiment.

***P. longum*** L. (Jaborandi Pepper, Long Pepper). Himalayas through India. The fruits (Long Pepper) are used as a condiment. A decoction of the roots is used locally as a diuretic and to induce sweating. The plant is cultivated.

***P. medium*** Jacq. Tropical America. In Costa Rica, the leaves are used as a dressing for snake-bites.

***P. mekongense*** C. DC. Indochina. The fruits are used locally as a condiment.

***P. methysticum*** Forst. (Kava Pepper). Pacific Islands. The roots are used to make the non-alcoholic but narcotic drink Kava, which has been used as a ceremonial drink among the Polynesians since ancient times. The dried roots and rhizomes are also used medicinally as a diuretic, antiseptic, expectorant and urino-genital stimulant.

***P. nigrum*** L. (Pepper). India, Malaysia. Cultivated throughout the Far East for the fruits (Pepper) which are much-prized as a condiment and spice. Pepper was known to the Greeks and Romans and the trade in pepper was one of the main reasons for the opening up of trade with the Far East and the voyages of discovery during the 15th and 16th centuries. Most of the world supply comes from Java and Sumatra. There is some 17 million pounds exported annually. The fruits are picked by hand and sun-dried (or sometimes over fires) when they ferment, changing from green to black. The whole fruits are ground to make Black Pepper. White pepper is made by grinding the dried seeds after removing the flesh from the fruit (peppercorn) by fermentation and washing in water. The plants are propagated by cutting and grown up supports, often shade trees. The pungency of the pepper is due to the alkaloid piperine of which there is some 4–8 per cent. An oil extracted from the seeds (Oil of Pepper) is also used to flavour foods; it is also used in perfumery. Medicinally the oil is used as a stimulant and to reduce fevers. There are several varieties of the plant and commercial grades of pepper including Alleppy Pepper, Madagascar Pepper, Madagascar White Pepper, Lampong Black Pepper, Muntok White Pepper, Penang Black Pepper, Saigon Pepper, Singapore Pepper and Tellicherry Pepper.

***P. officinarum*** C. DC. = P. chaba.

***P. pachyphyllum*** Baker. (Mahalatsaka, Voampirifery), Madagascar. The seeds are used locally as a condiment.

***P. palmeri*** C. DC. (Hachogue, Matico). Mexico. A decoction of the plant is used to treat skin diseases and digestive upsets.

***P. retrofractum*** Vahl. (Javanese Long Pepper). Malaysia. The fruits are used as a condiment, especially in curries and pickles. Cultivated.

***P. ribesioides*** Wall. Burma, Malaya. See P. cubeba.

***P. saigonense*** C. DC. (Lolo). Indochina. Cultivated locally for the fruits which are used as a condiment.

***P. sanctum*** (Miq.) Schlecht. Mexico. The leaves are used locally to flavour soups and a decoction of the leaves is used to treat stomach upsets.

***P. umbellatum*** L. = Pothomorpe umbellata.

Pipiloxhuitl – Cestrum nocturnum.

Pipiri – Pithecellobium jupunba.

PIPTADENIA Benth. Leguminosae. 11 spp. Mexico to tropical S.America. Trees.

***P. cebil*** Griseb. Argentine. Cebil Gum, containing some 80 per cent arabinose is extracted from the trunk.

***P. colubrina*** (Vel.) Benth. = Mimosa colubrina. Brazil, Peru, Bolivia. The ground seeds are used locally to make a highly narcotic snuff. In Brazil the astringent bark is used as a powder or decoction to treat lung diseases.

***P. macrocarpa*** Benth. (Cebil Colorado, Curupag). Argentine. The bark is used locally for tanning.

***P. peregrina*** (L.) Benth. (Cohoba Tree, Yoke). The heavy hard wood is used for railway sleepers. The ground seeds are used as a narcotic snuff (Cohoba Snuff, Coxoba Snuff). Its use induces frenzy.

***Piptostegia*** Hoffmgg. = Operculina.

PIPTURUS Wedd. Urticaceae. 40 spp. Mascarene Islands to Australia and Polynesia. Shrubby trees. The bark of the following species gives a fibre, used for sails, nets, mats etc.

***P. albidus*** Gray = P. gaudichaudianus.

***P. argenteus*** Wedd. = P. taitensis = Urtica argentea. Malaysia, Australia, Pacific. The bark yields a brown dye.

***P. gaudichaudianus*** Wedd. = P. albidus. Hawaii. The bark is used to make tapa (paper cloth).

***P. incanus*** Wedd. = P. velutinus.

***P. taitensis*** Wedd. = P. argenteus.

***P. velutinus*** Wedd. = P. incanus. Malaysia, New Caledonia. The bark fibre is used in New Caledonia.

Piralahi – Landolphia perieri.

***Piratinera guianensis*** Aubl. = Brosimum aubletii.

Pirazha – Spilanthes ocmella.

Piriguaua – Anchietae salutaris.

Piripiri – Holarrhena microteranthera.

Pirirma Oil – Syagrus cocoides.

Pisang bau – Musa balbisina.

Pisang jacki – Musa acuminata.

Pisa Palm – Areca hutchinsoniana.

PISCIDIA L. Leguminosae. 10 spp. Florida, Mexico, W.Indies. Trees.

***P. erythrina*** L. = Ichthyomethia piscipula. (Jamaican Dogwood, Jamaican Fishfuddle Tree). The hard, heavy, yellow wood is used for boat-building, fuel and charcoal. The tree contains the narcotic piscidin and the powdered bark and roots were used in the W.Indies to stupify fish.

PISOLITHUS Alb. and Schw. Sclerodermataceae. 2 spp. Widespread.

***P. tinctorius*** (Pers.) Coker and Couch. The fruitbodies give a brown dye used in S.Europe to colour silk.

PISONIA L. Nyctaginaceae. 50 spp. Tropics and subtropics. Shrubs or trees.

***P. aculeata*** L. (Garabato, Huele de noche, Uña de gato, Zarza). S.Florida to tropical S.America, W.Indies, S.Asia. A decoction of the bark and leaves is used in Jamaica to treat venereal diseases and rheumatism. The stems are used in Jamaica to hoop barrels.

***P. alba*** Span. (Lettuce Tree, Maluko). The young leaves are used locally as a vegetable.

***P. brunoniana*** Endl. = Ceodes brunoniana.

***P. capitata*** (S. Wats.) Stand. = Cryptocarpus capitatus. (Bainoro prieto, Garambulla, Vainoro prieto). Mexico. The fruits are used by the local Indians to treat fevers.

***P. grandis*** R. Brown. = Ceodes umbellifera (Cabbage Tree, Gatae, Puatea). Polynesia, Australia, New Guinea, Ceylon, Indonesia, E.Africa. The leaves are sometimes used as an emergency food.

***P. sylvestris*** Teijsm. and Binn. Moluccas. Cultivated throughout Indonesia for the leaves which are eaten as a vegetable.

Pistache, Chinese – Pistacia chinensis.

Pistache, Egyptian – Pistacia khinjuk.

Pistache, Lentisk – Pistacia lentiscus.

Pistache, Mount Atlas – Pistacia atlantica.

Pistache, Terebinth – Pistacia terebinthus.

Pistache, Turkish Terebinth – Pistacia mutica.

Pistachio (Nut) (Oil) – Pistacia vera.

PISTACIA L. Pistaciaceae. 10 spp. Mediterranean to Afghanistan, S.E.Asia to Malaya, S. U.S.A., Mexico, Guatemala. Trees.

***P. atlantica*** Desf. (Betoum, Mount Atlas Pistache). Sahara, Canary Islands. Damage by the fungus Polyporus tinctorius causes galls which are used for tanning and making ink and dyes. The fruits are eaten locally and the leaves are fed to livestock.

***P. cabulica*** Stocks. = P. mutica.

***P. chinensis*** Bunge. (Chinese Pistache). China. The young shoots (huang-nu ya-tzu) are eaten locally as a vegetable and the wood is used as rudder posts.

***P. integerrima*** Stewart. (Kattar). N.India, Himalayas. The hard, durable brown wood is attractively marked. It is used for furniture, ploughs and spinning wheels. Insect galls from the leaves are a commercial source of tannin.

***P. khinjuk*** Stocks. (Egyptian Pistache). W.Himalayas. The wood is used to make furniture. The leaves are used for livestock fodder and galls yield tannin. The fruits are used locally to flavour milk.

***P. lentiscus*** L. (Lentisk Pistache). Mediterranean. Cultivated locally for the resin (Chois Mastic) extracted from the bark. It is an expensive commodity, used by the ancient Egyptians for embalming and now used as a varnish for coating paintings, metals and in lithography. It is also used as a theatrical fixative and chewed in the East to sweeten the breath. Medicinally it is used as an aid to digestion, for plasters and temporary teeth fillings.

***P. mutica*** Fisch. and Mey=P. cabulica. (Turkish Terebinth Pistache). Central Asia. The trunk yields a mastic (Bombay Mastic) used in varnishes and chewed locally.

***P. terebinthus*** L. (Terebinth Pistache). Mediterranean. The plant was an important source of turpentine (Chian Turpentine). Insect galls (Carobbe di Guidea, Pistacia Galls) are used for tanning.

***P. vera*** L. (Pistachio). Mediterranean, Middle East. The tree has been cultivated locally since ancient times for the fruits. It is now cultivated commercially in Turkey, Iran, Italy, Syria, Afghanistan and California, with an annual production of about 50,000 tons of nuts a year. Turkey is the main external supplier to the U.S.A. The nuts are used mainly in confectionery and ice-cream and also yield an oil Pistachio Nut Oil. Galls on the plant (Bokhara Galls, Gul-i-pista) are produced in Iran and used for tanning. There are several commercial varieties. The plants are propagated by budding onto seedling rootstock, fruiting begins in the fifth year, but there is no commercial crop until the tenth year. The trees live to an immense age. The fruits are usually picked by hand and sun-dried when the outer parts of the fruit peel away to leave the seeds. This splits longitudinally to reveal the "nut". The marketable kernels contain about 20 per cent protein and 60 per cent oil.

Pistacia Galls – Pistacia terebinthus.

PISTIA L. Araceae. 1 sp. Tropics and subtropics. Floating water plant.

***P. stratiotes*** L. (Tropical Duckweed, Water Lettuce). The plant is grown in fishponds in Java to encourage the breeding of edible water shrimps, it is also mixed with rice to feed to ducks. The Chinese use it as fodder for pigs. The plant is of considerable nuisance as a waterweed in tropical waterways.

PISUM L. Leguminosae. 6 spp. Mediterranean, W.Asia. Herbs.

***P. abyssinicum*** A. Braun=P. sativum var. abyssinicum. Abyssinia, Arabia. Cultivated locally as a food crop.

***P. arvense*** L. (Field Pea). Mediterranean. Cultivated for the seeds which are used for human consumption and feeding livestock. It is also grown as a green manure. There are several varieties, including Bavarian Pea, Black-podded Pea, Königsberger Pea, East Prussian Pea, Sand Pea and Smyrna Pea.

***P. elatius*** M. Bieb. (Djelben, Hammez, Pea of the Oasis, Pois des Oasis). Sahara. Cultivated around oases for the seeds which are used dry or ground into a flour.

***P. sativum*** L. (Garden Pea). Mediterranean. The plant has been cultivated since ancient times, fossil remains having been found in Swiss Lake villages. It is eaten as a vegetable and grown commercially as a field or garden crop. Formerly most of the crop was dried and used as split peas as a vegetable, often in soups. With the advent of canning and later deep-freezing, a lot of the crop is now marketed in these forms. In latter years the harvesting of the crop has become increasingly mechanised. There are many varieties depending on plant size, shape and size of pods and time of maturing.

***P. sativum*** var. ***abyssinicum*** Alef.=P. abyssinicum.

Pitahay dulce – Lemaireocereus thunberi.

Pitahaya – Acanthocereus pentagonus.

Pitahaya – Echinocereus stramineus.

Pitahaya – Hylocereus undatus.

Pitahaya – Lemaireocereus queretaroensis.

Pitahaya – Lemaireocereus thurberi.

Pitahaya – Machaerocereus gummosus.

Pitahaya agria – Machaerocereus gummosus.

Pitahaya de Agosto – Echinocereus conglomeratus.

Pitahaya Naranjada – Acanthocereus pentagonus.

Pitanga – Eugenia uniflora.

Pitanga tuba – Eugenia selloi.

Pitch, Burgundy – Picea excelsa.

Pitch, Canada – Tsuga canadensis.
Pitch, Hemlock – Tsuga canadensis.
Pitch Pine – Pinus palustris.
Pitch Pine – Pinus rigida.
Pitcher Plant – Sarracenia purpurea.
Pitchery – Duboisia hopwoodii.
Piteira – Furcraea gigantea.
Piteira gigante – Furcraea gigantea.
Pith Plant – Aeschynomene aspera.
Pitjes, Makassar – Brucea sumatrana.

PITHECELLOBIUM Mart. Leguminosae. 200 spp. Tropics. Trees.

***P. albicans*** (Kunt.) Benth. = Acacia albicans. (Chucum, Huisache). Mexico. The trunk yields a gum used as an emulsifying agent.

***P. arboreum*** (L.) Urban. = P. filicifolium. = Mimosa arborea. (Cojolia, Colu de mico, Coralillo Frijolillo). Central America, W.Indies. The hard durable wood is used locally for house construction.

***P. auremontemo*** Mart. Brazil. Yields a gum similar to Gum Arabic.

***P. bigeminum*** Mart. (Aragvadha, Kachlora, Kalpakku). N.India. E.Himalayas. The seeds are used locally to treat diabetes; a decoction of the leaves is used as a hair tonic and a fish poison.

***P. brevifolium*** Benth. (Guajillo, Huajillo, Tenaza). Mexico. The foliage is fed to livestock and the wood is used locally for general carpentry.

***P. dulce*** Benth. (Manila Tamarind). Tropical America. Sometimes cultivated in Old and New World tropics for the fruit (arils) which are eaten raw or made into a drink. The bark gives a yellow dye used for tanning leather.

***P. filicifolium*** Benth. = P. arboreum.

***P. flexicaule*** (Benth.) Coult. = P. texense = Mimosa flexicaulis. Mexico, Texas. The seeds and green pods are eaten locally as a vegetable. The testa is used as a coffee substitute. The wood is used to make furniture and wagons.

***P. hymeneaefolia*** Benth. Tropical America. Yields a gum (Goma de Orore) similar to Gum Arabic.

***P. jiringa*** Prain. (Jering). Borneo. The fairly hard brown wood is durable and used for house interiors and furniture.

***P. jupunba*** (Willd.) Urb. (Pipiri). W.Indies, N. S.America. The wood (Bois Cicerou) is used locally for general carpentry. The leaves are used as a soap substitute; the bark is used to stupify fish and the water from the boiled wood to treat dysentery.

***P. lobatum*** Benth. Burma, Malaysia. The leaves, fruits and flowers are eaten as vegetables in Java, where the plant is cultivated.

***P. mexicanum*** Rose. (Chino, Palo chino). Mexico. The wood is used locally for general carpentry and making furniture.

***P. minahassae*** Koord. (Bowoi, Rajango, Merang). Indonesia. The wood is used for general carpentry and making furniture.

***P. saman*** Benth. (Cow Tamarind, Saman). Tropical America. The wood is hard, durable and polishes well to a dark brown. It is used to make furniture. The seeds are eaten locally and fed to livestock. The tree is grown for shade in cacao and coffee plantations in Guyana and the W.Indies.

***P. texense*** Coult. = P. flexicaule.

***P. unguis-cati*** Benth. Central America, W.Indies. The fruits are eaten locally.

PITHECOCTENIUM Mart. ex Meissn. Bignoniaceae. 7 spp. Mexico to tropical S.America, W.Indies. Shrubs.

***P. echinatum*** (Jacq.) Schum. = Bignonia echinata = P. muricatum. (Corneta, Hoja de Cucharilla, Mareposa, Palomitas, Petaquillas). The seeds are used in Mexico as a cure for headaches, by placing them on the forehead. A paste of the ground seeds and tallow is placed on the temples to soothe sore eyes. The fruits are used as back-scratchers and pin-cushions.

***Pithecolobium*** auctt. = Pithecellobium.

PITHOPHORA Wittrock. Cladophoraceae, Fresh-water green alga.

***P. affinis*** Nordst. Hawaii. The alga is eaten locally with fresh-water shrimps.

Pitomba – Eugenia luschnathiana.
Pitre – Fucraea hexapetala.

PITTOSPORUM Banks ex Soland. apud Gaerth. Pittosporaceae. 150 spp. Tropical and subtropical Africa, Asia, Australia, New Zealand, Pacific. Small trees or shrubs.

***P. angustifolium*** Lodd. = P. phillyraeoides.

***P. crassifolium*** Soland (Karo Pittosporum). New Zealand. The tough white wood is used for inlay work.

***P. phillyraeoides*** DC. = P. angustifolium. (Bitter Bush). Australia. The seeds are made into a flour by the aborigines.

***P. viridiflorum*** Sims. S.Africa. The roasted bark is used locally to treat dysentery.

Pitura Fibre – Furcraea gigantea.

Pituri – Duboisia hopwoodii.

PITYROGRAMMA Link. Gymnogrammaceae. 40 spp. Tropical America, Africa. Ferns.

***P. calomelanos*** (L.) Link. Tropical America. Has become a weed in tropical Asia. In the W.Indies the leaves are used on wounds to stop bleeding.

Piule – Rivea corymbosa.

PLAGIANTHUS J. R. and G. Forst. Malvaceae. 15 spp. Australia, New Zealand. Trees.

***P. betulinus*** A. Cunn. (Lowland Ribbon Wood). New Zealand. The bark (Lace Bark) yields a fibre used by the Maoris to make cordage and fishing nets; sometimes used as a substitute for raffia for tying-up plants.

Plains Bamboo – Bambusa balcooa.

Plaited Usnea – Usnea plicata.

PLANCHONELLA Pierre. Sapotaceae. 100 spp. Indonesian, Malaya, tropical Australia, New Zealand, Pacific, Seychelles; 2 spp. S.America. Trees.

***P. firma*** Dubard. = Sideroxylon firmum. (Kane polo, Njatoh labar). Malaysia. The wood is easy to work and beautifully grained. It is used especially by the Chinese for carving household utensils, doors and door-posts.

PLANCHONIA Blume. Barringtoniaceae. 8 spp. Andaman Islands to N. and N.E.Australia. Trees.

***P. sundaica*** Miq. = P. valida.

***P. valida*** Blume. = P. sundaica. (Darah, Pootah). Indonesia. The wood is much used in Java for making houses and the leaves are eaten as a vegetable with rice.

Plane (Tree Maple) – Acer pseudo-platanus.

Plane, Oriental – Plantanus orientalis.

PLANTAGO L. Plantaginaceae. 265 spp. Cosmopolitan. Herbs.

***P. amplexicaulis*** Chav. (Isafghol, Ispaghul). N.India, E. Pakistan. A decoction of the leaves is used by the Hindus to treat chest complaints, fevers and opthalmia.

***P. arenaria*** Waldst. and Kit. = P. ramosa. Central to S.Europe, through Caucasia to Siberia. The plant is cultivated for the seeds (French Psyllium Seed, Spanish Psyllium Seed). These are mucilagenous and when wet swell-up. They are used as a laxative.

***P. coronopus*** L. (Buckthorn Plantain, Crowfood Plantain). Europe, N.Africa, Asia, introduced to Australia and New Zealand. The leaves are sometimes used in salads.

***P. decipiens*** Barneoud. (Goose Tongue, Seaside Plantain). Saltmarshes of N.E.America. The leaves are used as a salad or vegetable by the local fishermen.

***P. decumbens*** Forsk. = P. ovata.

***P. ispaghula*** Roxb. = P. ovata.

***P. lanceolata*** L. (Rib Grass). Europe to temperate Asia, introduced to N.America. The leaves are sometimes eaten in salads. A possible emergency food.

***P. major*** L. (Plantain, Ripple Grass, Waybread). Europe, to temperate Asia, naturalized in America. The leaves are rubbed on bee, wasp, nettle etc. stings to relieve the pain. They are diuretic and are used also to relieve piles and a decoction is taken to treat diarrhoea. They are sometimes eaten in salads.

***P. major*** L. var. ***asiaticum*** Dcne. Temperate Asia. The Chinese use the seeds as a diuretic.

***P. maritima*** L. (Sea Plantain). Coastal areas of Europe and Russia. The leaves are used locally in salads and as a vegetable.

***P. ovata*** Forsk. = P. decumbens = P. ispaghula. Mediterranean, Central Asia, to India. The plant is cultivated in India for the mucilagenous seeds (Blond Plantago Seed, Indian Plantago Seed, Ispaghul, Spogel) which are used in India as a laxative and to control dysentery.

***P. psyllium*** L. (Psyllium). Mediterranean. Cultivated for the mucilagenous seeds (Flea Seed, Plantain Seed, Psyllium Seed) which are used medicinally as a laxative.

***P. ramosa*** (Gilib.) Asch. = P. arenaria.

***P. virginica*** L. (Dwarf Plantain). Rhode Island, to Pacific Coasts of U.S.A., Mexico. The local Indians used garlands of the leaves as health symbols during their ceremonies.

Plantago Seed, Blond – Plantago ovata.

Plantago Seed, Indian – Plantago ovata.

Plantain – Musa paradisiaca.

Plantain – Plantago major.

Plantain, Buckthorn – Plantago coronopus.

Plantain, Crowfoot – Plantago coronopus.

Plantain, Downy Rattlesnake – Goodyera pubescens.

Plantain, Dwarf – Plantago virginica.

Plantain, Sea – Plantago maritima.

Plantain, Seaside – Plantago decipiens.

Plantain, Seaside – Plantago maritima.

Plantain Seed – Plantago psyllium.
Plantain, Water – Alisma plantago.
PLATANTHERA Rich. Orchidaceae. 200 spp. Temperate and tropical Eurasia, N.Africa, N. to Central America. Herbs.
***P. bifolia*** (L.) Reichb. Europe to Asia Minor, N.Africa. See Orchis latifolia.
PLATANUS L. Platanaceae. 1 spp. S.E.Europe, Near East; 1 spp. Indochina; 1 spp. E. N.America, 7 spp. S.W. N.America, Mexico. Trees.
***P. mexicana*** Moric. (Mexican Sycamore). The wood is used for general carpentry and carving.
***P. occidentalis*** L. (Buttonwood, American Sycamore). E. N.America. The wood is tough and hard and is used for house interiors, furniture, crates etc. The hybrid P. occidentalis × P. orientalis (London Plane) is frequently planted as an ornamental tree in the streets of towns in U.K.
***P. orientalis*** L. (Buna, Doolb, Oriental Plane, Oriental Sycomore). S.E. Europe, Near East. The wood is heavy, difficult to work, but not durable. It is used for pulping and inlay work.
***Platea corniculata*** Becc. = Urandra corniculata.
PLATHYMERIA Benth. Leguminosae. 3 spp. Brazil, Argentine. Trees.
***P. reticulata*** Benth. (Vinhatico). Brazil. The yellow-orange wood is durable and easy to work. It is locally valued for furniture, flooring, interior work, ship-building.
PLATONIA Mart. Guttiferae. 1–2 spp. Brazil. Trees.
***P. grandiflora*** Planch. and Triana. (Pahoorie, Ubacury). Brazil, Guyana. The orange-brown wood is hard and coarse-grained. It is used for general construction work, crates etc., ship building and cooperage.
***P. insignis*** Mart. (Bacury). Brazil, Guyanas. The yellow-brown wood is fairly heavy and durable. It is used for general construction work, carving and flooring. The fruits are used in preserves and pastries, or eaten raw. They are sold locally. A non-drying oil (Caury Kernel Oil) is extracted from the seeds and used to make candles and soap.
PLATYCARYA Sieb. and Zucc. Juglandaceae. 2 spp. E.Asia. Trees.
***P. strobilacea*** Sieb. and Zucc. China, Japan. A black dye from the bark is used in Japan to dye fishing nets and a similar dye from the fruits is used in China to dye cotton.
PLATYCODON A. DC. Campanulaceae. 1 sp. N.E.Asia. Herb.
***P. grandiflorum*** DC. (Chinese Bellflower, Japanese Bellflower). The plant is cultivated, especially in China where the leaves (Chieh-keng) are used to cure stomach chills.
PLATYLOPHUS D. Don. Cunoniaceae. 1 sp. S.Africa. Tree.
***P. trifoliatus*** D. Don. (White Alder). The durable white wood is used locally for making furniture, boxes, wagons etc.
PLATYMISCIUM Vog. Leguminosae. 30 spp. Mexico to tropical S.America. Trees.
***P. dimosphandrum*** D. Don. (Yama Cocobola, Yama Rosewood). Central America. The heavy red-brown wood is used for making small articles, e.g. knife handles, billiard cue handles etc.
***P. duckei*** Dug. Nicaragua. See P. dimosphandrum.
***P. paraense*** Huber. = P. ulei.
***P. pinnatum*** (Jacq.) Dug. = P. polystachum. (Quira Macawood). Tropical America, especially the Caribbean area. The red-purple wood is decorative and strong. It is used locally for heavy construction work, furniture and railway sleepers. It is exported to U.S.A. where it is used for carving brush handles, tool handles etc.
***P. polystachum*** Benth. = P. pinnatum.
***P. ulei*** Harms. = P. paraense. (Macacauba). Lower Amazon. The wood is used locally for furniture and general construction work.
PLECTANEIA Thou. Apocynaceae. 14 spp. Madagascar. Shrubs.
***P. microphylla*** Jum. and Perr. (Mahavoahavana). Madagascar. Yields a latex, sometimes mixed with that from Landolphia.
PLECTOCOMIA Mart. and Blume. Palmae. 1 spp. Assam to S.China and W.Malaysia. Climbing Palm.
***P. griffithii*** Becc. The stems are used for basketry and chair-making.
PLECTRANTHUS L'Hérit. Labiatae. 250 spp. Tropical Africa to Japan, Malaysia, Australia, Pacific. The species mentioned are cultivated locally for the root tubers (Kaffir Potato, Umbondive) which are eaten as a vegetable.
***P. esculentus*** N. E. Br. Natal.
***P. floribundus*** N. E. Br. Natal, Rhodesia.

***P. rotundifolius*** Spreng.=Coleus rotundifolius.

***P. tuberosus*** Blume.=Coleus rotundifolius.

***P. tuberosus*** Roxb.=Orthosiphon rubicundus.

PLECTRIDIUM Stoerer. Bacteriaceae.

***P. pectinovorum*** Stoerer. An anaerobic bacterium which hydrolyses pectin. It is useful in the retting of hemp and flax.

***Plectronia*** L.=Canthium.

PLEIOGYNIUM Engl. Anacardiaceae. 2 spp. Philippines, Lesser Sundra Islands, New Guinea, Queensland. Trees.

***P. solandri*** Engl. (Burdekin Plum). Queensland. The seeds are pleasantly flavoured and are eaten locally.

PLEIONE D. Don. Orchidaceae. 10 spp. India to Formosa and Thailand. Herbs.

***P. pogonioides*** Rolfe. China. The pseudobulbs (Ch'uan peimu) are used locally in a cure of asthma and tuberculosis.

***Pleomele aurea*** (H. Mann.) N. E. Brown= Dracaena aurea.

PLEOPELTIS Humb. and Bonpl. Polypodiaceae. 40 spp. Throughout the tropics, few only in Malaysia and Polynesia. Ferns.

***P. longissima*** Moore. Malaysia. The shoots are used locally as a vegetable or eaten raw.

Pleurisy Root – Asclepias tuberosa.

PLEUROSPA Rafin. Araceae. 2 spp. Tropical America, W.Indies. Large herbs.

***P. arborescens*** Schott. (Aninga Uva). Tropical America, W.Indies, Central America. Some Indian tribes use the juice as a cure for boils.

***P. linifera*** (Arruda) Schott. (Anenga, Aninga Tree). Brazil. The juice is used by some of the natives to clean rusty utensils and weapons.

PLEUROTUS (Fr.) Quél. Agaricaceae. 100 spp. Cosmopolitan. The fruitbodies of the following species are edible.

***P. anas*** Van Overeem. Indonesia.

***P. eryngii*** (DC.) Quél. Eaten in Italy. Called Cardarellas.

***P. fissilis*** (Lev.) Sacc. Indonesia.

***P. flabellatus*** Berk.=Lentinus djamor.

***P. griseus*** Quél. Contains the antibiotic pleurotin which seems active against Gram positive bacteria.

***P. neapolitanus*** (Pers.) Sing.=Agaricus coffeae. Cultivated on coffee grounds in Italy.

***P. opuntiae*** (Dur. and Lev.) Sacc.=P. ostreatus.

***P. ostreatus*** (Jacq.) Quél.=P. opuntiae=P. yuccae=Agaricus ostreatus. (Mock Oyster). Popular in China, where it is imported from Russia. It is also cultivated on beech in Germany.

***P. ulmarius*** Bull.=Clitocybe tessulata.

***P. yuccae*** Mer.=Pleurotus ostreatus.

PLUCHEA Cass. Compositae. 50 spp. Tropics and subtropics. Shrubs.

***P. indica*** Less. Tropical Asia, Australia. The natives eat the leaves as a vegetable and use them to make a tea.

***P. odorata*** Nat. (L.) Cass.=Conyza odorata (Alinanche, Canelón, Chaché, Hierba de Santa Maria, Srup Bush). Florida to S.America, W.Indies. In Mexico an alcoholic infusion of the leaves is used as an embrocation for rheumatism and neuralgia.

***P. sericea*** (Nutt.) Cov. (Marsh Fleabane). S.W. U.S.A., Mexico. The flowers are a good source of honey. The Indians use the straight stems for arrows.

PLUKENETIA L. Euphorbiaceae. 10 spp. Warm America; 1 sp. Madagascar. Vines.

***P. conophora*** Muell. Arg Tropical Africa. The seeds yield a quick-drying oil (N'Ghat Oil) used locally for cooking. The plant is cultivated locally.

***P. corniculata*** Smith.=Hedriostylus corniculatus=Pterococcus corniculatus. (Paina-paina). S E.Asia. The sweet leaves are eaten locally as a vegetable.

***P. peruviana*** Muell. Arg.=P. volubilis.

***P. volubilis*** L.=P. peruviana. Brazil, W.Indies. The leaves are eaten as a vegetable locally and are fed to livestock in the W.Indies.

Plum – Prunus domestica.

Plum, American Wild – Prunus americana.

Plum, Apricot – Prunus simonii.

Plum, Beach – Prunus maritima.

Plum, Botoko – Flacourtia ramontchi.

Plum, Bullace – Prunus institia.

Plum, Burdekin – Pleiogynium soldri.

Plum, Cherry – Prunus cerasifera.

Plum, Chickasaw – Prunus angustifolia.

Plum, Chinese – Prunus salicina.

Plum, Chinese Date – Diospyros kaki.

Plum, Clamath – Prunus subcordata.

Plum, Coco – Chrysobalanus icaco.

Plum, Damson – Prunus instititia.

Plum, Darling – Reynosia septentrionalis.

Plum, Date – Diospyros lotus.

Plum, Downward – Bumelia spiniflora.

Plum, Fladwoods – Prunus umbellata.

Plum, Gingerbread – Parinari macrophyllum.
Plum, Governor's – Flacourtia ramontchi.
Plum, Greengage – Prunus instititia.
Plum, Hog – Prunus umbellata.
Plum, Hog – Spondias lutea.
Plum, Hog – Spondias tuberosa.
Plum, Hog – Symphonia globulifera.
Plum, Hortulana – Prunus hortulana.
Plum, Icaco – Chrysobalanus icaco.
Plum, Japanese – Eriobotrya japonica.
Plum, Japanese – Prunus salicina.
Plum, Java – Euglenia jambolana.
Plum, Jew's – Spondias dulcis.
Plum, Kaffir – Harpephyllum caffrum.
Plum, Kaffir – Lannea caffra.
Plum, Klamath – Prunus subcordata.
Plum, Madagascar – Flacourtia ramontchi.
Plum, Marmalade – Calocarpum mammosum.
Plum, Martinique – Flacourtia inermis.
Plum, Mirabella – Prunus instititia.
Plum, Mobola – Parinarium mobola.
Plum, Moxie – Chiogenes hispidula.
Plum, Natal – Carissa grandiflora.
Plum, Oklahoma – Prunus gracilis.
Plum, Pacific – Prunus subcordata.
Plum, Pigeon – Coccolobis laurifolia.
Plum, Reine Claude – Prunus instititia.
Plum, River – Prunus americana.
Plum, Runeala – Flacourtia cataphracta.
Plum, Saffron – Bumelia spiniflora.
Plum, Spanish – Spondias purpurea.
Plum, Tamarind – Dialium indum.
Plum, Wild Goose – Prunus munsoniana.

PLUMBAGO L. Plumbaginaceae. 12 spp. Warm. Woody Herbs. The juice from the following species causes blistering of the skin.

***P. angustifolia*** Spach. = P. europaea.

***P. auriculata*** Blume = P. zeylanica.

***P. europaea*** L. = P. angustifolia. (Plumbago). The root (Radix Dentariae) is used for treating toothache and swellings.

***P. rosea*** L. Middle East, India. Beggars use the juice to produce blisters on the body to solicit sympathy.

***P. scandens*** L. Tropical America. See P. rosea. The root is also used to treat toothache.

***P. zeylanica*** L. = P. auriculata = Thelá alba. India, Malaysia. A paste of the roots and leaves is used throughout the Far East to treat skin complaints. A similar mixture, with the mucilage from Hibiscus esculentus is used in Nigeria to treat leprosy. In India and Japan, the juice is considered an abortive.

Plumbago – Plumbago europaea.
Plumegrass, Longhair – Dichelachne crinata.
Plumegrass, Shorthair – Dichelachne sciurea.

PLUMERIA L. Apocynaceae. 7 spp. Warm America. Trees.

***P. acuminata*** Roxb. = P. acutifolia.

***P. acutifolia*** Poir. = P. acuminata = P. obtusa. Mexico, S.America. The plant is cultivated in the Old and New World tropics for the bark which is considered a good cure for gonorrhoea. It is also a purgative. The leaves are used locally as a poultice for bruises and the latex as an embrocation for rheumatism. The latex is a potential source of rubber.

***P. lanicfolia*** Muell. Arg. Brazil. The bark is used medicinally in the treatment of asthma and syphilis. It is also used to control menstruation and as a purgative. The bark is exported to Europe and U.S.A.

***P. obtusa*** Lour. = P. acutifolia.

PLUTEUS Fr. Agaricaceae. 75 spp. Widespread.

***P. cervinus*** Quél. = Hyporhodius cervinus. Widespread. The fruitbodies are edible.

POA L. Graminae. 300 spp. Cosmopolitan. Grasses.

***P. abyssinica*** Juss. = Eragrostis abyssinica. Abyssinia, E. Africa. Locally the seeds are eaten as grain. The straw is mixed with mud to make bricks; it is also fed to livestock.

***P. alpina*** L. Europe to Caucasus. A good mountain pasture grass.

***P. arachnifera*** Torr. (Texas Bluegrass). W. U.S.A. Used for winter pasture.

***P. compressa*** L. (Canada Bluegrass). Europe, N.America. Used as pasture in the Old and New World.

***P. cynosuroides*** Retz. = Eragrostis cynosuroides.

***P. epilis*** Scribn. (Skyline Bluegrass). W. U.S.A. A good mountain pasture.

***P. fendleriana*** (Steud.) Vasey. (Mutton Grass). W. U.S.A. to Mexico. A good mountain pasture.

***P. fertilis*** Host. = P. palustris.

***P. microstachya*** Link. = Eragrostis lugens.

***P. nemoralis*** L. = P. rigidula. (Wood Bluegrass). Europe to temperate Asia, N.America. Used for pasture and hay.

***P. palustris*** L.=P. fertilis=P. serotina. (Fowl Bluegrass). Europe, temperate Asia, N.America. It is used for pasture and hay.

***P. pratensis*** L. (Kentucky Bluegrass). S.E.Europe. Grown for pasture and hay. The grass had spread throughout Europe by the Middle Ages and was introduced to America as a pasture grass by the early colonists. It is much-used in America and is one of the best grasses to establish and protect the soil. Various cultivars are the most popular lawn grasses in America.

***P. rigidula*** Koch.=P. nemoralis.

***P. rupicola*** Nash. (Timberline Bluegrass). W. N.America. A good pasture for mountainous areas and poorer soils.

***P. scabrella*** (Thurb.) Benth. (Pine Bluegrass). W. U.S.A. to Mexico. An important fodder grass in the lower areas.

***P. serotina*** Ehrh.=P. palustris.

Poaya Blancha – Polygala angulata

Pochote – Ceiba aesculifolia.

Pó da Bahia – Vataireopsis speciosa.

Pó da Goa – Vatairieopsis speciosa.

Pod Mahogany – Afzelia cuanzensis.

PODACHAENIUM Benth. ex Oerst. Compositae. 2 spp. Mexico, Colombia. Shrubs.

***P. eminens*** (Lag.) Schultz.=P. paniculatum =Ferdinanda eminens. (Tacote, Tora). Mexico to Costa Rica. The leaves are used locally as a dressing for wounds.

***P. paniculatum*** Benth.=P. eminens.

Podanganari – Premna tormentosa.

PODAXIS Desv. Hymenogasterales. 1 sp. Tropics and subtropics.

***P. pistillaris*** (L. ex Pers.) Fr. (Soozan Katamoogoo). The fruitbodies are eaten in Africa and used to treat tumours. The spores are used by the Hottentots as face powder and to treat burns.

Podgrass, Shore – Triglochin maritima.

PODOCARPUS L'Hérit. ex pers. Podocarpaceae. 100 spp. Tropical to temperate S.Hemisphere, N. to Himalayas and Japan. Trees.

***P. agathifolia*** Blume.=P. blumei.

***P. amara*** Blume. Java. The wood is used locally for posts and house construction.

***P. blumei*** Endl.=P. wallichinanus=P. agathifolia. (Kim giao, Thong mu). Malaysia, Philippines. The beautifully grained wood is used for interior work.

***P. cupressinus*** R. Br. Mirb.=P. imbricatus.

***P. dacrydioides*** Rich. (Kahika). New Zealand. The light-coloured wood is used for house construction and interiors, furniture, boxes and boat-building. It is also used for paper pulp. The fruits are eaten locally.

***P. elata*** R. Br. (Goongum, She Pine, White Pine). Queensland, New South Wales. The light, close-grained wood is used for furniture making.

***P. elongata*** L'Hérit. (Quinteniqua Yellow Wood). S.Africa. The light brown, soft, durable wood is used for house-building, railway sleepers and furniture.

***P. ferrugineus*** Don. (Miro). New Zealand. The hard, tough wood is used for house-building and furniture.

***P hallii*** Kirk. New Zealand. The wood is resistant to teredo worms and is used for piers, wharves etc. and ship-building.

***P horsfieldii*** Wall.=P. imbricatus.

***P. imbricatus*** Blume=P. cupressinus=P. horsfieldii. (Bach tung, Thong-nang). Malaysia. The beautifully grained wood is used for interior work.

***P. madagascariensis*** Baker. Madagascar. The wood is used locally for general carpentry and house-building.

***P. montana*** (Willd.) Lodd.=P. taxifolia.

***P. nerifolia*** D. Don. New Guinea, through the Himalayas to China. The light, yellowish hard wood is used for general carpentry in Burma. The fruits are eaten locally in Nepal.

***P. oleifolius*** D. Don. Bolivia to Costa Rica. The yellowish wood is used locally for carving and general carpentry.

***P. rumphii*** Blume. Malaysia. The light yellow wood is easy to work and is used locally for houses, boats and turning.

***P. spicata*** R. Br. (Black Pine). New Zealand. An important commercial timber. It is used for house-building, bridges, ballroom floors and railway sleepers. The bark is used for tanning.

***P. spinulosa*** R. Br. Queensland. The fruits are used locally in jams and preserves.

***P. taxifolia*** H.B.K.=P. montana. S.America. The wood is used locally, especially in Colombia for furniture.

***P. thunbergii*** Hook. (Yellow Wood). The bright yellow wood is used for furniture, general building and coaches and wagons.

***P. totara*** Cunningh. (Totara). New Zealand. The deep-red wood is durable and teredo-resistant. It is used for marine work, building houses and bridges and telegraph poles.

***P. wallichinanus*** Ridl. non Presl. = P. blumei.

PODOPHYLLUM L. Podophyllaceae. 10 spp. Himalayas, E.Asia; 1 sp. E. N.America. Herbs. Both the species mentioned contain the drug podophyllin in the rootstock. This is used medicinally as a violent purgative and skin irritant. The fruits are edible.

***P. emodi*** Wall. (Himalayan Mayapple). Used medicinally by the Hindus since ancient times. The drug is sometimes extracted commercially.

***P. peltatum*** L. (Ground Lemon, Mandrake, Mayapple). E. N.America. The rootstock is used medicinally and was known to the Indians.

PODOSTEMUM Michx. Podostemaceae. 17 spp. Central and tropical S.America, tropical Africa, Asia. Aquatic herbs.

***P. minutiflorus*** Benth. Madagascar. The leaves are eaten locally in salads.

Poet's Narcissus – Narcissus poeticus.

POGA Pierre. Anisophylleaceae. 1 sp. W.Equatorial Africa. Tree.

***P. oleosa*** Pierre. (Ovoga). The seeds yield an edible oil (Oil of Inoy Kernel) and are used locally as a condiment.

POGOGYNE Benth. Labiatae. 5 spp. California, S.Oregon. Herbs.

***P. parviflora*** Benth. California. The leaves are used as a condiment by the local Indians.

POGOSTEMON Desf. Labiatae. 40 spp. China, Indomalaysia. Herbs.

***P. cablin*** Benth. = P. patchouli.

***P. heyneanus*** Benth. See P. patchouli.

***P. patchouli*** pellet. = P. cablin. (Patchouly). Philippines. The plant is cultivated in the Seychelles, Malaya, Sumatra, parts of Africa, Madagascar and Brazil for the oil (Oil of Patchouly) which is steam-distilled from the leaves. The young leaves yield the most oil (3–5 per cent of the dry weight of the leaves). The young leaves are harvested about once every six months and left overnight to slightly ferment before they are sun-dried and packed for shipment. Most of the oil is produced in the importing countries, e.g. U.S.A. and U.K. The world production is about 100 tons of oil a year. The plants are propagated from cuttings and continue to yield for about three years. The oil consists mainly of sesquiterpenes and has a cedar-like odour, often associated with fabrics imported from the East, where the plant is used as an insect repellent in cloth. It is one of the best fixatives for "heavy" perfumes and is widely used in perfumery and soap-manufacture.

Pohon sago – Metroxylon rumphii.

***Poiciana*** L. – Caesalpinia.

Pointleaf Manzanita – Arctostaphylos pungens.

Poinxter Flower – Rhododendron nudiflorum.

Pois de Oasis – Pisum elatius.

Poison Bulb – Urginea macrocentra.

Poison Hemlock – Conium maculatum.

Poison Ivy – Rhus toxicodendron.

Poison Nut – Strychnos castelnaei.

Poison Nut, Curare – Strychnos toxifera.

Poison Nut, St. Ignatius – Strychnos ignatii.

Poison Wood – Metopium toxiferum.

Poivre du kissi – Piper guineense.

Poivre lolet – Piper lolot.

Poke, Indian – Veratrum viride.

Pokea – Portulaca lutea.

Pokeberry, Common – Phytolacca decandra.

Pokeberry, Venezuela – Phytolacca rivinoides.

POLAKOWSKIA Pittier. Cucurbitaceae. 1 sp. Costa Rica. Vine.

***P. tacaco*** Pittier. (Tacaco). The fruits are eaten locally as a vegetable. The plant is cultivated.

Polamaria – Callophylum tacamahaca.

***Polanisia icosandra*** Wigh. and Arn. = Cleome viscosa.

Polar Plant – Silphium laciniatum.

POLEMONIUM L. Polemoniaceae. 50 spp. Temperate Eurasia, N.America, Mexico; 2 spp. Chile. Herbs.

***P. coeruleum*** L. (Greek Valerian, Jacob's Ladder). Temperate Europe, Asia, N.America. The leaves are used as an astringent and to induce sweating.

***P. reptans*** L. (Abscess Root). N.America, especially E.coast states of U.S.A. The rhizome is expectorant and induces sweating. A decoction of the rhizome is used in treating all forms of chest and lung diseases.

Poli (Oil) – Carthamus oxycanthus.

POLIANTHES L. Agavaceae. 13 spp. Mexico to Trinidad. Herbs.

***P. tuberosa*** L. (Tuberose). Mexico. The

plant is grown commercially in China and Java for the flowers from which Tuberose Flower Oil is extracted by solvent extraction. This is used in high-grade perfumery. The plant is also grown as an ornamental.

POLIOMINTHA A. Gray. Labiatae. 4 spp. S.W.United States of America, Mexico. Shrubs.

***P. incana*** (Torr.) Gray.=Hedeoma incana. S.W.U.S.A. The leaves are eaten as a vegetable by the local Indians, who use the flowers as a condiment.

Polish Wheat – Triticum polonicum.

***Pollindium binatum*** (Retz.) C. E. Hubbard= Eulaliopsis binata.

POLYADOA Stapf. Apocynaceae. 4 spp. Tropical Africa. Trees.

***P. elliottii*** Hook. Sierra Leone. The wood is used locally for making combs.

POLYALTHIA Blume. Annonaceae. 120 spp. Old World tropics, especially Indomalaya. Trees.

***P. aberans*** Maingay=Papowia aberrans.

***P. mortahani*** De Wild.=P. suaveolens.

***P. oliveri*** Engl. W.Africa. In Liberia the bark is used to combat intestinal worms and the stems are used as spear shafts in the Cameroons.

***P. siamensis*** Boerl.=Popowia aberrans.

***P. suaveolens*** Engl. and Diels=P. mortahani. S.Nigeria, Gabon. The hard wood is used locally for building houses.

***P. suberosa*** (Roxb.) Thw.=Uvaria suberosa =Phaeanthus cumingii. India, S.China, Malaysia, Philippines. In the Philippines, an extract of the root is used to induce abortions.

***P. thorelii*** Fin. and Gagnep. (Ngân chày). A decoction of the bark is used locally to treat stomach complaints and the wood for making charcoal, used in the manufacture of fireworks.

POLYCARPAEA Lam. Caryophyllaceae. 50 spp. Tropics and subtropics. Herbs.

***P. corymbosa*** Lam. (Pail l'ou weng). India, Indochina, China. The leaves and flowers are used in India, as a poultice and decoction against snakebite. In China they are used as an emollient and astringent.

***P. frankenioides*** Presl.=Mollugo oppositifolia.

POLYGALA L. Polygalaceae. 500–600 spp. Cosmopolitan, except New Zealand, Polynesia and Arctic. Herbs.

***P. amara*** L. Europe. The dried leaves (Herba Polygalae) are used in local (especially Germany) treatment of asthma and stomach pains.

***P. anatolica*** Boiss.=P. major.

***P. angulata*** DC. Brazil. Yields White Ipecac or Poaya Blanca, used as a substitute for Ipecacuanha.

***P. bakeriana*** Chod. (Tangaluhira). Tropical Africa. The leaves are used by the natives in remedies against intestinal worms.

***P. butyracea*** Heck. Tropical Africa. Cultivated very locally for the stem fibres which are used to make fabric and cordage. The parched ground seeds are eaten in soups.

***P. costaricensis*** Chodat. Central America. The plant is used as a substitute for Ipecacuanha.

***P. glomerata*** Lour. (Andong petooroon, Djookoot malela, Lidah a jam). Indonesia, Malaysia. A paste from the roots is used locally to treat asthma, fatigue and pains in the stomach from eating too much pineapple.

***P. javana*** DC.=P. tinctoria. E.India, Malaysia. The leaves yield a blue dye used locally for staining fabrics.

***P. major*** Jacq.=P. anatolica. Used as a substitute for P. senega.

***P. paniculata*** L.=P. variabilis (Djookoot rindik, Sasapooan, Totombe). Brazil. The plant has escaped to Java where the roots are used to scent linen and a decoction of the plant is used to treat gonorrhoea.

***P. pruinosa*** Boiss. Far East. Used as a substitute for P. senega.

***P. rugelii*** Shuttlw. (Yellow Bachelor's Button). Florida. A decoction of the leaves was used by the local Indians against snakebite.

***P. ruwenzoriensis*** Chod. (Lovishohe, Shere). Tropical Africa. The twigs are used locally to make arrows.

***P. senega*** L. (Seneca, Snake Root). E. N.America. The dried roots contain the glucoside senegin and are used medicinally as an expectorant stimulant and irritant. They were used by the local Indians to treat snakebite.

***P. serpentaria*** E. and Z. (Kafferslangen Wortel). S.Africa. The Zulus use the powdered root as a purgative for children and when mixed with milk as a mild enema.

***P. siberica*** L. Tropics and subtropics around Himalayas. The roots are used in Indochina as a diuretic and to treat amnesia, sexual impotency and bronchitis. They are used in China and Japan as a substitute for P. senega.

***P. theezans*** L. Java to Japan. The leaves are used locally to make a tea (Thé des carolines).

***P. tinctoria*** Vahl.=P. javana.

***P. ukirensis*** Gürke=P. usafuensis. Tropical Africa. The stem fibres are used locally to make light cordage.

***P. usafuensis*** Auct. non Gürke=P. ukirensis.

***P. usafuensis*** Gürke. See P. ukirensis.

***P. variabilis*** Hassk.=P. paniculata.

POLYGONATUM L. Polygonaceae. s. str. 50 spp. Cosmopolitan. Herbs.

***P. biflorum*** (Walt.) Ell. (Small Solomon's Seal). E. U.S.A. to Florida and Texas. The roots were eaten as a vegetable by the local Indians.

***P. canaliculatum*** Pursh.=P. giganteum.

***P. giganteum*** A. Dietr.=P. canaliculatum. Temperate Asia, W. N.America. The rhizomes are eaten in Japan as a vegetable, or preserved in syrup. They are also used as a source of starch.

***P. multiflorum*** Allem. (Solomon's Seal). Europe, temperate Asia, Japan, N.America. The rootstocks are demulcent and astringent. They are used in the treatment of lung complaints, including bleeding and as a poultice for bruises etc. and haemorrhoids.

***P. officinale*** Moench. (Drug Solomon's Seal). The young shoots can be eaten as a vegetable.

POLYGONUM L. Polygonaceae. s. lat. 300 spp., s. str. 50 spp. Cosmopolitan, especially temperate. Herbs.

***P. aviculare*** L. (Prostrate Knotweed). Temperate. The leaves are used to make an infusion (Hemero Tea) used in Europe for the treatment of lung complaints and asthma. They are also used in treating haemorrhoids and rheumatism.

***P. barbatum*** L. Tropics. The ground leaves are used to prevent bleeding from wounds.

***P. bistortam*** L. (Bistort, English Serpentary, Snake Root). Europe, Asia, N.America. The rhizomes are eaten by the Indians and Eskimos of N.America. They are used medicinally as a tonic.

***P. cuspidatum*** Sieb. and Zucc. (Japanese Fleeceflower). China, Japan. Cultivated locally for the root bark which yields a yellow dye and for the roots which are used medicinally as an emollient.

***P. douglasii*** Greene. (Douglas Knotweed). W. N.America. The seeds were used by the local Indians to make flour.

***P. fugax*** Small. Arctic. The roots are eaten by the Eskimos.

***P. maritimum*** L. N.Temperate. The leaves are used as a dressing for burns.

***P. maximowiczii*** Regel. Japan. Cultivated locally for the leaves which are eaten as a vegetable.

***P. muehlenbergi*** (Meisn.) S. Wat. N.America. The young shoots were eaten as a vegetable by the local Indians.

***P. ochreatum*** Houtt.=P. tomentosum.

***P. odoratum*** Lour. (Chi kras-sang tomhom, Rau râm). Indochina. The plant is cultivated locally for use as a condiment. The leaves are also used as a diuretic, to treat sickness and to reduce fevers.

***P. pulchrum*** Blume. Europe, Old and New World tropics. Used in Africa as a source of salt and the leaves are sometimes used as a vegetable. The leaves are eaten raw in Indochina.

***P. sachalinense*** F. Schmidt. (Sachalin Knotweed). Sachalin. The young shoots are eaten locally as a vegetable.

***P. salicifolium*** Brouss. ex Willd. (Mbwetomasa, Neninga). See P. pulchrum.

***P. senegalense*** Meisn. See P. pulchrum.

***P. strigosum*** R. Br. Tropical Africa, Asia, Australia. Used in Africa to manufacture salt.

***P. tomentosum*** Willd.=P. ochreatum. India and Far East. The leaves are eaten raw and cooked in Indochina and are used in Malaya as a tonic.

***P. vivparum*** L. Arctic. The roots are eaten as a vegetable by the Eskimos.

***P. weyrichii*** F. Schmidt. Japan, Sacalin. The seeds are eaten locally as a porridge or as a flour.

POLYMNIA L. Compositae. 20 spp. Warm America. Herbs.

***P. edulis*** Wedd. (Yacon Strawberry). S.America, especially Peru. The roots are eaten as a vegetable in Colombia and are fermented to make alcohol from the stored inulin. The whole plant is used as livestock feed.

***P. uvedalia*** L. (Yellow Leafcup). E. N.America. A decoction of the roots is used by the local Indians as a laxative, stimulant and to relieve pain.

Polynesian Chestnut – Inocarpus edulis.

Polynesian Ivorynut Palm – Coelococcus amicarum.

POLYPODIUM L. 75 spp. Cosmopolitan. Ferns.

***P. angustifolium*** Sw. S.Mexico to Peru. The rhizomes are used in Mexico to reduce fevers and treat chest complaint and in Peru to treat fevers and dropsy.

***P. aureum*** L.=Phlebodium aureum. Tropical and subtropical America. In Mexico the rhizomes (Calaguala) are used to treat coughs and fevers.

***P. furfuraceum*** Schlecht. Tropical America. The rhizome is used in El Salvador to reduce pain.

***P. lanceolatum*** L. var. ***trichophorum*** Weatherb. Mexico to Guatemala. The rhizome is used locally in the treatment of coughs and to reduce fevers.

***P. plebejum*** Schlecht. and Cham. S.Mexico to S.America. A decoction of the rhizome is used in Mexico to treat coughs and as a purgative.

***P. vulgare*** L. (Polypody Root, Rockbrake, Brake Root). Europe, N.America. A decoction of the rhizomes or leaves is used to treat coughs, to improve the appetite, to relieve dyspepsia and to treat skin diseases. A resin from the rhizome combats intestinal worms. The rhizome is also used to flavour liqueurs.

Polypody Root – Polypodium vulgare.

POLYPORUS. Mich ex Fr. Polyporaceae. 250 spp. Cosmopolitan.

***P. alveolaris*** (DC.) Bong. and Sing.= Hexagona mori. Asia, America, Europe. Used, particularly in Italy to make a brown dye used for staining fabrics.

***P. anthelminticus*** Berk. India, Burma. Parasitic on bamboo. The fruitbodies are used locally to treat intestinal worms in cattle.

***P. arcularius*** (Batsch.) Fr. Tropics. The fruitbodies are eaten in Malaya.

***P. australiensis*** Wakefield. Australia. A brown dye from the fruitbodies is used in Flinders Island for dyeing raffia.

***P. betulinus*** Bull.=Piptoporus betulinus. Temperate. On birch. The fruitbodies are used as razorstrops and the wood attacked by the fungus is used to burnish watches in Switzerland.

***P. cristata*** Pers. Temperate. The fruitbodies are eaten.

***P. farlowii*** Lloyd. S.W. U.S.A. The fruitbodies are eaten cooked by the local Indians.

***P. frondosus*** Dicks. Temperate. The fruitbodies are eaten.

***P. grammocephalus*** Berk. Tropical Asia. The young fruitbodies are eaten locally.

***P. hispidus*** Bull.=Inonotus hispidus. Temperate. The fruitbodies yield a brown dye used in parts of Europe and Asia to dye fabrics.

***P. nidulans*** Fr.=Haplopilus nidulans. Temperate. The fruitbodies are used in Russia and Scandinavia to make bottle stoppers.

***P. officinalis*** (Vill.) Fr.=Boletus laricis. Temperate. The fruitbodies (Fungus Chirurgorum) were used in surgery to absorb blood etc. during operations. They are sometimes used as a purgative.

***P. ovinus*** Schäff. Temperate. The fruitbodies are eaten.

***P. pes captae*** Pers. Temperate. The fruitbodies are eaten.

***P. ramosissimus*** Schäff. Temperate. The fruitbodies are eaten.

***P. sanguineus*** (L.) Meyer.=Pycnoporus sanguineus. Tropics. In the East Indies the fruitbodies are used to treat intestinal complaints, venereal diseases and eczema.

***P. seriales*** Fr.=Trametes seriales.

***P. squamosus*** (Huds.) Fr. Temperate. The young fruitbodies are eaten and the old ones used to make razor strops.

***P. sulfureus*** Bull. Cosmopolitan. The fruitbodies are eaten and used to make a brown dye.

***P. tinctorius*** Quél. Tropics and subtropics, parasitic on Pistacia atlantica. The fruitbodies (Serra) are used to make ink and a dye.

***P. tuckahoe*** (Gussow) Lloyd. N.W. N.America. The Indians eat the sclerotia (Tuckahoe).

***P. tuneanus*** (Pat.) Sacc. and Trott. Mediterranean. The fruitbodies are eaten.

***P. undus*** Jungh. Tropics. The young fruitbodies are eaten in Java.

***P. versipellis*** Fr.=Leccinum aurantiacum. Temperate. The fruitbodies are eaten in Europe. They are used instead of truffles in Germany, and pickled (Krassny Grib) in Russia.

***P. vibecinus*** Fr. Pacific. The fruitbodies are eaten locally.

***P. volvatus*** Peck. N.America, E.Asia. The fruitbodies are eaten raw in Tibet.

POLYSCIAS J. R. and G. Forst. Araliaceae. fruitbodies are eaten raw in Tibet.

POLYSCIAS J. R. and G. Forst. Araliaceae. 80 spp. Old World Tropics. Trees.

***P. fruticosa*** Harms. Java. The plant is cultivated locally for the aromatic leaves and roots which are used as a condiment with fish, meat and soups. They are diuretic.

***P. nodosa*** Seem. = Eupteron nodosus (Delag, Ki langil, Papaja ootan). Indonesia. The wood is used locally for making utensils and knife handles. The leaves are used to stupify fish.

***P. rumphiana*** Harms. Moluccas. The plant is cultivated and used locally as a vegetable.

POLYSTICTUS Fr. Polyporaceae. 100 spp. Cosmopolitan.

***P. sacer*** Fr. Old World Tropics. In parts of Africa and S.Asia the sclerotia are used to treat chest complaints, asthma and pulmonary tuberculosis.

POLYTOCA R. Br. Graminae. 6 spp. Indomalaya. Grasses.

***P. bracteata*** R. Br. Malaysia. The grass is fed to cattle.

***P. punctata*** Stapf. = Sclerachne punctata.

POLYTRICHUM Willd. Polytrichaceae. Moss.

***P. juniperinum*** Willd. (Bearsbed, Hair Cap Moss, Robins Rye). N.Hemisphere. A decoction of the plant is diuretic and is used to treat urinary complaints and dropsy.

POMADERRIS Labill. Rhamnaceae. 45 spp. Australia, New Zealand. Trees.

***P. elliptica*** Labill. Australia, New Zealand. In New Zealand, a decoction of the leaves is used to treat chest and kidney complaints indigestion, skin cancer and sores.

***P. zizyphoides*** Hook. and Arn. = Alphitonia zizyphoides.

Pomegranate – Puncia granatum.

Pomelo – Citrus grandis.

POMETIA J. R. and G. Forst. Sapindaceae. 10 spp. Indomalaya. Trees.

***P. eximea*** Bedd. = P. tomentosa. Java. The hard tough wood is used locally for building and heavy construction work.

***P. pinnata*** J. R. and G. Forst. Pacific. The red wood is hard, tough, but elastic. It is used for house-building, furniture, agricultural implements and cooperage. The fruits are eaten raw and the seeds are eaten roasted or boiled.

***P. tomentosa*** Teijsm. and Binn. = P. eximea.

Pomme Chique – Clusia plukenetii.

Pomolche – Jatropha gaumeri.

POMPHOLYX Corda. Sclerodermataceae. 2 spp. Europe, N.America.

***P. sapidum*** Corda. Central and E.Europe. The fruitbodies are eaten locally.

Pompon Lily – Lilium pomponicum.

Pompona Bova – Vanilla pompona.

PONCIRUS Rafin. Rutaceae. 1 spp. N.China. Tree.

***P. trifoliata*** Rafin. (Trifoliate Orange). Cold resistant. Crossed with Citrus sinensis gives the Citange, hardier than C. sinensis. The fruits are used for drinks and marmalades. Crossed with Citrus sinensis and Fortunella marganta gives the Citrangequat which is hardy, but of little commercial importance.

Pond Apple – Annone glabra

Pondlily, Yellow – Nuphar advena.

Pond Pine – Pinus serotina.

Ponderosa Pine – Pinus ponderosa.

Pong kapong – Voacanga foetida.

Pong tai hai – Scaphium affinis.

PONGAMIA Adans. mut Vent. Leguminosae. 1 spp. Indomalaya. Tree.

***P. glabra*** Vent. = P. pinnata.

***P. pinnata*** (L.) Merr. = P. glabra. The seeds yield an oil (Hongay Oil, Pongam Oil) which is used in Asia to treat skin diseases and for burning. It is also used to make candles and soap.

Pongam Oil – Pongamia pinnata.

PONTEDERIA L. Pontederiaceae. 4 spp. America. Aquatic Herbs.

***P. cordata*** L. (Pickerel Weed). Warm America. The fruits are edible and are sometimes used as an emergency food.

Pontianak Gutta – Dyera costulata.

Pontuik Copal – Agathis alba.

Pony Bee Balm – Monarda pectinata.

Poolai kapoor – Alstonia pneumatophora.

Poolai – Alstonia polyphylla.

Poole batoo – Alstonia acuminata.

Poolas laoot – Mischocarpus sundaicus.

Poon Spar Tree – Calophyllum tomentosum.

Poonac – Cocos nucifera.

Poonah – Tetramerista glabra.

Poondoong – Baccaurea javanica.

Poopoi – Sariotheca griffithii.

Pootah – Planchonia valida.

Pootat lemlik – Barringtonia spicata.
Popeal – Shorea harmandii.
Popindh – Ardisia crispa.
Popioe – Fagara chalybea.
Poplar, Arizona – Populus arizonica.
Poplar, Balsam – Populus balsamifera.
Poplar, Black – Populus heterophylla.
Poplar, Black – Populus nigra.
Poplar, Japanese – Populus maximowiczii.
Poplar, Native – Codonopsis cotinifolia.
Poplar, Pink – Euroschinus falcatus.
Poplar, Silver-leaf – Populus alba.
Poplar, White – Populus alba.
Poplar, Yellow – Liriodendron tulipfera.
Poplear thom – Grewia paniculata.
Popli – Osyris arborae.

POPOWIA Endl. Annonaceae. 90 spp. Tropical Africa to tropical Australia. Trees.

***P. aberrans*** Pierre = Polyalthia aberans = Polyathia siamensis. = Unona mesnyi. (Dok dâm duan, Rom doul). Cambodia, Laos, S.Vietnam. The flowers are mixed with beeswax and the oil from Sterculia lychnophra to make a scented lip paste.

***P. fornicata*** Baill. Coasts of Zanzibar. A decoction of leaves is used locally to treat snakebite.

Poppy, Californian – Eschscholtzia californica.
Poppy, Corn – Papaver rhoeas.
Poppy, Hawaiian – Argemone glauca.
Poppy, Horned – Glaucium flavum.
Poppy, Iceland – Papaver nudicaule.
Poppy, Mexican Prickly – Argemone mexicana.
Poppy, Opium – Papaver somniferum.
Poppy, Red – Papaver rhoeas.
Poppy Seed Oil – Papaver somniferum.

POPULUS L. Salicaceae. 35 spp. N.Temperate. Trees.

***P. alba*** L. (Abele, Silver-leaf Poplar, White Poplar). Europe through to Himalayas. The yellowish, coarse light wood is used for boxes, packing material, matches, shoes, railway carriage interiors and as a source of cellulose. A decoction of the bark is used medicinally as a tonic and to reduce fevers.

***P. arizonica*** Sarg. (Arizona Poplar). S.W. U.S.A., Mexico. The wood is used locally for carts, wheels and water troughs.

***P. balsamifera*** Moench. = P. tacamahaca (Balsam Poplar). N.America to Florida. The dried unopened buds (Balm of Gilead, Balsam Poplar Buds) contain a resin, gallic acid, malic acid, salicin, populin and an essential oil. They are used medicinally as a stimulant and expectorant. The light wood is used for making paper pulp, boxes and pails.

***P. ciliata*** Wall. (Bangikat, Falsch, Paluch). Himalayan Region. The Hindus use a decoction of the bark as a stimulant.

***P. euphratica*** Oliv. (Bahan, Hodung). Arabia to N.W.India. A decoction of the bark is used locally to combat intestinal worms.

***P. fremontii*** Wats. (Fremont's Cottonwood). S.W. U.S.A. The local Indians use the twigs to make baskets and the bark as an antiscorbutic.

***P. grandidentata*** Michx. (Great Aspen, Largetoothed Aspen). E. N.America. The soft wood is used for pulp, matches, veneer, carving etc. The inner bark is eaten as a vegetable by the local Indians.

***P. heterophylla*** L. (Black Cottonwood, Swamp Cottonwood). N.America to Southern States. The wood (Black Poplar) is used for house interiors locally.

***P. maximowiczii*** Henry. (Japanese Poplar). Pacific E.Asia, Japan and surrounding islands, Korea, Manchuria. The grey wood is used locally for matches, pulp and wooden utensils.

***P. nigra*** L. (Black Poplar). Europe, temperate Asia. The tree is cultivated for the coarse light wood which is used for matches, carving, small boxes, railway carriage interiors. The resin from the buds is used as a salve and to promote hair growth.

***P. tacamahaca*** Mill. = P. balsamifera.

***P. tremula*** L. (European Aspen). Throughout Europe, through to Japan, N.Africa. The soft wood is elastic but not durable out of doors. The wood is used for carving, shoes, boxes, interiors of railway carriages and the charcoal is used for gunpowder.

***P. tremuloides*** Michx. (Trembling Aspen). N.America to Mexico. The wood is used for lumber, matches, wood pulp and veneer. The inner bark is eaten as a vegetable by the local Indians and a decoction of the bark is a stimulant and diuretic used to treat debility, nerve conditions and urinary complaints.

***P. wislizeni*** (S. Wats.) Sarg. S.W. N.America. The flower heads were eaten by the local Indians.

PORCELIA Ruiz. and Pav. Annonaceae. 5 spp. Central and tropical S.America. Trees.

***P. saffordiana*** Rusby. N.Bolivia. The large fruits are edible and are much esteemed locally.

Porcupine Grass – Stipa vaseyi.

Porcupine Wood – Cocos nucifera.

PORIA Pers. ex S. F. Gray. Polyporaceae. 200 spp. Cosmopolitan.

***P. cocos*** Wolf. = Pachyma cocos. Temperate. The American Indians eat the sclerotia (Indian Bread, Tuckahoe). The fruitbodies are used by the Chinese to treat stomach ache.

***P. hoelen*** Fr. Cosmopolitan. Used in the Far East to treat stomach complaints and fevers (Foe lin, Hoelen, Tahi oeler sawah, Chinese Root). Exported from China.

PORLIERIA Ruiz. and Pav. Zygophyllaceae. 6 spp. Mexico to Andes. Shrubs or small trees.

***P. angustifolia*** (Engelm.) Gray. (Soap Bush). Mexico into Texas. The root bark is used by the local Indians as a soap substitute for washing wool.

***P. hygrometrica*** Ruiz. and Pav. Argentina, Chile, Peru. Gives a useful timber. A decoction of the leaves is used locally to treat rheumatism.

PORPHYRA Agardh. Bangiaceae. Marine Red Algae.

***P. denticulata*** Kjellm. (Gamet). Pacific. A popular food in the Philippines, where it is dried and sold in the markets.

***P. leucosticta*** Thurot. Mediterranean, Indian Ocean, Pacific. Eaten in Hawaii (Limu Luan) with shellfish.

***P. nereocystis*** Anders. Pacific. Eaten in China.

***P. perforata*** J. Ag. Pacific N.America. Eaten by the Indians of California and the Chinese. Exported to China.

***P. suborbiculata*** J. Ag. Pacific. Much prized by the Chinese (Tsu Choy) who eat it in soups, pickles and sweetmeats.

***P. tenera*** Kjellm. Coasts of China and Japan. Cultivated in Japan and eaten with soy (Amanori). Exported to America, where it is used to make seaweed soup in Chinese restaurants.

***P. umbilicalis*** (L.) J. Ag. (Slack, Sloke). N.Atlantic. Eaten in U.K., Eire and parts of Europe.

***P. vulgaris*** J. Ag. (Laver). Pacific and Atlantic. Eaten with oatmeal as laverbread in U.K. Also eaten in China and Japan, where it is cultivated and eaten in various dishes called Asakusanori, Sushi, Norino tsukudani. In China it is much used in soups.

***Porphyranthus zenkeri*** Engl. = Panda oleosa.

Portia Tree – Thespesia populnea.

Portland Arrowroot – Colocasia antiquorum.

***Portlandia pterosperma*** S. Wats. = Coutarea pterosperma.

Portuguese Ipecacuanha – Naregamia alata.

Portuguese Oak – Quercus lusitanica.

PORTULACA L. Portulacaceae. 200 spp. Tropics and subtropics. Herbs.

***P. lutea*** Forst. (Aturi, Pokea). Polynesia. The young shoots are used as a vegetable locally.

***P. meridiana*** Blanco = P. quadrufida.

***P. napiformis*** F. v. Muell. N.Australia, Queensland. The aborigines eat the tubers as a vegetable.

***P. oleracea*** L. (Purslane, Pursley). Europe, introduced to N.America. The plant is found widely in the more arid regions of the N.Hemisphere where it is frequently cultivated locally as a vegetable. In India and China the leaves are used for soothing poultices on wounds, swellings etc.

***P. quadrufida*** L. = P. meridiana. Tropical Asia, Africa, Polynesia. The plant is often cultivated. The Zulus eat the leaves as a vegetable and use them as a diuretic. In India the plant is used to treat skin diseases, urinary complaints and lung diseases.

***P. retusa*** Engelm. W. N.America. The plant is eaten as a vegetable by the local Indians.

***P. tosten*** Blanco. = Trianthema portulacastrum.

***P. triangularis*** Jacq. = Talinum triangulare.

PORTULACARIA Jacq. Portulacaceae. 2 spp. S.Africa. Herbs.

***P. afra*** Jacq. S.Africa. A valuable fodder in times of drought.

Posh té – Annona scleroderma.

POSIDONIA Koenig. Posidoniaceae. 2 spp. Mediterranean, Australia. Aquatic Herbs.

***P. australis*** Hook. f. Australia. The stems yield a fibre (Cellonia, Lanamar, Posidonia Fibre) used for coarse fabrics and sacks. It is sometimes mixed with wool.

Posidonia Fibre – Posidonia australis.

POSOQUERIA Aubl. Rubiaceae. 15 spp.

Tropical Central and S.America, W.Indies. Small trees.

***P. latifolia*** Roem. and Schult. (Brazilian Oak). E.Brazil. The wood is used for making walking sticks.

Post Oak – Quercus obtusiloba.

Pot Majoram – Origanum vulgare.

Pota-eta-tel Oil – Kakoona zeylandica.

POTAMOGETON L. Potamogetonaceae. 100 spp. Cosmopolitan. Aquatic herbs.

***P. natans*** L. Temperate and subtropics. The starchy rootstocks are sometimes used as an emergency food.

Potanin Sumach – Rhus potanini.

Potato – Solanum tuberosum.

Potato, Air – Dioscorea bulbifera.

Potato Bean – Apios americana.

Potato, Chinese – Dioscorea batatas.

Potato, Duck – Sagittaria latifolia.

Potato, Fendler – Solanum fendleri.

Potato, Hausa – Coleus rotundifolius.

Potato, Kaffir – Plectranthus esculentus.

Potato, Kaffir – Plectranthus floribundus.

Potato, Sweet – Ipomoea batatas.

Potato, Wild – Anemonella thalectroides.

Potato, Wild – Solanum fendleri.

Potato Yam – Dioscorea bulbifera.

Potato Yam – Dioscorea esculenta.

POTENTILLA L. Rosaceae. 500 spp. Nearly cosmopolitan, chiefly N.Temperate and Arctic. Herbs.

***P. anserina*** L. (Silverweed). Temperate. The leaves (Herba Anserinae) are used as a decoction or in wine to treat diarrhoea, kidney stones, arthritis, vaginal discharges and cramp. The roots are sometimes used as an emergency food.

***P. bicolor*** L. Europe to Himalayas and Japan. The Chinese eat the roots as a vegetable.

***P. erecta*** (L.) Hampe. = Tormentilla erecta. (Tormenetil). Europe, temperate Asia, introduced to America. The dried rhizome is used medicinally as a tonic and astringent. A tea from the rhizome is used to treat intestinal catarrh and an infusion in brandy to treat stomach complaints.

***P. fruticosa*** L. (Bush Cinquefoil). N.Hemisphere. The Eskimos use the leaves as tea.

***P. glandulosa*** Lindl. W. N.America. The local Indians use the leaves as a tea.

***P. multifida*** L. Europe to Japan and Himalayas. The roots are eaten as a vegetable in China.

***P. palustris*** Scop. = Comarum palustre. Europe, temperate Asia, introduced to N.America. The flowers yield a red dye, used locally in Europe for dyeing wool. The rhizomes are sometimes used for tanning.

***P. reptans*** L. = Cinquefoil, Fivefinger, Five-leaf Grass). Europe, temperate Asia, introduced to N.America. The leaves and roots are used to make an astringent lotion and a decoction is taken to reduce fevers.

***P. rupestris*** L. Temperate Europe, Asia. The leaves are used locally in Russia to make a tea (Siberian Tea).

Potherb Fame Flower – Talinium triangulare.

Potherb Mustard – Brassica japonica.

POTHOÏDIUM Schott. Araceae. 1 sp. Philippines, Celebes, Moluccas. Herb.

***P. lobbianum*** Schott. The stem fibres are used locally to make fish-baskets.

POTHOMORPHE Miq. Piperaceae. 10 spp. Tropics. Herbs.

***P. umbellata*** L. (Baquiña, Mano de Zopilote, Santilla de Culebra). Central to S.America. A decoction of the fruits or root is used locally to treat rheumatism and as a diuretic.

POTHOS L. Araceae. 75 spp. Madagascar, Indomalaya. Vines.

***P. cannaefolia*** Durand. = Spathiphyllum candicans.

***P. giganteus*** Roxb. = Epipremnum giganteum.

***P. scandens*** L. India, Malaya, Indonesia. A decoction of the leaves is used locally to treat fevers.

Potonxihuite – Cestrum dumetorum.

Poui, Yellow – Tecoma serratifolia.

Poulard Wheat – Triticum turgidum.

POULSENIA Eggers. Moraceae. 1 spp. Ecuador, to Central America. Tree.

***P. armata*** (Miq.) Standl. = Inophloeum aromaticum. The inner bark is used locally to make bark cloth, mats, hammocks etc.

Pounce – Tetraclinis articulata.

POUPARTIA Comm. ex Juss. Anacardiaceae. 12 spp. Tropics, especially Madagascar and Mascarene Islands. Trees.

***P. amazonica*** Ducke. Brazil. The fruits are eaten locally.

POUROUMA Aulb. Urticaceae. 50 spp. Central and tropical S.America. Trees.

***P. cecropiaefolia*** Mart. Brazil. Cultivated locally for the pleasantly flavoured fruits.

***Pourretia coarctata*** Ruiz. and Pav. = Puya chilensis.

POUTERIA Aubl. Sapotaceae. 50 spp. Tropical America. Trees.

***P. caimito*** (Ruiz. and Pav. Radlk. = Lucuma caimito.

***P. guyanensis*** Aubl Guyana. The hard, dark brown wood is used to make furniture.

***P. multiflora*** (A. DC.) Eyma. (Jácana). Tropical America, W.Indies. The hard, strong red-brown wood is used for heavy construction work, houses, furniture, flooring, agricultural implements and sports goods.

***P. tovarensis*** (Klotzsch and Karst.) Engl. (Mispero del Monte). Colombia. The fruits are eaten locally.

POUZOLZIA Gaudich. Urticaceae. 50 spp. Tropical America, tropical and S.Africa, Tropical Asia. Woody herbs.

***P. denudata*** De Wild. (Betaume, Detanve). Congo. The leaves are eaten locally as a vegetable and are used as a poultice for wounds.

***P. guinensis*** Benth. See P. denudata.

***P. occidentalis*** Wedd. (Yaquilla). Central America to N.E. S.America. The stem fibres are used locally to make cordage and coarse fabrics.

***P. pentandra*** Benn. Tropical Asia. The stem fibres are used locally to make sails.

***P. tuberosa*** Wight. India. The tubers are eaten locally.

***P. viminea*** Wedd. E.India to Malaysia. The stem fibres are used in Java to make fishing nets.

Poverty Seed – Iva axillaris.

Powder, Countess – Cinchona officinalis.

Powdered Swiss Lichen – Nephroma parilis.

Po-Yak Oil – Afrolicania elaeosperma.

Poya – Borreria capitata.

Pracachy (Oil) – Pentaclethra filamentosa.

Pracaxi – Pentaclethra filamentosa.

Pracaxy Oil – Pentaclethra filamentosa.

***Pradosia lactescens*** Radlk. = Lucuma glycyphloca.

***Prainea limpato*** Beumée = Artocarpus limpato.

Prairie Acacia – Acacia angustissima.

Prairie Cherry – Prunus gracilis.

Prairie Cordgrass – Spartina pectinata.

Prairie Larkspur – Delphinium virescens.

Prairie Turnip – Psoralea esculenta.

Pramdoring – Zanthoxylum capense.

Pranadjuva – Sterculia javanica.

Prandjiwa – Euchresta horsfieldii.

PRANGOS Lindl. Umbelliferae. 30 spp. Mediterranean to Central Asia. Herbs.

***P. pabularia*** Lindl. (Avipriva, Badiankohi, Konal). Iran to Pakistan and N.India. The fruits are used to treat stomach complaints and control menstruation and as a diuretic.

Prahout – Garcinia vilersiana.

Pravak – Strychnos nux-blanda.

Preas phnau – Terminalia nigrovenulosa.

PREMNA L. Verbenaceae. 200 spp. Tropics and subtropical Africa, Asia. Shrubs or small trees.

***P. arborea*** Roth. = Gmelina arborae.

***P. cordifolia*** Roxb. (Ambong-ambong laoot, Baroowas, Barooweh). Malaya. The wood is used to make rudders for native boats. A decoction of the leaves and roots is used locally to reduce fevers and of the roots alone to cure poor breathing.

***P. integrifolia*** L. (Headache Tree). Malaysia. The hard wood is beautifully grained and is used in Java to make knife-handles.

***P. nauseosa*** Blanco. Tropical Asia. The leaves are sometimes chewed with betel nuts instead of Piper betle.

***P. tomentosa*** Willd. (Ije, Narve, Podanganari). India. The hard light wood is yellow-brown. It is used for carving, turning and other fancy work. The leaves are used as plates.

PRENANTHES L. Compositae. 40 spp. N.America, Canaries, tropical Africa, temperate and tropical Asia. Herbs.

***P. altissimus*** L. (White Lettuce). E. N.America. The leaves are used to poultice sores, ulcers etc.

***P. serpentaria*** Pursh. (Gall of the Earth, Lion's Foot). E. N.America. The leaves (Herba prenanthes serpentariae) are used as a home remedy for snake-bite.

Prickly Ash – Aralia spinosa.

Prickly Ash – Zanthoxylum clava-herculis.

Prickly Ash, Martinique – Zanthoxylum martinicense.

Prickly Chaff Flower – Achyranthes aspera.

Prickly Comfrey – Symphytum asperrimum.

Prickly Comfrey – Symphytum uplandicum.

Prickly Currant – Ribes lacustre.

Prickly Elder – Aralia spinosa.

Prickly Gooseberry – Ribes cynosbati.

Prickly Juniper – Juniperus oxycedrus.

Prickly-leaved Elephant's Foot – Elephantopus scaber.
Prickly Lettuce – Lactuca scariola.
Prickly Pear – Opuntia ficus-indica.
Prickly Poppy, Mexican (Seed Oil) – Argemone mexicana.
Prickly Sago Palm – Metroxylon rumphii.
Prickly Turkey (Gum) – Acacia senegal.
Primavera – Cybistax donnell-smithii.
Primrose, Evening – Oenothera biennis.
Primrose, Malanga – Xanthosoma violaceum.

PRIMULA L. Primulaceae. 500 spp. N.Hemisphere. Herbs.

***P. auricula*** L. (Auricula). Alpine Europe. The leaves (Herba Auriculae muris) are used to treat headaches.

***P. officinalis*** (L.) Hill. = P. veris.

***P. reticulata*** Wall. (Bishkopra). Himalayas. The Hindus use a decoction of the leaves as a pain-reliever.

***P. veris*** L. = P. officinalis. (Cowslip). Europe to temperate Asia. A decoction of the flowers and roots is used in Europe to treat coughs. The leaves are used as a tea-substitute and as a nerve tonic.

Prince of Wales' Feathers – Brachystegia boehmii.
Princewood – Cordia gerascanthus.

PRINGLEA Anders ex Hook. Cruciferae. 1 spp. Anarctica. Herb.

***P. antiscorbutica*** R. Br. (Kerguelen Cabbage). The leaves are eaten locally, raw or as a vegetable. A valuable local preventitive against scurvy.

PRIORIA Griseb. Leguminosae. 1 sp. Panama, Jamaica. Tree.

***P. copaifera*** Griseb. (Cativo, Cautivo). The light brown wood is strong but not durable. It is used locally for rough furniture, crates etc. and exported to U.S.A. where it is used for veneer.

***Pritchardia filifera*** Lindl. = Washingtonia filifera.

***Pritzelia sanguinea*** Kl. = Begonia sanguinea.

Privet, European – Ligustrum vulgare.
Privet, Glossy – Ligustrum lucidum.
Privet, Ibota – Ligustrum ibota.
Privet, Japanese – Ligustrum japonicum.
Privet, Swamp – Forestiera acuminata.
Priya-darsa – Crossandra infunduliformis.
Promotion Nut – Anacardium occidentale.

***Porphyroscias decursioa*** Miq. = Peucedanum decursivum.

Proso Millet – Panicum miliaceum.

PROSOPIS L. Leguminosae. 40 spp. Warm America; 1 sp. Tropical Africa; 2 spp. Caucasus to W.India. Trees.

***P. africana*** (Guill. and Perr.) Taub. = P. oblonga. Tropical Africa. The hard, heavy red-brown wood is used for general carpentry, agricultural implements and boat-building. It is used locally for charcoal. The bark is used for tanning. The ground leaves are a local aphrodisiac and the pods are used to stupify fish.

***P. alba*** Griseb. (Algaroba blanca). Tropical America. The pods are eaten locally.

***P. algarobilla*** Griseb. Brazil. The hard wood is used locally for general construction work and posting.

***P. cineraria*** (L.) Druce = P. spicigera. Iran, India. The purple-brown wood is hard, but not durable. It is used locally for house construction, furniture, wagons, agricultural implements and fuel. The pods are eaten locally.

***P. juliflora*** DC. (Mesquite). Tropical America. The pods are an important local food for humans and stock in parts of Mexico. The gum (Mesquite Gum) is used as an emulsifying agent. The flowers are a good source of honey and the bark is used for tanning. The heavy red wood is used for railway sleepers, posts, fuel and charcoal.

***P. nigra*** Hieron. Brazil, Argentina. The fruits are used locally to make an alcoholic beverage. The wood is used for general carpentry.

***P. oblonga*** Benth. = P. africana.

***P. odorata*** Torr. and Frem. = Strombocarpa odorata. (Screw Bean, Tornillo). S.W. U.S.A. The sweet pods are eaten by the local Indians.

***P. pubescens*** Benth. S.W. U.S.A. to Mexico. The pods are fermented by the local Indians to make a beverage. The wood is used for tool handles.

***P. spicigera*** L. = P. cineraria.

***Prosorus indica*** Dalz. = Phyllanthus indica.

Prostrate Amaranth – Amaranthus blitoides.
Prostrate Knotweed – Polygonum aviculare.

PROTEA L. Proteaceae. 130 spp. Tropical and S.Africa. Small trees.

***P. angolensis*** Welw. (Mubondo, Sugar Bush). Central Africa. The wood is used to make charcoal.

***P. mellifera*** L. (Sugar Protea) S.Africa. The

flowers yield large amounts of nectar which makes them a good honey source; the local farmers use the nectar to make a cough syrup (Syrupus Protae).

Protea, Sugar – Protea mellifera.

PROTIUM Burm. f. Burseraceae. 90 spp. Madagascar to Malaysia, tropical America. Trees. Most species mentioned yield resins used in varnishes, as salves and for incense.

**P. *aracouchini*** March. S.America. The balsam is used as a salve for wounds.

**P. *attenuatum*** (Rose) Urban. W.Indies. Resin called Bois d'Encens, Gommier Blanc, Gommier l'Incense.

**P. *carana*** (Humb.) L. Tropical America. Resin called Carana, Hyowana Resin, Mararo.

**P. *chapelieri*** Engl. (Tsiramiramy). Madagascar. The resin (Ramy) is used locally as an incense during various ceremonies.

**P. *copal*** L. Central America. The resin has been used for religious ceremonies by the local Indians since ancient times.

**P. *guianense*** March. Tropical America. The resin (Elemi of Guiana) is used as incense and for lacquers.

**P. *heptaphyllum*** March. Tropical America. The resin (Hyawa Gum, Ronima Resin) is used as incense and in varnishes. It is sometimes mixed with clay and used locally for making pots.

**P. *icicariba*** March. Argentina, Paraguay, Uruguay. The resin (Elemi Occidental) is used for varnishes.

**P. *javanicum*** Burm. (Bernang, Gollang, Tanggooloon). Indonesia. The hard red-brown wood is difficult to work. It is used locally for the teeth of sugar mills. A paste of the leaves is used by the natives to treat intestinal complaints.

PROUSTIA Lag. Compositae. 15 spp. W.Indies, Andes, temperate S.America. Shrubs.

**P. *pungens*.** Poepp. Chile. A decoction of the roots and leaves is used locally in medicinal baths for the treatment of gout and rheumatism.

Provision Tree – Pachira aquatica.

Prune – Prunus domestica.

PRUNELLA L. Labiatae. 7 spp. Temperate Eurasia, N.W.Africa, N.America. Herbs.

**P. *grandiflora*** (L.) Jacq. (Bigflower Self Heal). Europe to temperate Asia. The leaves are occasionally used in salads.

**P. *vulgaris*** L. (Heal All, Self Heal). Europe, temperate Asia, N.America. A decoction of the leaves is used to treat sore throats, internal bleeding and vaginal discharges.

PRUNUS L. Rosaceae. sens. str. 36 spp. N.Temperate. Trees.

**P. *americana*** Marsh. (American Wild Plum). E. and S. N.America. Not as vigorous as the European spp. Cultivated locally, especially in Canada for the fruits which are used in jellies etc. or eaten raw or cooked.

**P. *amygdalus*** Batsch. = Amygdalus communis.

**P. *angustifolia*** Marsh. (Chickasaw Plum, Mountain Cherry). S.E. U.S.A. The fruits are used locally in jellies, jams etc. The plant is sometimes cultivated.

**P. *armeniaca*** L. = Armeniaca vulgaris.

**P. *avium*** L. (Dukes, Sweet Cherry, Mazzard). Europe, S.Asia Cultivated locally since ancient times and introduced to America by the early settlers. The trees are about 35 ft. high, bearing sweet fruits, yellow to black and about 1 inch in diameter. They are eaten raw, cooked in various ways, canned, dried. Dukes are a hybrid – P. avium × P. cerasus. var. ***duracina*** (L.) Koch is the Hard Cherry (Bigareaus). var. ***juliana*** (L.) Koch is the Heart Cherry (Gean Cherry). The species is diploid in contrast to the tetraploid P. cerasus. Cultivars of both spp. are propagated by grafting on to rootstocks. The world production of both types of cherry is approximately 1½ million tons a year.

**P. *besseyi*** Bailey. (Western Sand Cherry). N.America. A shrubby spp. The dark fruits are eaten locally.

**P. *brigantina*** Vill. = Armeniaca brigantina.

**P. *cantabrigiensis*** Stapf. = P. pseudocerasus. (Chinese Sour Cherry). China. Cultivated locally. The sour fruits are used for making jams etc.

**P. *capollin*** Zucc. (Capulin). Mexico to Peru. The fruits are eaten raw, cooked in preserves. The fruit juice mixed with cornmeal (Cupultamal), is eaten as a cake by the local Indians. A distilled decoction from the leaves is Cherry Water. The wood is used locally for making furniture and carpentry.

**P. *cerasifera*** Ehrh. = P. myrobalana. (Cherry Plum). Asia. The plant is cultivated, sometimes for the fruit, but mainly as a rootstock for grafting plum cultivars.

***P. cerasoides*** D. Don. = P. puddum.

***P. cerasus*** L. (Sour Cherry). S.E.Europe, W.Asia. See P. avium for cultural and general details. The fruits are dark red and smaller than P. avium. The trees are also smaller. The fruits are used mainly for cooking and preserves but are also used to make various liqueurs, e.g. Cherry Brandy, Cherry Heering, Marachino and Kirschwasser. A semi-drying oil (Cherry Kernel Oil) is derived from the seeds and is used refined as a salad oil and in cosmetics. The gum from the stem is sometimes used in cotton printing and the leaves can be used as a substitute for tea. The wood is hard, elastic, but difficult to split. It is used for turning, inlay, furniture and instruments.

***P. domestica*** L. (Plum). The plant has been grown since antiquity in Europe and W.Asia. It has been found in the Swiss Lake Dwellings and is mentioned in literature of 2000 B.C. It is one of the most climatically tolerant of the orchard crops. The fruits are eaten raw or cooked, but most of the commercial production is used for canning etc. or for making prunes. These are the sun-dried fruits of varieties with a high sugar content. Prunes are produced mainly in Yugoslavia, Rumania, Italy, California and Oregon. The fruits are also used to make various alcoholic beverages and liqueurs, e.g. Prunella, Zwetschenwasser (Slobovitz, Slivovica). The wood is sometimes used for turning. There are over 1000 cultivars.

***P. gracilis*** Engelm. and Gray. (Oklahoma Plum, Prairie Cherry). Oklahoma, Arkansas, Texas. The fruits are eaten by the local Indians, who also dry them for winter use.

***P. garyana*** Maxim. (Gray's Chokeberry). Japan, China. The salted flower buds are eaten in Japan. The hard compact wood is used in Japan for carving, turning, printing blocks, utensils, furniture etc.

***P. hortulana*** Bailey. (Hortulana Plum). Cultivated locally for the fruits which are made into preserves.

***P. ilicifolia*** (Nutt.) Walp. = Cerasus ilicifolia = Laurocerasus ilicifolia. California, Mexico. The fruits are eaten by the local Indians who also eat the seeds when ground into a meal.

***P. instititia*** L. (Bullace Plum, Damson). Europe. Cultivated on a small scale since ancient times. The fruits of all varieties are eaten raw or in preserves, or cooked. var. ***italica*** (Borkh) Aschers is the Greengage or Reine Claude, with green fruits and var. ***syriaca*** (Borkh.) Koehne. is the Mirabelle with yellow or red fruits.

***P. laurocerasus*** L. = Laurocerasus officinalis (Cherry Laurel). S.E.Europe, Asia Minor. The leaves are poisonous if used in large quantities, but an extract of them (Cherry Laurel Water) is used as a pain-reliever and sedative particularly in treating coughs. They are sometimes used as a substitute for bitter almond as a flavouring. The plant is cultivated as an ornamental.

***P. mahaleb*** L. (Mahaleb Cherry, St. Lucie Cherry). Europe, W. Asia. The leaves are used to flavour milk and as a tobacco substitute. The very hard wood (St. Lucie Wood) is scented and is used to make tobacco pipes, cigarette holders, small boxes, turning and walking sticks.

***P. maritima*** Wang. (Beach Plum). E. U.S.A. The fruits are used locally in preserves.

***P. maximowiczii*** Rupr. (Miyana Cherry). Japan, Manchuria, Korea. The hard red-brown wood is used in Japan for carving, furniture, utensils and sports equipment.

***P. melanocarpa*** (A. Nels.) Rydb. (Rocky Mountain Cherry). W. N.America. The fruits are eaten by the local Indians and are used locally for preserves.

***P. munsoniana*** Wight. and Hedrick. (Wild Goose Plum). N.America. The plant is cultivated locally for the fruits which are used for preserves etc.

***P. myrobalana*** Lois. = P. cerasifera.

***P. padus*** L. = Padus avium. (European Bird Cherry). Europe, through temperate Asia to Japan. The fruits are used locally to make preserves etc. The hard brown wood polishes well and is used for furniture, house interiors, boats and walking sticks.

***P. paniculata*** Thunb. = P. pseudocerasus. Japan. The double-flowered varieties are cultivated locally. The flowers do not fall readily and are collected, salted and made into a tea.

***P. persica*** (L.) Batsch. = Amygdalus persica.

***P. pseudocerasus*** Koids. = P. cantabrigiensis.

***P. pseudocerasus*** Lindl. = P. paniculata.

***P. puddum*** Roxb. = P. derasoides. Himalayas. The gum from the stem (Cherry

Gum) is used to adulterate gum tragacanth.

***P. salicina*** Lindl. = P. triflora (Chinese Plum, Japanese Plum). China. Cultivated in China, Japan and California for the fruits which are eaten raw or cooked. P. salicina × Amygdalus perica is the Plum-Peach with edible fruits and is sometimes cultivated. P. salicina × Armeniaca vulgaris is the Plumcot. The fruits are edible but of no commercial value.

***P. serotina*** Ehrh. (Rum Cherry, Wild Black Cherry). N.America. The fruits are used to flavour liqueurs and the bark is used medicinally as a tonic and sedative, especially in cough syrups. The hard, light brown wood is used for interiors of houses etc. and furniture making.

***P. simonii*** Carr. (Apricot Plum). N.China. The fruits are edible, but sometimes have a bitter almond flavour.

***P. spartoides*** (Spach.) Schneid. = Amygdalus spartiodes.

***P. spinosa*** L. (Blackthorn, Sloe). Europe, Mediterranean to W.Asia. The fruits are used to make the liqueur Sloe Gin. The dark brown wood is used for turning and walking sticks.

***P. ssiori*** F. Schmidt. Manchuria, China, Japan. Used in Japan. See P. maximowiczii.

***P. subcordata*** Benth. (Klamath Plum, Pacific Plum). California, Oregon. The fruits are eaten locally, raw or in preserves.

***P. tomentosa*** Thunb. (Manchu Cherry). Japan. The sweet fruits are eaten locally.

***P. triflora*** Roxb. = P. salicina.

***P. umbellata*** Ell. (Black Sloe, Fladwood's Plum, Hog Plum). Coasts of S.W. U.S.A. The fruits are used locally for preserves.

***P. virginiana*** L. (Choke Cherry). N.America. The fruits are used locally in preserves or cooked. The bark is sedative and astringent and is used in cough medicines.

***Psalliota*** (Fr.) Quél. = Agaricus.

***Psamma arenaria*** R. and Sch. = Ammophila arundinacea.

PSATHYRELLA (Fr.) Quél. Agaricaceae. 20 spp. Temperate.

***P. candolleana*** (Fr.) A. H. Sm. = Agaricus appendiculatus = Hypholoma candolleanum. Temperate. The fruitbodies are eaten occasionally in U.S.A.

PSEUDOCEDRELA Harms. Meliaceae. 6 spp. Tropical Africa. Trees.

***P. kotschyi*** (Schroeinf.) Harms. Tropical W.Africa. The trunk yields an inferior gum. The bark is used locally in the treatment of fevers and rheumatism and yields a brown dye, used locally for dyeing cloth.

PSEUDOCINCHONA A. Chev. ex E. Perrot. Rubiaceae. 4 spp. Tropical Africa. Trees.

***P. africana*** Chev. Tropical Africa. A decoction of the bark is used to reduce fevers.

PSEUDOCYMOPTERUS Coulter and Rose. Umbelliferae. 7 spp. S.W. United States of America. Herbs.

***P. aletifolius*** Rydb. Mexico. The leaves are eaten as a vegetable by the local Indians.

PSEUDOSPONDIAS Engl. Anacardiaceae. 2 spp. W. and C. Tropical Africa. Trees.

***P. microcarpa*** (A. Rich.) Engl. (Agopia, Bololo, Mubulu, Ngubi) Central and W.Central Africa. The bark is used in the Congo to treat dysentery.

PSEUDOSTACHYUM Munro. Graminae. 1 sp. E.Himalayas to Burma. Bamboo.

***P. polymorphum*** Munro. The stems are used in India to make baskets etc. and are pulped for paper manufacture.

PSEUDOSTUGA Carr. Pinaceae. 7 spp. E.Asia, W. N.America. Trees.

***P. douglasii*** Carr. = P. menziesii.

***P. menziesii*** (Mirb.) Franco. = P. douglasii = P. mucronata = P. taxifolia. (Douglas Fir, Douglas Spruce, Red Fir). W. N.America. One of the most important lumber trees of W. N.America. The wood is light yellow to red in colour and of variable hardness. It is used for general construction work, fuel, plywood and railway sleepers. A balsam (Oregon Balsam) is extracted from the trunk and the bark is used for tanning. A tea is sometimes made from the leaves.

PSIADIA Jacq. Compositae. 60 spp. Tropical Africa, Madagascar, St. Helena, Mascarene Islands, Socotra. Shrubs.

***P. dodoneafolia*** Steetz. Tropical Africa, Madagascar. The leaves are used to anneal water pitchers in Madagascar.

PSIDIUM L. Myrtaceae. 140 spp. Tropical America, W.Indies. Shrubs or small trees. The fruits of all the following species are edible. They are eaten raw, but mainly as jellies, pastes and jams. None, except P. guajava and P. cattleianum are grown commercially.

***P. araca*** Raddi.=P. guineense. (Araca, Brazilian Guava). Tropical S.America, especially Brazil and Cuba. The fruits are used locally for jellies and the plant is cultivated locally.

***P. cattleianum*** Sab. (Cattley Guava, Strawberry Guava). Brazil. Grown throughout the tropics. The fruits are used mainly for drinks, jellies etc. var. lucida is the Chilean Strawberry Guava. The plant is more hardy with fruit of a better flavour than P. guajava. Propagation is by seed.

***P. cinereum*** Mart. Brazil. The fruits are eaten locally and the leaves are used as an astringent.

***P. decaspermum*** L. f.=Decaspermum fruticosum.

***P. fluviatile*** Rich.=P. guineense. Brazil, Guyana. The fruits are eaten locally.

***P. friedrichsthalianum*** Ndz. (Costa Rican Guava). Central America.

***P. guajava*** L. (Guava). Mexico, Peru, W.Indies. Cultivated in Old and New World tropics, mainly as a garden crop, with some grown commercially in Florida. Used in jellies and preserves, especially Guava Cheese which is sold commercially. The preserved skins are eaten in W.Indies. The fruit is a good source of vitamin C. The plant is propagated by seed, not lending itself to vegetative propagation.

***P. guineense*** Pers.=P. fluviatile.

***P. guineense*** Sw.=P. araca.

***P. incanescens*** Mart. Brazil. The fruits are edible. The wood is used for agricultural implements etc. The leaves are astringent.

***P. microphyllum*** Britt. W.Indies. The fruits are edible.

***P. molle*** Bertol. Mexico, Central America. The fruits are edible. The plant is occasionally cultivated locally.

***P. oerstedeanum*** Berg. (Arrayán). Central America. The fruits are eaten locally.

***P. sartorianum*** (Berg.) Niedenzu.=Mitranthes sartoriana. (Arrayan, Guayabillo, Pichiché). Mexico. The fruits are edible. The bark is used for tanning and the leaves as an astringent. The plant is cultivated locally.

Psillium, French – Plantago arenaria.

Psillium Seed – Plantago psyllium.

Psillium, Spanish – Plantago arenaria.

PSILOSYBE (Fr.) Quél. Agaricaceae. 25 spp. Temperate. The Mexican spp. especially ***P. mexicana*** Heim. contain hallucinogenic and psychotropic drugs. They are used locally in various ceremonies. One of the drugs is psilocybine, which may be of value in psychological medicine.

PSILOPEGANUM Hemsl. Rutaceae. 1 sp. Central China. Shrub.

***P. sinense*** Hemsl. The Chinese use a decoction of the plant to treat dropsy.

PSITTACANTHUS Mart. Loranthaceae. 50 spp. Tropical America. Parasitic shrubs.

***P. calyculatus*** (DC.) Don.=Loranthus calyculata. Mexico, Central America. A decoction of the leaves is used locally as a salve for wounds and as a cosmetic.

***P. martinicensis*** Eichl. (Capitaine Bois, Roi Bois). W.Indies, especially Martinique. The leaves are used locally in medicinal baths.

PSOPHOCARPUS Neek. ex DC. Leguminosae. 10 spp. Tropical Africa, Mascarene Islands. Herbs.

***P. longepedunculatus*** Hassk.=P. palustris. Tropical Africa. Cultivated locally, especially in W.Africa, for the tuberous roots and pods which are eaten as a vegetable.

***P. palustris*** Desv.=P. longepedunculatus.

***P. tetragonolobus*** DC. (Asparagus Bean, Asparagus Pea, Goa Bean). Tropical Asia. Cultivated locally as a smallholding crop in Old and New World. The leaves, young shoots and immature pods are eaten in soups and as a vegetable. The dried seeds are roasted and eaten with rice in Java.

PSORALEA L. Leguminosae. 130 spp. Tropics and subtropics. Herbs.

***P. castorea*** S. Wats. (Beaverbread, Scurfpea). S. U.S.A. and Mexico. The local Indians use the roots as a vegetable and to make a porridge and flour.

***P. corylifolia*** L.=Trifolium unifolium. (Côt chu, Ku tzü, Phá côt chi). India and Far East. The leaves are used locally to treat stomach complaints and skin diseases. In Vietnam an alcoholic extract of the seeds is used to treat rheumatism and "female complaints".

***P. esculenta*** Pursh. (Breadroot, Prairie Turnip). E. and S. U.S.A. The large roots are eaten as a vegetable by the local Indians who also use them as a flour.

***P. glandulosa*** L. Chile, Peru. The leaves are used as a tea substitute in Chile.

***P. macrostachya*** DC. (Leatherroot Scurf pea). W. U.S.A. The local Indians used the bark fibres for making thread for sewing.

***P. mephitica*** S. Wats. S.W. U.S.A., Mexico. The local Indians used the roots as a vegetable and a source of flour and to make a yellow dye.

***P. pedunculata*** (Mill.) Vail. E. U.S.A. The local Indians used the leaves to make a bitter tonic.

***P. tenuiflora*** Pursh. (Few-flowered Psoralea). Central U.S.A. to Texas. The local Indians used the roots as a vegetable and to make flour.

Psoralea, Few-flowered – Psoralea tenuiflora.

PSOROSPERMUM Spach. Guttiferae. 40–45 spp. Tropical Africa, Madagascar. Shrubs.

***P. febrifugum*** Spach. Tropical Africa. In Angola the bark is used to treat leprosy and to reduce fevers.

PSYCHOTRIA L. Rubiaceae. 700 spp. Warm. Shrubs.

***P. cooperi*** Standl. S.America. A decoction of the plant is used locally in Colombia to treat rheumatism.

***P. emetica*** L. f.=Cephaëlis emetica.

***P. ipecacuanha*** Stokes=Cephaëlis ipecacuanha.

***P. jackii*** Hook. f. (Halan, Ruhac). Malaysia. A decoction of the plant is used locally to soothe snake bites and insect stings.

***P. luzoniensis*** (Cham. and Schlecht.) Vill. Philippines. A decoction of the roots is used locally to treat dysentery.

Psyllium (Seed) – Plantago psyllium.

Psyllium Seed, French – Plantago arenaria.

Psyllium Seed, Spanish – Plantago arenaria.

PTAEROXYLON Eck. and Zeyh. Ptaeroxylaceae. 1 sp. S.Africa. Tree.

***P. obliquum*** (Thunb.) Radlk.=P. utile.

***P. utile*** Eckl. and Zeyh.=P. obliquum. (Cape Mahogany, Sneezewood). The hard heavy tough wood is used for bridges and piers etc. It is scented like pepper and is used by the natives as a snuff to treat headaches and in cloth as a moth repellent.

PTELEA L. Rutaceae. 3 spp. S. U.S.A. Mexico. Shrubs.

***P. trifoliata*** L. (Hop Tree). U.S.A. The fruits are used as a substitute for hops in making beer and a decoction of the bark as a tonic.

PTERIDIUM Scop. Dennstaedtiaceae. 1 sp. Cosmopolitan. Fern.

***P. aquilinum*** (L.) Kuhn. sub sp. ***aquilinum***= Pteris aquilina. (Bracken, Eagle Fern). N.Hemisphere. The young fronds and stalks are eaten as a vegetable in Japan. The rhizomes are used as a vegetable in parts of the Pacific and the Indians of N.America used the rhizome to combat intestinal worms.

***P. aquilinum*** (L.) Kuhn. sub sp. ***caudatum*** (L.) Bonpl.=P. esculentum. New Zealand. The rhizome is used as an invalid food locally.

***P. esculentum*** Forst.=P. aquilinum sub sp. caudatum.

PTERIS L. Pteridaceae. 250 spp. Cosmopolitan. Ferns.

***P. aquilinia*** L.=Pteridium aquilinum sub sp. aquilinum.

***P. crenata*** Sw.=P. ensiformis.

***P. ensiformis*** Burm.=P. crenata.=Photolobus chinensis. Tropical Asia, Australia, Polynesia. The young fronds are eaten as a vegetable in India and the juice from the leaves is used in Malaya to clean the tongues of sick children.

***P. multifida*** Poiv.=P. serrulata. Indochina to China and Japan. A decoction of the rhizome is used in China to treat intestinal worms and to cure diarrhoea.

***P. serrulata*** L. f.=P. multifida.

PTEROCARPUS Jacq. Leguminosae. 100 spp. Tropics. Trees.

***P. angolensis*** DC. (Bloodwood, Mokula, Muklwa, Mubsamaropa). Tropical Africa, especially Congo and areas S. and E. The hard, red wood is used as a substitute for teak and is a valuable local timber, much is exported, especially from Katanga. Locally a red dye is prepared from the wood and the roots are used as an aphrodisiac.

***P. draco*** L.=P. officinalis. (Dragonsblood, Padauk). W.Indies, Guyana. The juice from the bark is hardened and used as West Indian Dragon's Blood. The light wood is used to float fishing nets.

***P. erinaceus*** Lam. (African Kino, African Rosewood). The red hard wood is much valued for making fine furniture. A red resin (Gum Kino, Sangue de Draco) is extracted from the stem. The resin is used locally as a poultice for wounds and is exported to U.K. and Portugal where it is used medicinally as an astringent. In Nigeria, particularly N., Camwood dye is made from the wood.

***P. grandiflorus*** Micheli. = Craiba grandiflora.

***P. lucens*** Guill. and Perr. Tropical Africa. The hard yellow-white wood is used for general construction work and furniture.

***P. indicus*** Willd. Tropical Asia. The beautifully grained wood is used to make furniture.

***P. macrocarpus*** Kurz. (Brown Padauk, Red Padauk). India, Burma. The red-brown hard wood is used for inlay work.

***P. marsupium*** Roxb. (Vengai Padauk). E.India to Ceylon. The hard durable wood is brown. It is used for house-building, railway sleepers, boats, carts, agricultural implements and furniture. It is an important lumber tree. The gum (East Indian Kino, Malabar Kino) is used medicinally as an astringent.

***P. officinalis*** Jacq. = P. draco.

***P. osun*** Craib. (Red Camwood). Tropical Africa. The heartwood, roots and bark are red and are ground into a paste used by the natives of S.Nigeria for colouring their skin. The tree yields an astringent kino.

***P. santalinus*** L. f. (Red Saunders, Red Sandalwood, Paduak Sandalwood, Lignum Santali Rubrum). E.India, Ceylon to Philippines. The wood is used to make furniture and yield the red dye Santalin which is used by the Hindus for caste marks and as an emetic.

***P. soyauxii*** Taub. (Baywood, African Paduak, Red Wood). Tropical W.Africa. The wood yields a red dye used locally for colouring the skin. The wood is used for canoes etc.

***P. tinctorius*** Welw. Tropical Africa. The wood yields a red-brown dye used in some parts to colour the feet and paint newly-born children.

PTEROCARYA Kunth. Juglandaceae. 10 spp. Caucasus to Japan. Trees. The fruits of the following species are edible.

***P. caucasica*** C. A. Mey. = P. fraxinifolia.

***P. fraxinifolia*** Spach. = P. caucasica. (Caucasian Wingnut). Caucasus to Persia. The soft light wood is used to make matches and wooden shoes.

***P. rhoifolia*** Sieb. and Zucc. (Japanese Wingnut). Japan. See P. fraxinifolia.

***P. stenoptera*** DC. China. The leaves and bark are used locally in a decoction to treat stomach upsets and intestinal worms.

PTEROCAULON Ell. Compositae. 25 spp. Madagascar, warm Asia, Australia, America. Herbs.

***P. pycnostachya*** (Michx.) Ell. = P. undulatum.

***P. undulatum*** (Walt.) C. Mohr. = P. pycnostachya. (Black Root, Indian Black Root). S. N.America. The root is used to prevent bleeding, to aid digestion, to control menstruation, induce abortion and reduce fevers.

PTEROCELASTRUS Meissn. Celastraceae. 6 spp. S.Africa. Trees.

***P. variabilis*** Sond. S.Africa. The leaves are used for tanning.

PTEROCLADIA J. Ag. Gelidiaceae. Marine Red Algae.

***P. lucida*** (R. Br.) J. Ag. S.Pacific. Eaten cooked in New Zealand. It is used in Japan and New Zealand as a source of agar.

***Pterococcus corniculatus*** Pax. and Hoffm. = Plukenetia corniculata.

PTEROGYNE Tul. Leguminosae. 1 sp. Brazil. Tree.

***P. nitens*** Tul. (Embiráro). The wood is hard, heavy, tough and strong, pinkish-brown, darkening on exposure. It is used locally for furniture, interiors of buildings, railway sleepers, wheel- and barrel-making.

PTEROLOBIUM R. Br. ex Wight and Arn. Leguminosae. 1 sp. Tropical Africa; 10 spp. tropical and subtropical Asia and Australia. Trees.

***P. lacerans*** R. Br. = Cantuffa exosa. Tropical E.Africa. The leaves are mixed with rust and used locally as a solution to dye leather black.

PTERORHACHIS Harms. Meliaceae. 1 sp. W.Equatorial Africa. Shrub.

***P. zenkeri*** Harms. The bark is used locally as an aphrodisiac.

PTEROSPERMUM Schreb. Sterculiaceae. 40 spp. E.Himalayas, S.E.Asia, W.Malaysia. Trees. The timber from the following species is strong and durable. It is used locally for bridge-building, boats, house-building etc.

***P. acerifolium*** Willd. E.India, Malaysia.

***P. celebicum*** Miq. (Wajoo). Celebes.

***P. javanicum*** Jungh. Java.

PTERYGOTA Schott and Endl. Sterculiaceae. 20 spp. Mainly Old World tropics. Trees.

***P. alata*** Br. (Bekaro, Budahanarikella, Tula). W.India, Andamas, Sylhet. The seeds are used as an opium substitute in Sylhet.

PTYCHOPETALUM Benth. Olacaceae. 2 spp. Tropical S.America; 5 spp. Tropical Africa. Trees. Both the species mentioned yield Muira Puama which is used medicinally in the treatment of rheumatism, paralysis, nerve complaints and dyspepsia.

***P. olacoides*** Benth. Brazil, Guyana.

***P. uncinatum*** E. Ansel. Brazil.

Pua – Osmanthus sandwicensis.
Puakala – Argemore glauca.
Puatea – Pisonia grandis.
Pubescent Oak – Quercus pubescens.

PUCCINELLIA Parl. Graminae. 100 spp. N.Temperate, S.Africa. Grasses.

***P. nuttalliana*** (Schult.) Hitch. (Nuttall Alkali Grass). W. U.S.A. A good forage grass, especially on alkaline soils.

Puchury Bean – Nectandra puchury-minor.
Pudding Pipe Tree – Cassia fistula.
Pudding Pipe Tree – Cassia obovata.

PUERARIA DC. Leguminosae. 35 spp. Himalayas to Japan, S.E.Asia, Malaysia, Pacific. Woody Vines.

***P. thunbergiana*** Benth. (Kudzu, Thunberg Kudzu Vine). Japan, China. Cultivated as a forage crop in warm regions of Old and New World. The stem is a source of a fibre and sometimes the roots are eaten as a vegetable. The starch (Japanese Arrowroot) is extracted in Japan.

Puerto Rican Hat Palm – Sabal causiarum.
Puff Balls – Lycoperdon spp.
Puffed Shield Lichen – Parmelia physodes.
Puget Balsamroot – Balsamorhiza deltoidea.

PUGIONUM Gaertn. Cruciferae. 5 spp. Mongolia. Herbs.

***P. cornutum*** Gaertn. (Sagai). Mongolia. The plant is cultivated locally and eaten as a vegetable.

Pulai – Alstonia scholaris.
Pulai paya – Alstonia spathulata.
Pulasan – Nephelium mutabile.
Pulguitas – Rhynchosia phaseoloides.

PULICARIA Gaertn. Compositae. 50–60 spp. Temperate and warm Eurasia, tropical and S.Africa. Herbs.

***P. crispa*** Sch. Tropical Africa, N.E.Africa to India. The dried leaves are used locally as a poultice for bruises.

PULMONARIA L. Boraginaceae. 10 spp. Europe. Herbs.

***P. officinalis*** L. (Lungwort) Europe to central and S.Russia. The dried leaves are used in decoctions to treat coughs, bronchitis etc.

Pulque – Agave atrovirens.
Pulque – Agave complicata.
Pulque – Agave crassispina.
Pulque – Agave gracilispina.
Pulque – Agave mapisaga.
Pulque – Agave melliflua.
Pulasan – Nephelium mutabile.
Pulsatilla – Anemone pulsatilla.
Pummelo – Citrus grandis.
Pumpkin – Cucurbita maxima.
Pumpkin – Cucurbita moschata.
Pumpkin Seed Oil – Cucurbita maxima.
Pun – Typha elephantinia.

PUNICA L. Punicaceae. 1 spp. Socotra; 1 sp. Balkans to Himalayas. Small trees.

***P. granatum*** L. (Pomegranate). Balkans to Himalayas. The plant is cultivated in the mild regions of the Old and New World, especially the Mediterranean and California. The juicy arils around the seeds are edible, but the skin of the fruit is tough and contains some 26 per cent tannin which is used commercially for making leather. The juice from the fruits is used to make the drink Grenadine. The dried bark (Pomegranate Bark, Granatum Bark) is used medicinally to treat intestinal worms, diarrhoea and fevers. The plant has been cultivated in the East since ancient times and is referred to in the Bible.

Purging Cassia – Cassia fistula.
Purging Croton – Croton tiglium.
Purging Flax – Linum catharticum.
Purging Nut – Jatropha curcas.
Purging Root – Euphorbia corollata.
Purgio – Manilkara bidentata.
Purple Angelica – Angelica atropurpurea.
Purple Arrowroot – Canna edulis.
Purple Cane Raspberry – Rubus neglectus.
Purple Cone Flower – Echinaceae pallida.
Purple Cone Spruce – Picea purpurea.
Purple Dead Nettle – Lamium purpureum.
Purple Grenadilla – Passiflora edulis.
Purple Heart – Peltogyne paniculata.
Purple Loosestrife – Lythrum salicaria.
Purple Medic – Medicago sativa.
Purple Osier Willow – Salix purpurea.
Purple Poppy Mallow – Callirhoë involucrata.
Purple Prairie Clover – Petalostemon purpureum.
Purple Sage – Salvia carnosa.

Purple Tephrosia – Tephrosia purpurea.
Purple Trillium – Trillium erectum.
Purple Vetch – Vicia benghalensis.
Purple Willow – Salix purpurea.
Purslane – Portulaca oleracea.
Purslane, Sea – Sesuvium portulacastrum.
Purslane, Surinam – Talinum triangulare.
Purslane, Winter – Claytonia perfoliata.
Pursley – Portulaca oleracea.
Pusa – Mouriri pusa.
Puso – Musa errans.
Pute – Boehmeria platyphylla.
***Putranjiva*** Wall. = Drypetes.
Putt's Pine – Flindersia acuminata.
Putty Root – Aplectrum hymale.

PUYA Molina. Bromeliaceae. 120 spp. Andes. Herbs.

***P. coarctata*** Fisch. = P. chilensis.

***P. chilensis*** Molina = P. coarctata = Pourretia coarctata. Chile. The leaves are a source of fibre used locally to make fishing nets. The plant also yields a gum (Chagual Gum, Maguey Gum).

PYCANTHEMUM Michx. Labiatae. 17 spp. N.America. Herbs.

***P. albescens*** Torr. and Gray. = Koellia albescens. N.America. The local Indians use a decoction of the leaves to induce sweating especially for treating colds.

***P. incanum*** (L.) Michx. = Koellia incanum. (Atlantic Mountain Mint). E. U.S.A. The local Indians use a decoction of the leaves as an astringent to stop nose bleeding.

***P. virginianum*** (L.) Durand and Jacks. (Virginia Mountain Mint). E. N.America. The local Indians use the flower buds and leaves as a condiment when cooking meat dishes.

PYCANTHUS Warb. Myristicaceae. 8 spp. Tropical Africa. Trees.

***P. kombo*** (Baill.) Warb. = Myristica kombo (Arbre à suif, Bohamba, Cashon, Etang, Kombo, Mulingo, Mutuge, Nkombo). W.Africa. The seeds yield a fat (Kombo Butter) which is aromatic, but is used locally for cooking etc.

PYCNARRHENA Miers. Menispermaceae. 25 spp. S.E.Asia, Indomalaya, Australia. Climbing shrubs.

***P. manilensis*** Vidal. (Ambal). Philippines. The powdered roots are used locally as a tonic and as a soothing poultice for wounds and snake-bites, it is supposed to be excellent for facilitating the healing of wounds.

PYCNOCOMA Benth. Euphorbiaceae. 14 spp. Tropical Africa, Madagascar, Mascarene Islands. Shrubs.

***P. macrophylla*** Benth. Tropical Africa. A decoction of the fruits is used for tanning in Natal.

PYCNONEURUM Decne. Asceliadaceae. 2 spp. Madagascar. Shrubs. The roots of both spp. are used locally as emergency food.

***P. junciforme*** Decne. (Kifoko, Kitsangana).

***P. sessiliflorum*** Decne.

***Pycnosporus sanguineus*** (L.) Murr. = Polyporus sangineus.

***Pycnothamnus*** Small = Satureja.

***Pygeum*** Gaertn. = Lauro-Cerasus.

Pyinma Andaman – Lagerstroemia hypoleuca.

***Pyracantha crenulata*** (Roxb.) Reom. = Crataegus crenulata.

***Pyrethrum balsamita*** (L.) Willd. = Chrysanthemum balsamita.

***Pyrethrum cinerariifolium*** Trév. = Chrysanthemum cinerariifolium.

***Pyrethrum roseum*** Lindl. = Chrysanthemum coccineum.

***Pyrethrum sinense*** DC. = Chrysanthemum sinense.

Pyrol, Woolly – Phaseolus mungo.

PYROLA L. Pyrolaceae. 20 spp. N.Temperate. Herbs.

***P. rotundifolia*** L. Europe to temperate Asia. In Central Europe the leaves are used to heal wounds.

***P. secunda*** L. N.America, Mexico, across Europe to Japan. The leaves (Herba Pirolae Secondae) are used locally in various parts to heal wounds.

***P. uniflora*** L. = Moneses uniflora = M. reticulata. W. N.America, Mexico, Japan, through Asia to Europe. The leaves (Herba Pirolae uniflorae) are used as a diuretic.

PYRRHOPAPPUS DC. Compositae. 8 spp. N.America. Herbs.

***P. carolinianus*** (Walt.) DC. = Leontodon carolinianum = Sitilias caroliniana (Leafy-stemmed False Dandelion). N.America. The roots were eaten as a vegetable by the local Indians.

PYRROSIA Mirbel. Polypodiaceae. 100 spp. Old World, tropics and temperate to N.E.Asia. Ferns.

***P. adnascens*** Desv. Tropical Africa through to Polynesia. A decoction of the rhizome is used in Malaysia to treat dysentery.

PYRULARIA Michx. Santalaceae. 4 spp. Himalayas, China, S.E. United States of America. Shrubs.

***P. pubera*** Michx. (Oil Nut, Buffalo Nut). E. N.America. The fruits are edible.

PYRUS L. Rosaceae. 30 spp. Temperate Eurasia. Trees. Distinguished from Malus, by having stone cells in the fruit.

***P. americana*** DC.=Sorbus americana.

***P. aucuparia*** Gaertn.=Sorbus aucuparia.

***P. angustifolia*** Ait.=Malus angustifolia.

***P. baccata*** L.=Malus baccata.

***P. communis*** L. (Pear). Europe, W.Asia. The plant has been grown since ancient times for the fruit. Homer mentions it in the tenth century B.C. It is now cultivated throughout the temperate areas of the world for the dessert fruit and for cooking and canning. The main canning variety is the Bartlett, while there are many dessert and culinary varieties in cultivation. Pears grow under moderate conditions of temperature but are less tolerant of extremes than the apple. The total world production is over 5 million tons annually and most of this comes from Italy, Germany, France, Switzerland and the Balkans. Pears are grown extensively in the U.S.A., mainly for canning, as they are in Australia, New Zealand, S.Africa, Canada, Chile and Japan. A sizeable proportion of the European crop is used to make the alcoholic drink, perry. Cultivars are propagated by grafting, sometimes on to quince stock. The heavy durable wood is sometimes used for turning and knife handles.

***P. coronaria*** L.=Malus coronaria.

***P. cydonia*** L.=Cydonia oblonga.

***P. germanica*** Hook. f.=Mespilus germanica.

***P. hupehensis*** Pamp.=Malus hupehensis.

***P. malus*** L.=Malus communis.

***P. praecox*** Pall.=Malus praecox.

***P. prunifolia*** Willd.=Malus prunifolia.

***P. ringo*** Wenzig. China, Japan. The fruits are eaten locally, raw or dried.

***P. rivularis*** Dougl.=Malus rivularis.

***P. serotina*** Rehd. Central China. One of the species from which the Chinese pears have been derived.

***P. serrulata*** Redh. See P. serotina.

***P. sieboldi*** Regel.=P. toringo Japan. The bark yields a yellow dye used locally.

***P. sinensis*** Lindl.=P. ussuriensis. China. Cultivated locally for the edible fruits. One of the species from which the Chinese pears have been derived.

***P. toringo*** Sieb.=P sieboldi.

***P. ussuriensis*** Maxim.=P. sinensis.

***P. torminalis*** Ehrh.=Sorbus torminalis.

# Q

Quackgrass – Agropyron repens.

Quaino – Urera baccifera.

Quamash – Camassia quamash.

Quamo – Inga leptoloba.

Quandong – Eucarya acuminatus.

Quandong, Silver – Elaeocarpus grandis.

Quanja peruan – Chenopodium canihua.

Quaoua – Rajania cordata.

QUARARIBEA Aubl. Bombacaceae. 50 spp. Central and tropical S.America. Trees.

***Q. fieldii*** Millsp. (Saha). Mexico. The flowers are used locally for flavouring chocolate.

***Q. turbinata*** (Sw.) Poir. (Swizzle-stick Tree). W.Indies, tropical America. The yellow wood has a spicy smell. The small branched twigs are cut to make swizzle sticks, used particularly in the W.Indies for mixing drinks, chocolate drinks, sauces and soap.

QUASSIA L. Simaroubaceae. 40 spp. Tropics, especially America. Trees.

***Q. amara*** L.=Q. officinalis. N.E. S.America. The wood (Lignum Quassiae Surinamense) contains the bitter principle quassin (about 0·1 per cent of the wood). This is extracted from the chipped wood and used medicinally as a gastric tonic and to treat intestinal nematodes. As the extract contains no tannin it can be used as a tonic in conjunction with iron salts.

***Q. excelsa*** Sw. W.Indies. An extract of the wood is used locally, especially in Jamaica to treat fevers and dysentery. It is also used as a substitute for hops in making beer and in aperitifs.

***Q. gabonensis*** Pierre. Tropical Africa. The fruits (Nzeng) are used locally, especially in Gabon as a condiment.

***Q. indica*** Gaertn. (Niepa Bark Tree). India to Polynesia. A decoction of the leaves is used in Indonesia to kill termites, to treat biliousness and as an emetic and purgative.

***Q. klaineana*** Pierre and Engl. Tropical W.Africa. The light wood is used locally to make canoes, light planking for houses and ceilings. A decoction of the bark is a local treatment for stomach complaints.

***Q. officinalis*** DC. = Q. amara.

Quassia, Jamaica – Picrasma excelsa.

Quassia, (Surinam) – Quassia amara.

Quaters – Thespesia garckeana.

Quauhchichic – Garrya longifolia.

Quê do – Cinnamomum tetragonum.

Quebec Oak – Quercus alba.

***Quebranchia*** Griseb. = Loxopterygium.

Quebrancho Blanco – Aspidosperma quebrancho blanco.

Quebrancho Colorado – Loxopterygium lorentzii.

Quebrancho Colorado – Schinopsis balansae.

Quebracho, White – Aspidosperma quebracho blanco.

Quebrancho, Willowleaf Red – Schinopsis balansae.

Quebracho, Woolly White – Aspidosperma tomentosum.

Queen Crapemyrtle – Lagerstroemia flosreginae.

Queen's Delight – Stillingia sylvatica.

Queen-of-the-Meadow – Filipendula ulmaria.

Queen-of-the-Meadow Root – Eupatorium purpureum.

Queen's Root – Stillingia sylvatica.

Queensland Arrowroot – Canna edulis.

Queensland Hemp – Sida rhombifolia.

Queensland Maple – Flindersia brayleyana.

Queensland Nut – Macadamia ternifolia.

Queensland Nut – Macamaia tetraphylla.

Queensland Raspberry – Rubus probus.

Queensland Walnut – Endiandra palmerstoni.

Queensland Wild Lime – Microcitrus inodora.

Quelite salado – Suaeda ramosissima.

Quemador – Urera caracasana.

Querá – Schrasdera marginata.

Quercitron Bark – Quercus tinctoria.

Queso de Tuna – Opuntia magacantha.

QUERCUS L. Fagaceae. 450 spp. N.America through temperate and subtropical Eurasia, N.Africa, W. tropical S.America. Trees.

***Q. aegilops*** L. E.Europe to W.Asia. The unripe fruits (valonia) are used locally for tanning.

***Q. agrifolia*** Nees. See Q. gambelii.

***Q. alba*** L. (Quebec Oak, White Oak). N.America through to Texas. The tough, strong, durable light-brown wood is used for general construction work, furniture, agricultural implements, railway sleepers, interiors of buildings, fencing, barrels and as a fuel. The wood is much esteemed for making barrels in which whisky is matured. The ground bark (White Oak Bark) contains tanning and is used medicinally as an astringent and tonic.

***Q. aliena*** Bl. China, Korea to Japan. The timber is of local value, especially for boat-building. The leaves are used to feed silk worms in China.

***Q. apennina*** Zucc. = Q. pubescens.

***Q. ballota*** Desf. = Q. ilex.

***Q. bicolor*** Willd. = Q. platanoides (Swamp White Oak). W. U.S.A. for use of timber. See Q. alba.

***Q. borealis*** Michx. f. = Q. rubra du Roi non L. (Red Oak). N.America. The light red-brown wood is hard and durable. It is much prized for house construction and furniture.

***Q. cambelii*** Nutt. = Q. undulata.

***Q. cerris*** L. = Q. vallonae (European Turkey Oak). Europe and Middle East. Yields a manna (Gaz, Gazu, Oak Manna) used in the Middle East to make a sweetmeat (Gazenjuben). This is sometimes boiled with the leaves of the tree to make a cake. The galls (Morea Galls) are used as an adulterant for Allepo Galls used in tanning.

***Q. chrysolepis*** Liebm. (Canyon Live Oak, Maul Oak). S.W. U.S.A., Mexico. The heavy tough wood is used to make agricultural implements and wagons.

***Q. coccifera*** L. (Kermes Oak). Mediterranean. Used as a food for the Kermes insect from which a red dye is extracted.

***Q. coccinea*** Wangenh. (Scarlet Oak). E. N.America. A rather coarse-grained timber. See Q. borealis.

***Q. costata*** Blume. = Pasania costata. Malaysia. The light brown wood is used locally for building.

***Q. crispula*** Blume = Q. monglica.

***Q. cuspidata*** Thunb. Japan. The roasted acorns are eaten locally.

***Q. cyclophora*** Endl. = Pasania cyclophora. Malaysia. The hard wood is used in

Sumatra for tool handles and rice mortars.

***Q. dentata*** Thunb. (Daimyo Oak). Korea, Manchuria, China, Formosa, Japan. The bark is used for tanning in Japan.

***Q. densiflora*** Sarg. = Lithocarpus densiflora.

***Q. digitata*** (Marsh) Sudw. (Spanish Oak, Red Oak). N.America. The hard wood is coarse-grained and not durable. It is used sometimes for construction work. The bark is used for tanning and medicinally as an astringent.

***Q. eduardii*** Trel. = Q. oligodenta. (Encina colorada). Mexico. The hard, red wood is much used for general carpentry.

***Q. emoryi*** Torr. (Emory Oak, Western Black Oak). S.W. U.S.A., Mexico. The acorns are eaten locally.

***Q. faberi*** Hance. China. The leaves are used locally to feed silkworms.

***Q. fontanesii*** Taub. See Q. suber.

***Q. gambelii*** Nutt. (Shin Oak). W. N.America. The ground, boiled acorns were used as flour by the local Indians.

***Q. garryana*** Dougl. See Q. gambelii.

***Q. garryana*** Hook. (Oregon White Oak). W. N.America. The light yellow-brown wood is tough and hard. It is used to make furniture, ships, barrels, rolling stock and as fuel.

***Q. glabra*** Thunb. Japan. The acorns are eaten, roasted, locally.

***Q. glandulifera*** Blume. (Gland-bearing Oak). Korea, China, Japan. The strong red-brown wood is used in Japan to make barrels, farm implements etc. The dead wood is used to grow the mushroom Cortinellus berkleyanum (Sitake).

***Q. glauca*** Thunb. (Blue Japanese Oak). Himalayas. The wood is used locally for general construction work and the acorns are eaten locally.

***Q. grisea*** Liebm. (Encina blanca, Encina prieta). Texas to Mexico. The wood is used locally as fuel.

***Q. ilex*** L. = Q. ballota (Holly Oak, Holm Oak). Mediterranean. The tree is cultivated. The heavy wood is strong and elastic; it is used for furniture. The acorns are eaten locally. Galls (Istrian Galls), formed on the leaves due to feeding by Cynips tinctoria are used for tanning.

***Q. imbricaria*** Michx. (Laurel Oak, Shingle Oak). E. N.America. The coarse, hard wood is used for interiors of buildings, furniture, roof shingles etc. Galls (American Nut Galls), formed due to damage by Cynips aciculata, are used for tanning.

***Q. incana*** Roxb. N.India, Himalayas, Indochina. The hard red-brown wood is used for furniture and agricultural implements. It is a valuable local fire wood and is used to make charcoal.

***Q. lanuginosa*** Thuill. = Q. pubescens.

***Q. lobata*** Nees. (California White Oak, Valley Oak, Western White Oak). See Q. gambelii.

***Q. lusitanica*** Lam. (Portuguese Oak, Lusitanian Oak). Mediterranean. Galls (Aleppo Galls, Smyrna Galls, White Galls, Gallae Turcicae) formed by Cynips gallae tinctoriae are used for tanning and making ink.

***Q. lyrata*** Walt. (Overcup Oak, Water White Oak). See Q. alba. A decoction of the bark is used by the local Indians to treat dysentery and stomach pains.

***Q. macrocarpa*** Michx. (Bur Oak, Mossy Cup Oak). E. N.America. See Q. alba.

***Q. manifera*** Lindl. See Q. cerris.

***Q. marylandica*** Du Roi. (Black Jack Oak, Jack Oak). E. U.S.A. The wood is used to make charcoal.

***Q. mongolica*** Fisch. = Q. crispula. (Mongolian Oak). Siberia through to N.China and Japan. The wood is used locally for building, furniture-making, barrels, railway sleepers and wagons.

***Q. nigra*** L. (Black Jack Oak, Water Oak). E. U.S.A. The acorns are eaten by the local Indians and the wood is used as fuel.

***Q. nitida*** Mart. and Sal. = Q. uruapanensis.

***Q. obtusiloba*** Michx. = Q. stellata. (Post Oak, Iron Oak). The heavy, hard, durable wood is used for railway sleepers, posts etc. and sometimes for construction work. The acorns are eaten by the local Indians, who also use the dried acorns to brew a coffee-like drink. They also use the leaves as cigarette wrappers.

***Q. occidentalis*** Gay. See Q. suber.

***Q. oligodenta*** Seem. = Q. eduardi.

***Q. pagoda*** Raf. = Q. pagodaefolia. (Spanish Oak, Swamp Spanish Oak). See Q. alba.

***Q. pagodaefolia*** (Ell.) Ashe = Q. pagoda.

***Q. palustris*** Du Roi. (Swamp Spanish Oak, Pin Oak). E. N.America. The heavy, strong, coarse-grained wood is used for house construction, roof shingles and barrels.

***Q. pedunculata*** Ehrh.=Q. robur.

***Q. persica*** Jaub. and Spach (Manna Oak). Persia. The acorns are eaten locally.

***Q. petraea*** (Mattuschka) Liebl=Q. sessiliflora. (Durmast Oak). Europe. See Q. robur.

***Q. phellos*** L. (Willow Oak). E. N.America. The wood is strong, but not hard; it is used for fuel, charcoal, clapboard and wheel felloes.

***Q. platanoides*** (Lam.) Sudw.=Q. bicolor.

***Q. prinus*** L. (Chestnut Oak). E. N.America. The hard, durable wood is used for railway sleepers, fencing and fuel. The bark is used for tanning.

***Q. pseudosuber*** Santi. See Q. suber.

***Q. pubescens*** Willd.=Q. apennina=Q. lanuginosa. (Pubescent Oak). S.Europe to Asia Minor. The wood is durable under water. See Q. robur.

***Q. reticulata*** H.B.K. (Netleaf Oak). Mexico. The acorns are used locally as a coffee substitute.

***Q. robur*** L.=Q. pedunculata (English Oak). Europe to Caucasus and Asia Minor, N.Africa. The yellow-brown wood is hard, heavy, elastic and durable even under water. Uses: see Q. alba. The bark is used for tanning, as are the leaf galls (Acorn Galls, Knoppern) caused by Cynips calcies. The acorns are used as a coffee substitute (Eichel Kaffee), as an emergency food and are an important local food for pigs. Acetic acid, tannic acid and charcoal are also made from the wood.

***Q. rubra*** Du Roi non L.=Q. borealis.

***Q. serrata*** Sieb. and Zucc. See Q. aliena.

***Q. serrata*** Car. See Q. faberi.

***Q. sessiliflora*** Salis.=Q. petraea.

***Q. stellata*** Wang.=Q. obtusiloba.

***Q. suber*** L. (Cork Oak). S.W Mediterranean. The bark is the main commercial source of cork. The tree is cultivated mainly in Portugal, yielding some 200,000 tons a year of cork. Cylinders of bark are stripped from the trees during the summer. These must be at least 3 cm. thick. A large tree may give 1 ton of cork in a season, but it takes about 10 years for it to regenerate. The bark is seasoned in stacks for a few weeks, boiled to soften it and the rough outer layers are removed. It is then stacked in sheets ready for export. Cork has been used since the time of the ancient Greeks. It is useful in being exceptionally light, long-lasting, difficult to burn, an excellent insulator and non-conductor of electricity. It is used as a flooring material, insulator for buildings, gasket seals, linoleum, floats and bottle stoppers.

***Q. sundaica*** Blume=Q. hystrix. Malaysia. The heavy, hard wood is used to make houses and bridges.

***Q. tauricola*** Kotschy. S.E.Europe to Central Asia. Leaf galls (Dead Sea Apple, Rove, Mala Insana, Mala Sodomitica) caused by Cynips insana are used for tanning and making the red dye Rouge d'Adrianople. See Q. cerris.

***Q. texana*** Buckl. (Southern Red Oak, Texas Oak). S. U.S.A. The hard, heavy wood is used locally for general construction work.

***Q. tinctoria*** Bartr.=Q. velutina. (Black Oak). N.America. The bark (Querciton Bark) is used for tanning and is an important source of the dye quercetrin. The dye is extracted from the inner bark by hot water under pressure. It is yellow, but is made red by acid treatment. The dyes are used to colour wool and calico.

***Q. undulata*** Torr.=Q. cambelii. (Shin Oak, Wavyleaf Oak). W. U.S.A. The wood is used for fuel; the bark is used for tanning; and the acorns are used as a flour by the local Indians.

***Q. uruapanensis*** Trel.=Q. nitida. (Encino colarada). Mexico. The wood is used locally for making furniture and general carpentry.

***Q. utahensis*** Rydb.=Q. gambelii.

***Q. vallonea*** Kotschy=Q. cerris.

***Q. variabilis*** Blume. (Oriental Oak). Korea, N.China, Japan. The cupules around the acorns yield a black dye, used in China to colour silk. The leaves are fed to silk worms.

***Q. velutina*** Lam.=Q. tinctoria.

***Q. virginiana*** Mill. (Live Oak). S. U.S.A. through to Central America and W.Indies. The hard, tough wood was used for building ships.

Quick-in-Hand – Impatiens noli-tangere.

Quihuicha – Amaranthus caudatus.

Quilite – Urechites karwinskii.

QUILLAJA Molina. Rosaceae. 3 spp. Temperate S.America. Small trees.

***Q. brasiliensis*** (St. Hil.) Mart.=Fontanella

brasiliensis. S.Brazil, E.Argentine, Uruguay, Chile. The bark is used as a soap for washing fine textiles.

***Q. saponaria*** Molina. (Soapbark Tree, Quillaja). Peru, Chile. The dried inner bark (Soap Bark, Soap Tree Bark, Panama Bark) is used for washing clothes and as an emulsifying agent in tars. It is also used for washing the hair. The bark is also an irritant and expectorant, but is also a heart depressant, so should not be used internally.

Quina (blanco) – Croton niveus.
Quina – Esenbeckia febrifuga.
Quina – Exostema sanctae-luciae.
Quina do Campo – Strychnos pseudoquina.
Quina do Matto – Cestrum pseudoquina.
Quince – Cydonia oblonga.
Quince, Bengal – Aegle marmelos.
Quince, Chinese – Chaenomeles sinensis.
Quinilla – Manilkara bidentata.
Quinine – Cinchona spp.
Quinine Bush – Garrya elliptica.
Quinine Bush – Garrya fremontii.
Quinoa – Amaranthus caudatus.
Quinoa – Chenopodium quinoa.
Quinquina du pays – Ochrosia borbonica.
Quinteniqua Yellowwood – Podocarpus elongata.
Quiote – Agave quiotifera.
Quira Macawood – Platymiscium pinnatum.
Quirote – Serjania mexicana.
Quisache cortẽno – Acacia cochliacantha.

QUISQUALIS L. Combretaceae. 17 spp. Tropical and S.Africa, Indomalaya. Vines.

***Q. indica*** L. (Rangoon Creeper). Indomalaya. The roots and fruits are used locally in the treatment of intestinal worms. The plant is cultivated.

Quiza – Tuber gennadii.

# R

Rabbit Bush – Chrysothamnus confinis.
Rabbit Bush – Chrysothamnus nauseosus.
Rabbit Bush – Chrysothamnus viscidiflorus.
Rabbit's Thorn – Lycium pallidum.
Rabo de Nico – Heliotropium angiospermum.

RADERMACHERA Zoll. and Mor. Bignoniaceae. 40 spp. India to China, Philippines, Celebes, Java. Trees.

***R. gigantea*** Miq.=Spathodea gigantea=Stereospermum hypostictum. (Bedali, Radja matan, Toowi). Indonesia, Java. The wood is strong and durable, but is attacked by termites and ants. It is used for general construction work, bridges and household utensils.

***R. xylocarpa*** K. Schum. (Bersinge, Hulava, Kursingh). Tropical Asia, especially India. The Hindus use the oil distilled from the wood to treat skin infections.

Radiate Pine – Pinus radiata.
Radish – Raphanus sativus.
Radish, Chinese – Raphanus sativus.
Radish, Daikon – Raphanus sativus,
Radish, Horse – Armoracia lapathifolia.
Radish, Lobak – Raphanus sativus.
Radish, Spanish – Raphanus sativus.
Radish Tree, Horse – Moringa oleifera.
Radix Acori palustris – Iris pseudacorus.
Radix Alkannae syricae – Macrotomia cephalotus.
Radix Althaeae – Althaea officinalis.
Radix Anthorea – Aconitum anthora.
Radix Aplectri – Aplectrum hymale.
Radix Bryoniae – Bryonia diocia.
Radix Chinae ponderosae – Smilax china.
Radix Columba – Jateorhiza miersii.
Radix Columba – Jaterorhiza palmata.
Radix Dentariae – Plumbago europaea.
Radix Filipendula – Filipendula hexapetala.
Radix Gingseng – Panax quinquefolia.
Radix Glycyrrhizae – Glycyrrhiza glabra.
Radix Hellebori Albi – Veratrum album.
Radix et Herba Cannabis aquatica – Eupatorium cannabinum.
Radix et Herba Lepidii – Lepidium latifolium.
Radix et Herba Virage aurea – Solidago virgaurea.
Radix, Herba et Semen Oreo seleni – Peucedanum oreoselinum.
Radix Ipecacuanha albae lignosae – Hybanthus ipecacuanha.
Radix Ipecacuanha striata minor – Manettia cordifolia.
Radix Iridis – Iris germanica.
Radix Lichtensteiniae – Lichtensteinia interrupta.
Radix muira-Puama – Liriosma ovata.
Radix Ninsi – Panax quinquefolia.
Radix Ogkert – Silene macrosolen.

Radix pareira brava – Chondrodendron tomentosum.
Radix Sanley vel Acori – Acorus gramineus.
Radix Saponariae albae – Gypsophila arrostii.
Radix Saponariae hispanicae – Gypsophila paniculata.
Radix Saponariae levanticae – Gypsophila paniculata.
Radix et Semen Carvariae – Peucedanum carvaria.
Radix Tocoyenae – Tocoyena longiflora.
Radix Urticae – Urtica dioica.
Radja matan – Radermachera gigantea.
RADLKOFERA Gilg. Sapindaceae. 1 sp. Tropical Africa, especially Congo. Tree.
***R. calodendron*** Gilg. (Bie bie, Dugulu, Wompologa). The wood is used locally to make mortars.
Raffia – Raphia gigantea.
Raffia – Raphia pedunculata.
Raffia Palm, Giant – Raphia gigantea.
Raffia Palm, Ivory Coast – Raphia hookeri.
Raffia Palm, Madagascar – Raphia pedunculata.
Raffia Palm, Wine – Raphia vinifera.
RAFFLESIA R. Br. Rafflesiaceae. 12 spp. W.Malaysia. Root parasite.
***R. patma*** Blume. Java. The flowers are used locally as an astringent.
RAFNIA Thunb. Leguminosae. 32 spp. S.Africa. Herbs.
***R. perfoliata*** E. Mey.=Vascoa perfoliata. S.Africa. An infusion of the leaves is used locally as a diuretic.
***Ragala sanguinolenta*** Pierre=Ecclinusa sanguinolenta.
Rage – Lobaria pulmonaria.
Rage, Oak – Lobaria scrobiculata.
Ragged Cup – Silphium perfoliatum.
Ragi – Eleusina coracana.
Ragweed (Oil) – Ambrosia artemisiifolia.
Ragweed, Western – Ambrosia psilotstachya.
Ragwort, Golden – Senecio aurens.
Ragwort, Tansy – Senecio jacobea.
Rai – Brassica integrifolia.
Rain Tree – Lonchocarpus capassa.
Rain Tree, Manaca – Brunfelsia hopeana.
Raisin – Vitis vinifera.
Raisin Tree, Japanese – Hovenia dulcis.
Raiz Doce – Periandra dulcis.
Raiz de José Domingo – Aristolochia brasiliensis.
Rajango – Pithecellobium minahassae.
RAJANIA L. Dioscoreaceae. 25 spp. W.Indies. Vines.
***R. cordata*** L. (Bihi, Quaoua). W.Indies. The rootstocks are eaten locally.
***Rajapa*** Singer=Termitomyces.
Rajmahal Hemp – Marsdenia tenacissima.
Rakaf – Leontice leontopetalum.
Rakkyo – Allium chinense.
Ram's Horn – Martynia probocida.
Ramalina, Ivory-like – Ramalina scopulorum.
Ramalina, Mealy – Ramalina farinacea.
RAMALINA Ach. Usneaceae. Lichens.
***R. calicaris*** (L.) Röhling=Lobaria calicaris. (Twig Lichen). Temperate. An extract of the plant is used in Europe to dye wool yellow. It was used as a powder to dye wigs.
***R. cuspidata*** (Ach.) Nyl. Temperate. Used in Europe to dye wool a light brown.
***R. farinacea*** (L.) Ach. (Mealy Ramalina). Temperate. Used in Europe as a light brown dye for wool.
***R. fraxinea*** (L.) Ach.=Lobaria fraxinea. (Ash Twig Lichen). Temperate, especially Europe. Used in the manufacture of perfumes and cosmetics.
***R. pollinaria*** (Westr.) Ach.=Parmelia pollinaria. Mountains of N.Hemisphere. Contains ramalinic acid which is a potential antibiotic.
***R. scopulorum*** Ach. (Ivory-like Ramalina). Temperate. An extract of the plant is used in Scotland as a yellow-brown dye for wool.
Raman – Bouea bunnanica.
Ramie – Boehmeria nivea.
Ramie Bukit – Alchornea villosa.
Ramie Sengat – Abroma augustum.
Ramin – Gonystylus bancanus.
Ramón – Trophis racemosa.
Ramon Breadnut – Brosimum alacastrum.
Ramoncillo – Trophis racemosa.
Ramona, Desert – Salvia carnosa.
Ramontschi – Flacourtia ramontchi.
Rampion – Campanula rapunculus.
Ramtilla Oil – Guizotia abyssinica·
Ramtulsi – Ochium gratissimum.
Ran ptanas – Artocarpus hirsuta.
RANDIA L. Rubiaceae. 200–300 spp. Tropics. Shrubs or small trees.
***R. aculeata*** L.=R. mitis.
***R. armata*** (Sw.) DC.=Basanacantha armata.
***R. densiflora*** Benth.=R. racemosa=Stylocoryne densiflora=Webera densiflora.

Indomalaya, S.Australia, Hong Kong. A decoction of the roots is used to treat fevers in Cambodia.
***R. dumetorum*** Lam.=R. floribunda. Tropical Asia. The cooked berries are eaten locally.
***R. floribunda*** DC.=R. dumetorum.
***R. latifolia*** Lam.=R. mitis.
***R. maculata*** DC.=Rothmannia longiflora =Gardenia stanleyana. Tropical Africa. The fruits yield a dye, used locally for tattooing.
***R. malleifera*** Benth. and Hook. Tropical Africa. The juice from the fruit is used as ink in the regions of the White Nile. The dark sap is used locally to stain the skin.
***R. mitis***=R. aculeata=R. latifolia. (Agalla de costa, Palo de cotorra, Tintillo). Mexico to N. S.America, W.Indies. The juice from the fruits is used in the W.Indies for treating dysentery.
***R. racemosa*** F. Vill.=R. densiflora.
***R. ruiziana*** DC. Tropical America. The fruit pulp is eaten locally.
***R. walkeri*** Pellegr. (Oyem). Tropical Africa. The plant is used locally for stupifying fish. It is cultivated.
***R. wallichii*** Hook.=Tarenna incerta.
RANDONIA Coss. Resedaceae. 3 spp. N.Africa, Somaliland, Arabia. Shrubs.
***R. africana*** Coss. N.Africa, Sahara. The plant is eaten by camels.
Ranek fait – Diplotaxis duveyrierana.
Rang cira – Croton poilanei.
Range Ratany – Krameria parvifolia.
Ranggitan – Anotis hirsuta.
Rangoon Creeper – Quisqualis indica.
Rangoon Rubber – Urceola esculenta.
Rango-rango – Voanga foetida.
Rangu – Koordersiodendron pinnatum.
Rantil – Guizotia abyssinica.
RANUNCULUS L. Ranunculaceae. 400 spp. Cosmopolitan, temperate and cold tropical mountains. Herbs.
***R. ficaria*** L. (Lesser Celandine). Europe, temperate Asia. The bleached stems and leaves are occasionally used as a vegetable in Europe.
***R. hietus*** Banks and Soland.=R. plebeius. Australia, New Zealand. The plant is used locally in New Zealand to treat eye inflammations and toothache.
***R. pallasii*** Schlecht. Arctic. The roots are eaten by the Eskimos as a vegetable.
***R. plebeius*** R. Br.=R. hietus.
***R. rivularis*** Banks and Soland ex DC. Australia, New Zealand. In New Zealand an infusion of the leaves is used to treat quinsy and the sap as a massage for rheumatism and other joint pains.
***R. sceleratus*** L. (Celery-leaved Crowfoot, Blister Buttercup). N.Temperate. The juice, rubbed on the skin causes blistering. This is used by beggars to simulate sores.
Rapa Bark – Ficus tinctoria.
Papa ruvo – Brassica ruvo.
Rape (Oil) – Brassica napus.
Rape, Bird – Brassica campestris.
RAPHANUS L. Cruciferae. 8 spp. W. and C. Europe, Mediterranean, through to Central Asia. Herbs.
***R. raphinastrum*** L. var. ***sativus*** L.=R. sativus.
***R. sativus*** L.=R. raphinastrum var. sativus. (Radish, Lobak, Daikon). Temperate Europe and Asia. There are several varieties, all of which are grown for the roots, which are eaten mainly as a salad, but sometimes cooked. The leaves are sometimes also eaten as a salad. The varieties grown in temperate areas are small rooted, while those grown in the tropics and far East are white rooted and large. These latter are sometimes pickled in brine.
RAPHIA Beauv. Palmae. 30 spp. Tropical and S.Africa, Madagascar. Palm trees. The leaves of the species mentioned except R. sassandrensis, are used to make Raffia fibre which is used for basket work and by gardeners and nurserymen for tying plants. It dyes easily with vegetable dyes.
***R. gigantea*** Chev. (Giant Raffia Palm). Tropical Africa. The leaves are used locally for thatching etc. and the juice from the stems is used to make palm wine in W.Africa.
***R. hookeri*** Mann and Wendl. (Ivory Coast Raffia Palm, Wine Palm). Tropical Africa. The fibres are used locally to make cloth. The leaves are used for roofing etc. A palm wine (Mimbo) is made from the juice extracted from the inflorescence.
***R. pedunculata*** Beauv.=R. ruffia (Madagascar Raffia Palm). Madagascar. One of the main sources of raffia. The juice from the stem is used to make a palm wine (Harafa) and the shells from the fruits are made into trinkets.

***R. ruffia*** Mart.=R. pedunculata.

***R. sassandrensis*** Chev. W.Africa. The fruits are used locally to stupify fish.

***R. vinifera*** Beauv. (Bamboo Palm, Wine Raffia Palm). Tropical Africa. Used locally for raffia and palm wine.

***R. welwitschi*** Wendl. See R. vinifera.

***Raphidophora*** Hassk=Rhaphidophora.

***Raphiolepis*** Lindl.=Rhaphiolepis.

RAPHIONACME Harv. Periplocaceae. 20 spp. Tropical and S.Africa.

***R. brownii*** Elliot. Tropical Africa. Herb. The tubers are eaten as a vegetable locally, especially in Sierra Leone.

***R. utilis*** Brown and Stapf. Tropical Africa. Tree. A potential source of commercial rubber (Ekanda Rubber).

RAPUTIA Aubl. Rutaceae. 10 spp. Tropical America. W.Indies. Small trees or shrubs.

***R. alba*** Engl.=Almeida alba. Brazil. A decoction of the bark is used to reduce fevers and locally it is used to stupify fish.

***R. aromatica*** Aubl.=Sciuris aromatica. N.E. S.America. A decoction of the bark is used to reduce fevers and to relieve stomach complaints.

***R. magnifica*** Engl. Brazil. A decoction of the bark is used locally to treat intestinal worms.

Rasamala – Canarium oleosum.

Rasamala Resin – Altingia excelsa.

Rasamala Wood Oil – Altingia excelsa.

Rasant – Berberis lycium.

Rasin Tree, Japanese – Hovenia dulcis.

Rasna – Vanda roxburghii.

Rasp Pad – Flindersia australis.

Raspa – Tetracera volubilis.

Raspberry Acacia – Acacia acuminata.

Raspberry, American Red – Rubus strigosus.

Raspberry, Black – Rubus occidentalis.

Raspberry, Ceylon – Rubus albescens.

Raspberry, European Red – Rubus idaeus.

Raspberry, Golden Evergreen – Rubus ellipticus.

Raspberry, Japanese – Rubus parvifolius.

Raspberry, Macrae – Rubus macraei.

Raspberry, Mauritius – Rubus rosaefolius.

Raspberry, Mysore – Rubus albescens.

Raspberry, Purple Cane – Rubus neglectus.

Raspberry, Queensland – Rubus probus.

Raspberry, Wine – Rubus phoenicolasius.

Raspberry, Yellow Himalayan – Rubus ellipticus.

Rata, Northern – Metrosideros robusta.

Rata Vine – Metrosideros scandens.

Ratan Cane – Calamus rotang.

Ratanhia, Brazilian – Krameria argentea.

Ratanhia, Para – Krameria argentea.

Ratanhia Root – Krameria triandra.

Ratanhia, Savanilla – Krameria tomentosa.

Ratany, Para – Krameria argentea.

Ratany, Range – Krameria parvifolia.

***Ratonia*** DC.=Matayba.

Rattan – Calamus aquatilis, C. barteri, C. inops, C. javanensis, C. minahassae, C. ovoideus, C. radiatus, C. rotang.

Rattan Datoo – Calamus minahassae.

Rattan manau – Calamus ornatus.

Rattle Weed – Baptisia tinctoria.

Rattlesnake Master – Eryngium yuccifolium.

Rattlesnake Weed – Daucus pussilus.

Rattlesnake Weed – Echinacea angustifolia.

Ratu – Cassia sieberiana.

Rau râm – Polygonum odoratum.

Raukawa – Panax edgerleyi.

Rauli – Nothofagus procera.

RAUVOLFIA L. Apocynaceae. 100 spp. Tropics. Trees or shrubs.

***R. canescens*** L. (Trinidad Devilpepper). Tropical Africa. The juice from the fruits is used locally as an ink.

***R. heterophylla*** Roem and Schutt. See R. canescens.

***R. perakensis*** King and Gamble. Borneo. Used as an adulterant for R. serpentina.

***R. serpentina*** (L.) Benth. ex Kurz. (Java Devilpepper). India, Ceylon, Burma, Thailand and Indonesia. The roots have been used in Indian medicine since ancient times. They contain the alkaloid reserpine which is used medicinally in the relief of hypertension by reducing blood pressure and as a sedative. It depresses the activity of the hypothalamus. The plant is now cultivated in its native countries and the root exported to the West. Plants are propagated by seeds, cuttings or splitting the roots. Those from seeds take three years before the roots are harvested. The yield is about 4 tons per acre of roots, which contain about 2 per cent reserpine.

***R. vomitoria*** Benth. (Swizzlestick). W.Africa. The bark and roots are used as a sedative to treat convulsions and as an aphrodisiac. The leaves and fruits are an emetic and the leaf latex is rubbed on skin diseases.

***Rauwolfia*** L.=Rauvilfia.

RAVENEA H. Wendl. Palmae. 9 spp. Madagascar, Comoroa Islands. Palm trees.

***R. robustior*** Jum. and Perr. (Anivo). Madagascar. The buds are eaten locally as a vegetable. The leaflets are used to weave hats.

RAVENSARA Sonnerat. Lauraceae. 18 spp. Madagascar. Trees.

***R. aromatica*** Gmel.=Agathophyllum aromaticum. (Madagascar Nutmeg). The seeds are used locally as a spice and the aromatic bark in the manufacture of rum.

Ravinson Oil – Brassica campestris.

Rayung-payungar – Tacca fatsiifolia.

Re pa – Kadsura coccinea.

***Reaumurea*** Steud.=Reaumuria.

REAUMURIA Hasselq. ex L. Tamaricaceae. 20 spp. E.Mediterranean to Central Asia and Baluchistan. Shrubs.

***R. hypericoides*** Willd. (Lanisah). Throughout range. The leaves are used by the Hindus to treat diseases causing itching of the skin.

Rebba bisah – Enicostema verticillatum.

RECORDOXYLON Ducke. Leguminosae. 2 spp. Brazil. Trees.

***R. amazonicum*** Ducke.=Melanoxylon amazonicum. Brazilian Amazon. The yellow brown wood is hard, heavy, durable and insect resistant. It is used locally for heavy construction work.

Red Acaroid – Xanthorrhoea australis.

Red Alder – Alnus rubra.

Red Alder – Cunonia capensis.

Red Ash – Fraxinus pubescens.

Red Ash – Orites excelsa.

Red Banana – Musa sapientum.

Red Bark – Cinchona succiruba.

Red Bay – Persea borbonia.

Red Beech – Flindersia brayleyana.

Red Beech – Northofagus fusca.

Red Beet – Beta vulgaris.

Red Brch – Betula nigra.

Red Biox – Eucalyptus polyanthemos.

Red Box – Tristania conferta.

Red Buckeye – Aesculus pavia.

Red Bud – Cercis canadensis.

Red Campion – Melandrium rubrum.

Red Camwood – Pterocarpus osun.

Red Canary Grass – Phalaris arundinacea.

Red Cardinal – Erythrina arborea.

Red Cedar – Acrocarpus fraxinifolius.

Red Cedar Wood Oil – Juniperus virginiana.

Red Clover – Trifolium pratense.

Red Currant – Ribes rubrum.

Red Elderberry – Sambucus callicarpa.

Red Elm – Ulmus elata.

Red Fescue – Festuca rubra.

Red Fir – Abies magnifica.

Red Fir – Pseudotsuga menziesii.

Red-flowered Mangrove – Lumnitzera coccinea.

Red Gum – Eucalyptus camaldulensis.

Red Gum – Eucalyptus tereticornus.

Red Gum – Liquidambar styraciflua.

Red Ironbark – Eucalyptus siderophloia.

Red Ironbark – Eucalyptus sideroxylon.

Red Ironwood – Reynosia septentrionalis.

Red Kano Rubber – Ficus platyphylla.

Red Mahogany – Eucalyptus resinifera.

Red Mahogany – Khaya nyasica.

Red Mahogany – Khaya ivorensis.

Red Mangrove – Rhizophora mangle.

Red Maple – Acer rubrum.

Red Mauritius Ebony – Diospyros rubra.

Red Mombin – Spondias purpurea.

Red Mountain Ash – Eucalyptus gigantea

Red Mulberry – Morus rubra.

Red Myrtle – Eugenia myrtifolia.

Red Oak – Quercus borealis.

Red Oak – Quercus difitata.

Red Osier Bark, American – Cornus amomum.

Red Padauk – Pterocarpus macrocarpus.

Red Pepper – Capsicum annuum.

Red Peruvian Cotton – Gossypium murocarpum.

Red Pine – Dacrydium cupressinum.

Red Pine – Pinus resinosa.

Red Poppy – Papaver rhoeas.

Red Root – Ceanothus americanum.

Red Root – Lachnanthes carolina.

Red Sandal Wood – Adenanthera pavonia.

Red Sandwort – Spergularia rubra.

Red Sandalwood – Pterocarpus santalinus.

Red Shank – Geranium robertianum.

Red Saunders – Pterocarpus santalinus.

Red Saunderswood – Pterocarpus santalinus.

Red Siris – Albizia toona.

Red Spruce – Picea rubia.

Red Stinkwood – Lauro-Cerasus africanum.

Red Stopper – Eugenia confusa.

Red Stopper – Eugenia rhombea.

Red Sumach – Rhus glabra.

Red Tulip Oak – Argyrodendron perabatal.

Redberry Bryony – Bryonia dioica.

Redberry Buckthorn – Rhamnus crocea.

Redroot Amaranth – Amaranthus retroflexus.

Redshank Chamise – Adenostoma sparsifolium.
Red Top – Agrostis alba.
Redwater Tree – Erythrophleum guineensis.
Redwood – Adina microcephala.
Redwood – Eucalyptus piperata.
Redwood – Pterocarpus soyauxii.
Redwood – Sequoia sempervirens.
Redwood, Brazilian – Brosimum paraense.
Redwood, Brazilian – Caesalpinia brasiliensis.
Redwood, Indian – Chukrasia tabularis.
Redwood, Indian – Soymida febrifuga.
Redwood, Sierra – Sequoia gigantea.
Redwood, Zambezi – Baikiaea plurijuga.
Reed Canary Grass – Phalaris arundinacea.
Reed, Giant – Arundo donax.
Reed, Giant Bur – Sparganium eurycarpum.
Reed Grass – Phragmites communis.
Reef Thatches – Thrinax parviflora.
REHMANNIA Libasch. ex Fisch. and Mey. Scrophulariaceae. 10 spp. E.Asia. Herbs.
***R. glutinosa*** Lib. N.China. A decoction of the plant is used locally as a treatment for fevers.
***R. lutea*** Maxim. China. A decoction of the plant is used locally as a diuretic.
Rehu – Cinnamomum cecicodaphne.
Reiabo Rubber – Landolphia sphaerocarpa.
Reindeer Moss – Cladonia rangiferina.
Reine Claude Plum – Prunus instititia.
Rekondidi – Impatiens irvingii.
Remanso Gum – Manihot dichotoma.
REMIJIA DC. Rubiaceae. 35 spp. Tropical S.America. Trees.
***R. pedunculata*** Flueck. Colombia. The bark (Cuprea Bark), is a source of quinine. It is exported to Europe.
Remir – Garcinia harmandii.
Rempoodong – Elaeocarpus valetonii.
Remy's Gardenia – Gardenia remyi.
RENANTHERA Lour. Orchidaceae. 13 spp. India, S.E.Asia, Madagascar, Malaysia, Solomon Islands. Herbs.
***R. moluccana*** Blume. Moluccas. The leaves, pickled in salt and vinegar are a local delicacy.
RENAELMIA L. f. Zingiberaceae. 75 spp. Tropical America, W.Indies, Tropical W.Africa. Herbs.
***R. domingensis*** Horan. = Alpinia aromatica. Tropical America. The seeds are used locally in Brazil to control menstruation. The juice from the plant was used by the Mayas as a cure for haemorrhoids.
Rengasz – Gluta rhengas.
Reniala Oil – Adansonia digitata.
Rennet, Indian – Withania coagulans.
REPTONIA A. DC. Sapotaceae. 2 spp. Arabia, to Afghanistan and N.W.India. Trees.
***R. buxifolia*** A. DC. India. The sweet fruits are eaten locally.
Resah gelingga – Vatica bancana.
Rescue Grass – Bromus catharticus.
RESEDA L. Resedaceae. 60 spp. Europe, Mediterranean to Central Asia. Herbs.
***R. lutea*** L. (Dyer's Weed). Near East, Mediterranean. The plant is the source of a deep yellow dye, used since ancient times to dye cloth. It was extensively cultivated.
***R. odorata*** L. (Mignonette). Mediterranean. The source of an essential oil (Reseda Oil) used in perfumery. The plant is cultivated for the oil and as an ornamental.
***R. phyteuma*** L. Mediterranean, Middle East. The leaves are sometimes used locally as a vegetable.
Reseda Oil – Reseda odorata.
Resha Katmi – Lavatera kashmiriana.
Resin, Abu Beka – Gardenia lutea.
Resin, Alriba – Canarium strictum.
Resin, Apitong – Dipterocarpus grandiflorus.
Resin, Baláu – Dipterocarpus grandiflorus.
Resin, Baláu – Dipterocarpus vernicifluus.
Resin, Black Dammar – Canarium bengalense.
Resin, Bolax – Azorella caespitosa.
Resin, Bulungu – Canarium edule.
Resin, Bursa Opopanax – Commiphora kataf.
Resin, Carana – Protium carana.
Resin, Combee – Gardenia lucida.
Resin, Conima – Protium guianense.
Resin, Dammar – Agathis lanceolata.
Resin, Dammar – Shorea Wiesneri.
Resin, Dikkamaly – Gardenia gummifera.
Resin, Dragon's Blood – Daemonorops draco.
Resin, Dragon's Blood – Dracaena cinnabari.
Resin, Dragon's Blood – Dracaena draco.
Resin, Dragon's Blood – Pergularia africana.
Resin, Elmira – Protium heptaphyllum.
Resin of Fiji – Agathis vitiensis.
Resin, Galadoopa – Pahudia galedupa.
Resin, Guaiac – Guaiacum sanctum.
Resin, Hal-dummala – Vateria acuminata.
Resin, Hyowana – Protium carana.

Resin, Incienso de los Criollos – Libothamnus nerifolius.
Resin, Kauri – Agathis lanceolata.
Resin, Levant Scammony – Convolvulus scammonia.
Resin, Maina – Calophyllum longifolium.
Resin, Maracaibo – Copaifera officinalis.
Resin, Maria – Rheedia acuminata.
Resin, Minyak Keruong – Dipterocarpus kerri.
Resin, Njating Mahabong – Hopea balageran.
Resin, Njating Matapoesa – Hopea sangal.
Resin, Pagsahingin – Canarium villosum.
Resin, Panau – Dipterocarpus vernicifluus.
Resin, Pagsahingin – Canarium villosum.
Resin, Rasamala – Altingia excelsa.
Resin, Rimu – Dacrydium cupressinum.
Resin, Ronima – Protium heptaphyllum.
Resin, Sagapen – Ferula szotitziana.
Resin, Sahing – Canarium villosum.
Resin, Sal – Shorea talura.
Resin, Sanderac – Tetraclinis articulata.
Resin, Saul – Shorea talura.
Resin, Siam Benzoin – Styrax tonkinense.
Resin, Socotra Dragon's Blood – Dracaena cinnabari.
Resin, Sumatra Benzoin – Styrax benzoin.
Resin, Tabonico – Dacryodes hexandra.
Resin, Tacamahaca – Calophyllum tacamahaca.
Resin, Tacamahaca – Calophyllum inophyllum.
Resin, Thapsia – Thapsia garganica.
Resin, Turpenthin – Operculina turpethum.
Resin, Yellow – Pinus palustris.
Resin, Yellow Acaroid – Xanthorrhoea hastilis.
Resin – See also Balsam, Balsamo, Copal, Getah.
Resina Acroides – Xanthorrhoea hastilis.
Resina de Cuapinole – Hymenaea courbaril.
Resina Draconis – Daemonorops draco.
Resina Laretiae – Laretia acoulis.
Resina Lutea – Xanthorrhoea australis.
Resina de Mammee – Mammea americana.
Resina Ocuje – Calophyllum calaba.
Resina Scammoniae – Convolvulus scammonia.
Rest Harrow – Ononis spinosa.
***Retama*** (Rafin.) Boiss. = Genista.
RETANILLA Brongn. Rhamnaceae. 2 spp. Peru, Chile. Shrubs.
***R. ephedra*** Brongn. Peru, Chile. A decoction of the leaves is used locally to treat indigestion.
Retemshrub – Retama roetam.
***Retinodendron rassak*** Korth. = Vatica rassak.
Retinol – Pinus palustris.
Reunja – Acacia leucophloea.
Revalenta Arabica – Lens esculenta.
Rewa – Knightia excelsa.
Rewarewa – Knightia excelsa.
Rewari – Abies pindrow.
REYNOLDSIA A. Gray. Araliaceae. 14 spp. Polynesia. Trees.
***R. marchionensis*** F. Br. (Pilohe pimata, Pimata omoa). Polynesia. The plant is used locally for scenting coconut oil, used as a cosmetic.
REYNOSIA Griseb. Rhamnaceae. 1 sp. Florida; 15 spp. W.Indies. Small trees.
***R. septentrionalis*** Urb. (Darling Plum, Red Ironwood). Florida to Bahamas. The purple-black fruits are edible with a pleasant flavour. The very hard wood is used for cabinet-making.
Rhal Dhooma – Shorea robusta.
RHAMNUS L. Rhamnaceae. 110 spp. Cosmopolitan. Shrubs.
***R. brandegiera*** Standl. = Krugeodendron ferreum.
***R. carolinianus*** Blanco = Colubrina asiatica.
***R. catharticus*** L. (Common Buckthorn). Europe, N.Africa, Asia. The berries, which contain rhamnin and rhamnetin, have been used since the Middle Ages to make a purgative syrup (Sirupus Rhamni Catharticae). The bark yields a yellow dye and the hard yellow wood is used for turning.
***R. crocea*** Nutt. (Redberry Buckthorn). S.W. U.S.A., Mexico. The berries are eaten by the local Indians.
***R. dahurica*** Pall. = R. utilis (Dahuruan Buckthorn). China. The leaves yield a green dye (China Green) used for art work and locally for dyeing fabrics.
***R. ferreus*** Vahl. = Krugeodendron ferreum.
***R. frangula*** L. = Frangula alnus. (Alder Buckthorn, Gloss Buckthorn). Europe, Asia, N.Africa. The bark contains the glucoside frangulin, which is used as a laxative. The wood is used for shoe lasts, veneer, wooden nails and charcoal.
***R. globosa*** Bunge. N.China. The leaves yield a green dye (Chinese Green, Vert de Chine) used in art work and locally for dyeing fabrics.
***R. globosus*** Sieb. and Zucc. = R. japonica.

***R. graecus*** Boiss. and Rent. Greece. The berries yield a yellow dye.

***R. infectorius*** L. S.Europe, Asia Minor. The unripe berries (French Berries, Fructus Rhamni, Graines d'Avignon, Persian Berries, Yellow Berries) were an important source of a yellow dye. The plant was grown commercially, especially in Turkey.

***R. japonica*** Maxim. = R. globosus. (Japanese Buckthorn). Japan. The bark and fruits are used locally as a laxative and the wood to make furniture.

***R. oleoides*** L. Mediterranean. The berries yield a yellow dye.

***R. prinoides*** L'Hérit. S.Africa. The leaves are used in Abyssinia to make a stimulating wine with honey (Tetschen) and a beer (Tallas).

***R. purshiana*** DC. (Bearberry, Bearwood, Buckthorn, Cascara Buckthorn). W. N.America. The bark, which contains cascarin, emodin and purshinianin, is the main source of the laxative Cascara Sagrada. The plant is cultivated in N.America and Kenya.

***R. saxatilis*** L. Central and S.Europe. The berries yield a yellow dye.

***R. theezans*** L. = Sageretia theezans.

***R. tinctoria*** Waldst. China. The leaves yield the green dye called China green, used in artwork and locally for dyeing fabrics.

***R. utilis*** Dcne. = R. dahurica.

RHAPHIDOPHORA Hassk. Araceae. 100 spp. Indomalaya, New Caledonia. Herbs.

***R. korthalsii*** Schott. Malaysia. The juice is used as an arrow poison in Perak.

***R. pertusa*** (Roxb.) Schott. Indomalaya. The juice, mixed with black pepper, is used locally to treat snakebites.

RHAPHIOLEPIS Lindl. corr Poir. Rosaceae. 15 spp. Subtropical, E.Asia. Trees.

***R. japonica*** Sieb. and Zucc. Japan. The bark yields a brown dye, used locally for dyeing cloth.

***Rhaponticum scariosum*** Lam. = Centaurea rhaponticum.

Rhatany, Para (Brazilian) – Krameria argentia.

RHAZYA Decne. Apocynaceae. 1 sp. Greece to N.W.Asia Minor; 1 sp. Arabia to N.W.India. Small trees.

***R. stricta*** Decne. (Hisawang, Sewar, Sihar). Himalayan region. A laqueur is derived from the trees and a wax from the surface of the fruit. The former is used for Japanese lacquer work and the latter for making candles.

Rhea – Boehmeria nivea.

RHEEDIA L. Guttiferae. 45 spp. Central and tropical S.America, W.Indies, Madagascar. Trees.

***R. acuminata*** Planch. and Triana. Tropical America. A greenish resin (Maria Balsam) is extracted from the trunk.

***R. brasiliensis*** Planch and Triana. (Bakupari). Brazil. The slightly acid fruits are used locally for making jams.

***R. edulis*** Planch and Triana. Tropical America. The edible acid fruits (Limao do Matto) are used locally for making jams. The plant is sometimes cultivated, especially in Brazil.

***R. lateriflora*** L. Tropical America, W.Indies. The plant yields a wax.

***R. macrophylla*** Planch and Triana. Brazil. The slightly acid fruits are used locally to make jams. The plant is sometimes cultivated.

***R. madruno*** (H.B.K.) Planch and Triana. Tropical America. The juicy, slightly acid and aromatic fruits are eaten raw, made into jams and drinks.

RHEUM L. Polygonaceae. 50 spp. Temperate and subtropical Asia. Herbs.

***R. embodi*** Wall. (Himalayan Rhubarb). Himalayas. Sold locally as a purgative.

***R. hybridum*** Murray. (Rhubarb). Probably from Mongolia. Possibly a hybrid – R. rhaponticum × R. palmatum. The petioles are used as a dessert and to make wine.

***R. officinale*** Baill. (Medicinal Rhubarb). China. The dried rhizomes are used as a purgative, laxative and tonic. They are usually mixed with ginger and magnesium oxide. The plant is cultivated and exported, especially from Shanghai. The active principles include glucogallin and tetrarin and calcium oxalate.

***R. palmatum*** L. (East Indian Rhubarb, Turkey Rhubarb, China Rhubarb). The dried rhizome is used as a tonic and to treat stomach upsets. The plant is cultivated and the medicinal product is known as Chinese Rhubarb, Shensi Rhubarb and Canton Rhubarb.

***R. rhaponticum*** L. (Garden Rhubarb). S.Siberia. The plant is cultivated in the Old and New World. Its cultivation is of some local commercial importance. The petioles are eaten as a dessert, stewed and in pies. It was grown in China before 2700 B.C. as a drug plant and was introduced to England in 1777 for the same purpose. The active principle is chrysarobin which is a purgative. The leaves contain a high percentage of calcium oxalate and are poisonous. Like other Rheum spp. propagation is by splitting the rhizome.

***R. ribes*** L. (Currant Rhubarb). Asia Minor. The roots are used locally to treat intestinal worms in horses.

***R. undulatum*** L. China. See R. hybridum.

Rheumatism Root – Jeffersonia diphylla.

RHINACANTHUS Nees. Acanthaceae. 15 spp. Tropical Africa, Madagascar, E.Asia, Indomalaya. Small shrubs.

***R. communis*** Nees. = R. nasuta.

***R. nasuta*** (L.) Kurz. = R. communis = Justicia nasuta. India. The Hindus used the crushed leaves, roots and seeds mixed with lime juice to treat skin diseases. The seeds are also used as an aphrodisiac.

***R. osmospermus*** Roj. (Voanalakoly). Madagascar. The plant is aromatic and the leaves are used locally to scent the hair and make scented sachets.

***Rhioicissus capensis*** (Burm. f.) Planch. = Vitis carpensis.

RHIPOGONUM J. R. and G. Forst. emend. Spreng. Liliaceae. 7 spp. E.New Guinea, E.Australia, New Zealand. Vines.

***R. scandens*** Forst. (Supplejack) New Zealand. A decoction of the roots is used locally to treat intestinal complaints, rheumatism and skin diseases and to induce abortions. The sap from the stem is used to cauterize wounds and the burnt stem is used for the same purpose. The stems are woven into baskets and sometimes used to make rope ladders.

RHIZOBIUM Frank. Bacteriaceae.

***R. leguminosarum*** Frank. = Bacterium radicicola. In association with nodules on the roots of legumes, this bacterium is capable of fixing atmospheric nitrogen, making it available to plants.

RHIZOCARPON L. Lecideaceae. Lichens.

***R. geographicum*** (L.) DC. (Map Lichen). Temperate. The source of a brown dye used in Scandinavia for dyeing wool.

Rhizoma Anemopaegimae – Anemopaegma mirandrum.

Rhizoma calami – Acorus calamus.

Rhizoma Caricis arenariae – Carex arenaria.

Rhizoma galangae – Alpinia officinarum.

Rhizoma Iridis – Iris germanica.

Rhizoma Urticae – Urtica dioica.

Rhizoma Veratri – Veratrum album.

Rhizoma Vincetoxici – Cyanchum vincitoxicum.

RHIZOPHORA L. Rhizophoraceae. 7 spp. Tropical Coasts. Trees.

***R. conjugata*** L. (Bakau akik, Bako, Bangka minjak). Malaysia, Australia. The brace roots are used locally in Malaya for making anchors and fishing nets.

***R. mangle*** L. (Red Mangrove). Tropical America. The bark is an important commercial source of tannin. It is also used locally as a treatment for haemorrhages and to reduce fevers. The hard, heavy wood is resistant to Teredo worms and is used to make wharves etc. The flowers are a good source of honey.

***R. mucronata*** Lam. Tropical Africa and Asia. The bark is used for tanning.

RHIZOPOGON Fr. Hymenogasteraceae. 15 spp. Widespread.

***R. luteolus*** Fr. and Nordh. Temperate, subtropical. Lives as a mycorrhiza with the roots of Pinus spp. and is consequently important in establishing their seedlings and in reforestation.

***R. rubescens*** Tul. Temperate. The fruit-bodies are eaten locally in Japan.

RHODAMNIA Jack. Myrtaceae. 20 spp. S.E.Asia Malaysia, E.Australia, New Caledonia. Trees.

***R. arborea*** Jack. = R. concolor = R. spectabilis. (Andong, Djending, Ki beusi). India to E.Australia. The wood is used, especially in Sumatra for making houses and agricultural implements. It is also burnt for charcoal. The plant is sometimes cultivated.

***R. concolor*** Miq. = R. arborea.

***R. spectabilis*** Blume. = R. arborea.

Rhode Island Bent – Agrostis vulgaris.

***Rhodea*** Endl. = Rohdea.

Rhodes Grass – Chloris gayana.

Rhodesian Chestnut – Baikiaea plurijuga.

Rhodesian Ironwood – Copaifera mopane.

Rhodesian Teak – Baikiaea plurijuga.

Rhodesian Tree Hibiscus – Thespesia garckeana.

Rhodesian Wistaria – Bolusanthus speciosus.

***Rhodiola integrifolia*** Raf. = Sedum roseum.

Rhodium, Oil of – Convolvulus floridus.

RHODOCOLEA Baill. Bignoniaceae. 6 spp. Madagascar. Trees.

***R. racemosa*** (Baill.) Perr. Madagascar. The wood (Green Ebony, Ebène vert) is green, hard, durable and very attractive. It is used for cabinet work etc.

***R. telfairiae*** (DC.) Perr. See R. racemosa.

RHODODENDRON L. Ericaceae. 500–600 spp. N.Temperate. Shrubs of varying size.

***R. chrysanthum*** Pall. India. A decoction of the leaves causes sweating and is used in the treatment of rheumatism and gout.

***R. ferrugineum*** L.=Chamaerhodendron ferrugineum. (Rusty-leaved Alprose). Alpine Europe. The leaves and galls are poisonous, but a decoction of the dried leaves is used to induce sweating in the treatment of rheumatism and as a diuretic.

***R. maximum*** L. (Great Laurel, Rose Bay). E. N.America. A decoction of the leaves is used locally to treat rheumatism. The hard, strong, brittle wood is used for tool handles and for carving. Tobacco pipes are carved from the roots.

***R. nudiflorum*** (L.) Nutt.=Azalea nudiflora. (Poinzter Flower). S.E. N.America. Galls (May Apples) on the stems and leaves are pickled and used as a condiment.

Rhododendron, Singapore – Melastroma malabathricum.

RHODOMYRTUS Reichb. Myrtaceae. 20 spp. S.India, Ceylon, Thailand, New Guinea, New Caledonia, Australia. Shrubs.

***R. tomentosa*** Wight. (Downy Rose Myrtle). S.India, Malaysia. Cultivated in Old and New World tropics. The fruits are sometimes eaten in pies.

***Rhodopaxillus*** Maire=Tricholoma.

RHODOSPHAERA Engl. Anacardiaceae. 1 sp. E.Australia. Tree.

***R. rhodanthema*** Engl.=Rhus rhodanthema. (Yellow Cedar). The soft, beautifully marked wood is used for cabinet work.

***Rhodostachys argentina*** Baker=Bromelia serra.

RHODYMENIA Grev. Rhodymeniaceae. Red Algae.

***R. palamata*** (L.) Grev. (Dulse). Pacific and Atlantic Coasts. The plant is eaten cooked, when fresh or after drying in New England, Scotland and Ireland. It is used as a condiment and in making ragouts. The natives of Kamschatka ferment it to make an alcoholic beverage.

***Rhoicissus capensis*** (Burm. f.) Planch.= Vitis capensis.

RHOPALOSTYLIS H. Wendl. and Drude. Palmea. 3 spp. New Zealand, Norfolk Islands, Kermadec Islands. Trees.

***R. sapida*** Wendl. and Drude. (Nikau Palm). New Zealand. The Maoris used the leaves for thatching.

Rhubarb – Rheum hybridum.

Rhubarb, China – Rheum palmatum.

Rhubarb, Current – Rheum ribes.

Rhubarb, East Indian – Rheum palmatum.

Rhubarb, Garden – Rheum rhaponticum.

Rhubarb, Himalaya – Rheum emodi.

Rhubarb, Indian – Saxifraga peltata.

Rhubarb, Medicinal – Rheum officinale.

Rhubarb, Monk's – Rumex alpinus.

Rhubarb, Turkey – Rheum palmatum.

RHUS L. 250 spp. Subtropics, warm temperate. Shrubs, small trees, sometimes vines.

***R. abyssinica*** Hochst. E.Africa. The bark is used locally for tanning.

***R. ailanthoides*** Bunge=Picrasma quassoides.

***R. albida*** Schousb.=R. dioica=R. oxycanthoides N.Africa. The bark is used locally for tanning sheep skins and the wood is used for charcoal.

***R. americanus*** Nutt.=R. cotinoides.

***R. aromatica*** Ait.=R. canadensis (Fragrant Sumach). E. N.America to Texas. The root bark is diuretic and is used in the treatment of kidney and bladder complaints, diarrhoea, dysentery and vaginal discharges. The soaked berries are used to make a drink and the split stems are used by the local Indians to make baskets.

***R. canadensis*** Marsh.=R. aromatica.

***R. chinensis*** Mill.=R. semialata.

***R. copallina*** L.=Schmaltzia copallina. (Shining Sumach, Dwarf Sumach). E. N.America. The leaves are used for tanning and dyeing, the fruits were used by the local Indians to make a drink and they used the roots as a treatment for dysentery.

***R. coriaria*** L. (Sicilian Sumach). Mediterranean, Asia Minor. The leaves contain between 20 and 35 per cent tannin and are an important source of this material for tanning leather.

***R. cotinoides*** (Nutt.) Britt.=R. americanus =Cotinus americanus. (American Smoke Tree). E. N.America. The wood yields an orange dye which is used locally for dyeing cloth.

***R. cotinus*** L. = Cotinus coggygria. (Common Smoke Tree). Mediterranean to China. The powdered leaves (Molito) are an important source of tannin (Ventilato).

***R. dioica*** Willd. = R. albida.

***R. glabra*** L. (Red Sumach, Smooth Sumach). E. N.America. The fruits are used locally and by the local Indians to make an astringent and refreshing drink. The Indians also use a decoction of the flowers to gargle for sore throats, and smoke the leaves, sometimes mixed with tobacco. The roots are the source of a yellow dye.

***R. hindsiana*** Engl. = R. integrifolia.

***R. hirta*** (L.) Sudw. = R. typhina.

***R. incana*** Auct. = R. vulgaris.

***R. integrifolia*** (Nutt.) Benth. and Hook. = R. hindsiana = Neotyphonia integrifolia. (Lemonade Berry). S.California, Mexico. The berries are used locally to make a refreshing drink.

***R. javanica*** Thunb. non L. = R. semialata.

***R. metopium*** L. = Metopium toxiferum.

***R. ovata*** S. Wats. = Neotyphonia ovata. S.W. U.S.A., Mexico. The flowers yield a good honey and the wax from the plant is eaten as a vegetable by the local Indians.

***R. oxycanthoides*** Dum. = R. albida.

***R. potanini*** Maxim. (Potanin Sumach). China. Galls (Ch'-pei-tzu) are used to make a black ink.

***R. rhodanthema*** F. v. Muell. = Rhodosphaera rhodantheme.

***R. semialata*** Murr. = R. chinensis = R. javanica. China, Japan. Galls (Chinese Galls, Japanese Galls, Wu-pei-tzu) are used for tanning and the source of a blue dye used for dyeing silk in China.

***R. sempervirens*** Scheele = R. virens. (Evergreen Sumach). S.W. U.S.A., Mexico. The local Indians smoke the leaves, sometimes mixed with tobacco.

***R. simarubaefolia*** Gray. = R. taitensis. S.Pacific. The wood is used to make canoes in Samoa.

***R. succedanea*** L. (Wax Tree). China, Japan. The plant is cultivated locally for the fruits from which a wax (Sumach Wax, Japan Tallow, Vegetable Wax) is extracted. The wax is used for making varnishes, floor polish, ointments and plasters. Galls on the leaves are used for tanning.

***R. sylvestris*** Sieb. and Zucc. China, Japan. See R. vernicifera.

***R. taitensis*** Guill. = R. simarubaefolia.

***R. toxicodendron*** L. (Poison Ivy). E. N.America. Poisonous. The leaves were used medicinally for treating rheumatism, fevers, eye diseases, neuralgia and hypertrophy of the heart. Some Indian tribes considered the plants poisonous, while others used the roots as a poultice to open swellings.

***R. typhina*** = R. hirta. (Lemonade Tree, Staghorn Sumach). E. N.America. The fruits are used to make a drink (Indian Lemonade).

***R. veitchiana*** Kellogg = Pachycormus discolor.

***R. vernicifera*** DC. = R. verniciflua. (Lacquer Tree, Varnish Tree). Japan, China. A varnish (Japanese Varnish) is extracted from lacerations in the bark. The exuded liquid is collected and allowed to oxidize in the air, when it turns black and hard. An oil is also extracted from the plant which is used locally for making candles.

***R. virens*** Lindh. = R. sempervirens.

***R. vulgaris*** Miekle = R. incana. (Abili, Akelwak, Mugengo, Kishanula). Central and N.E.Africa. The wood is used locally for general construction work and as firewood.

***R. wallichii*** Hook. f. (Akoria, Arkol, Kambul). Himalayan region. The wood is used for tool handles and musical instruments. The wax from the fruit is used to manufacture candles and a varnish is extracted from the trunk.

RHYNCHOSIA Lour. Leguminosae. 300 spp. Tropics and subtropics, especially America and Africa. Herbs or small shrubs.

***R. longeracemosa*** Mart. and Gal. = Dolichos longeracemosa. Mexico. The poisonous seeds are used locally as a narcotic.

***R. monobotrya*** Harms. (Itendale, Mugharokoko). Congo, Tanzania. The leaves are eaten locally as a vegetable.

***R. phaseoloides*** DC. = Dolicholus phaseoloides. (Ojitos de Picho, Ojo de cangnejo, Pulguitas). Tropical America. The poisonous seeds are used locally as beads.

***R. pyramidalis*** (Lam.) Urb. = Dolicos phaseoloides. (Colorines, Nahuatl). Mexico. See R. longeracemosa.

RHYTICOCOS Becc. Palmae. 1 sp. W.Indies. Palm tree.

***R. amara*** (Jacq.) Becc.=Syagrus amara. (Coco Nain, Overtop Palm). The leaves are used for thatching and a palm wine was made from the sap. A fermented beverage was also made from the fruits.

Rib Grass – Plantago lanceolata.

Ribbon Eucalyptus – Eucalyptus viminalis.

Ribbon Gum – Eucalyptus nitens.

Ribbon Kino – Eucalyptus salicifolia.

Ribbon Wood, Lowland – Plagianthus betulinus.

RIBES L. Grossulariaceae. 150 spp. N.Temperate, Andes. Shrubs. The fruits of the species mentioned below are eaten, rarely raw, but mainly in pies, jams, etc.

***R. americanum*** Mill.=R. floridum. (American Wild Black Currant). E. N.America.

***R. aureum*** Pursh (Buffalo Currant, Golden Currant, Missouri Currant). W. N.America. Also used by the local Indians to flavour pemmican made of dried buffalo meat, tallow and the berries of R. aureum.

***R. bracteosum*** Dougl. W. N.America. Eaten with cooked salmon roe by the Alaskan Eskimos.

***R. cynosbati*** L. (American Wild Gooseberry, Prickly Gooseberry). E. N.America. Sometimes cultivated.

***R. divaricatum*** Dougl. W. N.America. Eaten by the local Indians, fresh or dried as winter food.

***R. floridum*** L'Hérit.=R. americanum.

***R. grossularia*** L. (European Gooseberry). Europe, across to N.China, N.Africa. Widely cultivated in the Old World. The plant was first cultivated in U.K. in 1600. It is grown as a garden and market-garden crop. A wine is sometimes made from the fruit. Propagation is by cutting. Cultivation is not encouraged in U.S.A. due to the prevalence of Gooseberry Rust and Gooseberry Mildew.

***R. lacustre*** Boir. (Prickly Currant). N.America. The fruits are eaten by the Eskimos.

***R. laxiflorum*** Pursh. See R. divaricatum.

***R. longeracemosum*** French. W.China. Cultivated locally.

***R. nigrum*** L. (European Black Currant). Europe to Central Asia. Cultivated widely in the Old World as a garden and market-garden crop. The fruits are not eaten raw, but are used mainly for making jam and drinks with a high vitamin C content. They are also fermented to make a liqueur (Cassis). The dried leaves are used in home remedies for coughs. Propagation is by cutting.

***R. odoratum*** Wendl. (Buffalo Currant). U.S.A. East of the Rockies. The fruits are used locally to make jellies and are eaten raw by the local Indians.

***R. oxycanthoides*** L. (Smooth Gooseberry). E. N.America.

***R. rubrum*** L.=R. silvestre. (Red Currant). Europe, Asia. Cultivated as a garden and market-garden crop. The fruits are used to make jellies, jam sauces, wine etc. There are red and white fruited varieties. Bar-le-Duc Jelly is made from a white variety. Propagation is by cutting.

***R. sivestre*** Mert. and Koch.=R. rubrum.

***R. triste*** Pall. (American Red Currant, Swamp Red Currant). E. N.America.

Ribseed – Aulospermum longipes.

Rice – Oryza sativa.

Rice Bean – Phaseolus mungo.

Rice of the Earth – Fritillaria camschatcensis.

Rice, Indian – Zizania aquatica.

Rice, Jungle – Echinochloa colonum.

Rice Paper – Fatsia papyrifera.

Rice Root – Epicampes macroura.

Rice, Wild – Oryza barthii.

Rice, Wild – Zizania aquatica.

RICHARDIA L. Rubiaceae. 10 spp. Tropical S.America. Shrubs.

***R. pilosa*** H.B.K. Tropical S.America. Yields an ipecachuana (Farinaceous Ipecachuana, Undulated Ipecachuana).

***Richardsonia*** Kunth.=Richardia.

RICHERIA Vahl. Euphorbiaceae. 7 spp. Tropical S.America. Trees.

***R. grandis*** Vahl. W.Indies. The wood (Bois Bande) is used locally for boards. The bark is used locally as an aphrodisiac.

Richmond Finger Grass – Digitaria diversinervis.

RICINODENDRON Muell-Arg. Euphorbiaceae. 6 spp. Tropical and S.W.Africa. Trees.

***R. africanum*** Muell-Arg.=R. heudelotii.

***R. heudelotii*** Klotzsch ex Pax.=R. africanum=Jatropha heudelotii. (Essang). W.African Coasts. The nuts (Essang Nuts, Ojok Nuts, Sanga Nuts) are eaten locally and yield an oil (Essang Oil). The oil is used locally for cooking food and industrially as a drying oil.

***R. rautanenii*** Schinz. S.W.Africa. The seeds yield a drying oil (Manketti Nut Oil) used for paints, varnishes and food preparation.

RICINUS L. Euphorbiaceae. 1 sp. Tropical Africa and Asia. Tree.

***R. communis*** L. (Castor Oil Plant, Castor Bean, Palma Christi).The plant is grown in the Old and New World Tropics for the seeds from which Castor Oil is extracted. India and Brazil produce 55–65 per cent of the world crop, with Manchuria, China and Mexico producing lesser amounts. Mechanical harvesting is difficult due to the pods shattering, but new varieties are being bred which have non-shattering pods. The seeds contain 35–55 per cent oil and a toxin ricin, which makes the expressed cake unsuitable for stock feed. The oil is extracted by pressing and the residue is treated with solvents which extract the remains of the oil. The cake is then used as fertilizer. When the oil is sulphonated by treatment with sulphuric acid it is called Turkey Red Oil, which is used with alizarine for dyeing fabrics. When heated, or treated with alkali it is used in perfumery, synthetic resins and plasticizers. When dehydrated it is used extensively in paints and varnishes. Its other uses include greases, hydraulic fluids, dyes, textile finishes, plastic, printing inks, fungistatic preparations, cosmetics and hair oils. Medicinally it is used as a laxative. The oil has S.G. 0·958–0·968; Sap. Val. 177–178; Iod. No. 82–90, Unsap. 0·3–0·7 per cent. The stems are pulped and used as a source of cellulose for making newsprint, cardboard and wall-boards and an insecticide is extracted from the leaves. The plant is a large annual and is propagated from seed.

Rièng rùng – Globba panicoides.

Rimu Resin – Dacrydium cupressinum.

Ring Lichen – Parmelia centrifuga.

Ringwood Senna – Cassia alata.

Riñon – Annona cineraria.

RINOREA Aubl. Violaceae. 340 spp. Warm. Shrubs. The leaves of the following species are eaten by the Negroes of Brazil as a vegetable.

***R. castaneaefolia*** (Spreng.) O. Ktze. S.America.

***R. physiphora*** (Mart. and Zucc.) O. Ktze. S.America.

Rio Nunez Coffee – Coffea robusta.

Rio Rosewood – Dalbergia nigra.

Ripple Grass – Plantago major.

Risoong batoo – Triomma macrocarpa.

Ritha – Sapindus trifoliatus.

RIVEA Choisy. Convolvulaceae. 5 spp. India, S.E.Asia, Central America. Vines.

***R. campanulata*** (L.) House. Central America. The sap is used locally to coagulate the latex of Castilla elastica.

***R. corymbosa*** (L.) Hallier f.=Convolvulus corymbosa, C. sidaefolius=Turbina corymbosa. (Ololiuqui, Piule, Yerba de la Virga). Florida, Central America, W.Indies. The seeds are used as an hallucinatory drug in parts of Mexico.

River Bank Grape – Vitis vulpina.

River Birch – Betula nigra.

River Black Oak – Casuarina suberosa.

River Oak – Casuarina stritya.

River Oak – Casuarina torulosa.

River Plum – Prunus americana.

RIVINA L. Phytolaccaceae. 3 spp. Tropical America. Shrubs.

***R. humilis*** L. (Rouge Plant). Tropical America. A red dye is extracted from the fruits.

***R. octandra*** L.=Trichostigma octandrum.

Riwat – Celtis cinnamomea.

Robin's Rye – Polytrichum juniperinum.

ROBINIA L. Leguminosae. 20 spp. E. N.United States of America, Mexico. Trees.

***R. pseudacacia*** L. (Black Locust, Locust) E N.America. The wood is heavy, hard, strong and durable. It is used for general construction, ship-building, turning and as fuel. The seeds were eaten cooked by the local Indians. The plant is poisonous, acting as a purgative and emetic. The flowers are a good source of honey. The plant was introduced to Europe in 1636 and is grown there as an ornamental.

Roble – Tabebuia heterophylla.

Roble Pellin – Nothofagus obliqua.

Roblo Blanco – Tabebuia heterophylla.

Roblo Blanco – Tabebuia pentaphylla.

Robusta Coffee – Coffea canephora.

Robusta Coffee – Coffea robusta.

ROCCELLA L. Rocellaceae. Lichen.

***R. fuciformis*** (L.) Lam. (Angola Weed, Lima Weed, Orcella). Mediterranean, Africa. Yields a purple to red-yellow dye

(Orchil) used in France and England for dyeing silk, wool, wood and marble.

***R. montagnei*** Bél. Africa, Asia, Australia. Yields a blue dye, used in Italy and Germany.

***R. phycopsis*** Ach. (Archil). Asia. Used in U.K. for dyeing blue broadcloth and as a tincture in alcohol in thermometers.

***R. tinctoria*** C. Mediterranean, Africa, Asia, Australia. The plant has been used since before Roman times to dye wool and silk a purple colour. It is also used to colour wines. Its main use now is as a pH indicator, being red in acid and blue in alkali. The Netherlands is now the main centre of production. The dye for fabrics is called Orseille or Persio and as an indicator, Litmus.

Rock Balsam – Clusia plukenetii.

Rock Dammar – Hopea odorata.

Rock Elm – Ulmus thomasii.

Rock Mushroom – Gyrophora esculenta.

Rock Tripe – Gyrophora deusta.

Rock Tripe – Gyrophora muhlenbergii.

Rock Urceolaria – Urceolaria scruposa.

Rockbrake – Polypodium vulgare.

Rocket, American Sea – Cakile edentula.

Rocket Larkspur – Delphinium ajacis.

Rocket Salad – Eruca sativa.

Rocket, Yellow – Barbarea vulgaris.

Rocky Mountain Cherry – Prunus melanocarpus.

Rocky Mountain Fir – Abies lasiocarpa.

Rocky Mountain Flax – Linum lewisii.

Rocky Mountain White Pine – Pinus flexilis.

ROGERIA J. Gay. Pedaliaceae. 6 spp. Brazil, tropical and S.Africa. Herbs.

***R. adenophylla*** J. Gay. Red Sea region, Tropical Africa. The mucilage from the plant is used to treat diarrhoea and the seeds are used by the Moors to make a porridge.

Roghum Oil – Carthamus oxycanthus.

ROHDEA Roth. Liliaceae. 3 spp. E.Asia. Herbs.

***R. japonica*** Roth. and Kunth. Japan. The Chinese use the plant as a diuretic.

Rohituka Oil – Amoora rohituka.

Roho Mahogany – Dysoxylum fraseranum.

Roi Bois – Psittacanthus martinicensis.

ROLLINIA A. St. Hil. Annonaceae. 65 spp. Central America to Argentine; W.Indies. Trees. The fruits of all the species mentioned are sweet and edible. They are much esteemed locally.

***R. deliciosa*** Safford. (Biribá). Brazil. Fruits called Fructa da Condenssa.

***R. dolabripetala*** (Reddi) St. Hil. = R. longifolia.

***R. laurifolia*** Schlecht. (Araticú, Mirim). Brazil.

***R. longifolia*** St. Hil. = R. dolabripetala. Brazil. Sometimes cultivated.

***R. mucosa*** (Jacq.) Baill. (Wild Cachiman). Mexico to S.America, W.Indies.

***R. orthopetala*** A. DC. S.America.

***R. pulchrinervis*** A. DC. Guyana.

***R. sieberi*** Standl. (Cachiman Montagne). W.Indies.

ROLLINIOPSIS Safford. Annonaceae. 5 spp. Brazil. Small trees.

***R. discreta*** Safford. Brazil. The fruits (Fructa de Macaco) are eaten locally. The plant is sometimes cultivated.

Rom doul – Popowia aberrans.

Roman Chamomille – Anthemis nobilis.

Roman Fennel – Foeniculum vulgare.

Roman Woodworm – Artemisia pontica.

Romduol si phlé – Goniothalamus repevensis.

Romerito – Suaeda ramissima.

Ronima Resin – Protium heptaphyllum.

Roodjak – Sonneratia ovata.

Roddon – Tarenna incerta.

Roof House Leek – Sempervivum tectorum.

Rooiwortel – Bulbine aloides.

Root, American Ginseng – Panax quinquefolia.

Root, Baboon – Babiana plicata.

Root, Brazilian Rhatany – Krameria argentea.

Root, Briar – Erica arborea.

Root, Brier – Smilax china.

Root, Chay – Oldenlandia corymbosa.

Root, Colic – Aletrus farinosa.

Root, Coral – Corallorhiza odontorhiza.

Root, Emetic – Euphorbia corollata.

Root, Gentian – Gentiana lutea.

Root, Helonias – Chamaelirium luteum.

Root, Hound's Tongue – Gynoglossum officinale.

Root, Life – Senecio aureus.

Root, Lopez – Toddalia aculeata.

Root, Mandragora – Mandragora officinarum.

Root, Mush – Ferula sumbul.

Root, Nardus – Nardostachys jatamansi.

Root, Orange Milkweed – Asclepias tuberosa.

Root, Orris – Iris florentina.

Root, Palo Pangue – Gunnera chilensis.

Root, Papoose – Caulophyllum thalictroides.
Root, Pareira – Chondodendron tomentosum.
Root, Pleurisy – Asclepias tuberosa.
Root, Polypody – Polypodium vulgare.
Root, Purging – Euphorbia corollata.
Root, Radix Junci – Cyperus corollata.
Root, Red – Ceanothus americanus.
Root, Red – Lachnanthes carolina.
Root, Rice – Epicampes macroura.
Root, Savanilla Krameria – Krameria ixina.
Root, Savanilla Rhatany – Krameria triandra.
Root, Squaw – Caulophyllum thalictroides.
Root, Sweet Corn – Calathea alluia.
Root, Tari – Caesalpinia digyna.
Root, Unicorn – Aletris farinosa.
Root, Violet – Ferula sumbul.
Rootbark, Archietea – Archietea salutaris.
Roquette – Eruca sativa.
***Rorippa armoracia*** Scop. = Armoracia lapathifolia.
***Rorippa nasturtium-aquaticum*** (L.) v. Hayek. = Nasturtium officinale.
ROSA L. Rosaceae. 250 spp. N.Temperate, tropical mountains. Shrubs. Various cultivars are grown commercially and sold extensively as ornamentals.
***R. banksiae*** Ait. (Banks' Rose). China. The roots are used locally for tanning.
***R. californica*** Cham. and Schlecht. (Californian Rose). Oregon to California. The fruits (Macuatas) are eaten cooked or raw after being sweetened by the frost.
***R. canina*** L. (Dog Rose, Dog Hip). Europe to temperate Asia. The fruits are used as a diuretic and astringent. They are a good source of vitamin C and are collected for the extraction of Rose Hip Syrup during emergencies, e.g. war time. The leaves are used locally as a tea substitute.
***R. centrifolia*** L. = R. gallica var. centrifolia. E. Caucasus. Cultivated, especially in S.France and Morocco for the flowers from which a rose water (Aqua Rosae Fortior) is made by distillation with water. It is used in perfumery.
***R. cherokeensis*** Donn. = R. laevigata.
***R. damascena*** Mill. (Damask Rose). Balkans and Asia Minor. Cultivated, especially in Bulgaria for the flowers from which Attar (Otto) of Roses is made by steam distillation. In France, the oil (Oil of Roses) is extracted by solvent extraction. 1000 g. of flowers yields 0·5 g. of oil which is used in perfumery and for flavouring. The oil, which is extracted from the nearly opened buds contains citronello, geranol, nerol, linalol, etc.
***R. eglanteria*** L. = R. lutea. (Eglantine Rose). Iran to N.India, Asia Minor. The flowers mixed with honey (Gulangabin) are used in confectionery. In Persia they are used to treat stomach pains and diarrhoea.
***R. gallica*** L. (French Rose). Europe, W.Asia. The petals from the nearly opened buds are dried and used as a tonic and astringent. The plant is cultivated.
***R. gallica*** L. var. ***centrifolia*** Reg. = R. centrifolia.
***R. laevigata*** Mich. = R. cherokeensis. China, escaped to S. U.S.A. An infusion of the leaves is used in China to treat spermatorrhoea.
***R. lutea*** Mill. = R. eglanteria.
***R. pomifera*** Herrm. = R. villosa var. pomifera. Europe to W.Asia. The fruits (Rose Hips) are used locally, especially in W.Europe, in preserves, sauces and drinks (Rose Wine, Rose Honey). The leaves are used as a tea substitute (Deutscher Tee).
***R. roxburghii*** Tratt. China. The fruits are used locally as an infusion to treat dyspepsia.
***R. rugosa*** Thunb. (Rugosa Rose). China, Japan. The Rose Hips are used as R. pomifera and are eaten locally in Japan.
***R. villosa*** L. var. ***pomifera*** (Herm.) Crép. = R. pomifera.
Rosa Morada – Lonchocarpus hondurensis.
Rosa de San Juan – Macrosiphonia hypoleuca.
Rosada, Balata – Sideroxylon cyrtibotryum.
Rosada, Balata – Sideroxylon resiniferum.
Rosary Pea – Arbus precatorius.
Rose Apple – Eugenia jambos.
Rose Apple, Samarang – Eugenia javanica.
Rose Apple, Water – Eugenia aquea.
Rose, Atta of – Rosa damascena.
Rose, Banks' – Rosa banksiae.
Rose Bay – Rhododendron maximus.
Rose, Californian – Rosa californica.
Rose, Christmas – Helleborus niger.
Rose Clover – Trifolium hirtum.
Rose, Damask – Rosa damascena.
Rose Dammar – Vatica rassak.
Rose, Desert – Adenium obesum.
Rose, Dog – Rosa canina.
Rose, Eglantine – Rosa eglanteria.

Rose Elf – Claytonia virginiana.
Rose, French – Rosa gallica.
Rose Gentian, Square-Stemmed – Sabatia angularis.
Rose and Gold Lichen – Stricta aurata.
Rose, Guelder – Viburnum opulus.
Rose Hip – Rosa pomata.
Rose Hip – Rosa rugosa.
Rose Honey – Rosa pomifera.
Rose of Jericho – Anastatica hierochuntica.
Rose Leek – Allium canadense.
Rose Mahogany – Ceratopetalum apetalum.
Rose Maple – Cryptocarya erythroxylon.
Rose Marara – Weinmannia lachnocarpa.
Rose, Mountain – Antigonon leptopus.
Rose, Oil of – Rosa damascena.
Rose, Otto of – Rosa damascena.
Rose, Rugosa – Rosa rugosa.
Rose, Stock – Sparmannia africana.
Rose Wine – Rosa pomifera.
Rosebay Willowherb – Epilobium angustifolium.
Roselle – Hibiscus sabdariffa.
Rosemary (Oil) – Rosmarinus officinalis.
Rosemary, Bog – Andropogon aciculatus.
Rosemary, Bog – Andropogon polifolia.
Rosemary, Marsh – Statice limonium.
Rosenoble – Scrophularia urdosa.
Rosetta Rosewood – Dalbergia latifolia.
Rosewood – Amyris balsamifera.
Rosewood – Dysoxylum fraseranum.
Rosewood – Eremophila mitchelli.
Rosewood – Synoum glandulosum.
Rosewood, Black – Dalbergia latifolia.
Rosewood, Bahia – Dalbergia nigra.
Rosewood, Bombay – Dalbergia latifolia.
Rosewood, Brazilian – Dalbergia nigra.
Rosewood, Brazilian – Physocalymma scaberrimum.
Rosewood, East Indian – Dalbergia latifolia.
Rosewood of Guatemala – Dalbergia cubilquitzensis.
Rosewood of Lower Amazon – Dalbergia spruceana.
Rosewood, Malabar – Dalbergia latifolia.
Rosewood, Nicaragua – Dalbergia retusa.
Rosewood Oil – Aniba terminalis.
Rosewood, Rio – Dalbergia nigra.
Rosewood, Rosetta – Dalbergia latifolia.
Rosewood, Seychelles – Thespesia populnea.
Rosewood of Siam – Dalbergia cochinchinensis.
Rosewood of Southern India – Dalbergia latifolia.
Rosewood, White – Dalbergia nigra.
Rosewood, Yama – Platymiscium dimosphandrum.
Rosha Grass – Cymbopogon martini.
Rosilla – Helenium puberulum.
Rosin (Oil) – Pinus palustris.
Rosinol – Pinus palustris.
Rosinweed – Silphium laciniatum.

ROSMARINUS L. Labiatae. 3 spp. Mediterranean. Herbs.

***R. officinalis*** L. (Rosemary). Mediterranean. Cultivated locally for the leaves from which an essential oil (Oil of Rosemary, Oleum Rosemarini) is distilled. It is used in perfumery (in Eau de Cologne and Hungarian Water), in mouth washes and Vermouth; medicinally it is used in liniments and to relieve indigestion. The dried leaves are used for flavouring meat dishes and sausages, in an infusion to relieve headaches and smoked to relieve asthma and bronchitis. They are also burnt as a disinfectant. The flowers are a good source of honey.

ROSTKOVIA Desv. Juncaceae. 1 sp. New Zealand, Campbell Islands, Tierra del Fuego, Falkland Islands, S.Georgia; 1 sp. Tristan da Cunha. Herbs.

***R. grandiflora*** Hook. f. = Marsippospermum grandiflorum. Falkland Islands etc. into Chile. The leaves are used locally for thatching, basket-making etc.

Rosval – Croton dioidus.
Rotenone – Lonchocarpus nicou.

***Rothmannia longflora*** Sal. = Randia maculata.

Rotreps Oel – Hesperis matronalis.
Rotsife – Entandrophragma cylindricum.

ROTTBOELLIA L. f. Graminae. 4 spp. Tropics and subtropical Africa and Asia. Grasses. The following spp. are used locally as fodder for livestock.

***R. compressa*** L. f Tropical and subtropical Africa and Asia.

***R. exaltata*** L. f. = Ophiurus appendiculatus. Tropical and subtropical Africa and Asia.

***R. fasciculata*** Desf. Tropical and subtropical Africa and Asia.

***R. glandulosa*** Trin. Malaysia, Indonesia.

***Rottlera*** Willd. = Mallotus.

ROUCHERIA Planch. 8 spp. Tropical S.America, Malaysia. Climbers.

***R. griffithiana*** Planch. (Akar bidji, Garamgarum). Malaysia. A decoction of the bark is used locally as an arrow poison.

Roucou – Bixa orellanra.

Rouge de Congo – Landolphia owariensis.
Rouge de Kassai – Landolphia gentilii.
Rouge de Kassai – Landolphia oeariensis.
Rouge Plant – Rivina humilis.
Rough Pea – Lathyrus hirsutus.
Rough-leaved Dogwood – Cornus asperifolia.
Rough-leaved Fig – Ficus aspera.
Rough Tongue – Aster macrophyllus.
Roummef – Haloxylon tamaricifolium.
Round Cardamon – Amomum kepulaga.
Round Chinese Cardamon – Amomum globosum.
Round-leaved Cornel – Cornus circinnata.
ROUPALA Aubl. Proteaceae. 50 spp. Central and S.Tropical America. Trees. The wood of the following spp., except R. montana, is hard and tough. It is used for general carpentry and ship-building.
***R. brasiliensis*** Klotzsch. Brazil.
***R. complicata*** H.B.K. Brazil.
***R. elegans*** Pohl. Brazil.
***R. heterophylla*** Pohl. Brazil.
***R. macrophylla*** Schott. Brazil.
***R. montana*** Aubl. (Trinidad Roupala). Trinidad, N. S.America. A decoction of the bark is used as a nerve tonic in Trinidad.
Roupala, Trinidad – Roupala montana.
ROUREA Aubl. Connaraceae. 80–90 spp. Tropical America, Africa, Madagascar, S.E.Asia, Malaysia, tropical Australia, Pacific. Shrubs or woody vines.
***R. glabra*** H.B.K. Central America to Venezuela, W.Indies. A fibre from the roots is used locally for cordage. The bark is used for tanning, staining the hides purple. The poisonous seeds are used to kill coyotes and as a criminal poison.
***R. heterophylla*** Planch.=R. volubilis.
***R. mimosoides*** Planch.=Santaloides mimosoides. Borneo, Nicobar Islands, Sumatra. A decoction of the roots is used locally to treat stomach pains.
***R. volubilis*** (Blanco) Merr.=R. heterophylla=Cnestis volubilis. Philippines. The fruits are used locally to poison dogs.
ROUREOPSIS Planch. Connaraceae. 2 spp. Tropical Africa; 8 spp. S.E.Asia, Malaysia. Woody vines.
***R. obliquifoliata*** (Gilg.) Schellenb. (Agiti, Dimi, Lokaka). W. and Central Africa. A decoction of the bark and roots is used locally to treat elephantiasis.
Rove – Quercus tauricola.
Rowan – Sorbus aucuparia.
Royal Fern – Osmunda regalis.
Royal Palm – Roystonea regia.
Royal Salep – Allium macleanii.
***Royena*** L.=Diospyros.
ROYSTONEA O. F. Cook. Palmae. 17 spp. Florida, Central and tropical S.America, W.Indies. Palm trees.
***R. oleracea*** (Mart.) Cook.=Oreodoxa oleracea. (Cabbage Palm). W.Indies. A sago is extracted from the trunk and the buds are eaten locally as a vegetable.
***R. regia*** (H.B.K.) Cook.=Oreodoxa regia. (Royal Palm). S.Florida, Central America, W.Indies. Cultivated locally for the trunks which are used for general construction work and wharves. The fruits and young buds are eaten locally.
Rozemarijn-hout – Osmoxylon umbelliferum.
***Rozites caperata*** (Bers.) Karst.=Pholiota caperata.
Rubaldo – Croton diodus.
Rubber, Accra Paste – Carpodinus hirsuata.
Rubber, Almeidina – Euphorbia rhipsaloides.
Rubber, Assam – Ficus elastica.
Rubber, Avara Seringa – Micranda siphonoides.
Rubber, Bokalahy – Marsdenia verrucosa.
Rubber, Borneo – Willughbeia firma.
Rubber, Camota – Sapium tabura.
Rubber, Caucho Blanco – Sapium pavonianum.
Rubber, Caucho Blanco – Sapium stylare.
Rubber, Caucho Blanco – Sapium thomsoni.
Rubber, Caucho Mirando – Sapium stylare.
Rubber, Caucho Negro – Castillea elastica.
Rubber, Caucho Virgin – Sapium thomsoni.
Rubber, Ceara – Manihot glaziovii.
Rubber, Chilte Highland – Cnidoscolus elasticus.
Rubber, Chupire – Euphorbia calyculata.
Rubber, Colombia Scrap – Sapium thomsoni.
Rubber, Colombia Virgin – Sapium thomsoni.
Rubber, Dahomey – Ficus vogelii.
Rubber, Dantung – Payena stipularis.
Rubber, Ekanda – Raphionacme utilis.
Rubber, Fingotra – Landolphia crassipes.
Rubber, Getah dantong – Payena dantung.
Rubber, Getah doojan – Palaquium treubii.
Rubber, Getah kelapang – Palaquium clarkeanum.

Rubber, Guayule – Parthenium argentatum.
Rubber, Hevea – Hevea brasiliensis.
Rubber, Huemega – Micranda minor.
Rubber, India – Ficus elastica.
Rubber, India – Marsdenia tenacissima.
Rubber, Intisy – Euphorbia intisy.
Rubber, Iré – Funtumia elastica.
Rubber, Jequie – Manihot dichotoma.
Rubber, Kappa – Clitandra orientalis.
Rubber, Kok-saghyz – Taraxacum kok-saghyz.
Rubber, Kompitso – Kompitsia elastica.
Rubber, Krim Sagiz – Taraxacum hybernum.
Rubber, La Glu – Carpodinus hirsuta.
Rubber, Lagos Silk – Funtumia elastica.
Rubber, Madagascar Rouge – Landolphia madagascariensis.
Rubber, Majunga Noir – Mascarenhasia arborescens.
Rubber, Majunga Rouge – Landolphia perieri.
Rubber, Mangabeira – Hancornia speciosa.
Rubber, Mavokeli – Pentopetia elastica.
Rubber, Mgoa – Mascarenhasia elastica.
Rubber, Mozambique Blamc – Landolphia kirkii.
Rubber, Mozambique Rouge – Landolphia kirkii.
Rubber, Niger Gutta – Ficus platyphylla.
Rubber, Noire du Congo – Clitandra orientalis.
Rubber, Nyassa – Landolphia kirkii.
Rubber, Ogbagha – Ficus platyphylla.
Rubber, Orinoco Scrap – Sapium jenmani.
Rubber, Palay – Cryptostegia grandiflora.
Rubber, Palay – Willughbeia martabanca.
Rubber, Palo Amarillo – Euphorbia fulva.
Rubber, Para – Hevea brasiliensis.
Rubber, Piauhy – Manihot piauhyensis.
Rubber, Pine – Landolphia kirkii.
Rubber, Rangoon – Urceola esculenta.
Rubber, Red Kano – Ficus platyphylla.
Rubber, Reiabo – Landolphia sphaerocarpa.
Rubber, Rouge du Congo – Landolphia owariensis.
Rubber, Rouge du Kassi – Landolphia gentilii.
Rubber, Rouge du Kassi – Landolphia owariensis.
Rubber, São Francisco – Manihot heptaphylla.
Rubber, Serapat – Urceola pilosa.
Rubber, Sernamby – Sapium taburu.
Rubber, Silk (Lagos) – Funtumia elastica.
Rubber, Tapuru – Sapium taburu.
Rubber, Tau-Saghys – Scorzonera wulfsbergii.
Rubber Tree, Male – Holarrhena wulfsbergii.
Rubber, Vahimainty – Secamonopsis madagascariensis.

RUBIA L. Rubiaceae. 60 spp. W. Europe through temperate Asia to Himalayas, E. tropical and S.Africa, Mexico to tropical S.America. Herbs. The roots of the species mentioned, except R. peregrina are a source of the red dye alizarin (Turkey Red).

***R. cordifolia*** L. (Indian Madder). Tropical to temperate Asia, Africa. The leaves are also eaten as a vegetable with rice in Java.

***R. khasiana*** Kurz. (Khasia Madder).

***R. peregrina*** L. (Levant Madder). Mediterranean. The powdered roots are used locally as a diuretic, abortive and to control menstruation. A tea made from the flowers is used as an aphrodisiac.

***R. sikkimensis*** Kurz. (Sikkim Madder). Himalayas.

***R. tinctorium*** L. (Common Madder). S.Europe, Middle East. Previously, the plant was an important source of alizarin, to the advent of the analine dyes. It was introduced commercially into Europe during the late Middle Ages and made Holland one of the leading dye-manufacturing countries. The dye has been identified on Egyptian mummies. The roots of two-year-old plants are washed, dried and pulverised to give a commercial source of the dye.

RUBUS L. Rosaceae. 250 spp. Cosmopolitan, especially N.Temperate. Shrubs. The fruits of all the species mentioned are eaten locally and usually raw.

***R. adenotrichos*** Schlecht. (Mora Común). Mexico to Ecuador. The fruits are sold locally.

***R. albescens*** Roxb. (Ceylon Raspberry, Mysore Raspberry). India, Ceylon to Malaya and Indonesia. Sometimes cultivated in Puerto Rico. A possible plant for improvement as a species for cultivation in warm countries.

***R. alcaeafolius*** Poir. = R. moluccanus.

***R. allegheniensis*** Porter. = R. nigrobaccus. (Alleghany Blackberry, Mountain Blackberry). E. N.America. Berries made into jams etc.

***R. amabilis*** Focke. W.China. The berries are eaten locally.

***R. arcticus*** L. (Arctic Bramble). Arctic. The yellow berries are prized by the Eskimos. The leaves are used as a tea substitute in China.

***R. brasiliensis*** Mart. Brazil. Cultivated locally.

***R. caesius*** L. (European Dewberry). Europe. Fruits eaten and the leaves are sometimes used as a tea substitute.

***R. chamaemorus*** L. (Cloudberry, Salmonberry, Yellowberry). N.Temperate.

***R. cochinchinensis*** Tratt. China, Cochin-China.

***R. ellipticus*** Smith. (Yellow Himalayan Raspberry). E.India. Introduced to Jamaica, Florida and California. Fruits used in preserves (Golden Evergreen Raspberry).

***R. elmeri*** Focke. Philippines. Fruits orange.

***R. flagellaris*** Willd.=R. villosus. (Northern Dewberry). E. N.America. Cultivated. The fruits are eaten raw, as fruit juice, in wines and in pies and jams. The juice is sometimes used to flavour medicines.

***R. flagellaris*** Willd. var. ***roribaccus*** Bailey E. N.America. (Lucretia Dewberry).

***R. floribundus*** H.B.K. (Zarzamora). Ecuador.

***R. flosculosus*** Focke. China.

***R. fruticosus*** L. (European Blackberry, Bramble). Europe. This species is possibly composed of several closely related species. The fruits are made into fruit juice, pies, jams etc. and eaten raw. They are also used to make the drink (Brombeersaft) and the wines Brombeerwasser and Blackberry Liqueur. The leaves (Folia Rubi fruticosi) are rich in tannins and when young make a good tea substitute. The older leaves are used in the treatment of diarrhoea, coughs, reducing fevers and as a diuretic.

***R. geoides*** Sm. Falkland Islands and adjoining territories.

***R. glaucus*** Benth. (Mora de Castilla). Ecuador. Cultivated locally. Also used to make a syrup (Jaropa de Mora).

***R. hawaiiensis*** A. Gray. (Akala). Hawaii.

***R. ichangensis*** Hemsl. and Kuntze. China.

***R. idaeus*** L. var. ***strigosa*** Michx.=R. strigosa.

***R. idaeus*** L. (European Red Raspberry). Europe, Asia. The fruit is mentioned by Pliny. It is cultivated, especially in Scotland for the fruit which is used mainly to make jam. The fruit is eaten raw, in preserves and confectionery and to make a Raspberry Liqueur and vinegar. The dried leaves are used as a substitute for tea. There are several commercial varieties. Propagation is by cuttings, or splitting the root stock.

***R. innominatus*** S. Moore. China.

***R. loganbaccus*** Bailey. Appeared in California in 1881. Possibly a hybrid between the octoploid R. ursinus and a diploid red raspberry. This is disputed. The fruits are larger than the European Red Raspberry and the plant is cultivated, mainly as a garden crop.

***R. macraei*** A. Gray. (Akala, Macrae Raspberry). Hawaii.

***R. macrocarpus*** Benth. (Colombian Berry). Colombia, Ecuador. Cultivated locally.

***R. microphyllus*** L. f.=R. palmatus. China, Japan.

***R. moluccanus*** L.=R. alcaeafolius. Malaysia, Indochina.

***R. morifolius*** Sieb. Japan.

***R. myranthus*** Baker. Madagascar.

***R. neglectus*** Peck. (Purple Cane Raspberry) =R. strigosus×R. occidentalis. N.America. Sometimes cultivated.

***R. nigrobaccus*** Bailey.=R. allegheniensis.

***R. niveus*** Thunb. (Pilay). Central China to Philippines.

***R. occidentalis*** L. (Black Cap, Black Raspberry). E. N.America. Cultivated in U.S.A. There are several varieties. The fruits are also used for drinks, preserves, ice-creams etc.

***R. omeiensis*** Rolfe.=R. flosculosus.

***R. palmatus*** Thunb.=R. microphyllus.

***R. parvifolius*** L. (Japanese Raspberry). China, Japan.

***R. pectinellus*** Max. Japan, Philippines.

***R. phoenicolasius*** Max. (Wineberry, Wine Raspberry). Japan. Sometimes cultivated locally and in U.S.A.

***R. probus*** Bailey. (Queensland Raspberry). Australia. Possibly a hybrid of R. ellipticus×R. rosaefolius.

***R. rosaefolius*** Smith. (Bramble of the Cape, Mauritius Raspberry). Tropical Asia, widely introduced. Sometimes cultivated the fruits are rather insipid. A decoction of the roots is used to treat diarrhoea in parts of S.Africa.

***R. roseus*** Poir. (Mora de Rocoto, Huagra

Mora). Ecuador, Peru. Used locally to make a drink.

***R. setchuemis*** Bur. and Franch. = R. omeiensis. China.

***R. spectabilis*** Purs. (Salmon Berry). W. N.America to Alaska. The Eskimos eat the fruit mixed with seal oil.

***R. strigosus*** Michx. = R. idaeus var. strigosos. (American Red Raspberry). E. N.America to New Mexico and Arizona. There are several varieties and the plant is cultivated commercially. The fruits are eaten raw, in jams etc. ice-cream, sherbets and sold frozen. The Mayberry of Burbank may be a hybrid, R. strigosus × R. microphyllus.

***R. trivialis*** Michx. (Southern Dewberry). S.E. U.S.A. The plant is cultivated locally. There are several varieties.

***R. villosus*** Ait. = R. flagellaris.

***R. xanthocarpus*** Bur. and French. W.China.

Rue Anemone – Anemonella thalictroides.

Rue, Common – Ruta graveolens.

Rue, Fringed – Ruta chalepensis.

Rue, Goat's – Galega officinalis.

Rue, Goat's – Tephrosia virginiana.

Rue, Oil – Ruta graveolens.

Rue, Wild – Anemonella thalictroides.

Rugosa Rose – Rosa rugosa.

Rugosi – Celosia stuhlmanniana.

Ruhac – Psychotria jackii.

Rukam – Flacourtia rukam.

Rula Ixtle – Agave lophantha.

RULINGIA R. Br. Sterculiaceae. 20 spp. Madagascar, Australia. Trees.

***R. pannosa*** R. Br. = Commersonia dasyphylla = Buettneria dasyphylla. Australia. The bark yields a fibre used for cordage.

Rum – All forms are made from Saccharum officinale.

Rumberry – Eugenia floridunda.

Rum Cherry – Prunus serotina.

RUMEX L. Polygonaceae. 170 spp. (sensu stricta). Cosmopolitan, especially N.Temperate. Herbs. The leaves of all spp. mentioned, except R. nepalensis, are used locally as a vegetable.

***R. abyssinicus*** Jacq. (Spanish Rhubarb Dock). Abyssinia. Cultivated in the Congo for the roots which are powdered and added to butter to colour it red.

***R. acetosa*** L. (Garden Sorrel). Temperate Europe, Asia. Cultivated in the Old and New World for the leaves.

***R. alpestris*** Jacq. = R. scutatus.

***R. alpinus*** L. (Alpine Dock, Monk's Rhubarb). Mountains of Central Europe. The leaves are also used to stop butter souring during the summer.

***R. arcticus*** Trautr. Arctic Asia and America. Eaten by the Eskimos of Alaska, fresh or preserved in oil.

***R. berlandieri*** Meisn. W. N.America. The leaves are eaten by the local Indians.

***R. brasiliensis*** Link. Brazil. Also, the roots are used locally as a diuretic and to reduce fevers.

***R. crispus*** L. (Curled Dock). Europe, N.Asia, introduced to N.America, Mexico, Chile, New Zealand. Also a decoction of the roots is used as a purgative and tonic.

***R. ecklonianus*** Meisn. S.Africa. An infusion of the roots is used locally to treat intestinal worms.

***R. hastifolius*** M. Bieb. = R. scutatus.

***R. hymenosepalus*** Torr. (Canaigre). S.W. U.S.A., Mexico. The tuberous roots contain some 35 per cent tannin (Raiz del India), for which the plant is sometimes cultivated. It would lend itself to mechanised harvesting, if the need arose. The roots also yield a yellow dye, used by the local Indians to dye wool. The petioles are sometimes eaten in pies. Some Indian tribes used a decoction of the roots to treat sore throats.

***R. maderensis*** Lowe. (Kafundbafunda). Tropical Africa. In the Congo a decoction of the leaves is used to reduce fevers.

***R. nepalensis*** Spreng. Himalayan Region. A decoction of the roots is used locally as a purgative.

***R. obtusifolius*** L. (Bitter Dock, Broad-leaved Dock). Europe, temperate Asia. Sometimes cultivated.

***R. mexicanus*** Meisn. W. U.S.A., Mexico. The leaves are eaten by the local Indians.

***R. occidentalis*** S. Wats. S.W. U.S.A., Mexico. The leaves are eaten by the local Indians.

***R. paucifolius*** Nutt. (Mountain Sorrel). N.W. N.America. The leaves are eaten by the local Indians.

***R. pubescens*** C. Koch. = R. scutatus.

***R. scutatus*** L. = R. alpestris = R. hastifolius = R. pubescens. (French Sorrel). S.Europe through to India. Sometimes cultivated. The leaves (Herba Acetosa romana) were also used as a purgative.

***R. vescarius*** L. N.Africa. The leaves are eaten by the Bedouins.

Rumf Sago Palm – Metroxylon rumphii.
Rumput Roman – Artemisia capillaris.
Rumy – Protium chapilieri.
Runeala Plum – Flacourtia cataphracta.
Rupture Wort – Herniaria glabra.
RUSCUS L. Ruscaceae. 3 spp. Madeira, W. and Central Europe, Mediterranean to Persia. Small shrub.
***R. aculeatus*** L. (Butcher's Broom, Kneeholm). Europe, Persia, N.Africa. The young shoots are sometimes used as a vegetable. A decoction of the cladodes is used to induce sweating, to treat jaundice, gravels in the bladder and uterine complaints.
Rush – Juncus glaucus.
Rush, Club – Scirpus lacustris.
Rush, Flowering – Botomus umbellatus.
Rush Skeleton Weed – Lygodesmia juncea.
Russell River Lime – Microcitrus inodora.
Russet Buffalo Berry – Sheperdia canadensis.
Russian Liquorice – Glycyrrhiza glabra.
Russian Thistle – Salsola kali.
Russian Vetch – Vicia villosa.
Russian Wild Rye – Elymus junceus.
RUSSULA Pers. ex S. F. Gray. Agaricaceae. 150 spp. Cosmopolitan. The fruitbodies of the following spp. are eaten locally.
***R. alutacea*** (Pers.) Fr. Temperate.
***R. atrovirens*** Beeli. Tropical Africa.
***R. flava*** Rom. Temperate.
***R. foetens*** Pers. = Agaricus fastidiosus = A. incrassa. Temperate. Eaten in Java without treatment, but in Russia they are first boiled and salted to remove a poisonous principle.
***R. madagascariense*** Heim. Madagascar.
***R. mariae*** Peck. Temperate.
***R. vesca*** Fr. Temperate.
***R. virescens*** (Schäff.) Fr. Temperate.
Rusty-leaved Alprose – Rhododendron ferrugineum.
Rút – Ilex godajam.
RUTA L. Rutaceae. 60 spp. Mediterranean, temperate Asia. Shrubs or herbs.
***R. angustifolia*** Pers. (Fringed Rue). Mediterranean. The juice is used in Morocco to control menstruation, to induce abortions, to treat intestinal worms and as a balm for sore eyes.
***R. bracteosa*** DC. Mediterranean. An essential oil (Algerian Oil) distilled from the leaves is used in perfumery and as a flavouring.
***R. chalepensis*** L. = R. angustifolia.
***R. graveolens*** L. (Common Rue). Mediterranean. Cultivated in the Old and New World for the leaves which are used as a flavour for sauces, meats, beverages, vinegar etc. A decoction of the leaves is also used to control menstruation and to cause abortion. An oil (Oil of Rue) is distilled from the leaves and young shoots and used in perfumery and flavourings.
***R. montana*** L. See R. bracteosa.
***R. tuberculata*** Forsk. N.Africa, Arabia, Iran. An essential oil from the plant is used locally as a flavouring.
Rutabaga – Brassica napo-brassica.
Ruvo Kale – Brassica ruvo.
RYANIA Vahl. Flacourtiaceae. 8 spp. N. tropical S.America, Trinidad. Trees.
***R. speciosa*** Vahl. = Patrisia pyrifera. Brazil, Guyana, Colombia. An extract of the wood yields the insecticide Ryanex.
Ryanex – Ryania speciosa.
Rye – Secale cereale.
Rye, Canada Wild – Elymus canadensis.
Rye, Giant – Triticum polonicum.
Rye, Giant Wild – Elymus condensatus.
Rye Grass, Italian – Lolium italicum.
Rye Grass, Perennial – Lolium perenne.
Rye, Jerusalem – Triticum polonicum.
Rye, Mandan Wild – Elymus canadensis.
Rye, Mountain – Secale montanum.
Rye, Robin's – Polytrichum juniperinum.
Rye, Russian – Elymus junceus.
Rye, Siberian Wild – Elymus giganteus.
Rye, Wild Beardless – Elymus triticoides.
RYPAROSA Blume Flacourtiaceae. 18 spp. Andaman and Nicobar Islands, W.Malaysia, New Guinea. Trees.
***R. caesia*** Blume. Java. The hard durable wood is used locally for heavy construction work, bridges etc.
RYTIDOSTYLIS Hook. and Arn. Cucurbitaceae. 5 spp. Tropical America, W.Indies. Vines.
***R. ciliatum*** Cogn. Central America. The young fruits are eaten locally as a vegetable.
RYTIPHLOEA Ag. Rhodomelaceae. Red Alga.
***R. rigidula*** Kg. = R. tinctoria.
***R. semicristata*** Ag. = R. tinctoria.
***R. tinctoria*** (Clem.) Ag. = R. rigidula = R. semicristata. Mediterranean. A red dye extracted from the plant was used by the Romans as a cosmetic.

# S

***Sabadilla*** Brandt. and Ratzeb.=Schoenocaulon.

SABAL Adans. Palmae. 25 spp. Warm America, W.Indies. Palm trees.

***S. causiarum*** (C. F. Cook) Becc.=Inodes causiarum (Puerto Rico Hat Palm, Yaray Palm). Puerto Rico, Dominica. The split leaves are used to make hats, baskets etc.

***S. mexicana*** Mart.=Inodes mexicana. Mexico, Guatemala. The leaves are an important local thatching material.

***S. palmetto*** (Walt.) Todd. (Cabbage Palmetto). S.E. U.S.A., Gulf of Mexico, W.Indies. The leaves are used for thatching, baskets, etc. The young buds are eaten locally as a vegetable. The stems are used to make brushes. The fruits are edible and the flowers are a good source of honey. Large trunks are resistant to water and are used for piling and sometimes the cross sections are polished as tables.

***S. texana*** Becc.=Inodes texana. Texas into Mexico. The leaves are used for thatching etc.

SABATIA Adans. Gentianaceae. 2 spp. S.E. U.S.A., Mexico, W.Indies. Herbs. The leaves of the following species are used to make a tonic tea.

***S. angularis*** (L.) Pursh. (American Century, Squarestem Rosegentian). E. N.America to Florida and Louisiana.

***S. campestris*** Nutt. E. N.America.

***S. elliotii*** Steud. E. N.America.

Sabicu – Calliandra formosa.

Sabino, Laurel – Magnolia splendens.

Sabonero – Phyllostylon brasiliensis.

Sabre Bean – Canavalia ensiformis.

SACCHAROMYCES Meyen ex Hansen. Saccharomycetaceae. 30 spp. Cosmopolitan.

***S. anamensis*** Will. and Heinnick. Sometimes used in alcoholic fermentations.

***S. apiculatus*** Reess. Ferments a wide variety of fruit juices.

***S. caslsbergensis*** Hansen. A yeast which ferments at the bottom of the vat during the manufacture of Lagers. It is used particularly in Copenhagen.

***S. cerevisiae*** Hansen. (Brewers' Yeast, Bakers' Yeast). There are many strains giving the characteristic flavours etc. to various beers. Other strains are used for rizing bread during baking. The compressed yeast is sold dried (Cerevisiae Fermentum Compressum) as a source of vitamin B complex and as a laxative.

***S. ellipsoides*** Hansen. (Wine Yeast). One of the most important yeasts used in making wine. The various strains give the particular characteristics to the various wines. Some strains, grown in an alkaline medium are used in the production of glycerol.

***S. fragilis*** Jörg. A potential commercial source of the enzyme lactase.

***S. kefir*** Beijerink. Ferments milk of various kinds into small lumps (Kefir Grains) which can be dried and preserved. They are much used in the Caucasus.

***S. lactis*** Dombr. A potential source of the enzyme lactase.

***S. marxianus*** Hansen. Used in the assay of the vitamin nicotinic acid.

***S. monacensis*** Hansen. Sometimes used as a bottom yeast in the fermentation of beer.

***S. pasteurianus*** Hansen. (Wild Yeast). A contaminant, causing turbidity in beer.

***S. pyriformis*** Hansen. Used in symbiosis with Bacterium vermiforme to make ginger beer.

***S. sake*** Yabe. The most important yeast used in the manufacture of sake. The starch in steamed rice is first hydrolysed by Aspergillus oryzae and the sugars so produced are fermented by S. sake.

***S. secundus*** Groenew. Used in Java for fermentations.

***S. theobromae*** Preyer. Plays a part in the initial fermentation and curing of cacao beans.

***S. tokyo*** Yabe. See S. sake.

***S. tuac*** Vorderm. Plays a part in the fermentation of palm wine made from the coconut tree.

***S. turbitans*** Hansen. A contaminant causing turbidity in beer.

***S. yeddo*** Yabe. See S. sake.

SACCHARUM L. Graminae. 5 spp. Tropics and subtropics. Large grasses.

***S. arundinaceum*** Retz. Tropical Asia. The leaves are used locally for thatching, making paper and Munj Fibre for cordage. The thick flower stalks make light furniture, carts, boats, etc.

***S. munja*** Roxb. (Munj, Sarkhand, Sirki). N.India, Pakistan. The leaves (Sarr) are used for thatching and twisted to make waterproof ropes. They are also used to make Munj fibre, for cordage and mats etc. The stems (Kana, Sentha) are used for light furniture, screens etc. The plant is grown in some parts of India to prevent erosion of sand and the young leaves are fed to livestock.

***S. officinale*** L. (Sugar Cane). Not known in the wild, probably originated in S.E.Asia. The plant has been cultivated locally since ancient times for chewing. It was introduced to the Canary Islands and Mediterranean region for chewing, via the Eastern trade routes. From here it was introduced to the Americas during the late fifteenth century. Higher yielding cultivars from Java (P.O.J. 2878) were not introduced to America until 1921. It is only since the beginning of the eighteenth century that sugar has become an important part of the diet in the Old World. World production of cane sugar is now about 441 million tons a year. The main areas of production are the W.Indian Islands, especially Cuba, the Pacific Islands, especially Hawaii, Java and the Philippines, but some sugar is grown almost everywhere where the climate is suitable. Most sugar is now grown as a plantation crop and is heavily mechanised. It is grown as an annual, but new plants are not necessarily planted every year, the old ones being allowed to sprout again. Propagation is by splitting the culms. Yield from the more modern varieties, with modern cultivation techniques is about 15 tons of sugar per acre. The varieties now planted are polyploids with n=15, 20, 30, 40, 50, 56 and higher. Some cultivars, e.g. the P.O.J. ones have involved crosses with ***S. spontaneum*** and ***S. sinense.***

The cane is harvested by mechanical cutters and the sugar liquid extracted by successive pressings through large rollers. The solution is concentrated by boiling and the crystals of sugar removed by centifuging off the liquor. The liquor left after centifuging is molasses which is used in confectionery, animal feeds, for making industrial alcohol, synthetic rubber and explosives. Most of the molasses is fermented to make rum. There are as many types of rum, or more, as there are sugar-growing areas in the world. Their different flavours, bodies and aromas depend on the local methods of preparation and the soil and climate in which the original cane was grown. The bagasse is used to fire the sugar factory boilers, to make paper and cardboard, or mixed with molasses and fed to cattle as Molascuit. The hard wax from the stem is sometimes used for polishes, candles etc.

***S. sape*** St. Hil.=Imperata brasiliensis.

***S. spontaneum*** L. (Bagberi, Dharb, Khus, Thatch Grass). India, Pakistan. See S. munja. The plant is also grown to prevent erosion of sands and to feed buffalo. The young shoots are eaten as a vegetable in Java.

***Saccopetalum*** Bennett.=Miliusa.

Sachalin Cork Tree – Phellodendron sachalinense.

Sachalin Knotweed – Polygonum sachalinense.

SACOGLOTTIS Mart. Houmiriaceae. 8 spp. Central to tropical S.America; 1 sp. W.Africa. Shrubs.

***S. gabonensis*** Baill. Tropical Africa. The fruits are used in Gabon to make a fermented drink (Stouton).

Sacred Datura – Datura meteloides.

Sacred Fir – Abies religiosa.

Sacxin – Abutilon trisulcatum.

Sadjang – Canarium littorale.

SADLERIA Kaulf. Blechnaceae. 6 spp. Hawaii. Ferns.

***S. cyatheoides*** Kaulf. Hawaii. The starchy pith is eaten locally as a vegetable and the plant yields a red dye.

Safea babul – Acacia leucophloea.

Safflower – Carthamus tinctoria.

Safflower, Wild – Carthamus oxycanthus.

Safflor Oil – Carthamus tinctoria.

Saffraanbossie – Sutera atropurpurea.

Safrole – Cinnamomum camphora.

Safrole – Ocotea caudata.

Saffron, Cape – Sutera atropurpurea.

Saffron Crocus – Crocus sativus.

Saffron Heart – Halfordia scleroxyla.

Saffron, Meadow – Colchicum autumnale.

Saffron Plum – Bumelia spiniflora.

Saffron-yellow Solorina – Solorina crocea.

Sagamomi – Abies firma.

Sagapen Resin – Ferula szowitziana.

Sagapan Gum – Ferula persica.

Sage – Salvia officinalis.

Sage, Black – Artemisia tridentata.

Sage, Black – Salvia mellifera.

Sage, Blue – Salvia mellifera.

Sage Bush – Artemisia gnaphalodes.

Sage Bush – Artemisia tridentata.
Sage Bush, Mariposa – Calochortus gunnisonii.
Sage, Clary – Salvia sclarea.
Sage, Garden – Salvia officinalis.
Sage, Linden-leaf – Salvia tiliaefolia.
Sage, Lyre-leaved – Salvia lyrata.
Sage, Meadow – Salvia pratensis.
Sage, Oil – Salvia officinalis.
Sage, Purple – Salvia carnosa.
Sage, Thistle – Salvia carduacea.
Sage, White – Eurotia lanata.
SAGERETIA Brogn. Rhamnaceae. 33 spp. Asia Minor, Somalia to Formosa, S. U.S.A. to tropical S.America. Shrubs.
***S. brandrethiana*** Ait. Tropical Asia, especially N.E. India. The fruits are eaten in Afghanistan.
***S. theezans*** (L.) Brogn.=Rhamnus theezans. E.India, Burma, China. The leaves are used in Tonkin to make a tea.
Sageung – Pentace polyantha.
SAGITTARIA L. Alismataceae. 20 spp. Cosmopolitan, especially America. Aquatic herbs.
***S. latifolia*** Willd.=S. variabilis. (Arrowleaf, Duck Potato, Wapato). N. to Central America. The starchy rhizomes are eaten as a vegetable by the local Indians and immigrant Chinese.
***S. sagittifolia*** L. (Arrowhead). Europe to Asia. Cultivated in China and Japan for the starchy rhizomes which are eaten as a vegetable.
***S. variabilis*** Engelm.=S. latifolia.
Sago – Caryota mitis.
Sago – Cycas circinalis.
Sago – Cycas revoluta.
Sago – Metroxylon laeve.
Sago Lily – Calochortus nuttallii.
Sago Palm – Metroxylon rumphii.
Sago Palm – Metroxylon sagu.
Sago Palm, Prickly – Metroxylon rumphii.
Sago, Pohon – Metroxylon rumphii.
***Sagorium volubile*** Baill.=Plukenetia volubilis.
Sagouier – Metroxylon rumphii.
Saguaragy (Bark) – Colubrina rufa.
Saguaro – Carnegiea gigantea.
SAHAGUNIA Lieb. Moraceae. 3 spp. Tropical America. Trees.
***S. strepitans*** Engl. Brazil. The sap and fruit are eaten locally and the wood is used for general construction work.
Sahing Resin – Canarium villosum.
Saigon Cinnamon – Cinnamomum loureirii.
Sain – Terminalia tormentosa.
Sainfoin – Onobrychis viciifoilia.
St. Andrew's Cross – Ascyrum hypericoides.
St. Domingo Apricot – Mammea americana.
St. Ignatius Poison Nut – Strychnos ignatii.
St. John's Bread – Ceratonia siliqua.
St. John's Wort – Hypericum perforatum.
St. Lucie Cherry – Prunus mahaleb.
St. Lucie Wood – Prunus mahaleb.
St. Martha Wood – Caesalpinia echinata.
St. Thomas Bean – Entada scandens.
St. Vincent Arrowroot – Maranta arundinacea.
Sajor caju – Lentinus sajor-caju.
Sajor kalapa – Cycas rumphii.
Sajor pepe – Gymnema syringifolium.
Sakat bawang – Acriopsis javavica.
Sake – Oryza sativa.
Sakhalin Spruce – Picea glehnii.
Sal (Dammar) (Butter) – Shorea robusta.
Sal Nut – Ginkgo biloba.
Sal Resin – Shorea talura.
SALACIA L. Celastraceae. 200 spp. Tropics. Shrubs or climbers.
***S. buddinghii*** Scheff.=S. macrophylla.
***S. celebica*** Blume=S. macrophylla.
***S. grandiflora*** Kurz. (Ampedal ajam, Nasi sedjook). Malaya. The fruits are eaten locally.
***S. macrophylla*** Blume=S. buddinghii=S. celebica. (Katji pot). Malaya. The fruits are eaten locally. A paste of the leaves is used to treat skin diseases and abdominal pains. The twigs are used as cordage.
***S. prinoides*** (Willd.) DC.=S. sinensis=Tondelea prinoides. Tropical Asia into Australia. In the Philippines a decoction of the roots is used to stimulate menstruation and as an abortive.
***S. senegalensis*** DC.=Hippocratea senegalensis=Hippocratea verticillata. Tropical Africa. The fruits are eaten in Senegal.
***S. sinensis*** Blanco=S. prinoides.
***S. soyauxii*** Loes. (Lole, Nkunde). Congo. The tough branches are used locally to make elephant traps.
Salad, Rocket – Eruca sativa.
Salatoon – Hydrocotyle sibtorpiodes.
Salelimisri – Eryngium coeruleum.
Salep – Orchis latifolia.
Salep, Royal – Allium macleanii.
Salibmisri – Eulophia campestris.
SALICORNIA L. Chenopodiaceae. 35 spp. Temperate and tropical sea coasts. Herbs.

***S. australis*** Soland. Australia. The pickled shoots are eaten locally.
***S. fruticosa*** L. (Leadbush). Mediterranean. Used as camel food.
***S. herbacea*** L. (Marsh Samphire, Saltwort). Europe. The stems are eaten pickled or as a vegetable. They were used as a source of soda for soap and glass making.
Saligna Gum – Eucalyptus saligna.
SALACCA Reinw. Palmae. 10 spp. Indomalaya. Small palms.
***S. edulis*** Reinw. = Zalaccia edulis. Malaysia. The sweetish fruits are eaten locally and when unripe they are eaten pickled in brine and sugar. They are sold locally.
SALIX L. Salicaceae. 500 spp. Chiefly N.Temperate. Trees. The branches of the species mentioned are flexible and the young ones are used to some extent to make baskets etc.
***S. acmophylla*** Boiss (Bada, Gadhbhains). N.W.India. A decoction of the bark is used locally to reduce fevers.
***S. acutifolia*** Willd. (Caspic Willow). E.Russia, Turkestan, Manchuria. Cultivated for basket work.
***S. alba*** L. = S. aurea. (White Willow). Europe, N.Africa and temperate Asia. Cultivated for the supple wood which is used to make shoes, boxes, boats, tool-handles and paper pulp. The leaves are used as a tea substitute. The bark contains tannin and salicin and a decoction is used as an astringent. The bark is sometimes used for tying plants. ***S. alba*** L. var. ***caerulea*** (Sm.) Sm. is used to make cricket bats.
***S. amygdalina*** L. = S. triandra.
***S. amygdaloides*** Anders. = S. nigra.
***S. aurea*** Salisb. = S. alba.
***S. auriculata*** Mill. = S. triandra.
***S. capensis*** Thunb. (Cape Willow). S.Africa. The shoots contain salicin and are used by the natives in the treatment of rheumatic fever.
***S. caprea*** L. (Goat Willow, Common Willow). The wood is soft and elastic. It is used for baskets, rugs etc. The charcoal makes a good gunpowder. The tree is cultivated.
***S. daphnoides*** Vill. = S. pulchra. Temperate into Arctic. The young catkin-bearing shoots are eaten raw or with seal oil by the Eskimos, who also eat the inner bark.
***S. fragilis*** L. (Brittle Willow). Europe, introduced to N.America. Wood is used for charcoal.
***S. humboldtiana*** Willd. (Humboldt's Willow). Central to S.America.
***S. irrorata*** Anders. (Bluestem Willow). S.W. U.S.A., Mexico. The charcoal is used by the local Indians for war paint, etc.
***S. jessoensis*** v. Seem. (Yeddo Willow). Japan. The soft, light wood is used locally for clogs, tooth picks, matches, boxes etc.
***S. longifolia*** Lam. = S. viminalis.
***S. nigra*** Marsh. = S. amygdaloides. (Black Willow). N.America. The soft light wood is used for charcoal and burning and to some extent for paper pulp and to make artificial legs. The bitter bark is used in home treatments for fevers.
***S. piperi*** Bebb. W. N.America. Used by the local Indians to make baskets.
***S. pulchra*** Cham. Alaska. See S. daphnoides.
***S. pulchra*** Wimner = S. daphnoides.
***S. purpurea*** L. (Purple Osier Willow, Purple Willow). Europe, temperate Asia and Africa. Also a commercial source of salicin, which is used medicinally in the treatment of rheumatism.
***S. sasaf*** Forsk. N. and tropical Africa. The twigs are used locally to make baskets, while a black dye from the leaves is used to stain mats.
***S. sitchensis*** Brongn. W. N.America. Also, the bark is used by the local Indians as cordage and a decoction from it is used as a tonic.
***S. tetrasperma*** Roxb. (Bains, Bent, Indian Willow Tree, Jalama). Tropical and subtropical India and Pakistan. The soft reddish wood is used locally to make furniture, boards, posting and wells. The charcoal is used for gunpowder. The wood is sometimes used for making pencils. A decoction of the bark is used to reduce fevers.
***S. triandra*** L. = S. amygdalina, S. auriculata. (Almond-leaved Willow, French Willow). Europe through to Japan. Cultivated.
***S. viminalis*** L. = S. longifolia = S. virescens (Basket Willow, Osier Willow, Osier). Europe, through N.Asia. Cultivated.
***S. virescens*** Vill. = S. viminalis.
Sallal – Gaultheria shallon.
Sally, Black – Acacia melanoxylon.
***Salmalia malabarica*** (DC.) Schott. and Endl. = Bombax malabaricum.

SALMEA DC. Compositae. 12 spp. Mexico, Central America, W.Indies. Shrubs.

***S. eupatoria*** DC.=S. scandens=Bidens scandens. Mexico. The root is used as a fish poison. When chewed it has a local anaesthetic effect on the tongue and gums and is used locally to treat toothache.

***S. scandens*** (L.) DC.=S. eupatoria.

Salmonberry – Rubus chamaemorus.

Salmonberry – Rubus spectabilis.

SALPICHROA Miers. Solanaceae. 25 spp. Warm America. Climbers.

***S. rhomboidea*** Miers. Argentina. The fruits (Huevos de Gallo), are used locally in preserves etc. They are sold locally.

Salsalamagui – Caulerpa sertularioides.

Salsify – Tragopogon porrifolius.

Salsify, Black – Scorzonera hispanica.

SALSOLA L. Chenopodiaceae. 150 spp. Cosmopolitan. Halophytic herbs.

***S. aphylla*** L. S.Africa. Used locally to feed livestock.

***S. asparagoides*** Miq. China, Korea, Japan. The young plants are eaten as a vegetable in Japan. The plant is cultivated.

***S. atriplicifolia*** Spreng.=Cycloloma atriplicifolia.

***S. foetida*** Del. Sahara. Used as fodder for camels.

***S. kali*** L. (Russian Thistle). Temperate. The young plants are used locally as a vegetable and emergency food.

***S. soda*** L. Mediterranean, Asia. Eaten locally as a vegetable, the plant is cultivated in Japan. It was formerly used as a source of sodium carbonate (Barilla) in S.France and Italy. This was used to manufacture glass and soap.

***S. zeyheri*** Shinz. S.Africa. Used as fodder for livestock.

Saltbush – Chenopodium auricomum.

Saltbush – Kochia aphylla.

Saltbush – Salvadora persica.

Saltbush, Fourwinged – Atriplex canesens.

Saltbush, Mediterranean – Atriplex halimus.

Saltbush, Small – Atriplex campanulata.

Saltwort – Batis maritima.

Saltwort – Salicornia herbacea.

Salvador Henequen – Agave letonae.

Salvador Sisal – Agave letonae.

SALVADORA L. Salvadoraceae. 4–5 spp. Warm Africa, Asia. Shrubs.

***S. persica*** Garc. (Mustard Tree, Salt Bush, Tooth-brush Tree). Tropical Africa, Asia. The leaves are eaten locally raw and as camel fodder. The twigs are used as toothbrushes. Salt (Kegr) is derived from the ashes and a fat extracted from the seeds is used to make candles and as a skin oil. The wood is used locally for different purposes. The fruits and bark are bitter and are used in local medicines.

SALVIA L. Labiatae. 700 spp. Tropics and temperate. Herbs.

***S. apiana*** Jep. W. N.America. The flowers are a good source of honey and a drink is made from the seeds.

***S. calycina*** Sibth. and Sm. Greece, Turkey. The leaves are used in Greece to make a tea.

***S. carduacea*** Benth. (Thistle Sage). S.W. U.S.A. The local Indians used the seeds to make a flour and a refreshing drink. The flowers are a good source of honey.

***S. carnosa*** Dougl. (Desert Ramona, Purple Sage). W. N.America. A decoction of the leaves and stem is used by the local Indians to treat colds.

***S. chia*** Fern (Chia). Mexico. Locally a drink is made from the seeds. An oil (Chia Oil) extracted from the plant was used for painting.

***S. columbariae*** Benth. (Californian Chia). W. N.America). The local Indians make a drink from the seeds and use them parched as a flour mixed with maize or wheat. The flowers yield a good honey.

***S. divinorum*** Epling and Játiva. (Ška Pastora, Yerba de Maria) Mexico. The leaves are used by the Maztac Indians as a psychotropic and hallucinogenic drug. The leaves (Hojas de la Pastora, Hojas de Maríia Pastora), are given to a patient as an infusion. This causes a delirium in which the patient describes his symptoms and disease.

***S. lyrata*** L. (Lyre-leaved Sage). E. N.America. The local Indians use an ointment made from the roots to treat sores.

***S. mellifera*** Greene (Black Sage, Blue Sage). California. The flowers are an excellent source of honey.

***S. micrantha*** Vahl.=S. tenella.

***S. militiorrhiza*** Bunge. China. The roots are used in local medicines.

***S. occidentalis*** Swartz. Tropical and subtropical America. In S.Mexico, a decoction of the leaves is used to treat stomach pains and dysentery.

***S. officinalis*** L. (Garden Sage). Mediterranean, Asia Minor. The dried leaves and the oil from them (Oil of Sage) are used for flavouring soups and meat dishes and to treat indigestion. The plant is cultivated in the Old and New World.

***S. pratensis*** L. (Meadow Sage). Europe. The mucilage from the fruits is used locally to treat eye complaints.

***S. sclarea*** L. (Clary Wort, Clary Sage). Mediterranean, Central Europe to Iran. The dried leaves are used to treat digestive upsets and, when powdered, as a snuff. The oil (Oil of Clary) extracted from the leaves is used in perfumery and in flavouring such wines as Vermouth and Muscatel. The plant is cultivated for the oil.

***S. sonomensis*** Greene. California. An excellent source of honey.

***S. tenella*** Sw.=S. micrantha. S.Mexico The leaves are used locally to treat ear ache.

***S. tiliaefolia*** Vahl. (Lindenleaf Sage). Mexico to Central America. The seeds are used in Mexico to make a drink and the leaves are used in Guatemala to kill lice.

***S. triloba*** L. Greece, Turkey. The leaves are used to make a tea locally in Greece.

***S. viridis*** L. (Bluebeard). Mediterranean to Iran. The oil extracted from the leaves is used to flavour wines and beers. The flowers are a good source of honey. The plant is sometimes cultivated.

Salvia – Hyptis laniflora.

Salvia – Lippia umbellata.

Salwa – Shorea robusta.

Sâm nam – Disacus asper.

Sâm nam – Hibiscus sagittifolius.

***Samadera*** Gaertn=Quassia.

Samamgate – Echinochloa pyramidalis.

Saman – Pithecellobium saman.

***Samara lacta*** L.=Embelia laeta.

Samarang Rose Apple – Eugenia javanica.

Sambal – Litsea cubeba.

Sambas Copal – Agathis loranthifolia.

Sambagan – Vitis adnata.

SAMBUCUS L. Sambucaceae. 40 spp. Cosmopolitan, except Amazonia, Africa (but 1 sp. in E.Africa), India, W.Australia, Pacific. Shrubs or small trees.

***S. australis*** Cham. and Schlecht. Argentina, Peru, parts of Brazil. The berries are used locally in preserves and an infusion of the leaves to treat indigestion and as a diuretic.

***S. callicarpa*** Greene (Red Elderberry). W. N.America. The berries are eaten by the local Indians, often preserved for the winter.

***S. canadensis*** L. (American Elder). E. N.America to Texas. The fruits are eaten in pies, preserves etc. and made into wine. A decoction of the flowers is used to treat stomach upsets, as a diuretic and stimulant, in eye lotions and as a salve for bruises.

***S. coerulea*** Raf.=S. glauca. W. N.America. (Blueberry Elder). The fruits are eaten by the local Indians and the wood to make bows.

***S. ebulus*** L. (Dwarf Elder). Europe, W.Asia, N.Africa. The fruits yield a blue dye used to stain leather, cloth and to colour wine. The leaves are made into a tea (Kneipp Tea).

***S. gaudichaudiana*** DC. (Elderberry). Australia. The fruits are eaten by the aborigines.

***S. javanica*** Reinw.=S. chinensis=S. thunbergiana. Tropical and subtropical Asia. The fruits are made into a drink and preserved. They are also used locally as a purgative and diuretic.

***S. mexicana*** Presl. (Mexican Elder). S.W. U.S.A. to Central America. The fruits are eaten locally in pies and preserves.

***S. nigra*** L. (European Elder). Europe. A decoction of the flowers (Flores Sambuci) is used to treat catarrh and as a gargle. They are also made into a wine. The fruits are used to make wine and in a syrup as a purgative. The pith from the stems is used for holding specimens for cutting to make microscope slides.

***S. peruviana*** H.B.K. Bolivia, N.Argentina, Peru. A syrup from the fruits is used locally in Peru to treat sore throats and ulcers.

***S. thunbergiana*** Blume.=S. javanica.

***S. xanthocarpa*** F. v. Muell. Australia. The fruits are eaten by the aborigines.

SAMOLUS L. Primulaceae. 10–15 spp. Cosmopolitan, especially S.Hemisphere. Herbs.

***S. valerandi*** L. (Brookweed). N. and S.America, temperate Asia to Japan, S.W. Australia. The young leaves are eaten raw or as a vegetable. They are antiscorbutic.

Samp-ki-khumb – Arisaema speciosum.

Sampalnan – Licania splendens.
Samphire – Crithmum maritimum.
Samphire, Golden – Inula crithmoides.
Samphire, Marsh – Salicornia herbacea.
Sam-rong – Sterculia lychnophora.
SAMUELA Trelease. Agavaceae. 2 spp. S.United States of America to Mexico. Tree-like shrubs.
***S. carnerosana*** Trelease. (Palma Barreta, Palma istlé). Mexico. The leaves are the source of a fibre which is used locally for cordage, brushes and sacking. It is exported to Europe and U.S.A. for the same purposes. The unopened flower heads are eaten as a vegetable by the local Indians and the trunks are used to make houses.
San chi – Panax repens.
San Domingo Apricot – Mammea americana.
San Domingo Box Wood – Phyllostaylon brasiliensis.
San Domingo Mahogany – Swietenia mahogani.
San eh'i – Gynura pinnatifida.
San Hemp – Crotalaria juncea.
Sand Apple – Parinari mobola.
Sand Bluestem – Andropogon halii.
Sand Dropseed – Sporobolus cryptandrus.
Sand Pea – Pisum arvense.
Sand Pine – Pinus clausa.
Sand Spurrey – Spergularia rubra
Sandal Bead Tree – Adenanthera pavonica.
Sandal Oil – Aptandra spruceana.
Sandal Oil – Guarea guara.
Sandal Oil – Lecontea amazonica.
Sandalo Brasilerio – Aniba canelilla.
Sandalo do Pará – Guarea guara.
Sandalo Inglez – Calophyllum brasiliense.
Sandalwood – Eremophila mitchelli.
Sandalwood – Pterocarpus santalinus.
Sandalwood, African – Baphia nitida.
Sandalwood, Ambon – Osmoxylon umbelliferum.
Sandalwood, Australian – Santalum lanceolatum.
Sandalwood, Bastard – Eremophila mitchelli.
Sandalwood, Bastard – Myoporum sandwicensis.
Sandalwood, East African – Osyris tenuifolia.
Sandalwood, Fiji – Santalum yasi.
Sandalwood, Fragrant – Eucarya spicatus.
Sandalwood, Hawaiian – Santalum freycinethianum.
Sandalwood, Juan Fernandez – Santalum fernandezianum.
Sandalwood, Lanceleaf – Santalum lanceolatum.
Sandalwood, New Caledonian – Santalum caledonicum.
Sandalwood, Padauk – Santalum santalinus.
Sandalwood, Red – Adenanthera pavonina.
Sandalwood, Shrub – Exocarpus latifolia.
Sandalwood, South Australian – Eucarya acuminatus.
Sandalwood (Oil), West Australian – Eucarya spicata.
Sandalwood, West Indian – Amyris balsamifera.
Sandalwood, White – Santalum album.
Sandalwood, Yellow Fiji – Santalum freycinetianum.
Sandand oopool – Licuala pumila.
Sandarac – Tetraclinis articulata.
Sandarac, Australian – Callitris calcarata.
Sandbox Tree – Hura crepitans.
Sandel Oil – Calophyllum brasiliense.
Sanders, Yellow – Terminalia hilariana
Sanderswood, Red – Pterocarpus santalinus.
Sandirerbena, Yellow – Abronia latifolia.
Sando de Maranhào – Aptandra spruceana.
SANDORICUM Cav. Meliaceae. 10 spp. Mauritius, Indomalaya. Trees.
***S. emarginatum*** Hiern. (Ketjapi kera). E.Sumatra. The light, soft, orange wood is durable and is used locally for building houses and boats.
***S. indicum*** Cav.=S. koetjape.
***S. koetjape*** Merr.=S. indicum. Malaysia, Philippines. The fruits are eaten locally, fresh, dried, spiced or candied.
Sandpaper Tree – Curatella americana.
Sanduri – Triticum timopheevi.
Sandwort, Red – Spergularia rubra.
Sandwort, Sand beach – Arenaria peploides.
Sang de Dragon – Dracaena draco.
Sanga Nut – Ricinodendron heudelotii.
Sangari – Vitex trifolia.
Sangi – Canarium amboinense.
Sangkat – Moschosma polystachyum.
Sango – Euphorbia hermentiana.
Sangre de Chucho – Calderonia salvadoriensis.
Sangre de Drago – Croton draco.
Sangre de Drago – Otoba gordoniifolia.
Sangre de Grado – Jatropha spathulata.
Sangu – Cissus petiolata.
Sangue de Draco – Pterocarpus erinaceus.

SANGUINARIA L. Papaveraceae. 1 sp. Atlantic N.America. Herb.

***S. canadensis*** L. (Blood Root). The rhizome contains the alkaloids chelerythrine, protopine and sanguinarine and a red resin. The dried rhizome is used medicinally as an expectorant and emetic and was used for the same purpose by the early colonists. The local Indians use the rhizome as a source of red dye for painting their bodies.

Sanguis Draconis – Daemonorops draco.

Sanguis Draconis – Dracaena draco.

SANGUISORBA L. Rosaceae. 2–3 spp. N.Temperate. Herbs.

***S. officinalis*** L. (Garden Burnet). Europe, temperate Asia to Japan, N.America. The leaves are sometimes used in salads or eaten as a vegetable.

Sanicle – Sanicula marilandica.

Sanicle, European Wood – Sanicula europaea.

SANICULA L. Umbelliferae. 37 spp. Cosmopolitan except New Guinea and Australia. Herbs.

***S. europaea*** L. (European Wood Sanicle). Europe, through Asia Minor to temperate Asia. The astringent leaves were used to treat complaints of the stomach and lungs.

***S. marilandica*** L. (Sabicle, Black Snakeroot). E. N.America to the Rocky mountains. A decoction of the roots is used as a sedative and astringent.

Sanleo-Calamus – Acorus gramineus.

Sannan – Ougeinia dalbergioides.

Sansapote – Licania platypus.

Sansada – Buxus wallichiana.

SANSEVERINIA Pentagna. Agavaceae. 60 spp. Tropical and S.Africa, Madagascar, Arabia. Herbs. The leaves of the following spp. (Except S. thyrsiflora) yield Bowstring Hemp, which is used locally for nets, bowstrings, sails and sometimes paper-making. The spp. are cultivated on a small scale locally.

***S. abyssinica*** N.E. Br. E.Africa. The leaves are also used locally by the women to make skirts.

***S. liberica*** Ger. and Labr. Tropical Africa.

***S. senegambica*** Baker. (African Bowstring Hemp). Tropical Africa.

***S. thyrsiflora*** Thunb. S.Africa. A decoction of the roots is used locally to treat intestinal parasites.

***S. trifasciata*** Prain. Tropical Africa.

***S. zeylanica*** Willd. (Ceylon Bowstring Hemp). Ceylon.

***Sansevieria*** Thunb. = Sanserverinia.

Santal Oil – Santalum album.

Santal rouge de Cochinchina – Aglaia baillonii.

Santalwood (Red) – Pterocarpus santalinus.

Santalin – Pterocarpus santalinus.

***Santaloides mimosoides*** Kuntze = Rourea mimosoides.

SANTALUM L. Santalaceae. 25 spp. E.Malaysia, Australia to Polynesia, Juan Fernandez. Parasitic small trees. The timber from the following species is durable, yellowish and highly scented. It is valued for making chests, boxes and fancy articles.

***S. acuminatum*** A. DC. = Eucarya acuminatus.

***S. album*** L. = Sirium myrtifolium. (White Sandalwood). E.India. Also, a fragrant oil (Oil of Santal, Oleum Santali) is distilled from the heartwood for use in perfumery and cosmetics. It is a good fixative for other perfumes. The oil is also used in the treatment of urinary complaints and venereal diseases and as an expectorant. The Chinese make joss sticks from the wood.

***S. austro-caledonicum*** Vieill. (New Caledonian Sandalwood). New Caledonia.

***S. cunninghamii*** Hook. New Zealand.

***S. fernandezianum*** Phil. Juan Fernandez. The plant is now entirely eradicated.

***S. freycinetianum*** Gaud. = S. insulare. (Yellow Fiji Sandalwood, Hawaiian Sandalwood). Polynesia.

***S. hornei*** Seems. Eromanga.

***S. insulare*** Bert. = S. freycinetianum.

***S. lanceolatum*** R. Br. (Australian Sandalwood, Lanceleaf Sandalwood). Tropical Australia.

***S. marchionense*** Skottb. Polynesia. The wood is used ceremonially in the Marquesas, for burning and decoration. The oil (Scented Santalumwood Oil) is used as a body oil and for embalming.

***S. paniculatum*** Hook. and Arn. New Zealand.

***S. pilgeri*** Rock. New Zealand.

***S. preissianum*** Miq. = Eucarya acuminatus.

***S. spicatum*** DC. = Eucarya spicata.

***S. yasi*** Seem. (Fiji Sandalwood). Pacific. Also yields Fiji Sandalwood Oil. See S. album.

Santalum Scented Woodoil – Santalum marchionense.

Santilla de Culebra – Pothomorphe umbellata.

SANTIRIA Blume. Burseraceae. 6 spp. Tropical W.Africa; 17 spp. Malaysia. Trees.

***S. griffithii*** Hook. f.=Trigonochlamys griffithii. Tropical Asia, especially Malaysia. The cream-coloured, hard scented wood is used for house construction.

***S. laevigata*** Blume. (Keranthai merah). Malaysia. The light brown, heavy wood is used for furniture and gun stocks.

***S. tomentosa*** Blume. (Kaboo-kaboo, Pegah). Malaysia. The wood is easy to work, but rots in contact with the soil. It is used for boards. The fruits are eaten cooked. An oil is extracted from the sun-dried fruits by steaming and pressing. It is used for cooking.

SANTOLINA L. Compositae. 10 spp. W. Mediterranean. Shrubs.

***S. chamaecyparissus*** L. (Lavender Cotton). W.Mediterranean. The flowers are sometimes used to treat ringworm and the stems and flowers hung with clothes to repel moths. An essential oil extracted from the leaves is occasionally used in perfumery.

Santonica, Barbary – Artemisia ramosa.

Sanwa Millet – Echinochloa frumentacea.

São Francisco Rubber – Manihot heptaphylla.

Saootan – Adenia singaporeana.

Saoyoo – Euodia latifolia.

Sapagen (Resin) – Ferula szowitziana.

Sape – Macaranga diepenhorstii.

Sapelé – Entandrophragma cylindricum.

Sapelé, Heavy – Entandrophragma candollei.

Sapelé Mahogany – Entandrophragma candollei.

Sapia – Macaranga mappa.

SAPINDUS L. Sapindaceae. 13 spp. Tropical and subtropical Asia, America, Pacific (not Australia). Trees. The fruits of all the spp. mentioned are used locally as a soap substitute for washing fabrics and bathing.

***S. detergens*** Roxb.=S. mukorossi.

***S. drummondii*** Hook. and Arn.=S. marginatus.

***S. laurifolius*** Vahl.=S. trifoliatus.

***S. marginatus*** Willd.=S. drummondii. (Western Soapberry). S. and S.W. U.S.A. The wood is used for making baskets for gathering cotton and pack-saddle frames.

***S. mukorossi*** Gaertn.=S. detergens. (Chinese Soapberry, Soapnut Tree). E.Asia, Himalayas. The Buddhists use the seeds to make rosaries.

***S. oahuensis*** Hillebr. Hawaii. The seeds are used locally as a strong laxative.

***S. rarak*** DC.=Dittelasma rarak. Java, Malacca. The fruits are used locally to treat scabies and as an insecticide. They are also used to intoxicate fish.

***S. saponaria*** L. (Southern Soapberry). S.Florida, tropical America. In Guadeloupe and Martinique an oil is extracted from the seeds.

***S. trifoliatus*** L.=S. laurifolius. (Areeta, Ritha, Soapnut Tree). India, Pakistan. The fruits are also used by the Hindus as a purgative, emetic and expectorant and to treat epilepsy and fits, hysteria and asthma. The fruit and the bark are used as fish poisons. The strong hard wood is used for general construction work, carts and mills.

SAPIUM P. Br. Euphorbiaceae. 120 spp. Tropics and subtropics (in S.America as far as Patagonia). Shrubs or trees.

***S. aucuparium*** Jacq.=S. hippomane. (Lecheros, Milk Tree). Tropical America. The sap yields a good rubber and is used locally as bird lime.

***S. biloculare*** (S. Wats.) Pax. Mexico. The chopped branches are used locally as a fish poison and the juice as an arrow poison. The seeds are "jumping beans".

***S. grahami*** Prain. Tropical Africa, especially the W. coast. The bark and leaves yield an orange dye, used by the women for ornamentation.

***S. hippomane*** Mey.=S. aucuparium.

***S. indicum*** Willd.=Excoecaria indica. Malaysia. The fruits are used locally as a fish poison.

***S. jenmani*** Hemsl. Guyana. The latex yields a good rubber (Orinoco Scrap). The plant is sometimes cultivated.

***S. lauricerasus*** Desf. (Manzanillo, Tabaiba). Tropical America, W.Indies. The soft wood is easy to work but is attacked by insects and fungi. It is used for boxes, etc., interior work and paper pulp.

***S. madagascariensis*** Prain. Tropical Africa. The latex is used as an arrow poison in E.Africa.

***S. pavonianum*** (Muell. Arg.) Hub. (Palo de

Leche). Peru, Colombia. The latex yields a rubber (Gaucho Blanco).

***S. sebiferum*** Roxb. (Chinese Tallow Tree). The plant is cultivated for the fruits. The outer seed yields a fat with a low iodine value (Chinese Vegetable Tallow) used to make candles and soap and the seed kernel gives a drying oil with a high iodine value (Stillingia Oil).

***S. stylare*** Muell. Arg. (Gaucho Blanco, Gaucho Mirado). S.America. Yields a rubber.

***S. taburu*** Ule. (Seringierana, Tapuru). Peru, Ecuador, Brazil. The latex yields a rubber (Camota, Sernamby or Tapuru Rubber) which is sometimes mixed with Hevea Rubber.

***S. thomsoni*** God.=S. tolimense. Colombia, Ecuador. Yields a rubber (Colombia Virgin, Gaucho Blanco, Gaucho Virgin, Scrap).

***S. tolimense*** Muell Arg.=S. thomsoni.

Sapnit – Mezonevron latisiliquium.

Sapodilla – Achras sapota.

Sapomaldin – Citrus paradisici×C. mitis.

SAPONARIA L. Caryophyllaceae. 30 spp. Temperate Eurasia, mainly Mediterranean. Herbs.

***S. officinalis*** L. (Soaproot, Soapwort). Europe, Asia, introduced to N.America. The roots contain saponin and are sometimes used as a soap-substitute for washing clothes and medicinally as a laxative. The plant is occasionally cultivated.

Sapote – Calocarpum mammosum.

Sapote, Black – Diospyros ebenaster.

Sapote, Black – Diospyros ebenum.

Sapote, Green – Calocarpum viride.

Sapote Negro – Diospyros ebenum.

Sapote, Yellow – Lucuma salicifolia.

Sapote, White – Casimiroa edulis.

Sapotón – Pachira macrocarpa.

Sappan Wood – Caesalpinia sappan.

Sapreewood – Callitris cupressoides.

Sapucaja Nuts – Lecythis zabucajo.

Sapucaja Oil – Lecythis ollaria.

Saquen Saqui – Bombacopsis sepium.

SARACA L. Leguminosae. 20 spp. Tropical Asia. Trees.

***S. indica*** L.=Jonesia asoca=J. pinnata. (Si soup). India, Indonesia. In India the bark is used to stimulate menstruation and to treat uterine disorders.

Saracuramira – Ampelozizyphus amazonicus.

Sarawak Bean – Vigna hosei.

SARCOBATUS Nees. Chenopodiaceae. 1–2 spp. N.America. Shrubs.

***S. vermiculata*** (Hook.) Torr. (Greasewood, Mexican Greasewood). S.W. U.S.A., Mexico. Used as cattle food. The seeds are edible.

***Sarcocephalus*** Afzel. ex R. Br.=Nauclea.

SARCOCHLAMYS Gaudlich. Urticaceae. 1 sp. Indomalaya. Shrub.

***S. pulcherrima*** Gaudich. The stem fibres (Duggal Fibre) is sometimes used for cordage and sacking.

Sarcocolla (Gum) – Penaea sarcocolla.

Sarcocolla – Astragalus fasciculiformis.

SARCOLOBUS R. Br. Asclepiadaceae. 15 spp. S.E.Asia, Malaysia. Vines.

***S. globosus*** Wall. (Kambing-kambing). Malaya. A sweet candied peel is made locally by boiling the skins of the fruits in syrup. It is eaten as a side dish or dessert.

***S. narcoticus*** Miq. Malaysia. The bark and root are used in Java for killing tigers, wild dogs and pigs.

***S. spanoghei*** Miq. (Gonggom, Lakambing, Peler kambing). Malaysia, Indonesia. The bark and root are used locally as S. narcoticus.

***Sarcostemma arenarium*** Benth.=Funastrum cumanense.

***Sarcostemma cumanense*** H.B.K.=Funastrum cumanense.

Sarepta Mustard – Brassica besseriana.

SARCOTHECA Blume. Averrhoaceae. 11 spp. Malaysia, Celebes. Small trees.

***S. griffithii*** Planch. (Koopooi, Poopoi). The fruits, stewed in sugar, are eaten locally in soups and as a candy.

***S. monophylla*** Planch. (Belimbang kera). Malaya, Malacca. The fruits are eaten locally in soups.

Sarewow – Mischophloeus vestiaria.

SARGASSUM C. A. Agardh. Fucaceae. Brown Algae.

***S. echinocarpus*** J. Ag. Pacific. Eaten in Hawaii, fresh, cooked or stuffed with fish (Limu Kala).

***S. enerve*** J. Ag. Pacific. Eaten in Japan.

***S. fusiforme*** (Harv.) Stech. E.Asia, especially China and Japan. Eaten in Japan and used to make stews, soups and a tea. It is also used as a source of iodine, potash and algin. The Chinese also use the seaweed in the treatment of goitre, glandular and lymphatic complaints.

***S. hemiphyllum*** (Turn.) J. Ag. Pacific. Used in China as a fertilizer.
***S. horneri*** J. Ag. See S. hemiphyllum.
***S. siliquastrum*** J. Ag. Pacific. Eaten in Japan.
***S. thunbergii*** O. Kuntze. See S. hemiphyllum.
***S. vulgare*** J. Ag. Pacific and Atlantic. Eaten in the Philippines.
Sargent Lily – Lilium sargentiae.
SARGENTODOXA Rehder and Wilson. Sargentodoxaceae. 1 sp. Central China. Shrub.
***S. cuneata*** (Oliv.) Rehder and Wilson. The stems and roots are used locally in a decoction to treat rheumatism.
Sarkhand – Saccharum munja..
SAROTHAMNUS Wimm. Leguminosae. 1 sp. Europe to W.Siberia, Atlantic Islands. Shrub.
***S. scoparius*** (L.) Wimm.=S. vulgaris=Spartium scoparium. (Broom). The plant is grown as a sand binder and is fed on by sheep and goats. The flowers are a good source of honey. In Germany the leaves and buds are pickled in salt and vinegar (Brahm, Geiss Kappern). The twigs are used to make brooms and sometimes to flavour beer; they also yield a fibre which is sometimes used for cordage and sacking. A yellow-brown dye is extracted from the bark and used for colouring paper and cloth. The flowerheads contain sparteine sulphate and are used as a diuretic and heart tonic.
***S. vulgaris*** Wimm.=S. scoparius.
Sarr – Saccharum munja.
SARRACENIA L. Sarraceniaceae. 10 spp. Atlantic N.America. Herbs.
***S. flava*** L. (Trumpets). Atlantic N.America. A decoction of the leaves is used to treat dyspepsia, constipation, liver and kidney complaints.
***S. purpurea*** L. (Pitcherplant). See S. flava.
Sarsaparilla – Smilax spp. Many local types are prepared from a great number of S. spp.
Sarsaparilla – Tecoma crucigera.
Sarsaparilla, Brown – Smilax regelii.
Sarsaparilla de Lisboa – Smilax spruceana.
Sarsaparilla de Marahaó – Smilax spruceana.
Sarsaparilla, German – Carex arenaria.
Sarsaparilla, Honduras – Smilax regelii.
Sarsaparilla, Indian – Hemidesmus indicus.
Sarsaparilla, Jamaica – Smilax ornata.
Sarsaparilla, Mexican – Smilax aristolochiaefolia.
Sarsaparilla, Pará – Smilax spruceana.
Sarsari – Silene macrosolen.
SASA Makino and Shibata. Graminae. 200 spp. E.Asia. Bamboo-like grasses.
***S. palmata*** (Mitf.) E. G. Camus=Bambusa palmata. China, Japan. The culms are sometimes used for pulping to make cardboard.
Sasah – Aporusa frutescens.
Sasanqua Camellia – Camellia sasanqua.
Sasapooan – Polygala paniculata.
Sasktoon Serviceberry – Amelanchier aluifolia.
Sasooroo – Osmoxylon umbelliferum.
SASSAFRAS Trew. Lauraceae. 2 spp. China and Formosa; 1 sp. Canada to Florida. Trees.
***S. alibidum*** (Nutt.) Nees.=S. officinale=S. variifolium. Canada to Florida. (Sassafras, Augue tree). The light, brittle, soft wood is used for barrels, yokes, light boats. A scented, antiseptic oil (Oil of Sassafras) is distilled from the root bark. It is used for flavouring sweets, toothpastes, medicines and tobacco and as a scent in perfumes and soaps. It is also used to flavour Sarsaparilla, Root Beer etc. In dentistry it is used as an antiseptic and in home remedies to treat colds, high blood pressure and as a tonic. When mixed with opium it is sold as Godfrey's Cordial. The plant is cultivated as a commercial source of safrole which makes up 80 per cent of the essential oil.
***S. goesianum*** Teysm. and Binn.=Cryptocarya aromatica.
***S. officinale*** Nees and Eberm.=S. alibidum.
***S. tzumu*** Hemsl. (Chinese Sassafras). China. The light timber is used locally.
***S. variifolium*** Kuntze=S. alibidum.
Sassafras – Doryphora sassafras.
Sassafras, Black – Cinnamomum oliveri.
Sassafras, Chinese – Sassafras tzumu.
Sassafras, Common – Sassafras alibidum.
Sassafras, Oil of – Sassafras alibidum.
Sassafras, Southern – Atherosperma moschatum.
***Sassafridium veraguense*** Meissn.=Ocotea veraguense.
Sasswood – Erythrophleum guineënsis.
Sassy Bark – Erythrophleum guineënsis.
Satapashpi – Anethum sowa.
Satin Ash, Grey – Eugenia gustavioides.
Satin Walnut – Liquidambar styraciflua.

Satinwood – Chloroxylon sweitenia.
Satinwood – Murraya paniculata.
Satinwood – Parinari guayanensis.
Satinwood – Zanthoxylum flavum.
Satinwood, Brazilian – Euxylophora paraensis.
Satinwood, Scented – Ceratopetalum apetalum.
Satra – Hyphaene shatan.
Satrabe – Medemia nobilis.
Satramira – Hyphaene shatan.
Satranabe – Medemia nobilis.
Satruma Orange – Citrus reticulata var. unshiu.
Satsumelo – Citrus paradisi × C. reticulata var. unshiu.

SATUREJA L. Labiatae. sensu lat. 200 spp, sensu strict. 30 spp. Temperate and warm. Herbs.

***S. acinos*** (L.) Scheele. = Calamintha acinos = C. arvensis = Thymus acinos. (Spring Savory). Europe, Asia, introduced to N.America. The leaves are sometimes used to treat stomach complaints.

***S. calamintha*** (L.) Scheele. = Calamintha officinalis. (Basil Thyme, Calamint). Europe, Mediterranean to Asia Minor. A decoction of the leaves are used to induce sweating and as an expectorant.

***S. hortensis*** L. (Summer Savory). Central Europe to Mediterranean and Asia Minor to Siberia. Distillation of the leaves yields an essential oil (Oil of Savory) which is used for flavouring prepared meats, e.g. sausages and sauces.

***S. illyrica*** Host. = S. montana.

***S. montana*** L. = S. illyrica = S. obovata. (Winter Savory). Mediterranean, Asia Minor to Ukraine. Cultivated in S.Europe and Germany. See S. hortense.

***S. obovata*** Lag. = S. montana.

***S. rigidus*** (Bart.) Small. (Pennyroyal). Florida. The leaves are used to make a tea.

Sauce – Baccharis glutinosa.
Sauerkraut – Lactobacillus cucumeris.
Sauh – Manilkara fasciculata.
Saul Resin – Shorea talura.
Saule – Dicraeanthus africanus.
Saunders, Red – Pterocarpus santalinus.

SAURAUIA Willd. Actinidiaceae. 300 spp. Tropical Asia, America. Trees.

***S. aspera*** Turc. S.Mexico. The fruits are eaten locally, raw or cooked.

***S. conzatti*** Busc. Mexico. The fruits are eaten locally.

***S. napaulensis*** DC. China, Himalayas, Indonesia. The fruits are eaten locally, with sugar and sometimes used to adulterate honey.

***S. roxburghii*** Wall. China, Burma, Indonesia. The Cambodian women use the mucilage from the leaves as a hair dressing.

SAUROMATUM Schott. Araceae. 6 spp. Tropical Africa to W.Malaya. Herbs.

***S. nubicum*** Schott. Tropical Africa. The rhizome is eaten locally as a vegetable.

SAUROPUS Blume. Euphorbiaceae. 40 spp. S.E.Asia, Indomalaya. Shrubs.

***S. albicans*** Blume. = S. androgynus. Malaysia. In Java the leaves are eaten as a vegetable with rice or in soups.

***S. androgynus*** Merr. = S. albicans.

***S. quadrangularis*** Muell. Arg. (Surasaruni). Tropical Asia. The Hindus smoke the leaves to treat tonsilitis.

***S. rhamnoides*** Blume (Katook badak). Indonesia. The fruits are eaten locally.

***Sauruopsis chinensis*** Turez. = Saururus chinensis.

SAURURUS L. Saururaceae. 1 sp. E.Asia, Philippines; 1 sp. E.United States of America.

***S. cernuus*** L. (Lizardtail). E. U.S.A. The roots are used medicinally as a sedative and as a poultice for wounds by the local Indians.

***S. chinensis*** Baill. = S. loureiri = Saururopsis chinensis. E.Asia. Locally a poultice of the leaves is used to treat ulcers and wounds.

***S. loureiri*** Decne. = S. chinensis.

Sausage Tree, African – Kigelia africana.

SAUSSUREA DC. Compositae. 400 spp. temperate Asia; 1 sp. Europe; 1 spp. W. N.America; 1 sp. Australia. Herbs.

***S. lappa*** C. B. Clarke = Aucklandia costus. (Costus). Kashmir. The heavy scented oil (Costus Root Oil) extracted from the roots is used widely in oriental perfumes, particularly in China and Persia and the Middle East, where it is imported. The oil is also used to treat skin diseases, as an aphrodisiac, in incense and as a hair wash to darken grey hair.

Savan – Juniperus sabina.
Savanilla krameria – Krameria ixina.
Savin (Oil) – Juniperus sabina.
Savory Oil – Satureja hortensis.
Savory, Spring – Satureja acinos.
Savory, Summer – Satureja hortensis.

Savory, Winter – Satureja montana.
Saw Grass – Cladium effusium.
Saw Greenbrier – Smilax bona-nox.
Saw Palmetto – Sernoa serrulata.
Saw Wort – Serratula tinctoria.
Sawara Cypress – Chamaecyparis pisifera.
Sawarri Fat – Caryocar amygdaliferum.

SAXE-GOTHAEA Lindl. Podocarpaceae. 1 sp. South Andes. Tree.

***S. conspicua*** Lindl. (Mañio). The durable wood is easy to work and is used for general carpentry locally.

SAXIFRAGA L. Saxifragaceae. 370 spp. N.Temperate, Arctic, Andes. Herbs.

***S. erosa*** Pursh.=S. micranthidifolia. (Deer Tongue, Mountain Lettuce). E. U.S.A. The leaves are eaten locally in salads.

***S. micranthidifolia*** Steud.=S. erosa.

***S. peltata*** Torr.=Peltiphyllum peltatum. (Indian Rhubarb). S.W. U.S.A. The leafstalks are eaten by the local Indians, raw or as a vegetable.

***S. punctata*** L. (Dotted Saxifrage). W. N.America. The Eskimos of Alaska eat the leaves raw or soaked in oil.

Saxifrage, Dotted – Saxifraga punctata.
Scabwort – Inula helenium.

SCAEVOLA L. Goddeniaceae. 80–100 spp. Tropics and subtropics, especially Australia. Shrubs.

***S. koenighii*** Vahl. Malaysia, Australia. The hard wood is used locally as nails for building boats. The fruits are eaten by the aborigines of S.Australia.

***S. plumieri*** Vahl. Malaysia, Australia. The pith is flattened and dried to be used as rice paper.

***S. sericea*** Vahl. See S. plumieri.

Scammony (Levant) – Convolvulus scammomia.

Scammony, Mexican – Ipomoea orizabensis.

SCANDIX L. Umbelliferae. 15–20 spp. Europe, Mediterranean. Herbs. The leaves of the following spp. are used locally in salads.

***S. cerefolium*** L.=Anthriscus cerefolium.

***S. grandiflora*** L. Greece, Asia Minor.

***S. pecten-veneris*** L. (Lady's Comb, Venus' Comb). Temperate Europe, Asia, introduced to N.America.

SCAPHIUM Endl. Sterculiaceae. 6 spp. W.Malaysia. Trees.

***S. affinis*** Pierre.=Firmiana affinis=Sterculia affinis. Malaysia. The seeds soaked in water and eaten with sugar (Antehai, Pong tai hai) are used locally as a mild laxative.

Scarlet Gilla – Gilia aggregata.
Scarlet Oak – Quercus coccinea.
Scarlet Pimpernel – Anagallis arvensis.
Scarlet Runner Bean – Phaseolus coccineus.

SCELETIUM N. E. Brown. Aizoaceae. 20 spp. S.Africa. Herbs.

***S. anatomicum*** (Haw.) Bolus.=Mesembryanthemum anatomicum. (Kanna, Kon, Kougoed). S.Africa. The plant is slightly narcotic. The juice is given to native children to drug them to sleep. The fermented plant causes intoxication, but the fresh plant is chewed to relieve thirst.

Scented Mahogany – Entandrophragma candollei.

Scented Satinwood – Ceratopetalum apetalum.

SCHAEFFERIA Jacq. Celastraceae. 16 spp. S. United States of America to S.America and W.Indies. Trees.

***S. frutescens*** Jacq. (Boxwood, Florida Boxwood, Yellow Boxwood). S.Florida, W.Indies. The hard wood is sometimes used for carving.

SCHEELEA Karst. Palmae. 40 spp. Tropical America. Palm trees. The seeds of the following spp. are minor sources of edible oil.

***S. costaricensis*** Burret.

***S. excelsa*** Karst. (Corozo, Palma de Vino, Yagua).

***S. humboldtiana*** Spruce. (Yagua).

***S. liebmannii*** Becc. (Corozo, Coyal Real).

***S. lundellii*** Bartl. (Corozo).

***S. macrocarpa*** Karst. (Coroba).

***S. macrolepis*** Burret. (Coroba).

***S. preussii*** Burret. (Coquito, Corozo).

***S. zonensis*** L. H. Bailey. (Coroza zonza).

SCHEFFLERA J. R. and G. Forst. Araliaceae. 200 spp. Tropics and subtropics. Trees.

***S. aromatica*** Harms.=Heptapleurum aromaticum. (Djangkorang, Klanting, Lingkersap). Indonesia. The aromatic leaves are eaten as a side dish with rice etc.

***S. digitata*** Forst. New Zealand. The soft wood is used by the Maoris for rubbing to make a fire.

***Scheidweileria luxurians*** Kl.=Begonia luxurians.

SCHIMA Reinw. ex Blume. Theaceae. 15 spp. Himalayas to Formosa, Malaysia. Trees.

***S. bancana*** Miq. (Medang seroo, Seroo). W.Malaysia. The heavy red-brown wood is used in Java to build houses, bridges etc. The bark is used to stupify fish.

***S. noronhae*** Reinw. (Burma Guger Tree). Malaya. See S. bancana.

***S. stellata*** Pierre. Cochin China. The beautifully marked wood is used for carving etc.

***S. wallichi*** (DC.) Choisy. Himalayas to Sumatra. The wood is used locally for building.

SCHINOPSIS Engl. Anacardiaceae. 7 spp. S.America. Trees.

***S. balanse*** Engl. (Quebracho Colorado, Willowleaf Red Quebracho). W.Argentina. The wood is an important commercial source of tannin. The heartwood contains some 30 per cent tannin which is extracted by pressurized steam.

***S. lorentzii*** (Griseb.) Engl. = Loxopterygium lorentzii.

SCHINUS L. Anacardiaceae. 30 spp. Mexico to Argentina. Trees.

***S. aroeira*** Vell. = S. terebinthifolius.

***S. dependens*** Orteg = Amyris polygama = Duvaua dependens. (Peru Peppertree). Brazil, Chile, Peru. The leaves are used in Chile to treat rheumatism.

***S. latifolius*** Engl. = Litrea molle. (Chilean Peppertree). S.America. In Chile the fruits are used to make an alcoholic drink.

***S. molle*** L. (Brazilian Peppertree, Californian Peppertree). Andes. The plant is cultivated. In Mexico the fruits are used to make a drink. The seeds are an adulterant for pepper. The gum (American Mastic) from the trunk is chewed and the ground bark is used as a purgative for men and animals.

***S. terebinthifolius*** Raddi. = S. aroeira. Argentina, Brazil, etc. The leaves are used as a tonic. A resin (Balsamo de Misiones is extracted from the trunk. The plant is cultivated.

***Schizandra hanceana*** H. Bn. = Kadsura coccinea.

SCHIZOGLOSSUM E. Mey. Asclepiadaceae. 50 spp. Tropical and S.Africa. Shrubs.

***S. shirense*** N. E. Brown. Tropical E.Africa. A decoction of the leaves is used locally to treat dysentery and other stomach complaints and as an aphrodisiac.

SCHIZOMERIA D. Don. Cunoniaceae. 18 spp. Moluccas, New Guinea, Queensland. Trees.

***S. ovata*** D. Don. (Crab Apple, New South Wales White Ash, White Birch). Queensland, New South Wales. The pinkish, light-brown wood is used for turning, veneer, coffins, boxes, shelving etc.

SCHIZOSACCHAROMYCES Lindner. Saccharomycetaceae. Yeasts. Both spp. play a part in alcoholic fermentation.

***S. asporus*** Beijerinck. Particularly important in fermenting arak in Java.

***S. pombe*** Lindner.

SCHIZOSTACHYUM Nees. Graminae. 35 spp. E.Asia. Bamboos.

***S. abyssinicum*** Hochst. = Oxytenanthera abyssinica.

***S. dielsianum*** (Pilger) Merr. = Dinodielsiana. (Bikal-Bohoi). Philippines. A decoction of the rhizome is used locally as a drink and an extract of the young shoots to treat conjunctivitis.

SCHLEICHERA Willd. Sapindaceae. 1 sp. Indomalaya. Tree.

***S. oleosa*** Merr. = S. trijuga.

***S. trijuga*** Willd. and Klein. = S. oleosa. (Ceylon Oak, Kussum Tree, Malay Lacktree). The hard, heavy, red wood is used to make small boats and sugar mill rollers. The bark is used for tanning. The wood also makes a good charcoal. The seeds are extracted to produce a non-drying oil (Macassar Oil, Kussum Oil) which is used for lighting, making candles, ointments and hair-dressings, soap and is sometimes eaten. The unripe fruits are eaten as pickles and the young leaves as a vegetable with rice.

***Schmaltzia*** Steud. = Rhus.

SCHOENOCAULON A. Grey. Liliaceae. 10 spp. Florida to Peru. Herbs. The seeds of the spp. mentioned contain the poison veratrin.

***S. officinale*** Gray = Sabadilla officinalis. Central America. The seeds (Cavadilla Seed, Semen Sabadallae) are used locally to kill vermin and medicinally to make a salve to treat rheumatism and neuralgia. They are also used as an insecticide and in veterinary medicine.

***Schoenodum chilense*** Grey = Leptocarpus chilensis.

SCHOUTENIA Korth. Tiliaceae. 8 spp. Malaysia, Indochina. Trees.

***S. ovata*** Koch. = Actinophora fragrans. Java. The heavy reddish wood is used for wheels and tool-handles.

SCHRADERA Vahl. Rubiaceae. 15 spp. Tropical S.America, W.Indies. Vines.

***S. marginata*** Standl. (Querá) N.W. S.America. The local Indians chew the young shoots as a tooth preservative and tooth brush. The juice stains the teeth black.

SCHULZERIA Bres. Agaricaceae. 10 spp. Widespread. The fruitbodies of the following spp. are eaten locally.

***S. goosseniae*** Beeli. Tropical Africa.

***S. nivea*** Beeli. Tropical Africa.

***S. robusta*** Beeli. Tropical Africa.

SCHUMANNIOPHYTON Harms. Rubiaceae. 5 spp. Tropical Africa. Shrubs. The bark of the following spp. is used as a stimulant for travellers and hunting dogs. It is also used as an aphrodisiac and to stupify fish. The bark is eaten with meat or fish.

***S. arboreum*** Chev. Gabon.

***S. magnificum*** Marms. W.Africa.

Schwarze Schafgabe – Achillea atrata.

Schweinitz Cyperus – Cyperus schweinitzii.

Schweiztee – Dryas octopetala.

SCIADOPITYS Sieb. and Zucc. Taxodiaceae. 1 sp. Japan. Tree.

***S. verticillata*** Sieb. and Zucc. (Parasol Pine, Umbrella Pine). The soft orange-white wood is water-resistant and is used to make boats and in waterworks. An oil (Oleum Sciadopitys) distilled from the wood is used for varnishes and dyes.

SCILLA L. Liliaceae. 80 spp. Temperate Eurasia, S.Africa, tropical Africa. Herbs.

***S. lilio-hyacinthus*** L. Pyrenees. The bulbs are used locally as a purgative.

***S. non-scripta*** Hoffm. and Link.=S. nutans=Endymion nonscriptum=E. nutans=Hyacinthus nonscriptus. (Bluebell). W. to central Europe. The bulbs are poisonous, but the gum from them was used to fix feathers to arrows.

***S. nutans*** Sm.=S. non-scripta.

***S. ridigifolia*** Kunth. (Wild Squill). S.Africa. The bulbs are poisonous, but a decoction of them is used by the Zulus to treat rheumatic fever.

SCINDAPSUS Schott. Araceae. 40 spp. S.China, S.E.Asia, Indomalaya. Vines.

***S. officinalis*** (Roxb.) Schott. Tropical Asia, especially India and Burma. The dried fruits are eaten by the Hindus to treat intestinal worms, to induce sweating and as a stimulant.

SCIRPODENDRON Zippel ex Kurz. Cyperaceae. 1 sp. Indomalaya, Australia, Polynesia. Sedge.

***S. costatum*** Kurz. Cultivated in Sumatra for making mats and in the Philippines for making hats.

SCIRPUS L. Cyperaceae. 300 spp. Cosmopolitan. Sedges.

***S. acutis*** L.=S. lacustris.

***S. acutis*** Muhl.=S. occidentalis. (Tulu). W. N.America. The local Indians use the leaves to make mats, baskets etc.

***S. articulatus*** L.=S. incurvatus. Old World Tropics. The Hindus use a decoction of the plant as a purgative.

***S. corymbosus*** Roth. (Tarrada). Swamps of N.Africa. A decoction of the plant is used locally to treat chest complaints. The dried plant is used as tinder.

***S. grossus*** L. f.=S. kysoor. India, Malaysia, Indochina. The Hindus use a decoction of the plant to stop vomiting and diarrhoea.

***S. incurvatus*** Roxb.=S. articulatus.

***S. kysoor*** Roxb.=S. grossus.

***S. lacustris*** L.=S. acutis L.=S. validus. (Club Rush, Great Bulrush). Europe, N.America, to central America and W.Indies. The American Indians eat the young shoots as a vegetable and use the rhizomes to make a flour, or eat them raw.

***S. nevadensis*** S. Wats. W. N.America. The rhizomes are eaten raw by the local Indians.

***S. occidentalis*** (Wats.) Chase.=S. acutis.

***S. paludosus*** A. Nels. N.America. The local Indians eat the rhizomes raw, or make them into flour. They also mix the pollen with flour to make bread.

***S. totara*** (Nees and Mey.) Kunth. (Totara, Tulu). Tropical America. The local Indians use the stems to make small boats.

***S. tuberosus*** Roxb. China, Japan. The plants are grown in the rice fields for the starchy tubers which are eaten as a vegetable. In China a starch (Batei-fun) is extracted from the tubers.

***S. validus*** Vahl.=S. lacustris.

***Sciuris aromatica*** Gmel.=Ratutia aromatica

SCLERACHNE R. Br. Graminae. 1 sp. Java Timor. Grass.

***S. punctata*** R. Br.=Chionachne masii=Polytoca punctata. (Co gao). The seeds are used locally to make a flour, which is eaten with molasses (Bông lôc).

SCLEROCARYA Hochst. Anacardiaceae. 5 spp. Tropical and S.Africa. Trees. The fruits of the following spp. are eaten locally raw and are used to make fermented drinks.

***S. birraca*** Hochst. (Dineygama, Hari, Kunan). Sudan, Senegal.

***S. caffra*** Sond. (Kaffir Marvola Nut). S.Africa. Also used to make preserves.

***S. schweinfurthii*** Schinz. Tropical Africa.

***Scolochloa*** Mert. and Koch. = Arundo.

***Scolopendrium*** Adans. = Phyllitis.

SCOLOPIA Schreb. Flacourtiaceae. 45 spp. Tropical and S.Africa, Asia, Australia. Trees.

***S. gaertneri*** Thw. = S. schreberi.

***S. mundtii*** Harv. = Phoberos mundtii. (Red Pear). S.Africa. The hard, heavy wood is used for building wagons and wheels.

***S. schreberi*** J. F. Gmel. = S. gaertneri. Ceylon. The hard, heavy, red wood is used locally for posts and house-building.

***S. zeyheri*** Arn. = Phoberos zeyheri. (Thorn Pear). S.Africa. The heavy, hard wood is used to make wagons and wheels.

SCOLYMUS L. Compositae. 3 spp. Mediterranean. Herbs.

***S. hispanicus*** L. (Spanish Oyster Plant). Mediterranean. The roots are eaten locally as a vegetable. The plant is occasionally cultivated, but was more so.

SCOPARIA L. Scrophulariaceae. 20 spp. Tropical America. Shrubs.

***S. dulcis*** L. (Sweet Broom). Tropical America. In the W.Indies branches are sometimes put into wells to give the water a cool flavour.

SCOPOLIA Jacq. corr. Link. Solanaceae. 6 spp. Central and S. Europe to India and Japan. Herbs. The spp. mentioned contain atropine, scopolamine and hyoscyamine which are used as narcotics and to produce dilation of the pupils.

***S. anomala*** (Link and Otto) Airy-Shaw. = S. lurida. N.India. An extract from the leaves is used medicinally by the Hindus.

***S. carniolica*** Jacq. (Scopolia). S.E.Europe. Sometimes cultivated. The drug is extracted from the dried rhizomes.

***S. japonica*** Maxim. Japan. Used medicinally by the Chinese.

***S. lurida*** Duna. = S. anomala.

Scorbute Grass – Cochlearia officinalis.

SCORODOCARPUS Becc. Olacaeae. 1 sp. W.Malaysia. Tree.

***S. borneensis*** (Baill.) Becc. The heavy, hard, purplish wood (Koolim) is used locally for heavy construction work and boats.

SCORODOPHLOEUS Harms. Leguminosae. 2 spp. Tropical Africa. Trees.

***S. zenkeri*** Harms. (Bodfidji, Monajembe). Tropical Africa. The bark is used as a condiment in the Congo.

SCORZONERA L. Compositae. 150 spp. Mediterranean, Central Europe to Central Asia. Herbs.

***S. hispanica*** L. (Scorzonera, Black Salsify). Central Europe to Mediterranean. Cultivated for the roots which are eaten as a vegetable and used as a coffee substitute.

***S. mollis*** Bieb. = S. undulata Mediterranean. The flowers are sometimes eaten in salads.

***S. schweinfurthii*** Boiss. N.Africa. The roots are eaten as a vegetable locally.

***S. scapigera*** Forst. = Microseris forsteri.

***S. tau-saghys*** Lips. and Bosse. (Tau-Saghys). Russia. The latex contains 20–30 per cent rubber for which the plant has been grown commercially in Russia.

***S. undulata*** Vahl. = S. mollis.

Scorzonera – Scorzonera hispanica.

Scotch Lovage – Ligusticum scothicum.

Scotch Maple – Acer pseudo-platanus.

Scotch (Scot's) Pine – Pinus sylvestris.

Scrap – Sapium thomsoni.

Scrap, Orinoco – Sapium jenmani.

Scratch-coco – Colocasia esculenta.

Screw Bean – Prosopis odorata.

Screw Pines – Pandanus spp.

Screw Tree, East Indian – Helicteres isora.

Scribbly Gum – Eucalyptus haemastoma.

Scrofula Weed – Goodyera pubescens.

SCROPHULARIA L. Scrophulariaceae. 300 spp. Temperate Eurasia; 12 spp. N. and tropical America. Herbs.

***S. marilandica*** L. (Maryland Figwort). E. N.America. A decoction of the leaves is used to stimulate the appetite and control menstruation.

***S. nodosa*** L. (Figwort, Rosenoble, Throatwort). Europe. A decoction of the leaves is diuretic. They are used as a poultice for wounds, boils etc.

Scrub Myrtle – Backhousia citriodora.

Scrub Sandalwood – Exocarpus latifolia.

Scurfpea, Beaverbread – Psoralea castorea.

Scurfpea, Leatherroot – Psoralea macrostachys.

SCURRULA L. Loranthaceae. 50 spp. S.E.Asia, W.Malaysia. Parasitic herbs.

***S. gracilifolia*** Danser. = Loranthus chinensis = L. gracilifolius (Chùm goi cây dâu). Ceylon, Burma, Assam to Vietnam. In

Vietnam a decoction of the leaves is used to treat boils, as a mild tonic and a hair tonic. A decoction of the fruits is used to clear the vision.

Scurvy Grass – Barbarea praecox.

Scurvy Grass – Cochlearia officinalis.

SCUTELLARIA L. Labiatae. 300 spp. Cosmopolitan except S.Africa. The plant mentioned contains the alkaloids baicalin and scutellarin. They are used locally medicinally as mild sedatives and nerve tonics in treating hysteria, convulsions etc.

***S. baicalensis*** Georg. (Baical Skullcap). China.

***S. laterifolia*** L. (Madweed, Quaker Bonnet, Skullcap). N.America.

SCYPHIPHORA Gaertn. f. Rubiaceae. 1 sp. Coasts of Indomalaya and Australia. Small tree.

***S. hydrophyllacea*** Gaertn. f. (Doodook rajap, Pererat lanang). A decoction of the leaves is used in Indomalaya to treat intestinal complaints. The wood is used to make small spoons.

SCYPHOCEPHALIUM Warb. Myristicaceae. 3 spp. Tropical W.Africa. Trees.

***S. ochocoa*** Warb. = Ochocoa gabonii. (Ochoco). W.Africa. The fat from the seeds (Ochoco Butter) is used locally for cooking.

Sdau phnom – Aglaia baillonii.

Sea Beet – Beta maritima.

Sea Buckthorn – Hippophae rhamniodes.

Sea Burdock – Xanthium strumonium.

Sea Girdle – Laminaria digitata.

Sea Grape – Coccoloba uvifera.

Sea Holly – Acanthus ebracteatus.

Sea Holly – Eryngium maritimum.

Sea Island Cotton – Gossypium barbadense.

Sea Kale – Crambe maritima.

Sea Kale, Tatarian – Crambe tatarica.

Sea Lavender – Statice limonium.

Sea Lettuce – Ulva lactuca.

Sea Milkwort – Glaux maritima.

Sea Oats – Uniola paniculata.

Sea Onion – Urginea maritima.

Sea Pea – Lathyrus maritimus.

Sea Pink – Armeria maritima.

Sea Plantain – Plantago maritima.

Sea Purslane – Sesuvium portulacastrum.

Sea Rocket, American – Cakile edentula.

Sea Staff – Laminaria digitata.

Sea Ware – Fucus vesiculosus.

Sea Whistles – Ascophyllum nodosum.

Seabeach Sandwort – Arenaria peploides.

Seacoast Abronia – Abronia latifolia.

Seacoast Angelica – Archangelica actaeifolium.

Seacoast Bluestem – Andropogon littoralis.

***Seaforthia vestiaria*** Mart. = Mischophloeus vestaria.

Seaside Morning Glory – Ipomoea pes-caprae.

Seaside Plantain – Plantago decipiens.

Seaside Plantain – Plantago maritima.

Seatron – Nereocystis luetkeana.

SEBAEA Soland. ex R. Br. Gentianaceae. 100 spp. Warm Africa, Madagascar, India, Australia, New Zealand. Herbs.

***S. crassulaefolia*** Schlecht. S.Africa. The leaves are used locally to treat snakebites.

Sebari – Bauhinia elongata.

SEBASTIANA Spreng. Euphorbiaceae. 95 spp. Tropical America; 1 sp. Tropical W.Africa, India to S.China and Australia: 3 sp. W. Malaysia. Shrubs.

***S. pavoniana*** Muell. Arg. Mexico. The juice is used as an arrow poison. The fruits, invaded by the larvae of the butterfly Carpocapsa saltitans, are "Jumping Beans".

Sebestans – Cordia myxa.

Sebongang – Neesia malayana.

SECALE L. Graminae. 4 spp. Mediterranean, E.Europe to Central Asia; 1 sp. S.Africa. Grasses.

***S. africanum*** Stapf. S.Africa. The straw is used for roofing native huts.

***S. cereale*** L. (Rye). Probably originated in S.W. Asia. An important cereal crop, usually grown where conditions are unsuitable for wheat or barley. The world production is about 35 million tons a year, the main producers being the U.S.S.R., Poland and Germany. The grain contains some 13 per cent protein making it suitable for bread-making. The bread is usually a dark grey in colour and moist. It is sometimes made from a mixture of rye and wheat and sometimes flavoured with carroway seed. The grain is fermented and distilled to give vodka and in U.S.A and Canada it is used to make rye whisky. The grain is also used for stockfeed and when roasted as a coffee substitute. The plant is also grown for hay and silage, especially with red clover. The straw is used for bedding stock, paper-making, packing material, for mushroom-growing and archery targets. The plant is diploid ($2n=14$) and can yield 1·8 tons of grain an acre.

***S. fragile*** Marsch. = S. sylvestre.

***S. montanum*** Guss. (Mountain Rye). S.Europe, N.Africa, Asia Minor. Sometimes grown as a grain crop and for animal fodder.

***S. sylvestre*** Host. = S. fragile. Europe, tropical Asia. Grown as a grain crop in the mountains of China and Tibet.

Secale cornuti – Claviceps purpurea.

SEACAMONE R. Br. Asclepiadaceae. 100 spp. Old World Tropics. Shrubs.

***S. myrtifolia*** Benth. Tropical Africa. A decoction of the leaves is used locally as a purgative.

SECAMONOPSIS Jumelle. Asclepiadaceae. 1 sp. Madagascar. Shrub.

***S. madagascariensis*** Jum. The latex is the source of a good rubber.

SECHIUM P. Br. Cucurbitaceae. 1 sp. Tropical America. Vine.

***S. edule*** Swartz. = Chayota edulis. (Chayote, Guisquil). The fruits (Choco) are eaten locally as a vegetable or as a sweetmeat. The roots are eaten as a vegetable.

Sechsämtertropfen – Sorbus aucuparia.

SECONDATIA A. DC. Apocynaceae. 7 spp. Tropical America, Jamaica. Shrubs.

***S. stans*** (A. Gray) Standl. = Trachelospermum Stans. (Hierba de la cucararcha). Mexico. The leaves are used locally for poisoning cockroaches.

SECURIDACA L. Polygalaceae. 80 spp. Tropics except Australia. Climbers.

***S. corybmosa*** Turez. = S. philippensis.

***S. longipedunculata*** Fres. Tropical Africa. The stems yield a fibre (Buaze Fibre) which is used locally for cordage etc. The leaves are used as a snuff in Togoland and as a purgative in Abyssinia. In Angola they are used as an ordeal poison.

***S. philippensis*** Chodat. = S. xorybmosa. Philippines. The bark is used locally as a soap substitute.

***Securinga obovata*** Muell. Arg. = Flueggea microcarpa.

Sedge, Fen – Cladium mariscus.

Sedge Oil – Cyperus esculentus.

Sedge, Thatching – Cladium mariscus.

SEDUM L. Crassulaceae. 600 spp. N.Temperate; 1 sp. Peru. Herbs.

***S. acre*** L. (Goldmoss Stonecrop, Mossy Stonecrop). Europe, temperate Asia, introduced to N.America. The leaves are used as a laxative.

***S. album*** L. (Wall Pepper, White Stonecrop). Europe, temperate Asia, N.Africa. The leaves are eaten in salads.

***S. collinum*** Willd. = S. reflexum.

***S. dendroideum*** Moc. and Sessé. (Texiote, Texiotl, Semperviva). Mexico. The juice is used to harden the gums, to treat chilblains, haemerrhoids and dysentery.

***S. reflexum*** L. = S. collinum. (Jenny Stonecrop). Europe. The leaves are eaten in salads or soups.

***S. roseum*** Scop. = Rhodiola integrifolia. Europe, N.America. The leaves are eaten raw, or in oil by the Eskimos of Alaska.

***S. rupestre*** L. See S. reflexum.

***S. telephium*** L. See S. album.

Seemann Ivory Palm – Phytelephas seemanni.

Seepweed, Alkali – Suaeda fruticosa.

Seepweed, Desert – Suaeda suffrutescens.

Segal – Dillenia excelsa.

Sejal – Acacia tortilis.

Seje – Jessenia bataua.

Sekakul – Malabaila sekakul.

Sekar waron – Hibiscus venustus.

Sekua Fat – Fevillea cordifolia.

Selan merak – Swintonia glauca.

Selangan merak – Shorea kunstleri.

Selapan – Garcinia macrophylla.

SELAGINELLA Beauv. Selaginellaceae. 700 spp. Chiefly tropics. Herbs.

***S. lepidophylla*** Spreng. S. U.S.A., Mexico. The leaves are used in Mexico as a diuretic.

SELENICEREUS (A. Berger) Britton and Rose. Cactaceae. 25 spp. S. United States of America, Mexico and Central America; 1 sp. Uruguay and Argentina. Cacti.

***S. grandiflorus*** (L.) Britton and Rose. = Cereus grandiflorus. (Night-blooming Cereus.) Mexico. Contains an alkaloid like digitalin, used to treat rheumatism and in parts of S.America, dropsy. The plant is cultivated.

SELENIPEDIUM Reichb. f. Orchidaceae. 3 spp. Tropical S.America, W.Indies. Herbs.

***S. chica*** Reichb. f. (Vanilla Chica). Central to S.America. The psuedobulbs were used as a substitute for vanilla.

Self Heal – Prunella vulgaris.

Self Heal, Bigflower – Prunella grandiflora.

***Selinum cervaria*** L. = Peucedanum cervaria.

***Selinum officinale*** Vest. = Peucedanum officinale.

***Selinum ostruthium*** Wallr. = Peucedanum ostruthium.

***Selinum palustre*** L. = Peucedanum palustre.

***Selinum sylvestre*** L. = Peucedanum palustre.
***Selinum monnieri*** L. = Ligusticum monnieri.
***Selloa*** Spreng. = Gymnosperma.
Semamelwana – Heliophila suavissima.
Semba gona Gum – Bauhinia variegata.
SEMECARPUS L. f. Anacardiaceae. 50 spp. Indomalaya, Solomon Islands. Trees.
***S. anacardium*** L. f. = Anacardium orientale. (Cashew Marking Nut Tree, Oriental Cashew Nut). Tropical Asia to Australia. Cultivated locally. The oil from the nuts is used for preserve floors etc. from white ants. The nuts are also used for tanning. The juice mixed with lime water makes a marking ink for clothes.
***S. cassuvium*** Roxb. Malaysia. The fleshy receptacles are eaten locally. The plant is sometimes cultivated.
***S. cuneiformis*** Blanco. = S. microcarpa.
***S. microcarpa*** Wall. = S. cuneiformis. Philippines. The fruits are eaten locally.
Semen Ammeos vulgaris – Ammi majus.
Semen Amomi vulgaris – Sison amomum.
Semen Contra – Artemisia judaica
Semen Ignatii – Strychnos ignatii.
Semen Mataybae – Matayba purgans.
Semen Sabadallae – Schoenocaulon officinale.
Semen Thlaspeos – Thaspi arvense.
Semen Urticae – Urtica dioica.
SEMIARUNDINARIA Makino ex Nak. Gramineae. 20 spp. E.Asia. Bamboos.
***S. fastuosa*** (Marl.) Makino = Arundinaria fastuosa. (Narihira Bamboo, Narihiradake). Japan. The shoots are eaten locally as a vegetable.
Seminole Tea – Asimina reticulata.
Semki-gona Gum – Bauhinia purpurea.
Semperviva Sedum dendroideum.
SEMPERVIVUM L. Crassulariaceae. 25 spp. Mountains of S.Europe to Caucasus. Herbs.
***S. tectorum*** L. (Hen-and-Chickens, Roof Houseleek). S.Europe. The plant is grown on the roofs of houses to keep the slates in place. The leaves are chewed to relieve toothache and used as a poultice for bee stings. A decoction of the leaves is used to treat intestinal worms.
Sempoor – Dillenia aurea.
Sen do – Shorea harmandii.
Senaar Gum – Acacia seyal.
Senarr, White – Acacia senegal.
Senasena – Uvaria catacarpa.
***Senebiera*** DC. = Coronopus.
Seneca – Polygala senega.
SENECIO L. Compositae. 2000–3000 spp. Cosmopolitan. Herbs, shrubs or small trees.
***S. ambavilla*** Pers. = Hubertia ambavilla. Madagascar. A decoction of the leaves is used locally as a diuretic.
***S. ancanthifolius*** Hort. = S. cineraria.
***S. aureus*** L. (Golden Ragwort, Life Root, Squaw Weed). N.America. A decoction of the dried leaves is used medicinally as a diuretic, uterine stimulant and to control menstruation.
***S. axillaris*** Klatt. = S. salignus.
***S. biafrae*** Oliv. Tropical Africa. The leaves are eaten locally as a vegetable and used as a tea.
***S. chinensis*** DC. = S. scandens.
***S. cineraria*** DC. = S. ancanthifolius = Cineraria maritima. (Dusty Miller, Jakobee, Joggobee). Mediterranean. The juice is used to treat cataracts. The plant is grown as an ornamental.
***S. confusus*** Elm. = S. scandens.
***S. gabonicus*** Oliv. Tropical W.Africa. The leaves are used as a vegetable locally.
***S. jacobae*** L. (Tansy Ragwort). Europe to Siberia, Caucasus, N.Africa to Asia Minor. Introduced to N.America. A decoction of the leaves is used to treat coughs, colds, rheumatism, cramp and intestinal parasites. It is also used as a tonic.
***S. japonicus*** Thunb. = Gynura pinnatifida.
***S. kaempferi*** DC. Japan. The petioles are used locally as a vegetable.
***S. nikoënsis*** Miq. (Chihli, Shensi). Japan. A decoction of the leaves is used as a stimulant by the Chinese.
***S. palmatus*** Pall. Siberia, China, Japan. The leaves are eaten locally in Japan and used as a stimulant by the Chinese.
***S. platyphyllus*** DC. Caucasus. The plant contains the alkaloid platyphyllin which is similar to atropine. It is used medicinally in the Ukraine.
***S. pteroneurus*** Sch. = Kleinia pteroneura.
***S. rotundifolius*** Hook. f. New Zealand. Shrub. The silvery wood is used for turnery and inlay work.
***S. salignus*** DC. = S. axillaris = S. vernus = Cineraria salicifolia. (Chilca, Flor de dolores, Jarilla). Arizona to Guatemala. In Mexico a decoction of the leaves is used to treat fevers and they are used as a poultice for rheumatism.

***S. scandens*** Ham.=S. chinensis=S. confusus=Cineraria chinensis. Vine. India, S.China, Formosa, Philippines. The Chinese used the stalks and leaves in the treatment of eye diseases.

***S. vulgaris*** L. (Groundsel). Europe, N.America. The leaves (Herbs Senecionis vulgaris) are used medicinally to stimulate menstruation and to reduce blood pressure. In dentistry they are used to treat bleeding from swollen gums.

Senegal Ebony – Dalbergia melanoxylon.

Senegal Gum – Acacia senegal.

Senegal Khaya – Khaya senegalensis.

Senegal Prickly Ash – Zanthoxylum senegalense.

Seng muang – Gonocaryum subrostratum.

Sengajob – Bhesa paniculata.

Sengkawang pinang – Shorea singkowang.

Senn (Gum) – Bauhinia variegata.

Senna, Aden – Cassia holsericea.

Senna, Alexandrian – Cassia acutifolia.

Senna, American – Cassia marilandica.

Senna, Arabian – Cassia angustifolia.

Senna, Coffee – Cassia occidentalis.

Senna, Dog – Cassia obovata.

Senna, Italian – Cassia obovata.

Senna, Mecca – Cassia angustifolia.

Senna, Mogdad – Cassia occidentalis.

Senna, Ringworm – Cassia alata.

Senna, Sickle – Cassia tora.

Senna, Smooth – Cassia laevigata.

Senna, Tennevelley – Cassia angustifolia.

Sennar – Acacia seyal.

Sentha – Saccharum munja.

Sepe – Muncuna poggei.

SEQUOIA Endl. Taxodiaceae. 2 spp. W. N.America. Very large trees. Both spp. are important lumber trees. The timber is light, soft, easy to work and durable. It is used for general construction work, buildings, fences, railway sleepers, etc.

***S. gigantea*** (Lindl.) Decne.=S. wellingtonia=Wellingtonia gigantea=Sequoiadendron giganteum. (Big Tree, Mammoth Tree, Sequoia, Sierra Redwood). California.

***S. sempervirens*** (Lamb.) Endl. (Redwood). California, Oregon.

***S. wellingtonia*** Seem.=S. gigantea.

Sequoia – Sequoia gigantea.

***Sequoiadendron giganteum*** (Lindl.) Buchh. =Sequoia gigantea.

Seradella – Ornithopus sativus.

Serapat Rubber – Urceola acuto-acuminata.

Seraya, Borneo White – Parashorea malaanonan.

Seraya putch – Parashorea malaanonan.

Seredik itam – Licuala elegans.

SERENOA Hook. f. Palmae. 1 sp. S.E. United States of America. Small palm.

***S. repens*** (Bartr.) Small.=S. serrulata.

***S. serrulata*** (Michx.) Hook. f.=S. repens. (Saw Palmetto). The dried fruits are used medicinally as a diuretic, sedative and to treat catarrh. The flowers are a good source of honey and the seeds were eaten by the local Indians.

Serepat Rubber – Urceola auto-acuminata.

Serepat Rubber – Urceola pilosa.

SERIANTHES Benth. Leguminosae. 12 spp. Malaysia, Polynesia. Trees.

***S. myriadena*** Planch.=Acacia myriadena Polynesia. The strong flexible wood is used to make waggons in New Caledonia.

Seringeirana – Sapium tabura.

SERJANIA Mill. Sapindaceae. 25 spp. S.United States of America to tropical S.America. Shrubs.

***S. curassavica*** Radlk.=Paullinia pinnata.

***S. mexicana*** (L.) Willd.=Paullinia mexicana. (Barbasco, Diente de culebra, Quirote culebra). Central America and adjacent territories. The stems are used in Mexico as cordage, to stupify fish and a decoction from them is used to treat rheumatism.

Serkkom – Amomum gracile.

Sernamby Rubber – Sapium tabura.

Seroo – Schima bancana.

Serpent Kelp – Nercocystis luetkeana.

Serpent's Tongue – Erythronium americanum.

Serpentary, English – Polygonum bistorta.

Serratella – Ornithopus sativus.

SERRATULA L. Compositae. 70 spp. Europe to Japan. Herbs.

***S. tintcoria*** L. (Centaury? Saw Wort). Europe. The leaves yield a yellow dye used for dyeing wool, cotton, silk and linen. It is sometimes mixed with other dyes, e.g. indigo. The leaves and roots (Herba et Radix Serratulae) was used as a salve for wounds and haemorrhoids.

***Sersalisia pohlmannianum*** Fr.=Sideroxylon pohlmannianum.

Service Berry – Amelanchier canadensis.

Service Berry, Saskatoon – Amelanchier alnifolia.

Service Berry, Shadblow – Amelanchier canadensis.

Service Berry, Western – Amelanchier alnifolia.

Service Tree – Sorbus domestica.

Service Tree – Sorbus torminalis.

Sesame – Sesamum indicum.

SESAMUM L. Pedaliaceae. 30 spp. Tropical and S.Africa, Asia. Herbs.

***S. alatum*** Thonn. = S. sabulosum. (Tacoutta). Sudan and Sahara. See S. indicum, but used locally.

***S. angustifolium*** Engl. Tropical Africa, especially E. The seeds yield an oil which is potentially useful for soap-making and for cooking when refined.

***S. calycinum*** Welw. Tropical Africa. The leaves are eaten as a vegetable in E.Africa.

***S. indicum*** L. = S. orientale. (Sesame, Oriental Sesame, Bene). E.Africa. The plant is grown widely in the Old and New World tropics and subtropics giving a world production of 1–2 million tons of seeds a year. China and India are the main producing countries. The seeds yield a semi-drying oil (Bene Oil, Sesame Oil, Teel Oil, Gingli Oil) S.G. 0·928, Iod. No. 100·8, Sap. Val. 189·3, Unsap. 1·73 per cent, which is used mainly as a cooking or salad oil, but also for margarine, cosmetics, ointments etc. and locally for lighting. The whole seeds are used in cakes and confectionery. The seeds contain 55 per cent oil and 16–32 per cent protein. The best oil is obtained by expression, but the final extraction is by high pressure expression of the heated seeds. This oil has to be refined before use. The oil is very stable and will keep for years. The residue after expression is used as cattle cake or fertilizer and the stems are burnt as fuel, the ash being used as fertilizer.

***S. orientale*** L. = S. indicum.

***S. sabulosum*** A. Chev. = S. alatum.

***Seseli amomum*** Scop. = Sison amomum.

SESBANIA Adans. mut Scop. 50 spp. Tropics and subtropics. Herbs, shrubs or trees.

***S. aculeata*** Poir. = Aeschynomene spinulosa. Tropical Africa, India, Ceylon, China. The stems fibres (Dundee Fibre) are used to make nets and sails.

***S. aegyptiaca*** Poir. E.Africa, India. The bark fibres are used as cordage in India. The leaves are fed to livestock and used in Africa to make an insect repellent wash for livestock. They are also used as green manure. The wood is burnt for charcoal.

***S. cinerascens*** Welw. Tropical Africa. The leaves are used to feed livestock.

***S. grandiflora*** (L.) Poir. = Aeschynomene grandiflora (Agati Sesbania). E.India to Australia. Cultivated in Old and New World Tropics. The flowers and pods are eaten in salads or as a vegetable in S.Asia. A decoction of the leaves is used as a diuretic and laxative and a decoction of the bark is a tonic and a treatment of intestinal worms.

***S. macrocarpa*** Muhl. S. and S.W. U.S.A. The local Indians use the bark fibres to make fishing nets and lines.

***S. tetraptera*** Hochst. Tropical Africa. The leaves are eaten locally as a vegetable.

***Seseli carvi*** Lam. = Carum carvi.

SESUVIUM L. Aizoaceae. 8 spp. Tropics and subtropics. Coasts. Herbs.

***S. portulacastrum*** L. (Sea Purslane). Throughout the tropics. Eaten as a vegetable in the Far East. Sometimes cultivated.

SETARIA Beauv. Graminae. 140 spp. Tropics and warm temperate. Grasses.

***S. almaspicata*** de Wit. = S. sphacelata.

***S. flebelliformis*** de Wit. = S. sphacelata.

***S. galuca*** Beauv. = Pennisetum typhoideum.

***S. italica*** L. = Chaetochloa italica (Foxtail Millet, Italian Millet). E.Asia. Grown widely in warm temperate and tropical countries as a grain crop, surviving well in the more arid regions, e.g. hills of India. There are several varieties. The plant has been grown since ancient times in China and Japan. In China it was one of the five sacred plants (2700 B.C.). It is grown in U.S.A. as a hay crop.

***S. sphacelata*** (Schum.) Stapf. and Hubbard ex Moss. = S. flebelliformis = S. almaspicata. S.Africa. Grown locally for grazing hay and silage. There are several types.

***S. verticillata*** (L.) Beauv. (Kan Kalagoua). Oases of Sahara. The seeds are used locally for flour.

Seven Bark – Hydrangea arborescens.

Seville Orange – Citrus aurantium.

Sewar – Rhazya stricta.

Seyal Acacia – Acacia seyal.

Seychelles Rosewood – Thespesia populina.

Shad Bush – Amelanchier canadensis.

Shadblow Serviceberry – Amelanchier canadensis.
Shaddock – Citrus grandis.
Shagbark Hickory – Carya ovata.
Shaggy Mane – Coprinus comatus.
Shallon – Gaultheria shallon.
Shallot – Allium ascalonicum.
Shallu – Sorghum vulgare.
Shama Millet – Echinochloa colonum.
Shantal citrin – Dysoxylon lourieri.
Sharol – Alnus nitida.
She ch'uang tzu – Ligusticum monnieri.
She-p'ao-tzu – Duchesnea filipendula.
She Pine – Podocarpus elata.
Shea Butter – Butyrospermum parkii.
Sheep's Fescue – Festuca ovina.
Shellbark Hickory – Carya ovata.
Shemba – Acacia pinnata.
Shengali – Carallia brachiata.
Shen yeh – Panax repens.
Shepherd's Purse – Capsella bursa-pastoris.
Shepherd's Tree – Boscia albitrunca.
SHEPHERDIA Nutt. Elaeagnaceae. 3 spp. N.America. Shrubs.

***S. argentea*** Nutt. (Silver Buffalo Berry). N.America. The fruits are used to make jams etc. The Indians dry them for winter use.

***S. canadensis*** (L.) Nutt. = Lepargyrea canadensis. (Russet Buffalo Berry). The Eskimos dry the berries in cakes for winter use.

Sherere – Polygala ruwenzoriensis.
Shibo-shi – Hosta ovata.
Shield Lichen, Pale – Cetraria glauca.
Shiitake – Lentinus edodes.
Shikimi – Illicium anisatum.
Shin Oak – Quercus gambeli.
Shin Oak – Quercus undulata.
Shingle Oak – Casuarina stricta.
Shingle Oak – Quercus imbricata.
Shingle Tree – Acrocarpus fraxinifolius.
Shining Gum – Eucalyptus nitens.
Shining Sumach – Rhus copallina.
Shiny Asparagus – Asparagus lucidus.
Shir-Khist – Cotoneaster racemiflora.
Snola – Aeschynomene aspera.
Shooting Star, Henderson – Dodecatheon hendersonii.
Shore Grindelia – Grindelia robusta.
Shore Lupin – Lupinus littoralis.
Shore Podgrass – Triglochin maritima.

SHOREA Roxb. ex Gaertn. Dipterocarpaceae. 180 spp. Ceylon to S.China, W.Malaysia, Moluccas, Lesser Sunda Islands. Trees. Members of the genus give a dark red brown durable coarse timber, which is termite resistant. It is used for general construction work, railway sleepers, props, brake blocks and barrels. These are marked here as "timber". Some spp. yield a resin used for varnishes, wax polishes, carbon paper, typewriter ribbons and caulking boats. These are marked "dammar". The seeds of some spp. (Illipé Nuts) yield a brittle fat (Borneo Tallow, Sal Butter) which is used locally for cooking and commercially as a substitute for cacao butter in making chocolate, here marked "fat".

***S. acuminata*** (Dyer) Alan. Malaysia. Timber.

***S. albida*** Sym. Borneo. Dammar (Dammar empenit).

***S. aptera*** Burck. (Borneo Shorea). Malaysia, Borneo. Fat.

***S. atrinervosa*** Sym. Borneo. Timber.

***S. balangeran*** Burck (Yakal Shorea). Malaysia. Dammar. Fat, used for candles and soap.

***S. bracteolata*** Dyer. Malaya. Timber. Fat for candles and soap.

***S. cambodiana*** Pierre. Cambodia. Dammar. Scented and used locally to treat intestinal worms.

***S. cochinchinensis*** Pierre. Indochina, Malaysia. Dammar, used locally as chewing gum.

***S. collina*** Didley. (Balau bookit). Malaya. Timber, dammar.

***S. coriacea*** Burck. Malayan, Borneo. Timber.

***S. curtisii*** King. (Meranti lahi). Malaya. Timber. The wood is beautifully grained but lighter than some other spp. It is used for furniture and lighter general construction.

***S. eximea*** Scheffer = Vatica sublacunosa (Almond Shorea). India. Dammar (Dammar Kloekoep, Dammar Tubang), used locally for lighting.

***S. geniculata*** Sym. Borneo. Timber.

***S. glauca*** King. Penang. Timber. Dammar (Dammar Kumus).

***S. gysbertiana*** Burck. (Tengkawang lajar). Malaysia, Indonesia. Fat. Timber, but lighter and softer than other spp.

***S. harmandii*** Pierre. (Sendo, Popeal). Indochina. A decoction of the bark is used locally to treat dysentery and the bark is used to prevent the fermentation

of the sap of Borassus flabellifer when used as a source of sugar.
***S. havilandii*** Brandis. Borneo. Timber.
***S. hemsleyana*** King. Malaysia. Timber.
***S. hypochra*** Hance. (Temal Shorea). Cambodia Dammar (Dammar Temak).
***S. koordersii*** Brandis. Celebes, Moluccas. Dammar (Dammar Tenang).
***S. kunstleri*** King. (Selangan merah). Malaysia. Timber.
***S. laevis*** Ridl. (Gelangan kumus). Borneo. Timber, Dammar. Fat, used locally for lighting.
***S. lepidota*** Blume = S. nitens = S. palembancia = S. stipulosa (Kenooar, Melebekan, Tengkawang). Malaysia. Timber, fat.
***S. leprosula*** Miq. Malaysia. Timber. Dammar (Dammar Daging).
***S. lucida*** Miq. = Parashorea lucida.
***S. macrantha*** Brandis. Malaysia, Borneo. Fat.
***S. macrophylla*** (De Vries.) Ashton. Borneo. Fat.
***S. martiana*** Scheff. Malaya, Borneo. Fat.
***S. materialis*** Ridl. Tropical Asia, especially E.Malaya. Timber.
***S. maxwelliana*** King. Borneo. Timber.
***S. nitans*** Miq. = S. lepidota.
***S. pachyphylla*** Ridl. (Kerukup). Borneo Timber, especially for boat-building.
***S. palembanica*** Miq. = S. lepidota.
***S. pauciflora*** King. (Meranti cheriak). Malaysia. Timber.
***S. platyclados*** van Slooten. Malaysia. Timber.
***S. resinosa*** Foxw. Malaysia. Timber.
***S. ridleyana*** King. Malaysia. Timber.
***S. robusta*** Gaertn. f. (Sal Tree, Salwa). Himalayas into India. Timber. Dammar (Sal Dammar, Rhal Dhooma). Fat. The dammar is burnt as incense and when distilled yields an oil (Chua Oil) used in perfumery. The bark yields a black dye.
***S. roxburghii*** G. Don. = S. talura.
***S. selanica*** Blume. Moluccas. Dammar, used locally for torches.
***S. seminis*** (De Vries) Van Sloten. Malaya, Borneo. Fat.
***S. siamensis*** Miq. = Pantacme siamensis. Burma, Indochina. Timber.
***S. singkowang*** Miq. = Hopea singkowang = S. thiseltoni. (Sengkawang piang). Indonesia. Fat (Tengkawa Fat) used for candles and lighting.
***S. stenocarpa*** Burck. Malaya. Fat. Cultivated.
***S. stenoptera*** Burck. Malaya. Fat.
***S. stipulosa*** Burck. = S. lepidota.
***S. talura*** Roxb. = S. roxburghii = Vatica laccifera. Malaysia. Dammar (Sal Resin, Saul Resin).
***S. thiseltoni*** Kurz. = S. singkowang.
***S. tumbuggaia*** Roxb. India. Dammar.
***S. utilis*** King. Malaysia. Timber.
***S. vulgaris*** Pierre. Tropical Asia, especially Cochinchina. Dammar (Huile de Bois).
***S. wiesneri*** Schiff. Malaysia, Philippines. Dammar (Resin Dammar, Gum Dammar).
Short Buchu – Barosma betulina.
Short Buchu – Barosma crenulata.
Short-capsuled Mahogany – Entandrophragma utile.
Short-hair Plume Grass – Dichelanche crinata.
Short-leaf Fig – Ficus laevigata.
Short-leaved Pine – Pinus echinata.
Short Staple American Cotton – Gossypium herbaceum.
Shoti – Iris nepalensis.
Showy Milkweed – Asclepias speciosa.
Shrub Lespedeza – Lespedeza bicolor.
Shrub Myrtle – Backhousia citridora.
Shrub Sandalwood – Exocarpos latifolia.
Shurni – Picrasma quassioides.
Siak (Fat) – Palaquium oleiferum.
Siak Illipé Nut – Palaquium burckii.
Siam Benzoin – Styrax tonkinense.
Siam Cardamon – Elettaria cardamomum.
Siam Cotton – Gossypium nanking.
Siam Gamboge – Garcinia hanburyi.
Siam Rosewood – Dalbergia cochinchinensis.
Siamelo – Citrus paradisi × C. reticulata.
Siamor – Citrus sinensis × (C. paradisi × C. reticulata var. deliciosa).
Siberian Buckwheat – Fagopyrum tataricum.
Siberian Crab Apple – Malus baccata.
Siberian Filbert – Corylus heterophylla.
Siberian Fir – Abies sibirica.
Siberian Hazelnut – Corylus heterophylla.
Siberian Motherwort – Leonurus sibiricus.
Siberian Pea Shrub – Caragana arborescens.
Siberian Tea – Potentilla rupestris.
Siberian Wild Rye – Elymus giganteus.
Siboosook – Cassia nodosa.
SICANA Naud. Cucurbitaceae. 2 spp. Tropical America, W.Indies. Vines.
***S. odorifera*** Naud. (Casa Banana, Curuba).

Brazil, Peru. The scented fruits are used locally as a vegetable and in preserves. They are also used to scent clothes.
Sicilian Sumach – Rhus coriaria.
***Sickingia*** Willd. = Simira.
Sickle Leaf Acacia – Acacia parpophylla.
Sickle Senna – Cassia tora.
Sickle Wort – Prunella vulgaris.
SICYDIUM Schlechtd. Cucurbitaceae. 1 sp. Tropical America. Vine.
***S. monospermum*** Schlechtd. An oil extracted from the seeds is used locally as a violent laxative.
SIDA L. Malvaceae. 200 spp. Warm, especially America. Herbs.
***S. acuta*** Murm. Tropics. The stem yields a good cordage fibre (Queensland Hemp).
***S. canariensis*** Willd. = S. rhombifolia.
***S. cordifolia*** L. = S. rotundifolia. The stem yields a fibre used for cordage, rather like hemp. The Hindus use a decoction of the roots to treat stomach complaints, asthma and heart conditions.
***S. phillipica*** Blanco. = S. retusa.
***S. retusa*** L. = S. phillipica = S. truncatula. Tropical Asia. The Hindus use a decoction of the roots to treat rheumatism.
***S. rhombifolia*** L. = S. canariensis. (Broomjute Sida). Tropics. The stem yields a good cordage fibre (Queensland Hemp). The leaves are used to make a tea in the Canary Islands.
***S. rotundifolia*** Lam. = S. cordifolia.
***S. truncatula*** Blanco. = S. retusa.
Sida, Broomjute – Sida rhombifolia.
Siddhi – Cannabis sativa.
Side-Oat Grama – Bouteloua curtipendula.
SIDERITIS L. Labiatae. 100 spp. N.Temperate, Eurasia. Herbs. The flower heads and leaves of the following spp. are used locally to make an aromatic tea.
***S. peloponnensiaca*** Boiss. and Heldr. Greece.
***S. roesceri*** Boiss. and Heldr. Greece.
***S. theezans*** Boiss. and Heldr. Greece.
SIDEROXYLON L. Sapotaceae. 100 spp. Tropics. Trees.
***S. angustifolium*** Standl. Mexico. The bark is used locally for curdling milk.
***S. attenuatum*** A. DC. India, Philippines. Yields a guttapercha.
***S. capiri*** (A. DC.) Pittier. = S. mexicanum = S. petiolare = Lucuma capiri. (Capiro, Tempixque, Zapote de ave, Totozapotl). Mexico. The sweet fruits are eaten locally, raw or cooked.
***S. cyrtibotyrum*** Miq. Brazil. A gum (Balata Rosada) is extracted from the trunk.
***S. dulcificum*** A. DC. Tropical Africa. The ripe fruits are used locally to sweeten palm wine. If eaten immediately afterwards they remove acid, bitter or sour tastes, e.g. of vinegar, citrus fruits etc.
***S. firmum*** Pierre = Godoya firma.
***S. foetidissimum*** Jacq. = S. mastichodendron.
***S. glabrescens*** Miq. Malaysia. The gum from the trunk is used for chewing gum.
***S. kaernbachianum*** Engl. New Guinea. A Gutta Percha (Getah Nalu) is extracted from the trunk.
***S. mastichodendron*** Jacq. = S. foetidissimum Jacq. (Mastic, Wild Olive). Florida, W.Indies. The hard, heavy orange wood is used locally to make furniture and boats.
***S. mexicanum*** Hemsl. = S. capiri.
***S. petiolare*** A. Gray. = S. capiri.
***S. pohlmannianum*** Benth. and Hook. = Sersalisia pohlmannianum. (Yellow Boxwood). New South Wales, Queensland. The yellow-brown fine-grained wood is used for rulers, carving and turning, wood cuts, household utensils, croquet mallets, fishing rod handles and roller skates.
***S. resiniferum*** Ducke. Brazil. The trunk yields a gum (Balata Rosada).
Siebold Walnut – Juglans sieboldiana.
Siejas (Wax) – Langsdorffia hypogaea.
Sierra Juniper – Juniperus occidentalis.
Sierra Redwood – Sequoia gigantea.
Sierra Leone Arrowroot – Canna edulis.
Sierra Leone Butter – Pentadesmia butyracea.
Sierra Leone Copal – Copiafera guibourtiana.
Sierra Leone Copal – Copaifera salikounda.
Sieva Bean – Phaseolus lunatus.
***Sieversia paradoxa*** D. Don. = Fallugia paradoxa.
Sifu-sifu – Afzelia pachyloba.
SIGESBECKIA L. Compositae. 6 spp. Tropics and warm temperate. Herbs.
***S. orientalis*** L. = Minyranthus heterphylla. Old World Tropics. A tincture of the leaves is used to treat skin infections, e.g. ringworm, leprosy. The Hindus used a decoction of the plant to treat diseases of the urethra and the Chinese use it to treat rheumatism, kidney complaints and malaria.

Signel – Meum athamanticum.
Sigua canelo – Ocotea veraguensis.
Siguaraya – Trichilia havanensis.
Sihar – Rhazya stricta.
Sikalo – Hugonia platysepala.
Sikatan – Glycosmis cochinchinensis.
Sikay – Lagenaria leucantha.
Sikkim Madder – Rubia sikkimensis.
Sil – Echinochloa pyramidalis.
SILENE L. Caryophyllaceae. 500 spp. N.Temperate, especially Mediterranean. Herbs.
***S. acaulis*** L. (Moss Campion). Arctic, Alpine. The plant is eaten in Iceland, cooked with butter.
***S. cucubalus*** Wibel.=S. inflata.
***S. inflata*** Sm.=S. cucubalus=S. latifolia. (Bladeer Campion). Europe, introduced to N.America. The leaves are sometimes eaten as a vegetable.
***S. latifolia*** (Mill.) Britt. and Rendle.=S. inflata.
***S. macrostolon*** Steud. E.Africa. The roots (Radix Ogkert, Sarsari) are used in Abyssinia to treat intestinal worms.
Silenli – Sphenostylis briartii.
SILER Mill. Umbelliferae. 1 sp. Europe to Siberia. Herbs.
***S. montanum*** Crabtz.=Laserpitium siler.
***S. trilobum*** (L.) Borkh.=Laserpitium aquilegiaefolium=L. trilobum. The fruits (Fructus Sileris) are fed to livestock to relieve stomach upsets.
Silk Cotton Tree – Bombax buononpozence.
Silk Cotton Tree – Ceiba pentandra.
Silk Oak – Grevillea robusta.
Silk Rubber – Funtumia elastica.
Silkwood – Flindersia brayleyana.
Silkwood,Bolly – Cryptocarya oblata.
Silkwood, Maple – Flindersia brayleyana.
Silkwood, Silver – Flindersia acuminata.
Silkwood, Tarzali – Cryptocarya oblata.
Silkwood, White – Flindersia acuminata.
Silky Cornel – Cornus amomum.
Silky Oak – Cardwellia sublimis.
Silky Oak – Stenocarpus salignus.
Silky Sophora – Sophora sericea.
Siloki – Vatica simalurensis.
Silongkis – Vatica simalurensis.
SILPHIUM L. Compositae. 15 spp. E. United States of America. Herbs.
***S. laciniatum*** L. (Compass Plant, Polar Plant, Rosinweed). Prairies of N.America. The leaves and roots are used as an expectorant in treating coughs, asthma, bronchitis etc. They are also used as a diuretic, emetic and milk sedative. The gum from the roots is used as chewing gum by the local Indians.
***S. perfoliatum*** L. (Cup Plant, Indian Cup Plant, Ragged Cup). Prairies of N.America. The roots are used to make a tonic and are used to reduce fevers and treat diseases of the liver and spleen.
Silphium – Ferula narthax.
Silver Beech – Nothofagus menziesii.
Silver Berry – Elaeagnus argentea.
Silver Buffalo Berry – Shepherdia argentia.
Silver Fir – Abies amabalis.
Silver Fir – Abies pectinata.
Silver Grey Wood, Indian – Terminalia bialata.
Silver Maple – Acer saccharinum.
Silver Nailroot – Paronychia argentea.
Silver Pine Needle Oil – Abies pectinata.
Silver Quandong – Elaeocarpus grandis.
Silver Silkwood – Flindersia acuminata.
Silver Sycomore – Cryptocarya glaucescens.
Silver Tree – Leucadendron argenteum.
Silver Wattle – Acacia dealbata.
Silverleaf Nightshade – Solanum elaednifolium.
Silverleaf Poplar – Populus alba.
Silverweed – Potentilla anserina.
Sily Oak – Orites excelsa.
SILYBUM Adams. Compositae. 2 spp. Mediterranean. Herbs.
***S. marianum*** Gaertn.=Carduus marianum. (Holy Thistle, Milk Thistle). Mediterranean. The young leaves are eaten in salads and the older leaves are used in an infusion to treat dropsy, jaundice, coughs etc. and spleen complaints. The seeds are roasted as a coffee substitute. The plant is cultivated.
***Simaba*** Aubl.=Quassia.
Simba Bali – Combretum glutinosum.
***Simaruba*** Aubl.=Quassia.
SIMIRA Aubl. Rubiaceae. 35 spp. Mexico to tropical S.America. Trees.
***S. maxonii*** Standl. Costa Rica. A decoction of the bark is used locally to reduce fevers and as a purgative.
***S. ruba*** (Mart.) Schum. Brazil. The wood is used by the natives for building houses and carving. The bark yields a red dye.
***S. tinctoria*** (H.B.K.) Schum. Brazil. The wood yields a red dye.
***S. viridiflora*** Schum.=Arariba viridiflora. Brazil. The bark is used locally to reduce fevers.

SIMMONDSIA Nutt. Simmondsiaceae. 1 sp. California into Mexico. Small tree.

***S. californica*** Nutt.=S. chinensis=S. Pabulosa. (Pignut, Goatnut). The seeds are eaten by the local Indians and when roasted are used as a substitute for coffee, or drunk in a mixture with eggs and milk. They also yield a non-drying oil (Jojoba Oil) used in melting point apparatus.

Simon Bamboo – Arundinaria simonii.

Simpoh ayer – Dillenia subfruticosa.

Simson Oil – Trianthema salsoloides.

Simulo Caper – Capparis coriacea.

Sinaloa – Arthroenemum subterminale.

SINAPIS L. Cruciferae. 10 spp. Mediterranean, Europe. Herbs.

***S. alba*** L.=Brassica alba. (White Mustard). Europe. The ground seeds are White Mustard, used as a condiment. The oil from the seeds (White Mustard Seed Oil) is used as a lubricant and for lighting. The plant is cultivated in the temperate regions of the Old and New World.

***S. cernua*** Thunb.=Brassica nigra.

***S. incana*** L.=Brassica adpressa.

***S. integrifolia*** Willd.=Brassica juncea.

***S. juncea*** L.=Brassica juncea.

***Sinarundinaria*** Nak.=Phyllostachys.

SINDORA Miq. Leguminosae. 1 sp. Tropical Africa; 20 spp. S.E.Asia, W.Malaysia, Celebes, Hainan, Moluccas. Trees.

***S. inermis*** Merr. Philippines. An oil extracted from the wood (Kayu-galu Oil) is used locally to treat skin diseases and for lighting. It may have some use in perfumery.

***S. leiocarpa*** Backer. Sumatra. The soft, pinkish wood is used locally for carving small household articles.

***S. supa*** Merr. Phillipines. The heavy hard, dark golden wood is used locally for heavy construction work, brldges, furniture and house-building. An oil from the wood (Sapa Oil) is used locally for varnishes and paints, lighting, transparent paper, and to treat skin diseases.

Singapore Copal – Agathis alba.

Singapore Copal – Agathis loranthifolia.

Singapore Manila – Agathis alba.

Singapore Rhododendron – Melastroma malabathricum.

Singhara Nut – Trapa bispinosa.

Sinkwa Towel Gourd – Luffa acutangula.

Sinoarundinaria Nak.=Phyllostachys.

***Sinocalamus beecheyanus*** (Munro) McClure =Bambusa beechyana.

***Sinocalamus latiflorus*** (Munro) McClure= Dendrocalamus latiflorus.

Sintok (Bark)=Cinnamomum sintok.

Sioerno (Oil) – Xanthophyllum lanceolatum.

SIPARUNA Aubl. Atherospermataceae. 150 spp. Mexico to tropical S.America, W.Indies. Trees.

***S. cujabana*** DC. Brazil. A decoction of the bark is used locally to induce sweating and as an abortive.

***S. oligandra*** DC. See S. cujabana.

SIPHOCAMPYLUS Pohl. Campanulaceae. 215 spp. Tropical America, W.Indies. Shrubs. Both spp. mentionted yield a rubber.

***S. caoutchouc*** G. Don. Peru.

***S. giganteus*** G. Don. Andes.

SIPHONODON Griff. Siphonodontaceae. 5–6 spp. S.E.Asia, Malaysia, N.E.Australia. Trees.

***S. australe*** Benth. (Ivory Wood). S.E.Australia. The creamy-white wood has an even texture and is used to make rulers etc., inlay work, carving and turning.

Siricote – Cordia dodecandra.

Siricote Blanco – Cordia sebestina.

Sirih – Piper betel.

Sirih popar – Ficus septica.

Siriphal – Aegle marmelos.

Siris, Red – Albizia toona.

Sirki – Saccharum munja.

Sirpoon Tree – Calophyllum tormentosa.

Sirupus Mororum – Morus nigra.

Sisal Agave – Agave sisalana.

Sisal Hemp – Agave sisalana.

Sisal, Salvador – Agave letonae.

Sisal, Yucatan – Agave fourcroyoides.

Sisilan niboot – Ziziphus calophylla.

SISON L. Umbelliferae. 2 spp. Europe, Mediterranean. Herbs.

***S. amomum*** L.=Cicuta amomum=Seseli amomum=Sium amomum=Sium aromaticum. (Bastard Stone Parsley, Hedge Sison). Europe, Asia Minor to N.Africa. The fruits (Semen Amomi vulgaris) and leaves are used to relieve digestive complaints, as a diuretic and to induce sweating. They are sometimes used as a condiment. The roots are occasionally eaten as a vegetable.

Sison, Hedge – Sison amomum.

Sissoo of India – Dalbergia sissoo.

SISYMBRIUM L. Cruciferae. 90 spp. Temperate

Eurasia, Mediterranean, S.Africa, N.America, Andes. Herbs.

***S. alliaria*** Scop. = Alliaria officinalis. (Hedge Garlic). Eurasia. The leaves, which taste of garlic, are sometimes used as a condiment.

***S. apetalum*** Lour. = Nasturtium indicum var. apetala.

***S. canescens*** Nutt. N.America. The seeds are used, particularly in Mexico, to make a drink, mixed with wines, lime juice and sugar.

***S. officinale*** (L.) Scop. (Hedge Mustard). Temperate. The leaves (Herba Erygimi) were used as a diuretic, stimulant and antiscorbutic.

***S. sophia*** L. = Descurainia sophia.

SISYRINCHIUM L. Iridaceae. 100 spp. America, W.Indies. Herbs.

***S. acre*** Mann. Hawaii. The rhizome yields a dye used locally for painting the body.

***S. galaxoides*** Gomez. = Trimezia lurida.

***Sitilias caroliniana*** (Walt.) Raf. = Pyrrhopappus carolinianus.

Sitka Columbine – Aquilegia formosa.

Sitka Cypress – Chamaecyparis nootkatensis.

Sitka Spruce – Picea sitchensis.

SIUM L. Umbelliferae. 10–15 spp. Cosmopolitan, except S.America and Australia. Herbs.

***S. amomum*** Roth. = Sison amomum.

***S. aromaticum*** Lam. = Sison amomum.

***S. cicutaefolium*** Schrank. (Water Parsnip). E. N.America. The roots are eaten as a vegetable by the local Indians and they use the leaves as a condiment.

***S. sisarum*** L. (Chervin, Skirret). Mediterranean, E.Asia. The roots were used in salads and as a vegetable and when roasted as a coffee substitute. The plant was cultivated.

***S. thunbergii*** DC. (Water Parsnip, Tandpijnwortel). S.Africa. The roots are used locally to treat toothache.

***Siuris aromatica*** Gmel. = Raputia aromatica.

Six-rowed Barley – Hordeum hexastichum.

Ška Pastora – Salivia divinorum.

Skeleton Weed – Lygodesmia juncea.

SKIMMIA Thunb. Rutaceae. 7–8 spp. Himalayas, E.Asia, Philippines. Shrubs.

***S. laurifolia*** Sieb. and Zucc. ex Walp. (Nair, Ner). N.W. Himalayas. An essential oil distilled from the leaves is used for scenting soap. The leaves are burnt locally as incense.

Skirret – Sium sisarum.

Skullcap – Scutellaria laterifolia.

Skullcap, Baical – Scutellaria baicalensis.

Skunk Bush – Garrya fremontii.

Skunk Cabbage – Symplocarpus foetidus.

Skunkweed – Croton texensis.

Skyline Bluegrass – Poa elipis.

Sla condor – Pinanga duperreana.

Slack – Porphyra umbellicalis.

Slangenhout – Strychnos ligustrina.

Slangkop, Natal – Urginea macrocentra.

Slangkop, Transvaal – Urginea burkei.

Slash Pine – Pinus caribaea.

Sleeve Palm – Manicaria saccifera.

Slender Bamboo – Oxytenanthera stocksii.

Slender Parosela – Dalea ennandra.

Slender Prairie Clover – Petalostemon oligophyllum.

Slender Wheatgrass – Agropyron trachycaulum.

Slim Amaranth – Amaranthus hybridus.

Slippery Elm – Fremontodendron californica.

Slippery Elm – Ulmus fulva.

SLOANEA L. Elaeocarpaceae. 120 spp. Tropical Asia and America. Trees.

***S. berteriana*** Choisy. (Motillo). Tropical America, W.Indies. The hard, heavy wood varies from grey to brown in colour. It is not resistant to termites or marine borers and does not work easily. It is used for heavy construction work, farm implements, heavy flooring and pilings.

***S. woollsii*** F. v. Muell. (Grey Carabeen, Yellow Carabeen). Queensland, New South Wales. The brownish wood is soft and not durable. It is used for packing cases etc., toys, brush backs and interiors of railway carriages.

Sloe – Prunus spinosa.

Sloe, Black – Prunus umbellata.

Sloe Blackthorn – Prunus spinosa.

Sloke – Porphyra umbilicus.

Slough Grass, American – Beckmannia syzigachne.

Small Buffalo Grass – Panicum coloratum.

Small Cranberry – Vaccinium oxycoccus.

Small Flowered Clover – Trifolium parviflorum.

Small Galangal – Alpinia officinarum.

Small Leaved Linden – Tilia cordata.

Small Leaved Nettle – Laportea photiniphylla.

Small Millet – Panicum sumatrense.

Small Nettle – Urtica urens.
Small Saltbush – Atriplex campanulata.
Small Soapweed – Yucca glauca.
Small Solomon's Seal – Polygonatum biflorum.
SMILACINA Desf. Liliaceae. 25 spp. Himalayas and E.Asia to N. and Central America. Herbs.
***S. racemosa*** Desf. = Vagnera racemosa. (Solomon's Plumes, Wild Spikenard). N.America. The fruits are occasionally eaten locally.
SMILAX L. Smilacacaeae. 350 spp. Tropics and subtropics. Vines.
***S. aristolochiaefolia*** Mill. = S. medica. (Mexican Sarsaparilla). S.Mexico. The dried rhizomes are one of the main sources of Sarsaparilla which is used medicinally as a tonic. The product is known as Costa Rican, Grey, Guayquil, Lima, Mexican, Red Jamaican, Tampico, Vera Cruz or Virginian Sarsaparilla. The plant is cultivated. The local Indians use the rhizome as a treatment for digestive disorders, rheumatism, skin diseases, venereal diseases and kidney complaints.
***S. auriculata*** Walt. E. N.America. The local Indians use the rhizomes as a starchy food.
***S. beyrichii*** Kunth. S.E. U.S.A. The local Indians make a bread and soup from the rhizomes.
***S. bona-nox*** L. (Brisly Greenbrier, Saw Greenbrier). S.E. U.S.A. See S. beyrichii.
***S. calophylla*** Wall. Malaysia. The rhizomes are used locally to make an aphrodisiac and tonic.
***S. china*** L. (China Root, Greenbrier). China, Japan. A decoction of the dried root (China Root, Brier Root, Radix Chinae ponderosae, Pai-fu-long) has been used in China since ancient times for the treatment of gout. Sarsaparilla is made from it in India.
***S. corbularia*** Kunth. = S. pseudo-china Lour. non L. = S. hypoglauca (Dây kim cang). China, Indochina. The leaves are used in Indochina to make a tea.
***S. glauca*** Walt. See S. beyrichii.
***S. herbacea*** L. var. ***nipponicum*** Miq. Japan. The leaves are eaten locally as a vegetable. The young shoots are also used as a vegetable.
***S. hypoglauca*** Benth. = S. corbularia.
***S. lanceolata*** L. See S. auriculata.
***S. latifolia*** Blanco = S. leucophylla.
***S. laurifolia*** L. S.E. U.S.A. See S. beyrichii.
***S. leucophylla*** Blume = S. latifolia = S. vicaria. Malaysia. A decoction of the rhizomes is used in the Philippines to treat skin diseases, syphilis, rheumatism and to purify the blood.
***S. medica*** Schlecht. = S. aritolochiaefolia.
***S. megacarpa*** DC. India to Java. The rhizomes and preserved fruits are eaten in Java.
***S. myosotiflora*** DC. Malaysia, Java. The rhizomes are used locally as an aphrodisiac.
***S. oblongifolia*** Pohl. Brazil. A decoction of the roots is used locally as a tonic and laxative.
***S. ornata*** Lam. Central America. One of the more important sources of Sarsaparilla (Jamaican Sarsaparilla). Much is exported.
***S. papyracea*** Spruce = S. spruceana.
***S. pseudo-china*** L. (Long-stalked Greenbrier). E. U.S.A., W.Indies. See S. beyrichii.
***S. pseudo-china*** Lour. non L. = S. coebularia.
***S. relegi*** Killip. and Morton. Central America. Yields a Sarsaparilla (Brown Sarsaparilla, Honduras Sarsaparilla).
***S. rotundifolia*** L. (Cat Brier, Common Greenbrier, Horse Brier). N.Canada throughout E. U.S.A. See S. beyrichii.
***S. spruceana*** A. DC. = S. papyracea. (Pará Sarsaparilla). Formerly of commercial importance as a source of sarsaparilla (Sarsaparilla de Lisboa, Sarsparilla de Marahaõ).
***S. vivaria*** Kunth. = S. leucophylla.
Smoke Bush – Dalea polyandra.
Smoke Tree, American – Rhus cotinoides.
Smoke Tree, Common – Rhus cotinus.
Smokeweed – Eupatorium maculatum.
Smoky Shield Lichen – Parmelia omphalodes.
Smooth Alder – Alnus rugosa.
Smooth-barked African Mahogany – Khaya anthotheca.
Smooth Brome – Bromus inermis.
Smooth Gooseberry – Ribes oxycanthoides.
Smooth Holly – Hedycarya angustifolia.
Smooth Hydrangea – Hydrangea arborescens.
Smooth Senna – Cassia laevigata.
Smooth Sumach – Rhus glabra.
Smut, Corn – Ustilago maydis.
Smyrna Gall – Quercus lusitanica.

Smyrna Pea – Pisum arvense.
SMYRNIUM L. Umbelliferae. 8 spp. Europe, Mediterranean. Herbs.
***S. olus-atrum*** L. (Alexanders, Maceron). Mediterranean, Asia Minor, Canary Islands. The petioles were used as a vegetable. Previously cultivated but now replaced by celery.
Snail Clover – Medicago scutellata.
Snake Bark, West Indian – Colubrina ferruginea.
Snake Flower – Bulbine asphodeloides.
Snake Flower – Bulbine narcissifolia.
Snake Gourd, Edible – Trichosanthes anguinea.
Snake Gourd, Japanese – Trichosanthes cucumeriodes.
Snake Lily – Haemanthus natalensis.
Snakeroot – Eryngium campestre
Snakeroot – Polygonum bistorta.
Snakeroot – Strychnos colubrina.
Snakeroot, Black – Actaea spicata.
Snakeroot, Black – Cimicifuga racemosa.
Snakeroot, Black – Sanicula marilandica.
Snakeroot, Button – Eryngium yuccifolium.
Snake root, Button – Liatris spicata.
Snakeroot, Seneca – Polygala seneca.
Snakeroot, Texas – Aristolochia reticulata.
Snakeroot, Virginia – Aristolochia serpentaria.
Snakeroot, White – Eupatorium aromaticum.
Snakewood – Strychnos colubrina.
Sneezeweed, Bitter – Helenium tenuifolium.
Sneezeweed – Ptaeroxylon utile.
Sneezewort – Achillea ptarmica.
Sneng kou – Tabernaemontana stralensis.
Sner – Manilkara fasciculata.
Snow Lichen – Cetraria nivalis.
Snow-in-the-mountain – Euphorbia marginata.
Snow Trillium – Trillium grandiflorum.
Snowbread – Cyclamen europaeum.
Snowdrop – Erythronium americanum.
Snowdrop, Wood – Monotropa uniflora.
Snuff, Cohoba – Piptadenia peregrina.
Snuff, Coxoba – Piptadenia peregrina.
Snuff, Schneeberger – Asarum europaeum.
Snuff, Tobacco – Nicotiana tabacum.
Snuff, Ýa-ka – Virola calophylla.
Snuff, Ýa-ko – Virola calophylla.
Soap Aloe – Aloe saponaria.
Soapbark – Quillaja saponaria.
Soapberry, Chinese – Sapindus mukorossi.
Soapberry, Southern – Sapindus saponaria.
Soapberry, Western – Sapindus marginatus.
Soapnut – Sapindus mukorossi.
Soapnut – Sapindus trifoliatus.
Soap Plant, California – Chenopodium californicum.
Soap Pod – Acacia cocinna.
Soaproot – Saponaria officinalis.
Soaproot, Californian – Chlorogalum pomeridianum.
Soaproot, White – Gypsophila arrostii.
Soaptree Yucca – Yucca elata.
Soapweed, Small – Yucca glauca.
Soapwood – Caryocar glabrum.
Soapwort – Saponaria officinalis.
Sobheuri – Bauhinia elongata.
Socotra Aloe – Alöe perryi.
Socotra Dragon's Blood – Dracaena cinnabari.
Socotra Dragon's Blood – Dracaena schizantha.
SOCRATEA Karst. Palmae. 12 spp. N.tropical S.America. Palm trees.
***S. durissima*** (Oerst.) Wendl. = Iriartea durissima. Central America. The stems are used locally for thatching and the buds are eaten as a vegetable. The wood (Paxiuba Wood) is used for posts and fences.
***S. exorrhiza*** Wendl. (Paxiuba Palm). Tropical America. The wood, (Raxiuba Wood) is used locally for fencing, posts etc., casings for blow-pipes and trumpets.
Sodom, Apple of – Calotropis procera.
Soga – Peltophorum pterocarpum.
Sogo – Acroceras amplectens.
Soko – Acroceras amplectens.
***Soja hispida*** Moench = Glycine max.
SOLANUM L. Solonaceae. 1700 spp. Tropics and temperate. Herbs.
***S. aethiopicum*** L. Tropical Africa, Asia. Cultivated in N. and Central Africa for the fruits which are eaten.
***S. agrarium*** Sendt. Brazil. The fruits are eaten locally and a decoction of the leaves are used to treat gonorrhoea.
***S. alternato-pinnatum*** Steud. = S. oleraceum. Argentina to Brazil. The leaves are used locally as a diuretic and narcotic.
***S. andigenum*** Juz. and Buk. Peru to Colombia. The stem tubers are eaten locally. The plant is cultivated.
***S. anomalum*** Thonn. Tropical Africa. The fruits (Children's Tomatoes) are used locally as a condiment, fresh or dried.
***S. aviculare*** Forst. = S. laciniatum = S. vescum. (Kangaroo Apple). Australia.

The fruits are eaten locally, fresh or as a vegetable.

***S. blumei*** Nees. (Booloong). Java. The leaves are eaten locally as a vegetable, or raw.

***S. carolinense*** L. (Carolina Horse Nettle). N.America. The fruits are used medicinally as a mild sedative.

***S. coronopus*** Dunal. = Chamaesaracha coronopus.

***S. diversifolium*** Schlecht. = S. fendleri. Central America, W.Indies. The salted fruits are used locally as a condiment with fish.

***S. dubium*** Fresn. Arabia, E.Africa. The fruits, ground in water are used in the Sudan to remove hair from hides.

***S. dulcamara*** L. (Bittersweet). Europe, Asia, N.Africa, N.America. The plant contains the glucoside dulcamarin. The dried stems are used medicinally as a sedative and diuretic.

***S. duplosinuatum*** Klotzsch. = S. farini = S. kalimandschari. Tropical to S.Africa. The fruits are used locally in soups. Cultivated locally.

***S. elaeagnifolium*** Cav. (Silverleaf Nightshade). S. U.S.A. into tropical America. The local Indians use the seeds to curdle milk and mix them with brain tissue to tan skins.

***S. ellipticum*** R. Br. Australia. The fruits are eaten by the aborigines.

***S. farini*** Damann. = S. duplosinuatum.

***S. fendleri*** Gray. (Fendler Potato, Wild Potato). S.W. U.S.A. into Central America. The tubers are eaten by the New Mexican Indians.

***S. fendleri*** Van Heurck. and Muell. = S. diversifolium.

***S. ferox*** L. S.E.Asia, Malaysia. The fruits are used as a condiment in curries.

***S. inaequilaterale*** Merr. Philippines. The leaves are smoked locally as tobacco.

***S. incanum*** L. E.Africa, Asia. In the Sudan, the seeds are used to curdle milk.

***S. insidiosum*** Mart. (Jurubeba). Tropical America. A decoction of the roots is used in Brazil as a diuretic and to relieve stomach upsets.

***S. jamesii*** Torr. S.W. U.S.A. into Mexico. The tubers are eaten as a vegetable by the local Indians.

***S. kalimandschari*** Dammar = S. duplosinuatum.

***S. laciniatum*** Ait. = S. aviculare.

***S. lycopersicum*** L. = Lycopersicum esculentum.

***S. macrocarpon*** L. Origin unknown. Cultivated in W.Africa, for the fruits which are eaten as a vegetable.

***S. mammosum*** L. (Berenjene, Cichihua, Pichichio). Mexico into S.America, W.Indies. The leaves are used in Costa Rica to treat kidney and bladder ailments.

***S. melongena*** L. (Aubergine, Eggplant). Tropical Asia. Grown widely in Old and New World where the temperature is high enough. The fruits are used as a vegetable. The plant was introduced to Brazil by the Portuguese and is grown widely in S. U.S.A. and S.America. It is grown in more temperate climates and features frequently in French and Italian cooking. It is also an important vegetable in China and Japan. There are several varieties with fruits varying in colour from white to dark purple. Sometimes the plant is grown as an ornamental.

***S. muricatum*** Ait. (Pepino). Peru. Cultivated in Central and S.America for the fruits which a juicy and pleasantly flavoured. They are eaten raw.

***S. nigrum*** L. (Black Nightshade). Cosmopolitan. The fruits (Woderberries) are eaten in pies etc. The shoots and leaves are used as a vegetable.

***S. nodiflorum*** Jacq. Sahara, Nigeria. The leaves are used locally as a vegetable. Cultivated locally.

***S. oleraceum*** Vell. = S. alternato-pinnatum.

***S. olivare*** Paill. See S. macrocarpon.

***S. paniculatum*** L. Brazil. An alcoholic extract of the roots is used locally to treat liver, bladder and spleen complaints.

***S. pierreanum*** Bois. (Olombe). Tropical Africa. The fruits are eaten locally.

***S. piliferum*** Benth. Mexico. The pleasantly flavoured fruits are eaten locally.

***S. quitoense*** Lam. (Naranjillo). Ecuador, Peru. The fruits are slightly acid and make a refreshing drink. The plant is grown commercially in Ecuador.

***S. saniwongsei*** Craib. Tropical Asia. The fruits are eaten in Thailand and are used there to reduce the blood sugar content in cases of diabetes.

***S. saponaceum*** Dun. = S. scabrum. Peru. The fruits are used locally as a soap substitute for washing cloths.

***S. scabrum*** Ruiz. and Pav. = S. saponaceum.

***S. sodomeum*** L. (Apple of Sodom). Mediterranean, S.Africa. Poisonous. The leaves and fruits are used as a diuretic in treating cystitis and in local remedies for coughs.

***S. spirale*** Roxb. (Mak dit). India, Indonesia. The plant is used locally as a condiment.

***S. supinum*** Dun. (Bitter Apple). S.Africa. The natives use the fruits to curdle milk.

***S. topiro*** H.B.K. (Cocona). N. S.America. Probably a cultigen. The fruits are eaten locally and are used to make a slightly acid refreshing drink.

***S. torvum*** Sw. (Soushumber). Tropics. In Java the young shoots are eaten as a vegetable and in the W.Indies the unripe fruits are used as a condiment with fish.

***S. trifolorum*** Nutt. W. N.America to Mexico. The local Indians eat the fruits raw, as a vegetable or ground to make a flour.

***S. tuberosum*** L. (Potato). Temperate Andes. The underground stem tubers are starchy and are eaten as a vegetable. The plant is also grown for the commercial production of alcohol and starch. In the Scandinavian countries one of the most popular alcoholic drinks, Akvavit (Aquavit) is made from potatoes. There are hundreds of varieties. The potato was introduced into Europe in 1570 and into N.America from Europe in 1621. The crop is grown throughout the world where the climate is suitable but the main producing countries are Russia (30·4 per cent), Poland (12·8 per cent) and W.Germany (8·3 per cent) of a world production of some 3,114,250,000 tons a year. A large proportion of the crop in U.K. and U.S.A. is used for making chips, dehydrating and to make starch. Potatoes are also fed to livestock. The potato contains 8–28 per cent starch and 1–4 per cent protein. Propagation is by planting the stem tubers (seed potatoes). The crop grows best on a well-drained loam, with a high organic content, yielding about 10 tons of tubers an acre. Most of the crop is harvested mechanically after cutting the tops or removing them with chemicals, e.g. sulphuric acid. They store well and can be kept throughout the winter. S. tuberosum is an autotetraploid (n=48).

***S. uporo*** Dun. Polynesia. The fruits are eaten locally. The plant is cultivated in Fiji.

***S. vescum*** F. v. Muell. = S. aviculare.

***S. xanthocarpum*** Schrad. and Wendl. Tropics. The seeds are sometimes used as an expectorant.

Solazzi – Glycyrrhiza glabra.

SOLENOSTEMMA Hayne. Asclepiadaceae. 1 sp. Egypt, Arabia. Shrub.

***S. argel*** Hayne. A poultice of the leaves is used to heal wounds on camels.

SOLENOSTEMON Thonn. Labiatae. 10 spp. Tropical Africa. Herbs.

***S. ocymoides*** Schum. and Thonn. Tropical Africa. The leaves and tubers are used locally as vegetables, especially in W.Africa. The plant is sometimes cultivated.

SOLIDAGO L. Compositae. 100 spp. America; 1 sp. Europe. Herbs.

***S. altissima*** L. S.E. U.S.A. The latex is a potentially good source of rubber.

***S. californica*** Nutt. (Californian Golden Rod). California. The leaves are used as a poultice for wounds and sores.

***S. canadensis*** L. (Canada Golden Rod). E. N.America. The seeds are eaten by the local Indians.

***S. fistulosa*** Mill. See S. altissima.

***S. leavenworthii*** Torr. and Grey. (Leavenworth Golden Rod). See S. altissima.

***S. nana*** Nutt. See S. canadensis.

***S. odorata*** Ait. (Sweet Golden Rod). E. N.America. The leaves have an anise scented essential oil. They are also used to make a tea.

***S. rigida*** L. See S. altissima.

***S. spectabilis*** Gray. See S. canadensis.

***S. virgaurea*** L. (European Golden Rod). Europe, Asia, introduced to N.America. The roots and leaves (Radix et Herba Virgae aurea) were used medicinally as a mild sedative, diuretic and to treat digestive upsets. The leaves were sometimes used to make a tea and as a poultice for wounds.

Solimán – Asclepias linaria.

Solimán – Croton ciliato-glandulosus.

Solo – Euphorbia kamarunica.

Solomon's Plumes – Smilacina racemosa.

Solomon's Seal – Polygonatum multiflorum.

Solomon's Seal, Drug – Polygonatum officinale.

Solomon's Seal, Small – Polygonatum biflorum.

SOLORINA Ach. Peltigeraceae. Lichens.

***S. crocea*** (L.) Ach. (Saffron-yellow Solorina). Temperate. The yellow dye from the plant is used in Scotland for dyeing wool.

Solorina, Saffron-yellow – Solorina crocea.

Somali Gum – Acacia glaucophylla.

Som toum – Cissus repens.

Somangana – Dombeya condensata.

Sona – Bauhinia purpurea.

SONCHUS L. Compositae. 50 spp. Eurasia, Mediterranean, tropical Africa, Atlantic Islands. Herbs.

***S. oleraceus*** L. (Hare's Lettuce, Milkweed, Milky Tassel, Sow Thistle). Through range. The leaves are sometimes eaten as a vegetable and fed to stock. They are used in Switzerland to feed edible snails. The juice is sometimes used to treat liver and intestinal complaints.

***S. taraxcifolius*** Willd.=Lactuca taraxaciflora.

Soncoya – Annona purpurea.

Song – Homalium fagifolium.

Songga langit – Ochrosia oppositifolia.

Songo – Dillenia elliptica.

Songgom – Barringtonia insignis.

SONNERATIA L. f. Sonneratiaceae. 5 spp. Coasts of tropical E.Africa and Madagascar through to Malaysia, Polynesia and N. Australia. Trees, along coasts, mangrove swamps etc.

***S. acida*** L. f. Tropical Asia. The aerial roots are used as cork.

***S. alba*** Smith. Tropical Asia. The brown, heavy, tough wood is used to build bridges and houses and the leaves are eaten as a vegetable in Malaya.

***S. caseolaris*** (L.) Engl. (Perepat). Malaya, Borneo. The heavy, hard wood is used for general construction work, houses, furniture, bridges, piles, pavings and boat-building. The air-roots are used for floats and making razor hones. The fruits are eaten raw or as a vegetable and when unripe in chutneys and curries.

***S. ovata*** Backer. (Bogen, Kedaboo). Malaysia. The fruits are used locally in a syrup (Roodjak) to stimulate the appetite of invalids.

Sonora Ironwood – Olneya tesota.

Songsgaravi – Muncuna monosperma.

Sookmadiloowih – Gunnera macrophylla.

Soolaketan – Antidesma stipulare.

Soolangkar – Leea indica.

Sooling – Cassia nodosa.

Soolo – Chasmanthera welwitschii.

Soomaralah – Horsfieldia irija.

Soompoo roompooh – Ardisia colorata.

Soongkai – Heterophragma macrolobum.

Soong-kai – Peronema canescens.

Soonga karsik – Vatica songa.

Soongoong – Combretum tetralophum.

Soonhari – Actinorhytis calapparia.

Soopa loobang – Oudemansiella canarii.

Sooranto – Parashorea lucida.

Soorooghan – Morinda geminata.

Soozan katamoogoo – Podaxis pistillaris.

Sop, Sweet – Annona squamosa.

***Sophia*** Adans.=Descurainia.

SOPHORA L. Leguminosae. 50 spp. Tropics, warm temperate. Trees.

***S. angustifolia*** Sieb. and Zucc.=S. galegoides (Hoè giác, K'u shên). China. The leaves are an important local drug, used as a tonic, to relieve stomach upsets, chest complaints and as a diuretic.

***S. chrysophylla*** Seem.=Edwardsia chrysophylla. Hawaii. The hard wood is used locally for sledge runners.

***S. flavescens*** Ait. China. A decoction of the roots is used locally to treat dysentery.

***S. galegoides*** Pall.=S. angustifolia.

***S. heptaphylla*** Blanco=S. tomentosa.

***S. japonica*** L. (Japanese Pagoda Tree). N.China, Japan. A yellow dye from the pods is used to dye cloth. It can be mixed with indigo to give a green colour. The Chinese use an extract of the leaves and fruits to adulterate opium.

***S. secundiflora*** (Orteg.) Lag. ex DC. (Mescal Bean). S.W. U.S.A. to Mexico. The seeds (Mescal Beans) are used to make necklaces used in ceremonial dances. They are also used as an intoxicant when added to Agave.

***S. sericea*** Nutt. (Silky Sophora). N.America. to Mexico. The sweet roots are eaten by the local Indians.

***S. tetraptera*** Ait. (Fourwing Sophora). Chile, Juan Fernandez, New Zealand. The light brown wood is very heavy, strong and durable. It is used for furniture-making, turning and for bearings and shafts for machinery.

***S. tomentosa*** L.=S. heptaphylla. Throughout the tropics. In the Philippines the leaves and seeds are used to treat stomach upsets and in New Zealand the roots and seeds are used for the same purpose.

Sophora, Fourwing – Sophora teraptera.

Sophora, Silky – Sophora sericea.

SORBUS L. Rosaceae. 100 spp. N.Temperate. Trees.

***S. americana*** Marsch.=Pyrus americana. (Mountain Ash). E. N.America. The fruits are used in home remedies for colds etc.

***S. aria*** Crantz.=Pyrus aria=Mespilus aria (White Beam). Europe. The fruits are used to make beans and vinegar and when dried to treat coughs etc.

***S. aucuparia*** L.=Pyrus aucuparia=Mespilus aucuparia. (Rowan Tree, European Mountain Ash). Europe through to Siberia, Asia Minor. The fruits are used to make jellies etc. Used in brandy in Germany (Sechämtertropen) and in vodka in Russia. They are also used as a substitute for coffee. The leaves and flowers are used as a tea and as an adulterant for tea. The hard, tough, elastic wood is used for turning and building wagons.

***S. domestica*** L. (Service Tree). S.Europe into Asia, N.Africa. The heavy, fine-grained wood is used for making furniture, wine-presses and screws. The bark is used for tanning. The fruits are edible after being frosted and are used to make wine.

***S. torminalis*** (L.) Crantz.=Aria torminalis =Pyrus tominalis. (Checker Tree, Service Tree). Europe, Asia Minor, Caucasus, N.Africa. The fruits are used to make, wine, brandy and vinegar.

Sore Eye Flower – Boophone disticha.

SORGHASTUM Nash. Graminae. 12 spp. Warm and tropical America, tropical Africa. Grasses.

***S. nutans*** (L.) Nash. (Indian Grass). N.America. Used to stabilize inland sand dunes.

SORGHUM Moenxh. Graminae. 60 spp. Tropics and subtropics. Grasses.

***S. almum*** Parodi. (Columbus Grass). Argentina. A hybrid, with S. halepense as one parent. Used for livestock feed and silage, especially in low-rainfall areas.

***S. caudatum*** Stapf. W.Africa. A red dye is extracted from the stems and leaf-sheaths. The plant is cultivated locally.

***S. halepense*** (L.) Pers.=Holcus halepensis. (Johnson Grass). Mediterranean. Cultivated as cattle fodder in warm parts of Old and New World. An autotetraploid (n=20) probably derived from the diploid annuals.

***S. margaritiferum*** Stapf. Congo, Niger Basin. The seeds are used locally as grain.

***S. vulgare*** Pers.=Andropogon sorghum.= Sorghum, Sweet Sorghum, Great Millet, Durra, Guinea Corn). Nile Valley, Central India. Has been used as a grain crop in these areas since time immemorial. A diploid (n=10). It cross fertilises easily, leading to the production of a great number of varieties, with a wide range of environmental requirements. The main type grown is durra (Shallu) which is used in the arid regions of Africa and India, but others, e.g. Sweet Sorghum, is grown for sugar production and others for cattle feed either as forage or as seed. Durra grain contains no gluten, so does not make a bread. It is eaten as a porridge; it is also fermented to make an alcoholic drink.

Sorghum – Sorghum vulgare.

Sorghum, Sweet – Sorghum vulgare.

SORINDEIA Thou. Anacardiaceae. 50 spp. Tropical Africa, Madagascar. Trees.

***S. claessenensii*** De Wild. (Diloso, Uhambaka). Congo. The wood is used locally for making arrows and the roots for caulking boats.

***S. gilletii*** De Wild.=S. kimuenzae=S. maxima. (Biembie, Elokoloko, Okote). Gabon, Congo, Angola. The hard pinkish wood is used locally for general construction work.

***S. juglandifolia*** Planch. ex Oliv. Tropical Africa. The fruits (Damsons), are eaten locally in W.Africa.

***S. kimuenzae*** De Wild.=S. gilletii.

***S. madagascariensis*** DC. (Grape Mango). Madagascar. The fruits are eaten locally.

***S. maxima*** Verm.=S. gilletii.

***S. werneckei*** Engl. Tropical Africa. The fruits yield a blue dye used for painting the body.

Sorrel, French – Rumex scutans.

Sorrel, Garden – Rumex acetosa.

Sorrel, Jamaica – Hibiscus sabdariffa.

Sorrel, Mountain – Oxyria digyna.

Sorrel, Mountain – Rumex paucifolius.

Sorrel Tree – Oxydendrum arboreum.

Sorrel, Upright Yellow Wood – Oxalis stoicta.

Sorrel, Wood – Oxalis acetosella.

Sosa – Suaeda ramisissima.

Sosan – Iris nepalensis.

Sotol – Dasylirion simplex.

Sotol, Texas – Dasylirion texanum.

Sotol, Wheeler – Daslirion wheeleri.
Souari Fat – Caryocar amygdaliferum.
Souari Nut – Caryocar nucifera.
Souari Nut – Caryocar tomentosum.
Souari Tree – Caryocar tomentosum.
SOULAMEA Lam. Simaroubaceae. 10 spp. Borneo, Moluccas, New Guinea to Fiji. Trees.
***S. amara*** Lam. Malaysia. A decoction of the fruits and roots is used locally to treat stomach pains and pleurisy.
Soulkhir – Agriophyllum gobicum.
Soungax – Euphorbia hermentiana.
Sour Cherry – Prunus cerasus.
Sour Grass – Amphilophis pertusa.
Sour Gum – Nyssa multiflora.
Sour Orange – Citrus aurantium.
Sour Tupelo – Nyssa ogeche.
Soursop – Annona muricata.
Soursop, Mountain – Annona montana.
Sourwood – Oxydendrum arboreum.
Soushumber – Solanum torvum.
South African Doum Palm – Hyphaene crinita.
South Australian Sandalwood – Eucarya acuminatus.
South Cameroom Ebony – Diospyros bipindensis.
Southern Cane – Arundinaria gigantea.
Southern Cypress – Taxodium distichum.
Southern Dewberry – Rubus trivalis.
Southern Fox Grape – Vitis rotundifolia.
Southern Magnolia – Magnolia grandiflora.
Southern Pine – Pinus palustris.
Southern Red Cedar – Juniperus barbadense.
Southern Red Oak – Quercus texana.
Southern Sassafras – Atherosperma moschatum.
Southern Soapberry – Sapindus saponaria.
Southern White Cedar – Chamaecyparis thyoides.
Southwestern Condalia – Condalia lycioides.
Sow Thistle – Sonchus oleraceus.
Sowa – Anethum sowa.
Soybean (Oil) – Glycine max.
Soya Bean – Glycine max.
SOYMIDA A. Juss. Meliaceae. 1 sp. Indomalaya. Tree.
***S. febrifuga*** Juss. (Bastard Cedar, Indian Red Wood). The bark yields a strong cordage fibre and is used for tanning. The wood is used to make houses and furniture, for carving and household utensils.
Spangletop, Green – Leptochloa dubia.
Spanish Alfalfa – Medicago sativa.
Spanish Bayonet – Yucca aloifolia.
Spanish Broom – Spartium junceum.
Spanish Chestnut – Castanea sativa.
Spanish Dagger – Yucca aloifolia.
Spanish Espercet – Hedysarum coronarium.
Spanish Hops – Origanum creticum.
Spanish Hops – Origanum heraclecticum.
Spanish Lime – Melicoccus bijugatus.
Spanish Liquorice – Glycyrrhiza glabra.
Spanish Moss – Tillandsia usneoides.
Spanish Mustard – Brassica peroiridis.
Spanish Oak – Quercus digitata.
Spanish Oak – Quercus pagoda.
Spanish Origanum Oil – Thymus capitatus.
Spanish Oyster Plant – Scolymus hispanicus.
Spanish Plum – Spondias purpurea.
Spanish Psyllium Seed – Plantago arenaria.
Spanish Rhubarb Dock – Rumex abyssinicus.
Spanish Turpetroot – Thapsia garganica.
SPARASSIS Fr. Clavriaceae. 5 spp. N.Temperate. The fruitbodies of the following spp. are eaten locally.
***S. crispa*** (Wulf.) Fr.
***S. laminosa*** Fr.
SPARGANIUM L. Sparganiaceae. 20 spp. N.Temperate, New Zealand, Australia. Herbs.
***S. eurycarpum*** Engl. (Giant Bur Reed). N.America. The tubers were eaten as a vegetable by the local Indians.
***Sparganophorus*** Boehm. = Struchium.
SPARMANNIA L. f. Tiliaceae. 7 spp. Tropical and S.Africa, Madagascar. Small trees.
***S. africana*** L. f. (Stock Rose). S.Africa. A fibre is obtained from the bark.
SPARTINIA Schreb. Graminae. 16 spp. Mostly temperate America, some coastal Europe and Africa. Grasses.
***S. pectinata*** Link. (Prairie Cordgrass). N.America. The coarse grass is used for thatching hay stacks.
SPARTIUM L. Leguminosae. 1 sp. Mediterranean. Shrub.
***S. junceum*** L. (Spanish Broom, Weaver's Broom). The bark fibres are sometimes used for cordage, mats, filling pillows etc. and to make paper. The stems are used to make baskets. An essential oil obtained from the flowers is occasionally used in perfumery.
***S. scoparium*** L. = Sarothamnus scoparius.
***S. tinctorium*** Roth. = Genista tinctoria.

SPATHIPHYLLUM Schott. Araceae. 36 spp. Central and tropical S.America, Philippines, Celebes, Moluccas. Herbs.

***S. cannaefolium*** (Dryand) Schott. = S. candicans.

***S. candicans*** Poepp. and Endl. = S. cannaefolium = Pothos cannaefolia. Tropical America. The leaves are used locally in Peru, for flavouring tobacco.

***S. caudatum*** Porpp. and Endl. = Urospatha caudata.

***Spathodea gigantea*** Blume = Radermachera gigantea.

***Spathodea macroloba*** Miq. = Heterphragma macrolobum.

SPATHOGLOTTIS Blume. Orchidaceae. 46 spp. China, Indomalaya, Australia. Herbs.

***S. eburnea*** Gagnep. (Bai tre nep). Cambodia. The tubers are eaten locally as a vegetable.

***S. lilacina*** Griff. = S. plicata.

***S. plicata*** Blume = S. lilacina = Bletia angustata. Malaysia, Indonesia. In Perak, a poultice of the leaves is used to relieve rheumatic pains.

***Spathyema*** Rafin. = Symplocarpus.

Spearhead – Heteropogon hirtus.

Spearleaf Grass Tree – Xanthorrhoea hastilis.

Spearmint (Oil) – Mentha spicata.

Spearwood – Acacia doratoxylon.

Spearwood – Acacia homalophylla.

Speedwell, Drug – Veronica officinalis.

Speik – Achillea clavenae.

Spelt – Triticum spelta.

SPERGULARIA (Pers.) J. and C. Presl. Caryophyllaceae. 40 spp. Cosmopolitan. Herbs.

***S. arvensis*** L. (Corn Spurrey). Europe, introduced to N.America. The seeds, mixed with wheat are used as a famine food in Scandinavia. Sometimes cultivated for cattle food.

***S. rubra*** (L.) J. and C. Presl. = Arenaria rubra. (Red Sandwort, Sand Spurrey). S.Europe. The leaves (Herba Areneriae rubra, Herba Spergulae rubrae) are used medicinally to treat kidney stones and catarrh of the bladder.

***Spermolepis gummifera*** Grongn. and Gris. = Arillastrum gummiferum.

SPHAERALCEA A. St. Hil. Malvaceae. 6 spp. Warm America, S.Africa. Herbs.

***S. angustifolia*** (Cav.) G. Don. (Narrowleaf Globemallow). W. U.S.A. to Mexico. The stems are chewed by the local Indians.

SPHAERANTHUS L. Compositae. 40 spp. Africa, Madagascar, Asia Minor, through to W.Malaysia, Celebes, N.E.Australia. Herbs.

***S. hirtus*** Willd. = S. indicus.

***S. indicus*** L. = S. hirtus = S. mollis. India, Malaysia, Indonesia, Australia. In India the leaves are used to treat intestinal worms, the plant cooked in butter, flour and sugar is a tonic and the fried or boiled seeds are used as an aphrodisiac. In Java the plant is used as a diuretic.

***S. mollis*** Roxb. = S. indicus.

***S. senegalensis*** DC. (China Chenuet). Senegal, Mauritania. The leaves are used in N.W.Africa for stuffing pillows and treating boils.

SPHAGNUM L. Sphagnaceae. 320 spp. Cosmopolitan.

***S. cymbifolium*** Ehrh. (Sphagnum Moss). Temperate. Used horticulturally for lightening heavy soils and growing plants in hanging baskets etc. It can be used as a surgical dressing.

Sphagnum Moss – Sphagnum cymbifolium.

SPATHIPHYLLUM Schott. Araceae. 36 spp. Philippines, Celebes, Moluccas, tropical S.America. Herbs.

***S. cannifolium*** (Dryad) Schott. = Monstera cannaefolia. Tropical S.America. The dried leaves are used locally for scenting tobacco.

SPHENOCENTRUM Pierre. Menispermaceae. 1 sp. Tropical W.Africa. Shrub.

***S. jollyanum*** Pierre. The roots are used locally as a laxative and to treat stomach pains. They are also chewed and being bitter give food eaten subsequently a sweet taste.

SPHENOCLEA Gaertn. Sphenocleaceae. 1 sp. throughout the tropics; 1 sp. W.Africa. Herbs.

***S. zeyalica*** Gaertn. (Goonda). Malaya, Indonesia. A weed in rice; the leaves are eaten as a vegetable with rice, particularly in Java.

SPHENOSTYLIS E. Mey. Leguminosae. 16 spp. Africa. Vines. The seeds and tubers of the following spp. are eaten locally.

***S. briartii*** (De Wild.) Bak. f. (Sileli, Tulanda). Central Africa.

***S. schweinfurthii*** Harms. (Yam Bean). Tropical Africa. Cultivated.

***S. stenocarpa*** (Hochst.) Harms. Tropical Africa, especially Abyssinia and W.Coast. Cultivated.

Spice Bush – Lindera benzoin.

Spickled Alder – Alnus incana.

Spicy Cedar – Beilschmiedia mannii.

Spider Antelope Horn – Asclepias decumbens.

Spiderwort – Tradescantia virginica.

SPIGELIA L. Spigeliaceae. 50 spp. Warm America. Herbs.

***S. anthelmia*** L. (West Indian Spigelia). W.Indies, N. S.America. A decoction of the roots is used to treat intestinal worms and as a criminal poison.

***S. flemmingiana*** Cham. and Schlecht. Brazil. A decoction of the plant is used to induce sweating and treat intestinal worms.

***S. glabrata*** Mart. Brazil. See S. flemmingiana.

***S. marilandica*** L. (Pinkroot). E. N.America. The dried rhizomes are used medicinally to treat intestinal worms.

Spigelia, West Indian – Spigelia anthelmia.

Spignel, Broad-leaved – Peucedanum cervaria.

Spike Oil – Lavendula latifolia.

Spiked Loosestrife – Lythrum salicaria.

Spiked Millet – Pennisetum typhoideum.

Spikenard – Nardostachys jatamansi.

Spikenard, American – Aralia racemosa.

Spikenard, Wild – Smilacina racemosa.

SPILANTHES Jacq. Compositae. 60 spp. Tropical America, Africa, Malaysia, N.Australia. Herbs.

***S. acmella*** Murr. (Akarakara, Maratitige, Pará Cress, Pirazha). Throughout the tropics. The leaves are eaten raw or as a vegetable and when crushed to stupify fish. An extract of the flower heads placed in water kills mosquito larvae and relieves toothache.

***S. mutisii*** H.B.K. S.America. An extract of the flowers is used in Colombia to treat liver disorders.

SPINACEA L. Chenopodiaceae. 3 spp. E.Mediterranean to Central Asia and Afghanistan. Herbs.

***S. oleracea*** L. (Spinach) S.W.Asia. The plant has been grown locally as a salad and vegetable crop since ancient times. Later it was introduced to Europe and from there to America. It is grown widely in temperate regions of the Old and New World as a vegetable, for canning and freezing. There are several varieties.

Spinach – Spinacea oleracea.

Spinach, Mountain – Atriplex hortensis.

Spinach, New Zealand – Tetragonia tetragonioides.

Spindle Tree – Euonymus europaeus.

Spiny Bamboo – Bambusa arundinacea.

Spiny Vitis – Vitis davidi.

Spiny Yam – Dioscorea spinosa.

SPIRAEA L. Rosaceae. 100 spp. N.Temperate, S. to Himalayas and Mexico. Shrubs.

***S. blumei*** Don. China. The leaves are used locally as a tea.

***S. chinensis*** Maxim. See S. blumei.

***S. filipendula*** L. = Filipendula hexapetala.

***S. henryi*** Memsl. See S. blumei.

***S. ulmaria*** L. = Filipendula ulmaria.

SPIRAENTHEMUM A. Gray. Cunoniaceae. 20 spp. New Guinea, Polynesia. Small tree.

***S. samoense*** Gray. Polynesia. The wood is used locally to make canoes and clubs.

SPIRANTHES Rich. Orchidaceae. 25 spp. Cosmopolitan, except Central and tropical S.America, tropical and S.Africa and Mascarene Islands. Herbs.

***S. chiliensis*** A. Rich. = S. diuretica.

***S. diuretica*** Lindl. = S. chilensis = Neottia diuretica. Chile. A decoction of the leaves is used locally as a diuretic.

Spirit, Turpentine – Pinus palustris.

Spirit Weed – Lachnathes carolina.

Spogel – Plantago ovata.

SPONDIANTHUS Engl. Euphorbiaceae. 2 spp. Tropical Africa. Trees. The bark of both spp. is used in W.Africa as a rat poison.

***S. preussii*** Engl. = Megabarea trillesii.

***S. ugandensis*** Hutch.

SPONDIAS L. Anacardiaceae. 10–12 spp. Indomalaya, S.E.Asia, tropical America. Trees. The fruits of the spp. mentioned are eaten raw, cooked, or in preserves. They have a pleasant, slightly acid, spicy flavour.

***S. cytherea*** Sonner. = S. dulcis.

***S. dulcis*** Forst. f. = S. cytherea. (Ambarella, Jew Plum, Otaheite Apple). Tropics. Cultivated in Old and New World tropics.

***S. laosensis*** Pierre. Tropical Asia. Cultivated in Tonkin.

***S. lutea*** L. (Hog Plum, Jobo, Yellow Mombin). Tropics. Cultivated.

***S. mangifera*** Willd. = Evia amara. Tropical Asia. The flowers are sometimes eaten as a vegetable or in salads. Sometimes cultivated in Indo-China.

***S. mombin*** L. = S. purpurea.

***S. purpurea*** L. = S. mombin. (Ciruela,

Jacote, Red Mombin, Spanish Plum). Central America, Mexico.

***S. tuberosa*** Arruda. (Hog Plum, Imbu). N.E.Brazil. Sometimes cultivated locally and used to make a milk jelly Imbuzada.

***S. wirtgenii*** Hassk. Java. The trunk yields a gum.

Sponge, Vegetable – Luffa cylindrica.

Spoon Tree – Malouetia tamaquarina.

Spoonwort – Cochlearia officinalis.

SPOROBOLUS R. Br. Graminae. 150 spp. Tropics and warm temperate. Grasses.

***S. cryptandrus*** (Torr.) A. Gray. (Sand Dropseed). N.America into Mexico. The seeds are eaten by the local Indians. The grass is used for reclamation work.

***S. elongatus*** R. Br.=S. indicus.

***S. fimbriatus*** Nees. S.Africa. Used as feed for livestock and reclamation.

***S. flexuosus*** (Thurb.) Rydb. (Mesa Dropseed). N.America into Mexico. The seeds are eaten by the local Indians.

***S. indicus*** R. Br.=S. elongatus (Sweet Grass). Tropics. The leaves are used to make hats.

***S. lindeleyii*** Benth.=S. pallidus.

***S. pallidus*** Lindl.=S. lindelyii. Australia. The seeds are used as a bread flour by the aborigines.

Spotted Blue Gum – Eucalyptus maideni.

Spotted Bur Clover – Medicago arabica.

Spotted Gum – Eucalyptus citriodora.

Spotted Gum – Eucalyptus goniocalyx.

Spotted Gum – Eucalyptus maculata.

Spotted Wintergreen – Chimaphila umbellata.

Sprangletop, Green – Leptochloa dubia.

Spreading Dogbane – Apocynum androsaemifolium.

Spreading Pasqueflower – Anemone patens.

Spring Adonis – Adonis vernalis.

Spring Beauty – Claytonia sibirica.

Spring Beauty – Claytonia virginiana.

Spring Grass – Anthoxanthum odoratum.

Spring Savory – Satureja acinos.

Sprinkled Parmelia – Parmelia conspersa.

Sprout Lichen – Dermatocarpon miniatus.

Spruce, Alcock – Picea bicolor.

Spruce, Beer – Picea mariana.

Spruce, Black – Picea mariana.

Spruce, Bog – Picea mariana.

Spruce, Douglas – Pseudotsuga menziesii.

Spruce, Dragon – Picea asperata.

Spruce, Engelmann – Picea engelmanni.

Spruce Gum – Picea mariana.

Spruce Himalayan – Picea smithiana.

Spruce, Norway – Picea excelsa.

Spruce, Purple Cone – Picea purpurea.

Spruce, Red – Picea rubra.

Spruce, Sakhalin – Picea glehnii.

Spruce, Sitka – Picea sitchensis.

Spruce, White – Picea glauca.

Spruce, Yeddo – Picea jezoensis.

Spur Bush – Pluchea odorata.

Spurge, Balsam – Euphorbia balsamifera.

Spurge, Caper – Euphorbia lathyris.

Spurge, Flowering – Euphorbia corollata.

Spurge, Ipecac – Euphorbia ipecacuanha.

Spurge Olive – Cneorum tricoccum.

Spurgewort – Iris foetidissima.

Spurrey, Corn – Spergularia arvensis.

Spurrey, Sand – Spergularia rubra.

Squarestem Rosegentian – Sabatia angularis.

Squash – Cucurbita maxima.

Squash, Banana – Cucurbita maxima.

Squash, Hubbard – Cucurbita maxima.

Squash, Mammoth – Cucurbita maxima.

Squash, Turban – Cucurbita maxima.

Squaw Berry – Rhus trilobata.

Squaw Grass – Elymus triticoides.

Squaw Root – Caulophyllum thalictroides.

Squaw Vine – Mitchella repens.

Squaw Weed – Senecio aureus.

Squill – Urginea maritima.

Squill, Indian Drug – Urginea indica.

Squill, Indigo – Camassia esculenta.

Squill, Wild – Scilla rigidifolia.

Squirrel Corn – Dicentra canadensis.

Squirting Cucumber – Ecballium elaterium.

STACHYS L. Labiatae. 300 spp. N. and S. tropics and subtropics, except Australia and New Zealand. Herbs.

***S. affinus*** Bunge=S. sieboldi.

***S. betonica*** Crantz.=S. recta.

***S. bufonia*** Thuill.=S. recta.

***S. californica*** Benth. California. The leaves and stems are used as a poultice or infusion to dress sores and wounds.

***S. officinalis*** (L.) Trev. (Common Betony). Europe to Caucasus. Various extracts of the leaves are used in home remedies for coughs, stomach upsets and kidney, bladder and spleen complaints.

***S. palustris*** L. (Marsh Woundwort). Europe, N.America. The leaves are used to dress wounds and in an extract to treat internal bleeding, dysentery, cramp, gout, joint pains and vertigo. The tubers are occasionally eaten as a vegetable.

***S. recta*** L.=S. betonica=S. bufonia. Europe to Asia Minor. A decoction of the leaves is used in the Balkans to wash children to protect them from diseases.

***S. sieboldi*** Miq. = S. affinus = S. tuberifera. (Chinese Artichoke, Japanese Artichoke). China, Japan. Cultivated locally and in Belgium and France for the tubers which are eaten as a vegetable.

***S. tuberifera*** Naud. = S. sieboldi.

STACHYTARPHETA Vahl. Verbenaceae. 100 spp. America, some spp. are widely distributed in the tropics as weeds.

***S. angustifolia*** Vahl. Guiana. The roots are used by the Arabs to treat haemorrhoids.

***S. dichotoma*** Vahl. = S. jamaicensis = Valerianioides jamaicensis = Verbena jamaicense. (Bastard Vervain). Old and New World tropics and subtropics. A decoction of the leaves is used to treat dysentery, intestinal worms, erysepalis, dropsy, stomach complaints, ulcers and venereal diseases. The leaves (Brazilian Tea) are used in Brazil to adulterate tea. Sometimes exported.

***S. jamaicensis*** Vahl. = S. dichotoma.

Stag Bush – Viburnum pruifolium.

Staghorn Sumach – Rhus typhina.

STANLEYA Nutt. Cruciferae. 8 spp. W. United States of America. Herbs. The leaves of the following spp. were eaten as a vegetable by the local Indians.

***S. albescens*** Jones.

***S. elata*** Jones.

***S. pinnatifida*** Nutt.

STANLEYELLA Rydb. Cruciferae. 2 spp. S.W. United States of America, Mexico. Herbs.

***S. wrightii*** (Gray). Rydb. S.W. U.S.A., Mexico. A dye from the plant was used by the local Indians for colouring pottery.

Star Anise (Oil) – Illicium verum.

Star Apple – Chrysophyllum cainito.

Star Apple, African – Chrysophyllum africanum.

Star, Blazing – Chamaelirium luteum.

Star Grass – Aletris farinosa.

Star Grass – Cynodon plectostachyus.

Star Thistle – Centaurea nigra.

Star Tree, Ash-leaf – Astronium fraxifolium.

Starwort, Yellow – Inula helenium.

STATICE L. = Armeria Willd. + Limonium Mill. Plumbaginaceae, 380 spp. Mediterranean to central Asia, N.Temperate to Andes. Herbs.

***S. chilensis*** Phil. Chile. The leaves are used locally to treat ulcers, dysentery and scrofula.

***S. limonium*** L. (Inkroot, Marsh Rosemary, Sea Lavender). Europe, N.America. A decoction of the roots is used to treat sore throats, catarrh, mouth ulcers, bleeding in the chest, haemorrhoids, vaginal discharges.

***S. ornata*** Ball. Mediterranean. The leaves are eaten locally in salads.

***S. tataricum*** L. = Goniolimon tataricum. S.Europe to Siberia. The roots are used for tanning in Siberia.

***S. thouini*** Viv. See S. ornata.

STAUDTIA Warb. Myristicaceae. 2–3 spp. W.Africa. Trees.

***S. stipitata*** Warb. Tropical W.Africa. The hard, orange wood is used locally for building houses, turning, furniture, making walking sticks and pencils. The seeds are used as a poultice for skin diseases and as a bait for animals.

***S. kamerunensis*** Warb. Cameroons. The wood (Bosé Wood) is attractively coloured red and black. It is used for furniture, carving etc. The seeds yield a fat (Staudtia Butter) used locally for cooking.

Staudtia Butter – Staudtia kamerunensis.

STAUNTONIA DC. Lardizabalaceae. 15 spp. Burma to Formosa and Japan. Vines.

***S. hexaphylla*** DC. (Japanese Staunton Vine). Japan. The sweet juicy fruits are eaten locally. The plant is cultivated.

Staunton Vine, Japanese – Stauntonia hexaphylla.

STAUROGYNE Wall. Acanthaceae. 80 spp. Tropics especially W.Malaysia. Herbs.

***S. elongata*** Kunze. Malaysia. The leaves are eaten as a vegetable in Java.

Stavisacre – Delphinium staphisagria.

STEGNOSPERMA Benth. Stegnospermataceae. 3 spp. Lower California to Central America, W.Indies. Shrubs.

***S. halimifolium*** Benth. Throughout range. In Mexico the roots are used as a soap substitute.

Steinraute – Achillea clavenae.

STELECHOCARPUS Hook. f. and Thoms. Annonaceae. 5 spp. Thailand, Malaysia. Trees.

***S. burakol*** Hook. f. = Uvaria burakol. Malaysia. The fruits are eaten locally.

STELLARIA L. Caryophyllaceae. 120 spp. Cosmopolitan. Herbs.

***S. media*** (L.) Cyr. (Common Chickweed). Europe. The leaves are sometimes eaten as a vegetable.

STEMMADENIA Benth. Apocynaceae. 20 spp. Mexico to Ecuador. Shrubs.

***S. galeottiana*** (A. Rich.) Miers. Cuba, S.E.Mexico. The plant yields a latex which is used in Mexico as chewing gum.

STEMONA Lour. Roxburghiaceae. 25 spp. E.Asia, Indomalaya, N.Australia. Shrubs. The roots of the following spp. contain stemorin and are potentially useful as a source of insecticide.

***S. sessilifolia*** Franch. Tropical Asia.

***S. tuberosa*** Lour. Tropical Asia.

STENOCARPUS R. Br. Proteaceae. 25 spp. E.Australia, New Caledonia. Trees.

***S. daroides*** Brongn. and Gris. New Caledonia. The wood is used for making furniture.

***S. laurifolius*** Planch. and Sebert.=S. lourinus.

***S. lourinus*** Brongn. and Gris.=S. laurifolius. New Caledonia. The dark coloured wood is used to make furniture.

***S. salignus*** R. Br.=Embothrium rubicaule =Hakea rubicaulis (Silky Oak, Beef Wood, Willow Firewheel Tree). New South Wales. The hard reddish wood is used for furniture, veneers, picture frames and walking sticks.

STENOCHLAENA J. Sm. Blechnaceae. 5 spp. Africa to Pacific. Ferns. The leaves of the following spp. are eaten locally as a vegetable.

***S. palustris*** (L.) Mett. Malaysia. The leaves are eaten in Java where they are also used for cordage, anchor ropes and fish traps.

***S. tenuifolia*** Moore.=Acrostichum meyerianum. Madagascar, S.Africa, Mascarene Islands.

STENOCLINE DC. Compositae. 13 spp. Brazil, S.Africa, Madagascar. Shrubs.

***S. incana*** Baker. Madagascar. The leaves are used locally to make a tonic, to treat stomach disorders and as an aphrodisiac.

STENOTAPHRUM Trin. Graminae. 7 spp. Tropics and subtropics. Grasses.

***S. madagascariense*** A. Cam. (Ahidrendra). Madagascar. Used locally as livestock fodder.

STEPHANIA Lour. Menispermaceae. 40 spp. Tropical Africa, Asia, Australia. Vines.

***S. japonica*** (Thunb.) Miers=Cissampelos psilophylla=Menispermum japonicum. India, through to Formosa and Japan, Philippines. In the Philippines, the leaves are used as a salve for itches.

***S. rotunda*** Lour.=Clypea rotunda. (Cû môt, Ngâi tuong). Indochina, Vietnam. The ground root is used locally for treating dysentery, fevers, tuberculosis and asthma.

***Stephanoluma rugosa*** Bn.=Micropholis melinoniana.

***Stephegyne parvifolia*** Korth.=Mitragyna parvifolia.

STERCULIA L. Sterculiaceae. 300 spp. Tropics. Trees. Several of the spp. mentioned yield gums, which, apart from any local use are used as substitutes for or adulterants of Gum Tragacanth. (Astragalus gummifer) especially for dyeing.

***S. affinis*** Mast.=Scaphium affinis.

***S. apetala*** (Jacq.) Karst.=S. carthaginensis.

***S. carthaginensis*** Cav.=S. apetala (Jacq.) Karst. (Panama Tree). S.Mexico into S.America, W.Indies. The wood is used locally for general construction work.

***S. chicha*** St. Hil. Brazil, Guyana. The seeds (Castanha de Maranháo, Maranháo Nuts), are eaten locally. The oil extracted from them is used in paints and for lubricating watches.

***S. cineria*** G. and P. E.Africa. Yields a gum (Tartar Gum).

***S. cordifolia*** Cav. Tropical Africa. The fruit pulp is eaten locally and the bark yields a local cordage fibre.

***S. foetida*** L. Tropics, especially Old World. The seeds yield a non-drying oil (Kaloempang Oil) and are eaten locally.

***S. hypochra*** Pierre. (Tlon, Trom). S.Vietnam. The trunk yields a gum which is used locally as chewing gum.

***S. javanica*** R. Br. (Binong, Hantap badak, Kalong, Kalongan). Indonesia, Java. The seeds (Pranadjiwa) are used locally as a tonic.

***S. lanceaefolia*** Roxb. India, S.W.China. The seeds are eaten locally and are used by the Chinese as a condiment.

***S. luzonica*** Warb. Philippines. The bark fibre is used locally for cordage.

***S. lychnophora*** Hance. Tropical Asia, especially Cochin-China. The seeds are used in Cambodia to make a drink (Sam-rong).

***S. oblonga*** Mart.=Eriobroma klaineana. Tropical Africa. The heavy, hard, yellowish wood (Ekonge, Yellow Wood) is used locally for furniture, boat-building, railway sleepers and turning.

***S. plantanifolia*** L. f.=Firmiana simplex.

***S. rogersii*** N. E. Br. Central Africa. The bark fibre is used locally for fishing nets and mats.

***S. rupestris*** Benth. Australia. The trunk yields a gum.

***S. scaphigera*** Wall. Tropical Asia. The Chinese and Siamese eat a pulp of the seeds with sugar as a delicacy and use it to reduce temperatures.

***S. tomentosa*** Guill. and Perr. Tropical Africa. The trunk yields a gum (Gomme de Senegal, Icacia Gum).

***S. tragacantha*** Lindl. (African Tragacanth). Tropical Africa, especially W.coast. The trunk yields a gum. The wood is used locally for general construction work.

***S. urens*** Roxb. (Kateeragum Sterculia). Tropical Asia. The trunk yields a gum (Karaya Gum, Kateera Gum).

***S. villosa*** Roxb. (Udal, Vakenar). India, Pakistan. The bark yields a coarse fibre, used for heavy cordage, bags, etc. The soft, light wood is used for tea chests.

STEREOCAULON L. Cladoniaceae. Lichens.

***S. paschale*** (L.) Hoffm.=S. tomentosum. (Easter Lichen). Temperate. Used in Europe as the source of a grey-green dye, used for dyeing wool.

***S. tomentosum*** Fr.=S. paschale.

STEREOSPERMUM Cham. Bignoniaceae. 24 spp. Tropical Africa, Asia. Trees.

***S. chelonoides*** DC. Himalayas and S.China to Ceylon and Java. The hard, grey, elastic wood is used for house-building, furniture, canoes and tea-chests.

***S. hypostichum*** Miq.=Redermachera gigantea.

***S. kunthianum*** Engl. (Golombi). Tropical W.Africa. A decoction of the bark is used locally to treat dysentery and diarrhoea and as a lipstick by the girls.

STEVIA Cav. Compositae. 150 spp. Tropical and subtropical America. Herbs.

***S. rebaudiana*** Bertoni.=Eupatorium rebaudianum. S.America, especially Paraguay. The plant is a possible sugar substitute as it contains estevin which is 150 times sweeter than sugar weight for weight.

STICTA Ach. Lichen. Stictaceae.

***S. aurata*** Ach. (Lungwort, Rose and Gold Lichen). Temperate, subarctic. A brown dye from the plant is used to dye wool in U.K. and Scandinavia.

***S. crocata*** (L.) Ach. Gold Edge Lichen. See S. aurata.

***S. glomulerifera*** Del. Temperate. The plants were eaten by the American Indians.

***S. pulmonaria*** (L.) Schaer. (Lungwort, Oak Lichen). Temperate. The lichen looks like lung tissue and so was thought to be a cure for pulmonary tuberculosis. It was used locally in England as a brown dye for wool. It is an important food for moose and a possible famine food.

STIGEOCLONIUM Kg. Chaetophoraceae. Green Algae.

***S. amoenum*** Kg. Hawaii. Eaten locally with fresh-water shrimps.

STILLINGIA Garden ex L. Euphorbiaceae. 30 spp. Warm America; 1–2 spp. Mascarene Islands; 1 sp. E.Malaysia, Fiji. Herbs.

***S. sylvatica*** L. (Queen's Delight, Queen's Root, Yaw Root). S.E. U.S.A. The dried root is used medicinally as a laxative, tonic, diuretic and is used to treat syphilis and scrofula. It contains an essential oil, sylvacrol and a resin.

Stillingia Oil – Sapium sebiferum.

Stinging Nettle – Urtica dioca.

Stinkhorn – Phallus impudicus.

Stinking Cedar – Torreya taxifolia.

Stinking Gladwin – Iris foetidissima.

Stinking Ground Pine – Camphorosma monspeliaca.

Stinkweed, Red – Lauro-cerasus africanum.

Stinkwood – Celtis kraussiana.

Stinkwood, Black – Ocotea bullata.

Stinkwood, Red – Pygeum africanum.

STIPA L. Graminae. 300 spp. Temperate and dry tropics. Grasses.

***S. comata*** Trin. and Rupr. N.American prairies. An important pasture grass.

***S. hyalina*** Nees. S.America. A pasture grass in Argentina and Brazil.

***S. ichu*** Kunth.=S. jarava.

***S. jarava*** Beauv.=S. ichu. S.America into Mexico. A good fodder grass in arid areas.

***S. tenacissima*** L.=Macrochloa tenacissima. (Esparto, Halfa). Mediterranean. Used locally and exported to make fine paper. The grass is also used to make cordage, sails, mats etc.

***S. vaseyi*** Scribn. (Porcupine Grass). W. U.S.A. Mexico. Used locally to make brushes.

***S. viridula*** Trin. (Green Needle Grass).

Prairies of N.America. A good forage grass for livestock.
Stipites Laminariae – Laminaria cloustoni.
STIXIS Lour. Capparidaceae. 7 spp. E.Himalayas to Indochina, Haman, W.Malaysia, Lesser Sundra Islands. Vines.
***S. flavescens*** Pierre. (Mang ram bo). Vietnam, Laos. The roots are used locally to make a tea.
***Stizolobium*** P. Br.=Mucuna.
Stock Rose – Spamannia africana.
***Stoechas officinarum*** Moench.=Lavandula stoechas.
Stone Cotton – Gossypium brasiliense.
Stone Crottle – Parmelia caperata.
Stone Leek – Allium fistulosum.
Stone Mint – Cunila origanoides.
Stone Pine – Pinus pinea.
Stonecrop, Goldmoss – Sedum acre.
Stonecrop, Jenny – Sedum reflexum.
Stonecrop, Mossy – Sedum acre.
Stonecrop, White – Sedum album.
Stonedrop, Ditch – Penthorum sedioides.
Stonedrop, Virginia – Penthorum sedioides.
Stoneroot – Collinsonia canadensis.
Stoneseed – Lithospermum ruderale.
***Stophostyles*** Ell.=Phaseolus.
Stopper, Red – Eugenia confusa.
Stopper, Red – Eugenia rhombea.
Stopper, Wild – Eugenia axillaris.
Storax – Styrax officinale.
Storax, American – Liquidambar styraciflua.
Storax, Levant – Liquidambar orientalis.
Storax, Oriental – Liquidambar orientalis.
Stork's Bill – Erodium cicutarium.
Stouton – Aubrya gabonensis.
Stramonium – Datura stramonium.
Strassburg Turpentine – Abies pectinata.
STRATIOTES L. Hydrocharitaceae. 1 sp. Europe. Water plant.
***S. aloides*** L. (Crab's Claw, Water Soldier). Used locally as manure.
Straw Lily – Clintonia borealis.
Strawberry – Fragaria vesca.
Strawberry Bush – Euonymus americanus.
Strawberry Cactus – Echinocereus enneacanthus.
Strawberry, Chiloë – Fragaria chiloensis.
Strawberry Clover – Trifolium fragiferum.
Strawberry Guava – Psidium cattleianum.
Strawberry, Hautbois – Fragaria moschata.
Strawberry, Mexican – Echinocactus stramineus.
Strawberry Tomato – Physalis alkekengi.
Strawberry Tree – Arbutus unedo.
Strawberry, Virginian – Fragaria virginiana.
Strawberry, Yacon – Polymnia edulis.
STREBLUS Lour. Moraceae. 22 spp. Madagascar, S.E.Asia, Indo-Malaya. Small trees.
***S. asper*** Lour. Tropical Asia. The plant is used to make paper in Thailand. The rough leaves are used to polish ivory and are used locally in Burmese cheroots. A decoction of the bark is used in India to treat dysentery, diarrhoea and fevers.
***S. tonkinensis*** Dub. and Eberh. Indochina. Used locally to make paper. The latex is a potential source of a good rubber.
STREPTOCAULON Wight. and Arn. Periplocaceae. 5 spp. Indomalaya. Vines.
***S. baumii*** Decne.=S. obtusum=Periploca calumpitensis. (Hing-gui-Kalabau). Philippines. The latex is used locally as a plaster on wounds.
***S. obtusum*** Turcz.=S. baumii.
STREPTOCOCCUS Lactobacilliaceae. Bacteria. Ferment sugars to lactic acid. The spp. mentioned play a part in the ripening of cheeses.
***S. lactis*** Dombr. Cheddar cheese.
***S. thermophilus*** Orla-Jensen. Swiss cheeses.
STREPTOMYCES Waksm. and Henrici. Streptomycetaceae. 73 spp. Cosmopolitan.
***S. albidoflavus*** and several other spp. produce vitamin $B_{12}$ by fermentation.
***S. albus, S. antibiotica*** produce the antibiotic actinomycin.
***S. aureofaciens*** produces aureomycin.
***S. griseus*** produces streptomycin.
Striata Clover – Trifolium striatum.
Striated Ipecac – Cephaëlis emetica.
***Strigelia acuminata*** Miers.=Styrax acuminatum.
***Stringelia camporum*** Miers=Styrax camporum.
Stringybark – Eucalyptus macrorhynchia.
Stringybark – Eucalyptus obliqua.
Strinbybark, Brown – Eucalyptus capitellata.
Stringybark, Messmate – Eucalyptus obliqua.
Stringybark, White – Eucalyptus eugenoides.
Stringybark, White – Eucalyptus piperata.
Stringybark, Whitetop – Eucalyptus gigantea.
Stringybark, Yellow – Eucalyptus muelleriana.
Stringybark, Yellow – Eucalyptus triantha.

STROBILANTHES Blume. Acanthaceae. 250 spp. Madagascar, tropical Asia. Herbs.

***S. flaccidifolius*** Nees. India, China, Thailand. Cultivated locally, especially in S.W.China and N.Malaya for a blue dye, used for dyeing fabrics.

***Strombocarpa*** A. Gray=Prosopis.

STROMBOSIA Blume. Olacaceae. 17 spp. Tropical Africa, India, Ceylon, Burma, W.Malaya. Vines. The plant contains the drug strophanthin, which is used medicinally as a heart stimulant and diuretic. This is collected from the dried ripe fruits of ***S. kombe*** and ***S. hispidus.*** The juice or soaked seeds of all the spp. mentioned is used as an arrow poison locally.

***S. cumingi*** A. DC.=S. erectus. Philippines.

***S. eminii*** Asch. and Pax. Tropical Africa.

***S. erectus*** Merr.=S. cumingi.

***S. gratus*** Franch. Tropical Africa, especially W.Coast

***S. hispidus*** DC. S.Africa. Yields Brown Strophanthus.

***S. preussii*** Engl. and Pax. Tropical Africa, especially W.Coast. The juice is also used locally for coagulating latex from Funtumia elastica.

***S. sarmentosus*** DC. Tropical Africa, especially Nigeria and Congo.

***S. thollonii*** Franch. Central Africa

STROPHARIA (Fr.) Quél. Agaricaceae. 70 spp. Temperate.

***S. cubensis*** Earle. Mexico, W.Indies. Contains the hallucinogen psilocybine and is used in various ceremonies by the natives of S.Mexico.

STRUCHIUM P. Br. 1 sp. Tropical America, W.Indies, tropical Africa. Herb.

***S. vaillantii*** Crantz. The leaves are used as a flavouring in soups, particularly in W.Africa.

STRUTHANTHUS Mart. Loranthaceae. 75 spp. Tropical America. Parasitic shrubs.

***S. syringifolius*** Mart. Tropical America. A potential source of rubber.

Strychnii Semen – Strychnos nux-vomica.

Strychnine Tree – Strychnos nux-vomica.

STRYCHNOS L. Strychnaceae. 200 spp. Tropics. Climbers or trees. Members of the genus contain the alkaloids strychnine and brucine which are extremely poisonous. They are used medicinally as a nerve tonic but cause paralysis of the motor nerves in excess. Several spp. are used as arrow poisons (Curare poison).

***S. acuminata*** Vall.=S. nux-blanda.

***S. atherstonei*** Harv. (Cape Teak). S.Africa. The hard, heavy, red-brown wood is used locally to make barrels and kitchen utensils.

***S. castelnaei*** Wedd. (Amazon Poison Nut, Urari). S.America. The Orinoco Indians use the bark as a source of curare.

***S. colubrina*** L. W.India, Ceylon. (Snake root, Snakewood). A decoction of the wood is used locally to treat malaria and stomach upsets.

***S. dekindtiana*** Gilg. Central Africa. Used as an Ordeal and weapon poison.

***S. gaulthieriana*** Pierre.=S. malaccensis. (Hoang nan). Indonesia. The bark is used locally to treat leprosy and other skin diseases.

***S. guineensis*** (Aubl.) Baill. Guyana. Used as an arrow poison.

***S. henningsii*** Gilg. (Hard Pear). S.Africa. The hard wood is used for tool handles.

***S. icaja*** Baill. (Bengue, M'loundu). Tropical Africa. Used as an Ordeal and weapon poison.

***S. ignatii*** Berg.=Ignatia amara. (St. Ignatius Poison Nut). Malaysia. The seeds (St. Ignatius Beans, Faba St. Ignatii, Semen Ignatii) contain brucine and strychnine and are used medicinally.

***S. innocua*** Delile. E.Africa. The fruits are edible in moderation. The seeds are poisonous.

***S. jollyana*** Pierre ex Chev. Tropical W.Africa. The fruit pulp is used locally as a soap substitute and a single seed as a laxative.

***S. kipapa*** Gilg. Tropical Africa. The root bark is used as an ordeal and weapon poison.

***S. ligustrina*** Blume (Bidara goognong, Bidara iaoot, Dara laoot, Malia pootih). Indonesia. The wood, especially from the roots, in a decoction of arak is used as a tonic, to treat fevers, intestinal worms. The bark extracted in water is applied to snake bites and skin diseases.

***S. malaccensis*** Clarke=S. gaulthieriana.

***S. malaccensis*** Bull. (Malacca Poison Nut). Indochina. The bark contains brucine and strychnine. It is used medicinally (Hoang Nan) as a nerve tonic and as an arrow poison.

***S. melinoniana*** Baill. Guyana. The bark is used locally as an aphrodisiac.

***S. nux-blanda*** Hill.=S. acuminata=S. ranconensis. (Kok toung ka, Pravak).

Burma, Indonesia. The seeds are sometimes used with S. nux-vomica seed.

***S. nux-vomica*** L. (Strychnine Tree). East Indies, Ceylon, India. The seeds (Nux Vomica, Strychni Semen) are the main commercial source of strychnine. Most is extracted from seed collected in the wild, but some plants are grown commercially.

***S. odorata*** Chev. W.Africa. The local women rub the leaves on their skin as scent.

***S. potatorum*** L. f. (Clearing Nut Tree, Neimal, Nirmali). India, E.Pakistan. The hard, yellow wood is used for house-building, posts and agricultural implements. The fruit pulp is edible and the seeds are supposed to be non-poisonous. They are used as an emetic, to treat eye diseases and when ground they are added to muddy water to clear it.

***S. pseudoquina*** St. Hil. Brazil. The bark (Quina do Campo), is used locally to treat intestinal worms.

***S. pubescens*** C. B. Clarke. Malaysia. Used as dart poison.

***S. quadrangularis*** A. W. Hil. See S. pubescens.

***S. ranconensis*** Pierre.=S. nux-blanda.

***S. spinosa*** Lam.=Brehmia spinosa. Madagascar. The fruit pulp is edible. The seeds are poisonous.

***S. tieute*** Lesch. (Tjetok). Java. Used locally as a poison, all parts of the plant are extremely poisonous. A potential commercial source of strychnine.

***S. toxifera*** Benth. (Curare Poison Nut, Urari). Costa Rica to Brazil. Curare is extracted from the bark and roots.

***S. triclisioides*** Franch. (Bowihi). W.Africa. The fruit pulp is edible, but is used as an emetic for children.

***S. unguacha*** A. Rich. Tropical Africa, especially E. The fruit pulp is eaten locally.

***S. wallichiana*** Benth. Assam to Malaysia. Used as an arrow poison.

STRYPHODENDRON Mart. Leguminosae. 15 spp. Tropical America. Trees.

***S. barbatimam*** Mart.=Mimosa barbartimam. (Barbatimao). Brazil. The bark is an important source of tannin (Casea Virgindade). It is also used medicinally to treat diarrhoea and internal bleeding, especially of the uterus.

***S. guianense*** Benth.=Mimosa guianensis. Guyana. The wood is used locally to make furniture and the bark is used for tanning.

STYLOCERAS Juss. Stylocerataceae. 3 spp. Tropical Andes. Trees.

***S. laurifolium*** H.B.K. Andes. The cream-coloured wood is used locally for furniture and other joinery.

STYLOCHITON Lepr. Araceae. 21 spp. Tropical Africa. Herbs.

***S. warneckii*** Engl. Forests of tropical Africa. The rhizomes are eaten as an emergency food.

***Stylocoryne densiflora*** Wall. Randia densiflora.

***Stylocoryne weberi*** Miq.=Tarenna incerta.

STYLOSANTHES Sw. Leguminosae. 50 spp. Tropical and subtropical America, Africa and Asia. Herbs.

***S. guianensis*** Sw.=S. surinamensis. Guyana. Grown in Brazil and Australia as fodder for livestock.

***S. mucronata*** Willd. (Wild Lucerne). Tropical America and Africa. Used as a fodder for livestock.

***S. surinamensis*** Miq.=S. guianensis.

STYPHELIA (Soland. ex G. Forst.) Sm. Epacridaceae. 11 spp. Australia, Tasmania. Shrubs.

***S. glaucescens*** Sieb.=S. trifolia.

***S. trifolia*** Andr. New South Wales, Queensland. The sweet fruit-pulp is eaten by the aborigines.

STYRAX L. Styracaceae. 130 spp. Warm Eurasia, Malaysia, America. Trees.

***S. acuminatum*** Pohl.=Strigella acuminata. Brazil. The wood is used locally for general carpentry.

***S. argentum*** Presl. Central America, Mexico. The gum from the trunk is used in local churches as incense.

***S. benzoin*** Dryand. S.E.Asia, E.Indies. The trunk yields a gum (Benzoin Gum, Sumatra Benzoin) which is used medicinally as an expectorant, antiseptic and diuretic. It is one of the ingredients of Friar's Balsam. The plant is cultivated locally.

***S. camporum*** Pohl.=Strigelia camporum. Brazil. The resin from the trunk is used locally as incense.

***S. ferrugineum*** Nees. and Mart. E. S.America. See S. argentum.

***S. japonicum*** Sieb. and Zucc. Japan. The fine white wood is used locally for umbrella handles. An oil is extracted from the seeds.

***S. leprosum*** Hook. and Arn. See S. acuminatum.
***S. tomentosum*** Humb. and Bonpl. S.America. Yields a fragrant resin.
***S. tonkinense*** Craib. Malaya. The stem yields a resin (Siam Balsam) which is used medicinally as an antiseptic and in making perfumes and soap. It is also used in incense and to flavour chocolate. The resin is produced mainly in Cochin-China.
SUAEDA Forsk. ex Scop. Chenopodiaceae. 110 spp. Cosmopolitan on sea shores. Herbs.
***S. californica*** (S. Wats.) Heller S.W. U.S.A., Mexico. The local Indians use the plant as a vegetable and as the source of a black dye for dyeing baskets. The ashes are used to make soap.
***S. fruticosa*** Forsk. (Alkali Seepweed). N.Temperate, Mediterranean. The plant is fed to camels.
***S. ramosissima*** Standl. (Jaujá, Quelite salado, Romerito, Sosa). See S. californica.
***S. suffrutescens*** Wats. See S. californica.
Suakim Gum – Acacia seyal.
Suakwa – Luffa cylindrica.
Sub Clover – Trifolium subterraneum.
Succinite – Pinus succinifera.
SUCCISA Haller. Dipsacaceae. 1 sp. Europe, W.Siberia, N.W.Africa; 1 sp. N.W.Spain; 1 sp. W.Equatorial Africa. Herbs.
***S. pratensis*** Moench. = Scabiosa succisa. (Devil's Bit Ofbit). Europe, W.Siberia, N.W.Africa. An infusion of the leaves is used locally to treat fevers, coughs and internal inflammations.
Succory – Cichorium intibus.
Succus Coni – Conium maculatum.
Succus Hypocistidis – Cytinus hypocistis.
Sucrier Demontage – Tetragastris balsamifera.
Sudarsana – Tinospora malabarica.
Sudan Grass – Sorghum vulgare.
Sudan Teak – Cordia myxa.
Sugar Apple – Annona squamosa.
Sugar Beet – Beta vulgaris.
Sugar Berry – Celtis occidentalis.
Sugar Berry – Ehretia elliptica.
Sugar Bush – Protea angolensis.
Sugar Bush – Protea melliflora.
Sugar Cane – Saccharum officinale.
Sugar Grass – Glyceria fluitans.
Sugar Gum – Eucalyptus cladocalyx.
Sugar, Horse – Symplocos tinctoria.
Sugar Maple – Acer saccharum.
Sugar Palm – Arenga pinnata.
Sugar Palm – Borasus flabellifer.
Sugar Palm – Cocos nucifera.
Sugar Palm – Nipa frutescens.
Sugar Palm – Phoenix sylvestris.
Sugar Pine – Pinus lambertiana.
Sugar Protea – Protea mellifera
Suir Tree (Oil) – Xanthophyllum lanceolatum.
Sulphur Root – Peucedanum officinale.
Sulphur Root – Peucedanum ostruthium.
Sultana – Vitis vinifera.
Sumac – Sorghum vulgare.
Sumac, Coral – Metropium toxiferum.
Sumach, Chinese – Rhus semialata.
Sumach, Dwarf – Rhus copallina.
Sumach, Evergreen – Rhus sempervirens.
Sumach, Fragrant – Rhus aromatica.
Sumach, Lemonade – Rhus triloba.
Sumach, Pontanin – Rhus potanini.
Sumach, Red – Rhus glabra.
Sumach, Shining – Rhus copallina.
Sumach, Sicilian – Rhus coriaria.
Sumach, Smooth – Rhus glabra.
Sumach, Staghorn – Rhus typhina.
Sumach Wax – Rhus succedanea.
Sumatra Benzoin – Styrax benzoin.
Sumbak – Equisetum debile.
Summer Adonis – Adonis aestivalis.
Summer Cypress – Kochia scoparia.
Summer Haw – Crataegus flava.
Summer Savory – Satureja hortensis.
Sunberry – Physalis minima.
Sundew – Drosera rotundifolia.
Sundri Tree – Heritiera minor.
Sunflower – Helianthus annuus.
Sunflower, Oblongleaf – Helianthus doronicoides.
Sunflower, Oregon – Balsamorrhiza sagittata.
Sunflower Seed Oil – Helianthus annuus.
Sunganisrie – Eulophia campestris.
Sunn Hemp – Crotalaria juncea.
Supa Oil – Sindora supa.
Supplejack – Rhipogonum scandens.
Supti – Mundulea sericea.
Surasaruni – Sauropus quadrangularis.
Surat Cotton – Gossypium obtusifolium.
Surf Grass – Phyllospadix scouleri.
Surgi – Mammea longifolius.
Surian bawang – Melia excelsa.
Surinam Ant Tree – Triplaris surinamensis.
Surinam Cherry – Eugenia uniflora.
Surinam Greenheart – Tecoma leucoxylon.
Surinam Purslane – Talinum triangulare.

Surinam Quassia – Quassia amara.
Surungi – Mammea longifolius.
Sushi – Porphyra tenera.
SUTERA Roth. Scrophulariaceae. 130 spp. Tropical and S.Africa, Canary Islands. Shrubs.
***S. atropurpurea*** Hiern.=Lyperia atropurpurea=L. crocea. (Cape Saffron, Saffraanbossie). S.Africa. The yellow dye, saffron is extracted from the flowers.
SUTHERLANDIA R. Br. ex Ait. Leguminosae. 6 spp. S.Africa. Shrubs.
***S. frutescens*** R. Br. (Cancer Bush, Cancer Wort, Kanker Bos). A decoction of the leaves is used to treat coughs, stomach and intestinal complaints, uterine troubles and as a tonic.
Suto – Urginea burkei.
Suwa – Anethum sowa.
***Svenhedia minor*** Urb.=Talauma plumieri.
Swallow Root – Cynanchum vincetoxicum.
Swallow's Wort – Cyanchum vincetoxicum.
Swamp Bay – Magnolia glauca.
Swamp Blueberry – Vaccinium corymbosum.
Swamp Cabbage – Spathyema foetida.
Swamp Cottonwood – Populus heterophylla.
Swamp Cypress – Taxodium distichum.
Swamp Cypress – Taxodium mucronatum.
Swamp Ebony – Diospyros mespiliformis.
Swamp Gum – Eucalyptus regnans.
Swamp Gum – Eucalyptus viminalis.
Swamp Mahogany – Diospyros botryoides.
Swamp Mahogany – Diospyros mespiliformis.
Swamp Mallet – Eucalyptus spathulata.
Swamp Milkweed – Asclepias incarnata.
Swamp Oak – Casuarina equisetifolia.
Swamp Oak – Casuarina suberosa.
Swamp Pine – Pinus caribaea.
Swamp Privet – Forestiera acuminata.
Swamp Red Currant – Ribes triste.
Swamp Spanish Oak – Quercus pagoda.
Swamp Spanish Oak – Quercus palustris.
Swamp Valerian – Valeriana uliginosa.
Swamp White Oak – Quercus bicolor.
Swampwood – Dirca palustris.
SWARTZIA Schreb. Leguminosae. 100 spp. Tropical America, Africa. Trees.
***S. madagascariensis*** Desv. Madagascar, tropical Africa. The hard, heavy, red wood is used for heavy construction work and making furniture and pianos.
Swatow Grass – Boehmeria nivea.
Swaziland Finger Grass – Digitaria swazilandensis.
Swedish Coffee – Astragalus boëticus.
Swedish Shield Lichen – Cetraria fahlunensis.
Swedish Turnip – Brassica napo-brassica.
Sweet Acacia – Acacia farnesiana.
Sweet Almond (Oil) – Amygdalus communis.
Sweet Basil (Oil) – Ocimum basilicum.
Sweet Bay – Magnolia glauca.
Sweet Bay – Persea borbonia.
Sweet Bean – Gleditsia triacanthos.
Sweet Blueberry – Vaccinium pennsylvanicum.
Sweet Broom – Scoparia dulcis.
Sweet Buckeye – Aesculus octandra.
Sweet Calabash – Passiflora maliformis.
Sweet Cassava – Manihot esculenta.
Sweet Cherry – Prunus avium.
Sweet Cicily – Myrrhis odorata.
Sweet Clover, White – Melilotus alba.
Sweet Clover, Yellow – Melilotus officinalis.
Sweet Coltsfoot – Petasites palmata.
Sweet Corn Root – Calathea alliuia.
Sweet Elm – Ulmus fulva.
Sweet Fennel – Foeniculum vulgare.
Sweet Fern – Comptonia aspleniifolia.
Sweet Flag – Acorus calamus.
Sweet Gale – Myrica gale.
Sweet Goldenrod – Solidago odora.
Sweet Grass – Hierochloe odorata.
Sweet Grass – Sporolobus indicum.
Sweet Grenadilla – Passiflora lingularis.
Sweet Gum – Liquidambar styraciflua.
Sweet Gum, Chinese – Liquidambar formpsana.
Sweet Leaf – Symplocos racemosa.
Sweet Leaf – Symplocos tinctoria.
Sweet Lemon – Citrus limetta.
Sweet Lime – Citrus aurantifolia var. limetta.
Sweet Majoram – Majorana hortensis.
Sweet Marigold – Tagetes lucida.
Sweet Melon – Cucumis melo.
Sweet Orange – Citrus sinensis.
Sweet Orange Pip Oil – Citrus sinensis.
Sweet Pea – Lathyrus odoratus.
Sweet Potato – Ipomoea batatas.
Sweet Root – Acorus calamus.
Sweet Root – Osmorhiza occidentalis.
Sweet Scented Myrrh – Myrrhis odorata.
Sweet Scented Sumach – Rhus aromatica.
Sweetshoot Bamboo – Phyllostachys dulcis.
Sweet Sop – Annona squamosa.

Sweet Sorghum – Sorghum vulgare.
Sweet Tangle – Laminaria saccharina.
Sweet Thorn – Acacia horrida.
Sweet Trefoil – Trigonella coerulea.
Sweet Vernal Grass – Anthoxanthum odoratum.
Sweet Viburnum – Viburnum cassionoides.
Sweet Violet – Viola odorata.
Sweet Winter Grape – Vitis cinerea.
Sweet Woodruff – Asperula odorata.

SWEETIA Spreng. Leguminosae. 12 spp. S.America. Trees.

***S. elegans*** Benth. (Lapachillo, Perobinha). S.Brazil, Argentina. The strong wood is used locally for general carpentry and making charcoal.

***S. nitens*** (Vog.) Benth. Central and N. S.America. The heavy, durable wood is used locally for railway sleepers, heavy construction work and agricultural implements.

***S. panamensis*** Benth. (Billyweb Sweetia). Central and N. S.America. The hard, strong, durable wood is used locally for railway sleepers, heavy construction work and agricultural implements. An extract of the bark (Cascara Amarga) is used to treat scrofula.

Sweetia, Billyweb – Sweetia panamensis.

SWERTIA L. Gentianaceae. 100 spp. Eurasia, Africa, Madagascar. Herbs.

***S. chirata*** Buch. Ham. (Chirata). Himalayas. The dried leaves are used medicinally as a bitter tonic. Sold locally in bazaars.

SWIETENIA Jacq. Meliaceae. 7–8 spp. Tropical America, W.Indies. Trees. The true Mahoganies. The rich, red-brown woods darken on exposure. They are hard and work well. These woods are extremely popular for furniture, house interiors, carving, pianos and other musical instruments, coffins, interiors of ships. S. mahagoni was the first discovered. The oldest known mahogany carving is in the cathedral of Santo Domingo (1514–40). It was first used in England in Nottingham Castle (1680).

***S. candollei*** Pitt. (Venezuela Mahogany). Guyana, Venezuela. Used particularly in France.

***S. cirrhata*** Blake. (Vendillo). Mexico.

***S. humulis*** Zucc. (Mexican Mahogany, Palo Zopilote). Mexico to El Salvador.

***S. macrophylla*** King. (Caoba, Hondureña, Honduras Mahogany, Mexican Mahogany). Central America. The most important commercial mahogany.

***S. mahagoni*** Jacq. (Common Mahogany, West Indian Mahogany, Caoba Mahogany, Tabasa Mahogany, San Domingo Mahogany). Cuba, Bahamas, Jamaica, Hispaniola, S.Florida. The bark is sometimes used as a substitute for quinine.

SWINGLEA Merr. Rutaceae. 1 sp. Philippines. Small tree.

***S. glutinosa*** (Blanco). Merr. = Chaetospermum glutinosum = Feronia ternata. = Limonia glutinosa. (Malakabuyan, Taboc, Tabog). The juice (Kurad) is used locally to treat skin diseases and to remove ticks from dogs.

SWINTONIA Griff. Anacardiaceae. 15 spp. S.E.Asia, W.Malaysia. Trees.

***S. glauca*** King. (Selan merah). Malaysia. The heavy, fairly hard wood is used locally for general construction work and making matches.

***S. specifera*** Hook. See S. glauca.

Swiss Chard – Beta vulgaris.
Swiss Stone Pine – Pinus cembra.
Switchgrass – Panicum virgatum.
Switchsorrel – Dodonaea viscosa.
Swizzlestick – Rauvolfia vomitoria.
Swizzlestick Tree – Quararibea turbinata.
Sword Bean – Canavalia ensiformis.
Sword Bean – Canavalia gladiata.

SYAGRUS Mart. Palmae. 50 spp. Tropical America. Palm trees.

***S. amara*** Mart. = Rhyticocos amara.

***S. cocoides*** Mart. (Piririma). S.America. The seeds yield an edible oil (Piririma Oil).

***S. coronata*** Mart. = Cocos coronata. (Licuri, Nicuri, Ouricuru Palm). A non-drying oil from the seeds (Ouricuri Palm Kernel Oil) is produced commercially and exported for the manufacture of margarine. The fruit pulp is eaten locally and fed to livestock.

Sycamore – Plantanus occidentalis.
Sycamore, American – Plantanus occidentalis.
Sycamore Fig – Ficus sycomorus.
Sycamore, Mexican – Platanus mexicana.
Sycamore, Oriental – Platanus orientalis.
Sycamore, Silver – Cryptocarya glaucescens.
Sydney Blue Gum – Eucalyptus saligna.
Sydney Peppermint – Eucalyptus piperata.

SYMPHONIA L. f. Guttiferae. 2 spp. Colombia; 1 sp. Tropical America, Africa; 18 spp. Madagascar. Small trees.

***S. clusioides*** L. Madagascar. The stem yields a resin and the wood is used locally for general carpentry.

***S. globulifera*** L. (Doctor's Gum, Hog Plum, Karamanni Wax, Guiana Symphonia). Tropical Africa, America. The Indians of Guyana use the wax, mixed with beeswax and charcoal as a general household glue for sticking wood, arrow heads, caulking boats and making torches. An oil from the seeds is used in Brazil to treat skin diseases.

***S. microphylla*** R. E. Schultes. Colombia. Yields a gum. See S. globulifera.

***S. utilisima*** R. E. Schultes. Colombia. See S. globulifera.

Symphonia, Guiana – Symphonia globulifera.

SYMPHORICARPOS Duham. Caprifoliaceae. 1 sp. China; 17 spp. N.America. Shrubs.

***S. orbicularis*** Moench. = S. vulgaris. (Coralberry, Indian Currant). E. U.S.A. A decoction of the leaves is used by the local Indians as an eye salve.

***S. vulgaris*** Michx. = S. orbicularis.

SYMPHYTUM L. Boraginaceae. 25 spp. Europe to Mediterranean and Caucasus. Herbs.

***S. asperrimum*** Don. (Prickly Comfrey). Asia Minor, Caucasus. The plant is sometimes cultivated as food for pigs, rabbits and goats.

***S. officinale*** L. (Comfrey). Europe, introduced to N.America. The leaves are sometimes eaten as a vegetable. The leaves and roots form the basis of home remedies for coughs, bronchitis, etc. and the leaves are applied as a poultice to swellings to reduce inflammation.

***S. orientale*** auct. non L. = S. uplandicum.

***S. peregrinum*** Bot. Mag. non Ledeb. = S. uplandicum.

***S. uplandicum*** Nyman. = S. peregrinum = S. orientale. (Prickly Comfrey). Probably a hybrid S. asperum × S. officinale. Sometimes grown as a food for livestock.

SYMPLOCARPUS Salis. Araceae. 1 sp. N.E.Asia, Japan, Atlantic, N.America.

***S. foetidus*** (L.) Salis. = Spathyema foetida. (Skunk Cabbage, Swamp Cabbage). The cooked roots are used as an emergency food by the N.American Indians.

SYMPLOCOS Jacq. Symplocaceae. 350 spp. Tropical and subtropical Asia, Australia, Polynesia, America. Small trees.

***S. fasciculata*** Zoll. = Dicalyx tinctorius. Malaya. The bark yields a red dye.

***S. ferruginea*** Roxb. = S. javanica.

***S. hamiltoniana*** Wall. = S. racemosa.

***S. javanica*** Kurz. = S. ferruginea = S. lachnobotrya. (Djirak, Kandoeen). Java. A brown dye from the bark is used locally for dyeing cloth.

***S. lachnobotrya*** Miq. = S. javanica.

***S. odoratissima*** Choisy. = Dicalyx odoratissima. (Kisariawan). Java. The inner bark (Kajoo seriawan, Koolit Seriawan) is chewed locally as a cure for thrush.

***S. orogenes*** Brand. = S. racemosa.

***S. petelotii*** Merr. = S. racemosa.

***S. racemosa*** Roxb. = S. hamiltoniana = S. orogenes = S. petelotii. (China brasilensis, Lodebark, Sweetleaf). N.India. S.China, Indochina. A red dye is extracted from the bark which is also used medicinally by the Hindus for treating eye diseases, abdominal pains and ulcers. The fruits are eaten locally and the flowers are a good source of honey.

***S. sumuntia*** Buch.-Ham. (Lode Tree, Lodh). N.India. The bark and leaves yield a yellow dye, which can be boiled with sodium carbonate to form a red dye.

***S. tinctoria*** (L.) L'Hérit. Common Sweetleaf, Horse Sugar). E. N.America. The fruit and leaves yield a red dye. The roots are used as a tonic and the red-brown, soft wood is used for turning.

SYNCARPIA Tenore. Myrtaceae. 5 spp. Queensland. Trees.

***S. laurifolia*** Tenore. = Metrosideros glomulifera = M. procera. (Boorea, Turpentine Tree). The soft, dark brown wood is durable and difficult to burn. It is used for ship-building.

SYNEDRELLA Gaertn. Compositae. 50 spp. Warm America, tropical Africa, Madagascar, India. Herbs.

***S. nodiflora*** Gaertn. (Babadotan lalaki). Malaya. The Javanese eat the young leaves with rice.

SYNOUM A. Juss. Meliaceae. 2 spp. Australia. Trees.

***S. glandulosum*** Juss. = Trichilia glandulosa. (Brush Bloodwood, Rosewood). New South Wales, Queensland. The dark red, scented wood polishes well and is used to make furniture.

Syrian Alkanna – Macrotomia cephalotus.

Syrian Cotton – Gossypium herbaceum.

SYRINGA L. Oleaceae. 30 spp. S.E.Europe to E.Asia. Shrubs.

***S. vulgaris*** L. (Lilac). S.E.Europe. Cultivated widely in temperate regions as an ornamental. The flowers (Flores Syringae) are used to reduce fevers.

Syrup, Cane – Saccharum officinale.

Syrups Protae – Protea mellifera.

SYZGIUM Gaertn. Myriaceae. 500 spp. Old World tropics. Trees.

***S. guinense*** (Willd.) DC. (Mukute, Mushingu, Musombo, Water Berry). Tropical Africa. The very ripe fruits are eaten locally and are sometimes made into vinegar.

***S. jambolanum*** DC.=Eugenia jambolana.

***S. multipetalum*** Brongn. and Gris. New Caledonia. The light red, hard wood is used to make furniture.

***S. nodosum*** Miq.=Eugenia operculata.

***S. racemosum*** DC.=Eugenia jamboloides.

Szechuan – Cupressus funebris.

Szechuan – Fagus sinensis.

Szechuan – Koelreuteria paniculata.

# T

Ta hsiao kiai – Cirsium japonicum.

Tabaiba – Sapium laurocerasus.

Tabalooko – Pandanus papuanus.

Taban Merak Fat (Tallow) – Palaquium oleiferum.

Tabasco – Capsicum annuum.

TABEBUIA Gomes ex DC. Bignoniaceae. 100 spp. Mexico to Argentine, W.Indies. Trees.

***T. donnell-smithii*** Rose=Cybistax donnell-smithii.

***T. heterophylla*** (DC.) Britton. (Roble, Roble Blanco). W.Indies. The coarse greyish-yellow wood is strong, coarse, but polishes well. It is durable, but susceptible to insect and borer damage. The wood is used for veneers, planks, posts, boat decks, sports goods.

***T. leucoxyla*** DC. Brazil. The wood is used for wooden shoes and a decoction of the bark is used as a diuretic.

***T. pentaphylla*** (L.) Hemsl.=Bignonia pentaphylla (Roblo blanco, White Cedar, White Wood, West Indian Boxwood). W.Indies, Central America, N. S.America. The light yellow hard wood is used to build houses, furniture, wagons and to make oars. It is used in Europe as a substitute for Boxwood.

TABERNAEMONTANA L. Apocynaceae. 100 spp. Tropics. Trees or shrubs.

***T. alba*** Mill. Guyana. The latex is sometimes used to adulterate balata.

***T. amygdalaefolia*** Jacq. (Cojón de Berraco). Nicaragua. The fruits are a local source of rubber.

***T. angolensis*** Stapf. (Cata Grande). Angola. Yields an inferior rubber.

***T. bovina*** Lour. Cochin-China. A rubber is extracted from the roots.

***T. citrifolia*** L. (Cojón de Gato, Lecherillo). S.Mexico into S.America, W.Indies. The juice is applied to warts to destroy them.

***T. coronaria*** Willd.=Ervatamia coronaria.

***T. corymbosa*** Roxb. (Djelotong badak). Malaya. The bark is used as a local remedy for tertiary syphilis.

***T. divaricata*** R. Br.=Ervatamia coronaria.

***T. pachysiphon*** Stapf.=Conopharyngia pachysiphon.

***T. pandacaqui*** Poir. Philippines to New Guinea. The leaves are used in the Philippines for bleaching.

***T. stralensis*** Pierre. (Sneng kou, Motes prey). Indonesia. The roots are used locally to treat snake-bites.

TABERNANTHE Baill. Apocynaceae. 7 spp. Tropics. Shrubs.

***T. iboga*** Baill. Tropical Africa, especially Congo. A decoction of the bitter roots (Iboga Root) is used locally to treat fevers, as a tonic and aphrodisiac.

Table Mountain Pine – Pinus pungens.

Tablote – Guazuma ulmifolia.

Taboc – Swinglea glutinosa.

Tabog – Swinglea glutinosa.

Tabonico (Resin) – Dacryodes excelsa.

Tabonuco – Dacryodes excelsa.

Tabosa Mahogany – Swietenia mahagoni.

Tacamahaca – Bursera gummifera.

Tacamahaca Resin – Calophyllum tacamahaca.

TACCA J. R. and G. Forst. Taccaceae. 30 spp. Tropics, especially E.Asia. Herbs.

***T. fatsiifolia*** Warb. (Kanalong, Nonno, Rayung-payungar). Indonesia, Philippines. The plant is used locally as a poultice for snake-bites and wounds.

***T. hawaiiensis*** Limpr. f.=T. oceanica. (Hawaiian Arrowroot). The starch from the tubers was sold commercially as arrowroot.

***T. leontopetaloides*** (L.) O. Kuntze=T. pannatifida.

***T. montana*** Schultes=T. palmata.

***T. palmata*** Blume=T. montana=T. vesicaria. (Gadoong tikoos, Koomis ootjing, Temoo giling). Malaysia, Philippines. The tubers are bitter. They are chewed in the Philippines as a tonic and to relieve stomach pains. They are used as a poultice for boils and snake-bites.

***T. pinnatifida*** Forst.=T. leontopetaloides (Katjondong, Loki, Likir, Pia, Telo, E.Indian Arrowroot). Pacific Islands. Cultivated locally and on a small scale in E. and W.Africa. The tuber yields a rather indigestible arrowroot (Fiji Arrowroot, Tahiti Arrowroot, William's Arrowroot, Feculé de kabija, Feculé de Pia). A bitter principle has to be removed from the tubers by boiling before the starch is collected. The rhizomes are eaten locally as a vegetable and are used locally to treat dysentery and diarrhoea. The starch is used as flour and in laundry work.

***T. umbrarum*** Jum. and Perr. (Kobitsondolo). Madagascar. The tubers are eaten locally as a vegetable.

TACHIA Aubl. Gentianaceae. 4 spp. Guyana, Brazil. Shrubs.

***T. guianensis*** Aubl. Guyana, Brazil. The roots are used in Brazil to treat fevers.

TACHIADENUS Griseb. Gentianaceae. 7 spp. Madagascar. Herbs.

***T. longifolius*** Griseb. (Befelopela). Madagascar. A bitter extract from the plant is used locally to flavour beer.

Tachinote – Urera caracasana.

Tacoutta – Sesamum alatum.

Tacso – Passiflora molissima.

Tacso – Passiflora tripartita.

***Tacsonia van-volxemii*** Lem.=Passiflora van-voxvolexmii.

***Taetsia*** Medik.=Cordyline.

TAGETES L. Compositae. 50 spp. Warm America. Herbs.

***T. erecta*** L. (Big Marigold). Mexico. Cultivated widely in Old and New World. In India the flowers are used as the source of a yellow dye. More widely a decoction of the flowers and leaves is used to treat intestinal worms, stomach upsets and to control menstruation.

***T. filicfolia*** Lag.=T. multifida.

***T. glandulifera*** Schrank.=T. minuta.

***T. lucida*** Cav. (Sweet Marigold). Mexico, Central America. The flowers are used locally as a condiment.

***T. minuta*** L.=T. glandulifera. Tropical America. A decoction of the leaves is used to treat stomach upsets, as a diuretic and to induce sweating. An essential oil distilled from the leaves is used as an insect repellent.

***T. multifida*** DC.=T. filicifolia. Mexico, Central America. A decoction of the leaves is used locally as a diuretic.

Tagua Palm – Phytelephas macrocarpa.

Tahiti Arrowroot – Tacca pinnatifida.

Tahite Chestnut – Inocarpus edulis.

Tahitean Fig – Ficus prolixa.

Tainghe – Cananga latifolia.

Tak, Malacca – Afzelia palembanica.

Takamahaca Resin – Calophyllum inophyllum.

Takemomi – Abies homolepis.

Takut (Galls) – Tamarix articulata.

TALAUMA Juss. Magnoliaceae. 50 spp. E.Himalayas, S.E.Asia, Mexico to tropical S.America, W.Indies. Trees.

***T. mexicana*** (DC.) Don.=Magnolia mexicana. Mexico. A decoction of the bark is used locally to treat fevers, heart diseases, paralysis and epilepsy.

***T. minor*** Urb.=T. plumerieri.

***T. plumerieri*** Griseb.=T. minor=Svenhedina minor. (Maroñon de la Maestre). W.Indies. In Cuba an extract of the astringent bark and leaves in wine is used to treat stomach upsets. The flowers are used to flavour liqueurs.

Talawrinta – Manilkara kauki.

Talca – Acacia seyal.

Talda bans – Bambusa tulda.

Talha – Acacia seyal.

Talha – Acacia tortilis.

Tali sesawi – Lophopyxis pierrei.

TALINUM Adans. Portulacaceae. 50 spp. Warm America, except Mexico, Africa, Asia. Herbs.

***T. aurantiacum*** Engl. (Orange Fame Flower). S.W. U.S.A. The roots are eaten as a vegetable by the local Indians.

***T. cuneifolium*** (Vahl.) Willd.=T. indicum. (Aby, Muiki, Ndele, Mgottmerkua). Africa, Arabia, India. The leaves are eaten as a vegetable in parts of Africa and are used as an aphrodisiac in Tanzania.

***T. indicum*** Wight. and Arn. = T. cuneifolium.

***T. portulacifolium*** (Forsk.) Aschers. Tropical Africa, Asia. The plant is eaten as a vegetable in parts of Africa. It is sometimes cultivated.

***T. racemosum*** Rohrb. = T. triangulare.

***T. triangulare*** Willd. = T. racemosum = Portulaca triangularia (potherb Fame Flower, Surinam Purslane). Tropical America. The leaves are eaten locally as a vegetable.

Talipot Palm – Corypha umbraculifera.

Taliste – Willardia mexicana.

Talkeh – Astragalus echidnaeformis.

Talki Gum – Acacia seyal.

Tall Albizia – Albizia procera.

Tall Meadow Oat Grass – Arrhenatherum avenaceum.

Tallow, Borneo – Shorea apyera.

Tallow, Borneo – Shorea stenocarpa.

Tallow, Chinese Vegetable – Sapium sebiferum.

Tallow, Jaboty – Erisma calcaratum.

Tallow, Japan – Rhus succedanea.

Tallow, Mafura – Trichilia emetica.

Tallow, Malabar – Vateria indica.

Tallow Mowra Butter – Madhuca longifolia.

Tallow, Piney – Vateria indica.

Tallow, Rambutan – Nephelium lappaceum.

Tallow, Taban Merak – Palaquium oleiferum.

Tallow Tree – Pentadesma butyracea.

Tallow Tree, Chinese – Sapium sebiferum.

Tallow Tree, Malabar – Vateria indica.

Tallow Wood – Eucalyptus microcorys.

Tallow Wood – Ximenia americana.

Talok – Grewia celtidifolia.

Tam that – Panax repens.

Tamarack – Larix americana.

Tamarack – Larix occidentalis.

Tamaquare – Caraipe psidifolia.

Tamarind – Tamarindus indica.

Tamarind, Cow – Pithecolobium saman.

Tamarind, Manila – Pithecolobium dulce.

Tamarind Plum – Dialium indum.

Tamarind, Wild – Lysiloma latisiliqua.

TAMARINDUS L. Leguminosae. 1 sp. Tropical Africa. Tree.

***T. indica*** L. (Tamarind). The bitter-sweet pulp from the ripe pods is used to make drinks, chutneys, curries, etc. It was used by sailors instead of limes as a source of Vitamin C. The pulp is also used as a laxative. The plant is cultivated in the Old and New World.

Tamarisk – Tamarix gallica.

TAMARIX L. Tamaricaceae. 90 spp. W.Europe, Mediterranean through to India and N.China. Shrubs.

***T. articulata*** Vahl. N.Africa, Middle East. Galls (Takut, Tegant) from the plant are used by the Arabs for tanning and staining skins pink or purple. An infusion of the galls is used to treat stomach aches. The wood is used for small construction work and as fuel.

***T. chinensis*** Lour. (Tch'eng lieou). China, Japan, Indonesia, Indochina. The Chinese use the leaves as a poultice on wounds, to aid digestion and as a diuretic. In Indonesia they are put into baths to prevent children getting measles.

***T. dioica*** Roxb. (Jau, Pilchi). N.W.India. The galls, leaves and bark are used for tanning. The wood is used to make wheels, turning and as fuel. The twigs are made into baskets, brooms and huts.

***T. gallica*** L. (Jhau, Kori, Tamarisk). N.India, W.Himalayas. Galls (Sakum) on the plant are used for tanning. The wood is white, fairly hard, but not strong. It is used for general construction, turning, fancy articles, fuel and particularly for poles.

***T. mannifera*** Ehrbg. Arabia, Middle East. A sweet manna is exuded from the plant due to wounding by Coccus manniparus. Possibly the manna of the Bible.

Tambalang – Codium geppii.

Tambookie Grass – Hyparrhenia aucta.

Tamisan – Citrus longispina.

Tamparal – Osyris arborae.

Tampico Fibre – Agave falcata.

Tampico Fibre – Hesperryucca whipplei.

Tampico Jalap – Ipomoea simulans.

Tampico Sarsaparilla – Smilax aristolochiaefolia.

Tampoosing – Bruguiera eriopetala.

TAMUS L. Dioscoreaceae. 5 spp. Canaries, Madeira, Europe, Mediterranean. Vines.

***T. communis*** L. (Black Bindweed, Black Bryony, Lady's Seal). S.W.Europe. The leaves and shoots are sometimes eaten as a vegetable. The root is used as a diuretic.

Tan Bay – Gordonia lasianthus.

Tan chu – Lopatherum gracile.

TANACETUM L. Compositae 50–60 spp. N.Temperate. Herbs.

***T. multiflorum*** Thunb. = Matricaria multiflora.

***T. vulgare*** L. = Chrysanthemum tanacetum (Tansy). N.Temperate. The dried leaves and flowers and an oil distilled from them (Oil of Tansy) are used medicinally to treat intestinal worms, control menstruation and as a stimulant. It is also used as an insect repellent when applied to the skin. The oil contains borneol, thujone, camphor and can be dangerous if taken in excess.

Tanbark Oak – Lithocarpus densiflora.

Tandpijnwortel – Sium thunbergii.

Tang, Black – Fucus vesiculosus.

Tangahibira – Polygala bakeriana.

Tangedu Bark – Cassia auriculata.

Tangelo – Citrus paradisi × C. reticulata var. deliciosa.

Tangelolo – Citrus paradisi × (C. paradisi × C. reticulata var. deliciosa).

Tangerine (Oil) – Citrus reticulata var. deliciosa.

Tanggooli – Cassia javanica.

Tanggooloon – Protium javanicum.

Tanghin of Madagascar – Tanghinia venenifera.

TANGHINIA Thou. Apocynaceae. 1 sp. Madagascar. Tree.

***T. venenifera*** Poir. = Cerbera venenifera = C. tanghin. (Tanghin of Madagascar). The plant contains a heart poison and is used locally as an ordeal poison.

Tangier Pea – Lathyrus tingitanus.

Tangkal bintenoo – Melochia umbellata.

Tangkawang Fat – Shorea balangeran.

Tangle – Laminaria digitata.

Tangle, Sweet – Laminaria saccharina.

Tanglegrass – Heteropogon hirtus.

Tangloo – Aglaia odoratissima.

Tanias – Xanthosoma sagittifolium.

Tanonak – Rhus oxycanthoides.

Tansy (Oil) – Tanacetum vulgare.

Tansy Phacelia – Phacelia tanacetifolia.

Tansy Ragwort – Senecio jacobae.

Tao yi – Docynia delavayi.

Tapa – Ficus marquesensis.

Tapa – Ficus prolixa.

Tapa Cloth – Broussonetia papyrifera.

Tapana – Hieronima caribaea.

Tapang – Koompassia excelsa.

Tape Grass – Vallisneria americana.

Tapioca – Manihot esculenta.

TAPURA Aubl. Dichpetalaceae. 20 spp. Tropical S.America, W.Indies; 4 spp. Tropical Africa. Trees.

***T. antillana*** Gleason. (Bois Côte). W.Indies, especially the S. The wood is used locally for canoe masts and the buttress roots for paddles.

Taparu (Rubber) – Sapium taburu.

Tara – Actinidia callosa.

Tarakat – Grewia betulifolia.

***Taraktogenos*** Hassk. = Hydnocarpus.

Taramniya – Combretium hypotilinum.

Tarandjabine – Alhagi personum.

TARAXACUM Weber. Compositae. 60 spp. Mostly N.Temperate, 2 spp. Temperate S.America. Herbs.

***T. hybernum*** Steven. (Krim Sagiz). Italy, Balkans, Asia Minor, Ukraine. The latex is a commercial source of rubber. The plant is cultivated in Russia.

***T. koksaghyz*** Rodin. (Kok-saghyz). Turkestan. See T. hybernum.

***T. magellanicum*** Comm. ex Sch. New Zealand, S.Argentina. In New Zealand a decoction of the roots is used to stimulate the appetite.

***T. megalorrhizon*** (Forsk.) Handel-Mazz. See T. hybernum.

***T. officinale*** Weber. = Leontodon taraxacum. (Dandelion). Temperate. Sometimes cultivated for the leaves which are eaten in salads. There are several cultivars. The dried root is used medicinally as a tonic and diuretic and when roasted as a coffee substitute. The flowers are made into Dandelion Wine.

Tarbrush – Flourensia cernua.

TARCHONANTHUS L. Compositae. 2 spp. Mexico; 2 spp. S.Africa. Shrubs.

***T. camphoratus*** L. (Hottentot Tobacco). S.Africa. The heavy wood is used for fancy work and making musical instruments. The leaves, which taste of camphor are smoked by the Hottentots and chewed by the Mahommedans. A decoction of the leaves is used to treat asthma and to induce sweating.

Tare – Lolium temulentum.

TARENNA Gaertn. Rubiaceae. 370 spp. Tropical Africa, Madagascar, Seychelles, tropical Asia, Australia. Trees.

***T. incerta*** Kod. and Val. = Randia wallichii = Stylocoryne weberi. (Brasan, Kikeujeup, Roodon). Indonesia. The wood is used locally for house pillars.

Tarentang – Campnosperma brevipetiolata.

Taro – Calocasia antiquorum.

Taro – Calocasia esculenta.

Taro des Maraia – Cyrtosperma senegalense.
Taro, Giant – Alocasia cucullata.
Tarout – Cupressus dupreziana.
Tarrada – Scirpus corymbosus.
Tarragon – Artemisia dracunculus.
Tarragon, False – Artemisia dracunculoides.
Tarragon Oil – Artemisia dracunculus.
***Tarrietia*** Blume=Heritiera.
Tarweed – Madia glomerata.
Tarweed – Madia sativa.
Tatalencho – Gymnosperma glutinosum.
Tartar Gum – Sterculia cineria.
Tartuji Bianchi – Tuber gennadii.
Tarzali Silkwood – Cryptocarya oblata.
Tasco – Passiflora mollissima.
Tasco – Passiflora pinnatistipula.
Tasco – Passiflora tripartita.
Tasmanian Beech – Nothofagus cunninghami.
Tasmanian Blue Eucalyptus – Eucalyptus globosus.
Tasmanian Blue Gum – Eucalyptus globosus.
Tasmanian Myrtle – Nothofagus cunninghami.
Tatarian Sea Kale – Crambe tatarica.
Tatary Buckwheat – Fagopyrum tataricum.
Tau-Saghys – Scorzonera tau-saghys.
Tauary – Couratari tauari.
Tavolohaza – Dilobeia thouarsii.
Tawas – Wrightia pubescens.
Tawny Daylily – Hemerocallis fulva.

TAXODIUM Rich. Taxodiaceae. 3 spp. S.E. United States of America, Mexico. Trees. The Swamp Cupresses. The soft, light woods are decay-resistant.

***T. distichum*** (L.) Rich. (Bald Cypress, Southern Cypress). Swamps of S.E. U.S.A. The wood is used for general construction work, fences, barrels, railway sleepers.

***T. mucronatum*** Ten. Mexico. See T. distichum. A decoction of the bark is used locally as a diuretic and to control menstruation. The resin is used as a poultice for wounds and ulcers.

TAXUS L. Taxaceae. 10 spp. N.Temperate, S. to Himalayas, Philippines, Celebes and Mexico. Trees.

***T. baccata*** L. (English Yew). Europe to Asia Minor, N.Africa. The heavy, hard, elastic wood was used for longbows and is used for knife handles, comb-backs, alpine stocks and carving. It is sometimes called Deutsches Ebenholtz. A decoction of the leaves is used in Switzerland to treat insects on livestock.

***T. brevifolia*** Nutt. (Pacific Yew). Pacific N.America. The heavy, red, elastic wood is used by the local Indians to make bows, spear handles and paddles.

***T. canadensis*** Willd. (American Yew, Ground Hemlock). Shrub. The arils are edible but the seeds are poisonous.

***T. cuspidata*** Sieb. and Zucc. (Japanese Yew). China, Japan. The fine-grained wood polishes well and is used in Japan for furniture, interiors of houses, kitchen utensils, carving and turning. It is also used to make arrows. The heartwood yields a brown dye, used for dyeing cloth.

Tazargade – Artemisia maciverae.
Tch'eng lieou – Tamarix chinensis.
Tchianda – Encephalartos poppgei.
Tchidi douan – Euphorbia kerrii.
Tchitscha – Chenopodium quinoa.
Té de Pais – Lippia geminata.
Té de Pais – Lippia graveolens.
Té del Monte – Clinopodium macrostemum.
Tê tan – Asarum sieboldii.
Tê tan nam – Asarum balanse.
Tea – Camellia sinensis.
Tea, Amacha – Hydrangea thunbergii.
Tea, Angelica – Angelica archangelica.
Tea, Arabian – Catha edulis.
Tea, Balm – Melissa officinalis.
Tea, Betony – Stachys officinalis.
Tea, Black – Camellia sinensis.
Tea, Black – Melaleuca bracteata.
Tea, Blackberry-leaf – Rubus spp.
Tea, Bohemian – Lithospermum officinale.
Tea, Brazilian – Stachytarpheta dichotoma.
Tea, Broom – Leptospermum scoparium.
Tea, Bush – Cyclopia vogelii.
Tea, Bush Cinquefoil – Potentilla fruticens.
Tea Bush, Gambian – Lippia adoënsis.
Tea Bush, Wild – Leucas martinicensis.
Tea, Cape Barren – Correa alba.
Tea, Carolina – Ilex vomitoria.
Tea, Caspa – Cyclopia subterenata.
Tea, Caucasian – Arctostaphylos uva ursi.
Tea, Chamomille – Matricaria chamomilla.
Tea, Chaparral – Croton corybulosus.
Tea, Chinese – Camellia sinensis.
Tea, Croatian – Lithospermum officinale.
Tea, Desert – Ephedra trifuca.
Tea, deutscher Tee – Rosa pomifera.
Tea, False Paraguay – Viburnum cassinoides.

Tea, Germander – Teucrium chamaedrys.
Tea, Green – Camellia sinensis.
Tea, Hemero – Polygonum aviculare.
Tea, Hoarhound – Helichrysum serphylifolium.
Tea, Indian Kidney – Orthosiphon stamineus.
Tea, Japanese – Camellia sinensis.
Tea, Java – Orthosiphon stamineus.
Tea, Kaisertee – Dryas octopetala.
Tea, Kapporie – Epilobium angustifolium.
Tea, Kapor – Epilobium angustifolium.
Tea, Kneipp – Sambucus ebulus.
Tea, Kutai – Arctostaphylos uva ursi.
Tea, Labrador – Ledum groenlandicum.
Tea, Mallow – Malva sylvestris.
Tea, Maté – Ilex paraguensis.
Tea, Mexican – Chenopodium ambroisoides.
Tea, Mountain – Gaultheria procumbens.
Tea, New Jersey – Ceanothus americanus.
Tea, Oolong – Camellia sinensis.
Tea, Oswego – Monarda didyma.
Tea, Paraguay – Ilex paraguensis.
Tea, Pennyroyal – Pycnothamnus rigidus.
Tea, Rosemary – Rosmarinus officinalis.
Tea, Schweizertee – Dryas octopetala.
Tea Seed Oil – Camellia sasanqua.
Tea, Seminole – Asimina reticulata.
Tea, Siberian – Potentilla rupestris.
Tea, Stone Mint – Cunila origanoides.
Tea, Sweet – Hydrangea thunbergii.
Tea, Sweet Golden Rod – Solidago odora.
Tea, Sweet Trefoil – Trigonella coerulea.
Tea, Tartarian – Rhododendron chrysanthum.
Tea, Teamaster's – Ephedra trifurca.
Tea, Thé des Canares – Cedronella triphylla.
Tea, Thé de Carolines – Plygala theezans.
Tea, Thorn – Cliffordia ilicifolia.
Tea Tree – Melalenca uncinata.
Tea Tree, Broom – Leptospermum scoparium.
Tea Tree, Heath – Leptospermum ericoides.
Tea Tree, Yellow – Leptospermum flavescens.
Tea, Tschgorish – Bergemia crassifolia.
Tea, Warsaw – Ulmus campestris.
Tea, Yellow – Leptospermun flavescens.
Tea, Yerba Maté – Ilex paraguensis.
Tea, Yerba Santa – Eriodictyon californica.
Teaberry – Gaultheria procumbens.
Teak – Tectonia grandis.
Teak, African – Oldfieldia africana.
Teak, Bastard – Butea superba.
Teak, Borneo – Intsia palembanica.
Teak, Cape – Strychnos atherstonei.
Teak, Grey – Gmelina leichthardtii.
Teak, Khartoum – Cordia myxa.
Teak, Malacca – Afzelia palembanica.
Teak, Rhodesian – Baikiaea plurijuga.
Teak, Sudan – Cordia myxa.
Teamaster's Tea – Ephedra trifurca.
Teasel Clover – Trifolium parviflorum.
Teazel, Fuller's – Dipsacus fullonum spp. sativus.

TECOMA Juss. Bignoniaceae. 16 spp. Florida, W.Indies, Mexico to Argentine. Trees or vines.

***T. crucigera*** (L.) Bureau. = Bignonia capreolata.

***T. ipe*** Mart. = T. ochracea.

***T. leucoxylon*** (L.) Mart. (Greenhart Ebony, Green Ebony, Surinam Greenhart). S.America, W.Indies. The heavy, green-olive wood is used for general construction work, furniture and turning. It is not resistant to marine borers.

***T. ochracea*** Cham. = T. ipe. (Lapacho). S.Brazil into Argentine. The wood yields a yellow dye.

***T. radicans*** (L.) Juss. = Bignonia radicans. (Trumpet Creeper). E. N.America into Texas. A decoction of the root is used locally to induce sweating and as a salve for wounds.

***T. serratifolia*** G. Don. (Yellow Boui). N. S.America, Trinidad. The hard, durable wood is grey-green in colour and is used for general construction work.

***T. stans*** (L.) H.B.K. = Bignonia stans. Subtropical S.America. A decoction of the root is used locally as a diuretic and to combat intestinal worms.

TECTARIA Cav. Aspidiaceae. 200 spp. Throughout the tropics. Ferns.

***T. filix mas*** (Male Fern, Male Shield Fern). The dried rhizomes and fronds are used medicinally to treat tapeworms.

***T. spinulosum*** (O. F. Muell) Sw. Temperate Europe, Asia and N. America. The dried rhizomes (Rhizoma Filicis Maris) are used to treat intestinal worms. The rhizomes, when cooked on hot stones are eaten by the Eskimos.

TECTONA L. f. Verbenaceae. 3 spp. Indomalaya. Trees.

***T. grandis*** L. f. (Djati, Teak). E.India into Malaysia. A very valuable timber spp. which has been used locally for more than 2,000 years. The plant is cultivated, over

75 per cent of the world production coming from Burma. The wood is durable, strong, hard and easily worked. It is used for house construction, furniture, railway sleepers, bridges, boats and general construction work. A yellow dye from the bark is used locally for dyeing basket work.

Tee coma – Albizia procera.

Teel Oil – Sesamum indicum.

Teff – Poa abyssinica.

Taggant – Tamarix articulata.

TEIJMANNIODENDRON Koorders. Verbenaceae. 14 spp. Indochina, Malaysia except Java and Lesser Subdra Islands. Trees.

***T. pteropodum*** Bakh. = Vitex pteropoda. (Medang poodi, Sipanooh). W.Malaysia. The seeds are used locally to make an ointment to treat haemorrhoids.

TEINOSTACHYUM Munro. Graminae. 6 spp. India, Ceylon, Burma. Bamboos.

***T. dullooa*** Gamble. N.E. India. The stems are used locally to make baskets, mats, umbrellas, water-buckets and for light construction work.

***T. wightii*** Munro. India. Used locally to make light bridges.

***Telanthera*** R. Br. = Alternanthera.

TELECTADIUM Baill. Periplocaceae. 3 spp. Indochina. Shrubs.

***T. edule*** H.B. S.Vietnam, Laos. The bitter plant is eaten locally in Laos.

Telepathine – Banisteriopsis caapi.

TELFAIRIA Hook. Cucurbitaceae. 2 spp. Tropical Africa. Vines.

***T. occidentalis*** Hook. f. (Oyster Nut). Tropical Africa. The plant is cultivated locally for the seeds which are eaten cooked and the leaves that are used as a vegetable.

***T. pedata*** Hook. f. (Zanzibar Oil Vine). Tropical Africa. The seeds, which are edible, yield an oil (Castanah Oil) which is used to make soap and candles.

Telinga kerbaoo – Vanilla griffithii.

TELITOXICUM Moldenke. Menispermaceae. 6 spp. Tropical S.America. Vines. The roots of both spp. mentioned are used by the local Indians as a source of curare for arrow poisons.

***T. minutiflorum*** (Diels.) Mold. Brazil.

***T. peruvianum*** Mold. S.America.

Tellicherry Bark – Holarrhena antidysenterica.

Telo – Tacca pinnatifida.

***Teloschistes candelarius*** (L.) Fink. = Xanthoria candelaria.

TELOSMA Coville. Asclepiadaceae. 10 spp. Old World tropics. Vines.

***T. cordata*** Merr. = T. odoratissima = Pergularia minor. India, Indochina, China. The flowers and leaves are eaten as a vegetable in Thailand. The Chinese in Java make a confection from the roots.

***T. odoratissima*** Coville = T. odorata.

***T. procumbens*** Merr. Philippines. The fruits are eaten as a vegetable, locally.

Temal Shorea – Shorea hypochra.

Tembega – Calotropis gigantea.

Temiche Palm – Manicaria saccifera.

Tempixque – Sideroxylon capiri.

Temoo gihng – Tacca palmata.

Temple Mandarin – Citrus sinensis × C. reticulata.

Ten Month Yam – Dioscorea alata.

Tena – Elaeophorbia drupifera.

TENAGOCHARIS Hochst. Limnocharitaceae. 1 sp. Tropical Africa, India, S.E.Asia, Java, Madura, N. and N.E.Australia. Herb.

***T. emarginata*** Humb. and Bonpl. (Yellow Velvetleaf). In Java, the leaves and flowers are eaten as a vegetable.

Tenaza – Pithocellobium brevifolium.

Tendergreen – Brassica perviridis.

Tengkawang lajar – Shorea gysbertsiana.

Tengkawang madjau – Shorea lepidota.

Tenkwa fun – Trichosanthes japonica.

Tenio – Weinmannia trichosperma.

Tennevelley Senna – Cassia angustifolia.

Teodora botoan – Musa errans.

Teonancatl – Panaeolus campanulatus.

Teosinte – Fuchlaena mexicana.

Tep pirou – Cinnamomum cambodianum.

Tepalcahuite – Coccoloba schiedeana.

Tepary Bean – Phaseolus acutifolius.

Tepeacuilotl – Cornus excelsa.

TEPHROSIA Pers. Leguminosae. 300 spp. Tropics and subtropics. Herbs.

***T. candida*** DC. (White Tephrosia). Tropical Asia. Grown as a cover and green manure in coffee and rubber plantations.

***T. cinerea*** Pers. (Ashen Tephrosia). Tropical America. The roots are used in Guyana as a fish poison.

***T. densiflora*** Hook. f. = T. interrupta. Tropical W.Africa. Cultivated locally as a cover crop. The roots are used to stupify fish.

***T. interrupta*** A. Vhev. = T. densiflora.

***T. nitens*** Benth. Tropical America. In Brazil the leaves are used to stupify fish.

***T. periculosa*** Baker. Tropical E.Africa. The roots are used locally to stupify fish.

***T. purpurea*** Pers. (Purple Tephrosia). Throughout the Tropics. The roots are used widely to stupify fish. They are used in parts of Africa to flavour milk and a decoction of them is a local treatment for coughs, liver and kidney complaints and indigestion. They are also used as a salve for elephantiasis and boils. Some American Indians use a decoction of the plant to treat nose bleeding. The plant is grown as a green manure in rubber plantations.

***T. toxicaria*** (Sw.) Pers.=Galega toxicaria. (Fishdeath Tephrosia). Tropical and subtropical America. The roots are used as a fish poison. The plant is cultivated as the source of an insecticide.

***T. virginata*** (L.) Pers.=Cracca virginiana. (Devil's Shoestrings, Goat's Rue, Virginia Tephrosia). E. N.America. A decoction of the roots is used as a diuretic, to treat intestinal worms and to induce sweating. It is also used by the local Indians as a tonic for children. The roots are sometimes used to stupify fish.

***T. vogelii*** Hook. f. (Vogel Tephrosia). Tropical Africa. The plant is grown as a green manure and cover crop in plantations. The roots are used locally to stupify fish and as an insecticide.

Tepozán – Cestrum dumetorum.

Tequampatli – Marsdenia zimpanica.

Tequila – Agave tequilana.

Terai Bamboo – Melocanna bambusioides.

Terbulan – Endospermum malaccense.

Terebinth Pistache – Pistacia terebinthus.

Terebinthina Laricina – Larix decidua.

Terebinthina Veneta – Larix decidua.

Terentanabang – Campnosperma macrophylla.

Terentang – Campnosperma minor.

Terfaz – Terfezia boudieri.

TERFEZIA Tul. Tuberaceae. 12 spp. Warm. The fruitbodies of the following spp. are eaten locally.

***T. boudieri*** Chatin. (Terfaz, Torfaz). N.Africa, Asia Minor.

***T. claveryi*** Chatin.=T. hafizii=T. hanotauxii. (Caucasian Truffles). Asia Minor, Caucasus.

***T. gennadii*** Chatin=Tuber gennadii.

***T. hafizii*** Chatin=T. claveryi.

***T. hanotauxii*** Chatin=T. claveryi.

***T. magnusii*** Matt.=Choiromyces magnusii.

TERMINALIA L. Combretaceae. 250 spp. Tropics. Trees.

***T. arunja*** Wight. and Arn. (Arjan Terminalia). E.India to Malaysia. The hard, brown streaked wood is used for housebuilding, carts, agricultural implements and boat making. A decoction of the bark is used as a heart stimulant.

***T. avicennioides*** Guill. and Perr. (Ouôlo). Tropical Africa. A light, reddish, spongy resin (Ouôlo mogo), from the trunk is used locally as incense.

***T. bellerica*** Roxb. (Belleric Terminalia). E.India to Philippines. The plum-like acid-sweet fruits (Baleric Myrobalans) are eaten raw or in preserves. A decoction of the fruits is used as an eye-lotion. In India, the leaves are used for tanning.

***T. bialata*** Steud. (Indian Silver Grey Wood, White Chuglam). Andaman Islands. There are two types of heart wood, Indian Silver Grey Wood which is used for veneer, furniture, panelling, picture frames etc. and the non-decorative White Chuglam used for general construction work, oars, mathematical instruments spars etc. A decoction of the bark is used in Cambodia to treat dysentery.

***T. calamansanay*** (Blanco) Rolfe. Philippines. A decoction of the bark is used for treating stones in the bladder.

***T. capitata*** Sauv.=T. hilariana.

***T. catappa*** L. (Country Almond, Indian Almond, Tropical Almond, Myrobalan, Almendro). Tropical Asia. Cultivated in Old and New World, primarily for the oily seeds which are eaten. The fruits, bark and roots are used for tanning and the fruits are used in parts of E.India for staining the teeth black. The leaves are fed to silkworms.

***T. chebula*** Retz. Central Asia. The fruits are used for tanning and as a tonic to stimulate the appetite.

***T. comintana*** (Blanco.) Merr. (Binggas Terminalia). Philippines. A decoction of the fruits is used locally to treat diarrhoea and thrush.

***T. coriacea*** W. and A. (Banapu, Sadagam). Tropical Asia. A decoction of the bark is used by the Hindus as a heart stimulant.

***T. hilariana*** Steud.=T. capitata. (Yellow Sanders). Tropical America. The strong, satiny wood is used for general carpentry, posts etc.

***T. karnbachi*** Schum. New Guinea. The pleasantly flavoured fruits are eaten locally.

***T. macroptera*** Guill. and Perr. (Rebreb Terminalia). Tropical Africa. In the Sudan the bark is used as a source of perfume.

***T. mauritiana*** L. (Mauritius Terminalia). Mauritius, Réunion. The bark (Jamrosa Bark) is used for tanning, giving a light colour to the leather.

***T. myriocarpa*** Heurcl. and Muell. Arg. Tropical Asia. See T. coriacea.

***T. nigrovenulosa*** Pierre. (Chieu lieu, Preas phnau). Indochina. A decoction of the bark is used in Cambodia to treat diarrhoea.

***T. obovata*** Steud. (Nargusta Terminalia, Yellow Olivier). S.America, W.Indies. The strong, fairly hard wood is used locally for general construction work and furniture making.

***T. sericea*** Burch. (Assegai Wood, Mangwe, Yellow Wood). Central Africa. The wood is used locally to make furniture. The bark is used for tanning leather yellow.

***T. splendida*** Engl. and Diels. Tropical Africa. The powdered bark is used in the Sudan as snuff.

***T. sumatrana*** Miq. (Djaha keling, Kajoo koongit). Indonesia. A decoction of the bark is used locally to dye cotton an ochre colour.

***T. superba*** Engl. and Diels. Tropical W.Africa. An important local timber tree, sometimes exported. The wood is fairly hard, tough and works easily. It is used for general construction work, house-building, canoes, bowls, bridges etc.

***T. tomentosa*** W. and D. (Ain, Asna, Sain, Saj, Laurel). India. The dark brown wood is hard and durable. It is used for house-building, furniture-making, toys, agricultural implements, mills, engine brake-blocks and electrical casing. The bark is used locally for tanning, preserving fishing nets and the source of a black dye. A gum from the bark is used as incense and in cosmetics. The juice from the bark is used in the clarifying of palm sugar.

TERMITOMYCES fide Singer (is based on Schulzeria Bres). Agaricaceae. 10 spp. Widespread.

***T. eurhisa*** (Berk.) Sing. Tropical Asia, growing in termitaria. The fruitbodies (Djamoer rajap) are eaten in Malaya.

TERNSTROEMIA Mutis ex L. f. Theaceae. 100 spp. Tropics. Trees.

***T. robinsonii*** Merr. Amboina. The wood is used locally to make tool-handles and the bark to stupify fish.

Teroondjam – Calophyllum wallichianum.

Terote – Jatropha spathulata.

Terra Japonica – Acacia catechu.

Terra Japonica – Uncaria gambir.

Tesota – Olneya tesota.

TESSMANNIA Harms. Leguminosae. 12 spp. Tropical W.Africa. Trees.

***T. africana*** Harms. (Kakweshe, Kankonde). Congo. The wood is used for general carpentry, house-building and railway sleepers.

***T. anomala*** (Micheli) Harms. See T. africana.

***T. dewildemaniana*** Harms. (Musise). Congo. The wood is used locally for furniture-making.

Teta toondjoo – Canarium patentinervium.

Tetlate – Bursera bipinnata.

TETRACARPIDIUM Pax. Euphorbiaceae. 1 sp. W.Africa. Tree.

***T. conophorum*** Hutch. and Dalz. The seeds yield a very rapidly-drying oil which is of potential value in making paints and varnishes and soap. The seeds are eaten locally and the leaves as a vegetable with rice.

TETRACERA L. Dilleniaceae. 40 spp. Tropics. Scrambling shrubs or vines.

***T. assa*** DC.=T. indica=T. sericea=T. trigyna. (Chon qué, Giây chieu). Malaysia, Indochina. In Cambodia the twigs are used as cordage and sandpaper.

***T. alata*** Presl.=T. volubilis.

***T. erecta*** Sessé and Moc=T. volubilis.

***T. indica*** Merr.=T. assa.

***T. sarmentosa*** Vahl.=T. scandens=Delima sarmentosa. Malaysia. The leaves are used locally as sandpaper and in India for polishing metal. A poultice of the leaves is used as a dressing for boils and a decoction of them is a treatment for dysentery.

***T. scandens*** Merr.=T. sarmentosa.

***T. sessiflora*** Triana and Plach=Delima dioica=D. mexicana. (Bejuca de agua). S.Mexico, Central America. The leaves are used locally as sandpaper and the stems as cordage.

***T. trigyna*** Roxb.=T. assa.

***T. volubilis*** L.=T. alata=T. erecta. (Bejuco carey, Bejuco tomé, Hoja chegüe, Raspa). Tropical America. A decoction of the seeds and leaves is used locally to induce sweating and control fevers, as a cure for syphilis and a diuretic.

TETRACLINIS Mast. Cupressaceae. 1 sp. S.Spain to Tunisia and Malta. Tree.

***T. articulata*** (Vahl.) Mast.=Callitris quadrivalvis. The trunk yields a resin (Sandarac, Pounce) used to make a fine varnish for preserving paintings, lacquers, coating paper and metal, bookbinding and in dental cements. It is also burnt as incense and used medicinally as a stimulant. The wood (Arar Wood) is used for house- and bridge-building.

TETRAGASTRIS Gaertn. Burseraceae. 8 spp. Central America to N. S.America, W.Indies. Trees.

***T. balsamifera*** (Sw.) Kuntze. (Amacet, Bois Cochin, Masa, Pig's Balsam, Sucrier Demontage). W.Indies. The high quality timber is reddish to orange-brown and is used for furniture, cabinet work, panelling and oars. The wood is fairly hard, tough and compact, but susceptible to termites and marine borers. It is sometimes used for heavy construction work and flooring. A resin from the trunk is used to treat inflammation of the mucous membranes and to reduce fevers.

TETRAGONIA L. Tetragoniaceae. 50–60 spp. Africa, E.Asia, New Zealand, Australia, temperate S.America. Herbs.

***T. expansa*** Murr.=T. tetragonioïdes.

***T. tetragonioides*** (Pall.) Kuntze.=T. expansa. (New Zealand Spinach). New Zealand, Australia. The leaves are eaten as a vegetable. The plant is widely cultivated as a vegetable in warm countries.

***Tetrahit longiflorum*** Moench.=Galeopsis ocholeuca.

TETRAMELES R. Br. Tetramelaceae. 1 sp. Indochina, Indomalaya. Tree.

***T. nudiflora*** R. Br.=T. rufinervis. (Binong, Gangganan, Kajoo laboo). The wood is used to make matches and boxes; it is also used to make canoes.

TETRAMERISTA Miq. Tetrameristaceae. 2–3 spp. Sumatra, Malaysia, Borneo. Trees.

***T. glabra*** Miq. (Kajoo malaka, Poonah). An important local swamp timber, much used locally in Singapore for beams. It is subject to termite damage when in contact with the ground. The sub-acid fruits are eaten raw or drunk as a syrup (Roodjak).

***Tetramicra bicolor*** Rolfe=Leptotes bicolor.

***Tetranthera polyantha*** Wall.=Litsea cubeba.

***Tetrapanax papyrifera*** Koch=Fatsia papyrifera.

TETRAPLEURA Benth. Leguminosae. 2 spp. Tropical Africa. Trees.

***T. thonningii*** Benth. Tropical W.Africa. The pods are eaten locally as a soup and the wood is used to make native houses and furniture.

TETRAPTERYS Cav. emend A. Juss. Malpighiaceae. 80 spp. Mexico to tropical S.America, W.Indies. Vines.

***T. methystica*** R. E. Schultes. Amazon. The leaves and stems are used by the Indians to make an intoxicating drink.

TETRASTIGMA Planch. Vitidaceae. 90 spp. S.E.Asia, Indomalaya, Australia. Vines.

***T. harmandi*** Planch. Tropical Asia, especially Malaya. The fruits are used locally to make jellies.

***T. lanceolarium*** Planch.=Vitis lanceolaria.

***T. papillosum*** Planch.=Vitis papillosa.

TEUCRIUM L. Labiatae. 300 spp. Cosmopolitan, especially Mediterranean. Herbs.

***T. capense*** Thunb. (Aabeibossie, Drieblaar, Katjee, Koonboisie). S.Africa. An infusion of the leaves is used to treat haemorrhoides and an extract of the leaves in brandy is a tonic and relieves stomach upsets.

***T. chamaedrys*** L.=T. officinalis=Chamaedrys officinalis. (Common Germander, Chamaedrys Germander). Europe, W.Asia, N.Africa. A decoction of the leaves has been used as a medicine since ancient times to treat complaints of the spleen, gout, dropsy, stomach upsets and scrophula. It has also been used as a tonic and diuretic and to reduce fevers.

***T. cubense*** Jacq. Tropical America, W.Indies. A decoction of the leaves is used in Cuba to reduce fevers and as a tonic.

***T. marum*** L. W.Mediterranean. A decoction of the leaves (Herba Mari veri) is used as a tonic, diuretic and to relieve stomach upsets.

***T. officinale*** Lam.=T. chamaedrys.

***T. polium*** L. Mediterranean. A decoction of the leaves is used in India to treat abcesses and fungus diseases of the skin.

***T. scordium*** L.=Chamaedrys scordium.

(Water Germander) Europe, Asia, N.Africa. An extract of the leaves (Extractum Scordii Dialystum) is used to treat tuberculosis and fungal infections of the skin. In Central Europe the leaves are used to make a green dye for dyeing cloth.

***T. scorodonia*** L. (Wood Germander). Mediterranean. The leaves are used as a dressing for wounds.

***T. thea*** Lour. China. The leaves are used locally as a tea.

Texas Bluegrass – Poa arachnifera.

Texas Croton – Croton texensis.

Texas Millet – Panicum texanum.

Texas Oak – Quercus texana.

Texas Snakeroot – Aristolochia reticulata.

Texas Sotol – Dasylirion texanum.

Texas Walnut – Juglans rupestris.

Texiote – Sedum dendroideum.

Texiotl – Sedum dendroideum.

***Teysmanniodendron*** Koord. = Teijsmanniodendron.

THALICTRUM L. Ranunculaceae. 150 spp. N.Temperate, tropical S.America, tropical and S.Africa. Herbs.

***T. angustifolium*** L. E.Europe. A decoction of the roots and leaves is used in the Ukraine as a diuretic and the plant is fed to cows to stimulate milk production.

***T. aquifolium*** L. (Columbine Meadow Rue). Europe, N.Asia. The roots are eaten raw or as a vegetable by the Ainu.

***T. collinum*** Wallr. E.Europe. A decoction of the flowers and roots is used locally to treat stomach aches, scrofula, to clean wounds and as a diuretic.

***T. foliolosum*** DC. (Mamira, Mariran, Penjari). Himalayas. The Hindus use a decoction of the roots as a tonic, to treat stomach upsets, to reduce fevers, as a purgative and as a salve to treat opthalmia.

***T. hernandezii*** Tausch. = T. lasiostylum. (Alboquillo de Campo). Mexico to S.America. A decoction of the root is diuretic and purgative. It is used in Argentina as a treatment for rheumatism.

***T. lasiostylum*** Presl. = T. hernandezii.

THAMNOSMA Torr. and Frém. Rutaceae. 8–9 spp. S. United States of America, S.W.Africa, Socotra, Arabia. Shrubs.

***T. montana*** Torr. and Frém. (Cordocillo) S.W. U.S.A. A decoction of the leaves is used in Arizona as a tonic and by the local Indians to treat gonorrhoea.

Thanh thât – Ailanthus malabarica.

THAPSIA L. Umbelliferae. 6 spp. Mediterranean. Herbs.

***T. garganica*** L. (Gargan Death Carrot). Mediterranean. The root (Spanish Turpethroot) yields a resin (Thapsia Resin, used for plasters.

Thapsia Resin – Thapsia garganica.

Thatch Grass – Saccharum spontaneum.

Thatch Palm – Thrinax parviflora.

Thatch Screw Pine – Pandanus odoratissimus.

Thatching Sedge – Cladium mariscus.

Thé de Bourbon – Angraecum fragrans.

Thé des Canares – Cedronella triphylla.

Thé des Carolines – Polygala theezans.

***Thea*** L. = Camellia.

THECOSTELE Reichb. f. Orchidaceae. 5 spp. Burma to W.Malaysia. Herbs.

***T. porlana*** Gagnep. (Bâ chout). N.Vietnam. A mixture of the leaves, boiled with arsenic and rice is used locally to kill rats.

Theetsee – Melanorrhoea usitata.

***Thela alba*** Lour. = Plumbago zeylanica.

THELEPOGON Roth. ex Roem. and Schult. Graminae. 1 sp. Abyssinia to India. Grass.

***T. elegans*** Roth. The leaves are fed to horses as a tonic.

THELESPERMA Less. Compositae. 12 spp. Warm America. Herbs.

***T. gracile*** (Torz.) Gray. W. U.S.A. The local Indians use the leaves to make a drink.

***T. megapotanicum*** Spreng. W. U.S.A. The flowers yield a yellow dye and the rest of the plant a brown dye. Both were used by the local Indians to dye baskets and clothing.

***T. trifidum*** (Poir). Britt. See T. gracile.

THELIGONUM L. Theligonaceae. 3 spp. Canaries, Mediterranean, S.W.China, Japan. Herbs.

***T. cynocrambe*** L. Mediterranean. The young shoots are sometimes eaten as a vegetable and were used as a laxative.

THELOSCHISTES Muell. Arg. Theloschistaceae Lichens.

***T. flavicans*** (Sw.) Muell. Arg. (Yellow Borrera). Temperate. Used in Germany as the source of a yellow dye used to dye wool.

***T. parietinus*** (L.) Norm. Temperate. Used in England and Sweden as the source of a yellow dye, used to dye wool.

THELYCRANIA (Dumort.) Fourr. Cornaceae. 2 spp. Temp. Trees.

***T. sanguinea*** (L.) Fourr.=Cornus sanguinea. (Bloodtwig Dogwood, Dogberry). Europe to N. India. The hard, tough wood is used for turnery, tool handles etc.

***T. sericea*** (L.) Dandy=Cornus stolonifera. N.America. The inner bark is used by the local Indians as a tobacco substitute.

***Thelygonum*** Schreb.=Theligonum.

THELYPTERIS Schmid. Thelipteridaceae. 500 spp. Cosmopolitan. Ferns.

***T. anthamantica*** Moore. S.Africa. The Zulus use the powdered rhizome to treat intestinal worms.

THEMEDA Forsk. Graminae. 10 spp. Warm Africa, India, Australia. Grasses.

***T. ciliata*** L. (Kangaroo Grass). E.India, Africa. Planted as fodder for livestock in Australia.

***T. membranacea*** Lindl. (Barcoo Grass). Australia. An excellent pasture grass, especially used in Queensland.

***T. triandra*** Forsk. Africa, India, Australia. Grown as a fodder grass. The seeds are used as a famine food.

THEOBROMA L. Sterculiaceae. 30 spp. Tropical America. Trees. The seeds of the following spp. are all made into cacao, either locally or on a minor commercial scale. They are sometimes mixed with the seeds of T. cacao, the main cacao-producing spp. The seeds contain varying amounts of theobromine, which is used medicinally as a diuretic and stimulant and in so-called cola drinks.

***T. albiflora*** Goud.=Herrania albiflora. (Cacao Montaras, Cacao Simarron). Colombia.

***T. angustifolium*** DC.=T. speciosa. (Cacao mico, Cacao de Sonusco, Cacao silvestre). Mexico, Central America. An important local source of cacao and one of the types originally imported to Spain.

***T. bicolor*** Humb. and Bonpl. (Cacao Blanco, Nicaraguan Cacao, Patashti). Mexico through to S.America. The seeds have a high cacao butter content.

***T. cacao*** L. (Cacao). Mexico, Central America. Cultivated in the Old and New World tropics for the seeds (beans) which are the main source of cacao. The seeds were used as a beverage and as currency by the Aztecs. Chocolate as a drink was introduced to Spain by Colombus and into London in 1657 from France. The world production of beans now exceeds 1 million tons a year, 33 per cent coming from Ghana, 16 per cent from Nigeria and Brazil, and 7 per cent from the Ivory Coast and Cameroons. The first commercial production was in the W.Indies and was introduced to W.Africa and later to the Far East, where production is increasing. The African crop is produced on small farms while that in the other countries is from plantations. Each tree yields about 1–2 lb. of dried beans a year (about 350 lb. per acre) but the yield may be over 1,000 lb. using selected clones and proper management. The plant is usually propagated by seed, but the increasing use of disease-resistant and high-yielding clones has led to more propagation by cutting. The trees start bearing after 4–5 years and continue to bear for about 50 years. Cacao requires an average annual temperate of over 21°C and rainfall of 80 inches a year. There are two main types of cacao, Forastero and the less vigorous, thinner shelled Criollos. The Criollos produce a finer quality cacao, but now play little part in world trade. About 90 per cent of the world cacao comes from the sub-group Amazonian Forestero, which was the type introduced from America to W.Africa. The rest of the world supply coming mainly from the W.Indies, and from Trinitario cacao, which is considered to be a complex group resulting from hybridization between Forastero and Criollo. The cacao produced is of a finer quality than that of pure Amazonian Forestero.

The pods are harvested and the beans extracted. They are left in piles to ferment. This removes the pulp and develops the chocolate flavour. The beans are then sun-dried to 6–8 per cent water content and exported. The chocolate manufacturer roasts the beans, shells them and grinds them to a liquor. The beans contain 50 per cent fat (Cacao Butter). This may be pressed from the liquor and the resulting powder is the cocoa powder of commerce. If chocolate is being produced, cacao butter is added to the liquor as are milk powder, flavouring etc. This is allowed to set giving the block-chocolate. The cocoa powder is used as a drink, in confectionery and as a flavouring. Cacao butter (Sp. Gr. 0·8823–0·8830; Sap. Val. 192–198; Iod. No. 32–40; Unsap. 0·3–0·8 per cent) is used in the manufacture of confectionery, toilet preparations and pharmaceuticals. A liqueur (Crême de Cacao) is also made from cacao. Cocoa powder contains 40·3 per cent carbohydrate, 18·1 per cent protein, 22 per cent fat, 2·2 per cent theobromine and 0·1 per cent caffeine.

***T. glauca*** Karts. Tropical America.

***T. grandiflora*** Schum. = Guazuma grandiflora.

***T. guazuma*** L. = Guazuma ulmifolia.

***T. leiocarpa*** Bern. (Cacao Calabacillo). Tropical America.

***T. mariae*** Schum. = Abroma mariae = Herrania mariae. (Cacaoti). Amazon Basin.

***T. martiana*** Dietr. N.Brazil.

***T. microcarpa*** Mart. Brazil. Cultivated in Bahia.

***T. pentagona*** Bern. (Cacao Lagarto). Central America.

***T. purpureum*** Pitt. (Cacao de Mico). Central America. Used by the local Indians to make a drink.

***T. speciosa*** Willd. = T. angustifolium.

***T. sunincanum*** Mart. Guyana, Brazil, Peru.

THERMOPSIS R. Br. Leguminosae. 30 spp. Central Asia to Himalayas, E. United States of America. Herbs.

***T. lanceolata*** R. Br. Russia to Siberia. The leaves are used locally as an expectorant.

THESPESIA Soland. ex Corrêa. Malvaceae. 15 spp. Tropics. Shrubs.

***T. garckeana*** F. Hoffm. (Quaters, Rhodesian Tree Hibiscus). Central Africa. The fruits are eaten locally.

***T. lambas*** Dalz. and Gibs. (Kapas ooton, Kapasan) Malaysia. The bark fibres, removed after retting, are used in Java as cordage.

***T. populnea*** Soland. (Portia Tree, Tulip Tree). Small tree. Tropical Africa, Asia. The wood (Seychelles Rosewood) is very hard and keeps well under water. It is light brown, decorated with black streaks and is used for wheels, gunstocks, boats and carving etc.

THEVETIA (L.) Juss. ex. Endl. Apocynaceae. 9 spp. Tropical America, W.Indies. Shrubs. The genus is a potential source of balata.

***T. ahouai*** A. DC. Tropical S.America. A decoction of the bark is used to reduce fevers and ulcers. It causes vomiting and is laxative.

***T. nereifolia*** Juss. = T. peruviana (Exile Tree, Yellow Oleander). Tropical America, W.Indies. The plant is cultivated in the Old and New world tropics for the bark, which is used medicinally to reduce fevers. The seeds are used to produce Exile Oil in India and are considered lucky charms in the W.Indies (Lucky Bean, Lucky Seeds).

***T. peruviana*** Schum. = T. nereifolia.

***T. thevetioides*** H.B.K. = T. yecotli. = Cerbera thevetioides. (Codo de fraile, Cabrito, Yucucaca, Yo yotl). A decoction of the seeds causes violent sickness and paralysis of the heart. They are used as a substitute for digitalin and as a salve for haemorrhoids.

Thi – Diospyros decandra.

Thick-leaf Phlox – Phlox carolina.

Thistle, Black Ray – Centaurea nigra.

Thistle, Blessed – Cnicus benedictus.

Thistle, Canada – Cirsium arvense.

Thistle, Carline – Carlina acaulis.

Thistle, Holy – Silybum marianum.

Thistle, Milk – Silybum marianum.

Thistle, Musk – Carduus nutans.

Thistle, Russian – Salsola kali.

Thistle Sage – Salvia carduacea.

Thistle, Sow – Sonchus oleraceus.

Thistle, Star – Centaurea nigra.

Thistle, Yellow-spined – Cirsium ochrocentrum.

THLASPI L. Cruciferae. 60 spp. Mostly N.Temperate Eurasia, few N. and S.America. Herbs.

***T. arvense*** L. (Field Penny Cress, Penny Cress). Europe, N.Asia, introduced to N.America. The seeds (Semen Thlaspeos) were used medicinally as a diuretic, to treat rheumatism and as a general tonic.

Tho nan – Eurycoma longifolia.

Tho shui – Cananga latifolia.

Thông – Pinus tonkinensis.

Thong mu – Podocarpus blumei.

Thong-nang – Podocarpus imbricatus.

THONNINGIA Vahl. Balanophoraceae. 5 (or 1) spp. tropical Africa; 1 sp. Madagascar. Herbs parasitic on tree roots.

***T. sanguinea*** Vahl. Tropical Africa. A decoction of the flowers is used in Nigeria to treat sore throats and the underground parts are used by the Hausa as a condiment.

Thorn, Buffalo – Ziziphus mucronata.
Thorn, Cape – Ziziphus mucronata.
Thorn, Christ – Ziziphus spina-Christi.
Thorn, Cock-spur – Crataegus crus-galli.
Thorn, Knob – Acacia nigrescens.
Thorn Pear – Scolopia zeyheri.
Thorn, Sweet – Acacia horrida.
Thorn Tea – Clifforia ilicifolia.
Thornapple – Datura stramonium.

THRINAX L. f. ex Sw. Palmae. 12 spp. W.Indies. Palm trees.

***T. argentea*** Millsp.=T. wendlandiana.

***T. miraguana*** Mart. (Yuraguana). Especially Cuba. The leaf-fibres are used locally for cordage.

***T. parviflora*** Sw. (Reef Thatches, Thatch Palm). Florida, W.Indies. The dried leaves are exported for decoration.

***T. wendlandiana*** Becc.=T. argentea. W.Indies, Central America. The fruits are edible and a fibre from the trunk is used locally for stuffing pillows etc.

Throatwort – Scrophularia nodosa.

THUJA L. Cupressaceae. 5 spp. China, Japan, N.America. Trees.

***T. occidentalis*** L. (Eastern Arbor-Vitae, White Cedar). E. U.S.A. The coarse, soft, light wood is durable and is used for fencing, railway sleepers and roofing. An oil (Oil of Arbor Vitae, Oil of White Cedar, Oleum Cedri Folii) is taken to relieve rheumatism, to control menstruation and as an expectorant. Externally it is applied to skin diseases. It has some use in making perfumes.

***T. plicata*** Don. (Giant Arbor Vitae, Western Red Cedar). The brown wood is coarse-grained, soft, light and not strong. It is used for interiors of buildings, furniture, roofing and barrels. The Indians use it to make canoes, houses etc. The Indians also use the bark fibre for making blankets, ropes and for thatching.

***T. tetragona*** Hook.=Pilgerodendron uviferum.

THUJOPSIS Sieb. and Zucc. Cupressaceae. 1 sp. Japan. Tree.

***T. dolobrata*** Sieb. and Zucc.=T. hondai. The soft, durable wood is used locally for building, furniture, ships, bridges, railway sleepers, wooden pipes and barrels. The bark is used for caulking boards.

***T. hondai*** Henry=T. dolobrata.

Thunberg Kudzu Vine – Pueraria thunbergiana.
Thuôe sôt rét – Tinospora crispa.
Thurber Fescue – Festuca thurberi.
Thyme (Oil) – Thymus vulgaris.
Thyme, Basil – Satureja calamintha.
Thyme, Conehead – Thymus capitatus.
Thyme, Creeping – Thymus serpyllum.
Thyme, Mastic – Thymus mastichina.
Thyme, Oil of – Monarda didyma.

***Thymeleaea hirsuta*** Endl.=Passerina hirsuta.

THYMUS L. Labiatae. 300–400 spp. Temperate Eurasia. Small shrubs or herbs.

***T. acinos*** L.=Satureja acinos.

***T. capitatus*** Hoffm. and Link.=Coridothymus capitatus. (Conehead Thyme). Mediterranean. An essential oil (Spanish Origanum Oil) distilled from the plant is used for flavouring.

***T. chamissionis*** Benth.=Micromeria chamissonis.

***T. hirtus*** Willd. N.Africa, Spain. A decoction of the leaves is used to relieve stomach upsets in N.Africa and the ground leaves are used as a condiment.

***T. mastichina*** L. (Mastic Thyme, Spanish Marjoram). Mediterranean. An essential oil (Oil of Marjoram, Oil of Wild Marjoram) distilled from the leaves is used for flavouring.

***T. serpyllum*** L. (Creeping Thyme). Europe, temperate Asia, Africa, N.America. A decoction of the dried flowers and leaves is used to treat whooping cough. The plant is sometimes cultivated.

***T. vulgaris*** L. (Thyme). Mediterranean, introduced to N.America. The dried leaves are used as a condiment in flavouring meat dishes. An essential oil (Thyme Oil, Oleum Thymi) distilled from the leaves and flowers is used to relieve stomach upsets and as a stimulant. It is also used as a linament. Thyme oil is a commercial source of thymol.

THRYSOSTACHYS Gamble. Graminae. 2 spp. N.India, Burma, Thailand. Bamboos.

***T. siamensis*** Gamble. (Monastry Bamboo, Umbrella Handle Bamboo). Throughout range. Grown as an ornamental in the gardens of local monasteries and used to make umbrella handles.

THYSANOLAENA Nees. Graminae. 1 sp. Tropical Asia. Grasses.

***T. maxima*** Kuntze. The panicles are used for brooms.

Tiama Mahogany – Entandrophragma macrophyllum.

TIARELLA L. Saxifragaceae. 5 spp. Himalayas and E.Asia, Pacific and Atlantic N.America. Herbs.

***T. cordifolia*** L. (Coolwort, Miterwort). Atlantic N.America. A decoction of the leaves is diuretic and is used in the treatment of stones in the bladder, retention of urine and other urinary complaints.

Tidnefulu – Albizia versicolor.

TIEGHEMELLA Pierre. Sapotaceae. 2 spp. Tropical W.Africa. Trees.

***T. africana*** Pierre. The seeds yield an edible fat, used locally.

***T. heckeli*** Chev. The seeds yield an edible fat (Doumori Butter) which is used locally. The wood (African Mahogany) is red and beautifully grained. It is used for furniture-making and the interiors of railway carriages and motor cars. It is exported.

Tien ch'i – Gynura pinnatifida.

T'ien chi wang – Hypericum japonica.

Tiên dó – Vitis pentagona.

Tiger Grass – Thysanolaena maxima.

Tigerflower – Tigridia pavonia.

Tiger Lily – Lilium tigrinum.

Tiger Nut Oil – Cyperus esculentus.

Tigerwood Lovoa – Lovoa klaineana.

Tigli Oil – Croton tiglium.

TIGRIDIA Juss. Iridaceae. 12 spp. Mexico to Chile. Herbs.

***T. pavonia*** (L. f.) Ker-Gawl. (Tigerflower). Throughout range. The starchy bulbs are eaten roasted by the local Indians.

Tik – Curcuma angustifolia.

Tikan – Curcuma leucorhiza.

Tikitiki Extract – Oryza sativa.

Tikor – Curcuma angustifolia.

Tikul – Garcinia pedunculata.

Tikur – Curcuma angustifolia.

TILIA L. Tiliaceae. 50 spp. N.Temperate, S. to India, China and Mexico. Trees.

***T. americana*** L. (American Basswood, Linden). The light, soft, light brown wood is used for furniture, carriages, carving, panelling and inner parts of shoes. The bark fibres are used for cordage, making mats, etc. The flowers are a good source of honey.

***T. cordata*** Mill.=T. parvifolia=T. ulmifolia. (Lime, Small-leaved Linden). Europe to Caucasus. The light, soft wood is used for all types of carving, furniture, inlays, brakes, instruments. The charcoal is used for drawing and charcoal. The flowers give a good honey and are used medicinally as a tea to treat stomach upsets, cramp and as a mouthwash and gargle. The bark fibre is used to make mats (Muscovite Mats).

***T. japonica*** Simk. (Japanese Linden). China, Japan. The light, soft, cream-coloured wood is used locally for matches, boxes, etc. The bark fibres are used for cordage.

***T. parvifolia*** Ehrh.=T. cordata.

***T. platyphyllos*** Scop. See T. cordata.

***T. tuan*** Szysyl. (Tuan Linden). W. to Central China. The bark fibres are used locally to make sandals.

TILLIACORA Colebr. Menispermaceae. 25 spp. Tropical Africa, Indomalaya. Vines.

***T. bequaertii*** De Wild. (Inaola-a-liwye). Congo. The dried flowers are used locally as perfume.

TILLANDSIA L. Bromeliaceae. 500 spp. Warm America. Herbs.

***T. macrocnemis*** Griseb.=T. xiphoides.

***T. odorata*** Gill.=T. xiphoides.

***T. usneoides*** L.=Dendropogon usneoides. (Crin Végétal, Florida Moss, Louisiana Moss, Spanish Moss, Vegetable Hair, Wool Hair). S. U.S.A., W.Indies to Argentine. Epiphyte. The whole plant is used as a packing material and for stuffing pillows etc.

***T. xiphoides*** Ker.=T. macrocnemis=T. odorata E. S.America. A decoction of the flowers is used locally to treat chest complaints.

Timberline Bluegrass – Poa rupicola.

Timbo da Matta – Derris pterocarpus.

Timboüva – Enterolobium timboüva.

Timinche – Azorella yarata.

TIMONIUS DC. Rubiaceae. 150 spp. Malaysia, Australia, Pacific; 2 spp. Seychelles, Mauritius; 1 sp. Ceylon; 1 sp. Andaman Islands. Trees.

***T. rumphii*** DC. Malaysia, Queensland, S.Australia. The fruits are eaten locally.

Timor Chaste Tree – Vitex littoralis.

Timothy, Alpine – Phleum alpinum.

Timothy Grass – Phleum pratense.

Timpiute – Gilia aggregata.

Timpoot – Kurrimia paniculata.

Timpoot poolau – Gymnopetalum quinquelobacum.

Tripe de Roch – Gyrophora muhlenbergii.

Tincture of Cudbear – Ochrolechia tartarea.

Tingting – Combretium tetralophum.

Tinker's Weed – Triosteum perfoliatum.

Tinnus – Ougeinia dalbergioides.

TINOMISCIUM Miers. Menispermaceae. 8 spp. S.E.Asia, Indomalaya. Vines.

***T. philippense*** Diels. (Bayating). Philippines. The fruits are used locally as a fish poison.

***T. tonkinense*** Gagnep. (Var kan). N.Vietnam, N. Annam. The juice is used locally to treat tooth decay.

TINOSPORA Miers. Menispermaceae. 40 spp. Tropical Africa, S.E.Asia, Indomalaya, Australia. Vines.

***T. bakis*** Miers=Cocculus bakis. Tropical Africa. A decoction of the root is used in Senegal to treat fevers and as a diuretic.

***T. cordifolia*** Miers=T. tuberculata (Galancha Tinospora). Malaysia. A decoction of the stems, roots and leaves is used by the Hindus to reduce fevers and in India and Malaya as a salve for sore eyes.

***T. crispa*** F. Vill.=T. rumphii.

***T. crispa*** Miers=Cocculus crispus=C. tuberculatus=Menispermum crispum. (Bandoul pech, Thuôc sôt rét). India, Malaya, Indochina. In Vietnam a decoction of the stem is used to treat fevers, tetanus and to aid digestion. In Cambodia it is a treatment for rheumatism, while in India and Malaya the stem is used in treating syphilitic ulcers and the powdered leaves as a poultice for wounds and itch.

***malabarica*** (Lam.) Miers. (Sudarsana, ***T.***Gulvel, Gurch). Tropical Asia, especially Bengal and Tonkin. The leaves and ***T.***stems are used locally to treat rheumatism.

***rumphii*** Boerl.=T. crispa F. Vill.=Menispermum rimosum. Malaya, Philippines. The leaves are used locally as a general tonic, for treating diarrhoea and other stomach complaints. It is also used to treat tropical ulcers and drunk in cordials and cocktails.

***T. sinensis*** Merr.=T. tomentosa=Cocculus tomentosus. (Coeur kambaur, Xeng con thon). India, China, Indochina. The leaves and stems are used locally as a poultice to relieve rheumatism.

***T. tomentosa*** Miers.=T. sinensis.

***T. tuberculata*** Beumee.=T. cordifolia.

Tinospora, Galancha – Tinospora cordifolia.

Tintero – Basanacantha armata.

Tintillo – Randia mitis.

TIRMANIA Chatin. Tuberaceae. 1 sp. N.Africa. Fungus.

***T. africana*** Chatin. (Gros Terfaz Blanc). The fruitbodies are eaten locally.

Tirpu – Hopea parviflora.

TITHONIA Desf. ex Juss. Compositae. 10 spp. Mexico, Central America, W.Indies Shrubs

***T. platylepis*** Schultz.=T. scaberrima.

***T. scaberrima*** Benth.=T. platylepis. (Mirasol, Palagaste). Mexico to Nicaragua. A bath of the soaked leaves is used in El Salvador to treat fevers and colds.

Titi – Cliftonia monophylla.

Titi – Gmelina moluccana.

Titi, Black – Cyrilla racemiflora.

Titinga – Clerodendrum buchananii.

Tizon Orange – Citrus reticulata var. papillaris.

Tjangal padi – Vatica faginea.

Tjankring – Erythrina fuseda.

Tjempaka selong – Horsfieldia odorata.

Tjenderai ootan – Grewia fibrocarpa.

Tjetok – Strychnos tietue.

Tjeuri – Garcinia dioica.

Tjintjaco Idjo – Cyclea peltata.

Tjoomantoong merah – Elaeocarpus mastersii.

Tjoopa – Baccaurea dulcis.

Tjoro – Ficus ceriflua.

Tlaxisqui – Amelanchier denticulata.

Tlon – Sterculia hypochra.

Toad Tree – Conopharyngia elegans.

Toadflax, Bastard – Comandra umbellata.

Toadflax, Blue – Linaria canadensis.

Toadflax, Yellow – Linaria vulgaris.

Tobacco – Nicotiana tabacum.

Tobacco, Aztec – Nicotiana rustica.

Tobacco, Hottentot – Tarchonanthus camphoratus.

Tobacco, Indian – Lobelia inflata.

Tobacco, Tree – Nicotiana glauca.

Tobacco, Tree Wild – Acnistus arborescens.

Tobaquilla – Lippia umbellata.

Tobory – Treculia perrieri.

Tocote – Podachaenium eminens.

TOCOYENA Aubl. Rubiaceae. 20 spp. Mexico, tropical S.America, W.Indies. Shrubs.

***T. longiflora*** K. Schum. Brazil. The root (Radix Tocoyenae), are used to make an Ipecac.

TODDALIA Juss. Rutaceae. 1 sp. Tropical Africa, Madagascar, tropical Asia. Small tree.

***T. aculeata*** Pers.=T. lanceolata.=Paullinia asiatica. (Wild Orange Tree). The hard, white wood (White Iron Wood) is used in E.Africa for general carpentry.

The roots (Lopez Root) yield a yellow dye, used in India, where the rootbark is used as an infusion to treat intestinal worms, stomach upsets and as a tonic. In E.Africa, the fruits are used as a flavouring and as an antiseptic, to relieve stomach upsets, as a diuretic and expectorant.

***T. lanceolata*** Lam.=T. aculeata.

Toddy – Arenga pinnata.

Toddy – Cocos nucifera.

Toddy – Phoenix sylvatica.

Todjong pipit – Eugeissona minor.

Tofiri modjioo – Palaquium obtusifolium.

Togari Wood – Morinda trifolia.

Toho – Gmelina moluccana.

Tokodole – Laportea aestuans.

Tokooloo – Zanthoxylum celebicum.

Tolambo – Colocasia indica.

Toledo Wood – Lecythis longipes.

Tolu Balsam – Myroxylon balsamum.

Tomate de Campo – Physalis neo-mexicana.

Tómatl – Physalis ixocarpa.

Tomato – Lycopersicon lycopersicum.

Tomato, Childrens' – Solanum anomalum.

Tomato, Ground – Physalis neo-mexicana.

Tomato, Husk – Physalis ixocarpa.

Tomato, Seed Oil – Lycopersicon lycopersicum.

Tomato, Strawberry – Physalis peruviana.

Tomato, Strawberry – Physalis pubescens.

Tomato, Tree – Cyphomandra betacea.

Tombak-tombak – Blumea chinensis.

Tomgus – Marsdenia tenacissima.

***Tondelea prinoides*** Willd.=Salacia prinoides.

Tongolo voalovo – Dipcadi cowanii.

Tongue, Cow – Clintonia borealis.

Tonka (Bean) – Coumarouna odorata.

Tonka, Dutch – Coumarouna odorata.

Tonka, English – Coumarouna oppositifolia.

Tonkin Bamboo – Arundinaria amabilis.

Tonkin Cane – Arundinaria amabilis.

Toombawa Wood – Dyxoxylum densiflorum.

TOONA M. Roem. Meliaceae. 15 spp. Tropical Asia, Australia. Trees.

***T. sinensis*** M. Roem.=Cedrela sinensis.

***T. sureni*** Merr.=Cedrela febrifuga Blume non King. (Ingoo, Ki beureum, Kooroo, Laoot). Malaysia, Indonesia. The durable wood is suitable for interior work and is used to make cigar boxes. A decoction of the bark is used locally to treat fevers and dysentery.

Tooroo – Oenocarpus batana.

Toothache Tree – Zanthoxylum americanum.

Toothache Tree – Zanthoxylum clavaherculis.

Toothbrush Tree – Gouania lupuloides.

Toothbrush Tree – Salvadora persica.

Toothed Lavender – Lavendula dentata.

Tootoop antjoor – Macaranga triloba.

Toowi – Radermachera gigantea.

Toparejo – Inga ruiziana.

Topee tampo – Calathea allouia.

Topinambour – Helianthus tuberosus.

Topoza – Cornus excelsa.

Topozán – Buddleja americana.

Topozana – Lippia umbellata.

Toquilla – Carludovica palmata.

Tora – Podachaenium eminens.

Torbisco – Asclepias linaria.

Torchwood – Bikkia mariannensis.

TORDYLIUM L. Umbelliferae. 6 spp. Europe, N.Africa, S.W.Asia. Herbs.

***T. apulum*** L.=Concylocarpus apulus. Mediterranean. The young plants are eaten as a vegetable in Greece.

Torementi – Indiofera trita.

Torfez – Terfezia boudieri.

Tormentil – Potentilla erecta.

***Tormentilla erecta*** L.=Potentilla erecta.

Torhoi – Grewia laevigata.

Tornasolis – Chrozophora tinctoria.

Tornillo – Prosopis odorata.

Torote – Bursera microphylla.

***Torresea cearensis*** Allem.=Amburana cearensis.

***Torresia odorata*** Hitchc.=Hierochloë odorata.

Torrey's Pine – Pinus torreyana.

TORREYA Arn. Taxaceae. 6 spp. E.Asia, California, Florida. Trees.

***T. grandis*** Forst. (Chinese Torreya). China. The roasted seeds are eaten locally.

***T. nucifera*** Sieb. and Zucc. (Japanese Torreya). Japan. The seeds are eaten locally, raw or roasted or in confectionery. They also yield an edible oil.

***T. taxifolia*** Arn.=Tumion taxicolia. (Florida Torreya, Stinking Cedar). Florida. The wood is used for fencing.

Torreya, Chinese – Torreya grandis.

Torreya, Japanese – Torreya nucifera.

Tortora – Jatropha multifida.

TORULA Pers. ex Fr. Dermatiaceae. 25 spp. Cosmopolitan.

***T. cremoris*** Hammer and Cordes. Used in the production of lactose.

***T. pulcherrima*** Hammer and Cordes. Used to convert waste and low cost carbohydrate into animal fodder.

TORULOPSIS Berl. Pseudosaccharomycetaceae. 20 spp. Widespread.

***T. lipofera*** Dooren de Jong. Synthesizes 18–43 per cent of its dry weight in lipids and so is potentially useful as a means of producing fats.

***T. utilis*** (Hanneb.) Lodder. Useful in producing animal food of high protein and vitamin B-complex content from waste materials.

Tos lelake – Intsia plurijuga.

Tossa – Corchorus olitorius.

Totai Palm – Acrocomia totai.

Totara – Podocarpus totara.

Torara – Scirpus totara.

Totombe – Polygala paniculata.

Totozapotl – Sideroxylon capiri.

Totum – Fillaeopsis discophora.

TOUCHARDIA Gaudich. Urticaceae. 1 sp. Hawaii. Shrub.

***T. latifolia*** Gaudich. (Olana) The stem yields a strong, water-resistant fibre, used locally for fishing nets

Touch-me-not – Impatiens noli-tangere.

Touchwood – Amyris balsamifera.

Touloucouna (Oil) – Carapa guineensis.

Touloucouna Oil – Carapa procera.

TOURNEFORTIA L. Boraginaceae. 150 spp. Tropics and subtropics. Shrubs.

***T. argentea*** L. f. = Messerschmidia argentia. (Velvetleaf). Tropical Asia. The leaves are used as a tobacco substitute in the Seychelles. The plant is cultivated.

Tournesol – Chrozophora tincoria.

Towoonba Grass – Phalaris tuberosa.

***Toxicodendron*** auctt. = Hyaenanche.

TOXOCARPUS Wight. and Arn. Asclepiadaceae. 40 spp. Tropical Africa, Madagascar, Mascarene Islands, S.E.Asia, Indomalaya. Trees.

***T. decaryi*** Choux. (Voansifitra). W.Madagascar. The hairs from the fruits are used locally for stuffing pillows etc.

***T. tomentosus*** Jum. and Berr. See T. decaryi.

***Toxophoenix aculeatissima*** Schott. = Astrocaryum ayri.

***Toxylon pomiferum*** Raf. = Maclura aurantiaca.

Tozokamit – Farsetia ramosissima.

Trà ring – Acaypha evrardii.

***Trachalospermum stans*** A. Gray. = Secondatia stans.

TRACHYCARPUS H. Wendl. Palmae. 8 spp. Himalayas, E.Asia. Palms.

***T. excelsus*** Wendl. N.E.India, China, Japan. The bark fibre is used by the Chinese to make cordage and brushes. They use the leaves to make hats and rough coats.

TRACHYLOBIUM Hayne. Leguminosae. 1 sp. Tropical E.Africa, Madagascar, Mauritius. Tree.

***T. verrucosum*** Oliv. A copal (Zanzibar Copal, Gum Copal of Madagascar) is derived from the trunk and from the soil around the trees.

***Trachyspermum copticum*** Link. = Carum copticum.

TRADESCANTIA L. Commelinaceae. 60 spp. N.America to tropical S.America. Herbs.

***T. virginica*** L. (Spiderwort). S. U.S.A. introduced N. The young shoots and leaves are occasionally used in salads and as a vegetable.

Tradet – Vitis pentagona.

Tragacanth, African – Sterculia tragacantha.

Tragacanth Gum – Astragalus echidnaeformis.

Tragacanth Gum – Astragalus elymaiticus.

Tragacanth Gum – Astragalus globiflorus.

Tragacanth Gum – Astragalus gossypinus.

Tragacanth Gum – Astragalus gummifera.

Tragacanth Gum – Astragalus microcephalus.

Tragacanth Gum – Astragalus myriacanthus.

Tragacanth Gum – Astragalus prolixus.

Tragacanth, Indian – Astragalus heratensis.

Tragacanth, Morea – Astragalus cylleneus.

Tragacanth, Persian – Astragalus pycnocladus.

TRAGOPOGON L. Compositae. 50 spp. Temperate Eurasia; 1 sp. S.Africa. Herbs.

***T. porrifolius*** L. (Salsify, Oyster Plant, Vegetable Oyster). Mediterranean. Cultivated in Old and New World for the roots which are eaten as a vegetable. The root latex is used as chewing gum by the Indians of British Columbia.

Trailing Arbutus – Epigaea repens.

Trailing Wild Bean – Phaseolus helvola.

Trameng – Garcinia delpyana.

Transvaal Slangkop – Urginea burkei.

TRAPA L. Trapaceae. 30 spp. Central and S.E.Europe, temperate and tropical Asia and Africa. Aquatic herbs. The seeds of the following spp. are eaten locally.

***T. bicornis*** L. China. Cultivated in China, Japan and Korea. The seeds are eaten as a vegetable or preserved in honey or syrup.

***T. bispinosa*** Roxb. (Singhara Nut). Tropical Asia. Cultivated in India where the seeds are also used to make flour.

***T. incisa*** Sieb. and Zucc. Japan.

***T. natans*** L. (Water Caltrop, Water Chestnut). Mediterranean, Asia, introduced to N.America and Australia. Eaten in Central Europe.

Travancore Starch – Cucuma angustifolia.

Traveller's Tree, Madagascar – Ravenala madagascariensis.

Traver's Daisybush – Olearia traversii.

Trecul Yucca – Yucca treculeana.

TRECULIA Decne. ex Tréc. Moraceae. 12 spp. Tropical Africa, Madagascar. Trees. The seeds of the following spp. are eaten locally.

***T. africana*** Decne. (African Breadfruit, Okwa). The seeds are eaten raw or as a flour.

***T. madagascariensis*** Br. (Varonhyfotsymalailey). E.Madagascar.

***T. perrieri*** Jum. (Katoka, Tobory). W.Madagascar.

Tree Alfalfa – Cytisus pallilans.

Tree Cotton – Gossypium arboreum.

Tree Fern – Cyathea spp.

Tree of Heaven – Ailanthus glandulosa.

Tree Lucerne – Cytisus pallilans.

Tree Mallow – Lavatera plebeia.

Tree Peony – Paeonia moutan.

Tree Tobacco – Nicotiana glauca.

Tree Tomato – Cyphomandra betacea.

Tree, Wild Tobacco – Acnistus arborescens.

Trefoil, Bird's Foot – Lotus corniculatus.

Trefoil, Sweet – Trigonella coerula.

TREMA Lour. Ulmaceae. 30 spp. Tropics and subtropics. Trees.

***T. guineensis*** Ficalho. W. and E.Africa, Madagascar. The soft, light wood is used for general building purposes. In Nigeria a decoction of the roasted wood is used to treat dysentery.

***T. orientale*** Blume. Malaysia. The bark is used locally for cordage. It yields brown dye used for treating fishing nets.

Trembling Aspen – Populus tremuloides.

TREMELLA Dill. ex Fr. Tremellaceae. 40 spp. Cosmopolitan.

***T. fuciformis*** Berk. Tropics and subtropics. The fruitbodies are eaten by the Chinese and used by them in treating lung diseases, tuberculosis and stomach ulcers.

Trementina de Ocoto – Pinus teocote.

***Trevsia moluccana*** Miq. = Boerlagiodendron palmatum.

TREVOA Miers. ex Hook. Rhamnaceae. 6 spp. Andes. Shrubs.

***T. trinervia*** Miers = Colletia treba. (Chilean Trevoa). Chile. The bark is used locally as a salve for burns.

Trevoa, Chilean – Trevoa trinervia.

TREWIA L. Euphorbiaceae. 2 spp. W.Himalayas to Ceylon and S.E.Asia. Trees.

***T. nudiflora*** L. (Bhillaura, Gumhar). Himalayas, E.Pakistan, N.India. The light grey, soft wood is used locally for carving, making drums, boxes and tea-chests and agricultural implements.

TRIANTHEMA L. Aizoaceae. 20 spp. Tropical and subtropical Africa, Asia, Australia; 1 sp. tropical America. Herbs.

***T. monogyna*** L. = T. portulacastrum.

***T. pentandra*** L. Tropical Africa, Asia. The leaves are eaten as a famine food in India. The ashes are used as a source of salt in N.W.Africa and the plant is used as an astringent in treating abdominal pains in the Sudan.

***T. portulacastrum*** = T. monogyna = Portulaca tosten. Throughout Old World tropics. The leaves are eaten as a vegetable in India and Ghana. A decoction of the roots is used in the Philippines to control menstruation and as an abortive. When powdered with ginger the root is used as a purgative.

***T. salsoloides*** Fenzl. Tropical Africa. In the Sudan the ashes, mixed with Simson Oil are used to make soap.

TRIBULUS L. Zygophyllaceae. 20 spp. Tropics and subtropics. Herbs.

***T. terrestris*** L. Old and New World tropics. The fruits are used as a diuretic in the Sudan. The seeds are a potential source of oil.

TRICALYSIA A. Rich. Rubiaceae. 10 spp. Tropical Africa, Madagascar; few in Indomalaya. Trees.

***T. coffeoides*** Good. Congo. The roasted seeds are used locally as a coffee substitute.

TRICHILIA P. Br. Meliaceae. 300 spp. Mexico to tropical S.America, W.Indies, tropical Africa. Trees.

***T. capitata*** Klotzsch. Tropical Africa. The roots are used locally to treat snakebites.

***T. cedrata*** Chev. = Guarea cedrata.

***T. emetica*** Vahl. (Cape Mahogany, Mafura Bitterwood, Manubi Mahogany, Natal

Mahogany). Tropical Africa. The wood is used locally to make furniture, canoes and yokes. The poisonous fat from the seeds (Mafura Oil, Mafura Tallow) is used locally as an ointment for wounds etc. and commercially to make soap and candles.

***T. glabra*** L. Tropical America. The wood is used to make tool-handles.

***T. glandulosa*** Smith=Synoum glandulosum.

***T. guara*** L.=Guarea guara.

***T. havanensis*** Jacq. (Cucharillo, Limoncillo, Siguaraya). Mexico, Central America, W.Indies. The wood is used for carving small articles in Mexico.

***T. hirta*** L. (Napa huite). Mexico, Jamaica. The oil from the seeds is used locally as a hairdressing.

***T. spondioides*** Jacq. (White Bitterwood). W.Indies. The wood is used for tool handles and carving small articles.

TRICHOCEREUS (Berger) Riccob. Cactaceae. 40 spp. subtropical and temperate S.America. Cacti.

***T. chiloensis*** Colla=Cereus quisco. (Gardon de Candelabro). S.America. The fruits are eaten raw or as a brandy or syrup.

***T. pasacana*** (Weber) Britt. and Rose. Argentina, Bolivia. The trunks are used in Bolivia for making small huts and fencing for stock.

TRICHOCLADUS Pers. Hamamelidaceae. 6–8 spp. Tropical and S.Africa. Small trees.

***T. ellipticus*** Eckl. and Zeyh. ex Sond. Central and E.Africa, S.Africa. The wood is used locally for house interiors.

TRICHOLAENA Schrad. Graminae. 8 spp. Canaries, Mediterranean, Africa, Madagascar. Grasses.

***T. rosea*** Nees. (Natal Grass). S.Africa. Cultivated in warm countries of Old and New World as fodder. It makes a good hay.

***T. sphacelata*** Benth. Tropical Africa. A good fodder grass.

TRICHOLEPIS DC. Compositae. 15 spp. Central Asia, Himalayas, Burma. Herbs.

***T. glaberrimum*** DC. (Brahmadandi, Mitabid). Tropical Asia, especially India. The Hindus use the plant to treat skin diseases and as a nerve tonic and aphrodisiac.

TRICHOLOMA (Fr.) Quél. Agaricaceae. 140 spp. Widespread. The fruitbodies of the following spp. are eaten locally.

***T. amethystinus*** (Berk. and Br.) van Overbeen. E.Indies.

***T. caffrorum*** Kalckbr. and McOwan. S.Africa.

***T. cnista*** Fr. sensu Bres.=Melanoleuca evenosa. Alpine and subalpine Europe and Asia.

***T. colubretta*** (Fr.) Quél. Europe and Japan.

***T. equestre*** (L.) Quél. Europe and Japan.

***T. flavobrunneum*** (Fr.) Quél. Europe and Japan.

***T. gambosum*** Gil.=Agaricus graveolens=A. cerealis. Also produces an antibiotic of potential medical value.

***T. georgii*** (Fr.) Quél. Europe.

***T. mongolica*** Imai. (Pai ku mo). E.Asia, Mongolia and China.

***T. nudum*** Bull. Europe.

***T. personatum*** (Fr.) Quél. (Blewitt). Europe.

***T. rutinlans*** (Schäff.) Quél. Temperate. Various T. spp. are eaten in the Philippines under the name of Kabuteng higunte.

TRICHOSANTHES L. Cucurbitaceae. 15 spp. Indomalaya, Australia. Vines.

***T. anguina*** L. (Edible Snake Gourd). Tropical Asia. The fruits are eaten locally as a vegetable.

***T. cucmerina*** L. Tropical Asia, Australia. The fruits are eaten locally as a vegetable.

***T. cucmeroides*** Max. (Japanese Snake Gourd). Japan, China. The fruits are used locally as a soap substitute and the roots as a source of starch.

***T. japonica*** Regel. Japan. The roots are a local source of starch (Tenkwa-fun) and the fruits are eaten salted or in soy sauce.

***T. kadam*** Miq.=Hodgsonia macrocarpa.

***T. macrocarpa*** Blume=Hodsonia macrocarpa.

***T. oxigera*** Blume. (Are ujtiwook). Indonesia. The fruits are eaten locally as a vegetable.

***T. palmata*** Benth.=T. subvelutina.

***T. quinquangulata*** Gray. (Patolang-Gabat). Philippines. The ground seed, cooked in oil is used locally as an ointment for itch, or when mixed with wine relieves stomach upsets.

***T. subvelutina*** F. v. Muell.=T. palmata. Australia. The aborigines eat the roasted roots.

***T. tricuspidata*** Lour.=Anguina tricuspidata. The leaves are used locally as a vegetable and the sap is used as a salve for skin eruptions in Cambodia.

TRICHOSCYPHA Hook. f. Anacardiaceae. 50 spp. Tropical Africa. Trees.

***T. acuminata*** Engl.=T. braunii=T. congoensis. (Bondole, Butika). Congo, Angola. The fruits are eaten locally.

***T. braunii*** Engl.=T. acuminata.

***T. congoensis*** Engl.=T. acuminata.

***T. ealaensis*** Van der Veken. Congo. The hard, flexible wood is used locally to make bows and arrows.

TRICHOSPERMUM Blume. Tiliaceae. 20 spp. Malaysia, Pacific, Nicobar Islands. Trees.

***T. kurzii*** King. Malaysia, Nicobar Islands. The bark fibres are used locally for cordage.

TRICHOSTIGMA A. Rich. Phytolaccaceae. 4 spp. Tropical America. Shrubs.

***T. octandrum*** (L.) H. Walt.=Rivina octandra=Villamilla octandra. (Bojuco de paloma, Guacamayo, Guanique). Tropical America. The leaves are used locally in Colombia as a poultice for wounds. The stems are used in Jamaica for barrel hoops.

TRICLISIA Benth. Menispermaceae. 25 spp. Tropical Africa, Madagascar. Vine

***T. gelletii*** (De Wild.) Staner. (Bahotot, Efiri). Congo. The leaves are used locally as a poultice for boils, as an infusion against diarrhoea and as a spear poison. It is possibly an excellent antidote for malaria.

Tridarent – Cordia gharaf.

***Tridesmis formosa*** Korth.=Cratoxylum formosum.

TRIFOLIUM L. Leguminosae. 300 spp. Temperate and subtropical, not S.E.Asia or Australia. Herbs.

***T. alexandrinum*** L. (Egyptian Clover). Mediterranean An important pasture and fodder plant in Egypt and Mediterranean countries. It is also cultivated in India, Burma.

***T. amabile*** H.B.K. (Aztec Clover, Chicmu). Mexico and S.America. In Peru the leaves are mixed with maize to make a food called Chucan.

***T. ambignum*** Bieb. (Kara Clover). Caucasus. A good source of honey.

***T. baccarinii*** Chiov. Tropical Africa. A potentially good pasture spp.

***T. campestre*** Schreb.=T. pubescens. (Large Hop Clover). Europe, W.Asia, N.Africa, introduced to N.America. Recommended as a winter annual forage plant in moist S. U.S.A.

***T. ciliatum*** Nutt. W. U.S.A. The leaves and stems are eaten raw or as a vegetable by the local Indians.

***T. fibriatum*** Lindl. (Beach Clover). W. N.America. The roots are eaten as a vegetable by the local Indians.

***T. fragiferum*** L.=T. neglectum. (Strawberry Clover). Europe, W.Asia, N.Africa, Canary Islands. Grown as fodder for livestock, especially in wet soils. It is salt tolerant. Used in W. U.S.A.

***T. glomeratum*** L. (Cluster Clover). Mediterranean. Grown as a winter annual in S. U.S.A., acid tolerant, but not salt tolerant.

***T. gracilentum*** Torr. and Gray. See T. ciliatum.

***T. hirtum*** All.=T. hispidum. (Rose Clover). Mediterranean. Recommended as a winter annual on hill sites in California.

***T. hispidum*** Desf.=T. hirtum.

***T. hybridum*** L. (Alsike Clover). Europe, Caucasus, Asia Minor, N.Africa. Widely grown as an annual or biennial forage and hay crop.

***T. hybridum savinon*** L.=T. nigrescens.

***T. incarnatum*** L. (Crimson Clover). Central Europe to Balkans and N.Africa. Widely grown in Old and New World for fodder, hay and a green manure.

***T. lappaceum*** L.=T. nervosum (Lappa Clover). Mediterranean, Asia Minor. Grown as a winter annual for forage.

***T. luinaster*** L. Poland, Russia to Japan. Grown as a fodder crop and for hay.

***T. microcephalum*** Pursh. See T. campestre.

***T. neglectum*** Fisch. and Mey.=T. fragiferum.

***T. nervosum*** Presl.=T. lappaceum.

***T. nigrescens*** Viv.=T. hybridum savinon. (Ball Clover). W.Mediterranean. Naturalized widely, it is grown as a winter annual, especially in S. U.S.A. Used for fodder.

***T. pannonicum*** L. (Hungarian Clover). S.Europe to Balkans, Russia and Asia Minor. More drought resistant than Red Clover, it is often used for food for stock under these conditions.

***T. parviflorum*** Ehrh. (Small-flowered Clover, Teasel Clover). S. and S.E.Europe. Cultivated locally as fodder and hay.

***T. pratense*** L. (Red Clover). Temperate Europe. Cultivated widely as a pasture and hay, usually in perennial pasture with Rye Grass or Timothy. The flowers yield a yellow dye and a decoction of the dried flowers is sometimes used to improve the appetite and as a sedative.

***T. pubescens*** L.=T. campestre.

***T. repens*** L. (White Clover). Europe, temperate Asia, N.Africa. Widely cultivated in Old and New World for perennial pastures and hay. The flowers are an excellent source of honey and are used medicinally (Flores Trifolii albi) to treat swollen glands, gout and rheumatism. The improved white clover (Ladina Clover) from Italy is cultivated in U.S.A.

***T. resupinatum*** L.=T. suaveolens. (Persian Clover). Mediterranean to Afghanistan. Grown locally for fodder.

***T. reuppellianum*** Fres. Tropical Africa. Cultivated in Ethiopia for fodder.

***T. striatum*** L. (Striate Clover). Europe, Mediterranean, N.W.Africa, Madeira. Grown for winter forage, particularly in heavy limy soils of S. U.S.A.

***T. subterraneum*** L. (Sub Clover). Mediterranean, W.Europe, Canary Islands, W.Asia. Grown for grazing, especially in N.W. N.America.

***T. suaveolens*** Willd.=T. resupinatum.

***T. tridentatum*** Lindl. See T. ciliatum.

***T. unifolium*** Forsk.=Psoralea corylifolia.

***T. usambarense*** Taub. Tropical Africa. Used locally as fodder for livestock. A potentially valuable fodder spp. in the humid tropics.

TRIGLOCHIN L. Juncaginaceae. 15 spp. Cosmopolitan, especially Australia and temperate S.America. Herbs.

***T. maritima*** L. (Shore Podgrass). N.Temperate. The young leaves are sometimes eaten as a vegetable and as an emergency food.

***T. procerum*** R. Br. Australia. The rhizomes are eaten locally as a vegetable.

TRIGONELLA L. Leguminosae. 100 spp. Mediterranean, Europe, Asia, S.Africa, Australia. Herbs.

***T. coerulea*** (L.) Ser. (Sweet Trefoil). Mediterranean. The leaves are used, particularly in Switzerland for flavouring cheeses and bread. They are also used to flavour soups etc. and as a tea. The plant is cultivated.

***T. foenum-graecum*** L. (Fenugreek, Fenugrec). Mediterranean. The plant is cultivated as a fodder crop in S.Europe, N.Africa and India. The seeds are aromatic and contain about 23 per cent protein, 9 per cent oil and 10 per cent carbohydrate. They also contain trigonelline and choline, but their flavour is mainly due to the presence of coumarin. They are used in curries and sometimes eaten as a pulse crop and cattle food, a yellow dye is also extracted from them in India. In N.Africa the seeds are used for flavouring food and medicinally to relieve stomach upsets. They are used in Europe as a demulcent.

TRIGONIASTRUM Miq. Trigoniaceae. 1 sp. W.Malaysia. Tree.

***T. hypoleucum*** Miq. The hard, light yellow wood is used locally for making tables.

***Trigonochlamys griffithii*** Hook.=Santiria griffithii.

TRIGONOPLEURA Hook. f. Euphorbiaceae. 1 sp. W.Malaysia, except Java. Tree.

***T. malayana*** Hook. f. (Gambir ooran, Kajoo gambir). The leaves are eaten as a vegetable locally and are used for tanning and yield a black dye used for dyeing and printing.

TRILISA Cass. Compositae. 3 spp. Atlantic United States of America. Herbs.

***T. odoratissima*** (Walt.) Cass. (Deer's Tongue, Vanilla Trilisa). Throughout range. The leaves contain coumarin and are used to flavour tobacco.

Trilisa, Vanilla – Trilisa odoratissima.

TRILLIUM L. Trilliaceae. 30 spp. W.Himalayas to Japan, N.America. Herbs.

***T. erectum*** L. (Bethroot, Purple Trillium). E. N.America. The dried rhizome is used medicinally as a uterine stimulant.

***T. grandiflorum*** (Michx.) Salisb. (Snow Trillium, Wake Robin). E. N.America. The leaves are eaten as a vegetable.

Trillium, Purple – Trillium erectum.

Trillium, Snow – Trillium grandiflorum.

***Trimeza*** Salis.=Trimezia.

TRIMEZIA Salisb. ex Herb. Iridaceae. 5 spp. Mexico to tropical S.America, W.Indies. Herbs.

***T. lurida*** Salisb.=Sisrinchium galaxoides Tropical America A decoction of the plant is used in Colombia as a laxative

Trinidad Devilpepper – Rauvolfia canescens

Trinidad Roupala – Roupala montata

TRIOMMA Hook. f. Burseraceae. 1 sp. W.Malaysia. Tree.

***T. macrocarpa*** Backer=Arytera macrocarpa. (Damar Siap Mata, Risoong batoo). The soft, red wood is used locally for boards. A scented resin (Dammar asam) extracted from the wood is used locally for torches.

TRIOSTEUM L. Caprifoliaceae. 6 spp. Himalayas, E.Asia; 4 spp. N.America. Herbs. The berries of the following spp. were used by the early settlers in America as a coffee substitute.

***T. angustifolium*** L. E. N.America.

***T. aurantiacum*** Bickn. E. N.America.

***T. perfoliatum*** L. (Wild Coffee, Tinker's Weed).

Tripa de zopilote – Vitis sicyiodes.

TRIPETALUM K. Schum. Guttiferae. 1 sp. New Guinea. Tree.

***T. cymosum*** K. Schum. The juice from the fruits is used locally to colour the teeth black. The plant is cultivated.

TRIPHASIA Lour. Rutaceae. 2 spp. Tropical Asia, Philippines. Shrubs.

***T. aurantiola*** Lour.=T. trifoliata. Throughout range. The fruits are eaten locally, raw or cooked. In the Philippines the leaves are used for aromatic baths. The plant is cultivated locally.

***T. trifoliata*** DC.=T. aurantiola.

TRIPLARIS Loefl. ex L. Polygonaceae. 25 spp. Tropical S.America. Trees.

***T. surinamensis*** Cham. (Surinam Ant Tree). Guyana. The light, tough, elastic wood is used locally for boxes and interiors of houses.

TRIPSACUM L. Graminae. 7 spp. Warm America. Grasses. The following spp. are used as fodder grasses in warm countries.

***T. laxum*** Nash. (Guatemala Grass). Central America.

***T. dactyloides*** L. Tropical America.

TRIPTERYGIUM Hook. f. Celastraceae. 4–5 spp. E.Asia. Shrubs.

***T. wilfordii*** Hook. f. China, Japan. The powdered root is used as an insecticide in parts of China.

TRISTACHYA Nees. Graminae. 25 spp. Tropical America, Africa, Madagascar. Grasses.

***T. leiostachya*** Nees. Tropical America. Used as a fodder grass in S.America.

TRISTANIA R. Br. Myrtaceae. 50 spp. Malaysia, Queensland, New Caledonia, Fiji. Trees.

***T. conferta*** R. Br.=T. subverticillata=Lophostemon argorescens. (Brush Box, Red Box, White Box). Queensland. The strong, durable wood is used for marine works, bridges and ship-building.

***T. guillaini*** Vieill. New Caledonia. The hard, red wood is used for turning.

***T. obovata*** Benn. (Pelawan). Sumatra. The heavy, hard, dark red wood is used locally for charcoal and fuel.

***T. suaveolens*** Sm.=Melaleuca suaveolens. (Bastard Peppermint, Broad-leaved Water Gum). Queensland. The tough, strong durable, elastic wood is used for posts, wheels, cogs and tool handles.

***T. subverticillata*** Wendl.=T. conferta.

***T. sumatrana*** Miq.=T. whitiana.

***T. whitiana*** Griff.=T. sumatrana. Sumatra. The hard, heavy wood is used locally for house construction.

***Tristaniopsis*** Bong. and Gris.=Tristania.

TRITICUM L. Graminae. About 20 spp. Europe, Mediterranean, W.Asia. Grasses.

The genus is divided into three groups, depending on the chromosome number (n=7). (1) Einkorn -2n- small, bearded types which break so that the grain stays in the hull. The group includes T. boeoticum and T. monococcum. (2) Emmer -4n- most of which are bearded, including T. dicoccum, T. timopheevi, T. durum, T. turgidum, T. polonicum, T. turanicum, T. carthlicum. (3) Common Wheat. This is not known in the wild, the probable centre of origin is Afghanistan. It has probably arisen as a cross T. dicoccum × Aegilops squarrosa, while T. dicoccum is probably itself a cross between an unknown T. sp. and Aegilops speltoides.

***T. aestivum*** L. emend Foir. and Paol.=T. sativum=T. vulgare. (Common Wheat). Probably originated in Afghanistan (see above). The main cereal crop of the World. It is grown in temperate and subtropical countries throughout the world, the high gluten content of the grain making the flour excellent for bread making. The world production is about 2·7 million tons a year with the best yields being over 2 tons an acre. The U.S.S.R. produces about 23·7 per cent of the world yield, U.S.A. 15·1 per cent, Canada, France 4·8 per cent, India 4·4 per cent, Italy 3·8 per cent, Turkey 3·0 per cent, Argentina 2·5 per cent, Australia 2·5 per cent. There are numerous varieties divided into (1) Hard and Soft, (2) Spring and Winter, (3) Bearded and not Bearded, (4) Red and White. Wheat flour is used for making all types of bread, biscuits and confectionery. The grain is fermented to make alcoholic drinks and industrial alcohol and the starch extracted from the grain is used for stiffening cloth. The straw is used for feeding and bedding livestock and making hats, baskets, chair seats, beehives, etc. The crop is an annual and is highly mechanized.

***T. carthlicum*** Nevski.=T. persicum (Persian Wheat). Asia Minor The plant is grown locally as a grain crop.

***T. compactum*** Host. (Club Wheat). Europe. Cultivated in the Old and New World for a low-gluten flour used for making bread, biscuits and breakfast foods.

***T. dicoccum*** Schrank. (Emmer). Europe, temperate Asia. Cultivated for flour and feed for livestock. There are several varieties.

***T. durum*** Desf. (Durum Wheat, Macaroni Wheat). S.Russia, Mediterranean, Asia. A hard, semi-translucent wheat, grown particularly in Italy for making macaroni, spaghetti etc. Semolina and Farina are flours derived from the hulled wheat.

***T. geogicum*** Dekaprelew. (Kolchic Emmer). S.Russia. It has been grown as a flour wheat in Georgia since ancient times.

***T. macha*** Dekr. and Men. S.Russia. Sometimes cultivated locally.

***T. monococcum*** L. (Einkorn). Europe. Sometimes grown in Italy, Germany and Switzerland.

***T. orientale*** Percival. (Kharassan Wheat). Mediterranean, Near East. Sometimes grown locally as a grain crop.

***T. persicum*** Vav.=T. carthlicum.

***T. polonicum*** L. (Polish Wheat, Astrakan Wheat, Giant Rye, Jerusalem Rye). Central Europe. A hard-grained wheat, suitable for macaroni, but not bread.

***T. sativum*** Lam.=T. aestivum.

***T. spelta*** L. (Spelt). Europe. Sometimes cultivated locally as a bread wheat. There are several varieties.

***T. sphaerococcus*** Percival. (Indian Dwarf Wheat). Central India. Grown locally as a grain crop for flour.

***T. timpheevi*** Zhukowski. (Sanduri, Tschalta sanduri). S.Georgia, U.S.S.R. Grown locally as a grain crop.

***T. turgidum*** L. (Poulard Wheat). Asia Minor A branch-headed wheat. Grown in Europe and Mexico. The flour is suitable for macaroni etc. It is also fed to livestock.

***T. vulgare*** Vill.=T. aestivum

***Tritonia aurea*** Papp.=Crocosmia aurea.

TRIUMFETTA L. Tiliaceae. 150 spp. Tropics. Shrubs. The bark fibres of the following spp. are tough, soft and glossy. They are used locally for cordage and making fabrics.

***T. lappula*** L. (Cadillo, Mozote de caballo, Pega-pega). Mexico, Central America, W.Indies, W.Africa. Also, the mucilage from the plant is used to treat colds and clarify syrup in Costa Rica.

***T. rhomboidea*** Jacq. Tropics.

***T. semitriloba*** Jacq. Mexico, W.Indies, Central America, S.America.

***Trixis neriifolia*** Bonpl.=Espeletia nerifolius.

TROCHODENDRON Sieb. and Zucc. Trochodendraceae. 1 sp. Japan, Ryukyu Islands, Formosa. Tree.

***T. araloides*** Sieb. and Zucc. The ground bark is used locally as birdlime.

TROCHOMERIA Hook. f. Cucurbitaceae. 10 spp. Africa. Herbs.

***T. dalzielii*** E. G. Baker. Nigeria. The roots are eaten locally as a vegetable.

Troitskie Truffle – Choiromyces venosus.

Trom – Sterculia hypochra.

Tromeng – Garcinia oliveri.

Tronera del Monte – Buchnera lithospermifolia.

Tronodor – Abutilon trisulcatum.

TROPAEOLUM L. Tropaeolaceae. 90 spp. Mexico to temperate S.America. Climbing herbs.

***T. brasiliensis*** Casar. Brazil. The leaves are eaten locally in salads.

***T. edule*** Paxt. Chile, Peru. The tubers are eaten locally as a vegetable.

***T. majus*** L. (Nasturtium). S.America. Cultivated widely as an ornamental in Old and New World. The flower buds and seeds are used as a substitute for capers and to flavour vinegar. The plant has antibiotic properties towards aerobic spore-forming bacteria.

***T. patagonicum*** Speg. N.W. and W. S.America. The tubers are eaten as a vegetable. The plant is cultivated locally.

***T. tuberosum*** Ruiz. and Pav. See T. patagonicum.

TROPHIS P. Br. Moraceae. sensu lat. 11 spp. Madagascar, Malaysia, tropical America, W.Indies; sensu str. 4 spp. Mexico, Central America. Trees.

***T. americana*** L.=T. racemosa.

***T. racemosa*** (L.) Urban.=T. americana=Bucephalon racemosum. (Confitura, Leche Maria, Ramón, Ramoncillo). Mexico, Central America, W.Indies, Columbia. In Mexico the branches are used widely as fodder for stock. The bark contains tannin and is used as an astringent.
Tropical Almond – Terminalia catappa.
Tropical Duckweed – Pistia stratiotes.
True Bay – Laurus nobilis.
True Myall – Acacia pendula.
Truffe Blanche – Tuber aestivum.
Truffe Blonde – Tuber magnatum.
Truffe d'Été – Tuber aestivum.
Truffe de France – Tuber melanosporium.
Truffe de la Bourgogne – Tuber uncinatum.
Truffe Violette – Tuber brumale.
Truffe Vraie – Tuber melanosporum.
Truffle – Tuber aestivum.
Truffle, Caucasian – Terfezia claveryi.
Truffle, Hart's – Elaphomyces cervinus.
Truffle, Troitskie – Choiromyces venosus.
Truffle, White – Choiromyces venosus.
Trufos de Ciervo – Elaphomyces cervinus.
Trumpets – Sarracenia flava.
Trumpet Creeper – Tecoma radicans.
Trumpet, Desert – Eriogonum inflatum.
Trumpet Flower – Tecoma crucigera.
Trumpet Lichen – Cladonia fimbricata.
Truxillo Coca – Erythroxylum nova-granatense.
Tsa – Ziziphus mauritana.
Ts'ao K'ou jen – Alpina globosa.
Tsauri Grass, Oil of – Cymbopogon giganteus.
Tschalta sanduri – Triticum timopheevi.
Tschitscha – Chenopodium quinoa.
Tshihangi – Lychnodiscus cerospermum.
Tsi – Eutrema wasabi.
Tsi – Houttuynia cordata.
Tsiavango – Phylloxylon ensifolius.
Tsiramiramy – Protium chapelieri.
Tsiriky – Pholga polystachya.
Tsirohoroka – Musa perrieri.
Tsu Choy – Porphyra suborbiculata.
Tsubaki Oil – Camellia japonica.
TSUGA Carr. Pinaceae. 15 spp. Himalayas, E.Asia, N.America. Trees.

***T. canadensis*** (L ) Carr. (Eastern Hemlock). E. N.America. The light-brown, light, soft, coarse-grained wood is used as lumber and exteriors of buildings. An oleo-resin (Canada Pitch, Hemlock Pitch, Oil of Hemlock) is extracted from cuts in the trunk, or by boiling the wood and skimming of the oleo-resin from the surface of the water. The bark is used for tanning and the leaves are used to make a tea locally.

***T. chinensis*** Pritz. (Chinese Hemlock). China. The soft, durable wood is used locally for roofing.

***T. heterophylla*** (Raf.) Sarg. (Western Hemlock). W. N.America. The light brown, hard, light wood is used for house construction, mill-work, furniture, crates, laundry appliances, refrigerators, blinds and pulp-wood. The inner bark is eaten by the local Indians.

***T. yunnanensis*** (Franch.) Mast. (Yunnan Hemlock). W.China. The wood is used locally for boards and roofing.

Tu be rou – Cinnamomum tetragonum.
Tu-chung – Eucommia ulmoides.
Tuan Linden – Tilia tuan.
Tuart – Eucalyptus gomphocephala.
Tuba – Corpha elata.
TUBER Mich. ex Fr. Tuberaceae. 50 spp. Widespread. The underground fruitbodies are eaten. They are considered a delicacy and cooked in a wide variety of ways. The fruitbodies are scented-out by trained dogs or pigs.

***T. aestivum*** Vitt.=T. cibarium.=T. albidum. (Truffle, Truffe d'Été, Truffe Blanche). Europe, especially Mediterranean.

***T. albidum*** Fr.=T. aestivum.

***T. album*** With.=Choiromyces venosus.

***T. brumale*** Vitt.=T. cibarium Fr.=Truffe Violette. Central and S.Europe.

***T. cibarium*** Bull.=T. melanosporum.

***T. cibarium*** Fr.=T. brumale.

***T. excavatum*** Vittad.=Vittadinion montagnei. (Truffe Jaune). Central Europe, especially France and Italy, England.

***T. gennadii*** (Chatin.) Pat.=T. lacunosum=Terfezia gennadii. Mediterranean, Canary Islands. Fruitbodies called Quiza or Tartufi Biancho.

***T. griseum*** Fr.=T. magnatum.

***T. lacumosum*** Matt.=T. gennadii.

***T. magnatum*** (Pico) Vitt.=T. griseum. (Truffe Blonde, Truffe grise). France, N.Italy.

***T. melanosporum*** Vitt.=T. cibarium Bull. (Truffe de France, Truffe Franche, Truffe Vraie). France, N.Italy, Spain.

***T. uncinatum*** Chatin. (Truffe de la Bourgogne). Central and S.Europe.

Tuber Nasturtium – Tropaeolum tuberosum.

Tuberose (Flower Oil) – Polianthes tuberosa.

Tuberous Bitter Vetch – Lathyrus maritimus.

Tuberous-rooted Chinese Mustard – Brassica napiformis.

***Tubiflora squamosa*** (Jacq.) Kuntze = Elytraria tridentata.

Tuc doan – Dipsacus asper.

Tuckahoe – Polyporus kuckahoe.

Tuckahoe – Poria cocos.

Tuckahoe, Virginian – Peltandra virginica.

Tucuma (Oil) – Astrocaryum tucuma.

Tucuma (Oil) – Astrocaryum vulgare.

Tufted Bamboo – Oxytenanthera nigrociliata.

Tufted Hairgrass – Deschampsia caespitosa.

Tuhu – Coriaria sarmentosa

Tula – Pterygota alata.

Tulanda – Sphenostylis briartii.

TULBAGHIA L. corr. Giseke. Aliaceae. 26 spp. Tropical and S.Africa. Herbs. The bulbs of the following spp. are used locally to treat intestinal worms.

***T. alliacea*** L. S.Africa.

***T. cepacea*** L. f. S.Africa.

***T. violacea*** Harv. S.Africa.

Tulip, Cilician – Tulipa montana.

Tulip, Edible – Tulipa edulis.

Tulip, Garden – Tulipa gesneriana.

Tulip Tree – Liriodendron tulipifera.

Tulip Tree – Thespia populnea.

Tulip Wood – Harpullia pendula.

Tulip Wood – Physocalymma scaberrimum.

Tulip Wood, Brazil – Dalbergia cearensis.

Tulip Wood, Oil – Physocalymna scaberrimum.

TULIPA L. Liliaceae. 100 spp. Temperate Eurasia, especially steppes of Central Asia. Herbs. The bulbs of the following spp. are eaten.

***T. edulis*** Baker. = Orithia oxypetala. (Edible Tulip). China, Japan. The leaves are eaten as a vegetable in Japan, where the bulbs are used as a source of starch.

***T. gesneriana*** L. (Garden Tulip). The cultivated tulip, much grown as an ornamental. The bulbs have been eaten as an emergency food in the Netherlands.

***T. montana*** Lindl. (Cilician Tulip). Asia Minor, Afghanistan.

***Tulipastrum acuminata*** (L.) Small. = Magnolia acuminata.

Tulsi – Ocimum sanctius.

Tulu – Scirpus totara

Tuma – Poulsenia armata.

***Tumion taxicolia*** (Arn.) Green = Torreya taxifolia.

Tummy Wood – Careya arborea.

Tuna – Opuntia ficus-indica.

Tuna – Opuntia imbricata.

Tuna – Opuntia megacantha.

Tuna Cardona – Opuntia streptacantha.

Tuna de Castilla – Opuntia ficus-indica.

Tunbridge Ware – Chlorosplenium aeruginosum.

Tung Oil – Aleurites fordii.

Tung Oil, Japanese – Aleurites cordata.

Tunic Flower. Little Leaf – Dianthus prolifera.

***Tunica*** Ludw. = Dianthus.

Tupelo, Black – Nyssa sylvatica.

Tupelo Gum – Nyssa aquatica.

Tupelo Gum – Nyssa ogeche.

Tupelo, Sour – Nyssa ogeche.

Tupelo, Water – Nyssa biflora.

Tupelo, Water – Nyssa aquatica.

***Turbina corymbosa*** (L.) Raf. = Rivea corymbosa.

TURBINARIA J. Ag. Fucaceae. Marine Algae. Malaysia, India. The following spp. are eaten raw or pickled. They are also used as fertilizer around coconut trees.

***T. conoides*** J. Ag.

***T. ornata*** Kurz.

Turbuli – Oldenlandia umbellata.

Turka – Apocynum venetum.

Turkcap Lily – Lilium superbum.

Turkey Corn – Dicentra canadensis.

Turkey Foot – Andropogon halii.

Turkey Mullein – Croton setigerus.

Turkey Oak – Quercus cerris.

Turkey Red – Peganum harmala.

Turkey Red Oil – Agonandra brasilensis.

Turkey Red Oil – Ricinus communis.

Turkey Rhubarb – Rheum palmatum.

Turkish Alkanna – Macrotomia cephalotus.

Turkish Hazel – Corylus colurna.

Turkish Terebinth Pistache – Pistacia mutica.

Turmeric – Cucuma longa.

Turmeric, African – Canna speciosa.

Turmeric, Wild – Curcurma aromatica.

TURNERA L. Turneraceae. 60 spp. Tropical and subtropical America. Shrubs or small trees.

***T. angustifolia*** Mill. = T. ulmifolia.

***T. aphrodisiaca*** Ward. = T. diffusa.

***T. diffusa*** Willd. = T. aphrodisiaca = T. microphylla. T. humifusa, T. pringlei.

(Damiana, Hierba de la pastora, Hierba del venado, Pastorata, Xmisibcoc). Throughout range. The dried leaves are used locally as a laxative and are exported to U.S.A. for medicinal use. They are also used in Mexico as a tea substitute, to flavour wines etc. and as an aphrodisiac.

***T. humifusa*** Endl.=T. diffusa.

***T. pringlei*** Rose=T. diffusa.

***T. trioniflora*** Sims.=T. ulmifolia.

***T. ulmifolia*** L.=T. angustifolia=T. trioniflora. (Clavel de oro, Oreja de coyote). Throughout range, naturalized in Old World tropics. A decoction of the leaves is used in Mexico to treat indigestion and chest complaints.

Turnip – Brassica rapa.

Turnip, Indian – Arisaema triphyllum.

Turnip, Indian – Eriogonum longifolium.

Turnip, Prairie – Psoralea esculenta.

Turnip-rooted Chervil – Chaerophyllum bulbosum.

Turnip-rooted Parsley – Petroselinum hortense.

Turnip Swedish – Brassica napo-brassica.

Turnip Wood – Dysoxylon muelleri.

Turpenthin Resin – Operculina turpethum.

Turpentine (Gum) (Oil) (Spirit) – Pinus palustris.

Turpentine, Austrian – Pinus nigra.

Turpentine, Bordeaux – Pinus pinaster.

Turpentine, Canada – Abies balsamea.

Turpentine, Chian – Pistacia terebinthis.

Turpentine, French Oil – Pinus pinaster.

Turpentine, Greek – Pinus halepensis.

Turpentine, Hungarian – Pinus cembra.

Turpentine, Indian – Pinus longifolia.

Turpentine, Japanese – Larix dahurica.

Turpentine, Jura – Picea excelsa.

Turpentine, Strassburg – Abies pectinata.

Turpentine Tree – Bursera gummifera.

Turpentine Tree – Copaifera mopans.

Turpentine Tree – Syncarpia laurifolia.

Turpentine, Trementina de Ocote – Pinus teocote.

Turpentine, Venice – Larix decidua.

Turpeth Root – Operculina turpethum.

Turpethroot, Spanish – Thapsia garganica.

Turraie – Grenvillea striata.

Turury – Manicaria saccifera.

TUSSILAGO L. Compositae. 1 sp. Temperate Eurasia, except China and Japan; N.Africa, introduced to N.America. Herbs.

***T. farfara*** L. (Coltsfoot). The leaves and flowers are used in remedies for coughs, bronchitis etc. They are also smoked in herbal tobaccos for the relief of asthma. The leaves are also drunk as a tea and eaten as a vegetable.

Tussock Grass – Deschampsia caespitosa.

Tsubaki Oil – Camellia japonica.

Twelve month Yam – Dioscorea cayenensis.

Twinberry – Lonicera involucrata.

Twinleaf – Jeffersonia diphylla.

Twinvein Wattle – Acacia binervata.

Twisted-leaf Garlic – Allium obliquum.

Twitch Grass – Agropyron repens.

Two-leaved Vetch – Vicia unijuga.

Two-rowed Barley – Hordeum distichon.

TYLOPHORA R. Br. Asclepiadaceae. 50 spp. Old World tropics and S.Africa. Shrubs.

***T. brevipes*** (Turcz.) Vill. Philippines. The roots are used as a substitute for ipecacuanha. A decoction of the roots is used locally to treat stomach pains and to control menstruation.

***T. perrottetiana*** Decne. Philippines. The leaves are used locally to dress wounds.

***Tylostemon*** Engl.=Beilschmiedia.

TYPHA L. Typhaceae. 10–12 spp. Temperate and tropics. Herbs.

***T. angustifolia*** L. (Narrowleaf Cat Tail). Throughout range. The leaves are used for matting. The roots are eaten as an emergency food and the hairs from the fruits are used to stuff pillows etc.

***T. australis*** Schum. and Thonn. S.Africa. The rhizomes are eaten as an emergency food and the florets are used to stuff pillows etc.

***T. elephantina*** Roxb. (Elephant Grass, Hagla, Pun). Banks of rivers in India. The leaves and stems are used to make mats, baskets etc. and small boats. The pollen (Boor, Booratoo) is eaten locally. The plant is planted as a soil binder along the river banks.

TYPHONIUM Schott. Araceae. 30 spp. S.E.Asia, Indomalaya. Herbs.

***T. angustilobium*** F. v. Muell. Australia. The roasted corms are eaten by the aborigines.

***T. brownii*** Schott. See T. angustilobium.

***T. divaricatum*** (L.) Decne. S.Asia. The corms are used locally to treat diarrhoea.

***T. orixense*** (Roxb.) Schott. India. The rhizomes are used locally as a poultice for boils.

***T. trilobatum*** (L.) Schott. India. The roasted rhizomes are eaten locally.

TYPHONODORUM Schott. Araceae. 1 sp. E.Africa, Madagascar, Mascarene Islands. Herbs.

***T. lindleyanum*** Schott. (Viha). The rhizomes are eaten locally as an emergency food.

Tzouchu – Bambusa arundinacea.

# U

UAPACA Baill. Uapacaceae. 50 spp. Tropical Africa, Madagascar. Trees.

***U. clusoides*** Baker. Madagascar. The fruits are eaten locally.

***U. guineensis*** Muell. Agr. Tropical Africa, especially W.Coast and Uganda. The fruits are eaten locally, and the wood is used for general construction work.

***U. heudelotii*** Baill. See U. guineensis.

***U. kirkiana*** Muell. Arg. (Wild Loquat, Mahobohobo). See U. guineensis. The wood is also used for charcoal and the fruits made into wine.

***U. sansibarica*** Pax. Tropical Africa, Zanzibar. A blue dye from the roots is used to dye cloth in Zanzibar.

***U. staudtii*** Pax. Tropical Africa. The hard, durable wood is resistant to termites. It is used locally for general construction work and furniture.

Ubacury – Platonia grandiflora.

Ubi – Vitis sicyoides.

Ubussi (Oil) – Manicaria sacifera.

Ucahuba – Virola surinamensis.

Ucuiba Butter – Virola surinamensis.

Udal – Sterculia villosa.

Udo – Aralia cordata.

Udslap – Paeonia emodi.

Uganda Aloe – Aloe ferox.

Uhambaka – Sorindeia claessensii.

Uhiuhi – Mezonervron kavaiensis.

Uiltómatt – Physalis ixocarpa.

Ukiuki – Dianella lavarum.

ULEX L. Leguminosae. 20 spp. W.Europe, N.Africa. Shrubs.

***U. europaeus*** L. (Gorse). Europe, introduced to N.America. The young shoots growing after burning are used as pasture for hill sheep. The crushed branches are sometimes fed to cattle, but are poisonous after flowering.

Ulluco – Ullucus tuberosus.

ULLUCUS Caldas. Basellaceae. 1 sp. Andes. Herb.

***U. tuberosus*** Caldas. (Ulluco). The tubers are eaten as a vegetable by the local Indians.

***Ulmaria filipendula*** Kostel. = Filipendula hexapetala.

Ulmo – Eucryphia cordifolia.

ULMUS L. Ulmaceae. 45 spp. N.Temperate to Himalayas, Indochina and Mexico. Trees.

***U. alata*** Michx. (Red Elm, Winged Elm). The reddish wood is hard, heavy, but not strong. It is used for tool handles and wheel hubs. The bark is used for tying cotton bales.

***U. americana*** L. (American White Elm, White Elm). The hard, heavy brownish wood is used for boat-building, barrel making, flooring, tool handles, saddles and wheel hubs.

***U. campestris*** L. = U. glabra. (English Elm, European Elm). Central and S.Europe, Asia Minor to Himalayas and China. The reddish elastic wood is hard and durable. It is used to make furniture, water-pipes, rifle butts, tobacco pipes and ships. The leaves are sometimes used as a substitute for tea (Warsaw Tea). The bark is sometimes used as a diuretic and the ash as a source of potash.

***U. crassifolia*** Nutt (Cedar Elm). S.E. U.S.A., Mexico. The wood is used in Mexico to make furniture and wheel hubs.

***U. fulva*** Michx. (Indian Elm, Slippery Elm, Sweet Elm). E. N.America. The heavy, strong, red-brown wood is used for wheel hubs, railway sleepers, fencing and agricultural implements. The mucilage from the inner bark is used medicinally to treat diarrhoea, dysentery and urinary inflammations and as a poultice for boils etc. Mixed with milk, the bark is used as an invalid food, while the local Indians eat it cooked in buffalo fat.

***U. glabra*** Mill. = U. campestris.

***U. japonica*** Sarg. (Japanese Elm). Japan, China, Korea. The hard, heavy wood is used in Japan for wheel-hubs and turning.

***U. macrocarpa*** Hance. China. The mucilaginous seeds are used locally to treat intestinal worms.

***U. mexicana*** (Lieb.) Planch. = Chaetoptelea mexicana.

***U. montana*** With. (Mountain Elm). See U. campestris.

***U. thomasii*** Sarg. (Rock Elm). E. N.America. The strong, heavy red-brown wood is used for wheel-hubs, furnitures, agricultural implements, railway sleepers and tool handles.

ULVA (L.) J. Ag. Ulvaceae. Marine Green Algae. The following spp. are eaten locally.

***U. fasciata*** Delile. Pacific. Eaten with squid in Hawaii. The dish is called Limu Lipahapaha.

***U. lactuca*** J. Ag. (Lettuce Laver, Sea Lettuce). Cosmopolitan. Eaten in the Far East and fed to pigs.

***U. latissima*** L. See U. lactuca.

***U. nematoidea*** Kuetz. Eaten in New Caledonia.

***U. penniformis*** Mart. Siberia. Used medicinally as U. pertusa.

***U. pertusa*** Kjellm. Pacific. Eaten in China and used to reduce temperature.

Umagaqana – Bowiea volubilis.

Umar – Ficus glomerata.

Umari – Geoffroea superba.

UMBELLULARIA Nutt. Lauraceae. 1 sp. California. Tree.

***U. californica*** Nutt. (Californian Laurel, California Olive). The hard, strong, light brown wood is used for furniture and house interiors. An essential oil is produced from the leaves by steam-distillation.

UMBILICARIA Lichens.

***U. cylindrica*** Dub.=Gyrophora cylindrica.

***U. esculenta*** Hoffm.=Gyrophora esculenta.

***U. flocculosa*** Hoffm.=Gyrophora deusta.

***U. pustulata*** (L.) Hoffm. (Blistered Umbilicaria). Temperate. The lichen, extracted with ammonia was used in Norway and Germany to dye wool red, brown or purple.

Umbilicaria, Blistered – Umbilicaria pustulata.

Umbini – Garcinia gerrardi.

Umbondive – Plectranthus floribundus.

Umbrella Acacia – Acacia oswaldii.

Umbrella Grass – Panicum decompositum.

Umbrella-handled Bamboo – Thyrostachys siamensis.

Umbrella Pine – Sciadopitys verticillata.

Umbrella Tree – Cussonia kirkii.

Umbrella Tree – Melia azedarach.

Umgusi – Baikiaea plurijuga.

Umiviesa – Maytenus senegalensis.

Umkadhlu – Conopharyngia elegans.

Umkokolo – Doryalis caffra.

Umpetawhale – Markhamia acuminata.

Umumena Pappea ugandensis.

Umungwamagi – Allophylus subcoriaceus.

Umzingila – Phyllogeiton discolor.

Uña de gato – Acacia wrightii.

Uña de gato – Pisonia aculeata.

UNCARIA Schreb. Naucleaceae. 60 spp. Tropical. Vines.

***U. gambir*** Roxb. (Bengal Gambir). Tropical Asia. A tannin and dye (Black Gambir, Cube Gambir, Terra Japonica) is extracted from the leaves and twigs. It is used for tanning to give a soft leather, and for dyeing and printing. The tannin is sometimes used to clear beer. The leaves are chewed locally with betel.

***U. guianensis*** Gmel. Guyana. A decoction of the leaves is used to treat dysentery, locally.

***U. sclerophylea*** Roxb. (Akar kawil-kawil, Kait beusi). Malaysia. In Sumatra the bark is used as the source of a dark dye for dyeing cloth.

***U. sessilifructus*** Drake=U. tonkinensis.

***U. tonkinensis*** Havil.=U. sessilifructus. (Dây cau múc). Indochina. The bark is chewed locally instead of betel nuts.

UNDARIA Harv. Alariaceae. Brown Algae.

***U. pinnatifida*** (Harv.) Suring.=Ulopteryx pinnatifida. China, Japan. The seaweed is eaten locally, often cooked or with vinegar.

Undulated Ipecac – Richardsonia pilosa.

Unicorn Plant – Martynia probocidea.

Unicorn Root – Aletris farinosa.

Unicorn Root – Chamaelirium luteum.

UNIOLA L. Graminae. 10 spp. America. Grasses.

***U. paniculata*** L. (Sea Oats). Sand dunes of E. U.S.A. The plant is frequently grown as a sand-binder and makes a useful pasture under these conditions.

***U. racemiflora*** Trin. W.Indies. The leaves and stems are used locally to make paper.

***Unona*** L. f.=Xylopia, BUT:—

***U. brandesiana*** Pierre=Cananga latifolia.

***U. grandiflora*** DC.=Uvaria purpurea.

***U. latifolia*** Hook. f. and Thw.=Cananga latifolia.

***U. mesnyi*** Pierre.=Popowia aberrans.

UNONOPSIS R. E. Fries. Annonaceae. 27 spp. Mexico to tropical S.America, W.Indies.

***U. veneficiorum*** (Mart.) R. E. Fries.= Guatteria veneficiorum. Amazon. The plant is used locally as an arrow poison.

Unyoro Ebony – Dalbergia melanoxylon.

Upas Tree – Antiaria toxicaria.
Upo – Lagenaria leucantha.
Upright Yellow Wood Sorrel – Oxalis stricta.
UPUNA Symington. Dipterocarpaceae. 1 sp. Borneo. Tree.
***U. borneensis*** Symington. (Penyau). The very heavy, hard, dark brown wood is used locally for heavy construction work and building boats. It is attacked by insects and fungi.
***Uragoga ipecacuanha*** Baill. = Cephaelis ipecacuanha.
URANDRA Thw. Icacinaceae. 17 spp. Indomalaya. Trees.
***U. corniculata*** Foxw. = Platea corniculata. (Daru urandra). The heavy, scented wood is used locally for heavy construction work.
Urari – Strychnos castelnaei.
Urari – Strychnos toxifera.
URARIA Desv. Leguminosae. 20 spp. Tropical Africa, S.E.Asia, Formosa, Indomalaya, N.Australia, Pacific. Shrubs or herbs.
***U. crinata*** Desv. Old World tropics. In some parts the plant is grown as a green manure and cover crop. In Malaya the roots are used to treat diarrhoea and the leaves as a poultice to remove lice.
***U. picta*** Desv. Tropical Africa. The crushed leaves are used in Nigeria as a treatment for gonorrhoea and in S.India as an antidote for snake-bites.
***Urbanella procera*** Pierre. = Lucuma procera.
URCEOLA Roxb. Apocynaceae. 15 spp. Burma, W.Malaysia. Shrubs or vines. The following spp. are minor sources of rubber.
***U. acuto-acuminata*** Boerl. Malaysia (Serapat Rubber).
***U. brachysepala*** Hook. Malacca, Sumatra.
***U. elastica*** Roxb. Malaysia.
***U. esculenta*** Benth. India, Burma, Malaya. (Rangoon Rubber).
***U. maingayi*** Hook. f. Malaysia.
***U. pilosa*** Boerl. Borneo. (Serapat Rubber).
URCEOLARIA Ach. Lichens.
***U. calcarea*** Sommerf. (Limestone Urceolaria). Temperate. An extract of the lichen with ammonia (Cudbear) is used in U.K. to dye wool red.
***U. cinerea*** Ach. (Greywish Urceolaria). Temperate. See U. calcarea.
***U. scruposa*** (Schreb.) Ach. (Rock Urecolaria). Temperate. Used in U.K. as the source of a red-brown dye, used for dyeing wool.
Urceolaria, Greyish – Urceolaria cinerea.
Urceolaria, Limestone – Urceolaria calcarea.
Urceolaria, Rock – Urceolaria scruposa.
Urcurana – Hieronima alchorneoides.
Urd – Phaseolus mungo.
URECHITES Muell. Arg. Apocynaceae. 2 spp. Florida to Central America, W.Indies. Small shrubs.
***U. karwinskii*** Muell. Arg. = Mandavilla potosina. (Loroco, Quilite). S.Mexico, Central America. The flowers, shoots and young leaves are used locally as a vegetable.
***U. suberecta*** Muell. Arg. Tropical America. The plant yields a poison which is used locally to stupify fish (Wooraia) and medicinally to treat snake bites, warts, gonorrhoea and stomach upsets.
URENA L. Malvaceae. 6 spp. Tropics and subtropics. Shrubs. The bark fibres of the following spp. are similar to jute.
***U. heterophylla*** Smith. = U. sinuata.
***U. lappago*** Smith. = U. sinuata.
***U. lobata*** L. (Aramina, Candillo). F.orida, S.America, W.Indies, Africa, Asia. Sometimes cultivated.
***U. sinuata*** L. = U. heterophylla = U. lappago. Tropics.
URERA Gaud. Urticaceae. 35 spp. Warm America, S.Africa, Madagascar. Herbs.
***U. baccifera*** (L.) Gaud. = Urtica baccifera. (Chichicastle, Ortega de cabilla, Quaina). S.Mexico to S.America, W.Indies. In Cuba the stem fibres are used for cordage, while in Porto Rico a decoction of the roots is used to treat gonorrhoea.
***U. cameroonensis*** Wedd. (Amulu-bala, Obondji). Tropical Africa. The leaves are eaten as a vegetable. Cultivated locally.
***U. caracasana*** Jacq. = Urtica caracasana = Urtica chichicaztli. (Mal hombre, Mala mujer, Ortega, Quemador, Tachinote). Mexico to S.America, W.Indies. The leaves are used locally in Mexico as a treatment for syphilis.
***U. hypselodendron*** (Hochst.) Wedd. = U. cameroonensis.
***U. oligoloba*** Baker. (Amiana). E.Madagascar. The stem fibres are used locally to make cloth and bags.
URGINEA Steinh. Liliaceae. 100 spp. Mediterranean, Africa, India. Herbs.

***U. altissima*** Baker. Tropical Africa, S.Africa. See U. maritima.

***U. burkei*** Baker. (Gavu, Suto, Transvaal Slangop). S.Africa. The local natives of Rhodesia, rub the bulbs on the skin to cause paralysis and used a decoction of the bulb to induce abortion and treat circulatory ailments.

***U. indica*** Kunth. (Indian Drug Squill). Tropical Asia and Africa. Used as U. maritima.

***U. macrocentra*** Baker. (Natal Slangkop, Poison Bulb). S.Africa.

***U. maritima*** (L.) Baker=U. scilla. (Sea Onion, Squill). Mediterranean. The bulbs contain the glucosides scillaren-A and scillaren-B and are used medicinally as a heart tonic, diuretic, emetic and expectorant. Red quills, from the red varieties are used as rat poison.

***U. micrantha*** Solms. Red Sea. The starch from the bulbs is used locally in the Sudan to stiffen the hair.

***U. scilla*** Steinh=U. maritima.

UROCHLOA Beauv. Graminae. 25 spp. Tropical Africa, Asia. Grasses.

***U. mosambicensis*** (Hack.) Dandy. (Buffel Grass). Dry. S.Africa. A good pasture grass for these areas.

UROPHYLLUM Wall. Rubiaceae. 150 spp. Tropical Africa, Asia to Japan and New Guinea. Trees.

***U. arboreum*** Korth.=U. glabrum (Kitjenkeh). Malaysia. The scented leaves are used locally as a spice.

UROSPATHA Schott. Araceae. 20 spp. Central and tropical S.America. Herbs.

***U. antisylleptica*** R. E. Schultes. Colombia. The dried, ground, unripened spadix is used by some local tribes as an oral contraceptive.

***U. caudata*** (Poepp. and Endl.) Schott.= Spathiphyllum caudatum. (Apé, Caa Apé). Amazon region of Brazil. The roasted rhizome is eaten by the local Indians.

***Urostigma*** Gasp.=Ficus.

URTICA L. Urticaceae. 50 spp. Mostly N.Temperate, a few tropics and S.Temperate. Herbs.

***U. aestuans*** L.=Laportea aestuans.

***U. argentea*** Forst.=Pipturus argenteus.

***U. baccifera*** L.=Urera baccifera.

***U. breweri*** Wats. W. N.America. The stem fibres are used locally to make cordage and fishing nets.

***U. cannabina*** L. Siberia to Iran. The stem fibres are used locally to make cordage.

***U. capitellata*** Poir.=Leucosyke capitellata.

***U. caracasana*** Jacq.=Urera caracasana.

***U. chichicazthi*** Sessé and Moc.=Urera caracasana.

***U. dioica*** L. (Bigstring Nettle, Stinging Nettle). Temperate. The young shoots and leaves are eaten as a vegetable and used to make a beer. The stem yields a fibre. They (Folia et Herba Urticae) are also used in home remedies for rheumatism, gout, dropsy, skin troubles and rashes, burns and to wash the hair. A decoction of the roots (Nettle Root, Radix Urticae, Rhizoma Urticae) is used as an astringent and as a hair-wash. A decoction of the seeds (Nettle Seeds, Semen Urticae) is also used as a hair wash and tonic and to treat coughs, bronchitis etc. The plant is a commercial source of chlorophyll.

***U. ferox*** Forst. New Zealand. A decoction of the bark is used locally to treat eczema and venereal diseases.

***U. gigas*** Cuningh.=Laportea gigas.

***U. holosericea*** Nutt. W. N.America. The stem fibres are used locally to make cloth.

***U. lyallii*** Wats. W. N.America. The local Indians use the stem fibres for cordage and cloth.

***U. massaica*** Mildbr. (Kiotsenga, Igisura). Tropical Africa. The leaves are eaten locally as a vegetable.

***U. photiniphylla*** Cunninh.=Laportea photiniphylla.

***U. puya*** Wall.=Maoutia puya.

***U. thunbergiana*** Sieb. and Zucc. Japan. The leaves are eaten locally as a vegetable and the stem fibres are used for cordage.

***U. urens*** L. (Dog Nettle, Small Nettle). Cosmopolitan. The young leaves are eaten as a vegetable and a decoction of the leaves and flowers is used as a diuretic.

Urucurana – Hieronima alcorneoides.

USNEA L. Lichens.

***U. barbata*** Hoffm.=U. dasypoga (Bearded Usnea, Old Man's Beard). Temperate. Yields a red-brown dye, used for dyeing wool. The powdered plant (Cyprus Powder) was used in the 17th century as a toilet powder and for powdering wigs.

***U. dasypoga*** (Ach.) Nyl.=U. barbata.

***U. florida*** (L.) Hoffm. (Flowering Usnea). Temperate. Used in Europe as the source of a red-brown or light green dye, used for dyeing wool.

***U. philippina*** Waino. (Lumot Kahoi). Philippines. The red decoction of the plant is used locally to relieve stomach upsets.

***U. plicata*** Hoffm. (Plaited Usnea). Temperate. Yields a green or yellow dye, used for dyeing wool.

Usnea, Bearded – Usnea barbata.

Usnea, Flowering – Usnea florida.

Usnea, Plaited – Usnea plicata.

USTILAGO (Pers.) Roussel. Ustilaginaceae 300 spp. Cosmopolitan Parasitic Fungi.

***U. maydis*** (DC.) Tul. (Corn Ergot, Corn Smut). Parasitic on Maize. The fungus is used medicinally in cases of difficult labour, to treat bleeding from the lungs and bowels and to treat discharges of the urinogenital tract. In parts of Mexico it is eaten as a delicacy (Cuiteche).

Utah Juniper – Juniperus utahensis.

Utis – Alnus nitida.

Uva silvestre – Vitis caribaea.

Uvalha – Eugenia uvalha.

UVARIA L. Annonaceae. 150 spp. Tropical Africa, Madagascar, Indomalaya, Australia. Trees.

***U. aethiopica*** Guill. and Poir. = Xylopia aethiopica.

***U. buesgenii*** Diels. (Paddle Wood). Tropical Africa. The orange, dense, hard wood does not snap easily and is used for furniture, walking sticks and pencils.

***U. burakol*** Blume = Stelechocarpus burakol.

***U. calamistra*** Hance. N.Vietnam, Hongkong. The leaves are used in China to aid fermentation and are exported from Vietnam to China.

***U. catocarpa*** Baker. (Senasena). Madagascar. The seeds are used medicinally as a stimulant.

***U. chamae*** Beauv. Tropical Africa, especially W. The wood is used locally to make oars and a decoction of the leaves as an eye-wash.

***U. febrifuga*** Humb. and Bonpl. = Xylopia xylopiodes.

***U. grandiflora*** Wall. = U. purpurea.

***U. lanceolata*** Sw. = Oxandra lanceolata.

***U. leptocladon*** Oliv. E.African coasts, Zanzibar. A decoction of the roots is used locally in E.Africa to treat epilepsy, sunstroke and tonsilitis.

***U. purpurea*** Blume = U. grandiflora = Unona grandiflora. (Cuôi con chông). India, Indochina, Malaysia, China. The sweetish fruits are eaten locally.

***U. rufa*** Blume (Calabao). Malaysia. The fruits are eaten locally, especially in the Philippines.

***U. suberosa*** Roxb. = Polyalthia suberosa.

Uvero – Coccolobus schliedeana.

UVULARIA L. Liliaceae. 4 spp. E. N.America. Herbs.

***U. sessiliflora*** L. = Oaksia sessiliflora. (Bellwort). Throughout range. The young shoots and roots are eaten as vegetables, occasionally.

# V

VACCINIUM L. Ericaceae. 300–400 spp. N.Temperate, tropical mountains (except Africa), Andes, S.Africa, Madagascar. Small shrubs.

The fruits of the following spp. are eaten locally, especially in tarts and pies. They are sometimes sold locally, but a few spp. (as mentioned) are cultivated and sold over a wider area and canned.

***V. andringitrense*** Perr. Madagascar.

***V. arbuscula*** (A. Gray) Mart. W. N.America, Alaska.

***V. berberifolium*** (A. Gray) Skottsb. Hawaii.

***V. corymbosum*** L. (Blueberry, High Bush Blueberry, Swamp Blueberry). E. N.America. Cultivated. Canned etc. There are several varieties.

***V. dentatum*** J. Sm. = V. penduliflorum = Metagonia penduliflora. (Dentate Ohelo, Ohelo). Hawaii.

***V. erythrocarpum*** Michx. (Bear Berry, Dingleberry, Mountain Cranberry). S.E. N.America.

***V. floribundum*** H.B.K. (Andean Blueberry, Columbian Blueberry, Mortiño). Andes.

***V. leucanthum*** Schlecht. Mexico.

***V. macrocarpon*** Ait. = Oxycococcus macrocarpon. (Cranberry). E. N.America. Cultivated. The berries are eaten particularly as a sauce with turkey or chicken.

***V. meridionale*** Sw. (Billberry). Jamaica.

***V. mortinia*** Benth. (Mortiña). Andes.

***V. myrsinites*** Lam. (Ground Blueberry). N.America.

***V. myrtillus*** L. (Billberry, Blueberry, European Blueberry, Whortleberry). Europe. The fruits are also used in Central Europe to make a wine (Heidelbeersekt, Heidelbeerwein) which is sometimes distilled (Heidelbeergeist). They are sometimes used to colour wine red.

***V. myrtoides*** (Blume) Miq. Malaya. Used to make preserves in the Philippines.

***V. nitidum*** Andr. (Blueberry). E. N.America.

***V. oyxcoccus*** L.=Oxycoccus palustris. (Small Cranberry). N.Temperate and arctic.

***V. penduliflorum*** Gaud.=V. dentatum.

***V. pennsylvanicum*** Lam. (Early Sweet Blueberry, Low Sweet Blueberry). E. N.America.

***V. uliginosum*** L. (Bog Billberry). Europe. The fruits are also used, soaked in brandy to treat stomach and intestinal catarrh.

***V. vacillans*** Kalm. E. N.America.

***V. vitis-indaea*** L. (Cowberry, Foxberry). Temperate N.hemisphere.

***Vagnera*** Adans.=Smilacina.

Vahimainty – Secamonopsis madagascariensis.

Vaikantha – Epimeredi malabarica.

Vainora prieto – Pisonia capitata.

Vairoa – Boehmeria platyphylla.

Vakenar – Sterculia villosa.

Valerian – Valeriana officinalis.

Valerian, African – Fedia cornucopiae.

Valerian, Greek – Polemonium coeruleum.

Valerian, Swamp – Valeriana uliginosa.

VALERIANA L. Valerianaceae. 200+ spp. Eurasia, S.Africa, temperate N.America, Andes. Herbs.

***V. capensis*** Thunb. (Balderjan). S.Africa. A decoction of the leaves is used locally to treat nerve complaints including epilepsy and hysteria, intestinal worms and to induce sweating.

***V. celtica*** L. Alps. A decoction of the root is used locally as a nerve tonic.

***V. hardwickii*** Wall. India, China, Indonesia. The leaves are used locally to treat nerve complaints.

***V. mexicana*** DC. Mexico. See V. officinalis.

***V. officinalis*** L. (Valerian). Europe to temperate Asia. The plant is grown as an ornamental in U.K. and other parts of Europe. The dried roots contain bornyl valerate and bornyl formate and are used medicinally for nerve complaints and as a nerve tonic, also to relieve stomach pains and intestinal catarrh. The leaves were used as a condiment and as a perfume. They are still used as a perfume in parts of the East.

***V. phu*** L. Siberia and Urals. See V. officinalis.

***V. toluccana*** DC. See V. officinalis.

***V. uliginosa*** (Torr. and Gray) Rydb. (Swamp Valerian). E. N.America. The local Indians use a decoction of the roots to treat headaches, sore throats, coughs, etc. and cramp. They also use a poultice of the roots on wounds, etc. The juice is supposed to attract fish.

***V. wallichi*** DC. Himalayas. The oil from the rhizomes is used locally in perfumes.

VALERIANELLA Mill. Valerianaceae. 80 spp. Europe to Central Asia and Afghanistan. Herbs. The leaves of the spp. mentioned are eaten in salads and are cultivated.

***V. eriocarpa*** Desv. (Italian Corn Salad). Mediterranean.

***V. olitoria*** Pollich. (Corn Salad). Europe, Caucasus, Africa. Introduced to N.America.

***Valerianioides jamaicensis*** (L.) Medic.= Stachytarpheta jamaicensis.

VALLESIA Ruiz. and Pav. Apocynaceae. 10 spp. Florida to Argentine. Shrubs.

***V. glabra*** Cav. Throughout range. The fruits are edible. The juice from the plant is sometimes used to treat eye diseases.

Valley Oak – Quercus lobata.

VALLISNERIA L. Hydrocharitaceae. 6–10 spp. Tropics and sub-tropics. Aquatic herbs.

***V. americana*** Michx. (Eel Grass, Tape Grass). E. N.America. A good food for ducks.

***V. spiralis*** L. Europe, Asia, America. The leaves are eaten in salads in Japan. They are also eaten by ducks.

Valonia – Quercus aegilops.

Valuka – Aporusa lindleyana.

Valulrai – Celastrus paniculata.

Vanashempaga – Euodia lunu-ankenda.

VANDA R. Br. Orchidaceae. 60 spp. China, Indomalaya, Marianne Islands. Herbs.

***V. roxburghii*** R. Por.=V. tesselata. (Bandaneke, Rasna). Tropical Asia, especially India. The Hindus use a decoction of the roots to treat rheumatism.

***V. tessellata*** Hook. ex G. Don.=V. roxburghii.

Vandevenona – Parinari emirnensi.

Vandke Grass – Panicum flavidum.
Vàng nhura – Garcinia vilersiana.
VANGUERIA Comm. ex Juss. Rubiaceae. 27 spp. Tropical Africa, Madagascar. Small trees. The fruits of the following spp. are eaten locally, but must be over-ripe.
***V. apiculata*** Schum. (White Medlar). Tropical Africa.
***V. edulis*** Vahl. = V. madagascariensis.
***V. esculenta*** S. Moore. (Chirinda Medlar). Tropical Africa.
***V. madagascariensis*** J. F. Gmel. = V. edulis. Tropical Africa, Madagascar. The plant is cultivated in various parts of the tropics.
***V. munjiro*** S. Moore. (Common Munjiro). Tropical Africa.
VANGUERIOPSIS Robyns ex Good. Rubiaceae. 18 spp. Tropical Africa. Trees.
***V. lancifolia*** (Hiern.) Robyns. (Wild Medlar). Central Africa. The fruits are eaten locally.
***Vaniera cochinchinensis*** Lour. = Cudrania javanensis.
VANILLA Mill. Orchidaceae. 90 spp. Tropics and subtropics. Herbs. The spp. mentioned are sources of vanilla, with the exceptions of V. griffithii and V. decaryi. Over 90 per cent of the world's supply comes from Madagascar by the cultivation of V. planifolia. The plant was introduced to Java in 1819 from its native Mexico and from there to the other areas of production, including Seychelles, Fiji, Réunion, Tahiti, Hawaii and the West Indies. The method of production is the same for any of the spp. grown. The vines are grown on frames or in the shade of trees under which the plants are grown. The plants are propagated by stem cutting, which are topped at a suitable height to induce flowering. They begin bearing after three years. The fruits (beans) are collected before they are ripe and successively sun-dried and fermented for several days until the fruits become dark brown. It is during this process that the vanillin develops by the enzymatic conversion of a glucoside. After treatment the fruits contain 2–3 per cent vanillin, which is widely used in flavouring confectionery, chocolate and in perfumery.
***V. abundiflora*** J. J. Smith. (E. Borneo). Grown locally.
***V. decaryi*** Perr. Madagascar. The plant is used locally as an aphrodisiac.
***V. gardneri*** Rolf. Brazil. (Vanilla of Brazil, Vanilla of Bahia).
***V. grandiflora*** Lindl. = V. pompona.
***V. griffithii*** Reichb. (Telinga kerbaoo). India, Malaysia. The fruits are sweet but do not produce vanilla. They are eaten locally. The juice from the crushed flowers is rubbed on the body to reduce fevers and the juice from the leaves is used as a hair tonic.
***V. guianensis*** Splitgerb. Brazil, Guyana.
***V. phaeantha*** Reich. W.Indies.
***V. planifolia*** Andr. (Vanilla). Mexico. The main commercial source of vanilla, which is also used to make a liqueur Crême de Vanille.
***V. pompona*** Schiede. = V. grandiflora. Marinique, Guadeloupe. (Pompona Bova, Vanilla Bouffie). Cultivated locally.
Vanilla – Vanilla planifolia.
Vanilla Bouffie – Vanilla pompona.
Vanilla Chica – Selenipedium chica.
Vanilla of Bahia – Vanilla gardneri.
Vanilla of Brazil – Vanilla gardneri.
Vanilla Plant – Trilisa odoratissima.
Vanilla Trilisa – Trilisa odoratissima.
VANILLOSMOPSIS Sch. Bip. Compositae. 7 spp. Brazil. Trees.
***V. erythropappa*** Sch. Bip. = Albertinia candolleana. Brazil. The wood is used to make boats and telegraph poles.
Var kan – Tinomiscium tonkinense.
Vara blanca – Croton niveus.
Vare Goldies – Xyris ambigua.
VARIOLARIA Ach. Pertusariaceae. Lichens.
***V. orcina*** Ach. Temperate. Used as the source of a purple dye, used in France to dye wool.
Varitas de Apizaco – Amelanchier denticulata.
Varnish, Burmese – Melanorrhoea usitata.
Varnish, Piney – Vateria indica.
Varnish Tree – Rhus vernicifera.
***Vascoa perfoliata*** DC. = Rafnia perfoliata.
***Vasconella cestriflora*** DC. = Carica cestriflora.
Vasey Grass – Paspalum urvillei.
VATAIREA Aub. Leguminosae. 7 spp. Tropical S.America. Trees.
***V. guianensis*** Aub. Guyana, Brazil. The yellow-brown wood is used for furniture and ship-building. The sap is used locally as a salve for ringworm and a decoction of the bark for treating ulcers.
VATAIREOPSIS Ducke. Leguminosae. 3 spp. Brazil. Trees.

***V. speciosa*** Ducke. Throughout range. The wood is used locally for general construction work. The powdered bark (Araroba Powder, Pó da Bahia, Pó da Goa) is used medicinally.

VATERIA L. Dipterocarpaceae. 21 spp. Seychelles, S.India, Ceylon, Malaysia. Trees.

***V. acuminata*** Hayne. E.India, Ceylon. The trunk yields good varnish resin (Haldummala). The fruits are eaten locally and the wood is used to make tea-chests.

***V. indica*** L. (Indian Copal Tree, Malabar Tallow Tree, White Dammar). S.E.India. The trunk yields a resin (Dhupa Fat, Piney Tallow, Piney Varnish, White Dammar) which is used medicinally instead of Pine Resin and in varnishes. Locally it is used for cooking, making candles and burning for light.

VATICA L. Dipterocarpaceae. 76 spp. S.India, Ceylon, Siam, Indochina, Hainan, Malaysia. Trees.

***V. bancana*** Scheff. (Kajoo daging, Resak gelingga). The hard, heavy, termite-resistant wood is used locally for beams and posts.

***V. faginea*** Dyer. (Dammar baott, Tjangal padi). Malaysia. The trunk yields an inferior resin (Dammar rasak). The wood is used locally to make boats.

***V. laccifera*** Wight.=Shorea talura.

***V. leucocarpa*** Foxw. Borneo. The heavy, dark brown wood is termite-resistant and is used locally for general construction work.

***V. papuana*** Dyer. Borneo, Moluccas, New Guinea. The trunk yields a resin (Dammar Hiroe).

***V. rassak*** Blume.=Retinodendron rassak. Borneo, Malaysia. The trunk yields a resin (Rose Dammar) which is exported.

***V. simalurensis*** van Slooten. (Siloki, Silongkis). Indonesia. The hard, durable wood is used locally for general construction work.

***V. songa*** van Slooten. (Songa karsik). Indonesia. The hard, elastic, durable, dark red wood is used locally to make boats.

***V. sublacunosa*** Miq.=Shorea eximea.

***V. sumatrana*** van Slooten. Sumatra. The trunk yields a resin, used locally for lighting.

***V. teijsmanniana*** Burck. Sumatra. The hard, durable wood is resistant to termites and is used locally for house-building.

VAUPESIA R. E. Schultes. Euphorbiaceae. 1 sp. Colombia, W.Brazil. Tree.

***V. cataractarum*** R. E. Schultes. (Ma-há-wa-soo). The oil-rich seeds are eaten locally after cooking, which breaks down the poison they contain.

VAUQUELINIA Corrêa ex Humb. and Bonpl. Rosaceae. 8–10 spp. S.W. United States of America, Mexico.

***V. corymbosa*** Corrêa. Mexico. A yellow dye from the bark is used locally to stain goat skins.

Vegetable Beefsteak – Fistulina hepatica.

Vegetable Hair – Tillandsia usneoides.

Vegetable Marrow – Cucurbita pepo.

Vegetable Oyster – Tragopogon porrifolius.

Vegetable Sponge – Luffa cylindrica.

Vegetable Wax – Rhus succedanea.

Veitch's Silver Fir – Abies veitchii.

Vekharo – Indigofera trita.

Veldtgrass – Ehrharta calycina.

Velampisini Gum – Feronia limonia.

Velvet Bean – Muncuna nivea.

Velvet Grass – Holcus lanatus.

Velvetleaf – Tournefortia argentia.

Velvetleaf, Yellow – Limnocharis emarginata.

Velvet Weed – Abutilon avincennae.

Venadillo – Swietenia cirrhata.

Venenillo – Asclepias linaria.

Venezuela Mahogany – Sweitenia candolleri.

Venezuela Pokeberry – Phytolacca rivinoides.

Vengai Paduak – Pterocarpus marsupium.

Venice Turpentine – Larix decidua.

VENIDIUM Less. Compositae. 20–30 spp. S.Africa. Herbs.

***V. arctotoives*** Less. S.Africa. The leaves are used locally as a dressing for wounds.

Venkelwortel – Carum capense.

Ventilagin – Vantilago maderaspatana.

VENTILAGO Gaertn. Rhamnaceae. 35 spp. India, China to New Guinea, Australia and Pacific Islands; 1 sp. tropical Africa; 1 sp. Madagascar. Vines.

***V. maderaspatana*** Gaertn. Tropical Asia, especially India, Burma, Ceylon. The root bark yields the red dye Ventilagin of commerce which is used to dye wool.

Ventilato – Rhus cotinus.

Venus' Comb – Scandix pecten-veneris.

VERATRUM L. Liliaceae. 25 spp. N.Temperate. Herbs.

***V. album*** L. (White False Hellebore, White Hellebore). Europe. The roots (Radix Hellebori Albi, Rhizoma Veratri) contain the alkaloids jervin, pseudojervin, rubijervin and the glucoside veratramarin and are used medicinally as a sedative and cardio-vascular depressant.

***V. californicum*** Durand.=V. speciosum. (False Hellebore). W. N.America. The local Indians use a decoction of the roots as an oral contraceptive.

***V. speciosum*** Rydb.=V. californicum.

***V. viride*** Ait. (American False Hellebore, American White Hellebore, Indian Poke). E. N.America. The rhizome is used medicinally as V. album and as an insecticide.

VERBASCUM L. Scrophulariaceae. 300 spp. N.Temperate Eurasia. Herbs.

***V. lychnitis*** L. Europe, introduced to N.America. The flowers are sometimes used to repel mice.

***V. phlomoides*** L. (Clasping Mullein). Europe, introduced to N.America. The ground seeds are used to stupify fish and the whole plant to repel mice.

***V. thaspsus*** L. (Flannel Mullein). Central and S.Europe, to W.Asia, introduced to N.America. An infusion of the leaves is used medicinally as a demulcent and to treat lung complaints. They are also used in cigarettes, smoked to relieve asthma and as a poultice for wounds. The dried flowers are used similarly. The seeds are used to stupify fish.

VERBENA L. Verbenaceae. 250 spp. tropical and temperate America; 2–3 spp. Old World. Herbs.

***V. hastata*** L. (American Blue Vervain, Wild Hyssop). The dried leaves are used medicinally as an expectorant and to induce sweating.

***V. jamaicense*** L.=Stachytarpheta jamaicensis.

***V. officinalis*** L. (Vervain, Holy Wort). Mediterranean, but found throughout temperate regions of the World. The dried leaves are used medicinally (Herba Verbenae) as a diuretic, to induce sweating, to control menstruation and as a poultice for skin diseases and ulcers. The ancient Egyptians burnt the whole plant (Herba Sancta, Tears of Isis), during ceremonies.

Verbena Oil – Lippia citriodora.

VERBESINIA L. Compositae. 150 spp. Warm America. Herbs or shrubs.

***V. calendulacea*** L.=Wedelia calendulacea.

***V. fruticosa*** L.=Bidens fruticosa= Zexmenia frutescens. Mexico, Central America. The ash from the plant is used in El Salvador by the women when spinning to keep their fingers lubricated.

***V. virginiana*** L. (White Crown Beard). E. U.S.A. The local Indians used a decoction of the roots to treat fevers.

Vernal Grass, Sweet – Anthoxanthum odoratum.

VERNONIA Schreb. Compositae. 1,000 spp. America, Africa, Asia, Australia. Herbs, shrubs or trees.

***V. albicans*** DC.=V. chinensis.

***V. amygalina*** Delile. (Bitterleaf). Tropical Africa. Shrub. In W.Africa, the twigs are chewed to relieve stomach pains and stimulate the appetite.

***V. anthelmintica*** Willd. (Kinka Oil Ironweed). Tropical Asia. Herb. The leaves are used locally in India as a salve for leprosy and skin diseases and in a decoction as an abortive.

***V. arborea*** Hom.=V. javanica=Eupatorium celebicum. India, Indochina, Indonesia, Philippines. Trees. The light wood is used locally to make matches and an infusion of the leaves is used to reduce fevers in Malaya.

***V. aschenborniana*** Schauer. Mexico, Central America. Herb. The leaves are used in Guatemala to treat fevers.

***V. chinensis*** Less.=V. albicans=V. patula= Conyza chinensis. China, Indochina. Herb. The plant is used locally to treat diarrhoea and fevers.

***V. javanica*** DC.=V. arborea.

***V. menthaefolia*** Less. Cuba. The leaves are used locally in baths to relieve rheumatism, stomach pains and diarrhoea.

***V. merana*** Baker. Madagascar. Tree. The wood is used locally to build houses.

***V. missurica*** Raf. (Drummond's Ironweed). S. U.S.A. A purple dye is extracted from the flowers and a decoction of the flowers is used to treat dandruff.

***V. nigritiana*** Oliv. and Hiern. (Nigerian Ironweed). Shrub. Tropical W.Africa. A decoction of the roots is used locally to treat dysentery, constipation, as a diuretic and emetic. The leaves are used similarly to reduce fevers and stop internal bleeding.

***V. patula*** Merr. = V. chinesis.

***V. pectoralis*** Baker. Madagascar. Shrub. The leaves are used locally as a tonic.

***V. senegalensis*** Less. (Bitters Tree of Gambia). Tree. A decoction of the stem and root barks is used locally to treat diarrhoea, fevers and as a tonic. The leaves are chewed.

***V. woodii*** Hoff. S.Africa. Herb. A decoction of the leaves is used locally to relieve stomach aches.

Verona Orris Root – Iris germanica.

VERONICA L. Scrophulariaceae. 300 spp. Mostly N.Temperate, alpine, a few S.Temperate and tropical mountains. Herbs or shrubs.

***V. anagallis*** L. Temperate. The leaves are eaten as salads in parts of Japan. The plant is cultivated.

***V. beccabunga*** L. (Beccabunga Brook-lime). Europe through temperate Asia. The young shoots and leaves are eaten as a vegetable. A decoction of the leaves is diuretic and improves the appetite. It is sometimes used to treat scurvy.

***V. officinalis*** L. (Drug Speedwell). Temperate Europe, Asia and N.America. A decoction of the leaves is used locally to treat coughs, colds, bronchitis, catarrh of the bladder and skin diseases. The leaves are also used as a substitute for tea.

***V. salicifolia*** Forst. Shrub. New Zealand. A decoction of the leaves is used locally to treat dysentery and diarrhoea.

***V. virginica*** L. = Leptandra virginica. (Culver's Root). E. N.America to Texas. The roots are used medicinally as a violent purgative and emetic.

VERPA Sw. ex Fr. Helvellaceae. 5 spp. N.Temperate.

***V. bohemica*** (Krombh.) Schroet. Central Europe. The fruit-bodies are eaten locally.

Vervain – Verbena officinalis.

Vervain, American Blue – Verbena hastata.

Vervain, Bastard – Stachytarpheta jamaicensis.

Vetch, Bard – Vicia monantha.

Vetch, Bitter – Vicia ervilla.

Vetch, Canada Milk – Astragalus canadensis.

Vetch, Chickling – Lathyrus sativus.

Vetch, Common – Vicia sativa.

Vetch, Gerard – Vicia cracca.

Vetch, Hairy – Vicia villosa.

Vetch, Horseshoe – Hippocrepis comosoa.

Vetch, Hungarian – Vicia pannonica.

Vetch, Kidney – Anthyllis vulneraria.

Vetch, Leavenworth – Vicia leavenworthii.

Vetch, Milk – Astragalus glycyphyllos.

Vetch, Monantha – Vicia monanthos.

Vetch, Narbonne – Vicia narbonensis.

Vetch, Narrowleaved – Vicia articulata.

Vetch, One-flowered – Vicia articulata.

Vetch, Purple – Vicia benghalensis.

Vetch, Russian – Vicia villosa.

Vetch, Tuberous Bitter – Lathyrus montanus.

Vetch, Two-leaved – Vicia unijuga.

Vetch, Winter – Vicia villosa.

Vetch, Woollypod – Vicia dasycarpa.

Vetiver – Vetiveria zizanoides.

VETIVERIA Bory. Graminae. 10 spp. Tropical Africa, Asia, Australia. Grasses.

***V. nigritana*** Stapf. = Andropogon nigritanus. Sahara, Sudan. The scented rhizomes are used locally for perfuming clothes, etc.

***V. odorata*** Virey. = V. zizanoides.

***V. zizanoides*** Stapf. = V. odorata = Andropogon muricatus. (Khus-khus, Vetiver). Tropical Asia, especially India, Ceylon and Burma. A heavy essential oil is distilled from the rhizomes and is used in perfumery, mainly as a fixative. The rhizomes are also used to make perfumed mats, baskets, fans etc. The plant is cultivated.

VIBURNUM L. Caprifoliaceae. 200 spp. temperate and subtropics, especially Asia, N.America; 16 spp. Malaysia excluding New Guinea. Shrubs or small trees.

***V. acerifolium*** L. (Arrow Wood, Dockmackie). E. N.America. The local Indians use a decoction of the inner bark to treat cramp and stomach pains.

***V. cassinoides*** L. (Nanny Berry, Sweet Viburnum). E. N.America. The fruits are eaten locally and the leaves are used as a tea substitute (False Paraguay Tea).

***V. opulus*** L. (Cranberry Tree, Guelder Rose). N.America, Europe, temperate Asia. The fruits are used in jellies as a substitute for cranberries. The bark (Cramp Bark, High Bush Cranberry Bark) is used medicinally as a sedative and by the local Indians as a diuretic.

***V. pauciflorum*** La Pyalaie. (Mooseberry Viburnum). N.America. The berries are eaten by the Eskimos and preserved for winter use.

***V. prunifolium*** L. (Black Haw, Stag Bush).

N.America. The root bark contains the bitter resin viburnin which is used as a sedative, abortive, uterine relaxant and diuretic. The fruits are edible when frosted.

***V. setigerum*** Hance.=V. theiferum.

***V. theiferum*** Rehd.=V. setigerum. China. The leaves are used locally as a tea.

VICIA L. Leguminosae. 150 spp. N.Temperate, S.America. Herbs. The following spp. except V. faba, are grown as cover crops and for fodder.

***V. angustifolia*** L.=V. multicaulis=V. polymorpha. (Narrow-leaved Vetch). Europe, W.Asia, N.Africa. Used particularly in orchards of S. U.S.A.

***V. articulata*** Hormen. (One-flowered Vetch). Mediterranean, Asia Minor. Sometimes cultivated locally.

***V. atropurpurea*** Desf.=V. benghalensis.

***V. benghalensis*** L.=V. atropurpurea=V. broteriana. (Purple Vetch). S.Europe, N.Africa. Cultivated in mild temperate areas.

***V. broteriana*** Ser.=V. benghalensis.

***V. calcarata*** Sesf. (Demehi). Sahara. Cultivated for the seeds locally. These are eaten, particularly at oases.

***V. cracca*** L. (Gerrard Vetch). Europe, Asia. Cultivated locally.

***V. dasycarpa*** Ten. (Woollypod Vetch). Europe. Grown particularly in S. U.S.A. There are several varieties.

***V. ervilia*** (L.) Willd.=Ervilia sativa=Ervum ervilia. (Bitter Vetch, Ervil). Mediterranean. The seeds are fed to livestock and sometimes eaten in soups. The plant is cultivated.

***V. faba*** L.=Faba vulgaris. (Broad Bean, Horse Bean). Mediterranean. A sp. with an upright habit and no tendrils. It has been cultivated for centuries for the seeds which are eaten as a vegetable. It is grown widely in temperate areas of the Old and New World. The seeds are sometimes eaten roasted or ground with wheat flour. Some varieties are grown for feeding to livestock, particularly horses.

***V. leavenworthii*** Torr. and Gray. (Leavenworth Vetch). S. U.S.A., especially Texas. Grown locally.

***V. monanthos*** Desf. (Bard Vetch, Monantha Vetch). Mediterranean. Sometimes cultivated. The seeds are occasionally eaten in soups.

***V. multicaulis*** Wallr.=V. angustifolia.

***V. narbonensis*** L. (Narbonne Vetch). Mediterranean, N.Africa. Cultivated locally.

***V. pannonica*** Crantz. (Hungarian Vetch). Mediterranean. Cultivated in Caucasus and Balkans.

***V. polymorpha*** Godron.=V. angustifolia.

***V. sativa*** L. (Common Vetch). Mediterranean, W.Asia. Widely cultivated in Old and New World.

***V. unijuga*** A. Br.=Orobus lathyroides. (Two-leaved Vetch). E.Siberia to Japan. Cultivated locally.

***V. villosa*** Roth. (Hairy Vetch, Winter Vetch). Mediterranean, N.Africa, W.Asia. Sometimes grown locally.

VICTORIA Schomb. Euyralaceae. 2–3 spp. Tropical America. Large aquatic herbs.

***V. amazonica*** (poepp.) Sow.=V. regia. Amazon. The seeds are eaten locally.

***V. regia*** Lindl =V. amazonica.

Victoria Blue Gum – Eucalyptus globosus.

Vid silvestre – Vitis sicyoides.

***Vieillardia*** Brong. Gris=Calpidia.

VIGNA Savi. 80–100 spp. Tropics, especially Africa and Asia. Herbs.

***V. catjang*** Walp. (Black Eye Pea, Catjang, Hindu Cowpea). Old World tropics. Grown widely for the pods which are eaten as a vegetable and the whole plant used as forage. There are several varieties.

***V. haumaniana*** Wilczek. var. ***paniculata*** Wilczek. Tropical Africa. The roots are used locally as a purgative.

***V. hosei*** Baker.=Dolichos hosei. (Sarawak Bean). N.Borneo. Grown locally as a cover crop in rubber plantations.

***V. maranguensis*** (Taub.) Hrams. (Kaingo, Lushamla). Tropical Africa. Grown locally for forage.

***V. mungo*** (L.) Hepper.=Phaseolus mungo. The systematics is disputed.

***V. multiflora*** Hook. f. (Aluta, Kakakoro). Tropical Africa. A decoction of the roots is used locally to treat intestinal worms.

***V. sinensis*** Endl.=V. unguiculata=Dolichos sinensis.

***V. unguiculata*** L.=V. sinensis. (Cowpea, Horse Gram, Cherry Bean). Central Africa. Cultivated widely in Old and New World tropics. The immature pods are eaten as a vegetable, the leaves are eaten in salads and the ripe seeds are cooked in a variety of ways. The plant is a good cover crop and a good fodder crop. There are several varieties.

***Villamilla octandra*** Benth. and Hook. = Trichostigma octandrum.

***Villaresia*** Ruiz. and Pav. = Citronella.

***Villarsia cristata*** Spreng. = Nymphoides cristata.

VILLEBRUNEA Gaudlich. Urticaceae. 12 spp. Ceylon to Japan. Shrubs.

***V. integrifolia*** Gaudlich. N.E.India, tropical Asia. The stem fibres (Mesakhi Fibre) are used in India for cordage.

Vin de Cornoulle – Cornus mas.

VINCA L. Apocynaceae. 7 spp. Europe to W.Asia. Herbs or small shrubs.

***V. major*** L. (Periwinkle). Throughout range. A decoction of the leaves (Herba Vincea pervincae) is used to clear catarrh and to clean the blood. The leaves are also used as a poultice to stop bleeding and treat eczema.

***V. minor*** L. See V. major.

***V. rosea*** L. (Cape Periwinkle, Madagascar Periwinkle, Old Maid). Throughout tropics and subtropics, supposed to have escaped from Madagascar. The plant contains vincaleukoblastine (Velban) and inhibits the growth of some human tumours. The plant is cultivated.

***Vincetoxicum officinale*** Moench. = Cynanchum vincetoxicum.

Vine, Bean – Phaseolus polystachyus.

Vine, Coral – Antigonon leptopus.

Vine, Grape – Vitis vinifera.

Vine, Grecian Silk – Periploca graeca.

Vine, Kolomikta – Actinidia callosa.

Vine, Lemon – Pereskia aculeata.

Vine, Madeira – Andredera baselloides.

Vine Maple – Acer circinatum.

Vine Mesquite – Panicum obtusum.

Vine Mignonette – Anredera baselloides.

Vine, Rata – Metrosideros scandens.

Vine Spinach – Basella rubra.

Vine, Squaw – Mitchella repens.

Vinhatico – Plathymenia reticulata.

VIOLA L. Violaceae. 500 spp. Cosmopolitan, especially N.Temperate, many in the Andes. Herbs.

***V. arborea*** L. = Corynostylis arborea.

***V. cucllata*** Ait. E. N.America. The leaves and flowers are sometimes used to make soup, especially as an emergency food.

***V. odorata*** L. (Sweet Violet). Europe, Asia. Cultivated, mainly in S.France for the flowers from which an essential oil is extracted. This is used in perfumes and hairdressing and to make a liqueur (Crême de Violette, Parfait Amour). The flowers are also candied and used in confectionery. A decoction of the leaves is used to treat coughs.

***V. mirabilis*** L. (Wonder Violet). Europe to Japan). A decoction of the plant is used in the Ukraine to treat heart palpitations.

***V. palmata*** L. (Early Blue Violet, Palmate Violet). E. N.America. The plant (Wild Okra) is used by the Negroes to thicken soups.

***V. tricolor*** L. (Pansy). Europe, temperate Asia. Cultivated widely as an ornamental. A decoction of the leaves is used locally as an expectorant and the leaves are used to treat skin diseases.

Violet, Early Blue – Viola palmata.

Violet, Palmate – Viola palmata.

Violet Root – Ferula sumbul.

Violet, Sweet – Viola odorata.

Violet, Wonder – Viola mirabilis.

Violet Wood – Copaifera bracteata.

Violeta – Dalbergia cearensis.

Viper's Bugloss – Echium vulgaris.

VIRGILIA Poir. Leguminosae. 1–2 spp. S.Africa.

***V. capensis*** Lam. Throughout range. The light, soft wood is used locally for general construction work, yokes and fuel.

Virginian Bugle Weed – Lycopus virginicus.

Virginia Creeper – Parthenocissus quinquefolia.

Virginian Mountain Mint – Pycanthemum virginianum.

Virginian Snakeroot – Aristolochia serpentaria.

Virginian Strawberry – Fragaria virginiana.

Virginia Stonecrop – Penthorum sedoides.

Virginia Tephrosia – Tephrosia virginiana.

Virginian Tukkahoe – Peltandra virginica.

Virginian Waterleaf – Hydrophyllum virginicum.

VIROLA Aubl. Myristicaceae. 60 spp. Centra and tropical S.America. Trees.

***V. bicuhyba*** Warb. (Bicuhyba). Brazil. The seeds yield a fat (Bicuhyba Fat) used to make soap and candles.

***V. calophylla*** Warb. Amazon. The bark yields a narcotic snuff (Yá-ka, Yá-lo), used locally, especially by the witch-doctors.

***V. calophylloides*** Markgr. See V. calophylla.

***V. koschnyi*** Warb. Central America. The wood is used locally for general construction work.

***V. merendonis*** Pitt. (Merendon Virola). Central America. The wood is used in Guatemala for general construction work.

***V. micheli*** Heckel. N. S.America. The fat from the seeds is exported to U.S.A. and Europe to make soap and pleasantly-smelling candles.

***V. sebifera*** Aubl. See V. micheli The fat is called Virola Fat.

***V. surinamensis*** Warb. (Ucahuba). N.E. S.America. The fat from the seeds (Ucahuba Butter, Ucuiba Butter) is sometimes used as a substitute for cacao butter.

***V. venezuelensis*** Warb. (Cuojo). N. S.America. The fat from the seeds (Cuojo Fat) is used locally as a salve for rheumatism.

Virola Fat – Virola sebifera.

Virola, Merendon – Virola merendonis.

VISCUM L. Viscaceae. 60–70 spp. Warm Old World. Parasitic shrubs.

***V. album*** L. (Mistletoe). Europe to Asia Minor. The berries are used locally as bird-lime, while the Romans used them to treat epilepsy.

***V. orientale*** Willd. India to S.China, Philippines to Australia. Poisonous. The Hindus used the plant to cause vomiting.

***Visenis indica*** Miq. = Melochia umbellata.

VISMIA Vand. Guttiferae. 30 spp. Mexico to tropical S.America; 2–3 spp. W. Tropical Africa; 2 spp. E. Tropical Africa. Trees.

***V. cayenensis*** (L.) Pers. W.Indies, N.Brazil. Yields a resin.

***V. guianensis*** (Aubl.) Choisy. See V. cayenensis.

***V. leonensis*** Hook. f. Tropical Africa. The resin from the bark is used locally as a salve for skin diseases.

VISNEA L. f. Theaceae. 1 sp. Canaries. Small tree.

***V. mocanera*** L. The fruits (Mocanes) are eaten locally and are made into a syrup (Charcherquem, Lamedor de Moca) used to treat internal bleeding.

VITEX L. Verbenaceae. 250 spp. Tropics and temperate. Trees.

***V. agnus castus*** L. (Chaste Tree). Mediterranean. Cultivated in Old and New World. The leaves and fruits are used as a sexual suppressant and the fruits are used as a substitute for pepper. The twigs are used for basket making.

***V. celebica*** Koord. Celebes. The hard, elastic wood is used locally for general construction work and boat-building.

***V. cienkowkii*** Kotschy and Perr. Tropical Africa. The fruits are eaten locally and the leaves are used as a substitute for tea. The gum from the trunk is an antidote to arrow poisons and is used to make ink.

***V. cofassus*** Reinw. = V. punctata. Malaysia. The wood is used locally to make boats.

***V. divaricata*** Sw. (Guiana Chaste Tree). Guyana, W.Indies. The leaves are used locally for tanning.

***V. erioclona*** Lam. (Lako). Indonesia. The wood is used locally to build boats.

***V. glabrata*** R. Br. = V. minahassae. (Bihbool, Gentileng). Malaysia. The wood is used locally for general construction work and household utensils.

***V. heterophylla*** Roxb. = V. quinata.

***V. lasiophylla*** Benth. = V. mollis.

***V. latifolia*** Lam. = V. pubescens.

***V. littoralis*** Cunningh. (Puzizi, Timor Chaste Tree). The strong, durable, hard, dark brown wood is used for general construction work, bridges, machine bedding, blocks etc.

***V. loureiri*** Hook. and Arn. = V. quinata.

***V. minahassae*** Koord. = V. glabrata.

***V. mollis*** H.B.K. = V. lasiophylla. Mexico. The fruits are eaten locally and a decoction of the leaves is used to treat diarrhoea.

***V. moluccana*** Blume = Gmelina moluccana.

***V. parviflora*** Juss. (Molave Chaste Tree). Malaysia, Philippines. The hard, light-brown wood is used locally for furniture and turning.

***V. peduncularis*** Wall. India, Burma. The leaves are used locally to reduce fevers.

***V. pubescens*** Vahl. = V. latifolia. Tropical Asia. The hard, heavy, durable wood is used locally for general construction work, furniture and agricultural implements.

***V. punctata*** Schau. = V. cofassus.

***V. quinata*** Williams. = V. heterophylla = V. loureiri. (Chân chim, Man kinh). Vietnam. A decoction of the bark is used locally as a tonic.

***V. tirfolia*** L. (Gendavasi, Lagondi, Sangari). Malaysia, Indonesia. A paste of the leaves is used locally as a salve for bruises and with Guava and Sembog to treat beri-beri.

VITIS L. Vitidaceae. 60–70 spp. Mainly N.Hemisphere. Vines.

***V. adnata*** Wall.=V. cymosa=Cissus adnata =C. latifolia. (Akar gamih, Sambagan). S.E.Asia, Indonesia. A decoction of the leaves is used locally to treat coughs.

***V. aestivalis*** Michx. (Summer Grape). N.America. The fruits are edible and the plant is occasionally cultivated. It is resistant to Phylloxera disease and has been crossed with V. vinifera.

***V. amurensis*** Rupt.=V. thunbergii regel.= V. shiragni. China. The fruits are edible and are made into wine.

***V. arachnoidea*** Backer. (Ampelopsis arachnoidea=A. indica. (Areuj beungbeurootan, Bentangon, Ojoa ajer). S.E.Asia, Indonesia. The berries are fed to ducks to increase egg-production.

***V. aralioides*** Welw.=Cissus aralioides. Tropical Africa. The stems are used locally during the tanning of leather.

***V. arizonica*** Engl. (Cañon Grape). S.W. U.S.A., Mexico. The fruits are eaten by the local Indians.

***V. capensis*** Burm. f.=Rhoicissus capensis. S.Africa. The fruits are used locally to make jellies.

***V. caribaea*** DC. (Uva silvestre). Central America, W.Indies. The fruits are eaten locally.

***V. cinerea*** Engl. (Downy Grape, Sweet Winter Grape). S. and S.W. U.S.A., Mexico. The fruits are eaten by the local Indians and made into jellies.

***V. compressa*** Backer.=Cissus compressa. (Akar ritjak, Areuj ki berera). S.W.Asia, Indonesia. The juice from the plant is used locally as a salve for wounds.

***V. cymosa*** Blume=V. adnata.

***V. davidi*** (Roman) Foëx. (Spiny Vitis). W.China. The fruits are eaten locally. The plant is sometimes cultivated.

***V. discolor*** Dalz.=Cissus sicyoides. Tropical America. The stems are used locally for making baskets and as a soap substitute for washing clothes.

***V. geniculata*** Miq.=Cayratia geniculata= Cissus geniculata. (Areuj ki barera, Lakoom). Indonesia. The fruits are eaten locally in soups.

***V. labrusca*** L. (Fox Grape). E. N.America. The fruits are eaten raw, or pressed for grape-juice. They are also used to make sweets, ice-cream, jellies etc. The pressed fruits are fed to livestock. The seeds yield an oil and tannin and are used to make cream of tartar. The plant is resistant to Phylloxera disease. It is cultivated locally and there are several varieties.

***V. lanceolaria*** Wall.=Tetrastigma lanceolarium. (Akar darik-darik, Bantengan, Ojod, Opok). S.E.Asia, Indonesia. The leaves are eaten locally in salads and the juice from the plant is taken to relieve coughs.

***V. lawsoni*** King. Malaysia. The fruits are eaten locally.

***V. multistriata*** Baker.=V. pentaphylla. Tropical Africa. The fruits are eaten locally.

***V. pallida*** Wight. and Arn. Tropical Asia, Africa. In W.Africa the fruits are eaten raw and in soups, while the juice is used to adulterate honey.

***V. palmatifida*** Baker. See V. multistriata.

***V. papillosa*** Backer.=Cissus papillosa= Tetrastigma papillosum. (Akar benang tikkoos, Areuj ki barera). Indonesia. The stems are used locally for cordage and a paste from the leaves and aerial roots is used as a hair tonic for children.

***V. pentagona*** Dield. and Gilg.=V. thunbergii Sieb. and Zucc. var. yunnanensis Planch. (Nho, Tien dó, Tradet). China to Indochina. The fruits are eaten locally and used to make a vinegar. They make a poor wine.

***V. pentaphylla*** Guill. and Perr.=V. multistriata.

***V. quarangularia*** L. Tropical Africa. The juice is used by camel drivers to treat saddle sores on camels. Locally the juice is attributed with wide healing powers. It may be poisonous.

***V. repens*** Wight. and Arn.=Cissus glauca= C. repens. (Areuj harijang). S.E.Asia, Indonesia. The leaves are eaten locally in salads.

***V. rotundifolia*** Michx. (Muscadine Grape, Southern Fox Grape). E. U.S.A. Cultivated in S. of area. The fruits are edible, but rather thick-skinned.

***V. shiragai*** Makino.=V. amurensis.

***V. sicyoides*** L.=Cissus elliptica. (Bejuco iasú, Caro, Tripa de zopilote, Ubi, Vid silvestre). Tropical America. The stems are used locally for cordage and as a soap substitute for washing clothes. In Mexico a decoction of the stems is used as a treatment for rheumatism.

***V. thunbergii*** Regel=V. amurensis.

***V. thunberii*** Sieb. and Zucc. var. ***yunnanensis*** Planch = V. pentagona.

***V. vinifera*** L. (Grape, European Grape, California Grape, Grape Vine). Mediterranean, S.E.Europe, W.Asia. The plant has been grown since time immemorial for the fruits and is now grown throughout the world where the climate is suitable. The world production now amounts to some 50 million tons a year, mainly from France, Italy and Spain. 90 per cent of the crop is used for making wine, while the rest is eaten raw, or dried as raisins or currants, the latter being produced mainly in Greece. There are a great number of varieties used, depending on the use to which the grapes will be put. The plants are propagated by cuttings and more frequently by grafting on to rootstocks of N.American V. spp. that are Phylloxera-resistant. Cultivation involves a great deal of care and pruning etc. The N.American grape industry, mainly in California, was not firmly established until the 1850's with the introduction of V. vinifera from Europe into California. Earlier introductions met with little success.

Wine is essentially fermented grape juice, but the wide variety of wines depends on the type of grape used and the area in which the grapes are grown. The soil, climate and type of yeast used affected the characteristics of the wine considerably. Wines bottled before fermentation is complete result in the Champagne-type. Brandy is made by distilling wine and the fortified wines, port, sherry etc. are made by adding brandy to a completely fermented wine. Various other wines, e.g. Vermouth are produced by flavouring wine with herbs of various sorts. Red wines are made from red grapes and white wines from green or white grapes. Evaporated grape juice (Grape Honey, Turkish Honey, Dips) is a delicacy in the Near East. Currants, raisins and sultanas are the sun-dried whole fruits, of which there are several seedless varieties. The residue from pressing grapes for wine-making is fed to livestock or used to make tannin and cream of tartar. The seeds yield Grape Seed Oil, a drying oil used for lighting, paints and when refined for cooking.

***V. vulpina*** L. (Frost Grape, River Bank Grape). N.America. The fruits are edible and the plant is sometimes cultivated.

Vitis, Spring – Vitis davidii.

Vittil – Aporusa lindleyana.

Vivaona – Dilobeia thouarsii.

VOACANGA Thou. Apocynaceae. 25 spp. Tropical Africa, Madagascar, Malaysia. Shrubs.

***V. africana*** Stapf. W.Africa. Yields a latex used to adulterate rubber.

***V. foetida*** Rolf. = Orchipeda feotida. (Hamperoo, Pong kapong, Rango-rango). Malaysia. In Sumatra, the latex is used to heal wounds and the wood to make knife handles.

Voahandrintsahona – Garcinia cernua.

Voampirifery – Piper pachyphyllum.

Voanalakoly – Rhinacanthus osmospermus.

VOANDZEIA Thou. Leguminosae. 1 sp. Tropical Africa, Madagascar. Herb.

***V. poissoni*** Chev. = V. subterranea.

***V. subterranea*** Thon. = V. poissoni. (Congo Coober, Bambara Ground Nut). An important local pulse crop. Known in Central America, where it was probably introduced during the Slave Trade. The pods develop underground and the seeds are eaten cooked in a variety of ways. They contain some 4–6 per cent fat, 10 per cent protein and 50–60 per cent starch.

Voansifitra – Toxocarpus decaryi.

Voara – Ficus tiliaefolia.

Voaromainty – Rubus myrianthus.

Voavanga – Vangeria madagascariensis.

Voeur sanda – Cissus repens.

Vogel Trephrosia – Tephrisoa vogelii.

VOLVARIA (Fr.) Quél. Agaricaceae. 40 spp. Widespread.

***V. esculenta*** Mass. Tropics. The fungus grows in rice straw, banana waste etc. and is eaten in many tropical countries. It dries well for keeping. It is sometimes cultivated.

***V. volvacea*** (Bull.) Guel. See V. esculenta.

Vomitel – Cordia sebestina.

VONITRA Becc. Palmae. 5 spp. Madagascar. Palm trees.

***V. thouarsiana*** Becc. (Vonitra ambohitra). E.Madagascar. The fibres from the leaves are used locally for cordage, baskets, brushes etc.

***V. utilis*** Jum. (Vonitra, Vonitrandrano). E.Madagascar. The buds are eaten locally as a vegetable.

Vonitra – Vonitra utilis.
Vonitra ambohitra – Vonitra thouarsiana.
Vonitrandrano – Vonitra utilis.
Vov – Ficus benghalensis.
Vriadhatulasi – Ocium gratissimum.
Vulparaiso Weed – Roccella tinctoria.
Vung – Careya sphaerica.

# W

Wahoo Bark – Euonymus atropurpureus.
Wajoo – Pterospermum celebicum.
Wake Robin – Trillium grandiflorum.
Wala – Licuala rumphii.
Wale Robin – Arum maculatum.
Waldhaar – Carex brizoides.
***Walkeria*** A. Chev. = Nogo baehui.
Wallaba – Epeura falcata.
Wallflower – Cheiranthus cheiri.
Wallich's Crane's Bill – Geranium wallichianum.
Wall Pepper – Sedum album.
Walikookoon – Schoutenia burmani.
Walnut – Juglans regia.
Walnut, African – Lovea klaineana.
Walnut, Australian – Endiandra palmerstoni.
Walnut Bean – Endiandra palmerstonii.
Walnut, Black – Juglans nigra.
Walnut, Bolivian Black – Juglans boliviana.
Walnut, Brown – Lovoa klaineana.
Walnut, Cathay – Juglans cathayensis.
Walnut, Cordate – Juglans sieboldiana.
Walnut, East Indian – Albizia lebbek.
Walnut, English – Juglans regia.
Walnut, Guatemalan – Juglans mollis.
Walnut, Manchurian – Juglans mandschurica.
Walnut, Persian – Juglans regia.
Walnut, Queensland – Eudiandra palmerstoni.
Walnut, Satin – Liquidambar styraciflua.
Walnut, Siebold – Juglans sieboldiana.
Walnut, Texas Black – Juglans rupestris.
Walnut, Yellow – Beilchmiedia bancroftii.
Waloba – Eperua falcata.
WALTHERIA L. Sterculiaceae. 50 spp. Tropical America; 1 sp. Rhodesia; 1 sp. Madagascar; 1 sp. Malaysia; 1 sp. Formosa. Shrubs.

***W. americana*** L. = W. indica. Tropical America, introduced throughout the tropics. A decoction of the plant is used, in Surinam and Guadaloupe to reduce fevers, in W.Africa as a purgative and abortive, and in S.Africa to cure sterility.
***W. indica*** L. = W. americana.
Wampi – Clausena lansium.
Wandoo – Eucalyptus redunca.
Wane – Ocotea rubra.
Wapato – Sagittaria latifolia.
Wapo – Ecclinusa sanguinolenta.
Waras – Maughania conjesta.
Ware, Sea – Fucus vesiculosus.
Warragarra – Grenvillea robusta.
Warty Leather Lichen – Lobaria scrobiculata.
Warty Olive – Olea verrucosa.
***Wasabi pungens*** Matsum. = Eutrema wasabi.
Wasabi – Eutrema wasabi.
WASHINGTONIA H. Wendl. Palmae. 2 spp. S.California, Arizona, into Mexico. Palm trees.
***W. filifera*** (Lindl.) Wendl. = Neowashingtonia filifera = Pritcharia filifera. (Canon Palm). Throughout range. The local Indians use the leaves for making huts, baskets etc. They eat the fruits fresh or ground to a flour, and the buds when roasted.
***W. sonorea*** S. Wats. (Palma blanca, Palma colorado, Palma negra). Throughout range. The fruits are eaten by the local Indians.
Water Agrimony – Bidens tripartitus.
Water Berry – Syzygium guinense.
Water Caltrop – Trapa natans.
Water Chestnut – Eleocharis tuberosa.
Water Chinquapin – Nelumbo pentapetala.
Water Cress – Nasturtium officinale.
Water Dropwort – Oenanthe stolonifera.
Water Fennel – Oenanthe aquatica.
Water Germander – Teucrium scordium.
Water Gum – Nyssa biflora.
Water Hyacinth – Eichhornia crassipes.
Water Lemon – Passiflora laurifolia.
Water Lettuce – Pistia stratiotes.
Water Nut – Nelumbo pentapetala.
Water Oak – Quercus nigra.
Water, Omum – Carum copticum.
Water Parsley – Oenanthe sarmentosa.
Water Parsnip – Sium cicutaefolium.
Water Parsnip – Sium thunbergii.
Water Soldier – Stratiotes aloides.
Water Tree – Eugenia cordata.
Water Tree – Ilex carpensis.

Water Tupelo – Nyssa aquatica.
Water Tupelo – Nyssa biflora.
Water Yam – Dioscorea alata.
Waterhemp – Eupatorium cannabinum.
Waterleaf, Virginia – Hydrophyllum virginicum.
Waterlily, White – Nymphaea alba.
Watermelon – Citrullus vulgaris.
Watershield – Brasenia schreberi.
Watery Rose Apple – Eugenia aquea.
Wattle, Black – Acacia binervata.
Wattle, Black – Acacia decurrens.
Wattle, Golden – Acacia pycnantha.
Wattle, Graceful – Acacia decora.
Wattle Gum – Acacia pycnantha.
Wattle, Silver – Acacia dealbata.
Wattle, Twinvein – Acacia binervata.
Wattung Urree – Banksia serrata.
Wau Beech – Elmerrillia papuana.
Wavyleaf Oak – Quercus undulata.
Wax, Afridi – Carthamus tinctoria.
Wax Berry – Myrica cordifolia.
Wax Bush – Myrica cordifolia.
Wax, Candelilla – Euphorbia antisyphilitica.
Wax, Candelilla – Pedilanthus pavonis.
Wax, Caranday – Copernica australis.
Wax, Carnauba – Copernica cerifera.
Wax, Fig – Ficus ceriflua.
Wax, Gandong – Ficus ceriflua.
Wax, Gourd – Benincasa hispida.
Wax, Karamanni – Symphonia globulifera.
Wax Myrtle – Myrica cerifera.
Wax Palm – Ceroxylon andicola.
Wax Palm – Copernicia cerifera.
Wax, Roghun – Carthamus tinctoria.
Wax, Siejas – Langsdorffia hypogaea.
Wax, Sugar Cane – Saccharum officinale.
Wax, Sumach – Rhus succedanea.
Wax, Vegetable – Rhus succedanea.
Wayaka Yam Bean – Pachyrrhizus angulatus.
Waybread – Plantago major.
Weaver's Broom – Spartium junceum.
***Webera densiflora*** Wall. = Randia densiflora.
WEDELIA Jacq. Compositae. 70 spp. Tropics and warm temperate. Herbs.
***W. calendulacea*** Less. = W. zollingeriana. India, Far East. The dried plant is used, especially in India, for treating chronic intestinal complaints and constipation.
Weed, Scrofula – Goodyera pubescens.
Weed, Squaw – Senecio aureus.
Weeping Lovegrass – Eragrostis curvula.
WEINMANNIA L. Cunoniaceae. 170 spp. Madagascar, Mascarene Islands, Malaysia throughout Pacific to New Zealand, Andes to Mexico. Trees or shrubs.
***W. benthamii*** F. v. Muell. = Geissois benthami. (Brush Mahogany). New South Wales, Queensland. The pinkish-brown wood is used mainly for rough construction work, but is sometimes used for furniture, small carvings etc., turning, veneer and plywood.
***W. lachnocarpa*** Maid. = Giessois lachnocarpa. (Mararie, Rose Marana). Australia. The hard, pinkish wood is used for turning, printers' blocks, tool handles, bearings and coach-building.
***W. macrostachys*** DC. = W. tinctoria.
***W. pinnata*** L. (Oreganillo). Mexico to Brazil, W.Indies. A gum from the bark is used to adulterate quinine in parts of Cuba.
***W. racemosa*** L. (Kamahi). New Zealand. The bark is used locally for tanning.
***W. selloi*** Engl. S.America. The astringent bark is used locally in Brazil as a poultice for wounds.
***W. silvicola*** L. See W. racemosa.
***W. tinctoria*** Smith = W. macrocarpa Mauritius. A decoction of the bark is used locally to stain leather red.
***W. trichosperma*** Cav. (Tenio). Chile. The bark is used locally for tanning.
***Wellingtonia gigantea*** Lindl. = Sequoia gigantea.
Welsh Onion – Allium fistulosum.
WENDTIA Meyen. Ledocarpaceae. 3 spp. Chile, Argentina. Shrubs.
***W. calycina*** Griseb. Argentina. The leaves are used locally to treat indigestion.
Werinnua Oil – Guizotia abyssinica.
Wesnoo – Melochia umbellata.
West African Boxwood – Nauclea diderrichii.
West African Copaiba – Oxystigma mannii.
West African Ebony – Diospyros mesipiliformis.
West African Indigo – Lonchocarpus cyanescens.
West African Jute – Hibiscus quinquelobus.
West African Mahogany – Mitragyna macrophylla.
West Australian Sandalwood (Oil) – Eucalyptus spicata.
West Australian Sandalwood – Eucarya spicata.
West Indian Boxwood – Phyllostylon brasiliensis.

West Indian Boxwood – Tabebina pentaphylla.
West Indian Cedar – Cedrela odorata.
West Indian Cherry – Malpighia punicifolia.
West Indian Elemi – Bursera gummifera.
West Indian Gherkin – Cucumis anguria.
West Indian Gooseberry – Pereskia aculeata.
West Indian Lemon Grass Oil – Cymbopogon citratus.
West Indian Locust – Hymenaea courbaril.
West Indian Sandalwood (Oil) – Amyris balsamifera.
West Indian Snake Bark – Colubrina ferruginea.
West Indian Spigelia – Spigelia anthelmia.
Western Alder – Alnus rubra.
Western Australian Blackbutt – Eucalyptus patens.
Western Balm – Monardella parviflora.
Western Black Oak – Quercus emoryi.
Western Buckeye – Aesculus arguta.
Western Catalpa – Catalpa speciosa.
Western Flowering Dogwood – Cornus nuttallii.
Western Hackberry – Celtis reticulata.
Western Hemlock – Tsuga heterophylla.
Western Larch – Larix occidentale.
Western Mugwort – Artemisia gnaphalodes.
Western Ragweed – Ambrosia psilotstachya.
Western Red Cedar – Thuya plicata.
Western Sand Cherry – Prunus besseyi.
Western Service Berry – Amelanchier alnifolia.
Western Skunk Cabbage – Lysichitum americanum.
Western Squaw Lettuce – Hydrophyllum occidentale.
Western Soapberry – Sapindus marginatus.
Western Wheatgrass – Agropyron smithii.
Western White Oak – Quercus lobata.
Western White Pine – Pinus montana.
Western Yellow Pine – Pinus ponderosa.
Westland Pine – Dacrydium westlandicum.
Whangee Canes – Phyllostachys spp.
Wheat – Triticum aestivum.
Wheat, Astrakan – Triticum polonicum.
Wheat, Club – Triticum compactum.
Wheat, Durum – Triticum durum.
Wheat, Einkorn – Triticum monococcum.
Wheat, Emmer – Triticum dicoccum.
Wheat, Inca – Amaranthus caudatus.
Wheat, Indian – Fagopyrum tataricum.
Wheat, Indian Dwarf – Triticum sphaerococcus.
Wheat, Kharassan – Triticum orientale.
Wheat, Macaroni – Triticum durum.
Wheat, Persian – Triticum carthlicum.
Wheat, Polish – Triticum polonicum.
Wheat, Poulard – Triticum turgidum.
Wheat, Wild – Alymus triticoides.
Wheatgrass, Bluebunch – Agropyron spicatum.
Wheatgrass, Crested – Agropyrum cristaum.
Wheatgrass, Intermediate – Agropyron intermedium.
Wheatgrass, Slender – Agropyron trachycaulon.
Wheatgrass, Western – Agropyron smithii.
Wheeler Sotol – Dasylirion wheeleri.
Wingleaf Prickly Ash – Zanthoxylum alatum.
Whip Tree, Common – Luehea dicaricata.
Whish, Mexican – Epicampes macroura.
Whisky, Barley – Hordeum distichon.
Whisky, Bourbon – Zea mays.
Whisky, Canada – Secale cereale.
Whisky, Corn – Zea mays.
Whisky, Irish – Hordeum distichon.
Whisky, Rye – Secale cereale.
Whisky, Scotch – Hordeum distichon.
White Alder – Platylophus trifoliatus.
White Ash – Fraxinus americana.
White Asparagus – Asperagus albus.
White Avens – Geum canadense.
White Bark Pine – Pinus albicaulis.
White Beam – Sorbus aria.
White Beech – Gmelina leichthardtii.
White Bent – Agrostis alba.
White Birch – Schizomeria ovata.
White Bitterwood – Trichilia spondioides.
White Box – Eucalyptus hemipholia.
White Box – Tristania conferta.
White Brittlebush – Encelia farinosa.
White Brown Beard – Verbesina virginiana.
White Bryony – Bryonia alba.
White Buttonwood – Laguncularia racemosa.
White Campion – Melandrium album.
White Candytuft – Iberis amara.
White Cedar – Chamaecyparis thyoides.
White Cedar – Chukrasia tabularis.
White Cedar – Dysoxylum malabaricum.
White Cedar – Tabebuia pentaphylla.
White Cedar (Oil) – Thuja occidentalis.
White Chuglam – Terminalia bialata.
White Cinnamon – Canella alba.
White Clover – Trifolium repens.
White Crottle – Pertusaria corallina.

White Crown Beard – Verbesina virginiana.
White Cypress Pine – Callitris glauca.
White Dammar – Vateria indica.
White Dammar Pine – Agathis alba.
White Dead Nettle – Lamium album.
White Ebony – Diospyros malacapai.
White Elm – Ulmus americana.
White False Hellebore – Veratrum album.
White Fir – Abies amabilis.
White Fir – Abies concolor.
White Fringe Fungus – Fusarium aleyrodes.
White Gall – Quercus lusitanica.
White Gentian – Laserpitium latifolium.
White Gourd – Benincasa hispida.
White Gum – Eucalyptus gomphocephala.
White Gum – Eucalyptus leucoxylon.
White Gum – Eucalyptus redunca.
White Gum – Eucalyptus viminalis.
White Haiari – Lonchocarpus densiflorus.
White Hellebore – Veratrum album.
White Ipecac – Hybanthus ipecacuanha.
White Ipecac – Polygala angulata.
White Ironbark – Eucalyptus leucoxylon.
White Ironwood – Toddalia aculeata.
White Jute – Corchorus capsularis.
White Lettuce – Prenanthes altissimus.
White Lupin – Lupinus albus.
White Mahogany – Cybistax donnell-smithii.
White Mahogany – Eucalyptus robusta.
White Mahogany – Eucalyptus triantha.
White Mahogany – Khaya anthotheca.
White Mallet – Eucalyptus falcata.
White Mangrove – Avicennia officinalis.
White Mangrove – Laguncularia racemosa.
White Maple – Acer saccharinum.
White Medlar – Vangueria escylenta.
White Mulberry – Morus alba.
White Mushroom – Psalliota campestris.
White Mustard (Seed Oil) – Sinapis alba.
White Oak (Bark) – Quercus alba.
White Pear – Apodytes dimidiata.
White Pepper – Piper nigrum.
White Periera – Abuta candollei.
White Pine – Pinus albicalis.
White Pine – Podocarpus elata.
White Pine, Eastern – Pinus strobus.
White Pine, Western – Pinus monticola.
White Poplar – Populus alba.
White Quebracho – Aspidosperma quebracho.
White Rosewood – Dalbergia nigra.
White Sage – Eurotia lanata.
White Sally – Eucryphia moorei.
White Sandalwood – Santalum album.
White Sapote – Casiniroa edulis.
White Senaar – Acacia senegal.
White Silkwood – Flindersia acuminata.
White Snakeroot – Eupatorium aromaticum.
White Soap Root – Gypsophila arrostii.
White Spotted Aloe – Aloe saponaria.
White Spruce – Picea glauca.
White Stonecrop – Sedum album.
White Stopper – Eugenia axillaris.
White Stringybark – Eucalyptus eugenoides.
White Stringybark – Eucalyptus piperata.
White Tephrosia – Tephrosia candida.
Whitetop Stringybark – Eucalyptus gigantea.
White Truffle – Choiromyces venosus.
White Waterlily – Nymphaea alba.
White Willow – Salix alba.
White Wood – Atalaya hemiglauca.
White Wood – Tabebuia pentaphylla.
White Wood, African – Enantia chlorantha.
White Wood, Bahama – Canella alba.
White Yam – Dioscorea alata.
Whitefly Fungus, Brown – Aegerita webberi.
Whitespot Giant Arum – Amorphophallus campanulatus.

WHITFIELDIA Hook. f. Acanthaceae. 15 spp. Tropical Africa. Shrubs.

***W. longifolia*** T. And. Central Africa. The leaves yield a black dye, used locally.

WHITFORDIODENDRON Elmer. Leguminosae. 9 spp. Formosa, China, W.Malaysia. Trees or climbers.

***W. erianthum*** Dunn. = Adinobotrys erianthus = Milletia eriantha. Malaysia. The acid fruits are eaten locally as a vegetable.

***W. pubescens*** Burkill = Adinobotrys atropurpureus = Padbruggea pubescens. Throughout range. The heavy, soft wood is used in Burma for house construction.

Whitkowwort – Paronychia jamesii.
Whortleberry – Vaccinium myrtillus.

WIDDRINGTONIA Endl. Cupressaceae. 5 spp. Tropical and S.Africa. Trees.

***W. cupressoides*** (Schrad.) Endl. = Callitris cupressoides.

***W. juniperivides*** Endl. = Callitris arborea.

***W. whytei*** Rendle. (Mlanje Cedar). Central Africa. The pale, reddish wood is termite-resistant and is used for general construction work, furniture and particularly fencing. When burnt as fuel they give a pleasant smell.

WIKSTROEMIA Endl. Thymelaeaceae. 70 spp. S.China, Indochina to Pacific Islands and Australia. Shrubs.

***W. candolleana*** Meissn. (Bakoo bakooan).

Malaysia, Indonesia. A paste of the leaves is used in Madura to cure wounds on buffalo.

***W. canescens*** Meissn. (Gampi). Ceylon, Himalayas to China. The bark fibres are used in Japan to make paper.

***W. japonica*** Miq. Japan. See W. canescens.

***W. sikokiana*** Franch. and Sav. Japan. See W. canescens.

Wild Angelica – Angelica sylvestris.
Wild Basil – Calamintha clinopodium.
Wild Basil – Pycanthemum incanum.
Wild Bean – Phaseolus polystachyus.
Wild Beardless Rye – Elymus triticoides.
Wild Bergamot – Monarda fistula.
Wild Black Cherry – Prunus serotina.
Wild Breadfruit – Artocarpus elastica.
Wild Buckwheat – Eriogonum microthecum.
Wild Cabbage – Caulanthus glauca.
Wild Cachiman – Rollinia mucosa.
Wild Cane – Gynerium asgittatum.
Wild Cherimoya of Jaliso – Annona longiflora.
Wild Chestnut – Brabejum stellatifolium.
Wild Cinchona – Cinchona pubescens.
Wild Cinnamon – Canella alba.
Wild Coffee – Triosteum perfoliatum.
Wild Cucumber – Cucumis myriocarpus.
Wild Cucumber – Cucumis naudinianus.
Wild Date Palm – Phoenix sylvestris.
Wild Durian – Cullenia excelsa.
Wild Garlic – Allium canadense.
Wild Goose Plum – Prunus munsoniana.
Wild Gourd – Cucurbita foetidissima.
Wild Hoarhound – Eupatorium aromaticum.
Wild Hyacinth – Camassia esculenta.
Wild Hyssop – Verbena hastata.
Wild Ipecac – Apocynum androsaemifolium.
Wild Ipecac – Asclepias curassavica.
Wild Iris – Iris missouriensis.
Wild Kapiega – Bulbine asphodeloides.
Wild Loquat – Uapaca kirkiana.
Wild Leek – Allium tricoccum.
Wild Lettuce – Lactuca taraxaciflora.
Wild Mango – Irvingia gabonensis.
Wild Liquorice – Glycyrrhize lepidota.
Wild Lucerne – Stylosanthes mucronata.
Wild Marjoram, Oil of – Thymus mastichina.
Wild Medlar – Vangueriopsis lancifolia.
Wild Nutmeg – Myristica malabarica.
Wild Oat – Avena fatua.
Wild Oleander – Adina microcephala.
Wild Olive – Elaeocarpus serratus.
Wild Olive – Sideroxylon mastichodendron.
Wild Olive Seed Oil – Ximenia americana.
Wild Orange Tree – Toddalia aculeata.
Wild Potato – Anemonella thalictroides.
Wild Potato – Solanum fendleri.
Wild Rice – Oryza barthii.
Wild Rice – Zizania aquatica.
Wild Rue – Anemonella thalictroides.
Wild Safflower – Carthamus oxycanthus.
Wild Siamese Cardamon – Anomum xanthioides.
Wild Spikenard – Smilacina racemosa.
Wild Squill – Scilla rigidifolia.
Wild Tamarind – Lysiloma latisiliqua.
Wild Tea Bush – Leucas martinicensis.
Wild Turmeric – Curcuma aromatica.
Wild Wheat – Elymus triticoides.
Wild Wisteria – Bolusanthus speciosus.
Wild Yam – Discorea spinosa.
Wild Yam – Dioscorea villosa.
Wilde castanjes – Brabejum stellatifolium.
Wiliwili – Erthrina sandwicensis.

WILLARDIA Rose. Leguminosae. 6 spp. Mexico. Shrubs or small trees.

***W. mexicana*** (S. Wats.) Rose=Coursetia mexicana. (Nesco, Palo piojo, Taliste). Throughout range. A decoction of the bark is used locally to treat parasites on stock. The wood is used for mine props.

William's Arrowroot – Tacca pinnatifida.
Willow, Almond-leaved – Salix triandra.
Willow, Basket – Salix viminalis.
Willow, Black – Salix nigra.
Willow, Bluestem – Salix irrorata.
Willow, Brittle – Salix fragilis.
Willow, Cape – Salix capensis.
Willow, Caspic – Salix acutifolia.
Willow, Common – Salix caprea.
Willow, Desert – Chilopsis lineari.
Willow, Firewheel Tree – Stenocarpus salignus.
Willow, Flowering – Chilopsis lineari.
Willow, French – Epilobium angustifolium.
Willow, French – Salix triandra.
Willow, Goat – Salix caprea.
Willow, Humboldt's – Salix humboldtiana.
Willow, Indian – Salix tetrasperma.
Willow Oak – Quercus phellos.
Willow, Osier – Salix viminalis.
Willow, Purple – Salix pupurea.
Willow, White – Salix alba.
Willow, Yeddo – Salix jessoensis.
Willowherb, Rosebay – Epilobium angustifolium.

Willowleaf Eucalyptus – Eucalyptus salicifolia.
Willowleaf Red Quebrachó – Schinopsis balansae.
WILLUGHBEIA Roxb. Apocynaceae. 25 spp. Indomalaya. Vines.
***W. coriacea*** Wall. Malacca. Yields a poor quality rubber, used to adulterate others of better quality.
***W. edulis*** Roxb. India, Malaysia, Himalayas, Cochin-China. The fruits are eaten locally.
***W. firma*** Blume. Malaysia. Yields a rubber (Borneo Rubber, Getah Borneo, Getah Soesoe) which is coagulated by salt water.
***W. tenuiflora*** Dyer. (Dango, Djelo, Gitan palau ninjak). Indonesia. Yields an inferior rubber, which is coagulated by lemon juice and is plastic in hot water. It is used locally to treat skin diseases.
***W. martabanica*** Wall. Malaysia. Yields an inferior rubber (Palay Rubber).
Windflower, Grove – Anemone nemorosa.
Wine, Dandelion – Taraxacum officinale.
Wine Palm – Raphia hokkeri.
Wine Palm (of Chile) – Jubaea chilensis.
Wine Raffia Palm – Raphia vinifera.
Wine Raspberry – Rubus phoenicolasius.
Wine, Rice – Oryza sativa.
Wine, Rose – Rosa pomifera
Winged Elm – Ulmus alata.
Winged Pignut – Cycloloma atriplicifolium.
Wingleaf Prickly Ash – Zanthoxylum alatum.
Wingnut, Caucasian – Pterocarya fraxinifolia.
Wingnut, Japanese – Pterocarya rhoifolia.
Winter Berry – Gaultheria procumbens.
Winter Berry – Ilex verticillata.
Winter Cherry – Physalis alkekengi.
Winter Cress – Barbarea praecox.
Winter Crookneck – Cucurbita moschata.
Winter Fat – Eurotia lanata.
Winter Marjoram – Origanum heraclecticum.
Winter Purslane – Claytonia perfoliata.
Wintergreen (Oil) – Gaultheria procumbens.
Wintergreen, Indian – Gaultheria fragantissima.
Wintergreen, Spotted – Chimaphila umbellata.
Winter's Bark – Drimys winteri.
Winter's Bark Drimys – Drimys minteri.
Winter Vetch – Vicia villosa.
WISSADULA Medik. Malvaceae. 40 spp. Tropics especially America. Shrubs. Both spp. mentioned yield a bark fibre similar to jute.
***W. rostrata*** Planch. Tropics especially S.Africa.
***W. zeylanica*** Medic. Tropics.
WISTARIA Nutt. Leguminosae. 10 spp. E.Asia, E. N.America. Vines.
***W. brachybotrys*** Sieb. and Zucc. China, Japan. The bark fibres are used in Japan for cloth. The plant is cultivated locally.
Wisteria, Rhodesian – Bolusanthus speciosus.
Wisteria, Wild – Bolusanthrus speciosus.
Witch Hazel (Bark) (Extract) – Hamamelis virginana.
Witgatboom – Boscia albitrunca.
WITHANIA Pauquy. Solanaceae. 10 spp. S.America, S.Africa, Canaries, Mediterranean to India. Shrubs.
***W. coagulans*** Dunal. (Cheesemaker, Indian Rennet). E.India, Afghanistan. The fruits are used locally to coagulate milk.
***W. somnifera*** Dunal. Mediterranean, Africa to India. The plant is narcotic. The seeds are used in the Sudan to coagulate milk and a decoction of the roots is used by the Kaffirs to treat diseases of the rectum.
Witlook, Brussel – Cichorium intybus.
Wodierwood – Lannea grandis.
Wolf's Bane, Yellow – Aconitum lycoctonum.
Wolfberry, Anderson – Lycium andersonii.
Wolfberry, Arabian – Lycium arabicum.
Wolfberry, Chinese – Lycium chinense.
Wompologa – Radlkofera calodendron.
Wonderberry – Solanum nigrum.
Wonder Violet – Viola mirabilis.
Wong Chuk – Bambusa textilis.
Wood Anemone – Anemone nemorosa.
Wood Apple – Limonia acidissima.
Wood Bluegrass – Poa nemoralis.
Wood, Cocobola – Lecythis costaricensis.
Wood Crane's Bill – Geranium sylvaticum.
Wood, Devil – Osmanthus americana.
Wood Germander – Teucrium scorodonia.
Wood, Green – Halfordia scleroxyla.
Wood, Kerosene – Halfordia scleroxyla.
Wood Lily – Lilium philadelphicum.
Wood Nettle – Laportea canadensis.
Wood Pimpernel – Lysimachia nemorum.
Wood Snowdrop – Monotropa uniflora.
Wood Sorrel – Oxalis acetosella.
Wood, Toledo – Lecythis longipes.
Wood, Tummy – Careya arborea.

Woodland Angelica – Angelica sylvestris.
Woodruff, Dyer's – Gallium tincortia.
Woodruff, Sweet – Gallium odorata.
Woodworm, Roman – Artemisia ponyica.
Woodwort, Marsh – Stachys palustris.
Wool Crape – Tillandsia usneoides.
Wool Flower – Lachnanthes carolina.
Wool Spider – Harpagophytum procumbens.
Woolly Asiaglory – Argyreia speciosa.
Wooly Buckthorn – Bumelia lanuginosa.
Woollybutt – Eucalyptus gigantea.
Woolly Finger Grass – Digitaria pentzii.
Woolly Mint – Mentha rotundifolia.
Woolly Pyrol – Phaseolus mungo.
Woollypod Milkweed – Asclepias eriocarpa.
Wollypod Vetch – Vicia dasycarpa.
Woolly White Quebrancho – Aspidosperma tormentosum.
Wooraia, Poison – Urechites suberata.
Wooroogading – Nyssa javanica.
Wormbos – Dicoma anomala.
***Wormia excelsa*** Jack.=Dillenia excelsa.
***Wormia ochreata*** Miq.=Dillenia ochreata.
Wormseed, American – Chenopodium ambrosioides var. anthelminticus.
Wormseed – Artemisia cina.
Wormwood – Artemisia absinthium.
Wormwood – Chenopodium ambrosioides.
Wort, Cancer – Southlandia frutescens.
Wort, Holy – Verbena officinalis.
Wort, Rupture – Herniaria glabra.
Wort, Yellow – Blackstonia perfoliata.
Woundwort – Laserpitium latifolium.
Wrack, Bladder – Fucus vesiculosus.
Wrack, Knobbed – Ascophyllum nodosum.
Wrack, Lady – Fucus vesiculosus.
Wrack, Yellow – Ascophyllum nodosum.
Wray – Lolium temulentum.
WRIGHTIA R. Br. Apocynaceae. 23 spp. Tropical Africa, Asia, Australia. Small trees.
***W. javanica*** DC. (Benteli lalaki, Bintaos). Java. The fine, white, durable wood is used locally for carving.
***W. pubescens*** R. Br. (Dediteh, Mentaos, Tawas). The white, fine wood is used locally for carving, house-building, tool handles and other small articles.
***W. tinctoria*** Por. (Duahi, Kalakudi). India. The light yellow, hard wood is used for carving a wide variety of articles, e.g. combs, chessmen. The leaves yield a dark blue dye (Mysore Pala Indigo).
***W. tomentosa*** Roem. and Schult. (Daira, Duahi, Pala). India. For uses of the wood see W. tinctoria. The bark from the stem and the roots is used locally to treat snake bites and a yellow dye from the leaves is used in Nepal to stop bleeding.
Wrinkled Lichen – Parmelia caperala.
Wu-pei-tzu – Rhus semialata.
Wuda – Lannea grandis.
Wur – Ficus benghalensis.
WYETHIA Nutt. Compositae. 14 spp. W. N.America. Herbs.
***W. amplexicaulis*** Nutt. The heated roots are eaten by the local Indians after fermentation.
***W. longicaulis*** Gray. See W. amplexicaulis.
***W. robusta*** Nutt. Throughout range. The seeds are eaten by the local Indians.

# X

Xa meh – Dialium cochinchinensis.
Xa pho – Dillenia ovata.
Xà sàng – Ligusticum monnieri.
Xaché – Manihot carthaginensis.
XANTHIUM L. Compositae. 30 spp. Cosmopolitan. Herbs.
***X. canadense*** Mill. (Burweed) E. N.America. A decoction of the leaves was used to improve the appetite, and the leaves are used as an adulterant for Datura stramonium.
***X. spinosum*** L. (Cotweed). N.America. The leaves are used locally as a mild sedative, to induce sweating and as a diuretic. They are also used as an antidote to snake bite.
***X. stramonium*** L. (Broad Cocklebur, Burweed, Sea Burdock). Cosmopolitan. The leaves (Herba Lappae minores) are used to improve the appetite and as a diuretic.
XANTHOPHYLLUM Roxb. Xanthophyllaceae. 60 spp. Indomalaya. Trees.
***X. excelsum*** Miq. Java. The bark yields a yellow dye, used locally. It is also used as a decoction against stomach pains.
***X. glaucum*** Wall. Burma, Indochina. The leaves are used in Indochina to give a bitter flavour to beer.
***X. lanceolatum*** J. J. Sm. (Suir Tree). The seeds yield an oil (Suir Oil, Sioerno Oil)

used locally for cooking and exported to Europe to make soap and candles.

XANTHORIA Ach. Parmeliaceae. Lichens.

***X. candelaria*** (Ach.) Arn.=Teloschistes candelarius. Temperate and Arctic. In Sweden it is used as the source of a yellow dye, used to dye wool.

***Xanthorrhiza*** Marshall.=Zanthorhiza.

XANTHORRHOEA Sm. Xanthorroeaceae. 15 spp. Australia. Woody, tree-like monocotyledons.

***X. arborea*** R. Br. (Dackowar Grass Tree). Australia. The young leaf-bases are eaten as a vegetable by the aborigines.

***X. australis*** R. Br. (Australian Grass Tree). Australia. The trunk yields a gum (Grass Tree Gum, Red Acaroid Gum, Yakkagum Gum, Resina Lutea) used to make varnishes.

***X. hastilis*** R. Br. (Black Boy, Char, Spearleaf Grass Tree). Yields a gum (Botany Bay Gum, Grass Tree Gum, Gummi Acaroides, Resina Acaroides, Yellow Acaroid Resin), used to make varnish.

***X. preussii*** Endl. See X. hastilis.

***X. quadrangulata*** F. v. Muell. S.Australia. The gum from the trunk is used by the aborigines to glue weapon parts.

XANTHOSOMA Schott. Araceae. 45 spp. Mexico to tropical S.America, W.Indies. Herbs. The rhizomes of the following spp. are starchy and edible.

***X. atrovirens*** C. Koch. (Darkleaf Malange, Indian Kale). N. S.America. The rhizomes are eaten locally.

***X. brasiliense*** (Desf.) Engl. (Belembe). W.Indies, S.America. The leaves are eaten as a vegetable. The plant is cultivated locally.

***X. caracu*** Koch. and Bouché. (Caracu). Tropical S.America, W.Indies. The corms and leaves are eaten as vegetables. The plant is cultivated in W.Indies.

***X. jacquinii*** Schott. (Yautia Palma). W.Indies, N. S.America. Grown locally to feed to pigs.

***X. sagittifolium*** Schott. (Malanga amarilla, Malanga blanca, Tania, Yautio amarilla, Yellow Yautia). Tropical America. The plant is grown locally throughout the Old and New World tropics for the starchy rhizomes which are eaten as a vegetable. The leaves are also eaten cooked.

***X. violaceum*** Schott. (Indian Kale, Otó, Primrose Malanga). See X. sagittifolium.

Xcanol-ak – Bignonia unguis-cati.

Xeng con thon – Tinospora sinensis.

XEROCHLAMYS Baker. Sarcolaenaceae. 16 spp. Madagascar. Shrubs.

***X. pilosa*** Bak. (Antsikana, Hatsikana). Madagascar. The root bark is used locally to flavour liqueurs.

XEROPHYLLUM Michx. Liliaceae. 3 spp. N.America. Herbs.

***X. tenax*** (Pursh.) Nutt. (Common Bear Grass). W. N.America. The leaves are used by the local Indians to make watertight baskets.

XEROSPERMUM Blume. Sapindaceae. 20–25 spp. Assam, S.E.Asia, W.Malaysia. Trees.

***X. intermedium*** Radlk. Malaysia. The hard wood is used locally for general construction work.

***X. noronhianum*** Blume. (Aloore, Booron-dool). Indonesia. The hard wood is used locally for agricultural implements, pestles, etc.

XIMENIA L. Olacaceae 10–15 spp. Tropical America, tropical S.Africa, tropical Asia, Australia.

***X. americana*** L. (Tallow Wood). Tropics and subtropics. The fruits are eaten locally, raw or pickled. The seeds yield a non-drying oil (Wild Olive Seed Oil) used in India for cooking. The wood is used as a substitute for Sandalwood in the Sudan.

Xmisibcoc – Turnera diffusa.

Xpayumak – Capparis flexuosa.

Xunh xe – Kadsura coccinea.

XYLARIA Hill ex Grev. Xylariaceae. 100 spp. Cosmopolitan.

***X. obovata*** Berk.=Coelorhopalon obovatum. Malaya. The powdered sclerotia, mixed with coconut oil is a local treatment for burns.

XYLIA Benth. Leguminosae. 15 spp. Tropical Africa, Madagascar, tropical Asia. Trees.

***X. dolabriformis*** Benth.=X xylocarpa. Malaysia, Burma, Siam. The hard, reddish wood is used locally for house-building, railway sleepers, bridges and ship-building.

***X. xylocarpa*** Taub.=X. dolabriformis.

XYLINABARIA Pierre. Apocynaceae. 2 spp. Indochina, Java. Shrubs.

***X. reynaudi*** Jum. Tonkin. Yields a latex.

XYLOPIA L. Annonaceae. 100–150 spp. Tropics, especially Africa. Trees.

***X. aethiopica*** (Dun.) Rich.=Uvaria aethiopica. Tropical Africa. The seeds (African Grains of Selim, Guinea Pepper, Negro Pepper, Piper aethiopicum) are used locally as a condiment and were used to a great extent in Europe. The light yellow, elastic wood is used for masts of small boats, locally.

***X. antunesii*** Engl. and Diels. S.W.Africa. A decoction of the roots is used by the local natives to cure sterility.

***X. carminativa*** (Arunda) Fries.=Unonia carminativa. (Pao d'Embira, Pimenta de Macaco). Brazil to N.E. S.America. The fruits are used as a condiment and locally to relieve stomach complaints. The bark fibres are used for cordage.

***X. concolor*** Willd. Guyana. The fruits are used locally as a condiment.

***X. discolor*** Vahl. Tropical Asia. The fruits are used as a condiment. A decoction of the roots is used locally to treat dysentery. An essential oil from the flowers is used in perfumery.

***X. ferruginea*** Hook. f. and Th. (Kelili jongkong). Malaysia. The light, soft wood is used locally for general construction work and packing cases.

***X. frutescens*** Aubl. Tropical America. The fruit is used locally as a condiment and the bark fibres as cordage.

***X. grandiflora*** St. Hil. (Malagueto). Tropical America to Panama. An essential oil from the flowers is used in Brazil to treat digestive upsets and as a tonic.

***X. obtusifolia*** A. Rich.=Habzelia obtusifolia. (Guimba). Cuba. The yellow wood (Pico de Gallo) is used locally for furniture etc.

***X. parvifolia*** Hook. f.=Unona tripetalodea. Ceylon. The flowers, fruits and root bark are chewed locally as a substitute for betel and to cure ulcers.

***X. sericea*** St. Hil. Brazil. The fruits are used locally as a condiment.

***X. striata*** Engl. Cameroons. The wood (Bosé Wood) is used locally for furniture etc.

***X. undulata*** Beauv. Tropical Africa. The fruits are used locally as a condiment.

XYRIS Gronov. ex L. Xyridaceae. 250 spp. Tropics and subtropics. Herbs.

***X. ambigua*** Bayr. (Vare Goldies). Coasts of S.E. U.S.A. A decoction of the plant is used locally to treat colds, coughs, etc.

***X. caroliniana*** Walt. E. N.America. The leaves and roots are used locally as a poultice for skin diseases.

***X. communis*** Kunth.=X. indica=X. laxifolia. Tropical America. The roots are used in Brazil to treat skin diseases, including leprosy.

***X. indica*** Vell.=X. communis.

***X. laxifolia*** Mart.=X. communis.

***X. melanocephala*** Miq. Indonesia. The leaves are used locally to make mats.

XYSMALOBIUM R. Br. Asclepiadaceae. 1 sp. Tropical and S.Africa. Herbs.

***X. heudelotianum*** Decne. The roots (Yakhop) are eaten locally as a vegetable.

# Y

Yá-ka – Virola calophylla.
Yakhop – Xysmalobium hendelotianum.
Yá-lo – Virola calophylla.
Yacon Strawberry – Polymnia edulis.
Ya'nte – Conocybe siligineoides.
Yaga-na – Guaiacum coulteri.
Yagua – Scheelea humboldtiana.
Yajé – Banisteriopsis caapi.
Yakal Shorea – Shorea balangeran.
Yakasa – Ficus exasperata.
Yakka Gum – Xanthorrhoea australis.
Yam, Affun – Dioscorea cayenensis.
Yam, Akam – Dioscorea latifolia.
Yam, Atlantic – Dioscorea villosa.
Yam Bean – Pachyrrhizus angulatus.
Yam Bean – Pachyrrhizus bulbosus.
Yam Bean – Pachyrrhizus tuberosus.
Yam Bean – Sphenostylis schweinfurthii.
Yam, Chinese – Dioscorea batatas.
Yam, Dye – Dioscorea rhipogonoides.
Yam, Fancy – Dioscorea esculenta.
Yam, Greater – Dioscorea alata.
Yam, Guinea – Dioscorea cayenensis.
Yam, Lesser Asiatic – Dioscorea esculenta.
Yam, Malacca – Dioscorea atropurpurea.
Yam, Negro – Dioscorea cayenensis.
Yam, Potato – Dioscorea bulbifera.
Yam, Potato – Dioscorea esculenta.
Yam, Spiny – Dioscorea spinosa.

Yam, Ten-month – Dioscorea alata.
Yam, Twelve-month – Dioscorea cayenensis.
Yam, Water – Dioscorea alata.
Yam, White – Dioscorea alata.
Yam, Wild – Dioscorea spinosa.
Yam, Wild – Dioscorea villosa.
Yam, Yellow – Dioscorea cayenensis.
Yama Momo – Myrica nagi.
Yama Rosewood – Platymiscium dimosphandrum.
Yamagua – Guarea guara.
Yambi – Dioscorea trifida.
Yampa – Carum gairdneri.
Yan chuk – Bambusa pervariabilis.
Yana múco – Neea parviflora.
Yanga – Geonoma dominicana.
Yaquilla – Pouzolzia occidentalis.
Yareta – Azorella caespitosa.
Yareta – Azorella yareta.
Yaray Palm – Sabul causiarum.
Yarran, Curly – Acacia homalophylla.
Yarrow – Achillea lanulosa.
Yarrow – Achillea millefolium.
Yarrow, Musk – Achillea moschata.
Yaruru – Aspidosperma excelsum.
Yaupon – Ilex vomitoria.
Yautio amarilla – Xanthosoma sagittifolium.
Yautia Palma – Xanthosoma jacquinii.
Yautia, Yellow – Xanthosoma sagittifolium.
Yaw Root – Stillingia sylvatica.
Yaxci – Agave sisalana.
Yeburi – Atropa acuminata.
Yeddo Spruce – Picea jezoensis.
Yeddo Willow – Salix jessoensis.
Yeh-hua-tsen – Melilotus macrorhiza.
Yeheb – Cordeauxia edulis.
Yellow Acaroid Resin – Xanthorrhoea hastilis.
Yellow Adder's Tongue – Erythronium americanum.
Yellow Ash – Cladrastis tinctoria.
Yellow Avens – Geum macrophyllum.
Yellow Box – Eucalyptus hemiphloia.
Yellow Box – Eucalyptus melliodora.
Yellow Balsam – Impatiens noli-tangere.
Yellow Bark – Cinchona calisaya.
Yellow Bark – Cinchona ledgeriana.
Yellow Batchelor's Button – Polygala rugelii.
Yellow Berries – Rhamnus infectorius.
Yellow Birch – Betula lutea.
Yellow Borrera – Theloschistes flavicans.
Yellow Box Wood – Sideroxylon pohlmannianum.
Yellow Buckeye – Aesculus octandra.
Yellow Carabeen – Sloanea woollsii.
Yellow Cedar – Rhodosphaera rhodanthema.
Yellow Cirouaballi – Nectandra pisi.
Yellow Cypress – Chamaecyparis nootkatensis.
Yellow Elder – Tecoma stans.
Yellow Flag – Iris pseudacorus.
Yellow-flowered Alfalfa – Medicago falcata.
Yellow Fiji Sandalwood – Santalum freycinetianum.
Yellow Gentian – Gentiana lutea.
Yellow Groove Bamboo – Phyllostachys aureosulcata.
Yellow Heart Prickly Ash – Zanthoxylum flavum.
Yellow Himalaysian Raspberry – Rubus ellipticus.
Yellow Jessamine – Gelsemium sempervirens.
Yellow Leafcup – Polymnia uvedalia.
Yellow Mombin – Spondias lutea.
Yellow Nut Grass – Cyperus esculentus.
Yellow Oleander – Thevetia nereifolia.
Yellow Olivier – Terminalia obovata.
Yellow Pareira – Aristolochia glaucescens.
Yellow Pine – Pinus echinata.
Yellow Pond Lily – Nuphar advena.
Yellow Poplar – Liriodendron tulipifera.
Yellow Poui – Tecoma serratifolia.
Yellow Rocket – Barbarea vulgaris.
Yellow Resin – Pinus palustris.
Yellow Root – Zanthorhiza apiifolia.
Yellow Sand Verbena – Abroma latifolia.
Yellow Sanders – Terminalia hilariana.
Yellow Sapote – Lucuma salicifolia.
Yellow Silver Pine – Dacrydium intermedium.
Yellow Spined Thistle – Cirsium ochrocentrum.
Yellow Star Anise – Illicium parviflorum.
Yellow Starwort – Inula helenium.
Yellow Stringbark – Eucalyptus muelleriana.
Yellow Stringbark – Eucalyptus triantha.
Yellow Sweet Clover – Melilotus officinalis.
Yellow Tea Tree – Leptospermum flavescens.
Yellow Toadflax – Linaria vulgaris.
Yellow Velvetleaf – Limnocharis emarginata.
Yellow Yautia – Xanthosoma sagittifolium.
Yellow Walnut – Beilschmiedia bancroftii.
Yellow Wild Indigo – Baptisia tinctoria.

Yellow Wolf's Bane – Aconitum lycoctonum.
Yellow Wood – Cladrastis tinctoria.
Yellow Wood – Flindersia oxleyana.
Yellow Wood – Podocarpus thunbergii.
Yellow Wood – Schaefferia frutescens.
Yellow Wood – Sterculia oblonga.
Yellow Wood – Terminalia sericea.
Yellow Wood – Zanthoxylum flavum.
Yellow Wood Ash – Flindersia oxleyana.
Yellow Wood Extract – Chlorophora tinctoria.
Yellow Wood, Marianne – Ochrosia mariannensis.
Yellow Wort – Blackstonia perfoliata.
Yellow Wrack – Ascophyllum nodosum.
Yellow Yam – Dioscorea cayenensis.
Yellow Zedoary – Curcuma aromatica.
Yellowbark Cinchona – Cinchona calisaya.
Yellowberry – Rubus chamaemorus.
Yellowwood, Quinteniaqua – Podocarpus elongata.
Yerba Buena – Micromeria chamissonis.
Yerba Buena – Micromeria douglassi.
Yerba del Angel – Eupatorium collinum.
Yerba de la Sangre – Cordia globosa.
Yerba de la Virga – Rivea corymbosa.
Yerba de María – Salvia divinorum.
Yerba Maté – Ilex paraguensis.
Yerba Reuma – Frankenia grandifolia.
Yerba Santa – Eriodictyon californica.
Yew, American – Taxus canadensis.
Yew, English – Taxus baccata.
Yew, Japanese – Taxus cuspidata.
Yew, Pacific – Taxus brevifolia.
Yezomatsu – Picea jezoensis.
Ylang Ylang – Cananga odorata.
Ylang Ylang Oil – Canangium odoratum.
Ymira Piranga – Caesalpinia echinata.
Yo yotl – Thevetia thevetioides.
Yoco – Paullinia yoco.
Yoghurt – Lactobacillus bulgaricus.
Yohimbe Bark – Pausinystalia yohimba.
Yoke – Piptadenia peregrina.
Yolombo – Panopsis rubescens.
Yorkshire Fog – Holcus lanatus.
Yoruba Ebony – Diospyros mombuttensis.
Yoruba Indigo – Lonchocarpus cyanescens.
Ysaño – Tropaeolum tuberosum.
Yuca – Manihot esculenta.
Yuca del Monte – Manihot carthaginensis.
Yuca dulce – Manihot dulce.
Yucatan Sisal – Agave fourcroyoides.

YUCCA L. Agavaceae. 40 spp. S. United States of America, Mexico, W.Indies. Small trees. The leaves of all the spp. mentioned yield a fibre which is used locally for twine, cloth and baskets.

*Y. **aloifolia*** L. (Spanish Bayonet, Spanish Dagger). S.E. U.S.A., W.Indies.

*Y. **angustifolia*** Pursh.=Y. glauca.

*Y. **arborescens*** Trél. (Joshua Tree). S.W. U.S.A. The seeds are eaten by the local Indians and the wood is used to make small boxes etc.

*Y. **australis*** (Engl.) Trél. Mexico. The leaf fibres (Ixtle) are exported. The young stems and leaves are fermented to make alcohol and the pith from the stem is used to make mats and pads for pack animals.

*Y. **baccata*** Torr. (Dátil). S.E. U.S.A., Mexico. The fruits and flower buds are eaten locally. The fruits are dried for winter use.

*Y. **elata*** Engl. (Soap Tree Yucca). S. U.S.A. to Mexico. The roots are used locally as a soap substitute for washing clothes. They are also used to hold a foam on drinks.

*Y. **elephantipes*** Regel. (Itabo, Izote, Palmita). Mexico to Central America. The flowers are eaten locally in soups or with eggs. The plant is grown as terracing for coffee plantations.

*Y. **endlichiana*** Trél. Mexico.

*Y. **filamentosa*** L. (Adam's Needle, Eve's Needle, Common Yucca). S.E. U.S.A. The fruits are eaten locally and the roots are used as a soap substitute.

*Y. **funifera*** Koch.=Hesperoyuccua whipplei.

*Y. **glauca*** Nutt.=Y. angustifolia. (Small Soapwort). S.W. U.S.A. to Mexico. The flowers, stems and fruits are eaten as vegetables by the local Indians and they use the roots as a soap substitute.

*Y. **gloriosa*** L. S.E. U.S.A.

*Y. **macrocarpa*** (Torr.) Cav. (Palma criolla). S.W. U.S.A. to Mexico. The seeds are eaten by the local Indians.

*Y. **mohavensis*** Sarg. (Mohave Yucca). S.W. U.S.A. The fibres have been used commercially in war-time.

*Y. **recurvifolia*** Salisb. S.E. U.S.A.

*Y. **rostrata*** Engl. (Soyate). See Y. australis.

*Y. **treculeana*** Carr. (Trecul Yucca). Texas to Mexico.

*Y. **valida*** T. S. Brandeg. Mexico. The leaves are used for thatching and the trunks for walls of houses etc.

*Y. **whipplei*** Torr.=Hesperoyucca whipplei.

Yucca, Chaparral – Hesperoyucca whipplei.
Yucca cimarrona – Jatropha multifida.

Yucucaca – Thevetia thevetioides.
Yumete – Asclepias subulata.
Yunnan Hemlock – Tsuga yunnanensis.
Yuquilla – Manihot carthaginensis.
Yuraguana – Thrinax miraguana.

# Z

Za-bà-a – Dalea enneandra.
Zacate – Nolina longifolia.
Zachun Oil – Balanites roxburghii.
Zakana – Boswellia carteri.
Zakaton Grass – Epicampes macroura.
***Zalaccia*** Reinw. ex Bl. = Salacca.
Zamandoque (Fibre) – Hesperoyucca whipplei.
Zambezi Redwood – Baikiaea plurijuga.

ZAMIA L. Zamiaceae. 30–40 spp. Tropical America, W.Indies. Cycads.

***Z. floridana*** DC. = Z. integrifolia.

***Z. furfuracea*** Ait. Mexico to Central America. The starch from the roots is eaten by the local Indians, after heating to destroy the poisonous principle. The roots are poisonous, used as a criminal poison in Honduras.

***Z. integrifolia*** Ait. = Z. floridana. (Coontie). The starch from the stem is eaten by the local Indians and was sold as Florida Arrowroot.

***Z. latifolia*** Pren. Central America, W.Indies. The plant is poisonous, but the starch is eaten locally after heating. It is also used for laundering clothes.

***Z. lindenii*** Regel. Ecuador, Columbia. The seeds are eaten by the local Indians.

ZANTHOXYLUM L. (sensu Engl.) Rutaceae. 20–30 spp. Temperate and subtropical E.Asia, Philippines, E.Malaysia, N.America. Trees.

***Z. alatum*** Steud. = Z. planispinum Sieb. and Zucc. (Wingleaf Prickly Ash). China. The seeds are used in China to relieve stomach ache.

***Z. americanum*** Mill. (Northern Prickly Ash, Toothache Tree). E. N.America. The dried bark is used medicinally to relieve toothache and rheumatism.

***Z. avicennae*** (Lam.) DC. = Fagara avicennae = F. pterota. S.E.China, Philippines. A decoction of the stem is used locally to relieve stomach aches and as an antidote to snake bites.

***Z. budrunga*** Wall. = Fagara rhetsa.

***Z. bungei*** Planch. China. The seeds (Chinese Pepper) are used locally as a condiment. The plant is cultivated locally.

***Z. capense*** Harv. = Fagara capensis. (Fever Tree, Knobwood, Pramdoring). The fruits are used locally to treat stomach pains and flatulence. A decoction of the leaves is used to treat intestinal worms.

***Z. celebicum*** Koord. (Koomaroko, Tokooloo). The wood is used locally for general construction work and boat-building.

***Z. clava-herculis*** L. = Z. fraxineum (Hercules' Club Prickly Ash, Toothache Tree). S. U.S.A. A decoction of the bark is used medicinally to treat toothache and rheumatism. The berries are used as a tonic and stimulant.

***Z. coriaceum*** A. Rich. = Z. emarginatum.

***Z. emarginatum*** Sw. = Z. coriaceum = Fagara coriacea. W.Indies. The yellow wood is used locally for general construction work.

***Z. flavum*** Vahl. (Satin Wood, Yellow Heart Prickly Ash, Yellow Wood). The hard, heavy, brittle wood is bright orange in colour. It is used for furniture making, tool handles, small carvings etc.

***Z. fraxineum*** Willd. = Z. clava-herculis.

***Z. martinicense*** (Lam.) DC. (Espino Rubial, Martinique Prickly Ash). W.Indies. The light brown to greenish wood is hard, fairly heavy and works well. It is susceptible to dry-wood termites. The wood is used for furniture, boxes and window frames etc.

***Z. microcarpum*** Griseb. Central America. The wood is used for general construction work.

***Z. nitidum*** DC. = Fagara piperita. China, Cochin-China. The seeds are used locally as a condiment.

***Z. panemense*** Wilson. Central America. The wood is used for general construction work.

***Z. piperatum*** DC. (Japanese Prickly Ash). Japan. The seeds (Japan Pepper) are used locally as a condiment.

***Z. planispinum*** Sieb. and Zucc. = Z. alatum.

***Z. rhetsea*** DC. India, Indochina. The fruits are used locally as a condiment.

***Z. roxburgianum*** Cham. = Euodia lepta.

***Z. senegalense*** DC. (Senegal Prickly Ash).

Tropical Africa, mainly W. The seeds are used locally to treat rheumatism and the bark to induce sweating.
Zanzibar Aloë – Aloë perryi.
Zanzibar Copal – Trachylobium hornemannianum.
Zanzibar Copal – Trachylobium verrucosum.
Zanzibar Ebony – Diospyros mespiliformis.
Zanzibar Olive Vine – Telfairia pedata.
Zapatero – Tabebuia pentaphylla.
Zapote Amarillo – Couepia polyandra.
Zapote Amarillo – Lucuma salicifolia.
Zapote Blanco – Casimiroa edulis.
Zapote Cabillo – Licania platypus.
Zapote de ave – Sideroxylon capiri.
Zapote Negro – Diospyros ebenaster.
Zapotilla amarillo – Crataeva tapia.
Zapupe Azul – Agave zapupe.
Zapupe de Tartoyuca – Agave zapupe.
Zapupe de Tepetzintia – Agave lespinassei.
Zapupe Estopier – Agave zapupe.
Zapupe Verde – Agave zapupe.
Zapupe Vincent – Agave zapupe.
ZANTHORHIZA L.Hérit. Ranunculaceae. 1 sp. E. N.America.
***Z. apiifolia*** L. Hérit. (Yellow Root). The roots yield a yellow dye.
Zarza – Pisonia aculeata.
Zarzamora – Rubus floribundus.
ZEA L. Graminae. 1 sp. Tropical, Central and S.America. Grass.
***Z. mays*** L. (Maize, Corn, Indian Corn). The plant has been grown as a grain crop in Central America since at least 2000 B.C. It was introduced into Europe by Columbus in 1492 and from here spread rapidly to all areas of the world where it could be grown. The annual world production is about 200 million tons, over half of which is produced in U.S.A. Z. mays probably originated by a cross:—primitive Zea sp. × Manisurus sp. to give teosinte (Euchlaena mexicana) which back-crossed with a primitive Zea spp. 80 per cent of the U.S.A. crop is fed to livestock. The grain is used for flour which is made into a wide variety of starchy foods and breads throughout the world. They are also used for breakfast foods including corn flakes. The grain is also fermented to make whisky and industrial alcohol. The starch is used in cosmetics, adhesives and in the manufacture of glucose, syrup. The oil from the embryo (Corn Oil) (Sp. Gr. 0·921–0·927, Iod. No. 116–130, Sap. Val. 188–193, Unsap. 1·3–2·0 per cent) is used as salad oil, to make linoleum, paints, varnishes, soap and glycerine. There are a great number of cultivars which can be divided into 8 groups: (1) Dent Corn – white endosperm the main type grown in U.S.A. (2) Sweet Corn (Soft Corn) with a high sugar content – eaten as a vegetable, canned and frozen. (3) Flint Corn – hard endosperm, but lower starch content than dent, grown in N. U.S.A. and tropical lowlands. (4) Popcorn – no soft starch, used for confectionery, after the grain has been burst by heating. (5) Flour Corn (Soft Corn, Squaw Corn) soft-grained high starch. Easily ground and preferred by the Indians. It is not grown in U.S.A., but in the Andes. (6) Pod Corn – no economic importance. (7) Ceratina – with a waxy endosperm, probably originated as a mutation in the Far East where it is grown. (8) Starchy Sweet Corn – the lower part of the grain contains starch and the upper part sugar – of no economic importance.
Zebra Wood – Centrolobium robustum.
Zebra Wood – Diospyros kurzii.
Zebra Wood, Andaman – Diospyros kurzii.
Zebra Wood, Congo – Julbernardia sereti.
Zedoary (Zedoaria) – Curcuma zeodoaria.
Zedoary, Yellow – Curcuma aromatica.
ZEHNERIA Endl. Cucurbitaceae. 30 spp. Old World tropics. Vines.
***Z. connivens*** Miq. = Melothria heterophylla = Bryonia sagittata. (Kilong lagong). Indonesia. The unripe fruits are eaten locally as a vegetable.
ZELKOVA Spach. Ulmaceae. 6–7 spp. E.Mediterranean, Caucasus, E.Asia. Trees.
***Z. keaki*** Mayr. = Z. serrata.
***Z. serrata*** (Thunb.) Mak. = Z. keaki (Japanese Zelkova). Japan. The light yellow, tough wood is used locally for house-building, furniture, ships and tool handles.
Zelkova, Japanese – Zelkova serrata.
Zengakiveli – Craiba grandiflora.
ZEPHYRANTHES Herb. Amaryllidaceae. 35–40 spp. Warm America, W.Indies. Herbs.
***Z. atamasco*** Herb. (Atamasco Lily). S. U.S.A. The local Indians used the bulbs as an emergency food.
Zerumbet Ginger – Zingiber zerumbet.

ZEUXINE Lindl. Orchidaceae. 76 spp. Old World, tropics and subtropics. Herbs.

***Z. strateumatica*** (L.) Ames.=Orchis strateumatica. Tropical Asia. The tubers are used in India to make salep.

***Zexmenia frutescens*** (Mill.) Blake.=Verbesina fruticosa.

***Zeyhera*** DC.=Zeyheria.

ZEYHERIA Mart. Bignoniaceae. 2 spp. Brazil. Shrubs.

***Z. tuberculosa*** D. Don. Brazil. A non-drying oil (Zeyheria Oil) is expressed from the seeds.

ZILLA Forsk. Cruciferae. 3 spp. N.Africa to Arabia. Herbs.

***Z. myagroides*** Forsk. Throughout range. The leaves are eaten locally as a vegetable.

ZINGIBER Boehm. Zingiberaceae. 80–90 spp. E.Asia, Indomalaya, N.Australia. The rhizomes of all the spp. mentioned are used as a condiment or for flavouring.

***Z. amaricans*** Noron. Malaysia. Used locally.

***Z. cassumunar*** Roxb. (Cassumunar Ginger). China. Cultivated and used locally.

***Z. mioga*** (Thunb.) Rosc. (Japanese Ginger, Mioga Ginger). Japan. Cultivated and used locally. The flowers, fruits and shoots are eaten locally.

***Z. officinale*** Rosc. (Ginger). E.India. Cultivated in Old and New World tropics. It has been used as a condiment in the Far East since ancient times and was known as such to the Romans. Now most of the commercial ginger is grown in Jamaica. The rhizomes are prepared for market by peeling them, partly boiling them and then they are sun-dried before being sold in pieces or as a powder. Ginger is used as a flavouring, in confectionery, cakes etc., in wines and to make a beer. It is used medicinally to relieve stomach upsets. Green Ginger is produced mainly in China, where the rhizomes are prepared as above, but them boiled in sugar to preserve them. Ginger owes its particular flavour to the presence of zingiberine and gingerol.

***Z. zerumbet*** (L.) Sm. (Martinique Ginger, Zerumbet Ginger). Tropical Asia. Cultivated in the Far East and Martinique.

Zit-el-Harmel – Peganum harmala.

ZIZANIA Gronov. ex L. Graminae. 1 sp. N.E.India, Burma, E.Asia; 2 spp. N.America. Grasses.

***Z. aquatica*** L. (Indian Rice, Wild Rice). E. N.America. The seeds are eaten by the local Indians and used to attract gamebirds.

***Z. latifolia*** Turcz. China, Japan, Korea. The leaves are used locally as a vegetable and when older made into mats.

ZIZIPHUS Mill. Rhamnaceae. 100 spp. Tropical America, Africa, Mediterranean, Indomalaya, Australia. Shrubs or small trees.

***Z. abyssinica*** Hochst. ex A Rich. (Catchthorn, Kankona, Musau). Central Africa. The fruits are used in Malawi to make an alcoholic drink (Kachaso). A decoction of the roots is used to induce abortions and treat elephantiasis. The wood is used to make wagons.

***Z. calophylla*** Wall. (Dawai-dawai). Malaya. The fruits are eaten locally.

***Z. endlichii*** Loes. Mexico. The fruits are eaten and the bark is chewed to relieve toothache.

***Z. joageiro*** Mart. Brazil. The fruits are eaten locally and the plant is used as an emergency food for livestock. A decoction of the leaves is used to treat digestive upsets and a decoction of the bark to treat urinary complaints.

***Z. jujuba*** Lam =Z. mauritania.

***Z. jujuba*** Mill. non Lam.=Z. vulgaris. (Jujub, Chinese Jujub). E.India, Malaysia, China, Japan. The plant is cultivated locally and as far W. as the Mediterranean. The fruits are eaten raw, dried or stewed with millet or rice, or in a syrup of sugar or honey. The mucilage from the fruits (Pâté de Jujubes) was used to make bronchial pastilles. There are several varieties.

***Z. lotus*** Desf. (Lotus, Lotophages). N.Africa. The fruits are edible – the Lotus of the Ancient World.

***Z. lycioides*** Gray.=Condalia lycioides.

***Z. mauritiana*** Lam.=Z. jujuba Lam. non Mill.=Z. sosoria. Tropical Africa, Asia. The fruits (Chinese Dates, Tsa) are eaten locally. The leaves are used for tanning.

***Z. mexicana*** Rose. Mexico. The fruits are used locally as a soap substitute for washing clothes.

***Z. microphylla*** Roxb.=Z. nummularia.

***Z. mistol*** Griseb. (Argentine Jujub). Andes. The fruits are eaten locally and fermented to make a drink (Chicha).

***Z. mucronata*** Willd. (Buffalo Thorn, Cape Thorn). Tropical Africa. The fruits are eaten locally and a decoction of the bark is used to treat lumbago. The seeds are used as beads.

***Z. nummularia*** (Burm.) Wight. and Arn.= Z. microphylla=Z. rotundifolia Lam. non DC. N.India to Sahara. The fruits are eaten in the Sahara region and the wood is used for fences in India.

***Z. rotundifolia*** Lam. non DC.=Z. nummularia.

***Z. sonorensis*** S. Wats. Mexico. The fruits are used as a soap substitute.

***Z. sosoria*** Roem. and Schult.=Z. mauritiana.

***Z. spina-Christi*** Willd. (Christ Thorn). N.Africa, Arabia. The fruits are eaten locally.

***Z. xylocarpus*** Willd. (Bhander, Ghot her, Katti ber). India. The hard, durable, brown wood, is used locally to make houses, agricultural implements, carts and torches. The bark and fruits are used to tan leather black. The young shoots, leaves and fruits are fed to livestock.

ZIZYPHORA L. Labiatae. 25 spp. Mediterranean to Central Asia and Afghanistan. Herbs.

***Z. tenuior*** L. Throughout range. The leaves are used in Iran to make a drink and to relieve stomach upsets.

***Zizyphys*** Adans.=Ziziphus.

ZOLLERNIA Maximil. and Nees. Leguminosae. 10 spp. Central and tropical America. Trees.

***Z. ilicifolia*** Vog. (Mocitahyba). Brazil. The wood is used locally for furniture making, railway sleepers and beams. The bark is used for tanning.

***Z. paraensis*** Huber. (Muirapinima Preta). Amazon, Brazil. The wood is popular in U.S.A. and locally for furniture, knife handles, brush-backs, butts of billiard cues and turning.

Zompopo – Cordia globosa.

Zooagoo – Morinda lucida.

ZORNIA J. F. Gmel. Leguminosae. 75 spp. Tropics especially America. Herbs.

***Z. diphylla*** (L.) Pers. Tropics. The leaves are used locally as an emergency food for horses. A decoction of the leaves is used to treat dysentery.

Zorrillo – Cestrum lanatum.

Zorrillo – Petiveria alliacea.

ZOSTERA L. Zosteraceae. 10 spp. Subtropics, temperate, subarctic, subantarctic. Marine herbs.

***Z. marina*** L. (Alva marina, Grass Weed, Grass Wrack). Europe. The dried leaves and stems are used for stuffing pillows etc. and as a packing material.

ZSCHOKKEA Muell. Arg. Apocynaceae. 15 spp. Tropical S.America. Trees.

***Z. foxii*** Stapf. (Minyadotana). Bolivia. The stem yields a latex used to adulterate hevea rubber.

ZUELANIA A. Rich. Flacourtiaceae. 5 spp. Central America, W.Indies. Trees.

***Z. guidonia*** (Sw.) Britt. and Millsp. (Guaguasi). W.Indies. The leaves, bark and resin from the trunk are used, especially in Cuba as a diuretic and the resin as a purgative.

***Z. roussoviana*** Pitt. Central America. The gum from the stem is used locally to induce vomiting.

Zumbic Tree – Adenanthera pavonina.

ZYGOPHYLLUM L. Zygophyllaceae. 100 spp. Mediterranean to Central Asia, S.Africa, Australia. Xerophytes, herbs or shrubs.

***Z. album*** L. N.Africa, Canaries. An infusion of the dried flowers is used locally as a scent.

***Z. simplex*** L. Herb Tropical Africa, Red Sea. The plant is fed to camels.

ZYGOSACCHAROMYCES Barker. Saccharomycetaceae. 17 spp. Widespread.

***Z. acidifaciens*** Nickerson. The fungus has some potential importance in that it converts glucose to glycerol.

ZYGOTRITONIA Mildbraed. Iridaceae. 6–7 spp. Tropical Africa. Herbs.

***Z. crocea*** Stapf. W.Africa. The corms and fruits are used locally as an emergency food.